# Understanding Biology

# Understanding Biology

## S E C O N D ◆ E D I T I O N

### PETER H. RAVEN

Director, Missouri Botanical Garden;
Engelmann Professor of Botany
Washington University, St. Louis, Missouri


### GEORGE B. JOHNSON

Professor of Biology
Washington University, St. Louis, Missouri

 **Mosby**
**Year Book**

St. Louis   Baltimore   Boston   Chicago   London   Philadelphia   Sydney   Toronto

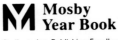

Publisher: EDWARD MURPHY
Developmental Editor: KATHLEEN SCOGNA
Editorial Assistant: RHONDA FOWLER
Project Manager: PATRICIA GAYLE MAY
Production Editor: DONNA WALLS

Art Director/Designer: ELIZABETH ROHNE RUDDER
Art Coordinator: NADINE B. SOKOL
Computer Art/Art Typesetting: PAGECRAFTERS, Inc.
Photo Researcher: MARY TEMPERELLI
Copyeditors: ELISABETH HEITZEBERG, CAROL BARRY
Artists: SCOTT BODELL, RAYCHEL CIEMMA, BARBARA COUSINS,
GEORGE KLATT, CHRISTY KRAMES, BILL OBER, LILI ROBINS,
KEVIN SOMERVILLE, NADINE B. SOKOL

Cover photo of African elephant bulls by Frans Lanting. These elephants were designated a "threatened" species by the World Wildlife Fund in January, 1991. By Fall, 1991, their status is expected to change to officially "endangered."

SECOND EDITION

Mosby-Year Book, Inc.
11830 Westline Industrial Drive
St. Louis, MO 63146

**Library of Congress Cataloging-in-Publication Data**

Raven, Peter H.
    Understanding biology / Peter H. Raven, George B. Johnson

    p.    cm.
    Includes index.
    ISBN 0-8016-2524-6
    1. Biology.   I. Johnson, George B. (George Brooks)
  II. Title.
QH308.2.R38      1991
574--dc20                                     90-22152
                                                CIP

C/CD/VH  9  8  7  6  5  4  3  2  1

# *Preface*

## BIOLOGY AND THE CITIZENS OF THE 90s

In the three years since the first edition of UNDERSTANDING BIOLOGY was published, 270 million people have been added to the world's population. At no other time in history has the world experienced such a rapid explosion in the human population, and as biologists prove and the media attest, solutions to the problems of a crowded planet must be found, and found quickly. As citizens of the Earth, it is important that you understand the biological aspects of this expanding world so that you may play a part in affecting change. We have made every effort in the second editon of UNDERSTANDING BIOLOGY to present these biological issues to you clearly. The decisions that you will make as citizens will shape your future and the future of the world.

In recent years, it has been fashionable to lament the passive attitude of students toward the challenges that await them. As educators in the university and in the public world outside the university, we have seen that this is not so. Students are profoundly interested in their planet and in the forces that threaten it, and they are making concrete efforts to contribute toward its preservation. If there is one trend that we have noted in students in the past few years, it is a hunger for practical knowledge that will assist them with the work that they will face in the coming years. It is a daunting challenge, as the media constantly reinforce the spectre of a dying Earth. Students see these images, but rather than throwing up their hands in despair, instead ask, "What can I do?"

Answering the challenges of the 1990s will not be easy. The problems we face are truly frightening. To name but one example, consider the spread of AIDS. In the past 3 years, 130,000 new cases of AIDS have been reported in the United States alone, and as we enter the 1990s the disease continues to spread. The World Health Organization reports that 700,000 people have developed AIDS worldwide and estimates that up to 8 million have contracted the AIDS virus. By the end of this decade, an estimated 6 million will be sick, and the total number infected may approach 20 million. AIDS is a problem that is not going to go away.

In the face of this challenge, the scientific community has been electrified by the crisis and has mounted a concerted effort to control the virus. The work that has been done to find a cure for this disease has opened up an astonishing understanding of the human immune system, and this understanding has implications for the defeat of other diseases, such as arthritis and cancer. We know more now about our bodies' defenses than ever before, thanks to the efforts of talented scientists all over the world combating AIDS. It is important for you as citizens to understand these efforts.

Another key area of concern is the environment. Rain forest destruction, acid rain, global warming, ozone depletion, and the growing roll call of endangered species all cast a pall on the future. The falling of the iron curtain in Eastern Europe was a wonderful welcome to the decade of the 1990s, but it has also revealed to Western eyes an environmental catastrophe. Communist governments have for the last 40 years permitted unrestrained industrial pollution, and only with monumental effort will it be possible to clean up. Solidifying ties with our friends in Eastern Europe should begin with contributing our expertise and technology, and most particularly our awareness, to the effort of restoring the land that was once behind the iron curtain. Today's students, we believe, will come to play a critical role in this task, an effort that will need to be carried on for many years.

While the study of biology will provide an important tool for students meeting the challenges of the 1990s, there are other tools that will also be needed, such as commitment and common sense. When all is said and done, however, understanding the problems that the world faces and the biological principles that underlie them must be the first step in eradicating these problems. Throughout UNDERSTANDING BIOLOGY, second edition we present basic biological principles in a simple, straightforward manner. Students need to know how viruses invade the body in order to understand AIDS. Knowledge of the delicate balance that exists in the water cycle is essential for realizing the impact of industrial pollution. Understanding is the first step to making informed decisions. It is our hope that UNDERSTANDING BIOLOGY, second edition will be a valuable resource for students as they face the future.

## HOW THIS BOOK IS ORGANIZED TO TEACH YOU BIOLOGY

Like the first edition, this second edition of UNDERSTANDING BIOLOGY is organized into nine parts that can be roughly ordered into three broad areas: basic biological principles (including ecology), ecology, and the structure and function of organisms (see diagram). This organization reflects what we believe to be the clearest way of teaching biology. First, by giving the student an overview of the principles of biological processes and the workings of the biosphere, it better prepares them to understand the form and function of organisms. Second, a full progression of principles that includes ecology makes logical sense: the cell and energy chapters describe the basic properties of all living things, and the three sections that follow present an ordered sequential view of how heredity shapes the world in which

we live. The genetics chapters are concerned with genes, the evolution chapters delineate how genes operate in populations, and the ecology section describes how evolution has shaped natural communities. A student moving through the book in this manner comes away with the understanding that ecology is the end result of evolution acting on genes.

The framework provided by these principles in turn makes possible a rich treatment of form and function in the latter half of the book, as these principles apply to every living creature. Those familiar with the first edition will know that evolution is the grounding feature of this text. Evolution is also the guiding perspective of the second edition, and our organization of the form and function chapters reflects this perspective.

## GOALS FOR THE SECOND EDITION

Although the first edition enjoyed great success and was heralded for its unique evolutionary approach, we decided that the second edition would benefit from a major revision effort. The importance of biology to students today cannot be overstressed, and every effort has been made in this edition to provide students with a readable, accessible, visually compelling text. Some of our goals for the second edition are listed below.

### Simplification of Presentation
Most of this book has been rewritten. In some cases, paragraphs have been rewritten for clarity; in other cases, whole chapters

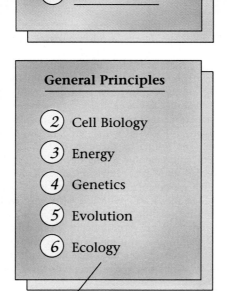

**1** Introduction

The text begins with an introduction to the nature of science.

**General Principles**

**2** Cell Biology

**3** Energy

**4** Genetics

**5** Evolution

**6** Ecology

The first half of the text is devoted to principles shared by all organisms.

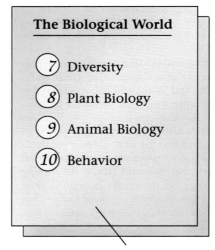

**The Biological World**

**7** Diversity

**8** Plant Biology

**9** Animal Biology

**10** Behavior

The second half of the text is devoted to particular organisms. Much of the emphasis is on vertebrate biology.

were discarded and rewritten from scratch. Language has been simplified in an effort to make the prose more friendly and readable. This type of revision sounds easy to effect, but in reality it takes countless hours of strenuous concentration. However, we believe that text that is easy to read and understand is the absolute *minimum* requirement of a biology textbook.

In addition, legends to the illustrations have been rewritten. Detail in the legends has been cut, and each legend now bears an illustration title.

## Reorganization of Presentation

An important facet of rewriting is reorganization of the internal structure of chapters. Within each chapter, the organization of topics was carefully scrutinized. Headings have been rearranged and rewritten to provide a better "blueprint" to the chapter material. This type of reorganization should aid the student in outlining the chapters and guide the student through the hierarchy of concepts he or she is expected to master for each chapter.

The order of chapters in the text has also been reorganized. Based on the suggestions of those who used the first edition, we have rearranged in several key ways. First, we have pulled together topics that were scattered over several chapters: all the material on DNA is now presented in Chapter 12, all the material on genetic engineering is now presented in Chapter 14, and so on. Second, several sections have been rearranged to provide a better flow of topics. The ecology section has been completely reorganized, and another chapter has been added. The animal biology section has also been reorganized, and it has also been augmented with an all-new chapter.

## Updating of Material

Three years is a century where biological advances are concerned. New, exciting discoveries such as the first tunneling microscopy photograph of DNA, the first use of gene therapy to combat cancer, and the discovery of the gene that determines human sex are but a few examples of biological progress that have been included in this edition.

## New Illustration Program

We are very proud to present an entirely new illustration program for the second edition of UNDERSTANDING BIOLOGY. Working closely with our designer, art coordinator, and developmental editor, we have revised and redrawn virtually every illustration in the book. Illustrations are an important tool for teaching complex concepts and today's students, versed as they are in the potential impact of visual images, are more likely to read and enjoy a text whose illustrations captivate the eye. We have spared no expense with this new illustration program, and we believe that the results speak for themselves.

## New Design

In keeping with the changes in text and art, we've introduced an entirely new design to complement the outstanding features of UNDERSTANDING BIOLOGY, second edition. You will notice that the typeface is bigger and bolder than that of the first edition, making the second edition much easier on the eyes. We've also increased the size of the figures an average of 33%, with the result that the art is much more prominent and compelling, as well as easier to comprehend. The new design of the end-of-chapter pedagogy makes this section easier to find and

refer to. And although we've added two completely new chapters to the text, increased the size of the typeface and art, and opened up the design of the pedagogy, we've increased the overall length of the text by just 14%. We thank our designer, Liz Rudder, for successfully undertaking this challenge and delivering a beautifully designed book.

## New Teaching Technology

Over the last several years the information revolution has caught up with college textbooks, many of which are now available with video disks and other novel ancillaries. We have joined that progress with great enthusiasm, presenting what we believe is the most user friendly—and useful—video disk on the market. It is keyed directly to the text with bar codes, and incorporates many of the key illustrations.

## WHAT'S NEW AND WHY

Given the fact that this revision of UNDERSTANDING BIOLOGY has focused intensely on both the text and the art, it has been necessary for us to garner help from a number of sources. We both began the revision process by collecting information from scientific literature, from colleagues in our respective fields, and from our own research efforts to use in the updating and upgrading of the text. A team of reviewers examined the first edition of the text and offered their suggestions for change, enhancement, and deletion. After a first draft was written, we convened two focus groups to "test" what we had accomplished. These focus groups, composed of experienced professors from across the country, focused separately on the principles section and the animal biology section. Based on focus group comments and comments from another team of reviewers, we wrote the second draft of the manuscript. Between the second and final drafts came numerous intermediate drafts, in which concepts and language were refined. This exhaustive process has yielded a text that is both readable and accurate.

While it is impossible to list every change that has taken place in the text, listed below are the major highlights of the revision:

1. Former Chapters 1 and 2 have been combined to focus more clearly on the scientific method. Following the suggestion of numerous reviewers, the discussion of creationism has been moved to Chapter 15 (The Evidence for Evolution)—students will understand the argument better if they have a background in the evidence for evolution. There are new boxes on scientific progress and how biologists do their work.

2. The chemistry chapter has been extensively revised with the introductory student in mind. Excess detail has been deleted, but information that introductory students need has been added—energy levels, building macromolecules, and nucleotides are all new additions.

3. The energetics section has been rethought and rewritten. Chapter 6 (Energy and Metabolism) has been augmented by new material on biochemical pathways and the evolution of metabolism. Chapter 7 (Cellular Respiration) has a new overview of the entire respiration process that gives the student a sound introduction before the details. Because of the comments of several reviewers, fermentation has been moved to precede oxidative respiration. There is a new boxed essay on metabolism, and

a new art piece on coupled reactions uses a simple spring analogy to teach this difficult concept. A miniglossary of terms pertinent to respiration appears in this chapter; miniglossaries are scattered throughout the text to help the student learn scientific vocabulary. We have added a new discussion of $C_3$ and $C_4$ photosynthesis to Chapter 8 (Photosynthesis).

4. The genetics section of the text (Chapters 9 through 14) has received very careful attention. Chapter 9 (How Cells Reproduce) has benefited from a lowering of the language level and the deletion of some advanced terms and concepts, all done at the suggestions of reviewers. There are new sections on the sexual life cycle, an introduction to sexual reproduction, and a discussion of the evolutionary consequences of sexual reproduction. The new illustration program for this chapter is, in short, stunning. New mitosis and meiosis illustrations clarify these difficult concepts. There is also a new illustration explaining exactly what a chromosome is composed of (chromatid, sister chromatid, homologous pair). Codominance, incomplete dominance, and pleiotropy have been moved from Chapter 10 (Mendelian Genetics) to Chapter 13 (Genes and How They Work). We added Rh blood groups and RFLPs to Chapter 11 (Human Genetics). All DNA information has been concentrated in Chapter 12 (DNA: The Genetic Material). In Chapter 13, the discussion of gene expression has been rewritten to improve precision and clarity. The art for Chapter 13 is another *tour de force*—three dimensional paintings bring the process of protein synthesis to life for students. Finally, Chapter 14 (Gene Technology)—a completely new chapter—discusses the fascinating topic of genetic engineering and incorporates diagrammatic illustrations of genetic engineering techniques.

5. The evolution chapters have benefited greatly from revision. The discussion of macroevolution and microevolution in Chapter 15 has been expanded and suffused throughout the chapter as a unifying theme. A helpful illustration comparing the two types of evolution also appears in this chapter. Hardy-Weinberg equilibrium has been clarified and expanded—more examples are used to teach the student this fundamental principle. A miniglossary appears in this chapter to aid students in learning unfamiliar vocabulary. In Chapter 16 (How Species Form), the explanations of isolating mechanisms have been completely rewritten; and please note the new illustration explaining the controversial debate between gradualism and punctuated equilibrium. Chapter 17 (The Evolution of Life on Earth) sports a new discussion of C-14 dating and an expanded discussion of continental drift, a specialty of one of the authors (Raven). Chapter 18 (How We Evolved: Vertebrate Evolution) gives a new, comprehensive review of the first vertebrates, emphasizing the significance of their appearance on earth.

6. The ecology chapters have been completely reorganized according to the following scheme:
*Chapter 19 (Population Dynamics) now covers the principles of population genetics.
*Chapter 20 (How Species Interact within Communities) covers community ecology—coevolution, predator-prey interactions, and mimicry.
*Chapter 21 (Dynamics of Ecosystems) covers biogeochemical cycles and succession. All the biogeochemical cycle illustrations have been reconceptualized to give a more accurate picture of the circulation of these chemicals throughout the environment.
*Chapter 22 (Atmosphere, Oceans, and Biomes) covers climate and biomes.
*Chapter 23 (Our Changing Environment) covers the future of

the biosphere and emphasizes solutions to environmental problems. This chapter benefits from both authors' expertise in and commitment to this area.

7. The chapters on animal and plant diversity (Chapters 24 through 31) have been improved by the addition of new photographs that more closely exemplify the featured organism. The discussion of plants incorporates the new division names of Hepatophyta, Bryophyta, and Anthocerophyta.

8. The chapters on Animal Biology (Chapters 32 through 44) have been extensively rewritten and reorganized. Our thanks to Warren Burrgren, Ph.D. of the University of Massachusetts, who read and reviewed the first edition chapters and suggested numerous changes, and to Charles Schauf, Ph.D. of the School of Science at Indianapolis and author of HUMAN PHYSIOLOGY, who contributed material to this section. We are also indebted to James Traniello, Ph.D. of Boston University, who made major contributions to the animal behavior chapter. Highlights of the revision to this section are as follows:
*The order of the chapters have been rearranged to provide a more logical flow of topics. The nervous system is presented first, followed by hormones, locomotion, digestion, respiration, the immune system, and water balance. Sex and reproduction and development follow, and the section ends with animal behavior.
*Chapter 32 (The Vertebrate Body) features an all-new discussion of homeostasis and feedback loops. This discussion introduces the student to the broad topic of homeostasis in preparation for the chapters that follow.
*Chapter 33 (How Animals Transmit Information) has been extensively revised. The entire chapter has been reorganized and rewritten and includes clearer definitions of the sodium-potassium pump and gated channels. A miniglossary is included in this chapter to assist the student in learning the terminology.
*Chapter 34 (The Nervous System) includes new information on the spinal cord and a clearer description of antagonistic controls.
*A discussion of the skeletal system has been added to Chapter 35 (How Animals Move). The chapter also includes a miniglossary. A rearrangement of topics that places the structure of the actin/myosin mechanism before its function makes this topic much easier to understand. The discussion of the function of the actin/myosin mechanism interaction has been clarified. The discussion of acetylcholine has been rewritten at a lower level.
*Chapter 36 (Hormones) is now a more cohesive chapter on hormones. Discussions of water balance and the control of digestion have been moved to more appropriate chapters (Chapters 41 and 37). There is a new introduction to receptors, target cells, and glands as well as extensive new coverage of the adrenal glands, thyroid gland, parathyroid gland, pancreatic islets, thymus gland, and pineal gland. Diabetes is discussed in more detail.
*Chapter 37 (How Animals Digest Food) features a new general introduction to the digestive process in addition to new information on the pancreas and liver. One new illustration in particular that should be pointed out is a figure on the fate of food (Figure 37-15) that depicts what happens to each class of nutrient in the digestive process.
*Partial pressure, a notoriously difficult topic for students to grasp, has been clarified in Chapter 38 (How Animals Capture Oxygen). The discussion of cross-current exchange has also benefited from rewriting. A discussion of the control of breath-

ing has been added, and the discussion of carbon dioxide transport has been expanded.

*Chapter 39 (Circulation) features a new box on high blood pressure and a new discussion of the clotting mechanism.

*Chapter 40 (How the Body Defends Itself) received much attention in the revision process. The chapter has been completely reorganized to provide a step-by-step description of this complex system. Appropriate introductory material, such as descriptions of the cells of the immune system, and an overview of specific and nonspecific defenses have been placed at the beginning of the chapter to acquaint the student with the overall process of the system before launching into the details. Spectacular full-color paintings of the humoral and cell-mediated responses are featured in this chapter, as well as a flowchart that outlines the entire immune response, from the invasion of an antigen to the proliferation of memory cells. James Smith, Ph.D. of California State University-Fullerton, should be acknowledged for his very detailed criticisms regarding the first edition chapter.

*New information on ADH, temperature regulation, and dialysis has been added to Chapter 41 (The Control of Water Balance).

*Chapter 42 (Sex and Reproduction) gives more attention to the role of homones in the reproduction cycles of both males and females. Updated material on birth control and AIDS appears in this chapter. In addition, the chapter was reorganized to place the structure of the male and female reproduction system before the function of the systems.

*Chapter 43 (Development) is a completely new chapter that covers comparative vertebrate embryology; the focus turns to human development in the second half of the chapter.

*Chapter 44 (Animal Behavior) has been extensively revised to include information on proximate and ultimate factors; the genetic and neural bases of behavior; learning; behavioral rhythms; animal communication; ecology and behavior; sociobiology; and insect and vertebrate societies.

## HELP FOR THE STUDENT

In our text we are highly sensitive to the fact that many of the students reading this book are novices to the subject of biology. To help the student master the material, we have included numerous teaching aids that are specifically designed to foster applications of key concepts instead of rote memorization.

1. An *Overview* prefaces each chapter. This overview briefly summarizes the content of the chapter for the student.

2. The *For Review* section, placed at the beginning of each chapter, lists terms that the student has encountered in previous chapters. The For Review list alerts students to terms that should be understood before proceeding further. The chapter in which each term appears is given for easy reference.

3. *Concept Summaries* are capsule summaries that appear throughout each chapter. These summaries provide a spot review of key concepts featured in each chapter.

4. The *Chapter Summary* at the end of each chapter provides additional reinforcement of key concepts.

5. *Review Questions* at the end of each chapter test student knowledge of the chapter content in a fill-in-the-blank format. The *Self Quiz* questions quiz students in a multiple choice format. *Thought Questions* ask students to apply the chapter's key

concepts to new and interesting situations not encountered in the chapter. Chapters 11 and 12 also include genetics problems. Answers to the Review and Self Quiz Questions are given in Appendix B. Extensive answers to the genetics problems are also given in Appendix B.

6. *Boxed essays* present interesting topics of social, medical, or environmental concern from a biological perspective. The research into new crops to feed a starving world (p. 553) and the health risks associated with smoking (pp. 334-335) are but two of the topics addressed in the boxed essays.

7. *Miniglossaries* are found in chapters that are particularly dense with vocabulary (such as Chapters 13 and 15). The miniglossaries list key vocabulary terms and their definitions within the chapter to provide the student with easy reference. Turn to pp. 325 and 373 for examples of miniglossaries.

8. The comprehensive glossary defines the bold-faced terms in the text and provides the derivation of each term.

## SUPPLEMENTS

We have developed a complete package of ancillaries to accompany the second edition of UNDERSTANDING BIOLOGY. These ancillaries will aid both student and instructor in managing what must sometimes seem an immense amount of material.

### For the Student

*UNDERSTANDING BIOLOGY Study Guide,* second edition, written by Ann Vernon of St. Charles Community College. This illustrated study guide is keyed directly to the text and provides students with significant additional study aids, including chapter overviews; chapter outlines; key terms exercises that reinforce vocabulary retention and understanding; "concept check" exercises that use a variety of techniques to enhance understanding of chapter topics; and mastery tests. Flash cards help students to conveniently review important concepts. Answers to all questions, including thought questions in the text, are included.

*UNDERSTANDING BIOLOGY Laboratory Manual* is a new addition to the ancillary package. This lab manual has been written specifically for this edition of UNDERSTANDING BIOLOGY by Jerry Davis, Richard Fletcher, Mark Sandheinrich, Martin Venneman and Alan Wortman of the University of Wisconsin-La Crosse. The 30 laboratory exercises in the manual, all class tested by the authors, take an investigative approach, and each exercise ends with a challenge to the student to apply what has been learned to new and different problems. Numerous illustrations, including many created specifically for the manual, provide extensive visual support.

### For the Instructor

*Mechanisms of Life: Stability and Change Videodisc* is an all-new, state of the art instructional medium for UNDERSTANDING BIOLOGY, second edition. The instructional value of this new videodisc stems from its versatility, compactness and ease of use. This customized resource combines high-resolution artwork from the text with film clips on biological processes to provide visual reinforcement for classroom presentation. Other outstanding features include the ability to quickly search and display images or animated sequences and extensive use of full motion along with still images. The videodisc is accompanied by the Instructor's Manual that contains all bar codes for player, as well as complete instructions.

*Instructor's Resource Guide,* written by Florence Ricciuti, of Albertus Magnus College, provides text adopters with substantial support in preparing for and teaching introductory biology with this text. The manual contains suggested course outlines, extensive sources of supplementary materials, and additional resources such as lists of audiovisuals and computer software; 100 overhead transparency masters; suggested learning objectives for each chapter; and chapter-by-chapter notes. Also included is a chapter for novice instructors, with down to earth suggestions for surviving the teaching experience.

*Overhead Transparency Acetates* of 102 of the text's most important four-color illustrations are available to instructors for use as teaching aids. The transparencies were selected with the assistance of a number of instructors of introductory biology. With this edition of acetates, we have greatly increased the size and boldness of the labels. The increased size and boldness of the acetate labels should make the acetates legible for the students sitting in the back of a large lecture hall.

A printed *Test Bank,* written and revised by Richard Van Norman of the University of Utah, provides an extensive battery of 2000 objective test items that may be used by instructors as a powerful instructional tool. Each chapter has between 40 and 50 questions, including multiple choice, short answer, and classification formats. For each question, in addition to the answer, we have identified the subject tested, given an approximate difficulty rating, and indicated the type of question (factual or conceptual) and the text page on which the question's information appears. New to this edition of the test bank are questions based on over 40 of the overhead transparency masters that appear in the *Instructor's Resource Guide.* These masters of important structures and diagrams are unlabeled for use with the test bank, or with questions developed by the instructor. Also new to the test bank is *"Proctor Practice,"* a separate bank of 1500 questions intended for use as practice exams.

A computerized version of the test bank, *Diploma II* is available in IBM and Apple versions. Diploma's EXAM allows instructors to add, edit, and delete questions and to print randomly—or manually—selected tests. GRADEBOOK records students grades, provides reports on individual or class performance, and graphically displays important information. PROCTOR (used with the separate *"Proctor Practice"*) disks included in the *Diploma II* package, allows instructors to provide students with practice exams based on a separate bank of test questions written by the author of the test bank. Scores on the *Proctor Practice* exams can be transferred to GRADEBOOK if desired.

*LXR Macintosh Computerized Test Bank* contains the same content as the printed and *Diploma II* test banks, but has different capabilities. This computerized test bank allows the instructor to select, edit, and delete questions to create custom tests and answer keys. Graphic images may be placed with the test questions, difficulty level and type of question can be controlled with a pop-up menu, and question information window allows simultaneous reference questions to the instructor's notes.

## ACKNOWLEDGMENTS

Anyone who has ever been involved with writing a text knows that the real authors are the reviewers and editors who shape the book. The hundreds of teachers who used the first edition and told us what they thought made a contribution to this edition as great as any author. We cannot name them all here, but we hope that they will recognize the improvements they initiated and appreciate the gratitude we feel. A few were particularly avid in detecting errors and inconsistencies, like Stephen Hedman of the University of Minnesota-Duluth and Clyde Bottrell of Tarrant County Junior College. The art benefited from the careful scrutiny of Steve Dina of St. Louis University and Ed Joern of the University of Missouri-St. Louis. Two teachers became even more deeply involved in our revision, offering criticisms and critiques so detailed that they totally reshaped particular segments of the text: Charles Schauf of the School of Science at Indianapolis put the entire eleven chapters of the animal biology section under a microscope, critically evaluating every statement and suggesting many improvements; and James Traniello of Boston University took the behavior chapter apart in a similar fashion, molding and changing it in many ways. We are deeply grateful for their interest and commitment to our text, which they greatly improved.

We are also indebted to an unusually creative editorial team. The authors write a book, but others create it. The beauty of this edition's art program, its lovely design and layout, the depth and clarity of its within-chapter topic reorganization, the innovative ancillaries—all these things are important aspects of the new edition, and they reflect the hard work of four people: a delightful chief artist and art coordinator, Nadine Sokol (who fit having a baby into a killer production schedule); a flexible book designer and page layout wiz, Liz Rudder, who never lost track of the fact that the text is a teaching tool; a perceptive, thorough, and incredibly hard-working developmental editor, Kathleen Scogna, who was fearless in her efforts to keep the authors focused on her vision of the book, but willing to accommodate ours; and a creative, risk-taking publisher, Ed Murphy, who shares with us a common vision of what an effective text ought to be. Many others helped too, more than we can name here. They will know what they did, and how grateful we are to each of them.

A particularly interesting contribution to this edition was made by the marketing staff, particularly the marketing manager Leslie Tinsley. It was Leslie who suggested that we might be willing to contribute to the international conservation efforts of the World Wildlife Fund, and Leslie who communicated to the sales staff what it was we were about.

This is the third text edition we have published together, and as in all the others we are deeply in debt to our families for their understanding of the demanding schedules we have had to keep, and forgiveness for the neglect that has often resulted. Loss of family time is one of the greatest hidden costs of text writing, and only the strong support of our wives Tamra and Barbara has made it possible to complete this revision.

Finally, a special word of thanks to Kathleen Scogna, for this edition is her creation as much as ours. This is her first book as a developmental editor. Look at it, and you will see why we suspect it will not be her last.

Thanks, all of you.

# A FINAL NOTE

Our commitment to educating citizens of the 1990s in important biological issues is deeply felt. It has shaped our personal and professional lives and forms the philosophical basis from which we teach. In keeping with our belief that students share our concern for the future of the earth and as a token of our commitment to education and conservation, we are contributing to the World Wildlife Fund one dollar from our royalties for every copy of the second edition sold. Buying this book may not save an elephant, but we hope it will help.

If you would like to make an individual contribution to the World Wildlife Fund, please write:

World Wildlife Fund
Membership Department ZB43
1250 24th Street N.W.
Washington, D.C. 20037

**Peter Raven**
**George Johnson**

## Focus Group Participants

**Focus Group I** (Chapters 1 through 31)
James Botsford, New Mexico State University
John Clamp, North Carolina Central University
Tom Emmel, University of Florida
Bill Glider, University of Nebraska-Lincoln
Stephen Hedman, University of Minnesota-Duluth

**Focus Group II** (Chapters 32 through 44)
Clyde Bottrell, Tarrant County Junior College
Judy Goodenough, University of Massachusetts-Amherst
Ed Joern, University of Missouri-St. Louis
James Smith, California State University-Fullerton
Rob Tyser, University of Wisconsin-La Crosse

## Art Reviewers
Steve Dina, St. Louis University
Ed Joern, University of Missouri-St. Louis

## Manuscript Reviewers
John Adler, Michigan Technological University
L. Rao Ayagari, Lindenwood College
Robert Beckman, North Carolina State University
Marlin Bolar, California State University-Sacramento
James Botsford, New Mexico State University
Clyde Bottrell, Tarrant County Junior College
David Bruck, San Jose State University
Warren Burrgren, University of Massachusetts-Amherst
John Clamp, North Carolina Central University
Roy Clarkson, University of West Virginia
Donald Collins, Orange Coast College
Roger Denome, University of North Dakota
Ronald Downey, Ohio University
Tom Emmel, University of Florida
Elizabeth Gardner, Pine Manor College
Gregory Grove, Pennsylvania State University
Elizabeth Gulotta, Nassau Community College
Joyce Hardin, Central State University
Holt Harner, Broward Community College
Stephen Hedman, University of Minnesota-Duluth
Robert Hersch, University of Kansas
George Hudock, Indiana University
Sylvia Hurd, Southwest Texas State University
Leonard Kass, University of Maine
Robert Kaul, University of Nebraska
Jay Kunkle, Ann Arundel Community College
Army Lester, Kennesaw State College
Ben Liles, University of Maine
Charles Lytle, North Carolina State University
Steven McCullagh, Kennesaw State College
Clifton Nauman, University of Tennessee-Knoxville
W. Brian O'Connor, University of Massachusetts-Amherst
Kevin Patton, St. Charles Community College
Jim Peck, University of Arkansas
Douglas Reynolds, Eastern Kentucky University
Michael Rourke, Bakersfield College
James Smith, California State University-Fullerton
Major James Swaby, United States Air Force Academy
Thomas Terry, University of Connecticut
Kathy Thompson, Louisiana State University
F. R. Trainor, University of Connecticut
James Traniello, Boston University
Rob Tyser, University of Wisconsin-La Crosse
Nancy Webster, Prince George's Community College
Dana Wrensch, Ohio State
John Zimmerman, Kansas State University
Steve Ziser, Austin Community College

# About the Authors

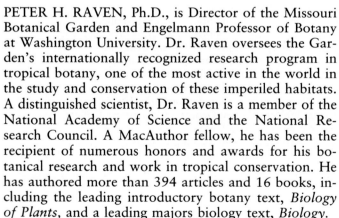

PETER H. RAVEN, Ph.D., is Director of the Missouri Botanical Garden and Engelmann Professor of Botany at Washington University. Dr. Raven oversees the Garden's internationally recognized research program in tropical botany, one of the most active in the world in the study and conservation of these imperiled habitats. A distinguished scientist, Dr. Raven is a member of the National Academy of Science and the National Research Council. A MacAuthor fellow, he has been the recipient of numerous honors and awards for his botanical research and work in tropical conservation. He has authored more than 394 articles and 16 books, including the leading introductory botany text, *Biology of Plants,* and a leading majors biology text, *Biology.*

GEORGE B. JOHNSON, Ph.D., is Professor of Biology at Washington University, where he teaches one of the university's largest courses, freshman biology for nonmajors. Also Professor of Genetics at the Washington University School of Medicine, he has authored more than 50 research publications and 7 books. A Guggenheim and Carnegie fellow, Dr. Johnson is a recognized authority on population genetics and evolution and is renowned for his pioneering studies on genetic variability. He is the founding Director of the St. Louis Zoo's trend-setting educational center, The Living World. Dr. Johnson is coauthor of *Biology* with Peter Raven. Old friends, they have known each other and collaborated together for over 20 years.

# Contents in Brief

# Contents

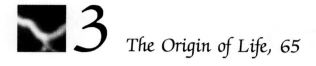

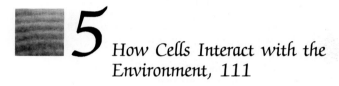

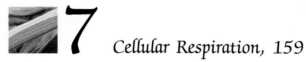

## 10 Mendelian Genetics, 233

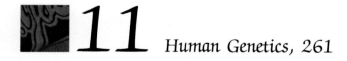

## 11 Human Genetics, 261

## 12 DNA: The Genetic Material, 287

## 13 Genes and How They Work, 311

 **14** *Gene Technology, 343*

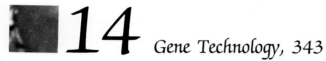

Part Five   Evolution

**15** *The Evidence for Evolution, 365*

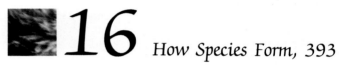 **16** *How Species Form, 393*

 **17** *The Evolution of Life on Earth, 411*

# 18 How We Evolved: Vertebrate Evolution, 433

## Part Six Ecology

# 19 Population Dynamics, 463

# 20 How Species Interact within Communities, 483

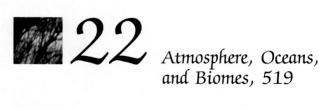

 **25** *The Invisible World: Bacteria and Viruses, 593*

 **26** *The Origins of Multicellularity: Protists and Fungi, 609*

 **27** *Plants, 629*

 **28** *Animals, 649*

## Part Eight    Plant Biology

# 29 The Structure of Plants, 681

# 30 Flowering Plant Reproduction, 701

# 31 How Plants Function, 717

## Part Nine    Animal Biology

## 32  The Vertebrate Body, 739

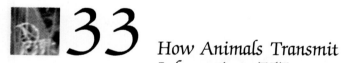

## 33  How Animals Transmit Information, 757

## 34  The Nervous System, 775

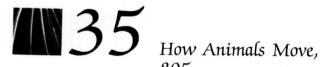

## 35  How Animals Move, 805

# 36 Hormones, 825

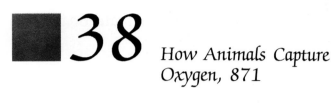

# 37 How Animals Digest Food, 847

# 38 How Animals Capture Oxygen, 871

# 39 Circulation, 889

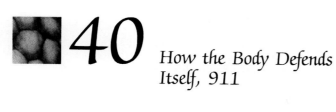

# 40 How the Body Defends Itself, 911

# 41 The Control of Water Balance, 937

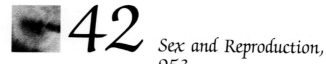

# 42 Sex and Reproduction, 953

# 43 Development, 977

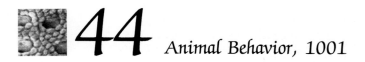

# 44 Animal Behavior, 1001

# Appendices

# A Classification of Organisms, A1

# B Answers, B1

## CHAPTER 1

# The Science of Biology

This cheetah couple faces an uncertain future. Their natural home, the African grasslands, is rapidly disappearing, and with it their ability to survive. The science of biology will play a crucial role in attempts to save them.

# THE SCIENCE OF BIOLOGY

## Overview

One of the most exciting aspects of studying biology is that biology is an active science, dynamic and ever-changing. Like all sciences, biology does not give us absolute truths, but rather "best guesses," which we call hypotheses. As we learn more about nature we sometimes revise our thinking, leading to the construction of new hypotheses. When hypotheses have been tested over a long period of time and still have not been shown to be false, they are called theories. Perhaps the most important theory in biology is the theory of evolution by natural selection. First proposed by Charles Darwin more than a hundred years ago, this theory has been widely examined and is now almost universally accepted.

The green turtle in Figure 1-1 has laid eggs during the night on the upper beach; as dawn breaks, she is returning to the sea, laboriously dragging her 400-pound body across the sand. She is on Ascension Island, in the middle of the Atlantic Ocean, far from her feeding grounds on the coast of Brazil, 1400 miles away. Every year great numbers of green turtles cross this great distance to Ascension Island, where they lay their eggs. Both males and females swim to the island, but the males remain offshore while the females scramble up on the beaches to lay their eggs. The journey lasts about 2 months: 2 months of churning through strong equatorial currents with no food. There are seven small sandy beaches on this tiny rocky island, and the turtles seek them out as if drawn by a beacon. By tagging the turtles, scientists have learned that individual females usually return faithfully to the same beach, or even the same stretch of beach, for nesting in successive years. Some evidence suggests that they swim back to the beaches where they were born. Once there, they lay their eggs and then turn their backs on the eggs to begin the long swim home.

In Brazil, where the green turtles feed and remain most of the year, there are countless miles of warm, sandy beaches. What is it that so draws the turtles to these isolated, cold strips of sand? How do they find Ascension Island, plowing head down through the waves for long days and weeks? The island is a mere speck in the ocean, hard for an airplane or ship to find even with modern navigational equipment. How can they know the island is out there, over the horizon, more than a thousand miles away? How do the newborn turtles then find their way back over open sea to a Brazil that they have never seen?

At first glance, the turtle in Figure 1-1 is just a turtle on a beach. It is only by observation and study that we learn enough to appreciate the mystery that is so much a part of its life. Similar mysteries are common in human life, and many of them have a biological basis: the more we know about the biology of other animals, the better we can understand ourselves. Biology is a science devoted to this larger vision. It provides knowledge about the living world of which we are a part,

**Figure 1-1**

**End of a long migration.** A female green turtle leaves Ascension Island in the South Atlantic Ocean after laying her eggs in the sand.

*Figure 1-2*

**Movement.** A graceful gull gliding through the air, a killer whale exploding from the sea—animals have evolved mechanisms that allow them to move about. Although not all kinds of organisms move from place to place, movement is one of the common properties of living things.

and, ceaselessly, it bombards us with questions. Slowly, a little at a time, it sometimes gives us answers. As we proceed through this text we will weave a tapestry of questions and answers, a picture of life on our planet. When we have completed our journey, we will ask again about the long voyage of the green turtles.

## BIOLOGY IS THE STUDY OF LIFE

Biology is a science that attempts to understand the teeming diversity of life on earth, a diversity of which we are a part. It is important that humans learn how to live in harmony with earth's other residents. The science of biology has much to contribute to this effort. A good way to start your study of biology is to focus for a moment on biology's subject, life. What is life? What do we mean when we use the term? This is not as simple a question as it appears, largely because life itself is not a simple concept. Pause for a moment and try to write a simple definition of "life." You will find that it is not an easy task. The problem is not your ignorance, but rather the loose manner in which the concept "life" is used. For example, imagine a situation in which two astronauts encounter a large, formless blob on the surface of some other planet. One might say to the other, "Is it alive?" We can try to answer the question of what life is by observing what the astronauts do to find out whether the blob has life. Probably they would first observe the blob to see whether it moves.

1. *Movement.* Most animals move about (Figure 1-2). A horse winning the Kentucky Derby, a dog chasing a car, you rolling over in bed—movement seems an integral part of living. However, movement from one place to another is not in itself a sure sign of life. Many animals, and most plants, do *not* move about, and many nonliving objects such as clouds can be observed to move. The criterion of movement is thus neither *necessary*—possessed by all life forms, nor *sufficient*—possessed only by life forms, even though it is a common attribute of many kinds of organisms.

The astronauts might prod the blob to see whether it responds and thus test for another criterion, sensitivity.

2. *Sensitivity*. Almost all living things respond to stimuli (Figure 1-3). Plants grow toward light, and animals retreat from fire. However, not all stimuli produce responses. Imagine kicking a redwood tree or singing to a mushroom. Different kinds of organisms often react to the same stimuli in different ways. This criterion, although superior to the first one, is still inadequate to define life.

The astronauts might watch the blob to see whether it changes and thus test yet another criterion, development.

3. *Development*. Most multicellular organisms exhibit development, an orderly progressive change in form and degree of specialization. You, for example, started life as a single cell. As you grew to be an embryo and then a baby, different cell types developed, giving rise to brain and lung and bone, and finally to your adult form. Without development, you would be simply a large blob of similar cells. But not all living things exhibit development. A single-celled bacterium, for example, does not develop from a simpler form; its parent cell simply divides into two identical daughter cells. Nor are all things that undergo progressive, orderly change alive. The progressive, orderly series of rocks that can be seen on the walls of the Grand Canyon does not indicate that the canyon was ever alive.

The astronauts might think that the motionless blob had once been alive, but is now dead.

4. *Death*. All living things die, whereas no inanimate objects do. Death is not the same as disorder. A car that breaks down does not die; we may say, "The car died on me," or "I killed the engine," but the now-broken car was never alive. Death is simply the termination of life. Unless one can detect life, death is a meaningless concept. Death is a terribly inadequate criterion.

Finally, the astronauts might attempt to pick up the blob and examine it more carefully, to see how complex it is.

5. *Complexity*. All living things are complex. Even the simplest bacterium contains a bewildering array of molecules organized into many complex structures. However, complexity is not diagnostic of life. A computer is also complex, but it is not alive. Complexity is a necessary condition of life, but not sufficient in itself to identify living things, because many complex things are not alive.

**Figure 1-3**

**Sensitivity.** The father is responding to a stimulus—he has just been bitten on the rump by his cub. As far as we know, all organisms respond to stimuli, although not always to the same ones.

**Figure 1-5**

**Metabolism.** The energy that this cedar wax wing chick will use to grow is obtained from the food it eats.

To determine whether the blob is alive, the astronauts must learn much more about it. The best thing they could do would be to examine it more carefully and determine the ways in which it resembles living organisms. All organisms that we know about share certain general properties, ones that we think must ultimately have been derived from the first organisms that evolved on earth. It is by these properties that we recognize other living things, and to a large degree these properties define what we mean by the process of life. Four fundamental properties shared by all organisms on earth are:

1. *Cellular organization.* All organisms are composed of one or more cells, complex organized assemblages of **molecules**—the smallest units of a chemical compound that still have the properties of that compound—surrounded by membranes (Figure 1-4). The simplest organisms possess only a single cell; your body contains about 100 trillion.

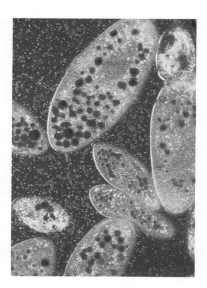

**Figure 1-4**

**Cellular organization.** This *Paramecium*, a complex protozoan, has just ingested several yeast cells, which are stained red in this photograph. The yeasts are enclosed within membrane vesicles called digestive vacuoles.

2. *Growth and metabolism.* All living things assimilate energy and use it to grow in a process called **metabolism** (Figure 1-5). Plants, algae, and some bacteria use the energy of sunlight to create the more complicated molecules that make up living organisms from carbon dioxide ($CO_2$) and water ($H_2O$). The process by which they do this is known as **photosynthesis**. Nearly all other organisms obtain their energy by consuming these photosynthetic organisms, or one another, in a constant ebb and flow of energy. All of the organisms you see about you drive the processes of life within themselves by using chemical energy first captured by photosynthesis. In all living things, this energy is transferred from one place to another by means of special, small, energy-carrying molecules called **ATP molecules.**

3. *Reproduction.* Some organisms live for a very long time. Some of the bristlecone pines *(Pinus longaeva)* growing near timberline in the western Great Basin of the United States have been alive for nearly 5000 years. But no organism lives forever, as far as we know. Because all individual organisms ultimately die, life as an ongoing process is impossible without reproduction (Figure 1-6).

4. *Homeostasis.* All living things maintain an internal environment quite different from their surroundings, with more of certain chemicals and less of others. Maintaining relatively stable internal conditions in an organism is called **homeostasis.** If someone were to grind you up into

**Figure 1-6**

**Reproduction.** These snakes hatching from their eggs in a Costa Rican rain forest represent successful reproduction. All organisms reproduce, although not all hatch from eggs. Some organisms reproduce several generations each hour; other only once in a thousand years.

**Figure 1-7**

Three generations in a family. Heredity ensures the transmission of characteristics from parent to offspring.

a soup, you would not be alive, even though all the same molecules would be present. The relationship between the molecules, which forms the stable internal environment necessary for you to live, would have been destroyed.

Are these properties adequate to define life? Is a membrane-enclosed entity that grows and reproduces alive? Not necessarily. Soap bubbles in water solution spontaneously form hollow spheres, membranes that enclose a small volume of air. These spheres may grow and subdivide and maintain an internal environment quite different from the water surrounding them. Despite these features, the soap bubbles are certainly not alive.

Therefore the four criteria just listed are necessary for life but are not sufficient to define life. One ingredient is missing: heredity, a mechanism for preserving features that determine the nature of an organism.

5. *Heredity.* All organisms on earth possess a "genetic" system that is based on the replication (duplication) of a complex linear molecule called **DNA.** The order of the subunits making up the DNA contains, in code, the information that determines what an individual organism will be like, just as the order of letters on this page determines the sense of what you are reading. Blocks of coded information in the DNA contain the directions for creating the molecules that determine what organisms are like. These subunits of DNA are called **genes.** Because DNA is copied faithfully from one generation to the next, any *change* in a gene is also preserved and passed to future generations.

To understand the role of heredity in our definition of life, let us return for a moment to soap bubbles.

When examining an individual bubble, we see it at that precise moment in time, but we learn nothing of its predecessors. It is likewise impossible to guess what future bubbles will be like. The bubbles are the passive prisoners of a changing environment, and it is in this sense that they are not alive. The essence of being alive is the ability to reproduce permanently the results of change.

**Heredity,** the transmission of characteristics from parent to offspring, therefore provides the basis for the great division between the living and the nonliving (Figure 1-7). A genetic system that enables this transmission to occur is the sufficient condition of life.

A natural consequence of heredity is adaptation. An **adaptation** is any peculiarity of structure, physiology (life processes), or behavior that promotes the likelihood of an organism's survival and reproduction in a particular environment. Organisms seem remarkably well suited to the environments in which they live. In the course of the progressive adaptation of organisms to the conditions of life on earth—a process known as **evolution**—those organisms which were less suited to particular places have not persisted. The ones that are found here today are the "winners," for the moment. When we look at any living organism, therefore, we see in its features a record of its history. Not only does life evolve, evolution is the very essence of life.

*All living things on earth are characterized by cellular organization, growth, reproduction, homeostasis, and heredity. These characteristics define the term "life."*

# BIOLOGY AS A SCIENCE

In its broadest sense, biology is the study of living things. Life, however, does not take the form of a uniform green slime covering the surface of the earth; rather, it consists of a diverse array of living forms. A biologist tries to understand the sources of this diversity, and in many cases to harness particular life forms to perform useful tasks. Even the narrowest study of a seemingly unimportant life form is a study in biological diversity.

As a science, biology is devoted to understanding biological diversity and its consequences. Like all other scientists, biologists achieve understanding by observing nature and drawing deductions from these observations. In doing so, biologists also attempt to explain the unity of structure and function that underlies the diversity, and the consequences of that unity. It is the special relationship between unity and diversity that underlies and distinguishes the science of biology.

# THE NATURE OF SCIENCE

What is "science"? The word conjures up images of people in white lab coats peering at instruments and shaking test tubes. What are they doing, and why?

Science is a particular way of investigating the world. Not all of the investigations we carry out are scientific. When you want to know how to get to Chicago from St Louis, for example, you do not carry out a scientific investigation—you look at a map to determine what turns to take. Making individual decisions by applying a "map" of general principles is called **deductive reasoning**. It is the reasoning of mathematics, of philosophy, of politics and ethics. It is the way in which a computer thinks and the way in which all of us make our everyday decisions about what to do. General principles are used as the basis for examining and evaluating specific decisions.

Where do the general principles come from? Religious and ethical principles often have a religious foundation; political principles reflect social systems. Some general principles, however, do not derive from religion or politics, but from observation of the physical world around us. If you drop an apple, it will fall whether or not you wish it to, despite any laws you may pass forbidding it to do so. Science is devoted to discovering the general principles that govern the operation of the physical world.

## How Scientists Work

How does a scientist discover such general principles? Where are they written? They are written in stone and air and fire, in a butterfly's wing and a tiger's stare. They are written wherever we look in the world around us. A scientist is above all an observer, someone who looks at the world in order to understand how it works. Said briefly, a scientist determines principles from observation.

This way of discovering general principles by the careful examination of specific cases is called **inductive reasoning**. It first became popular about 400 years ago, when Isaac Newton, Francis Bacon, and others began to carry out experiments, and from the results of particular experiments to infer general principles about how the world operates. The experiments were sometimes quite simple. Newton's consisted simply of releasing an apple from his hand. What happened? The apple fell to the ground. This simple result is the stuff of science, the sort of observation from which scientific principles are inferred. From a host of particular observations, each no more difficult to observe than the falling of an apple, Newton inferred a general principle— that all objects fall toward the center of the earth. What Newton did was to construct a mental model of how the world works, a family of working rules consistent with what he could observe. And, like Newton, that is what scientists do today. They are makers of mental models, and observations are the materials from which they build them.

## Testing Hypotheses

How do scientists learn which general principles are actually true, from among the many that might be true? They do this by attempting systematically to demonstrate that certain proposals are *not* valid—not consistent with what they learn from experimental observation. A great deal of careful and creative thinking is necessary for the construction of hypotheses that account for the facts, observations, and experiments that are available concerning a particular area of science. Proposals that scientists are not yet able to disprove they retain for the time being as useful, because they fit the known facts. Later, even these proposals might be rejected if, in the light of new information, they are found to be inconsistent with observation.

We call a proposal that might be true a **hypothesis**, and the test of a hypothesis an **experiment** (Figure 1-8). An experiment evaluates alternative hypotheses. Say for example that you face two closed doors. "There is a tiger behind the door on the left" is a hypothesis; an alternative hypothesis is "The door on the right has a tiger behind it"; a third alternative might be "There is no tiger behind either door." An experiment works by eliminating one or more of the hypotheses. To test these alternative hypotheses, you might open the door on the right. Let us say that when you do this,

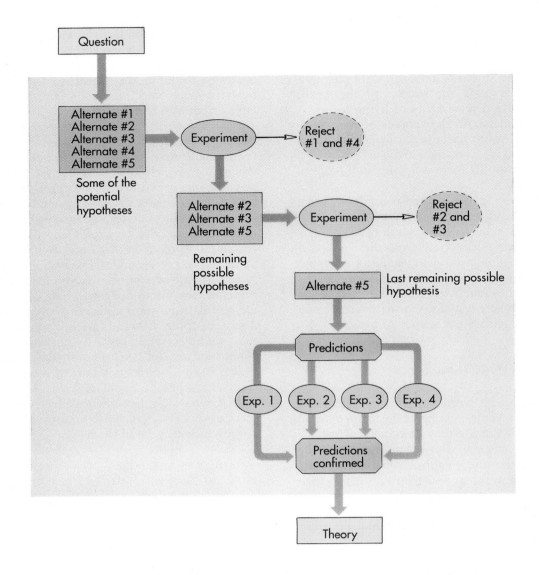

*Figure 1-8*

**The scientific method.** This diagram illustrates the way in which scientific investigations proceed. A number of potential explanations (hypotheses) are suggested in answer to a question; experiments are carried out in an attempt to eliminate one or more of these hypotheses; predictions are made based on these hypotheses; and further experiments are carried out to test these predictions. As a result of this process, the most likely hypothesis is selected. If it is validated by numerous experiments and stands the test of time, the hypothesis, together with other well-supported ones, may eventually lead to formation of a more general statement, a theory.

a tiger leaps out at you. The result of our experiment disproves the third hypothesis—it was clearly incorrect to say that there was no tiger behind either door. Note that a test such as this does not prove that only one alternative is true, but rather demonstrates that one of them is not. In this instance, the fact that a tiger is behind the door on the right does not rule out the possibility that a tiger also lurks behind the door on the left. A successful experiment is one in which one or more of the alternative hypotheses are demonstrated to be inconsistent with experimental observation and are thus rejected. Scientific progress is made in the same way a marble statue is, by chipping away unwanted bits.

> *The process of science consists of demonstrating that one or more hypotheses are not consistent with experimental observation.*

As you proceed through this text, you will encounter a great deal of information, often coupled with explanations. These explanations are hypotheses that have stood the test of experiment. Many will continue to do so; others will be revised. Biology, like all healthy sciences, is in a constant state of ferment, with new ideas bubbling up and replacing old ones.

## Controls

Often we are interested in learning about processes that are influenced by many factors. We call each factor that influences a process a **variable**. To evaluate alternative hypotheses about one variable, it is necessary to hold all the other variables constant so that we don't get misled or confused by these other influences. This is done by carrying out two experiments in parallel: in the first "experimental" test we alter one variable in a known way to test a particular hypothesis; in the second **control** experiment we do not alter that variable. In all other respects the two experiments are the same. We then ask "What difference is there between the outcomes of the two experiments?" Any difference that we see must result from the influence of the variable that we changed, as all other variables remained constant (the same in both experiments). Much of the challenge of experimental science lies in designing control experiments—in successfully isolating a particular variable from all other effects that might influence a process.

## The Importance of Prediction

A successful scientific hypothesis needs to be not only valid but useful—it needs to tell you something that you want to know. When is a hypothesis most useful? When it makes predictions. The predictions that a hypothesis makes provide a very important way to further test its validity. A hypothesis that your experiment does not reject, but which makes a prediction the experiment *does* reject, must itself be rejected. The more predictions a hypothesis makes that check out, the more demonstrably valid is the hypothesis. For example, Einstein's hypothesis of relativity was at first provisionally accepted because no one could think of an experiment that invalidates it. Acceptance soon became far stronger because the theory made a clear prediction, that the sun would bend the path of light passing by it. When this prediction was tested in a total eclipse, the light of background stars was indeed bent. Because the result was not known ahead of time, when Einstein's proposal was being formulated, it gave strong support to his hypothesis.

## Theories

Hypotheses that stand the test of time, their predictions often tested and never rejected, sometimes are combined together into general statements called **theories**. Thus we speak of the general principle first noted by Newton as the theory of gravity. Theories are the solid ground of science, that of which we are most certain. There is no absolute truth in science, however, only varying degrees of uncertainty. The possibility always remains that future evidence will cause a theory to be revised. A scientist's acceptance of a theory is always provisional.

The word "theory" is thus used very differently by scientists than by the general public. To a scientist, a theory represents that of which he or she is most certain; to the general public, the word *theory* implies a *lack* of knowledge, or a guess. As you can imagine, confusion often results. In this text, the word "theory" will always be used in its scientific sense, in reference to a generally accepted scientific principle.

> *A theory is a hypothesis that is supported by a great deal of evidence.*

Some theories are so strongly supported that the likelihood of their being rejected in the future is very small. Most of us would be willing to bet that the sun will rise in the east tomorrow, or that an apple, when dropped, will fall. In physics the theory of the atom is universally accepted, although until recently no one had ever seen one. In biology the theory of evolution by natural selection is so broadly supported by different lines of inquiry that biologists accept it with as much certainty as they do the theory of gravity. We will examine this particular theory later in this chapter as an example of how science is carried out. It is a particularly important theory to biologists because the theory of evolution provides the conceptual framework that unifies biology as a science.

# How Biologists Do Their Work

THE CONSENT

*Late in November, on a single night*
*Not even near to freezing, the ginkgo trees*
*That stand along the walk drop all their leaves.*
*In one consent, and neither to rain nor to wind*
*But as though to time alone: the golden and green*
*Leaves litter the lawn today, that yesterday*
*Had spread aloft their fluttering fans of light.*
*What signal from the stars? What senses took it in?*
*What in those wooden motives so decided*
*To strike their leaves, to down their leaves,*
*Rebellion or surrender? and if this*
*Can happen thus, what race shall be exempt?*
*What use to learn the lessons taught by time,*
*If a star at any time may tell us:* Now.
    Howard Nemerov

What is bothering the poet Howard Nemerov is that life is influenced by forces he cannot control, or even identify. It is the job of biologists to solve puzzles such as the one he poses—to identify and try to understand those things which influence life.

Nemerov asks "Why do ginkgo trees drop all their leaves at once?" To answer, scientists formulate *possible* answers, called **hypotheses,** and then try to see which answers are false:

> *Hypothesis 1:* Ginkgo trees possess an internal clock that times the release of leaves to match the season. On the day Nemerov describes, this clock sends a "drop" signal (perhaps a chemical) to all the leaves at the same time.
>
> *Hypothesis 2:* The individual leaves of ginkgo trees are able to sense day length, and when in the fall the days get short enough, each leaf responds independently by falling.
>
> *Hypothesis 3:* A strong wind arose the night before Nemerov made his observation, blowing all the leaves off the ginkgo trees.

Next the scientist attempts to eliminate one or more of the hypotheses by conducting an experiment. In this case one might cover some of the leaves so that leaves cannot use light to sense day length. If hypothesis 2 is true, then the covered leaves should not fall when the others do because they are not receiving the same information. Suppose, however, that despite the covering of some of the leaves, all the leaves fall together. This eliminates hypothesis 2 as a possibility. Either of the other hypotheses, and many others, remain possibilities.

This simple experiment with ginkgos serves to point out the essence of scientific progress: science does not prove that certain explanations are true, but rather that some possibilities are not. Hypotheses that are not consistent with experimental results are rejected. Hypotheses that are not proven false by an experiment or series of experiments are provisionally accepted—but may be rejected in the future when more information becomes available if they are not consistent with the new information. Just as a computer finds the correct path through a maze by trying and eliminating false paths, so a scientist gropes toward reality by eliminating false possibilities.

A hypothesis that stands the test of time, often tested and never rejected, is called a **theory.** In biology, the hypothesis that tiny organelles, called mitochondria, within your cells are the descendants of bacteria is a theory. It is not certain that this idea is correct, but the overwhelming weight of evidence supports the hypothesis, and most biologists accept it as "proven." However, there is no absolute truth in science, only varying degrees of uncertainty, and the possibility always remains that future evidence will cause a theory to be revised. A scientist's acceptance of a theory is always provisional.

# Scientific Progress

If there is one thing on which almost all scientists would agree, it is that science is a progressive enterprise. New observations and theories survive the scrutiny of scientists and earn a place in the edifice of scientific knowledge because they describe the physical or social world more completely or more accurately. Relativistic mechanics is a more thorough description of what we observe than is Newtonian mechanics. The DNA molecule is a double helix. Our apelike ancestors walked erect before brain sizes greatly increased.

Given the progressive nature of science, a logical question is whether scientists can ever establish that a particular theory describes the empirical world with complete accuracy. The notion is a tempting one, and a number of scientists have proclaimed the near completion of research in a particular discipline (occasionally with comical results when the foundations of that discipline shortly thereafter underwent a profound transformation). But the nature of scientific knowledge argues against our ever knowing that a given theory is the final word. The reason lies in the inherent limitations on verification. Scientists can verify a hypothesis by testing the validity of a consequence derived from that hypothesis. But verification can only increase confidence in a theory, never proving the theory completely, because a conflicting case can always turn up sometime in the future.

Because of the limits on verification, philosophers have suggested that a much stronger logical constraint on scientific theories is that they be falsifiable. In other words, theories must have the possibility of being proved wrong, because then they can be meaningfully tested against observation. This criterion of falsifiability is one way to distinguish scientific from nonscientific claims. In this light, the claims of astrologers or creationists cannot be scientific because these groups will not admit their ideas can be falsified.

Falsifiability is a stronger logical constraint than verifiability, but the basic problem remains. General statements about the world can never be absolutely confirmed on the basis of finite evidence, and all evidence is finite. Thus science is progressive, but it is an open-ended progression. Scientific theories are always capable of being reexamined and, if necessary, replaced. In this sense, any of today's most cherished theories may prove to be only limited descriptions of the empirical world and at least partially "erroneous."

From Committee on the Conduct of Science, National Academy of Sciences: *On Being a Scientist,* National Academy Press, Washington, D.C., 1989, pages 12-13.

## The Scientific Method

It used to be fashionable to speak of the "scientific method" as consisting of an orderly sequence of logical "either/or" steps, each step rejecting one of two mutually incompatible alternatives, as if trial-and-error testing would inevitably lead one through the maze of uncertainty that always slows scientific progress. If this were indeed true, a computer would make a good scientist—but science is not done this way. As British philosopher Karl Popper has pointed out, if you ask successful scientists how they do their work, you will discover that without exception they design their experiments with a pretty fair idea of how the experiments are going to come out—they use what Popper calls an "imaginative preconception" of what the truth might be. A hypothesis that a successful scientist tests is not just any hypothesis, but rather a "hunch" or educated guess in which the scientist integrates all that he or she knows, and also allows his or her imagination full play, in an attempt to get a sense of what *might* be true. It is because insight and imagination play such a large role in scientific progress that some scientists are so much better at science than others—for precisely the same reason that Beethoven and Mozart stand out above most other composers.

*The scientific method is the experimental testing of a hypothesis formulated after the systematic, objective collection of data. Hypotheses are not usually formulated simply by rejecting a series of alternative possibilities, but rather often involve creative insight.*

**Figure 1-9**

**Charles Darwin at the age of 29, 2 years after his return from the voyage of the *Beagle*.** Darwin had just been married to his cousin Emma Wedgewood and was hard at work studying the materials he had gathered on the voyage.

**Figure 1-10**

**A replica of the *Beagle* off the southern coast of South America.** Charles Darwin set forth on the H.M.S. *Beagle* in 1831 at the age of 22. During the 5 years of this voyage, which went around the world but mainly explored the coasts and coastal islands of South America, Darwin formulated and began to test his hypothesis of evolution by means of natural selection.

## HISTORY OF A BIOLOGICAL THEORY: DARWIN'S THEORY OF EVOLUTION

The idea of evolution—the notion that kinds of living things on earth change gradually from one form into another over the course of time—provides a good example of how an idea, an educated guess, is developed into a hypothesis, tested, and eventually accepted as a theory.

Charles Robert Darwin (1809-1882; Figure 1-9) was an English naturalist who, at the age of 50, after 30 years of study and observation, wrote one of the most famous and influential books of all time. The full title of this book, *On the Origin of Species by Means of Natural Selection, or the Preservation of Favoured Races in the Struggle for Life,* expressed both the nature of its subject and the way in which Darwin treated it. The book created a sensation when it was published in 1859, and the ideas expressed in it have played a

central role in the development of human thought ever since.

In Darwin's time, most philosophers believed that the various kinds of organisms and their individual structures resulted from the direct actions of the Creator. Species were held to be specially created and unchangeable over the course of time. In contrast, a number of other scholars, both before and during Darwin's time, held the view that living things must have changed during the course of the history of life on earth, and that some form of evolution from simple to complex forms must have taken place. Darwin, though, was the first to present a coherent, logical explanation for this process—natural selection—and was the first to bring the notion of evolution to wide attention. His book, as you can see from its title, presented a conclusion that differed sharply from conventional wisdom. Darwin argued that the evolution of organisms—the origin of the vast and diverse array of life on earth—followed a series of more or less orderly steps that could be studied and understood. These steps produced continual change and improvement. Although his theory did not challenge the existence of a divine creator, his views put Darwin at odds with most people of his time, who believed in a literal interpretation of the Bible and accepted the idea of a fixed and constant world. Darwin's theory was a revolutionary one that troubled many of his contemporaries, and Darwin himself, deeply.

The son of a wealthy father, Darwin had a troubled education, spending more time outdoors than in school. As a medical student in Edinburgh, for example, he cut lectures, spending the time collecting beetles! In desperation, his father sent him to Cambridge to train for the ministry. There, in 1831, when he was 22 years old, one of his professors recommended Darwin for a post as naturalist on a 5-year voyage around the coasts of South America on the *H.M.S. Beagle* (1831 to 1836; Figure 1-10). Darwin was offered the position. His father at first refused to allow him to go, and Darwin regretfully declined, but his girlfriend's father interceded at the last moment, and Darwin was off on a voyage that would change forever how we think of ourselves.

During his long journey (Figure 1-11), Darwin had the chance to study plants and animals on continents and islands and in far-flung seas. He was able to experience first-hand the biological richness of the tropical forests; the extraordinary fossils of huge extinct mammals in Patagonia, at the southern tip of South America (Figure 1-12); and the remarkable series of related but distinct forms of life on the Galapagos Islands, off the west coast of South America. Such an opportunity clearly played an important role in the development of his thoughts about the nature of life on earth.

When Darwin returned from the voyage, at the age of 27, he began a long life of study and contemplation. During the next 10 years he published important books on several different subjects, including the formation of oceanic islands from coral reefs and the geology of South America. He also devoted 8 years of study to barnacles, a group of marine animals, writing a four-volume work on their classification and natural history. In 1842, Darwin and his family moved a short distance out of London to a country home at Downe, in Kent. In these pleasant surroundings he lived, studied, and wrote for the next 40 years.

Darwin did not write *On The Origin of Species* as soon as he got back. Far from it. During the first few

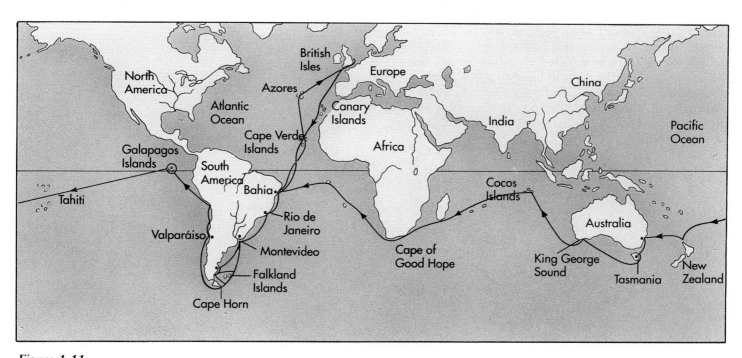

**Figure 1-11**

**The 5-year voyage of H.M.S. *Beagle*.** Most of the time was spent exploring the coasts and coastal islands of South America, such as the Galapagos Islands. Darwin's studies of the animals of the Galapagos Islands played a key role in his eventual development of the theory of evolution by means of natural selection.

years he formulated his ideas, presenting them in 1842 in a preliminary manuscript that he showed to trusted colleagues. This manuscript proposed, for the first time, a theory of evolution that explained *why* evolution occurred. Darwin presented his ideas in such convincing detail that they could logically be accepted as explaining the diversity of life on earth, the intricate adaptations of living things, and the ways in which they are related to one another. Then he put the manuscript aside and worked on other subjects for the next 17 years.

## DARWIN'S EVIDENCE

So much information had accumulated by 1859 that the acceptance of the theory of evolution now seems, in retrospect, to have been inevitable. Darwin was able to arrive at a successful theory, where many others had failed, because he rejected supernatural explanations for the phenomena that he was studying. One of the obstacles that blocked the acceptance of any natural theory of evolution was the incorrect notion, still widely believed at that time, that the earth was only a few thousand years old. However, the discoveries of thick layers of rocks, evidences of extensive and prolonged erosion, and the increasing numbers of diverse and unfamiliar fossils found during Darwin's time were making this assertion seem less and less likely. For example, the great geologist Charles Lyell (1797-1895), whose works Darwin read eagerly while he was sailing on the *Beagle*, outlined for the first time the story of an ancient world of plants and animals in flux. In this world, some species were constantly becoming extinct while others were emerging. It was this world that Darwin sought to explain.

**TABLE 1-1    DARWIN'S EVIDENCE THAT EVOLUTION OCCURS**

FOSSILS

1. Extinct species, such as the fossil armadillos of Figure 1-12 most closely resemble living ones in the same area, suggesting one had given rise to the other.
2. In rock strata (layers), progressive changes in characteristics can be seen in fossils from progressively older layers.

GEOGRAPHICAL DISTRIBUTION

3. Lands that have similar climates, such as Australia, South Africa, California, and Chile, have unrelated plants and animals, indicating that differences in environment are not creating the diversity directly.
4. The plants and animals of each continent are distinctive, although there is no reason why special creation should create this association: all South American rodents belong to a single group, structurally similar to the guinea pigs, whereas most of the rodents found elsewhere belong to other groups.

OCEANIC ISLANDS

5. Although oceanic islands have few species, those they have are very often unique ("endemic") and show relatedness to one another, such as the tortoises of the Galapagos (see Figure 1-14). This suggests that the tortoises and other groups of endemic species formed after their ancestors reached the islands and are therefore directly related to one another.
6. Species on oceanic islands show strong affinities to those on the nearest mainland. Thus the finches of the Galapagos (such as the one in Figure 1-13, *A*) closely resemble a finch seen on the western coast of South America (Figure 1-13, *B*). The Galapagos finches do *not* resemble birds of the Cape Verde Islands, islands in the Atlantic Ocean off Africa that are very similar to the Galapagos. Darwin visited the Cape Verde Islands and many other island groups personally and was able to make such comparisons on the basis of his own observations.

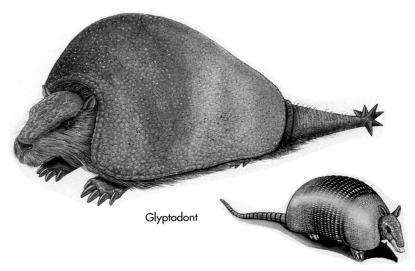

Glyptodont

Armadillo

*Figure 1-12*

**Reconstruction from a fossil of a glyptodont, a 2-ton South American armadillo, compared with a modern armadillo, which averages 10 pounds.** Finding the fossils of forms such as the glyptodonts and noting their similarity to living animals found in the same regions, Charles Darwin concluded that evolution had indeed taken place.

## What Darwin Saw

When the *Beagle* set sail, Darwin was fully convinced that species were unchanging and immutable. Indeed, he later wrote that it was not until 2 or 3 years after his return that he began to consider seriously the possibility that they could change. Nevertheless, during his 5 years on the ship, Darwin observed a number of phenomena that were of central importance to him in reaching his ultimate conclusion (Table 1-1). For example, in rich beds of fossils in southern South America, he observed fossils of extinct armadillos that were related directly to the armadillos that still lived in the same area (see Figure 1-12). Why would there be living

A

B

**Figure 1-14**

**Galapagos tortoises.** Tortoises with large, domed shells (**A**) are found in relatively moist habitats. The lower saddleback-type shells (**B**) in which the front of the shell is bent up, exposing the head and part of the neck, are found among the tortoises of dry habitats. Taken together, differences of these kinds make it possible to identify the races of tortoises that inhabit the different islands of the Galapagos.

A

B

**Figure 1-13**

**Darwin's finches.**
A One of Darwin's finches, the medium ground finch.
B The blue-black grassquit, which is found in grasslands along the Pacific Coast from Mexico to Chile. This bird may be the ancestor of Darwin's finches.

and fossil organisms, directly related to one another, in the same area, unless one had given rise to the other?

Repeatedly, Darwin saw that the characteristics of different species varied from place to place. These patterns suggested to him that organisms change gradually as they migrate from one area to another. On the Galapagos Islands, off the coast of Ecuador, Darwin encountered giant land tortoises. Surprisingly, these tortoises were not all identical. Indeed, local residents and the sailors who captured the tortoises for food could tell which island a particular animal had come from just by looking at it (Figure 1-14). This pattern of variation suggested that all of the tortoises were related, but had changed slightly in appearance after they had become isolated on the different islands.

In a more general sense, Darwin was struck by the fact that on these relatively young volcanic islands there was a profusion of living things, but that these plants and animals resembled those of the nearby coast of South America. If each one of these plants and animals had been created independently and simply placed on the Galapagos Islands, why did they not resemble the plants and animals of faraway Africa, for example? Why did they resemble those of the adjacent South American coast instead?

*The patterns of distribution and relationship of organisms that Darwin observed on the voyage of the Beagle ultimately made him certain that a process of evolution had been responsible for these patterns.*

### Darwin and Malthus

It is one thing to observe evolution, another to understand how it happens. Darwin's great achievement is his perception that evolution occurs because of natural selection. Of key importance to the development of Darwin's insight was his study of Thomas Malthus' *Essay on the Principles of Population*. In this book Malthus pointed out that populations of plants and animals, including humans, tend to increase geometrically, whereas, in our case, our food supply increases only arithmetically. A **geometric progression** is one in which the elements progress by a constant factor, as in 2, 6, 18, 54, and so forth; each number is three times the preceding one. An **arithmetic progression,** in contrast, is one in which the elements increase by a constant difference, as in 2, 6, 10, 14, and so forth. Each number is 4 more than the preceding one.

Virtually any kind of animal or plant, if it could reproduce unchecked, would cover the entire surface of the world within a surprisingly short period. In fact, this does not occur; instead, populations of species remain more or less constant year after year because death intervenes and limits population numbers. Malthus' conclusion was very important in the development of Darwin's ideas. It provided the key ingredient that was necessary to Darwin in developing the hypothesis that evolution occurs by natural selection.

*A key contribution to Darwin's thinking was Malthus' concept of geometric population growth. The fact that real populations do not expand at this rate implies that nature acts to limit population numbers.*

### Natural Selection

Sparked by Malthus' ideas, Darwin saw that, although every organism has the potential to produce more offspring than are able to survive, only a limited number of offspring actually survive and produce their own offspring. Combining this observation with what he had seen on the voyage of the *Beagle*, as well as with his own experiences in breeding domestic animals, Darwin made the key association: *those individuals which possess superior physical, behavioral, or other attributes are more likely to survive than those which are not so well endowed.* In surviving, these individuals have the opportunity to pass on their favorable characteristics to their offspring. Because these characteristics will increase in the population, the nature of the population as a whole will gradually change. Darwin called this process **natural selection** (Figure 1-15), and he referred to the driving force he had identified as the *survival of the fittest.*

*Natural selection is the increase in succeeding generations of the traits of those organisms which leave more offspring. Its operation depends on the traits being inherited. The nature of the population gradually changes as more and more individuals with those traits appear.*

Darwin was thoroughly familiar with variation in domesticated animals and began his *On the Origin of Species* with a detailed discussion of pigeon breeding. He knew that varieties of pigeons and other animals, such as dogs, could be selected to exhibit certain characteristics. Once this had been done, the animals would breed true for the characteristics that had been concentrated in them. Darwin had also observed that the differences that could be developed between domesticated races or breeds in this way were often greater than those which separated wild species. The breeds of domestic pigeon are much more different from one another in various ways than are all of the hundreds of wild species of pigeons found throughout the world. Such relationships suggested to Darwin that evolutionary change could occur very rapidly under the right circumstances.

## PUBLICATION OF DARWIN'S THEORY

Darwin drafted the overall argument for evolution by natural selection in 1842 and continued to enlarge and refine it for many years. The stimulus that finally brought it into print was an essay that he received in 1858. A young English naturalist named Alfred Russel Wallace (1823-1913; Figure 1-16) sent the essay to Darwin from Malaysia: it concisely set forth the theory of evolution by means of natural selection! Like Darwin, Wallace had been greatly influenced in his development of this theory by reading Malthus' 1798 essay. After receiving Wallace's essay, Darwin arranged for a joint presentation of their ideas at a seminar in London and proceeded to complete his own book, on which he had been working for so long, for publication in what he considered an abbreviated version.

Darwin's book appeared in November, 1859, and caused an immediate sensation (Figure 1-17). Many people were deeply disturbed by the idea that human beings were closely related to apes, for example. This idea was not discussed by Darwin in his book, but it followed directly from the principles that he did outline. It had long been accepted that humans closely resembled the apes in all of their characteristics, but the possibility that there might be a direct evolutionary relationship between them was unacceptable to many people. Darwin's arguments for the theory of evolution by natural selection were so compelling, however, that his views were almost completely accepted within the intellectual community of Britain after the 1860s.

**Figure 1-16**
Alfred Russel Wallace in 1902.

## EVOLUTION AFTER DARWIN: TESTING THE THEORY

Darwin did more than propose a mechanism that explains how evolution has generated the diversity of life on earth. He also assembled masses of facts, otherwise seemingly without logic, that began to make sense when they were viewed in the light of his theory. After

*"Can we doubt . . . that individuals having any advantage, however slight, over others, would have the best chance of surviving and of procreating their kind? On the other hand, we may feel sure that any variation in the least degree injurious would be rigidly destroyed. This preservation of favorable variations, I call Natural Selection."*

**Figure 1-15**
From Charles Darwin's *On the Origin of Species*.

publication of his book, other biologists continued this process, and it soon became evident that the theory of evolution was supported by a wide variety of biological information gathered by many investigators. These observations included the following three points: (1) the fact that members of different biological groups often share common features, (2) the ways in which embryos (organisms developing from fertilized eggs) are more similar at earlier embryo stages and diverge through later stages, and (3) the increasing complexity that is observed to develop in the fossil record through time. In the century since Darwin, evolution has become the main unifying theme of the biological sciences. It provides one of the most important insights that human beings have achieved into their own nature and that of the earth on which they have evolved.

More than a century has now elapsed since Charles Darwin's death in 1882. During this period, the evidence supporting his theory has grown progressively stronger. There have also been many significant advances in our understanding of how evolution works. These advances have not altered the basic structure of Darwin's theory, but they have taught us a great deal more about the mechanisms by which evolution occurs.

**Figure 1-17**

**Darwin greets his "monkey ancestor."** In his time, Darwin was often portrayed unsympathetically, as in this drawing from an 1874 publication.

## The Fossil Record

Darwin predicted that the fossil record should yield intermediate links (Figure 1-18) between the great groups of organisms—for example, between fishes and the amphibians thought to have arisen from them and between reptiles and birds. The fossil record is now known to a degree that would have been unthinkable in the nineteenth century. Recent discoveries of microscopic fossils have extended the known history of life on earth back to more than 3.5 billion years ago. The discovery of other fossils has shed light on the ways in which organisms have evolved from the simple to the complex over the course of this enormous time span. For vertebrate animals—those with backbones—especially, the fossil record is rich and exhibits a graded series of changes in form, with the evolutionary parade visible for all to see.

## The Age of the Earth

In Darwin's day, some physicists argued that the earth was only a few thousand years old. This bothered Darwin because the evolution of all living things from some single original ancestor would have required a great deal more time. Using evidence obtained by studying rates of radioactive decay, we now know that the earth formed some 4.5 billion years ago.

## The Mechanism of Heredity

It was in the area of heredity that Darwin received some of his sharpest criticism. Since at that time no one had any concept of genes, of how heredity works, it was not possible for Darwin to explain completely how evolution occurs. Theories of heredity current in Dar-

win's day seemed to rule out the possibility of genetic variation in nature, a critical requirement of Darwin's theory. Genetics was established as a science only at the start of the twentieth century, 40 years after the publication of Darwin's *On the Origin of Species*. When the laws of inheritance became understood, the problem with Darwin's theory vanished, because the laws of inheritance (discussed in Chapter 11) explain in a neat and orderly way the production of new variations in nature required by Darwin's theory.

## Comparing Different Kinds of Organisms

Comparative studies of organisms have provided strong evidence for Darwin's theory. As vertebrates have evolved, for example, the same bones sometimes get put to different uses—and yet they can still be seen, betraying their evolutionary past (Figure 1-19). Thus the forelimbs in Figure 1-20 are all constructed from the same basic array of bones, modified in one way in the wings of bats, in another way in the fins of porpoises, and in yet other ways in the legs of frogs, horses, and humans. The bones have the same evolu-

**Figure 1-18**

Fossil of an early bird, *Archaeopteryx*. A well-preserved fossil of this bird, about 150 million years old, was discovered within 2 years of the publication of *On the Origin of Species*. *Archaeopteryx* provides an indication of the evolutionary relationship that exists between birds and reptiles.

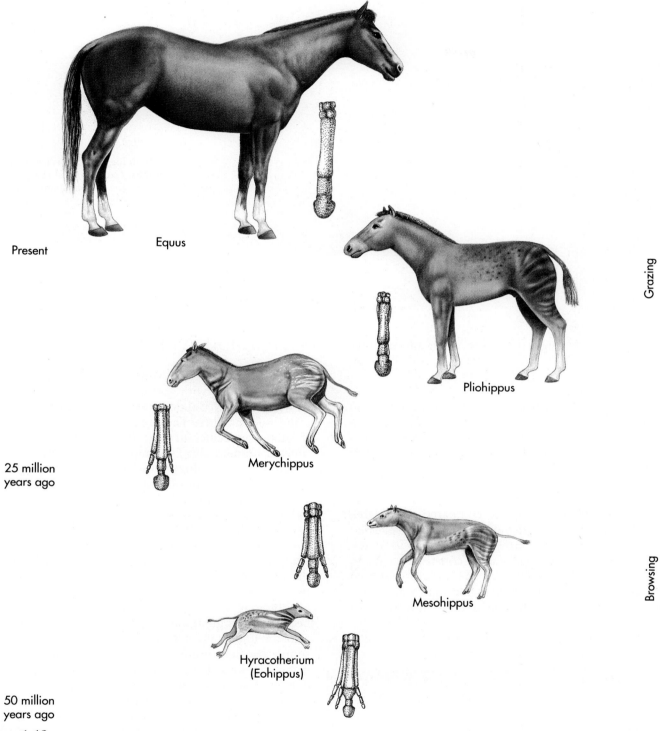

Present

Equus

Pliohippus

25 million
years ago

Merychippus

Mesohippus

Hyracotherium
(Eohippus)

50 million
years ago

### Figure 1-19

**Evolution of the horse.** Animals that included the earliest member of the evolutionary line *Hyracotherium* gave rise to several groups of mammals, including tapirs and rhinoceroses, in addition to the horses. As horses evolved, there was a progressive reduction in the number of toes.

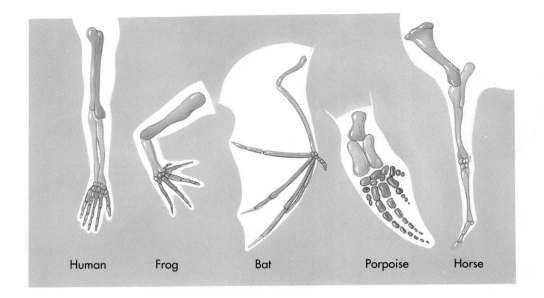

**Figure 1-20**

**Homology among vertebrate limbs.** Homologies between the forelimbs of four mammals and a frog show the ways in which the proportions of the bones have changed in relation to the particular way of life of the organism.

Human  Frog  Bat  Porpoise  Horse

tionary origin, although they now differ in structure and function.

## Molecular Biology

Biochemical tools have become of major importance in our efforts to reach a better understanding of how evolution has occurred. Within the last few years, for example, evolutionists have begun to "read" genes, much as you read this page, by recognizing in order the "letters" of the long DNA molecules that store genetic information. When the DNA sequences of different groups of animals or plants are compared, the degree of relationship between the groups can be specified more precisely than by any other means. In many cases, detailed "family trees" can be constructed. When study of the DNA that encodes the structure of two different molecules leads to the same family tree, this provides strong evidence that the tree is telling an accurate story of the evolutionary history of the group. Indeed, it is often possible to measure the rates at which evolution is occurring in the different groups.

*In the century since Darwin proposed his theory, a large body of evidence contributed by many branches of science has supported his view that evolution has occurred and that it has taken place by means of the mechanism of natural selection that he proposed.*

## WHY IS BIOLOGY IMPORTANT TO YOU?

Biologists do more than simply write books about evolution. They live with gorillas, collect fossils, and listen to whales. They isolate viruses, grow mushrooms, and grind up fruit flies. They read the message encoded in the long molecules of heredity and count how many times a hummingbird's wings beat each second. In its broadest sense, biology is the study of living organisms, of the diverse array of living things that blankets our earth. A biologist tries to understand the sources of this rich diversity of life and in many cases to harness particular life-forms to perform useful tasks. Even the narrowest study of a seemingly unimportant life-form is a study in biological diversity, one more brushstroke in the painting that biologists have labored over for centuries.

Biology is one of the most interesting of subjects because of its great variety. But not only is it fun, it is also an important subject for you and for everyone, simply because biology will affect your future in many ways. The knowledge that biologists are gaining is fundamental to our ability to manage the world's resources in a sustainable manner, prevent or cure diseases, and improve the quality of our lives and those of our children and grandchildren. Biologists are working on many problems that critically affect our lives (Figure 1-21), from dealing with the demands of the world's rapidly expanding population to attempts to find ways to prevent cancer and autoimmune deficiency syndrome (AIDS). Because the activities of biologists

# Biological Issues Today

### Human Population Growth

There were about 10 million people alive in the world when people first spread across North America; today there are over 5 billion, straining the earth's ability to support them.

### Infectious Disease

AIDS is a serious and growing health problem worldwide. Other diseases are even more serious: malaria will kill over 3 million people this year, most in the underdeveloped countries of the world. Scientists are struggling to produce vaccines against these scourges.

### Use of Addictive Drugs

The use of dangerous addictive drugs, particularly powerful heroin and cocaine derivatives, is creating a nightmare in many cities. Other dangerous but legal drugs are also in widespread use, such as cigarettes and other tobacco products that kill thousands by causing lung cancer.

### Industrial Alteration of the Environment

The destruction of the atmosphere's ozone, the creation of acid rain, the greenhouse effect caused by increasing concentrations of $CO_2$ in the atmosphere, the pollution of rivers and underground water supplies, the dilemma of how to dispose of radioactive wastes—all of these problems require urgent attention.

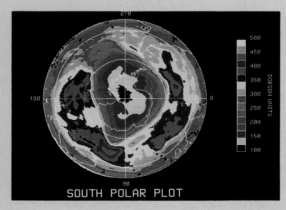

*Figure 1-21*

**Biological issues today.** Biologists face many urgent problems that will challenge their knowledge.

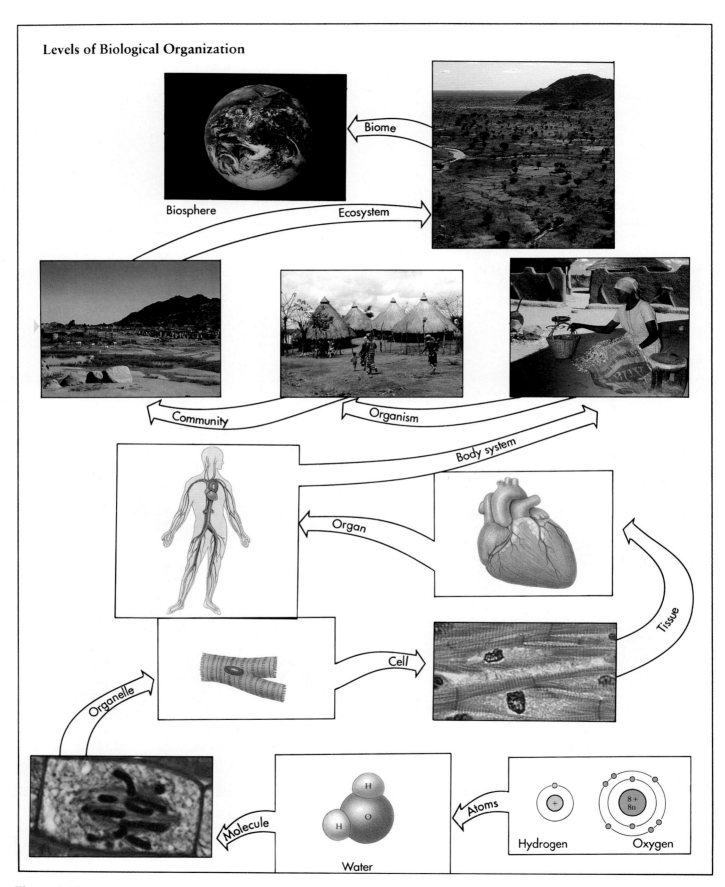

*Figure 1-22*

The levels of biological organization.

alter our lives in so many ways, an understanding of biology is becoming increasingly important for any educated person. It provides the key to the perpetuation of life on earth.

## HOW THIS TEXT IS ORGANIZED TO TEACH YOU BIOLOGY

In the century since Darwin's publication of *On the Origin of Species,* biology has exploded as a science, presenting today's student with a wealth of information and theory. There are many different ways in which a beginner can be introduced to biology and can see what is available to learn. In an introductory biology course you encounter a wealth of experiment and observation that you *could* learn, and from this you must select a small body of information that you *will* learn. Your target is the basic body of principles that unite biology as a science.

A very good way to begin your examination of basic biological principles is to focus on complexity. Arranging the wealth of information about biology in terms of the level of complexity leads to what has been called a "levels-of-organization" approach to the field (Figure 1-22).

The first half of this text is devoted to a description of the basic principles of biology. It uses a levels-of-organization framework to introduce different principles, each at the level where it is most easily understood. At the molecular, subcellular, and cellular levels of organization you will be introduced to the principles of **cell biology** and learn how cells are constructed and how they grow, divide, and communicate. At the organismal level you will learn the principles of **genetics,** which deals with the way in which the traits of an individual are transmitted from one generation to the next. At the population level you will study **evolution,** a field that is concerned with the nature of population changes from one generation to the next as a result of selection, and the way in which this has led to the biological diversity we see around us. Finally, at the community and global levels you will study **ecology,** which deals with how organisms interact with their environments and with one another to produce the complex communities characteristic of life on earth.

The second half of this book is devoted to an examination of organisms, the products of evolution. Organisms are classified into five kingdoms (Figure 1-23). Members of Monera—the bacteria—are prokaryotic, with relatively simple cellular structure; organisms in the other four kingdoms are eukaryotic, with a much more complex structure that we will examine in detail in the following chapters. Among eukaryotic organisms, Protists include a very diverse series of several dozen distinct evolutionary lines, most of them predominantly single-celled. From ancestors that would have been classified as protist, three primarily multicellular kingdoms—the plants (kingdom Plantae), the animals (kingdom Animalia), and the molds and mushrooms (kingdom Fungi)—have been derived independently. The diversity of living organisms is *incredible.* It is estimated that at least 10 million different kinds of plants, animals, and microorganisms exist. In the last section of the book we will take a detailed look at the vertebrates, the group of animals of which we are members. We will consider the vertebrate body and how it functions, focusing particularly on the human body, because this is information that is of interest and importance to us all.

As you proceed through this course, what you learn at one stage will give you tools to tackle the next. In the following chapter we will examine some simple chemistry. You are not subjected to chemistry first to torture you, but rather to make what comes later easier to comprehend. To understand lions and tigers and bears, you first need to know the basic chemistry that makes them tick, for they are chemical machines, as are you.

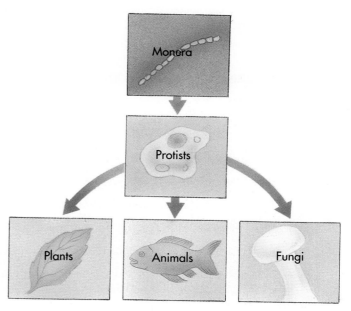

**Figure 1-23**
**The five kingdoms of life.**

## ■ SUMMARY

1. Biology is a science that attempts to describe and understand both the unity and the diversity of life.

2. Movement, sensitivity, development, death, and complexity are all characteristics of many living things, but they are not sufficient to define life, either singly or in combination.

3. Cellular organization, growth and metabolism, reproduction, homeostasis, and heredity characterize all living things, and together they define life. Of these, heredity is perhaps the key characteristic.

4. Science is the determination of general principles from observation and experiment.

5. Scientists test the validity of alternative hypotheses about a given phenomenon by attempting to reject some of the alternatives on the basis of experimentation.

6. Hypotheses that are supported by a large body of evidence are called theories. Unlike the everyday use of the word, the term "theory" in science refers to what we are most sure about.

7. Because even a theory is accepted only provisionally, there are no sure truths in science, no propositions that are not subject to change.

8. One of the most central theories of biology is Darwin's theory that evolution occurs by natural selection. Proposed over a hundred years ago, this theory has stood up well to a century of testing and questioning.

9. On an extended voyage, particularly while studying the animals and plants of oceanic islands, Darwin accumulated a wealth of evidence that evolution has occurred.

10. Sparked by Malthus' ideas, Darwin proposed that evolution occurs as a result of natural selection. Some individuals have heritable traits that let them produce more offspring in a given kind of environment than do other individuals who lack these traits. As a result of this process, the traits will increase in frequency through time.

11. A wealth of evidence since Darwin's time has supported his twin proposals: that evolution occurs and that its agent is natural selection. Together these two hypotheses are now usually referred to as Darwin's theory of evolution.

12. Biology may be considered at many levels of organization. This text treats basic principles first and then considers biological diversity.

## REVIEW

1. List six fundamental properties shared by all organisms on earth.

2. _____ is the complex linear molecule responsible for heredity.

3. A scientific _____ is a hypothesis that is supported by a great quantity of data.

4. Current evidence indicates that the earth is approximately _____ years old.

5. The arm of a man and the fin of a porpoise are considered to be _____ structures, made of the same bones, which have been modified in size and shape for different uses.

6. How many of the five kingdoms of living things contain multicellular organisms?

# SELF-QUIZ

1. It is easy to think of ways that animals respond to stimuli, but plants also respond to stimuli. Which of the following are examples of plants responding to a stimulus? (Give three.)
   (a) Plants grow toward light.
   (b) The roots of plants generally grow downward.
   (c) Cultivated plants sometimes escape from cultivation.
   (d) The trap of a Venus flytrap closes on a fly or other insect that has landed on it.
   (e) Plants die if they get no water.

2. A natural consequence of heredity is:
   (a) development.
   (b) growth.
   (c) adaptation.
   (d) death.
   (e) movement.

3. Hypotheses that are consistent with the available data are _____ accepted by scientists.
   (a) faithfully
   (b) provisionally
   (c) never
   (d) always
   (e) sometimes

4. Your dog suddenly drops to the ground in front of you. Five hypotheses suggest themselves to you: (1) your dog wants to play; (2) your dog has gone to sleep; (3) your dog is playing a trick on you; (4) your dog is chasing a bug; or (5) your dog just died. To find out what is going on, you try to eliminate one or more of these hypotheses by conducting an experiment. Which of the following experiments would eliminate at least one hypothesis, whatever the outcome?
   (a) Ask the dog why it dropped to the ground.
   (b) Push the the dog and see if it responds.
   (c) Look and see if you can detect a bug.
   (d) Wait and see if the dog moves.
   (e) Feed the dog.

5. Which of the following is *not* one of Darwin's pieces of evidence that evolution occurs?
   (a) The earth was created in 4004 BC.
   (b) Extinct species most closely resemble living species in the same area.
   (c) Species on oceanic islands show strong similarities to species on the nearest mainland.
   (d) The plants and animals of each continent are distinctive.
   (e) Progressive changes in characteristics of plants and animals can be seen in successive rock layers.

6. Darwin was greatly influenced by an essay written by _____, which pointed out that population growth is geometric, whereas increase in food is arithmetic.
   (a) Emma Wedgewood
   (b) Alfred R. Wallace
   (c) James Usher
   (d) Thomas Malthus
   (e) Charles Lyell

# THOUGHT QUESTIONS

1. It is sometimes argued that Darwin's reasoning is circular, that he first defined the "fittest" individuals as those which leave the most offspring and then turned around and said that the fittest survive preferentially (that is, leave the most offspring). Do you think this is a fair criticism of Darwin's theory as described in this chapter?

2. On the Galapagos Islands, Darwin saw a variety of different kinds of finches, but few other small birds. Imagine that you are visiting another group of islands about as far away from the South American mainland as are the Galapagos, but upwind, so that no birds travel between the two island groups. Do you expect that on your visit to this second island group you will find a variety of finches? (Comment on how your knowledge of the birds of the Galapagos, as discussed in this text, aids you in predicting what you will find on the second island group, if indeed it does.)

3. Malthus' *Essay on the Principles of Population* pointed out that populations of both plants and animals tend to increase geometrically. What is the next number in this geometric progression: 2, 8, 32, 128, _____?

# FOR FURTHER READING

ATTENBOROUGH, D.: *Life on Earth*, Little, Brown, & Co., Boston, 1979. The companion volume to a television series, a history of nature from the emergence of the first tiny, one-celled organisms to the appearance of upright human beings. Filled with interesting evolutionary stories and exceptional photographs.

BOWEN, B.W., A.B. MEYLAN, and J.C. AVISE: "An Odyssey of the Green Sea Turtle: Ascension Island Revisited," *Proceedings of the National Academy of Sciences USA*, vol. 86, 1989, pages 573-576. Although advanced, this article presents clearly the latest results of research on the fascinating annual migration of these animals.

DARWIN, C.R.: *On The Origin of Species by Means of Natural Selection, or the Preservation of Favoured Races in the Struggle for Life*, Cambridge University Press, New York, 1975 reprint. One of the most important scientific books of all time, Darwin's long essay is still comprehensible and interesting to modern readers.

DARWIN, C.R.: *The Voyage of the Beagle*, Natural History Press, Garden City, N.Y., 1962 reprint. Darwin's own account of his observations and adventures during his famous 5-year voyage.

FUTUYMA, D.: *Science on Trial: The Case for Evolution*, Pantheon Books, New York, 1983. An excellent exposition of the basic reasons that the creationist argument is flawed by serious errors.

GOULD, S.: "Darwinism Defined: The Difference Between Fact and Theory," *Discover*, January 1987, pages 64-70. A clear account of what biologists do and do not mean when they refer to the theory of evolution.

GOULD, S.: *Wonderful Life. The Burgess Shale and the Nature of History*, W.W. Norton & Company, Inc., New York, 1989. A marvelous book about the early evolution of animals, and about evolution in general.

IRVINE, W.: *Apes, Angels, and Victorians*, McGraw-Hill Book Co., New York, 1954. The story of Darwin and the early years of the theory of evolution; beautifully written.

MOORE, J.A.: "Science as a Way of Knowing—Evolutionary Biology," *American Zoologist*, vol. 23, 1983, pages 1-68. An outstanding exposition of the whole field of evolution.

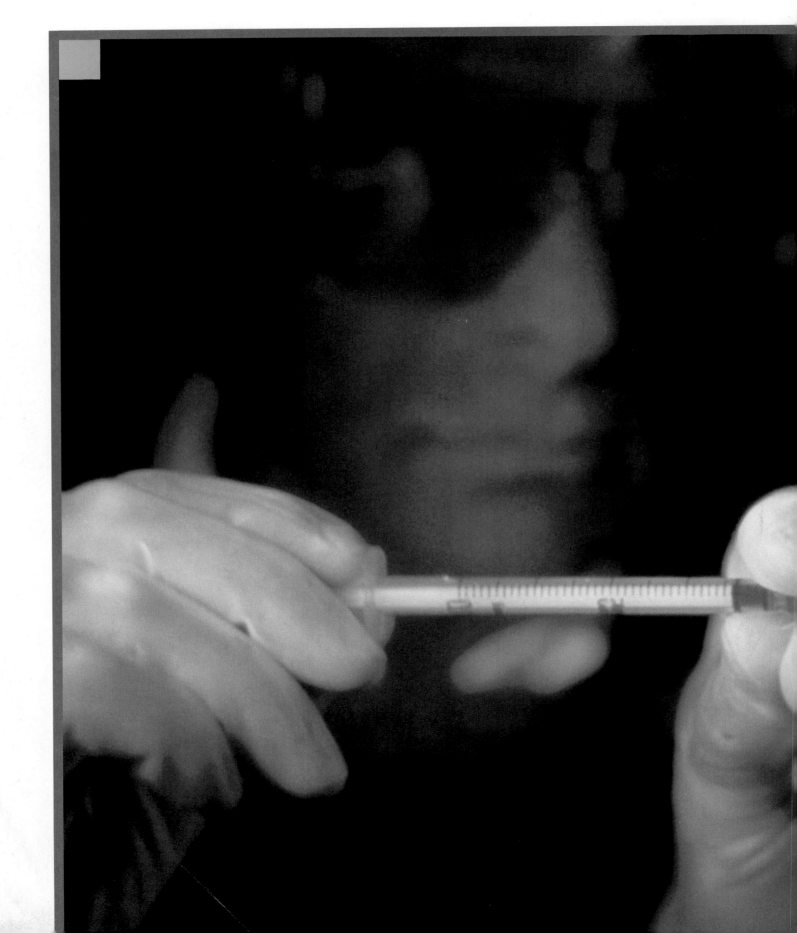

# The Chemistry of Life

This biologist is carefully extracting the flowing yellow material, a complex chemical called DNA, from the pink tube. The DNA he is recovering carries the genetic information of a plant resistant to antibiotics, and by studying it we may be able to greatly improve crop plants.

# THE CHEMISTRY OF LIFE

## Overview

You and I and all organisms are made up of molecules, which are collections of atoms bound to one another. Our bodies are built by assembling large molecules, much as a house is assembled from prefabricated building blocks. All of the molecules formed by living organisms contain the element *carbon*. Some of the molecular building blocks are long polymers, chains of similar units joined in a row. Among the polymers that make up the bodies of organisms are starches (used to store chemical energy), proteins (molecules that speed up specific chemical reactions), and nucleic acids (molecules in which hereditary information is stored).

Biology is the study of life, of ants and polar bears and roses. To start our study of biology, we begin with a brief look into chemistry—not a lot, just a taste. Organisms are chemical machines; to understand them, we must start by considering chemistry. Organisms are composed of molecules, which are collections of smaller units called atoms that are bound to one another. All atoms now in the universe are thought to have been formed long ago, as the universe itself evolved. Every carbon atom in your body was created in a star.

## ATOMS: THE STUFF OF LIFE

All matter is composed of small particles called **atoms** (Figure 2-1). Atoms are very small and hard to study,

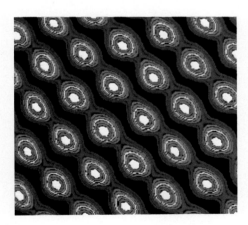

**Figure 2-1**

**Individual atoms on the surface of a silicon crystal.** A newly developed technique, tunneling microscopy, was used to take this remarkable picture.

and for a long time it was difficult for scientists to figure out their structure. It was only early in this century that experiments were carried out that suggested the first vague outlines of what an atom is like. We now know a great deal about the complexities of atomic structure, but the simple view proposed in 1913 by the Danish physicist Niels Bohr provides a good starting point (Figure 2-2). Bohr proposed that every atom possesses an orbiting cloud of tiny subatomic particles called **electrons** whizzing around the core like planets of a miniature solar system. At the center of each atom is a small, very dense nucleus formed of two other kinds of subatomic particles, **protons** and **neutrons.**

Within the nucleus, the cluster of protons and neutrons is held together by subatomic forces that work only over very short distances. Each proton carries a positive (+) charge. The number of charged protons (atomic number) determines the chemical character of the atom, because it dictates the number of electrons orbiting the nucleus and available for chemical activity: there is one electron for each proton. Neutrons are similar to protons in mass, but as their name implies, they are neutral and possess no charge. The atomic mass of an atom consists of the combined weight of all of its protons and neutrons. Atoms that occur naturally on earth contain from 1 to 92 protons and up to 146 neutrons (Figure 2-3).

### Isotopes

Different kinds of atoms are called elements. Formally speaking, an element is any substance that cannot be broken down to any other substance by ordinary chemical means. Each element is made up of one kind of atom, but may contain several versions of it.

Atoms that have the same number of protons but different numbers of neutrons are called **isotopes.** Isotopes of an atom differ in atomic mass but have similar

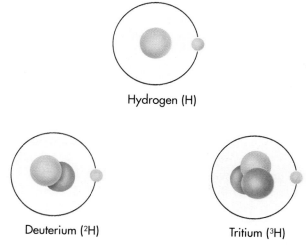

Hydrogen (H)

Deuterium (²H)                    Tritium (³H)

***Figure 2-2***

**Atoms.** The smallest atom is hydrogen (atomic mass, 1), whose nucleus consists of a single proton. Hydrogen also has two naturally occurring isotopic forms, which possess neutrons as well as the single proton in the nucleus: deuterium (one neutron) and tritium (two neutrons).

chemical properties. Most elements in nature exist as mixtures of different isotopes. There are, for example, three isotopes of the element carbon, all of which possess six protons (see Figure 2-3). The most common isotope of carbon has six neutrons. Because its total mass is 12 (6 protons plus 6 neutrons), it is referred to as carbon-12. Over 99% of the carbon in nature is carbon-12. Most of the rest is carbon-13, with 7 neutrons. The third isotope, carbon-14, is rare in nature. Unlike the other two isotopes of carbon, carbon-14 is unstable. Its nucleus tends to break up into elements with lower atomic numbers. This process is called radioac-

tive decay. Isotopes such as carbon-14 that decay in this fashion are said to be **radioactive**. By determining the ratios of the different isotopes of carbon and other elements in samples of biological origin and in rocks, scientists are able to make absolute determinations of the times when these materials formed. In addition, stable isotopes, introduced experimentally, have been used effectively to study dynamic processes in the physiology of individual organisms and in ecosystems.

## Electrons

The positive charges in the nucleus of an atom are counterbalanced by negatively (−) charged electrons orbiting the atomic nucleus at various distances. The negative charge of one electron exactly balances the positive charge of one proton. Thus atoms with the same number of protons and electrons have no net charge. An atom in which the number of protons in the nucleus is the same as the number of orbiting electrons is known as a neutral atom.

Electrons have very little mass (only $\frac{1}{1840}$ of the mass of a proton). Of all the mass contributing to your weight, the portion contributed by electrons is less than the mass of your eyelashes. Electrons stay in their orbits because they are attracted to the positive charge of the nucleus. This attraction is sometimes overcome by other forces, and one or more electrons fly off, lost to the atom. Sometimes atoms gain additional electrons. Atoms in which the number of electrons does not equal the number of protons are known as **ions,** and ions do carry an electrical charge. For example, an atom of sodium (Na) that has lost an electron becomes a positively charged sodium ion ($Na^+$) because the positive charge of one of the protons is not balanced by the negative charge of an electron.

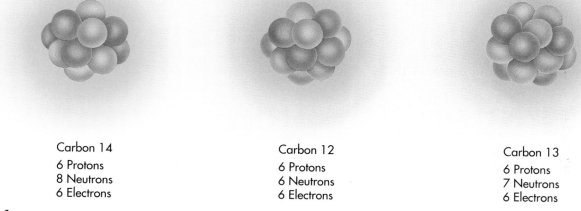

Carbon 14

6 Protons
8 Neutrons
6 Electrons

Carbon 12

6 Protons
6 Neutrons
6 Electrons

Carbon 13

6 Protons
7 Neutrons
6 Electrons

***Figure 2-3***

**The three most abundant isotopes of carbon.**

## Electrons Determine What Molecules are Like

The key to the chemical behavior of atoms lies in the arrangement of the electrons that spin around them. It is convenient to visualize individual electrons as spinning in well-defined circular orbits around a central nucleus, the way Earth, Mars, and Venus circle the sun. But such a simple picture is not realistic. Physicists have learned that it is impossible to locate precisely the position of any individual electron at any given time. In fact, theories predict that, at any given instant, a particular electron can be located anywhere from close to the nucleus to infinitely far away from it.

However, a particular electron is not equally likely to be located at all positions. Some locations are much more probable than others. For this reason, it is possible to say where an electron is *most likely* to be located. The volume of space around a nucleus where an electron is most likely to be found is called the **orbital** of that electron. Atoms can have many electron orbitals. Some are simple spheres enclosing the nucleus like a wrapper. Others resemble dumbbells and other complex shapes.

How far away from the nucleus are the orbiting electrons? Very far. Almost all the volume of an atom is empty space. If the nucleus of an atom were the size of an apple, the orbit of the nearest electron would be more than a mile out. It is for this reason that the elec-trons of an atom determine its chemical behavior—the nuclei of two atoms never come close enough to each other in nature to interact. That is why isotopes of an element, all of which have the same arrangement of electrons, behave the same way chemically.

## Energy within the Atom

Because electrons carry negative charges, they are attracted to the positively charged nucleus, and it takes work to keep them in orbit, just as it takes work to hold an apple in your hand when gravity is pulling the apple down toward the ground. The apple in your hand is said to possess energy, the ability to do work, because of its position—if you were to release it, the apple would fall. Similarly, electrons have energy in relation to their position; energy of this sort is called potential energy. It takes work to oppose the attraction of the nucleus and move the electron further out, and so moving an electron out to a more distant orbit requires an input of energy and results in an electron with greater potential energy. For the same reason, a bowling ball released from the top of a building hits with greater force than one dropped from only a meter. Moving an electron in toward the nucleus has the opposite effect; energy is released, and the electron ends up with less potential energy (Figure 2-4).

Sometimes an electron is transferred during a chemical reaction from one atom to another. The loss of an electron is called **oxidation;** the gain of an electron is called **reduction** (Figure 2-5). It is important to realize that when an electron is transferred in this way, it keeps its energy of position. In living organisms, chemical energy is stored in high-energy electrons that are frequently transferred from one atom to another.

In a schematic drawing of an atom (Figure 2-6), the nucleus is represented as a small circle with the

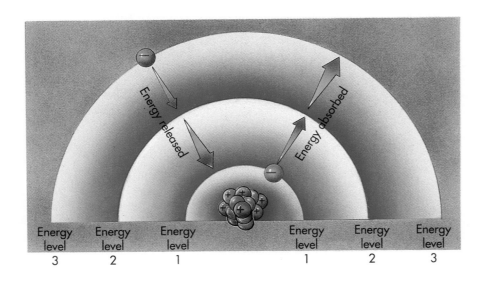

**Figure 2-4**

**Atomic energy levels.** When an electron absorbs energy, it moves to higher energy levels further from the nucleus. When an electron releases energy, it falls inward to lower energy levels.

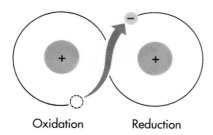

Oxidation          Reduction

*Figure 2-5*

**Oxidation is the loss of an electron; reduction is the gain of one.**

number of protons and neutrons indicated. The orbitals are depicted as concentric rings. In helium, for example, there is a single orbital containing two electrons. A given orbital may contain no more than two electrons. Such schematics are only diagrams; the actual orbitals have complex, three-dimensional shapes.

Most atoms have more than one orbital. When electrons from two different orbitals are the same distance from the nucleus, they are placed on the same ring by convention. When electrons in two orbitals are different distances from the nucleus, they are placed in separate concentric rings. These rings are called **energy levels,** or **shells.** The farther an electron is from the nucleus, the more energy it has, as if it were spinning faster.

## ELEMENTS AND MOLECULES

Much of the earth's core is thought to consist of atoms of iron, nickel, and other heavy elements. An **element** is defined as a substance that cannot be separated into

different substances by ordinary chemical methods. The composition of the earth's crust is quite different from the core, the crust being made up primarily of lighter elements (Table 2-1). Thus, by weight, 74.3% of the earth's crust (its land, oceans, and atmosphere) consists of oxygen and silicon. Most of these elements are combined, stable associations of atoms called **molecules.**

## CHEMICAL BONDS HOLD MOLECULES TOGETHER

A molecule is a group of atoms held together by energy. The energy acts as "glue," ensuring that the various atoms stick to one another. The force holding two atoms together is called a **chemical bond.** The force can result from the attraction of opposite charges, called an ionic bond, or from the sharing of one or more pairs of electrons, a covalent bond. Other, weaker kinds of bonds also occur.

### Ionic Bonds Form Crystals

**Ionic bonds** form when atoms are attracted to one another by opposite electrical charges. Common table salt, sodium chloride (NaCl), is a lattice of ions in which atoms are held together by ionic bonds. Sodium atoms (Na) have 11 electrons (Figure 2-7). Two of these are in the inner energy level, eight are at the next level, and one is at the outer energy level. The outer electron is unpaired ("free") and has a strong tendency to form a pair. A stable configuration is achieved if the outer electron is lost. The loss of this electron results in the formation of a positively charged sodium ion $(Na^+)$.

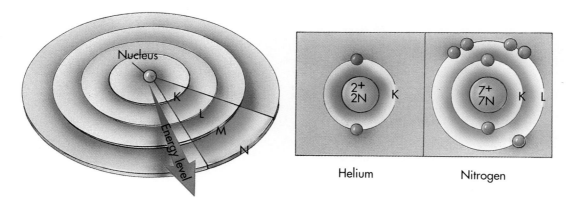

Helium          Nitrogen

*Figure 2-6*

**Electron energy levels for helium and nitrogen.** The electrons are indicated by orange dots. Each concentric circle represents a different distance from the nucleus and thus a different electron energy level.

**TABLE 2-1    THE MOST COMMON ELEMENTS ON EARTH AND THEIR DISTRIBUTION IN THE HUMAN BODY**

| ELEMENT | SYMBOL | ATOMIC NUMBER | APPROXIMATE PERCENT OF EARTH'S CRUST BY WEIGHT | PERCENT OF HUMAN BODY BY WEIGHT | IMPORTANCE OR FUNCTION |
|---|---|---|---|---|---|
| Oxygen | O | 8 | 46.6 | 65.0 | Required for cellular respiration; component of water |
| Silicon | Si | 14 | 27.7 | Trace | — |
| Aluminum | Al | 13 | 6.5 | Trace | — |
| Iron | Fe | 26 | 5.0 | Trace | Critical component of hemoglobin in the blood |
| Calcium | Ca | 20 | 3.6 | 1.5 | Component of bones and teeth; triggers muscle contraction |
| Sodium | Na | 11 | 2.8 | 0.2 | Principal positive ion bathing cells; important in nerve function |
| Potassium | K | 19 | 2.6 | 0.4 | Principal positive ion in cells; important in nerve function |
| Magnesium | Mg | 12 | 2.1 | 0.1 | Critical component of many energy-transferring enzymes |
| Hydrogen | H | 1 | 0.14 | 9.5 | Electron carrier; component of water and most organic molecules |
| Manganese | Mn | 25 | 0.1 | Trace | — |
| Fluorine | F | 9 | 0.07 | Trace | — |
| Phosphorus | P | 15 | 0.07 | 1.0 | Backbone of nucleic acids; important in energy transfer |
| Carbon | C | 6 | 0.03 | 18.5 | Backbone of organic molecules |
| Sulfur | S | 16 | 0.03 | 0.3 | Component of most proteins |
| Chlorine | Cl | 17 | 0.01 | 0.2 | Principal negative ion bathing cells |
| Vanadium | V | 23 | 0.01 | Trace | — |
| Chromium | Cr | 24 | 0.01 | Trace | — |
| Copper | Cu | 29 | 0.01 | Trace | Key component of many enzymes |
| Nitrogen | N | 7 | Trace | 3.3 | Component of all proteins and nucleic acids |
| Boron | B | 5 | Trace | Trace | — |
| Cobalt | Co | 27 | Trace | Trace | — |
| Zinc | Zn | 30 | Trace | Trace | Key component of some enzymes |
| Selenium | Se | 34 | Trace | Trace | — |
| Molybdenum | Mo | 42 | Trace | Trace | Key component of many enzymes |
| Tin | Sn | 50 | Trace | Trace | — |
| Iodine | I | 53 | Trace | Trace | Component of thyroid hormone |

The chlorine atom faces a similar dilemma. It has 17 electrons: 2 at the inner energy level, 8 at the next energy level, and 7 at the outer energy level. The outer energy level of the chlorine atom has an unpaired electron. The addition of an electron to the outer level causes the formation of a negatively charged chloride ion ($Cl^-$).

When placed together, metallic sodium and gaseous chlorine react swiftly and explosively, with the sodium atoms donating electrons to the chlorine atoms. The result is production of $Na^+$ and $Cl^-$ ions. Because opposite charges attract, (when all charges balance) electrostatic neutrality can be achieved by the association of these ions with each other. The ions aggregate, or come together, and form a crystal matrix, which has a precise geometry. Such aggregations are known as crystals of salt. If a salt such as $Na^+Cl^-$ is placed in water, the electrical attraction of the water molecules (for reasons we discuss later in this chapter) disrupts the forces holding the salt ions in their crystal matrix, causing the salt to dissolve into a roughly equal mixture of free $Na^+$ and free $Cl^-$ ions. Approximately 0.06% of the atoms in your body are free $Na^+$ or $Cl^-$ ions—about 40 grams, the weight of your fingernails.

*An ionic bond is an attraction between ions of opposite charge.*

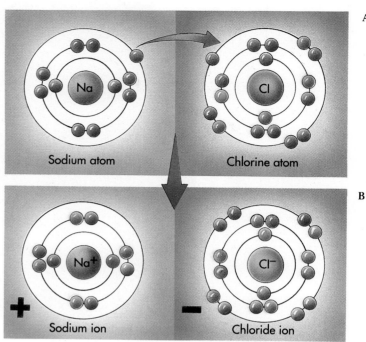

**Figure 2-7**

**A** The formation of ionic bonds in sodium chloride. When a sodium atom donates an electron to a chlorine atom, the sodium atom, lacking that electron, becomes a positively charged sodium ion; the chlorine atom, having gained an extra electron, becomes a negatively charged chloride ion.
**B** When water evaporates from solutions containing high concentrations of sodium chloride, a highly regular lattice of alternating $Na^+$ and $Cl^-$ ions forms—a crystal. You are familiar with these crystals as table salt.

## Covalent Bonds Build Stable Molecules

**Covalent bonds** form when two atoms share electrons. For example, each hydrogen atom has an unpaired electron. This is why the hydrogen atom is chemically unstable. When two hydrogen atoms are close enough to one another, however, each electron can orbit both nuclei. In effect, nuclei in close proximity are able to share their electrons. The result is a diatomic molecule (one with two atoms) of hydrogen gas ($H_2$) (Figure 2-8).

The diatomic hydrogen gas molecule that is formed as a result of this sharing of electrons is not charged. It still contains two protons and two electrons. Each hydrogen atom can be considered to have two orbiting electrons in the outer electron shell. Each shared outer-shell electron orbits both nuclei and therefore is included in the outer shell of *both* atoms. The relationship results in the pairing of the two free electrons. The two hydrogen atoms thus form a stable molecule. Note, however, that this stability is conferred by the electrons that orbit *both* nuclei; stability of this kind occurs only when the nuclei are very close. For this reason the strong chemical forces tending to pair electrons and to fill the outer energy level with the maximum number of electrons will act to keep the two hydrogen nuclei near one another. The bond between the two hydrogen atoms of diatomic hydrogen gas is an example of a covalent bond.

The bonds that hold the molecules of your body together are covalent bonds. Covalent bonds are far better suited to this task than are ionic ones because covalent bonds are *directional*—a covalent bond is formed between two specific atoms, whereas an ionic bond is formed between a charged atom and the electric field contributed by all nearby atoms of opposite charge. Ionic bonds can form regular crystals, but such crystals dissolve in water. Complex stable shapes require the more specific associations made possible by covalent bonds.

**Figure 2-8**

**Hydrogen gas.** Hydrogen gas in a diatomic molecule composed of two hydrogen atoms, each sharing its electron with the other. The flash of fire that consumed the *Hindenburg* occurred when the hydrogen gas used to inflate the airship combined explosively with oxygen gas in the air to form water.

*A covalent bond is a chemical bond formed by the sharing of one or more pairs of electrons.*

Covalent bonds can be very strong, that is, difficult to break. Covalent bonds that share *two* pairs of electrons, called **double bonds,** are stronger than covalent bonds sharing only one electron pair, called **single bonds.** Energy is required to form larger aggregations of atoms from smaller ones because of the need to establish new orbitals for the electrons. This energy is released when the bonds are broken. Covalent bonds are represented in chemical formulations as lines connecting atomic symbols. Each line between two bonded atoms represents the sharing of one pair of electrons. Hydrogen gas is thus symbolized H—H and oxygen gas, O═O.

Molecules are often made up of more than two atoms. One reason larger molecules may form is that a given atom is able to share electrons with more than one other atom. An atom that requires two, three, or four additional electrons to fill its outer energy level completely may acquire them by sharing its electrons with two or more other atoms. For example, carbon (C) atoms (atomic number 6) contain six electrons, two of them at the inner level and the other four in the outer shell. A carbon atom can form the equivalent of four covalent bonds. Because there are many ways that four covalent bonds may form, carbon atoms are able to participate in many different kinds of molecules, making carbon an ideal atom with which to construct the many molecules of living things.

Of the 92 kinds of atoms (elements) that form the crust of the earth, only 11 are common in living organisms. Table 2-1 lists the frequency with which various elements occur in the earth's crust and in the human body. Unlike the elements that occur most abundantly in the earth's crust, all of the elements common in living organisms are light. Each has an atomic number of less than 21 and thus a low mass. The great majority of the atoms in living things, for example, 99.4% of the atoms in the human body, are either nitrogen, oxygen, carbon, or hydrogen (Figure 2-9). You can remember these elements by their first letters, NOCH.

# THE CRADLE OF LIFE: WATER

If you were to count the atoms in your body, and then ask in what molecules those atoms are present, you would find that most of your atoms are parts of water molecules. The most common atoms in living things are oxygen and hydrogen atoms; the great majority of these are combined together in water molecules. Water has the chemical formula $H_2O$. As you will see, this seemingly simple molecule has many surprising properties. For example, of all the common molecules on earth, only water exists as a liquid at the relatively cool temperatures prevailing on the earth's surface (Figure 2-10). When life on earth was beginning, water, because it is a liquid at such temperatures, provided a medium in which other molecules could move around and interact without being bound by strong covalent or ionic bonds. Life evolved as a result of these interactions.

Life as it evolved on earth is inextricably tied to water (Figure 2-11). Three-fourths of the earth's surface is covered by water. You yourself are about two-thirds water, and you cannot exist long without it. All other organisms also require water. It is no accident that tropical rain forests are bursting with life, whereas deserts are almost lifeless except when water becomes temporarily plentiful, such as after a rainstorm. Farming is possible only in areas of the earth where rain is plentiful or water can be supplied by irrigation. No plant or animal can grow and reproduce in any but a water-rich environment.

The chemistry of life, then, is water chemistry. The way that life evolved was determined largely by the chemical properties of the water in which its evolution

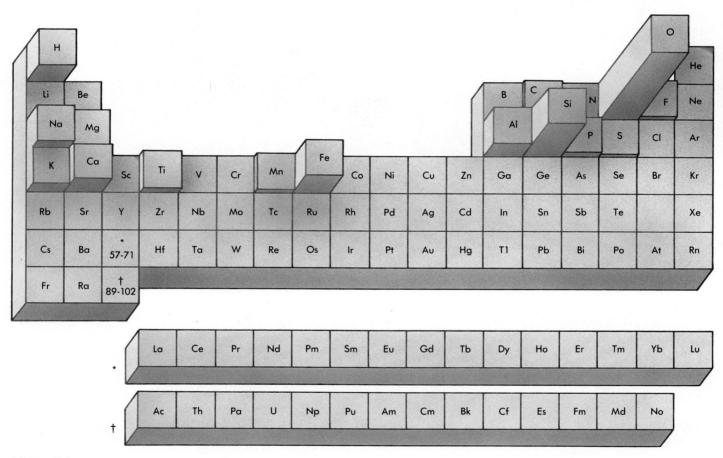

## Figure 2-9

**Periodic table of the elements.** In this representation, the frequency of elements that occur in the earth's crust in more than trace amounts is indicated in the vertical dimension. Elements found in significant amounts in living organisms are shaded in color. Many elements that are common on the surface of the earth, such as silicon and iron, are not present in living organisms in significant concentrations.

*Figure 2-10*

**Water takes many forms.** As a liquid, it fills our rivers and runs down over the land to the sea, sometimes falling in great cascades. The iceberg on which the penguins are holding their meeting was formed in Antarctica from huge blocks of ice breaking away into the ocean water. When water cools below 0° C, it forms beautiful crystals, familiar to us as snow and ice. However, water is not always plentiful. In a dry creek bed, there is no hint of water except for the broken patterns of dry mud.

**TABLE 2-2    THE PROPERTIES OF WATER**

| PROPERTY | EXPLANATION | EXAMPLE OF BENEFIT TO LIFE |
|---|---|---|
| High polarity | Polar water molecules are attracted to ions and polar compounds, making them soluble | Many kinds of molecules can move freely in cells, permitting a very diverse array of chemical reaction |
| High specific heat | Hydrogen bonds absorb heat when they break and release heat when they form, minimizing temperature changes | Water stabilizes body temperature, as well as that of the environment |
| High heat of vaporization | Many hydrogen bonds must be broken for water to evaporate | Evaporation of water cools body surfaces |
| Lower density of ice | Water molecules in an ice crystal are spaced relatively far apart because of hydrogen bonding | Because ice is less dense than water, lakes do not freeze solid, and they overturn in spring |
| Cohesion | Hydrogen bonds hold molecules of water together | Leaves pull water upward from roots; seeds swell and germinate |

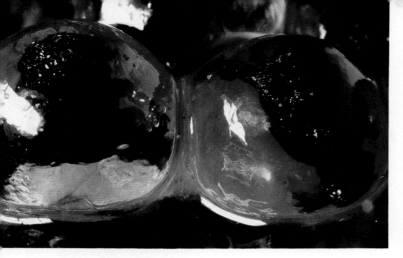

**Figure 2-11**

**Frogs about to hatch.** Life originated in the warm waters of the ancient earth, and many organisms, like these small frogs, seen through the walls of their eggs, begin life in water.

occurred (Table 2-2). The single most outstanding chemical property of water is its ability to form weak chemical associations with only 5% to 10% of the strength of covalent bonds. This one property of water, which derives directly from its structure, is responsible for much of the organization of living chemistry.

Water has a simple atomic structure—one oxygen atom bound by covalent single bonds to two hydrogen atoms. The resulting molecule is stable: it satisfies the tendency to fill the outer energy level with the maximum number of electrons, has no unpaired electrons, and does not carry a net electrostatic charge.

## Water Acts Like a Magnet

The remarkable story of the properties of water does not end here. The electron-attracting power of the oxygen atom (referred to by chemists as its electronegativity) is much greater than that of the hydrogen atom. As a result, the electron pair shared in each of the two single oxygen-hydrogen covalent bonds of a water molecule is more strongly attracted to the oxygen nucleus than to either of the hydrogen nuclei. Although electron orbitals encompass both the oxygen and hydrogen nuclei, the negatively charged electrons are far more likely, at any given moment, to be found near the oxygen nucleus rather than near one of the hydrogen nuclei. This relationship has a profoundly important result: the oxygen atom acquires a partial negative charge. It is as if the electron cloud were more dense in the neighborhood of the oxygen atom and less dense around the hydrogen atoms. This charge separation within the water molecule creates negative and positive electrical charges on the ends of the molecule (Figure 2-12). These partial charges are much less than the unit charges of ions.

The water molecule thus has distinct "ends," each with a partial charge, like the two poles of a magnet. Molecules such as water that exhibit charge separation are called **polar molecules** because of these magnetlike poles. Water is one of the most polar molecules known. *The polarity of water underlies its chemistry and thus the chemistry of life.*

*Much of the biologically important behavior of water results because the oxygen atom attracts electrons more strongly than do the hydrogen atoms, with the result that the water molecule has electron-rich (−) and electron-poor (+) regions, giving it magnetlike positive and negative poles.*

Polar molecules interact with one another. The partial negative charge at one end of a polar molecule is attracted to the partial positive charge of another polar molecule. This weak attraction is called a **hydrogen**

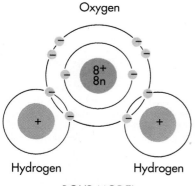

Oxygen

Hydrogen    Hydrogen

**BOHR MODEL**

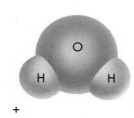

MOLECULAR MODEL

**Figure 2-12**

**Water.** Water has a simple molecular structure. Each molecule is composed of one oxygen atom and two hydrogen atoms. The oxygen atom shares a pair of electrons with each participating hydrogen atom, contributing one electron of each pair.

**bond.** Water forms a lattice of such hydrogen bonds. Each hydrogen bond is individually very weak and transient. A given bond lasts only 1/100,000,000,000 of a second. Although each bond is transient, a very large number of such hydrogen bonds can form, and the cumulative effects of very large numbers of these bonds can be enormous. The cumulative effect of very large numbers of hydrogen bonds is responsible for many of the important physical properties of water (Figure 2-13).

## Water is a Powerful Solvent

Water molecules gather closely around any molecule that exhibits an electrical charge, whether the molecule carries a full charge (ion) or a charge separation (polar molecule). For example, sucrose (table sugar) is com-

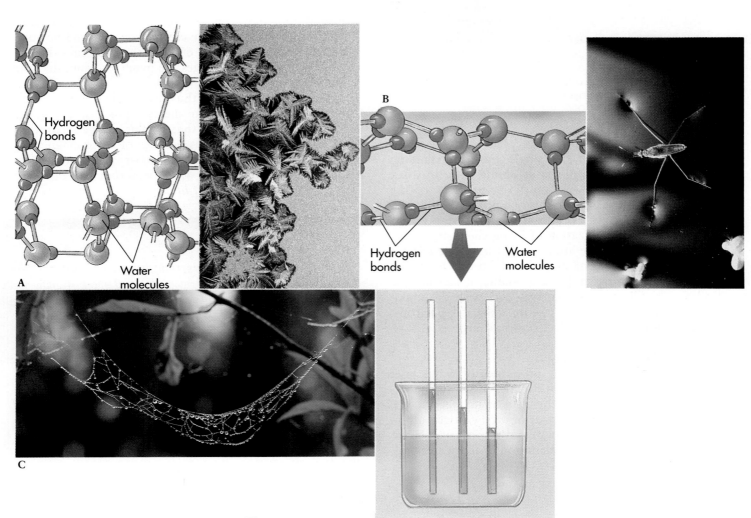

*Figure 2-13*

**Many of the physical properties of water depend on hydrogen bonding.**
**A** Ice formation. When water cools below 0° C, it forms a regular crystal structure in which the partial charges of each atom in the water molecule interact with opposite charges of atoms in other water molecules to form H bonds.
**B** Surface tension. Some insects, such as the water strider, literally walk on water. In this photograph you can see the dimpling its feet make on the water as its weight bears down on the surface. Because the surface tension of the water is greater than the force that one foot brings to bear, the water strider does not sink, but rather slides along.
**C** Adhesion. Capillary action will cause the water within a narrow tube to rise above the surrounding fluid; the adhesion of the water to the glass surface, drawing it upward, is stronger than the force of gravity, drawing it down. The narrower the tube, the greater the surface:volume ratio and the more the adhesion counteracts the force of gravity.

posed of molecules that contain slightly polar hydroxyl ($OH^-$) groups. A sugar crystal dissolves rapidly when it is placed in water (Figure 2-14) because water molecules can form hydrogen bonds with the polar hydroxyl groups of the sucrose molecules. Every time a sugar molecule dissociates or breaks away from the crystal, water molecules orient around it in a cloud. Such a **hydration shell,** formed by the water molecules, prevents every sucrose molecule from associating with other sucrose molecules. Similarly, hydration shells form around all polar molecules. Polar molecules that dissolve in water in this way are said to be **soluble** in water. Nonpolar molecules are not water-soluble. Oil is an example of a nonpolar molecule. Life originated in water not only because it is a liquid, but also because so many molecules are polar or ionized and thus are water-soluble.

## Water Organizes Nonpolar Molecules

Water molecules always tend to form the maximum number of hydrogen bonds possible. When nonpolar molecules, which do not form hydrogen bonds, are placed in liquid water, the water molecules act to exclude them. The water molecules preferentially form hydrogen bonds with other water molecules. The nonpolar molecules are forced to associate with one another, minimizing their disruption of the hydrogen bonding of water; they are thus crowded together. This is why oil and water do not mix but always separate. It seems almost as if the nonpolar compounds shrink from contact with the water, and for this reason they are called **hydrophobic** (Greek *hydros,* water + *phobos,* hating; "water-hated" might be a more apt description). The tendency for nonpolar molecules to band together in water solution is called **hydrophobic bonding.** Hydrophobic forces determine the three-dimensional shapes of many biological molecules, which are often surrounded by water within organisms.

## Water Ionizes

The covalent bonds of water sometimes break spontaneously. When this happens, one of the protons (hydrogen atom nuclei) dissociates from the molecule. Because the dissociated proton lacks the negatively charged electron that it had shared in the covalent bond with oxygen, its own positive charge is not counterbalanced; it is a positively charged hydrogen ion, $H^+$. The remaining bit of the water molecule retains the shared electron from the covalent bond and has one less proton to counterbalance it; it is a negatively charged hydroxyl ion ($OH^-$). This process of spontaneous ion formation is called **ionization.**

$$H_2O \rightarrow OH^- + H^+$$

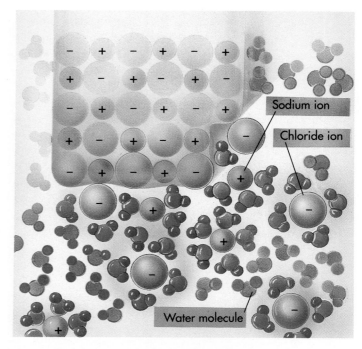

***Figure 2-14***

**Why salt dissolves in water.** When a crystal of table salt dissolves in water, individual $Na^+$ and $Cl^-$ ions break away from the salt lattice and become surrounded by water molecules. Surrounded by water in this way, $Na^+$ and $Cl^-$ ions never reenter the salt lattice.

Any substance that ionizes to form $H^+$ ions when dissolved in water is called an **acid.** A convenient indication of acid strength is the **pH scale** (Figure 2-15). pH is normally expressed as a positive number and is determined by taking the negative value of the exponent of the hydrogen concentration $H^+$. Water has a hydrogen concentration of $10^{-7}$ moles per liter (a mole is the number of grams of a substance that equals its molecular weight) and a pH of 7.

A mole is the atomic weight of a substance expressed in grams. In the case of $H^+$, the atomic weight equals 1, and a mole of $H^+$ ions would weigh 1 gram. The molar concentration of hydrogen ions in pure water, or number of moles per liter, 1/10,000,000 (more easily written as $10^{-7}$), results from the fact that roughly 1 out of each 550 million water molecules is ionized at any instant in time.

The stronger an acid is, the more $H^+$ ions it produces and the *lower* (smaller) its pH. Hydrochloric acid (HCl), which is abundant in your stomach, ionizes completely, so the molar concentration of $H^+$ in water containing $\frac{1}{10}$ of a mole of hydrochloric acid per liter

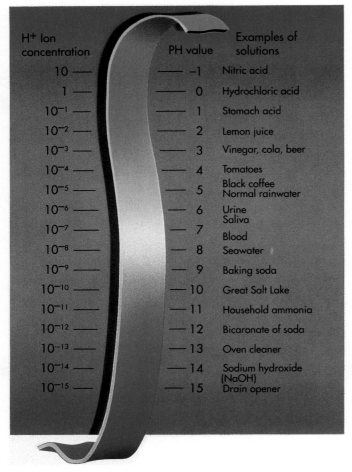

| H⁺ Ion concentration | PH value | Examples of solutions |
|---|---|---|
| 10 | -1 | Nitric acid |
| 1 | 0 | Hydrochloric acid |
| $10^{-1}$ | 1 | Stomach acid |
| $10^{-2}$ | 2 | Lemon juice |
| $10^{-3}$ | 3 | Vinegar, cola, beer |
| $10^{-4}$ | 4 | Tomatoes |
| $10^{-5}$ | 5 | Black coffee / Normal rainwater |
| $10^{-6}$ | 6 | Urine / Saliva |
| $10^{-7}$ | 7 | Blood |
| $10^{-8}$ | 8 | Seawater |
| $10^{-9}$ | 9 | Baking soda |
| $10^{-10}$ | 10 | Great Salt Lake |
| $10^{-11}$ | 11 | Household ammonia |
| $10^{-12}$ | 12 | Bicaronate of soda |
| $10^{-13}$ | 13 | Oven cleaner |
| $10^{-14}$ | 14 | Sodium hydroxide (NaOH) |
| $10^{-15}$ | 15 | Drain opener |

**Figure 2-15**

**The pH scale.** A fluid is assigned a value according to the number of hydrogen ions present in a liter of that fluid. The scale is logarithmic, so that a change of only 1 means a tenfold change in the concentration of hydrogen ions; thus lemon juice is 100 times more acidic than tomato juice, and seawater is 10 times more basic than pure water.

is $10^{-1}$ moles per liter, corresponding to a pH of 1. Some acids such as nitric acid are even stronger, although such very strong acids are rarely found in living systems. The pH of champagne, which bubbles because of the carbonic acid dissolved in it, is about 2.

*pH refers to the relative concentration of H⁺ ions in a solution. The numerical value of the pH is the negative of the exponent of the molar concentration. Low pH values indicate high concentrations of H⁺ ions (acids), and high pH values indicate low concentrations.*

Many of the processes on which our everyday lives depend form acids, often with destructive effects because acids are chemically reactive. For example, a lot of the energy that fuels industry is obtained by burning the plentiful form of coal that contains high concentrations of sulfur. The smoke from this burning is released from tall smokestacks so that the sulfur stench and pollution do not disturb the local inhabitants, but rather are blown great distances away. In the air, the $SO_2$ released from the coal combines with $H_2O$ to form $H_2SO_4$—sulfuric acid! Falling back to earth, this **acid rain** disrupts biological systems on a regional scale, a problem that is discussed further in Chapter 23.

Recall that H⁺ ions are not the only type of ion produced when water ionizes. Negatively charged OH⁻ ions are also produced in an equal concentration. Any substance that combines with H⁺ ions, as OH⁻ ions do, is said to be a **base.** In pure water the concentrations of H⁺ and OH⁻ ions are both $10^{-7}$ mole per liter, reflecting the spontaneous rate of dissociation of water. Thus pH 7 is neutral. Any increase in base concentration has the effect of lowering the H⁺ ion concentration, because base and H⁺ ions join spontaneously. Bases, therefore, have pH values above 7. Strong bases such as sodium hydroxide (NaOH) have pH values of 12 or more.

## THE CHEMICAL BUILDING BLOCKS OF LIFE

The basic chemical building blocks of organisms, the mortar and bricks used to assemble a cell, are made of molecules, just as a house is built of bricks. The molecules formed by living organisms, which contain carbon, are called **organic molecules.** Although many organic molecules are used in the construction of a cell, they need not confuse you. A good way to see through the tangle of different molecules is to focus on those bits of the molecules which are important, like focusing on who has the ball in a football game. Much of the complex structure of an organic molecule may have little to do with the biological process you are studying. It is often helpful to think of an organic molecule as a carbon-based core with special bits attached, groups of atoms with definite chemical properties. We refer to these groups of atoms as **functional groups.** For example, a hydrogen atom bonded to an oxygen atom, —OH, is a hydroxyl group. The most important functional groups are illustrated in Figure 2-16. Most chemical reactions that occur within organisms involve the transfer of a functional group from one molecule to another, or the breaking of a carbon-carbon bond. Proteins called kinases, for example, transfer phosphate groups from one kind of molecule to another.

| Compound | Examples | |
|---|---|---|
| Hydroxyl group | — OH | 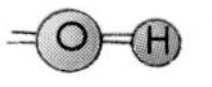 |
| Carbonyl group | $-\overset{\mid}{\underset{\mid\mid}{C}}-$  $O$ | 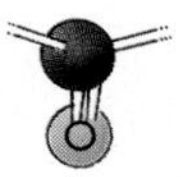 |
| Carboxyl group | $-C\overset{\displaystyle O}{\underset{\displaystyle OH}{}}$ | 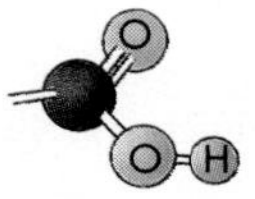 |
| Amino group | $-N\overset{\displaystyle H}{\underset{\displaystyle H}{}}$ | 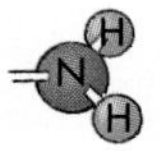 |
| Sulphydral | — S — H |  |
| Phosphate | $\overset{\displaystyle OH}{-O-\underset{\displaystyle O}{\overset{\mid}{\underset{\mid\mid}{P}}}-OH}$ | 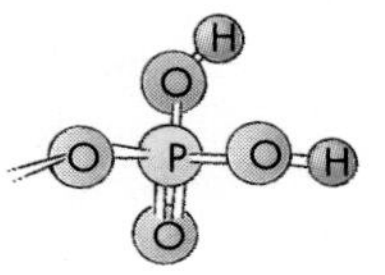 |

**Figure 2-16**

**The principal functional chemical groups.** These groups tend to act as units during chemical reactions and to confer specific chemical properties on the molecules that possess them. Hydroxyl groups make a molecule more basic, whereas carboxyl groups make a molecule more acidic.

Some molecules that occur in organisms are simple organic molecules, often with a single reactive functional group protruding from a carbon chain. Other molecules are far larger and are called **macromolecules.** This is particularly true of molecules that play a structural role in organisms or that store information. Most of these macromolecules are themselves composed of

simpler components, just as a wall is composed of individual bricks. Macromolecules fall into the following four classes (Table 2-3): carbohydrates, lipids, proteins, and nucleic acids.

*The molecules formed by living organisms, all of which contain carbon, are called organic molecules. Large organic molecules, or macromolecules, play a structural role or store information.*

Many macromolecules are polymers. A **polymer** is a molecule built of a long chain of similar molecules, like railway cars coupled together to form a train. Complex carbohydrates, for example, are polymers of simple ring molecules called sugars. Enzymes, membrane proteins, and other proteins are polymers of amino acids. DNA and RNA are two versions of a long-chain molecule called a nucleic acid, a polymer composed of a long series of molecules called nucleotides.

### Building Macromolecules

The units that are linked together to form a macromolecule are called subunits. Although the four different kinds of macromolecules are assembled from different kinds of subunits, they all put their subunits together in the same way: a covalent bond is formed between two subunit molecules in which a hydroxyl group (OH) is removed from one subunit and a hydrogen (H) is removed from the other (Figure 2-17). This process is called a **dehydration** (water-losing) **reaction** because in effect the removal of the OH and H groups constitutes removal of a molecule of water. In the synthesis of a polymer, one water molecule is removed for every link in the chain of subunits. Energy is required to break the chemical bonds when water is extracted from the subunits, and so cells must supply energy to assemble polymers. The process also requires that the two sub-

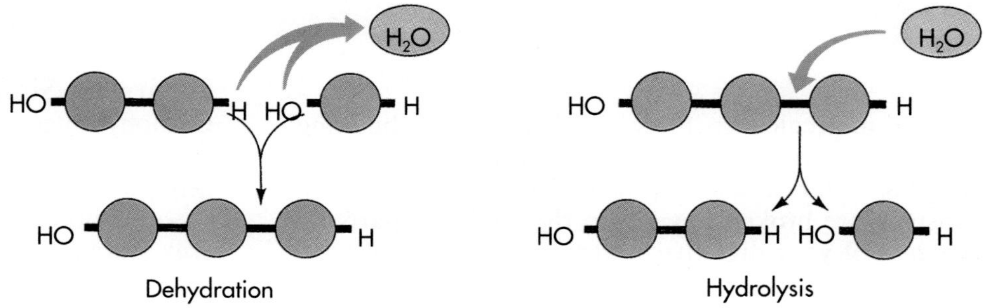

**Figure 2-17**

**Making (and breaking) macromolecules.** Biological macromolecules are formed by linking together subunits. The covalent bond between the subunits is formed in a dehydration reaction, in which a water molecule is eliminated. Breaking such a bond requires returning the water molecule, a hydrolysis reaction.

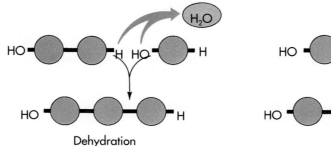

Dehydration                    Hydrolysis

**TABLE 2-3   MACROMOLECULES**

| MACROMOLECULE | SUBUNIT | FUNCTION | EXAMPLE |
|---|---|---|---|
| POLYSACCHARIDE CARBOHYDRATES | | | |
| Starch, glycogen | Glucose | Energy storage | Potatoes |
| Cellulose | Glucose | Cell walls | Paper |
| Chitin | Modified glucose | Cell walls | Crab shells |
| LIPIDS | | | |
| Fats | Glycerol + 3 fatty acids | Energy storage | Butter |
| Phospholipids | Glycerol + 2 fatty acids + phosphate | Cell membranes | Soap |
| Steroids | 4 carbon rings | Membranes; hormones | Cholesterol; estrogen |
| Terpenes | Long carbon chains | Pigments; structural | Chlorophyll; rubber |
| PROTEINS | | | |
| Globular | Amino acids | Enzyme catalysts | Hemoglobin |
| Structural | Amino acids | Support | Hair; silk |
| NUCLEIC ACIDS | | | |
| DNA | Nucleotides | Encodes genes | Chromosomes |
| RNA | Nucleotides | Operating blueprint of genes | Flu virus |

units be held close together and that the correct chemical bonds be stressed and broken. This process of positioning and stressing is called **catalysis**. In cells, catalysis is carried out by a special class of proteins called **enzymes**.

*Polymers are large molecules formed of long chains of similar molecules joined by dehydration, in which a hydroxyl (OH) group is removed from one subunit and a hydrogen (H) group is removed from the other. The process in which the subunits are held together and their bonds are stressed is called catalysis, which is carried out in organisms by enzymes, specialized proteins.*

No molecule lasts forever. When cells are building up some macromolecules, they are disassembling others. When you eat a steak, the protein and fat that you consume are broken down into their subunit parts by enzymes of your digestive system. The process of tearing down a polymer is essentially the reverse of dehydration—instead of removing a molecule of water, one is added. A hydrogen is attached to one subunit and a hydroxyl to the other, breaking the covalent bond. The breaking up of a polymer in this way is an example of a **hydrolysis reaction** (literally, "to break with water"—Greek *hydro*, water + *lyse*, break).

In discussing the polymers that make up the bodies of organisms (Figure 2-18), we will start with carbohydrates. Some carbohydrates are simple, small molecules. Others are long polymers. Carbohydrates are important as structural elements because of their role in energy storage. After we have discussed carbohydrates, we shall address lipids, amino acids and proteins, and nucleic acids.

## CARBOHYDRATES

### Sugars are Simple Carbohydrates

**Carbohydrates** are a loosely defined group of molecules that contain the elements carbon, hydrogen, and oxygen. Because they contain many carbon-hydrogen (C—H) bonds, carbohydrates are well suited for energy storage. Such C—H bonds are the ones most of-

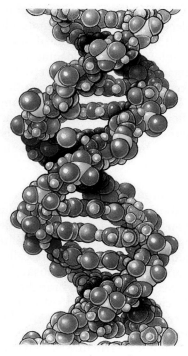

Nucleic acid

Protein

Carbohydrate

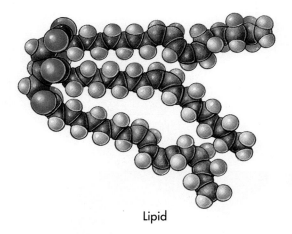

Lipid

**Figure 2-18**

The four fundamental kinds of biological molecules.

ten broken by organisms to obtain energy. Most carbohydrates contain the three elements carbon, hydrogen, and oxygen in the molar ratio of 1:2:1. A chemist would say that the empirical formula (a list of the atoms in a molecule with a subscript to indicate how many of each) is $(CH_2O)_n$, where $n$ is the number of carbon atoms. Among the simplest of the carbohydrates are the simple sugars or **monosaccharides** (Greek *monos*, single + *saccharon*, sweet). As their name implies, monosaccharides taste sweet. Simple sugars may have as few as three carbon atoms, but the molecules that play the central role in energy storage have six. They have the empirical formula:

$$C_6H_{12}O_6 \quad \text{or} \quad (CH_2O)_6$$

Sugars can exist in a straight-chain form, but in water solution they almost always form rings. The primary energy-storage molecule is glucose (Figure 2-19), a six-carbon sugar with seven energy-storing CH bonds.

*Among the most important energy-storage molecules in organisms are sugars. Many simple sugars contain six carbon atoms and seven energy-storing CH bonds.*

Many organisms transport sugars within their bodies. In human beings, glucose circulates in the blood. In many other organisms, glucose is converted to a **transport form** before it is moved from place to place. In transport form, glucose is less readily consumed (metabolized) while it is being moved. Transport forms of sugars are commonly formed by linking

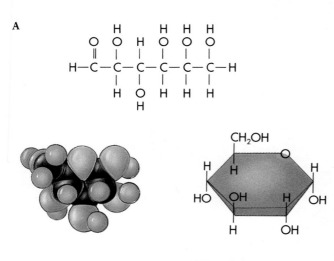

**Figure 2-19**

**A** Structure of the glucose molecule. Glucose is a linear six-carbon molecule that forms a ring in solution.
**B** Many animals consume sugar. This hungry butterfly has its uncoiled mouthparts extended down into a sugar solution, which it is sucking from the flowers of this wild relative of a sunflower in California.

two monosaccharide molecules to form a **disaccharide** (Greek *di,* two). Sucrose (table sugar) is a disaccharide formed by linking a molecule of glucose to a molecule of fructose. It is the common transport form of sugar in plants; much of the table sugar we consume is refined from the sap of sugar cane and sugar beets. If a glucose molecule is linked to galactose, the resulting disaccharide is lactose. Lactose is the molecule many mammals use to feed their babies. Maltose, two glucose subunits linked together, lends a sweet taste to the barley seeds that brewers ferment into alcohol in the process of making beer.

## Starches are Chains of Sugars

Organisms store the metabolic energy contained in glucose by converting it to an insoluble form and depositing it in specific storage areas. Sugars are made insoluble by joining them together into long polymers called **polysaccharides,** which are composed of monosaccharide sugar subunits. If the polymers are branched—that is, if they have side chains coming off of a main chain—the molecules are even less soluble. **Starches** are polysaccharides formed from glucose.

The starch with the simplest structure is amylose. Amylose is made up of many hundreds of glucose molecules linked together in long, unbranched chains. Potato starch is about 20% amylose. When amylose is digested by a sprouting potato plant (or by you), proteins called enzymes first break it into fragments of random length. These shorter fragments are soluble. Baking or boiling potatoes has the same effect.

Most plant starch, including 80% of potato starch, is a more complicated variant of amylose called amylopectin. **Pectins** are branched polysaccharides. Amylopectin is a form of amylose with short, linear amylose branches consisting of 20 to 30 glucose subunits. Most humans consume a great deal of plant starch. The seeds of rice, wheat, and corn supply about two-thirds of all the calories used by humankind.

Animals also store glucose in branched amylose chains. However, the average chain length is much longer in animals, and there are more branches. This results in a highly branched animal starch called **glycogen** (Figure 2-20).

*Starches are storage polysaccharides formed from glucose. Because they form long chains, starches are relatively insoluble and thus function well as storage.*

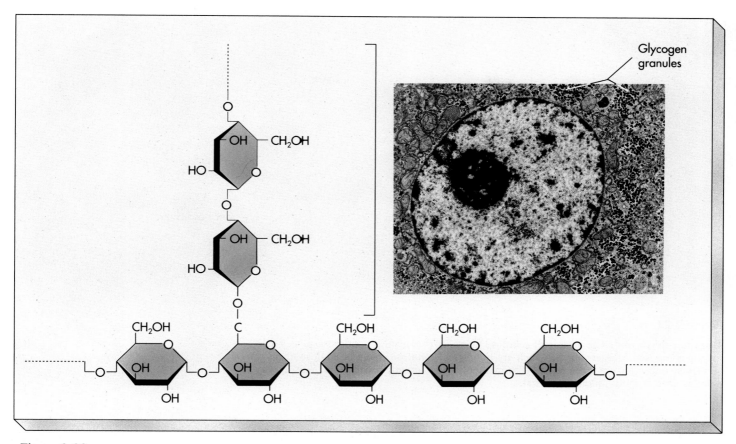

**Figure 2-20**

**Starch.** Storage polymers of glucose are called starches. The simplest starches are long chains of glucose called amylose. Most plants contain more complex starches called amylopectins, which contain branches. The nuclear membrane of the liver cell in the micrograph insert is surrounded by dense granules of animal starch called glycogen, which is even more highly branched.

## Cellulose is a Starch that is Hard to Digest

Imagine that you could draw a line down the central axis of a starch molecule, like threading a rope through a pipe. Because all the glucose subunits of the starch chain are joined in the same orientation, all the $CH_2OH$ groups would fall on the same side of the line (see Figure 2-20). There is another way to build a chain of glucose molecules, however, in which the glucose subunit orientations switch back and forth (the $CH_2OH$ groups alternate on opposite sides of the line). The resulting polysaccharide is **cellulose,** the chief component of plant cell walls. Cellulose is chemically similar to amylose, with one important difference (Figure 2-21): the starch-degrading enzymes that occur in most organisms cannot break the bond between two sugars in opposite orientation. It is not that the bond is stronger, but rather that its cleavage requires the aid of a different protein, one not usually present. Because cellulose cannot readily be broken down, it works well as a biological structural material and occurs widely in this role in plants. For those few animals able to break down cellulose, it provides a rich source of energy. Certain vertebrates, such as cows, which do not themselves produce the enzymes necessary to digest cellulose, do so by means of the bacteria and protists they harbor in their intestines. In humans, cellulose is a major component of dietary fiber, necessary for the proper functioning of the digestive system.

The structural material in insects, many fungi, and certain other organisms is **chitin** (Figure 2-22). Chitin is a modified form of cellulose in which a nitrogen group has been added to the glucose units. Chitin is a tough, resistant surface material. Few organisms are able to digest chitin.

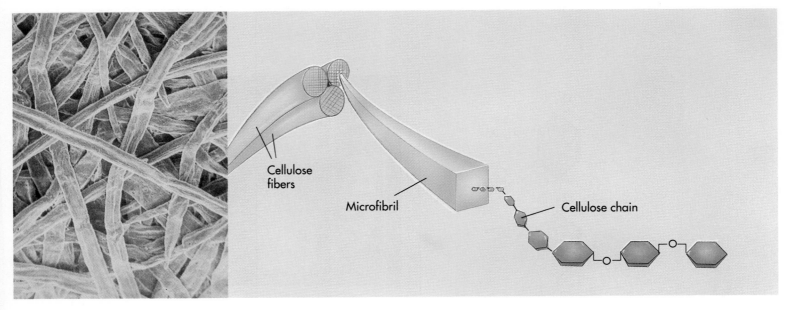

**Figure 2-21**

Cellulose fibers

Microfibril

Cellulose chain

**A journey into wood.** The jumble of cellulose fibers *(left)* is from a ponderosa pine. Each fiber is composed of microfibrils, which are bundles of cellulose chains. Cellulose fibers can be very strong; they are quite resistant to metabolic breakdown, which is one reason why wood is such a good building material.

# Nonfattening Sweets

Imagine a kind of table sugar that looks, tastes, and cooks like the real thing, but has no calories or harmful side effects. You could eat mountains of candy made from such a sweetener without gaining weight. As Louis Pasteur discovered in the late 1800s, most sugars are "right-handed" molecules, the hydroxyl group that binds a critical carbon atom being located on the right side. However, sugars can be readily made in the laboratory that are "left-handed"—the critical ion is on the left side, producing a mirror-image chemical twin of the natural form. The enzymes that break down sugars in the human digestive system can tell the difference. To digest a sugar molecule, an enzyme must first grasp it, much like a shoe fitting onto a foot, and all the body's enzymes are right-handed! A left-handed sugar just doesn't fit, any more than a right-handed shoe fits onto a left foot.

The Latin word for "left" is levo, and these new sugars are called levo- or l-sugars. These new laboratory-manufactured sweeteners, not yet on the market, do not occur in nature except as traces of red algae, snail eggs, and seaweed. Because they pass through the body without being used, they can let diet-conscious sweet lovers have their cake and eat it, too. Nor will they contribute to tooth decay—bacteria cannot metabolize them either.

**Figure 2-22**

**Chitin.** Chitin, which may be considered a modified form of cellulose with nitrogen groups added to the sugar subunits, is the principal structural element in the external skeletons of many invertebrates, such as this lobster, and in the cell walls of fungi.

## LIPIDS

### Fats Store Energy Very Efficiently

When organisms store glucose molecules for long periods, they usually convert the glucose into another kind of insoluble molecule that contains more C—H bonds than do carbohydrates. These storage molecules are called **fats.** The ratio of H to O in carbohydrates is 2:1, but in fat molecules it is much higher. Like starches, fats are insoluble and can therefore be deposited at specific storage locations within the organism. The insolubility of starches results from the fact that they are long polymers. The insolubility of fats, in contrast, arises from the fact that they are nonpolar. Unlike the H—O bonds of water, the C—H bonds of carbohydrates and fats are nonpolar and cannot form hydrogen bonds. Because fat molecules contain a large number of C—H bonds, they are hydrophobically excluded by water because water molecules tend to form hydrogen bonds with other water molecules. The result is that the fat molecules cluster together, insoluble in water.

Fats are one kind of **lipid,** a loosely defined group of molecules that are insoluble in water but soluble in oil. Oils such as olive oil, corn oil, and coconut oil are also lipids, as are waxes such as bee's wax and ear wax.

Fats are composite molecules; each molecule is built from two different kinds of subunits:

1. *Glycerol:* a three-carbon alcohol with each carbon bearing a hydroxyl (—OH) group. The three carbons form the backbone of the fat molecule, to which three fatty acids are attached.
2. *Fatty acids:* long **hydrocarbon** chains (chains consisting only of carbon and hydrogen atoms) ending in a carboxyl (—COOH) group. Three fatty acids are attached to each glycerol backbone.

The structure of an individual fat molecule, such as the one diagrammed in Figure 2-23, consists simply of a glycerol molecule with a fatty acid joined to each of its the three carbon atoms:

$$
\begin{array}{c}
\text{H} \\
| \\
\text{H—C—fatty acid} \\
| \\
\text{H—C—fatty acid} \\
| \\
\text{H—C—fatty acid} \\
| \\
\text{H}
\end{array}
$$

Because there are three fatty acids, the resulting fat molecule is called a triglyceride.

Fatty acids vary in length. The most common are even-numbered chains of 14 to 20 carbons. Fatty acids with all internal carbon atoms having two hydrogen side groups are called **saturated** because they contain the maximum number of hydrogen atoms possible. Some fatty acids have double bonds between one or more pairs of successive carbon atoms (Figure 2-24). Fats composed of fatty acids with double bonds are said to be **unsaturated** because the double bonds re-

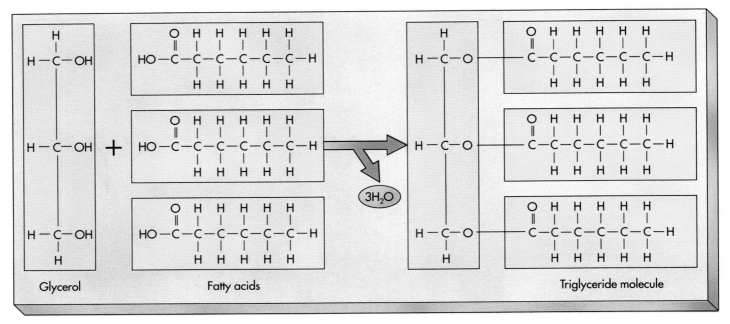

*Figure 2-23*

**Triglycerides are made from simple pieces.** Triglycerides are composite molecules made up
of three fatty acid molecules coupled to a single glycerol backbone.

place some of the hydrogen atoms. The fatty acids therefore contain fewer than the maximum number of hydrogen atoms. If a given fat has more than one double bond, it is said to be **polyunsaturated.**

Polyunsaturated fats have low melting points because their chains bend at the double bonds and the fat molecules cannot be aligned closely with one another. Consequently the fat may be fluid. A liquid fat is called an **oil.** Many plant fatty acids, such as oleic acid (a vegetable oil) and linolenic acid (a linseed oil) are unsaturated. Animal fats, in contrast, are often saturated and occur as hard fats.

It is possible to convert an oil into a fat by adding hydrogen. The peanut butter that you buy has usually been hydrogenated to convert the peanut fatty acids to hard fat. This prevents the fatty acids from separating out as oils while the jar sits on the store shelf.

Fats are very efficient energy-storage molecules because of their high concentration of C—H bonds. Most fats contain more than 40 carbon atoms. The ratio of energy-storing C—H bonds to carbon atoms is more than twice that of carbohydrates, making fats much more efficient vehicles for storing chemical energy. Fats usually yield about twice the amount of chemical energy per gram that carbohydrates yield. As you might expect, the more highly saturated fats are

richer in energy than are the less saturated ones. Animal fats contain more calories than do vegetable fats. Human diets with large amounts of saturated fats appear to upset the normal balance of fatty acids in the body, a situation that can lead to heart disease.

## There are Many Other Kinds of Lipids

Fats are just one example of the oily or waxy class of molecules called lipids. Your body contains many different kinds of lipids. The membranes of your cells are composed of a modified fat called a phospholipid. Membranes often also contain another kind of lipid called **steroids.** Many of the molecules that function as messengers that pass across cell membranes are steroids, such as the male and female sex hormones. Many cell membranes are rich in cholesterol, a steroid that makes them less rigid. In some people a diet high in saturated fats leads to a buildup of cholesterol deposits on the inner walls of blood vessels, producing elevated blood pressure and an increased risk of heart attack and stroke. Yet another kind of lipid forms many of the biologically important pigments, such as the photosynthetic pigment *chlorophyll* found in plants and the light-absorbing pigment *retinal* found in your eyes.

A saturated fat

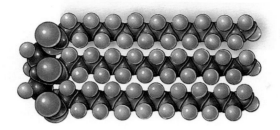

An unsaturated fat

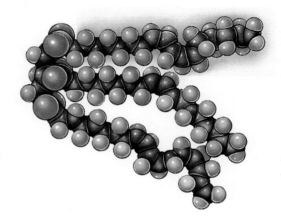

**Figure 2-24**

**Saturated and unsaturated fats.** Many animal triglyceride fats are saturated. Because their fatty acid chains can fit closely together, these triglycerides form immobile arrays called *hard fat.* Plant fats, in contrast, are typically unsaturated, and the many kinks that the double bonds introduce into the fatty acid chains prevent close association of the triglycerides and produce oils such as linseed oil, which is obtained from flax seed.

# PROTEINS

**Proteins** are the third major group of macromolecules that make up the bodies of organisms. Perhaps the most important proteins are **enzymes,** proteins capable of speeding up specific chemical reactions. Enzymes lower the energy required to activate or start the reaction, but are unaltered themselves in the process. Enzymes are biological **catalysts,** a more general term for substances that affect chemical reactions in this way. Other kinds of proteins also have important functions. Cartilage, bones, and tendons are made up of a protein called collagen. Keratin, another protein, forms both the horns of a rhinoceros and the feathers of a bird. The fluid within your eyeballs contains still other proteins. Short proteins called peptides are used as chemical messengers within your brain and throughout your body. Despite their diverse functions, all proteins have the same basic structure: a long polymer chain of amino acid subunits linked end to end.

## Amino Acids are the Building Blocks of Proteins

Amino acids are small molecules with a simple basic structure. An **amino acid** can be defined as a molecule containing an amino group ($-NH_2$), a carboxyl group ($-COOH$), a hydrogen atom, and a functional group designated R, all bonded to a central carbon atom:

$$H_2N-\underset{\underset{H}{|}}{\overset{\overset{R}{|}}{C}}-COOH$$

The identity and unique chemical properties of each amino acid are determined by the nature of the R group linked to the central carbon atom. An amino acid can potentially have any of a variety of different R groups, often called "side groups." Although many different amino acids occur in nature, only 20 are used in proteins. These 20 "common" amino acids and their side groups are illustrated in Figure 2-25. The different functional groups present on the side groups of the 20 amino acids give each amino acid distinctive chemical properties. For example, when the side group is $-H$, the amino acid (glycine) is polar, whereas when the side group is $-CH_3$, the amino acid (alanine) is nonpolar. The 20 amino acids that occur in proteins are commonly grouped into five chemical classes based on the chemical nature of their side groups.

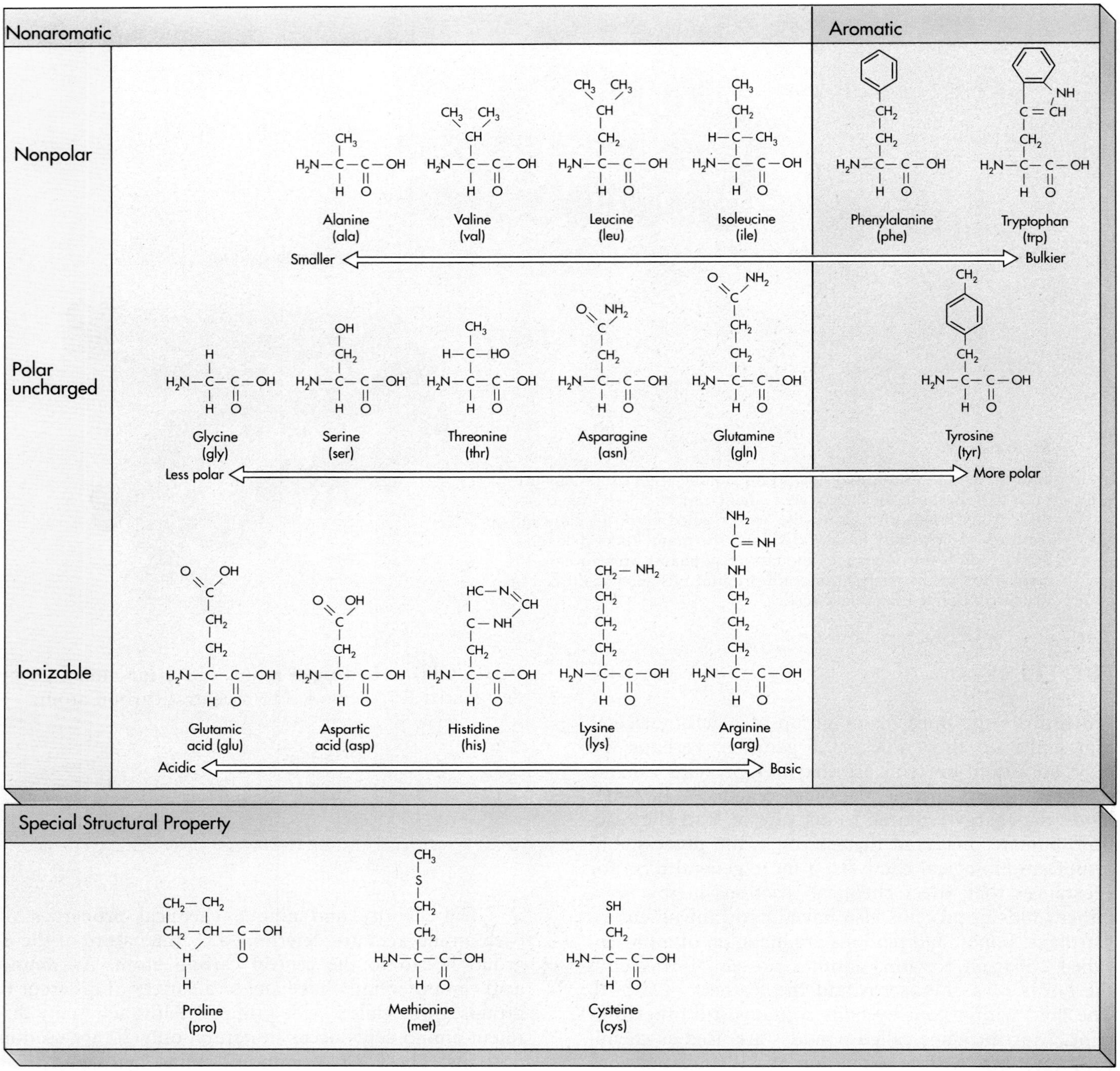

*Figure 2-25*

**The 20 common amino acids.** Each amino acid has the same chemical backbone but differs from the others in the side, or R, group that it possesses. Six of the amino acid R groups are nonpolar, some more bulky than others (particularly the ones containing ring structures, which are called the aromatic amino acids). Another six are polar but uncharged; these differ from one another in their polarity. Five more are polar and capable of ionizing to a charged form; under typical cell conditions some of these five are acids, others bases. The remaining three have special chemical properties that play important roles in forming links between protein chains or forming kinks in their shape.

The way that each amino acid affects the shape of a protein depends on the chemical nature of the amino acid's side group. Portions of a protein chain with many nonpolar amino acids tend to be shoved into the interior of the protein by hydrophobic interactions because polar water molecules tend to exclude nonpolar amino acid side groups.

*Proteins can contain up to 20 different kinds of amino acids. These amino acids fall into five chemical classes, which have properties quite different from one another. These differences determine what the proteins are like.*

Note that, in addition to its R group, each amino acid, when ionized, has a positive (amino, or $NH_3^+$) group at one end and a negative (carboxyl, or $COO^-$)

group at the other end. These two groups can undergo a chemical reaction, losing a molecule of water and forming a covalent bond between two amino acids (Figure 2-26). A covalent bond linking two amino acids is called a **peptide bond.**

## Polypeptides are Chains of Amino Acids

A protein, as just mentioned, is composed of a long chain of amino acids linked end to end by peptide bonds. The general term for chains of this kind is **polypeptide.** Proteins are therefore long, complex polypeptides. The sequence of amino acids that make up a particular polypeptide chain is termed its **primary structure** (Figure 2-27). Because the R groups that distinguish the various amino acids play no role in the peptide backbone of proteins, a protein can be composed of any sequence of amino acids. A protein made up of 100 amino acids linked together in a chain might have any of $20^{100}$ different amino acid sequences. That would be an enormous figure indeed. The great variability possible in the sequence of amino acids is perhaps the most important property of proteins, permit-

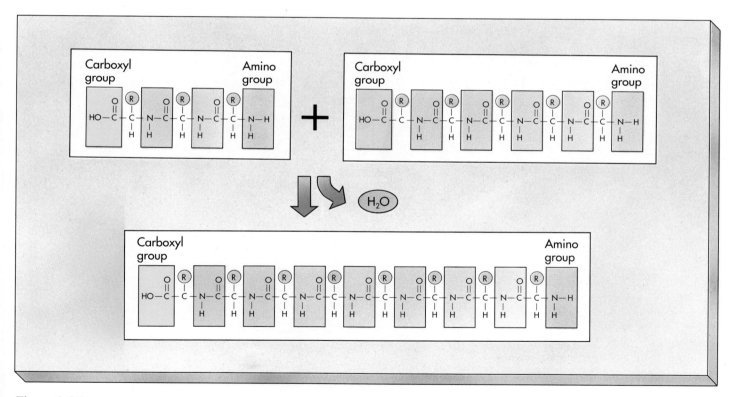

*Figure 2-26*

**The peptide bond.** Because of the partial double-bond nature of the C—N peptide bond, which forms when the —NH₂ end of one amino acid joins to the —COOH end of another, the resulting peptide chain cannot rotate freely around the peptide bond.

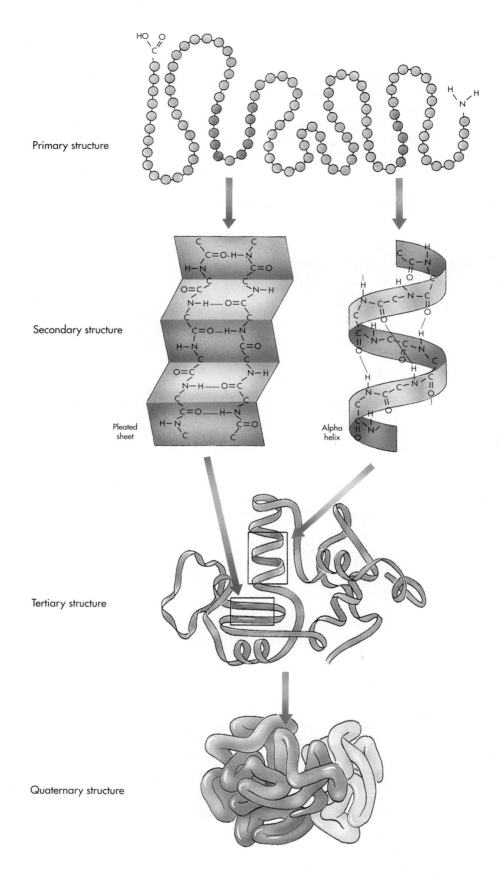

Primary structure

Secondary structure

Pleated sheet

Alpha helix

Tertiary structure

Quaternary structure

**Figure 2-27**

**How primary structure determines a protein's shape.** The amino acid sequence of the
enzyme protein lysozyme, called its *primary structure*, encourages the formation of
hydrogen bonds between nearby amino acids, producing coils and fold-backs called the
*secondary structure*. The lysozyme protein assumes a three-dimensional shape like a
cupped hand; this is called its *tertiary structure*. Many proteins (not lysozyme) aggregate in
clusters called the *quaternary structure* of the protein.

ting great diversity in the kinds and therefore functions of specific proteins.

Each amino acid of a polypeptide interacts with its neighbors, forming hydrogen bonds. Because of these near-neighbor interactions, polypeptide chains tend to fold spontaneously into sheets or wrap into coils. The form that a region of a polypeptide assumes is called its local **secondary structure.**

The three-dimensional shape, or **tertiary structure,** of a protein depends heavily on its secondary structure.

Proteins made up largely of sheets often form fibers that have a structural function (Figure 2-28), whereas proteins that have regions forming coils frequently fold into globular shapes. The shape of a globular protein is very sensitive to the order and nature of amino acids in the sequence. A change in the identity of a single amino acid can have either very subtle or very profound effects. We shall show in Chapter 6 that globular proteins are extremely effective biological catalysts because they can assume so many different shapes.

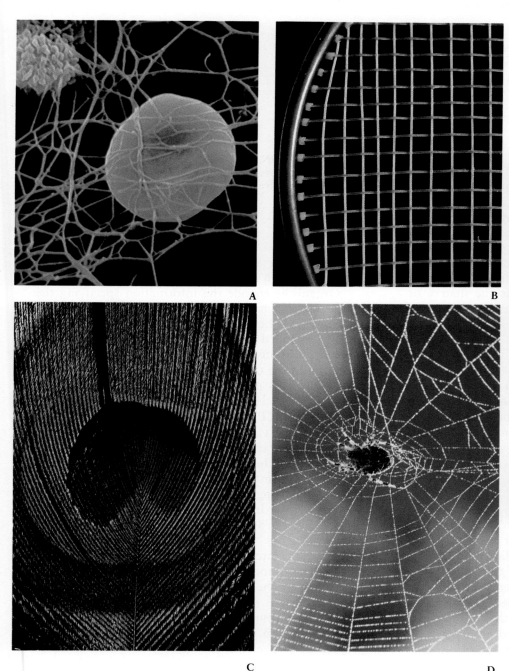

**Figure 2-28**

**Some of the more common structural proteins.**
A  Fibrin—electron micrograph of a blood clot.
B  Collagen—strings of a tennis racket.
C  Keratin—a peacock feather.
D  Silk—a spider's web.
E  Hair—a woman's hair.

When two protein chains associate to form a functional unit, the chains are termed subunits. Hemoglobin is a protein composed of four subunits. The subunits need not all be the same, although they can be the same. In hemoglobin molecules, there are two identical subunits of one kind and two identical subunits of a second kind. For proteins that consist of subunits, the way these subunits are assembled into a whole is called the **quaternary structure.** Proteins are discussed in detail in Chapter 6.

*The shape that proteins assume is determined by the sequence of amino acids in the polymer. Because different amino acid R groups have different chemical properties, the shape of a protein may be altered by a single amino acid change.*

Proteins perform many functions in your body. The thousands of different enzymes that carry out your body's chemical reactions are **globular proteins.** So are the antibodies that protect you from infection and cancer. **Fibrous proteins** play structural roles. The keratin in your hair is a fibrous protein, and so are the actin and myosin that make up your muscles. Indeed, the most abundant protein in your body is a fibrous protein—collagen, the protein that forms the matrix of your skin, ligaments, tendons, and bones.

## NUCLEIC ACIDS

All organisms store the information specifying the structures of their proteins in nucleic acids. *Nucleic acids* are long polymers of repeating subunits called *nucleotides.* Each nucleotide, the basic repeating unit, is a composite molecule made up of three smaller building blocks (Figure 2-29):

1. A five-carbon sugar
2. A phosphate group ($PO_4$)
3. An organic nitrogen-containing base

In the formation of a nucleic acid chain, the individual sugars are linked together in a line by the phosphate groups. The phosphate group of one sugar binds to the hydroxyl group of another, forming an —O—P—O bond. This bond is called a **phosphodiester** bond. A nucleic acid is simply a chain of five-carbon sugars (called ribose sugars) linked by phosphodiester bonds, with an organic base protruding from each

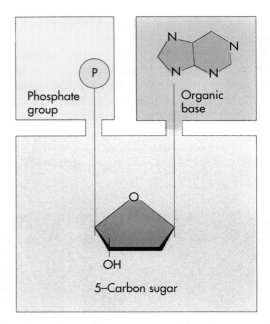

*Figure 2-29*

**The structure of DNA and RNA.** The nucleotide subunits of DNA and RNA have a composite structure; each is made up of three elements: a five-carbon sugar, an organic base, and a phosphate group.

sugar. Each of the repeating phosphate-sugar-base links in the chain is a nucleotide.

Organisms encode the information specifying the amino acid sequence of their proteins as sequences of nucleotides in the nucleic acid called DNA (Figure 2-A). This encoded information is used in the everyday metabolism of the organism as well as being stored and passed on to the organism's descendants.

Nucleotides play other critical roles in the life of the cell. The nucleotide adenine is a key component of the molecule **ATP,** the energy currency of the cell, and of the molecules **NAD** and **FAD,** which act to carry electrons whose energy is used to make ATP.

How does the structure of DNA permit it to store hereditary information? If DNA were a simple, monotonously repeating polymer, it could not encode the message of life. Imagine trying to write a story using only the letter E and no spaces or punctuation. All you could ever say is "EEEEEEE. . . ." You need more than one letter to write. We use 26 letters in the English alphabet. The Chinese use thousands of different characters to convey the same messages. You do not need so many individual symbols, of course, if the individual "letters" are grouped together into words. Morse code, which is used to transmit messages by telegraph, employs only two elements ("dot" and "dash"), as do

# Seeing Atoms

The beautiful pattern of yellow peaks you see in Figure 2-A is a DNA molecule as seen by a scanning-tunneling microscope, the bases following one another like marching soldiers. You cannot see DNA molecules with a light microscope, which cannot resolve anything smaller than 1000 atoms across. With an electron microscope you can image structures as small as a few dozen atoms across, but still cannot see the individual atoms that make up a DNA strand. This limitation has been overcome in the last few years with the introduction of the scanning-tunneling microscope.

out electrons instead of light). Another thing you could do would be to reach over and feel the chair's outline with your hands. You are "seeing" the chair by putting a probe (your hands) near its surface and measuring how far away the surface is. In a scanning-tunneling microscope, computers advance the probe over the surface of a molecule in tiny steps less than the diameter of an atom. Figure 2-B is the first photograph ever made of DNA. It shows the two strands magnified a million times! The strand is so slender it would take 50,000 of them to reach the diameter of a human hair.

A

What is so new? Imagine you are in a dark room and that there is a chair in the room, too. How would you know what the chair looked like? You could find out by shining a flashlight on it, letting the light bounce off the chair and form an image on your eye. That's what optical and electron microscopes do (in the latter, the "flashlight" shoots

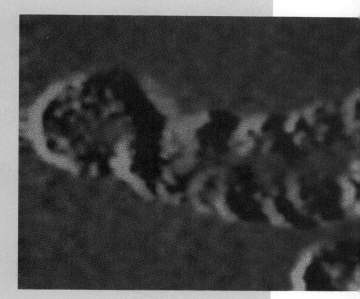

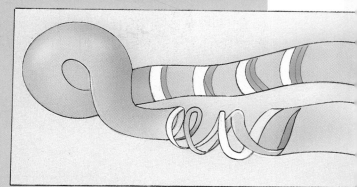

B

most modern computers (0 and 1). Nucleic acids can encode information because they contain more than one kind of organic base. Each sugar link in a nucleic acid chain can have any one of four different organic bases attached to it. Just as in the English language, the sequence of letters encodes the information. In nucleic acids there are not 26 letters, as in English, but only four letters, the four organic bases that occur in nucleic acids. (As you will see shortly, one of the four bases is present in different versions in the two principal forms of nucleic acid.)

*Organisms store and use hereditary information by encoding the sequence of the amino acids of each of their proteins as a sequence of nucleotides in nucleic acids.*

Organisms store hereditary information in two forms of nucleic acid. One form, **deoxyribonucleic acid (DNA)**, provides the basic storage vehicle, or master plan. The other form, **ribonucleic acid (RNA)**, is simi-lar in structure and is made as a template copy of portions of the DNA. This copy passes out into the rest of the cell, where it provides a blueprint specifying the amino acid sequence of proteins.

Two of the four organic bases that make up DNA and RNA (Figure 2-30), adenine and guanine, are large, double-ring compounds called **purines.** The other organic bases that occur in these molecules, cytosine (in both DNA and RNA), thymine (in DNA only), and uracil (in RNA only), are smaller, single-ring compounds called **pyrimidines.** In discussing the sequence of bases in RNA and DNA, the organic bases are usually referred to by their first initials: A, G, C, T, and U.

DNA chains in organisms exist not as single chains folded into complex shapes, as in proteins, but rather as double chains (Figure 2-31). Two of the polymers wind around each other like the outside and inside rails of a circular staircase. Such a winding shape is called a **helix,** and one composed of two molecules winding about one another, as in DNA, is called a **double helix.** The steps of the helical staircase are hydrogen bonds between the bases in one polymer chain and those opposite them in the other chain. These hydrogen bonds hold the two chains together as a duplex. Details of the structure of DNA, and how it interacts with RNA in the production of proteins, are presented in Chapter 13.

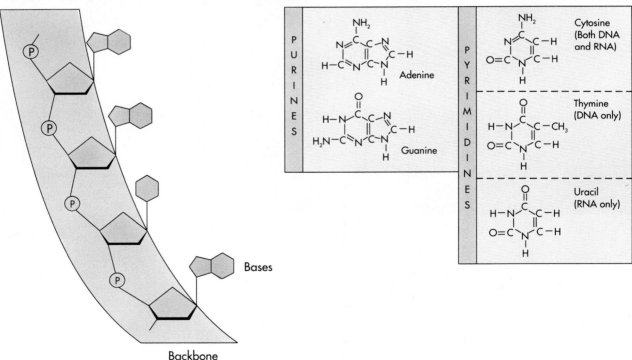

**Figure 2-30**

**The five nucleotide bases of nucleic acids.** In DNA the nucleotide thymine replaces the nucleotide uracil found in RNA. In a nucleotide chain, nucleotides are linked to one another via phosphodiester bonds, as shown on the left, which is one half of a DNA helix.

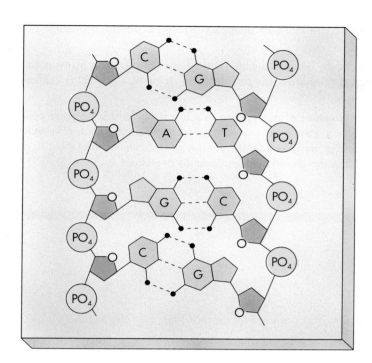
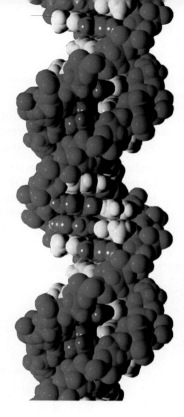

*Figure 2-31*

**Hydrogen bonds hold the DNA double helix together.** Hydrogen bond formation between the organic bases, called base pairing, causes the two chains of a DNA duplex to bind to each other. The molecule is not a straight chain, but rather a graceful double helix, rising like a circular staircase.

## ■ SUMMARY

1. The smallest stable particles of matter are protons, neutrons, and electrons; these particles associate, forming atoms. The core (nucleus) of an atom is composed of protons and neutrons; electrons orbit around the core in a cloud.

2. The chemical behavior of an atom is largely determined by the distribution of its electrons, particularly the number of electrons in its outermost level.

3. A molecule is a stable collection of atoms. The forces holding atoms together in a molecule are called chemical bonds.

4. The force of a chemical bond can result from the attraction of opposite charges, as in an ionic bond, or from the sharing of one or more pairs of electrons, as in a covalent bond.

5. The chemistry of life is the chemistry of water. In water molecules the oxygen atom more strongly attracts the electrons shared between oxygen and the hydrogen atoms. As a result, the oxygen atom is electron-rich (partial negative charge) and the hydrogen atoms are electron-poor (partial positive charge). This charge separation is like that of a magnet with positive and negative poles. Water is termed a "polar" molecule.

6. A hydrogen bond is formed by the attraction of the partial positive charge of one hydrogen atom of a water molecule with the partial negative charge of the oxygen atom of another. Water molecules tend to form the maximum number of hydrogen bonds and to exclude nonpolar molecules.

7. Organisms store energy in carbon-hydrogen (C—H) bonds. The most important of the energy-storing carbohydrates is glucose, a six-carbon sugar.

8. Excess energy resources may be stored in complex sugar polymers called starches, especially in plants. Glycogen, a comparable storage polymer occurring frequently in animals, is characterized by complex branching.

9. Fats are molecules containing many more C—H bonds than carbohydrates contain, providing more efficient energy storage.

10. Proteins are linear polymers of amino acids. Because the 20 amino acids that occur in proteins have side groups with very different chemical properties, the function and shape of a protein is critically affected by its particular sequence of amino acids.

11. Hereditary information is stored as a sequence of nucleotides in a linear nucleotide polymer called deoxyribonucleic acid, or DNA. DNA is a double helix. A second form of nucleic acid, ribonucleic acid, or RNA, is similar in structure and is made as a template copy of portions of the DNA. In the cell, RNA provides a blueprint specifying the amino acid sequence of proteins.

## REVIEW

1. Which molecule is most important to the evolution of life?

2. The three subatomic particles that make up atoms are _____, _____, and _____. Which two make up the nucleus of an atom?

3. Plants store glucose as _____. Animals store glucose in long, highly branched chains called _____.

4. The nucleic acid _____ is the molecule that encodes the information specifying the amino acid sequences for proteins and thus contains the information necessary to make organisms.

5. The bonds that hold the molecules of your body together are _____ bonds.

## SELF-QUIZ

1. Which is an example of a polar molecule that is soluble in water?
   (a) Water
   (b) Oxygen gas
   (c) Ammonia
   (d) Fats
   (e) Cellulose

2. The primary structure of a protein is
   (a) the sequence of amino acids that make up the chain.
   (b) usually a helical coil.
   (c) the shape of the protein.
   (d) influenced by the amino acid sequence.
   (e) when two proteins associate to form a functional unit.

3. A nucleotide, the repeating subunit that makes up a nucleic acid, is composed of which *three* smaller building blocks?
   (a) A glucose
   (b) A fatty acid
   (c) A ribose
   (d) A phosphate group
   (e) A nitrogen-containing base

4. Which of the following kinds of molecules are lipids?
   (a) Unsaturated fats
   (b) Waxes
   (c) Cholesterol
   (d) Chlorophyll
   (e) Chitin

5. Which of the following are characteristics of DNA?
   (a) Exists as single chains folded into complex forms
   (b) Serves as the natural catalyst of reactions in cells
   (c) Is a ready source of energy for metabolic reactions
   (d) Consists of a long chain of linked amino acids
   (e) Is a polymer

# THOUGHT QUESTIONS

1. Carbon (atomic number 6) and silicon (atomic number 14) both have four vacancies in their outer energy levels. Ammonia is even more polar than water. Why do you suppose life evolved on earth in the form of organisms made up of carbon chains in water solution rather than ones of silicon in ammonia?

2. Carbon atoms can share four electron pairs to form molecules. Why doesn't carbon form a bimolecular gas, in the way of hydrogen (one pair of shared electrons), oxygen (two pairs of shared electrons), and nitrogen (three pairs of shared electrons)?

3. How many different three-base DNA sequences are possible?

# FOR FURTHER READING

DOOLITTLE, R.: "Proteins," *Scientific American,* October 1985, pages 88-99. A good general description of protein primary, secondary, and tertiary structure, with emphasis on protein evolution.

FITZGERALD, J., and G. TAYLOR: "Carbon Fibres Stretch the Limits," *New Scientist,* May 1989, pages 48-53. A fascinating discussion of the ways in which the principles discussed in this chapter are used to produce modern, synthetic materials.

KARPLUS, M., and A. McCAMMON: "The Dynamics of Proteins," *Scientific American,* April 1986, pages 42-51. Proteins are not fixed in their shape like car parts, but rather are flexible. This article explains why flexibility is critical to protein function.

SHARON, N.: "Carbohydrates," *Scientific American,* November 1980, pages 90-116. An overview of the structures of carbohydrates and the diverse roles they assume in organisms.

SUTTON, C.: "Subatomic Forces," *New Scientist,* February 1989, supplement, pages 1-4. Contemporary review of the interactions between particles within the atom.

# The Origin of Life

Life evolved on earth over 3.5 billion years
ago. Our world was a different place then, with
no oxygen in the atmosphere, no ozone to protect its surface from the sun's radiation, and
violent electrical storms. Born in lightening, life
has since changed the earth and its atmosphere
in profound ways.

# THE ORIGIN OF LIFE

## Overview

Life originated on earth more than 3.5 billion years ago, within 1 billion years after our planet's formation. We do not know exactly how life originated, although the evidence is consistent with the hypothesis that it evolved from nonliving materials. All organisms are composed of one or more cells, which are the basic units of life. The cells of bacteria have little internal organization and were the first kind of cells to appear on earth. Later a new kind of cell evolved with a more elaborate internal organization; organisms that have such cells are called eukaryotes. Eventually multicellular organisms originated independently from a number of different groups of eukaryotes.

## For Review

*Here are some important terms and concepts that you will encounter in this chapter. If you are not familiar with them, you should review them before proceeding.*

**Properties of life** (Chapter 1)

**Amino acids** (Chapter 2)

**Lipids** (Chapter 2)

When we look around us and see a world teeming with life, it is difficult to imagine that there was a time when no life existed on earth, when no grass grew and no fish swam in the sea. However, the earth is much older than life. By studying the distribution of radioactive isotopes in ancient rocks, scientists have determined that the earth was formed about 4.5 billion years ago. At first the earth was molten, but soon a thin crust of rock, the shell on which we live, solidified over the hot core. Early earth was a land of molten rock and violent volcanic activity. The oldest rocks that have survived on earth are about 3.9 billion years in age. These ancient rocks contain no definite traces of life, or at least none that can be recognized with our current level of technology.

Today the world is very different, and life exists in every crack and crevice (Figure 3-1). Except for a blast furnace, you would be hard pressed to think of a place

**Figure 3-1**

**A fossil fish.** For a long time the earliest fossils known came from the Cambrian geological period, about 590 million years ago. Older fossils are simply too small to see with the naked eye. Now, with microscopes, we are able to trace the fossil record back 3.5 billion years.

anywhere on the surface of the earth where life does not exist in profusion. Where did all of this life come from?

It is not easy to answer this question. You cannot go back in time and see for yourself how life originated, nor are there any witnesses. We can learn something about what must have happened by studying the rocks of the earth, but the record of events is incomplete and often silent. Perhaps the most fundamental of these issues is the nature of the agency or force that led to the origin of life. In principle, there are at least three possibilities:

1. *Extraterrestrial origin.* Life may not have originated on earth at all but instead may have been carried to it, perhaps as an extraterrestrial infection of spores originating on a planet of a distant star. How life came to exist on *that* planet is a question we cannot hope to answer soon.

2. *Special creation.* Life forms may have been put on earth by supernatural or divine forces. This viewpoint, common to most Western religions, is the oldest hypothesis and is widely accepted. It forms the basis of the "scientific creationism" viewpoint discussed in Chapter 15.

3. *Evolution.* Life may have evolved from inanimate matter, with associations among molecules becoming more and more complex. In this view the force leading to life was selection; changes in molecules that increased their stability caused the molecules to persist longer.

In this book we deal only with the third possibility, by attempting to understand whether the forces of evolution could have led to the origin of life—and, if so, how the process might have occurred. However, showing that life could have originated in this way does not rule out the second possibility, which could be accepted only as a matter of faith, and also does not prove that life might not have originated on another planet. The evidence that we can accumulate can be taken merely as a test of the plausibility of the third possibility, which is the only readily testable hypothesis.

In our search for an understanding of the way in which life could have evolved on earth, we must look back to the time before life appeared, when the earth was just starting to cool. We must go back at least that far because fossils of bacteria exist in rocks that are about 3.5 billion years old, demonstrating the existence of life on earth within no more than 1 billion years of the origin of our planet. In attempting to determine how the first organisms originated, we must first consider the mode of origin of organic molecules, which are the building blocks of organisms. Then we shall

consider how organic molecules might have become organized into living cells.

*Life on earth may have evolved from inorganic substances, and the process by which this may have occurred can be studied directly. If life came to earth from another planet, it probably originated in some similar way. The suggestion that life was created by a supreme being lies beyond the scope of scientific investigation.*

## THE ORIGIN OF ORGANIC MOLECULES: CARBON POLYMERS

Scientists who study the conditions of the primitive earth are called geochemists. Geochemists believe that as the primitive earth cooled and its rocky crust formed, many gases were released from the molten core. It was a time of volcanoes, blasting enormous amounts of material skyward. These gases formed a cloud around the earth and were held as an atmosphere by the earth's gravity. The atmosphere we breathe now is very different from what it used to be; it has been changed by the activities of organisms, as we shall see later. Despite the changes, geochemists have been able to learn what the early atmosphere must have been like by studying the gases released by volcanoes and by deep sea vents in the earth's crust. Geochemists do not all agree on the exact composition of this original atmosphere, but they do agree that it was principally composed of nitrogen gas. It also contained significant amounts of carbon dioxide and water. It is probable, although not certain, that compounds in which hydrogen atoms were bonded to other light elements such as sulfur, nitrogen, and carbon were also present in the atmosphere of early earth. These compounds would have been hydrogen sulfide ($H_2S$), ammonia ($NH_3$), and methane ($CH_4$).

The atmosphere of early earth was probably rich in hydrogen, although there is debate on this point. We refer to such an atmosphere as a reducing one because of the ample availability of hydrogen atoms and associated electrons (as we saw in Chapter 2, in chemistry the donation of electrons to a molecule is called **reduction** and the removal of electrons is **oxidation**). Little if any oxygen gas was present. In such a reducing atmosphere it does not take much energy to form the carbon-rich molecules from which life evolved. Later, the earth's atmosphere changed as living organisms began to carry out photosynthesis, which involves harnessing

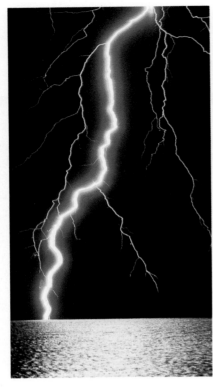

**Figure 3-2**

**Lightning.** Before life evolved, the simple molecules in the earth's atmosphere combined to form more complex molecules. The energy that drove some of these chemical reactions came from lightning and other forms of geothermal energy.

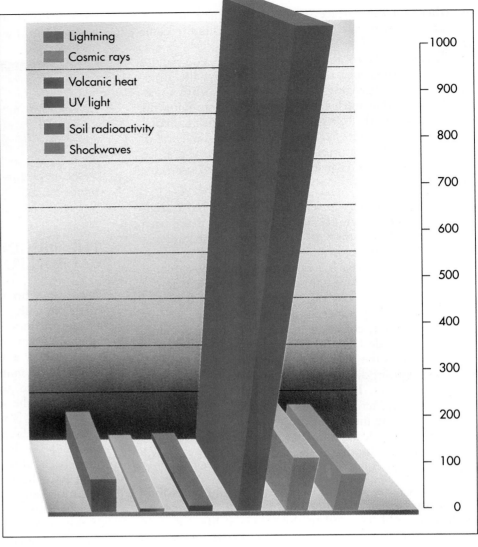

Lightning
Cosmic rays
Volcanic heat
UV light
Soil radioactivity
Shockwaves

1000
900
800
700
600
500
400
300
200
100
0

**Figure 3-3**

**Sources of energy for the synthesis of complex molecules in the atmosphere of the primitive earth.** Of the ultraviolet radiation, only the very short wavelengths (less than 100 nanometers) would have been effective in promoting chemical reactions. Electrical discharges are thought to have been more common on the primitive earth than they are now.

the energy in sunlight to split water molecules and form complex carbon-containing molecules, giving off gaseous oxygen molecules in the process. Our atmosphere is now approximately 21% oxygen. In the oxidizing atmosphere that exists today, the spontaneous formation of complex carbon-containing molecules cannot occur.

Therefore the first step in the evolution of life probably occurred in a reducing atmosphere that was devoid of gaseous oxygen and thus very different from the atmosphere that exists now. Those were violent times, and the earth was awash with energy (Figure 3-2): solar radiation, lightning from intense electrical storms, violent volcanic eruptions, and heat from radioactive decay. Living on earth today, shielded from the effects of solar ultraviolet radiation by a layer of ozone gas ($O_3$) in the upper atmosphere, most humans cannot imagine the enormous flux of ultraviolet energy to which the early earth's surface was exposed. Subjected to ultraviolet energy and to other sources of energy as well (Figure 3-3), the gases of the early earth's atmosphere underwent chemical reactions with each other and formed a complex assemblage of molecules. In the covalent bonds of these molecules, some of the abundant energy present in the atmosphere was captured as chemical energy.

What kinds of molecules might have been produced? One way to answer this question is to repeat the process: (1) assemble an atmosphere similar to the one thought to exist on early earth; (2) place this atmosphere over water, which was present on the surface of the cooling earth; (3) exclude gaseous oxygen from the atmosphere because none was present in the atmosphere of early earth; (4) maintain this mixture at a temperature somewhat below 100° C; and (5) bombard it with energy in the form of electrical sparks. When Harold C. Urey and his student Stanley L. Miller

performed this experiment in 1953, they found that, within 1 week, 15% of the carbon that was originally present as methane gas had been converted into more complex carbon-based molecules.

Among the first substances produced in the Miller-Urey experiment (Figure 3-4) were molecules derived from the breakdown of methane, including formaldehyde and hydrogen cyanide. These molecules then combined to form more complex molecules containing carbon-carbon bonds, including the amino acids *glycine* and *alanine*. Amino acids are important in understanding the origin of life because they are the basic building blocks of proteins, which are one of the major kinds of molecules of which organisms are composed. About 50% of the dry weight of each cell in your body

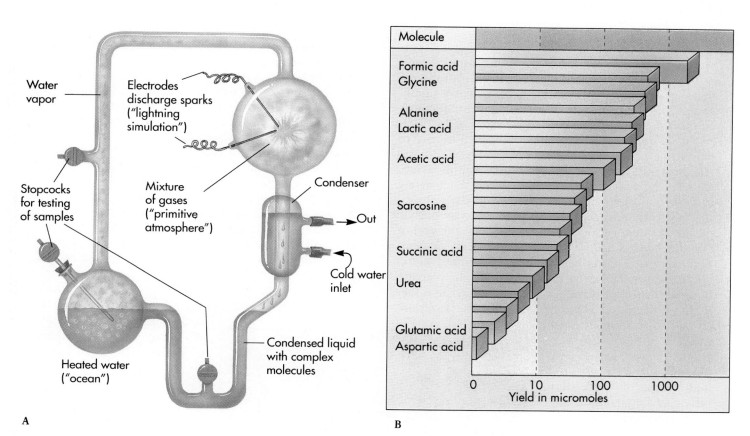

A

B

### Figure 3-4

**A** The Miller-Urey experiment. The apparatus consisted of a closed tube connecting two chambers. The upper chamber contained a mixture of gases thought to resemble the earth's atmosphere. Any complex molecules formed in the atmosphere chamber would be carried dissolved in these droplets to the lower "ocean" chamber, from which samples were withdrawn for analysis.

**B** Some of the 20 most common complex molecules detected in the original Miller-Urey experiments are indicated. Among these 20 molecules are four amino acids, the names of which are shown in blue type.

consists of amino acids—alone or linked together into protein chains.

In later experiments, more than 30 complex carbon-containing molecules were identified, including the amino acids *glycine, alanine, glutamic acid, valine, proline,* and *aspartic acid.* The production of amino acids indicates that proteins could have formed under conditions similar to those which probably existed on early earth. Other biologically important molecules were also formed in these later experiments, including the purine base *adenine,* which is a constituent of both DNA and RNA. Thus at least some of the key molecules from which life evolved were created in the atmosphere of early earth as a by-product of its birth.

*Among the molecules that form spontaneously under conditions thought to be similar to those of primitive earth are some of those molecules which form the building blocks of organisms.*

As the earth cooled, much of the water vapor present in the atmosphere condensed into liquid water and accumulated in the oceans. Judging from the results of the Miller-Urey experiments, we presume that the water droplets carried nucleotides, amino acids,

# The Puzzle of the Origin of Proteins

It is not clear how in the dilute, nonliving (**prevital**) soup of the early earth the more complex molecules characteristic of life were formed. Among the many questions that arise, one is particularly significant: *How did amino acids aggregate spontaneously to form the first protein?*

The question is a puzzle because it seems to defy what we know of the laws of chemistry. Biological proteins form by the joining of subunits into chains. The addition of each element to the growing chain requires the input of energy and the removal of a water molecule. This addition is accomplished by the removal of an —OH group from the end of the chain and an —H from the incoming element. Because this reaction is chemically reversible, an excess of water should, in principle, drive the reaction in the direction of breakdown of the molecule rather than synthesis. The puzzle is that these reactions critical to the evolution of life are thought to have taken place within the oceans and therefore in a high excess of water. It is difficult to imagine that the spontaneous formation of proteins could have occurred under these conditions.

It has been suggested that the first proteins may have formed on the surfaces present within silicate clays. The interior of clays such as kaolinite is made up of thin layers, only 0.71 nanometer apart, separated by water. A nanometer is a billionth of a meter ($10^{-9}$ meter). Thus a cube of kaolinite 1 centimeter to a side has a surface area of 2800 square meters! A 1-pound lump would have a surface greater than 50 football fields. Clay would thus have provided ample surfaces for protein formation to occur. In addition, there are many positive and negative charges on the surfaces of these layers, which would have facilitated the process. Supporting the hypothesis that clays may have played an important role in the origin of life, clays such as kaolinite have been shown to catalyze the formation of long polypeptide chains from amino acids when the amino acids are first joined to adenosine monophosphate (AMP), a molecule that produces the energy that makes the reaction possible. This same initial joining of amino acids to AMP is universally employed by all living systems in their synthesis of proteins.

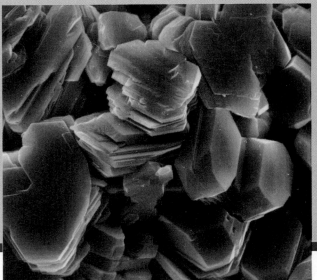

*Figure 3-A*

**A close look at clay.** Magnified nearly 10,000 times, the layers of clay are stacked, one on another, in chunks. At the molecular level, the layers provide enormous surface area and spaces for water and other substances.

and other compounds produced by chemical reactions in the atmosphere.

The primitive oceans must not have been pleasant places. It is odd to think of life originating from a dilute, hot, smelly soup of ammonia, formaldehyde, formic acid, cyanide, methane, hydrogen sulfide, and organic hydrocarbons. Yet within such oceans arose the organisms from which all later life forms were derived. One cannot escape a certain curiosity about the earliest steps that eventually led to the origin of all living things on earth, including ourselves. How did organisms evolve from complex molecules? What is the "origin of life"?

## ORIGIN OF THE FIRST CELLS

Many different kinds of molecules gather together or aggregate in water, much as people from the same foreign country tend to aggregate together in a large city. Sometimes the aggregation of molecules of one kind form a cluster big enough to see. If you shake up a bottle of oil-and-vinegar salad dressing, you can see this happen—small spherical bubbles of oil appear, grow in size, and fuse with one another. Small **coacervates** of this sort, spherical aggregations only 1 to 2 micrometers in diameter, form spontaneously from lipid molecules that are suspended in water (Figure 3-5). Similar coacervates that formed in the primeval soup may well have been the first step in the evolution of cellular organization. Coacervates have several remarkably cell-like properties:

1. Coacervates form an outer boundary that has two layers and thus resembles a biological membrane, as we shall see in Chapter 4.
2. Coacervates grow by accumulating more subunit molecules from the surrounding medium.
3. Coacervates form budlike projections and divide by pinching in two, as do bacteria.
4. Coacervates may contain amino acids and use them to facilitate several kinds of chemical reactions that are mainly found in living cells.

A process of chemical evolution involving coacervate microdrops of this sort may have taken place before the origin of life. The early oceans must have contained untold numbers of these microdrops—billions in a spoonful, each one forming spontaneously, persisting for a while, and then dispersing. Some of the droplets would by chance have contained amino acids with side groups that were better able than the others to catalyze growth-promoting reactions. These droplets would have survived longer than the others because the persistence of both protein and lipid coacervates is greatly increased when they carry out metabolic reac-

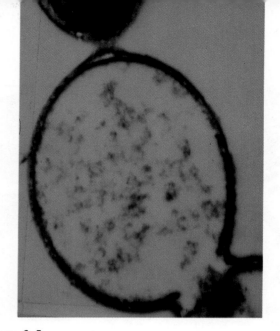

**Figure 3-5**

**Coacervate droplets.** These hollow microspheres possess many of the characteristics of living cells. If you look carefully, you can see that each is bounded by a bilayer membrane.

tions such as glucose degradation (breakdown) and when they are actively growing.

Over millions of years, those complex microdrops which were better able to incorporate molecules and energy from the lifeless oceans of early earth would have tended to persist more than the others. Also favored would have been those microdrops which could use such molecules to expand in size, growing large enough to divide into daughter microdrops with features similar to those of their "parent" microdrop. The daughter microdrops would have been able to use the same favorable combination of characteristics as their parents, and grow and divide as well. When a means occurred to facilitate this transfer of new ability from parent to offspring, heredity—and life—began.

There is considerable discussion among biologists as to how the first cells may have evolved. The recent discovery that RNA can act like an enzyme to assemble new RNA molecules on an RNA template has raised the interesting possibility that coacervates may not have been the first step in the evolution of life. Nucleotides were also produced in the Miller-Urey experiments. Perhaps the first macromolecules were RNA molecules, and the initial steps on the evolutionary journey were ones leading to more complex and stable RNA molecules. Later, stability might have been improved by surrounding the RNA within a coacervate. There is as yet no consensus among scientists studying this problem as to whether RNA evolved before a coacervate-like structure, or after.

|—20 nm—|

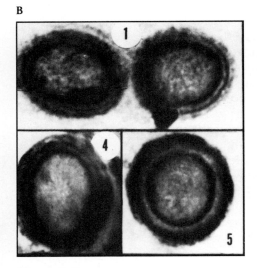

A B

**Figure 3-6**

**Fossil bacteria.**
**A** A scanning electron micrograph of one of the oldest fossils yet discovered, a bacterium from South African rocks from 3.4 billion years ago.
**B** Cross-sections through fossil bacteria from the Bitter Springs Formation of Australia, in which the cell walls are clearly visible. These fossils are about 850 million years old.

## THE EARLIEST CELLS

The fossils that we have found in ancient rocks (Figure 3-6) represent an obvious progression from simple to complex organisms during the vast period of time that began no more than 1 billion years after the origin of the earth. Living things may have been present earlier, but rocks of such great antiquity are rare, and fossils have not yet been found in them.

What do we know about these early life forms? We have learned from studying early microfossils that, for most of the history of life, all living organisms resembled living bacteria in their physical characteristics, although some ancient forms cannot be matched exactly. The ancient bacteria were small (1 to 2 micrometers in diameter) and single-celled, lacked external appendages, and had little evidence of internal structure.

We call these simple organisms with a body plan of this sort **prokaryotes,** from the Greek words for "before" and "kernel," or nucleus. The name reflects their lack of a **nucleus,** which is a spherical organelle (structure) characteristic of the more complex **eukaryotic** cells that evolved much later. We refer to the prokaryotes collectively as **bacteria.** Judging from the fossil record, eukaryotes did not appear until about 1.5 billion years ago. Therefore for at least 2 billion years— nearly half the age of the earth—bacteria were the only organisms that existed.

## Living Fossils

Most organisms living today resemble one another fundamentally, having the same kinds of membranes and hereditary systems and many similar aspects of metabolism. However, not all living organisms are exactly the same in these respects. If we look carefully in uncommon environments, we occasionally encounter organisms that are quite unusual, differing in form and metabolism from most other living things. Sheltered from evolutionary alteration in unchanging habitats which resemble those of earlier times, these living relics are the surviving representatives of the first ages of life on earth. In those ancient times, biochemical diversity was the rule, and living things did not resemble each other in their metabolic features as closely as they do today. In places such as the oxygenless depths of the Black Sea or the boiling waters of hot springs (see Figure 4-1), we can still find bacteria living without oxygen and displaying a bewildering array of metabolic strategies. Some of these bacteria have shapes similar to those of the fossils of bacteria that lived 2 or 3 billion years ago.

## Methane-Producing Bacteria

What were these early bacteria like? Among the most primitive ones that still exist today are the methane-producing bacteria. These organisms are typically simple in form and are able to grow only in an oxygen-free environment. For this reason they are said to grow "without air," or **anaerobically** (Greek *an,* without + *aer,* air + *bios,* life), and are poisoned by oxygen. The methane-producing bacteria convert $CO_2$ and $H_2$ into methane gas ($CH_4$). They resemble all other bacteria in that they possess hereditary machinery based on DNA, a cell membrane composed of lipid molecules, an exterior cell wall, and a metabolism based on an energy-carrying molecule called ATP. However, the resemblance ends at that point.

When the details of membrane and cell wall structure of the methane-producing bacteria are examined, they prove to be different from those of all other bacteria. There are also major differences in some of the fundamental biochemical processes of metabolism that are the same in all other bacteria. The methane-producing bacteria are survivors from an earlier time when there was considerable variation in the mechanisms of cell wall and membrane synthesis, in reading hereditary information, and in energy metabolism. They appear to represent a road not taken, a side branch of evolution.

## Photosynthetic Bacteria

One additional kind of bacteria deserves mention here: that which has the ability to capture the energy of light and transform it into the energy of chemical bonds within cells. These bacteria are photosynthetic, like plants and algae. The pigments used to capture light energy vary in different groups; when these bacteria are massed, they often color the earth, water, or other areas where they grow with characteristic hues (Figure 3-7).

One of the groups of photosynthetic bacteria that is very important in the history of life on earth is the **cyanobacteria,** sometimes called "blue-green algae." They have the same kind of chlorophyll pigment that is most abundant in plants and algae, plus other pigments that are blue or red. Cyanobacteria produce oxygen as a result of their photosynthetic activities, and when they appeared at least 3 billion years ago, they played the decisive role in increasing the concentration of free oxygen in the earth's atmosphere from below 1% to the current level of 21%. As the concentration of oxygen increased, so did the amount of ozone in the upper layers of the atmosphere, thus affording protection from most of the ultraviolet radiation from the sun—radiation that is highly destructive to proteins and nucleic acids. Certain cyanobacteria are also responsible for the accumulation of massive limestone deposits.

## The Origin of Modern Bacteria

The early stages of the history of life on earth seem to have been rife with evolutionary metabolic experimentation. Novelty abounded, and many biochemical pos-

***Figure 3-7***

**A look towards the past.** One can imagine that the earth looked something like this area in Yellowstone National Park as life began. The brownish streaks are masses of bacteria, indicating that the scene closely resembles conditions that occurred billions of years ago, soon after the origin of life.

sibilities were apparently represented among the organisms alive at that time. From the array of different early living forms, representing a variety of biochemical strategies, a very few became the ancestors of the great majority of organisms that are alive today. A few of the other "evolutionary experiments," such as the methane-producing bacteria, have survived locally or in unusual habitats, but most others became extinct millions or even billions of years ago.

*Most organisms now living are descendants of a few lines of early bacteria. Many other diverse forms have not survived.*

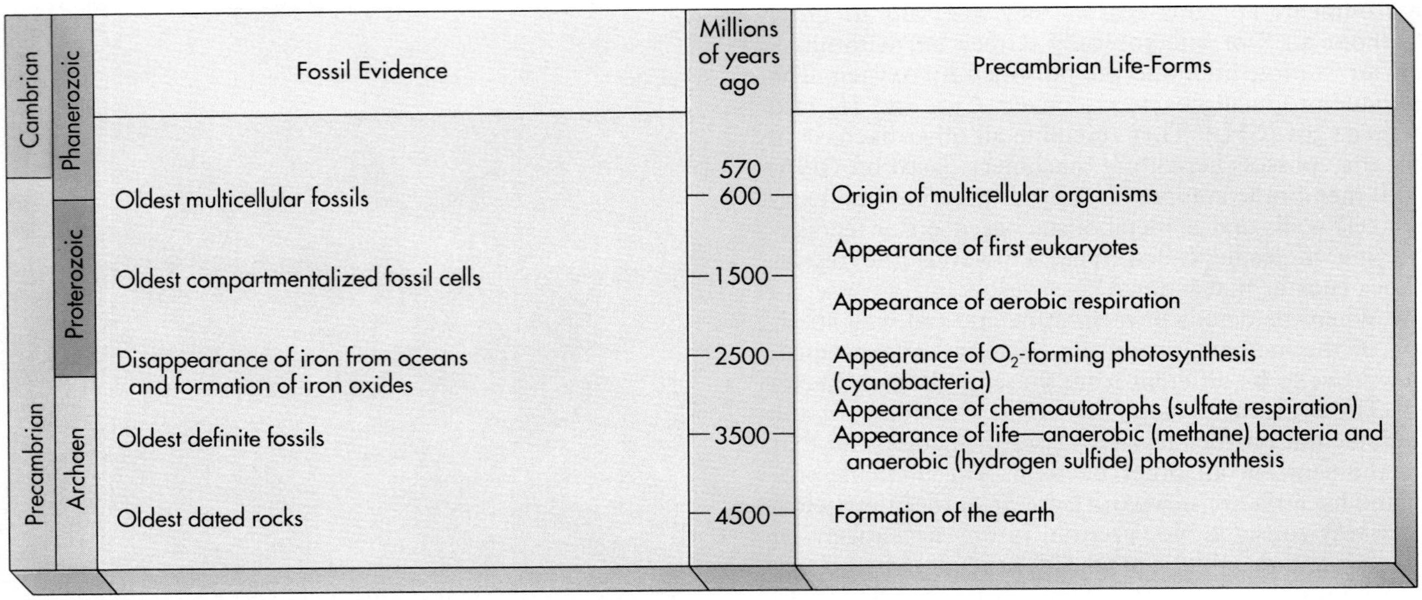

| | | Fossil Evidence | Millions of years ago | Precambrian Life-Forms |
|---|---|---|---|---|
| Cambrian / Phanerozoic | | | 570 | |
| Precambrian | Proterozoic | Oldest multicellular fossils | 600 | Origin of multicellular organisms |
| | | | | Appearance of first eukaryotes |
| | | Oldest compartmentalized fossil cells | 1500 | Appearance of aerobic respiration |
| | | Disappearance of iron from oceans and formation of iron oxides | 2500 | Appearance of $O_2$-forming photosynthesis (cyanobacteria) |
| | Archaen | | | Appearance of chemoautotrophs (sulfate respiration) |
| | | Oldest definite fossils | 3500 | Appearance of life—anaerobic (methane) bacteria and anaerobic (hydrogen sulfide) photosynthesis |
| | | Oldest dated rocks | 4500 | Formation of the earth |

**Figure 3-8**

**The geological time scale.** The periods refer to different stages in the evolution of life on earth. The time scale is calibrated by examining rocks containing particular kinds of fossils and determining the degree of spontaneous decay of radioactive isotopes locked within the rock when it was formed.

Modern bacteria, for the most part, seem to have stemmed from a tough, simple little cell; its hallmark was adaptability. For at least 2 billion years, bacteria were the only form of life on earth (Figure 3-8). All of the eukaryotes, including animals, plants, fungi, and protists, are their descendants.

## THE APPEARANCE OF EUKARYOTIC CELLS

All fossils that are more than 1.5 billion years old are generally similar to one another structurally. They are small, simple cells (Figure 3-9); most measure 0.5 to 2 micrometers in diameter, and none are more than about 6 micrometers thick.

In rocks about 1.5 billion years old we begin to see for the first time microfossils that are noticeably different in appearance from the earlier, simpler forms. These cells are much larger than bacteria and have internal membranes and thicker walls (Figure 3-10). Cells more than 10 micrometers in diameter rapidly increased in abundance. Some fossil cells that are 1.4 billion years old are as much as 60 micrometers in diameter. Others, 1.5 billion years old, contain what appear

to be small, membrane-bound structures. Many of these fossils have elaborate shapes, and some exhibit highly branched filaments, tetrahedral configurations, or spines.

These early fossil traces mark a major event in the evolution of life. A new kind of organism had appeared. These new cells are called **eukaryotes,** from the Greek words for "true" and "nucleus," because they possess an internal chamber called the cell nucleus. All organisms other than the bacteria are eukaryotes, and they rapidly evolved to produce all of the diverse organisms that inhabit the earth today, including ourselves (Figure 3-11). In the next chapter we shall explore in detail the structure of eukaryotes in relation to the factors involved in their origin.

*For at least the first 2 billion years of life on earth, all organisms were bacteria. About 1.5 billion years ago, the first eukaryotes appeared.*

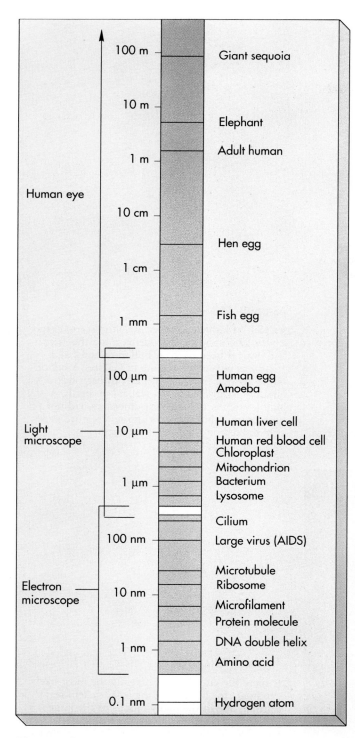

**Figure 3-9**

**The size of things.** Bacteria are generally 1 to 2 micrometers (μm) thick, and human cells are typically orders of magnitude larger. The scale goes from nanometers (nm) to micrometers (μm) to millimeters (mm) to centimeters (cm) and finally meters (m).

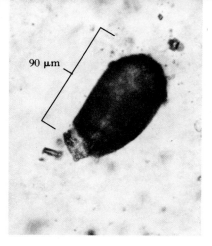

**Figure 3-10**

**Fossil unicellular eukaryote about 800 million years ago.** All life was unicellular until about the past 700 million years.

## IS THERE LIFE ON OTHER WORLDS?

The life forms that evolved on earth closely reflect the nature of this planet and its history. If the earth were farther from the sun, it would be colder, and chemical processes would be greatly slowed down. For example, water would be a solid, and many carbon compounds would be brittle. If the earth were closer to the sun, it would be warmer, chemical bonds would be less stable, and few carbon compounds would persist. Apparently the evolution of a carbon-based life form is possible only within the narrow range of temperatures that exists on earth, and this range of temperature is directly related to the distance from the sun.

The size of the earth has also played an important role because it has permitted a gaseous atmosphere. If the earth were smaller, it would not have a sufficient gravitational pull to hold an atmosphere. If it were larger, it might hold such a dense atmosphere that all solar radiation would be absorbed before it reached the surface of the earth.

Has life evolved on other worlds? In the universe there are undoubtedly many worlds with physical characteristics like those of our planet (Figure 3-12). The universe contains some $10^{20}$ stars with physical characteristics which resemble those of our sun; at least 10% of these stars are thought to have planetary systems. If only 1 in 10,000 planets is the right size and at the right distance from its star to duplicate the conditions in which life originated on earth, the "life experiment" will have been repeated $10^{15}$ times (that is, a million billion times). It seems likely that we are not alone.

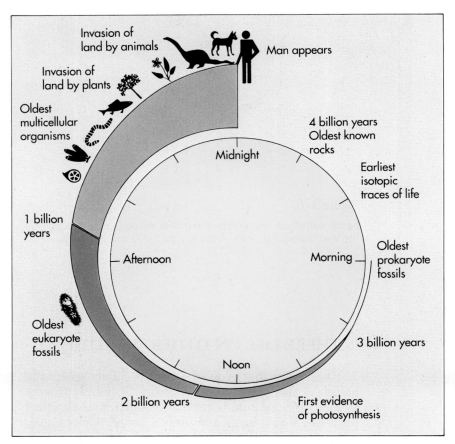

### Figure 3-11

**A clock of biological time.** A billion seconds ago it was 1957, and most students using this text had not yet been born. A billion minutes ago Jesus was alive and walking in Galilee. A billion hours ago the first human had not been born. A billion days ago no biped walked on earth. A billion months ago the first dinosaurs had not yet been born. A billion years ago no creature had ever walked on the face of the earth.

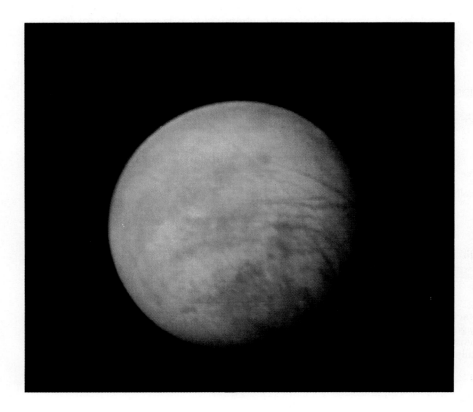

### Figure 3-12

**Is there life elsewhere?** Currently the most likely candidate for life within the solar system is Europa, one of the many moons of the large planet Jupiter. Although most of Jupiter's moons resemble earth's moon, pockmarked by meteors and devoid of life, Europa is covered with ice. Perhaps below its frozen surface the pressures of gravity create enough heat to maintain water in a liquid form. Life may have evolved in such an environment. The conditions on Europa now are far less hostile to life than the conditions that existed in the oceans of the primitive earth. We have not yet traveled to Europa to investigate.

# Is Anybody Out There?

In a heavily forested mountain valley outside the small town of Arecibo in Puerto Rico rests a huge white dish. One thousand feet across, it is the world's largest radio telescope, and it listens to the stars. Over the next few years NASA is proposing to aim it, and other smaller radio telescopes around the world, at one thousand of the stars nearest earth to listen for signals from other worlds. The Search for Extra-Terrestrial Intelligence (SETI) project will be the most comprehensive search ever made for signs that intelligent life is out there somewhere in the stars, trying to get in touch with us.

We have made earlier attempts to send messages telling of our presence. Every radio and television program ever broadcast travels outward, although the signals are very weak. Elvis Presley's appearance on the Ed Sullivan show in 1957 arrived at the star Zeta Herculis in 1989, having travelled 31 light years (1 light year = 5,878,000,000,000 miles). In 1974, a group of Cornell scientists led by Carl Sagan beamed a signal (much of it in image form) toward the Great Cluster, a massive aggregation of stars 25,000 light years away. We cannot expect a reply soon. If they respond the same year they receive our message, we will hear from them 50,000 years from now.

Rather than transmitting messages, the SETI project is designed to listen. For several years, an ongoing project called META (for Megachannel Extra-Terrestrial Assay) has already been listening in a narrow channel surrounding frequencies deemed likely for alien messages, with no success yet. The new NASA project dwarfs these attempts. Using new computer technology called *multi-channel spectrum analysis* it will simultaneously monitor millions of channels over a frequency range of 9 billion hertz (Hz), the entire range of clear radio frequencies that reach the earth. Static noise in our galaxy limits useful transmission by anybody to a "quiet window" of some 60 billion Hz, and all but frequencies between 1 billion and 10 billion Hz are obscured by water in the earth's atmosphere. Any future look at the rest of the 60 billion Hz window will have to be done from space. What does NASA intend to listen for? How will we know when we are being spoken to? The only approach that makes sense is to assume that the signals are designed to be easily detected and understood. What do you do when you want your dog's attention? You shout or clap your hands loudly—the sound you make is loud enough and different enough that it doesn't blend in with the rest of the background noise, and so its gets the dog's attention. An alien signal ought to be something similar, a patterned pulse, perhaps. Or maybe, like Carl Sagan, they will send a picture of themselves.

## SUMMARY

1. Evolution is a scientifically testable hypothesis concerning the origin of life on earth. It can be investigated experimentally. The religious belief that life on earth was specially created cannot be tested and rests on faith.

2. The experimental re-creation of atmospheres, energy sources, and temperatures similar to those thought to have existed on primitive earth leads to the spontaneous formation of amino acids and other biologically significant molecules.

3. The first cells are thought to have arisen by a process in which aggregations of molecules that were more stable persisted longer.

4. Microscopic fossils of bacteria are found continuously in the fossil record as far back as 3.5 billion years in the oldest rocks that are suitable for the preservation of organisms.

5. Bacteria were the only life form on earth for 2 billion years or more. The first eukaryotes appear in the fossil record about 1.5 billion years ago. All organisms other than bacteria are descendants of these first eukaryotes.

6. Bacteria are metabolically, but not structurally, diverse. Some distinctive ones, probably of ancient origin, survive in unusual habitats which may resemble those of the early earth.

## REVIEW

1. Which of the possible explanations for the origin of life permit the construction of testable hypotheses?

2. Fossils of bacteria have been found in rocks that are about _____ billion years old.

3. The Miller-Urey experiment demonstrated that a number of biologically important molecules could be produced from a mixture simulating the earth's early atmosphere, using _____ as an energy source. Name another energy source that could have been used.

4. _____, which form spontaneously in a mixture of lipids with water, display cell-like properties.

5. Eukaryotic cells—that is, cells that contain a nucleus and are found in complex organisms such as yourself first appeared about _____ years ago.

## SELF-QUIZ

1. According to geochemists, which of the following was *not* a major component of the earth's atmosphere when life began?
   (a) Nitrogen gas　　　　　(c) Carbon dioxide　　　　　(e) Oxygen gas
   (b) Methane　　　　　　　(d) Hydrogen

2. In which two locations would you *not* look for living organisms that might resemble some of the early bacteria?
   (a) In the hot springs in Yellowstone National Park.
   (b) In the depths of the Black Sea.
   (c) On some of the older items in your refrigerator.
   (d) In the floor of an old stable that has accumulated urine and manure for centuries.
   (e) In the Gunflint chert, an ancient rock found in Canada.

3. Organisms more complex than bacteria begin appearing as fossils in rocks that are about _____ years old.
   (a) 4.5 billion　　　　　(c) 1.5 billion　　　　　(e) 6000
   (b) 3.5 billion　　　　　(d) 600 million

4. The methane-producing bacteria resemble other prokaryotic organisms in that they (choose three)
   (a) can survive in the presence of oxygen gas.
   (b) use DNA as their hereditary information molecule.
   (c) have a cell membrane composed of lipid molecules.
   (d) use the molecule ATP to drive chemical reactions that require energy.
   (e) show the same biochemical processes when synthesizing cell walls and cell membranes.

5. In the Miller-Urey experiments, which of the following kinds of molecules were *not* produced:
   (a) Amino acids
   (b) Formaldehyde
   (c) Purine bases
   (d) Proteins
   (e) Nucleotides

# THOUGHT QUESTIONS

1. In Fred Hoyle's science fiction novel *The Black Cloud,* the earth is approached by a large interstellar cloud of gas. As the cloud orients around the sun, scientists discover that the cloud is feeding, absorbing the sun's energy through the excitation of electrons in the outer energy levels of cloud molecules, a process similar to the photosynthesis that occurs on earth. Different portions of the cloud are isolated from each other by associations of ions created by this excitation. Electron currents pass between these sectors, much as they do on the surface of the human brain, and endow the cloud with self-awareness, memory, and the ability to think. Using electricity produced by static discharges, the cloud is able to communicate with human beings and to describe its history, as well as to maintain a protective barrier around itself. The cloud tells our scientists that it once was smaller, having originated as a small extrusion from an ancestral cloud, but has grown by absorbing molecules and energy from stars such as our sun, on which it has been grazing. Eventually the cloud moves off in search of other stars. Is the cloud alive? Which of its features would you consider important in deciding whether or not the cloud is alive?

2. The nearest galaxy to ours is the spiral galaxy Andromeda. It contains millions of stars, many of which resemble our sun. The universe contains more than a billion galaxies. Each galaxy, like Andromeda, contains countless thousands of stars. It is interesting to speculate: on planets orbiting these stars, are there students speculating on *our* existence? If 1 in 10 of the stars that are like our sun has planets, if 1 in 10,000 of these planets is capable of supporting life, and if 1 in each million life-supporting planets evolves an intelligent life form, how many planets in the universe support intelligent life? Can you think of any objections to this estimate?

# FOR FURTHER READING

MARGULIS, L., and D. SAGAN: *Microcosmos: Four Billion Years of Evolution from Our Microbial Ancestors,* Summit Books, Simon & Schuster, Inc., New York, 1986. In a beautifully written essay, this mother-son team outlines the evolution of life on earth, showing how all the features that we see today are derived from the early evolution of bacteria. Highly recommended.

SCHOPF, J.W.: "The Evolution of the Earliest Cells," *Scientific American,* vol. 239 (3), 1978, pages 110-139. Discusses how the earliest cells gave rise to the oxygen present in the earth's atmosphere today.

TENNESEN, M.: "Mars. Remembrance of Life Past," *Discover,* July 1989, pages 82-88. A lively discussion of the way life may have originated, then disappeared, on another planet of our solar system.

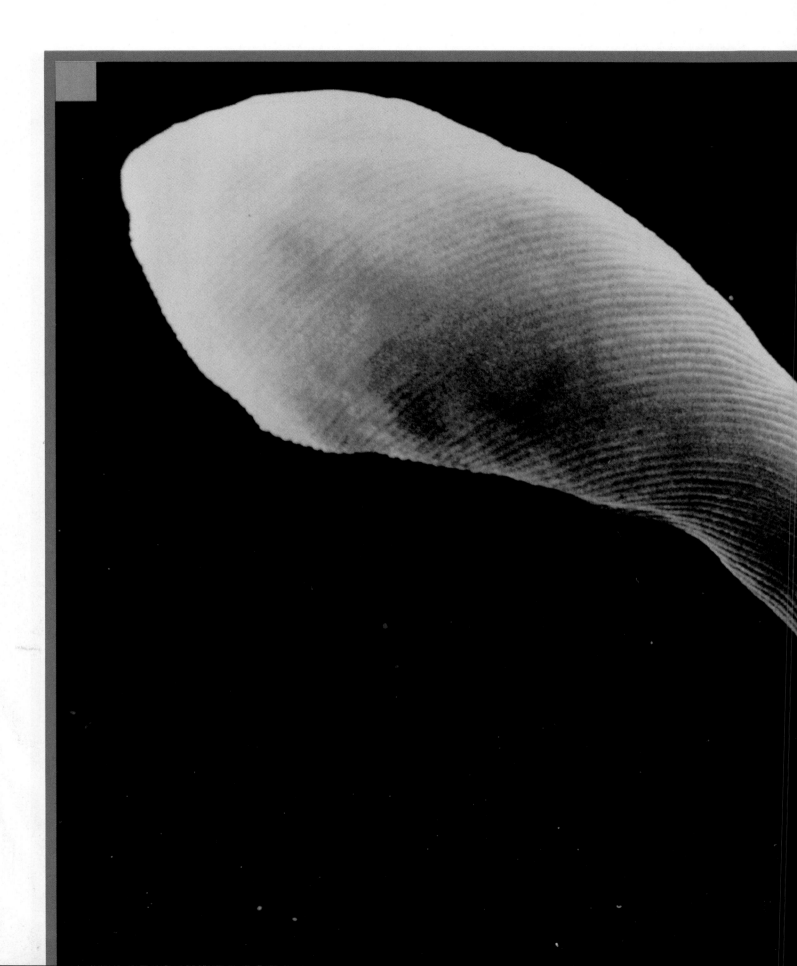

# Cells

This little organism, Euglena, is too little to see readily with the naked eye—it is only 60 micrometers long. A single cell, it nourishes itself by carrying out photosynthesis, and swims by waving its long, threadlike flagellum.

# CELLS

## Overview

The cells of bacteria are complex and efficient metabolic machines, bounded by a membrane, which in some kinds of bacteria is infolded into the interior of the cell. Eukaryotic cells are much more complex, with a considerably greater degree of internal organization, a dynamic system of membranes forming internal compartments, and one or more kinds of membrane-bound organelles. Some of the internal features of a eukaryotic cell are relatively permanent, such as the nucleus, in which the hereditary apparatus is isolated from the rest of the cell. Others are transient, such as the lysosomes that contain digestive enzymes. The partitioning of the cytoplasm into functional compartments is the most distinctive feature of the eukaryotic cell.

## For Review

*Here are some important terms and concepts that you will encounter in this chapter. If you are not familiar with them, you should review them before proceeding.*

**Proteins** (Chapter 2)

**Lipids** (Chapter 2)

**Distinction between prokaryotic and eukaryotic cells** (Chapter 3)

**Evolution of eukaryotes** (Chapter 3)

All organisms are composed of cells. Some are composed of a single cell (Figure 4-1), and some, such as us, are composed of many cells. The gossamer wing of a butterfly is a thin sheet of cells, and so is the glistening layer covering your eyes. The hamburger you eat is composed of cells, whose contents will soon become part of your cells. Your eyelashes and fingernails, orange juice, and the wood in your pencil—all were produced by or consist of cells. We cannot imagine an organism that is not cellular in nature. In this chapter we look more closely at cells and learn something of their internal structure. In the following chapters we focus on cells in action—how they communicate with their environment, grow, and reproduce.

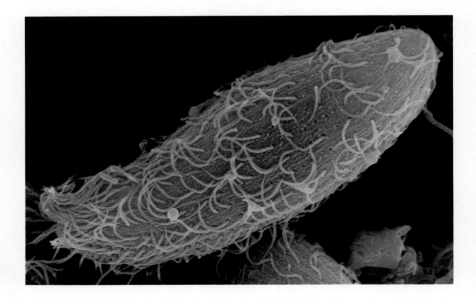

**Figure 4-1**

**Scanning electron micrograph (× 1000) of the single-celled protist *Dileptus anser*.** The hairlike projections that cover the surface are cilia, which the organism undulates to propel itself through the water.

# WHAT CELLS ARE LIKE

Before launching into a detailed examination of cell structure, it is useful first to gain an overview of what to expect. What would we find on the inside of a cell? What is a typical cell like?

A bacterial cell, with its prokaryotic organization, is enclosed by an outer cell wall that surrounds a **plasma membrane,** a lipid bilayer with embedded proteins that controls the permeability of the cell to water and dissolved substances. Within, bacterial cells are relatively uniform in appearance. Their **flagella,** if present, are uniform, threadlike protein structures that move the bacteria by rotating. Eukaryotic cells are far more complex; those of animals and a few protists lack cell walls. Within every eukaryotic cell is a **nucleus,** which is its control center. The power that drives a eukaryotic cell comes from internal bacteria-like inclusions called **mitochondria.** A eukaryotic cell is further subdivided into separate compartments by a winding membrane system called the **endoplasmic reticulum.** The flagella that propel motile eukaryotic cells are much more complex than those of bacteria and are not related to bacterial flagella in an evolutionary sense. They propel the cell through its medium by undulating rapidly.

1. A membrane surrounds every cell, isolating it from the outside. Chapter 5 describes the many different kinds of passageways and communication channels that span these membranes. They provide the only connection between the cell and the outside world.
2. The nuclear region directs the activities of the cell. In bacteria, most of the genetic material is included in a single, closed, circular molecule of DNA, which resides in a central portion of the cell that is not bounded by membranes. In eukaryotes, by contrast, a double membrane, the **nuclear envelope,** surrounds the nucleus, which contains the DNA.
3. A semifluid matrix called the **cytoplasm** fills bacterial cells; in eukaryotic cells, the cytoplasm occupies the volume between the nuclear region and the cell membrane. The cytoplasm contains the chemical wealth of the cell, the sugars and amino acids and proteins with which the cell carries out its everyday activities of growth and reproduction. In addition to these elements, the cytoplasm of a eukaryotic cell also contains numerous organized structures called **organelles,** or "little organs." Many of these organelles are created by the membranes of the endoplasmic reticulum, which close off compartments within which different activities occur. The cytoplasm

of eukaryotic cells also contains organelles that resemble bacteria; these organelles, called **mitochondria,** provide power.

All cells share this architecture. However, the general plan is modified in various ways in different classes of cells. For example, the cells of most kinds of organisms—plants, bacteria, fungi, protists—possess an outer cell wall that is rigid and provides structural strength; animal cells, and those of a few other kinds of organisms, do not. Eukaryotic cells generally possess a single nucleus, whereas the cells of fungi often have several nuclei, as do those of some other groups of organisms and those of particular kinds of tissues. Most cells derive all their power from the kind of organelles called mitochondria. But plant and algal (**algae** are photosynthetic protists) cells contain a second kind of bacteria-like powerhouse, **chloroplasts,** in addition to their mitochondria. Despite their differences, all cells are fundamentally similar.

*A cell is a membrane-bound unit containing hereditary machinery and other components, including enzymes. Because of these components, cells are able to metabolize, grow, and reproduce.*

## MOST CELLS ARE VERY SMALL

Sometimes important things seem so obvious that they are overlooked. In studying cells, for example, it is important that we do not overlook one of their most striking traits—their very small size. Cells are not like shoeboxes, big and easy to study. Instead, they are so small that you cannot see a single one of your body's cells with the naked eye. Your body contains about 100 trillion cells. If each cell were the size of a shoebox and they were lined up end-to-end, the line would extend to Mars and back, over 500 million kilometers.

### The Cell Theory
It is because cells are so small that they were not observed until microscopes were invented in the mid-seventeenth century. Cells were first described by Robert Hooke in 1665. Using a microscope he had built to examine a thin slice of cork, Hooke observed a honeycomb of tiny empty compartments. He called the compartments in the cork *cellulae,* using the Latin word for a small room. His term has come down to us as **cells.**

The first living cells were observed by the Dutch naturalist Antonie van Leeuwenhoek a few years later. van Leeuwenhoek called the tiny organisms that he observed "animalcules"—little animals (Figure 4-2). For another century and a half, however, the general importance of cells was not appreciated by biologists. In 1838 the German Matthias Schleiden, after a careful study of plant tissues, made the first statement of what we now call the cell theory. Schleiden stated that all plants "are aggregates of fully individualized, independent, separate beings, namely the cells themselves." The following year, Theodor Schwann reported that all animal tissues are also composed of individual cells.

The cell theory in its modern form includes four principles:

1. All organisms are composed of one or more cells, within which the life processes of metabolism and heredity occur.
2. Cells are the smallest living things, the basic unit of organization of all organisms.
3. Although life evolved spontaneously in the hydrogen-rich environment of early earth, biologists have concluded that additional cells are not originating at present. Rather, life on earth represents a continuous line of descent from those early cells.
4. Cells arise only by division of a previously existing cell.

**Figure 4-3**

**Acetabularia.** The marine green alga *Acetabularia*, a protist, is a large organism with clearly differentiated parts, such as the stalks and elaborate "hats" visible here. Each individual is actually a single cell several centimeters tall.

*All the organisms on earth are cells or aggregates of cells, and all of us are descendants from the first cells.*

## Why Aren't Cells Larger?

Cells are not all the same size. Individual cells of the marine alga *Acetabularia,* for example, are up to 5 centimeters long (Figure 4-3). In contrast, the cells of your body are typically from 5 to 20 micrometers in diameter. If a typical cell in your body were again the size of a shoebox, an *Acetabularia* cell to the same scale would be about 2 kilometers high! The cells of bacteria are much smaller, usually about 1 to 10 micrometers thick.

Why are our bodies made up of so many tiny cells? The centralized control of each cell is essential in maintaining it as a distinct, functional unit. Substances travel through a cell by diffusion, a relatively slow process of random molecular movement. If a cell were too large, it could not function efficiently. The nucleus must send commands to all parts of the cell, molecules that direct the synthesis of certain enzymes, the entry of ions from the exterior, and the assembly of organelles. These molecules must pass by diffusion from the nucleus to all parts of the cell, and it takes them a very long time to reach the periphery of a large cell. For this reason, an organism made up of relatively

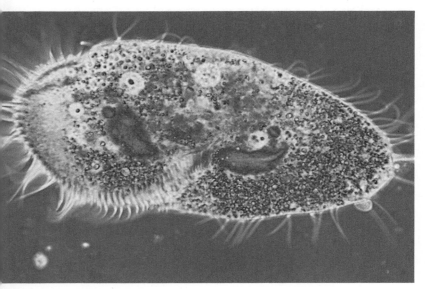

**Figure 4-2**

**Single-celled organisms are not simple.** Among the "animalcules" that can be seen with a microscope are individuals of *Paramecium*, which dash this way and that, engulfing smaller organisms, and clearly are vibrantly alive.

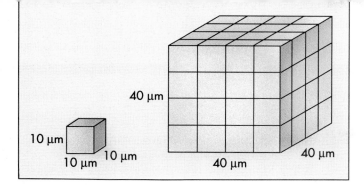

**Figure 4-4**

**Surface to volume ratio.** Since cells depend on their surface area to interact with their environment, large organisms are divided into smaller cells to provide the greatest surface area possible. The cube on the left symbolizes a single cell, the cube on the right a large organism. By dividing the larger cube into smaller cubes, surface area is maximized.

small cells is at an advantage over one composed of larger cells.

The advantage of small cell size is also seen in terms of what is called the **surface-to-volume ratio.** As cell size increases, the volume grows much more rapidly than does the surface area (Figure 4-4). For a round cell, surface area increases as the square of diameter, whereas volume increases as the cube. Thus a cell with 10 times greater diameter would have 10 squared, or 100 times, the surface area but 10 cubed, or 1000 times, the volume. A cell's surface provides its

only opportunity to interact with the environment, and large cells have far less surface-per-unit volume than do small ones. All substances must enter and exit from a cell via the plasma membrane, a structure of fundamental importance that is discussed in more detail in Chapter 5. The plasma membrane plays a key role in controlling cell function, something that is more effectively done when cells are relatively small.

*We have many small cells rather than few large ones because small cells can be commanded more efficiently and have a greater opportunity to communicate with their environment.*

## THE STRUCTURE OF SIMPLE CELLS: BACTERIA

Bacteria are the simplest cellular organisms. Over 2500 species are recognized, but doubtless many times that number actually exist and have not yet been described properly. Although these species are diverse in form (Figure 4-5), their organization is fundamentally simi-

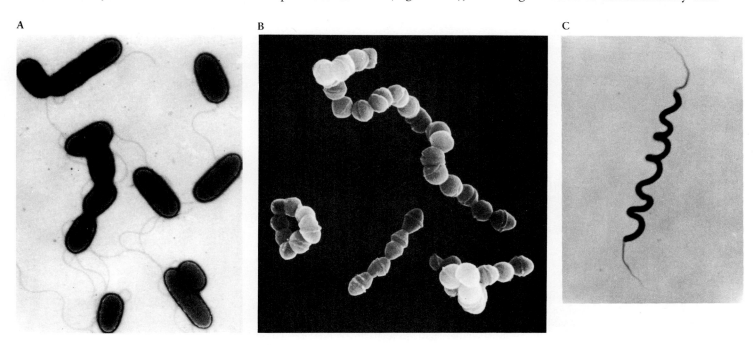

**Figure 4-5**

**Bacterial cells have several different shapes.**
**A** A rod-shaped bacterium, *Pseudomonas,* a type associated with many plant diseases.
**B** *Streptococcus,* a more or less spherical bacterium in which the individuals adhere in chains.
**C** *Spirillum,* a spiral bacterium. This large bacterium has a tuft of flagella at each end.

# Microscopes: Looking at What's Tiny

Cells are so small that you cannot see them with the naked eye. Most eukaryotic cells are between 10 and 30 micrometers in diameter. Why can't we see such small objects? Because when two objects are closer together than about 100 micrometers, the two light beams fall on the same "detector" cell at the rear of the eye. Only when two dots are farther apart than 100 micrometers will the beams fall on different cells, and only then can your eye tell that they are two objects and not one.

Robert Hooke and Antonie van Leeuwenhoek were able to see very small cells by magnifying their size, so that the cells appeared larger than the 100-micrometer limit imposed by the structure of the human eye. Hooke and van Leeuwenhoek accomplished this with simple microscopes, which magnified images of cells by bending light through a glass lens. To understand how such a single-lens microscope is able to magnify an image, examine Figure 4-A, *1*. The size of the image that falls on the screen of detector cells lining the back of your eye depends on how close the object is to your eye—the closer the object, the bigger the picture. However, your eye is unable to comfortably focus on an object closer than about 25 centimeters (Figure 4-A, *2*, top) because it is limited by the size and thickness of its lens. What Hooke and van Leeuwenhoek did was help the eye out by interposing a glass lens between the object and the eye (Figure 4-A, *2*, bottom). The glass lens added additional focusing power, producing an image of the close-up object on the back of the eye. But because the object is closer, the image on the back of the eye is bigger than it would have been had the object been 25 centimeters away from the eye. It is as big as a *much larger* object placed 25 centimeters away would have appeared without the lens. You perceive the object as magnified, or bigger.

The microscope used by van Leeuwenhoek consists of (1) a plate with a single lens, (2) a mounting pin that holds the specimen to be observed, (3) a focusing screw that moves the specimen nearer or farther from the eye, and (4) a specimen-centering screw.

van Leeuwenhoek's microscope, (Figure 4-B, *1*) although simple in construction, is very powerful. One of van Leeuwenhoek's original specimens, a thin slice of cork, was recently discovered among his papers. In the image of that section obtained with

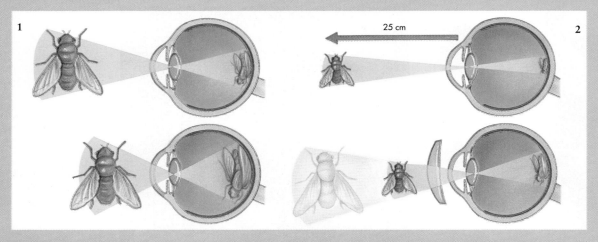

**Figure 4-A**

**How does a microscope magnify? 1,** The closer an object is to the eye, the larger the image that falls on the back of that eye, the rear surface of which is covered with cells called photoreceptors. **2,** The eye will not comfortably focus an object closer than about 25 centimeters because the lens of the eye must change shape to focus and cannot exceed this limit. The glass lens aids the eye in focusing the close object. Because the object is closer, it produces a larger image on the back of the eye and so appears "larger." A much larger object would have been required to produce an image of the same size without the lens.

van Leeuwenhoek's own microscope, the magnification is 266 times, as good as many modern microscopes. The finest structures visible are less than 1 micrometer (1000 nanometers) in diameter (Figure 4-B, *2*).

Modern microscopes use two magnifying lenses (and various correcting lenses) that act like back-to-back eyes. The first lens focuses the image of the object on the second lens. The image is then magnified again by the second lens, which focuses it on the back of the eye. Microscopes that magnify in stages by using several lenses are called **compound microscopes**. The finest structures visible with modern compound microscopes are about 200 nanometers in thickness. A contemporary light micrograph is shown in Figure 4-C, *1*.

Compound light microscopes are not powerful enough to resolve many structures within cells. A membrane, for example, is only 5 nanometers thick. Why not just add another magnifying lens to the microscope and so increase the **resolving power**, the ability to distinguish two lines as separate? This approach doesn't work because when two objects are closer than a few hundred nanometers, the light beams of the two images start to overlap. A light beam vibrates like a plucked string, and the only way two beams can get closer together and still be resolved is if the "wavelength" is shorter.

The wavelength of light ranges from about 0.4 micrometer for violet light to about 0.7 micrometer for red light. This limits the best light microscopes to a resolving power of 0.2 micrometer; they are thus able to improve on the naked eye about 500 times. They are able to distinguish the structures of eukaryotic cells and individual prokaryotic cells, but they are not able to visualize the internal structure of prokaryotic cells or of bacteria.

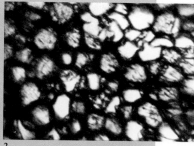

***Figure 4-B***

**1,** Antonie van Leeuwenhoek's microscope, with a model. **2,** An image of Leeuwenhoek's sample of cork, obtained with his microscope, which is preserved at Utrecht, in the Netherlands. In this image, the cork is modified × 266; it compares well with modern images of thin sections of cork. The finest structures that are visible are less than one micrometer across.

One way to accomplish greater magnifications is by using a beam of electrons rather than a light beam. Electrons have a much shorter wavelength, and a microscope employing electron beams has 400 times the resolving power of a light microscope. **Transmission electron microscopes** today are capable of resolving objects only 0.2 nanometer apart—just five times the diameter of a hydrogen atom (Figure 4-C, *2*).

The **scanning electron microscope** beams the electrons onto the surface of the specimen as a fine probe, which passes back and forth rapidly. The electrons that are reflected back from the surface of the specimen, together with other electrons that the specimen itself emits as a result of the bombardment, are amplified and transmitted to a television screen, where the image can be viewed and photographed. Scanning electron microscopy yields striking three-dimensional images and has proved to be very useful in understanding many biological and physical phenomena (Figure 4-C, *3*).

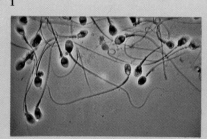

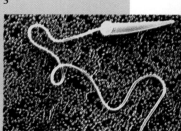

***Figure 4-C***

**1,** Image of sperm cell taken with a light microscope (× 400). **2,** Transmission electron micrograph of sperm cell (× 15,000). **3,** Scanning electron micrograph of sperm cell (× 8500).

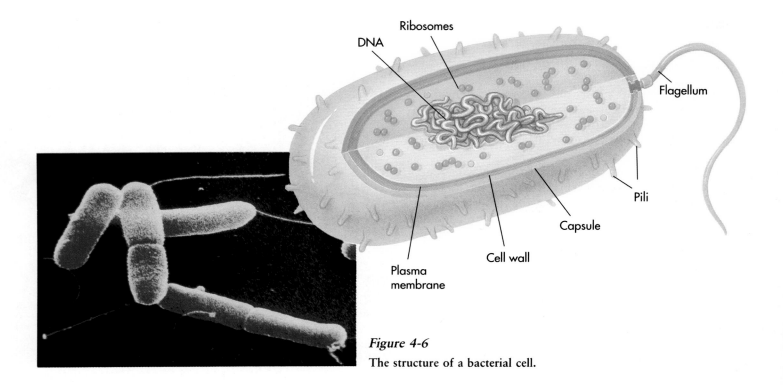

DNA

Ribosomes

Flagellum

Pili

Capsule

Cell wall

Plasma
membrane

**Figure 4-6**
The structure of a bacterial cell.

lar: small cells about 1 to 10 micrometers thick; enclosed, like all living cells, by a plasma membrane; and encased within a rigid cell wall, with no distinct interior compartments (Figure 4-6). Sometimes the cells of bacteria adhere in chains or masses, but fundamentally the individual cells are separate from one another.

*Compared with eukaryotic cells, which evolved from them, prokaryotic cells (bacteria) are smaller and lack interior organization.*

## Strong Cell Walls

Bacteria are encased by a strong **cell wall,** in which a framework made up of carbohydrates provides the basic structure. These carbohydrates are cross-linked into a rigid structure by short peptide units—chains of amino acids, like segments of proteins. No eukaryote possesses cell walls with this kind of structure.

## Simple Interior Organization

If you were to look at an electron micrograph of a thin section of a bacterial cell, you would be struck by its simple organization. There are no internal compartments bounded by membranes or membrane-bounded organelles—the kinds of distinct structures that are characteristic of eukaryotic cells. The entire cytoplasm of a bacterial cell is one unit, with no internal support structure; thus the strength of the cell comes primarily from its rigid wall.

The plasma membrane of bacterial cells often intrudes into the interior of the cell, where it may play an important role. In some photosynthetic bacteria, the cell membrane is extensively folded, with the folds extending into the cell's interior (Figure 4-7). The pigments connected with photosynthesis are located on these folded membranes.

Because there are no membrane-bounded compartments within a bacterial cell, both the DNA and the enzymes within such a cell have access to all parts of the cell. Reactions are not compartmentalized as they are in eukaryotic cells.

*Bacteria are encased by an exterior wall composed of carbohydrates cross-linked by short peptides. They lack a nucleus and other interior membrane-bound compartments.*

## SYMBIOSIS AND THE ORIGIN OF EUKARYOTES

The first eukaryotes had a cell structure that was radically different from that of all organisms that had existed earlier. Eukaryotic cells are far more complex internally than their bacterial ancestors. We know little about how this increase in internal organization first evolved, 1.5 billion years ago—with one exception. Within the cells of virtually all eukaryotes are organelles that resemble bacterial cells in both size and appearance. Most biologists interpret a wealth of evidence as indicating that that is just what they are, ancient bacteria that lived within ancestral eukaryote cells, where they provided (and still provide) their host the advantages associated with their special metabolic abilities. Mitochondria, the energy factories of eukaryotic cells, are organelles of this sort, and so are chloroplasts, the organelles in which photosynthesis takes place in eukaryotic cells. These critically important metabolic functions evolved in different groups of bacteria but are combined in individual eukaryotic cells through the process of **symbiosis,** the living together in close association of two or more dissimilar organisms.

## A COMPARISON OF BACTERIA AND EUKARYOTIC CELLS

Even a cursory examination of a eukaryotic cell (Figure 4-8) reveals it to be far more complex than a prokaryotic one. Compared with their bacterial ancestors, eukaryotic cells exhibit these significant differences:

1. Their DNA is packaged tightly into compact units called **chromosomes,** which contain both DNA and proteins. Except when they are dividing, these eukaryotic chromosomes are located within a separate organelle, the nucleus.
2. The interiors of eukaryotic cells are subdivided into membrane-bounded compartments, which permits one biochemical process to proceed independently of others that may be going on at the same time. The compartmentalization of biochemical activities in eukaryotes serves to increase the efficiency of the various processes.
3. The cells of animals and some protists lack cell walls.
4. The mature cells of plants often contain large, fluid-filled internal sacs called **central vacuoles** that are not present in animal cells (Figure 4-9).

*The interiors of eukaryotic cells are subdivided by a complex set of membranes. The DNA of the cell is present within the nucleus and is associated with protein in units called chromosomes. Most organisms possess cell walls, which are absent in animals and some protists.*

## THE INTERIOR OF EUKARYOTIC CELLS: AN OVERVIEW

Although eukaryotic cells are diverse in form and function, they share a basic architecture (Table 4-1). All such cells are bounded by a membrane called the **plasma membrane,** all contain a supporting matrix of protein called a **cytoskeleton,** and all possess numerous organelles. The organelles are of two general kinds: membranes or organelles derived from membranes (class 1), and DNA-containing organelles that resemble bacteria (class 2).

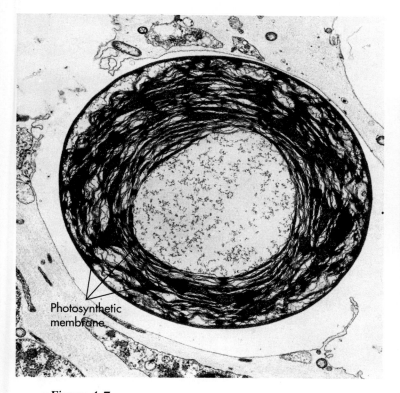

Photosynthetic membrane

**Figure 4-7**

**Electron micrograph of a photosynthetic bacterial cell,** ***Prochloron.*** Extensive folded photosynthetic membranes are visible. The cellular DNA is located in the clear area in the central region of the cell.

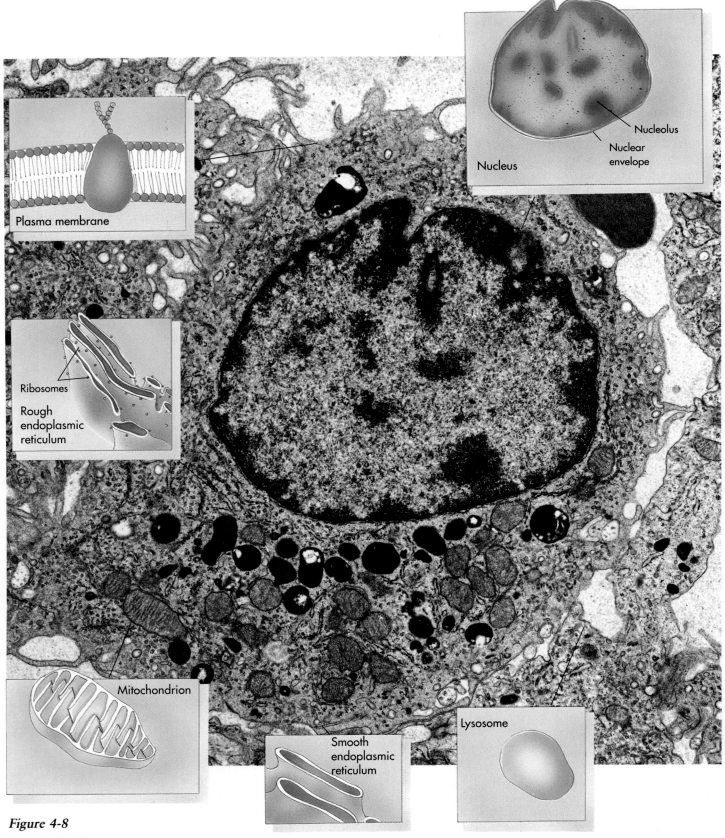

**Plasma membrane**

**Nucleus**

**Nucleolus**

**Nuclear envelope**

**Ribosomes**

**Rough endoplasmic reticulum**

**Mitochondrion**

**Smooth endoplasmic reticulum**

**Lysosome**

*Figure 4-8*
An animal cell.

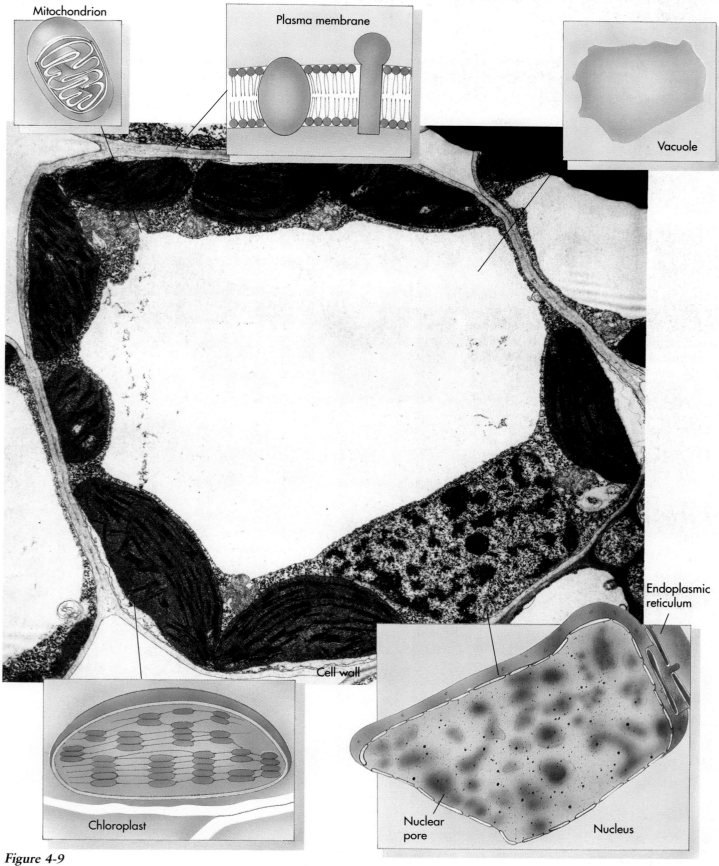

Mitochondrion

Plasma membrane

Vacuole

Cell wall

Endoplasmic reticulum

Chloroplast

Nuclear pore

Nucleus

*Figure 4-9*

A plant cell.

## TYPES OF ORGANELLES

| CLASS 1 | CLASS 2 |
|---|---|
| Endoplasmic reticulum | Mitochondria |
| Nucleus | Chloroplasts |
| Golgi bodies | Centrioles |
| Lysosomes | |
| Microbodies | |

## MEMBRANES

All biological membranes are phospholipid bilayers about 7 nanometers thick, with embedded proteins; viewed with the electron microscope in cross-section, such membranes appear as two dark lines separated by a lighter area (Figure 4-10). This distinctive appearance arises from the tail-to-tail packing of the phospholipid molecules that make up a membrane. The major proteins of a membrane are hydrophobic, therefore they associate with and become embedded in the phospholipid matrix. Very few molecules—water is one, as we shall see—are able to cross the lipid bilayer of a membrane; instead, their passage is selectively controlled by the proteins that lie embedded in this matrix, and these proteins also affect the properties of the individual kind of membrane. The properties of the plasma membrane and the various internal membranes of a eukaryotic cell will be discussed in relation to the various structures involved.

## A CELL'S BOUNDARY: THE PLASMA MEMBRANE

The different kinds of proteins that are embedded in the plasma membrane control the interactions of the

### TABLE 4-1 EUKARYOTIC CELL STRUCTURES AND THEIR FUNCTIONS

| STRUCTURE | DESCRIPTION | FUNCTION |
|---|---|---|
| **STRUCTURAL ELEMENTS** | | |
| Cell wall | Outer layer of cellulose or chitin, or absent | Protection; support |
| Plasma membrane | Lipid bilayer in which proteins are embedded | Regulates what passes in and out of cell; cell-to-cell recognition |
| Cytoskeleton | Network of protein filaments | Structural support; cell movement |
| Flagella (cilia) | Cellular extensions with 9 + 2 arrangement of pairs of microtubules | Motility or moving fluids over surfaces |
| **ORGANELLES** | | |
| Endoplasmic reticulum | Network of internal membranes | Forms compartments of vesicles |
| Ribosomes | Small, complex assemblies of protein and RNA, often bound to endoplasmic reticulum | Sites of protein synthesis |
| Nucleus | Spherical structure bounded by double membrane; contains chromosomes | Control center of cell; directs protein synthesis and cell reproduction |
| Chromosomes | Long threads of DNA associated with protein | Contain hereditary information |
| Nucleolus | Site on chromosome of rRNA synthesis | Assembles ribosomes |
| Golgi complex | Stacks of flattened vesicles | Packages proteins for export from cell; forms microbodies |
| Peroxisomes | Vesicles containing collections of oxidative and other enzymes | Isolate particular chemical activities from rest of cell |
| Lysosomes | Microbodies containing digestive enzymes | Digest worn-out mitochondria and cell debris; play role in cell death |
| **ENERGY-PRODUCING ORGANELLES** | | |
| Mitochondria | Bacteria-like elements with inner membrane highly folded | Power plant of cell; site of oxidative metabolism |
| Chloroplasts | Bacteria-like elements with vesicles containing chlorophyll | In plant cells; site of photosynthesis |

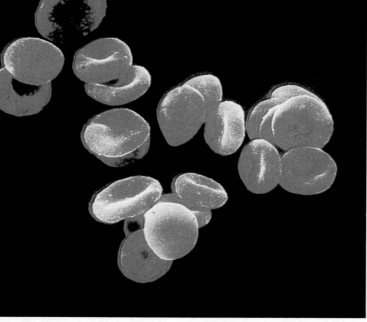

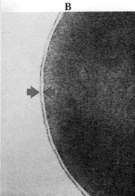

**Figure 4-10**

**A** A red blood cell. It is shaped like a flattened disk with a depressed center—not unlike a doughnut with a hole that doesn't go all the way through.
**B** This electron micrograph of a thin section of a red blood cell (× 200,000) shows clearly the double nature of the plasma membrane, which is indicated by arrows.

cell with its environment. The major kinds of proteins that occur in the plasma membrane are:

1. *Channels.* Some of these proteins, called channels, act as doors that admit specific molecules to the cell; for example, membranes possess specific channels for sodium ions and others for glucose.
2. *Receptors.* Other proteins that cross the membrane serve to transmit information rather than molecules. These proteins, called receptors, induce changes within the cell when they come in contact with particular molecules on the cell surface. Some hormones induce changes in cells by binding to such surface receptors.
3. *Markers.* A third class of proteins embedded within the membrane serves to identify a cell as being of a particular type. This is very important in a multicellular individual because cells must be able to recognize one another for tissues to form and function correctly.

Chapter 5 presents a more detailed look at the structure of cell membranes and how they function.

## THE ENDOPLASMIC RETICULUM: WALLS WITHIN THE CELL

As viewed with a light microscope, the interiors of eukaryotic cells exhibit a relatively featureless matrix within which various organelles are embedded. But with the advent of electron microscopes, a very striking difference became evident: the interior of a eukaryotic cell is packed with membranes. So thin that they are not visible with the relatively low resolving power of

light microscopes, these membranes fill the cell, dividing it into compartments, channeling the transport of molecules through the interior of the cell, and providing the surfaces on which enzymes act. This system of internal compartments created by these membranes in eukaryotic cells constitutes the most fundamental distinction between eukaryotes and prokaryotes.

The extensive system of internal membranes that exists within the cells of eukaryotic organisms is called the **endoplasmic reticulum,** often abbreviated ER (Figure 4-11). The term *endoplasmic* means "within the cytoplasm," and the term *reticulum* comes from a Latin word that means "a little net." Like the plasma membrane, the ER is composed of a double layer of lipid, with various enzymes attached to its surface. The ER, weaving in sheets through the interior of the cell, creates a series of channels and interconnections between its membranes that isolates some spaces as membrane-enclosed sacs called **vesicles.**

### Rough ER: Manufacturing for Export

The surface of the ER is the place where the cell manufactures proteins intended for export from the cell, such as enzymes secreted from the cell surface. The manufacture is carried out by **ribosomes,** organelles composed of protein and RNA, which translate RNA copies of genes into protein. As new proteins are made on the surface of the ER, they are passed out across the ER membrane (Figure 4-12) into the vesicle-forming system called the Golgi complex (discussed below). They then travel within vesicles to the inner surface of the cell, where they are released to the outside of the cell in which they were produced. From the time the protein is first synthesized on the ER-bound ribosome and crosses into these channels, it is, in a sense, already located outside the cell.

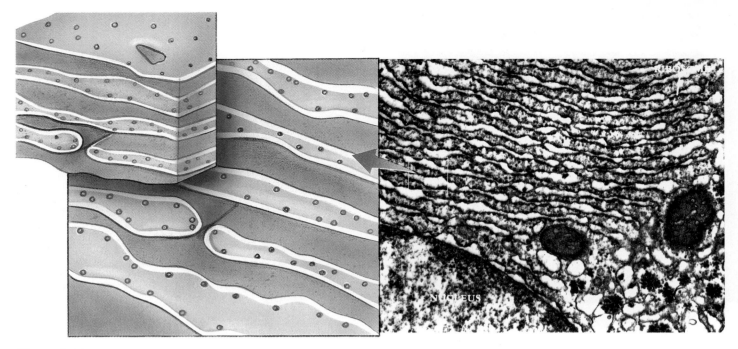

**Figure 4-11**

**Rough endoplasmic reticulum (ER).** The electron micrograph is of a rat liver cell, rich in ER-associated ribosomes. In the drawing, you can see that the ribosomes are associated with only one side of the rough ER; the other side is the boundary of a separate compartment within the cell into which the ribosomes extrude newly made proteins destined for secretion.

The surfaces of those regions of the ER devoted to the synthesis of such transported proteins are heavily studded with ribosomes. These membrane surfaces appear pebbly, like the surface of sandpaper, when visualized with an electron microscope. Because of this "rocky beach" appearance, the regions of ER that are rich in bound ribosomes are termed **rough ER.** Regions of the ER in which bound ribosomes are relatively scarce are correspondingly called **smooth ER.**

### Smooth ER: Organizing Internal Activities

Many of the cell's enzymes cannot function when they are free in the cytoplasm; they are active only when they are associated with a membrane. ER membranes contain many different enzymes embedded within them, and the enzymes carry out their functions from these positions. Enzymes anchored within the ER catalyze the synthesis of a variety of carbohydrates and lipids. In cells that carry out extensive lipid synthesis, such as the cells of the testicles, smooth ER is particularly abundant. Intestinal cells, which synthesize triglycerides, and the cells of the brain are also rich in smooth ER. In the liver, enzymes embedded within the smooth ER are involved in various detoxification processes. Drugs such as amphetamines, morphine, codeine, and phenobarbital are detoxified in the liver by components of the smooth ER.

*The endoplasmic reticulum (ER) is an extensive system of membranes that divides the interior of eukaryotic cells into compartments and channels. Rough ER functions mainly in the synthesis and transport of proteins across the ER membrane, whereas smooth ER organizes the synthesis of lipids and other molecules.*

## THE NUCLEUS: CONTROL CENTER OF THE CELL

The largest and most easily seen of the organelles within an eukaryotic cell is the **nucleus** (Figure 4-13), which was first described by the English botanist Robert Brown in 1831. The word "nucleus" is derived from the Greek word for a nut, an object that nuclei somewhat resemble. Nuclei are usually spherical; in animal cells, they generally occur near the center of the cell, sometimes appearing to be cradled in this position by a network of fine filaments. The nucleus is the repository of the genetic information that directs all the activities of a living cell. Some kinds of cells, such as

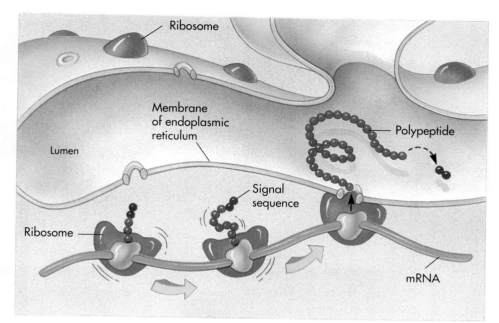

Ribosome

Membrane of endoplasmic reticulum

Lumen

Polypeptide

Signal sequence

Ribosome

mRNA

**Figure 4-12**

**How a protein enters the Golgi complex.** Many biologists now believe that short 20-amino–acid sequences at the tips of proteins direct their destinations in the cell. In this example, a sequence of hydrophobic amino acids on a secretory protein attaches them (and the ribosomes making them) to ER, injecting the leading end of the newly made protein across the ER into the lumen (space) of the Golgi complex. As the synthesis of the protein continues, it passes across the ER, "exported" into the lumen. The signal sequence is clipped off after the leading edge of the protein enters the lumen.

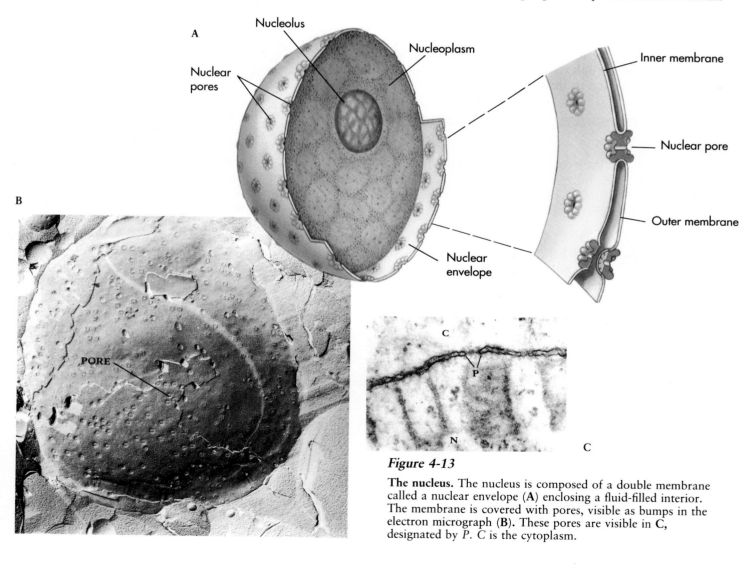

A

Nucleolus

Nucleoplasm

Nuclear pores

Inner membrane

Nuclear pore

Outer membrane

Nuclear envelope

B

PORE

C

P

N

C

**Figure 4-13**

**The nucleus.** The nucleus is composed of a double membrane called a nuclear envelope (**A**) enclosing a fluid-filled interior. The membrane is covered with pores, visible as bumps in the electron micrograph (**B**). These pores are visible in C, designated by *P*. *C* is the cytoplasm.

mature red blood cells, lose their nuclei at the final stage of their development. Once this loss has occurred, the cells lose all ability to grow, change, and divide and become merely vessels for the substances they contain.

## Getting In and Out: The Nuclear Envelope

The surface of the nucleus is bounded by *two* membranes, the outer and inner membranes of the **nuclear envelope** (see Figure 4-13). The ER is continuous with the outer layer of the nuclear envelope. Scattered over the surface of this envelope, like the craters of the moon, are shallow depressions called **nuclear pores.** These pores, 50 to 80 nanometers apart, form at locations where the two membrane layers of the nuclear envelope pinch together. A nuclear pore is not an empty opening like the hole in a doughnut; such a pore contains many embedded proteins that act as molecular channels, permitting certain molecules to pass into and out of the nucleus. Passage is restricted primarily to two kinds of molecules: (1) proteins moving into the nucleus, where they will be incorporated into nuclear structures or catalyze nuclear activities, and (2) RNA and protein-RNA complexes formed in the nucleus and subsequently exported to the cytoplasm.

*The nucleus of a eukaryotic cell contains the cell's hereditary apparatus, isolated from the rest of the cell.*

## The Chromosomes of Eukaryotes Are Complex

Both in bacteria and in eukaryotes, all the hereditary information specifying cell structure and function is encoded in DNA. However, unlike the DNA of bacteria, the DNA of eukaryotes is divided into several segments and associated with protein and RNA, forming **chromosomes** (Figure 4-14). Association with protein enables eukaryotic DNA to wind up into a highly condensed form during cell division. Under a light microscope, these condensed chromosomes are readily seen in dividing cells as densely staining rods. After cell division, eukaryotic chromosomes uncoil and can no longer be distinguished individually with a light microscope. Uncoiling the chromosomes into a more extended form permits the enzymes that make RNA copies of DNA to gain access to the DNA molecule. Only by means of these RNA copies can the hereditary information be used to direct the synthesis of enzymes.

**Figure 4-14**
Eukaryotic chromosomes *(arrow)* within an onion root-tip cell.

*A distinctive feature of eukaryotes is the organization of their DNA into chromosomes. Chromosomes can be condensed into compact structures when a eukaryotic cell divides, and later they can be unraveled so that the information the chromosomes contain can be used.*

## Ribosomes Also Occur in the Nucleus

To make proteins, a cell employs a special structure, the **ribosome,** which reads the RNA copy of a DNA gene and uses the information it finds there to direct the synthesis of a protein. Ribosomes are made up of several special forms of RNA bound within a complex of several dozen different proteins. When a lot of proteins are being made, a cell needs a lot of ribosomes to handle the work load and therefore needs to be able to make large numbers of ribosomes quickly. To do this, many thousands of copies of the portion of the DNA encoding the RNA components of ribosomes, called **ribosomal RNA** (rRNA), are clustered together on the chromosome. Copying RNA molecules from the cluster rapidly generates large numbers of the molecules needed to produce ribosomes.

## The Nucleolus Is Not a Structure

At any given moment, many rRNA molecules dangle from the chromosome at the sites of these clusters of RNA genes. The proteins that will later form part of the ribosome complex bind to the dangling rRNA molecules. These areas where ribosomes are being assembled on the chromosomes are easily visible within the nucleus as one or more dark-staining regions, called **nucleoli** (singular, *nucleolus*; Figure 4-15). The nucleoli can be seen under the light microscope even when the

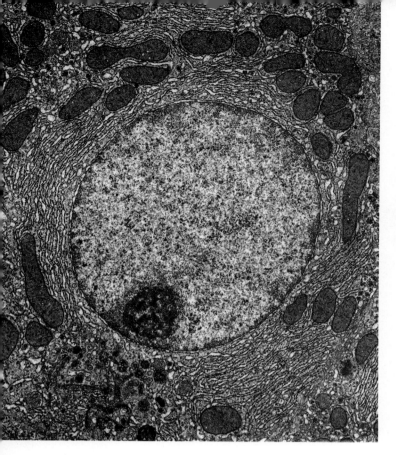

**Figure 4-15**

**The interior of a rat liver cell, magnified about 6000 times.** A single large nucleus occupies the lower center of the micrograph. The electron-dense area in the center of the nucleus is the nucleolus, the area where the major components of the ribosomes are produced. In the nucleoplasm around the nucleolus can be seen partly formed ribosomes.

chromosomes are extended, unlike the rest of the chromosomes, which are visible only when condensed. Because they are visible in nondividing cells, early scientists thought that the nucleoli were distinct cellular structures. We now know that this is not true. Nucleoli are, in fact, aggregations of rRNA and some ribosomal proteins which are transported into the nucleus from the rough ER and accumulate at those regions on the chromosomes where active synthesis of rRNA is taking place.

## THE GOLGI COMPLEX: THE DELIVERY SYSTEM OF THE CELL

At various locations in the cytoplasm are flattened stacks of membranes called **Golgi bodies** (Figure 4-16). These structures are named for Camillo Golgi, the nineteenth-century Italian physician who first called at-

tention to them. Animal cells contain 10 to 20 Golgi bodies each (they are especially abundant in glandular cells, which manufacture the substances that they secrete), whereas plant cells may contain several hundred. Collectively, the Golgi bodies are referred to as the **Golgi complex.**

Golgi bodies function in the collection, packaging, and distribution of molecules synthesized in the cell (Figure 4-17). The proteins and lipids that are manufactured on the rough and smooth ER membranes are transported through the channels of the ER, or as vesicles budded off from it, into the Golgi bodies. Within the Golgi bodies, many of these molecules are bound to polysaccharides, forming compound molecules. Among these are **glycoproteins,** consisting of a polysaccharide bound to a protein, and **glycolipids,** consisting of a polysaccharide bound to a lipid. The newly formed glycoproteins and glycolipids collect at the ends of the

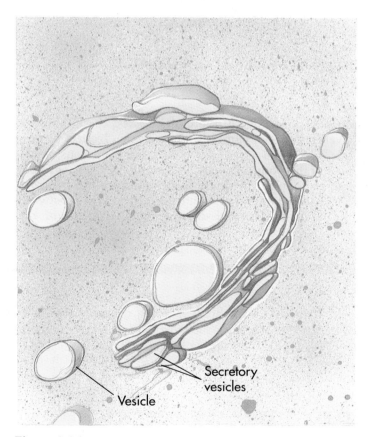

**Figure 4-16**

**A Golgi body in cross-section.** Within the Golgi body the endoplasmic reticulum is smooth, and it pinches together at its terminus to produce membrane-bounded vesicles. These vesicles contain whatever enzymes or other substances have been transported through the endoplasmic reticulum to the Golgi body.

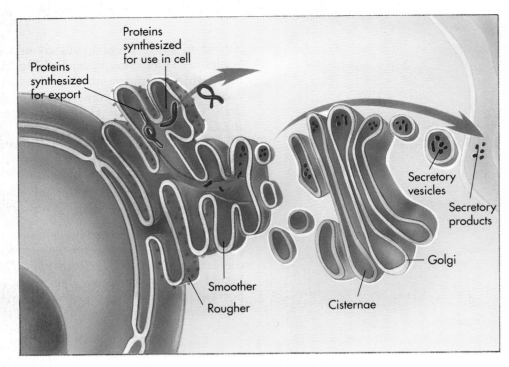

Proteins synthesized for export

Proteins synthesized for use in cell

Secretory vesicles

Secretory products

Golgi

Cisternae

Smoother

Rougher

**Figure 4-17**
How proteins are secreted across membranes.

membranous folds of the Golgi bodies; these folds are given the special name **cisternae** (Latin, "collecting vessels"). Around the Golgi bodies are numerous small, membrane-bounded vesicles containing glycoprotein and glycolipid molecules. These vesicles pinch off from the cisternae and are thought to carry material between the Golgi bodies and different compartments of the cell.

*The Golgi complex is the delivery system of the eukaryotic cell. It collects, modifies, packages, and distributes molecules that are synthesized at one location within the cell and used at another.*

## PEROXISOMES: CHEMICAL SPECIALITY SHOPS

**Peroxisomes,** membrane-bound spherical bodies 0.2 to 0.5 micrometers in diameter, are present in almost all eukaryotic cells. Apparently derived from the smooth ER, peroxisomes carry one set of enzymes active in converting fats to carbohydrates and another set that detoxifies various potentially harmful molecules—

strong oxidants—that form in cells. They do this by using molecular oxygen to remove hydrogen atoms from specific molecules.

## LYSOSOMES: RECYCLING CENTERS OF THE CELL

**Lysosomes** (Figure 4-18), another class of membrane-bound organelles, are about the same size as peroxisomes. They provide an impressive example of the metabolic compartmentalization achieved by the activity of the Golgi complex. They contain in a concentrated mix the digestive enzymes of the cell, enzymes that catalyze the rapid breakdown of proteins, nucleic acids, lipids, and carbohydrates. Lysosomes contain the enzymes that are responsible for the breakdown of macromolecules.

Lysosomes digest worn-out cellular components, making way for newly formed ones while recycling the materials locked up in the old ones. Cells can persist for a long time only if their components are constantly renewed. Otherwise the ravages of use and accident chip away at their metabolic capabilities and slowly degrade the cell's ability to survive. Cells age for the same reason that people do, because of a failure to renew themselves. Throughout the lives of eukaryotic cells, lysosomes break down the organelles and recycle their component proteins and other molecules at a fairly constant rate. As an example, mitochondria are replaced in some tissues every 10 days, with lysosomes digesting the old ones as new ones are produced.

Lysosomes that are actively engaged in digestive activities keep their battery of hydrolytic enzymes—those which catalyze the hydrolysis of molecules—fully active by maintaining a low internal pH—they pump protons into their interiors. Only at such acid pH values are the hydrolytic enzymes maximally active. Lysosomes that are not functioning actively do not maintain such an acid internal pH. A lysosome in such a "holding pattern" is called a **primary lysosome.** It is not until a primary lysosome fuses with a food vacuole or other organelle that its pH falls and the arsenal of hydrolytic enzymes is activated. When it becomes active, it is called a **secondary lysosome.**

It is not known what prevents lysosomes from digesting *themselves,* but the process requires energy, which is the reason that metabolically inactive eukaryotic cells die. Without a constant input of energy, the hydrolytic enzymes of lysosomes digest their membranes from within. When these membranes disintegrate, the digestive enzymes of the lysosome pour out into the cytoplasm of the cell and destroy it. In contrast to eukaryotes, bacteria do not possess lysosomes and do not die when they are metabolically inactive. Instead, they are able to remain quiet but alive until altered conditions restore their metabolic activity, a property that makes it possible for them to survive under unfavorable environmental conditions. For us, the very process that repairs the ravages of time within our cells eventually also leads to their destruction. Our dependency on a constant supply of energy is the price that we pay for our long lives.

*Lysosomes are membrane-bound organelles, about 0.2 to 0.5 micrometers in diameter and formed by the Golgi complex, which contain digestive enzymes. The isolation of these enzymes in lysosomes protects the rest of the cell from their digestive activity.*

## SOME DNA-CONTAINING ORGANELLES RESEMBLE BACTERIA

Most of the membrane-bound structures within eukaryotic cells are derived from the ER, but there are important exceptions. Eukaryotic cells also contain complex cell-like organelles that most biologists believe were derived from ancient symbiotic bacteria. An organism that is symbiotic within another is called an **endosymbiont.** The major endosymbionts that occur in eukaryotic cells are **mitochondria,** which occur in all but a very few eukaryotic organisms; **chloroplasts,** which occur in algae and plants; and **centrioles,** which occur in animals and most protists, but not in plants or fungi.

### Mitochondria: The Cell's Chemical Furnaces

The mitochondria that occur in all but a very few eukaryotic cells are thought by most biologists to have originated as symbiotic, **aerobic** (oxygen-requiring) bacteria. According to this theory, the bacteria that became mitochondria were engulfed by ancestral eukaryotic cells early in their evolutionary history. Before they

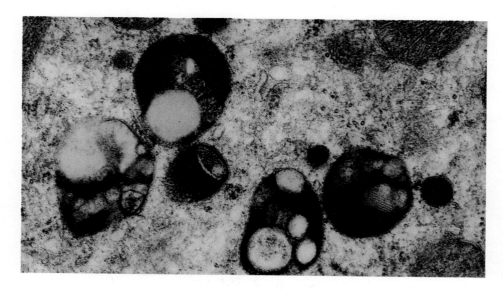

*Figure 4-18*

**Lysosomes.** These lysosomes (L) are within the cytoplasm of mouse kidney cells.

had acquired these bacteria, the host cells were unable to carry out the metabolic reactions necessary for living in an atmosphere that contained increasing amounts of oxygen. Metabolic reactions requiring oxygen are collectively called **oxidative metabolism,** a process that the symbiotic bacteria were able to carry out. Over the course of time, these bacteria became mitochondria.

Mitochondria (singular, *mitochondrion*) are tubular or sausage-shaped organelles 1 to 3 micrometers long (Figure 4-19); thus they are about the same size as most bacteria. Mitochondria are bounded by two membranes (Figure 4-20). The outer membrane is smooth and was apparently derived from the ER of the host cell, whereas the inner one, which was apparently the original plasma membrane of the bacterium that gave rise to the mitochondrion, is folded into numerous cristae that resemble the folded membranes that occur in various groups of bacteria. The cristae partition the mitochondrion into two compartments, an inner **matrix** and an **outer compartment.** On the surfaces of the membranes, and also submerged within them, are the proteins that carry out oxidative metabolism.

During the 1.5 billion years in which mitochondria have existed as endosymbionts in eukaryotic cells, most of their genes have been transferred to the chromosomes of their host cells. For example, the genes that produce the enzymes involved with the oxidative metabolism characteristic of mitochondria are located in the cell nucleus. But mitochondria still have some of their original genes, contained in a circular, closed mol-

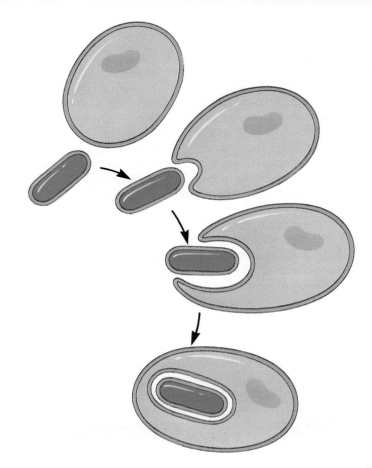

***Figure 4-20***

**Endosymbiosis.** This figure shows how a double membrane may have been created during the symbiotic origin of mitochondria.

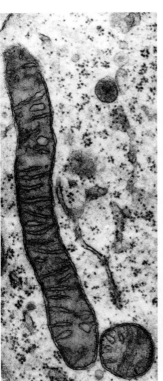

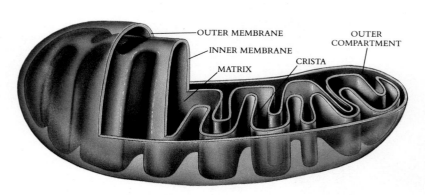

OUTER MEMBRANE
INNER MEMBRANE
OUTER COMPARTMENT
CRISTA
MATRIX

***Figure 4-19***

**Mitochondria.** These organelles are thought to have evolved from bacteria that long ago took up residence within the ancestors of present-day eukaryotes.

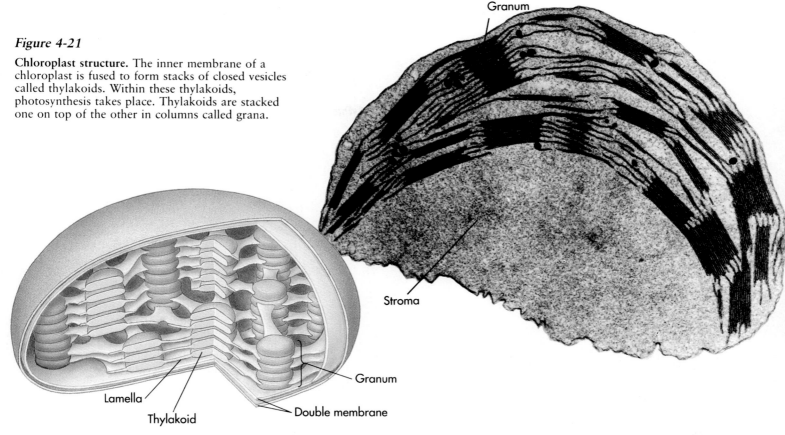

**Figure 4-21**

**Chloroplast structure.** The inner membrane of a chloroplast is fused to form stacks of closed vesicles called thylakoids. Within these thylakoids, photosynthesis takes place. Thylakoids are stacked one on top of the other in columns called grana.

Granum

Stroma

Granum

Lamella

Thylakoid

Double membrane

ecule of DNA which resembles those found in bacteria. On this mitochondrial DNA are located several genes that produce some of the proteins essential for the mitochondrion's role as the site of oxidative metabolism.

*Mitochondria are the chemical furnaces of the eukaryotic cell; within them occur the enzymes that carry out oxidative metabolism.*

Genes located on mitochondrial DNA are copied into RNA within the mitochondria and used there to make proteins. In this process the mitochondria use small RNA molecules and ribosomal components that are also encoded within the mitochondrial DNA. These ribosomes are smaller than those of eukaryotes in general, resembling bacterial ribosomes in size and structure.

When a eukaryotic cell divides, its mitochondria split into two by simple fission. The products of this division are partitioned between the new eukaryotic cells. All mitochondria are produced by the division of existing mitochondria in the same way that all bacteria are produced from existing bacteria. In both mitochondria and bacteria, the circular DNA molecule is replicated during the process of division.

## Chloroplasts: Where Photosynthesis Takes Place

Symbiotic events similar to those postulated for the origin of mitochondria also seem to have been involved in the origin of chloroplasts, which are characteristic of photosynthetic eukaryotes (algae and plants). Chloroplasts apparently were derived from symbiotic photosynthetic bacteria, which continue to carry out photosynthesis within the cells in which they occur. The advantage that chloroplasts bring to the organisms that possess them is obvious: these organisms can manufacture their own food. The photosynthetic endosymbionts that are thought to have given rise to chloroplasts were **anaerobic** (requiring no oxygen) photosynthetic bacteria.

Mitochondria apparently originated as endosymbiontic aerobic bacteria, whereas chloroplasts seem to have originated as endosymbiontic anaerobic photosynthetic bacteria.

A chloroplast is bounded, like a mitochondrion, by two membranes (Figure 4-21), the inner one apparently derived from the symbiotic bacterium and the outer one from the host cell's ER. Chloroplasts are larger than mitochondria, and their inner membranes

Microtubule triplet

A

B

## Figure 4-22

**Centrioles.**
**A** Electron micrograph of a pair of centrioles. The round shape is a centriole in cross section. The rectangular shape is a centriole in lingitudinal section. **B** Structure of a centriole, which is composed of nine triplets of microtubules.

have a more complex organization. They also have a larger circular DNA molecule than do mitochondria. Many of the genes that specify chloroplast components are located in the cell nucleus; they apparently were transferred from the chloroplasts to the nucleus over time, in a process similar to that which we have described for mitochondria. But the specific RNA and proteins necessary for photosynthesis are synthesized entirely within the chloroplast. Photosynthetic cells typically contain from one to several hundred chloroplasts, depending on the organism involved or the particular kind of cell. Neither mitochondria nor chloroplasts can be grown in a cell-free culture: they are dependent on the cells in which they occur for their survival.

*Both mitochondria and chloroplasts have transferred the bulk of their genetic material to the host chromosomes but retain certain specific genes related to their functions. Neither kind of organelle can be maintained in a cell-free culture.*

## Centrioles: Producing The Cytoskeleton
Centrioles are organelles associated with the assembly and organization of **microtubules** in the cells of ani-

mals and most protists. Microtubules are long, hollow cylinders about 25 nanometers in diameter, composed of the protein **tubulin**. They influence cell shape, move the chromosomes in cell division, and provide the functional internal structure of cilia and flagella, as we discuss below.

Centrioles occur in pairs within the cytoplasm of eukaryotic cells, usually located at right angles to one another near the nuclear envelope (Figure 4-22). They resemble tubes and are among the most structurally complex organelles of the cell. In cells that contain flagella or cilia, each flagellum is anchored by a form of centriole called a **basal body**. Most animal and protist cells have both centrioles and basal bodies. The cells of plants and fungi lack centrioles and basal bodies, and their microtubules are organized by amorphous structures.

In many respects, centrioles resemble spirochaete bacteria. For this reason it has been hypothesized, notably by Lynn Margulis of the University of Massachusetts, that centrioles actually originated as endosymbiotic spirochaetes. In 1989, David Luck and his coresearchers at the Rockefeller University demonstrated that at least some centrioles contain DNA, which apparently is involved in the production of their structural proteins. This discovery appears to lend strong support to Lynn Margulis' theory, which will doubtless become an area of even more active research in the future.

# THE CYTOSKELETON: INTERIOR FRAMEWORK OF THE CELL

The cytoplasm of all eukaryotic cells is crisscrossed by a network of protein fibers that support the shape of the cell and anchor organelles such as the nucleus to fixed locations (Figure 4-23). This network, called the **cytoskeleton,** cannot be seen with a light microscope because the fibers are single chains of protein, much too fine for microscopes to resolve. The fibers of the cytoskeleton are a dynamic system, constantly being formed and disassembled. Individual fibers form by **polymerization,** a process in which identical protein subunits are attracted to one another chemically and spontaneously assemble into long chains. Fibers are disassembled in the same way, by the removal from one end of first one subunit and then another.

Cells from plants and animals contain three different kinds of cytoskeleton fibers, each formed from a different kind of subunit (Figure 4-24): (1) long protein fibers called **actin filaments,** (2) hollow tubes called **microtubules,** and (3) ropes of protein called **intermediate fibers.**

Both actin filaments and intermediate fibers are anchored to proteins embedded within the plasma membrane; they provide the cell with mechanical support. Intermediate fibers act as intracellular tendons, preventing excessive stretching of cells, whereas actin filaments play a major role in determining the shape of cells. As we have seen, animal cells lack rigid cell walls; because actin filaments can form and dissolve readily, the shape of such cells can change rapidly. If you examine the surface of an animal cell with a microscope, you will find it alive with motion, projections shooting outward from the surface and then retracting, only to shoot out elsewhere moments later (Figure 4-25).

The cytoskeleton not only is responsible for the cell's shape, it also provides a scaffold on which the enzymes and other macromolecules are located in defined areas of the cytoplasm. Many of the enzymes involved in cell metabolism bind to actin filaments, as do the ribosomes that carry out protein synthesis. By anchoring particular enzymes near one another, the cytoskeleton participates with the ER in organizing the cell's activities.

Actin filaments and microtubules also play important roles in cell movement. Your own muscles use actin filaments to contract their cytoskeleton. Indeed, all cell motion is tied to these same processes. The fluttering of an eyelash, the flight of an eagle, and the awkward crawling of a baby all depend on the movements of actin filaments in the cytoskeletons of muscle cells.

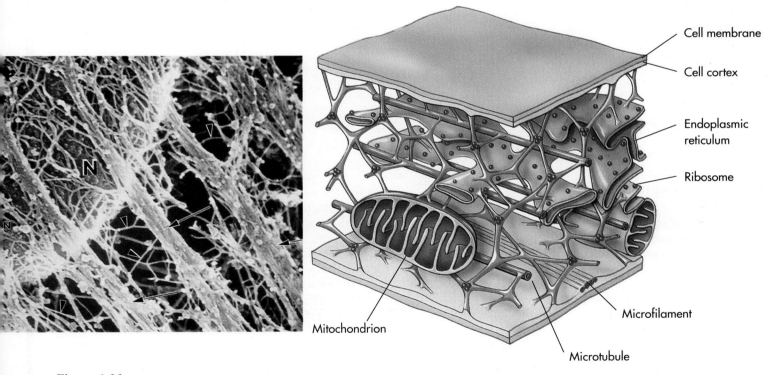

**Figure 4-23**

**The cytoskeleton.** In this diagrammatic cross-section of a eukaryotic cell, the mitochondria, ribosomes, and endoplasmic reticulum are all supported by a fine network of filaments, through which pass microtubules linking various portions of the cell.

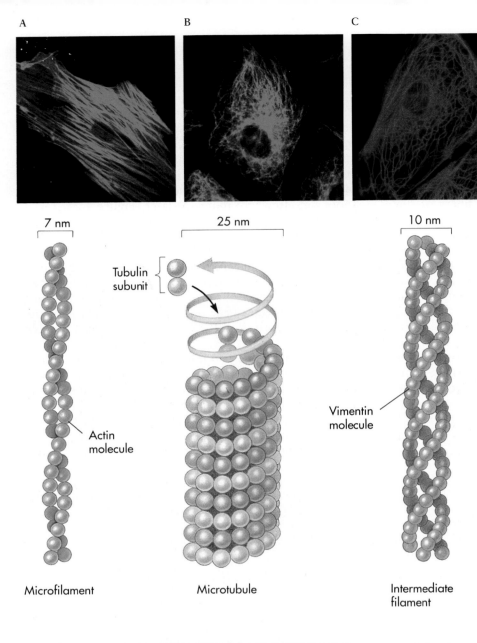

A           B           C

7 nm

Tubulin subunit

25 nm

Actin molecule

Microfilament

Microtubule

10 nm

Vimentin molecule

Intermediate filament

### Figure 4-24

**Molecules that make up the cytoskeleton.**
**A** Actin filaments. In the micrograph, actin filaments parallel the cell surface membrane in bundles which may have a contractile function.
**B** Microtubules. Each microtubule is composed of a spiral array of tubulin subunits. The microtubules visible in this micrograph radiate from an area near the nucleus (most heavily stained region). Microtubules act to organize metabolism and intracellular transport in the nondividing cell.
**C** Intermediate filaments. The best evidence suggests that three subunits are wound together in a coil, interrupted by uncoiled regions. In a skin cell, such as the one shown, intermediate filaments form thick, wavy bundles that probably provide structural reinforcement.

## FLAGELLA: SLENDER WAVING THREADS

**Flagella** (singular, *flagellum*) are fine, long, threadlike organelles protruding from the surface of cells; they are used in locomotion and feeding. The structures called flagella in bacteria are totally distinct in structure and origin from those which occur in eukaryotic organisms, but they have a similar function.

In the bacteria that possess them, flagella are long protein fibers. They rotate (Figure 4-26), and are so efficient that the bacteria in which they occur can move as much as 20 cell diameters per second. Imagine trying to run 20 body lengths per second! One or more flagella trail behind each swimming bacterial cell, depending on the species of bacterium. Each moves like a propeller, driven by a complex rotary "motor" embedded in the cell wall and membrane. Rotary motion is

mainly characteristic of bacteria; only a few eukaryotes have organs that rotate.

Eukaryotic cells have a completely different kind of flagellum, based on a cable made up of microtubules. Such flagella are increasingly called **undulipodia** to underscore their complete distinctiveness from bacterial flagella, which they resemble only to a limited degree in external form and function. Eukaryotic flagella, which arise from basal bodies, consist of a circle of nine microtubule pairs surrounding two central ones; they are called **9 + 2 flagella** (Figure 4-27). These complex structures are a fundamental feature of, and evidently evolved early in, the history of the eukaryotes. Even in cells that lack flagella, derived structures with the same 9 + 2 structure often occur, as we shall see.

The arrangement of flagella differs greatly in different eukaryotes. If the flagella are numerous and or-

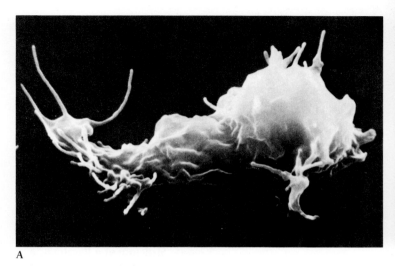
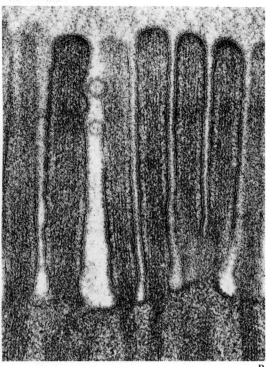

A                                                                                                          B

**Figure 4-25**

**The surfaces of animal cells are in constant motion.**
**A** This amoeba, a single-celled protist, is advancing toward you, its advancing edges extending projections outward.
**B** Animal cells often produce projections. This figure shows fingerlike projections, called microvilli, in the cells lining the human intestine. Microvilli can change their length quickly. They often appear to pop up almost instantaneously and to disappear just as quickly.

ganized in dense rows (see Figure 4-1), they are called **cilia** (singular, *cilium*), but cilia do not differ from flagella in their structure. In many multicellular organisms, cilia carry out tasks far removed from their original function of propelling cells through water. For example, in several kinds of human tissues, the beating of rows of cilia moves water over the tissue surface. The 9 + 2 arrangement of microtubules occurs in the sensory hairs of the human ear, where the bending of these hairs by pressure constitutes the initial sensory input of hearing. Throughout the evolution of eukaryotes, 9 + 2 flagella have been an element of central importance.

## AN OVERVIEW OF CELL STRUCTURE

The structure of eukaryotic cells is much more complicated and diverse than is the structure of bacterial cells (Table 4-2). The most distinctive difference between these two fundamentally distinct cell types is the exten-

*Flagella are slender, threadlike structures that aid cellular motion and other functions. Bacterial flagella, long protein fibers, are distinct from eukaryotic flagella (undulipodia), which are formed of microtubules in a characteristic 9 + 2 arrangement.*

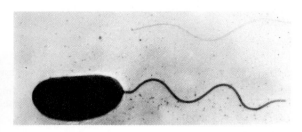

**Figure 4-26**
**Bacteria swim by rotating their flagella.**

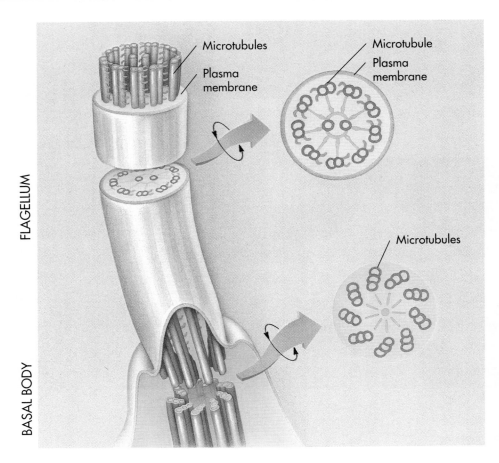

FLAGELLUM

BASAL BODY

Microtubules

Plasma membrane

Microtubule

Plasma membrane

Microtubules

**Figure 4-27**

**Structure of a flagellum.** A eukaryotic flagellum springs directly from a basal body and is composed of a ring of nine pairs of microtubules with two microtubules in its core.

**TABLE 4-2    A COMPARISON OF BACTERIAL, ANIMAL, AND PLANT CELLS**

|  | BACTERIUM | ANIMAL | PLANT |
|---|---|---|---|
| **EXTERIOR STRUCTURE** | | | |
| Cell wall | Present (protein-polysaccharide) | Absent | Present (cellulose) |
| Cell membrane | Present | Present | Present |
| Flagella | May be present (1 strand) | May be present | Absent except in sperm of a few species |
| **INTERIOR STRUCTURE** | | | |
| ER | Absent | Usually present | Usually present |
| Microtubules | Absent | Present | Present |
| Centrioles | Absent | Present | Absent |
| Golgi bodies | Absent | Present | Present |
| **ORGANELLES** | | | |
| Nucleus | Absent | Present | Present |
| Mitochondria | Absent | Present | Present |
| Chloroplasts | Absent | Absent | Present |
| Chromosomes | A single circle of naked DNA | Multiple units, DNA associated with protein | Multiple units, DNA associated with protein |
| Ribosomes | Present | Present | Present |
| Lyosomes | Absent | Usually present | Equivalent structures called "spherosomes" |
| Vacuoles | Absent | Absent or small | Usually a large single vacuole in mature cell |

sive subdivision of the interior of eukaryotic cells by membranes. The most visible of these membrane-bound compartments, the nucleus, gives eukaryotes their name. There are no equivalent membrane-bound compartments within prokaryotic cells. The membranes of some photosynthetic bacteria are extensively folded inwardly, but they do not isolate any one portion of the cell from any other portion. A molecule can travel unimpeded from any location in a bacterial cell to any other location.

In the following chapters we consider the consequences of these structural differences and how they influence the metabolism and biochemistry of eukaryotes. We shall see that the metabolic processes which occur within eukaryotic cells differ from those of bacteria and that these differences, like the structural ones discussed in this chapter, are substantial.

## ■ SUMMARY

1. The cell is the smallest unit of life. It is composed of a nuclear region that contains the hereditary apparatus within a larger volume called the cytoplasm, which performs the day-to-day functions of the cell under the supervision of the nuclear region. The cytoplasm is bounded by a lipid membrane in all cells.

2. The cells of bacteria, which are prokaryotic in structure, do not include membrane-bound organelles. Their DNA constitutes a single, closed, circular molecule.

3. The endoplasmic reticulum is a series of membranes that subdivides the interior of eukaryotic cells into separate compartments. It is the most distinctive feature of eukaryotic cells and one of the most important because it enables the cells to carry out different metabolic functions in separate compartments.

4. The nucleus is the largest of the closed compartments created by internal membranes. Within it, the cell's DNA is separated from the rest of the cytoplasm.

5. A third key membrane-associated organelle is the Golgi complex, which serves as a cellular "express package service," packaging molecules within special membrane vesicles and transporting them to various locations in the cell.

6. One class of vesicle created by the Golgi complex consists of the lysosomes, organelles that contain high concentrations of enzymes that constantly digest the cell's macromolecules and thus enable their renewal. A cell continuously expends energy to prevent the lysosomes from digesting themselves. When this activity ceases, the lysosomes soon burst, digesting and killing the cell.

7. Similar in size to lysosomes, peroxisomes contain enzymes that digest fats and reduce strong oxidants in cells.

8. Not all the eukaryotic cell's internal organelles are formed by internal membrane systems. Others— mitochondria, chloroplasts, and possibly centrioles—appear to have been derived from ancient symbiotic bacteria; they contain DNA, evidently derived from their ancestors.

9. Mitochondria are tubular or sausage-shaped organelles 1 to 3 micrometers long. Bounded by two membranes, they closely resemble the aerobic bacteria from which they were originally derived. As chemical furnaces of the cell, they carry out its oxidative metabolism.

10. Chloroplasts, which occur in plants and algae (photosynthetic protists), were apparently derived from anaerobic photosynthetic bacteria. They resemble mitochondria but are larger and have more DNA.

11. Many, but not all, of the genes that originally were present in mitochondrial and chloroplast DNA seem to have been transferred to, or had their functions taken over by, the DNA in the chromosomes of the host cell. However, both classes of organelles have retained the genes necessary to create the distinctive structures related to their particular functions.

12. Every eukaryotic cell includes a cytoskeleton of microtubules, which helps to determine its form and fixes the location of many internal components. In animals and most protists, but not in plants or fungi, the generation of microtubules is directed by specialized, paired, tubular organelles called centrioles.

13. Flagella are threadlike structures by virtue of which some cells move; in other kinds of cells, they acquired secondary functions during the course of evolution. Bacterial flagella are threads of protein; they propel the bacteria that possess them by rotating. Eukaryotic flagella, or undulipodia, have a distinctive and complex 9 + 2 structure of microtubules and are one of the characteristic features of eukaryotes.

## REVIEW

1. The photosynthetic pigments of photosynthetic bacteria are located on extensively convoluted infoldings of the _____.

2. The proteins embedded in eukaryotic plasma membranes perform three general classes of functions. They are (1) _____, (2) _____, and (3) _____.

3. _____ are the "garbage collectors/recycling centers" of eukaryotic cells, digesting worn-out cellular components and recycling their components. These "garbage collector" organelles are formed by what other organelle?

4. Which three organelles are thought to have originated as endosymbionts in an early cell?

5. DNA in animal cells occurs in the _____, the _____, and the _____.

## SELF-QUIZ

1. Which of the following statements is *not* included in what we call the cell theory?
   (a) All cells are so small that they cannot be seen with the naked eye.
   (b) All organisms are composed of one or more cells.
   (c) Cells are the smallest living things.
   (d) Hereditary information is contained within cells.
   (e) The cells that we see today are descended from early cells produced billions of years ago and are not produced spontaneously at the present time.

2. Which of the following are components of all eukaryotic cells (choose three)?
   (a) DNA and associated proteins found in a membrane-bound structure called the nucleus
   (b) Cytoplasm that contains numerous membrane-bound organelles
   (c) An intracellular matrix of protein called a cytoskeleton
   (d) Peptide cross-linkages in the cell wall
   (e) Chloroplasts

3. The DNA of eukaryotic cells is fragmented into several pieces called _____, unlike the DNA of bacteria, which is mostly contained in a single circular molecule.
   (a) ribosomes
   (b) peroxisomes
   (c) RNA
   (d) chromosomes
   (e) chloroplasts

4. Arrange in order the sequence of membrane-bound structures through which a protein that is going to be secreted by a cell must pass.
   (a) Golgi body
   (b) Secretory vesicle
   (c) Rough ER
   (d) Smooth ER

# THOUGHT QUESTIONS

1. Mitochondria are thought to be the evolutionary descendants of living cells, probably of symbiotic aerobic bacteria. Are mitochondria alive?

2. Some cells are very much larger than others. What would you expect the relationship to be between cell size and level of cell activity?

3. How does the Golgi complex know where to send what vesicle?

4. A cubical cell that is about 5 micrometers across contains what percentage of the volume of a cubical cell that is 20 micrometers across?

# FOR FURTHER READING

DE DUVE, C.: *A Guided Tour of the Living Cell*, vols. 1 and 2, Scientific American Books, New York, 1986. This review has excellent illustrations and presents a wide variety of the techniques used to study cells.

MARGULIS, L.: *Symbiosis in Cell Evolution*, W.H. Freeman Co., San Francisco, 1980. A broad-ranging exposition of the theory that mitochondria, chloroplasts, and other organelles were acquired by eukaryotes from symbiotes.

McDERMOTT, J.: "A Biologist Whose Heresy Redraws Earth's Tree of Life," *Smithsonian*, August 1989, pages 71-81. Engaging account of the scientific approach taken by Lynn Margulis, leading contemporary advocate of the endosymbiotic theory of the origin of some organelles.

SYMMONS, M., A. PRESCOTT, and R. WARN: "The Shifting Scaffolds of the Cell," *New Scientist*, February 1989, pages 44-47. New findings show that the microtubules of which our cells are built are constantly appearing and disappearing.

WEBER, K., and M. OSBORN: "The Molecules of the Cell Matrix," *Scientific American*, October 1985, pages 110-120. An up-to-date description of the many different molecules that make up the cell cytoskeleton.

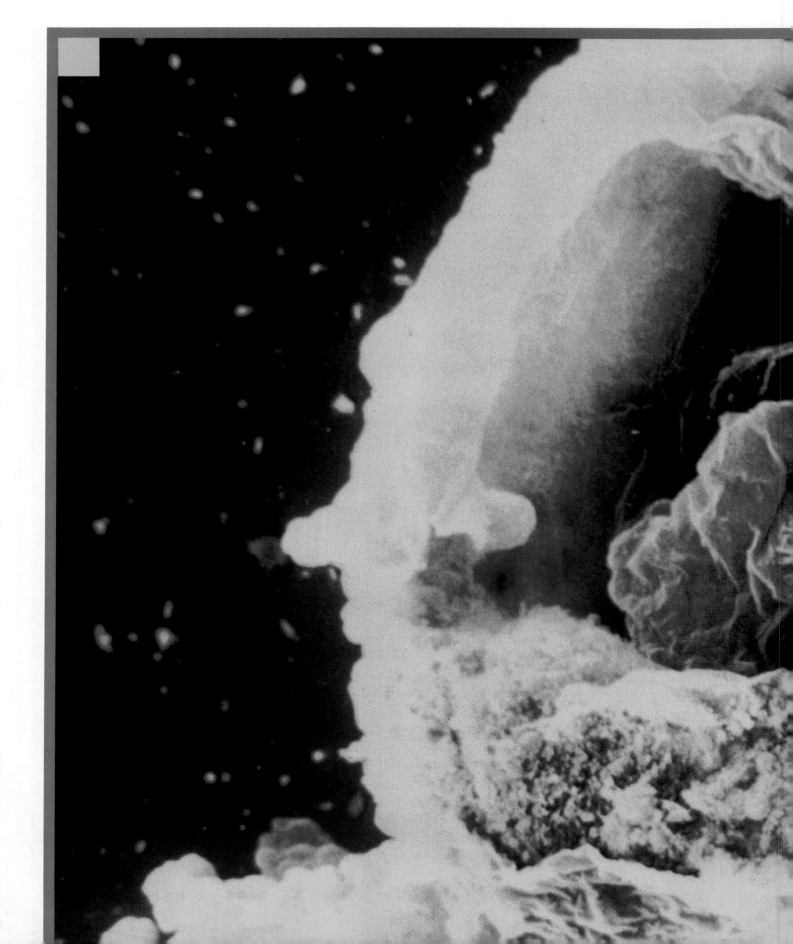

# *H*ow Cells Interact with the Environment

The massive cell in the foreground is an amoeba that has surrounded and is about to engulf its dinner, a ciliated protozoan. Being folded inside like this protozoan is one of many routes into a eukaryotic cell, although a little more dramatic than most.

# HOW CELLS INTERACT WITH THE ENVIRONMENT

## Overview

Every cell is enveloped within a liquid bilayer of lipid, the plasma membrane; this membrane is a fluid shell that isolates the cell's interior. Anchored within the lipids are a host of proteins that move about like boats on a lake. These proteins provide passage across the lipid layer (which is impermeable to most molecules) for selected molecules and information. They are the cell's only connection with the outer world.

## For Review
*Here are some important terms and concepts that you will encounter in this chapter. If you are not familiar with them, you should review them before proceeding.*

**Polar nature of water** (Chapter 2)

**Hydrogen bonds** (Chapter 2)

**Structure of fat molecules** (Chapter 2)

**Types of membrane proteins** (Chapter 4)

Among a cell's most important activities are its transactions with the environment, a give-and-take that never ceases. Cells are constantly feasting on food they encounter, ingesting molecules and sometimes entire cells. They dump their wastes back into the environment, together with many other kinds of molecules. Cells continuously garner information about the world around them, responding to a host of chemical clues and often passing on messages to other cells. This constant interplay with the environment is a fundamental characteristic of all cells. Without it, life could not exist (Figure 5-1).

Imagine that you were to coat a living cell in plastic, giving it a rock-hard, impermeable shell. All the cell's transactions with the environment would stop. No molecules could pass in or out, nor could the cell learn anything about the molecules around it. The cell might as well be a rock. Life in any meaningful sense would cease—unless, of course, you allowed for doors and windows in the shell.

That is, in fact, what living cells do. Every cell is encased within a lipid membrane, an impermeable shell through which no water-soluble molecules and little information (data about its surroundings) can pass, but the shell contains doors and windows made of protein. Molecules pass in and out of a cell through these passageways, and information passes in and out through the windows. A cell interacts with the world through a delicate skin of protein molecules embedded in a thin sheet of lipid. We call this assembly of lipid and protein a **plasma membrane**; alternative terms are *cell membrane* or *plasmalemma*. The structure and function of this membrane are the subject of this chapter.

**Figure 5-1**

**A foraminifera.** This beautiful foraminiferan carries out the same essential life processes that you do, and, as in your body, all these activities require that substances enter and leave the cell.

# THE LIPID FOUNDATION OF MEMBRANES

The plasma membranes that encase all living cells are sheets only a few molecules thick; it would take more than 10,000 of these sheets, which are about 7 nanometers thick, piled on one another to equal the thickness of this sheet of paper. But the sheets are not simple in structure, like a soap bubble. Rather, they are made up of diverse collections of proteins enmeshed in a lipid framework, like small boats bobbing on the surface of a pond. Regardless of the kind of cells or organelles that they enclose, all plasma membranes are similar in molecular structure.

## Phospholipids

As is true for all biological membranes, the lipid layer that forms the foundation of a plasma membrane is composed of molecules called **phospholipids**. Like the fat molecules you studied in Chapter 2, a phospholipid has a backbone derived from a three-carbon molecule called glycerol, with long chains of carbon atoms called fatty acids attached to this backbone. A fat molecule has three such chains, one attached to each carbon of the backbone; because these chains are nonpolar (do not form hydrogen bonds with water), the fat molecule is insoluble in water. A phospholipid, by contrast, has only two such chains attached to its backbone (Figure 5-2). The third position is occupied instead by a highly polar organic alcohol that readily forms hydrogen

bonds with water. Because this alcohol is attached to the glycerol by a phosphate group, the molecule is called a *phospho*lipid.

One end of a phospholipid molecule is therefore strongly nonpolar (water insoluble), whereas the other end is extremely polar (water soluble). The two nonpolar fatty acids extend in one direction, roughly parallel to each other, and the polar alcohol group points in the other direction. Because of this structure, phospholipids are often diagrammed as a (polar) ball with two dangling (nonpolar) tails (see Figure 5-2).

## Phospholipids Form Bilayer Sheets

Imagine what happens when a collection of phospholipid molecules is placed in water. The long nonpolar tails of phospholipid molecules are pushed away by the water molecules that surround them, shouldered aside as the water molecules seek partners that can form hydrogen bonds. Water molecules always tend to form the maximum number of such bonds; the long nonpolar chains that can't form hydrogen bonds get in the way, like too many chaperones at a party. The best way to rescue the party is to put all the chaperones together in a separate room, and that is what water molecules do—they shove all the long, nonpolar tails of the lipid molecules together, out of the way. The polar heads of the phospholipids are "welcomed," however, because they form good hydrogen bonds with water. What happens is that every phospholipid molecule orients so that its polar head faces water and its nonpolar

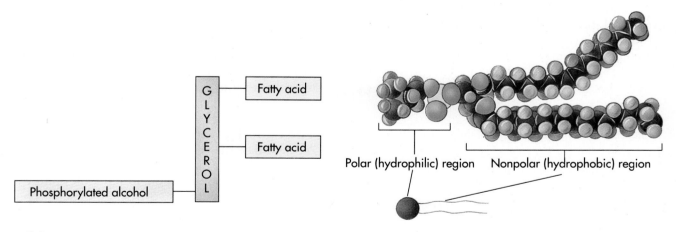

*Figure 5-2*

**Phospholipid structure.** A phospholipid is a composite molecule similar to a triglyceride, except in this case only two fatty acids are bound to the glycerol backbone, the third position being occupied by another kind of molecule called a phosphorylated alcohol. Because the phosphorylated alcohol usually extends from one end of the molecule and the two fatty acid chains from the other, phospholipids are often diagrammed as a polar head with two nonpolar hydrophobic tails.

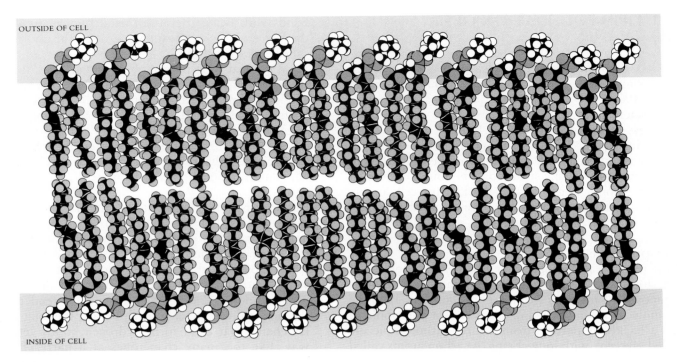

OUTSIDE OF CELL

INSIDE OF CELL

**Figure 5-3**

**A phospholipid bilayer.** The diagram running across the top of the page illustrates how the long nonpolar tails of the phospholipids orient toward one another. Because some of the tails contain double bonds, which introduce kinks in their shape, the tails do not align perfectly and the membrane is "fluid"—individual phospholipid molecules can move from one place to another in the membrane.

tails face away. Because there are *two* layers with the tails facing each other, no tails are ever in contact with water. The structure that results is called a **lipid bilayer** (Figure 5-3). Lipid bilayers form spontaneously, driven by the forceful way in which water tends to form hydrogen bonds.

*The basic foundation of all biological membranes is a lipid bilayer that forms spontaneously. In such a layer, the nonpolar tails of phospholipid molecules point inward, forming a nonpolar zone in the interior of the bilayer.*

Because the interior of a lipid bilayer is completely nonpolar, it repels any water-soluble molecules that attempt to pass through it (Figure 5-4), just as a layer of oil stops the passage of a drop of water (that's why ducks don't get wet). This barrier to the passage of wa-

ter-soluble molecules is the key biological property of the lipid bilayer. It means that a cell, if fully encased within a pure lipid bilayer, would be impermeable to water-soluble molecules such as sugars, amino acids, and proteins. But no cell is so imprisoned. In addition to the phospholipid molecules that make up the lipid bilayer, the membranes of every cell also contain proteins that extend across the lipid bilayer, providing passage across the membrane.

## The Lipid Bilayer is a Fluid

A lipid bilayer is very stable because the hunger of water for hydrogen bonding never stops. Although water continually urges phospholipid molecules into this orientation, it is indifferent to where individual phospholipid molecules are located. Water forms just as many hydrogen bonds when a particular phospholipid molecule is located here or there. As a result, individual lipid molecules are free to move about within the membrane (Figure 5-5). Because the individual molecules are free to move about, the lipid bilayer is not a solid, like a rubber balloon, but rather is a liquid, like the "shell" of a soap bubble. The bilayer itself is a fluid,

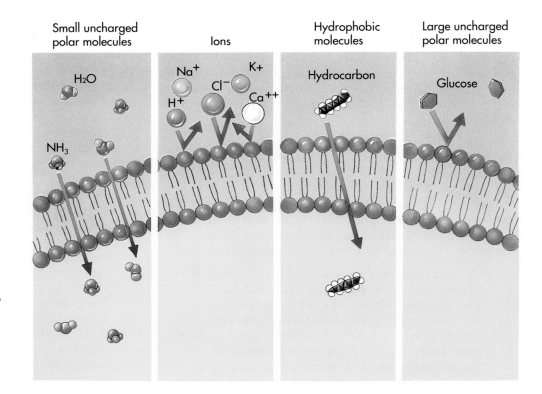

Small uncharged polar molecules

$H_2O$

$NH_3$

Ions

$Na^+$    $K^+$

$H^+$    $Cl^-$

$Ca^{++}$

Hydrophobic molecules

Hydrocarbon

Large uncharged polar molecules

Glucose

**Figure 5-4**

**Some molecules pass across bilayer membranes, others don't.** Lipid bilayer membranes are permeable to oxygen, lipids, and small uncharged molecules even if they are polar like water; they are not permeable to large molecules if they are polar, or to anything that is charged, such as ions or proteins.

with the viscosity of olive oil. Just as the surface tension holds the soap bubble together, even though it is made of a liquid, so the hydrogen bonding of water holds the membrane together.

Some membranes are more fluid than others. The tails of individual phospholipid molecules attract one another when they line up close together, stiffening the membrane because aligned molecules must pull apart from one another before they can move about in the membrane. The more alignment, the less fluid the membrane is. Some phospholipids have tails that don't align well because they contain one or more C=C

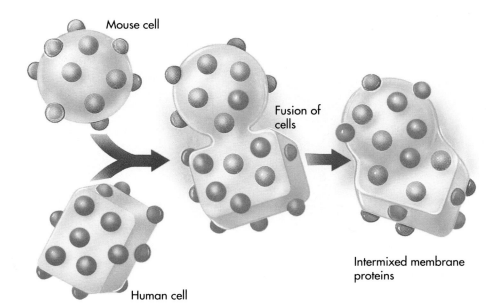

Mouse cell

Fusion of cells

Human cell

Intermixed membrane proteins

**Figure 5-5**

**Proteins in membranes move around.** Protein movement within membranes can be easily demonstrated by labeling the proteins of a mouse cell with fluorescent antibodies and then fusing that cell with a human cell. Within 1 hour, the labeled and unlabeled proteins are intermixed throughout the hybrid cell's membranes.

double bonds, which introduce kinks in the tail. Membranes containing phospholipids of these sorts are more fluid than those which lack them. Sometimes membranes contain other lipids, such as cholesterol, which prevent the phospholipid tails from coming into contact with one another and thus make the membrane more fluid.

## ARCHITECTURE OF THE PLASMA MEMBRANE

A eukaryotic cell contains several kinds of membranes that are similar in their lipid bilayers but that differ in the nature of the molecules embedded in them. All such membranes are assembled from four components:

1. *A lipid bilayer foundation.* Every biological plasma membrane has as its basic foundation a phospholipid bilayer. The other components of the membrane are enmeshed within the bilayer, which provides a flexible matrix and, at the same time, imposes a barrier to permeability.
2. *Membrane proteins.* A major component of every biological membrane is a collection of proteins that float within the lipid bilayer (Figure 5-6). In plasma membranes, these proteins provide channels through which molecules and information pass. Membrane proteins are not fixed in position; instead, they move about freely. Some membranes are crowded with proteins, side by side. In other membranes, the proteins are more sparsely distributed.
3. *Network of supporting fibers.* Biological membranes are structurally supported by proteins that reinforce the membrane's shape. This is why a red blood cell is shaped like a disk rather than being irregular. The membrane is held in that shape by a scaffold of protein on its inner surface (Figure 5-7). Membranes use networks of other proteins to control the lateral movements of some key membrane proteins, anchoring them to specific sites so that they do not simply drift away. Unanchored proteins have been observed to move as much as 10 micrometers in 1 minute.
4. *Exterior glycolipids.* In some kinds of biological membranes, and especially in plasma membranes, many carbohydrate and lipid molecules extend outward from the cell surface, like a thicket of brush. These chains act as cell iden-

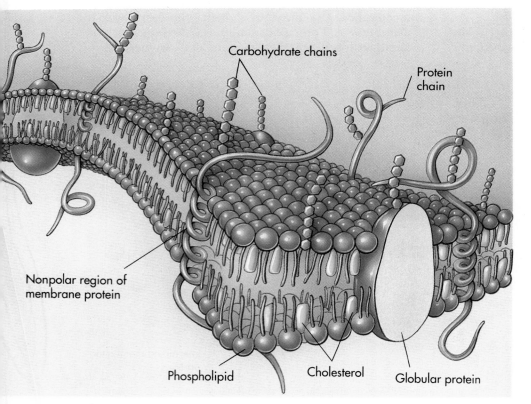

Carbohydrate chains

Protein chain

Nonpolar region of membrane protein

Phospholipid    Cholesterol    Globular protein

**Figure 5-6**

**Cells have complex surfaces.** A variety of proteins protrude through the plasma membrane, nonpolar regions of the proteins serving to tether them to the membrane's nonpolar interior. Carbohydrate chains (strings of sugar molecules) are often bound to these proteins and to lipids in the membrane itself as well. These chains serve as distinctive identification tags, unique to particular types of cells.

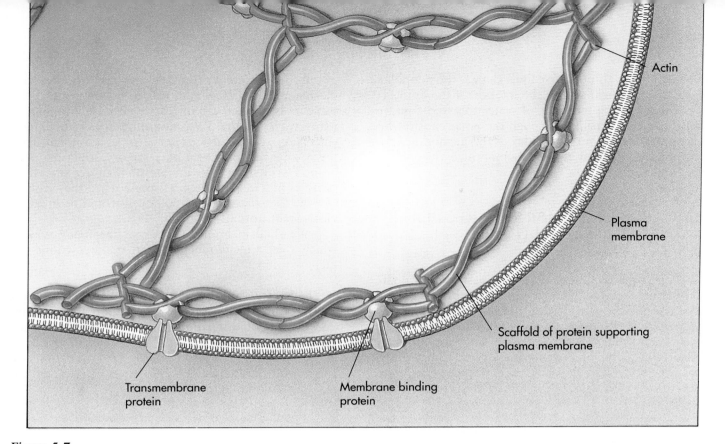

Actin

Plasma membrane

Scaffold of protein supporting plasma membrane

Transmembrane protein

Membrane binding protein

*Figure 5-7*

**Membrane support.** Cell membranes are reinforced by a network of protein fibers. Here we see a portion of the network, attached to the inside of the plasma membrane by proteins inserted into the lipid bilayer.

tity markers. Different cell types exhibit different kinds of carbohydrate chains on the surface of their plasma membranes.

*Plasma membranes are assembled from four components: a phospholipid bilayer foundation, membrane proteins, a network of supporting fibers, and exterior glycolipids.*

A membrane, then, is a sheet of lipid and protein that is supported by other proteins and to which carbohydrates are attached (Table 5-1). The key functional proteins act as passageways through the membrane, extending all the way across the bilayer. How do these transmembrane proteins manage to span the membrane, rather than just floating on the surface in the way that a drop of water floats on oil? The part of the protein that traverses the lipid bilayer is specially constructed, a spiral helix of nonpolar amino acids (Figure 5-8). Because water responds to nonpolar amino acids

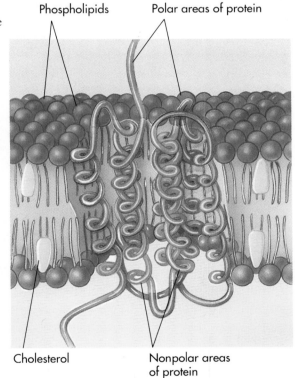

Phospholipids       Polar areas of protein

Cholesterol       Nonpolar areas of protein

*Figure 5-8*

**How proteins are anchored to membranes.** Membrane proteins have spiral regions containing many nonpolar amino acids that are embedded within the nonpolar interior of the bilayer.

much as it does to nonpolar lipid chains, the nonpolar helical spiral is held within the interior of the lipid bilayer by the strong tendency of water to avoid contact with these amino acids. Although the polar ends of the protein protrude from both sides of the membrane, the protein itself is locked into the membrane by its nonpolar helical segment.

## HOW A CELL'S MEMBRANES REGULATE INTERACTIONS WITH ITS ENVIRONMENT

The plasma membrane is a complex assembly of floating proteins loosely anchored to the membrane. This is probably not the sort of cell surface you or I would have designed. It is not strong, and its surface components are not precisely located. It is, however, a design of enormous flexibility, permitting a broad range of interactions with the environment. A plasma membrane is like a rack that can hold many different tools. A list of all the different kinds of proteins in one cell's plasma membrane would run to many pages. With these tools, the cell can interact with its environment in many ways (Figure 5-9), admitting a particular molecule here, sensing the presence of a hormonal signal there. Like the tools of a busy factory, a plasma membrane's proteins are in a state of high activity. If the proteins of a membrane made noise when they worked, like the tools in a factory do, a cell would be in a state of constant uproar.

We shall mention only six of the many ways in which a cell's membranes regulate its interactions with the environment:

1. *Passage of water.* Plasma membranes are freely permeable to water, but the spontaneous movement of water into and out of cells sometimes presents problems.
2. *Bulk passage into the cell.* Cells sometimes engulf other cells or large pieces of such cells, or gulp liquids.

**TABLE 5-1    COMPONENTS OF THE CELL MEMBRANE**

| COMPONENT | COMPOSITION | FUNCTION | HOW IT WORKS | EXAMPLE |
|---|---|---|---|---|
| Lipid foundation | Phospholipid bilayer | Permeability barrier matrix for proteins | Water-soluble molecules excluded from nonpolar interior of bilayer | Bilayer of cell is impermeable to water-soluble molecules |
| Transmembrane proteins | Single-coil channels | Transport of large molecules across membrane | Carrier "flip-flops" | Glycophorin channel for sugar transport |
| | Multicoil channels | Transport of small molecules across membrane | Create a tunnel that acts as a passage | Photoreceptor |
| | Receptors | Transmit information into cell | Bind to cell surface portion of protein; alter portion within cell, inducing activity | Peptide hormones; neurotransmitters |
| Cell surface markers | Glycoprotein | "Self" recognition | Shape of protein/carbohydrate chain is characteristic of individual | Major histocompatibility complex protein recognized by immune system |
| | Glycolipid | Tissue recognition | Shape of carbohydrate chain is characteristic of tissue | A, B, O blood group markers |
| Interior protein network | Spectrin | Determines shape of cell | Forms supporting scaffold beneath membrane, anchored to both membrane and cytoskeleton | Red blood cell |
| | Clathrins | Anchor certain proteins to specific sites | Form network above membrane to which proteins are anchored | Localization of low-density lipoprotein receptor within coated pits |

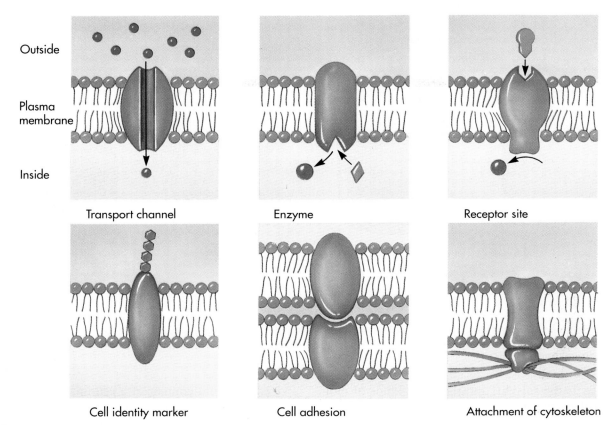

Outside

Plasma membrane

Inside

Transport channel

Enzyme

Receptor site

Cell identity marker

Cell adhesion

Attachment of cytoskeleton

**Figure 5-9**

Functions of plasma membrane proteins.

3. *Selective transport of molecules.* Plasma membranes are very picky about which molecules they allow to enter or leave the cell.
4. *Reception of information.* Plasma membranes can identify chemical messages with exquisite sensitivity.
5. *Expression of cell identity.* Plasma membranes carry molecular name tags that tell other cells who they are.
6. *Physical connection with other cells.* In forming tissues, plasma membranes make special connections with each other.

## THE PASSAGE OF WATER INTO AND OUT OF CELLS

Molecules dissolved in liquid are in constant motion, moving about randomly. This random motion causes a net movement of molecules toward zones where the concentration of the molecules is lower, a process called **diffusion** (Figure 5-10). Driven by random mo-

tion, molecules always "explore" the space around them, diffusing out until they fill it uniformly. A simple experiment will demonstrate this process. Take a small jar, fill it with ink, cap it, place it at the bottom of a full bucket, and remove the cap. The ink molecules will slowly diffuse out until there is a uniform concentration in the bucket and the jar.

*Diffusion is the net movement of molecules to regions of lower concentration as a result of random, spontaneous molecular motions. Diffusion tends to distribute molecules uniformly.*

The cytoplasm of a cell consists of molecules such as sugars, amino acids, and ions dissolved in water. The mixture of these molecules and water is called a **solution.** Water, the most common of the molecules in

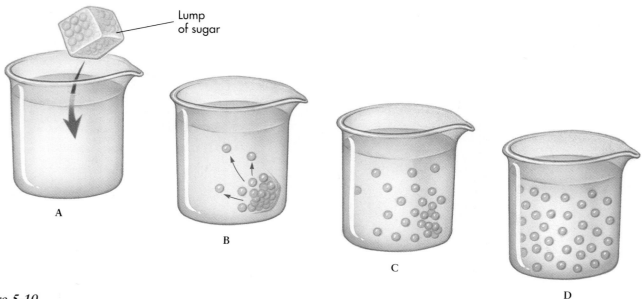

Lump
of sugar

A

B

C

D

*Figure 5-10*

**Diffusion.** If a lump of sugar is dropped into a beaker of water, its molecules dissolve (**A**) and diffuse (**B and C**). Eventually, diffusion results in an even distribution of sugar molecules throughout the water (**D**).

the mixture, is the **solvent,** and the other kinds of molecules dissolved in the water are **solutes.**

Because of diffusion, both solvent and solute molecules in a cell will move from regions where their concentration is greater to a region where their concentration is less. When two regions are separated by a membrane, what happens depends on whether or not the molecule can pass freely through that membrane; most kinds of solutes that occur in cells cannot do so. Sugars, amino acids, and other solutes are water soluble and not lipid soluble, and so are imprisoned within the cell: they are unable to cross the lipid bilayer of the membrane. Water molecules, in contrast, can pass through slight imperfections in the sheet of lipid molecules and so diffuse across the membrane into the cell. Water molecules stream into the cell across the membrane, thus diluting the high concentration of solutes within the cell so that it matches more and more closely the lower concentration in the outside solution. This form of net water movement into or out of a cell is called **osmosis** (Figure 5-11).

*Osmosis is the diffusion of water across a membrane that permits the free passage of water but not that of one or more solutes.*

The fluid content of a cell immersed in pure water is said to be **hypertonic** (Greek *hyper,* more than) with respect to its surrounding solution because it has a higher concentration of solutes than does the water. The surrounding solution, which has a lower concentration of solutes than does the cell, is said to be **hypotonic** (Greek *hypo,* less than) with respect to the cell. A cell with the same concentration of solutes as its environment is said to be **isotonic** (Greek *iso,* the same) (Figure 5-12).

Intuitively you might think that, as new water molecules diffuse inward, the pressure of the cytoplasm pushing out against the cell wall would build up, and this is indeed what happens. As water molecules continue to diffuse inward toward the area of lower *water* concentration (the concentration of the water is lower inside than outside the cell because of the dissolved solutes in the cell), the hydrostatic water pressure within the cell increases. We refer to pressure of this kind as **osmotic pressure** (Figure 5-13). Because osmotic pressure opposes the movement of water inward, such diffusion will not continue indefinitely. The cell will eventually reach an equilibrium—a point at which the osmotic pressure driving water inward is counterbalanced exactly by the hydrostatic pressure driving water out. In practice, the hydrostatic pressure at equilibrium is typically so high that an unsupported cell membrane cannot withstand it, and such an unsupported cell, suspended in water, will burst like an overinflated balloon. Cells whose membranes are surrounded by cell walls, in contrast, can withstand high internal hydrostatic pressures (Figure 5-14).

*Within the closed volume of a cell that is hypertonic to its surroundings, the movement of water inward, which tends to lower the relative concentration difference of water, will at the same time increase the internal hydrostatic pressure. The net movement of water stops when an equilibrium condition is reached or the cell bursts.*

## BULK PASSAGE INTO THE CELL

The lipid nature of plasma membranes raises a second problem for growing cells. The molecules required by cells as food are mostly polar molecules; they will not pass across the hydrophobic barrier interposed by a lipid bilayer. How then are organisms able to get food molecules into their cells? Particularly among single-celled eukaryotes, the dynamic cytoskeleton is employed to extend the cell membrane outward toward food particles such as bacteria. The membrane encircles and engulfs a food particle. Its edges eventually meet on the other side of the food particle, where, because of the fluid nature of the lipid bilayer, the mem-

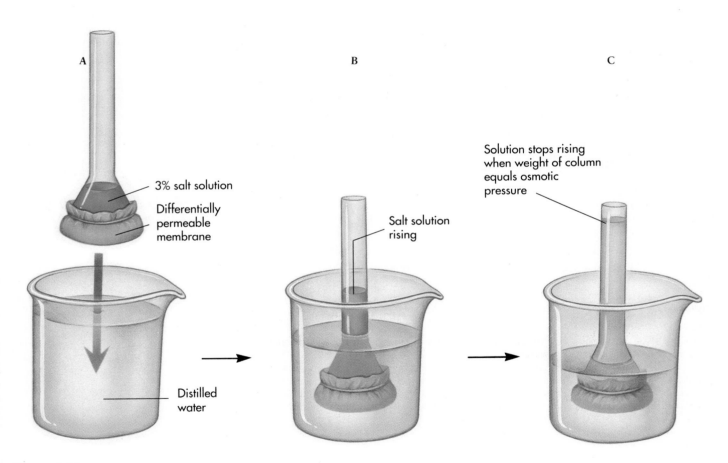

**Figure 5-11**

**An experiment demonstrating osmosis.**
**A** The end of a tube containing a 3% salt solution is closed by stretching a differentially permeable membrane across its face that will pass water molecules but not salt molecules.
**B** When this tube is immersed in a beaker of distilled water, the salt cannot cross the membrane; however, water can. The added water causes the salt solution to rise in the tube.
**C** Water will continue to enter the tube from the beaker until the weight of the column of water in the tube exerts a downward force equal to the force drawing water molecules upward into the tube. This force is referred to as osmotic pressure.

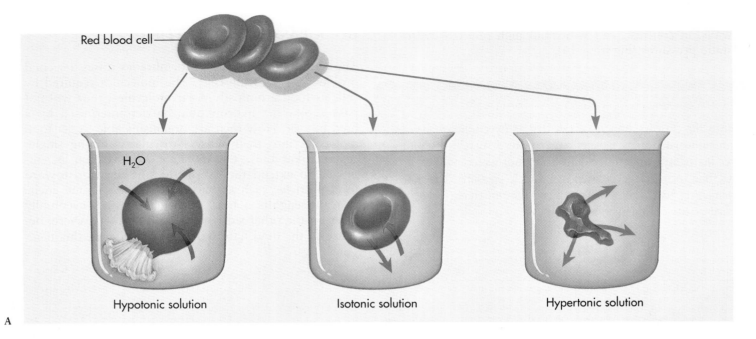

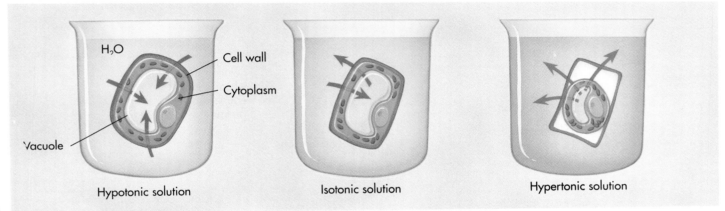

A

B

*Figure 5-12*

**Osmosis.**
A  Animal cells. When the outer solution is hypotonic with respect to the cell, water will move in; when it is hypertonic, water will move out.
B  Plant cells. In plant cells, the large central vacuole contains a high concentration of solutes, so water tends to diffuse inward, causing the cells to swell outward against their rigid cell walls. However, if a plant cell is immersed in a high-solute (hypertonic) solution, water will leave the cell, causing the cytoplasm to shrink and pull in from the cell wall.

branes fuse together, forming an enclosed chamber called a *vesicle* around it. This process is called **endocytosis.** Endocytosis involves the incorporation of a portion of the exterior medium into the cytoplasm of the cell by capturing it within a vesicle (Figure 5-15).

If the material that is brought into the cell is an organism (Figure 5-15, *C* and *D*) or some other fragment of organic matter, that particular kind of endocytosis is called **phagocytosis** (Greek *phagein*, to eat + *cytos*, cell). If the material brought into the cell is liquid and contains dissolved molecules, the endocytosis is referred to as **pinocytosis** (Greek *pinein*, to drink). Pinocytosis (Figure 5-15, *B*) is common among the cells of multicellular animals. Human egg cells, for example, are "nursed" by surrounding cells that secrete nutrients that the maturing egg cell takes up by pinocytosis.

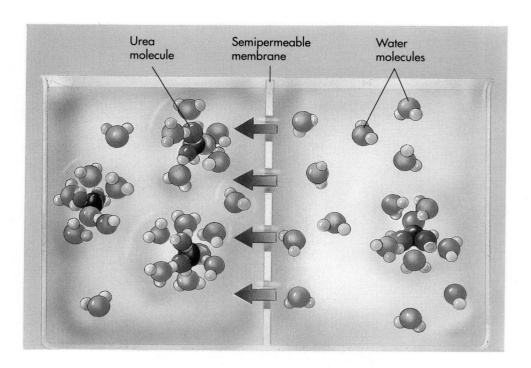

Urea molecule     Semipermeable membrane     Water molecules

**Figure 5-13**

**How solutes create osmotic pressure.** Charged or polar molecules such as urea are soluble in water because they form hydrogen bonds with water molecules clustered around them. When such a polar solute is added to one side of a membrane, the water molecules that gather around each urea molecule are no longer free to diffuse across the membrane—in effect, the polar solute has reduced the number of free water molecules on that side of the membrane. Because the other side of the membrane (on right, with less solute) has more unbound water molecules than the side with more solute, water moves by diffusion from the right to the left.

*Phagocytosis is a process in which cells engulf organisms or fragments of organisms, enfolding them within vesicles.*

Virtually all eukaryotic cells are constantly carrying out endocytosis, trapping extracellular fluid in vesicles and ingesting it. Rates of endocytosis vary from one cell type to another but can be surprisingly large. Some types of white blood cell ingest 25% of their cell volume each hour.

The reverse of endocytosis is **exocytosis,** the extrusion of material from a cell by discharging it from vesicles at the cell surface (Figure 5-16). In plants, vesicle discharge constitutes a major means of exporting the materials used in the construction of the cell wall through the plasma membrane. In animals, many cells are specialized for secretion using the mechanism of exocytosis.

## SELECTIVE TRANSPORT OF MOLECULES

Endocytosis is an awkward means of governing entrance to the interior of the cell. It is expensive from an energy standpoint because a considerable amount of

STOMA

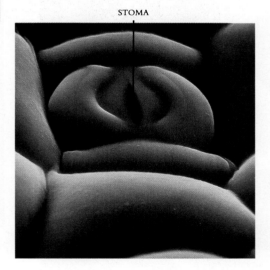

**Figure 5-14**

**Plants use osmosis to control water loss.** The cells surrounding this opening in a *Tradescantia* leaf are swollen by osmosis and have assumed a rigid, bowed shape. This creates an opening between them through which water vapor can pass. When relaxed, the guard cells rest against one another, closing the opening.

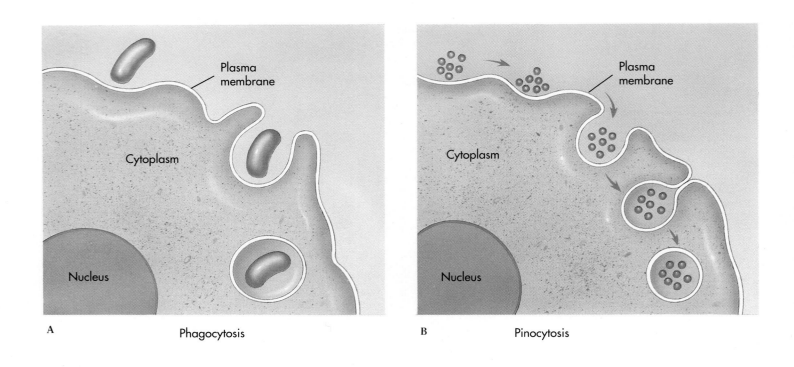

A            Phagocytosis           B            Pinocytosis

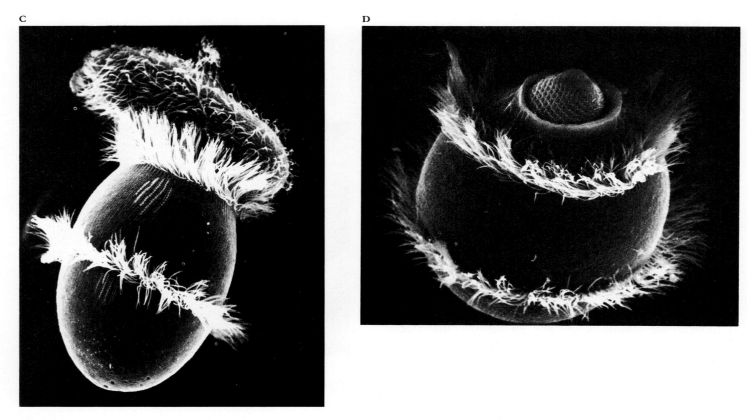

*Figure 5-15*

**Endocytosis.** Both phagocytosis (**A**) and pinocytosis (**B**) are forms of endocytosis. The large egg-shaped protist *Didinium nasutum* (**C**) has just begun eating the smaller protist *Paramecium;* (**D**) its meal is practically over.

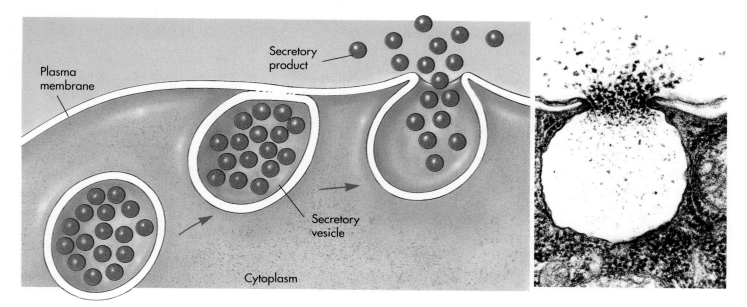

**Figure 5-16**

**Exocytosis.** Proteins and other molecules are secreted from cells in small pockets called vesicles, whose membranes fuse with the cell membrane, releasing their contents to the cell surface.

membrane movement is involved. More important, endocytosis is not selective. Particularly when the process is pinocytotic, it is difficult for the cell to discriminate between different solutes, admitting some kinds of molecules into the cell while excluding others.

However, no cell is limited to endocytosis. As we have seen, cell membranes are studded with proteins, which act as channels across the membrane. Because a given channel will transport only certain kinds of molecules, the cell membrane is **selectively permeable**; that is, it is permeable to some molecules and not to others. Different kinds of cells are permeable to different molecules because they have different protein channels embedded within their membranes.

*Selective permeability allows the passage across a membrane of some solutes but not others. Selective permeability of cell membranes is the result of specific protein channels extending across the membrane; some molecules can pass through a specific kind of channel, but others cannot do so.*

## The Importance of Selective Permeability

The most important property of any cell is that the cell constitutes an isolated compartment within which certain molecules can be concentrated and brought together in particular combinations. This essential isolation depends on allowing molecules to enter the cell selectively. Your home is private in the same sense—inoperable if no one, including you, can enter, but not private if everyone in the neighborhood is free to wander through at all times. The solution adopted by cells is the same one that most homeowners adopt: there are doors with keys, and only those possessing the proper keys can enter or leave. The channels through plasma membranes are the doors to cells. These doors are not open to any molecule that presents itself; only particular molecules can pass through a given kind of door. A cell is able to control the entry and exit of many kinds of molecules by possessing many kinds of doors. These channels through its membrane are among the most important functional feature of any cell.

## Facilitated Diffusion

Some of the most important channels in a plasma membrane are highly selective, facilitating *only* the passage of specific molecules or ions, but in either direction. An example is provided by the channel of vertebrate red blood cell membranes that transports negatively charged ions, or **anions**. This channel plays a key role in the oxygen-transporting function of these cells. The anion channels of the red blood cell membrane readily pass chloride ions ($Cl^-$) or carbonate ions ($HCO_3^-$) across the red blood cell membrane. We can easily demonstrate that the ions are moving by diffusion through the channels in the red blood cell membrane. If there are more $Cl^-$ ions within the cell, we will see that the net movement is outward, whereas if there are more $Cl^-$ ions outside the cell, then the net movement will be into the cell. Because the ion movement is always toward the direction of lower ion concentration, the transport process is one of diffusion.

Are these channels simply holes in the membrane, somehow specific for these anions? Not at all. If we repeat our hypothetical experiment, progressively increasing the concentration of $Cl^-$ ions outside the cell above that inside the cell, the rate of movement of $Cl^-$ ions into the cell increases only up to a certain point, after which it levels off and will proceed no faster despite increases in the concentration of exterior $Cl^-$ ions. The reason that the diffusion rate will increase no further is that the $Cl^-$ ions are being transported across the membrane by a protein "carrier," and all available carriers are in use: we have saturated the capacity of the carrier system. The transport of these anions is a diffusion process facilitated by a carrier. Transport processes of this kind are called **facilitated diffusion** (Figure 5-17).

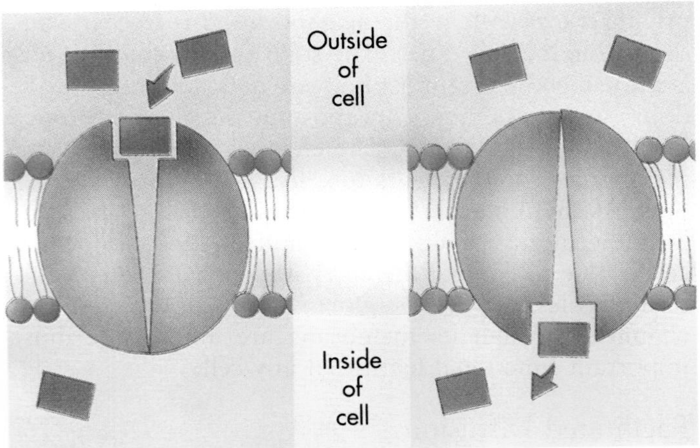

*Facilitated diffusion is the transport of molecules across a membrane by a carrier protein in the direction of lowest concentration.*

Facilitated diffusion provides the cell with a ready means of preventing the buildup of unwanted molecules within the cell, or of gleaning from the external medium the molecules that are present in high concentration. Facilitated diffusion has two essential characteristics: (1) it is *specific*, with only certain molecules being able to traverse a given channel, and (2) it is *passive*, the direction of net movement being determined by the relative concentrations of the transported molecule inside and outside the membrane.

## Active Transport

There are many molecules that the cell admits across its membrane that are maintained within the cell at a concentration different from that of the surrounding medium (Figure 5-18). In all such cases, the cell must expend energy to maintain the concentration difference. The kind of transport that requires the expenditure of energy is called **active transport**. Active transport may maintain molecules at a higher concentration inside the cell than outside it by expending energy to pump more molecules in than would enter by diffusion, or it may maintain molecules at a lower concentration by expending energy to pump them out actively.

*Active transport is the transport of a solute across a membrane to a region of higher concentration by the expenditure of chemical energy. The process uses protein carrier molecules.*

Active transport is one of the most important functions of any cell. It is by active transport that a cell is able to concentrate molecules. Without it, the cells of your body would be unable to harvest glucose—a major source of energy—from the blood because the concentration of glucose molecules is often already higher in the cells than it is in the blood from which it is extracted. Imagine how difficult it would be to survive as

**Figure 5-17**
Facilitated diffusion is a carrier-mediated transport process.

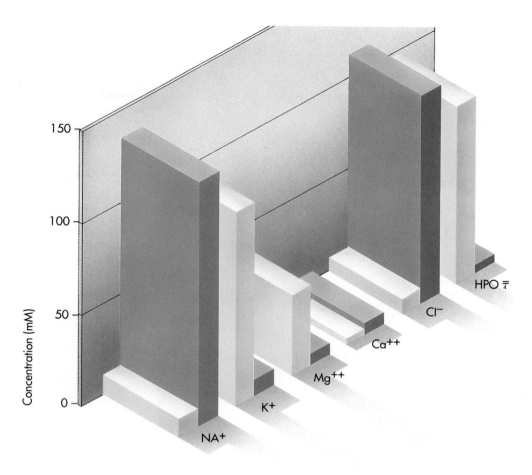

**Figure 5-18**

**The ionic composition of cells.** Many ions are present at far higher concentrations within cells *(gold)* than in the fluid surrounding the cells *(blue)*. These differences, maintained by active transport of ions, are responsible for many of the most important properties of cells.

a beggar if you could only obtain money from those who had less than you did! Active transport permits a cell, by expending energy, to take up additional molecules of a substance that is already present in its cytoplasm in concentrations higher than those found in the cell's environment.

There are many molecules that a cell takes up or eliminates against a concentration gradient. Some, such as sugars and amino acids, are simple metabolites that the cell extracts from its surroundings and adds to its internal stockpile. Others are ions such as sodium and potassium, that play a critical role in functions such as the conduction of nerve impulses. Still others are the nucleotides that the cell uses to synthesize DNA. These many kinds of molecules enter and leave cells by way of a wide variety of different kinds of selectively permeable transport channels. Some of the channels are permeable to one or a few sugars, others to a certain size of amino acid, and still others to a specific ion or nucleotide. You might suspect that active transport occurs at each of these channels, but you would be

wrong. There is *one* major active transport channel in plasma membranes that transports sodium and potassium ions; all the others work by tying their activity to this all-important channel. The many channels in the membrane that the cell uses to concentrate metabolites and ions are called **coupled channels.** We discuss the sodium-potassium channel first, and then these coupled channels.

***The Sodium-Potassium Pump.*** More than a third of all the energy expended by a cell that is not actively dividing is used to actively transport sodium ($Na^+$) and potassium ($K^+$) ions. The remarkable channel by which these two ions are transported across the plasma membrane is referred to as the **sodium-potassium pump** (Figure 5-19). Most animal cells have a low internal concentration of $Na^+$ ions and a high internal concentration of $K^+$ ions relative to their surroundings. They maintain these concentration differences by actively pumping $Na^+$ ions out of the cell and $K^+$ ions in. The transport of these ions is carried out by a highly spe-

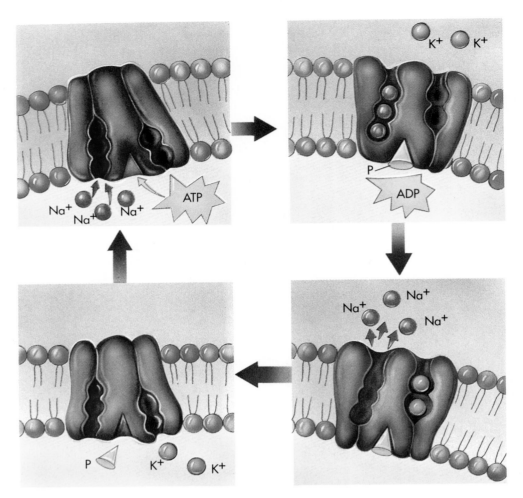

**Figure 5-19**
The sodium-potassium pump.

cific transmembrane protein channel. Passage through this channel entails changes in the shapes of the proteins within it.

The sodium-potassium pump is an active transport process, transporting $Na^+$ and $K^+$ ions from areas of low concentration to areas of high concentration. This transport into a zone of higher concentration is just the opposite of that which occurs spontaneously in diffusion; it is achieved only by the constant expenditure of metabolic energy. The energy used in the process is obtained from a molecule called **adenosine triphosphate** (**ATP**), the functioning of which is explained in Chapter 6. Some membranes contain large numbers of $Na^+-K^+$ channels, whereas others contain few. The changes in protein shape that go on within an individual channel are very rapid. Each channel is capable of transporting as many as 300 $Na^+$ ions per second when working full tilt.

*Coupled Channels.* The accumulation of many amino acids and sugars by cells is also driven against a concentration gradient: the molecules are harvested from a surrounding medium in which their concentration is much lower than it is inside the cell. The active transport of these molecules across the cell membrane takes place by coupling them with $Na^+$ ions, which pass simultaneously through a channel by facilitated diffusion. This process is driven by the very low concentration of $Na^+$ ions within the cell, maintained by active transport of these ions outward by the sodium-potassium pump. The special transmembrane protein of the **coupled channel** allows $Na^+$ back into the cell only when a sugar or amino acid is also bound to the protein's exterior surface (Figure 5-20). As a result, the facilitated diffusion inward of the $Na^+$ ion also results in the importation into the cell of a sugar or amino acid. In this process, the sugar or amino acid is literally dragged along osmotically because the Na⁺ ions are so much less concentrated within the cell than they are outside. Thus the transport of the sugar or amino acid inward to an area of its higher concentration, against the concentration gradient, occurs as a direct consequence of the sodium-potassium pump's activity.

***The Proton Pump.*** The sodium-potassium pump is of central importance because it drives the uptake of so many different molecules. A second channel of equal importance in the life of the cell is the **proton pump.**

***Figure 5-20***

**A coupled channel.** The sodium-potassium pump keeps the Na⁺ ion concentration higher outside the cell than inside. There is thus a strong tendency for Na⁺ ions to diffuse back in through the coupled channel—but their passage requires the simultaneous transport of a sugar molecule as well.

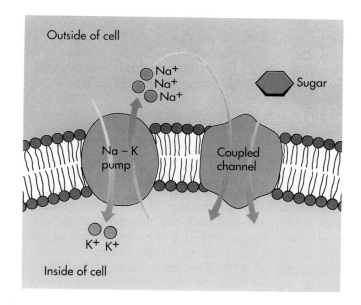

# Chloride Channels and Cystic Fibrosis

Cystic fibrosis is a fatal disease of human beings in which affected individuals secrete a thick mucus that clogs the airways of the lungs. These same secretions block the ducts of the pancreas and liver so that the few patients who do not die of lung disease die of liver failure. Cystic fibrosis is usually thought of as a children's disease because few affected individuals live long enough to become adults. There is no known cure.

Cystic fibrosis is a genetic disease resulting from a defect in a single gene that is passed down from parent to child. It is the most common fatal genetic disease of Caucasians. One in 20 individuals possesses at least one copy of the defective gene. Most carriers are not afflicted with the disease; only those children who inherit two copies of the defective gene, one from each parent, succumb to cystic fibrosis—about 1 in 1800 Caucasian children.

Cystic fibrosis has proved to be difficult to study. Many organs are affected, and until recently it was impossible to identify the nature of the defective gene responsible for the disease. In 1985 the first clear clue was obtained. An investigator, Paul Quinton, seized on a commonly observed characteristic of cystic fibrosis patients, that their sweat is abnormally salty, and performed the following experiment. He isolated a sweat duct from a small piece of skin and placed it in a solution of salt (NaCl) that was three times as concentrated as the NaCl inside the duct. He then monitored the movement of ions. Diffusion tends to drive both the Na⁺ and Cl⁻ ions into the duct because of the higher outer ion concentrations. In skin isolated from normal individuals, Na⁺ ions indeed entered the duct, transported by the sodium-potassium pump; Cl⁻ ions followed, passing through a passive channel. Both ions crossed the membrane easily. In skin isolated from individuals with cystic fibrosis, the sodium-potassium pump transported Na⁺ ions into the ducts, but no Cl⁻ ions entered. The passive chloride channels were not functioning in these individuals.

It appears that cystic fibrosis results from a defective channel within plasma membranes, one that transports Cl⁻ ions across the membranes of normal individuals but not across those of affected persons. It was learned in 1986 that the genetic defect is the result of an alteration in a protein regulating the activity of the channel, rather than in the transmembrane protein itself. The defective gene was isolated in 1987, and its position on a particular human chromosome was pinpointed in 1989. Now that scientists have finally identified the primary cause of the disease, they have much greater opportunities than were available earlier to find a cure for it.

The proton pump involves two special transmembrane protein channels: the first pumps protons ($H^+$ ions) out of the cell, using energy derived from energy-rich molecules or from photosynthesis to power the active transport. This creates a proton gradient in which the concentration of protons is higher outside the membrane than inside. As a result, diffusion acts as a force to drive protons back across the membrane toward a zone of lower proton concentration. But biological membranes are impermeable to protons, so the only way protons can diffuse back into the cell is through a second channel, which couples the transport of protons to the production of ATP. This process occurs within the cell and not primarily at its plasma membrane. The net result is the expenditure of energy derived from metabolism or photosynthesis and the production of ATP. This mechanism, called **chemiosmosis,** is responsible for the production of almost all the ATP that you harvest from food that you eat (see Figure 7-4) and for all the ATP produced by photosynthesis (see Figure 8-16). ATP provides the cell with a usable energy source that it can employ in its many activities.

## Transfer Between Cell Compartments

Relatively little is known about how molecules pass from one compartment of a cell to another. One process thought to be important is receptor-mediated endocytosis.

*Receptor-Mediated Endocytosis.* Within most eukaryotic cells there are vesicles that transport molecules from one place to another. In electron micrographs, some of these vesicles appear like membrane-bounded balloons coated with bristles on their interior (cytoplasmic) surfaces, as if they had a 2-day growth of beard. These coated vesicles appear to form by the pinching in of local regions of plasma membrane that are coated with a network of proteins called coated pits (Figure 5-21). Coated pits act like molecular mousetraps, closing over to form an internal vesicle when the right molecule enters the pit. The trigger that releases the mousetrap is a protein called a receptor, embedded within the pit, that detects the presence of a particular target molecule and reacts by initiating endocytosis of the coated pit. By doing so, it traps the target molecule within the new vesicle. This process is called **receptor-mediated endocytosis.** It is quite specific, so that transport across the membrane occurs *only* when the correct molecule is present and positioned for transport.

The mechanisms for transport across plasma membranes that we have considered in this chapter are summarized in Table 5-2.

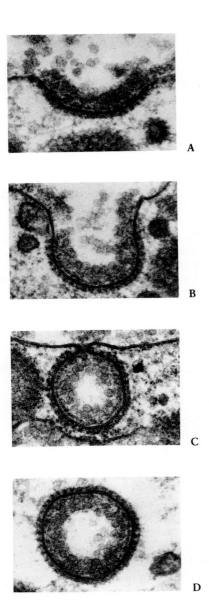

*Figure 5-21*

**How coated vesicles form.** A coated pit appears in the plasma membrane of a developing egg cell, covered with a layer of proteins (**A**). A coating of protein molecules can be seen just beneath the it, on the interior side of the membrane. When an appropriate collection of molecules gathers in the coated pit, the pit deepens (**B**) as the outer membrane of the cell closes in behind the pit (**C**), and the pit buds off to form a coated vesicle, which carries the molecules into the cell (**D**).

## RECEPTION OF INFORMATION

So far in this chapter we have focused on membrane proteins that act as channels—doors across a membrane through which only particular molecules may pass. But cells also interact with their environments in

**TABLE 5-2    MECHANISMS FOR TRANSPORT ACROSS CELL MEMBRANES**

| PROCESS | PASSAGE THROUGH MEMBRANE | HOW IT WORKS | EXAMPLE |
|---|---|---|---|
| **NONSPECIFIC PROCESS** | | | |
| Diffusion | Imperfections in lipid bilayer | Random molecular motion produces net migration of molecules toward region of lower concentration | Movement of oxygen into cells |
| Osmosis | Imperfections in lipid bilayer | Diffusion of water across differentially permeable membrane | Movement of water into cells placed in distilled water |
| **ENDOCYTOSIS** | | | |
| Phagocytosis | Membrane vesicle | Particle is engulfed by membrane, which folds around it, forming a vesicle | Ingestion of bacteria by white blood cells |
| Pinocytosis | Membrane vesicle | Fluid droplets are engulfed by membrane, which forms vesicles around them | Nursing of human egg cells |
| Exocytosis | Membrane vesicle | Vesicles fuse with plasma membrane and eject contents | Secretion of mucus |
| **SPECIFIC PROCESS** | | | |
| Facilitated diffusion | Protein channel | Molecule binds to carrier protein in membrane and is transported across; net movement is in direction of lowest concentration | Movement of glucose into cells |
| **ACTIVE TRANSPORT** | | | |
| Na-K pump | Protein channel | Carrier expends consumed energy to export Na ions against a concentration gradient | Coupled uptake of many molecules into cells against a concentration gradient |
| Proton pump | Protein channel | Carrier expends consumed energy to export protons against a concentration gradient | Chemiosmotic generation of ATP |
| Carrier-mediated endocytosis | Membrane vesicle | Endocytosis triggered by a specific receptor | Cholesterol uptake |

many ways that do not involve the passage of molecules across membranes. A second major class of transaction that cells carry out with their environments involves information. In examining receptor-mediated endocytosis, we encounter for the first time a second general class of membrane proteins, **cell surface receptors.** Recall that the receptor, a protein embedded within the membrane of a coated pit, binds specifically to particles but does *not* itself provide a transport channel for the particle; what the receptor transmits into the cell is information.

In general, a cell surface receptor is an information-transmitting protein that extends across a plasma membrane. The end of the receptor protein exposed on the cell surface has a shape that fits to specific hormones or other "signal" molecules, and when such molecules encounter the receptor on the cell surface, they bind to it. This binding produces a change in the shape of the other end of the receptor protein, the end protruding into the interior of the cell, and this change in shape in turn causes a change in cell activity in one of several ways.

Cell surface receptors play a very important role in the lives of multicellular animals. Among these receptors are the signals that pass from one nerve to another; the protein hormones such as adrenaline and insulin, which your body uses to regulate its metabolic level; and the growth factors such as epidermal growth factor, which regulate development. All of these substances act by binding to specific cell surface receptors. The antibodies that your body uses to defend itself against infection are themselves free forms of receptor proteins. Without receptor proteins, the cells of your body would be "blind," unable to detect the wealth of chemical signals that the tissues of your body use to communicate with one another.

*Cell surface receptors in membranes bind specifically to particles and transmit information about them into the cells that they surround. Among the information that they transmit are the presence of hormones and the signals that pass from one nerve to another. Antibodies are free forms of receptor proteins.*

## EXPRESSION OF CELL IDENTITY

In addition to passing molecules across its membranes and acquiring information about its surroundings, a third major class of cell interaction with the environment involves conveying information *to* the environment. To understand why this is important, let us consider multicellular animals like ourselves. One of their fundamental properties is the development and maintenance of highly specialized groups of cells called **tissues.** Your blood is a tissue, and so is your muscle. What is remarkable about having tissues is that each cell within a tissue performs the functions of a member of that tissue and not some other tissue, even though all the cells of the body have the same genetic complement of DNA and are derived from a single cell at conception. How does a cell "know" to which tissue it belongs? In the course of development in human beings and other vertebrates, some cells move over others, as if they were seeking particular collections of cells with which to develop. How do they sense where they are? Every day of your adult life, your immune system inspects the cells of your body, looking for cells infected by viruses. How does it recognize them? When foreign tissue is transplanted into your body, your immune system rejects it. How does it know that the transplanted tissue is foreign?

The answer to all these questions is the same: during development, every cell type in your body is given a banner proclaiming its identity, a set of proteins called **cell surface markers** that are unique to it alone. Cell surface markers are the tools a cell uses to signal to the environment what kind of cell it is.

Some cell surface markers are proteins anchored in plasma membranes. The immune system uses such marker proteins to identify "self." All the cells of a given individual have the same "self" marker, called a **major histocompatibility complex** (**MHC**) protein. Because practically every individual makes a different version of the MHC marker protein, these proteins serve as distinctive markers for that individual.

Other cell surface markers are **glycolipids,** lipids with carbohydrate tails. These are the cell surface markers that differentiate the various organs and tissues of the vertebrate body. The markers on the surfaces of red blood cells that distinguish different blood types, such as A, B, and O, are glycolipids. During the course of development, the cell population of glycolipids changes dramatically as the cells divide and differentiate.

*The cells that make up specific kinds of tissues are marked by proteins called cell surface markers. These are either proteins anchored in plasma membranes or glycolipids.*

## PHYSICAL CONNECTIONS BETWEEN CELLS

So far we have seen how cells pass molecules to and from the environment, how they acquire information from the environment, and how they convey information about their identity to the environment. A fourth major class of interaction with the environment concerns physical interactions with other cells. Most of the cells of multicellular organisms are in contact with other cells, usually as members of organized tissues in organs such as the lungs, heart, or gut. The immediate environment of many of these cells is the mass of other cells clustered around it. The nature of the physical connections between a cell and the other cells of a tissue largely determines that cell's contribution to what the tissue will be like.

The locations on the cell surface where cells of a tissue adhere to one another are called **cell junctions** (Figure 5-22). There are three general classes in animals (Table 5-3):

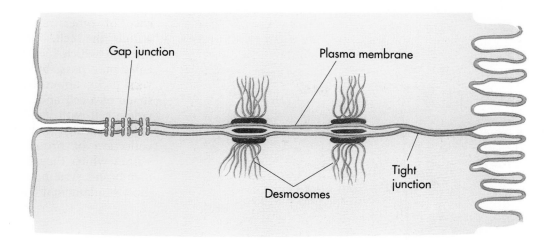

**Figure 5-22**

The three principal types of intercellular connection.

1. *Adhering junctions.* Adhering junctions called **desmosomes** hold cells together as if they were welds constructed of protein.
2. *Organizing junctions.* Organizing cell junctions partition the plasma membrane into separate compartments. **Tight junctions** are belts of protein that girdle each cell like the belt around a pair of jeans. These protein belts act like fences, preventing any membrane proteins afloat in the lipid bilayer from drifting across the boundary from one side of the cell to the other.
3. *Communicating junctions.* Communicating cell junctions called **gap junctions** (Figure 5-23) pass small molecules from one animal cell to another. Such passageways are large enough to permit the passage of small molecules, such as sugar molecules and amino acids, from one cell

to another, but small enough to prevent the passage of larger molecules, such as proteins. In plants the plasma membranes of adjacent cells come together through pairs of holes in the walls; the cytoplasmic connections that extend through such holes are called **plasmodesmata**.

## HOW A CELL COMMUNICATES WITH THE OUTSIDE WORLD

Every cell is a prisoner of its lipid envelope, unable to communicate with the outside world except by means of the proteins that traverse its lipid shell. Anchored within the lipid layer by nonpolar segments, these proteins are the "senses" of the cell. They detect the pres-

| | | CHARACTERISTIC SPECIALIZATIONS IN | WIDTH OF | |
|---|---|---|---|---|
| TYPE | NAME | CELL MEMBRANE | INTERCELLULAR SPACE | FUNCTION |
| Adhering junctions | Desmosomes | Buttonlike welds joining opposing cell membranes | Normal size 24 nm | Hold cells tightly together |
| Organizing junctions | Tight junctions | Belts of protein that isolate parts of plasma membrane | Intercellular space disappears as the two membranes are adjacent | Form a barrier separating surfaces of cell |
| Communicating junctions | Gap junctions | Channels or pores through the two cell membranes and across the intercellular space | Intercellular space greatly narrowed to 2 nm | Provide for electrical communication between cells and for flow of ions and small molecules |

**TABLE 5-3    TYPES OF INTERCELLULAR CONNECTIONS**

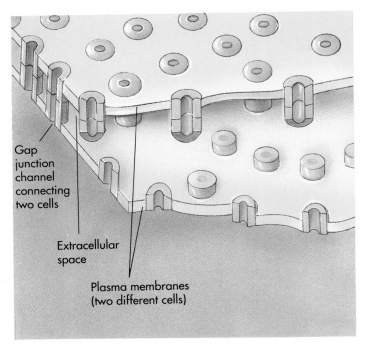

Gap
junction
channel
connecting
two cells

Extracellular
space

Plasma membranes
(two different cells)

ence of other molecules, often initiating responses within the cell. They provide doors into the cell through which food molecules, ions, and other molecules may pass, but like protective doormen they are very picky about whom they admit. They form the shape of a cell and bind one cell to another. They provide a cell with its identity, doing so by means of surface name tags that other cells can read. This diverse collection of proteins, together with the lipid shell within which they are embedded, constitute the cell's membrane system, a system that is among the cell's most fundamental features.

**Figure 5-23**

**Gap junctions.** Gap junctions are open channels connecting animal cells, like so many pipes between two rooms.

## ■ SUMMARY

1. Every cell is encased within a bilayer sheet of phospholipid, which exists as a liquid. This is the plasma membrane, which is similar in its basic structure to the other kinds of biological membranes.

2. Because cells contain significant concentrations of sugars, amino acids, and other solutes, and thus constitute a hypotonic envirnoment, water tends to diffuse into them. As it does so, a hydrostatic pressure builds that will rupture cells lacking a wall or other means of support.

3. Lipid bilayers are selectively permeable and do not permit the diffusion of water-soluble molecules into the cell. These molecules gain entry by crossing one of a variety of transmembrane proteins embedded within the membrane, with the proteins acting as transport channels.

4. Some channels involve carriers that transport molecules across the membrane much like cars of a train carry passengers across a bridge. This process is called facilitated diffusion because net movement is always in the direction of lowest concentration but cannot proceed any faster than the capacity of the number of carriers.

5. Some channels transport molecules against a concentration gradient by expending energy. The two most important ones transport sodium ions and protons. These channels create very low concentrations of sodium ions and protons (hydrogen ions) within the cell. These ions can diffuse back into the cell only through a second set of special channels, where their passage is often coupled to transport of another molecule inward or to the synthesis of ATP.

6. Passage across membranes from one cell compartment to another takes place when a particular molecule triggers a receptor in the membrane, initiating a form of endocytosis.

7. Many proteins embedded within the plasma membrane transmit information into the cell rather than transporting molecules. These proteins, called receptors, initiate chemical activity inside the cell in response to the binding of specific molecules on the cell surface.

8. Both proteins and glycolipids are used by cells as identification markers. These cell surface recognition markers permit cells of a given tissue to identify one another and also provide your body with a means of identifying foreign cells.

9. There are three general classes of cell junctions (connections between cells) in animal cells: (1) adhering junctions, or desmosomes, which hold cells together; (2) organizing junctions, which partition the plasma membrane into separate compartments; and (3) communicating junctions, which pass small molecules from one cell to another. In plants, the cell walls have openings that allow protoplasmic connections, called plasmodesmata, to connect adjacent cells.

## REVIEW

1. What three components are necessary to make a phospholipid?

2. Which class of cell surface markers function in "self" versus "nonself" recognition?

3. What physical process allows you to smell perfume when you are near someone wearing it?

4. What mechanism is responsible for the production of almost all the energy that you obtain from food?

5. Give an example of a substance that acts by binding to cell surface receptors.

## SELF-QUIZ

1. Phospholipids spontaneously form a lipid bilayer if placed in water. Why?
   (a) Because phospholipids, like fats, are repelled by the polar nature of water
   (b) Because phospholipids are polar molecules, so they mix easily with water
   (c) For the same reason that coacervates form spontaneously
   (d) They do not form a lipid bilayer spontaneously; proteins are needed to stabilize the phospholipids
   (e) Because part of the molecule, the alcohol end, is polar and has affinity for the water, whereas the two nonpolar fatty acids are repelled by water and thus associate with the fatty acid tails of the other phospholipids

2. If you put an animal cell into a hypotonic solution, it will pop. However, if you put a plant cell into the same hypotonic solution, it will almost never pop. Why?
   (a) Because the plant cell wall can withstand considerable hydrostatic pressure.
   (b) Because photosynthesis disrupts osmosis.
   (c) Because plant cells react oppositely from animal cells with regard to hypotonic solutions.
   (d) Because plant cells contain more water.
   (e) Because plant cells produce molecules that change the hypotonic solution into an isotonic one.

3. How many of the following means of moving things into or out of cells require an expenditure of energy?
   (a) Diffusion
   (b) Osmosis
   (c) Facilitated diffusion
   (d) Active transport
   (e) Sodium-potassium pump

4. How many of the following can move molecules *against* a concentration gradient?
   (a) Diffusion
   (b) Osmosis
   (c) Facilitated diffusion
   (d) Active transport
   (e) Sodium-potassium pump

5. How are large proteins transported into cells?
   (a) Diffusion
   (b) Facilitated diffusion
   (c) Active transport
   (d) The proton pump
   (e) Receptor-mediated endocytosis

# THOUGHT QUESTIONS

1. When a hypertonic cell is placed in water solution, water molecules move rapidly into the cell. However, if the introduced cell is hypotonic, water molecules leave the cell and enter the surrounding solution. How does an individual water molecule *know* what solutes are on the other side of the membrane?

2. Cells maintain many internal metabolite molecules at high concentrations by coupling their transport into the cell to the transport of sodium and potassium ions by the sodium-potassium pump. What happens to all the potassium ions that are constantly being pumped into the cell?

# FOR FURTHER READING

BRETSCHER, M.S.: "The Molecules of the Cell Membrane," *Scientific American*, October 1985, pages 100-108. A good description of the structure of the cell membrane and of how transmembrane proteins are anchored within the lipid bilayer.

HAKOMORI, S.: "Glycosphingolipids," *Scientific American*, May 1986, pages 44-53. Carbohydrate chains attached to lipid molecules play an important role in cell-to-cell recognition, an area of intensive present-day research.

STEAHLIN, L.A., and B.E. HULL: "Junctions Between Living Cells," *Scientific American*, May 1978, pages 140-152. Many passages exist between adjacent cells. They play important roles in the biology of organisms.

UNWIN, N., and R. HENDERSON: "The Structure of Proteins in Biological Membranes," *Scientific American*, February 1984, pages 78-94. A lucid account of how proteins are anchored within membranes by means of non-polar segments.

ENERGY

# Energy and Metabolism

Even a mother's love requires energy. These Japanese monkeys are huddled together sharing their body warmth, which they generate by converting body fat to ATP and then expending ATP to generate heat. Although their fur provides them with excellent insulation, they are constantly losing heat to the cold around them and would not survive long if they could not generate heat to replace that which is lost.

# ENERGY AND METABOLISM

## Overview

The life processes of every cell are driven by energy. From cell growth and movement to the transport of molecules across plasma membranes, all of a cell's activities require energy. Just as a car is powered by burning gasoline, so our bodies oxidize chemical fuel to obtain energy. Where is the energy in a chocolate bar? It is in the electrons, which spin in energetic orbitals about their atomic nuclei. Your cells strip these electrons away and use them to power their lives. Your every thought is fueled by the energy of electrons.

## For Review
*Here are some important terms and concepts that you will encounter in this chapter. If you are not familiar with them, you should review them before proceeding.*

**Mole** (Chapter 2)

**Nature of chemical bond** (Chapter 2)

**Nucleotides** (Chapter 2)

**Protein structure** (Chapter 2)

**Proton pump** (Chapter 5)

A well-known story tells of an extravagant man who purchased an expensive car, drove it until it ran out of gas and sputtered to a stop—and then went out and bought another! You and I and every other living creature are like this man's car—we cannot run without fuel. Fuel is a source of energy, and energy drives everything that we do. When you whistle or walk down the street, you use energy. When you plan or hope or dream as you stand in the shower or sleep in bed, you use energy. In fact, it takes more energy to power your brain when you are sound asleep than it does to power the 60-watt bulb that lights your bedroom.

You obtain your energy from the food you eat. If you stopped eating, you would soon begin to lose weight as your body used up its stored energy. If you were not supplied with some outside source of energy, you would eventually die. The same is true of all other living things. Deprived of a source of energy, life stops. Why does this happen? Why cannot life simply continue? The reason is that each of the significant properties by which we define life—growth, reproduction, and heredity—uses energy. And once this energy has been used, it is dissipated as heat and cannot be used again. To keep life going, more energy must be supplied, like putting more logs on a fire (Figure 6-1).

Life can be viewed as a constant flow of energy, which is channeled by organisms to do the work of living. This chapter focuses on energy—what it is, and how organisms capture, store, and use it. In the following two chapters we explore in more detail the energy-capturing and energy-using engines of cells, a network of chemical reactions that are the highway system for the energy of your body. This living chemistry—the total of all the chemical reactions that an organism performs—is called **metabolism.**

## WHAT IS ENERGY?

**Energy** is defined as the ability to bring about change or, more generally, as the capacity to do work. Instinctively we all know something about energy. It is "work," such as the force of a falling boulder, the pull of a locomotive, or the swift dash of a horse; it is also "heat," such as the blast from an explosion or the warmth of a fire. Energy can exist in many forms: as mechanical force, heat, sound, an electrical current, light, radioactivity, and the pull of a magnet. All are able to create change, to do work.

Energy exists in two states. Some energy is actively engaged in doing work, such as driving a speeding bul-

**Figure 6-1**

**All life's processes use energy.** The energy expended by these baby squirrels in keeping warm is supplied by the food they eat.

let or lifting a brick. This form of energy is called **kinetic energy,** or energy of motion. Other energy is not actively doing work but has the capacity to do so, just as a boulder perched on a hilltop has the capacity to roll downhill. This form of energy is called **potential energy,** or stored energy. Much of the work performed by living organisms involves the transformation of potential energy to kinetic energy (Figure 6-2).

*Energy is the capacity to bring about change. It can exist in many forms: some are actively engaged in doing work, and others store the energy for later use.*

Because energy exists in many forms, there are many ways to measure it. The most convenient way is in terms of heat because all other forms of energy can be converted into heat. Indeed, the study of energy is called **thermodynamics;** that is, "heat changes." The unit of heat most commonly employed in biology is the **kilocalorie (kcal),** which is equal to 1000 calories. A calorie is the heat required to raise the temperature of 1 gram of water 1 degree (from 14.5° to 15.5° C). It is important not to confuse the term *calorie* with a term often encountered in diets and discussions of nutrition, the Calorie (with a capital C), which is in fact another term for kilocalorie.

**Figure 6-2**

**The energy of motion.** Transforming potential into kinetic energy, this cheetah has just sprung into motion.

## THE LAWS OF THERMODYNAMICS DESCRIBE HOW ENERGY CHANGES

All of the changes in energy that take place in the universe, from nuclear explosions to the buzzing of a bee, are governed by two laws called the **laws of thermodynamics.** The **first law of thermodynamics** concerns amounts of energy. It states that energy can change from one form to another and can transform from potential energy to kinetic energy, but it can never be lost. Nor can any new energy be made. The total amount of energy in the universe remains constant.

*The first law of thermodynamics states that energy cannot be created or destroyed; it can only undergo conversion from one form to another.*

The ground squirrel that you see gnawing on a nut in Figure 6-3 is busy acquiring energy. He is not creating new energy, but rather is transferring the potential energy stored in the nut's tissues to his own body, where it will fuel running, digging, and all his other daily activities. Where is the potential energy in the nut's tissue? It is stored in chemical bonds. Recall the nature of covalent chemical bonds, which were discussed in Chapter 2. A covalent bond is created when two atomic nuclei share electrons, and breaking such a bond requires energy to pull the nuclei apart. Indeed, the strength of a covalent bond is measured by the amount of energy required to break it. For example, it takes 98.8 kilocalories of energy to break a mole (the atomic weight of a substance expressed in grams) of carbon-hydrogen (C—H) bonds.

What happens to the energy after the squirrel uses it? Some of it is transferred to other forms of potential energy—for example, stored as fat. Another portion accomplishes mechanical work such as bending blades of grass and running. Almost half is dissipated to the environment as heat, where it speeds up the random motions of molecules. This energy is not lost, but rather is converted to a nonuseful form, random molecular motion.

The **second law of thermodynamics** concerns this transformation of potential energy to the kinetic energy of random molecular motion; that is, to heat. It states that all objects in the universe tend to become more disordered and that the disorder in the universe is continuously increasing. The idea behind this law is easy to understand and part of everyone's experience. It is much more likely that a stack of six soda cans will tumble over than that six cans will spontaneously leap one onto another and form a stack. Stated simply, disorder is more likely than order. This is true of a child's room, of the desk where you study, of a waiting crowd of people—and of molecules.

*The second law of thermodynamics states that disorder in the universe constantly increases. Energy spontaneously converts to less organized forms.*

**Figure 6-3**

**An animal's body is an efficient furnace.** The oxidation of food that takes place during metabolism leads to the same chemical end-point, to $CO_2$ and water. An animal, however, siphons off much of the energy, while burning disperses the chemical-bond energy as heat.

As energy is transferred from one molecule to another, some always leaks away as kinetic energy of motion, thus increasing the random motion of molecules. At normal temperatures, all molecules dance about randomly, and this added energy increases the pace of the dance. We refer to this form of kinetic energy as heat energy.

With every transfer of energy, more potential energy is dissipated as heat. Although heat can be harnessed to do work when there is a gradient (that is how a steam engine works), it is generally not a useful source of energy for biological systems. Thus, from a biological point of view, the amount of useful energy in a system dissipates progressively. Although the total amount of energy does not change, the amount of **useful energy** available to do work decreases as progressively more energy is degraded to heat.

When energy becomes so randomized and uniform in a system that it is no longer available to do the work, the energy lost to disorder is referred to as **entropy**. Entropy is a measure of the disorder of a system. Sometimes the second law of thermodynamics is stated simply as "entropy increases." When the universe was formed about 14 billion years ago, it had all the potential energy it will ever have. It has been becoming progressively more disordered ever since, with every energy exchange frittering away useful energy and increasing entropy. Someday all the energy will be random and uniform in distribution; the universe will have wound down like an abandoned clock. No stars will shine, no waves will break upon a beach. But this final state will not occur soon. Scientists speculate it will be achieved perhaps 100 billion years from now. Our species has been around for less than 1 million years, and life on earth has existed for about 3.5 billion years. Clearly, many more immediate problems face us than the final end predicted by the second law of thermodynamics.

The first law of thermodynamics tells us that the universe as a whole is a closed system: energy doesn't come in or go out. The earth, however, is not a closed system; it is constantly receiving energy from the sun. It has been estimated that every year the earth receives in excess of $13 \times 10^{23}$ calories of energy from the sun, which is equal to 2 million trillion calories per second. Much of this energy heats up the oceans and continents, but some is captured by photosynthetic organisms: plants (Figure 6-4), algae (photosynthetic protists), and photosynthetic bacteria. In photosynthesis, energy acquired from sunlight is converted to chemical energy, combining small molecules (water and carbon dioxide) into ones that are more complex (sugars). The energy is stored as potential energy in the bonds of the

**Figure 6-4**

**All the energy that powers life is captured from sunlight.** These trees in a Michigan forest are the first of many plants to snare sunlight as it falls toward the forest floor.

sugar molecules. This energy then can be shifted to other molecules by forming different chemical bonds or can be converted into motion, light, electricity, and heat. As each shift and conversion takes place, more energy is dissipated as heat. Energy continuously flows through the biological world, new energy from the sun constantly flowing in to replace the energy that is dissipated as heat.

Life converts energy from the sun to other forms of energy that drive life processes. The energy is never lost, but as it is used, more and more is converted to heat energy, a form of energy that is not useful in performing biological work.

## OXIDATION-REDUCTION: THE FLOW OF ENERGY IN LIVING THINGS

The flow of energy into the biological world comes from the sun, which shines a constant beam of light on our earth and its moon. Life exists on earth because it

**Figure 6-5**

**A royal dinner.** The tissue this lion is consuming will undergo many changes on its metabolic journey, eventually becoming part of the lion and of the gases that the lion exhales, as well as solid waste products.

systems, oxygen, which strongly attracts electrons, is the most frequent electron acceptor (Figure 6-5).

When an atom or molecule gains an electron, it is **reduced,** and the process is called **reduction.** Oxidation and reduction always take place together because every electron that is lost by one atom (oxidation) is gained by some other atom (reduction) (Figure 6-6).

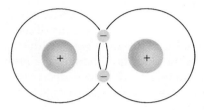

**Figure 6-6**

**Oxidation—reduction.** Oxidation is the loss of an electron; reduction is the gain of one.

is possible to capture some of that continual flow of energy and transform it into chemical energy that can be transferred from one organism to another and used to create cattle, and fleas, and you. Where is the energy in sunlight, and how is it captured? To understand the answers to these questions, we need to look more closely at the atoms on which the sunlight shines. As described in Chapter 2, an atom is composed of a central nucleus surrounded by one or more orbiting electrons, and different electrons possess different amounts of energy, depending on how far from the nucleus they are and how strongly they are attracted to it. Light (and other forms of energy) can boost an electron to a higher energy level. The effect of this is to store the added energy as potential energy; that is, as chemical energy that the atom can later release by dropping the electron back to its original energy level.

Energy stored in chemical bonds can be transferred to new chemical bonds, with the electrons shifting from one energy level to another. In some (but not all) of these chemical reactions, electrons actually pass from one atom or molecule to another. This class of chemical reaction is called an **oxidation-reduction reaction.** Oxidation-reduction (or redox) reactions are critically important to the flow of energy through living systems.

When an atom or molecule loses an electron, it is **oxidized,** and the process by which this occurs is called **oxidation.** The name reflects the fact that, in biological

*Oxidation is the loss of an electron; reduction is the gain of one.*

Oxidation-reduction reactions play a key role in energy flow through biological systems because the electrons that pass from one atom to another carry their own potential energy of position (that is, they maintain their distance from the nucleus). Energy that originally entered the system when light boosted the electrons in a photosynthetic pigment to higher energy levels is passed from one molecule to another by these electrons. Until the electrons return to their original low-energy level, they keep storing potential energy.

*Atoms can store potential energy by means of electrons that orbit at higher-than-usual energy levels. When such an energetic electron is removed from one atom (oxidation) and donated to another (reduction), it carries the energy with it and orbits the second atom's nucleus at the higher energy level.*

In biological systems, electrons often do not travel alone from one atom to another but rather in the company of a proton. Recall that a proton and an electron together make up a hydrogen atom. Thus oxidation-reduction in a chemical reaction usually involves the removal of hydrogen atoms from one molecule and the addition of hydrogen atoms to another molecule. For example, in photosynthesis, hydrogen atoms are transferred from water to carbon dioxide, reducing the carbon dioxide to form glucose:

$$6CO_2 + 6H_2O + energy \rightarrow C_6H_{12}O_6 + 6O_2$$

CARBON WATER        GLUCOSE OXYGEN
DIOXIDE

In this reaction, electrons move to higher energy levels. The conversion of carbon dioxide to form a mole of glucose stores 686 kilocalories of energy in the chemical bonds of the glucose.

The energy stored in a glucose molecule is released in a process called **cellular respiration,** in which the glucose is oxidized. Hydrogen atoms are lost by glucose and gained by oxygen:

$$C_6H_{12}O_6 + 6O_2 \rightarrow 6CO_2 + 6H_2O + energy$$

The oxidation of a mole of glucose releases 686 kilocalories of energy, the same amount that was stored in making it.

## ACTIVATION ENERGY: PREPARING MOLECULES FOR ACTION

As the laws of thermodynamics predict, all chemical reactions tend to proceed spontaneously toward a state of maximum disorder and minimum energy. Reactions in which the products contain less energy than do the reactants release the excess usable energy (called "free energy"). These reactions are spontaneous and are called **exergonic.** In contrast, reactions in which the products contain more energy than do the reactants require an input of usable energy from an outside source before they can proceed. These reactions are not spontaneous and are called **endergonic** (Figure 6-7).

*Any reaction producing products that contain less free energy than that possessed by the original reactants tends to proceed spontaneously.*

If all chemical reactions that release free energy tend to occur spontaneously, it is reasonable to ask

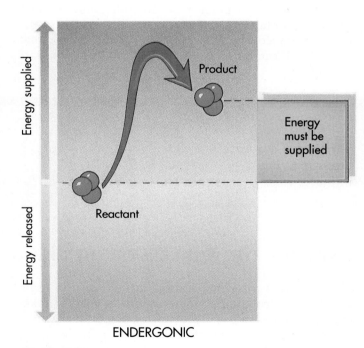

ENDERGONIC

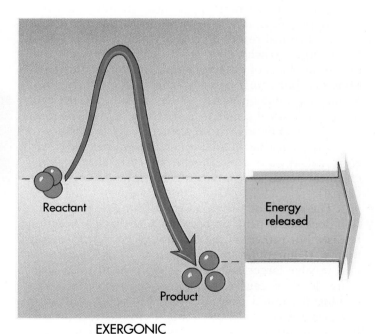

EXERGONIC

*Figure 6-7*

**Energy in chemical reactions.** In an *endergonic* reaction, the products of the reaction contain more energy than do the reactants, so the extra energy must be supplied for the reaction to proceed. In an *exergonic* reaction, the products contain less energy than do the reactants, and the excess energy is released.

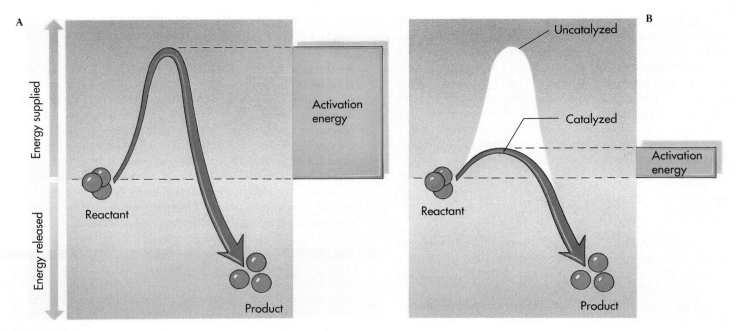

**Figure 6-8**

Exergonic reactions do not proceed spontaneously because it takes energy to get them going.
**A** Without an enzyme, energy must be supplied to destabilize existing chemical bonds.
**B** Enzymes catalyze particular reactions by lowering the amount of activation energy required to initiate the reaction.

why all such reactions have not already occurred. Clearly they have not. When gasoline is ignited, the resulting chemical reaction proceeds with a net release of free energy. So why doesn't all the gasoline within all the automobiles of the world and beneath all filling stations just burn up right now? Because most reactions require an input of energy to get started, like the heat from the flame of a match. Before it is possible to form new chemical bonds with less energy, it is necessary to break the existing bonds, and this requires extra energy, called *activation energy,* to destabilize existing chemical bonds and initiate a chemical reaction (Figure 6-8, *A*).

The speed, or reaction rate, of an exergonic reaction does not depend on how much energy the reaction releases but rather on the amount of activation energy that is required for the reaction to begin. Reactions that involve larger activation energies tend to proceed more slowly because fewer molecules succeed in overcoming the initial energy hurdle. However, activation energies are not fixed constants. Putting stress on particular chemical bonds can make them easier to break. The process of influencing chemical bonds in a way that lowers activation energies is called **catalysis,** and substances that perform catalysis are **catalysts.** Catalysts cannot violate the basic laws of thermodynamics;

for example, they cannot make an endergonic reaction proceed spontaneously. Only exergonic reactions proceed spontaneously, and catalysis cannot change that fact. What catalysts *can* do is make a reaction rate much faster.

*The speed of a reaction depends on the activation energy necessary to initiate it. Catalysts reduce the amount of activation energy required and thus speed reactions.*

## ENZYMES: THE WORKERS OF THE CELL

The chemistry of living things, metabolism, is organized by controlling the points at which catalysis takes place. Therefore life is a process regulated by **enzymes,** which are agents that perform catalysis in living organisms. Enzymes are globular proteins whose shapes are specialized to form temporary associations with the molecules that are reacting. By putting stress on particular chemical bonds, an enzyme lowers the amount of activation energy required for new bonds to form (Figure 6-8, *B*). The reaction thus proceeds much faster than it would otherwise. Because the enzyme itself is not changed, it can be used over and over.

To understand how an enzyme works, consider the joining of carbon dioxide ($CO_2$) and water ($H_2O$) to form carbonic acid ($H_2CO_3$):

$$CO_2 + H_2O \leftrightarrows H_2CO_3$$
CARBON DIOXIDE    WATER    CARBONIC ACID

This reaction can proceed in either direction, but in the absence of an enzyme it does so very slowly because there is a significant need for activation energy. Only about 200 molecules of carbonic acid form in 1 hour. Given the speed at which events usually occur within cells, this reaction rate is like a snail racing in the Indianapolis 500. Cells overcome this problem by employing an enzyme called carbonic anhydrase (enzymes are usually given names that end in *-ase*), which speeds the reaction dramatically. In the presence of carbonic anhydrase, an estimated 600,000 molecules of carbonic acid form every second! The enzyme has speeded the reaction rate about 10 million times.

Thousands of different kinds of enzymes have been described, each catalyzing a different chemical reaction. By facilitating particular chemical reactions, the enzymes in a cell determine the course of metabolism in that cell, much as traffic lights determine the flow of traffic in a city. Not all cells contain the same enzymes, which is why there is more than one type of cell. The chemical reactions within a red blood cell are very different from those going on within a nerve cell because the cytoplasm and membranes of red blood cells contain a different array of enzymes.

## How Enzymes Work

Enzymes are globular proteins with one or more pockets, or clefts, on their surface, which resemble deep creases in a prune (Figure 6-9). These surface depressions are called **active sites.** They are the locations at

A

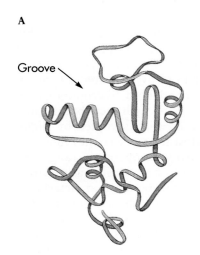

Groove

B

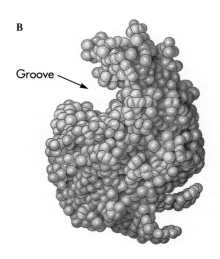

Groove

C

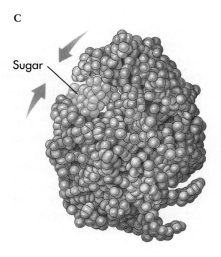

Sugar

### Figure 6-9

**How the enzyme lysozyme works.** The tertiary structure of the lysozyme, diagrammed as a ribbon in **A** and shown in a three-dimensional model in **B** and **C**, forms a groove through the middle of the protein. This groove fits the shape of the chains of sugars that make up bacterial cell walls. When such a chain of sugars, indicated in yellow in **C**, slides into the groove, its entry induces the protein to alter its shape slightly to embrace the substrate more intimately. This *induced fit* positions a glutamic acid in the protein right next to the bond between two adjacent sugars, and the glutamic acid "steals" an electron from the bond, causing it to break.

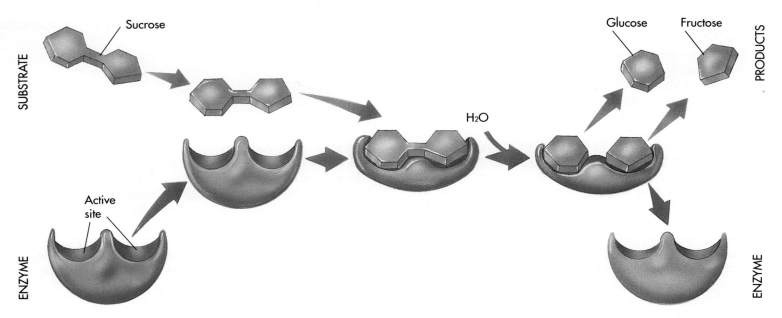

SUBSTRATE

Sucrose

Glucose    Fructose

PRODUCTS

H₂O

Active
site

ENZYME

ENZYME

*Figure 6-10*

**The catalytic cycle is an enzyme.** Enzymes increase the speed with which chemical reactions occur but are not altered themselves as they do this. In the reaction illustrated here, the enzyme is splitting the sugar sucrose (the sugar present in most candy) into its two parts, the simpler sugars glucose and fructose. After the enzyme releases the resulting glucose and fructose fragments, it is then ready to bind another molecule of sucrose and run through the catalytic cycle once again.

which catalysis occurs. For catalysis to occur, the molecule on which the enzyme acts, called a **substrate,** must fit precisely into the surface depression so that many of its atoms nudge up against atoms of the enzyme. The substrate fits into the active site of the enzyme like a foot into a tight-fitting shoe, in very close contact. Proteins are not rigid, however, and in some cases the binding of substrate may induce the protein to adjust its shape slightly, allowing a better fit.

When a substrate molecule binds to the active site of an enzyme, amino acid side groups of the enzyme are placed against certain bonds of the substrate, just as when you sit in a chair, certain parts of you press against the seat (Figure 6-10). These amino acid side groups chemically interact with the substrate, usually by stressing or distorting a particular bond, lowering the activation energy needed to break the bond.

Enzymes typically catalyze only one or a few different chemical reactions because they are very "picky" in their choice of substrate. The active site of each kind of enzyme is shaped so that only a certain substrate molecule will fit into it.

## Factors Affecting Enzyme Activity

The activity of an enzyme is affected by any change in conditions that alters its three-dimensional shape, in-

cluding changes in temperature or pH or the binding to the protein of specific chemicals that regulate the enzyme's activity.

*Temperature.* The shape of a protein is determined by hydrogen bonds that hold its arms in particular positions and also by the tendency of noncharged ("nonpolar") segments of the protein to avoid water. Chemists call interactions of this second kind **hydrophobic,** or water-hating, interactions. Both hydrogen bonds and hydrophobic interactions are easily disrupted by slight changes in temperature. Most human enzymes function best within a relatively narrow temperature range between 35° and 40° C (close to body temperature). Below this temperature level, the bonds that determine protein shape are not flexible enough to permit the induced change in shape that is necessary for catalysis; above this temperature, the bonds are too weak to hold the protein's arms in the proper position. In contrast, bacteria that live in hot springs have proteins with stronger bonding between their arms and therefore can function at temperatures of 70° C or higher (Figure 6-11, *A*).

*pH.* A third kind of bond that acts to hold the arms of proteins in position is the bond that forms between oppositely charged amino acids such as glutamic acid (−) and lysine (+). These bonds are sensitive to hydrogen ion concentration. The more hydrogen ions available in the solution, the fewer negative charges and the more positive charges that occur. For this reason, most enzymes have a pH optimum just as they have a temperature optimum; this optimum usually lies in the range of pH 6 to 8. Proteins that are able to function in very acidic environments have amino acid sequences that maintain their ionic and hydrogen bonds even in the presence of high levels of hydrogen ion (Figure 6-11, *B*). For example, the enzyme *pepsin* digests proteins in your stomach at pH = 2, a very acidic level.

## How Enzyme Activity is Regulated

The activity of an enzyme is sensitive not only to temperature and pH but also to the presence of specific chemicals that bind to the enzyme and cause changes in its shape. By means of these specific chemicals, a cell is able to regulate which enzymes are active and which are inactive at a particular time. When the binding of the chemical alters the shape of the protein and thus shuts off enzyme activity, the chemical is called an **inhibitor;** when the change in the enzyme's shape is necessary for catalysis to occur, the chemical is called an **activator.**

The change in shape that occurs when an activator or inhibitor binds to an enzyme is called an **allosteric** change (Greek *allos*, other + *steros*, shape). Enzymes usually have special binding sites for the activator and inhibitor molecules that affect them, and these binding sites are different from their active sites. The enzyme catalyzing the first step in a series of chemical reactions, or **biochemical pathway,** often has an inhibitor-binding site to which the molecule produced by the last step in the series binds. As the amount of this molecule builds up in the cell, it begins to bind to the initial enzyme in the biochemical pathway, thus inhibiting the activity of that enzyme. By this process the pathway is shut down when it is no longer needed. Such **end-product inhibition** is a good example of the way many enzyme-catalyzed processes within cells are self-regulating.

*The activity of enzymes is regulated by allosteric changes in enzyme shape; these changes result when specific, small molecules bind to the enzyme, molecules that are not substrates of that enzyme.*

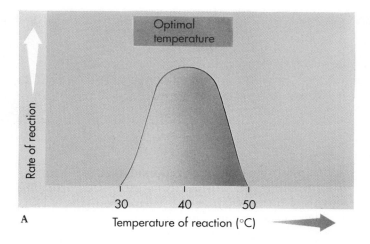

A

Temperature of reaction (°C)

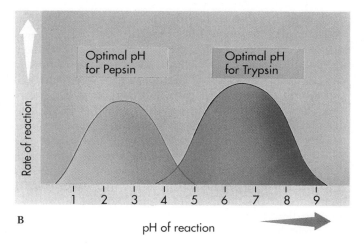

B

pH of reaction

*Figure 6-11*

**Enzymes are sensitive to their environment.** The activity of an enzyme is influenced both by temperature (**A**) and pH (**B**). Human enzymes tend to work best at temperatures of about 40° C, and within a pH range of 6 to 8.

## Coenzymes are Tools Enzymes Use to Aid Catalysis

Often enzymes use additional chemical components called **cofactors** as tools to aid catalysis. For example, many enzymes have metal ions locked into their active sites, and these ions help draw electrons from substrate molecules. The enzyme *carboxypeptidase* chops up proteins by using a zinc ion to draw electrons away from the bonds being broken. Many of the trace elements such as molybdenum and manganese, which are necessary for your health, also use metal ions in this way. A cofactor that is a nonprotein organic molecule is called a **coenzyme.** Many of the vitamins that your body requires are parts of coenzymes.

In many enzyme-catalyzed oxidation-reduction reactions, energy-bearing electrons are passed from the active site of the enzyme to a coenzyme that serves as the electron acceptor. The coenzyme then carries the electrons to a different enzyme that is catalyzing another reaction, releases the electrons (and the energy they bear) to that reaction, and then returns to the original enzyme for another load of electrons. In most cases the electrons are paired with protons as hydrogen atoms. Just as armored cars transport cash around a city, so coenzymes shuttle energy, in the form of hydrogen atoms, from one place to another in a cell.

One of the most important coenzymes is the hydrogen acceptor **nicotinamide adenine dinucleotide,** usually referred to by the abbreviation $NAD^+$ (Figure 6-12). When $NAD^+$ acquires a hydrogen atom from the active site of an enzyme, it becomes reduced as NADH. The energetic electron of the hydrogen atom is then carried by the NADH molecule, like money in your wallet. The oxidation of foodstuffs in your body, from which you get the energy to drive your life, takes place by the cell's stripping of electrons from food molecules and donating them to $NAD^+$, thus forming a wealth of NADH. This wealth is the principal energy income of your cells. However, much of the NADH is eventually converted to another currency.

## ATP: THE ENERGY CURRENCY OF LIFE

The chief energy currency of all cells is a molecule called **adenosine triphosphate,** or **ATP.** Just as the bulk of the energy that plants harvest during photosynthesis is channeled into production of ATP, so is most of the NADH that soaks up the energy resulting from the oxidation of your food. The energy stored in the chemical bonds of fat and starch is converted to ATP, as is the energy carried by the sugars circulating in your blood. Cells then use their supply of ATP to drive active transport across membranes, to power movement, to provide activation energy for chemical reactions, to grow—almost every energy-requiring process that cells perform is powered by ATP. Because ATP plays this central role in *all* organisms, it is clear that its role as the major energy currency of cells evolved early in the history of life.

Each ATP molecule (Figure 6-13) is composed of three subunits. The first subunit is a five-carbon sugar called **ribose,** which serves as the backbone to which the other two subunits are attached. The second subunit is **adenine,** an organic molecule composed of two carbon-nitrogen rings. Each of the nitrogen atoms in the ring has an unshared pair of electrons that weakly attract hydrogen ions. Adenine therefore acts as a chemical base and is usually referred to as a **nitrogenous base.** As described in Chapter 2, adenine plays another major role in the cell: it is one of the four nitrogenous bases that are the principal components of the genetic material DNA.

The third subunit is a **triphosphate group** (three phosphate groups linked in a chain). The covalent bonds linking these three phosphates are usually indicated by a squiggle ($\sim$) and are sometimes called **high-energy bonds.** When one is broken, slightly more than 7 kilocalories of energy are released per mole of ATP. These phosphate bonds possess what a chemist would call "high transfer potential;" that is, they are bonds that have a low activation energy and are broken easily, which releases their energy. In a typical energy transaction, only the outermost of the two high-energy bonds is broken, breaking off the phosphate group on the end. When this happens, ATP becomes **adenosine diphosphate** (ADP) and 7 kilocalories of energy are expended per mole of ATP.

Cells use ATP to drive endergonic reactions, reactions whose products possess more energy than their substrates. Such reactions will not proceed unless the reactants are supplied with the necessary energy, any more than a boulder will roll uphill. But as long as the cleavage of ATP's terminal high-energy bond is more exergonic than the other reaction is endergonic, the overall energy change of the two "coupled" reactions is exergonic, and the reaction will proceed. Because almost all endergonic reactions in the cell require less than 7 kilocalories of activation energy per mole, ATP is able to power all of the cell's activities. So although the high-energy bond of ATP is not highly energetic in an absolute sense (for example, it is not nearly as strong as a carbon-hydrogen bond), it is more energetic than the activation energies of almost all energy-requiring cell activities. It is for this reason that ATP is able to serve as a universal energy donor.

Cells contain a pool of ATP, ADP, and phosphate. ATP is constantly being cleaved into ADP plus phosphate to drive the endergonic, energy-requiring processes of the cell. An individual on a typical diet of 2000 Calories/day goes through about 125 pounds of ATP a day. But cells do not maintain large stockpiles of ATP, just as most people do not carry large amounts of cash with them. Instead, cells are constantly recycling their ADP, withdrawing from their energy reserves to rebuild more ATP. Using the energy derived from foodstuffs and from stored fats or starches (or, in the case of plants, from photosynthesis), ADP and phosphate are recombined to form ATP, with 7 kilocalories of energy per mole contributed to each newly

**Figure 6-12**

**NAD$^+$ and ATP have similar structures.** The cell's two most important cofactors, NAD$^+$ and ATP, are both nucleotides. A nucleotide (**A**) is simply a 5-carbon sugar with a phosphate group and an organic base attached to it. NAD$^+$ (**B**) is composed of two such nucleotides (nicotinamide and adenine), which is why it is called *nicotinamide adenine dinucleotide*. In ATP, the nicotinamide group is replaced with two phosphate groups. The detailed structures of NAD$^+$ are presented in **C**, of ATP in **D**.

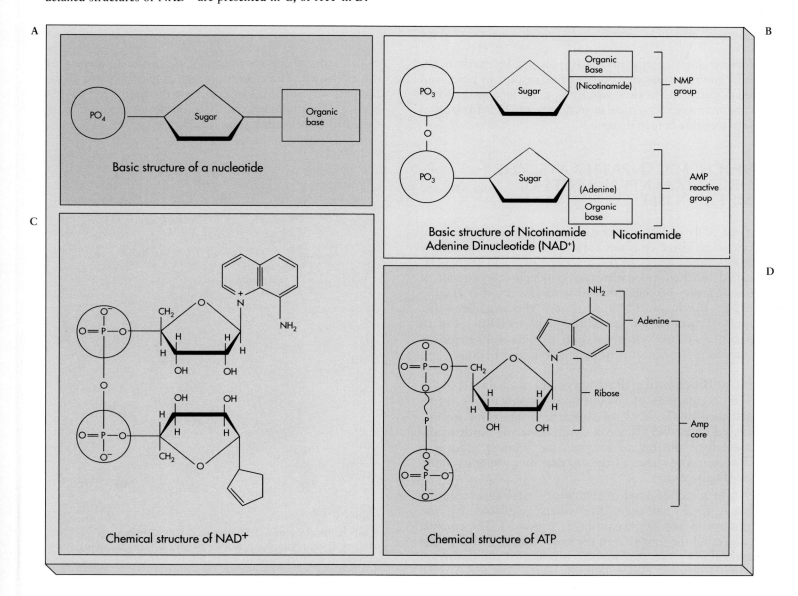

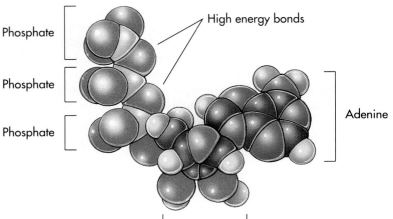

Phosphate

Phosphate

Phosphate

Ribose

High energy bonds

Adenine

**Figure 6-13**

A molecule of ATP, the primary energy currency of all cells.

formed high-energy bond. If you could mark every ATP molecule in your body at one instant in time, and then watch them, they would be gone in a flash. Most cells maintain a particular molecule of ATP for only a few seconds before using it.

## BIOCHEMICAL PATHWAYS: THE ORGANIZATIONAL UNITS OF METABOLISM

Your body contains over a thousand different kinds of enzymes. These enzymes catalyze a bewildering variety of reactions. Many of the reactions occur in sequences, called **biochemical pathways,** in which the product of one reaction becomes the substrate for another (Figure 6-14). Biochemical pathways are the organizational units of metabolism, just as the many metal parts of an automobile are organized into distinct subassemblies such as the carburetor, transmission, and brakes.

### How Biochemical Pathways are Regulated

The regulation of biochemical pathways is often achieved by the kinds of allosteric interactions we discussed on p. 149. When a signal molecule binds to the regulatory site on an enzyme, the binding induces a change in the shape of the enzyme that alters its activity (Figure 6-15).

In a second kind of regulatory mechanism, simple biochemical pathways are often regulated by the amount of product the pathway produces. The enzyme catalyzing the first step in the biochemical pathway has a regulatory site that fits the product of the pathway; when a product molecule binds to the site, the enzyme

changes shape and stops catalyzing the first reaction in the pathway. Thus, when levels of product are high, the pathway is shut down because the first step is inoperative (Figure 6-16). Such a mode of regulation is called **feedback inhibition**.

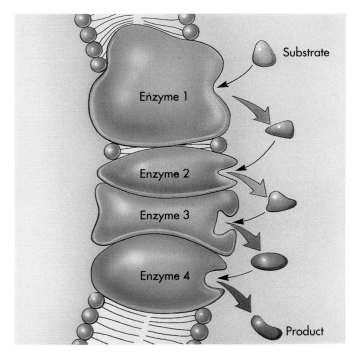

Substrate

Enzyme 1

Enzyme 2

Enzyme 3

Enzyme 4

Product

**Figure 6-14**

**A biochemical pathway.** The original substrate is acted on by enzyme 1, changing the substrate to a new form recognized by enzyme 2. Each enzyme in the pathway acts on the product of the previous stage.

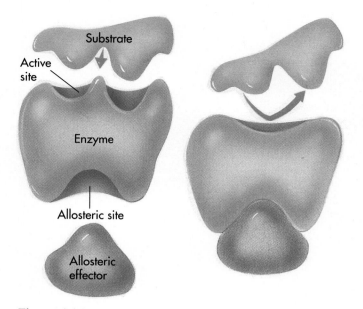

**Figure 6-15**

**Allosteric inhibition.** The binding to the enzyme of a particular metabolite, the "allosteric effector," causes the enzyme to alter its shape in such a way that the substrate no longer fits into the active site.

## THE EVOLUTION OF METABOLISM

The way in which contemporary organisms harvest energy and employ it to drive chemical reactions is different from the way in which the earliest organisms prob-

ably operated. For most of the period in which life has existed on earth, metabolic processes took place in what was essentially an oxygen-free, or **anaerobic**, environment. In fact, you will learn in the next chapter that, not only does anaerobic metabolism still exist, it remains the fundamental chemistry of all life processes. In this process, which is called cellular respiration, organic compounds are oxidized; it is the subject of Chapter 7. Respiration converts carbon compounds and $O_2$ gas into $CO_2$ and chemical energy (ATP).

The other half of the cycle, which is characteristic of our world today, is oxidative, or **aerobic**, metabolism, in which carbon atoms are cycled between organisms and the atmosphere (Figure 6-17). The portion of the cycle that aquires carbon atoms from atmospheric carbon dioxide and incorporates them into organisms is photosynthesis, the subject of Chapter 8. The photosynthesis that is carried out by today's plants, algae, and photosynthetic bacteria converts atmospheric $CO_2$ and the energy of sunlight into organic compounds with the release of oxygen gas ($O_2$).

This kind of chemical machinery, which is so characteristic of biological systems, is nevertheless a relatively recent development in life's evolutionary history. The more modern biochemical machinery of aerobic metabolism has simply been added on. In the next two chapters we examine respiration and photosynthesis and consider how evolution has shaped these two most fundamental metabolic processes.

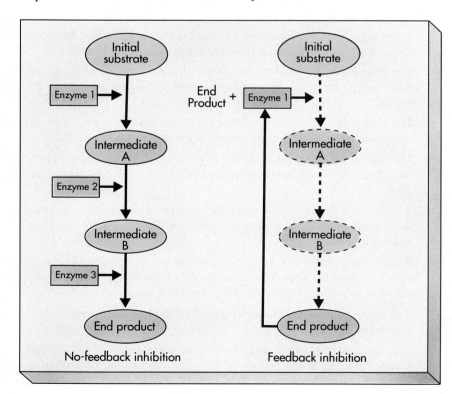

**Figure 6-16**

**Feedback inhibition.**

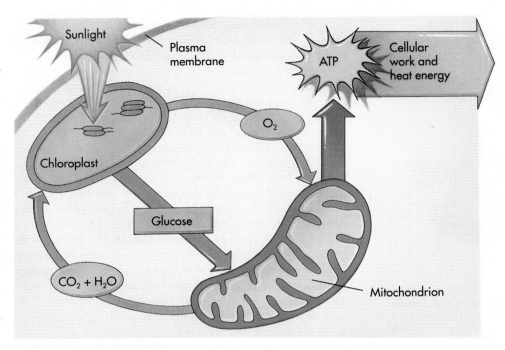

*Figure 6-17*
**An overview of aerobic metabolism.**

## ■ SUMMARY

1. Energy is the capacity to bring about change, to do work. Kinetic energy is actively engaged in doing work, whereas potential energy has the capacity to do work. Most energy transformations in living things involve conversion of potential energy to kinetic energy.

2. The first law of thermodynamics states that the amount of energy in the universe is constant and that energy cannot be destroyed or created. However, energy can be converted from one form to another.

3. The second law of thermodynamics says that disorder in the universe tends to increase. As a result, energy spontaneously converts to less organized forms.

4. The least organized form of energy is heat, which is random molecular motion. Heat energy can be used to do work when heat gradients exist, as in a steam engine, but cannot accomplish work in cells.

5. A redox reaction is one in which an electron is taken from one atom (oxidation) and donated to another (reduction). The electrons in redox reactions often travel in association with protons, as hydrogen atoms.

6. If the electron transferred in a redox reaction is an energetic electron, then the energy is transferred with the electron. The electron moves to a high-energy orbital in the reduced atom. Many energy transfers in cells take place by means of redox reactions.

7. Any chemical reaction whose products contain less free energy than the original reactants will tend to proceed spontaneously.

8. The speed of a reaction depends on the amount of activation energy required to break existing bonds. Catalysis is the process of lowering the amount of activation energies needed by stressing chemical bonds. Enzymes are the catalysts of cells.

9. Cells contain many different enzymes, each of which catalyzes a different reaction. A given enzyme is specific because its active site fits only one or a few potential substrate molecules.

10. Cells focus all of their energy resources on the manufacture of ATP from ADP and phosphate, which requires the cell to supply 7 kilocalories of energy obtained from photosynthesis or from electrons stripped from foodstuffs to form 1 mole of ATP. Cells then use this ATP to drive endergonic reactions.

## REVIEW

1. Energy is defined as the capacity to do _____.

2. When energy is converted from one form to another (an example is the conversion of chemical energy to mechanical energy in a car), a large percentage is lost as heat. Does this contradict the first law of thermodynamics?

3. The earth is constantly receiving energy. Where does this energy come from?

4. What cellular process releases the chemical energy stored in glucose for use by organisms?

5. The agents that perform catalysis in the human body are called _____. What class of organic molecules are they? (See Chapter 2.)

## SELF-QUIZ

1. Energy can exist in many forms. Which three of the following are forms of energy?
   (a) Heat
   (b) Sound
   (c) Smell
   (d) Diffusion
   (e) Light

2. When energy becomes so randomized and uniform in a system that it is no longer available to do work, it is referred to as
   (a) heat.
   (b) entropy.
   (c) enthalpy.
   (d) light.
   (e) chemical energy.

3. Light energy is captured and used by _____. Because of this ability to use light energy to produce chemical energy, these organisms allow most of the rest of life on earth to exist.
   (a) animals generally
   (b) mammals
   (c) plants and algae
   (d) fungi
   (e) most bacteria

4. Chemical reduction is
   (a) the gain of a proton.
   (b) the loss of a proton.
   (c) the gain of an electron.
   (d) the loss of an electron.
   (e) the exchange of an electron.

5. The molecule that is used by all cells to do work—that is, to drive chemical reactions by moving ions or molecules against a concentration gradient—is
   (a) ATP.
   (b) $NAD^+$.
   (c) NMP.
   (d) glucose.
   (e) water.

6. The fact that energy can be changed from one form to another, but never lost, is called
   (a) theory of energy conservation.
   (b) the second law of thermodynamics.
   (c) Avogadro's number.
   (d) the first law of thermodynamics.
   (e) the principle of energy-change equivalency.

7. Which of the following statements are true of enzymes generally?
   (a) They are short peptide chains.
   (b) They are globular proteins.
   (c) They work through allosteric interactions.
   (d) They may be either organic or inorganic.
   (e) Their catalysis takes place in strongly basic solutions.

8. Which of the following often facilitate the activities of enzymes?
   (a) Cofactors
   (b) Covalent bonds
   (c) Short peptide chains
   (d) Coenzymes
   (e) Vitamins

# THOUGHT QUESTIONS

1. Oxidation-reduction reactions occur between a wide variety of molecules. Why do you suppose reactions involving hydrogen and oxygen are the ones of paramount importance in biological systems?

2. On earth there are many more bears than there are lions and tigers. Why do you suppose this is so?

3. On summer nights in some parts of the country, one can often see fireflies glowing briefly in the dark. Do you suppose this light requires energy? If so, where does it come from?

4. If you eat a 1500-Calorie lunch consisting of a hamburger and some french fries, how many kilocalories have you consumed?

5. If you eat a mole of glucose, how many Calories do you consume?

# FOR FURTHER READING

HINKLE, P.C., and R.E. MacCARTY: "How Cells Make ATP," *Scientific American*, March 1978, pages 104-123. Describes how cells use electrons stripped from foodstuffs to make ATP.

LEHNINGER, A.L.: "How Cells Transform Energy," *Scientific American*, September 1961, pages 62-73. Written more than 25 years ago, this article still provides one of the clearest expositions of how cells channel energy through metabolism.

# CHAPTER 7

# *Cellular Respiration*

We are what we eat. These harvest mice are greeting each other after feasting on grains of wheat, the same food that you consume, in processed form, as bread. Like you, they "respire" their food, stripping high-energy electrons from it, using the energy, and ridding themselves of the spent electrons by combining them with oxygen gas.

# CELLULAR RESPIRATION

## Overview

All organisms fuel their metabolism with energy from ATP. The simplest processes for generating ATP, and the ones that evolved first, involve the rearrangement of chemical bonds. Later, organisms evolved far more efficient means of carrying out this process: plants, algae, and some bacteria obtain ATP by using energetic electrons to drive proton pumps, obtaining the electrons both by photosynthesis and by oxidizing sugars and fats. Animals, fungi, most protists, and most bacteria are heterotrophs; they use only the second oxidative process, which occurs within the symbiotically derived mitochondria that exist in all but a very few eukaryotic organisms.

## For Review    *Here are some important terms and concepts that you will encounter in this chapter. If you are not familiar with them, you should review them before proceeding.*

**Glucose** (Chapter 2)

**Chemiosmosis** (Chapter 5)

**Oxidation-reduction** (Chapter 6)

**ATP** (Chapter 6)

**Exergonic and endergonic reactions** (Chapter 6)

Life processes are driven by energy. All the activities that organisms perform—the swimming of bacteria, the purring of a cat, your reading these words—use energy. Even though the ways that organisms use energy are many and varied, all of life's energy ultimately has the same beginning: the sun. Plants, algae, and some bacteria harvest light energy from the sun by the process of photosynthesis, thus converting it to chemical energy (Figure 7-1). These organisms, along with a few others that use chemical energy in a similar way, are called **autotrophs,** or self-feeders. All organisms ultimately live on the energy produced by autotrophs. Organisms that do not have the ability to produce their own food are called **heterotrophs,** which means fed by others. At least 95% of the kinds of organisms on earth—all animals, all fungi, and most protists and bacteria—are heterotrophs; most of them live by feeding on the chemical energy fixed by photosynthesis. All of us, plants and bacteria, you and I, share the same ultimate dependency on the sun. We are all children of light.

In this chapter we discuss the processes that cells use to derive energy from organic molecules and to convert that energy to ATP. We consider photosynthesis in detail in Chapter 8. We treat the conversion of chemical energy to ATP first because all organisms—both photosynthesizers and the heterotrophs that feed on them—are capable of transforming energy in this way. In contrast, fewer than 1 in 20 of the kinds of organisms on earth is capable of photosynthesis. But as you will see, the two processes have much in common.

## USING CHEMICAL ENERGY TO DRIVE METABOLISM

Your body contains more than 2000 kinds of enzymes. These enzymes catalyze a bewildering variety of reactions. Many of the reactions occur in step-by-step sequences called **biochemical pathways,** which are the organizational units of metabolism and the pathways of energy and materials in the cell. Just as the many metal parts of an automobile are organized into distinct subassemblies such as the carburetor, transmission, and brakes, so the many enzyme-catalyzed reactions of an organism are organized into short biochemical pathways that accomplish particular metabolic jobs, such as extracting energy from a sugar molecule or building an amino acid.

In biochemical pathways, exergonic reactions—that is, those reactions which involve a release of free energy—can occur spontaneously; endergonic reactions—those reactions which require the addition of energy—cannot. To drive endergonic reactions, all organisms use the same mechanism: that of an energy-rich molecule such as ATP (Figure 7-2).

*Chemical energy powers metabolism by driving endergonic reactions. Chemical energy is used to create ATP, and the splitting of ATP is coupled to these reactions, providing the necessary energy.*

## HOW CELLS MAKE ATP

ATP is the energy currency of all living organisms. How do organisms make ATP? They do so in one of two ways:

1. *Substrate-level phosphorylation*. The enzymatic generation of ATP from ADP and inorganic phosphate ($P_i$) is called substrate-level phosphorylation (Figure 7-3). Because the formation of ATP from ADP plus $P_i$ requires an input of free energy, ATP formation is ender-

**Figure 7-1**

**Respiration picks up where photosynthesis leaves off.** All of the energy that drives our metabolism comes from plants or from animals that eat plants. This winter wheat, rich with the products of months of capturing energy from the sun through the process of photosynthesis, will soon become flour and provide energy for human activity.

**Figure 7-2**

**How coupled reactions work.** The conversion of compound A to compound B *(1)* requires the imput of energy—in this diagram, the left piston must be driven down to the same level as compound A *(2)*, and to do this, the left spring must be pulled, extending it. In cells, this extra energy is acquired from ATP. In this diagram, the cleavage of ATP drives the right piston down, compressing the right spring. *Because the two pistons are linked by a crossbar ("coupled"), the movement of the right piston pulls the left one down, too.* In cells, enzymes provide the physical link.

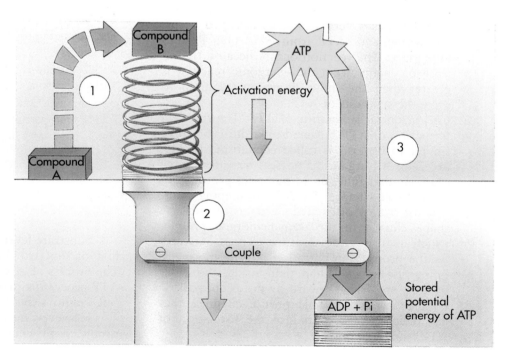

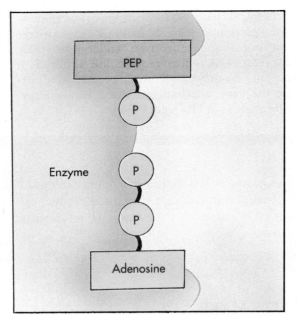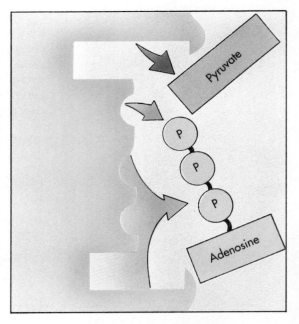

*Figure 7-3*

**Substrate-level phosphorylation.** Some molecules such as phosphoenolpyruvic acid (PEP) possess a high-energy phosphate bond similar to those of ATP. When the phosphate group of PEP is transferred enzymatically to ADP, the energy in the bond is conserved and ATP is created. The reaction takes place because the phosphate bond of PEP has higher energy than does the terminal phosphate bond of ATP.

gonic; that is, it does not occur spontaneously. However, when coupled with an exergonic reaction that has a very strong tendency to occur, the synthesis of ATP from ADP plus $P_i$ does take place. The reaction occurs because the release of energy from the exergonic reaction is greater than the input of energy required to drive the synthesis of ATP.

2. *Chemiosmotic generation of ATP.* The chemical formation of ATP powered by a diffusion force similar to osmosis is called chemiosmosis (Figure 7-4). All organisms possess transmembrane channels that function in pumping protons out of cells (see Chapter 5). Proton-pumping channels use a flow of excited electrons to induce a shape change in the transmembrane protein, which in turn causes protons to pass outward. As the proton concentration outside the membrane rises higher than that inside, the outer protons are driven inward by diffusion, passing back in through special proton channels that use their passage to induce the formation of ATP from ADP plus $P_i$.

*The harvesting of chemical energy takes place in one or both of two stages:*

*1. Substrate-level phosphorylation, which involves a reshuffling of chemical bonds to couple ATP formation to a highly exergonic reaction.*

*2. The transport of electrons to a membrane where they drive a proton pump and thus power the chemiosmotic synthesis of ATP.*

Substrate-level phosphorylation was probably the first of the two ATP-forming mechanisms to evolve. The process of glycolysis, which is the most basic of all ATP-generating processes and is present in every living cell, employs this mechanism.

However, most of the ATP that organisms make is produced by chemiosmosis. The electrons that drive proton-pumping channels involved in chemiosmosis are obtained by organisms from two sources:

1. *Light.* In photosynthetic organisms, light energy boosts electrons to higher energy levels, and these electrons are ferried to proton pumps.
2. *Chemical bonds.* In all organisms, photosynthetic and nonphotosynthetic alike, high-energy electrons are extracted from chemical bonds and carried by coenzymes to proton pumps.

The second of these two chemiosmotic processes, the extraction of electrons from chemical bonds, is an oxidation-reduction process. Oxidation is defined as the removal of electrons (see Chapter 6). When electrons are taken away from the chemical bonds of food molecules to drive proton pumps, the food molecules are being oxidized chemically. This electron-harvesting process is given a special name: **cellular respiration.** Because cellular respiration uses oxygen, it is also known as **oxidative respiration.** This is the term we use in this text. Oxidative respiration is the oxidation of food molecules to obtain energy. Do not confuse this with the breathing of oxygen gas that your body carries out, which is also called respiration.

## CELLULAR RESPIRATION

In most organisms, cellular respiration is carried out in three stages (Figure 7-5). The first stage of cellular respiration is a biochemical pathway called glycolysis. The enzymes that carry out glycolysis are present in the cytoplasm of the cell, not bound to any membrane or organelle. Some of these enzymes form ATP by substrate-level phosphorylation, whereas others strip away a few energetic electrons from food molecules and donate them to $NAD^+$ molecules to form NADH molecules. Because electrons are removed from them, the food molecules are said to be oxidized. In this process the electrons are accompanied by protons and thus travel as hydrogen atoms. These NADH molecules then donate the energetic electrons to proton-pumping channels to drive the chemiosmotic synthesis of ATP.

Glycolysis will produce ATP by substrate-level phosphorylation whether or not oxygen is present, but the use of energetic electrons to drive the chemiosmotic synthesis of ATP does not take place in the absence of oxygen because oxygen serves as the final acceptor of the electrons harvested from the food molecules:

$$C_6H_{12}O_6 + 6\ O_2 \rightarrow 6\ CO_2 + 6\ H_2O + \text{energy}$$

In the absence of oxygen, cells are restricted to the substrate-level phosphorylation reactions of glycolysis to obtain ATP.

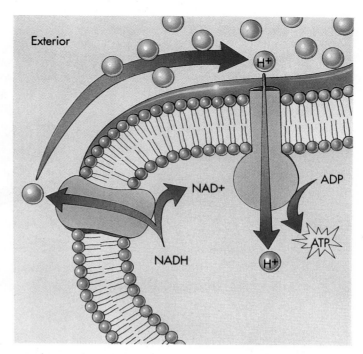

***Figure 7-4***

**Chemiosmosis.** High-energy electrons harvested from food molecules during metabolism are transported by NADH carrier molecules to "proton pumps," which use the energy to pump protons out across the membrane. As a result, the concentration of protons on the outside of the membrane rises, encouraging protons to diffuse back. They pass through the only channel open to them, a special ATP-forming protein complex. As each proton passes, an ATP molecule is formed.

All cells carry out glycolysis, but glycolysis is not efficient at extracting energy from food molecules—only a small percentage of the energy in the chemical bonds of food molecules is converted by glycolysis to ATP. In the presence of oxygen, two other pathways of cellular respiration also occur in your body, pathways that extract far more energy. The second pathway is the oxidation of pyruvate. The third pathway is called the citric acid cycle, after the six-carbon citric acid molecule formed in its first step. Alternatively, it is called the Krebs cycle, after British biochemist Sir Hans Krebs, who discovered it. (Less commonly, it is called the tricarboxylic acid cycle, because citric acid has three carboxyl groups.) In eukaryotes, the oxidation of pyruvate and the reactions of the citric acid cycle take place only within the cell organelles called mitochondria.

It is important to note that pyruvate oxidation and the reactions of the citric acid cycle take place in all organisms that possess mitochondria, as well as in many

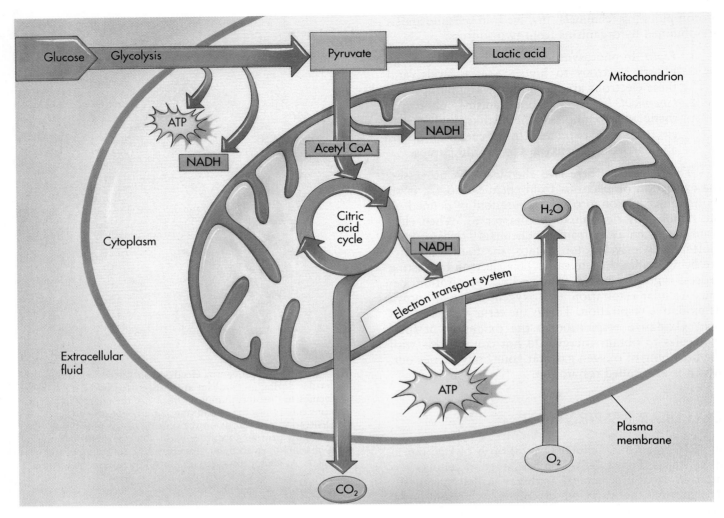

**Figure 7-5**
An overview of cellular respiration.

forms of bacteria (Chapter 3 recounts how mitochondria are thought to have evolved from bacteria). Thus plants as well as animals carry out oxidative respiration in this way. Because plants produce ATP by photosynthesis, it is sometimes easy to lose sight of the fact that they also produce ATP in this other oxidative fashion, just as you do.

*All cells make ATP by substrate-level phosphorylation and by oxidative respiration; some organisms also make ATP by photosynthesis.*

## THE FATE OF A CANDY BAR: CELLULAR RESPIRATION IN ACTION

When you metabolize food and thus obtain the ATP that powers your life, you employ both substrate-level phosphorylation and oxidative respiration. To understand what goes on better, it is instructive to follow the fate of something you eat and to see what happens to it. So let's eat a chocolate bar (Figure 7-6).

A chocolate bar, like many of the things that we consume, is a complex mixture of sugars, lipids, proteins, and other molecules. The first thing that happens in its journey toward ATP production is that the complex molecules are broken down to simple ones. Disaccharides such as sucrose are split into simple sugars, either glucose or sugars that are converted to glucose;

proteins are split up into amino acids; and complex lipids are broken into smaller bits. These initial steps usually yield no usable energy, but they serve to assemble the energy wealth of a diverse array of complex molecules into a small number of simple molecules such as glucose.

For simplicity, let's assume that our chocolate bar is entirely broken down to molecules of the six-carbon sugar glucose. Glucose occupies a central place in metabolism because many different foodstuffs are converted to it. Glucose is where the making of ATP begins.

The first stage of extracting energy from glucose is a 10-reaction biochemical pathway called **glycolysis.** In glycolysis, ATP is generated in two ways. For each glucose molecule, two ATP molecules are used up in preparing the glucose molecule, and four ATP molecules are formed by substrate-level phosphorylation, for a net yield of two ATP molecules. In addition, four electrons are harvested and used to form ATP molecules by oxidative respiration. But the total yield of ATP molecules is small. When the glycolytic process is completed, the two molecules of **pyruvate** that are left still contain most of the energy that was present in the original glucose molecule.

The second stage of extracting energy from glucose, after glycolysis, is a cycle of nine reactions, the citric acid cycle. The pyruvate left over from glycolysis is first converted to a two-carbon molecule, which feeds into the cycle. In the cycle, two more ATP molecules are extracted by substrate-level phosphorylation, and a large number of electrons are removed and donated to electron carriers. It is the harvesting of these electrons that leads to the greatest amount of ATP formation.

When this whole series of steps has been completed, the six-carbon glucose molecule has been divided into 6 molecules of $CO_2$, and 38 ATP molecules have been generated, 6 by substrate-level phosphorylation and 32 by oxidative respiration. Even though 2 of the ATP molecules must be expended to transport NADH out into the cytoplasm, the net yield of 36 ATP molecules is still a very good yield. The overall process for the catabolism of glucose can be summarized as:

| Substrate-level phosphorylation | + Oxidative respiration | − NADH transport | = Net ATP production |
|---|---|---|---|
| 6 | 32 | 2 | 36 |

Each ATP molecule represents the capture of 7.3 kilocalories of energy per mole of glucose; therefore 36 ATP molecules represent a total capture of 7.3 × 36 = 263 kilocalories per mole. The total energy content of the chemical bonds of glucose is only 686 kilocalories

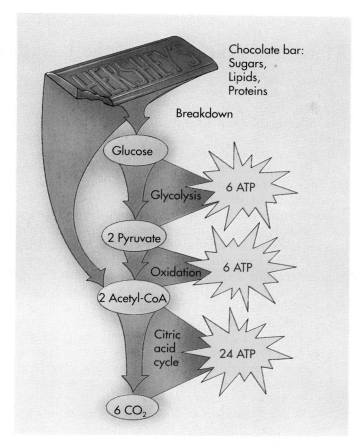

**Figure 7-6**

**The fate of a chocolate bar.** When you eat a chocolate bar, its complex molecules are first broken down to simple molecules: the sugars to glucose molecules, and the proteins, fats, and other lipids to 2-carbon molecules called acetyl-CoA. These breakdowns produce little or no energy, but they prepare the way for three major energy-producing processes: glycolysis, a process that converts glucose to two molecules of pyruvate; the oxidation of this pyruvate to two molecules of acetyl-CoA; and the oxidation of acetyl-CoA molecules in the citric acid cycle.

per mole; so we have succeeded in harvesting 38% of the available energy. By contrast, a car converts only about 25% of the energy in gasoline into useful energy.

This brief overview gives some sense of how cells organize their production of ATP from a food source such as a candy bar. We shall now examine glycolysis and the citric acid cycle as processes and study the ways in which they direct the flow of energy.

## GLYCOLYSIS

Among the many simple molecules that are available as a consequence of degradation, the metabolism of early bacteria clearly focused on the simple six-carbon sugar

# Metabolic Efficiency and the Length of Food Chains

The simplest heterotrophs are those anaerobic bacteria which first evolved glycolysis. These bacteria derive chemical energy by breaking carbon-hydrogen bonds, use the energy to promote ATP production through coupled reactions, and dispose of the hydrogen atoms by fermentations. The inefficiency of the process, which extracts only a small percentage of the chemical energy available in organic compounds, probably placed an important constraint on early organisms: most heterotrophs must have lived by consuming photosynthetic organisms rather than other heterotrophs. If a heterotroph preserves only 2% of the energy of a photosynthetic organism that it consumes, then any other population of heterotrophs that consumes this kind of heterotroph has only 2% of the original energy available to it and can harvest from this by glycolysis only 2% of *that* energy, or 0.04% of the original amount of available energy. Therefore a very large base of autotrophs would be needed to support a small number of heterotrophs.

When organisms became able to extract energy from organic molecules by oxidative metabolism, this constraint became far less severe because oxidative processes are much more efficient. The efficiency of oxidative respiration is 38%; therefore about two-thirds of the available energy is lost at each level of consumption. Thus animals that eat plants can obtain no more than approximately one-third of the energy in the plants, and other animals that eat the first animals obtain no more than a third of the energy in the original plant-eaters, or **herbivores**. But losses of this magnitude still result in the transmission of far more energy from one **trophic level** (de-

fined as a step in the movement of energy through an ecosystem) to another than does anaerobic glycolysis, in which more than 98% of the energy is lost at each step. Because of the improved efficiency of oxidative metabolism, it has been possible for **food chains** to evolve, in which some heterotrophs consume plants and others consume the plant-eaters. You will read more about food chains in Chapter 22.

The length of a food chain is limited by the efficiency of oxidative metabolism; most food chains involve only three and rarely four levels (Levels 1 through 4). Too much energy is lost at each transfer point to allow the chain to become much longer. You could not support a very large human population with the meat of lions captured from the Serengeti Plain of Africa—the amount of grass there will not support enough zebras to maintain a large population of lions. Thus the ecological complexity of our world is fixed in a fundamental way by the chemistry of oxidative respiration.

Level 1: *Photosynthesizer*

glucose, undoubtedly partly because glucose was a major constituent of the carbohydrates of the cell. Glucose molecules can be dismantled in many ways, but these early bacteria evolved the ability to do it in a way that includes reactions that release enough free energy to drive the synthesis of ATP in coupled reactions. The process involves a sequence of 10 reactions that convert glucose into 2 three-carbon molecules of pyruvate. For each molecule of glucose that passes through this

transformation, the cell acquires two ATP molecules. The overall process is called glycolysis.

## An Overview of Glycolysis

Glycolysis consists of two very different processes, one wedded to the other:

1. Glucose is converted to two molecules of the three-carbon compound **glyceraldehyde 3-phosphate** (G3P), with the expenditure of ATP.

Level 2: *Herbivore*

Level 3: *Carnivore*

Level 3: *Scavenger*

Level 4: *Refuse utilizer*

**Level 1:** *Photosynthesizer.* The grass growing under these palms grows actively during the hot, rainy season, and, capturing the energy of the sun abundantly, converts it into molecules of glucose that are stored in the grass plants as starch.

**Level 2:** *Herbivore.* These large antelopes, known as wildebeests, consume the grass and convert some of its stored energy into their own bodies.

**Level 3:** *Carnivore.* The lion feeds on wildebeests and other animals, converting part of their stored energy to the body of the lions.

**Level 3 (cont'd):** *Scavenger.* This hyena and the vulture occupy the same stage in the food chain as the lion. They are also consuming the body of the dead wildebeest, which has been abandoned by the lion.

**Level 4:** *Refuse utilizer.* These butterflies, mostly *Precis octavia*, are feeding on the material left in the hyena's dung after the material the hyena consumed has passed through its digestive tract. At each of these four levels, only about a third or less of energy present is utilized by the recipient.

2. ATP is generated from G3P.

The 10 reactions of glycolysis apparently proceed in four stages (Figure 7-7):

*Stage a:* The first stage is composed of three reactions that change glucose into a compound that can readily be split into three-carbon phosphorylated units. Two of these reactions require the cleavage of an ATP molecule, so this stage requires the investment by the cell of two ATP molecules.

*Stage b:* The second stage is **cleavage,** in which the six-carbon product of the first stage is split into two three-carbon molecules. One is G3P, and the other is converted to G3P by another reaction.

*Stage c:* The third stage is **oxidation,** in which a hydrogen atom carrying a pair of electrons is

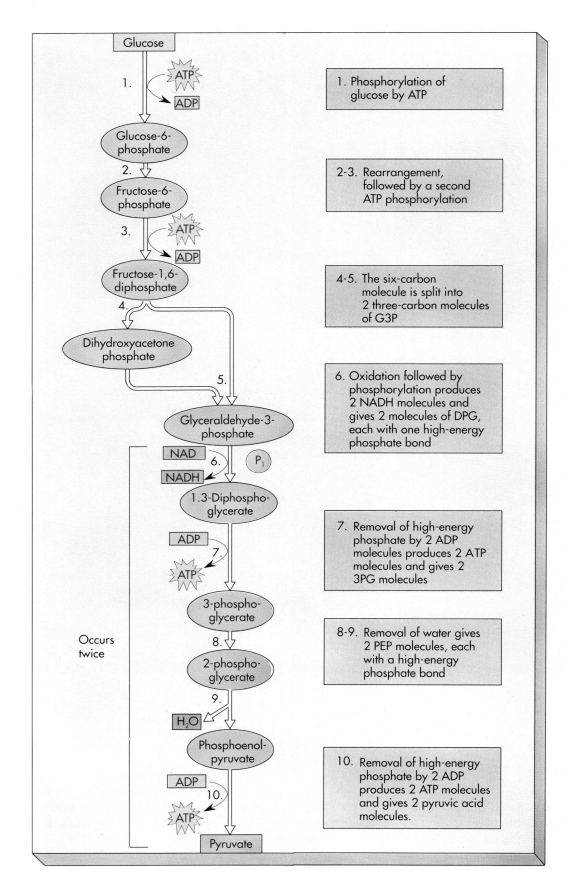

**Figure 7-7**

**The glycolytic pathway.** The first five reactions convert a molecule of glucose into two molecules of G3P. This process is endergonic and requires the expenditure of two ATP molecules to drive it. The next five reactions convert G3P molecules into pyruvate molecules and generate four molecules of ATP for each two molecules of G3P. These reactions also generate two molecules of NADH, which, as you will see, is further oxidized in your body to produce four more molecules of ATP. Subtracting the two ATP molecules expended in driving the initial endergonic reactions, the net yield in your body is thus six ATP molecules for each molecule of glucose. When cells are forced to operate without oxygen, the two molecules of NADH cannot be used to produce ATP, and the net yield is then only two molecules of ATP per molecule of glucose.

Within the figure:

Glucose

1. Phosphorylation of glucose by ATP

Glucose-6-phosphate

2-3. Rearrangement, followed by a second ATP phosphorylation

Fructose-6-phosphate

Fructose-1,6-diphosphate

4-5. The six-carbon molecule is split into 2 three-carbon molecules of G3P

Dihydroxyacetone phosphate

Glyceraldehyde-3-phosphate

NAD
NADH
$P_1$

6. Oxidation followed by phosphorylation produces 2 NADH molecules and gives 2 molecules of DPG, each with one high-energy phosphate bond

1.3-Diphospho-glycerate

ADP
ATP

7. Removal of high-energy phosphate by 2 ADP molecules produces 2 ATP molecules and gives 2 3PG molecules

3-phospho-glycerate

Occurs twice

2-phospho-glycerate

8-9. Removal of water gives 2 PEP molecules, each with a high-energy phosphate bond

$H_2O$

Phosphoenol-pyruvate

ADP
ATP

10. Removal of high-energy phosphate by 2 ADP produces 2 ATP molecules and gives 2 pyruvic acid molecules.

Pyruvate

removed from G3P and donated to $NAD^+$. $NAD^+$ acts as an electron carrier in the cell, in this case accepting the proton and two electrons from G3P to form NADH. Note that $NAD^+$ is an ion and that *both* electrons in the new covalent bond come from G3P.

*Stage d:* The final stage, called **ATP generation,** is a series of four reactions that convert G3P into another three-carbon molecule, **pyruvate,** and in the process generate two ATP molecules for each G3P.

Because each glucose molecule is split into *two* G3P molecules, the overall net reaction sequence yields two ATP molecules and two molecules of pyruvate:

$$\begin{array}{ll} -\ 2\text{ATP} & \text{Stage a} \\ +\ 2(2\text{ATP}) & \text{Stage d} \\ \hline +\ 2\text{ATP} & \end{array}$$

This is not a great amount of energy. When you consider that the free energy of the total oxidation of glucose to $CO_2$ and $H_2O$ is $-686$ kilocalories per mole (the sign is negative to indicate that the products of the oxidation have less energy), and that ATP's high-energy bonds have an energy content of $-7.3$ kilocalories per mole apiece, then the efficiency with which glycolysis harvests the chemical energy of glucose is only $^{14.6}/_{686}$, or 2%. Although far from ideal in terms of the amount of energy that it releases, glycolysis does generate ATP; and for more than a billion years during the long, anaerobic first stages of life on earth, this reaction sequence was the only way for heterotrophic organisms to generate ATP from organic molecules.

> *The glycolytic reaction sequence generates a small amount of ATP by reshuffling the bonds of glucose molecules. Glycolysis is a very inefficient process, capturing only about 2% of the available chemical energy of glucose.*

A **catabolic process** is one in which complex molecules are broken down into simpler ones; in contrast, an **anabolic process** is one in which more complex molecules are built up. Perhaps the original catabolic process involved only the ATP-yielding breakdown of G3P; the generation of G3P from glucose may have evolved later when alternative sources of G3P were depleted. Like many biochemical pathways, glycolysis

may have evolved backwards, the last stages in the process being the most ancient.

## All Cells Use Glycolysis

The glycolytic reaction sequence is thought to have been among the earliest of all biochemical processes to evolve. It uses no molecular oxygen and therefore occurs readily in an anaerobic environment. All of its reactions occur free in the cytoplasm; none is associated with any organelle or membrane structure. Every living creature is capable of carrying out the glycolytic sequence. However, most present-day organisms are able to extract considerably more energy from glucose molecules than does glycolysis. Of the 36 ATP molecules you obtain from each glucose molecule that you metabolize, only 2 are obtained by substrate-level phosphorylation during glycolysis. Why then is glycolysis still maintained even though its energy yield is comparatively meager?

This simple question has an important answer: evolution is an incremental process. Change occurs during evolution by improving on past successes. In catabolic metabolism, glycolysis satisfied the one essential evolutionary requirement: it was an improvement. Cells that could not carry out glycolysis were at a competitive disadvantage. By studying the metabolism of contemporary organisms, we can see that only those cells which were capable of glycolysis survived the early competition of life. Later improvements in catabolic metabolism built on this success. Glycolysis was not discarded during the course of evolution but rather was used as the starting point for the further extraction of chemical energy. Nature did not, so to speak, go back to the drawing board and design a different and better metabolism from scratch. Rather, catabolic metabolism evolved as one layer of reactions added to another, just as successive layers of paint can be found in an old apartment. We all carry glycolysis with us—a metabolic memory of our evolutionary past.

## THE NEED TO CLOSE THE METABOLIC CIRCLE

Inspect for a moment the net reaction of the glycolytic sequence:

Glucose + 2 ADP + 2 $P_i$ + 2 $NAD^+$ →
2 pyruvate + 2 ATP + 2 NADH + $2H^+$ + $2H_2O$

You can see that three changes occur in glycolysis: (1) glucose is converted to pyruvate, (2) ADP is converted to ATP, and (3) $NAD^+$ is converted to NADH.

As long as foodstuffs that can be converted to glucose are available, a cell can continually churn out ATP to drive its activities, except for one problem: the $NAD^+$. As a result of glycolysis, the cell accumulates NADH molecules at the expense of the pool of $NAD^+$ molecules. A cell does not contain a large amount of $NAD^+$, and for glycolysis to continue, it is necessary to recycle the NADH that it produces back to $NAD^+$. Some other home must be found for the hydrogen atom taken from G3P, some other molecule that will accept the hydrogen and be reduced.

What happens after glycolysis depends critically on the fate of this hydrogen atom. One of two things generally happens:

1. *Oxidative respiration.* Oxygen is an excellent electron acceptor, and in the presence of oxygen gas the hydrogen atom taken from G3P can be donated to oxygen, forming water. Because air is rich in oxygen, this process is referred to as **aerobic metabolism.**
2. *Fermentation.* When oxygen is not available, another organic molecule must accept the hydrogen atom instead. Such a process is called fermentation (Figure 7-8). This process, which is what happens when bacteria grow without oxygen, is referred to as **anaerobic metabolism,** or metabolism without oxygen.

## FERMENTATION

In the absence of oxygen, oxidative respiration cannot occur because of the lack of an electron acceptor. In such a situation a cell must rely on glycolysis to produce ATP. In the absence of oxygen, cells donate the hydrogen atoms generated by glycolysis to organic molecules during a process called **fermentation.**

Bacteria carry out many sorts of fermentations, each employing some form of carbohydrate molecule to accept the hydrogen atom from NADH and thus reform $NAD^+$:

Carbohydrate + NADH → Reduced carbohydrate + $NAD^+$

More than a dozen fermentation processes have evolved among bacteria, each using a different carbohydrate as the hydrogen acceptor. In organisms such as yeasts, the reduced compound is an alcohol.

Of these various bacterial fermentations, only a few occur among eukaryotes (Figure 7-9). In one fermentation process, which occurs in the single-celled fungi called yeasts, the carbohydrate that accepts the hydrogen from NADH is pyruvate, the end product of the glycolytic process. Yeasts remove a terminal $CO_2$ group from pyruvate, producing a toxic two-carbon molecule called **acetaldehyde,** and $CO_2$. Because of this $CO_2$ production, bread with yeast rises and "unleavened" bread—that is, bread without yeast—does not.

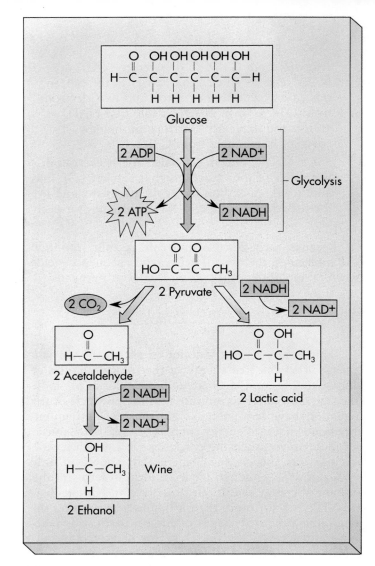

**Figure 7-8**

**How wine is made.** The conversion of pyruvate to ethanol is carried out by yeasts. It takes place naturally in grapes left to ferment on vines, as well as in fermentation vats of crushed grapes. When ethanol concentration reaches about 12%, the toxic effects of the alcohol kill the yeasts. What is left is wine. In your muscles, pyruvate is instead converted to lactic acid, which is less toxic than is ethanol.

Acetaldehyde then accepts the hydrogen from NADH, producing $NAD^+$ and ethyl alcohol, also called ethanol.

Most multicellular animals regenerate $NAD^+$ without removing a carbon atom. The processes that they use involve the production of by-products that are less toxic than alcohol. For example, your muscle cells use an enzyme called **lactate dehydrogenase** to add the hydrogen of the NADH produced by glycolysis back to the pyruvate that is the end product of glycolysis, converting pyruvate plus NADH into lactic acid plus $NAD^+$.

This process closes the metabolic circle, allowing glycolysis to continue for as long as the glucose holds out. Blood circulation removes excess lactic acid from muscles. When the lactic acid cannot be removed as

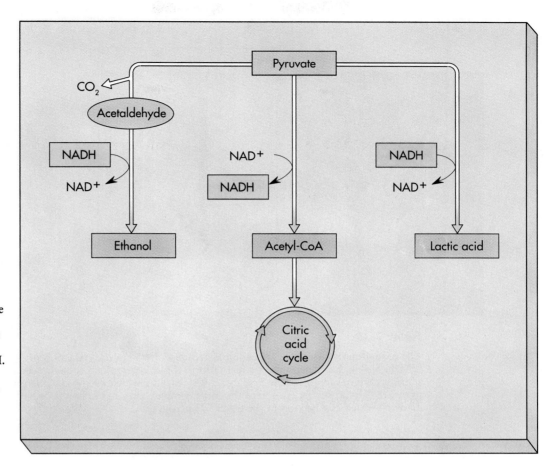

**Figure 7-9**

**What happens to pyruvate, the product of glycolysis?** In the presence of oxygen, pyruvate is oxidized to acetyl-CoA and enters the citric acid cycle. In the absence of oxygen, pyruvate instead is reduced, accepting the electrons extracted during glycolysis and carried by NADH. When pyruvate is reduced directly, as it is in your muscles, the product is lactic acid; when $CO_2$ is first removed from pyruvate and the remainder reduced, as it is in yeasts, the product is ethanol.

fast as it is produced, your muscles cease to work well. Subjectively they feel tired or leaden. Try raising and lowering your arm rapidly a hundred times and you will soon experience this sensation. A more efficient circulatory system developed through training will let you run longer before the accumulation of lactic acid becomes a problem—this is why people train before they run marathon races—but there is always a point at which lactic acid production exceeds lactic acid removal. It is chiefly because of this limit that the world record for running a mile is just under 4 minutes and not significantly less.

*In fermentations, which are anaerobic processes, the electron generated in the glycolytic breakdown of glucose is donated to an oxidized organic molecule. In contrast, in aerobic metabolism such electrons are transferred to oxygen, generating ATP in the process.*

## OXIDATIVE RESPIRATION

Even in aerobic organisms, not all cellular respiration is aerobic; glycolysis and fermentation play important roles in the metabolism of most organisms. But in all aerobic organisms the oxidation of glucose, which began in Stage c of glycolysis, is continued where glycolysis leaves off—with pyruvate. The evolution of this new biochemical process was conservative, as is almost always the case in evolution; the new process was simply tacked onto the old one. In eukaryotic organisms, oxidative respiration takes place exclusively in the mitochondria, which apparently originated as symbiotic bacteria. The oxidation of pyruvate takes place in two stages: the oxidation of pyruvate to form an intermediate product, acetyl-CoA, and the later oxidation of the acetyl-CoA. We shall consider them in turn.

### The Oxidation of Pyruvate

The first stage of oxidative respiration is a single oxidative reaction in which one of the three carbons of pyruvate is split off, departing as $CO_2$ (a reaction of this kind is called by chemists a **decarboxylation**) and leav-

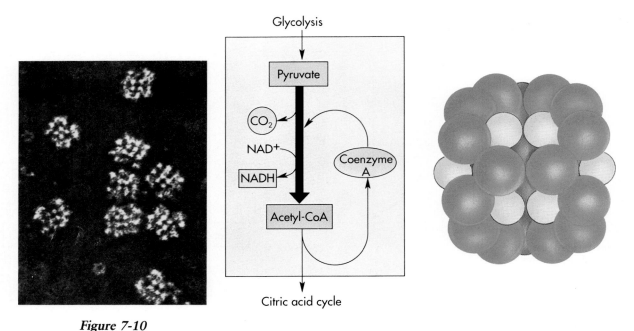

**Figure 7-10**

**The oxidation of pyruvate.** This complex reaction involves the reduction of $NAD^+$ to NADH and is thus a significant source of metabolic energy. Its product, acetyl-CoA, is the starting material for the citric acid cycle. The enzyme which carries out the reaction, pyruvate dehydrogenase, is one of the most complex enzymes known—it has 72 subunits, many of which can be seen in the electron micrograph.

ing behind two remnants (Figure 7-10): (1) a pair of electrons and their associated hydrogen, which reduce $NAD^+$ to NADH; and (2) a two-carbon fragment called an **acetyl group.**

This reaction is very complex, involving three intermediate stages, and is catalyzed by an assembly of enzymes called a **multi-enzyme complex.** Such multi-enzyme complexes serve to organize a series of reactions so that the chemical intermediates do not diffuse away or undergo other reactions. Within such a complex, component polypeptides pass the reacting substrate molecule from one enzyme to the next in line without ever letting go of it. The complex of enzymes called **pyruvate dehydrogenase** that removes the $CO_2$ from pyruvate is one of the largest enzymes known—it contains 48 polypeptide chains. In the course of the reaction, the two-carbon acetyl fragment removed from pyruvate is added to a carrier molecule called coenzyme A (CoA), forming a compound called **acetyl-CoA.**

$$\text{Pyruvate} + NAD^+ + \text{CoA} \rightarrow \text{Acetyl-CoA} + \text{NADH} + CO_2$$

This reaction produces a molecule of NADH, which is later used to produce ATP. Of far greater significance than the reduction of $NAD^+$ to NADH, how-

ever, is the residual fragment acetyl-CoA (Figure 7-11). Acetyl-CoA is important because it is produced, not only by the oxidation of the product of glycolysis, as previously outlined, but also by the metabolic breakdown of proteins, fats, and other lipids. Acetyl-CoA thus provides a single focus for the many catabolic processes of the eukaryotic cell, all the resources of which are channeled into this single molecule.

Although acetyl-CoA is formed by many catabolic processes in the cell, only a limited number of processes use acetyl-CoA. Most acetyl-CoA either is directed toward energy storage (it is used in lipid synthesis) or is oxidized to produce ATP. Which of these two processes occurs depends on the amount of ATP in the cell. When ATP levels are high, acetyl-CoA is channeled into making fatty acids, which is why people get fat when they eat too much. When ATP levels are low, the oxidative pathway is stimulated and acetyl-CoA flows into energy-producing oxidative metabolism.

## The Oxidation of Acetyl-CoA

The oxidation of acetyl-CoA begins with the binding of the acetyl group to a four-carbon carbohydrate. The resulting six-carbon molecule is then passed through a series of electron-yielding oxidation reactions, during which two $CO_2$ molecules are split off, regenerating the four-carbon carbohydrate, which is then free to bind another acetyl group. The process is a continuous cyclical flow of carbon. In each turn of the cycle a new acetyl group comes in to replace the two $CO_2$ molecules that are lost, and more electrons are extracted.

## The Reactions of the Citric Acid Cycle

The citric acid cycle, which oxidizes acetyl-CoA, consists of nine reactions, diagrammed in Figure 7-12. The cycle has two stages:

*Stage a:* Three **preparation reactions** set the scene. In the first reaction, acetyl-CoA joins the cycle, and in the other two reactions, chemical groups are rearranged.

*Stage b:* It is in the second stage that **energy extraction** occurs. Four of the six reactions are oxidations in which electrons are removed, and one generates an ATP equivalent directly by substrate-level phosphorylation.

Together the nine reactions constitute a cycle that begins and ends with oxaloacetate. At every turn of the cycle, acetyl-CoA enters and is oxidized to $CO_2$ and $H_2O$, and the electrons are channeled off to drive proton pumps that generate ATP.

## The Products of the Citric Acid Cycle

In the process of aerobic respiration the glucose molecule has been consumed entirely. Its six carbons were first split into three-carbon units during glycolysis. One of the carbons of each three-carbon unit was then lost as $CO_2$ in the conversion of pyruvate to acetyl-CoA, and the other two were lost during the oxidations of the citric acid cycle. All that is left to mark the passing of the glucose molecule is its energy, which is preserved in four ATP molecules and the reduced state of 12 electron carriers.

The oxidative metabolism of one molecule of glucose proceeds in three stages: glycolysis (which can be anaerobic), the oxidation of pyruvate, and the citric acid cycle. Both glycolysis and the citric acid cycle produce two ATP molecules by substrate-level phosphorylation, coupling ATP production to a reaction that involves a large release of free energy. All three processes involve oxidation reactions; that is, reactions that produce electrons. The electrons are donated to a carrier molecule, $NAD^+$ (or in one instance flavin adenine dinucleotide, $FAD^{++}$).

During the oxidation of glucose, many electron and some ATP molecules have been generated. The extracted electrons are temporarily housed within NADH molecules. In one reaction the extracted electrons are not energetic enough to reduce $NAD^+$, and a different

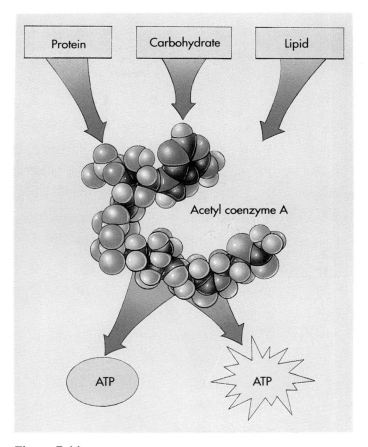

### Figure 7-11

**Acetyl-CoA is the central molecule of energy metabolism.** Almost all of the molecules that you use as foodstuffs are converted to acetyl-CoA when you metabolize them; the acetyl-CoA is then channeled into fat synthesis or into ATP production, depending on your body's energy requirements.

coenzyme called reduced **flavin adenine dinucleotide** (**$FADH_2$**) is used to carry these less energetic electrons. Let us count the number of molecules of ATP and of electron carriers that we have generated, starting with glucose (Table 7-1).

*The systematic oxidation of the pyruvate remaining after glycolysis generates two ATP molecules, which is as many as glycolysis produced. More important, this process harvests many energized electrons, which can then be directed to the chemiosmotic synthesis of ATP.*

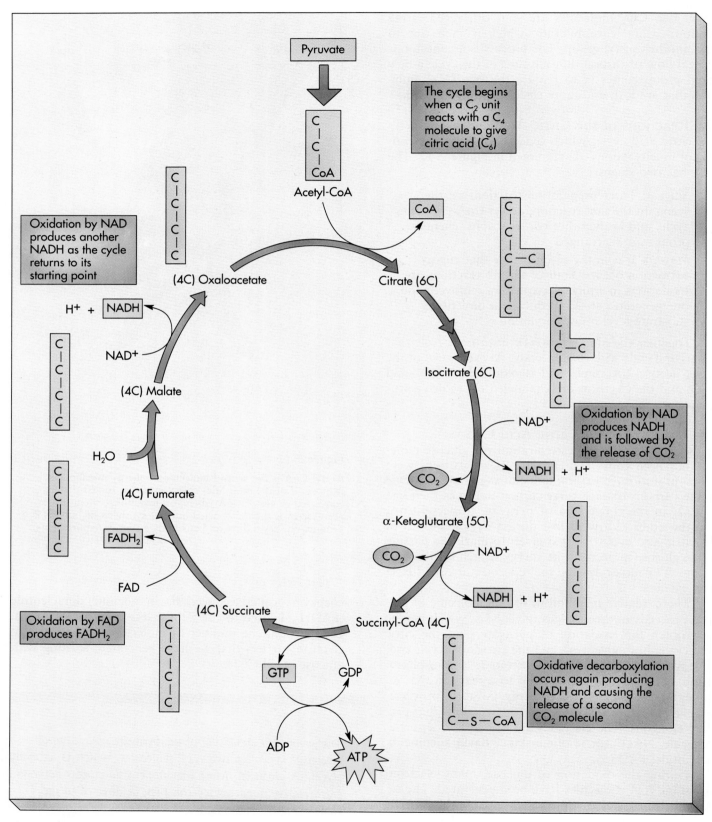

**Figure 7-12**

The citric acid cycle.

**TABLE 7-1    THE OUTPUT OF AEROBIC METABOLISM**

| METABOLIC PROCESS | SUBSTRATE-LEVEL PHOSPHORYLATION | OXIDATION | |
|---|---|---|---|
| Glycolysis | 2 ATP | 2 NADH | |
| Decarboxylation of pyruvate ($\times 2$) | | 2 NADH | |
| Citric acid cycle ($\times 2$) | 2 ATP | 6 NADH | 2 FADH$_2$ |
| TOTAL | 4 ATP | 10 NADH | 2 FADH$_2$ |

## USING THE ELECTRONS GENERATED BY THE CITRIC ACID CYCLE TO MAKE ATP

The NADH and FADH$_2$ molecules formed during glycolysis and the subsequent oxidation of pyruvate each contain a pair of electrons gained when NADH was formed from NAD$^+$ and when FADH$_2$ was formed from FAD$^+$. The NADH molecules carry their electrons to the inner membrane of the mitochondria (the FADH$_2$ is already attached to it), and there they transfer them to a complex membrane-embedded protein called **NADH dehydrogenase,** or, sometimes, ATPase. The electrons are then passed on to a series of respiratory proteins called **cytochromes** and other carrier molecules (Figure 7-13), one after the other, losing much of their energy in the process by driving several transmembrane proton pumps. This series of membrane-associated electron carriers is collectively called the **electron transport chain** (Figure 7-14). At the terminal step of the electron transport chain, the electrons are passed to the **cytochrome c oxidase complex,** which uses four of the electrons to reduce a molecule of oxygen gas and form water:

$$O_2 + 4H^+ + 4e^- \rightarrow 2H_2O$$

*The electron transport chain puts the electrons harvested from the oxidation of glucose to work driving proton-pumping channels. The ultimate acceptor of the electrons harvested from pyruvate is oxygen gas, which is reduced to form water.*

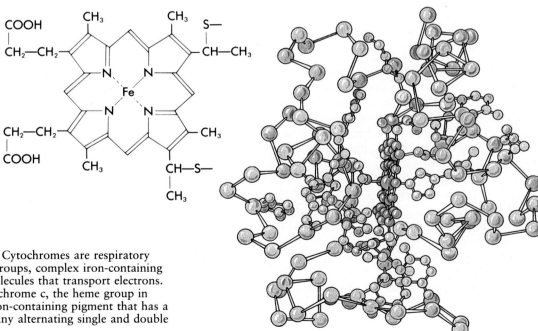

*Figure 7-13*

**The structure of cytochrome c.** Cytochromes are respiratory proteins that contain "heme" groups, complex iron-containing carbon rings often found in molecules that transport electrons. In the molecular model of cytochrome c, the heme group in indicated in red. Heme is an iron-containing pigment that has a complex ring structure with many alternating single and double bonds.

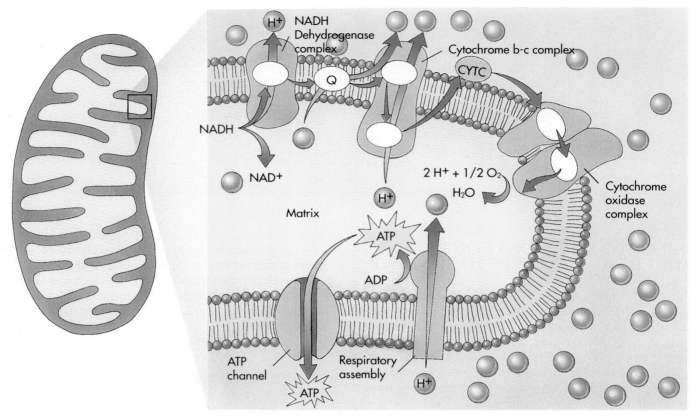

**Figure 7-14**

**The electron transport chain: how mitochondria use electrons to make ATP.** High-energy electrons harvested from food molecules *(red arrows)* are transported along a chain of membrane proteins, three of which use portions of the electron's energy to pump protons *(blue arrows)* out of the matrix into the other compartment. The electrons are finally donated to oxygen to form water. ATP is formed by chemiosmosis as protons reenter the matrix, driven by the high outside concentration.

It is the availability of a plentiful electron acceptor (that is, an oxidized molecule) that makes oxidative respiration possible. In the absence of such a molecule, oxidative respiration is not possible. The electron transport chain used in aerobic respiration is similar to, and probably evolved from, the one that is employed in aerobic photosynthesis.

Thus the final product of oxidative metabolism is water, which is not an impressive result in itself. But recall what happened in the process of forming that water. The electrons contributed by NADH molecules passed down the electron transport chain, activating three proton-pumping channels. Similarly, the passage of electrons contributed by $FADH_2$, which entered the chain later, activated two of these channels. The electrons harvested from the citric acid cycle have in this

way been used to pump a large number of protons out across the membrane, and *that* result is the payoff of oxidative respiration.

In eukaryotes, oxidative metabolism takes place within the mitochondria, which are present in virtually all eukaryotic cells. The internal compartment, or **matrix,** of a mitochondrion contains within it the enzymes that perform the reactions of the citric acid cycle. Electrons harvested there by oxidative respiration are used to pump protons out of the matrix into the outer compartments, the space between the two mitochondrial membranes. As a high concentration of protons builds up, protons cross the inner membrane back into the matrix, driven by diffusion. The only way they can get in is through special channels, which are visible in electron micrographs as projections on the inner surface of the membrane (see Figure 7-14). The proton channel is made up of four polypeptide chains that traverse the membrane. At the inner boundary of the membrane, the channel is linked by a stalk of several proteins to a large protein complex composed of five polypeptide chains. This protein complex synthesizes ATP within the matrix when protons travel inward through the channel (Figure 7-15). The ATP then passes out of the mitochondrion by facilitated diffusion and into the cell's cytoplasm.

*Electrons harvested from glucose and transported to the mitochondrial membrane by NADH drive protons out across the inner membrane. Return of the protons by diffusion generates ATP.*

## AN OVERVIEW OF GLUCOSE CATABOLISM: THE BALANCE SHEET

How much metabolic energy does the chemiosmotic synthesis of ATP produce? One ATP molecule is generated chemiosmotically for each activation of a proton pump by the electron transport chain. Thus each NADH molecule that the citric acid cycle produces ultimately causes the production of three ATP molecules because its electrons activate three pumps. Each $FADH_2$, which activates two of the pumps, leads to the production of two ATP molecules. However, because of an evolutionary development that we discuss later, eukaryotes carry out glycolysis in their cytoplasm and the citric acid cycle within their mitochondria. This separation of the two processes within the cell requires that the electrons of the NADH created during glycolysis be transported across the mitochondrial membrane, consuming one ATP molecule per NADH. Thus

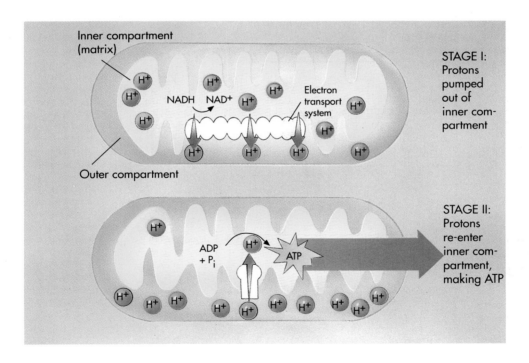

**Figure 7-15**

**How proton movement in the mitochondrion drives ATP production.** A summary of proton movement in mitochondria during chemiosmosis.
**Stage I** The electron transport system utilizes energetic electrons from NADH to drive the pumping of electrons out of the inner matrix compartment. The result is a deficit of protons in the matrix and an excess in the outer compartment.
**Stage II** Protons re-enter the matrix, driven by diffusion, through special channels that produce ATP. These ATP molecules diffuse out of the mitochondrion through passive channels *(red arrow).*

each glycolytic NADH produces only two ATP molecules in the final tally, instead of three.

The overall reaction for the catabolic metabolism of glucose is

$$C_6H_{12}O_6 + 10NAD^+ 2FAD^+ + 36ADP + 36P_i + 14H^+ + 6O_2 \rightarrow$$
$$6CO_2 + 36ATP + 6H_2O + 10NADH + 2FADH_2$$

As seen in Table 7-1, 4 of the 36 ATP molecules result from substrate-level phosphorylation; the remaining 32 ATP molecules result from chemiosmotic phosphorylation.

The overall efficiency of glucose catabolism is very high: the aerobic oxidation (combustion) of glucose yields 36 ATP molecules (a total of $-7.3 \times 36 =$ −263 kilocalories per mole) (Figure 7-16). The aerobic oxidation of glucose thus has an efficiency of $^{263}/_{686}$, or 38%. Compared with the two ATP molecules generated by glycolysis, aerobic oxidation is 18 times more efficient.

*Oxidative respiration is 18 times more efficient than is glycolysis at converting the chemical energy of glucose into ATP. It produces 36 molecules of ATP from each glucose molecule consumed, compared with the 2 ATP molecules that are produced by glycolysis.*

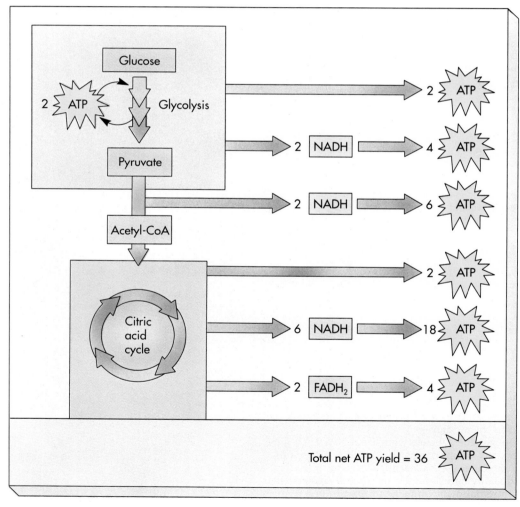

**Figure 7-16**

An overview of the energy extracted from the oxidation of glucose.

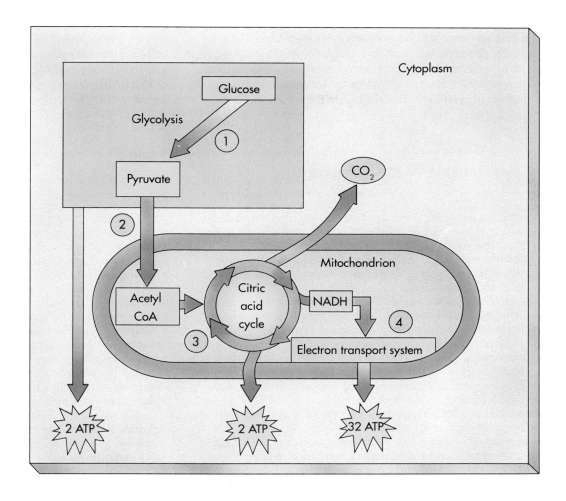

**Figure 7-17**

**An overview of oxidative respiration.** There are four basic processes:
1 The initial stage of glucose metabolism, glycolysis, does not require oxygen and occurs in the cytoplasm of cells; during it, each glucose molecule is converted to two molecules of pyruvic acid.
2 In eukaryotic cells, the pyruvic acid molecules enter the mitochondria, where the pyruvate molecules are converted to acetyl-CoA.
3 The acetyl-CoA enters the citric acid cycle and is broken down to carbon dioxide, harvesting chemical bond energy to drive the production of NADH.
4 The NADH in turn powers the fourth and final stage of respiration, the electron transport system that chemiosmotically produces ATP. Most of the ATP that results from oxidative respiration is at the final chemiosmotic stage.

Oxidative metabolism is so efficient at extracting energy from photosynthetically produced molecules that its development opened up an entirely new avenue of evolution. For the first time it became feasible for heterotrophic organisms, which derive their metabolic energy exclusively from the oxidative breakdown of other organisms, to evolve. As long as some organisms produced energy by photosynthesis, heterotrophs could exist solely by feeding on these autotrophs. The high efficiency of oxidative respiration was one of the key conditions that fostered the evolution of nonphotosyn-

thetic heterotrophic organisms. An overview of the oxidative respiration process is presented in Figure 7-17.

## REGULATION OF CELLULAR RESPIRATION

When cells possess plentiful amounts of ATP, the ATP molecules within the cell act to shut down key enzymes of glycolysis and the citric acid cycle and thus slow down ATP production. Conversely, when ATP levels in the cell are low, ADP molecules (left over after ATP is used by the cell) will activate enzymes in the pathway to stimulate more production of ATP. The regulation of these biochemical pathways by levels of ATP is an example of feedback inhibition, discussed in the previous chapter. Almost all of the biochemical pathways in cells are regulated by the sensitivity of key enzymes to particular metabolites. The cell's metabolic machinery operates like a well-designed automobile engine, every part working in concert.

---

A VOCABULARY OF ATP GENERATION

---

**aerobic respiration** That portion of cellular respiration which requires oxygen as an electron acceptor; includes the citric acid cycle and pyruvate oxidation.

**anaerobic respiration** That portion of cellular respiration which can occur in the absence of oxygen; includes glycolysis and fermentation.

**cellular respiration** The oxidation of organic molecules in cells; ATP is produced both by substrate-level phosphorylation and by chemiosmosis; includes glycolysis, pyruvate oxidation, and the citric acid cycle.

**chemiosmosis** The production of ATP by the transport of high-energy electrons to a membrane where they drive a proton pump and thus power the synthesis of ATP via the osmotic reentry of protons.

**fermentation** Respiration in which the final electron acceptor is an organic molecule.

**maximum efficiency** The maximum number of ATP molecules generated by oxidizing a substance, relative to its bond energy; in organisms, the realized efficiency is rarely maximal.

**oxidation** The loss of an electron. In cellular respiration, high-energy electrons are stripped from food molecules, oxidizing them.

**oxidative respiration** Respiration in which the final electron acceptor is molecular oxygen.

**photosynthesis** The energy to drive the chemiosmotic generation of ATP is provided by high-energy electrons as in cellular respiration, but in photosynthesis the energy is derived from light, whereas in cellular respiration it is provided by electrons stripped from chemical bonds.

**reduction** The gain of an electron. In cellular respiration, electrons from food molecules are added to oxygen molecules, reducing them to form water.

**substrate-level phosphorylation** The generation of ATP by coupling its synthesis to a strongly exergonic (energy-yielding) reaction.

---

## ■ SUMMARY

1. Metabolism is driven by the application of energy to endergonic reactions. Those reactions which require a net input of free energy are coupled to the cleavage of ATP, which involves a large release of free energy.

2. Organisms acquire ATP from photosynthesis or by harvesting the chemical energy in the bonds of organic molecules.

3. Energy can be harvested in two ways: substrate-level phosphorylation, in which some reactions involving a large decrease in free energy are coupled to ATP formation; or oxidative respiration, in which electrons are used to drive proton pumps, resulting in the chemiosmotic synthesis of ATP.

4. Glycolysis harvests chemical energy in the first of these two ways by rearranging the chemical bonds of glucose to form two molecules of pyruvate and two molecules of ATP.

5. Organisms living in anaerobic environments require a mechanism to dispose of the electron and the associated hydrogen produced in the oxidation step of glycolysis. They are donated to one of a number of carbohydrates in a process called fermentation.

6. Aerobic organisms direct the electron to the inner mitochondrial membrane, where it drives a proton pump. Pyruvate is further oxidized, yielding many additional electrons, all of which are also channeled to the mitochondrial membrane to drive proton pumps.

7. The excess protons in the outer mitochondrial compartment created by the proton pumps diffuse back across the inner mitochondrial membrane through special channels, and the passage of these protons drives the production of ATP.

8. In total, the oxidation of glucose results in the net production of 36 ATP molecules, all but 4 of them produced chemiosmotically.

9. Within all eukaryotic cells, oxidative respiration of pyruvate takes place within the matrix of mitochondria. The mitochondria act as closed osmotic compartments from which protons are pumped and into which protons pass by diffusion to create ATP. The ATP then leaves the mitochondria by facilitated diffusion.

10. Almost all of the biochemical pathways in cells are regulated by the sensitivity of key enzymes to particular metabolites.

## REVIEW

1. Proton-pumping channels are responsible for the chemiosmotic generation of _____.

2. Give, in order, the three stages of the oxidative metabolism of glucose.

3. The end result of glycolysis is to break a glucose molecule up into two molecules of _____ and generate a net gain of two ATP molecules.

4. In eukaryotic cells, oxidative metabolism takes place in which organelle?

5. What is the net production of ATP from a molecule of glucose?

## SELF-QUIZ

1. How do all cells drive thermodynamically unfavorable reactions?
   (a) By using the energy released from splitting ATP.
   (b) By using light energy.
   (c) These reactions occur only in warm-blooded organisms, and heat is used as the energy source.
   (d) Thermodynamically unfavorable reactions do not occur.
   (e) The information is coded in the DNA.

2. In glycolysis, in the absence of oxygen, ATP is generated by
   (a) electron transport chain.
   (b) substrate-level phosphorylation.
   (c) chemiosmotic generation of ATP.
   (d) citric acid cycle.
   (e) photosynthesis.

3. When oxygen is available as an electron acceptor, what happens after glycolysis?
   (a) Glycolysis occurs again
   (b) NADH is produced
   (c) Fermentation
   (d) Pyruvate is formed
   (e) Oxidative respiration

4. By the completion of oxidative metabolism, all of the carbons from glucose $(C_6H_{12}O_6)$ are gone. Where did they go?
   (a) Carbon dioxide
   (b) NADH
   (c) To make pyruvate
   (d) ATP
   (e) Water

5. What is the function of oxidative metabolism?
   (a) Carbon dioxide production
   (b) ATP production
   (c) Water production
   (d) $NAD^+$ production
   (e) Glucose utilization

# PROBLEM

1. Thirty-six ATP molecules are produced per molecule of glucose. How many ATP molecules are produced from a mole of glucose? As a helpful hint, remember that there are $6.023 \times 10^{23}$ molecules in a mole.

# THOUGHT QUESTIONS

1. If you poke a hole in a mitochondrion, can it still perform oxidative respiration? Can fragments of mitochondria perform oxidative respiration? Explain.

2. Why have eukaryotic cells not dispensed with mitochondria, by placing the mitochondrial genes in the nucleus and incorporating the mitochondrial enzyme ATPase and the proton pump into the cell's smooth endoplasmic reticulum?

# FOR FURTHER READING

DICKERSON, R.: "Cytochrome *c* and the Evolution of Energy Metabolism," *Scientific American*, March 1980, pages 136-154. A superb description of how the metabolism of modern organisms evolved.

HINKLE, P., and R. McCARTY: "How Cells Make ATP," *Scientific American*, March 1978, pages 104-125. A good summary of oxidative respiration, with a clear account of the events that happen at the mitochondrial membrane.

LEVINE, M., H. MUIRHEAD, D. STAMMERS, and D. STUART: "Structure of Pyruvate Kinase and Similarities with Other Enzymes: Possible Implications for Protein Taxonomy and Evolution," *Nature*, vol. 271, 1978, pages 626-630. An advanced article that recounts how the enzymes of oxidative metabolism may have evolved. Well worth the effort.

# Photosynthesis

Children of light, we and all other animals depend on photosynthesis for the energy we use. Plants literally fuel our lives, for only they can harvest energy from sunlight. Plants use these delicate beams of light to extract the forces that make life possible.

# PHOTOSYNTHESIS

## Overview

We all depend on the process of photosynthesis, which is the means by which the energy that ultimately builds our bodies is captured from sunlight. Photosynthesis is one of the oldest and most fundamental of life processes: the major types of photosynthesis first evolved billions of years ago among the bacteria. The first of these to evolve apparently used the energy obtained from sunlight to split hydrogen sulfide, generating sulfur as a by-product. Later, in the cyanobacteria, a system evolved in which water is split in the same way, yielding oxygen gas as a by-product. The outcome of this form of photosynthesis, performed for at least 3 billion years, has been the production of an atmosphere rich in oxygen; this atmosphere has set the stage for the evolution of all complex forms of life on earth, including ourselves.

## For Review    *Here are some important terms and concepts that you will encounter in this chapter. If you are not familiar with them, you should review them before proceeding.*

**Electron energy levels** (Chapter 2)

**Chemiosmosis** (Chapters 5 and 7)

**Oxidation-reduction** (Chapter 6)

**Glycolysis** (Chapter 7)

For all its size and diversity, our universe might never have spawned life except for one characteristic of overriding importance: it is awash in energy. Everywhere in the universe, matter is continually being converted to energy by thermonuclear processes. The energy resulting from those processes streams in all directions from the stars, including our sun. The total amount of radiant energy that reaches the earth from the sun each day is the equivalent of that of about 1 million Hiroshima-sized atomic bombs. Approximately a third of this energy is immediately radiated back into space, and most of the remainder is absorbed by the earth and converted to heat. Less than 1% of the energy that reaches the earth is captured in the process of photosynthesis and provides the energy that drives all the activities of life on earth (Figure 8-1). In this chapter we discuss how photosynthesis works and outline the major kinds of photosynthesis.

## THE BIOPHYSICS OF LIGHT

### The Photoelectric Effect

Where is the energy in light? What is there about sunlight that a plant can use to create chemical bonds? To answer these questions we need to begin by considering the physical nature of light itself. Perhaps the best place to start is in a laboratory in Germany in 1887 where a curious experiment was performed. A young physicist named Heinrich Hertz was attempting to verify a mathematical theory that predicted the existence of electromagnetic waves. To see whether such waves existed, Hertz constructed a spark generator in his laboratory—a machine composed of two shiny metal spheres standing near each other on slender rods. When a very high static electric charge built up on one sphere, sparks would jump across to the other sphere.

After Hertz had constructed this system, he proceeded to investigate whether the sparking would create invisible electromagnetic waves, as predicted by the mathematical theory. On the other side of the room he placed a thin metal hoop that was not quite a closed circle on top of an insulating stand. When he turned on the spark generator across the room, tiny sparks could be seen crossing the gap in the hoop! This was the first demonstration of radio waves. But Hertz noted a curious side effect as well. When light was shone on the ends of the hoop, the sparks crossed the gap more readily. This unexpected effect, called the **photoelectric effect,** puzzled investigators for many years.

**Figure 8-1**

**Light drives life.** These sunflowers, growing vigorously in the August sun, capture enough energy to power your body for hours.

Especially perplexing was the fact that the strength of the photoelectric effect depended not only on the brightness of the light shining on the gap in the hoop but also on its wavelength. Short wavelengths were much more effective than were long wavelengths in producing the photoelectric effect. This effect was finally explained by Albert Einstein as a natural consequence of the physical nature of light: the light was literally blasting electrons from the metal surface at the ends of the hoop, creating positive ions and thus facilitating the passage of the electronic spark induced by the radio waves. Light consists of units of energy called **photons,** and some of these photons were being absorbed by the metal atoms of the hoop. In this process, some of the electrons of the metal atoms were being boosted into higher energy levels and so ejected from the metal atoms into the gap.

## The Role of Photons

All photons do not possess the same amount of energy. Some contain a great deal of energy, others far less. Photons of short-wavelength light contain higher energy than photons of long-wavelength light. For this reason the photoelectric effect was more pronounced with light of a shorter wavelength: the metal atoms of the hoop were being bombarded with higher energy photons. Sunlight contains photons of many energy levels, only some of which our eyes perceive as visible light. The highest energy photons, which occur at the short-wavelength end of the **electromagnetic spectrum** (Figure 8-2), are gamma rays with wavelengths of less than 1 nanometer; the lowest energy photons, with wavelengths of thousands of meters, are radio waves.

**Figure 8-2**

**The electromagnetic spectrum.** Light is a form of electromagnetic energy and is conveniently thought of as a wave. The shorter the wavelength of light, the greater the energy. Visible light represents only a small part of the electromagnetic spectrum, that between 380 and 750 nanometers.

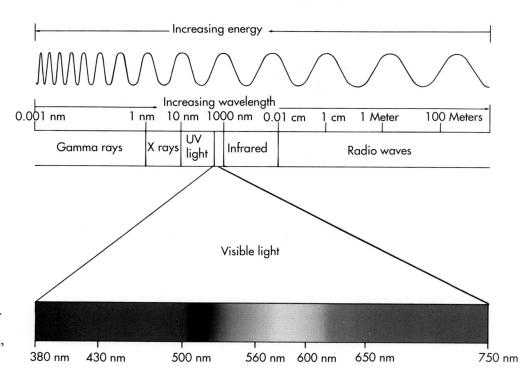

## Ultraviolet Light

Sunlight contains a significant amount of ultraviolet light, which possesses considerably more energy than does visible light because ultraviolet light has a shorter wavelength. Ultraviolet light is thought to have been an important source of energy on primitive earth when life originated, before the oxygen-rich atmosphere developed. Today's atmosphere contains ozone (derived from oxygen gas), which absorbs most of the ultraviolet-energy photons in sunlight, but anyone who has been sunburned knows that considerable quantities of ultraviolet-energy photons manage to penetrate the atmosphere. We discuss mysterious "holes" that have recently been discovered in the ozone layer in a later chapter; these holes, which have been caused by human activities, are associated with a general decrease in the amount of ozone in the upper layers of the atmosphere. Unless preventive measures are taken, this decrease will cause a significant increase in the incidence of skin cancers in human beings throughout the world.

As described in Chapter 2, electrons occupy distinct energy levels in their orbits around the atomic nucleus. Boosting an electron to a different energy level requires just the right amount of energy—no more and no less; similarly, when climbing a ladder, you must raise your foot just so far to climb a rung. Therefore specific atoms can absorb only certain photons of light—those which correspond to available energy levels. A given atom or molecule has a characteristic range, or **absorption spectrum,** of photons that it is capable of absorbing, depending on the electron energy levels that are available in it.

## CAPTURING LIGHT ENERGY IN CHEMICAL BONDS

The energy of light is "captured" by a molecule that absorbs it because the photon of energy is used to boost an electron of the molecule to a higher energy level. Molecules that absorb light are called **pigments,** and organisms have evolved a variety of such pigments. Two general sorts are of interest here:

1. *Carotenoids.* Carotenoids consist of carbon rings linked to chains in which single and dou-

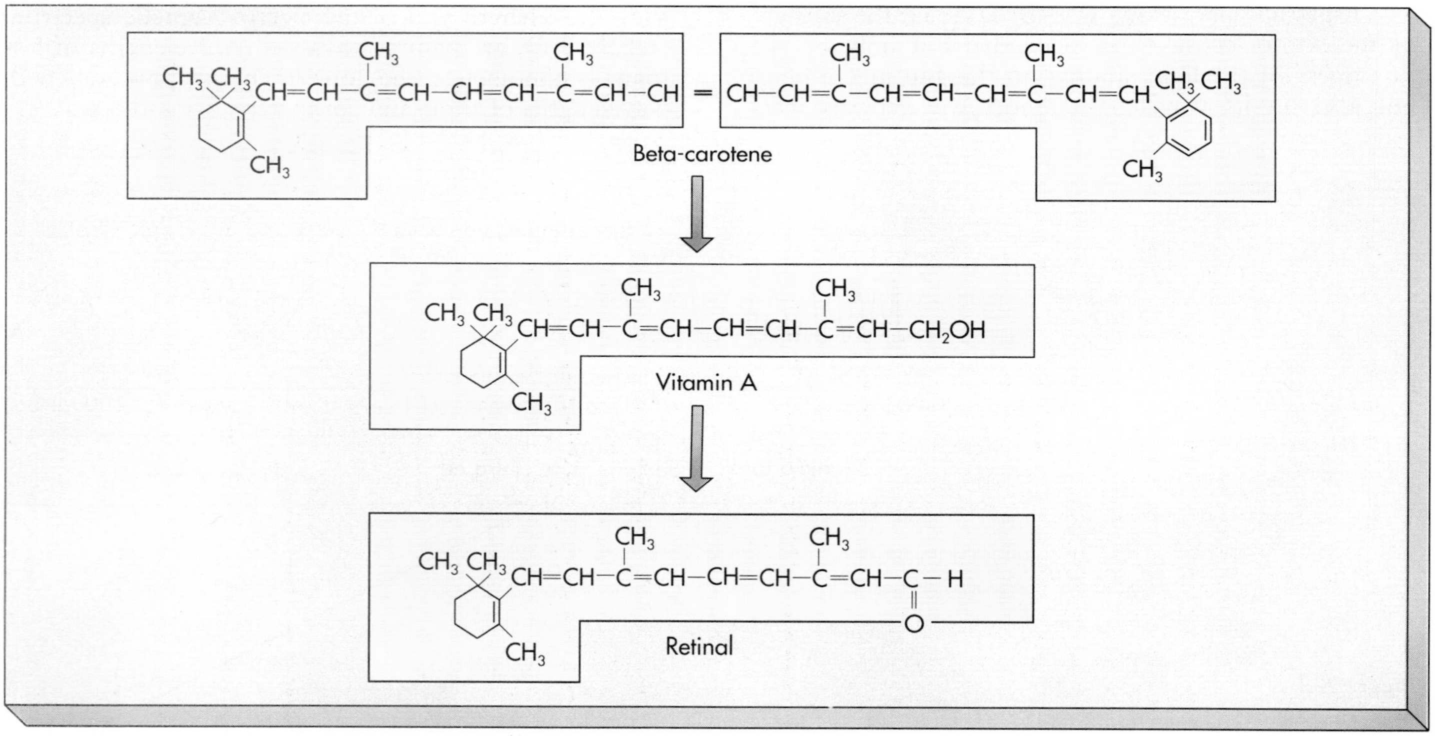

**Figure 8-3**

**What do carrots have to do with vision?** The pigment B-carotene is what makes carrots look orange. When the double bond that links the two halves of B-carotene is broken, two molecules of vitamin A are produced. Retinal, which functions as the key visual pigment in your eyes, is produced from vitamin A.

ble bonds alternate. Carotenoids can absorb photons of a wide range of energies, although not always with high efficiency. They are responsible for most of the yellow and orange colors seen in plants, where they play an important role in capturing light energy and transferring it to chlorophyll in the process of photosynthesis. A typical carotenoid is **beta-carotene** (Figure 8-3). In beta-carotene, 2 carbon rings are linked by a chain of 18 carbon atoms connected alternately by single and double bonds. Splitting a molecule of beta-carotene into equal halves results in the production of two molecules of **vitamin A.** When vitamin A is subsequently oxidized, **retinal,** the pigment used in human vision, is produced. Thus the claim that eating carrots (rich in beta-carotene) improves vision is based on fact. When retinal absorbs a photon of light, the resulting electron excitation causes a change in the shape of a pigment located in the membrane of certain cells in the eye and thus triggers a nerve impulse. Retinal absorbs photons that produce light ranging from violet (380 nanometers) to red (750 nanometers), and so determines the range of colors that we can see—**visible light.** Some other organisms use different light-absorbing pigments for vision and thus "see" a different portion of the electromagnetic spectrum. For example, most insects have eye pigments that absorb at lower wavelengths than does retinal. As a result, bees can perceive ultraviolet light, which is produced by photons with a shorter wavelength than violet has—carotenoids in plants are important in producing photons of such wavelengths—but cannot see red, which is produced by a photon with a relatively long wavelength.

2. *Chlorophylls.* Other biological pigments called chlorophylls absorb photons by means of an excitation process analogous to the photoelectric effect. These pigments use a metal atom (magnesium), which lies at the center of a complex ring structure called a **porphyrin ring.** This ring structure consists of alternating single and double carbon bonds. Photons absorbed by the pigment molecule excite electrons of the magnesium atom, which are then channeled away through the carbon-bond system. Several small side groups are attached outside the porphyrin ring. These side groups alter the absorption properties of the pigment in different kinds of chlorophyll.

Unlike retinal, the different kinds of chlorophyll absorb only photons of a narrow energy range. As you can see from their absorption spectra (Figure 8-4), the

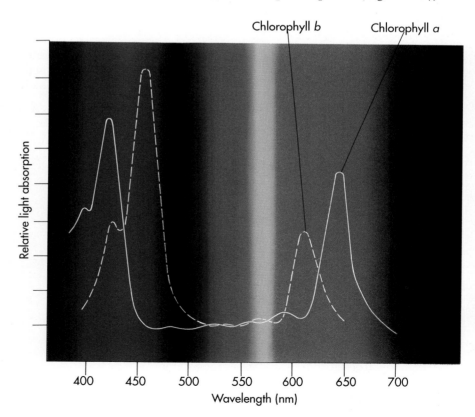

**Figure 8-4**

**Absorption spectra for chlorophylls *a* and *b.*** The cholorphylls absorb predominately violet-blue and red light in two narrow bands of the spectrum.

two kinds of chlorophyll that occur in plants—chlorophylls *a* and *b*—absorb primarily violet-blue and red light, respectively. The light between wavelengths of 500 and 600 nanometers is not absorbed by chlorophyll pigments and therefore is reflected by plants. The light reflected from a chlorophyll-containing plant has had all of its middle-energy photons absorbed by the chlorophyll except those in the 500- to 600-nanometer range. When these photons are subsequently absorbed by the retinal in our eyes, we perceive them as green.

*A pigment is a molecule that absorbs light. The wavelengths absorbed by a particular pigment depend on the energy levels available in that molecule to which light-excited electrons can be boosted.*

All plants and algae and all but one primitive group of photosynthetic bacteria use chlorophylls as their primary light-gatherers. Why don't these photosynthetic organisms use a pigment like retinal, which has a very broad absorption spectrum and can harvest light in the 500- to 600-nanometer wavelength range as well as at other wavelengths? The most likely hypothesis involves photoefficiency. Retinal absorbs a broad spectrum of light wavelengths but does so with relatively low efficiency; in contrast, chlorophyll absorbs in only two narrow bands, violet-blue and red, but does so with very high efficiency. Therefore, by using chlorophyll, plants and most other photosynthetic organisms achieve far higher overall photon capture

rates than would be possible with a pigment that allows a broader but less efficient spectrum of absorption.

# AN OUTLINE OF PHOTOSYNTHESIS

Photosynthesis is a single term describing a complex series of events that involves three kinds of chemical processes (Figure 8-5). The first process that occurs is the chemiosmotic generation of ATP by electrons, using energy captured from sunlight. The reactions involved in this process are called the **light reactions** of photosynthesis because the resultant synthesis of ATP takes place only in the presence of light. Second, the light reactions are followed by a series of enzyme-catalyzed reactions that use this newly generated ATP to drive the formation of organic molecules from atmospheric carbon dioxide. These reactions are called the **dark reactions** of photosynthesis because as long as ATP is available, they occur as readily in the absence of light as in its presence. Third, the pigment that absorbed the light in the first place is rejuvenated and made ready to initiate another light reaction.

## Absorbing Light Energy

The light reactions occur on **photosynthetic membranes**. In photosynthetic bacteria these membranes are the cell membrane itself; in plants and algae, photosynthesis is carried out by the symbiotic evolutionary descendants of photosynthetic bacteria, the **chloroplasts;** the photosynthetic membranes in plants and algae are contained within the chloroplasts.

The light reactions occur in three stages:

1. A photon of light is captured by a pigment. The

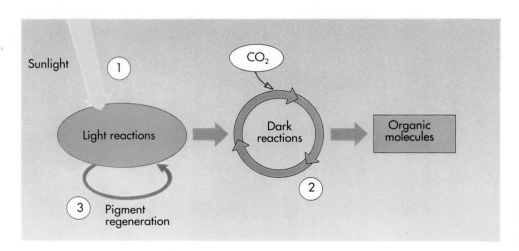

*Figure 8-5*

Overview of the three processes of photosynthesis.

result of this **primary photoevent** is the excitation of an electron within the pigment.

2. The excited electron is shuttled along a series of electron-carrier molecules embedded within the photosynthetic membrane to a transmembrane proton-pumping channel, where the arrival of the electron induces the transport of a proton inward across the membrane. The electron is passed on to an acceptor.

3. The later exit of protons drives the chemiosmotic synthesis of ATP, just as it does in aerobic respiration.

## Fixing Carbon

The chemical events that use the ATP generated by the light reactions take place readily in the dark. However, building an organism requires not only ATP but also a way of making organic molecules from carbon dioxide ($CO_2$). Organic molecules are "reduced" compared with $CO_2$ because they contain many C—H bonds (electrons and the associated hydrogens). To build an organism, a source of hydrogens and the electrons to bind them to carbon are needed. Photosynthesis uses the energy of light to extract hydrogen atoms (**reducing power**) from water in a way that we explain later. In the dark reactions overall, atmospheric $CO_2$ is incorporated into carbon-containing molecules through a process called **carbon fixation.**

## Replenishing the Pigment

The third kind of chemical event that occurs during continuous photosynthesis involves the electron that was stripped from the chlorophyll at the beginning of the light reactions. This electron must be returned to the pigment or another source of electrons must be used to replenish the supply of electrons in the pigment. Otherwise, continual electron removal would cause the pigment to become deficient in electrons (bleached), and the pigment could then no longer trap photon energy by electron excitation. As we shall see, various organisms have evolved different approaches to solving this problem during the course of their evolutionary history.

*Photosynthesis involves three processes: the use of light-ejected electrons to drive the chemiosmotic synthesis of ATP; the use of the ATP to fix carbon; and the replenishment of the photosynthetic pigment.*

The overall process of photosynthesis may be summarized by a simple oxidation-reduction equation:

$$6CO_2 + 12H_2^*O \xrightarrow{\text{light}} C_6H_{12}O_6 + 6H_2O + 6^*O_2$$

| Atmospheric carbon dioxide | Water vapor | | Sugar | Water | Oxygen gas from the original water molecules |

Although $H_2O$ appears on both sides of the equation, they are *not* the same water molecule. This can be demonstrated by carrying out photosynthesis with water vapor in which the oxygen atom is a heavy isotope. Most of the heavy oxygen atoms (indicated in the equation by the symbol $^*O$) end up in oxygen gas and not in water.

## HOW LIGHT DRIVES CHEMISTRY: THE LIGHT REACTIONS

Photosynthesis in plants, algae, and those bacteria in which it occurs is the result of a long evolutionary process. As this evolution progressed, new reactions were added to older ones, thus making the overall series of reactions more complex. Much of the evolution has centered on the light reactions, which apparently have changed considerably since they first evolved.

### Evolution of the Photocenter

In chloroplasts and all but one group of primitive bacteria, light is captured by a network of chlorophyll pigments working together. Each chlorophyll molecule within the network, which is called a **photocenter,** is capable of capturing photons efficiently. Pigment molecules are held on a protein matrix within the photocenter, and their arrangement permits the channeling of excitation energy from anywhere in the array to a central point. The assembly of chlorophyll molecules thus acts as a sensitive "antenna" to capture and focus photon energy. Its mode of operation is similar to the way a magnifying glass focusing light can generate enough heat energy at the point of focus to burn paper. Similarly, the photocenter channels the excitation energy gathered by any one of the pigment molecules to one called $P_{700}$ (Figure 8-6), which is associated with a membrane-bound protein called **ferredoxin.** The photocenter thus funnels to the ferredoxin many more electrons than would otherwise be possible.

When light of the proper wavelength strikes any pigment molecule of the photocenter, the resulting ex-

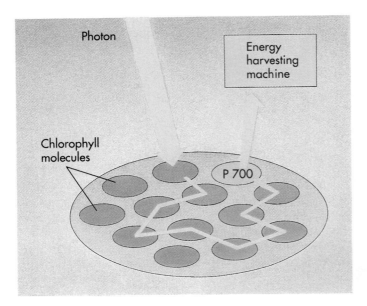

**Figure 8-6**

**How photocenters work.** When light of the proper wavelength strikes any chlorophyll molecule within a photocenter, the photon's energy is absorbed by it. The energy passes from one chlorophyll molecule to another until it encounters $P_{700}$, which channels the energy out of the photocenter to the energy-harvesting machinery.

citation passes from one chlorophyll molecule to another. The excited electron does not transfer physically from one pigment molecule to the next. Instead, the pigment passes the energy along to an adjacent molecule of the photocenter, after which the pigment's electron returns to the low-energy level it had before the photon was absorbed. A crude analogy to this form of energy transfer exists in the initial "break" in a game of pool. If the cue ball squarely hits the point of the triangular array of 15 pool balls, the 2 balls at the far corners of the triangle fly off and none of the central balls move at all. The energy is transferred through the central balls to the most distant ones. The protein matrix of the photocenters in which the molecules of chlorophyll are embedded serves as a sort of scaffold, holding individual pigment molecules in orientations that are optimal for energy transfer. In this way the process channels excitation energy to the membrane-bound ferredoxin in the form of electrons.

*A photocenter, which is an array of pigment molecules, acts as a light antenna, directing photon energy captured by any of its members toward a single pigment molecule and thus amplifying the light-gathering powers of the individual pigment molecules.*

## Where Does the Electron Go?

Photocenters probably evolved more than 3 billion years ago in bacteria similar to the group called green sulfur bacteria that exists today. In these bacteria the absorption of a photon of light by the photocenter results in the transmission of an electron from the pigment $P_{700}$ to ferredoxin. As in many oxidation-reduction processes, the electron is accompanied by a proton, traveling as a hydrogen atom. Where does the proton come from? It is extracted by green sulfur bacteria from hydrogen sulfide ($H_2S$) through a process that produces elemental sulfur as a residual by-product.

The ejection of an electron from $P_{700}$ and its donation to ferredoxin leaves $P_{700}$ short one electron. Before the photocenter of the green sulfur bacteria can function again, the electron must be returned. These bacteria channel the electron back to the pigment through an electron-transport system (see Chapter 7) in which the electron's passage drives a proton pump and thus promotes the chemiosmotic synthesis of ATP. Therefore the path of the electron originally extracted from $P_{700}$ is a circle (chemists call the process **cyclic photophosphorylation**) (Figure 8-7). However, the process is not a true circle. The electron that left $P_{700}$ was a high-energy−level electron, boosted to its high-en-

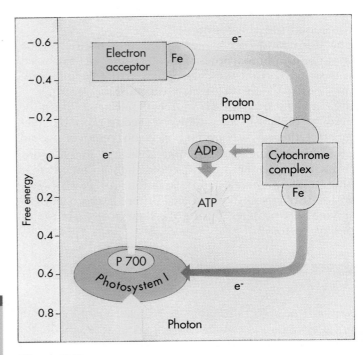

**Figure 8-7**

**The path of an electron in bacterial photosynthesis.** When an electron is ejected from the photocenter by a photon of light, its passage is a circle, the electron returning to the photocenter from which it was initially ejected.

ergy level by the absorption of a photon of energy; but the electron that returns has only as much energy left as it had before the photon absorption. The difference in the energy of that electron is the photosynthetic payoff; that is, the energy that is used to drive the proton pump.

*In bacteria, the electron ejected from the pigment by light travels a circular path in which light-excited electrons power a proton pump and then return to the photocenter where they originated.*

The light reactions of all other photosynthetic systems, including those in other groups of bacteria, evolved by adding to the simple cyclic photophosphorylation process of the green sulfur bacteria. Just as glycolysis is retained as a fundamental component of the respiratory metabolism of all organisms, so cyclic photophosphorylation remains a fundamental component of photosynthesis in the chloroplasts of all plants, algae, and most photosynthetic bacteria.

## LIGHT REACTIONS OF PLANTS

For more than 1 billion years, cyclic photophosphorylation was the only form of photosynthetic light reaction that organisms used. But it has a fundamental limitation: it is geared only toward energy production and not toward biosynthesis. To understand this point, consider for a moment that the ultimate point of photosynthesis is *not* to generate ATP but rather to fix carbon—to incorporate atmospheric carbon dioxide into new carbon compounds. Because the molecules (sugars) produced during carbon fixation are more reduced (have more hydrogen atoms) than their precursor ($CO_2$), a source of hydrogens must be provided. Cyclic photophosphorylation does not provide hydrogen atoms, and thus bacteria that use this process must scavenge hydrogens from other sources—a very inefficient undertaking.

### The Advent of Photosystem II
At some point after the appearance of the green sulfur bacteria, other kinds of bacteria evolved an improved version of the photocenter that solved the reducing power problem in a neat and simple way. These bacteria grafted a second, more powerful photosystem onto the original one, using a new form of chlorophyll called chlorophyll *a*. This great evolutionary advance

took place when the cyanobacteria originated, no less than 2.8 billion years ago.

In this second photosystem, called **photosystem II**, molecules of chlorophyll *a* are arranged with a different geometry in the photocenter so that more of the shorter-wavelength photons of higher energy are absorbed than in the more ancient bacterial photosystem (called **photosystem I** in algae and plants). As in the bacterial photosystem, energy is transmitted from one pigment molecule to another within the photocenter until it encounters a particular pigment molecule that is positioned near a strong membrane-bound electron acceptor. In photosystem II, the absorption peak of this pigment molecule is 680 nanometers; therefore the molecule is called $P_{680}$.

### How the Two Photosystems Work Together
Plants, algae, and cyanobacteria, as contrasted with other photosynthetic bacteria, use both photosystems—a two-stage photocenter (Figure 8-8). The new photosystem acts first. When a photon jolts an electron from photosystem II, the excited electron is donated to an electron transport chain, which passes it along to photosystem I. In its journey to photosystem I, each electron drives a proton pump and thus generates an ATP molecule chemiosmotically.

When the electron reaches photosystem I, it has already expended its excitation energy in driving the proton pump and thus contains only the same amount of energy as the other electrons of this photosystem. However, its arrival does give the photosystem an electron that it can afford to lose. Photosystem I now absorbs a photon, boosting one of its pigment electrons to a high-energy level. The electron is then channeled to ferredoxin and is used to generate reducing power. In plants, algae, and cyanobacteria, ferredoxin contributes two electrons to reduce $NADP^+$, generating NADPH. By using this molecule instead of the NAD used in oxidative respiration, these organisms keep the flow of electrons in the two processes separate (Figure 8-9).

*Plants, algae, and cyanobacteria use a two-stage photocenter. First, a photon is absorbed by photosystem II, which passes an electron to photosystem I. During this process the electron uses its photon-contributed energy to drive a proton pump and thus generate a molecule of ATP. Then another photon is absorbed, this time by photosystem I, which also passes on a photon-energized electron. This second electron is channeled away to provide reducing power.*

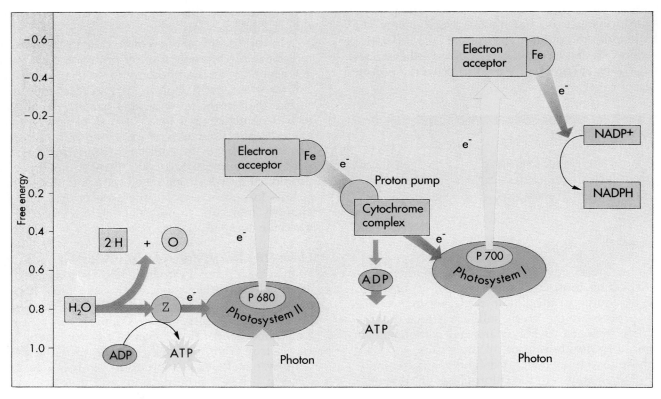

**Figure 8-8**

**The path of an electron in a chloroplast.** There are two photocenters that work one after the other. First, a photon of light ejects a high-energy electron from photosystem II that is used to pump a proton out across the membrane, contributing chemiosmotically to the production of a molecule of ATP. The ejected electron passes along a chain of cytochromes to photosystem I. When this photosystem in turn absorbs a photon of light, it ejects a high-energy electron to ferredoxin (Fe), which utilizes the electron's energy to drive the formation of the electron carrier NADPH.

Thus the energy produced in the first photoevent is spent in ATP synthesis; the energy of the second photoevent creates reducing power. These two processes together compose the light reactions of eukaryotic photocenters.

## The Formation of Oxygen Gas

By now you may wonder about the fate of $P_{680}$, the pigment of photosystem II that started the photosynthetic process by donating an electron. How does it make up for this loss if the electron is not returned but is instead expended to synthesize NADPH? As you might expect, an electron is obtained from another source. The loss of the excited electron from photosystem II converts $P_{680}$ into a powerful oxidant (electron-seeker), which obtains the required electron from a protein called Z. The removal of this electron from Z renders it a strong electron-acceptor in turn, and Z ob-

tains electrons from water to funnel to $P_{680}$. Z catalyzes a complex series of reactions in which water is split into electrons (passed to $P_{680}$), $H^+$ ions, and free $OH^-$ groups. The free $OH^-$ groups (chemists call them "free radicals") are collected and reassembled as water and oxygen gas. The hydrogen ions (protons) are exported across the membrane, thus augmenting the gradient in proton concentration that was established during the passage of electrons to photosystem I.

*The electrons and associated protons that oxygen-forming photosynthesis employs to form reduced organic molecules are obtained from water. The leftover oxygen atoms of the water molecules combine to form oxygen gas.*

The use of a two-stage photocenter containing both photosystems I and II thus solves in a simple way the evolutionary problem of how to obtain reducing power for biosynthetic reactions. Even though the cyclic photophosphorylation of green sulfur bacteria provides ATP by cyclic photophosphorylation, it does not provide a ready means of generating NADPH. Therefore organisms that use this form of photosynthesis must make NADPH in a roundabout way and, in doing so, expend a lot of ATP.

## Comparing Plant and Bacterial Light Reactions

It is useful to compare the two-stage $P_{680}/P_{700}$ photocenter with the $P_{700}$ photocenter from which it evolved. The removal of an electron from $P_{700}$ yields enough energy to extract hydrogen from $H_2S$ (78 kilocalories) but not from $H_2O$ (118 kilocalories). By contrast, the removal of an electron from $P_{680}$ yields considerably more energy, and that energy is adequate to split water molecules, producing gaseous oxygen as a by-product. In cyanobacteria, algae, and plants, all of which use this double photocenter, there is no cyclic flow of electrons. Instead, electrons and associated hydrogen atoms are extracted continually from water and are eventually used to reduce $NADP^+$ to NADPH. As a result of stripping hydrogens from water, oxygen gas is continuously generated as a product of the reaction. This photosynthetic process generates all of the oxygen in the air that we breathe (Figure 8-10).

*Every oxygen molecule in the air you breathe was once split from a water molecule by an organism carrying out oxygen-forming photosynthesis.*

## Accessory Pigments

As we saw earlier in this chapter, a simple chemical modification converts chlorophyll *a* into chlorophyll *b*. Chlorophyll *b* has an absorption spectrum shifted to-

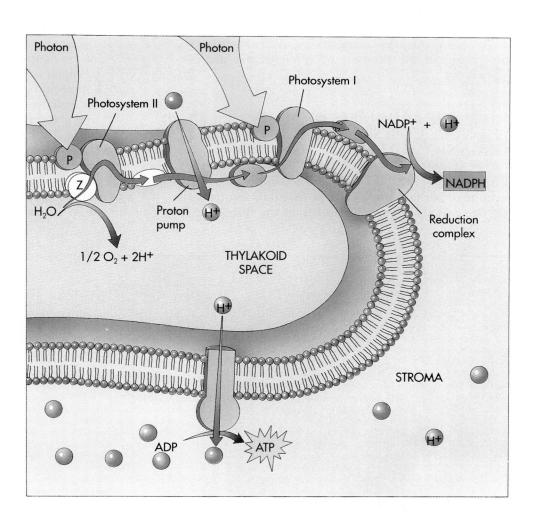

**Figure 8-9**

**The photosynthetic electron transport system.** When a photon of light strikes a pigment molecule *(P)* in photosystem II, it excites an electron. This electron is coupled to a proton stripped from water by a Z protein and passes along a chain of membrane-bound cytochrome electron carriers *(red arrow)* to a proton pump. There the energy supplied by the photon is used to transport a proton across the membrane into the thylakoid. The resulting proton gradient drives the chemosmotic synthesis of ATP. The spent electron then passes to photosystem I. When photosystem I absorbs a new photon of light, $P_{700}$ passes a second high-energy electron to a reduction complex, which drives NADPH generation.

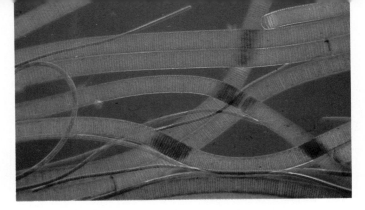

**Figure 8-10**           **A**

**All of the oxygen we breathe, indeed all the oxygen in the earth's atmosphere, has been generated by photosynthesis.** In the roughly 3 billion years between the advent of the cyanobacteria (**A**) and the cultivation of maize (**B**), enough oxygen has accumulated to account for 21% of the atomsphere.

          **B**

ward the green wavelengths. Like the carotenoids that we mentioned earlier in this chapter, chlorophyll *b*, which is located within the photocenter of plants, algae, and cyanobacteria, acts as an **accessory pigment,** one that is able to absorb photons that chlorophyll *a* cannot and subsequently transfer them to chlorophyll *a*. By doing this, the accessory pigments increase the percentage of the photons of sunlight that are harvested.

## HOW THE PRODUCTS OF PHOTOSYNTHESIS ARE USED TO BUILD ORGANIC MOLECULES FROM $CO_2$

The preceding section was concerned with the light reactions of photosynthesis. These reactions use light energy to produce metabolic energy in the form of ATP and to produce reducing power in the form of NADPH. But this is only half the story. Photosynthetic organisms employ the ATP and NADPH produced by the light reactions to build organic molecules from atmospheric carbon dioxide. This later phase of photosynthesis, which comprises the so-called dark reactions, is carried out by a series of enzymes.

Carbon fixation depends on the presence of a molecule to which $CO_2$ can be attached. The cell produces such a molecule by reassembling the bonds of two of the intermediates of glycolysis—fructose 6-phosphate (F6P) and glyceraldehyde 3-phosphate (G3P)—to form $CO_2$ and a 5-carbon sugar, **ribulose 1,5-bisphosphate (RuBP).**

The dark reactions of photosynthesis form a cycle of enzyme-catalyzed steps, just as the citric acid cycle is a circular biochemical pathway. In this cycle a carbon atom from atmospheric $CO_2$ is added to RuBP (Figure

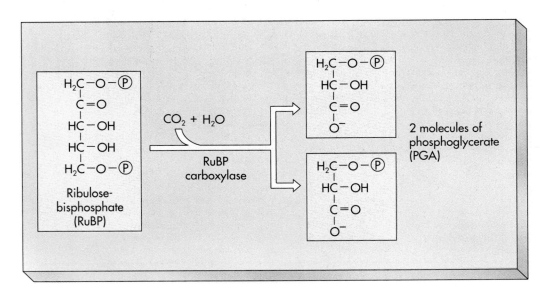

**Figure 8-11**

**The key step in carbon fixation.** Melvin Calvin and his coworkers at the University of California at Berkeley worked out the first step of what later became known as the Calvin cycle by exposing photosynthesizing algae to radioactive carbon dioxide ($^{14}CO_2$). Following the fate of the radioactive carbon atom (shown in color in these diagrams), they found that it is first bound to a molecule of ribulose bisphospate (RuBP), which splits immediately, forming two molecules of phosphoglycerate (PGA), one of which contains the radioactive carbon atom.

8-11), and the products run backwards through the glycolytic sequence to form F6P molecules, some of which are used to reconstitute RuBP. The remainder enters the cell's metabolism as newly fixed carbon in glucose. This cycle of reactions is called the **Calvin cycle,** after its discoverer Melvin Calvin of the University of California, Berkeley. The full Calvin cycle is diagrammed in Figure 8-12.

## THE CHLOROPLAST AS A PHOTOSYNTHETIC MACHINE

In eukaryotes, all photosynthesis takes place in chloroplasts (Figure 8-13). The internal membranes of chloroplasts are organized into flattened sacs called **thylakoids,** which are stacked on top of one another in arrangements called grana (Figure 8-14). The photo-

*Figure 8-12*

**The Calvin cycle.** For every three molecules of $CO_2$ that enter the cycle, one molecule of the 3-carbon compound glyceraldehyde phosphate is produced. Notice that the process requires energy, supplied as ATP and NADPH. This energy is generated by the light reactions, which generate ATP and NADPH.

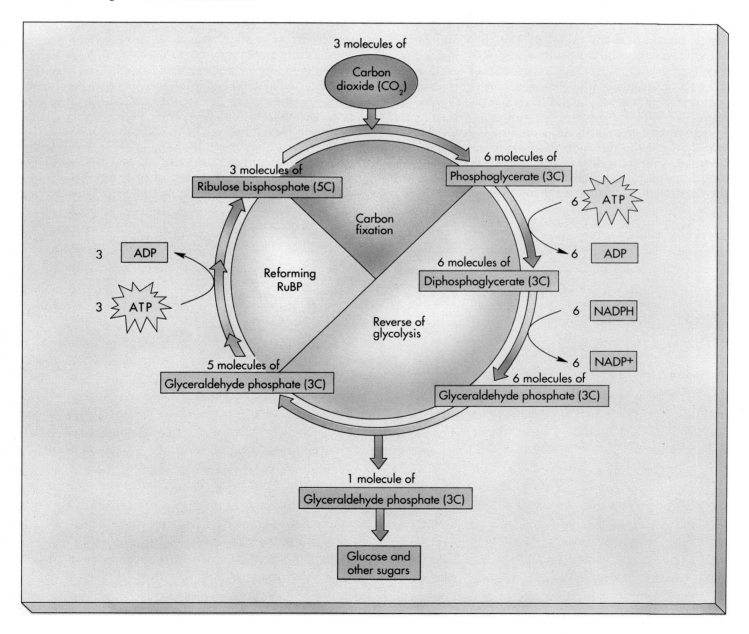

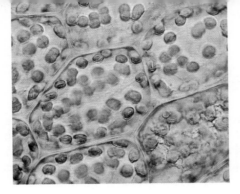

**Figure 8-13**

**Chloroplasts make plants green.** These moss cells are densely packed with bright-green chloroplasts.

synthetic pigments are bound to proteins embedded within the membranes of the thylakoids.

Each thylakoid is a closed compartment into which protons can be pumped. The thylakoid membrane is impermeable to protons and most molecules; therefore transit across it occurs almost exclusively through transmembrane channels. The diffusion of protons from the thylakoid interior takes place at distinctive ATP-synthesizing proton channels. This is similar to the process in mitochondria, except that in mitochondria the diffusion gradient is in the opposite direction—protons diffuse *into* rather than out of a mitochondrion. These channels protrude as knobs on the external surface of the thylakoid membrane, from

which ATP is released into the surrounding fluid. This fluid matrix inside the chloroplast, within which the thylakoids are embedded, is called the **stroma.** It contains the Calvin-cycle enzymes that catalyze the dark reactions of carbon fixation, using the ATP and NADPH that the photosynthetic activity of the thylakoids produces (Figure 8-15). Within the chloroplast the thylakoid membrane pumps protons from the stroma into the thylakoid compartment. As hydrogen ions pass back through the membrane by way of enzymes (ATPases), phosphorylation of ADP occurs on the stroma side of the membrane (Figure 8-16).

## PHOTOSYNTHESIS IS NOT PERFECT

One of the ironies of evolution is that processes evolve to be only as good as they need to be, rather than as good as they potentially might be. Evolution favors not optimum solutions but rather workable ones that can be derived from others that already exist. New reactions are often grafted onto old ones, as the citric acid cycle was grafted onto glycolysis. Photosynthesis is no exception. Several stages of the glycolytic pathway are used in the Calvin cycle, and one of the carry-over enzymes, RuBP carboxylase (the enzyme that catalyzes the key carbon-fixing reaction of photosynthesis), provides a decidedly nonoptimum solution. This enzyme

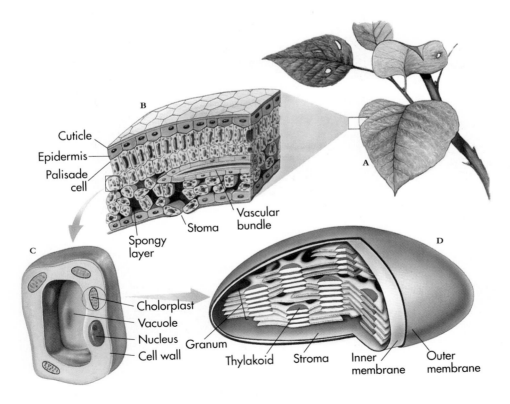

**Figure 8-14**

**Journey into a leaf.** The leaf shown (**A**) is from a plant that carries out the Calvin cycle. In cross-section (**B**), the leaf possesses a thick palisade layer whose cells (**C**) are rich in chloroplasts (**D**).

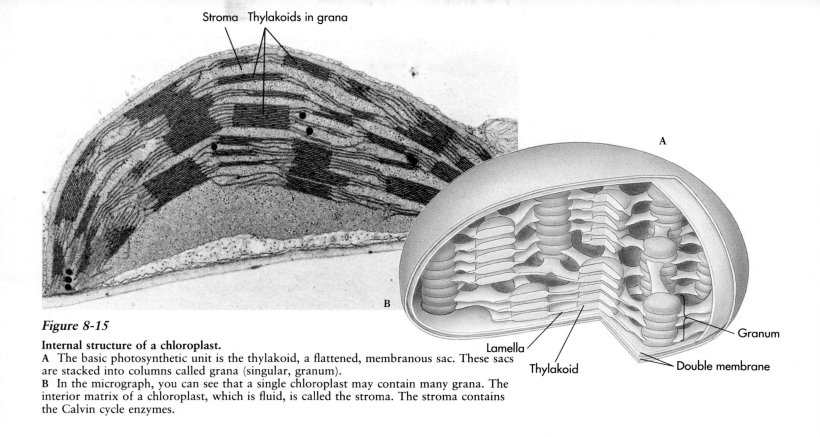

**Figure 8-15**

**Internal structure of a chloroplast.**
**A** The basic photosynthetic unit is the thylakoid, a flattened, membranous sac. These sacs are stacked into columns called grana (singular, granum).
**B** In the micrograph, you can see that a single chloroplast may contain many grana. The interior matrix of a chloroplast, which is fluid, is called the stroma. The stroma contains the Calvin cycle enzymes.

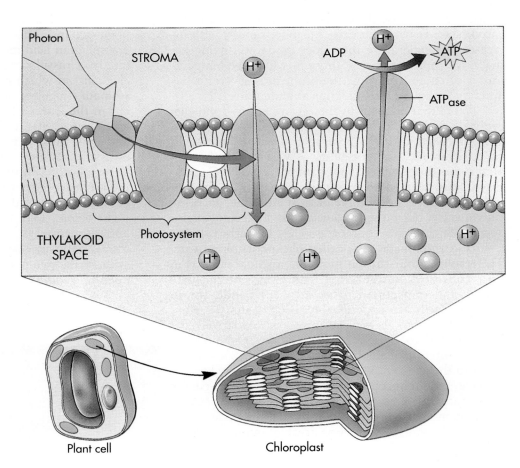

**Figure 8-16**

**Chemiosmosis in a chloroplast.**
Chloroplasts contain sacs called thylakoids. Proteins in the thylakoid membrane pump protons into the interior of the sac. ATP is produced on the outside surface of the membrane (stroma side) as protons pass back out of the sac through ATPase channels.

has a second activity that interferes with the successful performance of the Calvin cycle: RuBP carboxylase also initiates the *oxidation* of RuBP. In this process, called **photorespiration**, $CO_2$ is released without the production of ATP or NADPH. Because it produces neither ATP nor NADPH, photorespiration acts to undo the work of photosynthesis.

The undesirable oxidative reaction is initiated by RuBP carboxylase at the same active site that carries out the carbon-fixing carboxylation reaction so important to photosynthesis. When photosynthesis first evolved, there was little oxygen in the atmosphere and, because the decarboxylation reaction requires oxygen, there was little or no photorespiration. Under these conditions, the fact that the active site of RuBP carboxylase was capable of carrying out both reactions presented no problem. Only after millions of years of oxygen buildup in the atmosphere did the competition of $CO_2$ and $O_2$ for the same site lead to the problem that photorespiration now poses.

The loss of fixed carbon caused by photorespiration is not trivial. Plants that use the Calvin cycle to fix carbon are called **$C_3$ plants** and lose between a quarter and a half of their photosynthetically fixed carbon in this way. The extent of carbon loss depends largely on temperature because the oxidative activity of the RuBP carboxylase enzyme increases far more rapidly with increases in temperature than does its carbon-fixing activity. In tropical climates, those in which the temperature is often above 28° C, the problem is severe; it has a major limiting effect on tropical agriculture.

Plants that adapted to these warmer environments evolved two principal ways to deal with this problem. One approach, taken by a number of grasses, including corn, sugarcane, and sorghum, as well as members of about two dozen other plant groups, is called **$C_4$ photosynthesis**. In them, the first product of $CO_2$ fixation to be detected is not a three-carbon molecule, as in the Calvin cycle, but rather a four-carbon molecule, oxaloacetate. Such plants concentrate $CO_2$ by carboxylating a three-carbon metabolite called **phosphoenolpyruvate.** The resulting four-carbon molecule, oxaloacetate, is in turn converted to the citric acid cycle intermediate *malate* and transported to an adjacent **bundle-sheath cell.** Such cells are impermeable to $CO_2$ and therefore hold $CO_2$ within them. The malate is oxidatively decarboxylated to pyruvate, releasing a $CO_2$ molecule into the interior of the bundle-sheath cell. Pyruvate returns to the leaf cell, where an ATP is converted to adenosine monophosphate, or AMP (*two* high-energy bonds are split), in converting the pyruvate back to phosphoenolpyruvate, thus completing the cycle.

$C_4$ plants expend a considerable amount of ATP to concentrate $CO_2$ within the cells that carry out the Calvin cycle. Because $CO_2$ binds to the same place within the RuBP carboxylase active site that $O_2$ does, high concentrations of $CO_2$ act to commandeer the available enzyme for the $CO_2$-fixing reaction.

The path of carbon fixation in $C_4$ plants is diagrammed in Figure 8-17. The enzymes that carry out the Calvin cycle are located within the bundle-sheath cells, where the increased $CO_2$ concentration inhibits photorespiration. Because each $CO_2$ molecule is transported into the bundle-sheath cells at a cost of 2 high-energy ATP bonds, and because 6 carbons must be fixed to form a molecule of glucose, 12 additional molecules of ATP are required to form a molecule of glucose. The unique kind of leaf structure in which these processes occur is illustrated in Figure 8-18. In $C_4$ photosynthesis the energetic cost of forming glucose is almost doubled—from 18 to 30 molecules of ATP. However, in a hot climate in which photorespiration would otherwise remove more than half of the carbon fixed, it is the best compromise available. For this reason, $C_4$ plants are more abundant in warm regions than in cooler ones (Figure 8-19).

A second strategy to facilitate photosynthesis in hot regions has been adopted by many succulent (water-storing) plants such as cacti, pineapples (Figure 8-20), and some members of about two dozen other plant groups. This mode of carbon fixation is called crassulacean acid metabolism, or CAM, after the plant family Crassulaceae (the stonecrops, or hens-and-chickens) in which it was first discovered. In them, the **stomata** (singular, *stoma*), specialized openings that occur in the leaves of all plants and through which $CO_2$ enters and water vapor is lost, open during the night and close during the day (the reverse of how plants normally behave). Closing stomata during the day reduces photorespiration by preventing $CO_2$ from entering the leaves; the $CO_2$ necessary for producing sugars is instead provided from organic molecules made the night before. Like $C_4$ plants, these plants use both $C_4$ and $C_3$ pathways. They differ from $C_4$ plants in that the $C_4$ pathway operates at night and the time of operation of the two pathways differs, with the $C_3$ pathway operating *within the same cells* during the day. In $C_4$ plants, on the other hand, the two cycles take place in different, specialized cells.

*Photorespiration releases $CO_2$ without the production of ATP and so short-circuits photosynthesis. $C_4$ plants and CAM plants circumvent this waste by modificaitons of leaf architecture.*

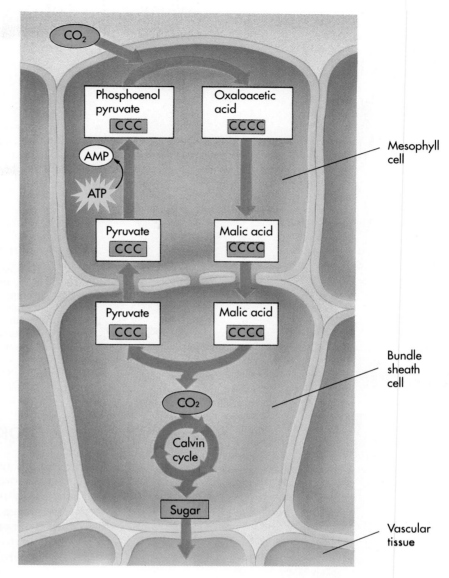

*Figure 8-17*

The path of carbon fixation in C₄ plants.

Mesophyll cell

Bundle sheath cell

Vascular tissue

*Figure 8-18*

**The leaf of a C₄ plant is very different from that of a C₃ leaf.** In a C₃ leaf, carbon dioxide is taken up by widely spaced cells (the spongy parenchyma), which pass it to the vascular bundles (veins). In a C₄ leaf, by contrast, $CO_2$ is taken up by mesophyll cells, which incorporate it into C₄ compounds. These C₄ compounds then pass inward to the bundle-sheath cells, which surround the veins, and release carbon dioxide there. Because these cells are relatively impermeable to carbon dioxide, this process can build up high carbon-dioxide concentrations in the bundle-sheath cells, which carry out the Calvin cycle.

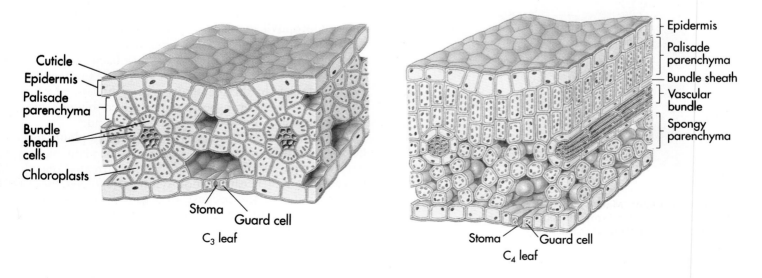

Cuticle
Epidermis
Palisade parenchyma
Bundle sheath cells
Chloroplasts
Stoma
Guard cell
C₃ leaf

Epidermis
Palisade parenchyma
Bundle sheath
Vascular bundle
Spongy parenchyma
Stoma
Guard cell
C₄ leaf

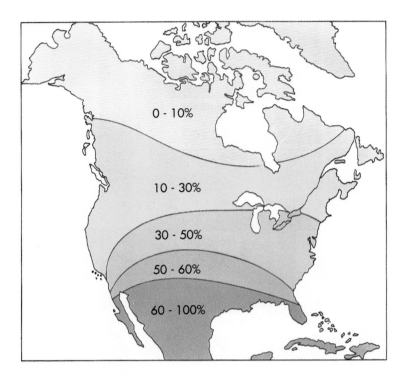

Figure 8-19

*Figure 8-19*

**The distribution of species of C₄ grasses in North America.** Many more $C_4$ species occur toward the south, where the average temperatures during the growing season are higher; at higher temperatures, photorespiration wastes more of the products of photosynthesis, and the ability of $C_4$ plants to counteract photorespiration is more of an advantage than it is in cooler regions, where $C_3$ grasses predominate.

*Figure 8-20*

**CAM photosynthesis.** Many CAM plants, including this pineapple, grow in warm tropical climates; far fewer grow in cooler northern temperate climates.

## A LOOK BACK

A cell's metabolism betrays its evolutionary past perhaps more than any other aspect of its life. This is particularly true of cells that perform photosynthesis. The two-stage photocenter of plants, algae, and cyanobacteria (Figure 8-21) has as its second stage a photosystem that evolved hundreds of millions of years earlier in anaerobic bacteria, which used hydrogen sulfide rather than water as a source of reducing hydrogen. The Calvin cycle uses part of the ancient glycolytic pathway, run in reverse, to produce glucose. The principal chlorophyll pigment of plants and algae evolved in cyanobacteria at least 2.8 billion years ago; they, in turn, are simple modifications of other bacterial chlorophylls that existed still earlier in anaerobic photosynthetic bacteria. In the metabolism of a modern plant, we can detect many aspects that evolved billions of years before any life whatever existed on land.

In Chapters 29 through 31, we examine plants in detail; photosynthesis is only one aspect of plant biology, although an important one. We have treated photosynthesis here because photosynthesis evolved long before plants did and because all organisms depend directly or indirectly on photosynthesis for the energy that powers their lives.

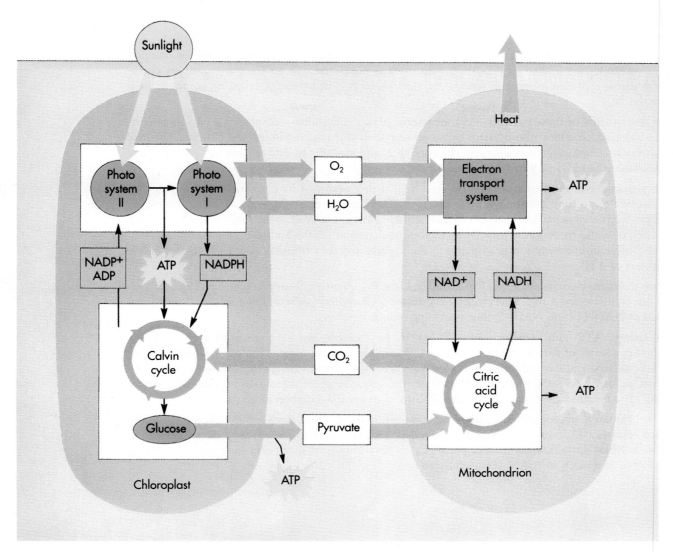

**Figure 8-21**

**The metabolic machine.** Within the chloroplast, sunlight drives the production of glucose, using up $CO_2$ and $H_2O$ while generating $O_2$. This $O_2$ is converted back to water in the mitochondrion by accepting electrons harvested from glucose molecules after the electrons have been used by the electron transport system to drive the synthesis of ATP. Water and oxygen gas thus cycle between chloroplasts and mitochondria within a plant cell, as do glucose and $CO_2$. Cells without chloroplasts, such as your own cells, require an outside source of glucose and oxygen and generate $CO_2$ and water.

■ **SUMMARY**

1. Light consists of energy packets called photons; the shorter the wavelength of light, the more energy in its photons.

2. When light strikes a pigment, photons are absorbed by boosting an electron to a higher energy level, thus causing an excited state. Most biological photopigments are carotenoids, such as the retinal of human eyes, or chlorophylls, such as those which make grass green.

3. Photosynthesis seems to have evolved in organisms similar to the green sulfur bacteria, which use a network of chlorophyll molecules (a photocenter) to channel photon excitation energy to one pigment molecule, referred to as $P_{700}$. $P_{700}$ then donates an electron to an electron transport chain, which drives a proton pump and returns the electron to $P_{700}$ in a process called cyclic photophosphorylation.

4. The descendants of these bacteria—cyanobacteria, algae, and plants—developed a two-stage photocenter in which a new photosystem called photosystem II was grafted onto the old one. The new photosystem employed a new pigment, chlorophyll *a*, and was able to generate enough energy to use $H_2O$ rather than $H_2S$ as a hydrogen source.

5. In organisms with the two-stage photocenter, light is first absorbed by photosystem II, which jolts an electron out of one chlorophyll molecule, $P_{680}$. This has two effects: (1) the absence of the high-energy electron causes photosystem II to seek another electron actively, which results in the eventual splitting of water to obtain it with $O_2$ as the by-product; and (2) the high-energy electron is passed to photosystem I, driving a proton pump in the process and thus bringing about the chemiosmotic synthesis of ATP.

6. When the spent electron arrives at $P_{700}$ from $P_{680}$, the $P_{700}$ pigment absorbs a photon of light, boosting one of its electrons out. This electron is directed to ferredoxin, where it is used to drive the synthesis of NADPH from $NADP^+$, thus providing reducing power.

7. The energy (ATP) and reducing power (NADPH) produced by the light reactions are used to fix carbon in a series of dark reactions called the Calvin cycle. In this process, ribulose 1,5-bisphosphate is carboxylated and the products run backwards through a series of reactions also found in the glycolytic sequence to form fructose 6-phosphate molecules, some of which are used to reconstitute RuBP. The remainder enters the cell's metabolism as newly fixed carbon in glucose.

8. In hot climates, photosynthetic carbon fixation tends to be short-circuited by photorespiration, which releases $CO_2$ instead of fixing it. Plants deal with this problem by concentrating $CO_2$ within the cells where carbon fixation takes place ($C_4$ photosynthesis) or by closing their stomates during the day to keep $CO_2$ in (CAM photosynthesis).

## REVIEW

1. Light energy is "captured" and used to boost an electron to a higher energy level in molecules called _____.

2. What is the point of photosynthesis? Is it to generate ATP or to fix carbon?

3. _____ is used as an electron carrier in the anabolic process of photosynthesis instead of _____, which is used in catabolic oxidative respiration.

4. All of the oxygen that you breathe has been produced by the splitting of water during _____.

5. The basic photosynthetic unit of a chloroplast is the _____, a flattened membranous sac that occurs in stacked columns called grana.

# SELF-QUIZ

1. Carbon fixation requires the expenditure of ATP molecules. How is this ATP generated?
   (a) By chemiosmotic synthesis during the light reactions.
   (b) By replenishment of the photosynthetic pigment.
   (c) By formation of glucose during the dark reactions.
   (d) By breaking the covalent bonds in carbon dioxide.
   (e) None of the above.

2. When $P_{700}$ receives photon-induced excitation energy, the excitation energy is channeled to membrane-bound ferredoxin. What is the form of this excitation energy?
   (a) Light energy
   (b) Protons
   (c) Neutrons
   (d) Electrons
   (e) Sex

3. Plants, algae, and cyanobacteria carry out oxygen-producing photosynthesis, employing both photosystem I and photosystem II. This means that _____ and _____ are generated, both of which are required to form organic molecules from atmospheric carbon dioxide.
   (a) water
   (b) ATP
   (c) ADP
   (d) NADH
   (e) NADPH

4. The rate of photorespiration in most plants increases at higher temperatures. Some plants have evolved a somewhat round-about system to deal with this problem. This series of reactions is called
   (a) the $C_3$ pathway.
   (b) the $C_4$ pathway.
   (c) the Calvin cycle.
   (d) photosystem II.
   (e) cyclic photophosphorylation.

5. A principal difference between a carotenoid and a chlorophyll is that
   (a) chlorophylls use a metal atom to snare electrons, whereas carotenoids do not.
   (b) chlorophylls absorb light over a much broader range than do carotenoids.
   (c) chlorophylls absorb light at much lower efficiency.
   (d) carotenoids contain carbon ring molecules with alternating single and double carbon bonds, whereas chlorophylls do not.
   (e) all of the above.

# THOUGHT QUESTIONS

1. What is the advantage of having many pigment molecules in each photocenter for every $P_{700}$? Why not couple *every* pigment molecule directly to an electron acceptor?

2. Why are plants that consume 30 ATP molecules to produce one molecule of glucose (rather than the usual 18 molecules of ATP per glucose molecule) favored in hot climates but not in cold climates? What role does temperature play?

# FOR FURTHER READING

BJORKMAN, O., and J. BERRY: "High Efficiency Photosynthesis," *Scientific American,* October 1973, pages 80-93. A description of $C_4$ photosynthesis and other strategies carried out by plants that live in Death Valley, where the problem posed by photorespiration is acute.

GOVINDJEE, and R. GOVINDJEE: "The Absorption of Light in Photosynthesis," *Scientific American,* December 1974, pages 69-87. A good account of how a photocenter works.

TING, I.P.: "Photosynthesis of Arid and Subtropical Succulent Plants," *Aliso,* vol. 12, 1989, pages 387-406. An excellent review of the occurrence and significance of CAM metabolism.

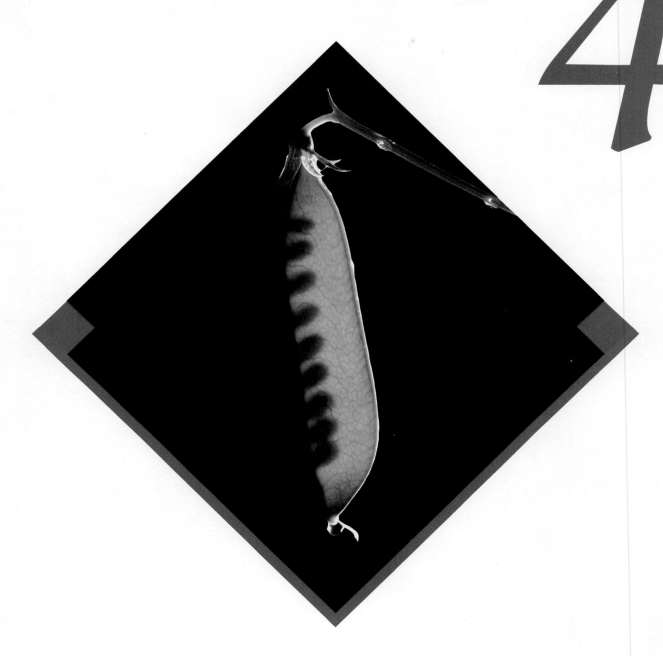

P A R T

4

G E N E T I C S

# $\mathcal{H}$ow Cells Reproduce

Just hatching from their eggs, peering out at the world they are about to enter, these baby hognose snakes are the product of many cell divisions since they were first initiated by the fusion of egg and sperm. Many more cell divisions lie ahead as they grow into 4-foot adult snakes.

# HOW CELLS REPRODUCE

## Overview

Your body consists of some hundred trillion cells, all derived from a single cell at the start of your life as a fertilized egg. Many millions of successful cell divisions, each resulting in genetically equivalent cells, occurred while your body was reaching its present form. In mitosis, the process by which this occurs, growing eukaryotic cells attach microtubules to each replicated chromosome and pull sister chromatids to opposite poles of the cell. Through meiosis, a second process of cell division, eukaryotic organisms produce their gametes, eggs and sperm. Meiosis includes two stages. In the first, the homologous chromosomes derived from the two parents pair longitudinally, exchange genetic material, and are pulled to the poles; in the second, these reorganized chromosomes split longitudinally into equal halves and move to new poles. As a result of each meiotic event, four cells—sperm, eggs, or undifferentiated gametes—are produced, each with half the chromosome number of the cell from which they were derived.

## For Review    *Here are some important terms and concepts that you will encounter in this chapter. If you are not familiar with them, you should review them before proceeding.*

**Bacterial cell division** (Chapter 4)

**Chromosomes** (Chapter 4)

**Nuclear envelope** (Chapter 4)

**Microtubules** (Chapter 4)

**Centriole** (Chapter 4)

**Figure 9-1**

**All reproduction of organisms depends on the reproduction of cells.** Like you, this turtle started life as a fertilized egg.

All living organisms grow and reproduce. Bacteria too small to see, alligators, the weeds growing in your lawn—from the smallest of creatures to the largest, all organisms produce offspring like themselves and pass on to them the hereditary information that makes them what they are (Figure 9-1). All reproduction of organisms depends on the reproduction of cells. In this chapter, we begin our consideration of heredity with an examination of how cells reproduce themselves. The ways in which cell reproduction is achieved, and their biological consequences, have changed significantly during the evolution of life on earth.

## CELL DIVISION IN BACTERIA

Among the bacteria, the process of cell division is simple. The genetic information, or **genome**, exists in bacteria as a single, circular, deoxyribonucleic acid (DNA) molecule, attached at one point to the interior surface of the cell membrane. This genome is replicated early in the life of the cell. At a special site on the chromo-

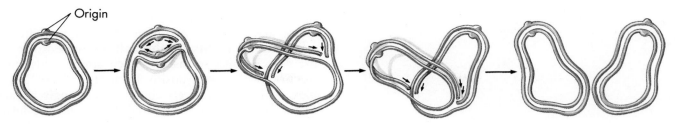

Origin

**Figure 9-2**

**How bacterial DNA replicates.** The circular DNA molecule that constitutes the genome of a bacterium initiates replication at a single site, moving out in both directions. When the two moving replication points meet on the far side of the molecule, its replication has been completed.

some called the **replication origin,** a battery of more than 20 different enzymes goes to work and starts to make a complete copy of the DNA molecule (Figure 9-2). When these enzymes have proceeded all the way around the circle of the DNA molecule, the cell then possesses two copies of the genome, attached side-by-side to the interior cell membrane.

The growth of a bacterial cell to an appropriate size induces cell division. First, new plasma membrane and cell wall materials are laid down in the zone between the attachment sites of the two "daughter" DNA genomes. This begins the process of binary fission. **Binary fission** is the division of a cell into two equal or nearly equal halves. As new material is added in the zone between the attachment sites, the growing plasma membrane pushes inward (invaginates), and the cell is progressively constricted (Figure 9-3). Initiating the constriction at a position between the membrane attachment sites of the two daughter DNA genomes provides a simple and effective mechanism for ensuring that each of the two new cells will contain one of the two identical genomes. Eventually the invaginating circle of membrane reaches all the way into the cell center, pinching the cell in two. A new cell wall forms around the new membrane, and what was originally one cell is now two.

*Bacteria divide by binary fission, a process in which a cell is pinched in two. The point where the constriction begins is located between the places where the two replicas of the chromosome are bound to the cell membrane, ensuring that one copy will end up in each daughter cell.*

# CELL DIVISION AMONG EUKARYOTES

The evolution of the eukaryotes introduced several additional factors into the process of cell division. Eukaryotic cells are much larger than bacteria, and they contain genomes with much larger quantities of DNA (Figure 9-4). This DNA is located in several individual chromosomes rather than in a single, circular molecule. In eukaryotic chromosomes, the DNA is associated with proteins and wound into tightly condensed coils (Figure 9-5). The eukaryotic chromosome is much more complex than the single, circular DNA molecule that plays the role of a chromosome in bacteria.

## The Structure of Eukaryotic Chromosomes

In the century since their discovery, we have learned a great deal about chromosomes, their structure, and

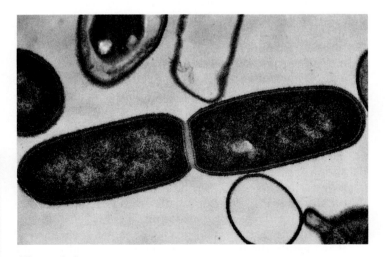

**Figure 9-3**

**Fission.** Bacteria divide by a process of simple cell fission. Note here the septum between the two daughter cells.

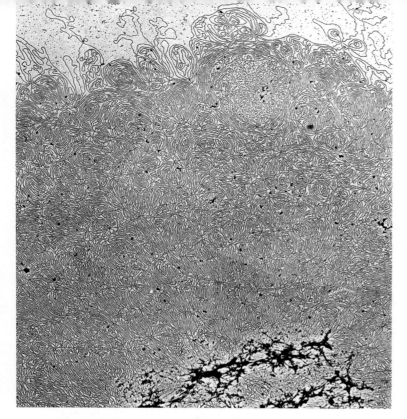

**Figure 9-4**

**A human chromosome contains an enormous amount of DNA.** The dark element at the bottom of the photograph is the protein matrix of a single chromosome. All of the surrounding material is the DNA of that chromosome.

how they function. Eukaryotic chromosomes are composed of a complex of DNA and protein. This complex is called **chromatin.** Most eukaryotic chromosomes are about 60% protein and 40% DNA. A significant amount of RNA is also associated with chromosomes because they are the sites of RNA synthesis. The DNA of a chromosome exists as one very long, double-stranded fiber, a **duplex,** which extends unbroken through the entire length of the chromosome. A typical human chromosome contains more than 300 million (3 $\times$ 10$^8$) nucleotides in its DNA fiber. If nucleotides were words, the amount of information one chromosome contains would fill about 600 books of 1000 pages each, assuming that each page had about 500 words on it. If the strand of DNA from a single eukaryotic chromosome were laid out in a straight line, it would be about 5 centimeters (2 inches) long. This is much too long to fit into a cell. So the DNA is coiled in the cell, thus fitting into a much smaller space than would be possible if it were not coiled.

How is the coiling of this long DNA fiber achieved? If we gently disrupt a eukaryotic nucleus and examine the DNA with an electron microscope, we find that it resembles a string of beads. Every 200 nucleotides, the DNA duplex is coiled around a complex of **histones,**

which are small, very basic polypeptides, rich in the amino acids arginine and lysine. Eight of these histones form the core of an assembly called a **nucleosome** (Figure 9-6). Because so many of their amino acids are basic, histones are very positively charged. The DNA duplex, which is negatively charged, is strongly attracted to the histones and wraps tightly around the histone core of each nucleosome. The core thus acts as a "form" that promotes and guides the coiling of the DNA. Further coiling of the DNA occurs when the string of nucleosomes wraps up into higher-order coils called **supercoils** (see Figure 9-5).

Condensed portions of the chromatin are called **heterochromatin.** Some remain condensed permanently, so their genes can never be used. The remainder of the chromosome, called **euchromatin,** is not condensed except during cell division, when the movement of the chromosomes is made easier by the compact packaging that occurs at that stage. At all other times, the euchromatin is present in an open configuration, and its genes are active.

Chromosomes may differ widely from one another in appearance. They vary in features such as the location of the **centromere,** the constricted region where the spindle fibers attach during cell division; the relative length of the two arms (regions on either side of the centromere); size; staining properties; and the position of constricted regions along the arms. The particular array of chromosomes that an individual possesses, called its **karyotype** (Figure 9-7), may differ greatly among species or sometimes even among individuals in a species.

To examine human chromosomes, investigators collect a blood sample, add chemicals that induce the cells in the blood sample to divide, and then add other chemicals that stop cell division at the stage when the chromosomes are most condensed. After arresting the process of cell division at this stage, the investigators break the cells to spread out their contents, including the chromosomes, and then stain and examine the chromosomes. To make the karyotype easier to examine, the chromosomes are usually photographed. Finally, the chromosomes are cut out of the photograph, like paper dolls, and arranged in order of their size (see Figure 9-7).

The karyotypes of individuals are often examined to detect genetic abnormalities, a number of which arise from extra or lost chromosomes. The human birth defect known as Down syndrome is associated with the presence of an extra copy of a particular chromosome, which can be recognized easily in photographs of the set of chromosomes. In this way, the presence of the defect can be detected by examining the karyotypes of fetal cells taken before birth.

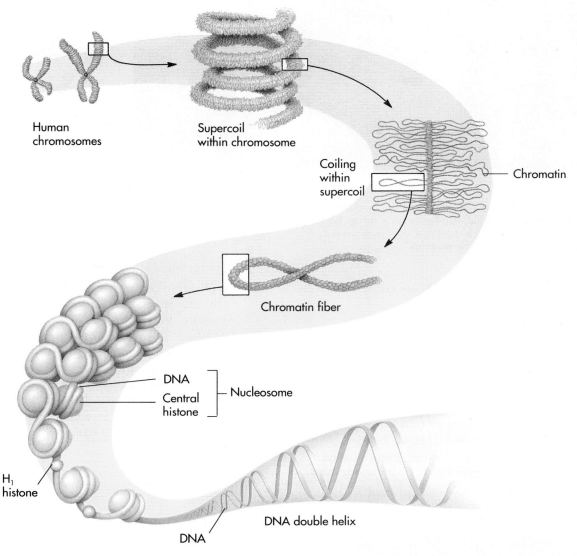

Human chromosomes

Supercoil within chromosome

Coiling within supercoil

Chromatin

Chromatin fiber

DNA

Central histone

Nucleosome

H₁ histone

DNA

DNA double helix

*Figure 9-5*

**Levels of chromosomal organization.**

## How Many Chromosomes are in a Cell?

Each of the cells in your body, except your gametes (sperm or eggs), contains 46 chromosomes, consisting of two nearly identical copies of each of the basic set of 23 chromosomes. This basic set of 23 chromosomes is present in all of your gametes, in eggs or sperm. The two nearly identical copies of each of the 23 different chromosomes are called homologous chromosomes, or homologues (from the Greek word *homologia,* agreement). Before cell division, each of the two homologues replicates, producing in each case two identical copies called **sister chromatids,** which remain joined together at the centromere. Thus at the beginning of cell division a body cell contains a total of 46 replicated chro-

mosomes (Figure 9-8), each composed of two sister chromatids joined by one centromere. By convention, each pair of sister chromatids is counted as a single chromosome as long as they remain joined.

## The Cell Cycle

The profound change in genome organization that occurred during the evolutionary transition from bacteria (a single circle of naked DNA) to eukaryotes (several segments of DNA packaged with protein) required radical changes in the way cells divide to partition two replicas of the genome accurately, one replica into each of the two daughter cells. The processes that occur dur-

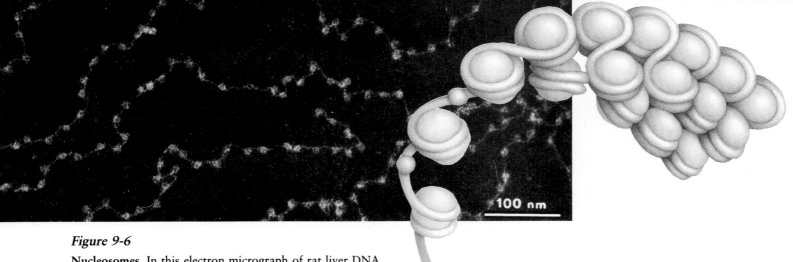

**Figure 9-6**

**Nucleosomes.** In this electron micrograph of rat liver DNA, rows of nucleosomes are visible, like beads on a string. Each nucleosome is a region in which the DNA duplex is wound tightly around a cluster of histone proteins.

ing the division of eukaryotic cells are conveniently diagrammed as a cell cycle (Figure 9-9):

$$G_1 \rightarrow S \rightarrow G_2 \rightarrow M \rightarrow C$$

$G_1$ *phase* The growth phase of the cell. For many organisms, this phase occupies the major portion of the cell's life span.

*S phase* The phase in which a replica of the genome is synthesized. The DNA duplicates, and the individual chromosomes now consist of two identical sister chromatids (Figure 9-10).

$G_2$ *phase* The stage in which preparations are made for genomic separation. This includes the replication of mitochrondia and other organelles, chromosome condensation, and the synthesis of microtubules.

*M phase* The phase in which the microtubular apparatus is assembled, binds to the chromosomes, and moves them apart. This phase, called **mitosis,** is the essential step in the separation of the two daughter genomes.

*C phase* The phase in which the cell itself divides, creating two daughter cells. This phase is called **cytokinesis.**

## MITOSIS

Mitosis and cytokinesis are the active phases in cell division; together, they often represent only a short portion of the cell cycle (see Figure 9-9). Both phases are of critical importance, and we discuss each of them in turn.

Mitosis (the M phase of cell division) has long fascinated biologists because of its central role in bringing about the proper division of the genetic material. During mitosis, the chromosomes undergo intricate movements as they separate. Our description of mitosis focuses on the process as it occurs in animals and plants; its details vary in other groups of organisms.

Mitosis is subdivided into four stages (Figure 9-11): prophase, metaphase, anaphase, and telophase. Such a subdivision is convenient, but the process is actually continuous, with the stages flowing smoothly one into another.

### Preparing the Scene for Mitosis: Interphase
Before the initiation of mitosis, events have occurred in the preceding **interphase** (that is, the $G_1$, S, and $G_2$

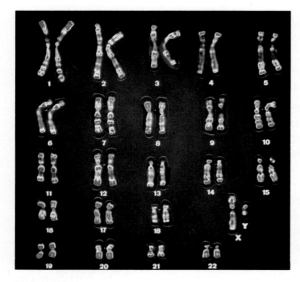

**Figure 9-7**

**A human karyotype.**

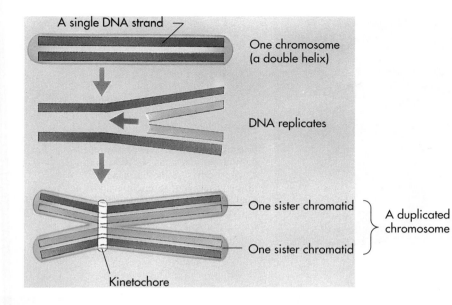

One chromosome (a double helix)

A single DNA strand

DNA replicates

One sister chromatid
One sister chromatid
A duplicated chromosome

Kinetochore

**Figure 9-8**

**A duplicated chromosome contains two identical "sister" chromatids.**

phases) that are essential for the successful completion of the mitotic process:

1. During the S phase, each chromosome replicates to produce two sister chromatids with identical DNA that remain attached to each other at the

centromere. The centromere includes a specific DNA sequence of about 220 nucleotides, to which is bound a disk of protein called the **kinetochore**. The centromere occurs at a specific site on any given chromosome, which is why, as we have seen, it is a useful characteristic in rec-

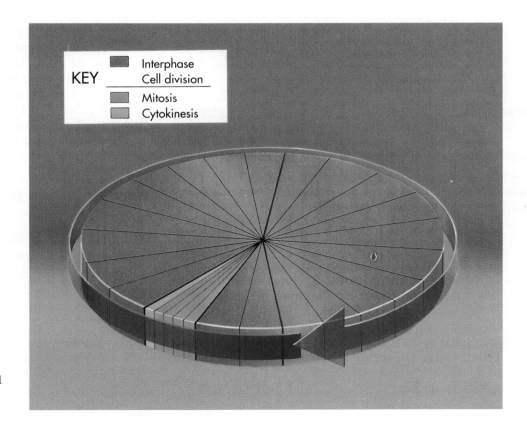

KEY

Interphase
Cell division
Mitosis
Cytokinesis

**Figure 9-9**

**The cell cycle.** Each wedge represents 1 hour of the 22-hour division cycle of human liver cells growing in culture.

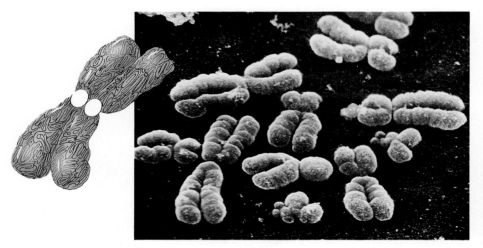

**Figure 9-10**

**Sister chromatids.** The photo and companion drawing show human chromosomes as they appear immediately before nuclear division. Each DNA strand has already been replicated, forming sister chromatids that are identical to each other, held together by the centromere.

ognizing the individual homologous chromosomes in a karyotype.

2. At the completion of the S phase, the replicated chromosomes remain fully extended and uncoiled; they are not visible under a light microscope. In the $G_2$ phase, the chromosomes begin the long process of **condensation,** coiling into more and more tightly compacted bodies.

3. During the $G_2$ phase, the cells begin to assemble the machinery that they will later use to move the chromosomes to opposite poles of the cell. In the cells of animals and most protists, the centrioles replicate. All cells undertake an extensive synthesis of tubulin, the protein of which microtubules are formed.

*Interphase is that portion of the cell cycle in which the chromosomes usually are not visible under a light microscope. It includes the $G_1$, S, and $G_2$ phases. In the $G_2$ phase, the cell mobilizes its resources for cell division.*

### Formation of the Mitotic Apparatus: Prophase

When the chromosome condensation begun in the $G_2$ phase reaches the point at which individual condensed chromosomes first become visible with a light microscope, the first stage of mitosis, **prophase,** has begun. This process of condensation continues throughout prophase, the individual chromosomes becoming progressively thicker.

*Prophase is the stage of mitosis at which the condensing chromosomes first become visible.*

As the chromosomes condense, the microtubular apparatus is assembled and the nuclear envelope breaks down. If centrioles are present, the pairs separate and move to the opposite poles of the cell, forming an axis of microtubules referred to as the **spindle fibers.** When the centrioles reach the poles of the cell, they radiate an array of microtubules called the **aster.** When centrioles are absent, as they are in plants, fungi, and some protists, the spindle fibers form anyway. The position of the spindle determines the plane in which the cell will divide, a plane that passes through the center of the cell at right angles to the spindle.

*During prophase, the nuclear envelope is reabsorbed, and a network of microtubules called the spindle forms between opposite poles of the cell.*

As prophase continues, a second group of microtubules grows out from the centromeres of the individual chromosomes to the poles of the spindles. Two of the secondary microtubules extend in opposite direc-

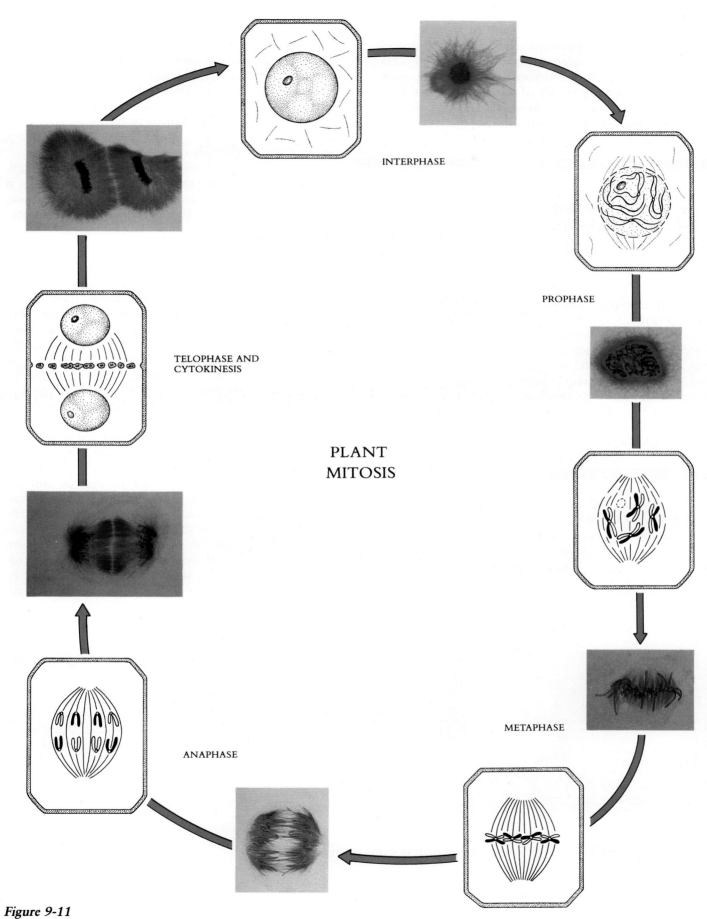

INTERPHASE

PROPHASE

TELOPHASE AND
CYTOKINESIS

PLANT
MITOSIS

ANAPHASE

METAPHASE

**Figure 9-11**

**Mitosis.** Chromosomes of the African blood lily, *Haemanthus katharinae,* are stained blue,
and microtubules are stained red.

tions from each chromosome, connecting the kinetochore of each of the sister chromatids to the two poles of the spindle. Because microtubules extending from each pole attach to their side of the centromere, the effect is to attach one sister chromatid to one pole and the other sister chromatid to the other pole.

*At the end of prophase, the centromere joining each pair of sister chromatids is attached by microtubules to the poles of the spindle.*

## The Alignment of the Chromosomes: Metaphase

The second phase of mitosis, **metaphase,** begins when the chromosomes, each consisting of a pair of chromatids, align in the center of the cell equidistant from the two poles. Viewed with a light microscope, the chromosomes appear to be lined up along the inner circumference of the cell in a circle perpendicular to the axis of the spindle (Figure 9-12). An imaginary plane passing through this circle is called the **metaphase plate.**

*Metaphase is the stage of mitosis characterized by the alignment of the chromosomes on a plane in the center of the cell. Each chromosome is drawn to that position by the microtubules extending from it to the two poles of the spindle.*

Each sister chromatid has one kinetochore, to which one or more microtubules are attached. At the end of metaphase, each centromere splits in two, freeing the two sister chromatids to be drawn to the opposite poles of the spindle in the next phase by the microtubules attached to their kinetochore.

### Separation of the Chromatids: Anaphase

**Anaphase,** the stage of mitosis during which the daughter chromosomes move rapidly to the opposite poles, is the shortest stage of mitosis. Two forms of movement take place simultaneously, each driven by microtubules:

1. *The poles move apart.* The microtubular spindle fibers slide past one another. Because the two members of each microtubule pair are physically anchored to opposite poles, their sliding past one another pushes the poles apart. Because the chromosomes are attached to these poles, they also move apart. In this process, the cell becomes visibly elongated if it lacks a rigid cell wall. Microtubular sliding is powered by adenosine triphosphate (ATP)-driven changes in the shape of proteins that bridge across pairs of microtubules.

2. *The centromeres move toward the poles.* The microtubules attached to the centromeres shorten. This shortening process is not a contraction because the microtubules do not get any thicker. Instead, tubulin subunits are continuously removed from the polar ends of the microtubules by the organizing center. As more and more subunits are removed, the progressive disassembly of the chromosome-bearing microtubule renders it shorter and shorter, pulling the chromosome ever closer to the poles of the cell.

*Figure 9-12*

**The metaphase plate.** In metaphase the chromosomes array themselves in a circle around the spindle midpoint.

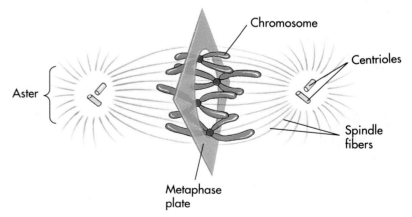

## Re-formation of the Nuclei: Telophase

With the movement of identical daughter chromosomes to the opposite poles in anaphase, the only tasks that remain in mitosis are to dismantle the stage and remove the props. The stage during which this occurs is called **telophase.** The spindle apparatus is disassembled, with the microtubules broken back down into tubulin monomers ready for use in constructing the cytoskeleton of the new cell. A nuclear envelope forms around each set of chromosomes while they begin to uncoil into the more extended form that permits genes to be used.

## CYTOKINESIS

At the end of telophase, mitosis is complete. The cell has divided its replicated genome into two nuclei, which are positioned at opposite ends of the cell. While this process has been going on, the cytoplasmic organelles, such as the mitochondria and, if present, the chloroplasts, have also been reassorted to the areas that will separate and become the daughter cells. Their replication occurs before cytokinesis, often in the S or $G_2$ stage.

But the process of cell division is still not complete. The division of the cell itself has not yet even begun. The stage of the cell cycle at which cytoplasmic division actually occurs is called **cytokinesis.** Cytokinesis generally involves the cleavage of the cell into roughly equal halves.

In animal cells, which lack cell walls, cytokinesis is achieved by pinching the cell in two with a contracting belt of microfilaments. As the contraction proceeds, a **cleavage furrow** becomes evident around its circumference (Figure 9-13)—an area where the cytoplasm is being progressively pinched inward by the decreasing diameter of the microfilament belt. As contraction proceeds, the cleavage furrow deepens until it eventually extends all the way into the residual spindle, and the cell is literally pinched into two.

Plant cells possess rigid cell walls that are far too strong to be deformed by microfilament contraction. A different approach to cell division has therefore evolved in plants. Plant cells assemble membrane components in their interior, at right angles to the spindle (Figure 9-14). This expanding partition is called a **cell plate.** It continues to grow outward until it reaches the interior surface of the cell membrane and fuses with it, at which point it has effectively divided the cell into two. Cellulose is then laid down on the new membranes, creating two new cells. The space between the two new cells becomes impregnated with pectins and is called a **middle lamella.**

## MEIOSIS

Most animals and plants reproduce sexually. In sexual reproduction, gametes of opposite sexes unite in the process of fertilization. The fusion of gametes forms a new cell, called a **zygote.** Because this fusion takes

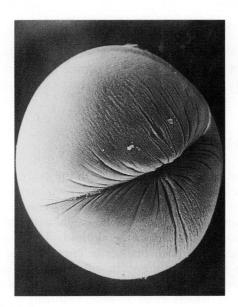

***Figure 9-13***

**Cytokinesis in animal cells.** A cleavage furrow is forming around this dividing sea urchin egg.

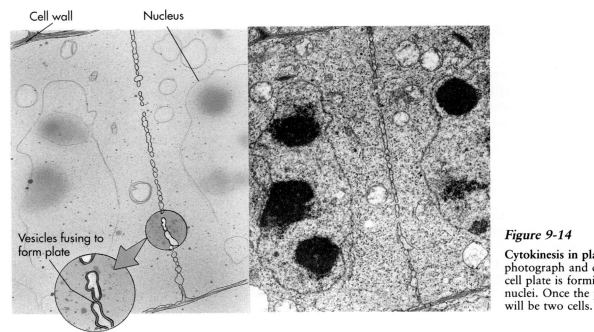

Cell wall    Nucleus

Vesicles fusing to form plate

**Figure 9-14**

**Cytokinesis in plant cells.** In this photograph and companion drawing, a cell plate is forming between daughter nuclei. Once the plate is complete there will be two cells.

place in every generation, it is clear that it must be balanced by some mechanism during the formation of gametes that reduces the number of chromosomes to half the number normally present in body cells. Otherwise the number of chromosomes would soon become impossibly large. The mechanism that reduces the number of chromosomes in gametes is a special form of cell division called **meiosis.**

---

*Meiosis is a process of nuclear division in which the number of chromosomes in cells is halved during gamete formation.*

---

## SEXUAL REPRODUCTION

The set of chromosomes present in the body or the vegetative cells of a plant or animal, as we have seen, contains two replicates of each individual chromosome. A cell, tissue, or individual with such a double set of chromosomes is said to be **diploid.** The gametes of these organisms, with one set of chromosomes each—half the number present in the body or vegetative cells—are said to be **haploid.** *Fertilization* is the process during which gametes fuse, restoring the diploid chromosome number, and *meiosis,* which alternates with fertilization in the cycle of reproduction, is the process by which the gametes, cells with the haploid

number of chromosomes, are produced. When the gametes are differentiated into smaller, motile ones and larger, nonmobile ones, as they are in humans and other animals, the former are called sperm and the latter are called eggs. Both of these types of gametes, and all other gametes, are haploid. In humans, the diploid body cells have 46 chromosomes—two sets of 23—and the haploid gametes have 23 chromosomes. When these gametes fuse, the diploid number of chromosomes, 46, is restored (Figure 9-15).

Reproduction that involves the alternation of fertilization and meiosis is called **sexual reproduction.** Its outstanding characteristic is that an individual offspring inherits genes from two parent individuals (Figure 9-16). You, for example, inherited genes from both your mother and your father, your mother's genes being contributed by the egg fertilized at your conception, and your father's by the sperm that fertilized that egg (Figure 9-17).

### Why Sex?

Sexual reproduction is not the only way that reproduction can occur. Consider sponges or dandelions. In different ways, they can reproduce simply by fragmenting their bodies. In such a process of **asexual reproduction,** a small portion of the organism divides from it and gives rise to a new individual. There is no alternation of haploid and diploid cells, no meiosis, and no gametes. Because these cell divisions are mitotic, every cell of the new individual has the same genetic makeup as the individual from which it arose. Asexual reproduction is widespread among different groups of or-

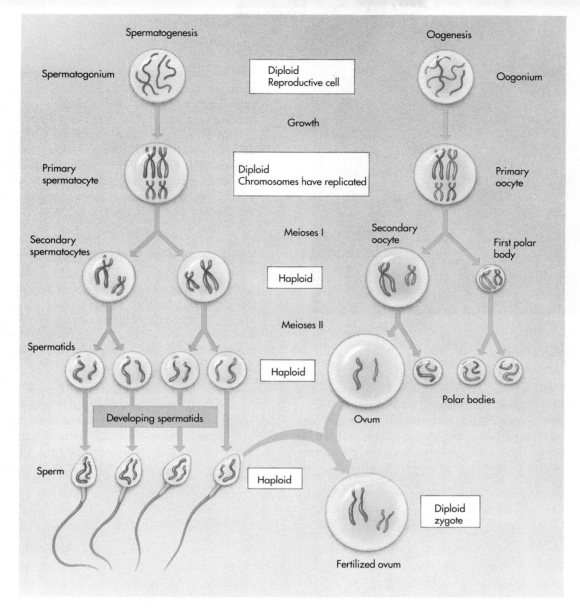

**Figure 9-15**

**The birth of a diploid zygote.** In diploid organisms, special lines of cells undergo meiosis to produce sperm in males and ova in females. The union of a haploid sperm with a haploid ovum produces the zygote, which is diploid. All of the billions of cells in your body are derived from a single such zygote cell.

ganisms. Its outstanding characteristic is that an individual offspring is *genetically identical* to its parent.

Even when meiosis and the production of gametes occur, reproduction may still occur without sex. The development of an adult from an unfertilized egg is called **parthenogenesis**—a common form of reproduction in insects. Among bees, the development of eggs into adults does not require fertilization. Fertilized eggs develop into diploid females, unfertilized eggs into haploid males. Parthenogenesis occurs even among the vertebrates. Some fishes, amphibians, and lizards are capable of reproducing in this way, their unfertilized eggs undergoing a mitotic division without cell cleavage to

produce a diploid cell, which then develops into a diploid adult.

## THE SEXUAL LIFE CYCLE

The life cycles of all sexually reproducing organisms follow the same basic pattern of alternating between the diploid and the haploid chromosome number (Figure 9-18). The first cell of a diploid individual is called the zygote. Dividing by mitosis, a human zygote eventually gives rise to an adult body with some 100 trillion

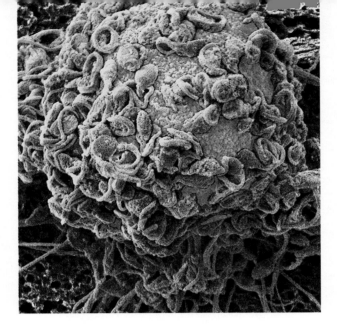

**Figure 9-16**

Sperm fertilizing an egg.

cells! Every one of your 100 trillion cells is genetically identical to the zygote from which you arose.

In animals the diploid cells that will eventually undergo meiosis to produce the gametes are known as **germ line cells.** These cells become obviously different from the **somatic cells**—all the other diploid body cells—early in the course of development. It is within germ line cells located in the sex organs of your body that meiosis occurs and gametes are produced (this process is described in detail in Chapter 42).

In plants the haploid cells that result from meiosis divide by mitosis, forming multicellular haploid individuals (see Figure 9-18). Certain cells of this haploid phase eventually differentiate into eggs or sperm, which, when they unite, form a zygote, the first cell of the diploid phase of the life cycle.

## THE STAGES OF MEIOSIS

Although meiosis, like mitosis, is a continuous process, we can describe this process most conveniently by dividing it into arbitrary stages. In a sense, meiosis consists of two rounds of nuclear division similar to that in mitosis, during the course of which two unique events occur:

1. In an early stage of the first of the two nuclear divisions, the two nearly identical versions of each chromosome, called **homologues,** pair with each other all along their length. Each chromosome has replicated during the S phase that precedes cell division and now consists of two

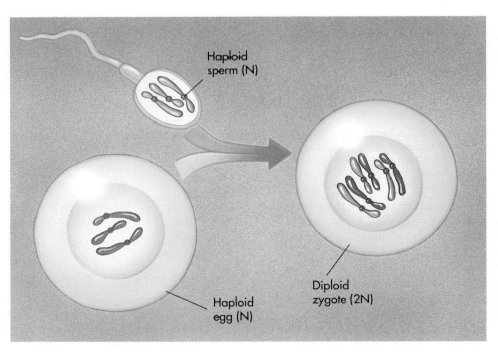

Haploid sperm (N)

Haploid egg (N)

Diploid zygote (2N)

**Figure 9-17**

**Diploid cells carry their parents' chromosomes.** A diploid cell contains two versions of each chromosome, one contributed by the haploid egg of the mother, the other by the haploid sperm of the father.

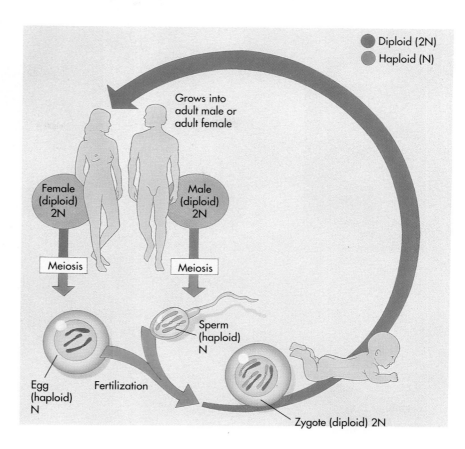

Grows into
adult male or
adult female

Female
(diploid)
2N

Male
(diploid)
2N

Meiosis

Meiosis

Sperm
(haploid)
N

Egg
(haploid)
N

Fertilization

Zygote (diploid) 2N

**Figure 9-18**

**The sexual life cycle.** In animals the vast majority of the life cycle is spent as a diploid organism, the completion of meiosis being soon followed by fertilization.

sister chromatids bound together by a single centromere. When chromosomes pair during meiosis, therefore, they unite four chromatids. While these four chromatids are held together, they exchange portions of DNA strands in a process called **crossing-over.** If the exchange takes place between chromatids of the different homologues, the two homologues will be held together, thus forming a pair of chromosomes that separates later in meiosis. When the two homologues do separate, their chromatids have exchanged genetic material, as we shall discuss in more detail later.

2. The second meiotic division is identical to a mitotic division, except that *the chromosomes do not replicate between the two divisions.*

*Two important features distinguish meiosis from mitosis:*
*1. In meiosis, homologous chromosomes pair lengthwise, and their chromatids exchange genetic material.*
*2. The sister chromatids of each homologue do not separate from each another in the first nuclear division during meiosis, and the chromosomes do not replicate between the two nuclear divisions.*

The two stages of meiosis (Figure 9-19) are called meiosis I and meiosis II—the first and second meiotic divisions. Each stage is further subdivided into prophase, metaphase, anaphase, and telophase, just as in mitosis. In meiosis, however, prophase is more complex than it is in mitosis.

## The First Meiotic Division

*Prophase I.* In prophase I, individual chromosomes first become visible, as viewed with a light microscope, as their DNA coils more and more tightly. Because the chromosomes (DNA) have replicated before the onset of meiosis, each of these threadlike chromosomes actually consists of two sister chromatids joined at their centromeres. The two homologous chromosomes then line up side-by-side, a process that is called **synapsis.** A lattice of protein and RNA is laid down between the chromatids of each pair of homologous chromosomes. This lattice holds the chromatids in precise relation with one another, each gene located directly across from its corresponding sister on the homologue. The effect is similar to zipping up a zipper. Within the lattice, the DNA duplexes of each chromatid unwind, and each strand of DNA pairs with a complementary strand *from the other homologous chromosome.* How many single strands of DNA are present in each lattice? Eight. Count them: 2 per DNA duplex molecule × 2 sister chromatids per homologue × 2 homologues.

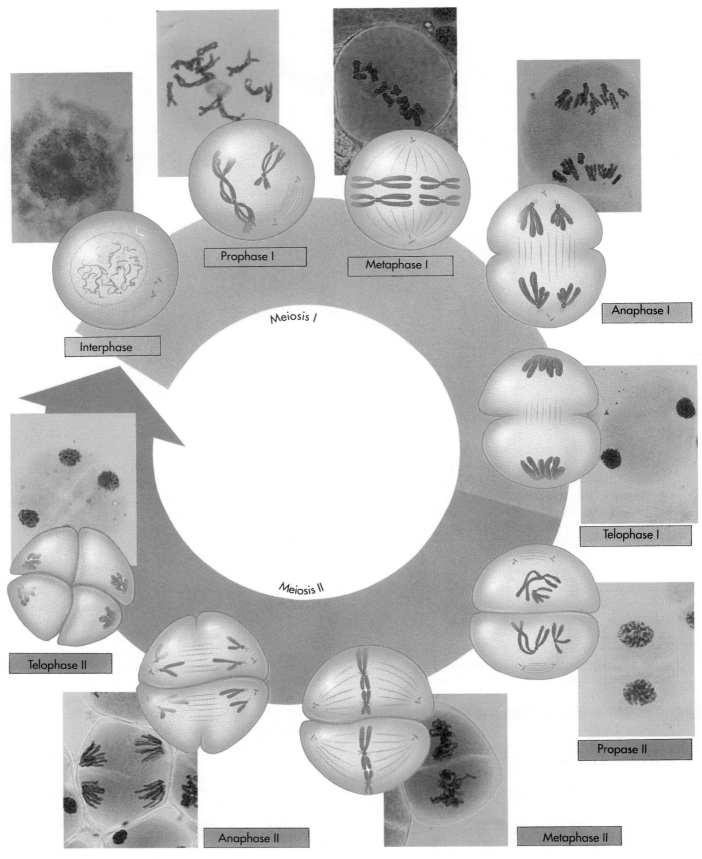

Meiosis I

Meiosis II

Interphase

Prophase I

Metaphase I

Anaphase I

Telophase I

Propase II

Metaphase II

Anaphase II

Telophase II

**Figure 9-19**
Meiosis.

*Synapsis is the close pairing of homologous chromosomes that occurs early in prophase I of meiosis. During synapsis, a molecular latticework aligns the genes of the two homologous chromosomes side-by-side. As a result, a DNA strand of one homologue can pair with the corresponding complementary DNA strand of the other.*

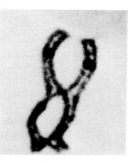

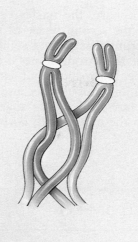

**Figure 9-21**
Chiasmata.

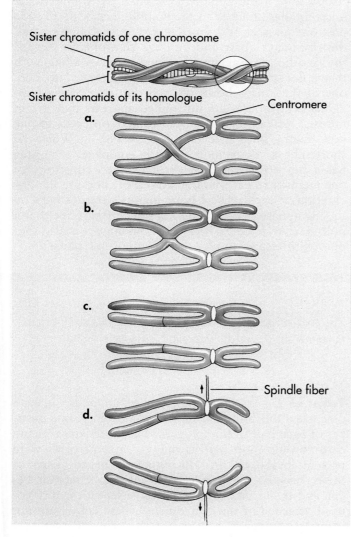

Sister chromatids of one chromosome

Sister chromatids of its homologue

Centromere

a.

b.

c.

Spindle fiber

d.

**Figure 9-20**

**Crossing over.** The open circle highlighting the upper chromosome indicates a region where homologous chromosomes are pairing. Crossing-over occurs when the paired chromosomes exchange arms.

The process of synapsis initiates a complex series of events, called crossing-over, in which DNA is exchanged between the two homologous chromosomes (Figure 9-20). Once the process of crossing-over is complete, the lattice breaks down. At that point, the nuclear envelope dissolves, and the chromatids begin to move apart. Each chromosome consists of four chromatids at this point: two homologous chromosomes, each replicated and so present twice as sister chromatids.

The points of crossing-over, where portions of chromosomes have been exchanged, can often be seen under a light microscope as X-shaped structures known as **chiasmata** (singular *chiasma*, the Greek word for cross) (Figure 9-21). The presence of a chiasma indicates that two of the four chromatids of paired homologous chromosomes have exchanged parts, one participant from each homologue.

*In prophase I, the individual DNA strands of the two homologues pair with one another. Crossing-over occurs between the DNA-paired strands, creating the chromosomal configurations known as chiasmata.*

*Metaphase I.* Late in prophase, the nuclear envelope disperses. In metaphase, the second phase of the first meiotic division, the microtubules form a spindle, just as in mitosis. But because the chromosomes are paired in meiosis I, one of the kinetochores of each centromere is inaccessible to microtubules because the homologous chromosomes are held tightly together by chiasmata. In mitosis, on the other hand, *both* kineto-

chores of a centromere are bound by microtubules, and the chromatids therefore separate from one another when anaphase occurs. The chromosomes line up double-file in meiosis and single-file in mitosis.

Because microtubules bind to only one kinetochore of each centromere, the centromere of one homologue becomes attached to microtubules extending to one pole, whereas the centromere of the other homologue becomes attached to microtubules extending to the other pole. Each joined pair of homologues then lines up on the metaphase plate. For each pair of homologues, the orientation on the spindle axis is random; which homologue is oriented toward which pole is a matter of chance.

*Anaphase I.* After spindle attachment is complete, the microtubules attached to the homologues begin to slide past one another and shorten. The microtubules break the chiasmata apart and pull the centromeres toward the two poles, dragging the chromosomes along with them. Because the microtubules are attached to only one of the kinetochores of each centromere, the individual centromeres are not pulled apart as they are in mitosis. Instead, the entire centromere proceeds to one pole, taking both sister chromatids with it. When the shortening of the spindle fibers is complete, each pole has a full complement of chromosomes, consisting of one member of each homologous pair. Because the orientation of each pair of homologous chromosomes on the metaphase plate is random, the chromosome that a pole receives from each pair of homologues is also random with respect to all other chromosome pairs.

## A VOCABULARY OF CELL DIVISION

**binary fission**  Asexual reproduction of a cell by division into two equal or nearly equal parts. Bacteria divide by binary fission.

**centromere**  A constricted region of the chromosome joining two sister chromatids, to which the kinetochore is attached. About 220 nucleotides in length, it is composed of highly repeated DNA sequences (satellite DNA).

**chromatid**  One of the two replicated strands of a duplicated chromosome, joined by a single centromere to the other replicated strand.

**chromatin**  The complex of DNA and proteins of which eukaryotic chromosomes are composed.

**chromosome**  The organelle within cells that carries the genes. In bacteria, the chromosome consists of a single naked circle of DNA; in eukaryotes, a chromosome consists of a single linear DNA molecule associated with proteins.

**cytokinesis**  Division of the cytoplasm of a cell after nuclear division.

**euchromatin**  Chromatin that is extended except during cell division, from which RNA is transcribed.

**heterochromatin**  That portion of a eukaryotic chromosome that remains permanently condensed and therefore is not transcribed into RNA. Most centromere regions are heterochromatic.

**homologues**  Homologous chromosomes. In diploid cells, one chromosome of a pair that carry equivalent genes.

**kinetochore**  A disk of protein bound to the centromere to which microtubules attach during mitosis, linking each chromatid to the spindle.

**microtubule**  A hollow protein cylinder, about 25 nanometers in diameter, composed of subunits of the protein tubulin. Microtubules grow in length by the addition of tubulin subunits to the end(s) and shorten by their removal.

**mitosis**  Nuclear division in which duplicated chromosomes separate to form two genetically identical daughter nuclei; usually accompanied by cytokinesis, producing two identical daughter cells.

**nucleosome**  The basic packaging unit of eukaryotic chromosomes, in which the DNA molecule is wound around a ball of histone proteins. Chromatin is composed of long strings of nucleosomes, like beads on a string.

*The two similar chromosomes of homologous pairs orient randomly on the metaphase plate.*

*Telophase I.* At the completion of anaphase I, each pole has a full complement of chromosomes: one member of each pair of homologues. There are half as many chromosomes and half as many centromeres as were present in the cell in which meiosis began. Each of these chromosomes replicated itself before meiosis began and thus contains two copies (sister chromatids) of itself attached at the centromere. These copies are not identical, however, because of the crossing-over that occurred in prophase I. The stage at which the two complements of chromosomes gather together at their respective poles to form two chromosome clusters is called telophase I. After an interval of variable length, the second meiotic division, meiosis II, occurs.

## The Second Meiotic Division

Meiosis II is simply a mitotic division involving the products of meiosis I, except that the sister chromatids are not genetically identical, as they are in mitosis, because of crossing-over. At the completion of anaphase I, each pole has a haploid complement of chromosomes, each of which is still composed of two sister chromatids attached at the centromere. Meiosis II occurs to separate these sister chromatids. At both poles of the original cell, the chromosomes divide mitotically. Spindle fibers bind each of the two kinetochores at the centromere region, which separate and move to opposite ends of each polar region. The result of this division is four haploid complements of chromosomes. At this point, the nuclei are reorganized and nuclear envelopes form around each of the four haploid complements of chromosomes. The cells that contain these haploid nuclei may function directly as gametes, as they do in animals, or they may divide again by mitosis, as they do in plants, fungi, and many protists, eventually producing gametes after further mitotic divisions.

## THE EVOLUTIONARY CONSEQUENCES OF SEX

The reassortment of genetic material that occurs during meiosis (Figure 9-22) is the principal factor that has made possible the evolution of eukaryotic organisms, in all their incredible diversity, over the past 1.5 billion years. Sexual reproduction represents an enormous advance in the ability of organisms to generate genetic variability. To understand why this is true, recall that most organisms have more than one chromosome. Human beings have 23 different pairs of homologous chromosomes. Each human gamete receives one of the two copies of each of the 23 different chromosomes, but which copy of a particular chromosome it receives is random. For example, the copy of chromosome number 14 that a particular human gamete receives has no influence on which copy of chromosome number 5 it will receive. Each of the 23 pairs of chromosomes goes through meiosis independently of all the others, so there are $2^{23}$ (more than 8 million) different possibilities for the kinds of gametes that can be produced, and no two of them are alike.

Because the zygote that forms a new individual is created by the fusion of *two* gametes, each produced independently, fertilization squares the number of possible outcomes (8 million $\times$ 8 million = 64 trillion). Fertilization therefore creates a unique individual, a new combination of the 23 chromosomes that probably has never occurred before and probably will never occur again.

The exchange that occurs as a result of crossing-over between the arms of homologous chromosomes adds even more possibilities to the random assortment of chromosomes discussed above. Thus the number of possible genetic combinations that can occur among gametes is virtually unlimited.

Few subjects have been so profoundly interesting to biologists as sex. As you might expect, the subject has proven quite controversial. For one thing, it is hard to understand how sex evolved in the first place. Evolutionary biologists think it may have arisen secondarily from selection for better repair of genetic damage.

The evolutionary consequences of sexual reproduction have been profound. No genetic process generates diversity more quickly, and genetic diversity is the raw material of evolution, the fuel that determines how far and how rapidly natural selection can change the characteristics of a population. Among bacteria and other groups of organisms in which genetic recombination is limited, the short generation time appears to play a similar role in enhancing the amount of genetic diversity available for selection during a given period

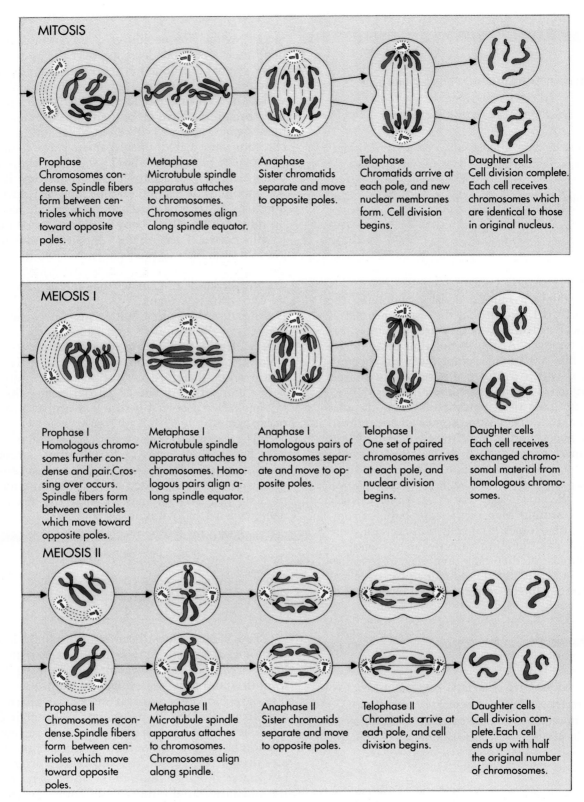

**MITOSIS**

Prophase
Chromosomes condense. Spindle fibers form between centrioles which move toward opposite poles.

Metaphase
Microtubule spindle apparatus attaches to chromosomes. Chromosomes align along spindle equator.

Anaphase
Sister chromatids separate and move to opposite poles.

Telophase
Chromatids arrive at each pole, and new nuclear membranes form. Cell division begins.

Daughter cells
Cell division complete. Each cell receives chromosomes which are identical to those in original nucleus.

**MEIOSIS I**

Prophase I
Homologous chromosomes further condense and pair. Crossing over occurs. Spindle fibers form between centrioles which move toward opposite poles.

Metaphase I
Microtubule spindle apparatus attaches to chromosomes. Homologous pairs align along spindle equator.

Anaphase I
Homologous pairs of chromosomes separate and move to opposite poles.

Telophase I
One set of paired chromosomes arrives at each pole, and nuclear division begins.

Daughter cells
Each cell receives exchanged chromosomal material from homologous chromosomes.

**MEIOSIS II**

Prophase II
Chromosomes recondense. Spindle fibers form between centrioles which move toward opposite poles.

Metaphase II
Microtubule spindle apparatus attaches to chromosomes. Chromosomes align along spindle.

Anaphase II
Sister chromatids separate and move to opposite poles.

Telophase II
Chromatids arrive at each pole, and cell division begins.

Daughter cells
Cell division complete. Each cell ends up with half the original number of chromosomes.

*Figure 9-22*

**A comparison of meiosis with mitosis.** Meiosis involves two serial nuclear divisions with no DNA replication between them and so produces four daughter cells, each with half the original amount of DNA. Crossing-over occurs in prophase I of meiosis.

of time. In many cases the pace of evolution appears to be geared to the level of genetic diversity: the greater the genetic diversity, the faster evolution proceeds. Programs for selecting larger size in domesticated cattle and sheep proceed rapidly at first, but then slow as all of the existing genetic combinations are exhausted; further progress must then await the generation of new gene combinations.

*The genetic recombination that is associated with sexual reproduction has had an enormous evolutionary impact because of the extensive variability that it can generate rapidly within the genome.*

## ■ SUMMARY

1. Bacterial cells divide by simple binary fission. The two replicated circular DNA molecules attach to the plasma membrane at different points, and cellular fission is initiated between the points to which they are attached.

2. The cells of eukaryotes contain much more DNA than do those of bacteria. Eukaryotic DNA is coiled around a framework of histone proteins and distributed among several chromosomes. Some of the DNA is permanently condensed into heterochromatin, whereas the rest becomes condensed only during cell division.

3. DNA replication is completed before mitosis begins. Immediately before the onset of mitosis, condensation of the chromosomes and synthesis of microtubules begin. The period preceding mitosis is called interphase.

4. During interphase, cells undergo a $G_1$ (growth) phase; an S phase, during which a replica of the genome is synthesized; and a $G_2$ phase, during which preparations are made for genomic separation, including replication of the organelles, synthesis of microtubules, and initiation of chromosome condensation. The subsequent M phase is called mitosis, and it is followed by a C phase, in which the cell itself divides.

5. The first stage of mitosis is prophase, during which the chromosomes become more condensed and the spindle fibers form. At the end of prophase, the nuclear envelope disassembles and microtubules form, connecting each pair of sister chromatids to the two poles of the cell.

6. The second stage of mitosis is metaphase, during which the chromosomes align around the periphery of a plane cutting through the center of the cell at right angles to the axis of the spindle. At the end of metaphase, the centromeres joining each pair of sister chromatids replicate, freeing each chromatid to be pulled to one of the poles of the cell by the microtubules attached to it.

7. The third stage of mitosis is anaphase, during which the chromatids physically separate, moving to opposite poles of the cell.

8. The fourth and final stage of mitosis is telophase, during which the mitotic apparatus is disassembled, the nuclear envelope re-forms, and the chromosomes uncoil.

9. Following mitosis, most cells undergo cytoplasmic division, or cytokinesis. In animal and many protist cells, which lack a cell wall, the cell is pinched in two by a belt of microtubules drawing inward around its midsection to form a cleavage furrow. In plant, fungal, and many protist cells, which are surrounded by a rigid cell wall, an expanding cell plate forms along the midline of the spindle.

10. Meiosis is a special form of nuclear division that precedes gamete formation in sexually reproducing eukaryotes. In meiosis, there is a single replication of the chromosomes and two chromosome separations. Meiosis results in the formation of four nuclei, each with half the original number of chromosomes. The cells that include these nuclei may serve as gametes directly, or they may undergo further mitotic divisions before the gametes differentiate.

11. Meiosis involves a pair of serial nuclear divisions. The two unique characteristics of meiosis are synapsis, which is the intimate pairing of homologous chromosomes, and the lack of chromosome replication before the second nuclear division.

12. Crossing-over is an essential element of meiosis. It occurs during prophase, when the chromosomes are so closely associated that the spindle fibers are able to bind to only one kinetochore of each homologous chromosome's centromere because the other kinetochore is covered by the opposite homologue. The physical crossing-over that occurs between homologous chromosomes slows down their separation at anaphase, as the chiasmata move to the ends of the paired chromosomes. The pull of the microtubules ultimately breaks the terminal chiasmata that still link them, and the chromosomes move to the poles of the cell.

13. At the end of meiosis I, one member of each pair of homologues is present at each of the two poles of the dividing nucleus. These chromosomes already consist of two chromatids, differing genetically from each other as a result of the crossing-over that occurred when the chromosomes were paired with their homologues. No further replication occurs before the next nuclear division, which is a normal mitosis that occurs at each of the two poles. The sister chromatids simply separate from each other. This results in the formation of four clusters of chromosomes, each with half the number of chromosomes that was present initially.

14. Meiotic recombination is the principal factor that has made possible the evolution of eukaryotic organisms. The number of possible genetic combinations that can occur among gametes is virtually unlimited because of meiosis and crossing-over.

## REVIEW QUESTIONS

1. After the naked DNA duplex is replicated and separated in bacteria, the cell divides by a simple process called _____, which is made easier by invaginations of the growing plasma membrane.

2. Which phase of the cell cycle is the essential step in the separation of the two daughter genomes?

3. What determines the plane in which a cell will divide?

4. The process by which homologous chromosomes come to line up side-by-side during prophase I of meiosis is called _____.

5. Meiotic recombination greatly enhances the ability of organisms to generate genetic _____.

## SELF-QUIZ

1. Eukaryotic DNA is coiled around _____, which are the first order of DNA packaging.
   (a) nucleosomes          (c) basic amino acids          (e) supercoils
   (b) histones             (d) chromosomes

2. Which of the following does *not* occur during interphase?
   (a) Each chromosome replicates.
   (b) The nucleolus disappears.
   (c) The sister chromosomes are attached to each other at the centromere.
   (d) The chromosomes begin the process of condensation.
   (e) Centrioles replicate.

3. The stage of the cell cycle during which the cell actually divides to form two cells is called
   (a) $G_1$
   (b) S
   (c) $G_2$
   (d) mitosis
   (e) cytokinesis

4. Sexual reproduction in plants and animals involves the production of gametes through a special type of cell division called _____, followed by the union of haploid gametes, which is called _____.
   (a) mitosis
   (b) fertilization
   (c) cytokinesis
   (d) chromatin
   (e) meiosis

5. Which of the following is *not* true of crossing-over?
   (a) It occurs during prophase I.
   (b) It occurs during prophase II.
   (c) It occurs between homologues.
   (d) It is seen as X-shaped structures called chiasmata.
   (e) Exchange of DNA occurs between homologous chromosomes at the points of crossing-over.

# THOUGHT QUESTIONS

1. In what way does the short generation time in bacteria produce an effect similar to that of the extensive recombination which occurs as a result of meiosis in eukaryotes?

2. Which do you think evolved first, meiosis or mitosis? What do you think may have been some of the stages in the evolution of one from the other?

3. The human genome contains more than 3 billion ($3 \times 10^9$) nucleotides; each nucleotide of the sugar phosphate backbone of DNA takes up about 0.34 nanometer (1000 nanometers = 1 micrometer; 1000 micrometers = 1 millimeter; 10 millimeters = 1 centimeter). If a human genome were fully extended, how long would it be?

4. Human beings have 23 pairs of chromosomes; 22 pairs that play no role in sex determination, and an XX (female) or XY (male) pair. Ignoring the effects of crossing-over, what proportion of your eggs or sperm contain all of the chromosomes you received from your mother?

# FOR FURTHER READING

KORNBERG, R.D., and A. KLUG: "The Nucleosome," *Scientific American,* February 1981, pages 52-64. This article describes how the DNA helix is wound on a series of protein spools.

SCIENCE: "The Cell Cycle," *Science,* vol. 246, no. 4930, November 3, 1989, pages 545-640. A comprehensive account of all aspects of modern research into the cell cycle.

SLOBODA, R.D.: "The Role of Microtubules in Cell Structure and Cell Division," *American Scientist,* vol. 68, 1980, pages 290-298. A good description of how the spindle apparatus works.

# *M*endelian Genetics

Human beings are extremely diverse in appearance. The differences between us are partly inherited and partly a result of the environmental factors that we encounter during the course of our lives.

233

# MENDELIAN GENETICS

## Overview

Why do members of a family tend to resemble one another more than they do members of other families? We have puzzled about heredity since before words were written down. Little progress in solving the puzzle was made, however, until Gregor Mendel conducted some simple but significant experiments in his monastery garden a century ago. What emerged from this obscure garden was nothing less than the key to understanding heredity, the pivotal piece of the puzzle. Mendel developed a set of rules—hypotheses—that predicted patterns of heredity accurately. When these rules became widely known, investigators avidly sought and soon found the physical mechanisms responsible. We now know that inherited traits are specified by genes, which are portions of the DNA molecules of chromosomes. Mendel's contribution is widely regarded as one of the greatest intellectual accomplishments in the history of science.

*For Review*   *Here are some important terms and concepts that you will encounter in this chapter. If you are not familiar with them, you should review them before proceeding.*

**DNA** (Chapter 2)

**Chromosomes** (Chapters 4, 9)

**Meiosis** (Chapter 9)

**Homologues** (Chapter 9)

**Chiasmata and crossing-over** (Chapter 9)

*Figure 10-1*

**Any group of people includes individuals of very different appearances.** These babies are each separate individuals, as a result of both heredity and environmental influences.

Some variation in physical characteristics is evident in all of us. Groups of people from different parts of the world often have distinct appearances. Within any of these groups, individual families often differ greatly from one another. Look around you in class; rarely will any two students resemble one another closely. Even our brothers and sisters are not exactly like us, unless they are our identical twins. Any group of people includes individuals of very different appearances (Figure 10-1). The reasons for these differences have always fascinated us, and human beings are not unique in this respect. Great differences in appearance also often exist within other species. Dogs are born in many varieties and breeds of every size and form. All of these are still dogs, however, able to breed with one another and produce puppies (see Figure 16-1).

Variation is not surprising in itself. Differences in diet during development can have great effects on adult appearance, as can variations in the environments that different individuals experience. Many arctic mammals develop white fur when they are exposed to the cold of winter and dark fur during the warm summer months. But one remarkable property in some patterns of vari-

**Figure 10-2**

**Many human traits are inherited within families.** This little redhead is exhibiting such a trait.

the progeny than would others. They also discovered that the less frequent forms would disappear in one generation, only to reappear unchanged in the next. In one such series of experiments in the 1790s, T.A. Knight crossed two true-breeding varieties of the garden pea, *Pisum sativum* (Figure 10-3). One of these varieties had purple flowers; the other, white flowers. All the progeny of the cross had purple flowers. Among the offspring of these hybrids, however, some plants had purple flowers, and others, occurring less frequently, had white ones. As had been observed earlier, a trait from one of the parents (white flowers) was hidden in one generation, only to reappear in the next.

ation has especially fascinated and puzzled us: some of the differences that we observe between individuals are inherited, passed down from parent to child.

As far back as there is a written record, such patterns of resemblance among the members of particular families have been noted and discussed. Some features shared by family members are unusual, such as the protruding lower lip of the Austrian royal family, the Hapsburgs, which is evident in pictures and descriptions of that family since the thirteenth century. Other characteristics are more familiar, such as the common occurrence of red-haired children within the families of red-haired parents (Figure 10-2). Such inherited features, the building blocks of evolution, are the focus of this chapter.

## EARLY IDEAS ABOUT HEREDITY: THE ROAD TO MENDEL

Two centuries ago, English gentleman farmers who were trying to improve varieties of agricultural plants carried out matings, called **crosses,** between different strains, selecting the most desirable of the offspring of each cross. Whereas the strains they used were **true-breeding,** producing offspring that resembled their parents in all respects, the set of **hybrids,** plants that resulted from crossing dissimilar parents, often was variable. The individual hybrid plants had different combinations of the features of their parents; obviously, not all offspring of a cross had received the same selection of inherited traits. The alternatives of a trait were "segregating" among the progeny; some progeny exhibited one alternative, others exhibited another.

In their crosses, the farmers found that some alternative forms of a trait would appear more often among

*Two centuries ago, plant breeders had learned that the hybrid plants derived from crosses between different strains presented variable combinations of the traits of their parents. One of the alternative states of a particular feature sometimes disappeared in the first hybrid generation, reappearing in its progeny.*

**Figure 10-3**

**The garden pea,** *Pisum sativum.* Easy to cultivate and with many distinctive varieties, the garden pea had been a popular choice as an experimental subject in investigations of heredity as much as a century before Gregor Mendel's investigations.

In these deceptively simple results were the makings of a scientific revolution. But it took another century before the process of **segregation of alternative traits** was understood. Why did it take so long? One reason was that some characteristics appeared to be **blended** in the offspring, so it was not clear that individual, distinct factors were involved. Another problem was that early workers did not quantify their results, and a numerical record of results proved to be crucial to understanding this process. For example, both Knight and later experimenters who carried out other crosses with pea plants noted that some traits had a "stronger tendency" to appear than others, but they did not record the actual numbers of the different classes of progeny. Science was young then, and it was not obvious that the numbers were important.

## MENDEL AND THE GARDEN PEA

The first quantitative studies of inheritance were carried out by an Austrian monk, Gregor Mendel (Figure 10-4). Born in 1822 to peasant parents, Mendel was educated in a monastery and went on to study science and mathematics at the University of Vienna, where he failed his examinations for a teaching certificate. Returning to the monastery, where he spent the rest of his life and eventually became abbot, Mendel initiated a series of experiments on plant heredity in the garden. The results of these experiments would ultimately change our views of heredity irrevocably.

For his experiments, Mendel chose the garden pea, the same plant that Knight and many others had stud-

**Figure 10-5**

**A Mendelian trait.** One of the differences among varieties of pea plants that Mendel studied affected the shape of the seed. In some varieties the seeds were round, whereas in others they were wrinkled. As you can see, wrinkled seeds look like dried-out, shrunken versions of the round ones.

ied earlier. The choice was a good one for several reasons:

1. Many earlier investigators already had produced hybrid peas by crossing different varieties. From their results, Mendel knew that he could expect to observe the segregation of alternative traits among the progeny of crosses. That is, Mendel knew that some of the progeny would exhibit the alternative characteristic of one variety, but other individuals would exhibit the alternative characteristic of the other variety.

2. Many varieties of peas were available. Mendel initially examined 32 varieties. Then, for further study, he selected lines that differed with respect to seven easily distinguishable traits, such as smooth versus wrinkled seeds (Figure 10-5) and purple versus white flowers (a characteristic that Knight had studied 60 years earlier).

3. Pea plants are small and easy to grow, produce large numbers of offspring, and mature quickly. Thus one can conduct experiments involving many plants and can obtain the results relatively soon.

4. The sexual organs of the pea are enclosed within the flower (Figure 10-6). The flowers of peas, like those of most flowering plants, are bisexual, containing both the structures in which

**Figure 10-4**
Gregor Johann Mendel.

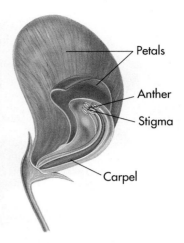

Petals

Anther

Stigma

Carpel

*Figure 10-6*

**In a pea plant flower the petals enclose the male and female parts, ensuring that self-fertilization will take place unless the flower is disturbed.**

male gametes are produced (the anthers) and those in which female gametes are produced (the carpels). **Self-fertilization** takes place automatically within an individual pea flower if it is not disturbed. As a result of this process, the offspring of garden peas are the progeny of a single individual. One can allow self-fertilization to occur within an individual flower, or one can remove its anthers before fertilization, introduce pollen (which is produced in the anthers and gives rise to the gametes) from a strain with alternative characteristics, and thus perform a controlled cross. **Self-pollination** leads directly to self-fertilization in peas.

*Gregor Mendel chose peas for his classical genetics experiments because they were variable in their features; the results of crosses in peas had been studied earlier; true-breeding strains are produced by self-pollination; the plants are relatively small, so many can be grown in a limited area; and the generation time is short.*

## MENDEL'S EXPERIMENTAL DESIGN

Mendel usually conducted his experiments in three stages:

1. He first allowed pea plants of a given variety to produce progeny by self-fertilization for several generations. Mendel was thus able to ensure that the characteristics he was studying were transmitted regularly

from generation to generation. Pea plants with white flowers, for example, produced only plants with white flowers among their progeny, regardless of the number of generations studied.

2. Mendel then conducted crosses between varieties exhibiting alternative traits (Figure 10-7). For example, he removed the male parts from a flower on a plant that produced white flowers and fertilized it with pollen from a purple-flowered plant. He would also carry out the reciprocal cross by reversing the procedure, using pollen from a white-flowered individual to fertilize a flower on a pea plant that produced purple flowers.

3. Finally, Mendel permitted the hybrid offspring produced by these crosses to self-pollinate for several generations. By doing so, he allowed the alternative traits to segregate among the progeny. That was the same experimental design that Knight and others had used much earlier. But Mendel added a new element: he counted the numbers of offspring in each class and in each succeeding generation. No one had ever done that before. The quantitative results that Mendel obtained were essential in helping him, and us, understand the process of heredity.

## WHAT MENDEL FOUND

When Mendel crossed two contrasting varieties, such as purple-flowered plants with white-flowered plants, the hybrid offspring that he obtained did not have an intermediate flower color. Instead, the hybrid offspring always resembled one of the parents. It is customary to refer to these hybrid offspring as the **first filial,** or $F_1$, generation. In a cross of white-flowered with purple-flowered plants, the $F_1$ offspring all had purple flowers,

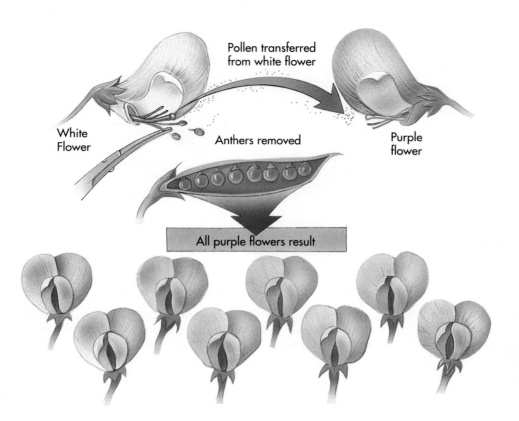

White Flower

Pollen transferred from white flower

Anthers removed

Purple flower

All purple flowers result

*Figure 10-7*

**How Mendel conducted his experiments.** Mendel pushed aside the petals of a white flower and cut off the anthers, where the male gametes are produced and enclosed in pollen grains. He then placed that pollen onto the female parts of a similarly castrated purple flower, where cross-fertilization took place. All of the seeds in the pod that resulted from this pollination were hybrids, with a white-flowered female parent and a purple-flowered male parent. Planting these seeds, Mendel observed what kinds of plants they produced. In this instance, all of them had purple flowers.

just as Knight and others had reported earlier. Mendel referred to the trait expressed in the $F_1$ plants as **dominant** and to the alternative trait, which was not expressed in the $F_1$ plants, as **recessive**. For each pair of contrasting traits that Mendel examined, one proved to be dominant and the other recessive.

After individual $F_1$ plants had been allowed to mature and self-pollinate, Mendel collected and planted the fertilized seed from each plant to see what the offspring in this **second filial**, or $F_2$, generation would look like. He found, just as Knight had earlier, that some $F_2$ plants exhibited the recessive trait. Latent in the $F_1$ generation, the recessive alternative reappeared among some $F_2$ individuals.

Mendel termed those features of the parental strains which occurred in all members of the $F_1$ generation as dominant traits and referred to their alternative features, which reappeared in some individuals of the $F_2$ and subsequent generations, as recessive traits.

It was at this stage that Mendel instituted his radical change in experimental design. He counted the numbers of each type among the $F_2$ progeny. Mendel was investigating whether the proportions of the $F_2$ types would provide some clue about the mechanism of heredity. He examined a total of 929 $F_2$ individuals in the cross between purple-flowered $F_1$ plants just described. Of these $F_2$ plants, 705 had purple flowers and 224 had white flowers. Almost exactly one-fourth of the $F_2$ individuals (24.1%) exhibited white flowers, the recessive trait.

Mendel examined each of the seven pairs of contrasting traits in this way (Figure 10-8). The numerical result was always the same: three-fourths of the $F_2$ individuals exhibited the dominant trait, and one-fourth displayed the recessive trait. The ratio of dominant to recessive among the $F_2$ plants was always $3:1$.

Mendel went on to examine how the $F_2$ plants behaved in later generations. He found that the one-fourth that were recessive were always true-breeding (that is, continued to exhibit the trait in all individuals of succeeding generations). In the cross of white-flowered with purple-flowered plants just described, the white-flowered $F_2$ individuals reliably produced white-flowered offspring when allowed to self-fertilize. By contrast, only one-third of the dominant $F_2$ individuals (one-fourth of the entire offspring) were true-breeding, whereas two-thirds were not. This last class of plants produced dominant and recessive $F_3$ individuals in the ratio 3:1. This result suggested that, for the entire sample, the 3:1 ratio Mendel observed in the $F_2$ generation was really a disguised 1:2:1 ratio, with one-fourth true-breeding dominant individuals to one-half not-true-breeding dominant individuals to one-fourth true-breeding recessive individuals.

## HOW MENDEL INTERPRETED HIS RESULTS

From these experiments, Mendel was able to learn four things about the nature of heredity.

1. The traits that Mendel studied did not produce intermediate types when plants that had them were crossed. Instead, the alternative features were inherited intact; they either appeared or

| Trait | Dominant vs recessive | $F_2$ generation results | | Ratio |
|---|---|---|---|---|
| | | Dominant form | Recessive form | |
| Flower color | 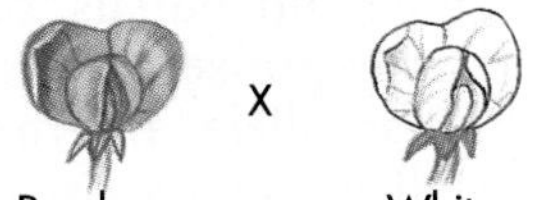 Purple X White | 705 | 224 | 3.15:1 |
| Seed color |  Yellow X Green | 6022 | 2001 | 3.01:1 |
| Seed shape |  Round X Wrinkled | 5474 | 1850 | 2.96:1 |
| Pod color | 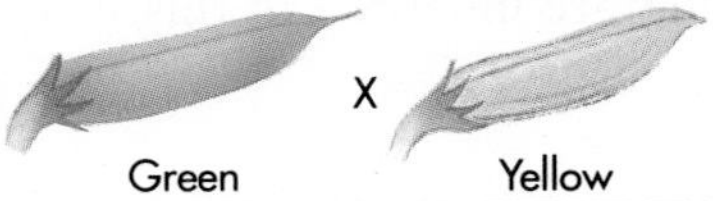 Green X Yellow | 428 | 152 | 2.82:1 |
| Pod shape | 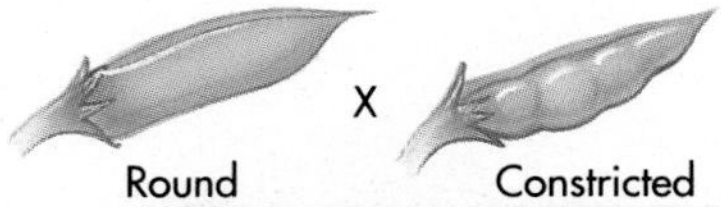 Round X Constricted | 882 | 299 | 2.95:1 |
| Flower position | 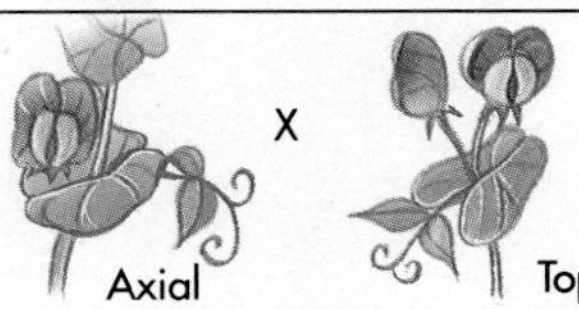 Axial X Top | 651 | 207 | 3.14:1 |
| Plant height | 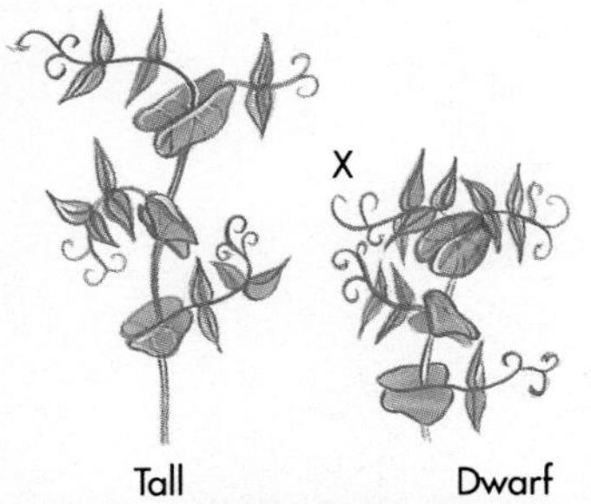 Tall X Dwarf | 787 | 277 | 2.84:1 |

*Figure 10-8*

**Mendel's experimental results.** The table illustrates the seven pairs of contrasting traits studied by Mendel in the garden pea and presents the data he obtained in crosses of these traits. Every one of the seven traits that Mendel studied yielded results very close to a theoretical 3:1 ratio.

did not appear in a particular generation.

2. For each pair of traits that Mendel examined, one alternative was not expressed in the $F_1$ hybrids, although it reappeared in some $F_2$ individuals. *The "invisible" trait must therefore have been latent (present but not expressed) in the $F_1$ individuals.*

3. The pairs of alternative traits that Mendel examined segregated among the progeny of a particular cross, some individuals exhibiting one trait, some the other.

4. Pairs of alternative traits were expressed in the $F_2$ generation in the ratio of three-fourths dominant to one-fourth recessive. This characteristic $3:1$ segregation ratio is often referred to as the **Mendelian ratio.**

*Mendel observed that alternative genetic traits were inherited intact and did not blend; that one of the alternatives was not expressed in the $F_1$ generation, but reappeared in the $F_2$ and subsequent generations; that the alternative traits segregated among the progeny of a particular cross; and that pairs of alternative traits were expressed in the $F_2$ generation in the ratio of three-fourths dominant, one-fourth recessive.*

To explain these results, Mendel proposed a set of hypotheses. These developed into one of the most famous theories in the history of science, containing simple assumptions and making clear predictions. Mendel's set of hypotheses has the following five elements, which are rephrased in modern terms when appropriate:

1. Parents do not transmit their features directly to their offspring; instead, they transmit encoded information about the features. Mendel called these bits of information, which act in the offspring to produce the trait, **factors.**

2. For every feature of a diploid organism, there are two factors, which may or may not be the same. If the factors differ from each other, the individual is said to be **heterozygous** for that characteristic. If they are the same, the individual is **homozygous.**

3. We now call Mendel's factors **genes.** Genes are composed of a sequence of nucleotides in a DNA molecule. Alternative forms of a particular gene are called **alleles,** and the position on a chromosome where a gene is located is called a **locus.**

4. The two alleles, one contributed by the male and one by the female parent, remain distinct; Mendel said they were "uncontaminated": alleles do not blend with one another or become altered in any other way. When an individual matures and produces its own gametes, these gametes include equal proportions of the elements that the individual received from its two parents.

5. The presence of a particular allele does not ensure that the encoded trait will actually be expressed in the individual that carries it. In heterozygous individuals, only one (the dominant) allele is expressed, and the features associated with that allele are the only ones seen. The other allele, the recessive one, although present, is unexpressed. Modern geneticists refer to the collection of alleles an individual possesses as its **genotype.** The physical appearance of an individual is its **phenotype.**

Many traits in humans are controlled by dominant or recessive alleles like those Mendel studied in peas (Figure 10-9). Table 10-1 lists a few of the human traits that are inherited in this way.

***Figure 10-9***

**A common recessive trait.** Blue eyes are considered a recessive trait in humans, although many genes influence the final color.

**TABLE 10-1  SOME DOMINANT AND RECESSIVE HUMAN TRAITS**

| TRAIT | PHENOTYPE |
| --- | --- |
| **RECESSIVE TRAITS** | |
| Common baldness | M-shaped hairline receding with age |
| Albinism | Lack of melanin pigmentation |
| Alkaptonuria | Inability to metabolize homogentisic acid |
| Red-green color blindnesses | Inability to detect red or green wavelengths of light |
| Cystic fibrosis | Failure of chloride ion transport system, leading to liver degeneration and lung failure |
| Duchenne muscular dystrophy | Wasting away of muscles during childhood |
| Hemophilia | Inability of blood to clot |
| Sickle cell anemia | Defective hemoglobin that collapses red blood cells |
| **DOMINANT TRAITS** | |
| Middigital hair | Presence of hair on middle segment of fingers |
| Brachydactyly | Short fingers |
| Huntington's disease | Degeneration of nervous system, starting in middle age |
| Phenylthiocarbamide (phenylthiourea) tasting | Ability to taste PTC as bitter |
| Camptodactyly | Inability to straighten the little finger |
| Hypercholesterolemia (the most common human Mendelian disorder, affecting one in every 500 persons) | Elevated levels of blood cholesterol and high risk of heart attack |
| Polydactyly | Extra fingers and toes |

*The phenotype of an individual is its physical appearance, the sum of all its features. The genotype, or genetic constitution, of the organism is the DNA blueprint for these features.*

The five elements just described constitute the set of hypotheses by which Mendel explained heredity. Do they predict the sort of result that Mendel actually obtained? Let us see.

## The F₁ Generation

Consider again Mendel's cross of purple-flowered with white-flowered plants. Let us assign the symbol $w$ to the recessive allele, associated with the production of white flowers. The symbol $W$ will refer to the dominant allele, associated with the production of purple flowers. As is conventional, the feature as a whole— purple versus white flowers, determined by a single gene—has been assigned a letter relating to its less common condition: thus, w for white flowers. The recessive allele, associated with white flowers, is indicated by lower case, as $w$; the alternative dominant allele, associated with purple flowers, is assigned the same symbol in upper case, W.

Using this system, the genotype of an individual that is true-breeding for the recessive white-flowered trait would be designated $ww$. A genotype is indicated by two letters to indicate the two alleles of a gene present at each locus. In such an individual, both copies of the allele specify white flowers. Similarly, the genotype of a true-breeding purple-flowered individual would be labeled $WW$. A heterozygote—heterozygous

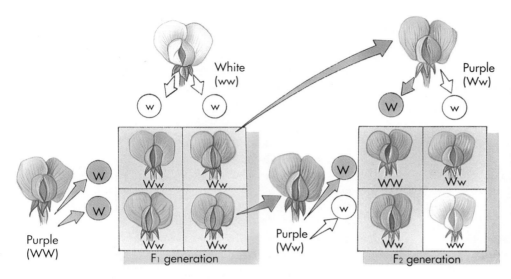

**Figure 10-10**

**Mendel's cross of peas differing in flower color.** The only possible offspring of the first cross are *Ww* heterozygotes, purple in color. These individuals are known as the $F_1$ generation. When two heterozygous $F_1$ individuals cross, three kinds of offspring are possible: *WW* homozygotes (purple); *Ww* heterozygotes (purple), which may form two ways; and *ww* homozygotes (white). Among these individuals, known as the $F_2$ generation, the ratio of dominant type to recessive type is 3:1.

individual—would be designated *Ww* (the dominant allele is usually written first). Using these conventions and a times sign (×) to denote a cross between two breeding lines, Mendel's original cross can be symbolized as *ww* × *WW* (Figure 10-10).

*In a cross between homozygous-dominant and homozygous-recessive individuals, all of the $F_1$ progeny will be heterozygous; they will all resemble the homozygous-dominant parent in their phenotype.*

Because the white-flowered parent can produce only *w*-bearing gametes and the purple-flowered parent can produce only W-bearing gametes, you can see that the union of an egg and a sperm from these parents can produce only heterozygous *Ww* offspring in the $F_1$ generation. Because the W allele is dominant, all these $F_1$ individuals are expected to have purple flowers. The *w* allele is present in these heterozygous individuals, but it is not phenotypically expressed.

### The $F_2$ Generation

When $F_1$ individuals are allowed to self-fertilize, the W and *w* alleles segregate at random during gamete formation. This occurs because of the properties of meiosis, discussed in Chapter 9; any gamete can have either the W allele or the *w* allele. The subsequent union of these gametes formed at fertilization to form $F_2$ individuals is also random; it is not influenced by the particular allele an individual gamete carries. What will the $F_2$ individuals look like? The possibilities may be visualized in a simple diagram called a **Punnett square** (Figure 10-11), named after its originator, the English geneticist Reginald Crundall Punnett. Mendel's model, analyzed in terms of a Punnett square, clearly predicts that in the $F_2$ generation, one should observe that three-fourths of the plants have purple flowers and one-fourth have white flowers, a phenotypic ratio of 3:1.

*In an $F_2$ generation derived from a heterozygous individual, any gamete can receive either allele. Their fusion following fertilization is random and is not influenced by the particular allele contained in a particular gamete. The outcome of such a cross can be illustrated graphically by a Punnett square.*

### Further Generations

As you can see from Figure 10-11, there are really three kinds of $F_2$ individuals. Of them, a fourth are true-breeding *ww* white-flowered individuals, half are heterozygous *Ww* purple-flowered individuals, and a fourth are true-breeding *WW* purple-flowered individuals. The 3:1 phenotypic ratio is the expression of the underlying 1:2:1 genotypic ratio because the heterozygotes are phenotypically indistinguishable from the homozygous purple-flowered (dominant) individuals.

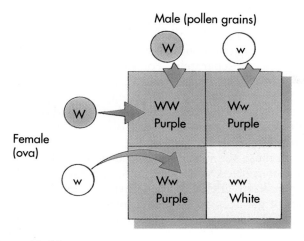

**Figure 10-11**

**A Punnett square.** To make a Punnett square, place the different possible types of female gametes along one axis of a square and the different possible types of male gametes along the other. Each potentially different kind of zygote can then be represented as the intersection of a vertical (male gamete) and horizontal (female gamete) line.

## THE TEST CROSS

To examine his hypotheses in more depth, Mendel devised a simple and powerful procedure called the **test cross.** Consider a purple-flowered individual: is it homozygous or heterozygous? It is impossible to tell simply by looking at it. To learn its genotype, Mendel crossed it with a homozygous recessive individual. His hypothesis predicted different results for homozygous and heterozygous test plants (Figure 10-12):

*Alternative 1* Test individual is homozygous. $WW \times ww$: all offspring have purple flowers *(Ww)*.

*Alternative 2* Test individual is heterozygous. $Ww \times ww$: half of the offspring have white flowers *(ww)*.

To test his hypothesis, Mendel crossed heterozygous $F_1$ individuals back to the parent homozygous for the recessive trait. In such a cross, a $Ww$ $F_1$ individual is crossed with its $ww$ parent. Mendel predicted that the dominant and recessive traits would appear in a 1:1 ratio:

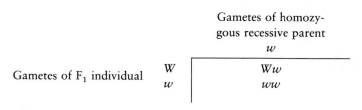

For each pair of alleles he investigated, Mendel observed phenotypic test-cross ratios very close to 1:1, just as he had predicted.

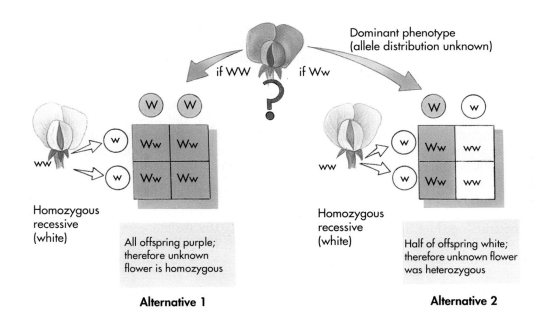

**Figure 10-12**

**A test cross.** To determine whether an individual exhibiting a dominant phenotype, such as purple flowers, is homozygous or heterozygous for the dominant allele, Mendel crossed the individual in question to a plant he knew to be double recessive, in this case a plant with white flowers.

Mendel's set of hypotheses thus accounted in a neat and satisfying way for the segregation ratios he had observed. Its central premises—(1) that alleles do not blend in heterozygotes; (2) that alleles segregate in heterozygous individuals; and (3) that they have an equal probability of being included in either gamete—have since been verified in countless other organisms. These points are usually referred to as **Mendel's First Law,** or the **Law of Segregation.** As you will see later in this chapter, the segregational behavior of alleles has a simple physical basis in meiosis, but that process was unknown to Mendel.

*In modern terms, Mendel's First Law states that:*
*1. Alleles do not blend in heterozygotes.*
*2. In meiosis, alleles segregate from one another; when they are present in a heterozygous condition, the results of the segregation can be observed.*
*3. Each gamete has an equal probability of possessing either member of an allele pair.*

## INDEPENDENT ASSORTMENT

After Mendel had demonstrated the segregation of the different alleles of individual genes, he went on to investigate whether different *genes* segregate independently of one another or as a unit. For example, would the particular alleles of a gene for seed shape that a gamete possessed have any influence on which allele the gamete had for a gene affecting seed color?

To answer this question, Mendel first established a series of true-breeding lines of peas that differed from one another with respect to two of the seven pairs of characteristics he had studied. Second, he crossed contrasting pairs of the true-breeding lines. In a cross involving different seed-shape alleles (round, $W$, and wrinkled, $w$) and different seed-color alleles (yellow, $G$, and green, $g$), all the $F_1$ individuals were identical, each being heterozygous for both seed shape $(Ww)$ and seed color $(Gg)$. The $F_1$ individuals of such a cross are said to be dihybrid individuals. A **dihybrid** is an individual heterozygous for two genes.

The third step in Mendel's analysis was to allow the dihybrid individuals to self-fertilize. If the segregation of alleles affecting seed shape and seed color were independent, the probability that a particular pair of seed-shape alleles would occur together with a particular pair of seed-color alleles would be simply the product of the two individual probabilities that each pair would occur separately. For example, the probability of an individual with wrinkled, green seeds appearing in the $F_2$ generation would be equal to the probability of observing an individual with wrinkled seeds ($\frac{1}{4}$) multiplied by the probability of observing an individual with green seeds ($\frac{1}{4}$), or $\frac{1}{16}$.

Because the genes concerned with seed shape and those concerned with seed color are each represented by a pair of alleles in the dihybrid individuals, four types of gametes are expected in a dihybrid: $WG$, $Wg$, $wG$, and $wg$. In the $F_2$ generation following the self-pollination of a dihybrid or a cross between two dihybrids, there are 16 possible combinations of alleles, each of them equally probable (Figure 10-13). Of the 16 combinations, 9 possess at least one dominant allele for each gene (the second allele is usually signified with a dash, $W—G—$, which means that the second allele at the w locus and the second allele at the g locus may be either dominant or recessive) and thus should have round, yellow seeds. Three possess at least one dominant $W$ allele but are homozygous recessive for color $(W—gg)$. Three other combinations possess at least one dominant $G$ allele but are homozygous recessive for shape $(wwG—)$. One combination among the 16 is homozygous recessive for both genes $(wwgg)$. In summary, the hypothesis that color and shape genes assort independently thus predicts that the $F_2$ generation of this dihybrid cross will display this ratio: nine individuals with round, yellow seeds to three individuals with round, green seeds to three individuals with wrinkled, yellow seeds to one with wrinkled, green seeds—a $9:3:3:1$ ratio.

What did Mendel actually observe? He examined a total of 556 seeds from dihybrid plants that had been allowed to self-fertilize, and he obtained the following results:

| Phenotype | Genotype |
|---|---|
| 315 Round, yellow | $W—G—$ |
| 108 Round, green | $W—gg$ |
| 101 Wrinkled, yellow | $wwG—$ |
| 32 Wrinkled, green | $wwgg$ |

This was very close to a theoretical $9:3:3:1$ ratio, which would have been $313:104:104:35$. Thus the two genes appeared to assort completely independently of one another. At the same time, round and wrinkled seeds still occur approximately in the ratio $3:1$ ($423:133$), as do yellow and green seeds ($416:140$). Those features were segregating normally, as well as independently of one another. Mendel strengthened his

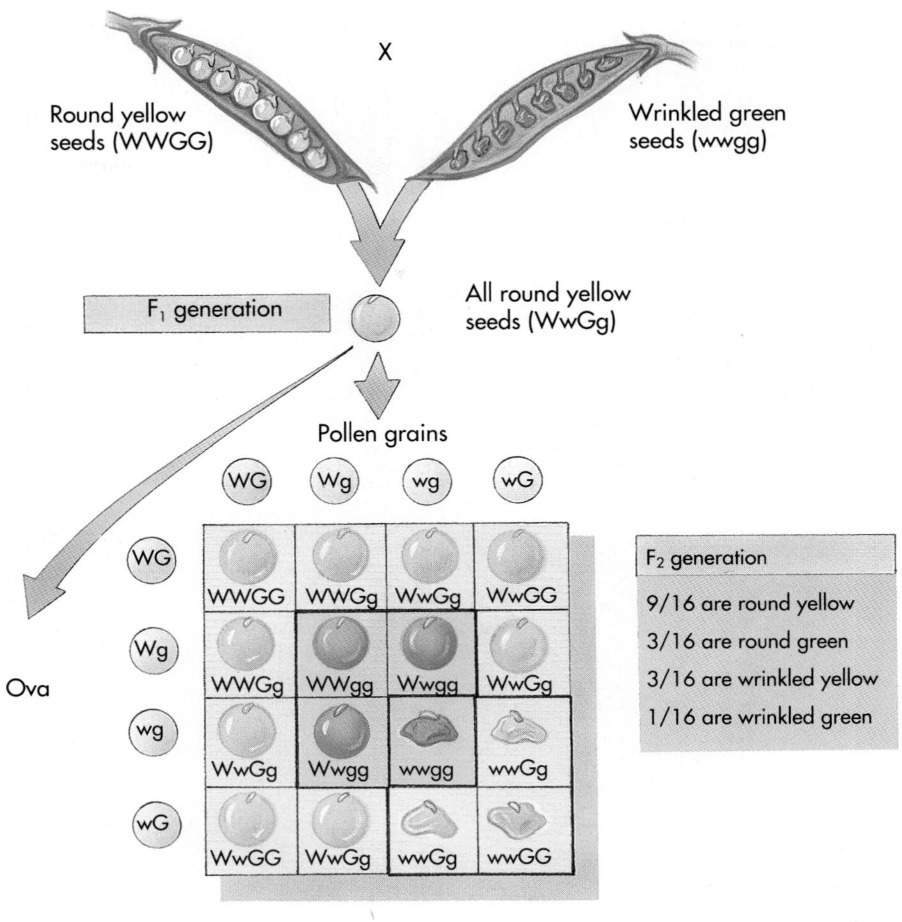
X
Round yellow seeds (WWGG)
Wrinkled green seeds (wwgg)
F1 generation
All round yellow seeds (WwGg)
Pollen grains
WG
Wg
wg
wG
Ova
WG
Wg
wg
wG
WWGG
WWGg
WwGg
WwGG
WWGg
WWgg
Wwgg
WwGg
WwGg
Wwgg
wwgg
wwGg
WwGG
WwGg
wwGg
wwGG
F2 generation
9/16 are round yellow
3/16 are round green
3/16 are wrinkled yellow
1/16 are wrinkled green

*Figure 10-13*

**A Punnet square explaining the results of Mendel's dihybrid cross of round vs wrinkled seeds and yellow vs green seeds.** The ratio of the four possible combinations of phenotypes is predicted to be 9:3:3:1, the ratio that Mendel found.

conclusions by obtaining similar results for other pairs of traits.

Mendel's conclusions are often referred to as **Mendel's Second Law** or the **Law of Independent Assortment.** As you will see, genes that assort independently of one another often do so because they are located on different chromosomes. Individual pairs of chromosomes orient independently of one another on the metaphase I plate in meiosis, which is why the genes which occur on those particular chromosomes also segregate independently (Figure 10-14). A modern restatement of Mendel's Second Law is that *genes located on different chromosomes assort independently during meiosis.*

*Mendel's Second Law states that genes located on different chromosomes assort independently of one another.*

Mendel's original paper describing his experiments, published in 1866, remains interesting reading today. His explanations are clear, and the logic of his arguments is presented in lucid detail. Unfortunately, Mendel failed to arouse much interest in his findings, which were published in the journal of the local natural history society. Only 115 copies of the journal were sent out, in addition to 40 reprints, which Mendel distributed himself. Although Mendel's results did not receive much notice during his lifetime, in 1900, 16 years after his death, 3 different investigators independently rediscovered his pioneering paper. They came across it while searching the literature in preparation for publishing their own findings, which were similar to those Mendel had quietly presented more than 3 decades earlier.

## CHROMOSOMES: THE VEHICLES OF INHERITANCE

Chromosomes are not the only kinds of organelles that segregate regularly when eukaryotic cells divide. Cen-

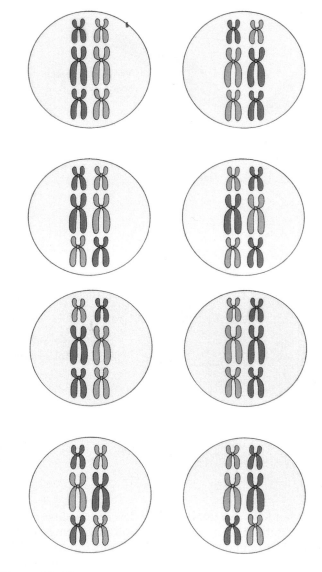

**Figure 10-14**

**Independent assortment.** Independent assortment occurs because the orientation of chromosomes on the metaphase plate is random. In this hypothetical 3-chromosome cell, 4 of the 8 possible orientations ($2^3$) are illustrated; the others are mirror images. Each orientation results in gametes with different combinations of parental chromosomes.

trioles also divide and segregate in a regular fashion, as do the mitochondria and chloroplasts in the cytoplasm. In the early years of this century, it was not obvious that chromosomes were the vehicles for the information of heredity. The German geneticist Carl Correns first suggested a central role for chromosomes in 1900 in one of the papers announcing the rediscovery of Mendel's work. Soon after, observations that homologous chromosomes paired with each another in the process of meiosis led directly to the **chromosomal theory of inheritance**, first formulated by the American Walter Sutton in 1902. Sutton argued as follows:

1. Reproduction involves the initial union of only two cells, egg and sperm. If Mendel's model is correct, then these two gametes must make equal hereditary contributions. Sperm, however, contain little cytoplasm. Therefore the hereditary material must reside within the nuclei of the gametes.

2. The two homologous chromosomes of each pair segregate during meiosis in a fashion similar to that exhibited by the alleles Mendel studied.

3. Gametes have a copy of one member of each pair of homologous chromosomes; diploid individuals have a copy of both members of each pair. Similarly, Mendel found that gametes have one allele of a gene and that diploid individuals have two.

4. During meiosis, each pair of homologous chromosomes orients on the metaphase plate independently of any other pair. This independent assortment of chromosomes is a process similar to the independent assortment of alleles that Mendel studied.

The proof that the genes were located on chromosomes was provided by a single small fly. In 1910, the American geneticist Thomas Hunt Morgan, studying the fly *Drosophila melanogaster,* detected a **mutant** fly, a male fly that differed strikingly from normal flies of the same species. In this fly, the eyes were white instead of the normal red (Figure 10-15).

Morgan immediately set out to determine if this new trait would be inherited in a Mendelian fashion. He first crossed the mutant male to a normal female to see whether white eyes were dominant or recessive. All $F_1$ progeny had red eyes, and Morgan therefore concluded that red eye color was dominant over white. Following the experimental procedure that Mendel had established long ago, Morgan then crossed flies from the $F_1$ generation with each other. Eye color did indeed segregate among the $F_2$ progeny, as predicted by Mendel's theory. Of 4252 $F_2$ progeny that Morgan examined, 782 had white eyes—an imperfect $3:1$ ratio, but one that nevertheless provided clear evidence of segregation. Something was strange about Morgan's result, however, something that was totally unpredicted by Mendel's theory: all the white-eyed $F_2$ flies were males!

How could this strange result be explained? Perhaps it was not possible to be a white-eyed female fly; such individuals might not be viable for some unknown reason. To test this idea, Morgan test-crossed one of the red-eyed $F_1$ female progeny back to the original white-eyed male and obtained white-eyed and red-eyed males and females. So a female could have white eyes. Why then were there no white-eyed females among the progeny of the original cross?

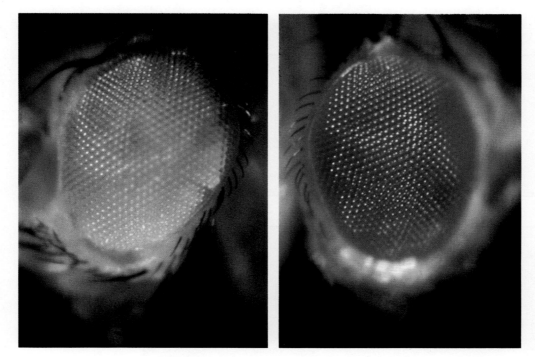

**Figure 10-15**

**White-eyed (mutant) and red-eyed (normal) *Drosophila.*** The white-eyed defect in eye color
is hereditary, the result of a mutation in a gene located on the sex-determining X
chromosome. By studying this mutation, Morgan first demonstrated that genes are on
chromosomes.

The solution to this puzzle proved to involve chromosomes. In *Drosophila,* the sex of an individual is influenced by the number of copies of a particular chromosome, the **X chromosome,** that an individual possesses. An individual with two X chromosomes is a female. An individual with only one X chromosome, which pairs in meiosis with a larger, dissimilar partner called the **Y chromosome,** is a male. Thus the female produces only X gametes, whereas the male produces both X and Y gametes. When fertilization involves an X sperm, the result is an XX zygote, which develops into a female. When fertilization involves a Y sperm, the result is an XY zygote, which develops into a male.

The solution to Morgan's puzzle lies in the fact that in *Drosophila* the white-eye trait resides on the X chromosome but is absent from the Y chromosome. We now know that the Y chromosome carries almost no functional genes, and there is no corresponding locus for the white-eye trait on it. The eye color of males, therefore, is determined by whichever allele is present on their X chromosome. Females that are homozygous recessive for the white-eye allele would have white eyes, but a cross involving a male with red eyes could not produce a progeny that included white-eyed females. A characteristic that is determined by genes located on the sex chromosomes is said to be **sex linked.** Knowing that the white-eye trait is recessive to the red-eye trait, we can now see that Morgan's result was a natural consequence of the Mendelian assortment of chromosomes (Figure 10-16).

Morgan's experiment is one of the most important in the history of genetics because it presented the first clear evidence that Sutton was right and that the factors determining Mendelian traits do indeed reside on the chromosomes. The segregation of the white-eye trait, evident in the eye color of the flies, has a one-to-one correspondence with the segregation of the X chromosome, evident from the sexes of the flies.

The white-eye trait behaves exactly as if it were located on an X chromosome, which is indeed the case. The gene that specifies eye color in *Drosophila* is carried through meiosis as part of an X chromosome. In other words, Mendelian traits such as eye color in *Drosophila* assort independently because chromosomes do. When Mendel observed the segregation of alleles in pea plants, he was observing a reflection of the meiotic segregation of chromosomes.

*Mendelian traits assort independently because they are determined by genes located on chromosomes, which also assort independently in meiosis.*

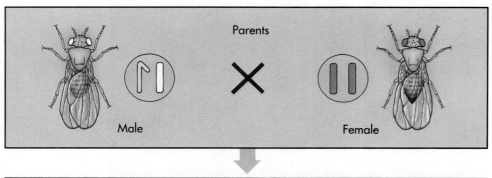

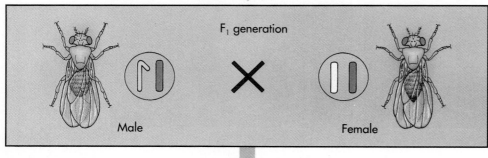

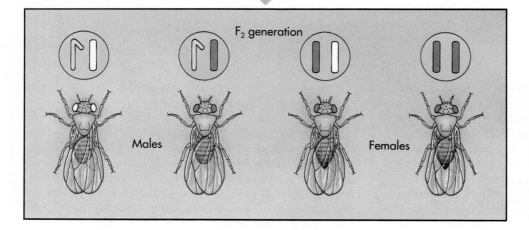

**Figure 10-16**

**Morgan's experiment demonstrating the chromosomal basis of sex linkage in *Drosophila*.** The white-eyed mutant male fly was crossed to a normal female. The $F_1$-generation flies all exhibited red eyes, as expected for flies heterozygous for a recessive white-eye allele. In the $F_2$ generation, all the white-eyed $F_2$-generation flies were male.

## CROSSING-OVER

Morgan's results led to the general acceptance of Sutton's chromosomal theory of inheritance. Scientists then attempted to find out why there are more genes that assort independently than there are chromosomes. In 1931, Curt Stern, an American geneticist, published the experimental proof that the formation of chiasmata between paired chromosomes (see Chapter 9) provided a means by which genes on the same chromosome could reassort into different combinations than those which occur originally. Stern's experiment is diagrammed in Figure 10-17.

Stern studied two sex-linked eye traits in strains of *Drosophila* whose X chromosomes were visibly abnor-

mal at both ends. He first examined many flies in crosses involving strains with different eye traits (one strain had two visible mutations; the other was normal) and identified those in which an exchange had occurred—a fly exhibiting only one of the mutant traits. Stern then studied the chromosomes of these flies to see if their X chromosomes had exchanged arms. He found that all the individuals exhibiting one trait also possessed chromosomes whose abnormal ends could be seen to have exchanged (Figure 10-18). The conclusion was clear: genetic exchanges of traits on a chromosome, such as eye color, involve crossing-over, the physical exchange of chromosome arms.

The patterns of assortment of genes that are **linked,** or located on the same chromosome, indicate

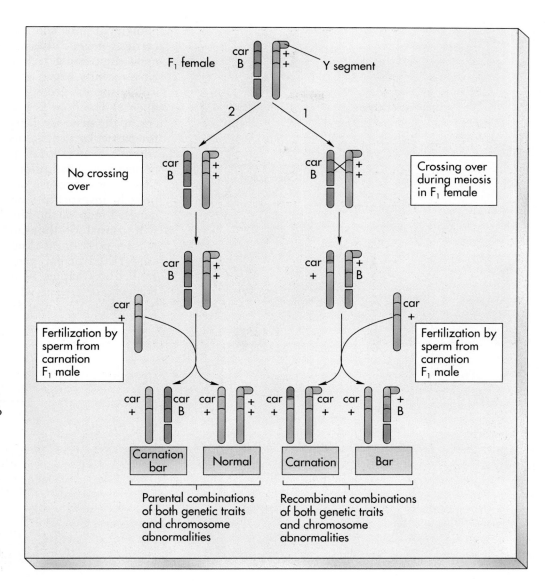

**Figure 10-17**

**Curt Stern's experiment demonstrating that a physical exchange of chromosomal arms occurs during crossing-over.** Stern monitored cross-over between two genes, recessive carnation eye color *(car)* and dominant Bar-shaped eye *(B)*, on chromosomes with physical peculiarities that he could see under a microscope. Whenever genes recombined through crossing-over, chromosomes recombined as well. The recombination of genes reflects a physical exchange of chromosome arms.

Within the figure:

F₁ female car B / + + Y segment

2 — No crossing over

1 — Crossing over during meiosis in F₁ female

Fertilization by sperm from carnation F₁ male

Carnation bar | Normal | Carnation | Bar

Parental combinations of both genetic traits and chromosome abnormalities

Recombinant combinations of both genetic traits and chromosome abnormalities

---

an important fact: in general, if the genes are located relatively far apart, crossing-over is more likely to occur between them than if they are located close together. If the genes are located far enough apart, they segregate independently.

## Genetic Maps

Because crossing-over occurs more often between two genes relatively far apart than between another set of two genes relatively close together, the frequency of crossing-over can be used to map the relative positions of genes on chromosomes. In a cross, the proportion of progeny derived from a gamete in which an exchange has occurred between two genes is a measure of how

frequently cross-over events occur between them and thus of the distance separating them.

A **genetic map** is a diagram showing the relative positions of genes. Genetic maps can be constructed by ordering fragments of DNA, studying chromosomal alterations, or performing crosses and seeing how frequently crossing-over occurs between pairs of genes. When the frequencies of crossing-over events in crosses are used to construct a genetic map, it is called a **recombination map.** Distances are measured in terms of the frequency of crossing-over. On such a map, one "map unit" is defined as the distance within which a cross-over event is expected to occur, on the average, in 1 of 100 gametes. This unit, 1% recombination, is now called a **centimorgan,** in honor of Morgan.

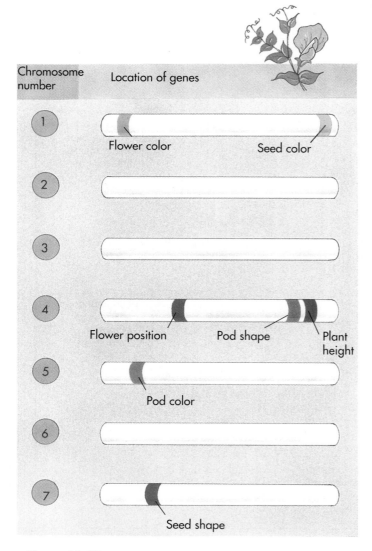

**Figure 10-18**

**The chromosomal locations of the seven genes studied by Mendel in the garden pea.** The genes for plant height and pod shape are very close to one another and do not recombine freely. Pod shape and plant height were not among the pairs of traits that Mendel examined in dihybrid crosses. One wonders what Mendel would have made of the linkage he would surely have detected had he tested this pair of traits.

In constructing a genetic map, one simultaneously monitors recombination among three or more genes that are located on the same chromosome. When such genes are close enough together, they do not segregate independently.

Genetic maps of human chromosomes (Figure 10-19) are of great importance to our own welfare. Knowledge of where particular genes are located on

human chromosomes can often be used to tell if a fetus carries a genetic disorder for which it is at risk. The genetic engineering techniques described in Chapter 14 have recently begun to permit investigators to actually isolate specific genes and determine their nucleotide sequence. The hope is that knowledge of what is different at the gene level may suggest a successful therapy for particular genetic disorders, and that knowledge of where a gene is located on the chromosome will soon permit the substitution of normal genes for dysfunctional ones. Because of the great potential of this approach, investigators are working hard to assemble a detailed map of the human genome. Initially this map will consist of thousands of small lengths of DNA whose position is known. Investigators wishing to study a particular gene will first use techniques described in Chapter 14 to screen this "library" of fragments to determine which one carries their gene. They can then analyze that fragment in detail.

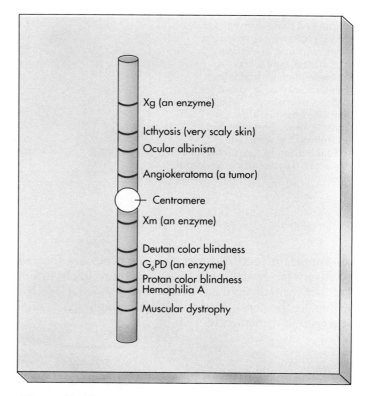

**Figure 10-19**

**A recombination map of the human X chromosome.** At least 93 human genes have been shown to be located on the X chromosome, and preliminary evidence indicates that almost as many others also belong on this chromosome.

# FROM GENOTYPE TO PHENOTYPE: HOW GENES INTERACT

It is important to keep in mind that the "gene" of Mendelian genetics is an abstract concept that Mendel used to explain the results of crosses. Later workers determined that these elements are located on the chromosomes. In fact, as we will learn in Chapter 12, the Mendelian gene is no more than a segment of a DNA molecule. Genes act in ways Mendel did not understand to produce differences among progeny of crosses. How genes work is the subject of Chapter 13.

The relationship between chromosomal genes and the phenotype that Mendelian traits exhibit is not always a simple one to understand, even with the extensive knowledge of DNA that we have today. Mendel was lucky in his choice of traits. Often genes reveal more complex patterns of inheritance than simple 3:1 ratios, including the following:

1. *Multiple alleles.* Although a diploid individual may possess no more than two alleles at one time, this does not mean that only two allele alternatives are possible for a given gene in the entire population. On the contrary, almost all genes that have been studied exhibit several different alleles. The gene that determines the human ABO blood group, for example, has three common alleles (Chapter 11).

2. *Gene interaction.* Few phenotypes are the result of the action of only one gene. Most traits reflect the action of many genes that act sequentially or jointly. When genes act sequentially, as in a biochemical pathway, an allele expressed as a defective enzyme early in the pathway blocks the flow of material through the pathway and thus makes it impossible to judge whether the later steps of the pathway are functioning properly. Such interactions between genes are the basis of the phenomenon called **epistasis.**

3. *Epistasis* is an interaction between the products of two genes in which one of them modifies the phenotypic expression produced by the other. For example, in a two-step biochemical pathway where the gene governing the second step has two alleles yielding black (dominant) or blonde (recessive) hair color, it is impossible to deduce which of these two alleles is present in individuals whose alleles in the first step are nonfunctional—the hair is white, no matter which allele governs step two. Epistatic interactions between genes make the interpretation of particular patterns of inheritance very difficult.

4. *Continuous variation.* When multiple genes act jointly to influence a trait such as height or weight, the contribution caused by the segregation of the alleles of one particular gene is difficult to monitor, just as it is difficult to follow the flight of one bee within a swarm. Because all of the genes that play a role in determining phenotypes such as height or weight are segregating independently of one another, one sees a graduation in degree of difference when many individuals are examined (Figure 10-20).

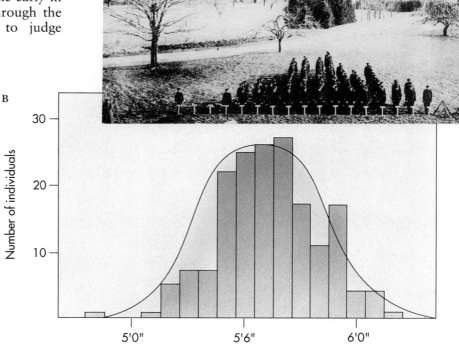

## Figure 10-20

**Height is a continuously varying trait.**
**A** Variation in height among students of the 1914 class of the Connecticut Agricultural College. Because many genes contribute to height and tend to segregate independently of one another, there are many possible combinations.
**B** The cumulative contribution of different combinations of alleles to height forms a continuous spectrum of possible heights—a random distribution, in which the extremes are much rarer than the intermediate values.

5. *Pleiotropy.* Often an individual allele will have more than one effect on the phenotype. Such an allele is said to be **pleiotropic.** Thus, when the pioneering French geneticist Lucien Cuenot studied yellow fur in mice, a dominant trait, he was unable to obtain a true-breeding yellow strain by crossing individual yellow mice with one another—individuals that were homozygous for the yellow allele died. The yellow allele was pleiotropic: one effect was yellow color, another effect was a lethal developmental defect. A pleiotropic gene alteration may be dominant with respect to one phenotypic consequence (yellow fur) and recessive with respect to another (lethal developmental defect). Pleiotropic relationships occur because the characteristics of organisms result from the interactions of products made by genes. These products often also perform other functions about which we may be ignorant.

6. *Incomplete dominance.* Not all alternative alleles are fully dominant or recessive in heterozygotes. Sometimes heterozygous individuals do not resemble one parent precisely. Some pairs of alleles produce instead a heterozygous phenotype that (1) is intermediate between the parents (intermediate or incomplete dominance, Figure 10-21), (2) resembles one allele closely but can be distinguished from it (partial dominance), or (3) is one in which both parental phenotypes can be distinguished in the heterozygote (co-dominance).

7. *Environmental effects.* The degree to which many alleles are expressed depends on the environment. Some alleles encode an enzyme whose activity is more sensitive to conditions such as heat (Figure 10-22) or light than are other alleles.

## Modified Mendelian Ratios

When individuals heterozygous for two different genes mate (a dihybrid cross), four different phenotypes are possible among the progeny: the dominant phenotype of both genes is displayed, one of the two dominant phenotypes is displayed, or neither dominant phenotype is displayed. Mendelian assortment predicts that these four possibilities will occur in the proportions $9:3:3:1$. But sometimes it is not possible for an investigator to successfully identify each of the four phenotypic classes because two or more of the classes look alike.

One example of such difficulty in identification is seen in the analysis of particular varieties of corn, *Zea mays.* Some commercial varieties exhibit a purple pigment called anthocyanin in their seed coats, whereas others do not. When in 1918 the geneticist R.A. Emerson crossed two pure-breeding corn varieties, neither of which typically exhibits any anthocyanin pigment, he obtained a suprising result: all of the $F_1$ plants produced purple seeds. The two white varieties, which had

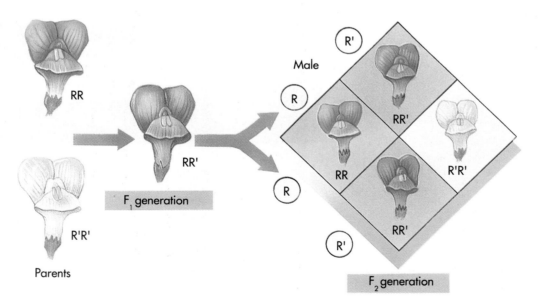

*Figure 10-21*

**Incomplete dominance.** A cross between a pink-flowered snapdragon, which has the genotype RR, and a white-flowered one (R'R'). Neither allele is dominant, and the heterozygotes have purple flowers and the genotype RR'.

never been observed to make pigment, would, when crossed, produce progeny that uniformly make the pigment.

When two of these pigment-producing $F_1$ plants were crossed to produce an $F_2$ generation, 56% were pigment producers and 44% were not. What was happening? Emerson correctly deduced that *two* genes were involved in the pigment-producing process and that the second cross had thus been a dihybrid cross as described by Mendel. Mendel predicted 16 possible genotypes in equal proportions (9 + 3 + 3 + 1 = 16), suggesting to Emerson that the total number of genotypes in his experiment was also 16.

How many of these were in each of the two phenotypic classes Emerson observed? He multiplied the fraction that were pigment producers—.56 × 16 = 9—and multiplied the fraction that were not—.44 × 16 = 7. Thus Emerson had in fact a **modified ratio** of 9:7 instead of the usual 9:3:3:1 ratio. Such a ratio illustrates epistasis, the kind of interaction between two genes that was discussed on page 251.

In this example, the pigment anthocyanin is produced from a colorless molecule by two enzymes that work one after the other. In other words, the pigment is the product of a two-step biochemical pathway:

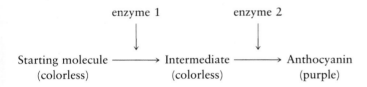

For pigment to be produced, a plant must possess at least one good copy of each enzyme. The dominant alleles encode functional enzymes; the recessive alleles, defective, nonfunctional ones. Of the 16 genotypes predicted by random assortment, 9 contain at least one dominant allele of both genes—these are the purple progeny. The 9:7 ratio that Emerson observed resulted from the pooling of the three phenotypic classes that lack dominant alleles at either or both loci (3 + 3 + 1 = 7), so all looked the same, nonpigmented.

## THE SCIENCE OF GENETICS

After many centuries of speculation about heredity, the puzzle was finally solved in the space of a few generations. Guided by the work of Mendel and a generation of investigators determined to explain his results, the basic outline of how heredity works soon became clear. Hereditary traits are specified by genes, which are integral parts of chromosomes, and the movements of chromosomes during meiosis produce the patterns of segregation and independent assortment that Mendel reported. Two of the most important discoveries were that (1) chromosomes exchange genes during meiosis, and (2) genes located far apart on chromosomes are more likely to have an exchange occur between them. These findings allowed investigators to learn how genes are distributed on chromosomes long before we knew enough to isolate and study them.

This core of knowledge, this basic outline of heredity, has led to a long chain of investigation and questions. What is the physical nature of a gene? How is information encoded within genes? How do genes change, and why? How does a gene create a phenotype? The people who ask such questions are called geneticists, and the body of what they have learned and are learning is called genetics. Genetics is one of the most active subdisciplines of biology. The next four chapters will answer the questions just posed and many others.

*Figure 10-22*

**The environment often influences how genes are expressed.** An arctic fox in winter has a coat that is almost white, so it is difficult to see the fox against a snowy background. In summer the fox's fur darkens to a red brown, in which condition it resembles the color of the tundra over which it runs.

**allele**   One of two or more alternative states of a gene.

**diploid**   Having two sets of chromosomes, referred to as *homologues*. Animals, the dominant phase in the life cycle of plants, and some protists are diploid.

**dominant allele**   An allele that dictates the appearance of heterozygotes. One allele is said to be dominant over another if an individual heterozygous for that allele has the same appearance as an individual homozygous for it.

**gene**   The basic unit of heredity. A sequence of DNA nucleotides on a chromosome that encodes a polypeptide or RNA molecule and so determines the nature of an individual's inherited traits.

**genotype**   The total set of genes present in the cells of an organism. This term is also often used to refer to the set of alleles at a single gene locus.

**haploid**   Having only one set of chromosomes. Gametes, certain protists, and certain stages in the life cycle of plants are haploid.

**heterozygote**   A diploid individual carrying two different alleles of a gene on its two homologous chromosomes. Most human beings are heterozygous for many genes.

**homozygote**   A diploid individual whose two copies of a gene are the same. An individual carrying identical alleles on both homologous chromosomes is said to be *homozygous* for that gene.

**locus**   The location of a gene on a chromosome.

**phenotype**   The realized expression of the genotype. The phenotype is the observable expression of a trait (affecting an individual's structure, physiology, or behavior), which results from the biological activity of proteins or RNA molecules transcribed from the DNA.

**recessive allele**   An allele whose phenotypic effects is masked in heterozygotes by the presence of a dominant allele.

## ■ SUMMARY

1. Knight and others noted the basic facts of heredity a century before Mendel. They found that alternative traits segregate in crosses and may mask each other's appearance. Mendel, however, was the first to quantify his data, counting the numbers of each alternative type among the progeny of crosses.

2. From counting progeny classes in crosses involving single traits, Mendel learned that the alternatives that were masked in hybrids appeared only 25% of the time when they subsequently segregated in the $F_2$ generation. This finding, which led directly to one of Mendel's most significant hypotheses concerning heredity, is usually referred to as the Mendelian ratio of $3:1$, the ratio of dominant to recessive traits.

3. Mendel deduced from the 3:1 ratio that traits are specified by what he called discrete "factors," which do not blend. We refer to Mendel's factors as genes and to his alternative factors as alleles. Often one member of a pair of alleles is dominant, the other recessive.

4. When two heterozygous individuals mate, the probability of obtaining two dominant alleles is $0.5 \times 0.5 = 0.25$, or 25%, and the probability of obtaining two recessive alleles is the same. The rest of the progeny, half, are heterozygotes; their phenotypes resemble the homozygous dominants. For this reason, the progeny thus appear as three-fourths dominant, one-fourth recessive, a ratio of $3:1$ dominant to recessive.

5. When two genes are located on nonhomologous chromosomes, the alleles included in an individual gamete are selected at random. The allele selected for one gene has no influence on which allele is selected for the other gene. Such genes are said to assort independently.

6. Thomas Hunt Morgan provided the first clear evidence that genes reside on chromosomes. Morgan demonstrated that the segregation of the white-eye trait in *Drosophila* flies was associated with the segregation of the X chromosome, the one responsible for sex determination. Genetic traits located on the sex chromosomes are said to be sex linked.

7. Curt Stern reported the first evidence that crossing-over occurs between chromosomes. He showed that when two Mendelian traits exchange during a cross, visible abnormalities on the ends of the chromosomes bearing the traits also exchange.

8. The frequency of crossing-over between genes can be used to construct genetic maps. Such maps are representations of the physical locations of genes on chromosomes, inferred from the degree of crossing-over between particular pairs of genes.

9. Traits are often influenced by the action of several genes. Sometimes a defect in one gene will mask your ability to detect the alleles of another gene ("epistasis"), and often the segregation of such traits will not exhibit simple Mendelian ratios.

# REVIEW

1. The set of alleles that an individual contains is called its _____, whereas the physical appearance of that individual is its _____.

2. A garden pea plant that is homozygous for the recessive white-flowered trait will produce gametes of what kind?

3. A purple-flowered garden pea can have one of two genotypes. How would you unambiguously determine the genotype of one individual?

4. Which of Mendel's laws is demonstrated by the 9:3:3:1 ratio observed in the $F_2$ generation of a dihybrid cross?

5. Because _____ occurs more frequently between genes far from each other on a chromosome and less frequently between genes close together, this information can be used to determine the relative position of genes on a chromosome.

6. In a cross of purple- and white-flowered garden peas, the $F_1$ offspring observed by Mendel were all purple. A cross of two of these purple plants yielded $F_2$ purple- and white-flowered plants in the ratio of _____.

7. When two heterozygous individuals are crossed, the percentage of the progeny that exhibit the recessive trait is _____.

8. The exchange of segments of chromosomes during meiosis is called _____.

9. In Sweden, many more people possess blue eyes than eyes of all other colors combined. It follows that the allele determining blue eyes that occurs in Sweden is dominant to most other eye color alleles. True or false?

10. The closer two genes are on a chromosome, the less likely that crossing-over will occur between them. True or false?

11. When two heterozygous individuals are crossed, the proportion of the progeny that are true-breeding for the dominant trait is _____.

12. Under what circumstances does Mendel's Second Law (independent assortment) not hold?

# SELF-QUIZ

1. Several people had carried out plant crosses before Mendel, but none is credited with being the founder of the science of genetics. What did Mendel do that was different?
   (a) He used garden peas.
   (b) He did specific crosses.
   (c) He kept track of the numbers of individuals of the different types produced from his crosses.
   (d) He removed the male parts from flowers so that self-fertilization could not occur.
   (e) Both a and b.

2. In all of Mendel's crosses, the $F_2$ plants displayed a 3:1 ratio of dominant to recessive traits. Of those showing the dominant trait, what proportion were true-breeding?
   (a) One-fourth
   (b) One-third
   (c) One-half
   (d) Two-thirds
   (e) All

3. Which of the following is *not* one of the five elements of Mendel's model?
   (a) Parents transmit discrete "factors" of information that will act in the offspring to produce specific traits.
   (b) The presence of any particular element does not guarantee its expression.
   (c) The two elements contributed, one by each parent, do not influence each other.
   (d) All copies of "factors" are identical.
   (e) Each parent, with respect to each trait, contains two factors, one of which will be packaged at random into the gametes.

4. The white color of a homozygous recessive plant is its
   (a) phenotype.
   (b) genotype.

5. Of the following crosses, which is a test cross?
   (a) $WW \times WW$  (c) $Ww \times ww$
   (b) $WW \times Ww$  (d) $Ww \times W$

6. Morgan's experiments on white-eyed *Drosophila* clearly showed that
   (a) white-eyed flies see better than normal flies.
   (b) chromosomes are the carriers of genetic information.
   (c) only female flies can have white eyes.
   (d) diploid individuals have two copies of each trait.
   (e) because eggs carry not only the genetic factors but large amounts of cytoplasm, the female has a greater influence on the genotype of the offspring.

7. Of the seven genes that Mendel studied, only two are located by themselves on a particular chromosome. The others are on a chromosome with another of Mendel's traits. In view of this, how could Mendel have come to the conclusion that genes assort independently of one another?
   (a) The data are not clear on this point.
   (b) They did not assort independently.
   (c) Mendel erred.
   (d) The traits were far enough apart that, because of the high probability of crossing-over between them, they behaved as if they were on different chromosomes.
   (e) Mendel knew that the traits were on the same chromosome and took this into account.

8. What is "independent assortment"?
   (a) Different genes on the same chromosome segregate independently of one another in genetic crosses.
   (b) Different alleles of the same gene segregate independently in genetic crosses.
   (c) Different genes on different chromosomes segregate independently of one another in genetic crosses.

9. How many alleles of a given gene are possible?
   (a) One
   (b) Two
   (c) Three
   (d) Four
   (e) Any number

# THOUGHT QUESTIONS

1. Many sexually reproducing lizard species are able to generate local populations that reproduce asexually. What do you imagine the sex of the local asexual populations would be, male or female or neuter? Explain your reasoning.

2. The plant *Haplopappus gracilis* has only 2 chromosomes, whereas the adder's tongue fern has 1,262. Can you suggest a reason for this wide variation in chromosome number? There is much less variation among mammals. Can you propose an explanation?

3. Imagine you were constructing an artificial chromosome. What elements would you introduce into it, at a *minimum*, so that it would function normally in mitosis?

# MENDELIAN GENETICS PROBLEMS

1. Why did Mendel observe only two alleles of any given trait in the crosses that he carried out?

2. The illustration at right describes Mendel's cross of *wrinkled* and *round* seed characters. What is wrong with this diagram?

3. The annual plant *Haplopappus gracilis* has two pairs of chromosomes 1 and 2. In this species, the probability that two traits *a* and *b* selected at random will be on the same chromosome of *H. gracilis* is the probability that they will both be on chromosome 1 ($\frac{1}{2}$), multiplied by the probability that they will both be on chromosome 2 (also $\frac{1}{2}$): $\frac{1}{2} \times \frac{1}{2} = \frac{1}{4}$, or 25%. This is often symbolized

   $$\frac{1}{2}^{(\# \text{ of pairs of chromosomes})}$$

   Human beings have 23 pairs of chromosomes. What is the probability that any two human traits selected at random will be on the same chromosome?

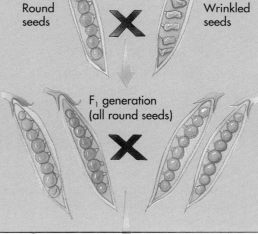

P generation

Round seeds    X    Wrinkled seeds

F₁ generation (all round seeds)

X

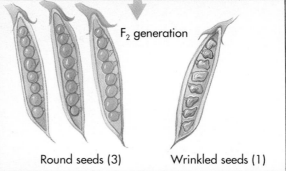

F₂ generation

Round seeds (3)    Wrinkled seeds (1)

4. Among Hereford cattle there is a dominant allele called *polled;* the individuals that have this allele lack horns. After college, you become a cattle baron and stock your spread entirely with polled cattle. You personally make sure that each cow has no horns. Among the calves that year, however, some grow horns. Angrily you dispose of them and make certain that no horned adult has gotten into your pasture. The next year, however, more horned calves are born. What is the source of your problem? What should you do to rectify it?

5. An inherited trait among human beings in Norway causes affected individuals to have very wavy hair, not unlike that of a sheep. The trait is called *woolly*. The trait is very evident when it occurs in families; no child possesses woolly hair unless at least one parent does as well. Imagine you are a Norwegian judge and that you have before you a woolly-haired man suing his normal-haired wife for divorce because their first child has woolly hair but their second child has normal long, blond hair. The husband claims this constitutes evidence of infidelity on the part of his wife. Do you accept his claim? Justify your decision.

6. In human beings, Down syndrome, a serious developmental abnormality, results from the presence of three copies of chromosome 21 rather than the usual two copies. If a female exhibiting Down syndrome mates with a normal male, what proportion of their offspring would be expected to be affected?

7. Many animals and plants bear recessive alleles for *albinism,* a condition in which homozygous individuals completely lack any pigments. An albino plant lacks chlorophyll and is white. An albino person lacks any melanin pigment. If two normally pigmented persons heterozygous for the same albinism allele mate, what proportion of their children would be expected to be albino?

8. Your uncle dies and leaves you his race horse, Dingleberry. To obtain some money from your inheritance, you decide to put the horse out to stud. In looking over the stud book, however, you discover that Dingleberry's grandfather exhibited a rare clinical disorder that leads to brittle bones. The disorder is hereditary and results from homozygosity for a recessive allele. If Dingleberry is heterozygous for the allele, it will not be possible to use him for stud because the genetic defect may be passed on. How would you go about determining whether Dingleberry carries this allele?

9. In the fly *Drosophila,* the allele for dumpy wings (symbolized $d$) is recessive to the normal long-wing allele (symbolized $D$). The allele for white eye (symbolized $w$) is recessive to the normal red-eye allele (symbolized $W$). In a cross of $DDWw \times Ddww$, what proportion of the offspring are expected to be "normal" (long wing, red eye)? What proportion "dumpy, white"?

10. As a reward for being a good student, your instructor presents you with a *Drosophila* named Oscar. Oscar has red eyes, the same color that normal flies possess. You add Oscar to your fly collection, which also contains Heidi and Siegfried, flies with white eyes, and Dominque and Ronald, which are from a long line of red-eye flies. Your previous work has shown that the white-eye trait exhibited by Heidi and Siegfried is caused by their being homozygous for a recessive allele. How would you determine whether or not Oscar was heterozygous for this allele?

11. In some families, children are born who exhibit recessive traits (and therefore must be homozygous for the recessive allele specifying the trait), even though one or both of the parents do not exhibit the trait. What can account for this occurrence?

12. You collect in your backyard two individuals of *Drosophila melanogaster*, one a young male and the other a young, unmated female. Both are normal in appearance, with the typical vivid red eyes of *D. melanogaster*. You keep the two flies in the same vial, where they mate. Two weeks later, hundreds of little offspring are flying around in the vial. They all have normal red eyes. From among them, you select 100 individuals, some male and some female. You cross each individually to a fly you know to be homozygous for a recessive allele called "sepia," which leads to black eyes when homozygous (these flies thus have the genotype *se/se*). Examining the results of your 100 crosses, you observe that in about half of them, only normal red-eyed progeny flies are produced. In the other half, however, the progeny are about 50% red-eyed and 50% black-eyed. What must have been the genotypes of your original backyard flies?

13. Hemophilia is a recessive sex-linked human blood disease that leads to failure of blood to clot normally. One form of hemophilia has been traced to the royal family of England, from which it spread throughout the royal families of Europe. For the purposes of this problem, assume that it originated as a mutation either in Prince Albert or in his wife, Queen Victoria (the actual explanation is in Chapter 12).

    (a) Prince Albert did not himself have hempophilia. If the disease is a sex-linked recessive abnormality, how could it have originated in Prince Albert (who was male and therefore would have been expected to exhibit recessive sex-linked traits), as he did not suffer from hemophilia?

    (b) Alexis, the son of Czar Nicholas II of Russia and Empress Alexandra (a granddaughter of Victoria), had hemophilia, but their daughter Anastasia did not. Anastasia died, a victim of the Russian revolution, before she had any children. Can we assume that Anastasia would have been a carrier of the disease? How is your answer influenced if the disease originated in Nicholas II or in Alexandra?

# FOR FURTHER READING

BLIXT, S.: "Why Didn't Gregor Mendel Find Linkage?" *Nature*, vol. 256, 1975, page 206. An interesting examination of the pea strains first studied by Mendel.
MORGAN, T.H.: "Sex-limited Inheritance in *Drosophila*," *Science*, vol. 32, 1910, pages 120-122. Morgan's original account of his famous analysis of the inheritance of the white-eye trait.
SUTTON, W.S.: "The Chromosomes of Heredity," *Biological Bulletin*, vol. 4, 1903, pages 213-251. The original statement of the chromosomal theory of heredity.

# Human Genetics

This child is competing in the Special Olympics. Many of the participants in the Special Olympics are afflicted with Down syndrome, a condition caused by an extra copy of chromosome 21. Although research is currently being conducted on the nature of Down syndrome, no means of prevention has yet been discovered.

# HUMAN GENETICS

## Overview

The principles of genetics apply not only to pea plants and *Drosophila*, but also to you. How closely you resemble your father or mother was largely established before your birth by the chromosomes that you received from them, just as meiosis in peas determined the segregation of Mendel's traits. But many of the alleles segregating within human populations demand more serious concern than the color of a pea. Some of the most devastating human disorders result from alleles specifying defective forms of proteins with important functions in our bodies. These disorders are passed from parent to child. By studying human heredity, we are beginning to learn how to limit the misery these disorders bring to so many human families.

## For Review
*Here are some important terms and concepts that you will encounter in this chapter. If you are not familiar with them, you should review them before proceeding.*

**Cell-surface proteins** (Chapter 5)

**Transport channels** (Chapter 5)

**Cholesterol uptake** (Chapter 5)

**Cystic fibrosis** (Chapter 5)

**Karyotypes** (Chapter 9)

**Sex linkage and sex chromosomes** (Chapter 10)

**Dominant and recessive traits** (Chapter 10)

Biologists often investigate heredity by studying odd creatures such as the *Drosophila* fly, organisms that we may care about very little. But the knowledge that biologists acquire by studying other organisms often tells us a great deal about ourselves. Nowhere is this more true than in the study of genetics, the study of the ways in which individual traits are transmitted from one generation to the next. When biologists studying pea plants and *Drosophila* established that patterns of heredity reflect the segregation of chromosomes in meiosis, their discovery opened the door to the study of human heredity. By studying human chromosomes, it is possible to learn why some hereditary disorders seem to leap generations and why others affect only males. Scientists are more able to predict which disorders parents might expect to pass on to their children, and with what probabilities (Figure 11-1).

Why do we devote a special chapter to human heredity? Because we are interested in ourselves. Although we humans pass our genes to the next generation in much the same way that other organisms do, we naturally have a special curiosity about ourselves.

Most of us know someone, whether a relative or a friend's relative, who suffers from a condition that might be hereditary. If a member of your family has had a stroke, it is difficult not to worry about your own future health, knowing that the propensity to suffer strokes can be hereditary. Few parents have babies without worrying about the possibility that their child may have some defect. Genes are also clearly involved in such conditions as adult diabetes, manic depression, and alcoholism, and the ways in which these genes interact with the environment to produce individuals with specific characteristics are the subject of continuing, intensive study. Because of the importance of genes in determining the course of our lives, we are all human geneticists, interested in what the laws of genetics reveal about ourselves and our families.

## HUMAN CHROMOSOMES

Although chromosomes were discovered more than a century ago, the exact number of chromosomes that

**Figure 11-1**

**Czar Nicholas II of Russia, the last of the czars, and his wife Alexandra with their children: Olga, Tatiana, Maria, Anastasia, and Alexis.** Alexandra was a carrier of the genetic disease hemophilia, a disorder in which the blood does not clot properly, and she passed it on to her son Alexis.

humans possess was not established accurately until 1956. By that time, techniques had been developed that allowed investigators to determine accurately the number and form of human and other mammalian chromosomes. As a result of the application of these techniques, we now know that each typical human cell has 46 chromosomes.

How do biologists examine human chromosomes? They collect a blood sample, add chemicals that induce the blood cells in the sample to divide, and then add other chemicals that arrest cell division at metaphase. Metaphase is the stage of mitosis when the chromosomes are most condensed and thus most easily distinguished from one another. The cells are then flattened, spreading out their contents and separating the individual chromosomes for examination. The chromosomes are stained and photographed, and the karyotype is prepared as we discussed in Chapter 9, with the photographs of the individual chromosomes arranged in order of descending size (Figure 11-2).

Of the 23 pairs of human chromosomes, 22 consist of members that are similar in size and morphology in both males and females; these chromosomes are called **autosomes.** The two members of the remaining pair, the sex chromosomes, are unlike each other in males and similar in females. As in *Drosophila*, females are designated XX and males are designated XY; the Y chromosome, in both organisms, bears few functional genes. In humans the Y chromosome is much smaller than the X chromosome, whereas in *Drosophila*, as we saw in Chapter 10, it is larger. Conventionally, the 23 pairs of human chromosomes, including the sex chromosomes, are arranged by their outlines into seven groups, designated A through G.

## Down Syndrome

Individuals with karyotypes that differ from the normal condition often are abnormal, sometimes seriously so. The karyotype seen in Figure 11-2 is characteristic of a very high proportion of all male human cells. Almost everyone has the same karyotype for the same reason that almost all cars have engines, transmissions, and wheels: other arrangements don't work well. Human individuals who are missing even one autosome (called **monosomics**) do not survive embryonic development. In all but a few cases, those who have received

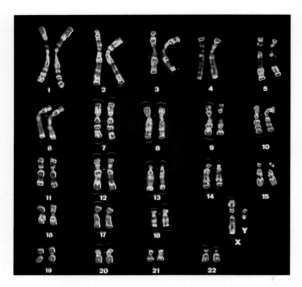

**Figure 11-2**

**A human karyotype.** In a karyotype such as the one shown here, the two members of each chromosome pair are grouped together, and the autosomes are arranged in order of descending size. Those chromosomes in which the centromere is near the middle are placed first within a given size class; chromosomes with the centromere located nearer one end are placed after them. A normal male has an X chromosome and a much smaller Y chromosome, as shown, and a normal female has two X chromosomes.

an extra autosome (called **trisomics**) also do not survive. Five of the smallest chromosomes, those numbered 13, 15, 18, 21, and 22, can be present in human beings as 3 copies and still allow the individual to survive for a time. Individuals with an extra chromosome 13, 15, or 18 undergo severe developmental defects, and infants with such a genetic constitution die within 3 months. In contrast, individuals who have an extra copy of chromosome 21 (Figure 11-3), and more rarely those who have an extra copy of chromosome 22, usually do survive to become adults. In such individuals, the maturation of the skeletal system is delayed, so generally they are short and have poor muscle tone. Their mental development is also affected, and children with trisomy 21 or trisomy 22 are always mentally retarded. The developmental defect produced by trisomy 21 was first described in 1866 by J. Langdon Down; thus it is called *Down syndrome* (formerly "Down's syndrome").

Down syndrome occurs frequently in all human racial groups, with an approximate incidence of 1 in every 750 children. Similar conditions also appear in chimpanzees and other related primates. In humans the defect is associated with a particular, small portion of chromosome 21. When this chromosomal segment is present in three copies instead of two, Down syndrome

results. In 97% of the human cases examined, all of chromosome 21 is present in three copies. In the other 3%, a small part of chromosome 21 containing the critical segment has been added to another chromosome, in addition to the normal two copies of chromosome 21.

The developmental role of the genes whose duplication produces Down syndrome is not known in detail, although clues are beginning to emerge. When human cancer-causing genes (see Chapter 13) were identified and localized on the chromosomes in 1985, one of them turned out to be located on chromosome 21 at precisely the location of the segment associated with Down syndrome. Cancer is indeed more common in children with Down syndrome. The incidence of leukemia is 11 times higher in children with Down syndrome than in children of the same age without Down syndrome.

What causes Down syndrome in the first place? In humans, it results almost exclusively from **primary nondisjunction** (the failure of homologous chromosomes to separate in meiosis I) of chromosome 21 in the meiotic event that leads to the formation of the egg. The cause of these failures to separate is not known, but, as with cancer, their incidence increases with age (Figure 11-4). In mothers less than 20 years of age, the incidence of Down syndrome is only about 1 per 1700 births; in mothers 20 to 30 years old, the risk is only slightly greater, about 1 per 1400. In mothers 30 to 35 years old, however, the risk doubles, to 1 per 750. In mothers older than 45, the risk is as high as 1 in 16 births.

Primary nondisjunctions occur much more often in women than in men because *all* the eggs that a woman will ever produce are present in her ovaries by the time she is born. In males, the development of new sperm cells continues throughout adult life. A much greater chance thus exists for problems of various kinds, including primary nondisjunction, to accumulate over time in the gametes of women than in those of men. For this reason, the age of the mother is more critical than that of the father when a couple contemplates childbirth.

## Sex Chromosomes

As you have seen, humans possess 22 pairs of autosomes that are the same in both sexes and one pair of sex chromosomes that differs between males and females. Females are designated XX because they have two sex chromosomes that are similar in size and appearance, although not necessarily in their genetic content, and males are designated XY because their sex chromosomes differ. The difference between them can be seen readily in Figure 11-3. The Y chromosome is

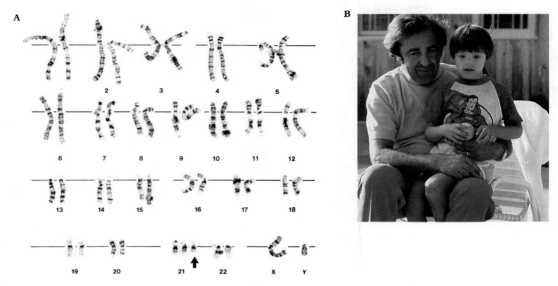

*Figure 11-3*

**Down syndrome.**
**A** As shown in this karyotype of a male, Down syndrome is usually associated with trisomy of chromosome 21.
**B** A child with Down syndrome sitting on his father's knee.

highly condensed, and few of its genes are used. For this reason, recessive alleles on the single X chromosome of males have no counterpart on the Y chromosome, or at least no active counterpart. The characteristics of such alleles are often expressed, just as if they were present in a homozygous condition in a female. However, the Y chromosome is not completely inert genetically: it does possess some active genes. For example, the genes that cause a zygote to develop into a male are located on the Y chromosome: any individual containing at least one Y chromosome is a male, and any individual containing no Y chromosome is a female. The number of X chromosomes an individual possesses does not determine the individual's sex.

Because a female has two copies of the X chromosome and a male only one, you might think that female cells would produce twice as much of the proteins encoded by genes on the X chromosome. This does not occur because in females one of the X chromosomes is inactivated shortly after sex determination, early in embryonic development. Such inactivation is called **Lyonization.** The inactivated chromosome can be seen as a deeply staining body, the **Barr body,** which remains attached to the nuclear membrane.

Individuals who lose a copy of the X chromosome or gain an extra one are not subject to the severe developmental abnormalities usually associated with similar changes in autosomes. When primary nondisjunction

results in the acquisition of an extra X chromosome in a particular individual or in a female with only one X chromosome, the resulting individuals may become

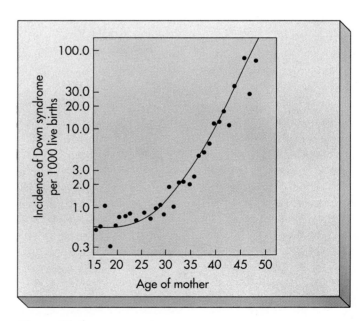

*Figure 11-4*

**As women age, the chance they will bear a child with Down syndrome increases.** After age 35, the frequency of nondisjunction of chromosome 21 increases rapidly.

mature, as discussed next, although they have somewhat abnormal features.

***The X Chromosome.*** Variation in the numbers of X chromosomes that are present is also associated with genetic abnormalities. These conditions come about because of the occasional failure of the X chromosome to separate, which yields some gametes that are XX and others that have no sex chromosome (designated O). If the XX gamete joins an X gamete to form an XXX zygote, the zygote that results develops into a female individual with one functional X chromosome and two Barr bodies. She is sterile but usually normal in other respects. If the XX gamete instead joins a Y gamete, the effects are more serious. The resulting XXY zygote develops into a sterile male who has many female body characteristics and, in some cases, diminished mental capacity. This condition, called **Klinefelter syndrome,** occurs in about 1 out of every 1000 male births.

If an O gamete, produced when the X chromosomes fail to separate, fuses with a Y gamete, the resulting OY zygote is nonviable and fails to develop further. Human beings cannot survive without any of the genes on the X chromosome. If, on the other hand, the O gamete fuses with an X gamete to form an XO zygote, the result is a sterile female of short stature, a webbed neck, and sex organs which resemble those of an infant. The mental abilities of an XO individual are in the low-normal range. This condition, called **Turner syndrome,** occurs roughly once in every 2000 female births. Figure 11-5 diagrams the ways in which nondisjunction can result in abnormalities in the number of sex chromosomes.

***The Y Chromosome.*** The Y chromosome may also fail to separate in anaphase II of meiosis. This leads to the formation of YY gametes and viable XYY zygotes, which develop into fertile males of normal appearance. The frequency of XYY among newborn males is about 1 per 1000. Interestingly, the frequency of XYY males in penal and mental institutions has been reported to be approximately 2% (that is, 20 per 1000), which is about 20 times the frequency of such individuals in the population at large. This observation has led to the suggestion that XYY males may be inherently antisocial, a suggestion that has proved highly controversial. The observation is confirmed in some studies but not in others. In any case, most XYY males do not appear to develop patterns of antisocial behavior.

## PATTERNS OF INHERITANCE

Imagine that you wanted to learn about an inherited trait that was present in your family. How would you find out if the trait is dominant or recessive, how many genes contribute to it, and how likely you might be to transmit it to your future children? If you wanted to study such a trait in *Drosophila,* you could conduct crosses, occasionally squashing a fly to examine its chromosomes. Studying your own heredity requires a more indirect approach.

To study human heredity, biologists look at the results of crosses that have already been made—they study family trees, called **pedigrees.** By identifying which relatives exhibit a trait, it is often possible to determine whether the gene producing the trait is sex-linked or autosomal and whether the trait's phenotype is dominant or recessive. Often one can infer which individuals are homozygous and which are heterozygous for the allele specifying the trait.

### A Pedigree: Albinism
Albino individuals lack all pigmentation, so that their hair and skin are completely white. In the United States, about 1 Caucasian person in 38,000 and 1 black person in 22,000 are albinos. In the pedigree of albinism presented in Figure 11-6, each symbol represents one individual in the family history, with the circles representing females and the squares, males. In

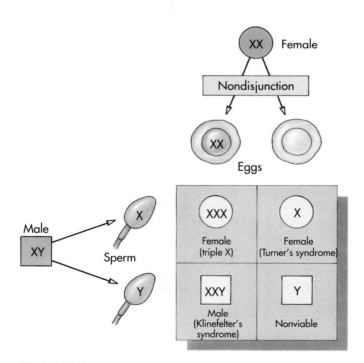

***Figure 11-5***

**How nondisjunction can result in abnormalities in the number of sex chromosomes.**

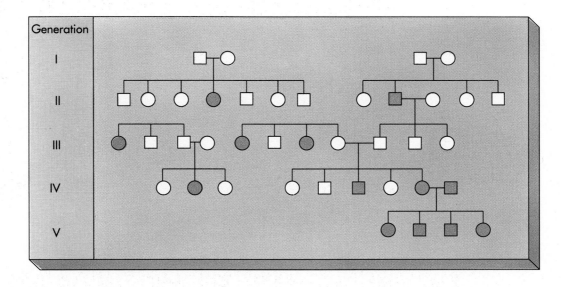

*Figure 11-6*

**A pedigree typical of albinism.** In pedigrees, males are conventionally shown as squares, females as circles, and marriages as horizontal lines connecting them, with the offspring shown below. The individuals who exhibit the trait being considered are indicated by solid symbols.

such a pedigree, individuals that exhibit a trait being studied, in this case albinism, are indicated by solid black symbols. Marriages are represented by horizontal lines connecting a circle and a square, from which a cluster of vertical lines indicate the children, arranged in order of their birth.

How does one analyze the pedigree in Figure 11-6?

1. First, let us inquire whether albinism is sex-linked or autosomal. If the trait is sex-linked, it is usually seen only in males; if it is autosomal, it will appear in both sexes equally. In Figure 11-6, the proportion of affected males ($^{5}/_{13}$, or 39%) is similar to the proportion of affected females ($^{8}/_{21}$, or 37%). Thus we can conclude that the trait is autosomal rather than sex-linked.

2. Second, let us ask whether albinism is dominant or recessive. If the trait is dominant, every albino child will have an albino parent; if recessive, an albino child's parents can appear normal (both parents may be heterozygous). In Figure 11-6, parents of most of the albino children do not exhibit the trait, which indicates that albinism is recessive. Four children in one family *do* have albino parents because the allele is very common among the Hopi Indians, from which this pedigree was derived. Thus, homozygous individuals, such as these albino parents, are present among the Hopis in sufficient numbers

that they sometimes marry. Note that in this marriage, *both* parents are albino and *all* four children are albino, as you would expect if the trait were recessive and both parents were therefore homozygous for this allele.

3. Third, let us see if the trait is determined by a single gene or by several. If the trait is determined by a single gene, then albinos born to heterozygous parents should occur in families in 3:1 proportions, reflecting Mendelian segregation in a cross. Thus you would expect that about 25% of the children should be albinos. Conversely, if the trait were determined by several genes, the proportion of albinos would be much lower, only a few percent. In this case, $^{9}/_{34}$ (you don't count the four children of the marriage between two homozygous individuals because this is not a cross between heterozygotes), or 27%, of the children are albinos, strongly suggesting that only one gene is segregating in these crosses.

This discussion indicates how pedigree analysis is done. Looking at the pattern of inheritance of albinism, we were able to learn that albinism is an autosomal recessive trait controlled by a single gene. Other traits are studied in a similar way. As a second example, consider red-green color blindness, a rare, inherited trait in humans. In the pedigree shown in Figure

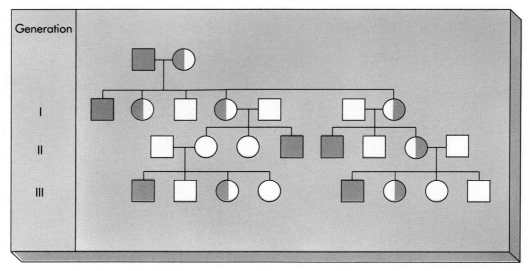

**Figure 11-7**

**A pedigree of color blindness.** Squares indicate males, circles indicate females. Solid symbols indicate a color-blind individual; half-filled symbols indicate a heterozygous individual who carries the trait but does not express it.

11-7, a color-blind man has five children with a woman who is heterozygous for the allele.

1. Is red-green color blindness sex-linked or autosomal? Of the seven affected individuals, all are male. The trait is clearly sex-linked.

2. Is red-green color blindness dominant or recessive? If the trait is dominant, then every color-blind child should have a color-blind parent—but in this pedigree that is not true in any family after that of the original male. The trait must be recessive.

3. Is the trait determined by a single gene? If so, then children born to heterozygous parents should be color-blind in about 25% of cases, reflecting a 3:1 Mendelian segregation of the trait. In this case $4/14$, or 28%, of the children of heterozygous mothers are color-blind, indicating that a single gene is segregating.

*In pedigree analysis, one acquires information about the phenotypes of family members to infer the genetic nature of a trait from the pattern of its inheritance.*

## MULTIPLE ALLELES

Mendel studied pairs of contrasting traits in pea plants. His plants were either tall or short, and their flowers were either purple or white. Similarly, Morgan's *Drosophila* flies had eyes that were either white or red (the normal color). Many human genes also exhibit two alternative alleles. For example, an individual may be either albino or pigmented. But remember that many genes possess more than two possible alleles. One such gene encodes an enzyme that adds sugar molecules to lipids on the surface of our body's blood cells. These sugars act as recognition markers in our immune system and are called **cell surface antigens.** The gene encoding the enzyme is designated "I" and possesses three common alleles: (1) allele $I^B$, which adds the sugar galactose; (2) allele $I^A$, which adds a modified form of the sugar, galactosamine; and (3) allele i, which does not add a sugar.

When more than one allele occurs, which allele is dominant? Often, no one allele is dominant. Instead, each allele has its own effect. Thus an individual heterozygous for the $I^A$ and $I^B$ alleles of the I gene produces both forms of the enzyme and adds both galactose and galactosamine to lipids on the cell surface. The cell surfaces of this individual's blood cells thus possess antigens with both kinds of sugar attached to them. Because both alleles are expressed simultaneously in heterozygotes, the $I^A$ and $I^B$ alleles are said to be **co-dominant.** Either is dominant over the i allele because in heterozygotes the $I^A$ or $I^B$ allele leads to sugar addition and the i allele does not.

*Many genes possess multiple alleles, several of which may be common within populations.*

## ABO Blood Groups

Different combinations of the three possible I gene alleles occur in different individuals because each person possesses two copies of the chromosome bearing the I gene and may be homozygous for any allele or heterozygous for any two. The different combinations of the three alleles produce four different phenotypes:

1. Persons who add only galactosamine are called **type A** individuals. They are either $I^A I^A$ homozygotes or $I^A i$ heterozygotes.
2. Persons who add only galactose are called **type B** individuals. They are either $I^B I^B$ homozygotes or $I^B i$ heterozygotes.
3. Persons who add both sugars are called **type AB** individuals. They are, as we have seen, $I^A I^B$ heterozygotes.
4. Persons who add neither sugar are called **type O** individuals. They are $I i I i$ homozygotes.

These four different cell-surface phenotypes are called the ABO blood groups or, less often, the Landsteiner blood groups after the man who first described them. As Landsteiner first noted, your immune system can tell the difference between these four phenotypes. If a type A individual receives a transfusion of type B blood, the recipient's immune system will recognize that the type B blood cells possess a "foreign" antigen (galactose) and attack the donated blood cells. If the donated blood is type AB, this will also happen. However, if the donated blood is type O, no attack will occur because no foreign galactose antigens are present on the surfaces of blood cells produced by the type O donor. In general, any individual's immune system will tolerate a transfusion of type O blood, so type O individuals are called **universal donors.** Because neither galactose nor galactosamine is foreign to type AB individuals (they add both to their red blood cells), an AB individual may receive any type of blood and is called a **universal recipient.** Figure 11-8 shows the combinations in which agglutination of blood cells will occur.

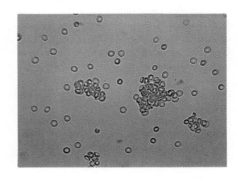

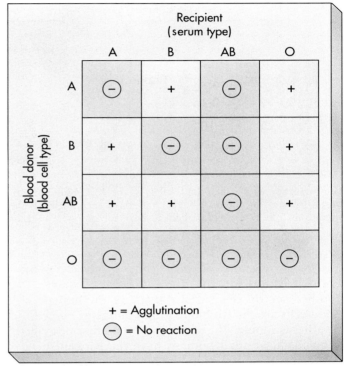

*Figure 11-8*

**The agglutination reaction.** Type A red blood cells agglutinate in type O serum. The unclumped red blood cells are type O cells. The diagram shows the combinations in which agglutination will occur.

In human populations, some of the ABO blood group phenotypes are more common than are others (Table 11-1). In general, type O individuals are the most common, and type AB individuals the least common. However, human populations differ greatly from one another. Among North American Indians the frequency of type A individuals is 31%, whereas among South American Indians, it is only 4%. Figure 11-9 illustrates the frequency with which the $I^B$ allele occurs in different parts of the world. This sort of genetic variation is an important property of genes in human and most other populations.

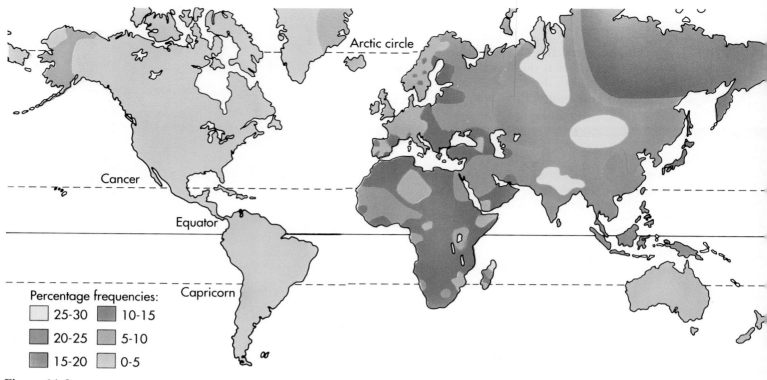

**Percentage frequencies:**

- 25-30
- 20-25
- 15-20
- 10-15
- 5-10
- 0-5

**Figure 11-9**

**The frequency with which the B allele of the ABO blood group occurs throughout the world.** By studying the distribution pattern of such alleles, biologists can often trace past migrations. The B allele is thought to have spread out from central Asia by migration.

### The Rh Blood Group

Another set of cell surface markers on human red blood cells are the **Rh blood group** antigens, named for the Rhesus monkey in which they were first described. About 85% of adult humans have the Rh cell surface marker on their red blood cells and so are called Rh positive (Rh+). Rh-negative (Rh−) persons lack this cell surface marker because they are homozygous recessive for the gene encoding it.

If an Rh− person is exposed to Rh+ blood, the Rh surface antigens of that blood are treated like foreign invaders by the Rh− person's immune system, which proceeds to make antibodies directed against the Rh antigens. This most commonly happens when an Rh− woman gives birth to an Rh+ child (the father being Rh+). Some fetal red blood cells cross the placental barrier and enter the mother's bloodstream, where they induce the production of "anti-Rh" antibodies by the mother. These antibodies persist for many years in the mother's bloodstream, and in later pregnancies can attack a new fetus and cause its red blood cells to clump, a potentially fatal condition called **erythroblastosis fetalis**.

It is important that prospective parents know their Rh blood type because therapy is available for couples comprising an Rh+ male and an Rh− female. In such cases the mother is injected with antibodies directed against the Rh blood group antigen (called Rho gammagobulins, or **Rhogam**) at the birth of the first child. These antibodies bind to the Rh antigens on any fetal blood cells that enter the mother's circulation, covering them so that her own immune system does not detect the Rh antigens and make antibodies directed toward them. A future fetus is safe because there are no anti-Rh antibodies in the mother's blood waiting to attack it.

### GENETIC DISORDERS

Humans differ greatly from one another genetically, and yet, for most loci, one allele is much more common than its alternatives. For most genes, variant alleles are rare, although there are several loci—such as those determining the blood types which we just considered—for which two or more alleles are common among humans. But why should one allele of most genes be overwhelmingly the most common?

Part of the answer seems to lie in the fact that the proteins encoded by most genes must function in a very precise fashion for the many complex processes of development and the regulation of bodily functions to occur properly. Alternative alleles may lead to the production of proteins that do not function properly, if they function at all. Chapter 13 discusses the process, called mutation, that is responsible for the production of such alternative alleles. Here we note only that mutation is a random process; changing a gene randomly rarely improves the functioning of its encoded protein, any more than randomly changing a wire in a computer is likely to improve the computer's functioning. Genetic disorders and other unfavorable conditions may be seen as a necessary consequence of this relationship. On the other hand, without mutation there would be no variation; and without variation there would be no natural selection and, consequently, no evolutionary change. Mutation is a necessary component in the progressive change and adaptation of life on earth, even though most individual mutations are harmful to the organisms that possess them.

You might expect that the alternative alleles with detrimental effects would be rare in human populations, and usually, but not always, they are rare. Sometimes a detrimental allele is common. This can happen in small, isolated communities, where an individual who happens to carry the allele is one of the few members. In many cases, we don't know why the allele is common. But whatever the reason for its being well represented, a common allele that results in unfavorable characteristics can have disastrous effects on the group of humans in which it occurs. Most alleles that lead to the production of abnormal phenotypes are recessive; such recessive alleles are more easily retained in a population than are dominant ones because they are not expressed in most individuals (heterozygotes) in which they occur. In populations where such alleles occur at high frequencies, the chance of two heterozygous individuals mating and producing abnormal, homozygous individuals as a quarter of their progeny is correspondingly great. Tragedies stemming from such causes affect many families, even in cases where the recessive allele is relatively rare. Learning how to avoid them is one of the principal goals of human genetics.

When a detrimental allele occurs at a significant frequency in human populations, the harmful effect that it produces is called a **genetic disorder.** Table 11-2 lists some of the most important human genetic disorders. We know a great deal about some of them, but much less about many others.

## Cystic Fibrosis

As you learned in Chapter 5, cystic fibrosis is the most common fatal genetic disorder among Caucasians, among whom about 1 in 20 individuals has a copy of the defective gene but shows no symptoms. Homozygous recessive individuals account for about 1 in 1800 children. In individuals afflicted by cystic fibrosis, the transmembrane channels that normally transport chloride ions into cells do not function, and water is prevented from passing from their bloodstream into the passages of their lungs. As a result, the mucus that is a normal component of the lung's inner surface becomes too thick. This mucus clogs the airways of their lungs and the passages of their pancreas and liver, and they inevitably die. The genetic and cellular bases of cystic fibrosis are now reasonably well understood.

TABLE 11-1   DISTRIBUTION OF ABO BLOOD GROUPS IN SOME HUMAN POPULATIONS

| POPULATION | PHENOTYPE FREQUENCY (%) | | | |
|---|---|---|---|---|
| | A | B | AB | O |
| U.S. whites | 39.7 | 10.6 | 3.4 | 46.3 |
| U.S. blacks | 26.5 | 20.1 | 4.3 | 49.1 |
| African (Bantu) | 25.0 | 19.7 | 3.7 | 51.7 |
| Amerindians (Navaho) | 30.6 | 0.2 | 0.0 | 69.1 |
| Amerindians (Ecuador) | 4.0 | 1.5 | 0.1 | 99.4 |
| Japanese | 38.4 | 21.9 | 9.7 | 30.1 |
| Russians | 34.6 | 24.2 | 7.2 | 34.0 |
| French | 45.6 | 8.3 | 3.3 | 42.7 |

**TABLE 11-2    SOME IMPORTANT GENETIC DISORDERS**

| DISORDER | SYMPTOM | DEFECT | DOMINANT/ RECESSIVE | FREQUENCY AMONG HUMAN BIRTHS |
|---|---|---|---|---|
| Cystic fibrosis | Mucus clogging lungs, liver, and pancreas | Failure of chloride ion transport mechanism | Recessive | $1/1800$ (whites) |
| Sickle cell anemia | Poor blood circulation | Abnormal hemoglobin molecules | Co-dominant | $1/1600$ (U.S. blacks) |
| Tay-Sachs disease | Deterioration of central nervous system while person is young | Defective form of enzyme hexosaminidase A | Recessive | $1/1600$ (Jews) |
| Phenylketonuria | Failure of brain to develop in infants | Defective form of enzyme phenylalanine hydroxylase | Recessive | $1/18,000$ |
| Hemophilia (Royal) | Failure of blood to clot | Defective form of blood clotting factor IX | Sex-linked recessive | $1/7000$ |
| Huntington's disease | Gradual deterioration of brain tissue in middle age | Production of an inhibitor of brain cell metabolism | Dominant | $1/10,000$ |
| Muscular dystrophy (Duchenne) | Wasting away of muscles | Degradation of myelin coating of nerves stimulating muscles | Sex-linked recessive | $1/10,000$ |
| Hypercholesterolemia | Excessive cholesterol levels in blood, leading to heart disease | Abnormal form of cholesterol cell-surface receptor | Dominant | $1/500$ |

Cystic fibrosis occurs when an individual is homozygous for the allele encoding the defective version of the protein regulating the chloride-transport channel. This allele is recessive to the normal-functioning version of the regulating protein. Thus the chloride channels of heterozygous individuals function normally, and such persons do not develop cystic fibrosis.

### Sickle-Cell Anemia

**Sickle-cell anemia** is a genetic disorder in which the affected individuals cannot transport oxygen to their tissues properly because the molecules within red blood cells that carry oxygen, **hemoglobin** proteins, are defective. When oxygen is scarce, these defective hemoglobin molecules become insoluble and combine with one another, forming stiff, rodlike structures. Surprisingly, the defective hemoglobin that causes sickle-cell anemia differs from normal hemoglobin in only 1 out of a total of about 300 amino acid molecules. In it, one molecule of valine occurs in place of the glutamic acid located in the same position in normal hemoglobin. Red blood cells that contain large proportions of defective molecules become sickle-shaped and stiff; normal red blood cells are disk-shaped and much more flexible (Figure 11-10). Because of their stiffness and irregular shape, the sickle-shaped red blood cells are able to move through the smallest blood vessels only with great difficulty. For the same reason, they also tend to accumulate in the blood vessels, forming clots. As a result, people who have large proportions of sickle-shaped red blood cells tend to have intermittent illness and a far shorter life span than those who have normal red blood cells.

Individuals homozygous for the sickle-cell allele show the characteristics just mentioned; those who are heterozygous for the allele are generally indistinguishable from normal persons. But in the blood of people who are heterozygous for this trait, some of the red cells show the sickling characteristic when they are exposed to low levels of oxygen. The allele responsible for the sickle-cell characteristic is thus said to be codominant. It is particularly common among blacks. In the United States, about 9% of blacks are heterozygous for this allele, and about 0.2% are homozygous and

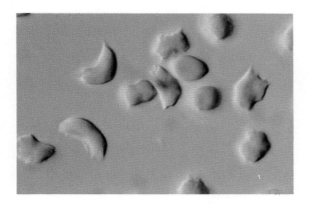

**Figure 11-10**

"Sickled" red blood cells. In individuals who are homozygous for the sickle-cell trait, many of the red blood cells have such shapes.

therefore have sickle-cell anemia. In some groups of people in Africa, up to 45% of the individuals are heterozygous for this allele.

People who are homozygous for the sickle-cell allele almost never reproduce because they usually die too young. This fact raises the question as to why the sickle-cell allele has not been eliminated from all populations, rather than being maintained at high levels in some. The answer has proved much easier to find than has the parallel question of why cystic fibrosis is so common. People who are heterozygous for the sickle-cell allele are much less susceptible to falciparum malaria, which is one of the leading causes of illness and death, especially among young children, in the areas where the allele is common. In addition, for reasons not understood, women who are heterozygous for this allele are more fertile than are those who lack it. Consequently, even though most people who are homozygous recessive, and thus develop sickle-cell anemia, die before they have any children, the sickle-cell allele is maintained at high levels in populations where falciparum malaria is common (Figure 11-11).

## Tay-Sachs Disease

**Tay-Sachs disease** is an incurable genetic disorder in which the brain deteriorates. Affected children appear normal at birth and usually do not develop symptoms until about the eighth month, when signs of mental deterioration become evident. Within a year after birth, affected children are blind; they rarely live past their fifth year (Figure 11-12).

Tay-Sachs disease is a rare disorder in most human populations, occurring in 1 in 300,000 births. However, it has a high incidence among Jews of Eastern and Central Europe (Ashkenazim) and among American Jews (90% of whom are descendants of Eastern and Central European ancestors). Approximately 1 in 3600 such Jewish infants have this genetic

disorder. Because it is a recessive condition, most people who carry the defective allele do not themselves develop the characteristic symptoms. An estimated 1 in

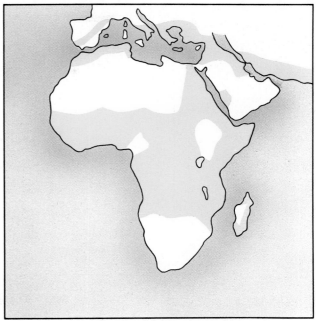

Distribution of *falciparum* malaria

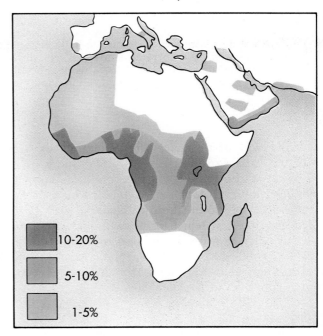

10-20%

5-10%

1-5%

Frequency of sickle cell gene

**Figure 11-11**

**Relationship between the frequency of sickle-cell trait and the distribution of falciparum malaria.** Falciparum malaria is one of the most devastating kinds of the disease; its distribution in Africa is closely correlated with that of the allele for the sickle-cell characteristic.

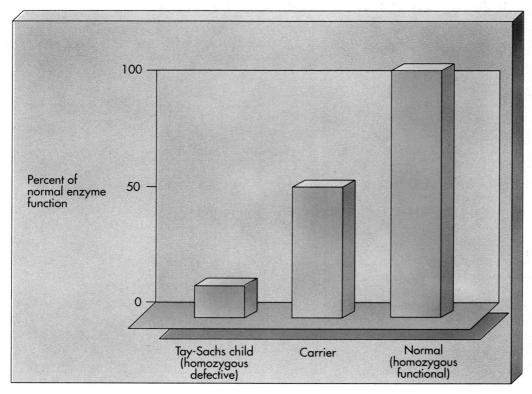

**Figure 11-12**

**Tay-Sachs disease.** Tay-Sachs disease is a genetic disorder in which an enzyme critical to lipid metabolism does not function, leading to harmful accumulations of fatty acids in the lysosomes of brain cells. Homozygous individuals typically have less than 10% of normal levels of the enzyme hexosaminidase A *(left)*, whereas heterozygous individuals have about 50% of normal levels, enough to prevent deterioration of the central nervous system.

28 individuals in these Jewish populations is a heterozygous carrier of the allele.

What is responsible for the condition known as Tay-Sachs disease? Homozygous individuals lack an enzyme necessary to break down a special class of lipid called **gangliosides,** which occur within the lysosomes of brain cells. As a result, the lysosomes fill with gangliosides, swell, and eventually burst, releasing oxidative enzymes that kill the brain cells. There is no known cure for this genetic disorder.

## Phenylketonuria

Many other hereditary disorders are not as common as cystic fibrosis in Caucasians, sickle-cell anemia in blacks, or Tay-Sachs disease in Central and Eastern European Jews. But even though they occur less frequently, some genetic disorders do affect significant numbers of people. A good example of a relatively infrequent genetic disorder is **phenylketonuria (PKU),** a condition in which the affected individuals cannot metabolize the amino acid phenylalanine normally. Because phenylalanine is necessary for the synthesis of many proteins, it is an essential ingredient in our diets. Individuals with PKU lack sufficient enzymatic activity to convert phenylalanine to tyrosine. Most of us obtain the tyrosine we need for protein synthesis through the metabolism of phenylalanine and do not require tyrosine in our diets. For PKU individuals, tyrosine becomes an essential dietary component. If individuals with PKU consume more than the amount of phenylalanine required for protein synthesis, the excess phenylalanine is converted by alternative metabolic pathways into abnormal products. These molecules are evidently associated with the extreme mental retardation seen in untreated cases of PKU; untreated individuals rarely live beyond the age of 30. Our understanding of these relationships enables us to design appropriate diets for individuals with PKU, provided that the condition is detected early enough, and thus allows them to avoid the deleterious consequences. Phenylalanine is not harmful to adults with PKU, presumably because their brain cells have already completed their development. Consequently, afflicted individuals who receive an appropriate diet when young mature into normal, healthy adults.

PKU is a recessive disorder caused by a mutant allele of the gene encoding the enzyme that normally breaks down phenylalanine. Only individuals homozygous for the mutant allele (in the United States, about 1 in every 15,000 infants) develop the disorder.

## Hemophilia

**Hemophilia** is a genetic condition in which the blood is slow to clot or does not clot at all. The reason you do not bleed to death when you cut your finger is that the blood in the immediate area of the cut is solidified and seals the cut, in a process similar to that which occurs in a puncture-proof tire. A blood clot forms from the polymerization of several kinds of protein fibers that circulate in the blood, all of them necessary for the proper formation of a blood clot. A mutation causing the loss of activity of any of the necessary proteins leads to a form of hemophilia.

Hemophilias are recessive disorders, expressed only when an individual does not possess at least one copy of the gene that is normal and so cannot produce one of the proteins necessary for clotting. Individuals homozygous for a mutant allele do not produce any active version of the affected clotting protein and thus cannot clot blood. Most of the dozen protein-clotting genes are on autosomes, but two (designated VIII and IX) are known to be located on the X chromosome. In the case of these particular protein-clotting genes, any male who inherits a mutant allele will develop hemophilia because his other sex chromosome is the Y (which is not expressed), and so he lacks a functioning allele of the protein-clotting gene.

A mutation in factor IX occurred in one of the parents of Queen Victoria of England (1819-1901) (Figure 11-13). In the 5 generations since Queen Victoria, 10 of her male descendants have had hemophilia. The British royal family escaped the disorder, often called the **Royal hemophilia,** because Queen Victoria's son King Edward VII did not inherit the defective factor IX allele. Three of Victoria's nine children did receive the defective allele and carried it by marriage into many of the royal families of Europe (Figure 11-14). It is still being transmitted to future generations among these family lines, except in Russia, where the five children of Alexandra, Victoria's granddaughter, were killed in the turbulent times soon after the Russian Revolution (see Figure 11-1).

## Huntington's Disease

Not all genetic disorders are recessive. Huntington's disease is associated with a dominant allele that causes the progressive deterioration of brain cells (Figure 11-15). This disorder killed folksinger and songwriter Woody Guthrie. Perhaps 1 in 10,000 individuals develop the disorder. Because Huntington's disease is a

*Figure 11-13*

**The Royal hemophilia.** In this photograph taken in 1894, Queen Victoria of England is surrounded by some of her descendants. Of Victoria's four daughters who lived to bear children, two, Alice and Beatrice, were carriers of Royal hemophilia. Two of Alice's daughters are standing behind Victoria (wearing feathered boas); to Victoria's right is Princess Irene of Prussia; to her left, Alexandra, who would soon become Tsarina of Russia (see Figure 11-1). Both Irene and Alexandra were also carriers of hemophilia.

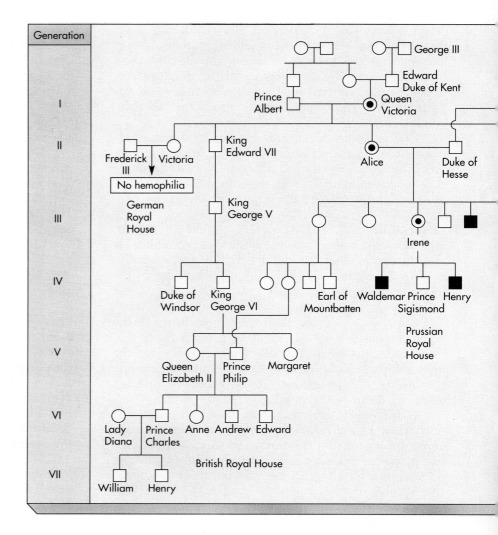

**Figure 11-14**

**The Royal hemophilia pedigree.** From Queen Victoria's daughter Alice, the disorder was introduced into the Russian and Austrian royal houses, and from her daughter Beatrice, it was introduced into the Spanish royal house. Victoria's son Leopold, himself a victim, also transmitted the disorder in a third line of descent.

dominant condition, every individual who carries an allele expresses it. You might then ask why the genetic disorder doesn't die out. The answer is that symptoms of Huntington's disease do not usually develop until the individuals are more than 30 years old, by which time most of them have had children. Thus the allele is transmitted before the lethal condition develops.

## GENETIC COUNSELING

Although most genetic disorders cannot yet be cured, we are learning much about them, and progress toward successful therapy is being made in many cases. But in the absence of a cure, the only recourse is to try to avoid producing children subject to these conditions. The process of identifying parents at risk for producing children with genetic defects, and of assessing the genetic state of early embryos, is called **genetic counseling.**

You might ask, "If the genetic defect is a recessive allele, how do potential parents *know* they carry the allele?" Pedigree analysis is often employed as an aid in genetic counseling. If one of your relatives has been afflicted with a recessive genetic disorder such as cystic fibrosis, there is a possibility that you are a carrier of the trait; in other words, you may carry the recessive allele in the heterozygous state. By analyzing your pedigree, it is often possible to estimate the likelihood of you being a carrier. When a couple is expecting a child and pedigree analysis indicates that both parents have a significant probability of being heterozygous carriers of a recessive allele responsible for a serious genetic disorder, the pregnancy is said to be a **high-risk pregnancy.** In such a pregnancy, there is a significant probability that the child will exhibit the clinical disorder.

Another class of high-risk pregnancies involves mothers who are more than 35 years old, as the frequency of Down syndrome increases dramatically after that age (see Figure 11-4).

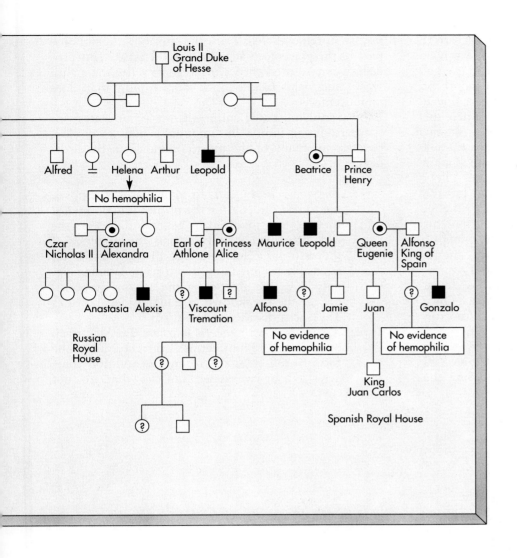

Louis II
Grand Duke
of Hesse

Alfred  Helena  Arthur  Leopold      Beatrice  Prince Henry

No hemophilia

Czar      Czarina    Earl of    Princess  Maurice  Leopold   Queen    Alfonso
Nicholas II  Alexandra  Athlone    Alice                        Eugenie   King of
                                                                          Spain

Anastasia  Alexis   Viscount   Alfonso  Jamie   Juan   Gonzalo
                     Tremation

Russian
Royal
House

No evidence
of hemophilia

No evidence
of hemophilia

King
Juan Carlos

Spanish Royal House

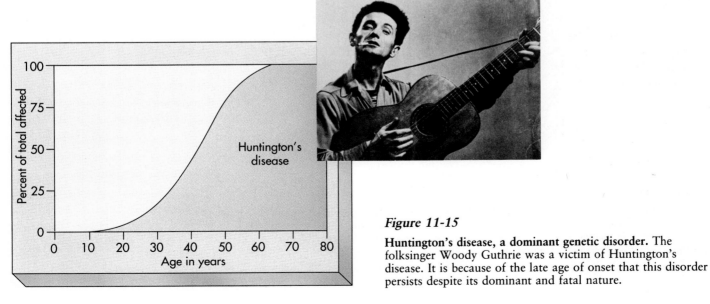

*Figure 11-15*

**Huntington's disease, a dominant genetic disorder.** The folksinger Woody Guthrie was a victim of Huntington's disease. It is because of the late age of onset that this disorder persists despite its dominant and fatal nature.

100
75
50      Huntington's
        disease
25
0
0  10  20  30  40  50  60  70  80

Percent of total affected

Age in years

When a pregnancy is diagnosed as being high-risk, many women elect to undergo **amniocentesis,** a procedure that permits the prenatal diagnosis of many genetic disorders (Figure 11-16). In the fourth month of pregnancy, a small sample of amniotic fluid is removed by means of a sterile hypodermic needle. When the needle is inserted into the expanded uterus of the mother, its position and that of the fetus are usually observed simultaneously by means of a technique called **ultrasound** (Figure 11-17). Sound waves allow an image of the fetus to be seen, as do x-rays, but the sound waves are not damaging to the mother or the fetus. Because ultrasound allows the position of the fetus to be determined, the person withdrawing the amniotic fluid can avoid damaging the fetus. In addition, the fetus can be examined for the presence of major abnormalities. The amniotic fluid, which bathes the fetus, contains free-floating cells derived from the fetus. Once removed, these cells can be grown as tissue cultures in the laboratory.

In the last few years, physicians have increasingly turned to a new, less invasive procedure, **chorionic villi sampling,** for genetic screening. In this procedure the physician removes cells from the chorion, a membrane part of the placenta that nourishes the fetus. This procedure can be used earlier in pregnancy (by the eighth week) and yields results much more rapidly than does amniocentesis.

By studying tissue cultures from amniocentesis or tissue from chorionic villi sampling, genetic counselors can test for many of the most common genetic disorders.

1. *Enzyme activity tests.* In many cases it is possible to test directly for the proper functioning of the enzymes involved in genetic disorders; the lack of proper activity signals the presence of the disorder. Thus the lack of the enzyme responsible for breaking down phenylalanine signals PKU, the absence of the enzyme responsible for the breakdown of gangliosides indicates Tay-Sachs disease, and so forth.
2. *Association with genetic markers.* For sickle-cell hemoglobin, Huntington's disease, and one form of muscular dystrophy (a condition characterized by weakened muscles that do not

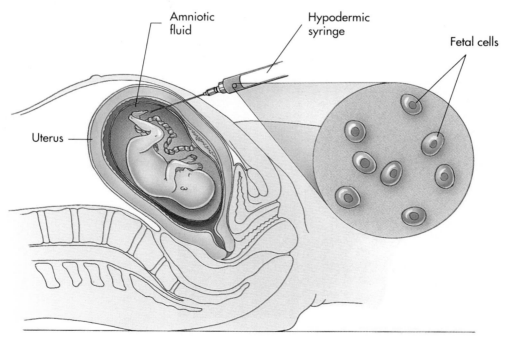

Amniotic fluid · Hypodermic syringe · Fetal cells · Uterus

*Figure 11-16*

**Amniocentesis.** A needle is inserted into the amniotic cavity, and a sample of amniotic fluid, containing some free cells derived from the embryo, is withdrawn into a syringe. The fetal cells are then grown in tissue culture so that their karyotype and many of their metabolic functions can be examined.

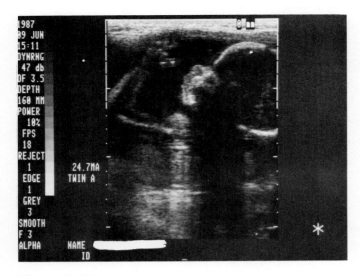

**Figure 11-17**

**The appearance of a fetus revealed by ultrasound.** During the fourth month of pregnancy, when amniocentesis is normally performed, the fetus usually moves about actively.

function normally), investigators have searched and found other mutations on the same chromosome that, by chance, occurred at about the same place as the disorder-causing mutation. By testing for the presence of the second mutation, an investigator can identify individuals with a high probability of possessing the disorder-causing defect. Identifying such mutations in the first place is a little like searching for a needle in a haystack, but persistent efforts have proved successful in these cases. The associated mutations are detected because they alter the length of DNA segments produced by enzymes that cut strands of DNA at particular places. Such enzymes, called **restriction enzymes,** are discussed in Chapter 14. The mutations are called restriction fragment—length polymorphisms, or **RFLPs.**

3. *Identification of heterozygotes.* As demonstrated in Figure 11-18, heterozygous individuals can often be detected during genetic counseling and can plan accordingly.

## Genetic Therapy

When the analysis of amniotic fluid indicates a severe genetic disorder in the fetus, the parents may consider terminating the pregnancy by means of therapeutic abortion. In some instances, other options are available. For example, if PKU is diagnosed, it is possible to avoid the defects of the disorder by placing the mother on a low-phenylalanine diet. This provides the mother and her unborn baby with enough phenylalanine to make proteins, but not enough to lead to the buildup of damaging molecules. After birth, the child is maintained on a low-phenylalanine diet until 6 years of age. At that age, the child's brain is fully developed, and PKU is no longer a potential health problem.

Listed below are some of the organizations studying the major genetic disorders. Their support is helping to find new and better ways to manage the individual conditions and may eventually lead to the detection of ways to alleviate them completely.

Cystic Fibrosis Foundation
3379 Peachtree Rd., N.E.
Atlanta, GA 30326

National Hemophilia Foundation
25 West 39th St.
New York, NY 10018

Committee to Combat Huntington's Disease, Inc.
250 West 57th St., Suite 2016
New York, NY 10107

Muscular Dystrophy Association
810 Seventh Ave.
New York, NY 10019

National Association of Sickle-Cell Disease, Inc.
945 South Western Ave.
Los Angeles, CA 90006

National Tay-Sachs and Allied Diseases Association
122 East 42nd St.
New York, NY 10017

Advances in gene technology are making it possible in some cases to correct undesirable genetic conditions directly by transferring genes from the cells of healthy individuals to those of individuals with faulty versions. The first such use of gene therapy in humans was approved in August, 1990. Patients suffering from a lethal form of skin cancer will have a potent cancer-fighting gene added to the cells of their immune system (this recent and very important advance in genetic therapy is described in Chapter 40). So far, gene transfers have not been carried out *in vitro* (in a test tube) with embryos recently fertilized there; in such a transfer, all the cells of an individual would be affected. As we learn more, it seems increasingly likely that this will occur.

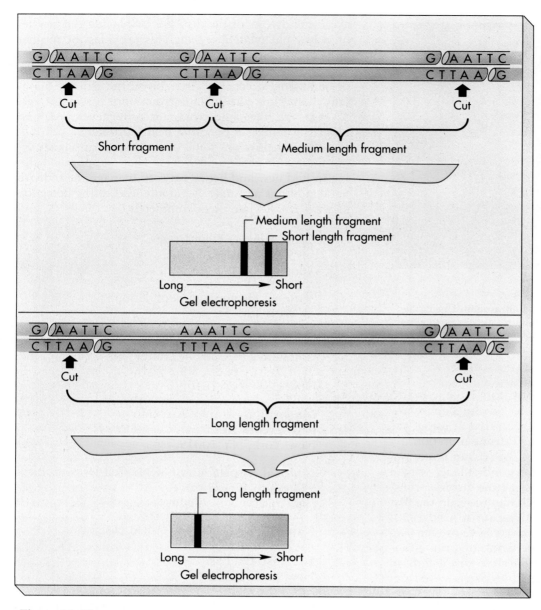

**Figure 11-18**

**RFLPs.** Restriction fragment length polymorphisms (RFLPs) are playing an increasingly important role in genetic identification. Here you can see how the mutation of a single nucleotide from G to A alters a restriction endonuclease cutting site so that the enzyme is no longer cut there. As a result, a single long fragment is obtained, rather than two shorter ones. Such a change is easy to detect by gel electrophoresis of the fragments.

## ■ SUMMARY

1. Human body cells contain 46 chromosomes: 44 autosomes and 2 sex chromosomes. The autosomes form 22 pairs of homologous chromosomes in meiosis. The two sex chromosomes may be similar in size and appearance, as occurs in females, where they are designated XX. The human Y chromosome is much smaller than the X chromosome. XY individuals are male because the genes that initiate the development of male characteristics are located on the Y chromosome.

2. In humans, the loss of an autosome is invariably fatal. Gaining an extra autosome, which leads to a condition called trisomy, is also fatal, with a few exceptions. The chromosomes that can be present in an extra copy and still allow that individual to survive and become an adult are chromosomes 21 and 22. Individuals with an extra copy of chromosome 21 (three copies in all) are retarded and have Down syndrome. As occurs in all trisomic persons, this trisomy arises from the chromosomes of that particular pair not separating during meiosis, and it usually occurs in mothers over 35 years of age.

3. Patterns of inheritance observed in family histories, called pedigrees, can be used to determine the mode of inheritance of a particular trait. By such analysis, one can often determine whether a trait is associated with a dominant or a recessive allele, if the gene determining the trait is located on the X chromosome (sex-linked), and if the trait is specified by more than one gene.

4. Many human genes possess more than two common alleles. An example is the ABO blood group gene. This gene encodes an enzyme that adds sugars to the surfaces of blood cells. The A and B alleles add different sugars, and the O allele adds none. Human populations vary in the proportions of these three alleles which they possess.

5. Genetic disorders are often caused by alleles that encode abnormal proteins; the effects of these proteins lead to serious health problems. Some genetic disorders are relatively common in human populations, whereas others are rare. Many of the most important genetic disorders are associated with recessive alleles, the functioning of which may lead to the production of defective versions of enzymes that normally perform critical functions. Because such traits are determined by recessive alleles and therefore are expressed only in homozygotes, the alleles are not eliminated from the human population, even though their effects in homozygotes may be lethal. Dominant alleles that lead to severe genetic disorders are less common. In some of those occurring more frequently, the expression of the alleles does not occur until after the affected individuals are in their reproductive years.

6. Parents who suspect their children may express a genetic disorder, such as Down syndrome, may elect to undergo amniocentesis. In this procedure, a sample of fetal cells obtained from amniotic fluid is used to establish a tissue culture, which can then be checked for the presence of various genetic disorders.

7. For a few genetic disorders, such as phenylketonuria, it is possible to initiate therapy if the disorder is diagnosed during pregnancy and thus avoid the detrimental effects. For most genetic disorders, we do not know how to accomplish such a result. The direct transfer of normal human genes to the chromosomes of individuals who would otherwise suffer from a particular genetic disorder may prove possible in the future, although this has not yet been accomplished. Meanwhile, although no cures exist for any genetic disorder, many of the conditions can be managed with increasingly positive results.

## REVIEW

1. What is the human diploid number of chromosomes? How many kinds of autosomes do you have?

2. The inactive X chromosome seen in the cells of human females is called a _____.

3. What percentage of Caucasians possess at least one copy of the allele for cystic fibrosis?

4. Nondisjunction of an X chromosome during meiosis can lead to an individual with an XXX genotype or to an individual with an XXY genotype. The XXY genotype leads to a condition called _____ and occurs in about 1 in 1000 male births.

5. _____ is a procedure that permits the prenatal diagnosis of many genetic disorders.

# SELF-QUIZ

1. Down syndrome results from having three copies of chromosome
   (a) 13.
   (b) 18.
   (c) 21.
   (d) 22.
   (e) both c and d.

2. Why are sex-linked traits expressed much more frequently in males than in females?
   (a) Because males have only one copy of the X chromosome.
   (b) Because males have only one copy of the Y chromosome.
   (c) Because males are more developmentally fragile than females.
   (d) Sex-linked traits are carried on the X chromosome and therefore occur with equal frequency in males and females.
   (e) These males lack an autosome.

3. Primary nondisjunction of the chromosomes
   (a) is associated with Huntington's disease.
   (b) occurs when chromatids do not separate during the second phase of meiosis.
   (c) sometimes results in babies with Down syndrome.
   (d) is a principal mechanism for meiosis.
   (e) occurs more frequently in the production of male gametes than in the production of female gametes.

4. Individuals with AB type blood
   (a) may receive transfusions successfully only from other AB individuals.
   (b) are known as universal donors.
   (c) may receive either A or B type blood.
   (d) often have trouble with blood clotting.
   (e) are predominant among the original inhabitants of the Western Hemisphere.

5. Which of the following statements is true of cystic fibrosis? More than one of the answers may be correct.
   (a) It is the disease that brought about Woody Guthrie's death.
   (b) It is caused by a gene that has been located on a particular chromosome.
   (c) It dries up the lungs by preventing the formation of mucus.
   (d) It is preserved in populations because of its associated disease resistance.
   (e) It can be corrected by appropriate dietary practices.

6. Which of the following genetic diseases is sex-linked?
   (a) Royal hemophilia
   (b) Tay-Sachs disease
   (c) Cystic fibrosis
   (d) Hypercholesterolemia
   (e) None of the above

7. A detailed pedigree of a family with a history of a particular genetic disorder can help determine
   (a) whether the disorder is sex-linked.
   (b) whether the disorder is autosomal.
   (c) whether the disorder is dominant.
   (d) whether the disorder is determined by a single gene or by several genes.
   (e) all of the above.

8. The sickle-cell trait is fatal when homozygous, yet the allele is retained at fairly high levels in some areas of the world. Why?
   (a) Because dominant alleles are difficult to purge from a population even if they are detrimental.
   (b) Because it has no effect on heterozygous individuals.
   (c) Because heterozygous individuals are resistant to falciparum malaria.
   (d) Because individuals homozygous for the trait generally live past their reproductive years.
   (e) None of the above.

# THOUGHT QUESTIONS

1. As you can see in Table 11-1, both North American and South American Indians exhibit very low frequencies of ABO blood group allele B. Can you think of a reason why they should differ from other human populations in this regard?

2. Schizophrenia is a serious mental disorder in which mental contact is lost with the environment. To assess whether schizophrenia is hereditary, twins who had been reared apart were studied; all they had in common was their being twins. Two kinds of twins were compared: those who were identical (monozygotic) and those who were not (dizygotic). Monozygotic twins are genetically identical, whereas dizygotic twins are normal brothers and/or sisters who happen to be born at the same time. Investigators asked if twins reared apart develop schizophrenia more often if they are genetically identical. Here is what they found, using data collected over 40 years: of 289 sets of monozygotic twins studied, both twins developed the disorder if one did in 51% of the cases; of 398 sets of dizygotic twins studied, both developed the disorder if one did in 10% of cases. Do you think these results suggest that schizophrenia is a hereditary disorder?

3. How do we know that the mutation to Royal hemophilia did not occur in one of Queen Victoria's own ova?

# HUMAN GENETICS PROBLEMS

1. George has Royal hemophilia and marries his mother's sister's daughter Patricia. His maternal grandfather also had hemophilia. George and Patricia have five children: two daughters are normal, and two sons and one daughter develop hemophilia. Draw the pedigree.

2. A couple with a newborn baby are troubled that the child does not appear to resemble either of them. Suspecting that a mix-up occurred at the hospital, they check the blood type of the infant. It is type O. Because the father is type A and the mother is type B, they conclude that a mistake must have been made. Are they correct?

3. Mabel's sister dies as a child from cystic fibrosis. Mabel herself is healthy, as are her parents. Mabel is pregnant with her first child. If she were to consult you as a genetic counselor, wishing to know the probability of her child developing cystic fibrosis, what would you tell her?

4. How many chromosomes would one expect to find in the karyotype of a person with Turner syndrome?

5. A woman is married for the second time. Her first husband was ABO blood type A, and her child by that marriage was type O. Her new husband is type B, and their child is type AB. What is the woman's ABO genotype and blood type?

6. Two bald parents have five children, three of whom eventually become bald. Assuming that this trait is governed by a single pair of alleles, is this baldness best explained as an example of dominant or recessive inheritance?

7. In 1986, *National Geographic* magazine conducted a survey of its readers' abilities to detect odors. About 7% of Caucasians in the United States could not smell the odor of musk. If both parents cannot smell musk, then none of their children is able to smell it. Conversely, two parents who can smell musk generally have children who can also smell it; only a few children in each family are unable to smell it. Assuming that a single pair of alleles governs this trait, is the ability to smell musk best explained as an example of dominant or recessive inheritance?

8. Total color blindness is a rare hereditary disorder among humans in which no color is seen, only shades of gray. It occurs in individuals homozygous for a recessive allele and is not sex-linked. A man whose father is totally color blind intends to marry a woman whose mother was totally color blind. What are the chances that they will produce offspring who are totally color blind?

9. A normally pigmented man marries an albino woman. They have three children, one of whom is an albino. What is the genotype of the father?

10. Four babies are born within a few minutes of each other in a large hospital, when suddenly an explosion occurs. All four babies are found alive among the rubble. None had yet been given identification bracelets. The babies prove to be of four different blood groups: A, B, AB, and O. The four pairs of parents have the following pairs of blood groups: A and B, O and O, AB and O, and B and B. Which babies belong to which parents?

11. This pedigree is of a rare trait in which children have extra fingers and toes. Which if any of the following patterns of inheritance is consistent with this pedigree?
    a. Autosomal recessive
    b. Autosomal dominant
    c. Sex-linked recessive
    d. Sex-linked dominant
    e. Y-linkage

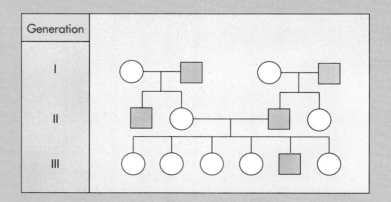

12. This pedigree is for a common form of inherited color blindness. What pattern of inheritance best accounts for this pedigree?

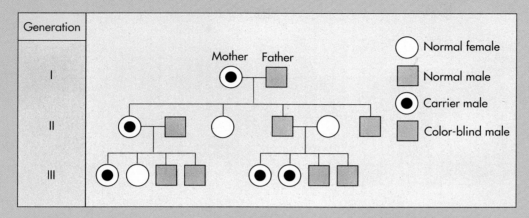

13. Of the 43 people in the 5 generations of this family, over a third exhibit an inherited mental disorder. Is the trait dominant or recessive?

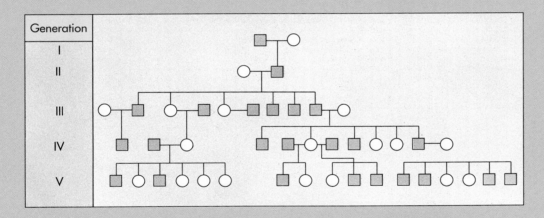

# FOR FURTHER READING

DIAMOND, J.: "Blood, Genes, and Malaria," *Natural History,* February 1989. A lucid account of the evolutionary history of sickle-cell anemia.

FRIEDMANN, T.: "Progress Toward Human Gene Therapy," *Scientific American,* June 1989. A discussion of the new gene therapy—approaches to genetic disease treatment by attacks directly on mutant genes.

GOULD, S.J.: "Dr. Down's Syndrome," *Natural History,* vol. 89, 1980, pages 142-148. An account of the history of Down syndrome, a relatively common chromosomal abnormality that results in severe mental retardation.

HOOK, E.B.: "Behavioral Implications of the Human XYY Genotype," *Science,* vol. 179, 1973, pages 139-150. A review of some of the conflicting data on this controversial subject.

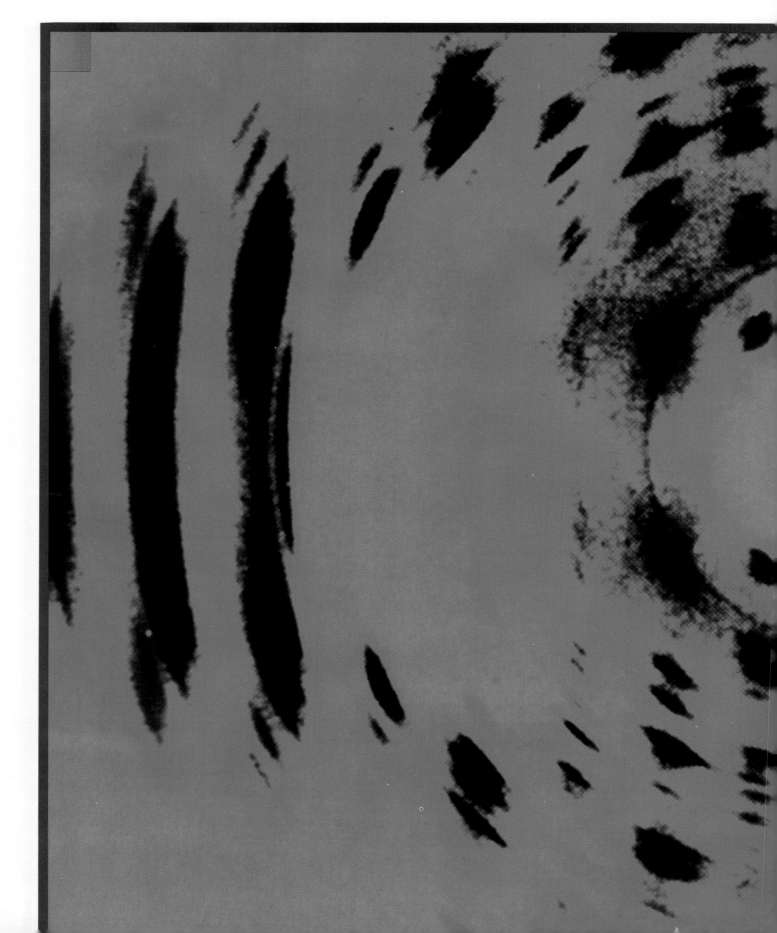

# $\mathcal{D}$NA: The Genetic Material

This is an actual X-ray of the DNA double helix, obtained from a crystallized form of DNA. The discovery of DNA's helical structure in 1953 by James Watson and Francis Crick was greatly helped by just such an X-ray, which clearly shows a helical pattern.

# DNA: THE GENETIC MATERIAL

## Overview

Mendel's laws of heredity and the chromosomal theory of inheritance that explained them answered the age-old question of how traits are inherited, only to ask another: How does it work? What is the mechanism that a cell uses to specify the nature of its descendants? We now know that cells store this information in DNA, copies of which are faithfully made and passed down from generation to generation. The experiments that uncovered this mechanism are among the most elegant in biology.

## For Review    *Here are some important terms and concepts that you will encounter in this chapter. If you are not familiar with them, you should review them before proceeding.*

**Nature of scientific experiments** (Chapter 1)

**Structure of DNA** (Chapter 3)

**Nucleotides** (Chapter 3)

**Structure of chromosomes** (Chapter 10)

The realization, reached at the beginning of this century, that patterns of heredity can be explained by the segregation of chromosomes in meiosis was one of the most important advances in human thought. Not only did it lead directly to formulation of the science of genetics and thus to great progress in agriculture and medicine, but also it profoundly influenced the way in which we think about ourselves. By removing the mystery from the process of heredity, it made the biological nature of human beings seem much more approachable (Figure 12-1). It also raised the question that was to occupy biologists for more than half a century: what is the exact nature of the connection between hereditary traits and the chromosomes?

We now are able to answer this question. We understand in considerable detail the mechanism by which the information on the chromosomes is converted into organisms with eyes, arms, and minds. Our understanding was not deduced in a single flash of insight, but was developed slowly over many years by a succession of investigators. How they did this and what they learned are the subjects of this chapter.

## WHERE DO CELLS STORE HEREDITARY INFORMATION?

Perhaps the most basic question one can ask about hereditary information is where it is stored in the cell. Of the many approaches one might take to answer this question, let's start with a simple one: cut a cell into pieces and see which of the pieces is or are able to express hereditary information. For this experiment we need a single-celled organism that is large enough to operate on conveniently and differentiated enough that the pieces can be distinguished from each other.

**Figure 12-1**

**Heredity shapes all of us.** Some of what this boy will be as an adult will be influenced by what he learns from his grandfather, but much will reflect the genes he has inherited from him.

## Figure 12-2

**The marine green alga *Acetabularia*.** Although *Acetabularia* is a large organism with clearly differentiated parts, such as the stalks and elaborate caps visible here, individuals are actually single cells.

An experiment of this kind was performed by the Danish biologist Joachim Hammerling in 1943. As an experimental subject, Hammerling chose the large unicellular green alga *Acetabularia* (Figure 12-2). Individ-

uals of this genus have distinct base, stalk, and cap regions, all of which are differentiated parts of a single cell. The nucleus of this cell is located in the base. As a preliminary experiment, Hammerling tried amputating the caps or bases of individual cells. He found that when the cap is amputated, a new one regenerates from the remaining portions of the cell. When the base is amputated and discarded, however, no new base is regenerated. Hammerling concluded that the hereditary information resided within the base of the giant cells of *Acetabularia*.

To test his hypothesis, Hammerling selected individuals from two species of *Acetabularia* in which the caps are very different: *A. mediterranea*, which has a disk-shaped cap, and *A. crenulata*, which has a branched, flowerlike cap. Hammerling cut the stalk and cap away from an individual of *A. mediterranea*; to the remaining base he grafted a stalk cut from a cell of *A. crenulata* (Figure 12-3). The cap that formed looked something like the flower-shaped one characteristic of *A. crenulata*, although it was not exactly the same.

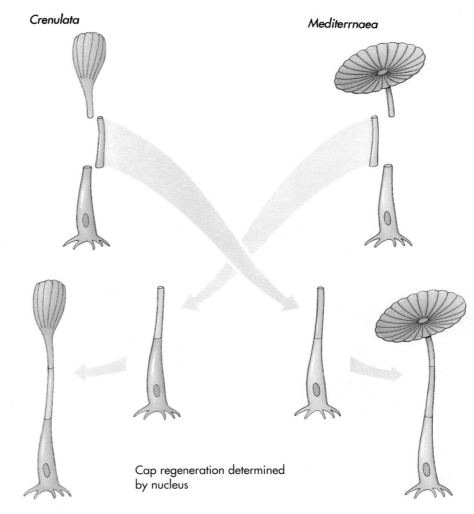

*Crenulata*

*Mediterrnaea*

Cap regeneration determined by nucleus

## Figure 12-3

**Hammerling's *Acetabularia* reciprocal graft experiment.** To the base of each species he grafted a stalk of the other. In each case, the cap that eventually developed was dictated by the base and not the stalk.

Hammerling then cut off this regenerated cap and found that a disk-shaped one exactly like that of *A. mediterranea* formed in the second regeneration and in every regeneration thereafter. This experiment strengthened Hammerling's earlier conclusion that the instructions specifying the kind of cap that is produced are stored in the base of the cell and probably therefore in the nucleus—and that these instructions must pass from the base through the stalk to the cap. In his regeneration experiment, the initial flower-shaped cap was formed as a result of the instructions that were already present in the transplanted stalk when it was excised from the original *A. crenulata* cell. In contrast, all subsequent caps used new information, derived from the base of the *A. mediterranea* cell onto which the stalk had been grafted. In some unknown way the original instructions that had been present in the stalk were eventually "used up" and replaced by new instructions from the base.

Hammerling's experiments identified the nucleus as the likely repository of the hereditary information but did not prove definitely that this was the case. To do that, isolated nuclei had to be transplanted from one cell to another. Such an experiment was carried out in 1952 by American embryologists Robert Briggs and Thomas King. Using a glass pipette drawn to a fine tip and working under a microscope (Figure 12-4), Briggs and King removed the nucleus from a frog egg; without a nucleus, the egg would not develop. They then replaced the absent nucleus with one they isolated from a cell of a young frog embryo. The diploid cell derived from that embryo took over directing the development of the original egg cell, which ultimately grew into an adult frog. Clearly, the nucleus was the important element in bringing about this result.

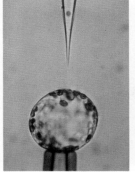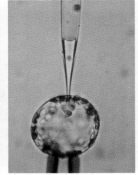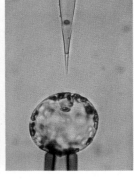

### Figure 12-4

**Microinjection.** Working under a microscope, it is possible to pierce a cell with the fine tip of a glass micropipette without rupturing the cell. These micrographs show microinjection into the protoplast of a cell from the leaf of a tobacco plant (the protoplast is about 50 micrometers in diameter).

## WHICH COMPONENT OF THE CHROMOSOMES CONTAINS THE HEREDITARY INFORMATION?

The identification of the nucleus as the source of hereditary information focused attention on the chromosomes, which had already been shown to be the vehicles of Mendelian inheritance. Specifically, biologists wondered how the actual hereditary information was arranged in the chromosomes. It was known that chromosomes contain both protein and DNA. On which of these was the hereditary information written?

Over a period of about 30 years, starting in the late 1920s, a number of groups of investigators addressed this issue, and ultimately their collective results resolved the problem clearly. Below are described three very different experiments, each of which yields a clear answer in a simple and straightforward manner.

### The Griffith-Avery Experiments: Transforming Principle is DNA

As early as 1928, a British microbiologist, Fred Griffith, made a series of unexpected observations while experimenting with a bacterium then called pneumococcus and now called *Streptococcus pneumoniae*. This bacterium had been recognized in 1886 as a primary cause of pneumonia in humans. Griffith was carrying out his experiments with mice.

Normally, cells of *S. pneumoniae* are enclosed by a polysaccharide capsule. Such cells are highly pathogenic (disease-causing); if even a few are injected into a mouse, the animal will die in a day or two as a result of the infection. However, mutant strains of the bacterium that lack capsules also exist; even if a large quantity of such mutant cells are injected into a mouse, the animal will remain healthy. If cells of the normal, pathogenic strain are killed by exposing them to high temperatures before injecting them into the mice, they have no effect, indicating that living bacteria are necessary to produce the harmful effects associated with this strain.

Building on these results, Griffith achieved a surprising result. He mixed heat-killed, pathogenic bacteria, which had capsules, with living, mutant, capsuleless bacteria and injected them into healthy mice (Figure 12-5). Unexpectedly, the injected mice developed disease symptoms, and many of them died. The blood of the dead mice was found to contain high levels of

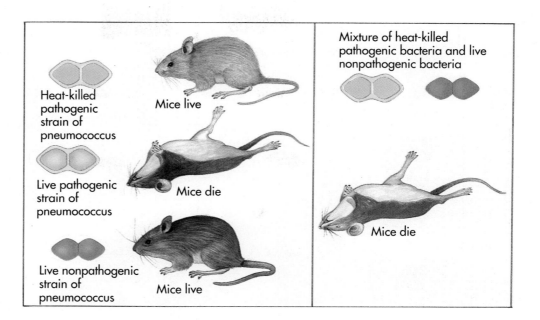

**Figure 12-5**
Griffith's discovery of transformation.

normal pathogenic *S. pneumoniae*. Somehow the information specifying the polysaccharide capsule had passed from the dead bacteria to the live but capsuleless ones in the control mixture, **transforming** them into normal virulent bacteria that infected and killed the mice.

The agent responsible for transformation in *S. pneumoniae* was discovered in a classic series of experiments conducted by Oswald Avery and his colleagues at Rockefeller University in New York during the 1930s; they were summarized in one of the classical papers of molecular biology, published in 1944. These scientists succeeded in characterizing what they referred to as the "transforming principle." Its properties resembled those of DNA rather than of protein: the activity of the transforming principle was not affected by protein-destroying enzymes but was lost completely in the presence of the DNA-destroying enzyme DNAse.

When the transforming principle was purified, it indeed consisted predominantly of DNA. Later, it was shown that all but trace amounts of protein (0.02%) could be removed without reducing the transforming activity. The conclusion was inescapable: DNA is the hereditary material in bacteria. It has since proved possible to use purified DNA to change the genetic characteristics of eukaryotic cells in tissue culture and even possible to inject pure DNA into fertilized *Drosophila* eggs and thereby alter the genetic characteristics of the resulting adult.

## The Hershey-Chase Experiment: Bacterial Viruses Direct Their Heredity with DNA

Avery's results were not widely appreciated at first. Many biologists preferred to believe that proteins were the source of the hereditary information. However, another very convincing experiment was soon performed that was difficult to ignore. It was done in a simple system—viruses—so that a direct experimental question could be asked. Viruses consist of either RNA or DNA with a protein coat; they are described in more detail in Chapter 25. These new experiments focused on bacteriophages, which are viruses that infect bacteria. When a bacteriophage infects a bacterial cell, it first binds to the cell's outer surface, and then injects its hereditary information into the cell. There the hereditary information directs the production of thousands of new virus particles within the cell. The host bacterial cell eventually falls apart, or lyses, releasing the newly made viruses.

In 1952, Alfred Hershey and Martha Chase set out to identify the material that was injected into the bacterial cell at the start of an infection. They used a strain of bacteriophage known as T2, which contains DNA rather than RNA, and designed an experiment to distinguish between the alternative hypotheses that the genetic material was DNA or that it was protein (Figure 12-6). Hershey and Chase used a technique known as labeling, which is the introduction of a radioactive isotope of a particular element into a molecule so that the source of that molecule can be identified and its natural pathways can be followed. They labeled the DNA of some of the bacteriophages with a radioactive isotope of phosphorus, 32P, and the protein coats of other viruses with an isotope of sulfur, 35S. The labeled viruses were permitted to infect bacteria. The bacterial cells were then agitated violently to shake the protein coats of the infecting viruses loose from the bacterial surfaces to which they were attached. They

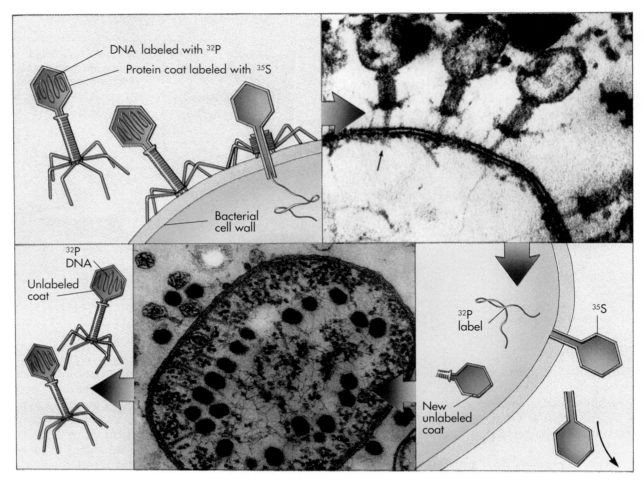

**Figure 12-6**
The Hershey-Chase experiment.

spun the cells at high speed so that they were pulled from solution by the gravitational forces; in this procedure, called centrifugation, the cells experience the same forces you do when pressed against the floor of a rapidly rising elevator.

Using this technique, Hershey and Chase found that, in their experiments involving labeled protein coats, the 35S label (and thus the virus protein) was now predominantly in solution with the dissociated virus particles. In the other set of experiments, involving the 32P label (and thus the DNA), they found by observing the patterns of radiation that the DNA had transferred to the interior of the cells. The viruses subsequently released from these infected bacteria contained the 32P label. The hereditary information injected into the bacteria that specified the new generation of virus particles was DNA and not protein.

### The Fraenkel-Conrat Experiment: Some Viruses Direct Their Heredity with RNA

Even with the evidence provided by the experiments we have just reviewed, one objection could still be raised

about accepting the hypothesis that DNA was the genetic material. Some viruses contain *no* DNA and yet manage to reproduce themselves quite satisfactorily. What was their genetic material?

In 1957, Heinz Fraenkel-Conrat and coworkers isolated tobacco mosaic virus (TMV) from tobacco leaves. From ribgrass (*Plantago lanceolata*), a common weed, they isolated a second, rather similar kind of virus, Holmes ribgrass virus (HRV). In both TMV and HRV the viruses consist of protein and a single strand of RNA. When they had isolated these viruses, the scientists broke them up, separating their protein from their RNA. By putting the protein component of one virus with the RNA of another, they were able to produce hybrid virus particles.

The payoff of the experiment was in its next step. To choose between the alternative hypotheses "the genetic material of viruses is protein" and "the genetic material of viruses is RNA," Fraenkel-Conrat now infected healthy tobacco plants with a hybrid virus composed of TMV protein capsules and HRV RNA, being careful not to include any nonhybrid virus particles

(Figure 12-7). The tobacco leaves that were infected with the hybrid virus particles developed the lesions that were characteristic of HRV and that normally formed on infected ribgrass. Clearly, the hereditary properties of the virus were determined by the nucleic acid in its core and not by the protein in its coat.

Later studies have shown that many kinds of viruses have as their genetic material RNA, rather than the DNA that is found universally in cellular organisms. When these viruses infect a cell, they may either multiply themselves directly or, in the case of the group called **retroviruses,** make DNA copies of themselves, which can then be inserted into the cellular DNA as if they were cellular genes.

*DNA is the genetic material for all cellular organisms and most viruses, although many viruses use RNA.*

## THE CHEMICAL NATURE OF NUCLEIC ACIDS

DNA was discovered only 4 years after the publication of Mendel's work. In 1869 a German chemist,

Friedrich Miescher, extracted a white substance from the cell nuclei of human pus and from fish sperm nuclei. The proportion of nitrogen and phosphorous was very different from that of any other known constituent of cells, which convinced Miescher he had discovered a new biological substance. He called this substance "nuclein" because it seemed to be associated specifically with the cell nucleus.

Because Miescher's nuclein was slightly acidic, it came to be called nucleic acid. For 50 years little work was done on it by biologists because nothing was known of its function in cells and there seemed little to recommend it to investigators. In the 1920s the basic chemistry of nucleic acids was worked out by the biochemist P.A. Levene. There were two sorts of nucleic acid, ribonucleic acid, or RNA, which had a hydroxyl group attached to a particular carbon atom, and deoxyribonucleic acid, or DNA, which did not. Levene found that DNA contained three basic components (Figure 12-8):

1. Phosphate ($PO_4$) groups
2. 5-carbon sugars, called ribose sugars
3. Four nitrogen-containing bases: adenine (A) and guanine (G) (double-ring compounds called purines), and thymine (T) and cytosine (C) (single-ring compounds called pyrimidines); RNA contained the pyrimidine uracil (U) in place of thymine.

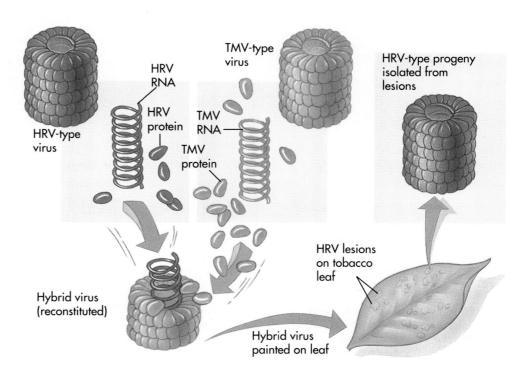

*Figure 12-7*

Fraenkel-Conrat's virus-reconstitution experiment.

From the roughly equal proportions of the three elements, Levene concluded correctly that DNA and RNA molecules are composed of units of these three elements, strung one after another in a long chain. Each component, a 5-carbon (ribose) sugar to which is attached a phosphate group and a nitrogen-containing base, is called a nucleotide. The identity of the base is the only factor that distinguishes one nucleotide in a nucleic acid from another.

The presence of the phosphate and hydroxyl groups is what allows DNA and RNA to form a polymer of nucleotides: these two groups can react chemically with one another. The chemical reaction between the phosphate group of one unit and the OH group of another causes the elimination of a water molecule and the formation of a covalent bond linking the two groups together (Figure 12-9). The linkage is called a **phosphodiester** bond because the phosphate group is now linked to the two sugars by means of two ester (—O—) bonds. Further linking up can occur in the same way: the two-unit polymer resulting from the elimination reaction we have just described still has a free phosphate group at one end and a free hydroxyl group at the other. In this way, many thousands of nucleotides can be linked together in long chains.

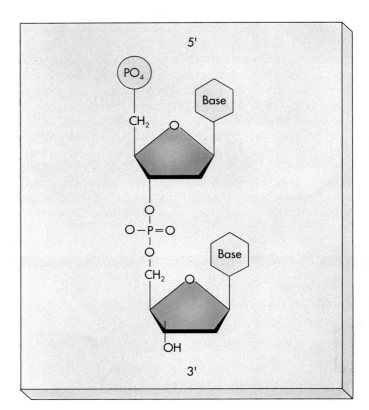

*Figure 12-9*

**A phosphodiester bond.**

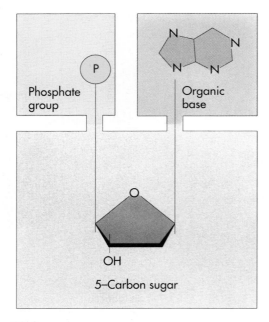

*Figure 12-8*

**Nucleotide subunit of DNA.** The nucleotide subunits of DNA are composed of three elements: an organic base, a phosphate group, and a 5-carbon sugar.

*A strand of DNA or RNA is a long chain of nucleotide subunits joined together like cars in a train.*

Levene's early studies indicated that all four types of DNA nucleotides were present in roughly equal amounts. This result, which was later proved an error, led to the mistaken belief that DNA was a simple repeating polymer in which the four nucleotides occurred together in a long series of identical units (for instance, ...GCAT...GCAT...GCAT...GCAT...). In the absence of sequence variation in such a repeating chain, it was difficult to see how DNA might contain the hereditary information necessary to specify even the simplest organism. The conclusions drawn from Avery's experiments, although they were crystal clear, were not readily accepted at first. It seemed more plausible to some biologists that DNA was no more than a structural element of the chromosomes, with proteins playing the central genetic role.

The key advance came after the World War II, when Levene's chemical analysis of DNA was repeated using more accurate techniques than had previously been available. Quite a different result was obtained. The four nucleotide bases were *not* present in equal proportions in DNA molecules after all. A careful study carried out by Erwin Chargaff showed that differences did exist in DNA nucleotide composition. Chargaff found that the nucleotide composition of DNA molecules varied in complex ways, depending on the source of the DNA. This strongly suggested that DNA was not a simple repeating polymer and that it might have the information-encoding properties required of genetic material after all. Despite DNA's complexity, Chargaff observed an important underlying regularity: the amount of adenine present in DNA molecules was always equal to the amount of thymine, and the amount of guanine was always equal to the amount of cytosine. Chargaff's finding is commonly referred to as **Chargaff's rule**.

*Chargaff pointed out that in all natural DNA molecules, the amount of A = T and the amount of G = C.*

## THE THREE-DIMENSIONAL STRUCTURE OF DNA

As it became clear that the DNA molecule was the repository of hereditary information, investigators began to puzzle over how such a seemingly simple molecule could carry out such a complex function. The significance of the regularities pointed out by Chargaff were not immediately obvious but soon became clear. A British chemist, Rosalind Franklin, had carried out x-ray crystallographic analysis of fibers of DNA. In this process the DNA molecule is bombarded with an x-ray beam. When individual rays encounter atoms, their path is bent, or diffracted; the pattern created by the total of all these diffractions can be captured on a piece of photographic film. Such a pattern resembles the ripples created by tossing a rock into a smooth lake. By carefully analyzing the diffraction pattern, it is possible to develop a three-dimensional image of the molecule.

Franklin's studies were severely handicapped by the fact that it had not proved possible to obtain true crystals of natural DNA, so she had to work with DNA in the form of fibers. Although the DNA molecules in a fiber are all aligned with one another, they do not form the perfectly regular crystalline array required to take full advantage of x-ray diffraction. One of Franklin's colleagues was the biochemist Maurice Wilkins, who was able to prepare more uniformly oriented DNA fibers than had been possible previously. Using these fibers, Franklin was able to obtain crude diffraction information on natural DNA. The diffraction patterns she obtained (Figure 12-10) suggested that the DNA molecule was a helical coil, a springlike spiral. From her photographs of the diffraction pattern, it was also possible to determine some of the basic structural parameters of the molecule; the pattern indicated that the helix had a diameter of about 2 nanometers and made a complete spiral turn every 3.4 nanometers.

Two young investigators at Cambridge University, Francis Crick and James Watson, learned about Franklin's results before they were published, and quickly worked out a likely structure of the DNA molecule (Figure 12-11), which we now know to be substantially correct. They analyzed the problem deductively: first they built models of the nucleotides, and then they tested how these could be assembled into a molecule that fit what they knew about the structure of DNA.

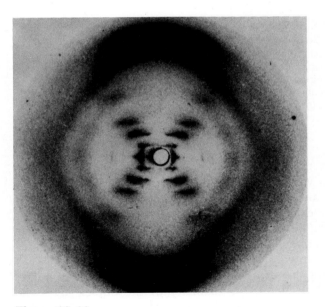

*Figure 12-10*

**The essential clue.** This x-ray diffraction photograph of fibers of DNA was made in 1953 by Rosalind Franklin in the laboratory of Maurice Wilkins. It suggested to Watson and Crick that the DNA molecule was a helix, like a winding staircase.

purine adenine (A) will not form proper hydrogen bonds in this structure with cytosine (C), but will with thymine (T), every A is paired to a T. Similarly, the purine guanine (G) will not form proper hydrogen bonds with thymine but will with cytosine, so every G is paired with C.

> The Watson and Crick structure of DNA is a "double helix," a spiral staircase composed of two polynucleotide chains hydrogen-bonded to each other, wrapped around a central axis.

### Figure 12-11

**The discovery of DNA's structure.** A key breakthrough in our understanding of heredity occurred in 1953, when James Watson, a young American postdoctoral student (he is the one peering up as if afraid their homemade model of the DNA molecule will topple over), and the English scientist Francis Crick (pointing) deduced the structure of DNA, the molecule that stores the hereditary information.

They tried various possibilities, first assembling molecules with three strands of nucleotides wound around one another to stabilize the helical shape. None of these early efforts proved satisfactory. They finally hit on the idea that the molecule might be a simple double helix, one in which the bases of two strands pointed inward toward one another (Figure 12-12). By always pairing a purine, which is large, with a pyrimidine, which is small, the diameter of the duplex stays the same, 2 nanometers. Because hydrogen bonds can form between the two strands, the helical form (Figure 12-13) is stabilized.

It immediately became apparent why Chargaff had obtained the results he had (Figure 12-14): because the

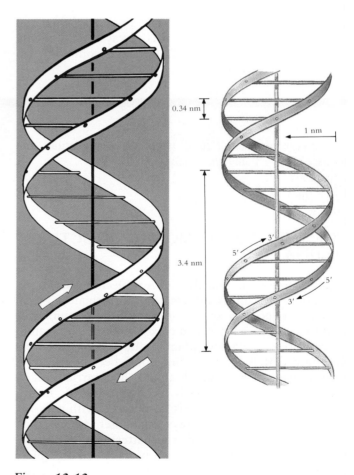

### Figure 12-12

**DNA is a double helix.** On the left is a reproduction of the original diagram of the double-helical structure of DNA, presented in a 1953 paper by Watson and Crick. The diagram above presents the exact dimensions of the double helix.

# HOW DNA REPLICATES

The Watson-Crick model immediately suggested that the basis for copying the genetic information is **complementarity.** One chain of the DNA molecule may have any conceivable base sequence, but this sequence completely determines that of its partner in the duplex. If the sequence of one chain is ATTGCAT, the sequence of its partner in the duplex *must* be TAACGTA. Each chain in the duplex is a complementary mirror image of the other. To copy the DNA molecule, one need only "unzip" it and construct a new complementary chain along each naked single strand.

## Replication is Semiconservative

The form of DNA replication suggested by the Watson-Crick model is called **semiconservative** because, after one round of replication, the original duplex is not conserved; instead, each strand of the duplex becomes part of another duplex.

The complementary nature of the DNA duplex provides a ready means of duplicating the molecule. If one were to unzip the molecule, one would need only to assemble the appropriate complementary nucleotides on the exposed single strands to form two daughter duplexes of the same sequence. This prediction of the Watson-Crick model was tested in 1958 by Matthew Meselson and Frank Stahl of the California Institute of Technology (Figure 12-15). These two scientists grew bacteria for several generations in a medium containing the heavy isotope of nitrogen 15N, so the DNA of their bacteria was eventually denser than normal. They then transferred the growing cells to a new medium containing the lighter isotope 14N and harvested the DNA at various intervals.

At first the DNA that the bacteria manufactured was all heavy. But as the new DNA that was being formed incorporated the lighter nitrogen isotope, DNA density fell. After one round of DNA replication was complete, the density of the bacterial DNA had decreased to a value intermediate between all-light isotope and all-heavy isotope DNA. After another round of replication, centrifugation was used to separate two density classes, one intermediate and the other light, corresponding to DNA that included none of the heavy isotope. These results indicate that after one round of replication, each daughter DNA duplex possessed one of the labeled strands of the parent molecule. When this hybrid duplex replicated, it contributed one heavy strand to form another hybrid duplex and one light strand to form a light duplex. Meselson and Stahl's experiment thus confirmed the prediction of the Watson-Crick model that DNA replicates semiconservatively.

*Figure 12-13*
**A three-dimensional model of DNA.**

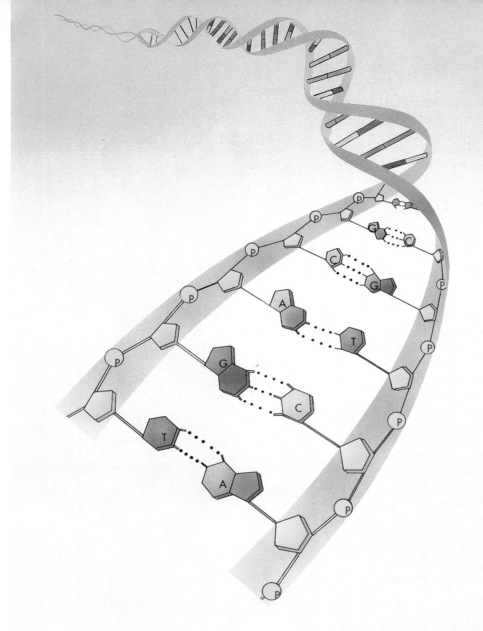

**Figure 12-14**

**Base pairing.** In a DNA duplex molecule, only two base pairs are possible: adenine (A) with thymine (T), and guanine (G) with cytosine (C). A G-C base pair has three hydrogen bonds, an A-T base pair only two.

*The basis for the great accuracy of DNA replication is complementarity. A DNA molecule is a duplex, containing two strands that are mirror images of each other, so either one can be used as a template to reconstruct the other.*

## How DNA Unzips to Copy Itself

When a DNA molecule replicates, the double-stranded DNA molecule separates at one end, forming a **replication fork** (Figure 12-16), and each separated strand serves as a template for the synthesis of a new complementary strand. Indeed, electron micrographs reveal Y-shaped DNA molecules at the point of replication, just as the model predicts.

To the surprise of those investigating the way in which DNA molecules replicate, it turned out that the two new daughter strands are synthesized on their templates differently and in different directions from the replication fork. One strand is built by simply adding nucleotides to its end. This strand grows inward toward the junction of the Y as the duplex unzips. Because this strand ends with an OH group attached to the third carbon of the ribose sugar, the strand is said to grow from its OH end. The enzyme that catalyzes this process is called **DNA polymerase.**

However, when investigators searched for a corresponding enzyme that added nucleotides to the other strand (which ends with a $PO_4$ group attached to the

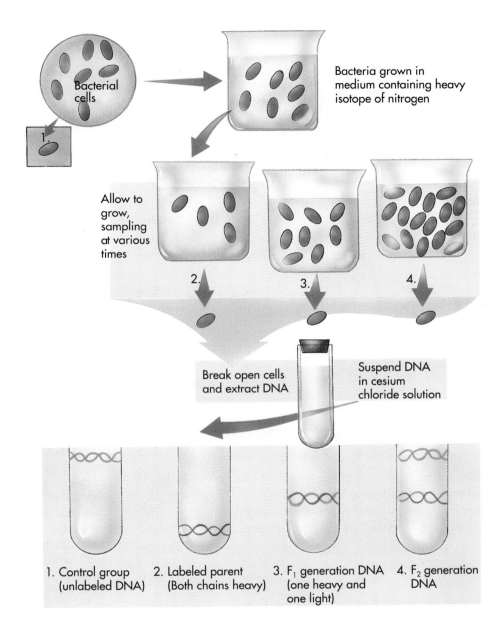

**Figure 12-15**

**The Meselson and Stahl experiment.**
Bacterial cells containing heavy
nitrogen isotopes were transferred to
a new medium containing only light
nitrogen isotope. At various times
thereafter, samples were taken and
the DNA centrifuged in a cesium
chloride solution. Because the cesium
ion is so massive, the cesium chloride
tends to settle in the rapidly spinning
tube, establishing a gradient of
cesium concentration. DNA
molecules sink in the gradient until
they reach a place where the cesium
concentration has the same density as
the density of the DNA; the DNA
then "floats" at that position.
Because DNA built with heavy
nitrogen isotopes is denser than
normal DNA, it sinks to a lower
position on the cesium gradient.

Bacterial cells

1.

Bacteria grown in
medium containing heavy
isotope of nitrogen

Allow to
grow,
sampling
at various
times

2.    3.    4.

Break open cells
and extract DNA

Suspend DNA
in cesium
chloride solution

1. Control group
(unlabeled DNA)

2. Labeled parent
(Both chains heavy)

3. $F_1$ generation DNA
(one heavy and
one light)

4. $F_2$ generation
DNA

fifth carbon of the ribose sugar and is called the phosphate end), they were unable to find one. Nor has anyone ever found one. DNA polymerases add only to the
OH ends of DNA strands.

How does the polymerase build the phosphate
strand? Along this strand, the chain is also formed in
the OH direction, the polymerase jumping ahead and
filling in backward (see Figure 12- 16). The DNA polymerase starts a burst of synthesis at the point of the
replication fork and moves outward, adding nucleotides to the OH end of a short new chain until this
new segment fills in a gap of 1000 to 2000 nucleotides

between the replication fork and the end of the growing chain to which the previous segment was added.
The short new chain is then added to the growing
chain and the polymerase jumps ahead again to fill in
another gap. In effect, it copies the template strand in
segments about 1000 nucleotides long and stitches
each new fragment to the end of the growing chain.
This mode of replication is referred to as **discontinuous
synthesis.** If one looks carefully at electron micrographs showing DNA replication in progress, one of
the daughter strands behind the polymerase appears
single-stranded for about 1000 nucleotides.

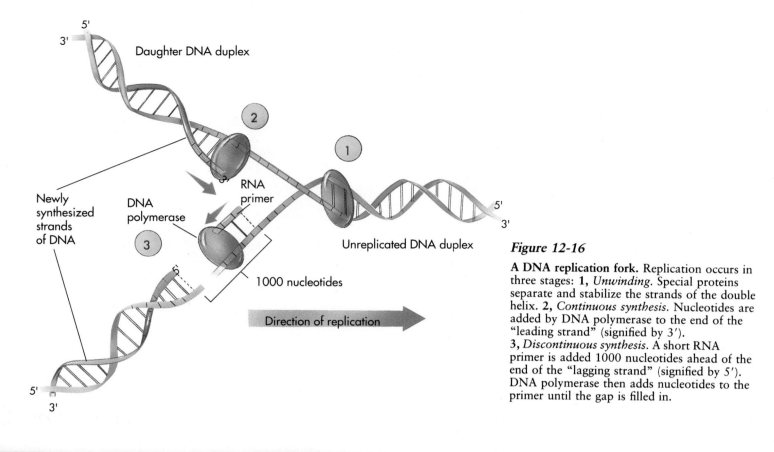

5'
3'
Daughter DNA duplex

2

1

3'
RNA primer

Newly synthesized strands of DNA

DNA polymerase

3

1000 nucleotides

Unreplicated DNA duplex

5'
3'

5'
3'

Direction of replication

*Figure 12-16*

**A DNA replication fork.** Replication occurs in three stages: **1,** *Unwinding.* Special proteins separate and stabilize the strands of the double helix. **2,** *Continuous synthesis.* Nucleotides are added by DNA polymerase to the end of the "leading strand" (signified by 3'). **3,** *Discontinuous synthesis.* A short RNA primer is added 1000 nucleotides ahead of the end of the "lagging strand" (signified by 5'). DNA polymerase then adds nucleotides to the primer until the gap is filled in.

*A DNA molecule copies (replicates) itself by separating its two strands and using each as a template to assemble a new complementary strand, thus forming two daughter duplexes. The two strands are assembled in different directions.*

### How Bacteria Replicate Their DNA

The genetic material of bacteria is organized as a single, circular molecule of DNA. The replication of this molecule is complex and involves many enzymes; it is not, however, difficult to visualize the overall process. The entire genome can be replicated simply by nicking the DNA duplex (that is, breaking one of the two strands) at one site and displacing the strand on one or both sides of the nick, creating one or two replication forks. These replication forks then proceed around the circle, creating a new daughter duplex as they go. In Figure 12-17 you can see a partially replicated circular DNA molecule, the displaced strand forming a loop. The mitochondria and chloroplasts of eukaryotic cells contain similar circular molecules of DNA, which replicate in the same way as as is seen in the bacteria from which they evolved.

### How a Eukaryote Replicates its Chromosome

The DNA within a eukaryotic chromosome is not naked, as in bacteria; it is clothed with a protein wrapping that makes it much easier to pack compactly (Figure 12-18). If you were to look at a eukaryotic chromosome under the electron microscope, you would see numerous replication forks spaced along the chromosome, rather than a single replication fork as in bacterial chromosomes. Each individual zone of the chromosome replicates as a discrete unit, called a **replication unit.** Replication units vary in length from 10,000 to 1 million base pairs; most are about 100,000 base pairs in length. They have been described for many different eukaryotes. Because each chromosome of a eukaryote possesses so much DNA (Figure 12-19), the orderly replication of DNA in eukaryotes undoubtedly requires sophisticated controls, which are as yet largely unknown.

### GENES: THE UNITS OF HEREDITARY INFORMATION

In 1902 a British physician, Archibald Garrod, noted that certain diseases were prevalent in particular families. Indeed, when he examined several generations within such families, he found that some of these disorders behaved as if they were controlled by simple reces-

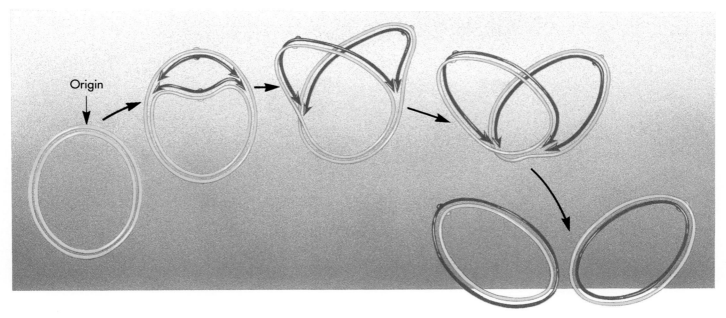

**Figure 12-17**

**How a bacterial chromosome replicates.** The circular chromosome of a bacterium initiates replication at a single site, moving out in both directions. When the two moving replication forks meet on the far side of the chromosome, the circular chromosome has been duplicated.

sive alleles. Garrod concluded that these disorders were Mendelian traits and that they had resulted from changes in the hereditary information that had occurred in an ancestor of the affected families.

Garrod examined several of these disorders in detail. In one, **alkaptonuria,** the patients passed urine that rapidly turned black on exposure to air. Such urine contains homogentisic acid (alkapton), which air oxidizes. In normal individuals, homogentisic acid is broken down into simpler substances, but the affected patients are unable to carry out that breakdown. With considerable insight, Garrod concluded that the patients suffering from alkaptonuria lacked the enzyme necessary to catalyze this breakdown and, more generally, that many inherited disorders might reflect enzyme deficiencies.

## The One Gene–One Enzyme Hypothesis

From Garrod's finding it is but a short leap of intuition to surmise that the information encoded within the DNA of chromosomes is used to specify particular enzymes. But this point was not actually established until 1941, when a series of experiments by Stanford University geneticists George Beadle and Edward Tatum finally provided definitive evidence. Beadle and Tatum deliberately set out to create Mendelian mutations in the chromosomes; they then studied the effects of these mutations on the organism.

*Creating Genetic Differences.* One of the reasons that Beadle and Tatum's experiments produced clear-cut results—and one of the characteristics of most successful laboratory experiments in biology—is that they made an excellent choice of experimental organism. They chose the bread mold *Neurospora,* a fungus that can be readily grown in the laboratory on a **defined medium** (a medium that contains only known substances such as glucose and sodium chloride, rather than some uncharacterized cell extract such as ground-up yeasts). To increase the numbers of mutations, Beadle and Tatum induced mutations by exposing the spores to x-rays. They then allowed the progeny to grow on a **complete medium** (a medium that contained all necessary metabolites and would therefore supply them to the growing fungi, whether or not individual strains could manufacture the metabolites for themselves). In this way the investigators were able to preserve strains that, as a result of the earlier irradiation, had experienced damage to their DNA in a region encoding the ability to make one or more of the compounds that the fungus needed for normal growth. Change of this kind is called **mutation,** and the strains that have lost the ability to use one or more compounds are called **mutant strains.**

*Identifying Mutant Strains.* The next step was to test the progeny of the irradiated spores to see if any mutations leading to metabolic deficiency actually had been cre-

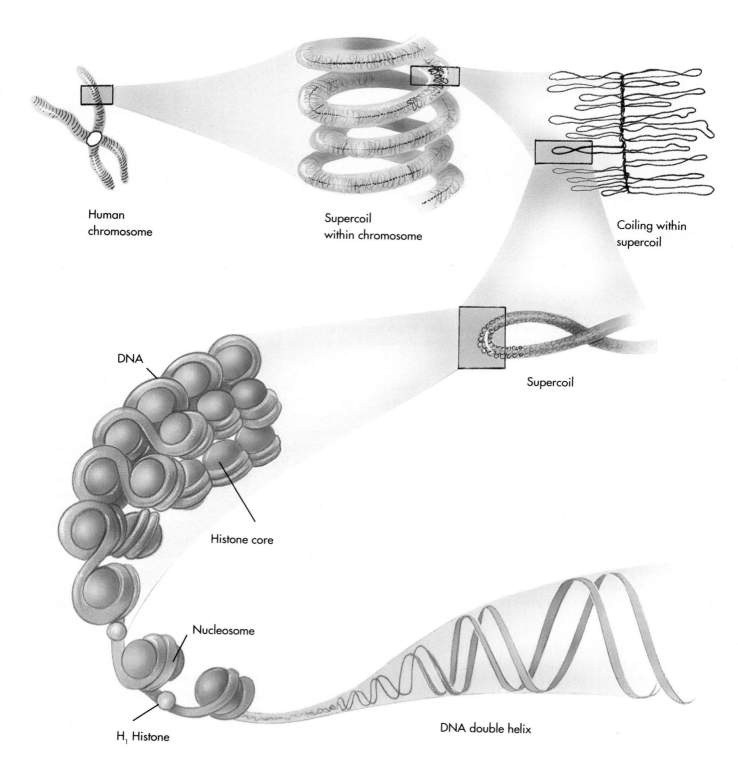

Human chromosome

Supercoil within chromosome

Coiling within supercoil

DNA

Histone core

Supercoil

Nucleosome

H₁ Histone

DNA double helix

**Figure 12-18**

A simplified scheme of the levels of chromosomal organizations.

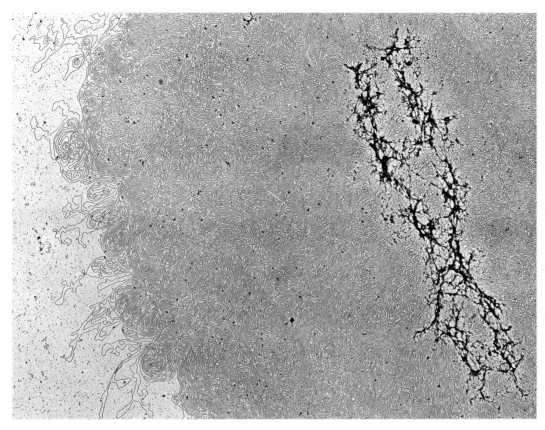

*Figure 12-19*

**A human chromosome contains an enormous amount of DNA.** The dark element at the right in the photograph is the protein matrix of a single chromosome. All the surrounding material is the DNA of that chromosome.

ated by the x-ray treatment. Beadle and Tatum did this by attempting to grow subdivisions of individual fungal strains on a **minimal medium,** which contained only sugar, ammonia, salts, a few vitamins, and water. A cell that had lost the ability to make a necessary metabolite would not grow on such a medium. Using this approach, Beadle and Tatum succeeded in identifying and isolating many deficient mutants.

*Pinpointing the Problem.* To determine the nature of each deficiency, Beadle and Tatum tried adding various chemicals to the minimal medium in an attempt to find one that would make it possible for a given strain to grow (Figure 12-20). In this way they were able to pinpoint the nature of the biochemical problems that many of their mutants had developed. Many of the mutants proved unable to synthesize a particular vitamin or amino acid. The addition of arginine, for example, permitted the growth of a group of mutant strains, dubbed *arg* mutants. When the chromosomal position

of each mutant *arg* gene was located, they were found to cluster in three areas (Figure 12-21).

For each enzyme in the arginine biosynthetic pathway, Beadle and Tatum were able to isolate a mutant strain with a defective form of that enzyme. The mutation always proved to be located at *one* specific chromosomal site, a different site for each enzyme. Thus each of the mutants that Beadle and Tatum examined could be explained in terms of a defect in one (and only one) enzyme, which could be localized at a single site on one chromosome. The geneticists concluded that genes produce their effects by specifying the structure of enzymes, and that each gene encodes the structure of a single enzyme. They called this relationship the **one gene–one enzyme hypothesis.**

Enzymes are responsible for catalyzing the synthesis of all components of the cell. They mediate the assembly of themselves and other proteins, as well as that of nucleic acids, and the synthesis of carbohydrates, fats, and lipids. From the hair on your head to the toe-

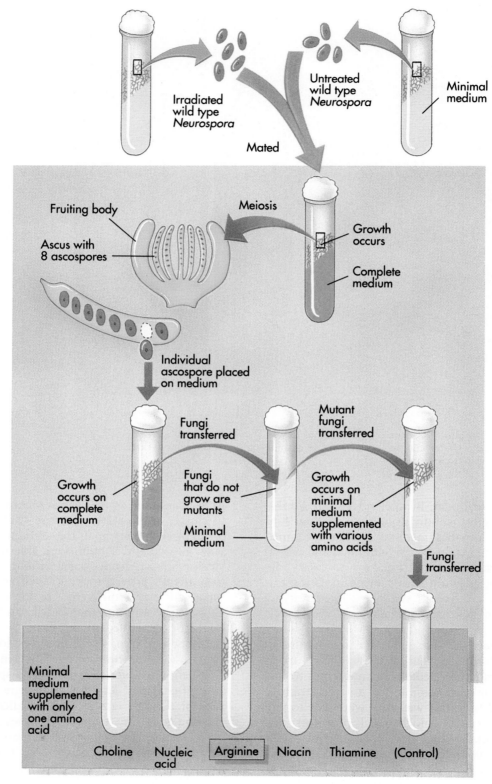

**Figure 12-20**

Beadle and Tatum's procedure for isolating nutritional mutations in *Neurospora*.

Labels within the figure:

Irradiated wild type *Neurospora*

Untreated wild type *Neurospora*

Minimal medium

Mated

Fruiting body

Ascus with 8 ascospores

Meiosis

Growth occurs

Complete medium

Individual ascospore placed on medium

Growth occurs on complete medium

Fungi transferred

Fungi that do not grow are mutants

Minimal medium

Mutant fungi transferred

Growth occurs on minimal medium supplemented with various amino acids

Fungi transferred

Minimal medium supplemented with only one amino acid

Choline | Nucleic acid | Arginine | Niacin | Thiamine | (Control)

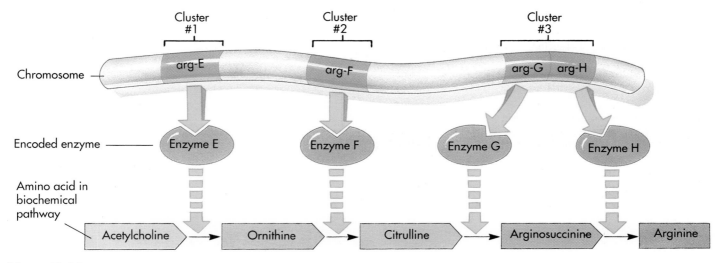

**Figure 12-21**

**Evidence for the "one gene-one enzyme" hypothesis.** The chromosomal locations of the many arginine mutations isolated by Beadle and Tatum cluster around three locations, corresponding to the locations of the genes encoding the enzymes that carry out arginine biosynthesis.

nails of your feet, all of you represents the product of enzyme-directed chemical reactions. By specifying your enzymes, DNA specifies you.

*Genetic traits are expressed largely as a result of the activities of enzymes. Organisms store hereditary information by encoding the structures of enzymes in the DNA of their chromosomes.*

## How DNA Encodes Proteins

What kind of information must a gene contain to specify a protein? For some time, the answer was not clear because protein structure seemed to be impossibly complex. For example, it was not evident whether or not a particular kind of protein had a consistent sequence of amino acids that would be the same in individual molecules in the same kinds of protein. The picture changed in 1953, the same year in which Watson and Crick unraveled the structure of DNA. The great English biochemist Frederick Sanger, after many years of work, announced the complete sequence of amino acids in the protein insulin. Sanger's achievement was extremely significant because it demonstrated for the first time that proteins consist of definable sequences of

amino acids instead of random or meaningless ones. For any given form of insulin, each molecule has the same amino acid sequence as every other, and this sequence can be learned and written down. All enzymes and other proteins are strings of amino acids arranged in a certain, definite order. The information necessary to specify an enzyme, therefore, is an ordered list of amino acids.

In 1956, Sanger's pioneering work was followed by Vernon Ingram's analysis of the molecular basis of sickle-cell anemia, a protein defect inherited as a Mendelian disorder. By analyzing the structure of normal and sickle-cell hemoglobin (Figure 12-22), Ingram, working at Cambridge University, showed that sickle-cell anemia was caused by the change of the amino acid from glutamate to valine at a single position in the protein. The alleles of the gene encoding hemoglobin differed only in their specification of this one amino acid in the hemoglobin amino acid chain.

These experiments, and other related ones, have finally brought us to a clear understanding of the nature of the unit of heredity. Like the dots and dashes of Morse code, the *sequence* of nucleotides in DNA is a code. The sequence provides the information that specifies the identity and order of amino acids in a protein. The sequence of nucleotides that encodes this information is called a **gene**. Although most genes encode proteins, there are also genes devoted to the production of special forms of RNA, many of which play important roles in protein synthesis.

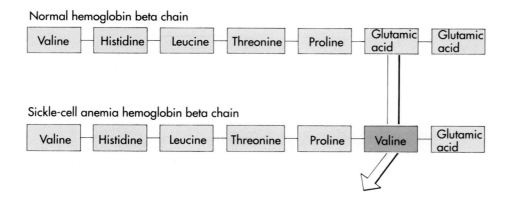

Normal hemoglobin beta chain

| Valine | Histidine | Leucine | Threonine | Proline | Glutamic acid | Glutamic acid |

Sickle-cell anemia hemoglobin beta chain

| Valine | Histidine | Leucine | Threonine | Proline | Valine | Glutamic acid |

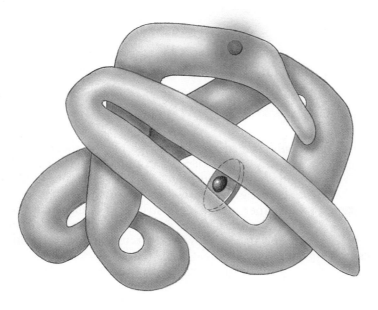

**Figure 12-22**

**The molecular basis of a hereditary disease.** Sickle-cell hemoglobin is produced by a recessive allele of the gene encoding the beta chain of hemoglobin. It represents a single amino-acid change from glutamic acid to valine at the sixth position in the chain.

*The amino acid sequence of a particular protein is specified by a corresponding sequence of nucleotides in the DNA. This nucleotide sequence is called a gene.*

In the next two chapters we focus on the molecular nature of genes, exploring how they work and what happens when one is changed. As we shall see, cancer is one example of what can happen when a gene is altered. Other deliberate changes induced by genetic engineers have proved beneficial; most of the insulin used today by diabetics is the product of a human gene introduced into bacterial cells. Our understanding of genes as the units of heredity, the foundations of which you have encountered in this chapter, represents one of the high-water marks of biology as a science. The intellectual path biologists have followed in their pursuit of this understanding has not always been a straight one, the best questions not always obvious. But however erratic and lurching the experimental journey, our picture of heredity has become progressively clearer and more sharply defined.

# ■ SUMMARY

1. Eukaryotic cells store hereditary information within the nucleus. When one transplants the nucleus, one also transplants the hereditary specifications of an organism.

2. In viruses, bacteria, and eukaryotes, the hereditary information resides in nucleic acids. The transfer of pure nucleic acid can lead to the transfer of hereditary traits. In all cellular organisms, the genetic material is DNA, although in some viruses, the genetic material is RNA.

3. A DNA molecule is a long chain made up of repeating units. Each unit is called a nucleotide and comprises a 5-carbon sugar, a phosphate group, and one of four nitrogen-containing bases (adenine, guanine, cytosine, and thymine).

4. DNA has the structure of a double helix, two chains held to each other by hydrogen bonds between nucleotides. The nucleotide adenine bonds with thymine, and guanine bonds with cytosine.

5. As a part of the process of cell division, the hereditary message is duplicated with great accuracy. The mechanism that achieves this high degree of accuracy is complementarity: the DNA molecule is organized as a double helix with two strands that are mirror images of one another. By adding complementary bases, either one of the strands can be used to re-form the helix.

6. DNA is replicated by a battery of enzymes, which include a DNA polymerase and a variety of other proteins. DNA is replicated by unwinding the duplex and using each of the single strands as a template to assemble a complementary new strand. The two strands are assembled in different directions, one by the continuous addition of nucleotides to the growing end, the other in the opposite direction in 1000-nucleotide segments, which are then joined to the end of the growing strand.

7. Most hereditary traits reflect the actions of enzymes. The traits are hereditary because the information necessary to specify these enzymes is encoded within the DNA.

8. Enzymes are encoded within DNA by specifying their amino acid sequence with a sequence of nucleotides.

9. Changes in individual nucleotides can alter the identity of a particular amino acid in a protein. When such an alteration in DNA, called a mutation, results in the production of a protein with altered biological activity, or no activity at all, the mutation may cause a significant change in appearance or abilities.

# REVIEW

1. Where do cells store their hereditary information?

2. The transforming principle in the Griffith-Avery experiments proved to be _____.

3. Erwin Chargaff observed that the amount of adenine in all DNA molecules is equal to the amount of _____, and that the amount of guanine is equal to the amount of _____.

4. Genetic traits are expressed largely as a result of the activities of proteins called _____, which are necessary for biosynthetic processes to take place.

5. In the Hershey-Chase experiments with bacteriophages, the genetic material was found to be in the _____.

# SELF-QUIZ

1. The Watson-Crick model of DNA structure suggested that the basis for the faithful copying of the genetic material is complementarity. This means that if you know the base sequence of one strand is AATTCG, the sequence of the other strand must be
   - (a) AATTCG.
   - (b) TTGGAC.
   - (c) TTAACG.
   - (d) TTAAGC.
   - (e) do not have enough information.

2. The sequence of DNA nucleotides that encodes the amino acid sequence of an enzyme is
   - (a) a gene.
   - (b) a polypeptide.
   - (c) a protein.
   - (d) a phenotype.
   - (e) a dominant trait.

3. The Hershey-Chase experiment demonstrated that
   - (a) virus DNA injected into bacteria cells is apparently the factor involved in directing the production of new virus particles.
   - (b) virus protein injected into bacterial cells is apparently the factor involved in directing the production of new virus particles.
   - (c) $^{12}P$ labeled protein is injected into bacterial cells by viruses.
   - (d) the transforming principle is the DNA and not the polysaccharide capsule.
   - (e) RNA is the genetic material of some viruses.

4. Some viruses direct their heredity with
   - (a) proteins.
   - (b) smooth capsular coats.
   - (c) RNA.
   - (d) guanine-cytosine bonds.
   - (e) none of the above.

5. The critical experiments in demonstrating that DNA is replicated semiconservatively were performed by
   - (a) Watson and Crick.
   - (b) Griffith and Avery.
   - (c) Meselson and Stahl.
   - (d) Chargaff.
   - (e) none of the above.

6. The presence of a single replication fork is associated with the replication of DNA in (more than one choice is possible)
   - (a) humans.
   - (b) bacteria.
   - (c) *Neurospora*.
   - (d) yeasts.
   - (e) chloroplasts.

7. The principal reason that Frederick Sanger's determination of the amino acid sequence in insulin was important is
   - (a) insulin could be produced on an industrial scale.
   - (b) it provided the critical information to develop a cure for sickle-cell anemia.
   - (c) it showed that insulin is produced on a DNA template.
   - (d) it demonstrated for the first time that proteins consist of definable sequences of amino acids.
   - (e) it provided a key to the structure and specialized activity of proteins.

# THOUGHT QUESTIONS

1. In Hammerling's experiments with *Acetabularia*, the grafting of a nucleus-containing basal portion of an *A. mediterranea* individual (with a disk-shaped cap) to the stalk of an *A. crenulata* individual (with a flower-shaped cap) resulted in the formation of a cap above the grafted stalk. The cap that formed was flower-shaped, like that of *A. crenulata*. When he removed that cap, another one formed, but this one was disk-shaped, like that of *A. mediterranea*. What was responsible for the change in morphology of these two regenerated caps?

2. From an extract of human cells growing in tissue culture, you obtain a white fibrous substance. How would you distinguish whether it was DNA, RNA, or protein?

3. In analyzing DNA obtained from your own cells, you are able to determine that 15% of the nucleotide bases it contains are thymine. What percentage of the bases are cytosine?

4. From a hospital patient afflicted with a mysterious illness, you isolate and culture cells and then purify DNA from the culture. You find that the DNA sample obtained from the culture contains two quite different kinds of DNA; one is double-stranded human DNA and the other is single-stranded virus DNA. You analyze the base composition of the two purified DNA preparations, with the following results:
Tube 1: 22.1% A : 27.9% C : 27.9% G : 22.1% T
Tube 2: 31.3% A : 31.3% C : 18.7% G : 18.7% T
Which of the two tubes contains single-stranded virus DNA?

# FOR FURTHER READING

CRICK, F.H.C.: "The Discovery of the Double Helix Was a Matter of Selecting the Right Problem and Sticking to It," *The Chronicle of Higher Education*, October 5, 1988, pages 1-9. Francis Crick's own recollections of the hectic days when he and James Watson deduced that the structure of DNA is a double helix.

HALL, S.S.: "James Watson and the Search for Biology's 'Holy Grail'," *Smithsonian*, February 1990, pages 41-49. The effort to provide a complete sequence of base pairs for the human genome is now underway.

JUDSON, H.F.: *The Eighth Day of Creation*, Simon & Schuster, Inc., New York, 1979. The definitive historical account of the experimental unraveling of the mechanism of heredity, based on personal interviews with the participants. This book is full of the feel of how science is really conducted.

SCHLIEF, R.: "DNA Binding by Proteins," *Science*, vol. 241, 1988, pages 1182-1187. An account of the different ways in which proteins bind to the DNA double helix.

WATSON, J.D.: *The Double Helix*, Atheneum Publishing Company, Inc., New York, 1968. A lively, often irreverent account of what it was like to discover the structure of DNA, recounted by someone in a position to know.

WATSON, J.D., and F.H.C. CRICK: "A Structure for Deoxyribose Nucleic Acid," *Nature*, vol. 171, 1953, page 737. The original report of the double helical structure of DNA. Only one page long, this paper marks the birth of molecular genetics.

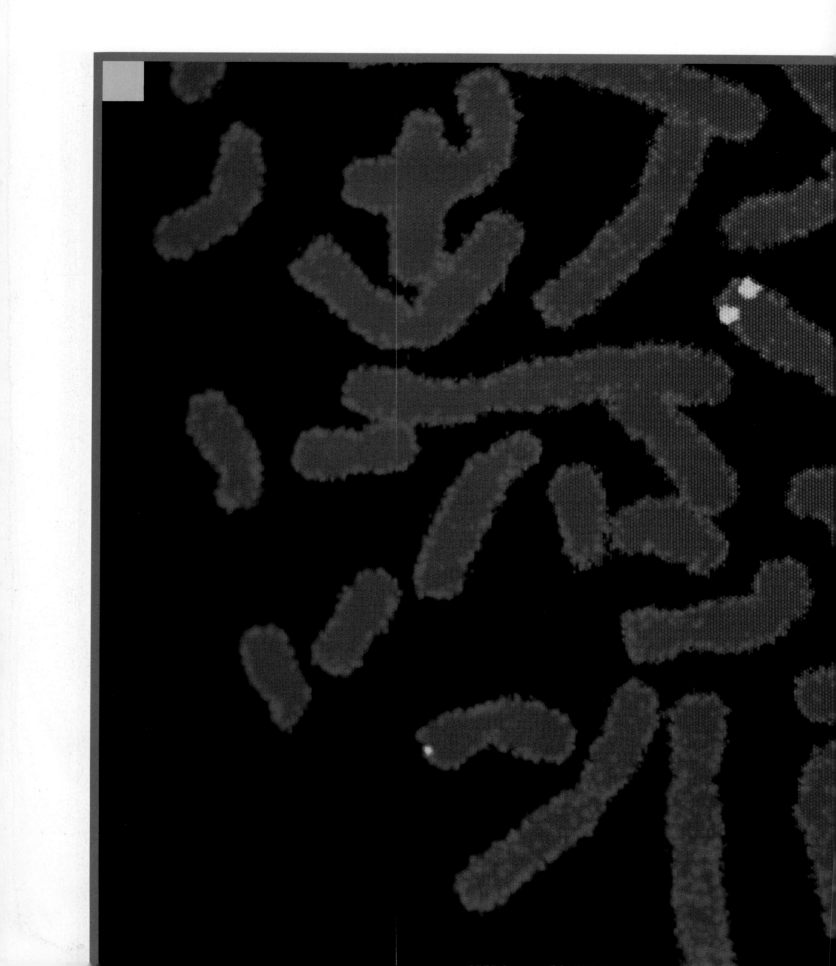

# Genes and How They Work

These chromosomes of the fruitfly *Drosophila melanogaster* are composed of genes strung one after another, like a beaded Indian necklace. Active genes appear as brighter bands. At any one time only a fraction of the many genes on a chromosome are actively producing RNA molecules.

# GENES AND HOW THEY WORK

## Overview

The discovery that genes were composed of DNA within chromosomes, and that genes acted by directing the production of specific proteins, left unanswered the question of how this is accomplished. We now know that special proteins bind to the DNA at the beginning of genes and then move through them, making an RNA copy of the gene as they go. It is this RNA copy that cells use to produce the polypeptides specified by the genetic information. Cells often control when a gene is "turned on" by controlling production of these RNA copies. Changes in genes often result in altered proteins that function abnormally. Such changes, which can be as simple as substituting a single nucleotide for another, are the raw material of evolution.

## For Review

*Here are some important terms and concepts that you will encounter in this chapter. If you are not familiar with them, you should review them before proceeding.*

**Ribosomes** (Chapter 4)

**Structure of DNA and RNA** (Chapters 2, 12)

**Enzymes and enzyme activity** (Chapter 6)

**Structure of eukaryotic chromosomes** (Chapter 9)

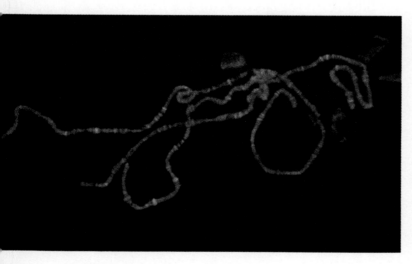

**Figure 13-1**

**Genes on a chromosome.** In this giant chromosome of the fly *Drosophila melanogaster*, active genes appear as brighter bands. At any one time, only a fraction of a chromosome's many genes are actively being transcribed to produce RNA molecules.

If the total DNA from a single cell of your body were stretched out, with all 46 chromosomes lined up, one after the other, the DNA would extend as high as you are tall. All the DNA in your body, stretched end-to-end, would extend about 200 billion kilometers! This enormous library of DNA (Figure 13-1) contains more information than is stored in this book. Its instructions specify that you will have arms and not fins, hair and not feathers, two eyes and not one. The color of your eyes, the texture of your fingernails, whether you dream in color—all of the many traits that you receive from your parents are recorded in the DNA present in every cell of your body.

The information in DNA is arrayed in little blocks like entries in a dictionary, each block a gene specifying a particular polypeptide. Some of these polypeptides function as entire proteins directly when they are formed, whereas many other proteins are formed of two or more polypeptides that are produced individually by separate genes. Proteins are the tools of heredity. Many of them are enzymes that carry out reactions within cells: what you are is the result of what they do.

Mistakes sometimes happen when genes are copied during chromosome duplication, producing genes with "typos"—mistakenly altered sets of instructions. Damage to the DNA often also produces changes in the instructions, just as shooting a bullet through a computer can change its performance. These changes in genes result in proteins that function differently.

The essence of heredity is the ability of a cell to faithfully copy its meters of DNA-encoded instructions, making few errors, and to use these instructions to bring about the production of particular polypeptides and so affect what the cell will be like. In this chapter we examine how this happens.

## CELLS USE RNA TO MAKE PROTEIN

To find out how a cell uses its DNA to direct the production of particular proteins, perhaps the simplest question you might ask is: "Where in the cell are proteins made?" You can answer this question by placing cells in a medium containing radioactive amino acids. The places where these labeled amino acids are assembled into proteins can then be determined. Using this method, cell biologists demonstrated that proteins are assembled not in the nucleus, where the chromosomal DNA resides, but rather in the cytoplasm, on large clusters of proteins called ribosomes (Figure 13-2). These polypeptide-making factories proved to be very complex, containing over 50 different proteins as well as RNA. As you will recall from Chapter 2, RNA is very similar to DNA (Figure 13-3), and its presence in ribosomes hints at the important role that it plays in polypeptide synthesis.

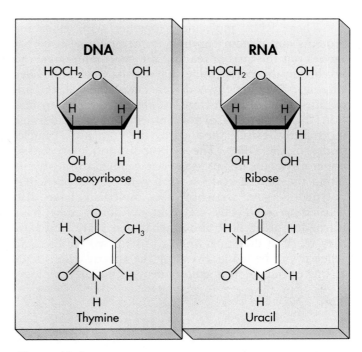

### Figure 13-3

**How RNA is different from DNA.** There are two important differences: first, in place of the sugar deoxyribose, RNA contains ribose, which has an additional oxygen atom. Second, RNA contains the pyrimidine uracil (U) instead of thymine (T). Another difference is that RNA does not have a regular helical structure and is usually single-stranded.

A cell contains many kinds of RNA. There are three major classes:

1. *Ribosomal RNA.* The class of RNA that is found in ribosomes, where it occurs with characteristic proteins, is called **ribosomal RNA,** or **rRNA.** During polypeptide synthesis, rRNA molecules provide the site on the ribosome where the polypeptide is assembled.

2. *Transfer RNA.* A second class of RNA, called **transfer RNA,** or **tRNA,** occurs as much smaller molecules than rRNA. Cells contain more than 60 kinds of tRNA molecules, which float free in the cytoplasm. During polypeptide synthesis, tRNA molecules transport the amino acids to the ribosome for use in building the polypeptide and position each amino acid at the correct place on the elongating polypeptide chain.

3. *Messenger RNA.* A third class of RNA is **messenger RNA,** or **mRNA.** Each mRNA molecule is a long, single strand of RNA that passes from the nucleus to the cytoplasm. During polypeptide synthesis, mRNA molecules bring information from the chromosomes to the ribosomes to direct which polypeptide is assembled.

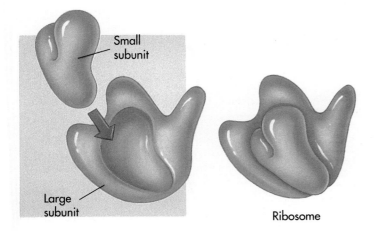

### Figure 13-2

**A ribosome is composed of two subunits.** The smaller subunit fits into a depression on the surface of the larger one.

These RNA molecules, together with ribosome proteins and certain enzymes, constitute a system that carries out the task of reading the genetic message and producing the polypeptide that the particular message specifies. They are the code-readers of the cell, the apparatus a cell uses to translate its hereditary information. You can think of this information as a message written in the code specified by the sequence of nucleotides in the DNA. The cell's polypeptide-producing apparatus reads this message one gene at a time, translating the genetic code of each gene into a particular polypeptide. As we shall see, biologists have also learned to read this code. In so doing, they have learned a great deal about what genes are and how they work in dictating what a protein will be like and when it will be made. Breaking the genetic code stands as one of the greatest achievements of modern biology.

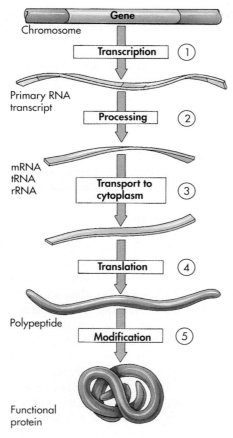

**Figure 13-4**

The 5 stages of gene expression.

# AN OVERVIEW OF GENE EXPRESSION

The hereditary apparatus of your body operates basically the same way as it does in bacteria. An RNA copy of each active gene is made, and the RNA copy directs the sequential assembly of a chain of amino acids at a ribosome. This whole process is known as **gene expression.** There are many minor differences in the details of gene expression between bacteria and eukaryotes and a single major difference that we discuss later in this chapter. But the basic apparatus used in gene expression appears to be the same in all organisms (Figure 13-4); it apparently has persisted virtually unchanged since very early in the history of life. The process of gene expression occurs in two phases (Figure 13-5), which are called transcription and translation.

## Transcription

The first stage of gene expression is the production of an RNA copy of the gene, called messenger RNA or mRNA (Figure 13-6). Like all classes of RNA that occur in cells, mRNA is formed on a DNA template. The production of mRNA is called **transcription;** the mRNA molecule is said to have been **transcribed** from the DNA. Transcription is initiated when a special enzyme, called an **RNA polymerase,** binds to a particular sequence of nucleotides on one of the DNA strands, a sequence located at the edge of a gene. Starting at that end of the gene, the RNA polymerase proceeds to assemble a single strand of mRNA with a nucleotide sequence **complementary** to that of the DNA strand it has bound. **Complementarity** refers to the way in which the two single strands of DNA that form a double helix relate to one another, with A (adenine) pair-

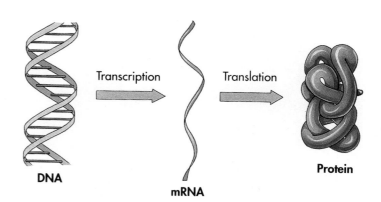

**Figure 13-5**

The central dogma of gene expression.

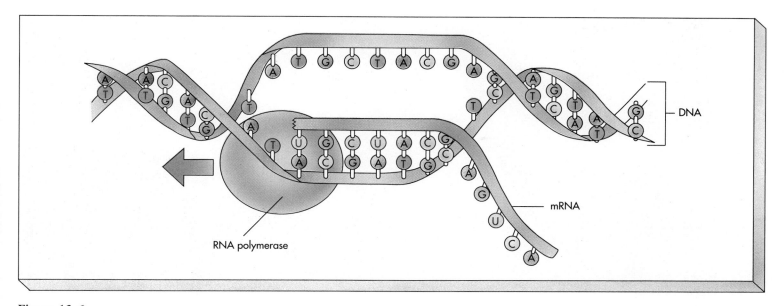

*Figure 13-6*

**Transcription.** One of the strands of DNA functions as a template, on which nucleotide building blocks are assembled by RNA polymerase into RNA.

ing with T (thymine) and G (guanine) pairing with C (cytosine). An RNA strand complementary to a DNA strand has the same relationships, but with U (uracil) in place of thymine.

As the RNA polymerase moves along the gene, encountering each DNA nucleotide in turn, it adds the corresponding complementary RNA nucleotide to the growing mRNA strand. When the enzyme arrives at the far end of the gene, it disengages from the DNA, releasing the newly assembled mRNA chain. This chain is complementary to the DNA strand from which the polymerase assembled it; thus it is an **RNA transcript** (copy), called the **primary mRNA transcript,** of the nucleotide sequence of the gene.

## Translation

The second stage of gene expression is the synthesis of a polypeptide by ribosomes, which use the information contained on an mRNA molecule to direct the choice of amino acids. This process is called **translation** because nucleotide-sequence information is translated into amino acid–sequence information. Translation begins when an rRNA molecule within the ribosomes binds to one end of an mRNA molecule. The ribosome then proceeds to move down it in steps of three nucleotides each. At each step, the ribosome adds an amino acid to a growing polypeptide chain. It continues to do this until it encounters a "stop" signal, which indicates the end of the polypeptide. The ribosome then disengages from the mRNA and releases the newly assem-

bled polypeptide. An overview of protein synthesis is presented in Figure 13-7.

*The information encoded in genes is expressed in two stages: transcription, in which a polymerase enzyme assembles an mRNA molecule whose sequence is complementary to the DNA; and translation, in which a ribosome assembles a polypeptide, using the mRNA to specify the amino acids.*

## HOW ARE GENES ENCODED?

The essential question of gene expression is: how does the *order* of nucleotides in a DNA molecule encode the information that specifies the order of amino acids in a protein? What, in other words, is the nature of the DNA genetic code? The answer came in 1961 as the result of an experiment led by Francis Crick. By treating virus DNA with chemicals that add or delete single nucleotides, he was able to show that genes are read in increments of three consecutive nucleotides. Each block of three nucleotides, a **codon,** codes for one amino acid.

Just which three-nucleotide code words correspond to which amino acids (the **genetic code**) was

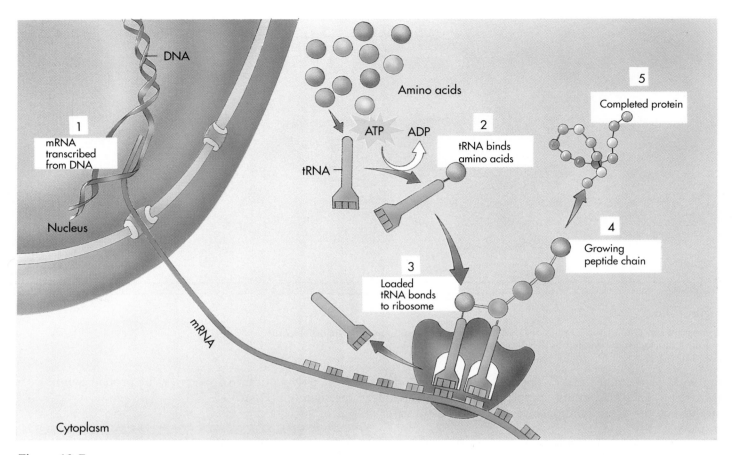

**Figure 13-7**

An overview of protein synthesis.

soon worked out, the first results coming within a year of Crick's experiment. Researchers had developed mixtures of RNA and protein isolated from ruptured cells ("cell-free systems") that would synthesize proteins in a test tube. To determine which three-nucleotide sequences specify which amino acids, researchers added artificial RNA molecules to these cell-free systems and then looked to see what proteins were made. For example, when Marshall Nierenberg of the National Institutes of Health added polyU (an RNA molecule consisting of a string of uracil nucleotides) to such a system, it proceeded to synthesize polyphenylalanine (a protein consisting of a string of phenylalanine amino acids). This result indicated that the three-nucleotide codon specifying phenylalanine was UUU. In this and other ways, all 64 possible triplets were examined, and the full genetic code was determined.

## THE GENETIC CODE

Working out the genetic code was a great step forward in removing the mystery from the process of gene ex-

pression. However, it left an important question unanswered. How is the information stored in a sequence of nucleotides, such as UUU, used to identify a specific amino acid such as phenylalanine? How is the genetic code deciphered?

To answer this question, we must look more carefully at the first events of the translation process. Translation occurs on the ribosomes. First, the initial portion of the mRNA transcribed from a gene binds to an rRNA molecule interwoven in the ribosome (Figure 13-8). The mRNA lies on the ribosome in such a way that only a three-nucleotide portion of the mRNA molecule—the codon—is exposed at the polypeptide-making site. As each bit of the mRNA message is exposed in turn, a molecule of tRNA with the complementary three-nucleotide sequence, or **anticodon**, binds to it (Figure 13-9). Because this tRNA molecule carries a particular amino acid, that amino acid and no other is added to the polypeptide in that position. Protein synthesis occurs as a series of tRNA molecules bind one after another to the exposed portion of the mRNA molecule as it moves through the ribosome. Each of these tRNA molecules has attached to it an amino acid,

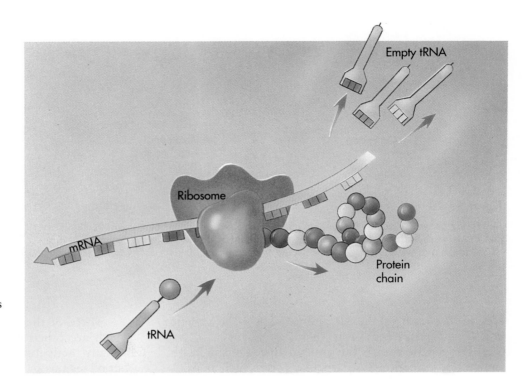

**Figure 13-8**

**Translation.** The mRNA strand acts as a template for tRNA molecules. The appropriate tRNA is selected and positioned by the ribosome, which moves along the mRNA in three-nucleotide steps.

and the amino acids are added, one after another, to the end of a growing polypeptide chain.

How does a particular tRNA molecule come to possess the amino acid that it does, and not just any amino acid? The correct amino acid is placed on each tRNA molecule by a collection of 20 enzymes called **activating enzymes.** There is one activating enzyme for each of the 20 common amino acids. An activating enzyme binds the amino acid that it recognizes (Figure 13-10) to a tRNA molecule. If one considers the nucleotide sequence of mRNA to be a coded message, then the 20 activating enzymes are the code books of the cell—the instructions for decoding the message. An activating enzyme recognizes both nucleotide-sequence information (a specific anticodon sequence of a tRNA molecule) and protein-sequence information (a particular amino acid).

The code word recognized by an activating enzyme is three nucleotides long. Because there are four kinds of nucleotides in mRNA (cytosine, guanine, adenine, and uracil instead of thymine), there are $4^3$ or 64 different three-letter code words, or codons, possible. Some of the activating enzymes recognize only one tRNA molecule, corresponding to one of these code words; others recognize two, three, four, or six different tRNA molecules, each containing a different anticodon. The base sequences of the tRNA anticodons are complementary to the associated sequences of mRNA and relate to the same amino acid as do those of their partner. The list of different mRNA codons

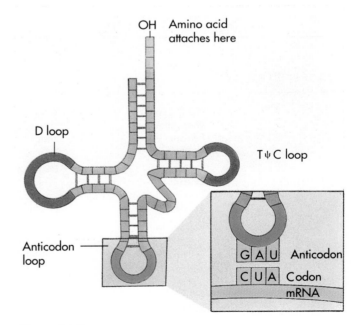

**Figure 13-9**

**Structure of a tRNA molecule.** The TψC loop and the D loop function in binding to the ribosomes during polypeptide synthesis. The third loop contains the anticodon sequence. The activating enzyme adds an amino acid to the free single-stranded —OH end.

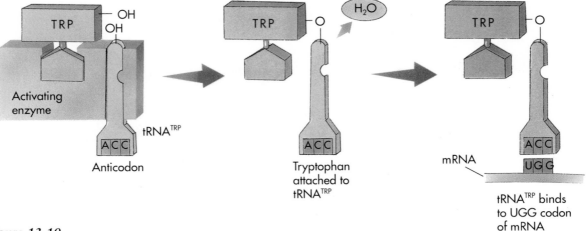

**Figure 13-10**

**Activating enzymes "read" the genetic code.** Each kind of activating enzyme recognizes and binds a specific amino acid, such as tryptophan, and also recognizes and binds the tRNA molecules with anticodons specifying that amino acid, such as ACC for tryptophan.

specific for each of the 20 amino acids—the genetic code—is presented in Table 13-1.

The genetic code is the same in all organisms, with only a few exceptions. A particular codon such as AGA corresponds to the same amino acid (arginine) in bacteria as in humans. The universality of the genetic code is among the strongest evidence that all living things share a common evolutionary heritage.

Three of the 64 codons (UAA, UAG, and UGA), which are called **nonsense codons,** are not recognized by any activating enzyme. These codons serve as "stop" signals in the mRNA message, marking the end of a polypeptide. The "start" signal, which marks the beginning of a polypeptide amino-acid sequence within a mRNA message, is the codon AUG, a three-base sequence that also encodes the amino acid methionine.

**TABLE 13-1    THE GENETIC CODE**

| FIRST LETTER | SECOND LETTER | | | | THIRD LETTER |
| --- | --- | --- | --- | --- | --- |
| | U | C | A | G | |
| U | Phenylalanine | Serine | Tyrosine | Cysteine | U |
| | Phenylalanine | Serine | Tyrosine | Cysteine | C |
| | Leucine | Serine | Stop | Stop | A |
| | Leucine | Serine | Stop | Tryptophan | G |
| C | Leucine | Proline | Histidine | Arginine | U |
| | Leucine | Proline | Histidine | Arginine | C |
| | Leucine | Proline | Glutamine | Arginine | A |
| | Leucine | Proline | Glutamine | Arginine | G |
| A | Isoleucine | Threonine | Asparagine | Serine | U |
| | Isoleucine | Threonine | Asparagine | Serine | C |
| | Isoleucine | Threonine | Lysine | Arginine | A |
| | (Start); Methionine | Threonine | Lysine | Arginine | G |
| G | Valine | Alanine | Aspartate | Glycine | U |
| | Valine | Alanine | Aspartate | Glycine | C |
| | Valine | Alanine | Glutamate | Glycine | A |
| | Valine | Alanine | Glutamate | Glycine | G |

A codon consists of three nucleotides read in the sequence shown above. For example, ACU codes threonine. The first letter, A, is read in the first column; the second letter, C, from the second letter columns; and the third letter, U, from the third letter column. Each of the codons is recognized by a corresponding anticodon sequence on a tRNA molecule. Some tRNA molecules recognize more than one codon sequence but always for the same amino acid. Most amino acids are encoded by more than one codon. For example, threonine is encoded by four codons (ACU, ACC, ACA, and ACG), which differ from one another only in the third position.

The ribosome uses the first AUG that it encounters in the mRNA message to signal the start of its translation.

*All organisms possess a battery of 20 enzymes, called activating enzymes, one or more of which recognizes a particular three-base anticodon sequence in a tRNA molecule. The mRNA codons specific for the 20 common amino acids constitute the genetic code.*

with which the nucleotide sequence will be translated into a polypeptide. This initiation complex, guided by another initiation factor, then binds to mRNA. It is important that the complex bind to the beginning of a gene, so that all of the gene will be translated. In bacteria, the beginning of each gene is marked by a sequence that is complementary to one of the rRNA molecules on the ribosome. This ensures that genes are read from the beginning; each mRNA binds to the ribosomes that read it by base-pairing between the sequence at its beginning and the complementary sequence on the rRNA, which is a part of the ribosome.

## THE MECHANISM OF PROTEIN SYNTHESIS

Polypeptide synthesis begins with the formation of an **initiation complex** (Figure 13-11). Special proteins called **initiation factors** position the initial tRNA on the ribosomal surface. The proper positioning of this first amino acid is critical because it determines the reading frame—the particular groups of three bases

*An initiation complex consists of a ribosome, mRNA, and a tRNA molecule.*

After the initiation complex has been formed, the synthesis of the polypeptide proceeds as follows (Figure 13-12):

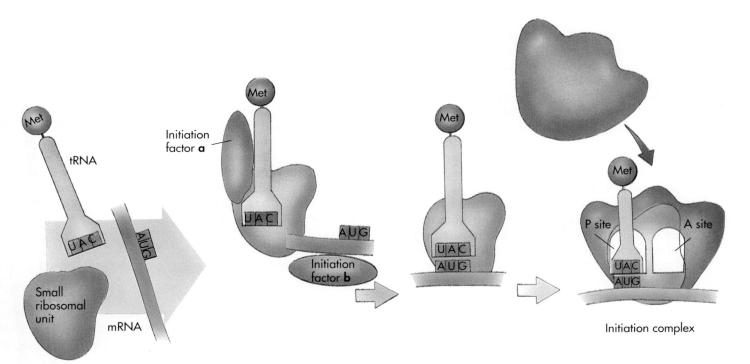

**Figure 13-11**

**Formation of the initiation complex.** Proteins called initiation factors play key roles in positioning the small ribosomal subunit and the met-tRNA molecule at the beginning of the mRNA message. When the met-tRNA is positioned over the first AUG codon sequence of the mRNA, the large ribosomal subunit binds, forming the A and P sites, and polypeptide synthesis begins.

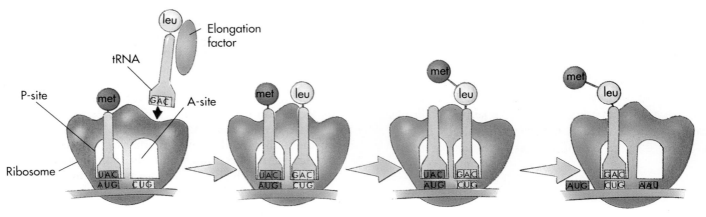

*Figure 13-12*

**How polypeptide synthesis proceeds.** The A site becomes occupied by a tRNA with an anticodon complementary to the mRNA codon exposed there. The growing polypeptide chain (here, met) is transferred to this incoming amino acid (here, leucine), and the ribosome moves three nucleotides to the right.

1. The ribosome exposes the codon on the mRNA next to the initiating AUG codon, positioning it for interaction with another incoming tRNA molecule. When a tRNA molecule with the appropriate anticodon appears, this tRNA binds to the mRNA molecule at the exposed codon position. Special proteins called **elongation factors** (because they aid in making the mRNA molecule longer) help to position the incoming tRNA. Binding the incoming tRNA to the ribosome in this fashion places the amino acid held by the incoming tRNA molecule next to the initial amino acid, which is held by the initiating tRNA molecule still bound to the ribosome.

2. The two amino acids undergo a chemical reaction in which the initial amino acid is released from its tRNA and is attached instead by a peptide bond to the adjacent incoming amino acid. The abandoned tRNA falls from its site on the ribosome, leaving that site vacant.

3. In a process called **elongation** (see Figure 13-12), the ribosome now moves along the mRNA molecule a distance corresponding to three nucleotides, guided by other elongation factors. This movement repositions the growing chain, at this point containing two amino acids, and exposes the next codon of the mRNA. This is the same situation that existed in step 1. When a tRNA molecule that recognizes this next codon appears, the anticodon of the incoming tRNA binds this codon, placing a new amino acid adjacent to the growing chain. The grow-

ing chain transfers to the incoming amino acid, as in step 2, and the elongation process continues (Figure 13-13).

4. When a stop codon is encountered, no tRNA exists to bind to it. Instead, it is recognized by special **release factors,** proteins that release the newly made polypeptide from the ribosome.

The sites on the ribosome where mRNA codons are exposed play critical roles in protein synthesis. The site where incoming tRNA molecules first bind to the mRNA is called the attachment site, or *A-site*. The site next to it, to which the growing chain of amino acids is attached, is called the polypeptide chain site, or *P-site*. Recent data suggest the existence of a third release site, or *R-site*, past the P-site that functions in releasing the tRNA after it has donated its amino acid to the growing chain.

*Protein synthesis is carried out by ribosomes, which bind to sites at one end of the mRNA and then move down the mRNA in increments of three nucleotides. Each step of the ribosome's progress exposes a three-base sequence to binding by a tRNA molecule with the complementary nucleotide sequence. Ultimately, the amino acid carried by that tRNA molecule is added to the end of the growing polypeptide chain.*

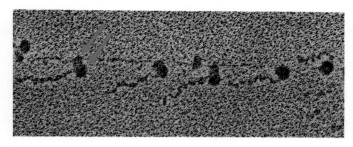

**Figure 13-13**

**Translation in action.** These ribosomes are reading along an mRNA molecule of the fly *Chironomus tentans* from left to right, assembling polypeptides that dangle behind them like the tail of a tadpole. Clearly visible are the two subunits *(arrows)* of each ribosome translating the mRNA.

## PROTEIN SYNTHESIS IN EUKARYOTES

Protein synthesis occurs in a similar way in both bacteria and eukaryotes, although there are differences. One difference is of particular importance. Unlike bacterial genes, most eukaryotic genes are much larger than they need to be, containing long stretches of nucleotides that are cut out of the mRNA transcript before it is used in polypeptide synthesis. Because these sequences are removed from the mRNA transcript before it is used, they are not translated into polypeptide. Intervening (lying between) the polypeptide-specifying portions of the gene (Figure 13-14), they are called **introns**.

The remaining segments of the gene—the nucleotide sequences that encode the amino-acid sequence of the polypeptide—are called **exons**. Exons are scattered among larger noncoding intron sequences. In a typical human gene the nontranslated (intron) portion of a gene is 10 to 30 times larger than the coding (exon) portion. Despite these differences, eukaryotes make polypeptides in much the same way as do bacteria.

## REGULATING GENE EXPRESSION

An organism must be able to control which of its genes are being transcribed, and when. There is little point for a cell to produce an enzyme when its substrate, the target of its activity, is not present in the cell. It costs the cell a great deal of energy to produce enzymes and other proteins, energy that can be saved if the enzyme is not produced very much until the appropriate substrate is encountered and the enzyme's activity will be of use to the cell.

From a broader perspective, the growth and development of many multicellular organisms, including human beings, entails a long series of biochemical reactions, each of which is delicately tuned to achieve a precise effect. Specific enzyme activities are called into play and used to bring about a particular developmental change. Once this change has occurred, those particular enzyme activities cease, lest they disrupt other activities which follow. During development, genes are transcribed in a carefully prescribed order, each gene for a specified period. The hereditary message is played like a piece of music on a grand organ, in which particular proteins are the notes and the hereditary information that regulates their expression is the score.

Organisms control the expression of their genes largely by controlling when the transcription of individual genes begins (Figure 13-15). Most genes possess special nucleotide sequences called **regulatory sites,** which act as points of control (Figure 13-16). These nucleotide sequences are recognized by specific regulatory proteins which bind to the sites.

### Negative Control

Regulatory sites often function to shut off transcription, a process called **negative control.** In these cases, the site at which the regulatory protein binds to the DNA is located between the site at which the polymerase binds and the beginning edge of the gene that the polymerase is to transcribe. When the regulatory protein is bound to its regulatory site, its presence there blocks the movement of the polymerase toward the gene. To understand this more clearly, imagine that you are shooting a cue ball at the eight ball on a pool table and that someone places a brick on the table between the cue ball and the eight ball. Functionally, this brick is like the regulatory protein that binds to the DNA: its placement blocks movement of the cue ball to the eight ball, just as placement of the regulatory protein between polymerase and gene blocks movement of the polymerase to the gene. The process of blocking transcription in this way is called **repression,** and the regulatory protein that is responsible for the blockage is called a **repressor protein.**

### Positive Control

Regulatory sites may also serve to turn on transcription, a process called **positive control.** In these situations, the binding of a regulatory protein to the DNA is necessary before the transcription of a particular gene can begin. The regulatory protein whose binding turns on transcription is called an **activator protein,** and this way of turning on the transcription of specific genes is called **activation.** Activation can be achieved by various mechanisms. In some cases, the activator protein's

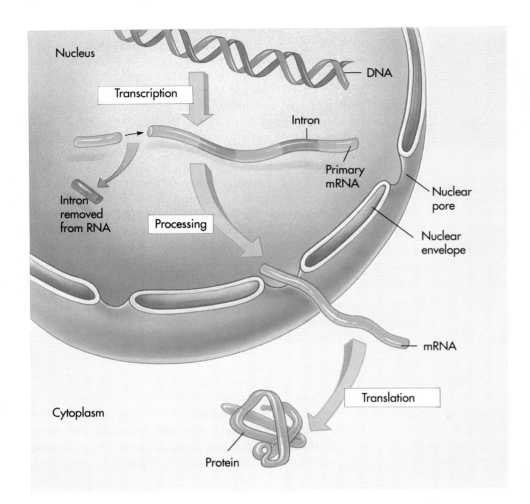

**Figure 13-14**
Gene processing.

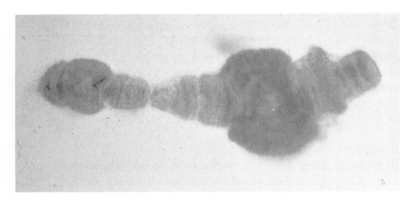

**Figure 13-15**

**Chromosome puffs.** In giant salivary chromosomes of the fly *Drosophila melanogaster,* the individual strands of DNA are replicated many times in parallel. For this reason, the activity of individual genes can readily be detected as "puffs."

binding promotes the unwinding of the DNA duplex. This facilitates the production of an mRNA transcript of a gene because the polymerase, although it can bind to a double-stranded DNA duplex, cannot produce an mRNA transcript from such a duplex: mRNA is transcribed from a single strand of the duplex.

## How Regulatory Proteins Work

How does the cell use regulatory sites to control which genes are transcribed? By influencing the shape of the regulatory proteins. Regulatory proteins possess binding sites not only for DNA but also for specific, small molecules within the cell. The binding of one of these small molecules can change the shape of a regulatory protein and thus destroy or enhance its ability to bind DNA. In some cases, the protein in its new shape may no longer recognize the regulatory site on the gene. In other cases, the recontoured regulatory protein may begin to recognize a regulatory site that it had previously ignored.

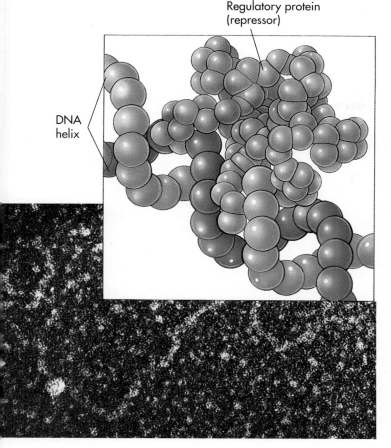

Regulatory protein
(repressor)

DNA
helix

**Figure 13-16**

**Repression.** The regulatory protein controlling transcription of the lactose genes is the large whitish sphere bound to the DNA strand at the lower left. Because the repressor protein fills the groove of the DNA double helix, the RNA polymerase cannot attach there, and the repressor blocks transcription.

The cell thus uses the presence of particular "signal" molecules within the cell to incapacitate particular regulatory proteins or to mobilize them for action. These regulatory proteins, in turn, repress or activate the transcription of particular genes. The pattern of metabolites in the cell sets "on/off" protein regulatory switches and, by doing so, achieves a proper configuration of gene expression.

*Organisms control the expression of their hereditary information by selectively inhibiting the transcription of some genes and facilitating the transcription of others. Control over transcription is exercised by modifying the shape of regulatory proteins and thus influencing their tendency to bind to sites on the DNA that influences the initiation of transcription.*

## THE ARCHITECTURE OF A GENE

Now that we have surveyed in general terms how specific polypeptides are assembled by ribosomes from mRNA copies of genes and how the production of these mRNA copies is regulated, we will examine in more detail a specific example and trace how the structure of a particular set of genes achieves the precise and timed production of the proteins it encodes. The set of genes we will examine is the *lac* system (Figure 13-17), a cluster of genes encoding three proteins that bacteria use to obtain energy from the sugar lactose. These proteins include two enzymes and a membrane-bound transport protein (a permease). Researchers have found this cluster to be typical of how genes are organized in bacteria. Within the cluster are five different regions:

1. *Coding sequence.* Three coding sequences specify the three lactose-utilizing enzymes. All three sequences are transcribed onto the same piece of mRNA and constitute part of an operational unit called an **operon.** An operon consists of one or more structural genes and the associated regulatory elements, the operator and the promoter, which are discussed below. This particular operon is called the *lac* operon because the three genes that it includes are all involved in lactose utilization. Such a pattern of clustering of coding sequences onto a single transcription unit, the operon, is common among bacteria but is rare in eukaryotes.

2. *Ribosome recognition site.* In front of the three coding sequences is the binding site for the ribosome, a series of nucleotides that lies within an initial untranslated portion of the mRNA. Each mRNA molecule transcribed from the cluster is composed of this series of nucleotides and the three coding sequences, transcribed in order.

3. *RNA polymerase-binding site.* Still farther in front of the coding sequences specifying the three lactose-utilizing enzymes is a specific DNA nucleotide sequence that the polymerase recognizes and to which it binds. Such polymerase-recognition sites are called **promoters** because they promote transcription.

4. *Regulatory protein-binding site.* Between the promoter and the ribosome recognition site is a regulatory site, the **operator,** where a repressor protein binds to block transcription.

5. *CAP site.* In front of the promoter is another regulatory site, the **CAP** (catabolite activator protein) site, where this activator protein binds. This in turn facilitates the unwinding of the DNA duplex and so enables the polymerase to bind to the nearby promoter.

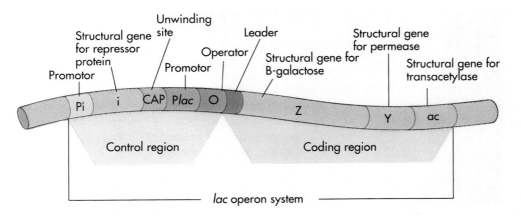

Figure 13-17

The *lac* region of the *Escherichia coli* chromosome.

Genes encoding enzymes also possess regulatory regions. The segment that is transcribed into mRNA is called a **transcription unit;** it consists of the elements that are involved in the translation of the mRNA—the ribosome-binding site and the coding sequences. In front of the transcription unit on the DNA are the elements involved in regulating its transcription—binding sites for the polymerase and for regulatory proteins.

This complex system of regulatory sites works to ensure that mRNA is copied from the three structural genes only when the cell can effectively utilize the proteins they encode: when there is lactose present, and when the cell requires the energy that would result from lactose breakdown. The regulatory region of the lac operon controls when the lac genes are transcribed in several interactive ways.

## Activation

A special activator protein, CAP, stimulates the transcription of the *lac* operon when the cell is low in energy. As a result of this CAP activation system, the enzymes needed for the metabolism of lactose are produced only when the cell requires the energy that lactose would provide.

## Repression

The sugar lactose is encountered by bacteria only occasionally, so the enzymes that metabolize lactose are not usually able to function, for lack of a substrate on which to act. Bacteria do not produce the enzymes under these conditions. The *lac* repressor protein blocks the proper functioning of RNA polymerase in this sys-

tem under most circumstances; cells in this condition are said to be **repressed** with respect to *lac* operon transcription (Figure 13-18, *A*).

Like the activator protein, the *lac* repressor protein is capable of changes in shape. When lactose binds to the repressor protein, the protein assumes a different shape, one that does not recognize the operator sequence! Therefore if the cell contains much lactose, the lac repressor proteins become inactive. This removes the block from in front of the polymerase (Figure 13-18, *B*) and so permits transcription of the *lac* genes to

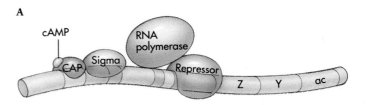

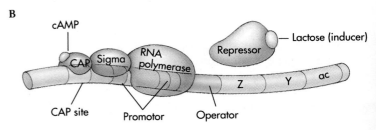

Figure 13-18

How *lac* operon works. A, The *lac* operon is shut down ("repressed") when the repressor protein is bound to the operator site. Because promotor and operator sites overlap, polymerase and repressor cannot bind at the same time, any more than two people can sit in one chair. B, The *lac* operon is transcribed ("induced") when lactose binding to the repressor protein changes its shape so that it can no longer sit on the operator site and block polymerase binding.

**anticodon**  The three-nucleotide sequence at the tip of a transfer RNA molecule that is complementary to, and base pairs with, an amino acid–specifying codon in messenger RNA.

**codon**  The basic unit of the genetic code; a sequence of three adjacent nucleotides in DNA or mRNA that code for one amino acid or for polypeptide termination.

**exon**  A segment of DNA that is both transcribed into mRNA and translated into protein; contrasts with introns. In eukaryotic genes, exons are typically scattered within much longer stretches of nontranslated intron sequences.

**intron**  A segment of DNA that is transcribed into mRNA but removed before translation. These untranscribed regions make up the bulk of most eukaryotic genes.

**nonsense codon**  A chain-terminating codon; a codon for which there is no tRNA with a complementary anticodon. There are three: UAA, UAG, and UGA.

**operator**  A site of negative gene regulation; a sequence of nucleotides that may overlap the promoter, which is recognized by a repressor protein. Binding of the repressor protein to the operator prevents binding of the polymerase to the promoter (just as two people cannot sit in one chair) and so blocks transcription of the structural genes of an operon.

**operon**  A cluster of functionally related genes transcribed onto a single mRNA molecule. A common mode of gene regulation in prokaryotes, it is rare in eukaryotes other than fungi.

**promoter**  An RNA polymerase binding site; the nucleotide sequence at the 5′ end of a gene to which RNA polymerase attaches to initiate transcription of mRNA.

**repressor**  A protein that regulates transcription of mRNA from DNA by binding to the operator and so preventing RNA polymerase from attaching to the promoter.

**RNA polymerase**  The enzyme that transcribes RNA from DNA.

**transcription**  The polymerase-catalyzed assembly of an RNA molecule complementary to a strand of DNA.

**translation**  The assembly of a protein on the ribosomes, using mRNA to direct the order of amino acids.

---

begin. It is for this reason that addition of lactose to a growing bacterial culture causes a burst of synthesis of the lactose-utilizing enzymes (Figure 13-19). The transcription of the enzymes is said to have been **induced** by the lactose. This element of the control system ensures that the *lac* operon is transcribed only in the presence of lactose.

The *lac* operon is thus controlled at two levels: (1) the lactose-utilizing enzymes are not produced unless the sugar lactose is available, and (2) even if lactose is available, the enzymes are not produced unless the cell has need of the energy. Similarly precise control mechanisms are known in eukaryotes, but the example provided here illustrates the way in which cellular control of protein synthesis operates.

## GENE MUTATION

There is an enormous amount of DNA within the cells of your body. This DNA represents a long series of DNA replications, starting with the DNA of a single cell, the fertilized egg. The replication of DNA that took place in producing the cells of your body is equivalent to producing a length of DNA nearly $97 \times 10^9$ kilometers long from an original 0.9-meter piece. Living cells have evolved many mechanisms to avoid errors during DNA replication and to preserve the DNA from damage. These mechanisms ensure the accurate replication of DNA duplexes by "proofreading" the strands of each daughter cell against one another for accuracy and correcting any mistakes. But the proofreading is not perfect. If it were, no mistakes would occur, no variation in gene sequence would result, and evolution would come to a halt.

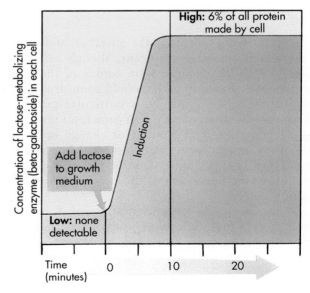

**Figure 13-19**

**Induction of enzyme synthesis.** Bacterial cells growing in the absence of lactose do not contain measurable amounts of the enzyme beta-galactosidase, which splits lactose into glucose and galactose, which bacteria can use as food. When lactose is added to a solution containing growing bacteria, the bacteria begin to produce large amounts of the enzyme. Within 10 minutes after lactose is added, 6% of all the protein made by a cell will be the enzyme beta-galactosidase.

**Figure 13-20**

**Mutation.** Fruit flies normally have one pair of wings, extending from the thorax. This fly is a mutant, *bithorax*. Due to a mutation in a gene regulating a critical stage at development, it possesses *two* thorax segments and thus two sets of wings.

In fact, living cells do make mistakes, and changes in the genetic message do occur, though only rarely (Figure 13-20). If changes were common, the genetic instructions encoded in DNA would soon degrade into meaningless noise. Typically, a particular gene is altered in only one out of a million gametes. Limited as it might seem, this small amount of change is the very stuff of evolution. Every single difference between the genetic message specifying you and the one specifying your cat, or the fleas on your cat, arose as the result of genetic change.

A change in the genetic message of a cell is referred to as a **mutation.** Some mutational changes affect the message itself, producing alterations in the sequence of DNA nucleotides. These alterations in the coding sequence are called **point mutations** because they usually involve only one or a few nucleotides. Other classes of mutation involve changes in the way in which the genetic message is organized. In both bacteria and eukaryotes, individual genes may move from one place on the chromosomes to another by a process called **transposition.** When a particular gene moves to a different location, there is often an alteration in its expression or in that of the neighboring genes. In eukaryotes, large segments of chromosomes may change

their relative location or undergo duplication. Such **chromosomal rearrangement** often has drastic effects on the expression of the genetic message. Sources and types of mutations are summarized in Table 13-2.

Point mutations, involving only one or a few nucleotides, result from (1) either chemical or physical damage to the DNA or from (2) spontaneous pairing errors that occur during DNA replication. The first class of mutation is of particular practical importance because modern industrial societies produce and release into the environment many chemicals capable of damaging DNA, chemicals called **mutagens.**

*Point mutations are changes in the hereditary message of an organism. They may result from physical or chemical damage to the DNA or from spontaneous errors during DNA replication.*

## DNA Damage

Although there are many ways in which a DNA duplex can be damaged, three are of major importance: (1) ionizing radiation, (2) ultraviolet radiation, and (3) chemical mutagens.

*Ionizing Radiation.* High-energy radiation such as x-rays and gamma rays is highly mutagenic. When such radiation reaches a cell, it is absorbed by the atoms that it encounters, imparting energy to the electrons of their outer shells and causing these electrons to be ejected from the atoms. The ejected electrons leave behind ionized atoms with unpaired electrons, called **free radicals.** Because most of a cell's atoms are present as water molecules and not as DNA, the great majority of free radicals created by **ionizing radiation** are produced from water molecules and not DNA.

Thus most of the damage that the DNA suffers is indirect. It occurs because free radicals are highly reactive chemically, reacting violently with the other molecules of the cell, including DNA. The action of free radicals on a chromosome is like that of shrapnel from a grenade blast tearing into a human body.

The locations on the DNA where damage occurs are random, and the damage is often severe. Cells can repair some of this damage. Chemical changes can be repaired by excising altered nucleotides, and single ruptured bonds can be re-formed. However, this **mutational repair** is not always accurate, and some of the mistakes are incorporated into the genetic message.

**TABLE 13-2   SOURCES AND TYPES OF MUTATION**

| SOURCE | PRIMARY EFFECT | TYPE OF MUTATION |
|---|---|---|
| **MUTATIONAL** | | |
| Ionizing radiation | Two-strand breaks in DNA | Deletions, translocations |
| Ultraviolet radiation | Pyrimidine dimers | Errors in nucleotide choice during repair |
| Chemical mutagens | Base analogue mispairing | Single nucleotide substitution |
| | Modification of a base leads to mispairing | Single nucleotide substitution |
| Spontaneous | Isomerization of a base | Single nucleotide substitution |
| | Slipped mispairing | Frameshift, short deletion |
| **RECOMBINATIONAL** | | |
| Transposition | Insertion of transposon into gene | Insertional inactivation |
| Mispairing of repeated sequences | Unequal crossing-over | Deletions, addition, inversions |
| Homologue pairing | Gene conversion | Single nucleotide substitution |

*Ultraviolet Radiation.* Ultraviolet (UV) radiation, the component of sunlight that leads to suntan (and sunburn), is much lower in energy than are x-rays. In DNA the principal absorption of UV radiation is by the pyrimidine bases thymine and cytosine. When these bases absorb UV energy, the electrons in their outer shells become reactive. If one of the nucleotides on either side of the absorbing pyrimidine is also a pyrimidine, a double covalent bond is formed between the two pyrimidines. This type of cross-link between adjacent bases of the DNA strand is called a **pyrimidine dimer** (Figure 13-21). If such a cross-link is left unre-paired, it can block DNA replication, in which case the damage would be lethal.

Fortunately, your body has very efficient systems for detecting and repairing DNA cross-links. Without a mechanism for repairing the damage to DNA caused by UV radiation, sunlight would wreak havoc on the cells of your skin. There is a rare hereditary disorder among humans called **xeroderma pigmentosum** (Figure 13-22) in which just this problem occurs. Individuals homozygous for a mutation that destroys the ability of the body's cells to repair UV damage develop extensive skin tumors after exposure to sunlight.

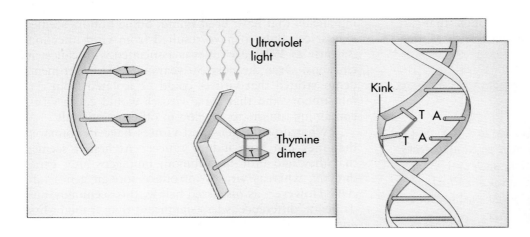

**Figure 13-21**

**Making a thymine dimer.** When two thymines are adjacent to one another in a DNA strand, the absorption of UV radiation can cause the formation of a covalent bond between them—a thymine dimer. Such a dimer introduces a "kink" into the double helix, which prevents replication of the duplex by DNA polymerase.

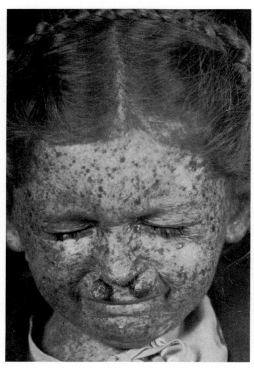

**Figure 13-22**

**The inherited genetic disorder xeroderma pigmentosum.** Those who have the disease develop extensive malignant skin tumors after exposure to sunlight.

*Chemical Mutagens.* Many mutations result from the direct chemical modification of the DNA bases. The chemicals that act on DNA fall into three classes: (1) chemicals that look like DNA nucleotides but pair incorrectly when they are incorporated into DNA; (2) chemicals that remove the amino group from adenine or cytosine, causing them to mispair; and (3) chemicals that add hydrocarbon groups to nucleotide bases, also causing them to mispair. This last group includes many particularly potent mutagens that are commonly used in the laboratory, as well as compounds that are sometimes released into the environment, such as mustard gas.

*The three major sources of mutational damage to DNA are (1) high-energy radiation, such as x-rays, which physically breaks the DNA strands; (2) low-energy radiation, such as UV light, which creates DNA cross-links whose removal often leads to errors in base selection; and (3) chemicals that modify DNA bases and thus alter their base-pairing behavior.*

# THE RIDDLE OF CANCER

Of all the diseases to which humans are susceptible, none is more feared than cancer. Practically everyone that reads this knew someone who died of cancer. Fully a third of the students who use this text will die of cancer. Finding a cure for this killer is a major goal of modern biology. We have learned a great deal, but much remains unclear.

Cancer is a growth disorder of cells. It starts when an apparently normal cell begins to grow in an uncontrolled and invasive way (Figure 13-23). The result is a ball of cells, called a **tumor,** that constantly expands in size. When this ball remains a hard mass, it is called a **sarcoma** (if connective tissue, such as muscle, is involved) or a **carcinoma** (if epithelial tissue, such as skin, is involved). Often cancer cells develop abnormally and can no longer recognize the tissue where they originated; such cells leave the tumor mass and spread throughout the body (Figure 13-24), forming new tumors at distant sites. The spreading cells are called **metastases.** Cancer is perhaps the most pernicious human disease. Of the children born this year, a third will contract cancer; a fourth of the males and fully a third of the females will someday die of the disease. Each of us has had family or friends affected by cancer, and many of us will die of it. Not surprisingly, a great deal of effort has been expended to learn the cause of this disease. Using the new techniques of molecular biology, much progress has been made, and the rough outlines of understanding are now emerging.

The search for a cause of cancer has focused in part on those environmental factors which were considered potential agents in causing the disease (Figure 13-25). Many have been found, including ionizing radiation, such as x-rays, and various chemicals. Many of these cancer-causing agents, or **carcinogens,** have in common the property of being potent mutagens. This observation led to the suspicion that cancer might be caused, at least in part, by the creation of mutations.

However, mutagens are not the only cancer-causing agents that have been found. Some tumors seemed almost certainly to have resulted from viral infection. As early as 1910, a virus was associated with cancer in chickens. Over the next 50 years, several experiments demonstrated that viruses could be isolated from certain tumors and that these viruses would cause virus-containing tumors to develop in other individuals.

What do mutagens and viruses have in common, that they can both induce cancer? At first it seemed that they had little in common: mutagens alter genes directly, whereas viruses introduce foreign genes into cells. However, as discussed below, this seemingly fundamental difference is less significant than it might first appear, and the two causes of cancer have in fact proved to be one and the same.

*Figure 13-23*

**Lung cancer cells.** These cells are from a tumor located in the alveolus of a lung.

## THE STORY OF CHICKEN SARCOMA

In 1910, American medical researcher Peyton Rous reported the presence of a virus, Rous avian sarcoma virus (RSV), that was associated with chicken sarcomas; he was awarded the 1966 Nobel Prize in physiology/medicine for his discovery. RSV proved to be a particular kind of RNA virus, a **retrovirus.** Among the RNA viruses, retroviruses are unusual in the way in which they replicate themselves. When they infect a cell (Figure 13-26), they make a DNA copy of their RNA—a copy that can be inserted into the animal DNA! The RSV virus was able to initiate cancer in chicken fibroblast (connective tissue) cells growing in culture when they were infected with the virus; from these **transformed** cells, more virus could be isolated.

How does RSV act to initiate cancer? When RSV was compared with a closely related virus, RAV-O, which was not able to transform chicken cells into cancer cells, the two viruses proved to be identical, except for one gene that was present in the RSV virus but absent from the RAV-O virus. The cancer-causing gene was called the *src* gene (for sarcoma) (Figure 13-27).

These and other results led to the hypothesis that cancer results from the action of a specific tumor-inducing *onc* gene, a hypothesis called the **oncogene theory.** The term **oncogene** was derived from the Greek word *oncos,* which means "cancer." This theory soon proved to be true for a broad range of tumor types. The RSV *src* gene is but one example of such *onc* genes. For example, seemingly spontaneous leukemia in mice is inherited in a Mendelian manner and can be mapped to specific sites on the mouse chromosomes.

*Figure 13-24*

**Portrait of a cancer.** The ball of cells is a carcinoma, developing from epithelial cells lining the interior surface of a human lung. As the mass of cells grows, it invades surrounding tissues, eventually penetrating into lymphatic vessels and blood vessels, both of which are plentiful within the lung. These vessels carry metastatic cancer cells throughout the body, where they lodge and grow, forming new centers of cancerous tissue.

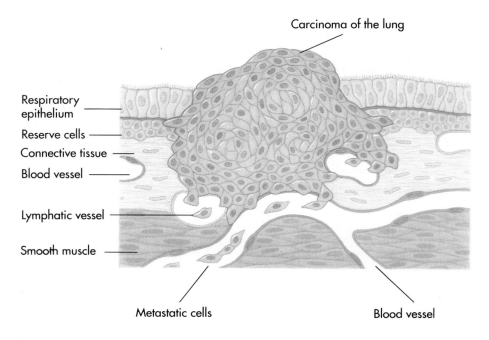

Carcinoma of the lung

Respiratory epithelium

Reserve cells

Connective tissue

Blood vessel

Lymphatic vessel

Smooth muscle

Metastatic cells

Blood vessel

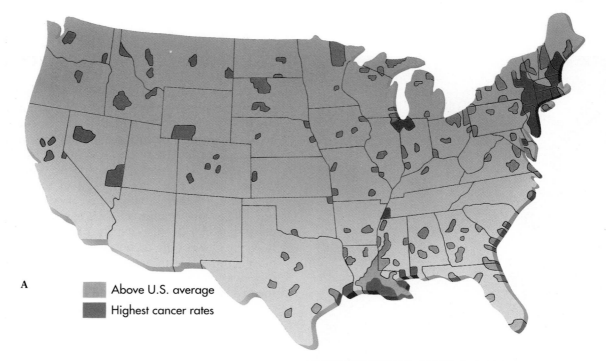

A

■ Above U.S. average

■ Highest cancer rates

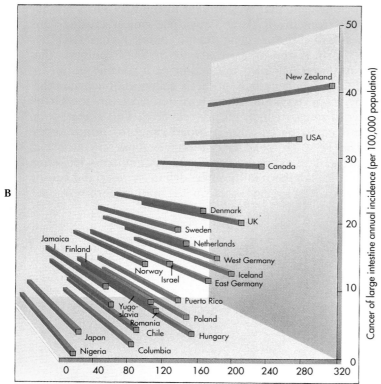

B

Cancer of large intestine annual incidence (per 100,000 population)

New Zealand — 40

USA — 30

Canada — 30

Denmark — 20

UK — 20

Sweden

Netherlands

West Germany

Norway    Iceland

Israel    East Germany

Jamaica — 10

Finland

Puerto Rico

Yugo-slavia    Poland

Romania    Chile

Japan    Hungary

Nigeria    Columbia — 0

Meat consumption (grams per person per day)

0    40    80    120    160    200    240    280    320

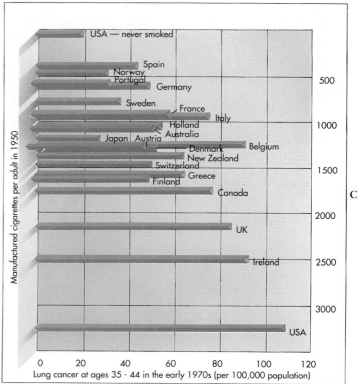

C

Manufactured cigarettes per adult in 1950

USA — never smoked

Spain — 500

Norway    Portugal

Germany

Sweden    France

Holland    Italy — 1000

Japan    Austria    Australia

Denmark    Belgium

New Zealand

Switzerland — 1500

Finland    Greece

Canada

UK — 2000

Ireland — 2500

USA — 3000

Lung cancer at ages 35 - 44 in the early 1970s (per 100,000 population)

0    20    40    60    80    100    120

## Figure 13-25

**Potential agents in causing cancer.**
A  The incidence of cancer per 1000 people is not uniform throughout the United States. Rather, it is centered in cities and in the Mississippi Delta. This suggests that pollution and pesticide runoff may contribute to cancer.
B  One of the most deadly cancers in the United States, that of the large intestine, is not at all common in many other countries, such as Japan. Its incidence appears to be related to the amount of meat an individual consumes—a high meat diet slows passage of food through the intestine, prolonging the exposure of the intestinal wall to digestive wastes.
C  The biggest killer among cancers is lung cancer, and the most important environmental agent producing lung cancer is cigarettes. When levels of lung cancer in many countries are compared, the incidence of lung cancer among adult males between 40 and 50 years of age is strongly correlated with the cigarette consumption in that country 20 years earlier.

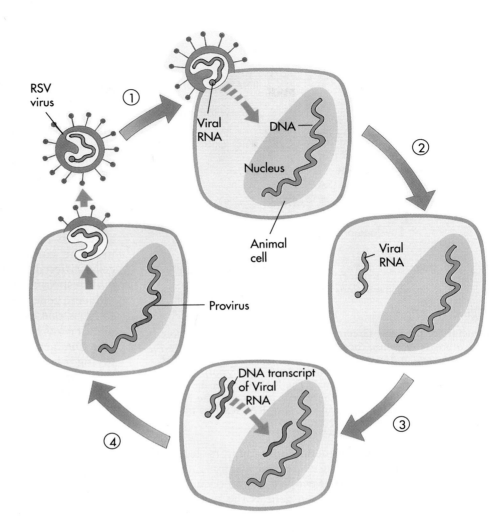

**Figure 13-26**

**Life cycle of the RSV virus. 1,** The genome of RSV is composed of RNA; that of animal cells is composed of DNA. **2,** On infecting an animal cell, the RSV RNA is added to the animal cell. **3,** There a DNA copy of it is made. **4,** Some or all of this DNA copy can become incorporated into the animal cell's DNA.

Labels on figure: RSV virus; Viral RNA; DNA; Nucleus; Animal cell; Viral RNA; DNA transcript of Viral RNA; Provirus; ①②③④

When investigated in detail, these sites proved to be sites of integrated virus genomes.

What might be the nature of a virus gene that causes cancer? An essential clue came in 1970, when RSV mutants were isolated that were temperature-sensitive. These mutants would transform tissue culture cells into cancer cells at 35° C, but not at 41° C. Temperature sensitivity of this kind is almost always associated with proteins. Therefore it seemed very likely that the *src* gene was actively transcribed by the cell, rather than serving as a recognition site for some unknown regulatory protein. This was a very exciting result because it suggested that the protein specified by this *onc* gene could be isolated and its properties studied.

The *src* protein was first isolated by Raymond Ericson and coworkers in 1977. It proved to be an enzyme of moderate size that acts to phosphorylate (add a phosphate group to) the tyrosine amino acids of proteins; such an enzyme is called a **tyrosine kinase.** Tyrosine kinases are not common in animal cells. Among the few that are known to occur in cells is **epidermal growth factor,** a protein that signals the initiation of cell division by binding to a special receptor site on the plasma membrane and phosphorylating key protein components. This fact raised the exciting possibility that RSV perhaps causes cancer by introducing into cells an active form of a normally quiescent enzyme. This indeed proved to be the case.

Does the *src* gene actually integrate into the host chromosome with the RSV virus? One way of investigating this question is to prepare a radioactive version of the *src* gene. We can then permit this radioactive *src* DNA to bind to complementary sequences on the chicken genome and examine where the chicken chromosomes become radioactive. Sites of radioactivity are sites where a sequence complementary to *src* occurs. As expected, radioactive *src* DNA binds to the site where RSV is inserted into the chicken genome—but unexpectedly it also binds to a second site, where there is no RSV.

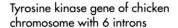

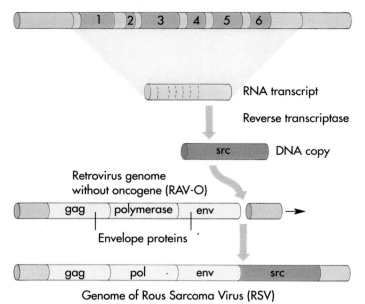

Tyrosine kinase gene of chicken
chromosome with 6 introns

RNA transcript

Reverse transcriptase

src DNA copy

Retrovirus genome
without oncogene (RAV-O)

gag polymerase env

Envelope proteins

gag pol · env src

Genome of Rous Sarcoma Virus (RSV)

**Figure 13-27**

**Structure of the Rous sarcoma virus (RSV).** The virus contains only a few virus genes, encoding the virus protein envelope (called the *env* and *gag* genes), and reverse transcriptase, which produces a DNA copy of its RNA genome (called *pol*). It also contains a gene, *src* (for sarcoma). The RAV-O virus shown here lacks this *src* gene, but otherwise is identical to the RSV retrovirus. RSV causes cancer in chickens, RAV-O does not.

From similar experiments, investigators learned that the *src* gene is not a virus gene at all, but rather a growth-promoting gene that evolved in and occurs normally in chickens. This chicken gene is the second site where *src* binds to chicken DNA. Somehow a copy of the normal chicken gene was picked up by an ancestor of the RSV virus in some past infection. Now part of the virus, the gene is active in a pernicious new way, escaping the normal regulatory controls of the chicken genome; its transcription is governed instead by virus promoters, which are actively transcribed during infection. Thus the study of RSV has shown that cancer results from the inappropriate activity of growth-promoting genes that are normally less active or completely inactive.

*Sarcoma in chickens results from the inappropriate activity of a normal chicken gene. Taken up by the RSV virus, it escapes the genetic regulation that is imposed by the chicken genome and induces cancer by being active when it should be inactive.*

## THE MOLECULAR BASIS OF CANCER

Not all cancer is associated with viruses. Other kinds of tumors have been studied by using techniques different from those which we have just explored. The most important of these techniques, called **transfection,** consists of (1) the isolation of the nuclear DNA from human cells that have been isolated from tumors, (2) its cleavage into random fragments by using appropriate enzymes, and (3) the testing of the fragments individually for the ability of any particular fragment to induce cancer in the cells that assimilate it.

By using transfection techniques, researchers found that a single gene isolated from a cancer cell is all that is needed to transform cells growing normally in tissue culture into cancerous ones. The cancerous cells differed from the normally growing ones only with respect to this one gene. In some cases the cancer-inducing gene identified by transfection proved to be the same as one of the *onc* genes that had previously been identified as being included in a cancer-causing virus.

*Onc* genes are normal genes gone wrong. By identifying *onc* genes by transfection and then isolating them, investigators have been able to compare them with their normal counterparts. In this way, researchers have been able to study why normal genes are converted into cancer-causing ones. The analysis of a number of *onc* genes associated with cancers of various tissues has led to the following conclusions:

1. The induction of many cancers involves changes in cellular activities that occur at the inner surface of the plasma membrane. In a normal cell, these activities are associated with the initiation of cell division (Figure 13-28): a cellular regulatory protein, epidermal growth factor (EGF), binds to a specific receptor on the inner plasma membrane, acts as a tyrosine kinase, and triggers cell division. In this process a protein encoded by a gene called *ras* is associated with the membrane EGF receptor site and acts to determine what cellular levels of EGF are adequate to initiate cell division. In several forms of human cancer, the cancerous, or *onc*, version of the *ras*-encoded protein activates the receptor site in response to much lower levels of EGF than does the normal version of the protein.

2. The difference between a normal gene encoding the proteins that carry out cell division and a cancer-inducing *onc* version need only be a single-point mutation in the DNA. For example, in a human bladder carcinoma induced by *ras*, a

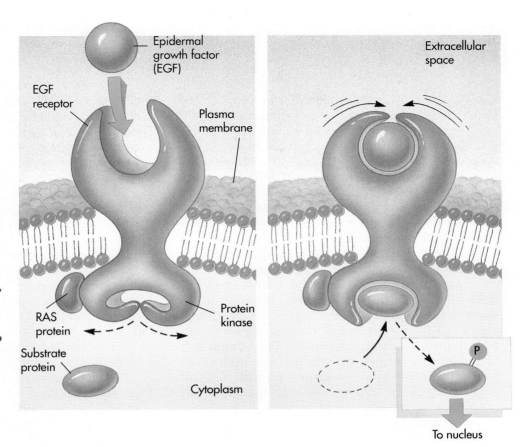

**Figure 13-28**

**How a mutation can cause cancer.** Mutations that lead to cancer often involve proteins of the plasma membrane associated with cell division. In a normal cell, cell division is triggered by a protein called epidermal growth factor (EGF), which binds to an EGF receptor protein on the exterior surface of the cell. This alters the shape of the portion of the receptor protuding into the cell, initiating a signal that passes to the cell nucleus and initiates cell division. The level of EGF necessary to start this process is affected by another protein, *ras*. A mutation increasing receptor efficiency or *ras* signal facilitation may trigger more frequent cell division—cancer.

single nucleotide alteration from G to T converts a glycine in the normal *ras* protein into a valine in the cancer-causing protein. There is no other difference between the normal and cancer-inducing forms of the *ras* gene.

3. The mutation of a gene such as *ras* to an *onc* form in different tissues can lead to different forms of cancer. There are probably no more than a few dozen genes whose mutation can lead to cancer, and there may be far fewer.

4. The induction of many cancers involves the action of two or more different *onc* genes. Particular *onc* genes can be isolated by transfection because, in the strains being studied, the other *onc* genes necessary for cancer are already turned to the "cancer" mode. The initiation of cancer may require changes both at the plasma membrane and in the nucleus. This may be the reason why most cancers occur in people over 40 years old (Figure 13-29). It is as if human cells accumulate mutational changes, and time is required for several such mutations to occur in the same cells.

The emerging picture of cancer is one that involves aborted regulation of the genes that normally signal the onset of cell proliferation. Cancer seems to occur when several of the controls that cells normally impose on their own growth and division become inoperative. Among the examples that have been studied in detail, the specific means whereby these controls are evaded vary, but many of them involve one of two general causes. In the first case, cancer may result from the mutation of a cellular gene with a regulatory function, as in the case of the *ras* gene, whose mutant *onc* form induces human bladder carcinoma. In the second case, cancer may result from the introduction of a regulatory gene into a cell as part of a virus, in which case the regulatory gene may escape the cell's control and so act abnormally. Thus in experiments in which a normal counterpart to an *onc* gene is carried into a cell by a virus, but is not changed in any other way, the counterpart acts as an *onc* gene, inducing cancer; the increased level of transcription of the gene under these conditions is the only difference. Again, the cancer is associated with an increase in the level of activity of a normal protein.

# How to Catch Cancer

Of the 1.7 million people who died in the United States last year, a fourth died of cancer and a third of these—140,000 people—died of lung cancer. About 160,000 cases of lung cancer were diagnosed each year in the 1980s, and 90% of these persons died or will die within 3 years. Of those who die, 96% will be cigarette smokers.

Smoking is a popular pastime in the United States. A third of the U.S. population smokes. American smokers consumed 443 billion cigarettes in 1983. These cigarettes emit in their tobacco smoke some 3000 chemical components, among them vinyl chloride, benzo *(a)* pyrenes, and nitroso-*nor*-nicotine, all of them potent mutagens. Smoking introduces these mutagens to the tissues of your lungs. Figure 13-A, a photograph of lung cancer in an adult human, illustrates the result. The bottom half of the lung is normal; the top half has been taken over completely by a cancerous growth. As you might imagine, a lung in this condition does not function well, but difficulty in breathing is rarely the cause of death in lung cancer. As the cancer grows within the lung, its cells invade the surrounding tissues and eventually break through into the lymph and blood vessels. Once the cancer cells have done this, they spread rapidly through the body, lodging and growing at many locations, particularly in the brain. Death soon follows.

Among cigarette manufacturers, it has been popular to argue that the causal connection between smoking and cancer has not been proved, that somehow the relationship is coincidental. Look carefully at the data presented in Figure 13-B and see if you agree. The upper graph presents data collected for American males, including the incidence of smoking from the turn of the century until now and the incidence of lung cancer over the same period. Note that as late as 1920, lung cancer was a rare disease. With a lag of some 20 years behind the increase in smoking, it became progressively more common.

Now look at the lower graph, which presents data on American females. Because of social mores, significant numbers of American females did not smoke until after World War II, when many social conventions changed. As late as 1963, when lung cancer among males was near current levels, this disease was still rare in females. In the United States that year, only 6588 females died of lung cancer. But as their smoking increased, so did their incidence of lung cancer, with the same inexorable lag of about 20 years. American females today have achieved equality with their male counterparts in the numbers of cigarettes that they smoke—and their lung cancer death rates are now approaching those for males. In 1983 more than 36,000 females died of lung cancer in the United States.

Among smokers, the current rate of deaths resulting from lung cancer is 180 per 100,000, or about 2 of each 1000 smokers *each year*. Smoking is very much like going into a totally dark room, standing still, and then calling in someone with a gun and closing the door. The person with the gun cannot see you, does not know where you are, and so just shoots once in a random direction and leaves the room. Every time an individual smokes a cigarette, he or she is "shooting" mutagens at their genes. Just as in the dark room, a hit is unlikely, and most shots will miss potential *onc* genes. As one keeps shooting, however, the odds of eventually scoring a hit get shorter and shorter. Nor do statistics protect any one individual; nothing says the first shot will not hit. Older people are not the only ones to die of lung cancer.

Except for eating powerful radioisotopes, there is probably no more certain way to catch cancer than to smoke.

---

*Cancer is a growth disease of cells in which the controls that normally restrict cell proliferation do not operate, transforming cells to a state of cancerous growth.*

## SMOKING AND CANCER

How can we prevent cancer? The most obvious strategy is to minimize mutational insult to our genes. Anything we do to increase our exposure to mutagens will result in an increased incidence of cancer for the unavoidable reason that such exposure increases the

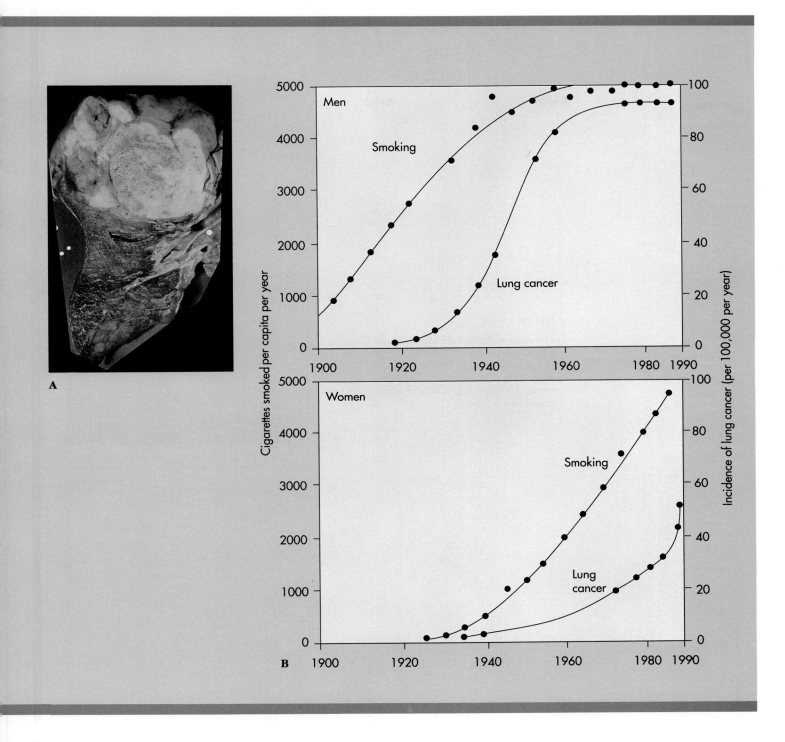

probability of mutating a potential *onc* gene. It is no accident that the most reliable tests for carcinogenic substances are those which measure their mutagenic ability.

Of all the environmental mutagens to which we are exposed, perhaps the most tragic are cigarettes. They are tragic because the cancers they cause are largely preventable. About a third of *all* cases of cancer in the United States can be attributed directly to cigarette smoking. The association is particularly striking for lung cancer. The dose-response curve obtained for male smokers (Figure 13-30) shows a highly positive correlation, with the risk of lung cancer increasing with increasing amounts of smoking. For those smoking two

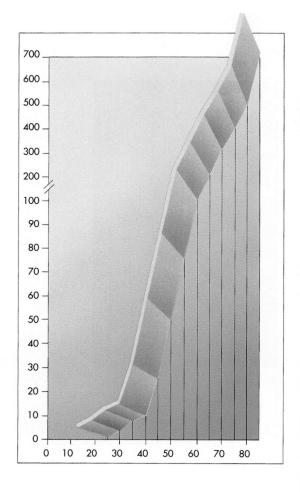

**Figure 13-29**

**The annual death rate from cancer is a function of age.** A logarithmic plot of age versus death rate is linear, suggesting that several independent events are required to give rise to cancer.

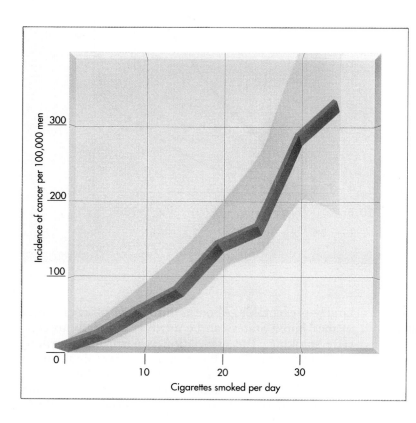

**Figure 13-30**

**Smoking causes cancer.** The annual incidence of lung cancer per 100,000 men is clearly related to the number of cigarettes smoked per day. Of every 100 American college students who smoke cigarettes regularly, 1 will be murdered, about 2 will be killed on the roads, and about 25 will be killed by tobacco.

or more packs a day, the risk of contracting lung cancer is 40 times or more greater than it is for nonsmokers. Note that the curve extrapolates back approximately to zero. Clearly, an effective way to avoid lung cancer is to not smoke. Life insurance companies have computed that, on a statistical basis, smoking a single cigarette lowers your life expectancy 10.7 minutes. (That is more than the time it takes to smoke the cigarette!) Every pack of 20 cigarettes bears an unwritten label:

"The price of smoking this pack of cigarettes is 3½ hours of your life."

Do we have any idea how to cure cancer, once it has started? Although surgery, radiotherapy, and chemotherapy can in some instances lead to remission, there is still no general "cure" for cancer. The path to a general cure lies in a better understanding of the thwarted cellular controls that are responsible for cancerous growth so that therapies can be devised to reinstitute appropriate cellular regulation within cancerous tissue. As you can imagine, this is an area of intense research activity.

## GENETIC SHORTHAND

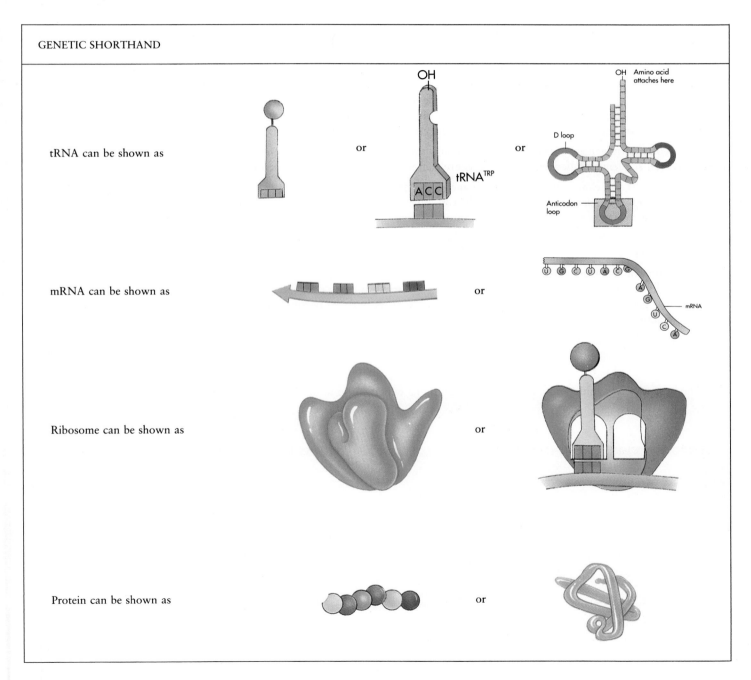

# ■ SUMMARY

1. The expression of hereditary information in all organisms takes place in two stages. First, an mRNA molecule with a nucleotide sequence that is complementary to a particular segment of the DNA is synthesized by the enzyme RNA polymerase. Second, an amino-acid chain is assembled by a ribosome, using the mRNA sequence to direct its choice of amino acids. The first process is called transcription, the second, translation.

2. The control of transcription is achieved largely by controlling the ability of RNA polymerase molecules to initiate transcription. In some instances, a nucleotide sequence located between the promoter and the transcription unit acts as a binding site for a repressor protein. When bound to the site, this repressor protein prevents the read-through of the RNA polymerase, just as a tree fallen across a road blocks traffic.

3. In other instances, nucleotide sequences located near the promoter site act as binding sites for activator proteins. The binding of the activator protein to the promoter site causes the unwinding of the DNA duplex. Because RNA polymerase can translate only a single strand of DNA, this forced unwinding greatly facilitates the translation process.

4. The coding sequence of a gene is read in increments of three nucleotides from the mRNA by a ribosome. The ribosome positions the three-nucleotide segments of the message so that a tRNA molecule with the complementary base sequence can bind to it. Attached to the other end of the tRNA molecule is an amino acid, which is added to the end of the growing polypeptide chain.

5. Most eukaryotic genes contain, embedded within the coding sequences of their transcription units, additional sequences called introns. Introns are removed from the transcript before it is translated.

6. A mutation is any change in the hereditary message. Mutations that change one or a few nucleotides are called point mutations. They may result from physical damage from ionizing radiation, chemical damage, or mistakes made during correction of damage caused by absorption of ultraviolet light.

7. Cancer is a growth disease of cells in which the regulatory controls that usually restrain cell division do not work.

8. The testing of DNA fragments from tumor cells to identify those fragments capable of inducing cancer has led to the isolation of various cancer-causing genes. In every case the cancer-causing gene is a normal gene whose product has a role in cell proliferation but that becomes active when it should not be active.

9. In some cases, cancer induction results from a single nucleotide mutation in a normal gene. In other cases, a normal gene is acquired by a virus and is transcribed at the high rate characteristic of the virus.

10. The best way to avoid cancer is to avoid those things which cause mutation, notably smoking.

## REVIEW

1. Strings of DNA nucleotides arranged in groups that specify the structure of a protein are called _____.

2. After mRNA is transcribed from the DNA, it is _____ into a polypeptide at a ribosome.

3. How many nucleotides does it take to specify an amino acid?

4. A codon is part of a(n) _____ molecule that specifies a particular amino acid. An anticodon is the corresponding portion of a(n) _____ molecule.

5. A regulatory protein responsible for blockage of DNA transcription is called a _____.

6. A chemical capable of damaging DNA is called a _____.

7. Retroviruses use _____ instead of DNA as their hereditary information molecule.

8. Cancer-causing genes are called _____, from the Greek word for cancer.

# SELF-QUIZ

1. RNA is a nucleic acid like DNA, but there are important differences. Choose these three differences from the following list.
   (a) The sugar in RNA is ribose instead of deoxyribose.
   (b) RNA contains uracil instead of thymine.
   (c) RNA is a single-stranded molecule.
   (d) RNA is a double-stranded molecule.
   (e) RNA is necessary for protein synthesis.

2. What is the first step in RNA transcription?
   (a) Nucleotides are assembled into RNA.
   (b) The DNA strands open up.
   (c) The hydrogen bonds of the DNA re-form.
   (d) RNA polymerase attaches to the DNA.
   (e) The DNA replicates.

3. A particular activating enzyme exists for
   (a) each of the 20 common amino acids.
   (b) each distinct codon.
   (c) the particular mRNA that recognizes each amino acid.
   (d) each of the four kinds of nucleotides in tRNA.
   (e) the three codons that signify the end of transcription.

4. Ultraviolet radiation damages cells by
   (a) burning their plasma membranes.
   (b) breaking DNA molecules into nonfunctional segments.
   (c) forming pyrimidine dimers that can block DNA replication.
   (d) causing xeroderma pigmentosum.
   (e) causing the cells to proliferate and become cancerous.

5. Which amino acid is specified by the codon CUC? (Consult Table 13-1.)
   (a) Lysine
   (b) Alanine
   (c) Glutamate
   (d) Proline
   (e) Leucine

6. Two key differences between bacterial and eukaryotic mRNA are
   (a) bacterial mRNA molecules do not need ribosomes for transcription.
   (b) eukaryotic mRNA molecules rarely contain more than one gene, whereas bacterial mRNAs often do.
   (c) ribosomes of eukaryotes are much larger than those of bacteria.
   (d) the primary mRNA transcripts of eukaryotes contain introns that are cut out before translation.
   (e) in bacterial mRNA the nucleotide sequence is not arranged in codons.

7. In the *lac* region of the bacterium *Escherichia coli's* chromosome, where does the repressor protein bind?
   (a) At the activator protein site.
   (b) At the promoter site.
   (c) At the operator site.
   (d) At the transcription unit.
   (e) At the permease gene site.

# THOUGHT QUESTIONS

1. You are provided with a sample of aardvark DNA, obtained at great risk to life. As part of your investigation of this DNA, you transcribe messenger RNA from the DNA and purify it. You then separate the two strands of the DNA and analyze the base composition of each strand, and then analyze that of the mRNA. You obtain the following results:

   |  | A | G | C | T | U |
   |---|---|---|---|---|---|
   | DNA strand 1 | 19.1 | 26.0 | 31.0 | 23.9 | 0 |
   | DNA strand 2 | 24.2 | 30.8 | 25.7 | 19.3 | 0 |
   | mRNA | 19.0 | 25.9 | 30.8 | 0 | 24.3 |

   Which strand of the DNA is the "sense" strand, serving as the template for mRNA synthesis?

2. In medical research, mice are often used as model systems in which to study the immune system and other physiological systems important to human health. Many medical centers maintain large colonies of mice for these studies. In one such colony under your supervision, a hairless mouse is born. What minimum evidence would you accept that this variant represents a genetic mutation?

3. The evidence associating lung cancer with smoking is overwhelming, and, as you have learned in this chapter, we now know in considerable detail the mechanism whereby smoking induces cancer. It introduces powerful mutagens into the lungs, which cause mutations to occur; when a growth-regulating gene is mutated by chance, cancer results. In light of this, why do you think that cigarette smoking is not illegal?

4. In a study of the segregation of leukemia in several generations of laboratory mice, a line selected for high incidence of leukemia was crossed with a line that never develops leukemia (it presumably lacks a leukemia-inducing oncogene). All of the $F_1$ progeny exhibited a high incidence of leukemia. Among the progeny resulting from a test cross of these $F_1$ individuals to the leukemia-free parent line, 25% were healthy and leukemia-free; the other 75% were leukemic. The result is not a 1:1 ratio, but rather a 3:1 ratio. How would you interpret this result?

# FOR FURTHER READING

BISHOP, J.M.: "Oncogenes," *Scientific American*, March 1982, pages 82-92. The story of *src* and how it causes cancer, by the man who first identified the protein product of the *src* gene. This article is particularly good at showing the chain of reasoning that underlies a major scientific advance.

DICKERSON, R.E.: "The DNA Helix and How It is Read," *Scientific American*, December 1983, pages 94-112. X-ray analysis of DNA molecules of different sequence shows that sequence differences create subtle differences in shape that may influence DNA site recognition by proteins. A very clear exposition of a complex subject.

DOLL, R.: "An Epidemiological Perspective of the Biology of Cancer," *Cancer Research*, vol. 38, 1978, pages 3573-3583. An advanced article that presents data on who contracts cancer and at what age. The data constitute a damning indictment of smoking as the leading cause of cancer.

FELDMAN, M., and L. Eisenbach: "What Makes a Tumor Cell Metastic?" *Scientific American*, November 1989, pages 60-85. Study of the molecules that stud the surface of cancer cells is beginning to enable cancer researchers to convert malignant cells into benign ones.

KARTNER, N., and V. Ling: "Multidrug Resistance in Cancer," *Scientific American*, March 1989, pages 44-51. Cancers resistant to many forms of chemotherapy seem to result from defects in a pump that flushes toxins out of cells.

PTASHNE, M.: "How Gene Activators Work," *Scientific American*, January 1989, pages 41-47. A lucid overview of how genes are turned on and off.

ROSS, J.: "The Turnover of Messenger RNA," *Scientific American*, April 1989, pages 48-55. The level of many proteins in the body is determined by how fast the messenger RNA encoding them is broken down, rather than by how speedily new transcripts are churned out.

WEINBERG, R.: "Finding the Anti-Oncogene," *Scientific American*, September 19, pages 44-51. This important paper by a developer of the transfection approach describes how certain growth-suppressing genes in a mutated form confer susceptibility to cancer.

# Gene Technology

Evolution created the plants we see in nature, but genetic engineers are attempting to improve on the design. This scientist is screening plant tissue culture, selecting those which respond best to treatment. The plants that result will be more resistant to chemicals and pests, more nutritious, and grow faster.

# GENE TECHNOLOGY

## Overview

Recent advances in molecular biology have given us powerful new tools with which to investigate genetics at the molecular level. The newly found ability to isolate individual genes and transfer them from one kind of organism to another has revolutionized our ability to improve the characteristics of useful plants and animals and to avoid disease. These techniques offer enormous potential for the agriculture and medicine of the future.

**For Review** *Here are some important terms and concepts that you will encounter in this chapter. If you are not familiar with them, you should review them before proceeding.*

**Enzymes** (Chapter 6)

**DNA** (Chapters 2 and 12)

**Point mutation** (Chapter 13)

During the last decade, a revolution in genetics has taken place because of the new and powerful techniques now available for the study and manipulation of DNA. These techniques have allowed biologists for the first time to intervene directly in the genetic fate of organisms (Figure 14-1). In this chapter we discuss these techniques and consider their application to specific problems of agriculture and medicine. Few areas of biology will have as great an impact on our lives, and the potential advances are just starting to be understood.

## PLASMIDS AND THE NEW GENETICS

In 1980, geneticists succeeded for the first time in introducing a human gene, the one that encodes the protein interferon, into a bacterial cell. Interferon is a rare protein, difficult to purify in any appreciable amounts, that increases human resistance to viral infection. It may prove to be effective in improving our resistance to cancer. At first this possibility was difficult to explore because purification of the substantial amounts of interferon required for large-scale testing was prohibitively expensive. But recently it has become possible to produce large amounts of interferon cheaply by introducing the gene that is responsible for its production into a bacterial cell.

Within the bacterial cell, the gene directs the production of copious amounts of interferon. Meanwhile, the bacterial cell with the introduced gene continues to grow and divide. Soon there are many millions of such cells in the culture, all descended from the original altered bacterial cell, and all producing human interferon. This procedure, called **cloning,** succeeds in making every cell in the culture a miniature factory for the production of human interferon. In a similar way, the

**Figure 14-1**

**Larger than life—a product of genetic engineering.** These two mice are genetically identical, except that the large one has one extra gene—the gene encoding a potent growth hormone not normally present in mice. The gene was added to the mouse genome by human genetic engineers and is now a stable part of the mouse's genetic endowment.

successful cloning of bacteria able to produce insulin, a polypeptide, has led to the development of new and relatively inexpensive ways to produce large amounts of the hormone. Cloning also has proved a valuable technique in molecular biology by allowing the multiplication of particular genes and gene products that are being studied. Collectively, these techniques of transferring genes from one kind of organism to another and multiplying them are termed **genetic engineering.**

Genetic engineering is based on the ability to cut up DNA into recognizable pieces and to rearrange these pieces in different ways. In the first interferon experiment the DNA segment carrying the gene was inserted into a **plasmid.** Plasmids are small fragments of DNA that replicate independently outside of the main bacterial chromosome; they are usually circular. They make up about 5% of the DNA of many bacteria but are rare in eukaryotic cells. In genetic engineering, plasmids or viruses are usually employed to carry a particular gene into a recipient cell.

The success of the initial step in a genetic engineering experiment is the key to the whole procedure. As you might expect, success depends on being able to cut up the source DNA (for example, human DNA in the interferon experiment), and the plasmid DNA so that the desired fragment of source DNA can be spliced permanently into the plasmid genome. This cutting is performed by a special kind of enzyme called a **restriction endonuclease.** These enzymes are able to recognize and cleave specific sequences of nucleotides in a DNA molecule; they are the basic tools of genetic engineering.

## RESTRICTION ENZYMES

Scientific discoveries often have their origins in odd little crannies—seemingly unimportant areas that receive little attention by researchers before their general significance is appreciated. One such obscure topic that turned out to have important implications was the warfare that takes place between bacteria and viruses.

Most organisms in nature eventually evolve means of defending themselves from predators, and bacteria are no exception to this rule. Among the natural enemies of bacteria are the viruses called **bacteriophages,** which infect bacteria, multiply within their cells, and eventually burst those cells open, releasing large numbers of new bacteriophages. For a bacterium, bacteriophages are lethal adversaries and act as agents of natural selection. Those bacteria which have the ability to resist viral infection survive in higher numbers than those which lack this ability, generally leaving more offspring than those which lack resistance. Some bacteria have powerful weapons against viruses: restriction

endonucleases, which chop up viral DNA as soon as it enters the bacterial cell but which leave the bacterial DNA unharmed. When viruses insert their DNA into a bacterial cell that is protected in this way, the viral DNA is immediately degraded by the restriction endonucleases. Why is the DNA of the bacteria not also degraded by the restriction enzymes? Because the bacterial cell has **modified** its own DNA in such a way that the restriction enzymes do not recognize it as DNA.

Restriction endonucleases recognize specific nucleotide sequences within a DNA strand, bind to DNA strands at sites where these sequences occur, and cut a bound strand of DNA at a specific place in the recognition sequence. Other bacterial enzymes called **methylases** recognize the same sequences in bacterial DNA, bind to them, and add methyl ($CH_3$) groups to the nucleotides. When the recognition sites of bacterial DNA have been modified with methyl groups in this way, they are no longer recognized by the restriction enzymes. Consequently the bacterial DNA is protected from being degraded. Viral DNA is not protected because it has not been methylated.

The sequences that restriction enzymes recognize are typically four to six nucleotides long and are symmetrical. Their symmetry is of a special kind, called **twofold rotational symmetry** (Figure 14-2). The nucleotides at one end of the recognition sequence are complementary to those at the other end so that the two strands of the DNA duplex have the same nucleotide sequence running in opposite directions for the length of the recognition sequence. This arrangement has two consequences. The first is of great importance to the bacteria; the second is of little significance in the bacterial systems, but of paramount importance to us:

1. Because the same recognition sequence occurs on both strands of the DNA duplex (running in opposite directions), the restriction enzyme is able to recognize and cleave both strands of the duplex, effectively cutting the DNA duplex in half.
2. The sites at which the two strands of a duplex are cut are offset from one another. This occurs because the position of the bond cleaved by a particular restriction enzyme typically is not in the center of the recognition sequence to which it binds. The sequence runs in opposite directions on the two strands, and the cleavage sites are therefore offset. Because of this, after cleavage, the two fragments of DNA duplex each possess a short, single strand a few nucleotides long dangling from the end. Look carefully at Figure 14-3. The two single-stranded tails are complementary to one another.

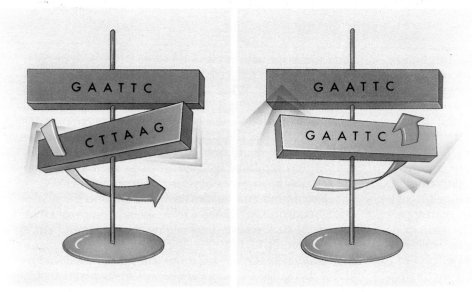

**Figure 14-2**

**Restriction sites have internal symmetry.** The nucleotide sequences recognized by restriction enzymes have twofold rotational symmetry—the second half of the sequence is the complementary mirror of the first. As a result, if the sequence CTTAAG is rotated through the plane of paper as illustrated here, it becomes its complementary sequence, GAATTC. The importance of twofold rotational symmetry is that both DNA strands have the same sequence, read in opposite directions.

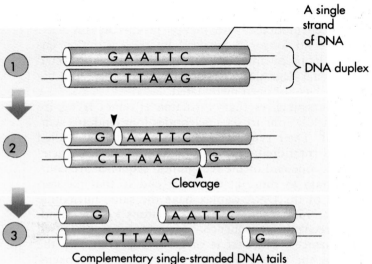

**Figure 14-3**

**The method in which restriction enzymes produce DNA fragments with "sticky ends."** The restriction enzyme EcoRI always cleaves sequence GAATTC at the same spot, after the first G. Because the same sequence occurs on both strands, both are cut. But the position of the G is not the same on the two strands; the sequence runs the opposite way on the other strand. As a result, single-stranded tails are produced. Because of the twofold rotational symmetry of the sequence, the single-stranded tails are complementary to each other, or "sticky."

There are hundreds of bacterial restriction enzymes recognizing a variety of four- to six-nucleotide sequences. Six-nucleotide sequences are the most common. Every cleavage by a given kind of restriction enzyme takes place at the same recognition sequence. By chance, this sequence will probably occur somewhere in any given sample of DNA so that a restriction endonuclease will cut DNA from any source into fragments.

The shorter the sequence, the more often it will arise by chance within a genome. Each of these fragments will have the dangling sets of complementary nucleotides (sometimes called "sticky ends") characteristic of that endonuclease. Because the two single-stranded ends produced at a cleavage site are complementary, they can pair with each other. Once they have done so, the two strands can then be joined back together with the aid of a sealing enzyme called a **ligase,** which reforms the phosphodiester bonds—the bonds between the sugars and phosphates of DNA. This latter property makes restriction endonucleases the invaluable tools of the genetic engineer: *any* two fragments produced by the same restriction enzyme can be joined together. Fragments of elephant and ostrich DNA cleaved by the same bacterial restriction enzyme can be joined to one another just as readily as can two bacterial fragments because they have the same complementary sequences at their ends.

*A restriction enzyme cleaves DNA at specific sites, generating in each case two fragments whose ends have one strand of the duplex longer than the other. Because the trailing strands of the two cleavage fragments are complementary in nucleotide sequence, any pair of fragments produced by the same enzyme, from any DNA source, can be joined together.*

# CONSTRUCTING CHIMERIC GENOMES

A chimera is a mythical creature with the head of a lion, the body of a goat, and the tail of a serpent. No chimera ever existed in nature. But human beings have made them—not the lion-goat-snake variety, but chimeras of much greater importance for our future.

The first actual chimera was a bacterial plasmid that American geneticists Stanley Cohen and Herbert Boyer made in 1973. Cohen and Boyer used a restriction endonuclease to cut up a large bacterial plasmid called a resistance transfer factor. From the resulting fragments of DNA, they isolated a fragment 9000 nucleotides long that contained both the sequence necessary for replicating the plasmid—the **replication origin**—and a gene that conferred resistance to an antibiotic, tetracycline.

Because both ends of this fragment were cut by the same restriction enzyme (called *Escherichia coli*–restriction endonuclease 1, or **Eco R1**), the ends could join together to form a circle, a small plasmid that Cohen dubbed pSC101 (Figure 14-4). Cohen and Boyer used the same restriction enzyme, Eco R1, to cut up DNA that they had isolated from an adult amphibian, the African clawed toad *Xenopus laevis*. They then mixed the toad DNA fragments with open-circle molecules of pSC101, allowed bacterial cells to take up DNA from the mixture, and selected bacterial cells that had become resistant to tetracycline (Figure 14-5). From among these pSC101-containing cells they were able to isolate ones containing the toad ribosomal RNA gene. These versions of pSC101 had the toad gene spliced in at the Eco R1 site. Instead of joining to

one another, the two ends of the pSC101 plasmid had joined to the two ends of the toad DNA fragment that contained the ribosomal RNA gene.

The pSC101 containing the toad ribosomal RNA gene is a true chimera. It is an entirely new kind of molecule that never existed in nature and would never have evolved there. It is a form of **recombinant DNA,** a DNA molecule created in the laboratory by molecular geneticists who joined together bits of several genomes into a novel combination.

*In 1973, Stanley Cohen and Herbert Boyer inserted an amphibian ribosomal RNA gene into a bacterial plasmid. By doing so, they initiated the age of genetic engineering.*

Inserting fragments of foreign DNA into bacterial cells by carrying them into the cells in plasmids or viruses is now a technique that scientists use frequently. Newer model plasmids with exotic names, such as pBR322, can be induced to make many hundreds of copies of themselves and thus of the foreign genes that are included in them within bacterial cells. Even easier entry into bacterial cells can be achieved by inserting the foreign DNA fragment into the genome of a bacterial virus, such as the lambda virus, instead of into a plasmid. The infective genome that harbors the foreign DNA and carries it into the target cell is called a **vector.** Not all vectors have bacterial targets. Animal viruses have been used as vectors to carry bacterial genes into monkey cells. Animal genes have even been carried into plant cells by using these methods.

There has been considerable discussion about the potential danger of inadvertently creating an undesirable life form in the course of a recombinant DNA experiment. What if one fragmented the DNA of a cancer cell and then incorporated the fragments at random into viruses that are propagated within bacterial cells? Might there not be a danger that one of the resulting bacteria could be capable of causing an infective form of cancer?

Even though most recombinant DNA experiments are not dangerous, such concerns should be taken seriously. Both scientists and governments monitor experiments in genetic engineering to detect and forestall any such hazard. In addition, researchers have established appropriate experimental safeguards. For example, the bacteria used in many recombinant DNA experiments

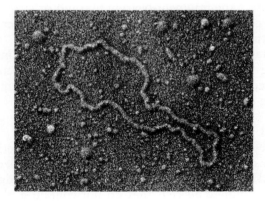

***Figure 14-4***

**A famous plasmid.** The circular molecule in the electron micrograph was the first plasmid used successfully to clone a vertebrate gene, pSC101. Its name refers to it being the hundred-and-first plasmid isolated by Stanley Cohen.

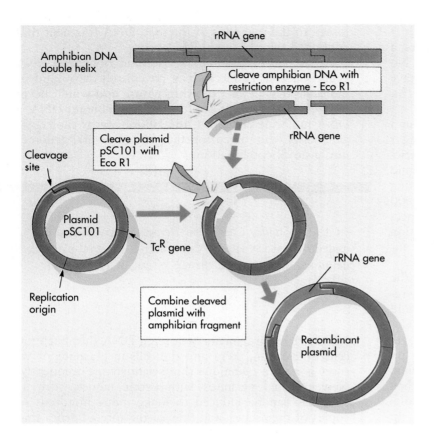

Amphibian DNA
double helix

rRNA gene

Cleave amphibian DNA with
restriction enzyme - Eco R1

rRNA gene

Cleave plasmid
pSC101 with
Eco R1

Cleavage
site

Plasmid
pSC101

Tc^R gene

Replication
origin

Combine cleaved
plasmid with
amphibian fragment

rRNA gene

Recombinant
plasmid

**Figure 14-5**

**The first genetic engineering experiment.** This diagram illustrates how Cohen and Boyer inserted an amphibian gene encoding rRNA into a bacterial plasmid. pSC101 contains a single site cleaved by the restriction enzyme EcoR1. The rRNA-encoding region was inserted into pSC101 at that site by cleaving the rRNA region with EcoR1 and allowing the complementary sequences to pair.

are unable to live outside of laboratory conditions; many of them can live only in an atmosphere that is free of oxygen and cannot survive in the earth's atmosphere. Experiments such as the cancer cell experiment described above, which is clearly dangerous, are prohibited.

## GENETIC ENGINEERING

The movement of genes from one organism to another is often termed *recombinant DNA technology,* or *genetic engineering.* Each experiment in this field presents unique problems, but all share four distinct stages:

Stage 1. *Cleavage.* The first stage of any genetic engineering experiment is the generation of specific DNA fragments by cleavage of a genome with restriction endonuclease enzymes. Different "libraries" of fragments may be obtained by using enzymes that recognize different sequences. The fragments are usually separated by electrophoresis, which permits their relative size to be estimated (Figure 14-6).

Stage 2. *Recombinant DNA.* Fragments of DNA are incorporated in plasmid or virus "vehicles," which allow the fragments to be replicated as part of the plas-

mid or virus genome. It is usually necessary at this stage to eliminate those plasmids or viruses which do *not* contain the fragment of DNA being moved.

Stage 3. *Cloning.* The fragment-containing plasmids or viruses are introduced into bacterial cells. Each such cell then reproduces, forming a clone of cells that all contain the fragment-bearing plasmid or virus. Each of the cell lines is maintained separately; together, they constitute what is called a **clone library.**

Stage 4. *Screening.* From the many clonal lines resulting from such an experiment, those which contain the specific fragment of interest, often a fragment containing a particular gene, are identified.

### Initial Screening of Clones

The key to a successful genetic engineering experiment lies in the strategy adopted to identify and select the desired fragment. To make this job easier, investigators try to eliminate all cells that do not contain a plasmid or virus and any that, by chance, do not contain a fragment from the original library. The investigators accomplish this by using genes that make the bacteria in which they occur resistant to antibiotics (Figure 14-7) (for example, penicillin, tetracycline, or ampicillin) that have the ability to block their growth.

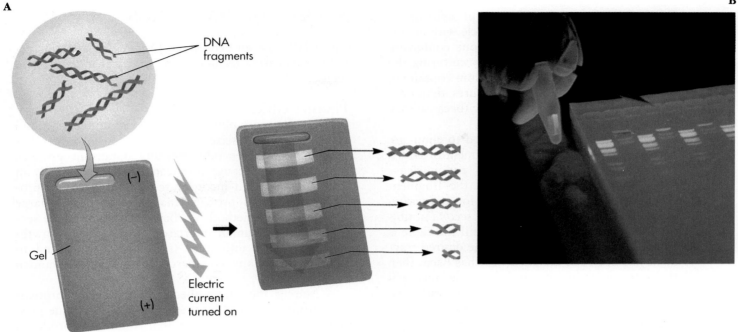

A

B

**Figure 14-6**

**Gel electrophoresis.** This process separates DNA or protein fragments (**A**), causing them to migrate within a gel in response to an electric field. The fragments migrate according to size and can be visualized easily (**B**), as the migrating bands glow in fluorescent light.

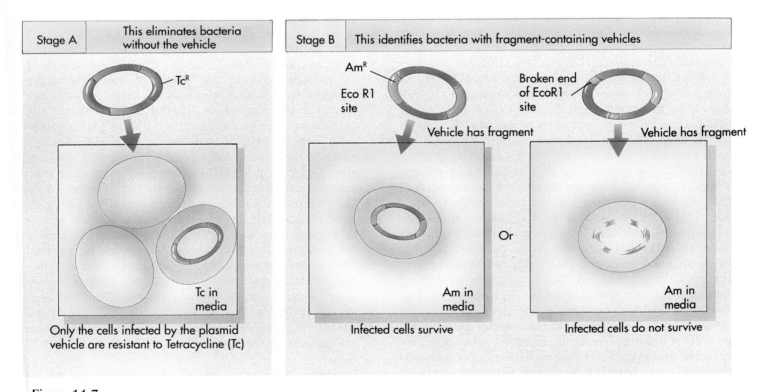

| Stage A | This eliminates bacteria without the vehicle |
|---|---|

TcR

Tc in media

Only the cells infected by the plasmid vehicle are resistant to Tetracycline (Tc)

| Stage B | This identifies bacteria with fragment-containing vehicles |
|---|---|

AmR

Eco R1 site

Broken end of EcoR1 site

Vehicle has fragment

Vehicle has fragment

Or

Am in media

Am in media

Infected cells survive

Infected cells do not survive

**Figure 14-7**

Screening restriction fragment clones.

1. To eliminate bacteria without a plasmid or virus vehicle (stage 2 above), a vehicle with an antibiotic resistance gene, such as one conferring tetracycline resistance, is used. By culturing the clones of stage 4 above on a medium containing tetracycline, the investigator ensures that only bacteria resistant to this antibiotic (because they contain the vehicle) will survive.

2. A more complex technique is used to eliminate bacteria that lack the desired fragment of DNA. A single site for the restriction endonuclease that is employed is included in the fragment that is being transferred, and it is located *within* a gene that confers resistance to a *second* antibiotic, such as ampicillin. Therefore if the clone is still resistant to that antibiotic, the desired fragment has not been included; if it has been, then the gene conferring the antibiotic resistance will have been broken up by the newly incorporated fragment, and the bacterial clone will not be resistant to the second antibiotic. By the technique of **replicate plating,** in which samples from a given clone are placed on separate media that do and do not contain an antibiotic such as ampicillin, the susceptible strains can be identified while other portions of the same clone are kept growing.

## Finding the Gene You Want

A library of restriction fragments may contain from a few dozen to many thousand individual members, each representing a different fragment of genomic DNA. A complete *Drosophila* library contains more than 40,000 different clones; a complete human library of fragments 20 kilobases long would contain 150,000 clones. To identify a single fragment containing a particular gene from within that immense clone library often requires ingenuity, and many different approaches have been used successfully.

One of the most useful procedures for identifying a specific gene has been the Southern blot, a procedure that employs a DNA copy of the gene's mRNA (called copy DNA or cDNA), as a "probe"; among thousands of fragments, only that fragment containing the proper gene will hybridize with the probe because only that fragment has a nucleotide sequence complementary to the cDNA's sequence. In such a screening procedure, the fragments are spread apart by electrophoresis (Figure 14-8), and a radioactive probe (synthesized in the presence of $^{32}P$, particular radioactive isotope of phorphorus) is "blotted" onto the resulting gel pattern, incubated, and then the excess washed off. Because only the gene-containing fragment will base pair with the probe, only that fragment produces a radioactive clear band on film left beneath the gel.

## A SCIENTIFIC REVOLUTION

The 1980s saw an explosion of interest in applying genetic engineering techniques to practical human problems.

### Pharmaceuticals

Perhaps the most obvious commercial application and the one that was seized on first was to introduce human genes that encode clinically important proteins into bacteria. Because they can be grown cheaply in bulk, bacteria that incorporate human genes can produce the human proteins that they specify in large amounts. This method has been used to produce several forms of human interferon, to place rat growth-hormone genes into mice (see Figure 14-1), and to manufacture many commercially valuable nonhuman enzymes.

Among the many medically important proteins now being produced by these techniques is tissue plasminogen activator, a protein that is normally produced by the body in minute amounts; it causes blood clots to dissolve. Plasminogen activator is now being produced in quantities large enough to test its effectiveness in preventing heart attacks and strokes. Also being produced are atrial peptides, small proteins produced in the hearts of mammals that regulate blood pressure and kidney function; genetically engineered versions are being tested as possible new ways to treat high blood pressure or kidney failure.

### Agriculture

A second major area of genetic engineering activity is the manipulation of the genes of crop plants. In broadleaf plants such as tomatoes, tobacco, and soybeans, a plasmid of the bacterium *Agrobacterium tumefaciens,* which causes crown gall in plants, is the main vehicle that has been used to introduce foreign genes. A part of this **Ti** plasmid integrates into the plant DNA, carrying with it whatever genes have been incorporated by earlier manipulation (Figure 14-9). The characteristics of a number of plants have been altered through the use of this technique, which will be valuable in improving crops and forest trees. Among the features that scientists would like to effect are resistance to disease, frost, and other forms of stress; nutritional balance and protein content; and herbicide resistance. Unfortunately, *Agrobacterium* generally does not infect the cereals such as corn, rice, and wheat, and alternative methods are being developed to introduce new genes into them.

***Herbicide Resistance.*** Genetically engineered plants have been produced that are resistant to the herbicide glyphosate, the active ingredient in Roundup, a power-

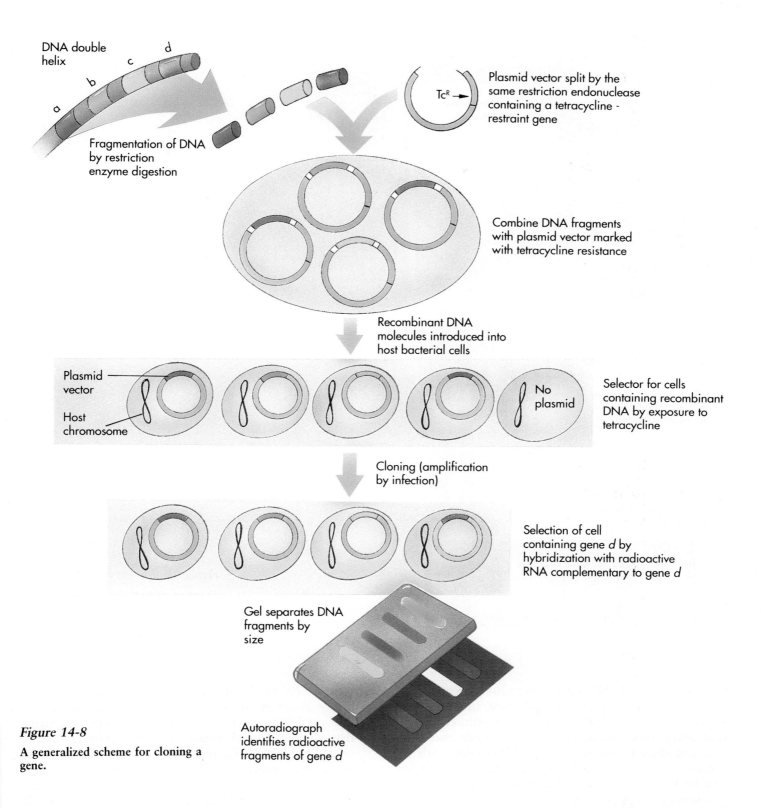

DNA double helix

Fragmentation of DNA by restriction enzyme digestion

Plasmid vector split by the same restriction endonuclease containing a tetracycline - restraint gene

Tc$^R$

Combine DNA fragments with plasmid vector marked with tetracycline resistance

Recombinant DNA molecules introduced into host bacterial cells

Plasmid vector

Host chromosome

No plasmid

Selector for cells containing recombinant DNA by exposure to tetracycline

Cloning (amplification by infection)

Selection of cell containing gene *d* by hybridization with radioactive RNA complementary to gene *d*

Gel separates DNA fragments by size

Autoradiograph identifies radioactive fragments of gene *d*

*Figure 14-8*

**A generalized scheme for cloning a gene.**

ful, biodegradable herbicide that kills most actively growing plants (Figure 14-10). Glyphosate kills plants by inhibiting an enzyme called *EPSP synthetase,* which is required by plants to produce certain amino acids. Humans do not make these amino acids but obtain them from food, and so are unaffected by glyphosate.

To obtain plants resistant to Roundup, agricultural scientists first inserted extra copies of the EPSP genes into plants, carrying them in on Ti plasmids. The engineered plants "overproduced" the enzyme, producing 20 times the normal levels, and so were able to withstand EPSP suppression caused by glyphosate—the ex-

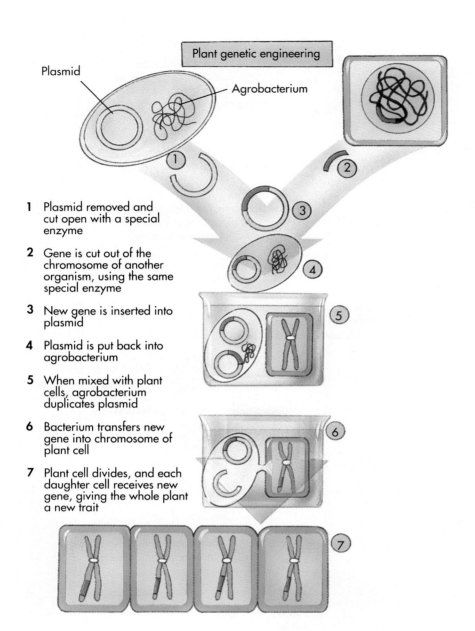

Plant genetic engineering

Plasmid

Agrobacterium

1 Plasmid removed and cut open with a special enzyme

2 Gene is cut out of the chromosome of another organism, using the same special enzyme

3 New gene is inserted into plasmid

4 Plasmid is put back into agrobacterium

5 When mixed with plant cells, agrobacterium duplicates plasmid

6 Bacterium transfers new gene into chromosome of plant cell

7 Plant cell divides, and each daughter cell receives new gene, giving the whole plant a new trait

**Figure 14-9**

The use of the T$_i$ plasmid of *Agrobacterium tumefaciens* in plant genetic engineering.

tra amounts of EPSP enzyme permitted them to still carry out protein production and grow. In later experiments, a bacterial form of the EPSP synthetase gene that differs from the plant form by a single nucleotide was introduced into plants via Ti plasmids with the same result—the plants became resistant to Roundup. The bacterial form of EPSP synthetase proved to be active within the plant cells.

This advance is of great interest to farmers because a crop resistant to Roundup would never have to be weeded if the field were simply treated with the herbicide. And because Roundup kills most growing plants, farmers would no longer need to employ a variety of herbicides on a single crop (most herbicides kill

only a few kinds of weeds). When natural selection brings about the development of glyphosate resistance in a particular species of weed, as inevitably happens, that weed can be eliminated by the short-term use of another herbicide. Glyphosate is readily broken down in the environment, which means that its use is a great improvement over many commercial herbicides that are commonly used in agriculture. A plasmid that will permit the introduction of EPSP genes into cereal crops, making them Roundup-resistant, is being actively sought.

**Virus Resistance.** Ti plasmids have been used to introduce a variety of other genes into broadleaf crop

**Figure 14-10**

**Genetically engineered herbicide resistance.** All four of these petunia plants were exposed to equal doses of the potent herbicide Roundup. The two on top were genetically engineered to be resistant to glyphosate, the active ingredient of Roundup, whereas the dead ones were not.

plants. In one of the most interesting experiments, Roger Beachy of Washington University in St. Louis made plants immune to a virus (Figure 14-11). Unlike humans, plants have no immune system of their own. Beachy introduced the tobacco mosaic virus (TMV) gene encoding the virus protein coat into a tobacco cell chromosome, using the Ti plasmid-transfer system. He then grew the plasmid-infected plant cell in tissue culture, and from the culture grew a whole tobacco plant. Every cell of this new plant was derived from the Ti-infected cell and contained the TMV gene. When the genetically engineered tobacco plant was reinfected with TMV, it did not contract the disease! This exciting avenue is being vigorously pursued because it offers hope of engineering resistance to many viral diseases of commercial crops.

***Immunity to Insects.*** Many commercially important plants are attacked by insects. Over 40% of the chemical insecticides used today are employed to kill boll weevil, boll worm, and other insects that eat cotton plants (Figure 14-12). Using genetic engineering, we may be able to produce cotton plants that are naturally resistant to these insect pests, and so remove the need for chemical insecticides. Reducing the massive doses

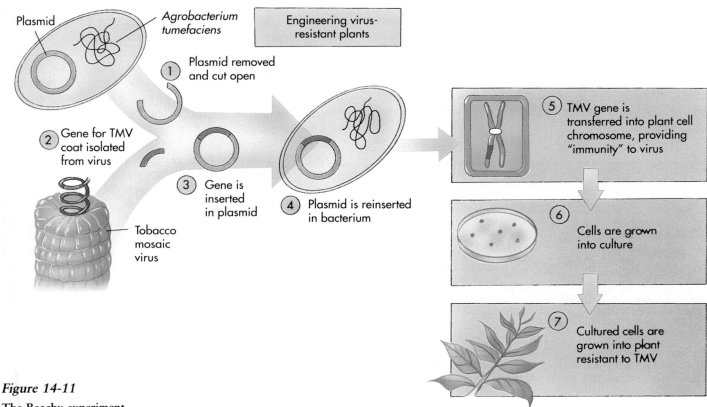

**Figure 14-11**

**The Beachy experiment.**

Plasmid

*Agrobacterium tumefaciens*

Engineering virus-resistant plants

① Plasmid removed and cut open

② Gene for TMV coat isolated from virus

Tobacco mosaic virus

③ Gene is inserted in plasmid

④ Plasmid is reinserted in bacterium

⑤ TMV gene is transferred into plant cell chromosome, providing "immunity" to virus

⑥ Cells are grown into culture

⑦ Cultured cells are grown into plant resistant to TMV

**Figure 14-12**

**An agricultural use of genetic engineering.** Over 40% of the chemical insecticides used in the world in 1987 were used to control insects on cotton. Using genetic engineering, scientists may be able to produce cotton plants that are naturally resistant to insect pests.

of chemical insecticides that are applied to cotton fields from the environment would be a very positive step because the insecticides affect a number of organisms other than those which they are intended to control.

One success makes use of the bacterium *Bacillus thuringiensis* (Figure 14-13), which normally lives in soil and produces enzymes that are toxic to the larvae (caterpillars) of moths and butterflies but not to other organisms. When the genes that produce these enzymes

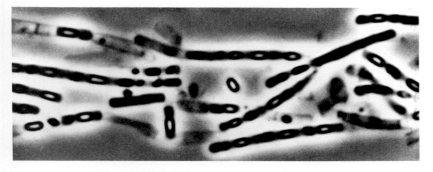

**Figure 14-13**

***Bacillus thuringiensis,* a bacterium that produces a protein toxic to many caterpillars.** The white spots within the bacteria are protein crystals that are transformed into a toxin by enzymes in the caterpillar's stomach.

are transferred via the Ti plasmid to plants such as tomatoes and tobacco, the plants become toxic to serious pests such as the tomato hornworms (Figure 14-14), which die after eating them. Plants containing such genes have been tested in the field and have been found to be protected from insects that normally attack them (Figure 14-15).

Many important plant pests attack roots, and to counter this threat biologists have introduced the *B. thuringiensis* insecticidal protein gene into root-colonizing bacteria, especially strains of *Pseudomonas. B. thuringiensis* does not normally colonize plant roots as does *Pseudomonas,* so this gene transfer affords a novel form of protection. The field testing of this promising approach is currently on hold, awaiting approval by the Environmental Protection Agency, which is assessing the possibility that other soil bacteria might be affected by the altered form of *Pseudomonas.* A variety of novel approaches are being employed to check this unlikely but important possibility (Figure 14-16).

*Nitrogen Fixation.* A long-range goal of agricultural genetic engineering is to increase yield and plant size in crop plants; however, we do not yet clearly know which genes are responsible for these complex characteristics. A long-range goal is to introduce into key crop plants the genes that enable symbiotic bacteria living in nodules on the roots of soybeans, other legumes, and certain other plants to "fix" nitrogen. Nitrogen fixation, which we discuss in detail in Chapter 21, is the process by which such bacteria obtain nitrogen from the atmosphere, where it is abundant, and convert it into a form that living organisms can use in their metabolism—a form that is scarce. Most plants do not form associations with nitrogen-fixing bacteria and must obtain their nitrogen from the soil; farmland, in which crops are grown repeatedly, soon becomes depleted of nitrogen unless fertilizers containing nitrogen are applied. Worldwide, farmers applied over 60 million metric tonnes of nitrogen fertilizers in 1987, an expensive undertaking. Farming would be much cheaper if major crops such as wheat and corn could be engineered to carry out nitrogen fixation. However, it has so far proved difficult to introduce the nitrogen-fixing genes from bacteria into plants because these genes do not seem to function properly in their new eukaryotic environment. Many experiments are being performed in the attempt to find a way around this difficulty. Various nitrogen-fixing bacteria are being tested to see whether their genes might function well in plants.

*Farm Animals.* One of the first genes to be cloned successfully was that for the growth hormone somatotropin. Instead of extracting the hormone from cow pituitaries at great expense, bovine somatotropin is produced in large amounts by a gene introduced into bacteria. The hormone, used as a supplement to a dairy cow's diet, improves the animal's milk production efficiency (Figure 14-17). It has no effect on humans because it is a protein and is digested in the stomach like all the other proteins we eat. Other experiments that use somatotropin to increase the weight of cattle and pigs are under way (Figure 14-18). In humans, the human version of this same growth hormone is currently being tested as a potential treatment for disorders in which the pituitary gland fails to make adequate levels of somatotropin, producing dwarfism.

## Probing the Human Genome

Genetic engineering techniques are revealing a great deal about the human genome. Any cloned gene can now be traced to a specific chromosomal location by seeing where a radioactive version at the cloned gene sticks on the chromosome; the radioactive label accumulates only at the site where the probe binds to the chromosome DNA. Several clonal libraries of the human genome have been assembled, using large restriction fragments. Maps indicating the location of these fragments on the chromosomes are being constructed, and a detailed map of the human genome will soon result. These attempts are significant because many genetic diseases can be shown to be associated with specific restriction fragments. When the genes associated with particular diseases have been cloned, they can be used as probes to screen the clone library and identify the appropriate restriction fragment. Mutations often

*Figure 14-15*

**A successful experiment in pest resistance.** Both of these tomato plants were exposed to destructive caterpillars under laboratory conditions. The nonengineered plant on the left has been completely eaten, whereas the engineered plant shows virtually no signs of damage.

affect the way in which the fragment moves on an electrophoresis gel, and they can always be detected as a change in nucleotide sequence of the fragment. Genetic screening of restriction fragments is a real possibility in the near future.

Efforts to sequence the entire human genome, which contains some $3 \times 10^9$ nucleotide base pairs, are under way. The analysis of one restriction fragment at a time will eventually reveal the whole sequence (Figure 14-19). Clearly the information gained will be of great value to medicine as well as to science.

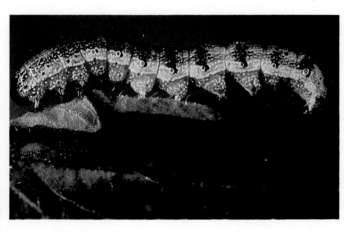

*Figure 14-14*

**A target for future genetic engineering.** Crop plants have been engineered to produce a naturally occurring protein that kills certain insect pests. Advances like this could reduce the farmer's dependence on chemical insecticides.

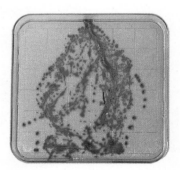

*Figure 14-16*

**Keeping track of genetically altered bacteria.** To monitor the location of genetically engineered *Pseudomonas* during and after a field test, scientists have introduced into them the bacterial *lac* genes. In the presence of the sugar lactose, which natural *Pseudomonas* ignore, the genetically engineered *Pseudomonas* break down the lactose and, as a result, turn a detector dye blue.

# Rape and DNA Fingerprints: The Molecular Witness

The 27-year-old woman never saw her assailant. She was asleep on February 22, 1987, when a man broke into her house in Orlando, Florida, covered her head with a sleeping bag, and raped her in bed. On November 3, 1987 a man named Tommie Lee Andrews was placed on trial for this rape. The case against him did not at first seem strong. There were two indistinct fingerprints on a window screen that resembled his, but the woman could not identify Andrews as her attacker—her head covered, she had never seen the man who raped her. Andrews' girlfriend and sister swore he never left home that night. Like many rape cases, this one presented the jury with conflicting testimony and evidence open to more than one interpretation.

Then the prosecuting attorney introduced a new line of evidence: there was a witness. The man who raped this woman could be clearly identified, without ambiguity and beyond the shadow of a doubt. The witness was not human, someone who saw what happened. Such a witness could always be doubted, as people do make mistakes in identification. The witness was DNA.

The chromosomes of every human cell contain scattered through their DNA short, highly repeated 15-nucleotide segments called "mini-satellites." The locations and number-of-repeats of any particular mini-satellite are so highly variable that no two people are alike. The probability of two unrelated individuals having the same pattern of location and repeat-number of mini-satellite is 1 in 10 billion—with a world population of just over 5 billion, no two people are ever alike. If one analyzes 2 mini-satellites to be sure, the probability is a minuscule $5 \times 10^{-19}$, which for all practical purposes is zero.

In Figure 14-A you can see the evidence presented by the prosecuting attorney. It consisted of autoradiographs, parallel bars on x-ray film resembling the line patterns of the universal price code found on groceries. Each bar represents the position of a mini-satellite, detected by techniques similar to those described in Figure 14-6. A vaginal swab had been taken from the victim within hours of her attack. From it, semen was collected, and the semen DNA analyzed for its mini-satellite patterns. These mini-satellite patterns, labelled "semen" in Figure 14-A, are beyond question those of the rapist.

Now compare a mini-satellite pattern of the rapist to that of the suspect Andrews, lining the two samples up with a standardized control lane that has many bands: you can see that the suspect's pattern is identical to that of the rapist (and not at all like that of the victim). And all the patterns are similarly the same. Clearly the semen collected from the rape victim and the blood collected from Tommie Lee Andrews came from the same person.

On November 6 the jury returned a verdict of guilty. Andrews became the first man in the United States to be convicted of a crime based on DNA evidence.

Since the Andrews verdict, over 2 dozen rape and murder trails in the United States have resulted in convictions based on DNA evidence. Just as fingerprinting revolutionized forensic evidence in the early 1900s, so DNA fingerprinting is revolutionizing it today. A hair, a minute speck of blood, a drop of semen—they all can serve as sources of DNA, to damn or clear a suspect. As the man who analyzed Andrews' DNA said, "It's like leaving your name, address, and social security number at the scene of the crime. It's that precise."

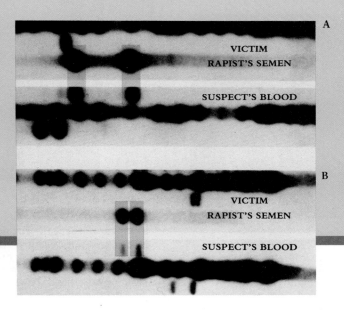

VICTIM
RAPIST'S SEMEN

SUSPECT'S BLOOD

A

B

VICTIM
RAPIST'S SEMEN

SUSPECT'S BLOOD

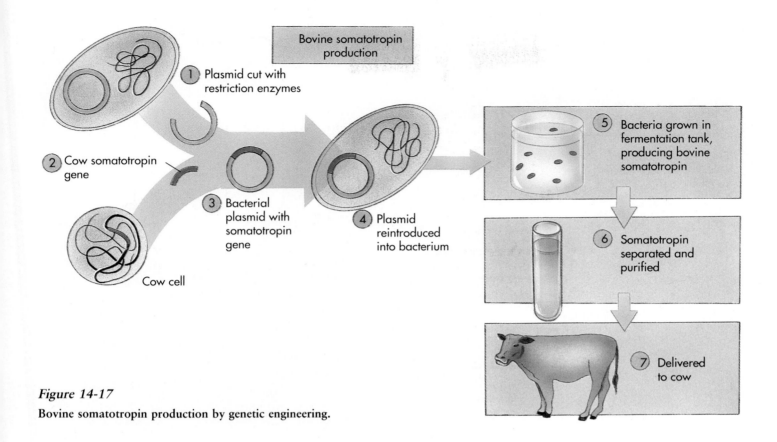

**Figure 14-17**

Bovine somatotropin production by genetic engineering.

## Piggyback Vaccines

A third area of potential significance involves using genetic engineering techniques to help produce important vaccines. Vaccines against herpesvirus and hepatitis viruses have recently been manufactured by using gene-splicing techniques. Genes specifying the protein-polysaccharide coat of the herpes simplex virus or hepatitis B virus have been spliced into a fragment of the DNA genome of the cowpox, or vaccinia, virus—the same virus that British physician Edward Jenner used almost 200 years ago in his pioneering vaccinations against smallpox.

Live vaccinia virus is introduced into a mammalian cell culture (Figure 14-20) along with spliced fragments of vaccinia virus DNA into which have been incorporated genes encoding the protein-polysaccharide coat of the herpes simplex virus genome (or, in other experiments, of the hepatitis B virus genome). Under such circumstances, the spliced herpes simplex gene is able to recombine into the vaccinia virus genome, producing a recombinant vaccinia virus. When this recombinant virus is injected into a mouse or rabbit, it dictates the production not only of the proteins that the

**Figure 14-18**

**A fatter pig.** Research is currently being conducted that fattens pigs by the introduction of human growth hormone.

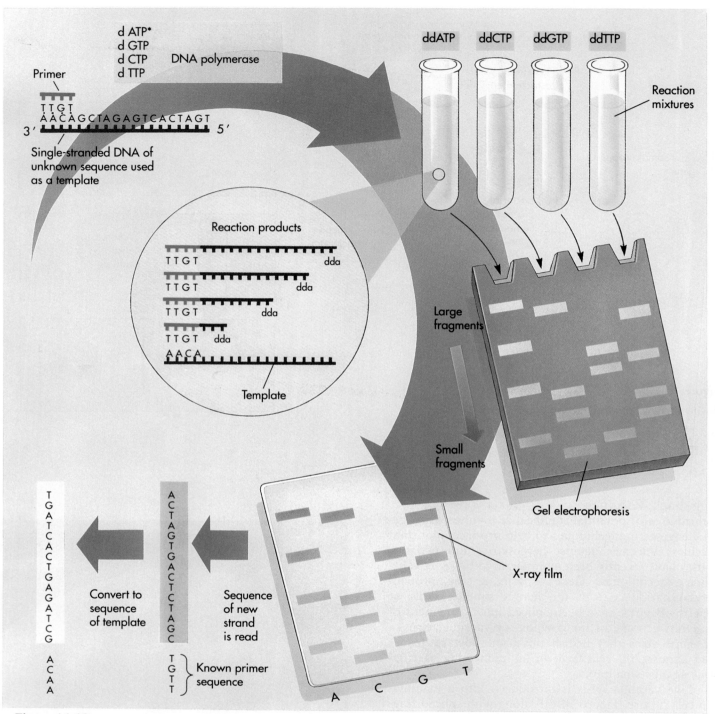

**Figure 14-19**

**Steps in sequencing DNA.** Most DNA sequencing is currently done using the technique of "primer extension." A short two-stranded primer is added to the end of a single-stranded fragment to be sequenced. This provides a 3-inch end for DNA polymerase to add to. The primed fragment is placed in four in vitro synthesis tubes, each containing a different synthesis-stopping dd-nucleotide. The first tube, for example, contains ddATP, and synthesis stops whenever ddATP is incorporated instead of dATP. Thus this tube will come to contain a series of fragments, corresponding to the different lengths the polymerase can travel from the primer before A nucleotides are encountered (ddATP is added at random and will not always hit every site). Electrophoresis separated these fragments according to size. A radioactive label (here dATP*) allows the fragments to be visualized on X-ray film, and the newly made sequence can be read directly. Try it. The fragment you started with has the complementary sequence.

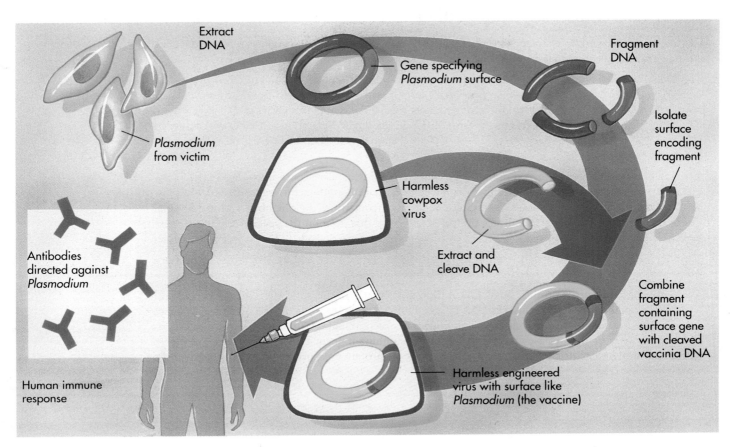

**Figure 14-20**

Constructing a "subunit" vaccine against malaria.

vaccinia virus genome specifies, but also of the protein-polysaccharide coat specified by the herpes genes that it carries. The viruses that it produces are thus somewhat like a sheep wearing a wolf's clothing—they have the interior of a benign vaccinia virus and the exterior surface of a herpes virus. Infected individuals contract cowpox instead of herpes, and cowpox is relatively harmless. At the same time, contact with the coat of the herpesvirus causes infected individuals to make antibodies against herpes and so become immune to it. Vaccines produced in this way are harmless because only a small fragment of the DNA of the disease-associated virus is introduced via the recombinant virus.

Similar "piggy-back" approaches are being used to correct genetic defects in humans—**gene therapy.** On September 14, 1990, a 4-year-old girl became the first human being to undergo gene therapy. She suffered from a rare genetic disorder called "severe combined immune deficiency." Certain critical white blood cells of her immune system called T-cells were not working for lack of a necessary enzyme called ADA (adenosine deaminase), leaving her no defense against infection. Doctors removed some of her blood and from it cultured T-cells. They then added to the cell culture viruses that possessed a good working copy of the human ADA gene that the girl lacked. These viruses carried the ADA gene with them as they infected the T-cells and spliced it along with their own genes into T-cell chromosomes. Doctors grew these altered T-cells in the laboratory until they numbered in the billions, and then injected a billion into a vein in the child's hand. Over the next year the genetically repaired cells are expected to multiply, populating the girl's immune system with gene-corrected T-cells that produce the necessary enzyme. If successful, she will have been the first human "cured" of a genetic disorder.

# ■ SUMMARY

1. Genetic engineering consists of the isolation of specific genes and their transfer to new genomes.

2. The key to genetic engineering technology is a special class of enzymes called restriction endonucleases, which cleave DNA molecules into fragments. These DNA fragments always have short, one-stranded tails that are complementary in nucleotide sequence to each other.

3. The sequences of the tails of the two DNA fragments resulting from the cleavage of DNA by restriction enzymes are complementary; therefore they can pair with each other and rejoin, whatever their source. DNA fragments from very different genomes can be combined in this way.

4. The first combination of DNA fragments from different genomes (recombinant DNA) was achieved by Stanley Cohen and Herbert Boyer in 1973. They inserted a ribosomal RNA gene from a toad into a bacterial plasmid.

5. Cloning is a technique in which fragment-containing plasmids or viruses are introduced into bacterial cells, which then reproduce, forming a clone of cells that all contain the fragment-bearing plasmid. A collection of such cell lines, maintained separately, constitutes a clone library of the original genome. Screening this library identifies the particular lines that contain the specific fragment of interest.

6. Cloning and screening are widely used in the pharmaceutical industry and in agriculture to produce large amounts of particular molecules that have important uses.

7. Gene splicing holds great promise as a clinical tool, particularly in the prevention of disease. Several vaccines are being prepared by inserting genes specifying the protein coats of disease-causing viruses or cells into the genomes of harmless viruses to create a recombinant form that is harmless and has the outside coat of the disease-causing form. The recombinant form can be used as a vaccine to induce antibody formation against the disease because its protein coat produces a response from the immune system.

## REVIEW QUESTIONS

1. The three kinds of genetic recombination that occur naturally in eukaryotes are _____, _____, and _____.

2. What single gene must every plasmid possess?

3. The passage of the F plasmid from one bacterial cell to another is called _____.

4. At what point in the cell replication cycle does crossing-over between homologous chromosomes occur?

5. What is the most numerous of the six classes of eukaryotic DNA sequences?

6. Which are more alike, the many copies of a tandem cluster, or the members of a multigene family?

## SELF-QUIZ

1. In nature, restriction endonucleases
   (a) transfer DNA between organisms.
   (b) protect bacteria from viruses.
   (c) enable viruses to invade nuclei.
   (d) restrict the ability of bacteria to form clones.
   (e) promote cell division in restricted bacteria.

2. Restriction endonucleases are an invaluable tool for genetic engineering because
   (a) they cut both strands of DNA at specific sites.
   (b) "sticky," or complementary, ends of the cleaved DNA are produced.
   (c) sticky ends of DNA from very different DNA sources can easily be joined together.
   (d) they restrict the movement of genes from nuclei.
   (e) a, b, and c.

3. A clone library is
   (a) a collection of different lines of bacterial cells, each containing a plasmid carrying one of a series of different DNA fragments.
   (b) a collection of books about clones.
   (c) the particular bacterial line that contains a DNA fragment of interest.
   (d) a series of restriction enzymes that can be used to produce different clones.
   (e) none of the above.

4. Screening of clones is accomplished by
   (a) using restriction enzymes.
   (b) using viruses that have known biochemical characteristics.
   (c) putting bacterial lines in a series of selective environments and observing them.
   (d) mapping genes.
   (e) estimating the sizes of linkage groups.

5. Nitrogen fixation is carried out by
   (a) all bacteria.       (c) legumes.       (e) genetically engineered plants.
   (b) only bacteria.      (d) cereals.

6. To increase the efficiency of herbicides, scientists have
   (a) produced more susceptible weeds.
   (b) caused genetically engineered viruses to attack weeds.
   (c) produced resistant crop plants.
   (d) generated large amounts of hormones that make the plants grow more rapidly and thus avoid being killed.
   (e) used glyphosate (Roundup) widely to increase competition.

7. A cloned gene can be traced to a specific area of a particular chromosome by
   (a) using a radioactive probe to detect in situ hybridization.
   (b) mapping the genome with crosses and comparing patterns.
   (c) using restriction enzymes to find fragments of comparable size.
   (d) testing its susceptibility to antibiotics known to attract particular chromosomes.
   (e) electrophoresis of break-down products.

8. Virus-resistant vaccines can be produced by
   (a) engineering viruses that produce the coat, but none of the other features, of a disease-causing virus.
   (b) engineering viruses that produce the DNA, but not the coat, of disease-causing viruses.
   (c) killing lethal viruses and letting them form clones, from which the vaccine can be extracted.
   (d) using bacteriophage to change the features of viruses.
   (e) understanding the infective features of malaria better.

# THOUGHT QUESTIONS

1. You are having a genetic engineering nightmare. In your dream, a well-meaning student cleaves the DNA of reticulocytes (red blood cells) isolated from a man dying of leukemia with the restriction enzyme Eco R1. The student then mixes the resulting fragments with the Eco R1–treated plasmid pSC101 and infects a growing culture of *Escherichia coli* bacteria with the resulting mix of normal and chimeric plasmids. From this mix, the student selects for cells containing plasmids by applying the antibiotic tetracycline to the liquid culture of growing bacteria. Many cells continue to grow. The student, in an excess of joy at this intimation of success, accidently drops the tube in the sink. It breaks, and the liquid flows down the drain. Knowing that *E. coli* is a common inhabitant of the human intestinal tract, the student breaks out into a cold sweat—and you wake up. Was the student right to be scared? Why?

2. A major focus of genetic engineering has been the attempt to produce large quantities of hormones and other molecules that are scarce in humans by placing the appropriate human gene into bacteria. Because prodigious numbers of bacteria can be produced rapidly and cheaply, large amounts of the desired molecule can be obtained readily. Human insulin is now manufactured in this way. However, if one attempts to use this approach to produce human hemoglobin (beta-globin), the experiment doesn't work: if an experimenter first identifies the proper clone from a clone library by using an appropriate radioactive probe, inserts the fragment containing the beta-globin gene into a plasmid, and infects bacterial cells with the chimeric plasmid, no human hemoglobin is produced by the infected cells. This negative result is obtained despite the fact that the proper fragment clone was chosen, the beta-globin gene was successfully incorporated into the plasmid, and the chimeric plasmid did successfully infect the *E. coli* cells. Why isn't this experiment working?

# FOR FURTHER READING

CHILTON, M. "A Vector for Introducing New Genes into Plants," *Scientific American,* June 1983, pages 50-60. A description of the only successful plant genetic engineering vector developed to date, important because of its agricultural implications.

DiLISI, C.: "The Human Genome Project," *American Scientist,* volume 76, 1988, pages 438-439. Mapping and deciphering the complete sequence of human DNA will stimulate research in fields ranging from computer technology to theoretical chemistry.

GODSON, G.: "Molecular Approaches to Malaria Vaccines," *Scientific American,* May 1985, pages 52-59. A revealing look at how the new techniques of molecular engineering are being used to defeat an ancient scourge.

HOFFMAN, P.: "The Human Mouse," *Discover,* August 1989, pages 4, 48-55. To defeat AIDS, an innovative young researcher, using the methods discussed in this chapter, has created remarkable new hybrids of mice and men.

NEUFELD, P.J. and N. COLMAN: "When Science Takes the Stand," *Scientific American,* May 1990, pages 46-53. DNA and other evidence is increasingly being applied to the solution of criminal cases but must be used with caution.

VERMA, I.M.: "Gene Therapy," *Scientific American,* November 1990, pages 68-84. Healing genetic inherited diseases by inserting healthy genes into patients is gene technology's newest advance.

E   V   O   L   U   T   I   O   N

# The Evidence for Evolution

There is something about our species that loves beauty. In caves in France, or here on the rocks of Kakadu in Australia, primitive man has left art, breathtaking and exquisite.

# THE EVIDENCE FOR EVOLUTION

## Overview

The theory of evolution by natural selection, which many find controversial, is the cornerstone of modern biology. That evolution has occurred is regarded as a demonstrated fact by biologists, most of whom also agree with Darwin's explanation of natural selection as the way in which evolution occurs. During the period of well over a century since Darwin first suggested this explanation for evolution, we have learned a great deal about the mechanisms that bring about evolution. Even though all available evidence points to the fact that evolution has occurred, the validity of this fact has been challenged recently in the United States, on the basis of a set of religious beliefs labeled "scientific creationism."

## For Review    *Here are some important terms and concepts that you will encounter in this chapter. If you are not familiar with them, you should review them before proceeding.*

**Darwin's theory of natural selection** (Chapter 1)

**Origin of life** (Chapter 3)

**Alleles** (Chapter 10)

**Heterozygosity and homozygosity** (Chapter 10)

**Punnett Square** (Chapter 11)

**Mutation** (Chapter 13)

Of all the major ideas of biology, evolution is perhaps the best known to the general public. This is not because the basic facts of evolution are well understood by the average person, but rather because many people mistakenly believe that it represents a challenge to their religious beliefs. Within the last few years, controversial attempts have been made to require the teaching of the set of religious dogmas known as scientific creationism in science classes. Similar highly publicized criticisms of evolution have occurred ever since the time of Darwin, and it is likely that others will occur in the future. For this reason it is important that, during the course of your study of biology, you address the issue squarely. Just what *is* the evidence for evolution?

As we saw in the first chapter, Darwin deduced that evolution, the progressive change that occurs in the characteristics of organisms through time, could best be explained by natural selection. Only a fraction of the individuals produced in each generation survive, and those individuals have genetic attributes which give them features promoting their survival. Survival is not random; rather, it is related to the genes of the individuals who do and do not survive. The resulting change in allelic frequencies from generation to generation is called natural selection.

In this chapter, we first ask, "What factors are thought to bring about evolutionary change?" Then we look at individual cases in which evolutionary change in natural populations can be seen to be adaptive, a key tenet of Darwin's theory of evolution by means of natural selection. Finally, we review the evidence that supports the general validity of Darwin's proposal.

When the word *evolution* is mentioned, it is difficult not to conjure up images of dinosaurs, woolly mammoths frozen in blocks of ice, or Darwin confronting a monkey. Traces of ancient forms of life, now extinct, survive as fossils that help us piece together the history of life on earth. With such a background, we usually think of evolution as meaning changes in the kinds of animals and plants on earth, changes that take place over long periods, with new forms replacing old ones. This kind of evolution is called **macroevolution.** Macroevolution is evolutionary change on a grand scale (Figure 15-1), encompassing the origin of novel designs, evolutionary trends, new kinds of organisms

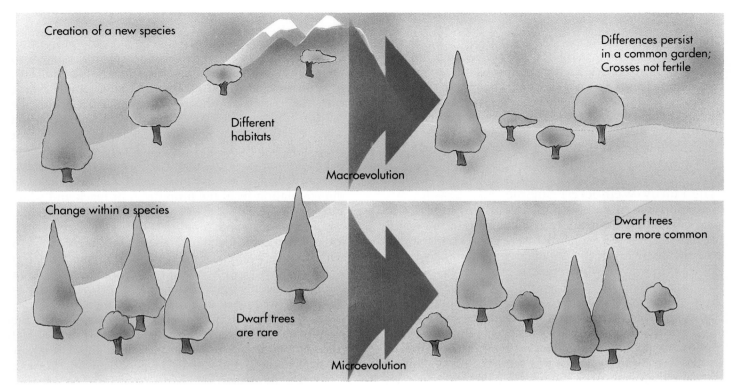

**Figure 15-1**

**Evolution from two perspectives.** Evolution within a species is called microevolution, whereas evolution between species (that is, the creation of new species) is called macroevolution.

penetrating new habitats, and major episodes of extinction.

Much of the focus of Darwin's theory of evolution, which we studied in Chapter 1, is directed not at the way in which new species are formed from old ones, but rather at the way that changes occur within species. Natural selection, the explanation that Darwin proposed for evolutionary change, is the process whereby some individuals in a population produce more surviving offspring than do others that lack their characteristics. As a result, the population gradually will come to include more and more individuals with the characteristics that provide an advantage to the individuals in which they occur, generation after generation, assuming that the characteristics have a genetic basis. In this way the population evolves. Change of this sort within populations—progressive change in allelic frequencies—is called **microevolution**. Natural selection is the process by which microevolutionary change occurs; the result of the process, the features that promote the likelihood of survival and reproduction by an organism in a particular environment, is called **adaptation**. The essence of Darwin's explanation of evolution is that progressive adaptation by natural selection is responsible for evolutionary changes *within*

a species, changes which, when they accumulate, lead to the creation of new kinds of organisms, new species. We shall return to the ways in which new species originate in Chapter 16.

## GENES WITHIN POPULATIONS

When considering Darwin's theory that evolution is a progressive series of adaptive changes brought about by natural selection, it is best to start by looking at the raw material available for the selective process—the genetic variation present among individuals within a species from which natural selection chooses the best-suited alleles. In the early part of this century the science of genetics—the study of genes—contributed relatively little to our understanding of the process of evolution. At first, geneticists were involved primarily with understanding the actions of individual genes. Evolutionists in turn could not understand how such observations would have a bearing on the evolution of a complex structure, such as an eye.

The gap between the geneticists and evolutionists finally started to close with the development of the field of **population genetics**—simply defined as the study of the properties of genes in populations—from the 1920s onward. At that time, scientists began to formulate a comprehensive theory of how alleles behave in populations and the ways in which changes in their frequencies lead to evolutionary change. The most fundamental model in the field of population genetics, the Hardy-Weinberg principle, was developed in the early years of this century; all other aspects of this field can be viewed in relation to it.

## The Hardy-Weinberg Principle

Variation within natural populations was a puzzle to Darwin and his contemporaries. The patterns by which certain features were inherited were poorly understood, and the discovery of the principles of genetics lay in the future. The way in which meiosis produces genetic segregation among the progeny of a hybrid had not yet been discovered, and the concept of blending inheritance, by which the features of organisms could merge with one another, was widely accepted. Even when the nature of dominant and recessive alleles was understood, in the early years of this century, it was thought that the dominant ones should drive the recessive ones out of populations, with selection favoring an optimal form.

The solution to the puzzle of why genetic variation persists was developed independently and published almost simultaneously in 1908 by G.H. Hardy, an English mathematician, and G. Weinberg, a German physician. They pointed out that in a large population in which there is random mating, and in the absence of forces that change the proportions of the alleles at a given locus (these forces are discussed below), the original proportions of the genotypes will remain constant from generation to generation. Dominant alleles do not, in fact, replace recessive ones. Because their proportions don't change, the genotypes are said to be in **Hardy-Weinberg equilibrium.**

In algebraic terms the Hardy-Weinberg principle is written as an equation. Its form is what is known as a binomial expansion. For a gene with two alternative alleles, which we shall call A and a, the frequency of allele A can be expressed as $p$ and that of the alternative allele a as $q$. By convention, the more common of the two alleles is designated $p$; the rarer allele, $q$. Because there are only two alleles, $p + q$ must always equal 1. The equation looks like this:

$$(p + q)^2 = \quad p^2 \quad + \quad 2pq \quad + \quad q^2$$

| Individuals homozygous for allele A | Individuals heterozygous with alleles A & a | Individuals homozygous for allele a |
|---|---|---|

In statistics, **frequency** is defined as the proportion of individuals falling in a certain category, relative to the total number of individuals being considered. Thus, in a population of 10 cats with 7 white and 3 black cats, the respective phenotypic frequencies would be 0.7 (or 70%) and 0.3 (or 30%). If we assume that the white cats were homozygous recessive for a gene that we will designate b, and the black cats were therefore either homozygous dominant BB or heterozygous Bb, we can calculate the frequencies of the two alleles in the population from the proportion of black and white cats.

In another population of 100 cats, with the genetics of coat color as stated above, we might find 16 white cats and 84 black ones, the phenotypic ratio. If $q$ is the frequency of the allele b, we can calculate from the Hardy-Weinberg equation that $q^2 = 0.16$ and $q = 0.4$. $p$, the frequency of allele B, must therefore be $1 - 0.4 = 0.6$. In addition to the 16 white cats, which have a bb genotype, there are $2pq$, or $2 \times 0.6 \times 0.4 \times 100$ (the number of individuals in the whole population), or 48 heterozygous individuals of genotype Bb. We can also calculate easily that there are $p^2 = (0.6)^2 \times 100$, or 36 homozygous dominant BB individuals.

Figure 15-2 allows you to trace genetic reassortment during sexual reproduction and see how it affects the frequencies of the B and b alleles during the next generation. In the lower element of Figure 15-2, a Punnett square, we have assumed that the union of sperm and egg in these cats is random, so all combinations of B and b alleles are equally likely. For this reason, the alleles are, in effect, mixed randomly and are represented in the next generation in proportion to their original representation; there is no inherent reason for them to change in frequency from one generation to the next. Each individual in each generation has a 0.6 chance of receiving a B allele and a 0.4 chance of receiving a b allele.

In the next generation, and for each succeeding generation, the chance of combining two B alleles is 0.36 (that is, $0.6 \times 0.6$), and on the average 36% of the individuals in the population will continue to have the BB genotype. The frequency of bb individuals ($0.4 \times 0.4$) will continue to be about 16%, and the frequency of Bb individuals will be $2 \times 0.6 \times 0.4$, or 48%.

This simple relationship has proven extraordinarily useful to biologists because it permits an investigator to predict readily what proportion of a population will be homozygous for a given allele and what proportion heterozygous, simply from a knowledge of that allele's frequency. As an example, consider the recessive allele that is responsible for cystic fibrosis. This allele is present in North American whites at a frequency of

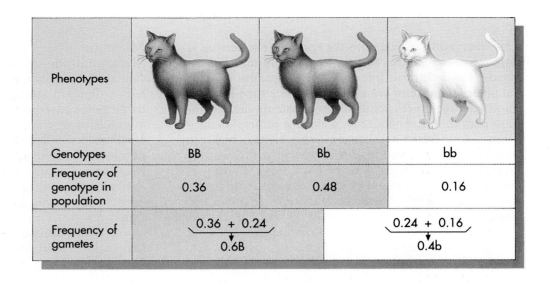

| Phenotypes | | | |
|---|---|---|---|
| Genotypes | BB | Bb | bb |
| Frequency of genotype in population | 0.36 | 0.48 | 0.16 |
| Frequency of gametes | 0.36 + 0.24 → 0.6B | | 0.24 + 0.16 → 0.4b |

B          b          Eggs

Sperm

B
p = 0.6

b
q = 0.4

| | B (Eggs) | b (Eggs) |
|---|---|---|
| B (p = 0.6) | BB (p²) = 0.36 | Bb (pq) = 0.24 |
| b (q = 0.4) | bB (pq) = 0.24 | bb (q)² = 0.16 |

**Figure 15-2**

**The Hardy-Weinberg equilibrium. In** the absence of factors that alter them, the proportions of alleles, genotypes, and phenotypes remain constant generation after generation.

about 50 per 1000 individuals, or 0.050. So what proportion of North American white individuals is expected to express this trait? The frequency of homozygous recessive individuals $(q^2)$ is expected to be 0.050 × 0.050, or 1 in every 400 individuals. What proportion is expected to be heterozygous carriers? If the frequency of the recessive allele $q$ is 0.050, then the frequency of the dominant allele $p$ must be $1 - 0.050$, or 0.950. The frequency of heterozygous individuals $(2pq)$ is thus expected to be 2 × 0.950 × 0.050, or 95 in every 1000 individuals.

How valid are these calculated predictions? For many genes, they prove to be very accurate. Most human populations are large and randomly mating and so are similar to the "ideal" population envisioned by Hardy and Weinberg. As we will see, however, the calculated predictions do *not* match the actual values for some genes. The reasons they do not do so tell us a great deal about evolution.

> The Hardy-Weinberg principle states that in a large population mating at random and in the absence of other forces that would change the proportions of the different alleles at a given locus, the process of sexual reproduction (meiosis and fertilization) alone will not change these proportions.

## WHY DO ALLELE FREQUENCIES CHANGE?

According to the Hardy-Weinberg principle, the proportions of homozygotes and heterozygotes in a large, random-mating population will remain constant, as long as the individual allele frequencies do not change.

This reservation tacked onto the end of the statement is important. In fact, it is the key to the importance of the Hardy-Weinberg principle for biology because individual allele frequencies *are* changing all the time in natural populations, with some alleles becoming more common than others. The Hardy-Weinberg principle establishes a convenient baseline against which to measure such changes. By looking at the ways in which various factors alter the proportions of homozygotes and heterozygotes in populations, we can identify those forces which are affecting particular situations that we observe.

Many factors can alter allele frequencies. But only five alter the proportions of homozygotes and heterozygotes enough to produce significant deviations from the proportions predicted by the Hardy-Weinberg principle (Table 15-1). These are (1) mutation; (2) migration (including both immigration into and emigration out of a given population); (3) genetic drift (random loss of alleles, which is more likely to occur in small populations), (4) nonrandom mating; and (5) selection. Of these, only the last factor, selection, produces adaptive evolutionary change because only in selection does the result depend on the nature of the environment. The other factors operate independently of the environment, so the changes they produce are not shaped by environmental demands.

*Five factors can bring about a deviation from the genotype frequencies predicted by the Hardy-Weinberg principle: mutation, migration, genetic drift (random loss of alleles), nonrandom mating, and selection.*

## Mutation

Mutation from one allele to another obviously can change the proportions of particular alleles in a population. But mutation rates are generally so low that they cannot alter Hardy-Weinberg proportions of common alleles very much. Many genes mutate about 1 to 10 times per 100,000 cell divisions, although a mutation in *any* gene of an individual will occur much more frequently. Because most environments are constantly changing, populations that are stable enough to accumulate differences in allele frequency produced this slowly are very rare. Nonetheless, as we have seen, mutation is the ultimate source of genetic variation and thus makes evolution possible.

## Migration

Migration is defined in genetic terms as the movement of individuals from one population into another. It can be a powerful force in upsetting the genetic stability of natural populations. Sometimes migration is obvious, as when an animal moves from one place to another. If the characteristics of the newly arrived animal differ from those already there, the genetic composition of the receiving population may be altered if the newly arrived individual or individuals are well-enough adapted to survive in the new area and mate successfully. Other important kinds of migration are not as obvious to the observer. These subtler movements include the drifting of gametes or immature stages of marine animals or plants from one place to another (Figure 15-3). The male gametes of flowering plants are often carried great distances by insects and other animals that visit their flowers. Their seeds may also be blown in the wind or carried by animals or other agents to new populations far from their place of origin. However it occurs, migration can alter the genetic characteristics of populations and prevent them from maintaining

| FACTOR | DESCRIPTION |
|---|---|
| Mutation | The ultimate source of all change; individual mutations occur so rarely that mutation alone does not change allele frequency much. |
| Migration | A very potent agent of change; migration locally acts to prevent evolutionary change by preventing populations that exchange members from diverging from one another. |
| Genetic drift | Statistical accidents; usually occurs only in very small populations. |
| Nonrandom mating | Inbreeding is the most common form; does not alter allele frequency, but lessens the proportions of heterozygotes (less than 2 pq). |
| Selection | The only form that produces *adaptive* evolutionary changes, only rapid for allele frequency $>0.01$. |

**TABLE 15-1   AGENTS OF EVOLUTIONARY CHANGE**

**Figure 15-3**

The yellowish-green cloud around these Monterey pines, *Pinus radiata,* is pollen being dispersed by the wind. The male gametes within the pollen reach the egg cells of the pine passively in this way. In genetic terms, such dispersal is a form of migration.

Hardy-Weinberg equilibrium. However, its evolutionary role is more difficult to assess, and it depends heavily on the selective forces prevailing at the different places where the species occurs.

## Genetic Drift

In small populations, the frequencies of particular alleles may be changed drastically by chance alone. The individual alleles of a given gene are all represented in few individuals, and some of them may be lost from the population if those individuals fail to reproduce. Because these changes in allele frequency appear to occur randomly, as if the frequencies were drifting, it is known as **genetic drift.** A series of small populations that are isolated from one another may come to differ strongly as a result of genetic drift.

*Genetic drift is the change in frequency of alleles at a locus that occurs by chance. In small populations, such fluctuations may lead to the loss of particular alleles.*

Sometimes one or a few individuals are dispersed and become the founders of a new, isolated population at some distance from their place of origin. When this occurs, the alleles that they carry are of special significance. Even if these alleles are rare in the source population, in their new area they will be a significant fraction of the whole population's genetic endowment. This effect—by which rare alleles and combinations of alleles may be enhanced in the new populations—is

called the **founder principle.** It is a particularly important factor in the evolution of organisms on islands, such as the Galapagos Islands, which Darwin visited. Most of the kinds of organisms that occur in such areas were probably derived from one or a few initial "founders." In a similar way, isolated human populations are often dominated by the genetic features that were characteristic of their founders, if only a few individuals were involved initially. Because humans have lived in small groups throughout much of their early history, genetic drift may have significantly affected the course of human evolution.

## Nonrandom Mating

Individuals with certain genotypes sometimes mate with one another more commonly than would be expected on a random basis, a phenomenon known as **nonrandom mating.** Inbreeding (mating with relatives), a type of nonrandom mating that is characteristic of many groups of organisms, causes the frequencies of particular genotypes to differ greatly from those predicted by the Hardy-Weinberg principle. Inbreeding does not change the frequency of the alleles, but rather the proportion of individuals that are homozygous. In inbred populations, there are more homozygous individuals than predicted by the Hardy-Weinberg principle. It is for this reason that populations of self-fertilizing plants consist primarily of homozygous individuals, whereas "outcrossing" plants, which interbreed with individuals different from themselves, have a higher proportion of heterozygous individuals (Figure 15-4).

Because inbreeding increases the proportion of homozygous individuals in a population, it tends to promote the occurrence of homozygous recessive combinations. It is for this reason that marriage between relatives is discouraged—it greatly increases the possibility

**Figure 15-4**

**Bees help flowers find distant mates.** This wild bee *(Ptilothrix)*, loading bristly areas on its hind legs with pollen from the desert "poppy" *(Kalstroemia)* in eastern Arizona, illustrates the way that flowering plants have co-opted animals to the task of dispersing their pollen precisely from plant to plant. Pollen grains, adhering to the hairy body of the bee, will have a good chance of reaching the female parts of other flowers of this same species.

of producing children homozygous for an allele associated with one or more of the genetic disorders discussed in Chapter 11. A dramatic example of this can be seen in Figure 15-5, which compares the incidence of five important genetic diseases in the United States, where inbreeding is rare, with the incidence in Japan, where it is more common. Even though alleles specifying some disorders, such as Tay-Sachs disease, are rarer in Japan, there is a much greater likelihood that a marriage between first cousins will produce affected children there because the Japanese population is more homozygous in general.

## Selection

As Darwin pointed out, some individuals leave behind more progeny than others, and the rate at which they do so is affected by their inherited characteristics. We describe the results of this process as **selection,** and speak both of **artificial selection** and of **natural selection.** In artificial selection, the breeder selects for the desired characteristics. In natural selection, the environment plays this role, with conditions in nature determining which kinds of individuals in a population are the most fit and so affecting the proportions of genes among individuals of future populations. This is

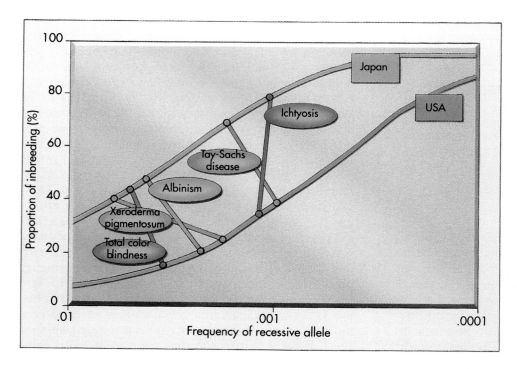

**Figure 15-5**

**Inbreeding often has detrimental effects.** In Japan it is common for first cousins to marry, and the Japanese population as a result is more homozygous than that of the United States, where such marriages are generally forbidden. Because of this increased homozygosity, recessive alleles tend to be expressed more often in Japan. For each of the five genetic disorders illustrated here, affected children in Japan are far more likely to have parents who are first cousins.

## A VOCABULARY OF GENETIC CHANGE

**adaptation**   A change in structure, physiology, or behavior that promotes the likelihood of an organism's survival and reproduction in a particular environment.

**allele**   One of two or more alternative versions of a gene.

**allele frequency**   The relative proportion of a particular allele among individuals of a population. Not equivalent to *gene frequency*, although the two terms are sometimes confused.

**evolution**   Genetic change in a population of organisms over time (generations). Darwin proposed that natural selection was the mechanism of evolution.

**fitness**   The genetic contribution of an individual to succeeding generations, relative to the contributions of other individuals in the population.

**gene**   The basic unit of heredity; a sequence of DNA nucleotides on a chromosome that encodes a protein or RNA molecule or regulates the transcription of such a sequence.

**gene frequency**   The frequency with which individuals in a population possess a particular gene. Often confused with *allele frequency*.

**genetic drift**   Random fluctuations in allele frequencies over time.

**macroevolution**   The creation of new species and the extinction of old ones.

**microevolution**   Evolution within a species.

**mutation**   A permanent change in a gene, such as an alteration of its nucleotide sequence.

**natural selection**   The differential reproduction of genotypes in response to factors in the environment; artificial selection, by contrast, is in response to demands imposed by human intervention.

**polymorphism**   The presence in a population of more than one allele of a gene at a frequency greater than that of newly arising mutations.

**population**   Any group of individuals, usually of a single species, occupying a given area at the same time.

**species**   A kind or organism. In nature, individuals of one species usually do not interbreed with individuals of other species.

rect: selection acts directly on the phenotype, which is determined by the interaction of the genotype and the environment, and the linkage between particular alleles and particular characteristics of the phenotype is less direct for some features than for others.

Although selection is perhaps the most powerful of the five principal agents of genetic change, there are limits to what it can accomplish. These limits arise because alternative alleles may interact in different ways with other genes. These interactions tend to set limits on how much a phenotype can be altered. For example, selecting for large clutch size in barnyard chickens eventually leads to eggs with thinner shells that break more often than did the shells of previous eggs. Because of the limits imposed by gene interactions, strong selection is apt to result in rapid change initially, but the change soon comes to a halt as the interactions between genes increase. For this reason, we do not have gigantic cattle that yield twice as much meat as our leading strains, chickens that lay twice as many eggs as the best layers do now, or corn with an ear at the base of every leaf, instead of just one or a few.

There is a second factor that limits what selection can accomplish: selection acts only on phenotypes. Only those characteristics which are expressed in the phenotype of an organism can affect the ability of that organism to produce progeny. For this reason, selection does not operate efficiently on rare, recessive alleles simply because they do not often come together as homozygotes and there is no way of selecting them unless they do come together. For example, when a recessive allele $a$ is present at a frequency $q$ equal to 0.1, 10% of the alleles for that particular gene will be $a$, but only 1 out of 100 individuals $(q^2)$ will be homozygous recessive and so display the phenotype associated with this allele. For lower allele frequencies, the effect is even more dramatic: if the frequency in the population of the recessive allele $q = 0.01$, the frequency of homozygotes in that population will be only 1 in 10,000.

the key point in Darwin's proposal that evolution occurs because of natural selection: the environment imposes the conditions that determine the results of selection and thus the direction of evolution.

Like mutation, migration, and genetic drift, selection causes deviations from Hardy-Weinberg proportions by directly altering the frequencies of the alleles. Darwin argued that the more successful reproduction of particular genotypes, which is how he defined selection, is the primary force that shapes the pattern of life on earth. But the selection of these genotypes is indi-

*Selection, the differential reproduction of genotypes as a result of the way that their phenotypic characteristics lead to reproductive success in the environment, is a powerful mechanism for producing deviations from Hardy-Weinberg equilibrium.*

What this means is that selection against undesirable genetic traits in humans or domesticated animals is very difficult unless the heterozygotes can also be detected. For example, if, for a particular undesirable recessive allele $r$, $q = 0.01$, and none of the homozygotes for this allele were allowed to breed, it would take 1000 generations, or about 25,000 years in humans, to lower the allelic frequency by half to 0.005. At this point, after 25,000 years of work, the frequency of homozygotes would still be 1 in 40,000, or 25% of what it was initially. This is the basic reason that few geneticists advocate **eugenics,** the field that deals with efforts to change the genetic characteristics of human beings by artificial selection. Aside from the moral implications, such efforts are essentially doomed to failure by the sheer difficulty of producing the desired results within any plausible human time frame.

## SELECTION IN ACTION

Selection operates in natural populations of a species something like skill does in football games; it is difficult to predict the winner in any individual game because chance can play an important role in the outcome; but, over a long season, the teams with the most skillful players usually win the most games. In nature, too, those individuals best suited to their environment tend to win the evolutionary game by leaving the most offspring, although chance can play a major role in the life of any one individual. Selection, you can see, is a statistical concept, just as is betting. Although you cannot predict the fate of any one individual, or any one coin toss, it *is* possible to predict which kind of individ-

ual will tend to become more common in populations of a species (Figure 15-6).

## Forms of Selection

In nature, many traits, perhaps most, are affected by more than one gene. The interactions between genes are typically complex, as we saw in Chapter 10. For example, alleles of many different genes play a role in determining human height (see Figure 10-20). In such cases, selection operates on all the genes, influencing most strongly those which make the greatest contribution to the phenotype. How selection changes the population depends on which genotypes are favored.

*Directional Selection.* When selection acts to eliminate one extreme from an array of phenotypes (Figure 15-7, *A*), the genes promoting this extreme become less frequent in the population. Thus in the *Drosophila* population illustrated in Figure 15-8, the elimination by the investigator of flies that move toward light causes the population to contain fewer individuals with alleles promoting such behavior. The result is that if you were to pick an individual at random from the new fly population, there is a lesser chance it would spontaneously move toward light than if you had selected a fly from the old population. The population has been changed by selection in the direction of lower light attraction. This form of selection is called **directional selection.**

*Stabilizing Selection.* When selection acts to eliminate *both* extremes from an array of phenotypes (Figure 15-

**Figure 15-6**

**Race horses.** Race horses are bred for one specific trait, speed.

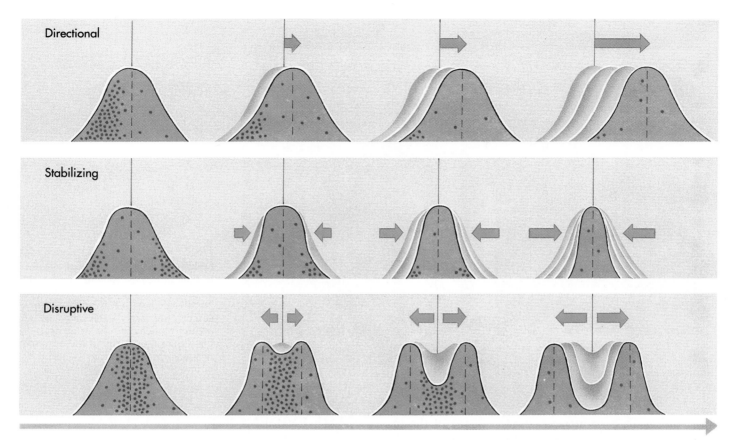

**Figure 15-7**

**Schematic representation of the three different kinds of natural selection acting on a trait, such as height, that varies in a population.** In this diagram, dots represent individuals that do not contribute to the next generation. The curves represent measurements from the trait taken on each individual in the population. All three kinds of natural selection have the same starting point, and the three series show the way that selection alters the distribution of the characteristics as time passes, moving to the right.

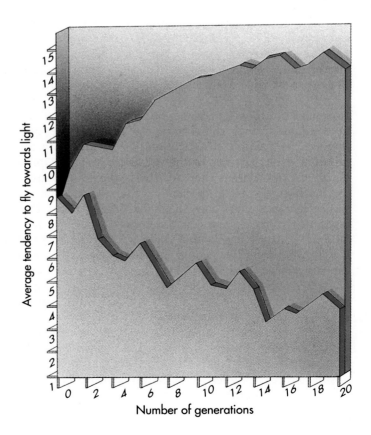

**Figure 15-8**

**Microevolution in action.** In generation after generation (expressed along the bottom of the graph), individuals of the fly *Drosophila* were selected for their tendency to fly toward light more strongly than usual (top curve); the bottom curve represent flies selected *against* the tendency to fly toward light. When flies with a strong tendency to fly toward light were used as the parents for the next generation, their offspring had a greater tendency to fly toward light.

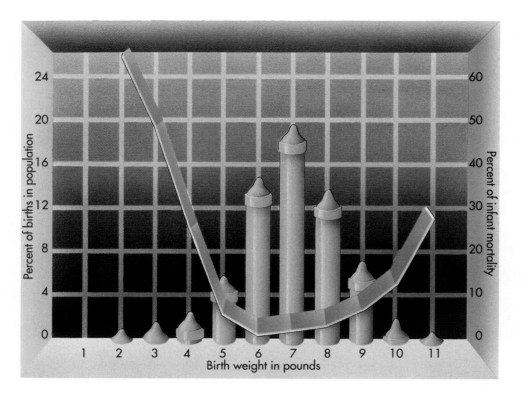

*Figure 15-9*

**Stabilizing selection for birth weight in human beings.** As you can see, the death rate among babies is lowest at an intermediate birth weight; both smaller and larger babies have a greater tendency to die at or near birth than those around the optimum weight, which is between 7 and 8 pounds.

7, *B*), the result is to increase the frequency of the intermediate type, which is already the most common. In effect, selection is operating to prevent change away from this middle range of values. In humans, infants with intermediate weight at birth have the highest survival rate (Figure 15-9). In ducks and chickens, eggs of intermediate weight have the highest hatching success. This form of selection is called **stabilizing selection.**

*Disruptive Selection.* In some situations, selection acts to eliminate rather than favor the intermediate type (Figure 15-7, *C*). A clear example is provided by the different color patterns of the African butterfly *Papilio dardanus.* In different parts of Africa, the color pattern of this butterfly is dramatically different, in each instance being a close copy of some other species that birds don't like to eat (*P. dardanus* is said to be a "mimic"). Any patterns that did not resemble those of distasteful butterflies would be detected readily and eaten by birds, and so any intermediate patterns that occur are eliminated. In this case, selection is acting to eliminate the intermediate phenotypes. This form of selection is called **disruptive selection.**

   Selection often acts against extreme phenotypes and operates on all genes influencing a phenotype; selection against one extreme moves the trait in the opposite direction, whereas selection against both extremes reinforces the existing phenotype.

## THE EVIDENCE THAT NATURAL SELECTION EXPLAINS MICROEVOLUTION

For a century, biologists have studied the genetic variation that occurs in nature to see if it is adaptive, as Darwin's theory would suggest. Much of it clearly is adaptive. Two of the best-studied examples are variation in human hemoglobin and variation in the darkness of moths.

### Sickle-Cell Anemia

Sickle-cell anemia is a hereditary disease affecting hemoglobin molecules in the blood. Sickle-cell anemia was first reported by a Chicago physician, James B. Herrick, in 1910. A West Indies black student exhibited symptoms of severe anemia that appeared related to abnormal red blood cells: "The shape of the red cells was very irregular, but what especially attracted attention was the large number of thin, elongated, sickle-shaped and crescent-shaped forms." The disease was soon found to be common among American blacks. In Chapter 11, we noted that this disorder, which affects roughly 2 American blacks out of every 1000, is associated with a particular recessive allele. Using the Hardy-Weinberg equation, you can calculate the frequency with which the sickle-cell allele occurs in the American black population; this frequency is the

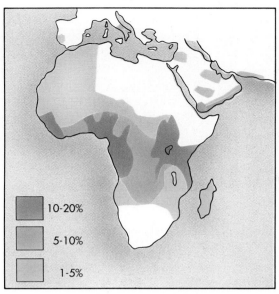

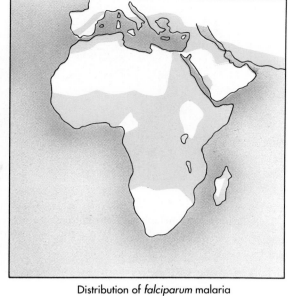

**Figure 15-10**

**Frequency of sickle-cell allele** *(left),* **and distribution of** **falciparan malaria** *(right).* Falciparan malaria is one of the most devastating kinds of the disease; its distribution in Africa is closely correlated with that of the allele for the sickle-cell characteristic.

10-20%

5-10%

1-5%

Frequency of sickle cell gene

Distribution of *falciparum* malaria

square root of 0.002, or approximately 0.045. In contrast, the frequency of the allele among American whites is only about 0.001.

Sickle-cell anemia is usually fatal. The disease occurs because of a single amino acid change. In this change, a valine is substituted for the usual glutamic acid at a location on the surface of hemoglobin near the oxygen-binding site. Unlike glutamic acid, valine is nonpolar (water-hating), and its presence on the surface of the molecule creates a "sticky" patch that will try to escape from the polar water environment by binding to another similar patch. As long as oxygen is bound to the hemoglobin molecule, there is no problem, because the oxygen atoms shield the critical area. But when oxygen levels fall, such as after exercise, then oxygen is not so readily bound to hemoglobin, and the exposed sticky patch binds to similar patches on other molecules, eventually producing long, fibrous clumps. The result is a sickle-shaped red blood cell.

Individuals who are heterozygous for the dominant, valine-specifying allele (designated allele *S*) are said to possess the sickle-cell trait. They produce some sickle-shaped red blood cells, but only 2% of the level seen in homozygous individuals.

The average incidence of the *S* allele in West Africa is about 0.12, a far higher value than that found among American blacks. From the Hardy-Weinberg principle, you can calculate that 1 in 5 Central African individuals are heterozygous at the *S* allele, and 1 in 100 develops the fatal form of the disorder. People who are homozygous for the sickle-cell allele almost never reproduce because they usually die before they reach reproductive age. Why is the *S* allele not eliminated from Central Africa by selection, rather than being maintained at such high levels? Because people who are heterozygous for the sickle-cell allele are much less

susceptible to malaria, which is one of the leading causes of illness and death in areas where the allele is common. Also, for reasons that are not understood, women who are heterozygous for this allele are more fertile than are those who lack it. Consequently, even though most people who are homozygous-recessive die before they have children, the sickle-cell allele is maintained at high levels in these populations because of its role in resistance to malaria in heterozygotes and its association with increased fertility in female heterozygotes.

As Darwin's theory predicts, it is the environment that acts to maintain the sickle-cell allele at high frequency. In this case the characteristic of the environment of Central Africa that is exercising selection is the presence of malaria. For the people living in areas where malaria is frequent, maintaining a certain level of the sickle-cell allele in the populations has adaptive value (Figure 15-10). Among American blacks, many of whom have lived for some 15 generations in a country where malaria is now essentially absent, the environment does not place a premium on resistance to malaria, so there is no adaptive value to counterbalance the ill effects of the disease; in this nonmalaria environment, selection is acting to eliminate the *S* allele.

## Peppered Moths and Industrial Melanism

The peppered moth, *Biston betularia,* is a European moth that rests on the trunks of trees during the day (Figure 15-11). Until the mid-nineteenth century, almost every individual of this species that was captured had light-colored wings. From that time on, individuals with dark-colored wings increased in frequency in the moth populations near industrialized centers until they composed almost 100% of those populations. The black individuals have a dominant allele that was present in populations before 1850, although it was

*Figure 15-11*

**Industrial melanism.** The photos show color variants of the peppered moth, *Biston betularia*. The dark moth is more easily seen in the photo on the left. The light moth in the photo on the right is more easily seen on the darker bark, which has been affected by industrial pollution.

very rare then. Biologists soon noticed that in industrialized regions where the dark moths were common, the tree trunks were darkened almost black by the soot of pollution, and the dark moths were much less conspicuous resting on them than were the light moths. In addition, the air pollution that was spreading in the industrialized regions had killed many of the light-colored lichens that had occurred on the trunks of the trees earlier, thus making these trunks even darker than they would have been otherwise.

Can Darwin's theory explain the increase in frequency of the dark allele? Why was it an advantage for the dark moths to be less conspicuous? Although initially there was no evidence, the ecologist H.B.D. Kettlewell hypothesized that the moths were eaten by birds while they were resting on the trunks of the trees during the day. He tested the hypothesis by rearing populations of peppered moths in which dark and light individuals were evenly mixed. Kettlewell then released these populations into two sets of woods: one, near Birmingham, was heavily polluted; the other, in Dorset, was unpolluted. Kettlewell set up rings of traps around the woods to see how many of both kinds of moths survived (Figure 15-12). To be able to evaluate his results, he had marked the moths that he had released with a dot of paint on the underside of their wings, where it could not be seen by the birds.

In the polluted area near Birmingham, Kettlewell trapped 19% of the light moths and 40% of the dark ones. This indicated that the dark moths had had a far better chance of surviving in the polluted woods, where the tree trunks were dark. In the relatively unpolluted Dorset woods, Kettlewell recovered 12.5% of the light moths but only 6% of the dark ones. These results indicated that where the trunks of the trees were still light-colored, the light moths had a better chance of survival than the dark ones. He later solidified his argument by placing hidden blinds in the woods and filming birds eating the moths. The birds so observed sometimes passed right over a moth that was of the "correct" color and thus well concealed.

**Industrial melanism** is a phrase used to describe the evolutionary process in which initially light-colored organisms become dark as a result of natural selection. The process, which is common among moths that rest on tree trunks, takes place because the dark organisms are better concealed from their predators in habitats that have been darkened by soot and other forms of industrial pollution.

Dozens of other species of moths have changed in the same way as the peppered moth in industrialized areas throughout Eurasia and North America, with dark forms becoming more common from the mid-nineteenth century onward as industrialization spread. In the second half of the twentieth century, with the widespread implementation of pollution controls, the trends are being reversed, not only for the peppered moth in many areas in England, but also for many other species of moths throughout the northern continents. Such examples provide some of the best documented instances of changes in allelic frequencies of natural populations because of natural selection in relation to specific factors in the environment.

## An Overview of Microevolution

These case histories of sickle-cell anemia and industrial melanism are but two of many well-documented cases of adaptation that provide clear evidence that microevolutionary changes can be produced by natural selection. All of these cases share the same fundamental characteristic: changes occur in the frequencies of alleles in a population that alters the characteristics of the population to make it better adapted to the environment in which it is living. In every case, it is the *environment* that dictates the direction and extent of the change. Just as Darwin's theory demands, it is the nature of the environment that leads to natural selection and so determines the direction of evolutionary change.

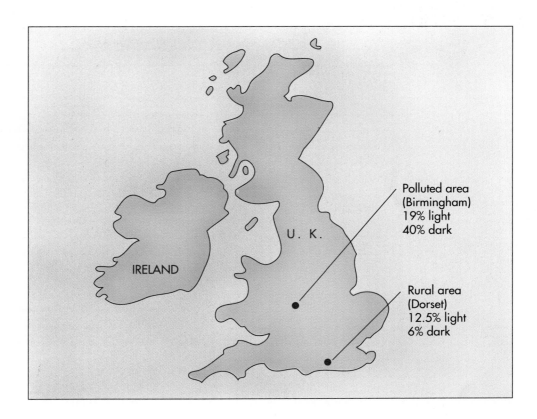

**Figure 15-12**

Kettlewell's industrial melanism experiment.

The map shows:

Polluted area (Birmingham)
19% light
40% dark

Rural area (Dorset)
12.5% light
6% dark

IRELAND

U. K.

# THE EVIDENCE FOR MACROEVOLUTION

Adaptation within natural populations such as those we have just described constitutes strong evidence that Darwin was right in arguing that selection could bring about genetic change within populations (Table 15-2). The examples we reviewed offer direct and compelling evidence of microevolutionary change. Because we can see the evolution as it occurs, we know that Darwin was correct. But what of macroevolution? What is the evidence that macroevolution has led to the diversity of life on earth?

In Chapter 1 we outlined the evidence that Darwin presented in favor of the hypothesis that macroevolution had occurred, a view that had become generally accepted by the time later scientists carried out his studies, but for which he was the first to provide an explanation of the mechanism. A great deal of additional evidence has accumulated since then, much of it far stronger than that available to Darwin and his contemporaries. Among the many lines of available evidence, we review seven.

## The Fossil Record

The most direct evidence of macroevolution is to be found in the fossil record; we now have a far more complete understanding of this record than was available in Darwin's time. Fossils are created when organisms become buried in sediment, the calcium in bone and other hard tissue is mineralized, and the sediment eventually is converted to rock. Sedimentary layers of rock reveal a history of life on earth in the fossils they contain.

By dating the rocks in which fossils occur, we can get a very accurate idea of how old the fossils are. In Darwin's day, rocks were dated solely by their position with respect to one another; rocks in deeper strata are generally older. Knowing the relative positions of sedimentary rocks, and the rates of erosion of different kinds of sedimentary rocks in different environments, geologists of the nineteenth century had achieved a fairly accurate idea of the relative ages of rocks.

More recently, much more accurate ways of dating rocks have been derived, and these provide dates that are absolute rather than relative. Currently, a rock is dated by measuring the degree to which certain radioisotopes that it contains have decayed since the rock formed; the older the rock, the more its isotopes will have decayed. Because radioactive isotopes decay at a constant rate that is not altered by temperature or pressure, the isotopes in a rock act as an internal clock, measuring the time since the rock was formed. For fossils less than 30,000 years old, the decay of carbon 14 (half-life—the time it takes for half of a sample of a

**TABLE 15-2    EXAMPLES OF THE EVIDENCE FOR EVOLUTION**

| | |
|---|---|
| The fossil record | When fossils are arrayed in the order of their age, a progressive series of changes is seen. |
| The molecular record | The longer organisms have been separated according to the fossil record, the more differences are seen in their DNAs. |
| Homology | All vertebrates contain a similar pattern of organs, suggesting that they are related to one another. |
| Development | During development, humans exhibit characteristics of other vertebrates, which suggests that humans are related to the other forms. |
| Vestigial structures | Many vertebrates contain structures that have no function but that resemble functional structures of other vertebrates; this suggests that the structures are inherited from a common ancestor. |
| Parallel adaptation | The marsupials in Australia closely resemble the placental mammals of the rest of the world, which suggests that parallel selection has occurred. |
| Patterns of distribution | Inhabitants of ocean islands resemble forms of the nearest mainland but show some differences, which suggests that they have evolved from mainland migrants. |

given size to change—of 5568 years) is used, whereas for older fossils, investigators examine the decay of radioactive potassium 40 into argon and calcium (half-life of 1.3 billion years). For very old fossils, a third isotope measure, the decay of uranium 238 into lead (half-life 4.5 billion years) is used. An investigator has only to measure the proportion of uranium 238 to lead in the rock to estimate its age.

When fossils are arrayed according to their age, from oldest to youngest, they provide evidence of progressive evolutionary change in the direction of greater complexity. Among the hoofed mammals illustrated in Figure 15-13, small, bony bumps on the nose can be seen to have changed progressively until they became large, blunt horns. In the evolution of horses (see Figure 1-19), the number of toes on the front foot has gradually reduced from four to one. About 200 million years ago, oysters underwent a change from small, curved shells to larger, flatter ones, with progressively flatter fossils being seen in the fossil record over a period of 12 million years (Figure 15-14). A host of other examples are known, all illustrating a record of progressive change. The demonstration of this progressive change is one of the strongest lines of evidence that evolution has occurred.

### The Molecular Record

If you think about it, the fact that organisms have evolved progressively from relatively simple ancestors implies that a record of evolutionary change is present in the cells of each of us, in our DNA. According to

evolutionary theory, every evolutionary change involves the substitution of new versions of genes for old ones, the new arising from the old by mutation and coming to predominance because of favorable selection. Thus a series of evolutionary changes involves a progressive accumulation of genetic change in the DNA. Organisms that are more distantly related will accumulate a greater number of evolutionary differences. This is indeed what is seen when DNA sequences are compared between various organisms. For example, the longer the time since the organisms diverged, the greater the number of differences in the nucleotide sequence of the gene for cytochrome *c*, a protein that plays a key role in oxidative metabolism (Figure 15-15). Again we see that evolutionary history involves a pattern of progressive change.

Some genes, such as the ones specifying the protein hemoglobin, have been particularly well studied, and the entire time course of their evolution can be laid out with confidence by tracing the origin of particular substitutions in their nucleotide sequences (Figure 15-16). The pattern of descent that is obtained is called a *phylogenetic tree*. It represents the evolutionary history of the gene. Note that the progressive changes seen in the hemoglobin molecule produce a tree that reflects precisely the evolutionary relationships predicted by study of anatomy. Whales, dolphins, and porpoises cluster together, as do the primates and the hoofed animals. The pattern of progressive change seen in the molecular record constitutes very strong direct evidence for macroevolution.

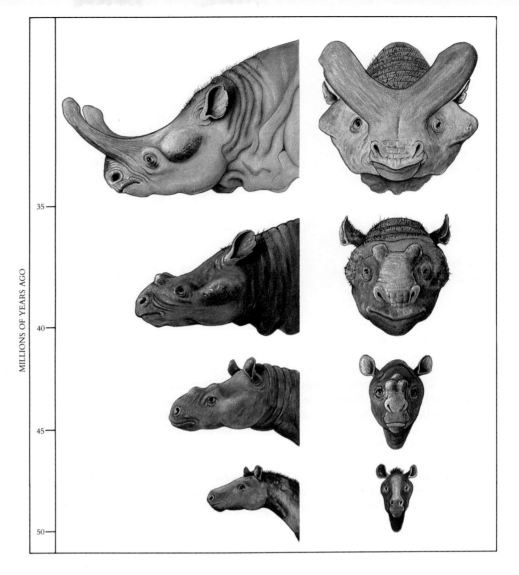

MILLIONS OF YEARS AGO

35

40

45

50

**Figure 15-13**

**Macroevolution.** Evolution in a group of hoofed mammals known as titanotheres between the Early Eocene Epoch (about 50 million years ago) and the Early Oligocene Epoch (about 36 million years ago). During this period of time, small, bony protuberances evolved into relatively large, blunt horns.

## Homology

A third demonstration of the process of macroevolution lies in the fact that many organisms exhibit structures that appear to have been derived from a common ancestral form. The forelimbs of all mammals contain the same pattern of bones, although they now carry out a variety of functions (see Figure 1-20). All vertebrates have the same pattern of bones, muscles, nerves, blood circulation, and organs, the pattern becoming gradually more complex as one moves from the fishes to amphibians to reptiles to mammals. It is difficult to avoid the conclusion that progressive change is taking place.

## Development

In many cases the evolutionary history of an organism can be seen to unfold during its development, with the embryo exhibiting characteristics of the embryos of its ancestors. For example, early in their development human fetuses possess gill slits like those of fish and later exhibit a tail, the vestige of which we carry to adulthood as the coccyx at the end of our spine. Human embryos even possess a fine fur (called *lanugo*) during the fifth month of development. These relict developmental features argue strongly that our development has evolved, with new instructions being layered on top of old ones and the overall developmental program getting progressively longer. Some vertebrate embryos are shown in Figure 15-17.

## Vestigial Structures

Many organisms possess structures with no apparent function that resemble the structures of presumed ancestors. You possess a complete set of muscles for wiggling your ears, just as a coyote does. Figure 15-18 illustrates the skeleton of a baleen whale, a representative of the group that contains the largest living mammals. It contains a pelvic bone just as do other mammals, even though such a bone serves no known function in the whale. Figure 20-9, *B* illustrates the hu-

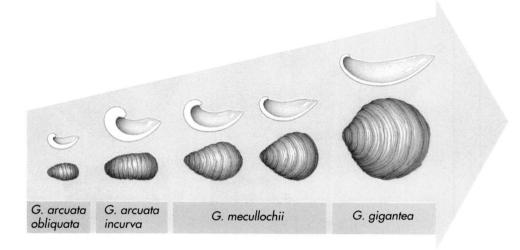

G. arcuata
obliquata | G. arcuata
incurva | G. mecullochii | G. gigantea

**Figure 15-14**

**Evolution of shell shape in oysters.** The apparently gradual evolution of a group of coiled oysters during a portion of the Early Jurassic Period, which lasted about 12 million years. During this interval, the shell became larger, thinner, and flatter. These animals rested on the ocean floor in the position known as the "life position," and it may be that the larger, flatter shells were more stable against potentially disruptive water movements.

man vermiform appendix, a hollow, wormlike appendage of the caecum, or sac, in which the large intestine begins. The vermiform appendix apparently is vestigial and represents the degenerate terminal part of the caecum. Although some suggestions have been made, it is difficult to assign any current function to the vermiform appendix. In many respects, it is a dangerous organ: quite often it becomes infected, leading to an inflammation called *appendicitis*. If it is not removed surgically when inflamed, the vermiform appendix may burst, allowing the contents of the gut to come in con-

tact with the lining of the body cavity. This condition can be fatal if unchecked. It is difficult to understand vestigial structures such as these in any way other than as evolutionary relics, holdovers from the evolutionary past. They argue strongly for the common ancestry of the members of the groups that share them, regardless of how different they have become subsequently.

## Parallel Adaptation

The plant and animal communities that occur in widely separated areas are often similar, especially if the areas

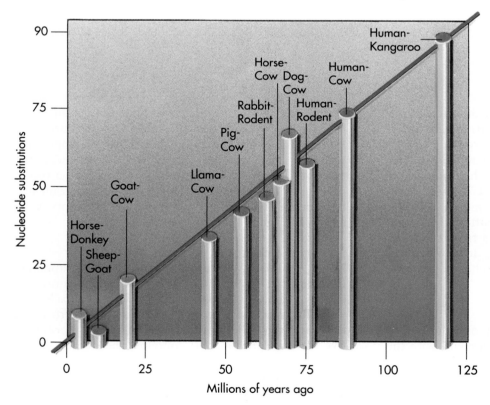

**Figure 15-15**

**The evolution of cytochrome *c*.** Comparing various pairs of organisms, investigators have counted the numbers of nucleotides in the cytochrome *c* genes that are not the same. When the number of such "substitutions" is plotted against the time evolutionists believe has elapsed since the pair of organisms diverged, a straight line is obtained. This constant rate of nucleotide substitution suggests that the cytochrome c gene is evolving at a constant rate.

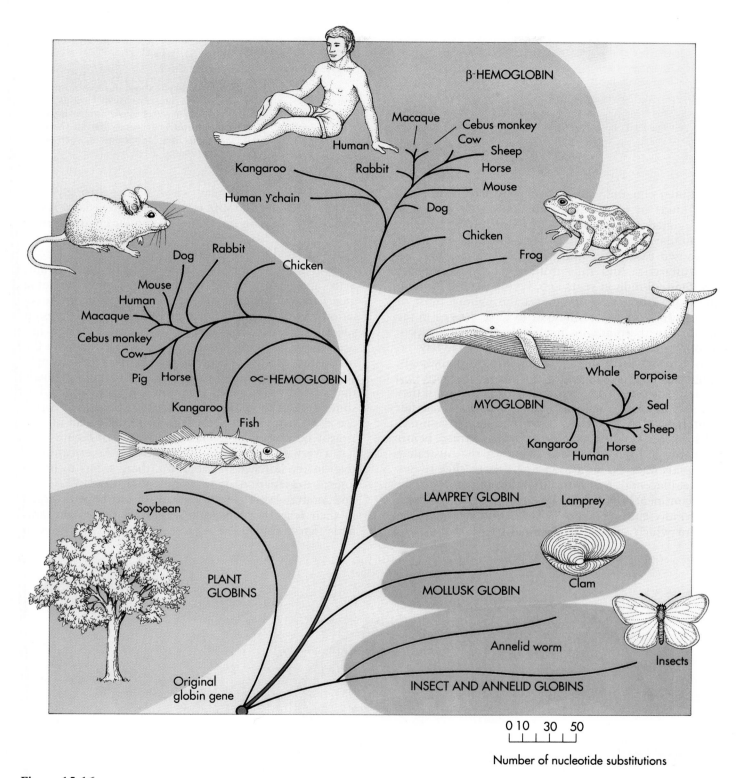

**Figure 15-16**

**Evolution of the globin gene.** The length of the various lines is relative to the number of nucleotide substitutions in the gene.

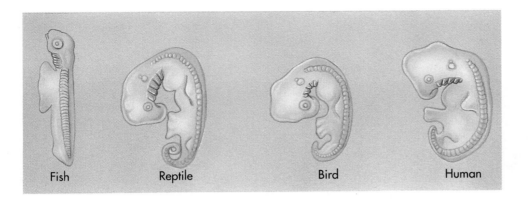

Fish　　　　Reptile　　　　Bird　　　　Human

**Figure 15-17**

**Our embryos show our evolutionary history.** The embryos of various groups of vertebrate animals showing the primitive features that all share early in their development, such as gill slits and a tail.

have similar climates. This may be true even if the individual kinds of plants and animals may be only distantly related to one other. It is difficult to explain these similarities as resulting from coincidence. In the best-known case, the continent of Australia separated from Antarctica and thus from the other continents more than 50 million years ago, before placental mammals—the group that dominates throughout most of the modern world—are thought to have arrived in the area. Today the placental mammals of Australia consist only of a few groups that arrived relatively recently. Most Australian mammals are marsupials, members of a group in which the young are born in a very immature condition and held in a pouch until they are ready to emerge into the outside world. Marsupials might have evolved earlier than placental mammals, and seem certainly to have arrived in Australia before its separation from Antarctica. What are the Australian marsupials like? To an astonishing degree, they resemble the placental mammals that are present on the other continents (Figure 15-19). The similarity of some of the individual members of these two sets of mammals, in which specific kinds have similar habits and find their food in similar ways, argues strongly that they have evolved in different, isolated areas as a result of natural selection in relation to similar environments.

## Patterns of Distribution

Darwin was the first to present evidence that the animals and plants living on oceanic islands resemble most closely the forms of the nearest continent—a relationship that would not make sense if they were all specially created. This kind of relationship, which has been observed many times since Darwin's time with the increasing exploration of the earth's surface, strongly suggests that the island forms evolved from individuals that came to the islands from the adjacent mainland at some time in the past. In many cases, the island forms are *not* identical with those which still occur on the nearby continents. The Galapagos finch of Figure 1-23 has a very different beak than its South American relative. In the absence of evolution, there seems to be no logical explanation of why individual kinds of plants and animals were clearly related to, but have diverged in their features from, other kinds of plants and animals that occur on the adjacent mainlands. As Darwin

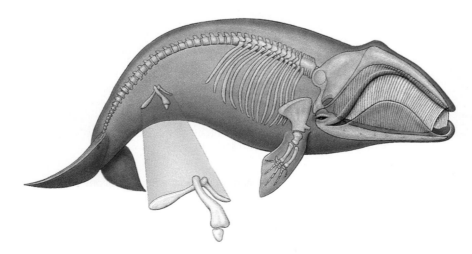

**Figure 15-18**

**Vestigial features.** The skeleton of a baleen whale, a representative of the group of mammals that contains the largest living species. The enlargement shows the pelvic bones, which resemble those of other mammals, but which are only weakly developed in the whale and have no apparent function.

Placental mammals

Marsupial mammals

Mole

Marsupial mole

Anteater

Numbat (anteater)

Mouse

Marsupial mouse

Lemur

Spotted cuscus

Flying squirrel

Flying phalanger

Bobcat

Tasmanian "tiger cat"

Wolf

Tasmanian wolf

**Figure 15-19**

The parallel adaptation of marsupials in Australia and placental mammals in the rest of the world.

pointed out, this relationship provides strong evidence that macroevolution has occurred.

In sum total, the evidence for macroevolution is overwhelming. Almost all biologists would agree both that (1) macroevolutionary changes have occurred, and (2) microevolutionary changes result from natural selection. In the next chapter, we consider Darwin's proposal that microevolutionary changes have led directly to macroevolutionary ones, the key argument in his theory that evolution occurs by natural selection.

## THE TEMPO AND MODE OF EVOLUTION

On the basis of a relatively complete fossil record, it has been estimated that most living species of mammals go back at least 100,000 years, almost none a million years. A good average value for the life of a mammal species might be about 200,000 years. Mammal genera, on the other hand, seem to persist several million years on the average. The American paleontologist George Gaylord Simpson pointed out that certain other groups, such as the lungfishes (Figure 15-20) are apparently evolving much more slowly than are the mammals. In fact, Simpson estimated that there have been few evolutionary changes among the lungfishes over the past 150 million years, and even slower rates of evolution are known in certain other groups.

Not only does the tempo, or rate, of evolution differ greatly from group to group, but groups apparently have rapid and relatively slow periods in their evolution. The fossil record provides evidence for such variability in evolutionary rate, and evolutionists are anxious to understand the environmental and other factors that account for it. In 1972 the paleontologists Niles Eldredge of the American Museum of Natural History in New York and Stephen Jay Gould of Harvard University proposed that it was the norm of evolution to proceed in spurts. They claimed that the process of evolution includes a series of **punctuated equilibria.** Evolutionary innovations would occur and give rise to new lines, then these lines might persist unchanged for a very long time. Eventually there would be a new spurt of evolution, creating a "punctuation" in the fossil record (Figure 15-21).

Eldredge and Gould proposed that rapid evolution would usually occur when populations were small, possibly different from their parent populations as a result of the founder effect, and still local enough for rapid adaptation to novel ecological circumstances. In contrast, **stasis,** or lack of evolutionary change, would be expected for large populations under diverse and conflicting selective pressures.

Unfortunately, the distinctions are not as clear-cut as implied by this discussion. The fossil record is seriously incomplete because of changes in the conditions under which fossils are deposited, and the interpretation of many of its "gaps" is problematical. Notwithstanding these difficulties, the punctuated equilibrium model has provided a very useful perspective for considering the mode of evolution and will continue to do so in the future. Eldredge and Gould contrasted their theory of punctuated equilibrium with that of **gradualism,** or gradual evolutionary change, which they claimed was what Darwin and most earlier students of evolution had considered normal. Whether or not Darwin and other evolutionists actually embraced gradualism is debatable.

*The punctuated equilibrium model assumes that evolution occurs in spurts, between which there are long periods in which there is little evolutionary change. The gradualism model assumes that evolution proceeds gradually, with progressive change in a given evolutionary line.*

**Figure 15-20**

**A lungfish.** The lungfish is a member of a group of vertebrates that has changed very little during the past 150 million years.

## SCIENTIFIC CREATIONISM

In the century since he proposed it, Darwin's theory of evolution by natural selection has become almost universally accepted by biologists as the best available explanation of biological diversity. Its predictions have

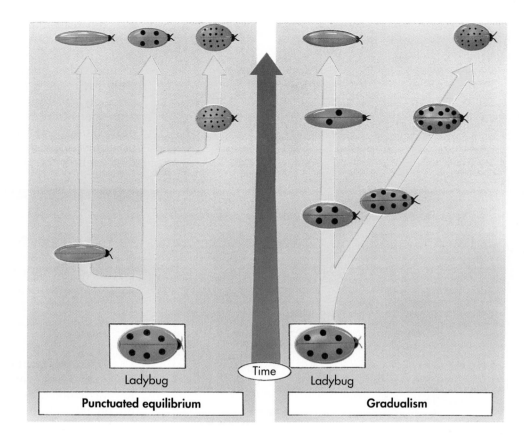

**Figure 15-21**

**Two views of the pace of macroevolution.** Punctuated equilibrium surmises that species formation occurs in bursts, separated by long periods of quiet, whereas gradualism surmises that species formation is constantly occurring.

been supported by the experiments and observations of generations of scientists. The fact that evolution has occurred and is occurring is as well accepted as is the existence of gravity, and its operation no more doubted than predictions that a dropped apple will fall or that the sun will rise tomorrow morning.

But evolution is not the only way in which the diversity of life on earth can be explained. The argument advanced in Darwin's day, that the Bible provides the correct and literal explanation of biological diversity, is still widely accepted by many people today, people who prefer a religious to a scientific explanation. Now, as in Darwin's time, the scientific perspective is only one of many worldviews. It is important in this regard to understand the limits of science. What science provides is a coherent means of organizing observations, of making predictions about how the world is going to behave. It is not a substitute for religion, which addresses a different arena of human concerns: questions of ethics and of ultimate causes. Religion and science do not preclude one another, but are regarded by many as complementary ways of viewing the world.

The clear distinction between science and religion sometimes gets muddled (Figure 15-22). Thus a number of individuals, mainly in the United States and starting largely in the 1970s, have put forward a view

they title **scientific creationism,** which holds that the biblical account of the origin of the earth is literally true, that the earth is much younger than most scientists believe, and that all species of organisms were individually created just as they are today. Scientific creationists are arguing in the courts that their views should be taught alongside evolution in classrooms. If both evolution and scientific creationism provide scientific explanations of biological diversity, they argue, then teachers have an obligation to present *both* views, taught side-by-side, so that students can choose knowledgeably between them.

This does not seem to be a bad argument if you accept the premise, which is that the view of the scientific creationists is indeed scientific. The confusion is not in the beliefs of the scientific creationists, which are religious beliefs that many people hold, but rather in their labeling of these beliefs as scientific. There is no scientific evidence to support the hypothesis that the earth is only a few thousand years old and none that indicates that every species of organism was created separately. These conclusions can be reached only on the basis of arbitrary faith; they are untestable, and, as such, they lie outside of the realm of science.

Science, as represented by the observations of scientists, has come to different conclusions. There are

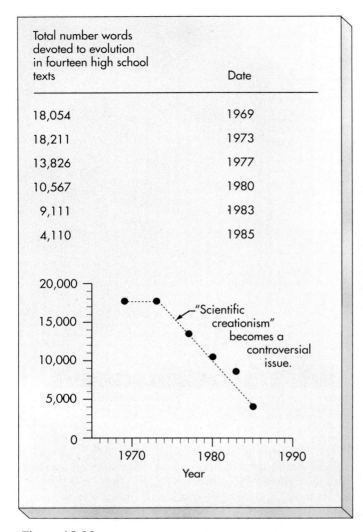

| Total number words devoted to evolution in fourteen high school texts | Date |
|---|---|
| 18,054 | 1969 |
| 18,211 | 1973 |
| 13,826 | 1977 |
| 10,567 | 1980 |
| 9,111 | 1983 |
| 4,110 | 1985 |

"Scientific creationism" becomes a controversial issue.

**Figure 15-22**

**Creationism loses the battle in courts, but wins in the schools.** Textbook publishers, wary of controversy, have been removing evolution from the schoolbooks.

virtually no differences of opinion among biologists, and indeed nearly all scientists, on the following major points: (1) the earth is about 4.5 billion years old; (2) organisms have inhabited it for the greater part of that time; and (3) all living things, including human beings, have arisen from earlier, simpler living things. The antiquity of the earth and the role the process of evolution played in the production of all organisms, living and extinct, is accepted by virtually all scientists.

In addition to the fact that it contradicts the considered judgment of almost all scientists, there is an even more fundamental problem associated with labeling scientific creationism as science: scientific creationism implicitly denies the intellectual basis of science it-

self, the reasoning on which the operation of science depends. Science insists on acceptance of the most predictive explanation of biological diversity, which is evolution, whereas scientific creationism says, "Yes, but God just made it look that way." Perhaps so, but this is simply substituting a religious argument for a scientific one. Science consists of inferring principle from observation, and when faith is substituted for observation, the conclusion is not science. It is in just this sense that scientific creationism is not science.

*Scientific creationism should not be labeled science for three reasons: (1) it is not supported by any scientific observations; (2) it does not infer its principles from observation, as does all science; and (3) its assumptions lead to no testable and falsifiable hypotheses.*

Scientific creationism implicitly denies the whole intellectual basis for the set of facts than human beings have assembled over the centuries about the nature of life on earth. It implies that a creator has made a world in which all phenomena are deceptively consistent with a great age for the earth and an evolutionary relationship between organisms. Evolutionary theory, in contrast, provides a coherent scientific explanation for the nature of the world in which we live and enables us to make predictions about that world. Certainly there is ample controversy among serious students as to the details of how evolution has occurred, just as there is controversy in every active scientific field. But there is no controversy about Darwin's basic finding that natural selection has played and is continuing to play the central role in the process of evolution.

The future of the human race depends largely on our collective ability to deal with the science of biology and all the phenomena that it comprises. We need the information that we have gained to deal with the problems, challenges, and uncertainties of the world in an appropriate way. We cannot afford to discard the advantages that this knowledge gives us because some people wish to do so as an act of what they construe as religious faith. Instead, we must use all of the knowledge that we are able to gain for our common benefit. With its help, we can come to understand ourselves and our potentials better. In no way should such rational behavior be taken as denial of the existence of a supreme being; it should rather be considered by those who do have religious faith as a sign that they are using their God-given gifts to reason and to understand.

# ■ SUMMARY

1. Macroevolution describes the grand outlines of evolution, which is based on the evolution of species. Microevolution, also called adaptation, refers to the evolutionary process itself. Adaptation leads to species formation and thus ultimately to macroevolution.

2. Population genetics is the branch of genetics that deals with the behavior of genes in populations. The Hardy-Weinberg principle, the baseline for all population genetic theory, illustrates the fact that in large populations with random mating, allele and genotype frequencies—and consequently phenotype frequencies—will remain constant indefinitely, provided that selection, net mutation or migration in one direction, inbreeding, and genetic drift do not occur.

3. Some organisms leave more offspring than do competing ones, and their genetic traits tend to appear in greater proportions among members of succeeding generations than do the traits of those individuals which leave fewer offspring. This process is called selection.

4. There is clear evidence of microevolutionary change in natural populations. For example, the disease called sickle-cell anemia occurs when an altered hemoglobin allele is present in homozygous form; it is almost invariably lethal. However, if the allele is present in heterozygous form, it not only does not produce anemia, but it also confers resistance to malaria and increased female fertility on heterozygous persons. For these reasons the allele has reached high frequencies in African populations that live in areas where malaria occurs frequently.

5. The two direct lines of evidence that argue that macroevolution has occurred are (1) the fossil record, which exhibits a record of progressive change correlated with age, and (2) the molecular record, which exhibits a record of accumulated changes, the amount of change correlated with age as determined in the fossil record.

6. Several indirect lines of evidence argue that macroevolution has occurred, including progressive changes in homologous structures, relict and vestigial developmental structures, parallel patterns of evolution, and patterns of biogeographic distribution.

# REVIEW

1. Genetic change within populations, which is the result of adaptation, is called _____.

2. Small, randomly mating populations of a species that are isolated from one another can come to differ greatly in their characteristics. In the absence of net immigration or mutation, this may occur as a result of _____ or of _____.

3. Which of the five mechanisms for producing deviations from Hardy-Weinberg allele frequencies is the most important in evolution?

4. _____ is a term used to describe the evolutionary process in which a population of predominantly light-colored individuals is gradually dominated by dark-colored individuals as a result of natural selection. The best-studied example of this is the peppered moth, *Biston betularia*.

5. What kind of evidence for evolution is illustrated by the striking similarity of the fossil marsupial sabertooth and the placental sabertooth?

# SELF-QUIZ

1. Which of the following is *not* a factor that causes change in the proportions of homozygous and heterozygous individuals in a population?
   - (a) Mutation
   - (b) Migration
   - (c) Genetic drift
   - (d) Random mating
   - (e) Selection against a specific allele

2. If you came across a population of plants and discovered a surprisingly high level of heterozygosity, what would you predict about their mating system?
   - (a) Their pollen is dispersed by wind.
   - (b) Their pollen is spread by insects that visit the flowers.
   - (c) They probably reproduce asexually.
   - (d) They are predominantly out-crossing.
   - (e) They are predominantly self-fertilizing.

3. The sickle-cell *(S)* allele has a relatively high frequency (0.12) in West Africa, even though individuals homozygous for this allele usually die before they reach reproductive age. Why has this allele persisted in the population in high frequencies when there appears to be such strong natural selection against it?
   - (a) Because heterozygous individuals are resistant to malaria.
   - (b) Because individuals homozygous for this characteristic are resistant to malaria.
   - (c) Because females heterozygous for this allele are more fertile than are those who lack it.
   - (d) Because females homozygous for this allele are more fertile than are those who lack it.
   - (e) Both a and c.

4. Successful adaptation is defined simply as
   - (a) the outcome of selection.
   - (b) moving to a new location.
   - (c) producing offspring.
   - (d) evolving new traits.
   - (e) living longer.

5. If the population of moths in a woods contained higher and higher proportions of light-colored individuals in subsequent years, which of the following could be considered plausible explanations?
   - (a) The woods were becoming less polluted.
   - (b) The allele or alleles associated with the lighter color were linked with other alleles that brought about pesticide resistance.
   - (c) Birds were becoming less frequent, so the color didn't matter.
   - (d) a and b.
   - (e) All of the above.

# THOUGHT QUESTIONS

1. In Central Africa there exists in low frequency a third hemoglobin allele called *C*, in addition to the *A* and *S* alleles discussed in this chapter. Individuals that are heterozygous for *C* and the normal allele *A* are susceptible to malaria just as are *AA* homozygotes, but *CC* individuals are resistant to malaria—and do *not* develop anemia! Assuming that the Bantu people entered Central Africa relatively recently from land where malaria is not common (we think this is what happened), and that among the original settlers both *C* and *S* alleles were rare, can you suggest a reason that *CC* individuals have not become predominant?

2. The North American human population is similar to the ideal Hardy-Weinberg population in that it is very large (over 270 million people in the United States and Canada) and generally randomly mating. Although mutation occurs, it does not lead by itself to great changes in allele frequencies. However, migration from Latin American and Asian countries occurs at relatively high levels—perhaps 1% per year. The following data were obtained in 1976 by the geneticist A.E. Mourant about relative numbers of individuals bearing the two alleles of what is known as the *MN* blood group:

|  | MM | MN | NN | Total |
|---|---|---|---|---|
| Observed number of individuals | 1787 | 3037 | 1305 | 6129 |

Do these data suggest that migration, selection, or some other factor is acting to perturb the Hardy-Weinberg proportions of the three genotypes?

3. In a large, randomly mating population with no forces acting to change gene frequencies, the frequency of homozygous-recessive individuals for the characteristic of extra-long eyelashes is 90 per 1000, or .09. What percent of the population carries this very desirable trait but displays the dominant phenotype, short eyelashes? Would the frequency of the extra-long–lash allele increase, decrease, or remain the same if long-lashed individuals preferentially mated with each other and no one else?

4. If you found that a series of small islands were each occupied by a distinctive population of land snails, what kind of evidence would you gather to attempt to determine whether the differences had resulted from natural selection or from genetic drift?

# FOR FURTHER READING

CARSON, H.L.: "The Process Whereby Species Originate," *BioScience*, vol. 37, 1987, pages 715-720. One of the modern masters of evolution discusses a fascinating aspect of the origin of species.

FUTUYMA, D.: *Science on Trial: The Case for Evolution*, Pantheon Books, New York, 1983. An excellent exposition of the basic reasons that the creationist argument is flawed by serious errors.

GILKEY, L.: *Creationism on Trial: Evolution and God at Little Rock*, Winston Press, Minneapolis, 1985. Excellent book, written by a theologian, outlining the case for creationism as argued in the courts at Little Rock.

GOULD, S.J.: "Darwinism Defined: The Difference between Fact and Theory," *Discover*, January 1987, pages 64-70. A clear account of what biologists do and do not mean when they refer to the theory of evolution by natural selection.

SIBLEY, C., and J. ALQUIST: "Reconstructing Bird Phylogenies by Comparing DNAs," *Scientific American*, February 1986, pages 82-92. A good introduction to the way in which biologists are beginning to use the tools of molecular biology to answer questions about evolution.

SIMPSON, G.G.: *Fossils and the History of Life*, Scientific American Library, New York, 1983. A short and beautifully illustrated account of how fossils are used to learn about life's evolutionary past by a master of the field.

# *H*ow Species Form

These Alaskan grizzly cubs are watching their mother fish for dinner, a skill each will quickly learn. Other species of bears have different specialities, each suited to where they live.

# HOW SPECIES FORM

*Overview*

To establish that Darwin's theory of evolution is correct, it is not enough to show that evolution has occurred, that the dinosaurs' ancestors were fishes, and that our ancestors were apelike primates. To be certain, it is necessary to *see* evolution in action. Although the replacement of one kind of organism with another usually occurs too slowly to observe within one human lifetime, less drastic changes within individual species can be observed. Many of these changes, when examined by biologists, turn out to reflect the selection of better-adapted kinds of organisms, just as Darwin proposed. The key issue, then, is the evidence that such changes *within* species lead to differences *between* species, to the formation of *new* species.

*For Review*     *Here are some important terms and concepts that you will encounter in this chapter. If you are not familiar with them, you should review them before proceeding.*

**Darwin's studies** (Chapters 1 and 15)

**Meiosis** (Chapter 9)

**Selection** (Chapter 15)

**Adaptation in populations** (Chapter 15)

If he were alive today, Darwin would be amazed at the quantity of the evidence that has accumulated to support his theory that natural selection is the primary mechanism responsible for evolution. He would not have needed convincing; the case he built in his *Origin of Species* is a very strong one. The case for evolution now rests solidly on three pillars of evidence, which respectively establish that (1) natural selection is the agent of microevolutionary change—change within populations; (2) microevolutionary change leads to macroevolutionary change—change in the kinds of plants and animals that occur on earth; and (3) macroevolution has indeed occurred.

In the preceding chapter, we considered the evidence that macroevolution has occurred and that selection is the agent of microevolutionary change. We saw that natural selection drives evolution by favoring genetic variations that better enable organisms to survive and reproduce. Different environments favor different responses from the organisms that survive in them, and there is a continuing process of change in relation to these environments. This is the key lesson of the previous chapter: evolution is not blind; instead, it is directed by the environment. Just as a football coach has the team try a variety of plays but keeps in the team's game plan only those plays which work, so a population of organisms keeps only those changes which "work." The population doesn't decide which changes to keep, any more than the football coach does. The pattern of success determines the outcome.

In this chapter we consider the third body of evidence on which Darwin's theory rests: the evidence that microevolutionary change leads to macroevolutionary change, that adaptive changes *within* a species convert to differences *between* species. This third point is the crux of Darwin's evolutionary argument.

## THE NATURE OF SPECIES

How do adaptive changes in natural populations lead to the origin of species? Darwin was extremely interested in this question because he considered species to be the most important evolutionary units. As we begin to think about the problem, we must first examine what the concept *species* means and how this concept has changed through the years.

John Ray (1627-1705), an English clergyman and scientist, was one of the first to propose a general definition of species. In about 1700 he pointed out how a species could be recognized: all the individuals that belonged to it could breed with one another and produce

**Figure 16-1**

**Artificial selection in action.** All of the different breeds of dogs are members of the same species, *Canis familaris*, and all can breed successfully with one another. The great differences between breeds of dogs were produced by artificial selection carried out by dog breeders.

progeny that were still of that species (Figure 16-1). Even if two different-looking individuals appeared among the progeny of a single mating, they were still considered to belong to the same species. All dogs were one species, all pigeons, and so on; carp, however, were not the same species as goldfish, nor horses the same species as donkeys (Figure 16-2), and so forth.

In an informal way, people had always recognized species. Indeed, the word *species* is simply Latin for *kind*. However, with Ray's observation, the species began to be regarded as an important biological unit that could be catalogued and understood. With other scientists of his time, Ray believed that species were individually created by a supreme being and did not change, a view that was widely held until it was challenged by Darwin in 1859. Darwin was interested in those situations in which it was not clear whether or not he was dealing with distinct species. He considered the fact that individuals that were intermediate in their features between two different species did occur—the species were then said to *intergrade*—to be important evidence

in support of his theory of evolution by natural selection. Darwin explained the relative constancy of species by saying that each had its own distinctive role in nature, a role that we would in modern terms call a *niche*. Thus each species occurs in a particular kind of place, displays different activities at different times of the year, has different habitats, and so forth.

From the 1920s onward, with the emergence of population genetics, there was a desire to define the category species more precisely. The definition that began to emerge was stated by the American evolutionist Ernst Mayr as follows: species were "groups of actually or potentially interbreeding natural populations which are reproductively isolated from other such groups." In other words, hybrids between species occur rarely in nature, whereas individuals that belong to the same species are able to interbreed freely. In fact, there are essentially no barriers to hybridization between the species in many groups of organisms, and strong barriers to hybridization between the species of other groups. In practice, scientists recognize species in dif-

**Figure 16-2**

**The mule, a sterile hybrid between a female horse and a male donkey.** By any standard, the horse and donkey are distinct species; mules, of course, cannot interbreed with either.

ferent groups primarily because they differ from one another in their visible features (Figure 16-3). As a general definition, it has not been possible to improve on the early view that species are kinds of organisms.

Within the units that are classified as species, the populations that occur in different places may be more or less distinct from one another (Figure 16-4). But when such populations occur in the same areas, individuals with intermediate or mixed characteristics usually are frequent; in other words, the distinct-appearing populations within a species usually intergrade with one another when they occur together. In areas where

these different-looking races, which may be classified as subspecies or varieties, approach one another, many individuals may occur in which the distinct features characteristic of each of the intergrading races are combined. Many of these individuals may not match either race in their characteristics. In contrast, when *species* occur together, they usually do not intergrade, although they may hybridize occasionally.

In some groups of organisms, even local races are not capable of interbreeding with one another. This pattern occurs in many annual plants. In contrast, species of trees, some groups of mammals, and fishes generally *are* able to form fertile hybrids with one another, even though they may not do so in nature. For still other kinds of plants and animals, we do not know whether the species can form hybrids. We define a species, therefore, as a group of organisms that is unlike other such groups of organisms and does not intergrade extensively with them in nature.

*Species are groups of organisms that differ in one or more characteristics and do not intergrade extensively if they occur together in nature.*

A          B          C

D          E          F

**Figure 16-3**

**Butterflies.** Distinct species in one genus. (A) American painted lady, *Vanessa virginiensis.* (B) Western painted lady, *Vanessa anabela.* (C) Red admiral, *Vanessa atalanta.* **Flowers.** (D) *Clarkia concinna.* (E) *Clarkia speciosa.* (F) *Clarkia rubicunda.*

**Figure 16-4**

**Extinction in our own backyard.** Subspecies of the seaside sparrow, *Ammodramus maritimus*, are quite local in distribution, and some of them are in danger of extinction because of the alteration of their habitats. The widespread subspecies *Ammodramus maritimus maritimus* (adult, *1*, and juvenile, *2*) is the most common; *A. m. fisheri (3)* occurs along the Gulf Coast. The Cape sable seaside sparrow, *A. m. mirabilis (4)*, occurs in a small area of southwestern Florida. The last subspecies, the dusky seaside sparrow *(5)*, *A. m. nigrescens*, occurred only near Titusville, Florida. The last individual, a male, died in captivity in 1987.

## THE DIVERGENCE OF POPULATIONS

Local populations are usually more or less separated geographically, and the conditions in which they occur are dissimilar. Populations of interbreeding individuals are often extremely small (Figure 16-5), and there may be very limited exchange of individuals, and thus of genetic material, between such populations. For these reasons, local populations often are able to adjust individually and effectively to the demands of their particular environment. As they do so, their characteristics change. The rate at which they change depends primarily on the selective forces to which they are responding. If these forces are strong, the populations will change rapidly; the populations certainly do not need to be isolated on islands to be able to become distinct species.

*The characteristics of populations tend to diverge more or less rapidly depending on the features of their particular environment. Because there is only a limited exchange of genetic material between populations, even if they are geographically close, all populations tend to become increasingly divergent from one another in their characteristics over time.*

Many genes interact in the production of most phenotypic characteristics. Furthermore, all of the developmental processes that occur within a single organism are closely integrated with one another. In addition, every population begins with a different endowment of alleles. Because of these three factors, the response of a given population of organisms to a combination of selective factors tends to be predictable. Even if two populations respond to similar selective forces, and even though they may be geographically close to one another, they still tend to change and to differ more and more from one another over time. If their environments are dissimilar enough or change rapidly enough, the populations may diverge rapidly and strikingly, and soon come to look very different from one another.

## ECOLOGICAL RACES

What kinds of patterns can be expected as a result of the differentiation of populations? One consequence is that the individuals of a species that occur in one part of its range often look different from those which occur elsewhere (see Figure 16-4). Such groups of distinctive individuals are informally called races, and they may, as we have mentioned, be classified as subspecies or varieties. The existence of such races proved fascinating to Darwin because he considered them to be an intermediate stage in the evolution of species. The kinds of ecologically defined races that we discuss first may change over time to the clusters of species that we consider next. Both provide important examples of the evolution of populations in nature.

Ecological races were first studied in detail in plants. As every gardener knows, the same plants may differ greatly in appearance depending on the places where they are grown. This is true even for genetically identical divisions of the same plant—clones. The part of an individual plant that is usually in the sun often produces leaves that are unlike those which the same plant produces in the shade. For example, shade leaves are usually thinner and broader and have more internal air spaces than do sun leaves. Thus it seemed to many botanists in the nineteenth century that environmental factors, rather than genetic differences, might account for many of the differences between races and even between species of plants.

## Ecotypes in Plants

In the 1920s and 1930s the Swedish botanist Göte Turesson perfomed a series of experiments that were designed to test whether differences between the races of plants were largely genetically determined or mainly caused by environmental factors. Turesson observed that many plants had distinctive races that grew in different habitats. These races differed from one another in characteristics such as height, leaf size and shape, degree of hairiness, flowering time, and branching pattern. Turesson dug up individuals representing these races and cultivated them together in his experimental garden at Lund, Sweden. In nearly every case he found that the unique features of individual races were maintained when the plants were grown in a common environment. Therefore most of the characteristics that he observed had a genetic basis; a few of them were environmental. Turesson called the ecological races that he studied and showed to have a genetic basis *ecotypes*.

Ultimately the kinds of studies first put on a scientific basis by Turesson and carried on by others led to the conclusion that most of the differences between individuals, populations, races, and species of plants are genetically determined. Despite the fact that plants change their characteristics in relation to the environments in which they grow, the differences between them are usually fixed genetically in the course of their evolution.

## Ecological Races in Animals

Similar patterns of variation are also found in animals (see Figure 16-4). The differences may be morphological or physiological, and they have the same basis as do ecotypes in plants. The differences between subspecies may be striking. For example, the larger races of some species of birds may often consist of individuals that weigh three or four times as much as individuals of the smaller races. Races may differ from one another in their tolerance to different temperatures, in the speed of their larval development, in their behavioral characteristics—in short, in virtually any feature that can be measured and studied. Their features are almost always genetically determined.

## BARRIERS TO HYBRIDIZATION

As isolated populations become more different from one another in their overall characteristics, they may eventually occupy different niches; in other words, they may exploit different resources in different ways. If such differentiated populations ever migrate back into contact with one another, they may still remain distinct in their characteristics. The populations may occur in different habitats, have different feeding habits, or otherwise be separated by differences that arose in them while they were apart (Figure 16-5). For these or other reasons the individuals of the differentiated populations may not hybridize with each other; if they do, functional hybrid individuals may not be formed, or the hybrids that are formed may be sterile. For animals that choose their mates, the differences between the populations that have originated in isolation may be so great that they may choose mates of their own kind, rather than those of the other, formerly isolated population. In other words, and for a variety of possible reasons, the populations may have become species.

*Populations of organisms tend to become increasingly different from one another in all of their characteristics. If the process continues long enough, or the seletive forces are strong enough, the populations may become so different that they are considered distinct species.*

Once species have formed, how do they keep their identity? The reasons that they do retain their identity may be grouped into two categories: (1) prezygotic isolating mechanisms, those preventing the formation of zygotes; and (2) postzygotic isolating mechanisms, those preventing the proper functioning of zygotes after they are formed. Some of these isolating mechanisms also occur, although often in a less marked form,

**Figure 16-5**

**Some animal populations are very localized.** The butterfly *Euphydryas editha* (**A**) occurs on Jasper Ridge (**B**), a biological preserve in the foothills south of San Francisco (see also Figure 25-10). The butterflies apparently constitute one continuous population throughout the open grassland, which occurs as an "island" surrounded by oak forest and chaparral. But extensive studies by Paul Ehrlich and colleagues over more than 30 years, involving marking butterflies with ink dots on their wings, releasing them, and recapturing them, have revealed that this species actually exists as a series of discontinuous populations between which individuals seldom move. Each of these populations, separated by dashed lines on the maps (**C**), is free to respond to the selective forces characteristic of its part of the ridge. Changes in the density of the populations, which have remained essentially constant in overall distribution since the late 1950s, are shown in the maps.

within species. When they do, they illustrate stages in the evolution of new species. In the following sections, we discuss various isolating mechanisms in these two categories and offer examples that illustrate how the isolating mechanisms operate to help species retain their identity.

## Prezygotic Isolating Mechanisms

*Geographical Isolation.* Most species simply do not exist together in the same places. Species are generally adapted to different climates or to different habitats. If this is the case, there is no possibility of natural hybridization between them. But they may hybridize if they are brought together in zoos, parks, or botanical gardens.

As an example of geographical isolation, the English oak, *Quercus robur,* occurs throughout those areas of Europe which have a relatively mild, oceanic climate. In its characteristics it is quite similar to the valley oak, *Q. lobata,* of California, and quite different from the scrub oak, *Q. dumosa,* also of California and adjacent Baja California (Figure 16-6). All of these species can hybridize with one another and form fertile

hybrids. But the English oak does not hybridize with the others in nature, simply because its geographical range does not overlap with theirs. Similarly, although lions, *Panthera leo,* and tigers, *P. tigris,* do not now occur together in nature, they do mate and produce hybrids in zoos. The hybrids in which the tiger is the father, called "tiglons," are viable and fertile; less is known about "ligers," hybrids in which the lion is the father.

*Ecological Isolation.* Even if two species occur in the same area, they may occur in different habitats and thus may not hybridize with each another. On the other hand, if they do hybridize with each another, the hybrids may not be well represented in the overall population because they may not be as fit in the habitat of either of their parents. In the latter case, one would speak of a postzygotic isolating mechanism of the kind that is discussed in the next section.

For example, in India the ranges of lions and tigers overlapped until about 150 years ago. But even when they did, there were no records of natural hybrids. Lions stayed mainly in the open grassland and hunted in groups called *prides;* tigers tended to be solitary crea-

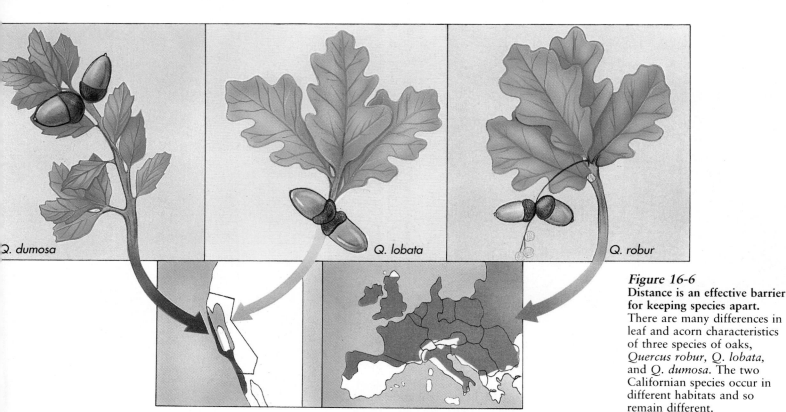

Q. dumosa    Q. lobata    Q. robur

**Figure 16-6**
**Distance is an effective barrier for keeping species apart.** There are many differences in leaf and acorn characteristics of three species of oaks, *Quercus robur, Q. lobata,* and *Q. dumosa.* The two Californian species occur in different habitats and so remain different.

tures of the forest. Because of the ecological and behavioral differences between them, lions and tigers rarely came into direct contact with one another, even though their ranges overlapped over thousands of square kilometers.

Similar situations occur among plants. We have already mentioned two species of oaks that occur in California: the valley oak, *Q. lobata,* and the scrub oak, *Q. dumosa* (Figures 16-6 and 16-7). Valley oak, a graceful deciduous tree, which can be as much as 35 meters tall, occurs in the fertile soils of open grassland on gentle slopes and valley floors in central and southern California. In contrast, scrub oak, *Q. dumosa,* is an evergreen shrub, usually only 1 to 3 meters tall, which often forms the kind of dense scrub known as chaparral. The scrub oak is found on steep slopes in less fertile soils. Hybrids between these very different oaks do occur, and they are fully fertile, but they are rare. The sharply distinct habitats of their parents limit their occurrence together, and there is no intermediate habitat where the hybrids might flourish.

*Temporal Isolation.* *Lactuca graminifolia* and *L. canadensis,* two species of wild lettuce, grow together along roadsides throughout the southeastern United States. Hybrids between these two species can easily be made experimentally and are completely fertile. But such hybrids are rare in nature because *L. graminifolia* flowers in early spring and *L. canadensis* flowers in summer. When their blooming periods overlap, as they do occasionally, the two species do form hybrids, which may even be abundant locally.

Many species of birds and amphibians that are closely related have different breeding seasons. Differences of this kind prevent hybridization between such species. For example, five species of frogs of the genus *Rana* occur together in most of the eastern United States. The peak time of breeding is different for each of them; because of this difference, hybrids are rare. In insects such as termites and ants, mating occurs when winged, reproductive individuals swarm from the nest. Species often differ in their swarming times, which eliminates hybridization between them.

*Behavioral Isolation.* In Chapter 44 we shall consider the often elaborate courtship and mating rituals of some groups of animals. Related species of organisms such as birds often differ in their mating rituals, which tend to keep these species distinct in nature even if they do occur in the same places. Indeed, much animal communication is related to the selection of mates.

In the Hawaiian Islands there are more than 500 species of flies of the genus *Drosophila.* This is one of the most remarkable concentrations of species in a single animal genus found anywhere. Many of these flies differ greatly from other species of *Drosophila,* exhibiting characteristics that can only be described as bizarre (Figure 16-8). The genus occurs throughout the world, but nowhere are the flies more diverse in their external appearance or behavior than in Hawaii.

The Hawaiian species of *Drosophila* are long-lived and often very large as compared with their relatives on the mainland. The females are more uniform than the males, which are often bizarre and highly distinctive. The males display complex territorial behavior and elaborate courtship rituals. Some of these are shown in Figure 16-8, *A* to *C*; another kind of courtship behavior is exhibited by *D. clavisetae* (Figure 16-8, *D*).

The patterns of mating behavior among the Hawaiian species of *Drosophila* are of great importance in maintaining the distinctiveness of the individual species. Despite the great differences between them, which are so evident in Figure 16-8, *D. heteroneura* and *D. silvestris* are very closely related. Hybrids between them are fully fertile. They occur together over a wide area on the island of Hawaii, and yet hybridization has been observed at only one locality. The very different and complex behavioral characteristics of these flies obviously play the major role in maintaining their distinctiveness.

*Mechanical Isolation.* Structural differences between some related species of animals prevent mating. Aside from such obvious features as size, the structure of the male and female copulatory organs may be so incompatible that mating cannot occur. In many insect and other arthropod groups, the sexual organs, particularly those of the male, are so diverse that they are used as a primary basis for classification. This diversity in structure is generally presumed by evolutionary biologists to have some importance in maintaining differences between species.

Similarly, the flowers of related species of plants often differ significantly in their proportions and structures. Some of these differences limit the transfer of pollen from one plant species to another. Bees may pick up the pollen of one species on one place on their bodies, and this area may not come into contact with the respective structures of the flowers of another plant species, so the pollen is not transferred. This difference would then decrease the frequency of hybridization between them.

*Prevention of Gamete Fusion.* In animals that simply shed their gametes into water, the eggs and sperm derived from different species may not attract one another. Many hybrid combinations involving land animals are not realized because the sperm of one species may function so poorly within the reproductive tract of another that fertilization never takes place. The growth of the pollen tubes may be impeded in hybrids between different species of plants. In both plants and animals the operation of such isolating mechanisms may prevent the union of gametes even following sucessful mating. The prevention of fusion between gametes is the last kind of prezygotic isolating mechanism possible before hybrids are formed.

A

B

*Figure 16-7*

**Closely related species may look very different.**
A *Quercus lobata* in Yosemite Valley.
B *Quercus dumosa* in the Coast Ranges south of San Francisco.

**Figure 16-8**

**Evolution among the Hawaiian *Drosophila*.** Males and females of closely related species of Hawaiian flies of the genus *Drosophila*, engaging in territorial defense and courtship.
**A** *Drosophila silvestris*. After approaching the female from the rear, the male lunges forward while vibrating his wings, with his head under the wings of the female, and raises his forelegs up and over the female's abdomen. Specialized hairs on the dorsal surface of one of the leg segments are then "drummed" over the dorsal surface of female's abdomen.
**B** *Drosophila heteroneura*. A male with extended wings approaching a female in typical courtship posture.
**C** *Drosophila heteroneura*. Two males lock antennae as part of the aggressive behavior involved in territorial defense.
**D** *Drosophila davisetae*. In this species, the males raise their abdomens up over their backs and spray a chemical signal over the female.

Prezygotic isolating mechanisms lead to reproductive isolation by preventing the formation of hybrid zygotes. The principal isolating mechanisms are geographical, ecological, seasonal, behavioral, and mechanical isolation and the prevention of gamete fusion.

## Postzygotic Isolating Mechanisms

All of the factors that we have discussed up to this point tend to prevent hybridization. But if hybridization does occur and zygotes are produced, there are still many factors that may prevent those zygotes from developing into normal, functional, fertile F$_1$ individuals. Development in any species is a complex process. In hybrids the genetic complements of two species may be so different that they cannot function together normally in embryonic development. For example, hybridization between sheep and goats usually produces embryos that die in the earliest developmental stages.

The leopard frog *(R. pipiens complex)* of the eastern United States is a series of very similar species between some of which it is difficult or impossible to produce viable hybrids. Before the nature of these species was understood, it was assumed that there were simply difficulties in producing hybrids between these frogs because they came from very different regions. The species are very similar, and hybrids between them are usually rare (Figure 16-9).

Many examples of this kind, in which similar species initially have been distinguished only as a result of hybridization experiments, are known in plants. Sometimes the hybrid embryos can be removed at an early stage and grown in an appropriate medium. When these hybrids are supplied with extra nutrients or other growth requirements that compensate for their weakness or inviability, they may complete their development normally.

Even if the hybrids survive the embryo stage, they may not develop normally. If the hybrids are weaker than their parents, they will almost certainly be eliminated in nature. Even if they are vigorous and strong, as in the case of the mule—a hybrid of the horse and the donkey (see Figure 16-2)—they may still be sterile and thus incapable of contributing to succeeding generations. The development of sex organs in hybrids may be abnormal, the chromosomes derived from the respective parents may not pair properly, or their fertility may simply be lower than normal for other reasons.

*Postzygotic isolating mechanisms are those in which hybrid zygotes develop abnormally or fail to develop entirely or in which hybrids cannot become established in nature.*

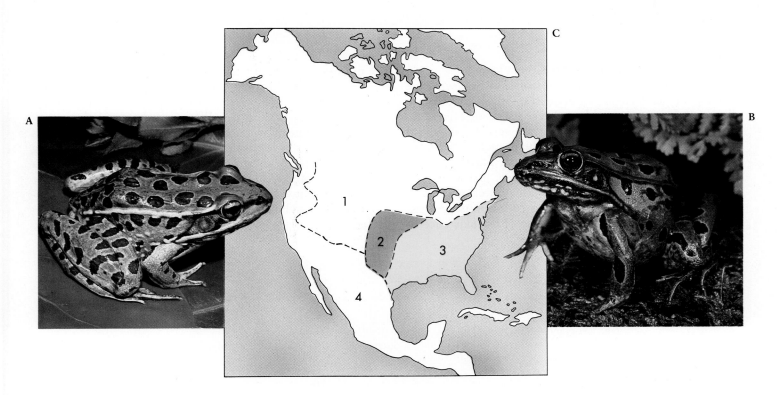

**Figure 16-9**

**Sibling species of leopard frogs.**
A The southern leopard frog, *R. berlandieri,* in California.
B The leopard frog, *Rana pipiens,* in Wisconsin.
C Numbers indicate the following species in the geographical ranges shown: 1, *R. pipiens;*
2, *R. blairi;* 3, *R. sphenocephala;* 4, *R. berlandieri.*

# Extinction

As a result of human population pressures, thousands of kinds of plants, animals, and microorganisms all over the world are facing extinction. One of these was the dusky seaside sparrow. A population greater than 6000 of these canary-sized birds used to inhabit the cordgrass flats along the St. Johns River and on Merritt Island in Brevard County, central Florida, but the last survivor (Figure 16-A) died in June, 1987. It had been living in a large cage in a secluded area on Discovery Island, an accredited zoo especially concerned with the preservation of selected species of birds, in Disney World near Orlando. The wild birds were driven near extinction by mosquito-abatement efforts near the Kennedy Space Center—heavy spraying with insecticides and then flooding the marshes—and other intensive development in the region, followed by wildfires. By the late 1960s, it was evident that the sparrows faced extinction. The last known nesting in the wild occurred in 1975, and the six remaining birds were captured in 1980.

But the last few dusky seaside sparrows that had been captured in nature were systematically backcrossed with members of the Gulf Coast subspecies of the seaside sparrow. So while the dusky seaside sparrow is now extinct, much of its genetic constitution has been preserved by these backcrossing efforts. The future release of these backcross progeny may serve to introduce a new species to the region with much of the evolutionary heritage of the one we have lost.

## Reproductive Isolation: An Overview

All of the kinds of reproductive isolation that we have discussed (Table 16-1) can arise in populations that are diverging from one another. These kinds of reproductive isolation all occur within certain species, as would be expected from the way in which such differences evolve. Therefore some populations of particular species cannot hybridize with one another, just as if they were distinct species as judged by this criterion.

The formation of species is a continuous process, one that we can understand because of the existence of intermediate stages at all levels of differentiation. If populations that are partly differentiated come into contact with one another, they may still be able to interbreed freely, and the differences between them may then disappear over the course of time. If their hybrids are partly sterile, or not as well adapted to the existing habitats as are their parents, these hybrids will be at a disadvantage. As a result, there may be selection for factors that limit the ability of the differentiated populations to hybridize. If hybrids are sterile or not as successful as their parents, individual plants or animals

---

**TABLE 16-1    REPRODUCTIVE ISOLATING MECHANISMS**

1. Prezygotic mechanisms: prevent the formation of zygotes.
   a. Geographical isolation. The species occur in different areas.
   b. Ecological isolation. The species live in different habitats and do not meet; there may be no habitat suitable for their hybrids.
   c. Temporal isolation. The species reproduce at different seasons or different times of day.
   d. Behavioral isolation. The behavior of the species may differ so that there is little or no attraction between them.
   e. Mechanical isolation. Structural differences between species may prevent mating.
   f. Prevention of gamete fusion. The gametes may not fuse, usually because of chemical factors.

2. Postzygotic mechanisms: prevent the proper functioning of zygotes once they are formed. These include the inviability or sterility of the hybrids or their irregular development.

which do not hybridize may be more fit than those which do hybridize.

Most species are separated by combinations of the isolating mechanisms that we have just discussed. For example, two related species may occur in different habitats, produce their gametes at different times of the year, have different behavioral patterns, and produce inviable embryos even if hybridization does take place. Such patterns, in which more than one factor functions in limiting the frequency of hybrids between two species, presumably arise for two reasons.

1. The factors that limit hybridization arise primarily as by-products of adaptive change in populations. Consequently, several factors that limit hybridization often emerge simultaneously and may characterize the differentiated populations.

2. If differentiated populations do come into contact with one another, natural selection may strengthen the isolating mechanisms that are already present. For example, if the hybrids do not complete their development beyond the embryo stage, then any factor that limits the hybridization that produces them (prezygotic isolating mechanisms) would be an advantage. Individuals that form hybrids that do not function well in nature waste reproductive energy by doing so and are less fit than are individuals that do not form such hybrids.

Both kinds of forces that limit hybridization are built up as a part of the overall process of change in populations that are isolated from one another, although they may sometimes be strengthened when and if the differentiated populations migrate into contact with one another.

*As a result of the way in which they originate, species are often separated by more than one factor. Some of these factors prevent the species from hybridizing at all; others limit the success of the interspecific hybrids once they are formed.*

## CLUSTERS OF SPECIES

One of the visible manifestations of species formation is the existence of groups of closely related species in certain locations. These species often have evolved relatively recently from a common ancestor. Such clusters are particularly impressive on groups of islands, in se-

ries of lakes, or in other sharply discontinuous habitats. The existence of these clusters makes sense only in the context of their arising by microevolutionary divergence from an ancestral form occupying diverse habitats. One of the best-known clusters of species is a group of birds, finches, found on the Galapagos Islands, where it was studied by Charles Darwin.

Thirteen species of Darwin's finches occur on the Galapagos Islands. A fourteenth species lives on Cocos Island, which lies about 1000 kilometers to the north. This group of birds provides one of the most striking and best-studied examples of species formation.

The Galapagos Islands are a remarkable natural laboratory of evolution. The islands are all relatively young in geological terms, and they have never been connected with the adjacent mainland of South America or with any other source area. The lowlands of the Galapagos Islands are covered with thorn scrub. At higher elevations, which are attained only on the larger islands, there are moist, dense forests. All of the organisms that occur on these islands have reached them by crossing the sea as a result of chance dispersal in the water, by wind, or by transport via another organism.

On oceanic islands there is often a disproportionate representation of certain groups of organisms. For example, in addition to the 13 species of Darwin's finches on the Galapagos Islands, there are only 14 other species of resident land birds. Perhaps the ancestor of Darwin's finches reached these islands earlier. In that case, all the habitats where birds occur on the mainland would have been unoccupied, and the ancestor of Darwin's finches would have been able to take advantage of them all. As new arrivals moved into these vacant niches, adopting new lifestyles, they were subjected to diverse sets of selective pressures. Under these circumstances, the ancestral finches split into a series of diverse populations, and some of these eventually became species.

The descendants of the original finches that reached the Galapagos Islands now occupy many different habitats (Figure 16-10). These habitats encompass a distinct group of birds on the mainland. Among the 13 species of Darwin's finches that inhabit the Galapagos, there are three main groups:

1. *Ground finches,* which feed on seeds of different sizes. The size of their bills is related to the size of the seeds on which the birds feed.
2. *Tree finches,* which have bills suitable for feeding on insects. One of the tree finches has a parrotlike beak and feeds on buds and fruit in trees. Another has a chisel-like beak with which it carries around a twig or cactus spine, which it uses to probe for insects in crevices. It is an extraordinary example of a bird that uses a tool.

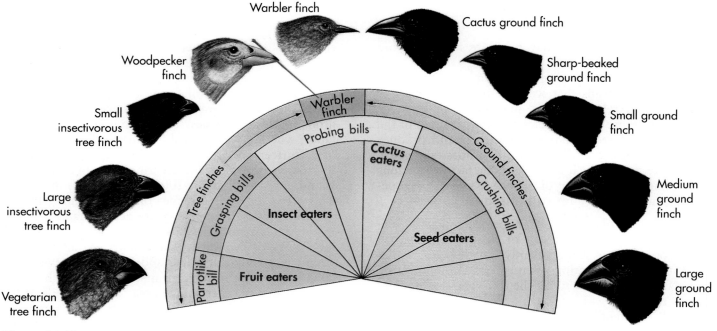

*Figure 16-10*

**Darwin's finches.** Ten species of Darwin's finches from Indefatigable Island, one of the Galapagos Islands, showing differences in bills and feeding habits. The bills of several of these species resemble those of different, distinct families of birds on the mainland. All of these birds are thought to have been derived from a single common ancestor.

3. *Warbler finches,* unusual birds that play the same ecological role in the Galapagos woods that warblers play on the mainland, searching continually over the leaves and branches for insects.

The evolution of Darwin's finches on the Galapagos Islands (Figure 16-11) and Cocos Island provides a classical example of species formation. It illustrates how adaptation to local conditions produces the divergence that is the heart of species formation. By the same sorts of processes, species are originating continuously in all groups of organisms.

The evolution of organisms on islands differs from that in mainland areas only in the degree of opportunity afforded for *rapid* evolution. Because evolution on islands is often rapid, and therefore recent, Darwin and other investigators have found the study of plants and animals on islands especially informative.

## THE EVIDENCE FOR EVOLUTION: AN OVERVIEW

In this and the preceding chapter we have considered the body of evidence that supports the three central tenets of evolution. The evidence that macroevolution has occurred is so strong as to be beyond further question. Although scientists are currently engaged in lively discussions about the rate at which macroevolution proceeds, and about whether it proceeds gradually or in spurts, there is essentially no disagreement among practicing biologists about the fact of macroevolutionary change.

The evidence for microevolution depends partly on observations of evolutionary change now in progress. In many cases, it has been possible to observe evolutionary change within populations of a single species in which selection has proven to be a powerful agent of adaptive change, just as Darwin proposed in 1859. Many factors cause natural populations to change progressively over time, including both environmentally guided selection and random factors such as genetic drift.

The evidence that microevolutionary change leads to macroevolutionary change is extensive. Biologists have observed the stages of the species-forming process in many different plants, animals, and microorganisms. These stages are indicated by such features as genetic variation in individual populations, genetic differences between different populations, divergence of populations from one another in response to different adaptive pressures, ecological races, progressively stronger reproductive isolation, and clusters of related species, each species occupying an ecologically distinct, but often adjacent, habitat. These patterns are so frequently observed in nature that there can be no doubt of the way in which species normally originate. Microevolution drives macroevolution, just as Darwin proposed.

*Fig. 16-11*
**A view of the Galapagos Islands.**

## ■ SUMMARY

1. The theory of evolution is based on three lines of evidence: (1) that macroevolution occurs, (2) that natural selection is responsible for the adaptive nature of microevolution, and (3) that microevolutionary change (evolution within a species) leads to macroevolutionary change (evolution of new species).

2. Species are kinds of organisms that differ from one another in one or more characteristics and that do not normally hybridize freely with other species when they come into contact in nature. They often cannot hybridize with one another at all. Individuals within a given species, on the other hand, usually are able to interbreed freely.

3. Populations change as they adjust to the demands of their environments. Even populations of a given species that are close to one another geographically are normally effectively isolated. Such populations are free to diverge in ways that are responsive to the needs of their particular environment.

4. In general, if the selective forces bearing on populations differ greatly, the populations will diverge rapidly; if these selective forces are similar, the populations will diverge slowly.

5. Among the factors that separate populations, and species, from one another are geographical, ecological, seasonal, behavioral, and mechanical isolation, as well as factors that inhibit the fusion of gametes or the normal development of the hybrid organism. Reproductive isolation between species arises as a normal by-product of the progressive differentiation of populations.

6. The evidence that microevolutionary change leads to macroevolutionary change is that it is possible to observe the intermediate stages of the process directly.

7. Ecological races and subspecies differentiate within species but often still intergrade with one another. The differences between races, in both plants and animals, are mostly determined by genetic factors.

8. Clusters of species arise when the differentation of a series of populations proceeds further, past race formation. On islands, such differentiation is often rapid because numerous open habitats are available. In many continental areas, differentiation is not as rapid, but there are local situations, such as those where many different habitats are developing close to one another, where differentiation may be rapid.

# REVIEW

1. Groups of organisms that differ from each other in one or more characteristics and do not hybridize extensively in nature are called _____.

2. The adaptive molding of local populations of a species, which leads to divergence of those populations, is called _____. This process, if it proceeds far enough, results in _____.

3. Lions and tigers will mate if brought together, and the hybrids are fertile. Which isolating mechanism is responsible for the lack of hybrids between these species in nature?

4. Mechanisms such as the prevention of gamete fusion so that hybrids are not formed are called _____ mechanisms. These mechanisms can be contrasted with _____ mechanisms, which act after fertilization has occurred.

5. Is the evolution of Darwin's finches on the Galapagos Islands an example of microevolution or of macroevolution?

◆

# SELF-QUIZ

1. Choose the phrase(s) that best complete this statement. Local populations of a species
   (a) respond to environmental conditions in their particular habitat.
   (b) tend to mate more frequently with other individuals in their population.
   (c) are groups of individuals that live in a particular place.
   (d) can be genetically different from other populations of the species.
   (e) all of the above.

2. Which of the following is *not* a prezygotic isolating mechanism?
   (a) Geographical isolation  (c) Seasonal isolation  (e) Mechanical isolation
   (b) Ecological isolation  (d) Hybrid sterility

3. Complete this sentence with the most appropriate choice. Göte Turesson performed a series of experiments to determine what caused the observed differences between races of plants. When plants were grown in a common garden, he found that
   (a) plants maintained their differences, and he concluded that the differences had a genetic basis.
   (b) plants maintained their differences, and he concluded that the differences were environmental.

4. On the Galapagos Islands, the descendants of a South American finch have produced a cluster of species. Darwin's finches fill a variety of niches. Which niche is *not* filled by any of these finches?
   (a) Seed eater  (c) Fish eater  (e) Fruit eater
   (b) Cactus eater  (d) Insect eater

5. In continental areas, as on islands, evolution sometimes results in clusters of species. In what kind of continental area would you expect to find species clusters?
   (a) An area of diverse habitats  (c) An alpine area  (e) A grassland
   (b) An arid area  (d) A tropical area

6. Which of the following is the most realistic definition of species?
   (a) A kind of organism.
   (b) A kind of organism that consists of one of more populations within which the individuals are interfertile with one another and which do not interbreed with other species.
   (c) A distinct kind of organism consisting of individuals that usually resemble one another more closely than they do other kinds of organisms and which do not normally interbreed with related kinds of organisms when they occur together in nature.
   (d) A geographical race.
   (e) The result of punctuated equilibrium.

7. Scrub oak and valley oak, both found in California, do not form many hybrids in nature because
   (a) they cannot hybridize for genetic reasons.
   (b) they occur in different habitats, and the hybrids are not as well suited to these habitats as are the parental species.
   (c) they are pollinated by different insects.
   (d) they produce flowers in different seasons.
   (e) they do not occur together.

# THOUGHT QUESTION

1. In the fall of 1986 the Supreme Court of the United States heard arguments for and against a Louisiana law requiring that "creation science" be taught in public schools on an equal footing with evolution. One of the arguments advanced in support of the law was that evolution is as much a religion as creationism, reflecting the beliefs of scientists and their faith in a particular worldview rather than the certain knowledge of objective reality that they claim. In June of 1987, the decision of the Court was announced: by 7 to 2, the justices struck down the law, rejecting the "creation science" argument. How would you have voted?

# FOR FURTHER READING

GOULD, S.J.: *Ever Since Darwin*, W.W. Norton & Company, New York, 1977. An entertaining collection of essays on evolution and Darwinism.

HITCHING, F.H.: *The Neck of the Giraffe or Where Darwin Went Wrong*, Chaucer Press (Pan), London, 1982. An entertaining and informal presentation of all the arguments currently being advanced *against* Darwin's theory.

GOSLOW, G.E., JR, K.P. DIAL, and F.A. JENKINS, JR: "Bird Flight: Insights and Considerations," *BioScience*, vol. 40 (2), 1990, pages 108-115. New techniques show that more than the wing participates in flying.

RYAN, M.J.: "Signals, Species, and Sexual Selection," *American Scientist*, vol. 78(1), 1990, pages 46-52. Studies of mate recognition in frogs and fishes reveal preferences for individuals, populations, and even members of closely related species.

# The Evolution of Life on Earth

These fossils of organisms known as crinoids date from the Cretaceous period, between 144 and 65 million years ago. With the help of fossils such as these we are learning much about the evolution of life on earth.

# THE EVOLUTION OF LIFE ON EARTH

## Overview

The central lesson of evolution is that the many kinds of plants, animals, and microorganisms now living on earth are the end result of a long history of success and failure, as organisms have for billions of years striven to survive and reproduce. Although much of what has happened is hidden from us now, traces of the evolutionary journey to the present remain frozen in rock as fossils. By studying fossils, we have been able to get a fairly clear picture of how life has evolved on earth since the first cells formed more than 3 billion years ago.

## For Review

*Here are some important terms and concepts that you will encounter in this chapter. If you are not familiar with them, you should review them before proceeding.*

**Kingdoms** (Chapter 1)

**The origin of life** (Chapter 3)

**The origin of species** (Chapter 16)

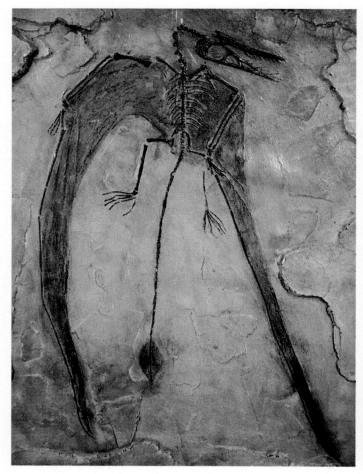

*Figure 17-1*

**Flight has evolved three separate times among the vertebrates.** Birds, bats, and Pterosaurs, a diverse group of flying reptiles that became extinct with the dinosaurs about 65 million years ago. The wing membrane stretched along the greatly extended fourth finger of this fossil pterosaur.

All of the many kinds of organisms now living on earth represent only a small fraction of the total that have lived. Every living creature occupies a place once taken by a different kind of animal or plant. Mammals such as you and me play an ecological role once filled by dinosaurs. Dinosaurs were once the dominant land vertebrates (Figure 17-1), but today no dinosaurs walk the earth. The trilobites, sea organisms related to the living horseshoe crabs, dominated the ocean floor in the Cambrian Period more than 500 million years ago (Figure 17-2), but they are gone now; so are the ammonites, octopuses with shells that were one of the dominant forms of ocean life in the Cretaceous Period 100 million years ago. As children, few of us escape fascination with dinosaurs; they appear so different that it is as if they inhabited a different world. And yet dinosaurs are the ancestors of the turkey you eat at Thanksgiving and are relatives of the ancestors of the rattlesnakes and turtles of the modern world. We know a great deal about dinosaurs and the many other forms of life that have preceded us in the evolutionary parade of life on earth because we have found their fossils. A fossil allows us to visualize the body of an organism that lived in the past because certain structures are mineralized and so preserved in rock. We don't have fossils of all the kinds of organisms that have lived before us, but we have found enough to piece together a fairly complete picture of the past. We review that picture in this chapter. First, a few words about fossils.

**Figure 17-2**

**Life in the Cambrian.**
A A reconstruction of a community of marine organisms in the Cambrian Period, 500 to 550 million years ago. The swimming animals, are trilobites, early members of the arthropod phylum. On the sea floor is a colony of sponges, members of another ancient animal phylum.
B A fossil trilobite.
C This assemblage of tribolites was found in an abandoned Ohio cement quarry.

A

B

C

# FOSSILS

A fossil is any record of an extinct organism; fossils may be nearly complete impressions of organisms, or merely burrows, tracks, molecules, or traces of their existence. Only fossils provide definite evidence about what extinct kinds of organisms looked like, including the ancestors of those which are living now. Fossils provide an actual record of organisms that once lived, an accurate understanding of where and when they lived, and some appreciation of the environment in which they lived. Fossils enable us to trace lines of evolutionary progression among individual groups of organisms, provided that enough fossils of various ages are available.

Only a minute fraction of the organisms living at any one time are preserved as fossils. Most of the fossils that we do have are preserved in **sedimentary** rocks. Sedimentary rocks are formed of particles of other rocks that are weathered off, deposited at the bottom of bodies of water or accumulated by wind, and then hardened into strata of the sort that Darwin found so filled with intriguing fossils in South America. During the formation of sedimentary rocks, dead organisms are sometimes washed down along with inorganic debris, such as mud or sand, and eventually reach the bottom of some pond, lake, or the ocean itself. In exceptional circumstances, fossils may be preserved in an organic substance such as tar: the La Brea Tar Pits in Los Angeles (Figure 17-3) illustrate this phenomenon. In other instances, organisms themselves may form the whole deposit, as in the formation of coal or oil deposits. However, many organisms do not live in places where their remains are likely to be preserved in sedimentary rocks; for this reason they are unlikely to be preserved as fossils.

For specific organisms to become well preserved as fossils, they must usually be buried before their decay is complete and before they are torn apart by scavengers. In addition, the process of decay, for at least some portions of the organism, must stop after they are buried. Because the soft parts usually decay rapidly, normally only structures such as bones, teeth, shells, scales, leaves, and wood are available in the fossil record. If fossils are exposed, they often disintegrate quickly; therefore most of those we find were exposed recently. When fossils are formed, the actual parts of the organisms are almost always replaced by minerals; a fossil rarely contains any of the material that made up the body of the organism originally, but rather is a mineralized replica of that body part.

In interpreting the structure of fossil organisms, we must normally use the hard parts to try to interpret what the soft parts looked like. For organisms such as worms, which have no hard parts, fossils are rare. Even though soft-bodied animals undoubtedly evolved before their hard-bodied counterparts, we have little evidence of their history in the fossil record. Sometimes soft-bodied animals are preserved in exceptionally fine-grained muds, in conditions under which the supply of oxygen was poor while the muds were being deposited and deterioration was therefore slowed.

**Figure 17-3**
The La Brea Tar Pit.

Fossils provide the concrete means whereby we can judge our deductions about the history of particular groups of organisms. Fossils are found mainly in sedimentary rocks.

## Dating Fossils

The direct methods of dating rocks and fossils that we mentioned in Chapter 15 first became available in the late 1940s. Naturally occurring radioactive isotopes of certain elements are employed in this process. Such isotopes are unstable and decay over the course of time at a steady rate, producing other isotopes. One of the most widely employed methods of dating, the **carbon-14** ($^{14}C$) method, employs estimates of the different isotopes present in samples of carbon.

Most carbon atoms have an atomic weight of 12; the symbol of this particular isotope of carbon is $^{12}C$. But a fixed proportion of the atoms in a given sample of carbon consists of carbon with an atomic weight of 14 ($^{14}C$), an isotope that has two more neutrons than does $^{12}C$. $^{14}C$ is produced from $^{12}C$ as a result of bombardment by particles from space. The carbon that is incorporated into the bodies of living organisms consists of the same fixed proportion of $^{14}C$ and $^{12}C$ that occurs generally. But after an organism dies and is no longer incorporating carbon, the $^{14}C$ in it gradually decays over time, by the loss of neutrons, to $^{12}C$. It takes 5600 years for half of the $^{14}C$ present in a sample to be converted to $^{12}C$ by this process; this length of time is called the **half-life** of the $^{14}C$ isotope. The relationships just outlined indicate that a sample that had a quarter of its original proportion of $^{14}C$ remaining would be approximately 11,200 years old.

Thus it is possible to date accurately fossil material that contains carbon, as all organic material does, provided that the fossil material is less than about 50,000 years old. This can be done by measuring as accurately as possible the proportion of $^{14}C$ that is still present in the carbon (Figure 17-4). For older fossils, the amount of $^{14}C$ remaining is so small that it is not possible to measure it precisely enough to provide accurate estimates of age.

To determine the ages of older samples, one can sometimes study the decay of other radioactive isotopes. For example, $^{40}K$ (potassium-40) decays very slowly into $^{40}Ar$ (argon-40) and can be used to date much older rocks. The half-life of $^{40}K$ is 1.3 billion years; other isotopes have even longer half-lives. With the use of such dating methods, our knowledge of the ages of various rocks has become more precise. Unfortunately, for sedimentary rocks, which incorporate pieces of other rocks, such dating methods can establish only the ages of the materials and not the time when they were incorporated into the sedimentary rocks—the time when the fossils were formed. But these methods do allow us to estimate the ages of all of the rocks on earth, regardless of how old they may be.

*The ages of fossils may be estimated by determining the proportions of different isotopes of carbon in the organic material preserved in them or the proportions of different isotopes in the rocks where they are found. Radioactive isotopes change from one form to another over time, so the isotope proportions in a given sample will provide an absolute date for the age of that sample.*

## DRIFTING CONTINENTS

As you saw in Chapter 3, the planets of our solar system, including the earth, began to coalesce approximately 4.6 billion years ago. The accretion (pulling together of fragments by gravity) of the early earth seems to have been completed about 4.5 billion years ago. After several hundred million years, the continents had formed. The oldest known rocks on earth are 3.8 billion years old; by the time they were consolidated, some portions of the earth's crust had begun to move. As these sections moved, they became thicker.

## The Discovery of a Living Coelacanth

New discoveries relating to the history of life on earth are not always made in the fossil record. Thus in 1938 scientists were surprised by the announcement that a trawler fishing in the Indian Ocean off the coast of South Africa had landed a large, very strange fish (Figure 17-A). The specimen, which was about 2 meters long, became rotten before it was seen by an ichthyologist (a scientist who specializes in the study of fish) and was skinned so that some of it could be saved. Nevertheless, as soon as J.L.B. Smith of the University of Grahamstown in South Africa saw it, he recognized it as a member of an ancient group of "lobe-finned" fishes (fishes in which the fins more closely resemble the limbs of amphibians and other land vertebrates than is usual). These fishes had been described from fossils more than a century earlier and had been thought to have been extinct for about 70 million years! They were called coelacanths, and the newly discovered living fish was dubbed *Latimeria chalumnae*.

Coelacanths are well represented in the fossil record for more than 300 million years, from about 390 million years to about 70 million years ago. Until *Latimeria* was discovered alive, it was assumed that the group, which was somewhat similar to the ancestors of the terrestrial vertebrates, had become extinct long ago. But here was indisputable evidence that one of their descendants was still alive, swimming the warm waters of the western Indian Ocean.

Because the first specimen had been skinned, scientists had no opportunity at first to learn about is internal parts, features that were of great interest in terms of its relationship to terrestrial vertebrates and other fishes. Leaflets and posters were distributed among the fishing communities of southern and eastern Africa, of fering rewards for another specimen of this remarkable fish. But it was not until 1952 that a second coelacanth was brought to the attention of scientists. It was landed in the Comoro Islands, about 3000 kilometers northeast of the place where the first specimen was caught. Coelacanths were landed occasionally in the Comoros and were well known to local fishermen.

Living mostly at depths of 150 to 300 meters in the sea, *Latimeria* is a very strange animal. Its features mark it as a member of the evolutionary line that gave rise to the terrestrial tetrapods. By studying the dozens of specimens that have been landed since 1952, scientists have been able to shed additional light on the nature of this ancient and archaic group of vertebrates.

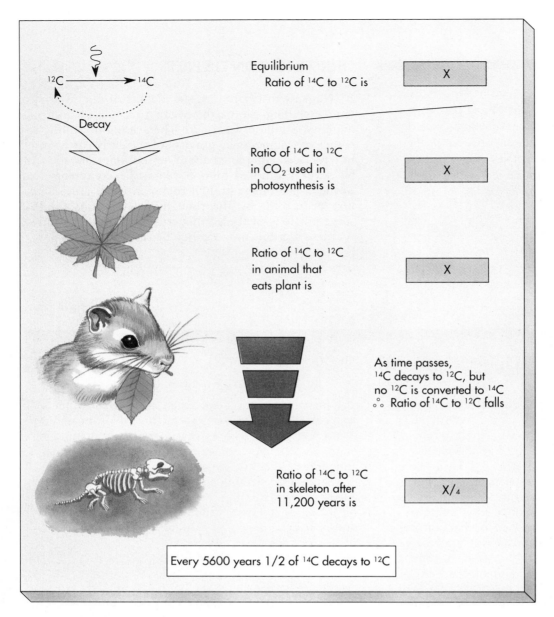

$^{12}C \longrightarrow ^{14}C$

Decay

Equilibrium
Ratio of $^{14}C$ to $^{12}C$ is

X

Ratio of $^{14}C$ to $^{12}C$
in $CO_2$ used in
photosynthesis is

X

Ratio of $^{14}C$ to $^{12}C$
in animal that
eats plant is

X

As time passes,
$^{14}C$ decays to $^{12}C$, but
no $^{12}C$ is converted to $^{14}C$
∴ Ratio of $^{14}C$ to $^{12}C$ falls

Ratio of $^{14}C$ to $^{12}C$
in skeleton after
11,200 years is

X/4

Every 5600 years 1/2 of $^{14}C$ decays to $^{12}C$

*Figure 17-4*
C-14 dating.

## Discovery of Continental Movement

About 200 million years ago, in the early Jurassic Period, the major continents, following a long history of earlier movements, were all together in one great supercontinent (Figure 17-5). Alfred Wegener, the German scientist who in 1915 first proposed the idea of continental movement in his book, *The Origin of Continents and Oceans,* called this giant land mass **Pangaea.** Because the specific mechanism Wegener proposed to account for continental movement was not feasible, his theory fell into discredit with most geologists and biologists. Indeed, although it is now generally accepted, Wegener's theory was denied by most scientists for nearly half a century.

In the early 1960s, new evidence developed that provided a mechanism for continental movement. The theory that emerged visualized heavy, basaltic, ocean-floor rocks moving like a conveyor belt away from the mid-ocean ridges where they were formed. In the process, they carried the lighter continental rocks along with them. For example, the mid-Atlantic Ridge is an enormous zone of upwelling basaltic lava. It generates basaltic ocean-floor rocks that move out from the Ridge both toward Europe and Africa and toward North and South America. As these rocks continue to be formed, both pairs of continents ride farther and farther apart from one another.

The earth's crust and associated upper mantle, which together are about 100 to 150 kilometers thick, are divided into plates. There are seven enormous ones and a series of smaller ones that lie between them. Because of the existence of these plates, the theory that explains the movement of continents in this way is called **plate tectonics.** Earthquakes, in general, are caused by the relative movement of these plates. Thus an earthquake along the San Andreas Fault in California, such as the one that destroyed San Francisco in 1906 or the large ones that occurred in Los Angeles in October, 1987, and the San Francisco Bay region in October, 1989, results from the relative movement of two gigantic segments of the earth's crust and mantle—the far edges of the plates that caused these earthquakes are in Japan and Iceland, respectively. Most mountains are thrust up by plate movements; the Himalayas have been thrust up to the highest elevations on earth as a result of the grinding, prolonged collision of the Indian subcontinent with Asia.

*Lava flowing out of fissures in the earth's crust, such as the mid-oceanic ridges, form great plates of relatively heavy rocks 100 to 150 kilometers thick. These rocks move, carrying the lighter continents along with them. There are seven huge plates, all moving in relation to one another, and a number of smaller plates.*

## A History of Continental Movements

The continents have gradually moved apart from their positions as parts of Pangaea 200 million years ago. The opening of the South Atlantic Ocean began about 125 to 130 million years ago, with Africa and South America having been directly connected earlier. Subsequently, South America has moved slowly toward North America, with which it eventually became connected by a land bridge, the Isthmus of Panama, between 3.1 and 3.6 million years ago. When the bridge was complete, many South American plants and animals, such as the opossum (Figure 17-6) and the armadillo, migrated overland into North America. At the same time, numerous North American plants and animals such as oaks, deer, and bears moved into South America for the first time. Changes on a greater or lesser scale occurred in the positions of all the continents, playing a major role in the patterns of distribution of organisms that we see today. For example, the ratite birds—ostriches, emus, rheas, cassowaries, and kiwis—now thought to have diverged very early in the course of their evolution from all other living groups of birds, undoubtedly migrated between southern continents and islands, now widely separated from one another, at a time when direct overland migration was possible: the Mesozoic Era. The explanations of present-day distributions that these continental movements indicate form a large part of **biogeography,** the study that is concerned with where different kinds of organisms occur and how they got there.

Australia provides a striking example of the way in which continental movements have affected the nature and distribution of organisms. Until about 53 million years ago, Australia was joined with a much warmer Antarctica, as was South America. Marsupials, which are best represented in Australia and South America at present, seem clearly to have moved overland between these continents via Antarctica when this was still possible. Isolated from the placental mammals that had become abundant and diverse in other parts of the world, marsupials underwent a major episode of evolutionary spread in Australia. As Australia moved northward toward the tropical islands fringing Southeast Asia, other groups of plants and animals did likewise.

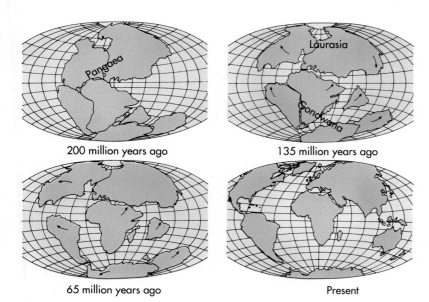

200 million years ago

135 million years ago

65 million years ago

Present

*Figure 17-5*

**How the continents have moved.** The positions of the continents at 200 million, 135 million, and 65 million years ago and at present.

*Figure 17-6*

**A North American opossum,** *Didelphis virginiana,* **with young.**
Marsupials, which once lived in Europe, Africa, and Antarctica,
are now best represented in Australia and South America. They
formerly occurred in North America but became extinct about
40 million years ago. This particular marsupial, the opossum, is
a mammal that migrated into North America from South
America once the Isthmus of Panama had been uplifted above
the sea, an event that occurred several million years ago.

As Australia moved northward, the Antarctic
Ocean opened and the cold Circumantarctic Current
began to flow around the Southern Hemisphere. A land
connection between Australia and Antarctica, through
the present position of Tasmania, seems to have persisted until about 38 million years ago. On the other
side of the world, the connection of southern South
America with West Antarctica lasted until about 23
million years ago. With the warmer climates of the
early Cenozoic Era, more-or-less direct overland migration between Australia and South America, via Antarctica, was possible until about 40 million years ago. The
formation of the Circumantarctic Current led directly
to the development of the Antarctic ice sheet, which
reached its full size by about 10 million years ago. Ultimately, the cold temperatures associated with the formation of that enormous mass of ice triggered the onset of widespread continental glaciation (the Ice Age) in
the Northern Hemisphere during the past few million
years.

*The movements of continents over the past 200 million years
have profoundly affected the distribution of organisms on earth.
Some of the major events include the linkage of South America
with North America about 3.1 to 3.6 million years ago and
the separation of Australia and South America from Antarctica,
which triggered the formation of the southern and ultimately the
northern ice sheets.*

## THE EARLY HISTORY OF LIFE ON EARTH

The earth itself is about 4.6 billion years old, and the
oldest rocks that have persisted in recognizable form
are about 3.8 billion years old. For many years, scientists believed that there were no fossils in such ancient
rocks, but we now know that the fossils were simply
too small to be seen clearly without an electron microscope. The earliest fossils found so far, all of them bacteria, are about 3.5 billion years old.

Massive limestone deposits called **stromatolites**
(Figure 17-7) became frequent in the fossil record
about 2.8 billion years ago. Produced by cyanobacteria, stromatolites were abundant in virtually all freshwater and marine communities until about 1.6 billion
years ago. Today stromatolites are still being formed,
but only under conditions of high salinity, aridity, and
high light intensities. About 2 billion years ago, many
kinds of bacteria existed, including single, rounded
cells; filaments apparently divided by cross-walls; tubular structures; branching filaments; and several unusual
forms that do not fit well into any of these categories.
For most of the time in which life has existed on earth,
the only organisms in existence were bacteria. Although the most conspicuous structures they formed
were the stromatolites, bacteria were clearly everywhere. After at least 2 billion years of bacterial domination, fossils that definitely seem to represent unicellular protists, the first eukaryotes, first appear about 1.5
billion years ago.

*The oldest fossils are bacteria; they date from about 3.5 billion
years ago. The oldest eukaryotic fossils, unicellular protists, are
from about 1.5 billion years ago.*

We find the first known fossils of multicellular organisms in rocks about 630 million years old from
southern Australia. Their appearance marks the onset
of a major period of the earth's history, **Phanerozoic
time.** Within Phanerozoic time, we begin to speak in
terms of geological eras, periods, epochs, and ages. For
many years the oldest known fossils were those from
the Cambrian Period (590 to 505 million years ago),
the first period of the Paleozoic Era. The geological
eras, with their dates in millions of years before the
present, are as follows:

1. *Paleozoic Era.* 590 to 248 million years ago.
   The name of this era is derived from the Greek
   words *paleos,* "old," + *zoos,* "life." Until the

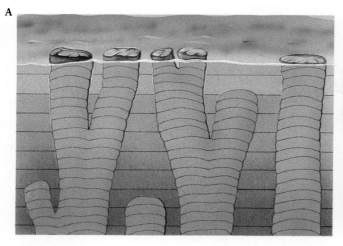

**Figure 17-7**

**Stromatolites.**
**A** This diagram shows how aggregations of the mineral calcium carbonate build up around massive colonies of cyanobacteria. Such deposits are known as stromatolites.
**B** Stromatolites in the intertidal zone at Shark Bay, Western Australia. The largest structures are about 1.5 meters across. These stromatolites formed during a time of slightly higher sea level, perhaps 1000 to 2000 years ago.

discoveries outlined above were made, this was the oldest period from which fossils were known.

2. *Mesozoic Era.* 248 to 65 million years ago. Greek *mesos,* "middle."
3. *Cenozoic Era.* 65 million years ago to the present. Greek *coenos,* "recent."

All of the strata older than the Cambrian Period, and thus older than the Paleozoic Era, are classified as Precambrian. To earlier scientists there seemed to be no fossils at all in Precambrian rocks, and their absence was regarded as a great mystery. Now we know that fossil organisms were not detected simply because they were unicellular and therefore so small that they were difficult to observe. With the evolution of external skeletons (that is, hard shells) visible fossils became abundant (Figure 17-8). Those multicellular animals which lived before the evolution of external skeletons in many cases were not well-preserved; traces of such organisms up to 700 million years old exist. The evolution of nearly all of the larger groups of organisms occurred during the Cambrian Period.

*Multicellular fossils first appear about 630 million years ago, with traces of multicellular animals preserved in rocks up to 700 million years old.*

## ORIGINS OF MAJOR GROUPS OF ORGANISMS

The Paleozoic Era was the time of origin and early diversification for virtually all of the major groups of organisms that survive at the present time, except for the plants (see inside back cover). Its beginning, the Cambrian Period (590 to 505 million years ago), was the period when multicellular animals became more diverse more rapidly (in geological terms) than was ever the case later. **Phyla** (singular **phylum**), the major groups into which kingdoms are divided (such as mollusks, sponges, or flatworms), appeared mainly at this time and exclusively in the sea. In contrast, plants originated on land, although they were derived from aquatic ancestors, the green algae. The basic record of the diversification of animal life on earth is a marine record, and the fossils that we have from the Paleozoic Era all originated in the sea.

Many of the multicellular animals that occurred in the early Paleozoic Era have no living relatives. In studying their remains in the rocks, one senses that this was a period of experimentation with different body forms and ways of life (Figure 17-9). Multicellular organisms seem to have diversified along many new evolutionary pathways, some of which ultimately led to the contemporary phyla of animals, and others to extinction. For example, the trilobites (see Figure 17-2) appear to be the ancestor of at least one living group, the horseshoe crabs, whereas the ammonites, abundant 100 million years ago, have no surviving descendants.

**Figure 17-8**

**Fossils found in South Australia.** All three are metazoans, or simple sponges.
**A** *Dickinsonia.*
**B** *Spriggina.*
**C** *Mawsonites.*

**Figure 17-9**

**Much of life's diversity is now extinct.** About 530 million years ago, the early Cambrian seafloor was inhabited by bizarre creatures such as *Hallucigenia* (**A**). The fossil shown here (**B**) is 12.5 millimeters long. This fossil and over a hundred other kinds are found together in the Burgess Shale, a formation that was formed from a fine-grained mud in the sea but has been uplifted to high elevations in the Rocky Mountains of British Columbia. Many of the fossils represent unique phyla. Only a few of the phyla present then still survive today.

*The early Paleozoic Era was a time of extensive diversification for marine animals. Many new kinds of animals appeared, and some of them have persisted to the present.*

Until the close of the first period of the Paleozoic, the Cambrian Period, all of the animals in the sea fed on unicellular organisms that floated freely in the water. During the following period, the Ordovician, true predators appeared, and the diversity of animals became even more extensive. The first corals also originated during the Ordovician Period and began to change the structure of marine communities permanently. By the end of this period, practically every mode of life that has ever existed had already evolved: for example, bottom feeders, scavengers, carnivores, colonies, and drifting multicellular forms. With all of the adaptive zones in which these animals occur having been filled, later opportunities for the additional evolution of novel forms have been much more limited.

## THE INVASION OF THE LAND

Only a few phyla, or groups, of organisms have invaded the land successfully; most others have remained exclusively marine. The first organisms that colonized the land were the plants, and they did so about 410 million years ago. The features of plants evolved in relation to their colonization of the land and were those which made them best suited for such colonization. Among these features are cuticle and bark, water-resistant outer coverings with specialized openings for gas exchange penetrating them (see Chapter 31); tissues for conducting water and nutrients; and drought-resistant pollen grains and spores. The ancestors of plants were specialized members of a group of photosynthetic protists known as the green algae. Although most algae are aquatic, the immediate ancestors of plants might themselves have been semiterrestrial. But it was with the plants themselves that the occupation of the land truly began (Figure 17-10, *A*).

The second major invasion of the land, and perhaps the most successful, was by the **arthropods,** a phylum of hard-shelled animals with jointed legs and a segmented body (Figure 17-10, *B*). Arthropods were originally marine organisms; the trilobites shown in Figure 17-2 were members of this phylum. Among the descendants of the first arthropods to invade the land are the insects, which have wings (Figure 17-11). This second invasion of the land occurred at about the same.

**Figure 17-10**

**Fossils from the Silurian and Devonian Periods.**
**A** This fragment of an unknown plant from the Silurian period, about 423 million years ago, shows the earliest evidence of an apparently well-organized conducting strand, consisting of banded, elongated conducting cells surrounded by unbanded, smooth-walled cells. It cannot be determined from the material found so far whether this organism was terrestrial or aquatic, or whether it was actually a plant, a large alga, or a transitional form.
**B** A mite (phylum Arthropoda) from the Devonian Period, about 376 to 379 million years ago, of what is now New York State. Land animals appeared about 40 million years earlier than the time this mite lived, but they were initially very scarce.

time as the evolution of the plants, about 410 million years ago. The body plan of arthropods has proved so well adapted to life on land that insects and other classes of this phylum now represent a large majority of all species of organisms. Among the features that are important to the success of the arthropods on land are their drought-resistant cuticles and efficient structures for conserving water and exchanging gases with the atmosphere (see Chapter 27). It seems certain that the plants colonized the land before arthropods did because plants would have been essential sources of food and shelter for the first arthropods that emerged from the sea.

The third major invasion of the land was by the vertebrates (Figure 17-12). Vertebrates are members of a phylum of animals called the **chordates** (see Chapter 28), and they include our own ancestors. The first of the vertebrates were the amphibians, represented today by such animals as frogs, toads, and salamanders. The earliest amphibians known are from the end of the Devonian Period, appearing just over 360 million years ago (Figure 17-13). Among their descendants on land

**Figure 17-11**

**Basic body patterns were established long ago.** The fossil dragonfly, which is about 170 million years old (Jurassic Period), closely resembles its modern counterpart. It dramatically illustrates the establishment of modern groups of organisms in the Mesozoic Era; many of these have persisted to the present day.

are the reptiles. Different groups of reptiles, in turn, ultimately became the ancestors of the dinosaurs (Figure 17-14), the birds, and the mammals, as we shall discuss further in Chapter 18. The air-breathing lungs of vertebrates; their scales, fur, and feathers, which regulate heat loss; their efficient circulatory and waste-removal systems; and their internal fertilization are all traits that admirably fit vertebrates to life on land.

The fact that all three of these major groups of organisms—plants, arthropods, and vertebrates—colonized the land within a few tens of millions of years of one another is probably related to the development of suitable environmental conditions, such as the formation of a layer of ozone in the atmosphere, which blocked ultraviolet radiation. Ozone ($O_3$) forms in equilibrium with oxygen ($O_2$) and thus was not abundant until the activities of photosynthetic bacteria had elevated the level of oxygen in the atmosphere sufficiently. These conditions have allowed the existence of multicellular organisms in terrestrial habitats for more than 400 million years, about a tenth of the age of the earth.

In addition to the plants, arthropods, and vertebrates, a fourth large group that has colonized the land

| | |
|---|---|
| | Neogene |
| | 24 million years ago |
| | Paleogene |
| | 63 |
| Age of mammals | |
| | Cretaceous |
| | 138 |
| Age of dinosaurs | |
| | Jurassic |
| | 205 |
| Early mammals | Triassic |
| | 240 |
| | Permian |
| | 290 |
| Early reptiles | |
| | Carboniferous |
| Early amphibians | 360 |
| | Devonian |
| Early coelacanths | 410 |
| | Silurian |
| Early fish | 435 |
| | Ordovician |
| | 500 |
| Age of marine invertebrates | |
| | Cambrian |
| | 570 million years ago |
| | Precambrian |

**Figure 17-12**

**Evolutionary history of the vertebrates.** Successive groups of vertebrates have flourished for more than 400 million years of earth history.

**Figure 17-13**

**Amphibians were the first vertebrates to walk on land.** Reconstruction of *Ichthyostega*, one of the first amphibians with teleologically efficient limbs used for crawling on land and a relatively advanced ear structure for picking up airborne sounds. Despite these features, *Ichthyostega*, which lived about 350 million years ago, was still quite fishlike in overall appearance.

consists of the fungi, which constitute a distinct kingdom of organisms. The success of the fungi on land, where they probably have been present for as long as any group of organisms, might be related to the structure of their cell walls, which are rich in chitin, a water-impermeable substance that also forms the external skeletons of arthropods.

*Plants and arthropods colonized the land about 410 million years ago; amphibians, the first terrestrial vertebrates, arrived about 50 million years later. Fungi may also have colonized the land at about the same time as plants. Earlier, there were no terrestrial, multicellular organisms.*

## MASS EXTINCTIONS

One of the most prominent features of the history of life on earth has been the periodic occurrence of major episodes of extinction. During the course of geological time there have been five such events. In each of them a large proportion of the organisms on earth at that time became extinct. Four of these events occurred during the Paleozoic Era, the first of them near the end of the Cambrian Period (about 505 million years ago). At that time most of the existing families of trilobites (see Figure 17-2) became extinct. A second major extinction marked the close of the Ordovician Period, about 438 million years ago, and a third occurred at the close of the Devonian Period, about 360 million years ago.

The fourth and most drastic extinction event in the history of life on earth happened during the last 10

**Figure 17-14**

**Dinosaurs.** Some of the remarkable diversity of dinosaurs, as shown in a reconstruction from the Peabody Museum, Yale University.

million years of the Permian Period, which ended the Paleozoic Era, about 238 to 248 million years ago. It is estimated that approximately 96% of all species of marine animals that were living at that time may have become extinct! The last of the trilobites and many other groups of organisms disappeared forever. The fifth and most recent major extinction event occurred at the close of the Mesozoic Era, 65 million years ago. It is familiar to most of us as the time when dinosaurs became extinct.

*In the history of life on earth there have been five major extinction events. The extinction event that took place at the end of the Permian Period, when about 96% of the species of marine animals became extinct, was the most drastic.*

Large-scale extinction events other than the five major ones summarized above also have occurred. One surprising correlation that was reported in 1983 by John Sepkoski and David M. Raup of the University of Chicago is that such events appear to occur regularly every 26 to 28 million years. The most recent such event occurred about 11 million years ago. The search for the sort of cosmic event, such as periodic comet showers, that could cause major events of extinction to occur with such regularity is now under way, as other scientists attempt to test the validity of the correlation made by Sepkoski and Raup. It has been hypothesized that the major extinction event that ended the Mesozoic Era is correlated with the impact of a large meteorite, as we discuss below. Changes in the relative positions of the continents and oceans also have played an important role in extinctions on a regional scale. Numerous other hypotheses have been advanced in explanation of major extinction events, but there is no general agreement concerning which explanation is true. The extinctions almost certainly had multiple causes. Within our own lives, the activities of human beings are bringing about an episode of mass extinction fully comparable to anything that has occurred in the past (see Chapter 23).

## DAWN OF THE MESOZOIC ERA

The Mesozoic Era, which began about 248 million years ago and ended about 65 million years ago, was a time very different from the present and one of intensive evolution of terrestrial plants and animals. The major evolutionary lines on land had been established during the mid-Paleozoic Era, but the evolutionary expansion of these lines—a radiation of novel forms that led to the establishment of the major groups of organisms living today—took place in the Mesozoic Era. In tracing the evolution of these lines, we need to first consider the events that happened just before the Mesozoic Era, in the Permian Period (286 to 248 million years ago), a time of drought and extensive glaciation that concluded the Paleozoic Era.

The Permian Period, the last period of the Paleozoic Era, ended with the greatest wave of extinction in the history of life on earth. In the sea, only about 4% of the species survived. The Mesozoic and Paleozoic Eras were initially recognized as distinct from each other because of the effects of this major extinction event: the marine animals of the Paleozoic Era can be recognized instantly as different from those of the Mesozoic Era. The few kinds of marine organisms that survived into the Mesozoic Era, including gastropods (a group of mollusks that includes snails and their relatives) and bivalves (mollusks such as oysters and clams), crustaceans (a group of arthropods that includes crabs, shrimp, and lobsters), fishes (aquatic chordates), and echinoderms (a phylum that includes starfish and sea urchins) began to evolve rapidly during the Mesozoic Era, producing many new kinds of organisms. Some of these had ways of living that were radically different from those of their ancestors. For example, the first efficient burrowers appeared among the echinoderms.

Both on land and in the sea, the number of species of almost all groups of organisms has been climbing steadily since the Permian extinction 250 million years ago and is now at an all-time high. Even though the evolutionary radiation of marine organisms during this period has been spectacular, the story of the evolution of life on land during the Mesozoic Era is of even greater interest for us, for we are products of that history. Many of the significant events in the history of the vertebrates took place during the Mesozoic Era.

## THE HISTORY OF PLANTS

The earliest known fossil plants are from about 410 million years ago. By the close of the Paleozoic Era, plants had become abundant and diverse. Shrubs and then trees evolved and came to form forests. These "Carboniferous forests" in turn formed many of the great coal deposits that we are now consuming. Much of the land was low and swampy at this time and provided excellent conditions for the preservation of plant remains. In these coal deposits, we have a relatively complete record of the horsetails, ferns, and primitive seed-bearing plants that made up these ancient forests.

The last part of the Paleozoic Era was cool and dry. The swamps of the Carboniferous forests, which existed at a time of worldwide moist and warm climates, largely disappeared. The end of the Paleozoic Era seems to have been one of ecological stress, during which many new life forms originated. One of these groups was the conifers, a group of seed-bearing plants that is represented today by pines, spruces, firs, and similar trees and shrubs. Today the descendants of these conifers still form extensive forests in many temperate and subtropical areas. Seed-bearing plants with featherlike leaves, similar to the living group called cycads, became abundant in the Mesozoic Era and helped to give that period its nickname, "the age of dinosaurs and cycads." But ultimately, the flowering plants, which apparently originated during the second half of the Mesozoic Era, became the dominant group of plants on the land.

The oldest fossils definitely known to be flowering plants are about 127 million years old. It seems likely that the group actually originated somewhat earlier, but no one is certain how much earlier. Like the mammals, the flowering plants were for a long time a minor group; they have been more abundant than any other group of plants for about 100 million years.

The evolution of the flowering plants, which began in the Mesozoic Era, has continued strongly to the present. Today there are about 240,000 species of this large group, which greatly outnumbers all other kinds of plants. As the flowering plants became more diverse, so did the insects, which had feeding habits that were closely linked with the characteristics of the flowering plants (Figure 17-15); the two groups have evolved to-

**Figure 17-15**

**Plants and insects coevolved.** A blister beetle, *Pyrota concinna*, eating the petals of a daisy in Zacatecas, Mexico. Beetles were among the first visitors to the flowers of the early flowering plants, spreading pollen and thus bringing about cross-fertilization as the beetles flew from flower to flower.

gether. Indeed, all groups of terrestrial organisms, including mammals, birds, and fungi, have characteristics that are largely related to those of the flowering plants. These groups now dominate life on the land, literally making our world look the way it does look.

*The earliest fossil plants, about 410 million years old, evolved into others that formed extensive forests within 50 million years of their appearance. Conifers evolved during the Permian Period (286 to 248 million years ago). Together with the flowering plants, which appeared in the fossil record about 127 million years ago, conifers have come to dominate the modern landscape.*

## EXTINCTION OF THE DINOSAURS

Everyone is generally familiar with the disappearance of the dinosaurs, an event of global importance that took place about 65 million years ago at the end of the Cretaceous Period. Less discussed, but actually of more fundamental importance, was the disappearance of many other kinds of organisms at about the same time. Among the **plankton**—free-drifting protists and other organisms that are still abundant in the sea—many of the larger (but still microscopic) forms suddenly disappeared about 65 million years ago, and a much lower number of smaller ones took their place. The same rapid changes occurred in at least some nonplanktonic marine animal groups such as bivalves. The ammonites, a large and diverse group of relatives of octopuses with shells, abruptly disappeared. In 1980 a group of distinguished scientists, headed by physicist Luis W. Alvarez of the University of California, Berkeley, presented a controversial hypothesis about the reasons for this drastic change.

Alvarez and his associates discovered that the usually rare element iridium was abundant in a thin layer that marked the end of the Cretaceous Period in many parts of the world (Figure 17-16). Iridium is rare on earth but common in meteorites. Alvarez and his colleagues proposed that if a large meteorite, or asteroid, had struck the surface of the earth then, a dense cloud would have been thrown up. The cloud would have been rich in iridium, and as its particles settled, the iridium would have been incorporated in the layers of sedimentary rock that were being deposited at that time. By darkening the world, the cloud would have greatly slowed or temporarily halted photosynthesis and driven many kinds of organisms to extinction (Figure 17-17).

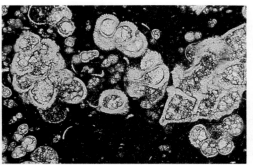
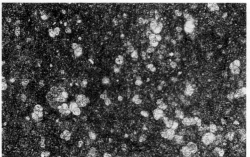

*Figure 17-16*

**The end of the Cretaceous Period can be seen in rocks.**
A The white limestone was deposited under the sea in the closing years of the Cretaceous Period, and the red limestone above was deposited during the first years of the Tertiary Period. They are separated by a layer of clay about 1 centimeter thick, in which iridium is abundant.
B Large, ornamented, diverse forams in sediments deposited during the late Cretaceous Period.
C Small, relatively unornamented, much less diverse forams deposited a few million years later, during the early years of the Tertiary Period. They are about one sixth the diameter of those that existed before the formation of the iridium layer.

Calculations have shown that the cloud produced by a meteorite about 20 kilometers in diameter would certainly have caused these effects to occur. Daytime conditions would have resembled those on a moonless night—below the amount of light required for photosynthesis—for several months. In biological communities such as the marine plankton, which are based directly on the continuous production of food by photosynthesis, such an effect would have had serious disruptive effects and could have produced the sudden change seen in the fossil record. Disruption of photosynthesis may also have been responsible for the extinction of certain other kinds of organisms—but is it reasonable to assume that it would have done away with the dinosaurs?

The Alvarez hypothesis has been controversial. It is not clear that the dinosaurs became extinct suddenly, as they should have if driven to extinction by a meteorite collision. Also, it is not clear that other kinds of animals and plants show the types of patterns that would have been predicted from these effects. Whether or not a meteorite impact caused widespread extinction 65 million years ago is a hypothesis that is still under active consideration.

*The occurrence of a worldwide layer rich in iridium 65 million years old suggests that a giant meteorite struck the earth at that time and threw up a huge cloud. The role of this cloud in the extinction of organisms is under active investigation.*

## THE CENOZOIC ERA: THE WORLD WE KNOW

We conclude this chapter with a brief account of some of the major evolutionary changes that have occurred during the past 65 million years, changes that have resulted in the conditions we now experience. The relatively warm and moist climates of the early Cenozoic Era have gradually given way to today's climates. By making the establishment of the Circumantarctic Current possible, the final separation of South America and Antarctica about 27 million years ago set the stage for worldwide glaciation. The ice mass that has been formed as a result of this glaciation has made the climate cooler near the poles, warmer near the equator, and drier in the middle latitudes than ever before.

In general, forests covered most of the land area of the continents, except for Antarctica, until about 15 million years ago, when they began to recede rapidly. During the time when these forests were receding and modern plant communities were appearing, some of the continents were approaching one another again after having been widely separated during most of the Cenozoic Era. The organisms in Australia and South America, particularly, evolved in isolation from all those in the rest of the world, mainly during the Cenozoic Era. Evolution in such isolated regions has been responsible for the distinctive characteristics of the groups of plants and animals found in different regions of the world. During the past several million years, the formation of extensive deserts in northern Africa, the Middle East, and India made migration between Africa and Asia very difficult for the organisms of tropical

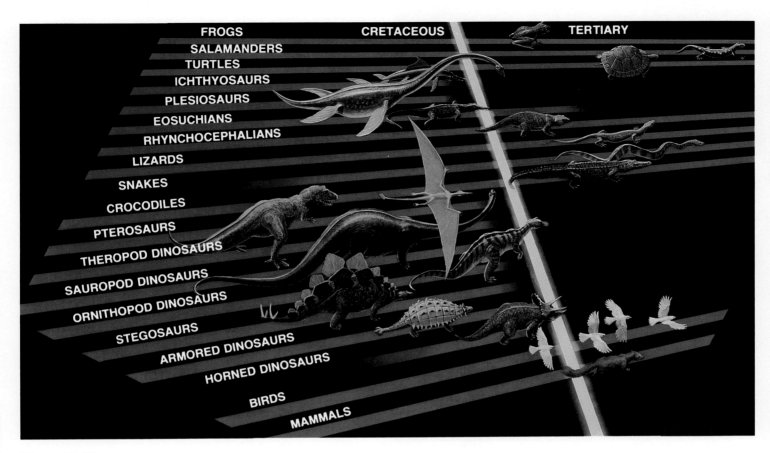

FROGS  CRETACEOUS  TERTIARY
SALAMANDERS
TURTLES
ICHTHYOSAURS
PLESIOSAURS
EOSUCHIANS
RHYNCHOCEPHALIANS
LIZARDS
SNAKES
CROCODILES
PTEROSAURS
THEROPOD DINOSAURS
SAUROPOD DINOSAURS
ORNITHOPOD DINOSAURS
STEGOSAURS
ARMORED DINOSAURS
HORNED DINOSAURS
BIRDS
MAMMALS

*Figure 17-17*

**Extinction of the dinosaurs.** The dinosaurs became extinct 65 million years ago *(yellow line)* in a major extinction event that also eliminated all the great marine reptiles (plesiosaurs and ichthyosaurs), as well as the largest of the primitive land mammals. The birds and smaller mammals survived and went on to occupy the aerial and terrestrial niches left vacant by the dinosaurs. Crocodiles, small lizards, and turtles also survived, but reptiles never again achieved the diversity of the Cretaceous Period.

forests. This formation of desert barriers, in turn, provided further opportunities for evolution in isolation. In general, the overall character of the Cenozoic Era has been set by a deteriorating climate, sharp differences in habitats even within small areas, and the regional evolution of distinct groups of plants and animals. These factors have enhanced the opportunities for the rapid formation of many new species by the processes outlined in Chapter 16.

*Throughout the 65 million years of the Cenozoic Era, the world climate has deteriorated steadily, and the distributions of organisms have become more and more regional in character.*

## LOOKING BACK

In concluding this chapter, it is useful to reflect on just how recent most of the evolutionary processes we have considered actually have occurred. If, as illustrated in Figure 4-9, we view the history of life on earth as having taken place during a single 24-hour day, then eukaryotes originated at about 4 PM, multicellular organisms at about 8 PM, and the invasion of the land just before 10 PM. Dinosaurs walked the earth from about 10:52 PM until about 11:40 PM, and the whole age of mammals, of which we are now the dominant product, has taken place in the last 20 minutes of our 24-hour earth history. Humans have been present on our 24-hour earth for only 40 seconds!

For most of the history of life on earth, bacteria were the only organisms present. The changes in the atmosphere that they brought about, and the symbiotic

events in which they participated, made possible the evolution of the protists—the first eukaryotes—and eventually the animals, plants, and fungi. Overall, there has been a great increase in structural complexity and in the number of species of organisms on earth (Figure 17-18), but most of the metabolic diversity that exists today was already present in our bacterial ancestors 2 billion years ago.

*Figure 17-18*

**A close relative?** In evolutionary terms, this orangutan is thought to be a close relative of ours. The possibility that this might be true has disturbed many people since Darwin first established a logical explanation for evolution, namely natural selection, in 1859. Although the theory of evolution by natural selection is now accepted by almost all biologists and indeed forms the foundation of modern biology, it still stirs heated debate in nonscientific circles.

## ■ SUMMARY

1. Fossils provide a record of life in the past. They occur largely in sedimentary rocks, with the actual organic remains gradually being replaced by minerals.

2. The earth originated about 4.6 billion years ago, and the oldest rocks are about 3.8 billion years old. Fossil bacteria about 3.5 billion years old are the oldest direct evidence of life on earth. Stromatolites, massive deposits of limestone formed by cyanobacteria, appear in the fossil record starting about 2.8 billion years ago.

3. The first unicellular eukaryotes appeared about 1.5 billion years ago; all earlier forms of life were bacteria. Multicellular animals appeared about 700 million years ago; the earliest, soft-bodied forms are poorly represented in the fossil record.

4. During Phanerozoic time, which started about 630 million years ago, fossils of multicellular organisms become frequent. Phanerozoic time is divided into three eras: the Paleozoic Era, 590 to 248 million years ago; the Mesozoic Era, 248 to 65 million years ago; and the Cenozoic Era, 65 million years ago to the present.

5. Judged primarily from the fossil record, all extant phyla of organisms except plants seem to have evolved during the Cambrian Period (590 to 505 million years ago).

6. The evolution of hard skeletons, including shells and similar structures, about 570 million years ago was correlated with the appearance of many diverse modes of life among animals.

7. Plants and terrestrial arthropods appeared about 410 million years ago; terrestrial vertebrates (amphibians) appeared about 360 million years ago. These groups of organisms, together with the fungi, have dominated life on the land since then.

8. The Paleozoic Era ended with the Permian Period (286 to 248 million years ago), a cold and dry period during which conifers and reptiles appeared.

9. The Mesozoic Era was a time when the outlines of life on earth as we know it were established. Flowering plants had become dominant by the end of the era; at this time, insects, mammals, birds, and other groups had begun to evolve in relation to the diversity of these plants.

10. There have been five major episodes of mass extinction during the history of life on earth. The most drastic was that at the end of the Permian Period, about 248 million years ago, when some 96% of marine animals became extinct. At the end of the Paleozoic Era, 65 million years ago, the dinosaurs and many other kinds of organisms disappeared.

11. The diversity of living things, measured in terms of species number, has been increasing steadily for the past 250 million years and is now at an all-time high. The wide separation of the continents and their arrangement into novel configurations probably has played a role in this trend.

## REVIEW

1. The giant land mass that was formed when all the continents united is called _____.

2. In biological classification, major groups into which kingdoms are divided are called _____.

3. The oldest fossils, which are bacteria, are about _____ years old.

4. What four major groups of organisms colonized the land from the sea?

5. The most drastic extinction event in the history of life on earth happened at the end of the _____ period.

## SELF-QUIZ

1. Place the following periods and eras in order, oldest first.
   (a) Permian
   (b) Cambrian
   (c) Cretaceous
   (d) Cenozoic
   (e) Devonian

2. Which of the following invaded land *first?*
   (a) Fungi
   (b) Arthropods
   (c) Plants
   (d) Vertebrates
   (e) We don't know

3. A stromatolite is a
   (a) colony of fossil cyanobacteria.
   (b) limestone deposit.
   (c) kind of fossil coral reef.
   (d) coelacanth.
   (e) mass of compacted seashells.

4. During the extinction that marked the end of the Cretaceous Period, which of the following forms did *not* become extinct?
   (a) Flowering plants
   (b) Smaller plankton
   (c) Ammonites
   (d) Mammals
   (e) Fishes

5. Which of the following continents approached one another most recently?
   (a) Australia—Antarctica
   (b) Europe—North America
   (c) India—Asia
   (d) North America—South America
   (e) South America—Antarctica

6. Before 1.5 billion years ago, the only forms of life on earth were
   (a) marine organisms.
   (b) cyanobacteria.
   (c) bacteria.
   (d) viruses.
   (e) soft-bodied.

7. The most drastic extinction event in the history of life on earth occurred during the
   (a) Cretaceous Period.
   (b) 20th century.
   (c) Permian Period.
   (d) Cambrian Period.
   (e) Devonian Period.

8. The most plausible explanation of the disappearance of the dinosaurs appears to be
   (a) that a large meteorite struck the earth.
   (b) that volcanoes threw up large clouds of smoke and darkened the sky.
   (c) that plants evolved defenses against them and became unpalatable.
   (d) that the climate changed, putting them at a disadvantage.
   (e) that the mammals took over, forcing them into extinction.

# THOUGHT QUESTIONS

1. Dinosaurs and mammals both lived throughout the Mesozoic Era, a period of more than 150 million years; all this time the dinosaurs were the dominant form, mammals being a minor group. Both mammals and small reptiles survived the Cretaceous extinction. Why do you suppose reptiles did not go on to become dominant again, rather than mammals?

2. There seems to be no iridium layer preserved in the rocks from four of the five major periods when mass extinction occurred. What are some of the ways in which one could account for these four extinction events?

3. Separation of Australia and South America from Antarctica led to the formation of a huge sheet of ice over Antarctica. Do you think that a land bridge connecting Alaska to Siberia would have had a similar effect, forming a huge Arctic ice mass over the North Pole?

# FOR FURTHER READING

ALVAREZ, W. and ASARO, F.: "What caused the mass extinction?—an extra-terrestrial impact," *Scientific American,* October 1990, pages 78-84. A debate on the cause of the extinction of the dinosaurs. The discoverers of iridium deposited at the time argue that a giant meteorite collided with the earth.

CORTILLOT, V.: "What caused the mass extinction?—a volcanic eruption," *Scientific American,* October 1990, pages 85-92. In a companion article, geologists argue that the iridium came not from a meteorite but from within the earth, blown into the sky by enormous volcanoes produced when plumes in the earth's mantel reached the surface 65 million years ago.

GORE, R.: "Extinctions," *National Geographic,* June 1989, pages 662-699. A superbly illustrated article about the events that have changed the nature of life in the past and are continuing at present.

GOULD, S.J.: *Wonderful Life. The Burgess Shale and the Nature of History,* W.W. Norton and Company, New York, 1989. A marvelous account of the discovery, study, and interpretation of the world's oldest fossil animals.

LESSEM, D.: "Secrets of the Gobi Desert," *Discover,* June 1989, pages 40-46. New fossil discoveries provide evidence about biological interchange between continents in ancient times.

RAUP, D.M.: *The Nemesis Affair,* W.W. Norton Co., New York, 1985. An account by one of the co-discoverers of the 26 million-year extinction cycle of how they discovered it and how it was treated in the media. Fascinating reading.

RICHARDSON, J.: "Brachiopods," *Scientific American,* September 1986, pages 100-106. A fascinating group of marine organisms, well represented in the fossil record, illustrates many of the kinds of changes that have occurred during the history of life on earth.

# How We Evolved: Vertebrate Evolution

Gorillas, chimpanzees, and humans are so similar that were they any other group scientists would probably call them the same genus. It is a tragedy that all wild populations of primates are in imminent danger of extinction.

# HOW WE EVOLVED: VERTEBRATE EVOLUTION

## Overview

The group of animals to which we belong, the vertebrates, originated in the sea as jawless fishes more than half a billion years ago. The story of their subsequent evolution—the intricate path by which we came to be what we are—is often surprising, but always of great interest to us because we are products of that evolution. Among living vertebrates, bony fishes are the most plentiful today in terms of species number, but it is the evolution of land vertebrates, which began about 350 million years ago, that captures our attention. The sequence leads from partial land-dwellers, the amphibians, to truly terrestrial animals, the reptiles. Reptiles, in turn, independently gave rise to the birds and the mammals, ultimately including human beings. Apes have existed for at least 36 million years, and our immediate ancestor, *Australopithecus,* was derived from this evolutionary line in Africa perhaps 5 million years ago. The genus *Homo* has existed for about 2 million years, *Homo sapiens,* our species, for perhaps 500,000.

## For Review

*Here are some important terms and concepts that you will encounter in this chapter. If you are not familiar with them, you should review them before proceeding.*

**Theory of evolution** (Chapter 15)

**How species evolve** (Chapter 16)

**Major features of evolutionary history** (Chapter 17)

**Origin of mammals** (Chapter 17)

The first vertebrates were jawless fishes that swam in the seas some 550 million years ago. None of these early fishes live today, but they represent a hallmark in evolutionary history, for from them evolved all of the large animals, such as lions and sharks and eagles. Most living vertebrates—over half of the 42,500 species—are fishes. The first vertebrates to walk on land were amphibians, ancestors of today's frogs and toads. Although they evolved from bony fishes about 350 million years ago, they have never fully adapted to life away from water. The first vertebrates to conquer the land completely were their descendants, the reptiles. In the milder climates of 65 million years ago, reptiles were the most successful of all terrestrial vertebrates, dominating the air, land, and sea. From different groups of reptiles, in turn, were derived the two most successful terrestrial groups of vertebrates today, the mammals and birds. In this chapter we review the evolutionary journey from jawless fishes to mammals such as ourselves (Figure 18-1).

Figure 18-2 shows the evolutionary relationship between the seven major groups, or **classes,** of vertebrates that have living representatives. Phyla are divided into classes, which in turn are divided into

### Figure 18-1

**The trail of our ancestors.** These fossil footprints were made in Africa 3.7 million years ago. A mother and child walking on the beach might leave such tracks. But these tracks are not human. Preserved in volcanic ash, these tracks record the passage of two individuals of the genus *Australopithecus,* the group from which our genus, *Homo,* evolved.

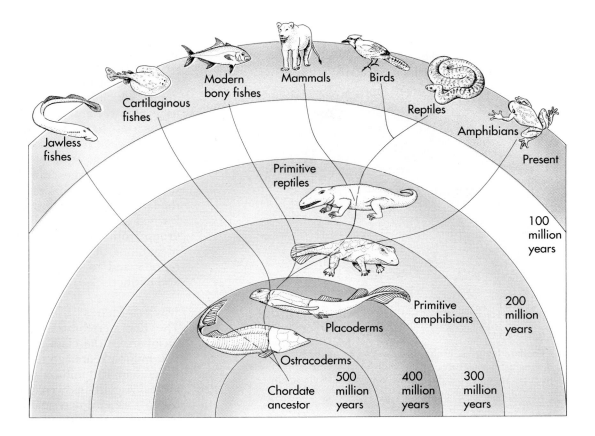

**Figure 18-2**

**The seven classes of vertebrates.**

smaller categories, **orders,** such as rodents, carnivores, and bats among mammals. These orders in turn consist of one or more **families,** such as cats and weasels among the carnivores. In turn, families include **genera** and **species.** Such a system of classification, in which successively smaller units of classification are included within one another, like boxes within boxes, is called a **hierarchical** system. It is illustrated further in Chapter 24.

## GENERAL CHARACTERISTICS OF VERTEBRATES

### Vertebrates are Chordates

Vertebrates are members of the phylum Chordata, characterized by a flexible rod that develops along the back in embryos of this class. The simplest chordates are tunicates, marine animals that are anchored to one spot as adults and look a bit like jugs. Adult tunicates develop from tadpolelike larvae with a notochord.

The approximately 42,500 species of chordates, including tunicates, birds, reptiles, amphibians, fishes, and mammals, are distinguished by three principal features, not all necessarily present in adult animals, but present during the course of development: (1) a single, dorsal (along the back), hollow **nerve cord,** or main trunk, to which the nerves that reach the different parts of the body are connected; (2) a rod-shaped **notochord,** which forms between the nerve cord and the developing gut (which becomes the stomach and intestines) in the early embryo; and (3) **pharyngeal slits.** The **pharynx** is a muscular tube that connects the mouth cavity and the esophagus. It serves as the gateway to the digestive tract and to the windpipe, or **trachea.**

### Vertebrates Have a Backbone

With the exception of tunicates and a small group of fishlike marine animals, the lancelets, which are considered in Chapter 28, all chordates are vertebrates. In vertebrates, the notochord becomes surrounded and then replaced during the course of the embryo's development by a bony **vertebral column,** a tube of hollow bones called vertebrae, which encloses the dorsal nerve cord like a sleeve and protects it. In addition, vertebrates (except for the agnathans) have a distinct and well-differentiated head; as a result, they are sometimes called the **craniate** chordates (Greek *kranion,* skull). Most vertebrates have a bony skeleton, although the living members of two of the classes of fishes, Agnatha (lampreys and hagfishes) and Chondrichthyes (sharks and rays), have a cartilaginous one. In vertebrates the notochord becomes surrounded and then replaced during the course of embryological development by the

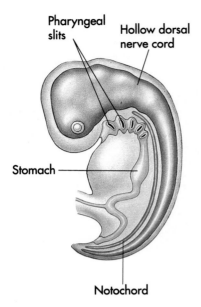

Pharyngeal slits

Hollow dorsal nerve cord

Stomach

Notochord

**Figure 18-3**

**Chordate features.** Embryos reveal some of the principle distinguishing features of the chordates, even though these features may be lost during the course of development.

vertebral column. But all of the features of chordates are evident in their embryos, even among the most advanced members of the group (Figure 18-3).

> *Vertebrates are a group of chordates characterized by a vertebral column surrounding a dorsal nerve cord.*

## THE FIRST VERTEBRATES

The first vertebrates were fishes. They evolved in the ocean some 550 million years ago, when seas covered much of what is now land. Their fossils are found on every continent, and their descendents, the modern fishes, are the most numerous and diverse of the vertebrates. Fishes live in almost every imaginable aquatic habitat, including small puddles, torrents, and in the perpetual dark tens of thousands of meters beneath the surface of the ocean. Fishes can endure temperatures varying from hot springs at nearly 40° C to near-freezing polar waters.

One of evolution's great successes, fishes have had a long history of change (Figure 18-4). Ancestral fishes,

which were jawless and lacked paired fins, eventually gave rise to five major evolutionary lines, all possessing jaws. Two of these groups are now extinct: the placoderms, which had extensive body armor, and the spiny fishes, bizarre animals with spiny fins. The third group consists of sharks and rays, whose skeletons are made of cartilage, rather than bone, as is found in the remaining two groups, the bony fishes. Most living fishes are ray-finned fishes, but there are a few survivors of a second, ancient group, the lobe-finned fishes. Among the latter are the coelacanth and the lungfishes that you learned about in Chapter 17; they are members of the group from which amphibians were derived hundreds of millions of years ago, ultimately giving rise to the reptiles and to all other terrestrial vertebrates, living and extinct.

### Jawless Fishes

The first vertebrates to evolve were jawless fishes, members of the class Agnatha, which appeared about 550 million years ago in the mid-Cambrian Period. For more than 100 million years, they were the *only* vertebrates. Agnathans were small and are thought to have fed in a head-down position, their fins acting as stabilizers while their small mouths sucked up organic particles from the bottom.

Many of the major groups of agnathans are extinct. The living ones, the lampreys and hagfishes, belong to a single, fairly uniform group of scaleless, eel-like fishes. In these living agnathans, the notochord persists throughout the life of the animal, and the body is supported by an internal skeleton made of cartilage. But we know from their fossils that the ancestral agnathans had bony skeletons and therefore that skeletons of this kind were lost during the course of evolution of the living ones.

Lampreys have round mouths that function like suction cups (Figure 18-5). Using these specialized mouths, lampreys attach themselves to the outsides of bony fishes. Once attached, they rasp through the skin of the fish with their tongues, which are covered with sharp spines, sucking out the blood of the fish through the hole. Sometimes lampreys can be so abundant as to constitute a serious threat to commercial fisheries.

> *Jawless fishes, the agnathans, were the first vertebrates; in fact, for more than 100 million years they were the only vertebrates. Living representatives are the lampreys and hagfishes.*

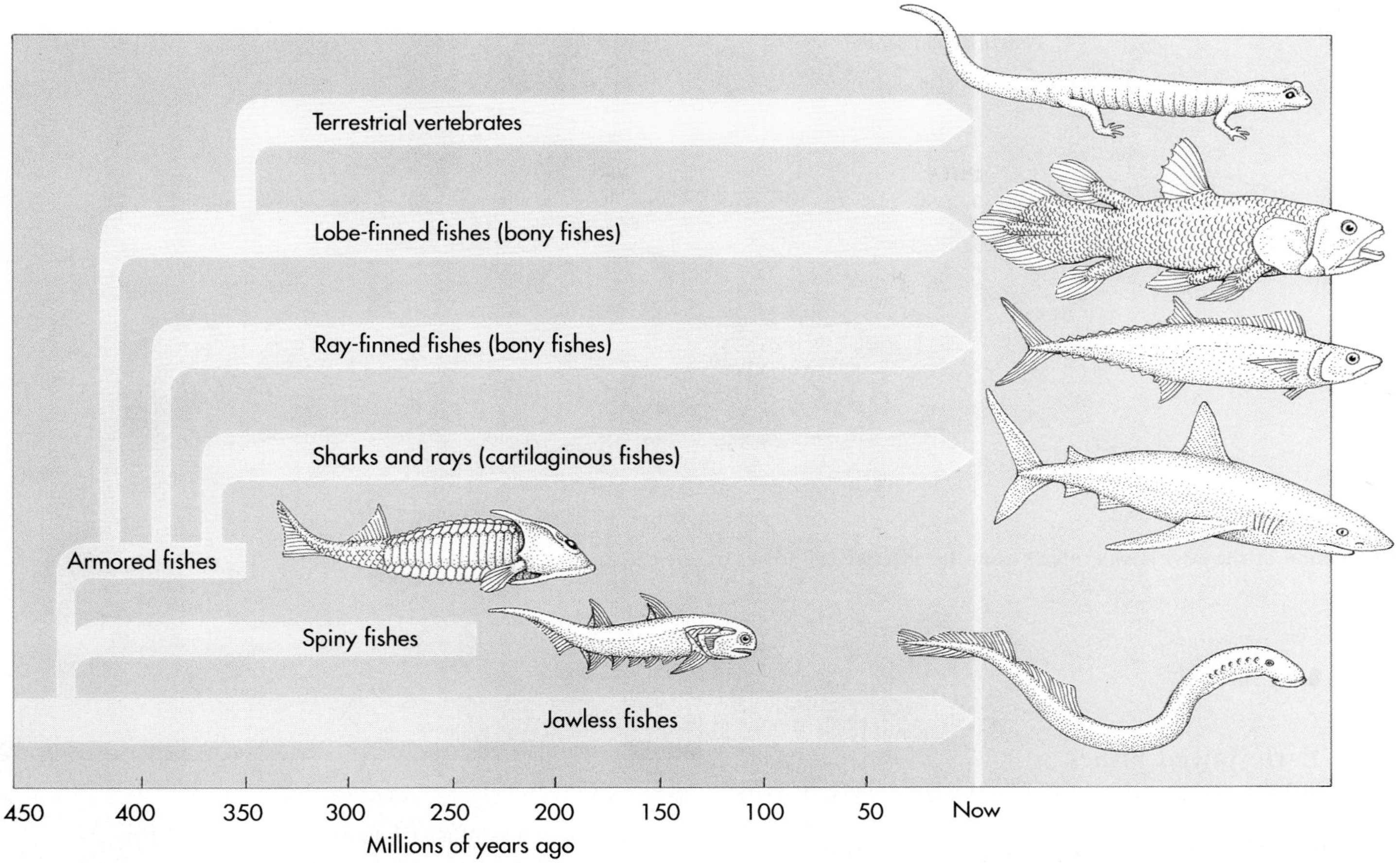
Terrestrial vertebrates
Lobe-finned fishes (bony fishes)
Ray-finned fishes (bony fishes)
Sharks and rays (cartilaginous fishes)
Armored fishes
Spiny fishes
Jawless fishes
450
400
350
300
250
200
150
100
50
Now
Millions of years ago

*Figure 18-4*

**Evolution of the fishes.**

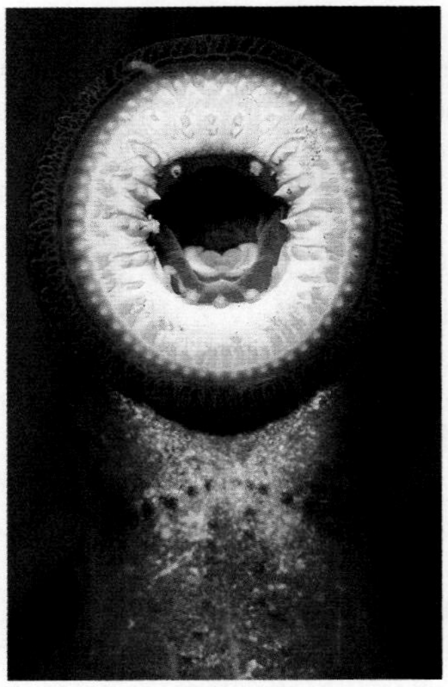

*Figure 18-5*

**A lamprey.** Lampreys attach themselves to fishes by means of their suckerlike mouths, rasp a hole in the body cavity, and suck out the blood and other fluids within.

## THE APPEARANCE OF JAWED FISHES

### The Evolution of Jaws

Jaws first developed among vertebrates that lived about 410 million years ago, toward the close of the Silurian Period. The biting jaws characteristic of modern vertebrates were produced by the evolutionary modification of one or more of the gill arches, originally the areas between the gill slits (Figure 18-6). Jaws allowed ancient fishes to become proficient predators, to defend themselves, and to collect food from places and in ways that had not been possible earlier. Fishes were able to bite and chew their food instead of sucking it or filtering it in as did all of the more primitive chordates. Because of these features, the early jawed fishes largely replaced their jawless ancestors over the course of time. Thus there are only about 63 living species of agnathans, whereas bony fishes have become dominant throughout the world's waters. Of the approximately 42,500 species of living chordates, about half are fishes.

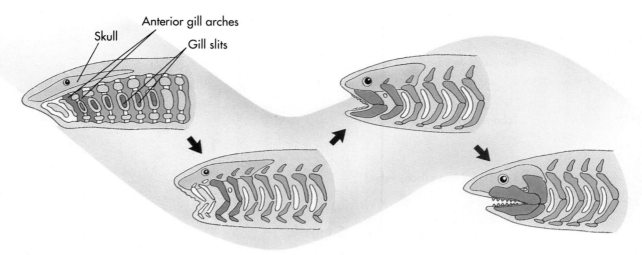

**Figure 18-6**

**Evolution of the jaw.** Jaws evolved from the anterior gill arches of ancient, jawless fishes.

## The Early Jawed Fishes

The earliest jawed fishes, the placoderms, were heavily armored and much larger than agnathans—the largest approached 10 meters in length. Placoderms dominated the seas for 50 million years and then became extinct 345 million years ago, replaced by sharks and bony fishes. A second group that developed jaws were the spiny fishes. They had strong spines along the edges of their fins and a torpedo-shaped body with numerous paired fins. All are now extinct.

## Sharks and Rays

Sharks appeared about 390 million years ago; they probably evolved from placoderm ancestors. Sharks are members of the class Chondrichthyes, which also includes dogfish, skates, and rays (Figure 18-7). Many hundreds of extinct species of Chondrichthyes are known from the fossil record, but there are only about 850 living ones. These fishes are mostly scavengers and carnivores. In the living species, the internal skeleton is composed of cartilage—a soft, light, and elastic material. Such skeletons are lighter and more buoyant than those of the kinds of fishes that the Chondrichthyes replaced. In addition to the evolutionary advance represented by such a skeleton, the Chondrichthyes were the first vertebrates to develop large, strong, mobile fins, which enabled them to pursue moving prey by quick adjustment of their motion through the water. With these advantages, they soon largely replaced placoderms throughout the world.

*The sharks and other Chondrichthyes were the first jawed vertebrates and the first to develop efficient fins. They evolved from placoderms that had bony skeletons, but soon developed cartilaginous ones.*

The skin of Chondrichthyes is covered everywhere with small, pointed **denticles,** similar in structure to the teeth of other vertebrates. As a result, the skin of these fishes has a rough, sandpaper-like texture. Their teeth, which are abundant in the fossil record, are enlarged versions of these denticles.

Sharks drive themselves through the water by sinuous motions of the whole body and by their thrashing tails. Such motion tends to drive the sharks downward, but this tendency is corrected by their two spreading pectoral fins. In the skates and rays, these pectoral fins have become enlarged and are undulated when these fishes move, which gives these animals a very characteristic appearance.

For their supply of oxygen, many sharks (and those bony fishes which swim constantly) depend on a constant stream of water that is forced over the gills and brings dissolved oxygen with it; others are able to pump water through their gills while they are stationary. Fishes that obtain their oxygen by moving con-

**Figure 18-7**

**Sharks and rays.**
A  Blue shark.
B  Manta ray.
C  Diamond sting ray.

stantly can literally drown if they are prevented from swimming. Drowned sharks are often found trapped in the nets used to protect Australian beaches.

## BONY FISHES

The vast majority of the more than 18,000 known species of fishes are bony fishes (class Osteichthyes). The first bony fishes were lobe-finned. This group was once quite abundant, but today only four genera survive, including the lungfishes and coelacanth, which we discussed in Chapter 17. It is from ancient members of this group of fishes that amphibians evolved.

The earliest ray-finned fishes, which appeared more than 300 million years ago, had a strong body armor consisting of heavy, overlapping scales. Fishes of this kind were dominant for 200 million years but have been replaced for the last 100 million years by **teleost** fishes, which evolved a unique new way to stay stationary in the water with no effort. Teleost fishes possess a **swim bladder** (Figure 18-8). A swim bladder is a gas-filled sac that allows the fishes to regulate their buoyant density; for this reason, they can remain suspended at any depth in the water without moving their fins. Swim bladders apparently evolved as outpocketings of the pharynx, specialized for respiration, in fishes of the Devonian Period.

Stationary bony fishes do not drown because they use muscles to draw water through their gills by way of their gill slits. These gill slits are simply the pharyngeal slits that are characteristic of all chordates.

Bony fishes are the most diverse of all groups of vertebrates. About half of all species of living vertebrates are bony fishes. The first bony fishes evolved in

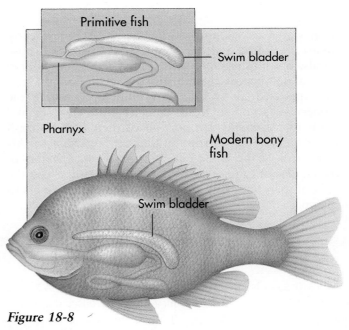

**Figure 18-8**

**Diagram of a swim bladder.** This structure, which evolved from an outpocketing of the pharynx, is used by the bony fishes to control their buoyancy in water.

**Figure 18-9**

**The bony fishes.** The bony fishes (class Osteichthyes) are extremely diverse.
**A** Koran angelfish, *Pomacanthus semicircularis*, in Fiji.
**B** Puffer. Puffers, which are also marine fishes, avoid being eaten by inflating themselves when attacked, thus projecting their spines outward.
**C** A sea horse. Sea horses move slowly and are difficult to see among marine algae and sea grasses.
**D** Moray eel, Rangiroa, French Polynesia.

fresh water, as we know from the places where early fossils are found and from the characteristics of those fossils. Bony fishes apparently entered the oceans only after a number of their distinctive evolutionary lines had appeared. Bony fishes are abundant both in the modern seas and in fresh water (Figure 18-9). Many kinds of fishes spend a portion of their lives in fresh water and another portion in the sea. For example, Atlantic and Pacific salmon are born in the headwaters of rivers, descend the rivers to the sea where they lead the bulk of their lives, and then return to the same rivers to spawn.

*Bony fishes (class Osteichthyes) possess swim bladders, which enable them to increase their buoyancy and thus offset the greater weight of a bony skeleton. They account for about half of all living vertebrate species.*

## THE INVASION OF THE LAND

About 350 million years ago, one of the lobe-finned fishes took the first steps in the evolutionary journey onto land. It certainly did not resemble closely the living lobe-finned fishes, the coelacanth and lungfishes. This ancient, ancestral fish, which was eventually to become the ancestor of all **tetrapods** (terrestrial, basically four-limbed vertebrates), perhaps began to explore an existence on land at the margins of freshwater ponds or swamps.

### Key Adaptations to Survival on Land

In the animals which ultimately became the earliest amphibians, primitive lungs gradually evolved into the efficient air-breathing organs seen in modern tetrapods. These lungs were originally supplementary organs, making it possible to breathe air directly when this was necessary; lungs of this sort still occur today in the lungfishes.

A second critical change occurred in the circulation system, where a new blood vessel now carried

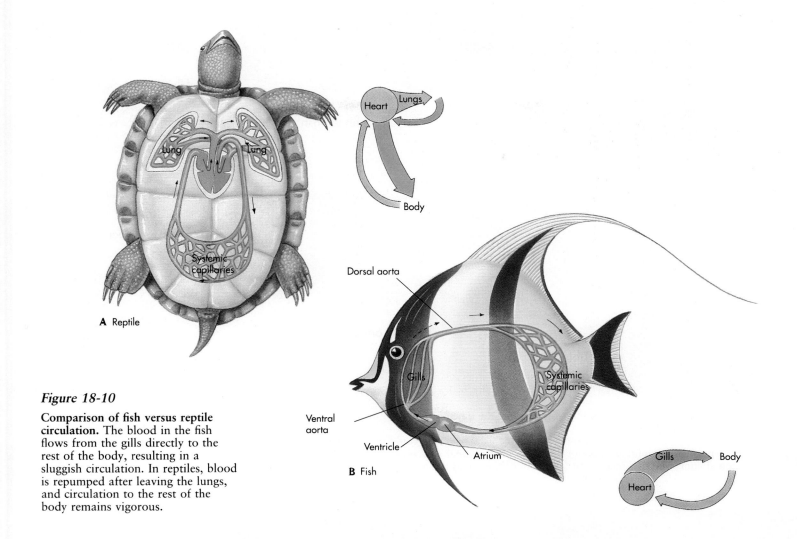

**A** Reptile

**B** Fish

*Figure 18-10*

**Comparison of fish versus reptile circulation.** The blood in the fish flows from the gills directly to the rest of the body, resulting in a sluggish circulation. In reptiles, blood is repumped after leaving the lungs, and circulation to the rest of the body remains vigorous.

blood from the lung *back to the heart,* where it was pumped again to circulate it through the body (Figure 18-10, *A*). This solved a critical problem in fish blood circulation—passage of blood through fish gills slows its flow, so circulation to the rest of the fish body is sluggish (Figure 18-10, *B*).

A third critical change was the evolution of walking legs. In these early terrestrial vertebrates, efficient locomotion on land was gradually made possible by the evolution of strong skeletal supports in the **thoracic** portion of the body, that part just behind the head (Figure 18-11). Such supports provided a more rigid base for the limbs, which were derived from the kinds of fins found in the lobe-finned fishes.

These three evolutionary advances—that is, the abilities to use gaseous oxygen for respiration, to circulate it efficiently through the body, and to move from one place to another on land—became the basis of the success and later evolutionary radiation of vertebrates on land.

## AMPHIBIANS

The first land vertebrates were amphibians (Figure 18-12), members of a class that first became abundant about 300 million years ago during the Carboniferous Period. The early amphibians had rather fishlike bodies, short stubby legs, and lungs; ultimately, they gave rise to all of the other tetrapods. As a group, amphibians, unlike reptiles, birds, and mammals, depend on the availability of water during their early stages of development. Many amphibians live in moist places even when they are mature. There are about 4200 species. Members of this class were the dominant vertebrates on land for about 100 million years, until the reptiles gradually replaced them.

The two most familiar orders of amphibians are (1) those which possess tails—the salamanders, mud puppies, and newts, which compose the order Caudata, with about 369 species; and (2) those which do not have tails as adults—the frogs and toads, which com-

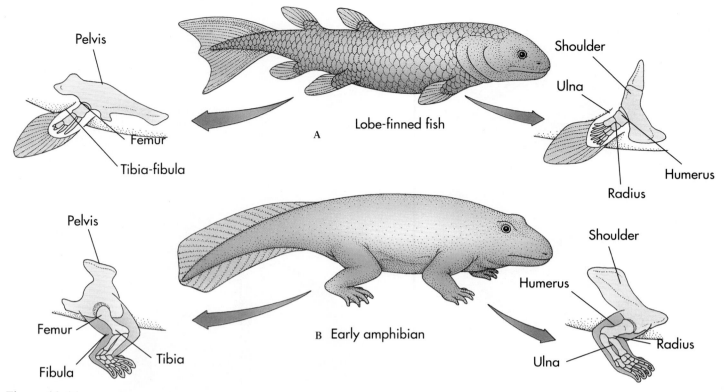

**Figure 18-11**

**Legs evolved from fins.** A comparison between the limbs of a lobe-finned fish and those of a primitive amphibian.
**A** A lobe-finned fish. Some of the animals of this appearance could move out onto land.
**B** A primitive amphibian. As illustrated by their skeletal structure, the legs of such an animal could clearly function on land better than could the fins of the lobe-finned fish.

**Figure 18-12**

**Amphibians.**
**A** Giant toad, *Bufo marinus*.
**B** Red-eyed tree frog, *Agalychnis callidryas*.
**C** Tiger salamander, *Ambystoma tigrinum*.
**D** Tennessee cave salamander, *Gyrinophilus palleucus*.

pose the order Anura, with about 3680 species. The first true lungs evolved in amphibians, but these organs were relatively inefficient. Respiration also takes place through their moist, glandular skin and through the lining of their mouth. The constant loss of water through the skins of amphibians is one of the reasons that these animals generally must remain in moist habitats. Amphibian larvae and adults of those species which remain permanently in the water respire by means of gills.

Amphibians lay their eggs, which lack water-retaining external membranes and shells and dry out rapidly, directly in water or in moist places. Anuran larvae are tadpoles, which usually live in the water, where they feed on minute algae. The adults, which are highly specialized for jumping and very different from the larvae in appearance, are carnivorous. In contrast to the anurans, young salamanders are carnivorous like the adults and look like small versions of them. Many salamanders swim efficiently and return to water to breed.

> *Amphibians are terrestrial but still depend on a moist environment; their eggs are laid in water, and the development of their larvae takes place there.*

Although amphibians appear primitive, they are in fact members of a successful group of animals, one that has survived over 300 million years. The amphibians evolved long before the dinosaurs and have, thus far, outlasted them by 65 million years.

## REPTILES

Reptiles (class Reptilia) have a dry skin, covered with scales, that efficiently retards water loss. As a result, the members of this large class, comprising nearly 6000 living species, are independent of water, unlike their ancestors. Nonetheless, of the the major orders of reptiles that have living representatives (Figure 18-13), the Crocodilia (crocodiles, alligators, and caimans) live in water, as do most of the members of the order Chelonia (turtles and tortoises). Members of a third, and by far the largest, order, Squamata (lizards, snakes, and other reptiles) are almost entirely terrestrial. Snakes evolved from lizards through the loss of their limbs and other evolutionary modifications.

The first reptiles appeared during the age of amphibian dominance, the Carboniferous Period, about

**Figure 18-13**

**A** River crocodile, *Crocodilus acutus*. Crocodiles, like birds, are related to dinosaurs.
**B** An Australian skink, *Sphenomorphus*. The snakes evolved from one line of legless lizards.
**C** Red-bellied turtles, *Pseudemys rubriventris*. This attractive turtle occurs frequently in the northeastern United States.
**D** Smooth green snake, *Opheodrys vernalis*. This harmless snake is common throughout the United States.

300 million years ago. Reptiles became abundant over the next 50 million years, gradually replacing amphibians. Immediately following the Carboniferous Period was the Permian Period (286 to 248 million years ago), a period of widespread glaciation and drought. The special adaptations of reptiles, including their water-re-

sistant skin and more efficient lungs, probably gave them an advantage as arid climates became widespread, and over the next 50 million years reptiles gradually replaced amphibians. The dinosaurs, by far the best-known group of fossil organisms, appeared near the start of the Mesozoic Era, at least 225 million years ago. They dominated the land for most of the following 180 million years, until the start of the Tertiary Period (Figure 18-14).

*Reptiles were the first truly terrestrial vertebrates. They have more efficient lungs than do amphibians and so do not need to breathe through their skin, which is dry and retards water loss efficiently.*

One of the most critical adaptations of reptiles in relation to their life on land is the evolution of the **amniotic egg**, which protects the embryo from drying out, nourishes it, and enables it to develop outside of water (Figure 18-15). Amniotic eggs, characteristic of reptiles and birds (and of the very few egg-laying mammals) retain their own water. They contain a large **yolk**, the primary food supply for the embryo, and abundant **albumen**, or egg white, which provides additional nutrients and water. An egg membrane, called the **amnion**, surrounds the developing embryo, enclosing a liquid-filled space within which the embryo develops. Blood vessels grow out of the embryo through the membranes of the egg to its surface, where they take in oxygen and release carbon dioxide. The egg is more easily permeable to these gases than to water. In most reptiles, the egg shell is leathery. In addition to the amniotic egg, another adaptation of reptiles to a terrestrial existence is the presence of copulatory organs, which the male inserts into the female. The presence of copulatory organs protects the eggs and sperm from drying out before their fusion.

*The key to successful invasion of the land by vertebrates, as it was by arthropods, was the elaboration of ways to avoid desiccation in dry air. In addition to a dry skin that retained water, a second pivotal innovation of reptiles was the watertight amniotic egg, which contains embryonic nutrients and is permeable to gases but not to water.*

*Figure 18-14*

**Early in the age of dinosaurs.** The reptiles on the left, from the Carboniferous Period, are not dinosaurs. The first dinosaurs evolved in the early Triassic (the center of the scene).

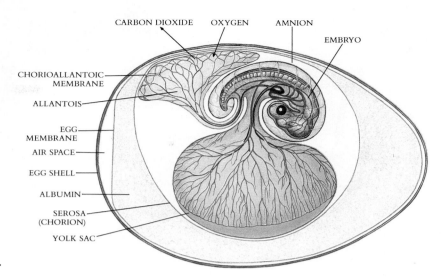

CARBON DIOXIDE   OXYGEN   AMNION

EMBRYO

CHORIOALLANTOIC
MEMBRANE

ALLANTOIS

EGG
MEMBRANE

AIR SPACE

EGG SHELL

ALBUMIN

SEROSA
(CHORION)

YOLK SAC

**Figure 18-15**

**Evolution of a water-tight egg.** The amniotic egg is perhaps the most important feature that allows reptiles to live in a wide variety of terrestrial habitats.

## TEMPERATURE CONTROL IN LAND ANIMALS

Reptiles, like amphibians and fishes but unlike birds and mammals, are **ectothermic** (from the Greek words *ectos*, "outside" + *thermos*, "heat"), regulating their body temperatures by taking in heat from the environment. They have developed a wide array of behavioral mechanisms that enable them to control their internal temperatures with remarkable precision. Even though ectothermic animals are called "cold-blooded," they often may actually be able to maintain bodily temperatures much warmer than their surroundings as a result of their behavior. Several scientists have postulated that at least some dinosaurs, like the birds and mammals, were **endothermic** (Greek *endos*, "inside"), capable of regulating their body temperatures internally. Endothermic animals are also sometimes called "warm-blooded" and **homeotherms** (Greek *homios*, similar).

The fact that birds (class Aves) and mammals (class Mammalia) are endothermic makes them unique among living organisms. Consequently, the members of these two classes are able to remain active at night, even if temperatures are cool. They are also able to live at higher elevations and farther north and south than most reptiles and amphibians. Endotherms can maintain their internal organs within a very narrow range of temperature more or less independently of external conditions, an adaptation that makes the maintenance and functioning of these animals more efficient than that of the ectotherms, even though more energy is required to maintain an endothermal system. About 80% of the calories in the food of endotherms goes to maintain their body temperature; a reptile that is the same

size as a mammal can therefore subsist on about a tenth of the amount of food required by the mammal. For this reason, reptiles are able to live in places such as some extreme deserts where food may be too scarce to support many birds and mammals.

## BIRDS

Birds are evolutionary descendants of the dinosaurs. The earliest known bird, *Protoavis* (Figure 18-16), which lived in west Texas about 225 million years ago, was about the size of a chicken. It had a number of distinctly dinosaurian features, including clawed fingers, a tail, and teeth in the front part of its jaw; on the other hand, it had a wishbone and a narrow shoulder girdle like that of modern birds. Even more telling, the skull, like that of all modern birds, had lost the upper arch behind the orbit of the eye; all dinosaurs retain this arch, which prevents them from raising the upper jaw to increase the gape as do birds.

A better-known fossil bird, *Archaeopteryx* (Figure 18-17) is about 150 million years old; the seven known fossils were all found in a single quarry in southern Germany. In many ways, *Archaeopteryx* is little more than a reptile with feathers; presumably its ability to fly was somewhat limited, and it may have been a glider, assisted in gliding by its feathers. Ultimately, these feathers made possible the subsequent evolution of birds that *could* fly efficiently, and the group began to become large and diverse about 90 to 65 million years ago. There are approximately 9000 species of living birds (class Aves; Figure 18-18).

The modern descendants of *Archaeopteryx* and other ancient birds have become true masters of the

**Figure 18-16**

**Protoavis, a 225-million-year-old bird about the size of a chicken.** This remarkable find, made in west Texas in 1986, extends the history of birds by 75 million years.

**Figure 18-17**

**Archaeopteryx.** Artist's reconstruction of *Archaeopteryx*, an early bird about the size of a crow. Closely related to its ancestors among the bipedal dinosaurs, *Archaeopteryx* lived in the forests of central Europe 150 million years ago. The teeth and long, jointed tail are features not found in any modern birds. Discovered in 1862, *Archaeopteryx* was cited by Darwin in support of his theory of evolution. The true feather colors of *Archaeopteryx* are not known.

air. In them, the wings are forearms, modified in the course of evolution. The scaly skin of birds has feathers, flexible, strong organs that can be replaced and that make up an excellent airfoil—a surface, such as a wing or rudder, that obtains resistance on its surfaces from the air through which it moves—for flying. The membranes that make up the flying surface of a bat or one of the extinct groups of flying reptiles can be severely damaged by a single rip. In contrast, the feathers of birds are individually replaceable. Even after the loss of many individual feathers, a bird can keep flying. Several hundred microscopic hooks along the sides of the individual barbules of a feather attach the barbules to one another. These hooks unite the feather so that in a marvelous way, unique to birds, a feather is magnificently adapted for flight (Figure 18-19). Overall, flight in birds is made possible by light, hollow bones, the replacement of scales with feathers, and the development of highly efficient lungs that supply the large amounts of oxygen necessary to sustain muscle contraction during prolonged flight.

Flight has evolved four times since animals first invaded the land. The first to fly were the insects, and they remain by far the most numerous fliers today. They were followed by the pterosaurs, flying relatives of the dinosaurs. The pterosaurs were a diverse and very successful group. Pterosaurs flew for millions of years, longer than birds have been flying. Birds and bats are late-comers to the air and fill the ecological niche once occupied by pterosaurs.

In birds, the eggs have hard shells and are highly resistant to drying out. Fertilization is direct, although copulatory organs are found only in a few groups, including ducks and geese.

*Birds evolved from reptiles and, with the bats, are one of the two living vertebrate groups that have achieved a full mastery of the air.*

A

B

C

D

E

## Figure 18-18

**Birds.**

**A** Ostrich, *Struthio camelus*. Wingless birds include the ostrich (Africa), rhea (South America), emu (Australia), and kiwi (New Zealand).

**B** A bee-eater, *Merops*, in Tanzania. The bee-eaters nest communally, with many adult individuals participating in the care of the young.

**C** Purple finch, *Carpodacus purpureus*, one of the very large group of birds that includes Darwin's finches, sparrows, warblers, chickadees, and many other familiar birds.

**D** Northern saw-whet owl, *Aegolius acadius*. Owls, hawks, eagles, and similar predatory birds are called raptors.

**E** A pair of wood ducks, *Aix sponsa*, showing the marked difference between males and females.

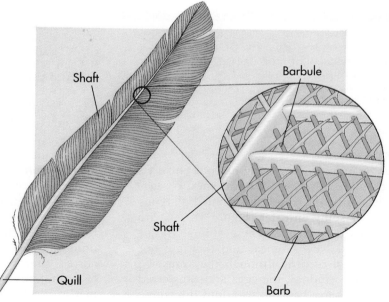

## Figure 18-19

**Feathers are more complex than they seem.** A feather, showing the way in which the vanes—secondary branches—are linked together by microscopic barbules.

# MAMMALS

The first known mammals are from the early Mesozoic Era, about 200 million years ago. They were clearly derived from reptiles; in fact, a number of the early fossil organisms are difficult to classify satisfactorily either as reptiles or as mammals. The first mammals probably were small; fed on insects, as suggested by their teeth; and may have been nocturnal, as suggested by their large eyes. Mammals remained secondary to the reptiles, both in absolute numbers and in numbers of species, until about 65 million years ago. After the extinction of the dinosaurs at that time (see Chapter 17), mammals became abundant.

## Characteristics of Mammals

There are about 4500 species of living mammals (class Mammalia), including human beings. Like birds, mammals are homeotherms. Their skin is covered with hair during at least some stage of their life cycle. Like birds and crocodiles, mammals have a four-chambered heart with complete double circulation: separate systems in which oxygen-rich and oxygen-poor blood circulate. Mammals have copulatory organs like their reptilian ancestors, and fertilization is internal. Mammals nourish their young with milk, a nutritious substance produced in the mammary glands of the mother. Their locomotion is advanced over that of the reptiles, which in turn is advanced over that of the amphibians; the legs of mammals are positioned much farther under the body than those of reptiles and are suspended from limb girdles, which permit greater leg mobility.

*Mammals nourish their young with milk and are covered with hair rather than scales or feathers.*

## Monotremes

In all but a few mammals, the young are not enclosed in eggs when they are born. The **monotremes**, or egg-laying mammals, consisting of the duck-billed platypus and the echidna, or spiny anteater (Figure 18-20), are the only exceptions to this rule. These animals occur only in Australia and New Guinea, and they are not known from the fossil record, although they must be an ancient group. In echidnas, the eggs are transferred to a special pouch, where they are held until the young hatch.

**Figure 18-20**

**Monotremes.**
A  A duck-billed platypus, *Ornithorhynchus anatinus*, at the edge of a stream in Australia.
B  Spiny echidna, *Tachyglossus aculeatus*, a desert dweller unrelated to the porcupine that it resembles.

## Marsupials

The **marsupials** are mammals in which the young are born early in their development, sometimes as soon as 8 days after fertilization, and are retained in a pouch, or **marsupium**. Living marsupials are found only in Australia, where they are abundant and diverse (Figures 18-21), and in North and South America. Ancient fossil marsupials, 40 to 100 million years old, are found in North America, where they became extinct until their geologically recent reintroduction, which occurred after North and South America were linked 3 to 4 million years ago. The opossum, a familiar mammal in North America, arrived from the south in just this way.

**Figure 18-21**

**A marsupial.** Kangaroo with young in its pouch.

## Placental Mammals

Most modern mammals are **placental mammals.** The first organ to form during the course of their embryonic development is the placenta. The **placenta** is a specialized organ, held within the womb in the mother, across which she supplies the offspring with food, water, and oxygen and through which she removes wastes. Both fetal and maternal blood vessels are abundant in the placenta, and substances can thus be exchanged efficiently between the bloodstreams of the mother and her offspring. In placental mammals, unlike marsupials, the young undergo a considerable period of development before they are born.

*Some mammals (monotremes) lay eggs; others (marsupials) give birth to embryos that continue their development in pouches (as do the monotremes); and still others (placental mammals) nourish their developing embryos within the body of the mother by means of a placenta until development is almost complete.*

Placental mammals are extraordinarily diverse; representatives of several orders are shown in Figure 18-22. One evolutionary line of mammals, the bats, has joined insects and birds as the only group of living animals that truly flies; another, the Cetacea (whales and porpoises), has reverted to an aquatic habitat like that from which the ancestors of mammals came hundreds of millions of years ago.

## THE EVOLUTION OF PRIMATES

The small Mesozoic mammals that ultimately gave rise to the primates were rare for much of their history. They were rare when other kinds of mammals became diverse very rapidly after the extinction of the dinosaurs. All of them were probably **nocturnal** (active primarily at night), judging from their large eyes and small size (Figure 18-23). The earliest fossils that resemble any living primates belonged to the group known as **prosimians.** By the end of the Eocene Epoch (38 million years ago), prosimians were abundant in North America and Eurasia and were probably present in Africa also. Their descendants now live only in tropical Asia and adjacent islands, in Africa, and especially on the island of Madagascar, which is about twice the size of the state of Arizona and lies in the Indian Ocean about 325 kilometers off the east coast of Africa (Figure 18-24).

### Anthropoids

The ancestors of monkeys, apes, and humans appear for the first time in the fossil record about 36 million years ago in the early Oligocene Epoch. These **anthropoid** primates—monkeys, apes, and human beings and their direct ancestors—seem to have replaced prosimians rather rapidly. Monkeys and apes are almost all **diurnal**; that is, active during the day. They are agile animals that feed mainly on fruits and leaves, and most of them live in trees. Monkeys and apes differ markedly in these respects from the smaller, mostly nocturnal prosimians. Their well-developed **binocular** vision (vision in which both eyes are located at the front of the head and function together to provide a three-dimensional view of the object being examined) has evolved in relation to a reduction in their snouts and is a great advantage in leaping through the treetops. This evolutionary trend, in turn, resulted in the flat faces that are characteristic of most members of the order.

The anthropoid primates have larger brains, and their braincase forms a larger portion of the head than it does in the prosimians and other mammals. The thumb of anthropoids is opposable: it stands out at an

A

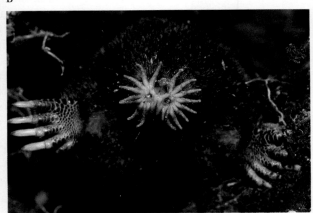

B

C

D

## Figure 18-22

**Placental mammals.**
**A** Snow leopard, *Uncia*, a cat (order Carnivora).
**B** Starnosed mole, *Condylura cristata*, a burrowing insectivore (order Insectivora).
**C** White-tailed deer, *Odocoileus virginianus*, abundant in eastern and central temperate North America (order Artiodactyla).
**D** Orca, *Orcinus orca*, a carnivorous whale (order Cetacea).

## Figure 18-23

**Where primates came from.** Reconstruction of an insect-eating mammal from the early Cenozoic Era, a mouse-sized animal that probably resembled the common ancestor of the primates and insectivores, two contemporary orders of mammals.

angle to the other digits and can be bent back against them, as when the animal grasps an object. Some of the advanced anthropoids have developed an extraordinary ability to manipulate objects by using this opposable thumb. Monkeys and apes live in groups in which complex social interaction occurs. In addition, they tend to care for their young for prolonged periods. Such a pattern of care makes possible the prolonged period of learning that is characteristic of these animals. The maternal care given the young of anthropoids seems to be related in some way to their large brains and the long periods that are necessary for their development.

*Contemporary primates are characterized by the expansion and elaboration of their brains; the shortening of their snouts, together with well-developed binocular vision; the free mobility of the fingers and toes, with the forefinger and thumb opposable; and an increasing complexity and quantity of social behavior.*

**Figure 18-24**

**Prosimians.**
**A** Ringtail lemur, *Lemur catta*. All living lemurs are restricted to the island of Madagascar, and all are in danger of extinction as the rain forests there are destroyed.
**B** Tarsier, *Tarsius syrichta*, tropical Asia. Note the binocular vision and large eyes of the tarsier, adaptations to nocturnal living in the trees.

A                                    B

*The Evolution of Hominoids.* The apes, together with the **hominids** (human beings and their direct ancestors), make up a group called the **hominoids.** The earliest known fossils of hominoids are from the Early Miocene Epoch (20 to 25 million years ago) and occur in the Old World. The living apes are usually classified as members of two families, the Pongidae (chimpanzees, gorillas, and orangutans) and the Hylobatidae (gibbons) (Figure 18-25). The living apes are confined to relatively small areas in Africa and Asia; no apes ever occurred in the New World.

Apes have larger brains than do monkeys. Apes exhibit the most adaptable behavior of all mammals except for human beings. All apes lack tails. With the exception of the gibbons, which are relatively small, all apes are larger than any surviving monkey.

Although the fossil evidence is still being accumulated, biochemical studies have told us a great deal about our relationships to the apes and their relationships to one another (Figure 18-26). Based on the rate with which differences seem to to have accumulated in the proteins they have studied, investigators have calculated that the evolutionary line leading to gibbons diverged from that leading to the other apes about 10 million years ago; the line leading to orangutans split off about 8 million years ago; and the split between hominids and the line leading to chimpanzees and gorillas occurred about 4 million years ago—although fossil evidence suggests that this last split, and therefore the appearance of *Australopithecus,* may actually have occurred somewhat earlier, perhaps about 5 million years ago. The time of divergence between chimpanzees and gorillas appears to have been a little less

than 3.2 million years ago, as determined by molecular analysis. The relationship between chimpanzees, gorillas, and human beings is so close that if they were members of any other group of organisms, they would almost certainly be classified as members of the same genus.

Molecular and other evidence suggests that chimpanzees and gorillas diverged from one another a little less than 3.2 million years ago, hominids from that line about 5 million years ago, orangutans from all of them about 8 million years ago, and gibbons from the rest about 10 million years ago.

## THE APPEARANCE OF HOMINIDS

### The Australopithecines

The two critical steps in the evolution of humans were the evolution of bipedalism and the enlargement of the brain. The earliest definite hominids, which share many of the characteristics that we regard as distinctively human, are assigned to the genus *Australopithecus* (Figure 18-27). Among their human features were bipedal (two-legged, erect) locomotion and rounded jaws. The australopithecines lived on the ground in the open grasslands of eastern and southern Africa, weighed upwards of 20 kilograms, and were up to 1.3 meters tall. Their brains were about the size of those of present-

**Figure 18-25**

A Gorilla, *Gorilla gorilla*.
B Mueller gibbon, *Hylobates muelleri*.
C Chimpanzee, *Pan troglodytus*.
D Orangutan, *Pongo pygmaeus*.

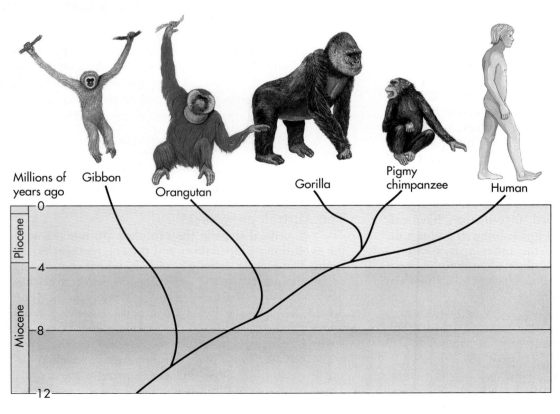

Millions of years ago

Gibbon

Orangutan

Gorilla

Pigmy chimpanzee

Human

Pliocene

Miocene

0

4

8

12

**Figure 18-26**

The evolution of living hominoids, the apes and human beings.

**Figure 18-27**

**"Lucy," from Ethiopia, is the most complete skeleton of *Australopithecus* discovered so far.** The reconstruction was made by a careful study of muscle attachments on the skull.

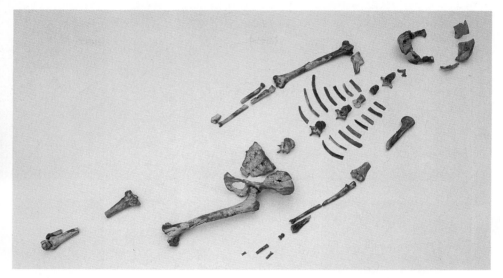

day gorillas, about 350 to 450 cubic centimeters in volume, as compared with an average volume of about 1450 cubic centimeters in modern humans. Individuals of *Australopithecus* appear to have eaten many kinds of plant and animal food, judging from the characteristics of their teeth and jaws. Although there is no general agreement on this point, there may have been four or more species of the genus *Australopithecus* (Figure 18-28). Their characteristics indicate clearly that this genus was ancestral to our genus, *Homo*.

The oldest known fossils that may represent *Australopithecus* are 5 to 5.5 million years old. They thus date from close to the postulated time of divergence of the hominid and chimpanzee-gorilla evolutionary lines. Other fossils of *Australopithecus* are commonly found from 4 million years ago onward at several localities in East and South Africa.

*The first definite hominids, Australopithecus, appeared about 5 million years ago in Africa. They lived on the ground, weighed less than 20 kilograms, and were less than 1.3 meters tall. They disappeared about 1.3 million years ago.*

## The Use of Tools: *Homo habilis*

Starting about 2 million years ago, tools appear in the sedimentary beds of a wide area of East and South Af-

rica. Such tools have been associated with the fossils of the first known humans, the earliest members of the genus *Homo*. The fossils and tools were discovered in the Olduvai Gorge, Tanzania. Because these people used tools, they were named *Homo habilis* (Latin *habilis*, able). *H. habilis* appeared about 2 million years ago and persisted for about 500,000 years (Figure 18-29).

*Homo habilis is the first member of our own genus and the first primate to use tools consistently. This species appeared in East Africa about 2 million years ago and survived for about 500,000 years. In this species, which certainly evolved from Australopithecus, fully developed human characteristics first emerged.*

Judged from its physical characteristics, *H. habilis* was certainly derived from *Australopithecus*. As the use of tools was developed in its earliest populations, the characteristics of the genus *Homo* may have begun to emerge. Judged from the structure of its hands, *H. habilis* regularly climbed trees, like *Australopithecus*, although it mainly stayed on the ground and walked erect on its two legs. Skeletons found in 1987 indicate that *H. habilis* was small in stature, like *Australopithecus*.

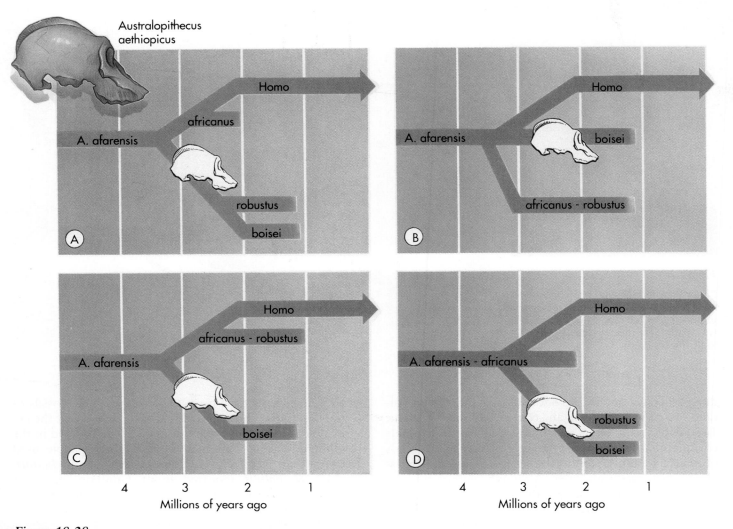

**Figure 18-28**

**Four possible human family trees.** Version **B** is the most popular among investigators of human fossils, but the evidence now available does not clearly eliminate any of the other three. The illustrated skull is of a new find, tentatively called *Australopithecus aethiopicus*, which appears to be an ancestor of *A. boisei*.

## Homo erectus

All of the early evolution of the genus *Homo* seems to have taken place in Africa. There, fossils belonging to the second and only other extinct species of *Homo*, *H. erectus*, are widespread and abundant from 1.7 million until about 500,000 years ago. By 1 million years ago, *H. erectus* had migrated into Asia and Europe. These large humans, who were about the size of our species, *H. sapiens*, had brains that were roughly twice as large as those of their ancestors, prominent brow ridges, rounded jaws, and massive teeth. *H. erectus* survived in Asia until about 250,000 years ago, much longer than in Africa or Europe.

The wide distribution of tools made by *H. erectus* suggests that their widely dispersed bands may have been in communication with one another. The first evidence of the use of fire by humans occurs at the camp-sites of this species in the Rift Valley of Kenya at least 1.4 million years ago, and fire was characteristically associated with populations of *H. erectus* from that time onward.

*Homo erectus, the second species of our genus to evolve, appeared in Africa about 1.7 million years ago and migrated throughout Eurasia by 1 million years ago. It disappeared in Africa about 500,000 years ago and in Asia about 250,000 years ago. H. erectus used fire, starting at least 1.4 million years ago, and made characteristic stone tools.*

## Modern Humans: *Homo sapiens*

Many scientists place the first appearance of our own species at about 500,000 years ago because of the larger brains, reduced teeth, and enlargement of the rear of some skulls that date from that time, as compared with those of *H. erectus*. *H. sapiens* appeared in Europe as *H. erectus* was becoming rarer. It appears almost certain that *H. sapiens* represents a derivative of *H. erectus* with certain "modern" features.

Recently, scientists analyzing mitochondrial DNA have calculated that all the races of living humans might have .originated in Africa, perhaps about 200,000 years ago. Because the DNA within mitochondria is transmitted only from the egg and not the sperm, lineages do not get rearranged by Mendelian re-

assortment and so it is possible to trace the parentage of humans back, from mother to grandmother to great grandmother. When this is done, scientists find that all humans alive today share mitochondrial DNA derived from a single, hypothetical, ancestral female, named "Eve," who lived about 200,000 years ago in Central Africa. Other sequences of mitochondrial DNA occur in living humans, but those derived from Eve are the only sequences present in all of us, as far as we have been able to determine. For this reason, Eve appears in some sense to be our common mother.

The fact that Eve's mitochondrial DNA is found throughout the world, and in all races of humans, provides powerful evidence that all individuals of *H. sapiens* are derived from a single common ancestor, and

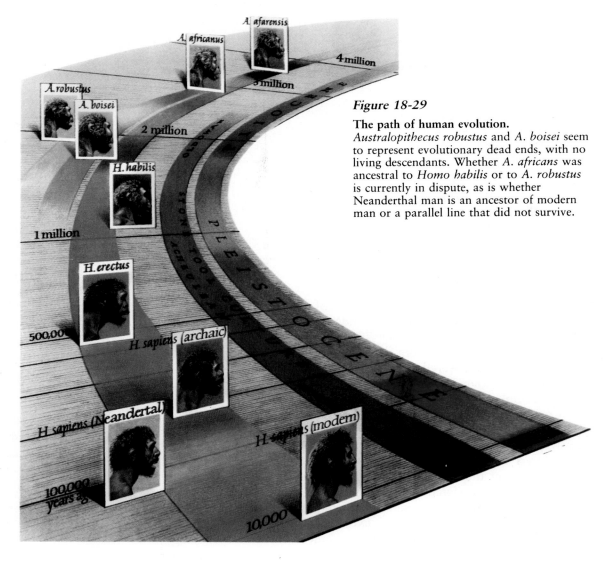

**Figure 18-29**

**The path of human evolution.**
*Australopithecus robustus* and *A. boisei* seem to represent evolutionary dead ends, with no living descendants. Whether *A. africans* was ancestral to *Homo habilis* or to *A. robustus* is currently in dispute, as is whether Neanderthal man is an ancestor of modern man or a parallel line that did not survive.

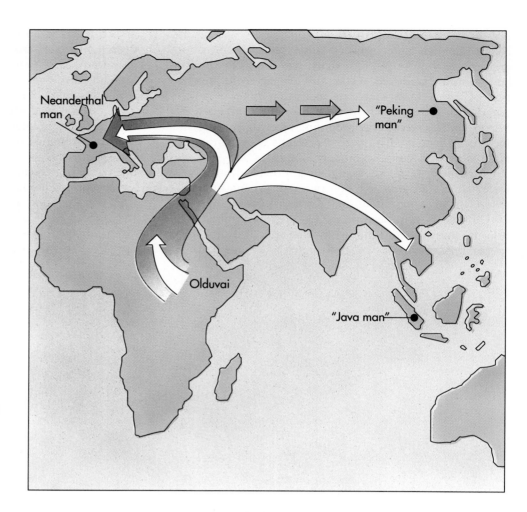

**Figure 18-30**

**Out of Africa—twice.** Evidence now supports the idea that our species first evolved in Africa. *Homo erectus* thus spread from Africa to Europe and Asia twice. First, *Homo* spread as far as Java and China. Later, it was replaced by *H. sapiens* in a second wave of immigration.

not independently from separate populations of *H. erectus*. The origin of our species seems clearly to have taken place in Africa and from there spread to all parts of the world, supplanting everywhere *H. erectus* populations, which had migrated out of Africa much earlier (Figure 18-30).

**Neanderthals.** Populations of *H. sapiens* of the sort that are called Neanderthals were abundant in Europe and western Asia between about 70,000 and about 32,000 years ago, judging from the fossil record. Such fossils were first discovered in the valley of the Neander River in Germany in 1856, and the name given to humans of this sort is derived from that of the valley.

Compared with ourselves, Neanderthals were powerfully built, short, and stocky. Their skulls were massive, with protruding faces, projecting noses, and rather heavy bony ridges over the brows. Their brains were even larger than those of modern humans, a fact that may have been related to their heavy, large bodies. The Neanderthals made diverse tools, including scrapers, borers, spearheads, and hand axes. Some of the Neanderthal tools were used for scraping hides, which

they used for clothing. They lived in hutlike structures or in caves. Neanderthals took care of their injured and sick and commonly buried their dead, often placing food and weapons, and perhaps even flowers, with the bodies. Such attention to the dead suggests strongly that they believed in a life after death. For the first time, the kinds of thought processes that are characteristic of modern *H. sapiens*, including symbolic thought, are evident in these acts.

*Our species, Homo sapiens, may be as much as 500,000 years old; it has been abundant for the past 150,000 years. Neanderthals were abundant in Europe and western Asia from about 70,000 to about 32,000 years ago.*

**Cro-Magnons.** About 34,000 years ago the European Neanderthals were abruptly replaced by people of essentially modern character, the Cro-Magnons. We can only speculate about why this sudden replacement occurred, but it was complete all over Europe in a short period. There is some evidence that the Cro-Magnons came from Africa, where fossils of essentially modern aspect, but as much as 100,000 years old, have been found. They seem to have replaced the Neanderthals completely in southwest Asia by 40,000 years ago, and then spread across Europe during the next few thousand years, coexisting and possibly even interbreeding with the Neanderthals for several thousand years.

The Cro-Magnons used sophisticated stone tools that rapidly became more diverse and thus more useful for many purposes. They also made a wide variety of tools out of bone, ivory, and antler, materials that had not been used earlier. The Cro-Magnons were hunters who killed game by using complex tools. In addition, their evident social organization suggests that they may have been the first people to have fully modern language capabilities. Soon after Cro-Magnons appeared, they began to make what were apparently ritual paintings in caves (Figure 18-31). The culture associated with these impressive and mysterious works of art existed during a period that was cooler than the present. Such conditions prevailed at the time of the last great expansion of continental ice. During that period, grasslands inhabited by large herds of grazing animals occurred across Europe. The animals that occurred in such habitats are often depicted in the cave paintings.

Humans of modern appearance eventually spread across Siberia to North America, which they reached at least 12,000 to 13,000 years ago, after the ice had begun to retreat at the end of the last glacial period. As

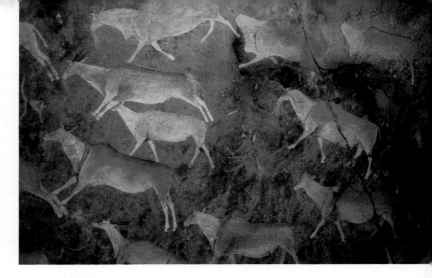

**Figure 18-31**

**Cave painting.** Paintings, almost always showing animals and sometimes hunters, were made by Cro-Magnon people, our immediate ancestors, in caves, especially in Europe. They were made for a period of about 20,000 years, until about 8000 to 10,000 years ago.

people spread throughout the world, they were apparently responsible for the extinction of many populations of animals and plants, a process that has been accelerated recently. Human remains are often associated with the bones of the large animals they hunted, and these animals often disappeared promptly from different parts of the world shortly after humans arrived for the first time. By the end of the Pleistocene Epoch, about 10,000 years ago, there were only about 10 million people throughout the entire world (compared with more than 5.3 billion now).

## SUMMARY

1. Chordates are characterized by a single dorsal hollow nerve cord; a flexible rod, the notochord, which forms on the dorsal side of the gut in the early embryo; and pharyngeal slits.

2. Vertebrates compose most groups of the phylum Chordata. They differ from other chordates in that they usually possess a vertebral column, a distinct and well-differentiated head, and a bony skeleton.

3. Members of the class Agnatha differ from the other vertebrates in that they lack jaws. Once abundant and diverse, they are represented among the living vertebrates only by the lampreys, which attach themselves to other fishes and suck their blood, and the hagfishes.

4. The two classes of fishes other than Agnatha consist of animals that have jaws, as do the members of the other four classes of vertebrates. Jawed fishes constitute about half of the estimated 42,500 species of vertebrates and are dominant in fresh and salt waters everywhere. Of the two classes of jawed fishes, the Chondrichthyes, or cartilaginous fishes, consist of about 850 species of sharks, rays, and skates; the Osteichthyes, or bony fishes, consist of about 18,000 species.

5. The first land vertebrates were the amphibians, one of the four classes of tetrapods (basically four-limbed vertebrates). Amphibians depend on water and lay their eggs in moist places. In many species the larvae, and in some the adults also, live in water. The amphibians probably gave rise to the reptiles, which otherwise may have been derived directly from the fishes.

6. The reptiles were the first vertebrates that were fully adapted to terrestrial habitats. Amniotic eggs, which evolved in this group but are also characteristic of the birds and the very few egg-laying mammals, represent a significant adaptation to the dry conditions that are widespread on land.

7. The birds and mammals were derived from the reptiles and are now the dominant groups of vertebrates on land in many areas. The members of these two classes have independently become endothermic (homeothermic), capable of regulating their own body temperatures. With few exceptions, all other living animals are ectothermic; their body temperatures are set by external influences.

8. Primates first appeared 55 to 38 million years ago, with the first anthropoid primates (a group that now includes monkeys, apes, and humans) appearing about 30 million years ago. Primates have proportionately large brains, binocular vision, and five digits, including an opposable thumb, and they exhibit complex social interactions.

9. The apes, which appeared 25 to 20 million years ago, gave rise to the gibbons, orangutans, and then to the African apes (chimpanzees and gorillas) and hominids (humans and their direct ancestors), starting more than 10 million years ago. Apes and hominids collectively are termed hominoids.

10. The earliest hominids, family Hominidae, belong to the genus *Australopithecus*. The members of this genus were the direct ancestors of humans. They appeared in Africa about 5 million years ago and were small, up to about 1.3 meters tall and about 20 kilograms in weight, with brains about 450 cubic centimeters in volume.

11. About 2 million years ago, with an enlargement of brain size that was perhaps associated with the increased use of tools, the genus *Australopithecus* gave rise to humans belonging to the genus *Homo*. The first species of this genus, *Homo habilis*, appeared in Africa about 2 million years ago. It became extinct about 1.5 million years ago.

12. The second species of *Homo, Homo erectus,* appeared in Africa at least 1.7 million years ago. *Homo erectus* migrated from Africa to Eurasia about 1 million years ago. They were replaced by our species, *Homo sapiens,* about 500,000 years ago in Africa and about 250,000 years ago in Asia.

## REVIEW

1. One of the pivotal innovations that led to the success of reptiles is the watertight egg, called a(n) _____ egg.

2. No mammals lay eggs. True or false?

3. Of the five classes of vertebrates, which has the most species?

4. The first hominids appeared about 5 million years ago. They belonged to the genus _____.

5. We are members of the genus *Homo*. How many species have existed in this genus?

# SELF-QUIZ

1. Which of the following are *not* distinguishing features of chordates?
   (a) Dorsal nerve cord
   (b) Spine
   (c) Notochord
   (d) Gills
   (e) Tail

2. Arrange the following animals in the order in which they are thought to have evolved.
   (a) Amphibians
   (b) Bony fishes
   (c) Lampreys
   (d) Reptiles
   (e) Sharks

3. Arrange the following hominids in order of their evolutionary distance from modern humans, the most distant first.
   (a) *Australopithecus*
   (b) Cro-Magnons
   (c) *Homo erectus*
   (d) *Homo habilis*
   (e) Neanderthals

4. Arrange the following animals in order of their evolutionary distance from humans, the most distant first.
   (a) Gibbons
   (b) Gorillas
   (c) Prosimians
   (d) Monkeys
   (e) Orangutans

5. Arrange the following animals in the order of number of species, the one with the greatest number of species first.
   (a) Amphibians
   (b) Birds
   (c) Bony fishes
   (d) Mammals
   (e) Reptiles

6. The living lobed-finned fishes include
   (a) hagfishes.
   (b) coelacanth.
   (c) lamprey.
   (d) mudskipper.
   (e) teleosts.

7. Ectothermic animals
   (a) control their temperature through their behavior.
   (b) control their temperature by expending energy.
   (c) have internal regulation of their body temperatures.
   (d) are more independent of the environment than are endothermic animals.
   (e) more than one of the above; list your choices.

# THOUGHT QUESTIONS

1. Of the 42,500 species of living chordates, twice as many (about 19,000 fishes) live in the sea as on the surface of the land (about 6000 reptiles and 4500 mammals). Why do you think there are so many more species of chordates in the sea than on land?

2. Our species, *Homo sapiens,* evolved from *Homo erectus* less than 1 million years ago. Do you think that evolution of the genus *Homo* is over, or might another species of man evolve within the next million years? Do you think this would involve the extinction of *H. sapiens*?

3. Flying reptiles were a very diverse group during the age of dinosaurs. However, although the present age might be considered the age of mammals, bats (the only flying mammals) exhibit far less diversity than did dinosaurs. Can you suggest an explanation for this?

4. Studies of the DNA of primates have revealed that humans differ from gorillas in only 1.4% of the DNA nucleotide sequences, and from chimpanzees in only 1.2%. This degree of genetic similarity (about 1%) is the same as is usually seen among "sibling" species (that is, species that have only recently evolved from a common ancestor). Yet humans are assigned not only to a different genus but to a different family! Do you think this is legitimate, or are humans just a rather unusual kind of African ape?

# FOR FURTHER READING

BAKKER, R.T.: *The Dinosaur Heresies,* William Morrow & Co., Inc., New York, 1986. A fine discussion of the controversy about whether dinosaurs could control their own temperatures internally; well-written and beautifully illustrated.

EASTMAN, J., and A. DE VRIES: "Antarctic Fishes," *Scientific American,* November 1986, pages 106-114. Did you ever wonder why fishes don't freeze in the Antarctic seas, even though ice forms in the oceans in which they swim? These biologists showed that the fishes produce their own potent antifreeze.

O'BRIEN, J.: "The Ancestry of the Giant Panda," *Scientific American,* November 1987, pages 102-107. A clear account of how modern molecular techniques are being used to solve a famous evolutionary puzzle: is the panda a raccoon or a bear?

WALKER, A. and M. TEAFORD: "The Hunt for Proconsul," *Scientific American,* January 1989, pages 76-82. An exciting account of the last common ancestor of great apes and human beings.

WALLACE, J.: "New Discoveries About Dinosaurs," *Science Year,* 1989, pages 42-76. Annual supplement to World Book. New evidence has convinced most scientists that many dinosaurs once considered dull and plodding were actually agile, fast-moving, and even social.

WELLNHOFER, P.: "Archaeopteryx," *Scientific American,* May 1990, pages 70-77. New information is continuing to provide insight into the evolution of flight in birds.

6

## CHAPTER 19

# Population Dynamics

A green blanket of Kudzu vine envelopes this entire South Carolina community. An introduced species, the vine overgrows everything as it greedily seeks sunlight. Natural communities, complexly organized by centuries of coevolution, are often vulnerable to the aggressive assault of outsiders.

# POPULATION DYNAMICS

## Overview

Ecology is the study of the relationships of organisms with one another and with their environment; as such, ecology is concerned with populations of organisms. Natural populations are affected by the ways in which they interact with other populations and with nonliving factors. Each population grows in size until it eventually reaches the limits of its habitat to support it; resources are always limiting. Some, but not all, of the limits to the growth of a population are related to the density of the individuals that constitute that population. Competition between populations of any two species limits their coexistence. Predation and other forms of interaction among populations also play an important role in limiting population size.

## For Review

*Here are some important terms and concepts that you will encounter in this chapter. If you are not familiar with them, you should review them before proceeding.*

**Adaptation** (Chapter 15)

**How species originate** (Chapter 16)

**Major features of evolution** (Chapter 17)

Stepping back and viewing the broad panorama of evolution, as we did in the preceding section, reveals the unity that underlies the diversity of life. All of us, camels and elephants and humans, share a common history of adaptation and change. But it is important not to limit our view to looking backward. Right now, all the organisms on earth are involved in a day-to-day struggle to survive and reproduce. Each living thing on earth is interacting with its physical environment and with the organisms around it in a journey toward the future we all share (Figure 19-1).

## THE HOUSE WE LIVE IN

In 1866 the German biologist Ernst Haeckel first gave a name to the study of how organisms interact with their environment, both with temperature and moisture and soil, and also with other organisms. He called this study of how organisms fit into their environment **ecology**, from the Greek words *oikos* (house, place where one lives) and *logos* (study of). Our study of ecology, then, is a study of the house in which we live.

Ecology is a complex but fascinating area of biology that has many important implications for each of us. Every principle of ecology would hold true even if there were no human beings in the world, but the operation of these principles is now being affected pro-

**Figure 19-1**

**Competition in the forest.** Each plant in this forest competes with all the individuals around it for light, soil, nutrients, and moisture. The competition for available resources is fierce, especially during the relatively brief period in spring before trees have so many leaves that photosynthesis is limited.

foundly by human activities. A human population that is growing rapidly, and has now attained the unprecedented level of more than 5.3 billion people, is severely straining the earth's capacity to sustain us all. In the face of this situation the principles of ecology assume a crucial importance and may enable us to chart a sound future.

Ecology is an area of scientific knowledge that is concerned with the most complex level of biological integration. Ecology attempts to tell us why particular kinds of organisms can be found living in one place and not another—the physical and biological variables that govern their distribution; the factors that control the numbers of particular kinds of organisms and maintain them at certain levels; and the principles that may allow us to predict the future behavior of assemblages of organisms.

*Ecology is the study of the relationships of organisms with one another and with their environment.*

Ecologists consider groups of different organisms at four progressively more inclusive levels of organization:

1. Individuals of a species that live together are called **population.** The place in which these individuals live is called their **habitat.**
2. Populations of different species that live together are called **communities.**
3. A community, together with the nonliving factors with which it interacts, is called an **ecosystem.** An ecosystem regulates the flow of energy, ultimately derived from the sun, and the cycling of the essential elements on which the lives of its plants, animals, and other organisms depend.
4. **Biomes** are major terrestrial assemblages of plants, animals, and microorganisms that occur over wide geographical areas and have distinct characteristics. They include deserts, tropical forests, and grasslands.

In our exploration of ecological principles, we first consider, in this chapter, the properties of populations, emphasizing their dynamics. In Chapter 20 we discuss communities and the sorts of interactions that occur in them. In Chapter 21 we move on to the dynamics of ecosystems, and then in Chapter 22 consider the properties of biomes and similar major aggregations in the seas. Finally, in Chapter 23, we consider the future of the biosphere and attempt to assess the impact of the

many problems that plague it. Its future will depend on the ecological principles presented in this section.

## POPULATIONS

A population consists of the individuals of a given species occurring together at one place and at one time. This is a flexible definition that allows us to speak in similar terms of the world's human population, the population of protozoa in the gut of an individual termite, or the deer that inhabit a forest.

All of these specific populations have characteristic features such as size, density, dispersion, and demography. Each occupies a particular place and plays a particular role in its ecosystem; that role is defined as the **niche** of that population. Although the properties of populations are ultimately determined by the genetic properties of the individuals they include, a population is fundamentally an operating system with its own distinct properties. Understanding these properties is crucial for understanding the nature of life on earth.

*A population is a group of individuals of a particular species that live together.*

One of the critical features of any population is its size. Population size has a direct bearing on the ability of a given population to survive. Random events or natural disturbances may endanger a few individuals more than they endanger many, and inbreeding can also be a negative factor in population survival if few potential mates are available. Not only does inbreeding often lead directly to a lowering of vigor, that is, of hardiness and the ability to meet the everyday challenges of living, but the reduced levels of variability that result from inbreeding are likely to detract from the population's ability to adjust to changing future conditions. If an entire species consists of only one or a few small populations, that species is likely to become extinct, especially if it occurs in areas that have been or are being changed radically (Figure 19-2).

In addition to population size, its **density**—the number of individuals that occur in a given area—is also highly significant. If the individuals that compose the population are widely spaced, they may rarely, if ever, encounter one another, and the reproductive capabilities—and therefore the future—of the population may be very limited, even if the absolute numbers of individuals over a wide area are relatively high.

B

C

**Figure 19-2**

A

**Highly endangered.** These animals exist in very small populations and are in danger of extinction unless the activities that are threatening them are brought under control.
A The Sumatran rhinoceros, *Didermoceros sumatrensis*, reduced to a few hundred individuals on the island of Sumatra.
B Mediterranean monk seal, *Monachus monachus*. Fewer than 500 individuals of this species are believed to exist, confined to remote cliffbound coasts and islands in the Mediterranean.
C The kagu, *Rhynochetus jubatus*, the only member of a family of birds that is restricted to the island of New Caledonia in the southwestern Pacific Ocean. It has been brought to the brink of extinction by introduced dogs that have run wild on the island.

## POPULATION GROWTH

A third key characteristic of any population is its capacity to grow. Most populations tend to remain relatively constant in number, regardless of how many offspring the individuals produce. As you saw in Chapter 1, Darwin partly based his theory of natural selection on this seeming contradiction. Many calculations have been done, showing that houseflies, bacteria, or even elephants would soon cover the world if their reproduction were unchecked. Under certain circumstances, a population size can increase very rapidly for a time (Figure 19-3). To understand how nature operates to limit population growth, we first consider situations in which populations do grow very rapidly for a time, and we then ask what brings their growth to a halt.

### Biotic Potential

The rate at which a population will increase when there are no limits of any sort on its rate of growth is called its **innate capacity for increase,** or **biotic potential.** But this theoretical rate is almost always impossible to calculate because there usually are limits to

growth. What biologists actually tend to calculate is the **actual rate of population increase,** which is defined as the difference between the birth rate and the death rate per given number of individuals per unit of time. The actual rate of increase, *r,* may also be affected by the movement of individuals into or out of the area. For example, the actual rate of increase for the human population of the United States during the closing decades of the twentieth century is more than half made up of immigrants to the country, less than half from the growth in numbers of the people already living in it.

The innate capacity for growth for any population is constant, determined largely by the physiology of the organism. The actual growth rate, on the other hand, depends on the death rate as well as the birth rate and so changes as the population increases in size. In general, as the population grows more crowded and begins to exhaust its resources, the death rate rises. If you were to keep track of the number of individuals in a population from its beginning, you would see that the population would grow rapidly at first, the number of individuals increasing exponentially (2, 4, 8, 16, 32,

**Figure 19-3**

**A marsh weed invades from Europe.** European purple loostrife, *Lythrum salicaria,* was introduced some time before 1860 and became naturalized over thousands of square miles of marshes and other wet places in North America.

64, 128, . . . ). Soon, however, the rate of increase would slow as the death rate began to rise. Eventually, just as many individuals would be dying as were being born. The early, rapid phase of population growth (Figure 19-4) lasts only for a short period, usually when an organism reaches a new habitat in which it has abundant resources. In this connection, one may think of such examples as the marsh weed called purple loosestrife (see Figure 19-3) reaching the wetlands of North America from Europe for the first time; of algae colonizing a newly formed pond; or of the first terrestrial immigrants that arrive on an island recently thrust up from the sea.

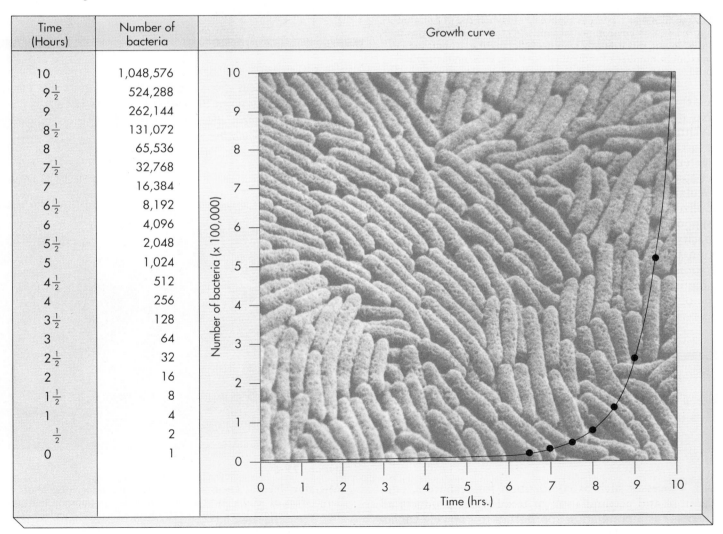

| Time (Hours) | Number of bacteria |
|---|---|
| 10 | 1,048,576 |
| $9\frac{1}{2}$ | 524,288 |
| 9 | 262,144 |
| $8\frac{1}{2}$ | 131,072 |
| 8 | 65,536 |
| $7\frac{1}{2}$ | 32,768 |
| 7 | 16,384 |
| $6\frac{1}{2}$ | 8,192 |
| 6 | 4,096 |
| $5\frac{1}{2}$ | 2,048 |
| 5 | 1,024 |
| $4\frac{1}{2}$ | 512 |
| 4 | 256 |
| $3\frac{1}{2}$ | 128 |
| 3 | 64 |
| $2\frac{1}{2}$ | 32 |
| 2 | 16 |
| $1\frac{1}{2}$ | 8 |
| 1 | 4 |
| $\frac{1}{2}$ | 2 |
| 0 | 1 |

Growth curve

**Figure 19-4**

Exponential growth in a population of bacteria.

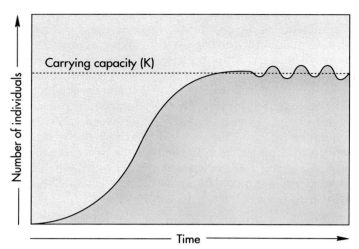

**Figure 19-5**

**The growth curve.** The sigmoid growth curve, characteristic of biological populations, begins with a period of exponential growth like that shown in Figure 19-4; when the population approaches its environmental limits (K), the growth begins to slow down, and it finally stabilizes, fluctuating around the maximum number of individuals that the environment will hold.

## Carrying Capacity

No matter how rapidly new populations grow, they eventually reach some environmental limit imposed by shortages of an important factor such as space, light, water, or nutrients. A population ultimately stabilizes at a certain size, called the **carrying capacity** of the particular place where it lives. The carrying capacity, symbolized by $K$, is the number of individuals that can be supported at that place indefinitely. As the carrying capacity is approached, the rate of growth of the population slows down greatly because, in effect, there is less "room" for each individual. Graphically, this relationship is the S-shaped **sigmoid growth curve**, characteristic of biological populations (Figure 19-5).

*The size at which a population stabilizes in a particular place is defined as the carrying capacity of that place for that species.*

Processes such as competition for resources, emigration, and accumulation of toxic waste products all increase as a population approaches its carrying capacity for a particular habitat. The resources for which the members of the population must increasingly compete may include food, shelter, light, mating sites, or any other factor necessary for them to carry out their life cycle and reproduce.

## Density-Dependent and Density-Independent Factors

Effects such as those we have just discussed regulate the growth of populations and are called **density-dependent effects.** Among animals, they are often accompanied by hormonal changes that may bring about alterations in behavior that directly affect the ultimate size of that population. One striking example occurs in migratory locusts ("short-horned" grasshoppers), which, when they are crowded, produce hormones that cause them to enter a new, migratory phase: the locusts take off as a swarm and fly long distances to new habitats (Figure 19-6). In contrast, **density-independent effects** are caused by factors such as the weather and physical disruption of the habitat that operate regardless of population size.

*Density-dependent effects are controlled by factors that come into play particularly when the population size is larger; density-independent effects are controlled by factors that operate regardless of population size.*

**Figure 19-6**

**Migratory locusts,** *Locusta migratoria,* **a legendary plague of large areas of Africa and Eurasia.** When the population density reaches a certain level, the locusts change their characteristics and take off as a swarm. The most serious infestation of locusts in 30 years occurred in North Africa in 1988. These Moroccan villagers are gathering locusts killed in 1988 spraying operations to burn them.

**Figure 19-7**

**Fishing off the coast of New England in the northeastern United States.** Intensive fishing here has driven the populations of commercially valuable fish species such as cod and halibut below the point of optimal yield. As increasingly sophisticated mehtods are used for fishing, the danger of such consequences increases.

Modern agricultural practice makes use of the characteristics of the sigmoid growth curve. Early in the history of a population, resources are not yet limiting the growth of individuals and net productivity is highest. For best yields, this is the best time to "harvest" a population. Commercial fisheries attempt to operate so that fish populations are always harvested in the steep, rapidly growing parts of the curve. The point of **optimal yield** (maximum sustainable catch from the population) lies partway up the sigmoid curve. Harvesting the population of an economically desirable species near this point will result in much better yields than can be obtained either when the population approaches the carrying capacity of its habitat or when it is very small (Figure 19-7). Overharvesting a population that is smaller than this critical size can destroy its productivity for many years, as evidently happened to the Peruvian anchovy fishery after the population had been depressed by the 1972 El Nino weather disturbance. Although it is often difficult to determine the population levels of commercially valuable species that are most suitable for long-term, productive harvesting, such estimates are the subject of much study because they are so important from both a commercial and an ecological point of view. In forestry, the harvesting of mature trees for plywood, pulp, or similar products tends to exploit the upper part of the sigmoid curve; it is very important to understand the characteristics of the populations being harvested for proper management.

## r Strategists and K Strategists

Many species, such as annual plants, a number of insects, and bacteria, have very fast rates of population growth. An individual bacterium can reproduce in less than 20 minutes. Growth that is this rapid cannot be controlled effectively by reducing population sizes. In such species, small surviving populations will soon enter an exponential pattern of growth and regain their original sizes. In contrast, a comparable reduction in population size among relatively slow-breeding organisms, such as whales, rhinoceroses, California redwoods, or most tropical rain forest trees, can lead directly to their extinction (see Figure 19-2).

Slow-growing populations of organisms tend to be limited in number by the carrying capacity of the environment and are sometimes called **K strategists.** Such organisms, including the whales and redwoods just mentioned, usually live in habitats that are fairly stable and predictable. In contrast, species whose populations are characterized by very rapid growth followed by sudden crashes in population size tend to live in unpredictable and rapidly changing environments. These organisms have a high intrinsic rate of increase, or $r$, and are called **r strategists.** Many organisms are neither "pure" $r$ strategists nor "pure" $K$ strategists. Rather, their reproductive strategies lie somewhere between these two extremes or change from one extreme to the other under certain environmental circumstances.

In general, $r$ strategists have many offspring (Figure 19-8). Their offspring are small, mature rapidly, and receive little or no parental care. Many offspring are produced as a result of each reproductive event. Such organisms include cockroaches and mice.

In contrast, $K$ strategists leave few offspring. These offspring are large, mature slowly, and often receive parental care. This group consists of organisms such as coconuts, whales (Figure 19-9), and humans. Many of them are in danger of extinction.

**Figure 19-8**

**A cockroach explosion.** The German cockroach (*Blatella germanica*), a major household pest, produces 80 young every 6 months. If every cockroach that hatched survived, kitchens might look like this. One pair of cockroaches could produce this horde of over 130,000 individuals in just three generations. The exhibit shown here is in the Smithsonian Institution's National Museum of Natural History.

## Human Populations

Human beings use their culture to avoid or postpone the effects of their population size, which is currently increasing at the rate of about 1.8% per year. Some 95 million people are added annually to a total population that already numbers more than 5.3 billion people. That's three every second. At this rate, the world population will double in less than 40 years. As we discuss in Chapter 22, such a rate of growth has potential consequences for our future that are extremely grave. By reducing the population sizes of most other species as our own enormous populations grow so rapidly, we are driving to extinction those relatively large and slow-breeding species that we view as desirable while favoring those weed and pest species that compete directly with us for food and other critical resources.

*The rapidly growing human population is tending to exterminate K strategists, such as whales, elephants, and forest trees, while favoring r strategists, including houseflies, cockroaches, and dandelions.*

## MORTALITY AND SURVIVORSHIP

A population's intrinsic rate of increase depends on the ages of the organisms in it and the reproductive performance of the individuals in the various age groups. Very young and very old individuals obviously are not

**Figure 19-9**

**A humpbacked whale, a K strategist.** Most whales have one calf at a time. Roger Paynes, who has contributed to our understanding of whales, has written, "In our long, sad history, we have brought hundreds of species to extinction. Unless we change our priorities, we will soon eradicate hundreds of thousands more. There is a curious fact, however; among all of the species we have destroyed, there is not one that occurred worldwide. . . . The closest that we have ever come to taking that fatally insane step is with the [humpbacked] whale—a species that once occurred off the western and eastern shores of every continent. We came so close to destroying that species that it really is something of a miracle that it survived."

as reproductively active as the rest of the population. The speed with which a population can grow depends critically on how many of its members are reproducing and thus on the **age distribution** of the population. A population with many reproducing individuals will grow much faster than one with a few reproducing individuals. Age distributions differ greatly from species to species and even, to some extent, from place to place within a given species.

One way to express the age distribution of a population is the survivorship curve. **Survivorship** is defined as the percentage of an original population that is living at a given age. Samples of different kinds of survivorship curves are shown in Figure 19-10. In the hydra, individuals are equally likely to die at any age, as indicated by the straight-line survivorship curve. But oysters produce vast numbers of offspring, only a few of which live to reproduce. Once they become established and grow into reproductive individuals, their rate of death, or **mortality,** is extremely low. Even though human infants are susceptible to death at relatively high rates, the highest mortality rates in people occur later in life, in their postreproductive years.

By convention, life cycles are designated type I, the type found in humans, where a large proportion of the individuals appear to reach their physiologically determined maximum age; type II, the situation found in hydra, in which the mortality rate remains more or less constant at all ages; and type III, the kind of life cycle found in oysters, in which mortality is high in the early stages but then declines. Many animal and protist populations in nature probably have survivorship curves that lie somewhere between those characteristic of type II and type III, and many plant populations, with high mortality at the seed and seedling stages, are probably closer to type III. Humans have probably approached type I more and more closely through the years, with the birth rate remaining relatively constant or declining somewhat but the death rate dropping markedly.

*In type I survivorship curves, a large number of individuals reach their maximum theoretical age. In type II curves, mortality is largely independent of age. In type III curves, mortality is highest during the young stages. These curves are convenient ways to describe the different kinds of life histories.*

## DEMOGRAPHY

**Demography** is the statistical study of populations. The term comes from two Greek words, *demos,* "the people," (the same root we see in the word "democracy") and *graphos,* "measurement." It therefore means measurement of people, or, by extension, of the characteristics of populations. Demography is the science that enables us to predict the ways in which the sizes of populations will alter in the future. It takes into account the age distribution of the population and its changing size through time.

A population whose size remains the same through time is called a **stable population.** In such a population, births plus immigration must exactly balance deaths plus emigration. In such a population, the size not only remains constant, but so does the age structure.

The characteristics of a population can be illustrated graphically by means of a **population pyramid**—a bar graph using 5-year age categories (Figure 19-11). In a population pyramid, males are conventionally placed to the left of the vertical age axis, and females to the right. A population pyramid thus shows the composition of a population by age and sex.

Examples of population pyramids for Mexico, the United States, and West Germany in 1980 are shown in Figure 19-11. In the population pyramid for the United States, the number of cohorts (groups of individuals) 45 to 49 years old is smaller in size than those preceding and following (these people were born during the depression years). The cohorts 20 to 24 and 25 to 29 years old represent the "baby boom."

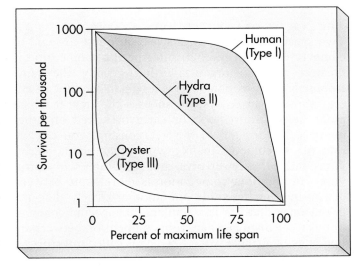

*Figure 19-10*

**Survivorship curves.** The shapes of the respective curves are determined by the percentages of individuals in populations that die at different ages.

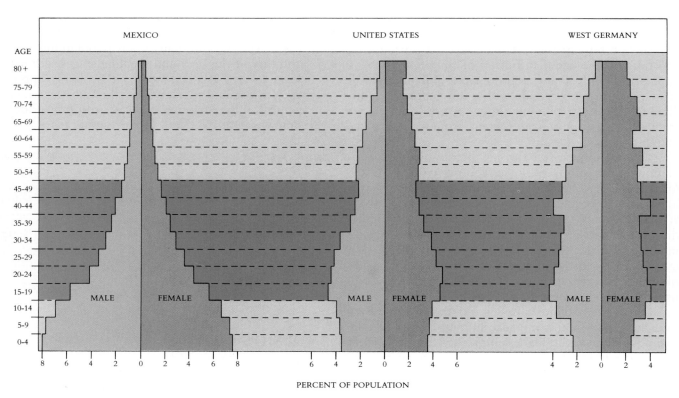

**Figure 19-11**

**Why Mexico's population is exploding.** Shown here is the age distribution of human males and females in 1980 in the populations of Mexico, the United States, and West Germany. Current rates of growth for these countries are approximately 2.4% (Mexico), 0.7% (United States), and −0.1% (West Germany). The shape of the Mexican population pyramid is characteristic of developing countries, where the number of children less than 15 years old often exceeds 40%.

The shape of the Mexican population pyramid in Figure 19-11 is characteristic of developing countries, where the number of children less than 15 years old often exceeds 40% of the total population; such individuals have yet to reach the child-bearing age, but when they do, the size of the population will grow explosively. Even if the people in Mexico were to limit their family size to two children each, effective immediately (half of the present average), population growth in the country would not level off until the middle of the next century.

## COMPETITION

Competition between individuals of two or more species for the same limiting resources, as well as the direct effects of predation or parasitism of one species on another, are among the kinds of interspecific interaction that may limit population size.

### The Nature of Competition

Two species that both use a scarce resource are said to **compete** with one another. Interspecific competition is greatest between organisms that occupy the same **trophic level**; a trophic level is a step in the flow of energy through an ecosystem, such as the step at which plants manufacture food or carnivores feed on other animals. In the light of this relationship, you can see that plants compete mainly with other plants, herbivores with other herbivores, and carnivores with carnivores. In addition, competition is more acute between organisms that resemble one another closely than between ones that are less similar. Interspecific competition is to be distinguished from competition between individuals of a single species, which is called **intraspecific competition.**

More than 50 years ago, the Soviet ecologist G.F. Gause formulated what he called the **principle of competitive exclusion.** This principle states that if two species are competing with one another for the same lim-

**Figure 19-12**

**The flour beetle *Tribolium confusum*, showing adults, larvae, and pupae.** This species, and the closely similar *T. castaneum*, were used in experiments that test what happens when two species of organisms compete with one another.

ited resource, then one of the species will be able to use that resource more efficiently than will the other, and the former will therefore eventually eliminate the latter locally.

*The principle of competitive exclusion states that if two species are competing with one another for the same limited resource in the same place, one of them will be able to use that resource more efficiently than will the other and eventually will drive that second species to extinction locally.*

In subsequent experiments the results of growing individuals of two species together in the laboratory have not always been easily predictable. For example, when Thomas Park and his colleagues at the University of Chicago grew two species of flour beetle, *Tribolium* (Figure 19-12), together in the same container of flour, one always became extinct. But the species that survived varied. *T. castaneum* usually won under relatively hot and damp conditions, whereas *T. confusum* won under cooler, drier conditions. Subsequent experiments with these species demonstrated that a genetic component was also involved in the unpredictability of the outcome. Some strains of one species would win over some—but not all—strains of the other under a given set of conditions.

## Competition in Nature

Examples demonstrating the principle of competitive exclusion can be found and studied, both by observation and by experiment, in nature. The competitive interactions between two species of barnacles that grew together on the same rocks along the coast of Scotland have been investigated by J.H. Connell of the Univer-

sity of California, Santa Barbara. Barnacles are marine animals (crustaceans) that have free-swimming larvae which settle down, cement themselves to rocks, and then remain permanently attached to that point. Of the two species Connell studied, *Chthamalus stellatus* lives in shallower water, where it is often exposed to air by tidal action, and *Balanus balanoides* occurs lower down, where it is rarely exposed to the atmosphere (Figure 19-13). In this deeper zone, *Balanus* could always outcompete *Chthamalus* by crowding it off the rocks, undercutting it, and replacing it even where it had begun to grow. But when Connell removed *Balanus* from the area, *Chthamalus* was easily able to occupy the deeper zone, indicating that no physiological or other general obstacles prevented it from becoming established in that area. In contrast, *Balanus* cannot survive in the shallow-water habitats where *Chthamalus* normally occurs. It evidently does not have the special physiological and morphological adaptations that allow *Chthamalus* to occupy this zone.

In another set of field observations, the late Princeton ecologist Robert MacArthur studied five species of warblers—small, insect-eating birds that coexist during part of the year in the forests of the northeastern United States and adjacent Canada. Although they all appeared to be competing for the same resources, MacArthur found that each species actually spent most of its time feeding in different parts of the trees, so each ate different subsets of insects in those trees. Some of the species of warblers fed on insects near the ends of the branches while others regularly penetrated well into the foliage, and some stayed high on the trees while others fed on the lower branches; these patterns were recombined in different ways characteristic of each of the warbler species (Figure 19-14). As a result of these different feeding habits, each species of warbler actually occupied a different niche; in other words, they had different ways of utilizing the resources of the environment and thus were not really in direct competition with one another for limited resources.

*Species of barnacles, warblers, and many other kinds of organisms coexist because the niches of the species involved differ and thus competitive exclusion does not occur.*

During the past decade there has been a lively debate about the real importance of competition in nature in relation to community structure. Environmental changes greatly affect the outcome of competitive situ-

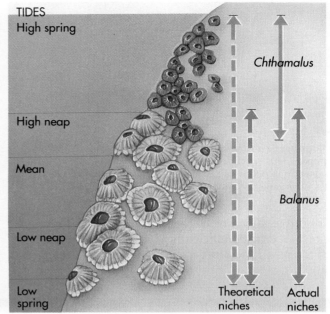

*Figure 19-13*

**Competition can limit niche use.**
**A** *Chthamalus* (the smaller, smooth barnacles) and *Balanus* (the larger, ridged barnacles) growing together on a rock along the coast of Scotland.
**B** The distribution of the two species with respect to different water levels and tidal effects is shown at the right. The neap tide is the extreme tide that occurs in each lunar month when the sun and the moon are lined up with one another.

ations. Furthermore, in the kinds of highly diverse situations that occur in nature, which are really mosaics of small patches of suitable environment, there are opportunities for many species to coexist in the same environment, so it is often very difficult to evaluate the competitive interactions that may be occurring.

## THE NICHE

Studies of the principle of competitive exclusion led to the development of the concept of the niche. A **niche** may be defined most simply as the role an organism plays in an ecosystem. It may be described in terms of space, food, temperature, appropriate conditions for mating, requirements for moisture, and so on. A full portrait of an organism's niche would also include its behavior and the ways in which this behavior changes at different seasons and different times of the day. In other words, *niche* is not synonymous with *habitat:* the habitat of an organism, the place where it lives, is defined by some, but by no means all, of the factors that make up its niche. The factors that make up an organism's niche determine whether it can exist in a given ecosystem and also how many species can exist there together.

Niche, in the sense we have just been discussing, is sometimes called the **actual niche** of an organism—the role the organism actually plays in a particular ecosystem. It is thus distinguished from the niche that it might occupy if competitors were not present, called the **theoretical niche.** Thus the theoretical niche of the barnacle *Chthamalus* in Connell's experiments in Scotland included that of *Balanus,* but its actual niche was much narrowed because *Chthamalus* was outcompeted by *Balanus* in its actual niche.

> *A niche may be defined as the role that an organism plays in the environment. The theoretical niche of an organism is the niche that it would occupy if competitors were not present. The actual niche is the niche that it actually occupies under natural circumstances.*

Another interesting example of some of the properties of a niche involves the flour beetles of genus *Tri-*

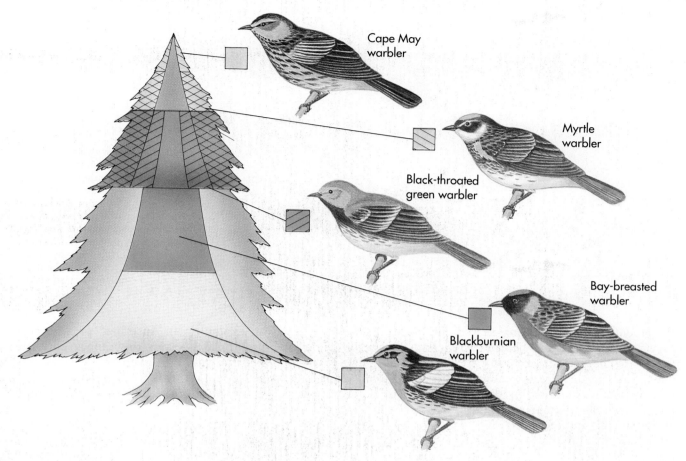

Cape May
warbler

Myrtle
warbler

Black-throated
green warbler

Bay-breasted
warbler

Blackburnian
warbler

**Figure 19-14**

**Carving up a niche avoids competition.** Although all five species of warbler shown here feed on insects in the same spruce trees at the same time, they mainly feed in different parts of the tree and in different ways. The shaded areas indicate areas where the given species spends at least half of its feeding time. In this way, competition is reduced.

*bolium* that we considered earlier. If *Tribolium* is grown in pure flour along with beetles of a second genus, *Oryzaephilus*, it will drive *Oryzaephilus* to extinction by means that are only partly understood. But if small pieces of glass tubing are added to the flour, providing refuges for *Oryzaephilus*, then both kinds of flour beetle will coexist indefinitely. In a sense, this experiment suggests why so many kinds of organisms can coexist in a really complex ecosystem, such as a tropical rain forest, and why competition is more direct in an ecosystem with fewer species, such as agricultural fields.

Gause's principle of competitive exclusion can be restated, in terms of niches, as follows: *No two species can occupy the same niche.* Certainly species can and do coexist while competing for the same resources, and we have just seen some examples of such relationships. Nevertheless, Gause's theory predicts that when two species do coexist on a long-term basis, one or more features of their niches will always differ; otherwise the extinction of one species will inevitably result.

## PREDATION

**Predation,** like competition, is a factor that may limit the size of populations. In this sense, predation includes everything from one kind of animal capturing and eating another to **parasitism,** the condition of an organism living in or on another organism, at whose expense the parasite is maintained. Predation and parasitism are two ends of a biological spectrum between which there is no clearly marked distinction. They are governed by similar principles.

When experimental populations are set up under very simple circumstances in the laboratory—as Gause did with *Paramecium* and its predator protozoan *Didinium*—the predator often exterminates its prey and then becomes extinct itself, having nothing to eat. If refuges are provided for the prey, its populations will be driven to low levels but can recover; one would then expect the populations of predators and prey to follow a cyclical pattern. The low population levels of the prey species provide scant food for the predators; then

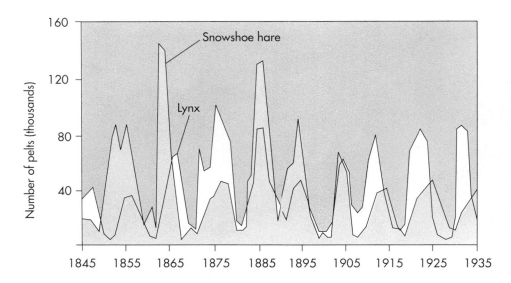

**Figure 19-15**

**A predator-prey cycle.** The numbers of lynxes and snowshoe hares oscillate in tune with each other in Northern Canada. The data are based on numbers of animal pelts from 1845 to 1935. As the number of hares grows, so does the number of lynxes; the cycle repeats about every 9 years. The number of hares is controlled not by predators, but by available resources; the number of lynxes is controlled by the availability of prey, hares.

the predators in turn become scarce. As this occurs the prey recover and again become abundant. Population cycles are characteristic of some species of small mammals, such as lemmings, and may be stimulated, at least in some situations, by their predators. An example of such fluctuating populations is expressed diagrammatically in Figure 19-15.

Relationships of this sort are of the greatest importance for biological control. If the prey becomes so rare that it is an infrequent event for the predator species to encounter it, the predator itself may become extinct. Ideally, in the case of deliberate biological control, the prey will survive in small numbers, thus making it possible for small populations of the predator to survive also and for the prey species to be controlled indefinitely.

For example, in Australia, prickly pear cactus *(Opuntia)* once overran the ranges and became so abundant that vast areas were effectively closed to cattle grazing (Figure 19-16). The situation was changed dramatically with the introduction of the moth *Cactoblastis* (Figure 19-17). The larvae of the moth feed on the pads of the cactus and rapidly destroy the plants. Within relatively few years, the moth had reduced the cactus to the status of a rare species in many regions where it was formerly abundant. It is now exceptional to find an individual of *Cactoblastis*, but the moth is still present and evidently keeps the cactus in check.

The future is more problematic for the American chestnut *(Castanea americana)*, a species of tree that has virtually been driven to extinction by the accidental introduction of an Asiatic fungus, *Cryptonectria (En-*

A

B

**Figure 19-16**

**Biological control of pests.** After an initial period in which prickly pear cacti *(Opuntia)*, introduced from Latin America choked many of the pastures of Australia with their rampant growth, they were controlled by the introduction of a cactus-feeding moth, *Cactoblastis*, from the areas where the cacti were native.
**A** An infestation of prickly pear cacti in scrub in Queensland, Australia, in October, 1926.
**B** The same view in October, 1929, after the introduction of *Cactoblastis*.

**Figure 19-17**

**The conquering hero.** *Cactoblastis*, a moth introduced from Latin America, still controls prickly pear cacti in Australia effectively, even though the moths are rarely encountered today.
**A** Larvae of *Cactoblastis* on a cactus pad.
**B** Adult *Cactoblastis* on a cactus spine.

*dothia) parasitica.* The chestnut used to be a dominant or codominant tree throughout much of the forests of eastern and central temperate North America. Chestnut blight, the disease caused by the fungus, was first seen in North America in New York State in 1904; it killed most of the chestnut trees in North America during the following 30 years. Today the American chestnut survives largely as sprouts that grow every year from the trunks of trees that were killed decades ago. A few populations of chestnuts appear to be isolated from the chestnut blight in remote areas, whereas others apparently are genetically resistant.

Many additional examples of the ways in which predator-prey relationships operate are familiar. Organisms that cause diseases which completely kill their host species are not successful because once they have done so they have also eliminated their own source of food. Thus those strains of the disease-causing organism which are less virulent will be favored by natural selection and will survive.

The history of the viral disease myxomatosis, which is caused by a virus that was introduced into Australia and New Zealand to control rabbits, pro-

vides an instructive example of this principle. The rabbits were brought to these countries as a convenient source of meat, but they soon ran wild, with devastating effects on the countryside (Figure 19-18). When the virus causing myxomatosis was introduced, most of the rabbits soon died. The most virulent strains of the virus disappeared with the dead rabbits, and less lethal strains became apparent in the remaining rabbit populations. At the same time, strains of rabbits that were resistant to the disease began to appear. Now the populations of both organisms have achieved the equilibrium relationship in which they can coexist indefinitely.

*An organism that causes a disease that always kills its host will die with it; one that produces sublethal effects will have the opportunity of spreading to another host.*

The relationships between large carnivores and grazing mammals are a subject of great interest and importance in many parts of the world. Appearances are sometimes deceiving. On Isle Royale in Lake Superior, moose reached the island and multiplied freely there in

**Figure 19-18**

**Introduced species often explode in numbers.** This photo shows rabbits crowding around a water hole in Australia. Imported from Europe as a source of meat, rabbits ran wild in Australia, until eventually controlled by a lethal virus that caused the disease myxomatosis.

**Figure 19-19**

**Predation often hits the old and infirm.**
On Isle Royale, Michigan, a large pack
of wolves is in unsuccessful pursuit of a
moose. This moose was chased for
almost 2 kilometers; it then turned and
faced the wolves, who by that time were
exhausted from running through
chest-deep snow. The wolves lay down
and the moose walked away.

isolation (Figure 19-19). When wolves later reached the island by crossing over the ice in winter as the moose had done earlier, it was widely assumed at first that they were playing the determining role in controlling the moose populations. But more careful studies have demonstrated that this is not the case. The moose that the wolves eat are mostly old and diseased and would not have survived long anyway. In general the moose are controlled by the amount of food available to them, their diseases, and many factors other than the wolves.

Wolves have been returning slowly to remnants of their former habitat in the United States outside of Alaska, where there are fewer than 1300 individuals living now. Only 2 of the 24 subspecies that originally roamed North America currently survive in this area. Because wolves kill a small amount of livestock, their presence in farming areas is controversial. Efforts to re-

introduce them into their former range in Yellowstone National Park, where biologists have set a goal of 10 breeding pairs, have likewise met with opposition from neighboring ranchers. On the other hand, there is no known case in North America of injury to a human being by a wild, nonrabid wolf, sensational stories to the contrary.

By controlling the levels of some species, predators make possible the continued existence of others—sometimes many others—in a given community. By keeping the numbers of individuals of some of the competing species low, the predators may prevent or greatly reduce competitive exclusion. Such patterns are particularly characteristic of biological communities in intertidal marine habitats. In many instances, they appear to play the key role in allowing species to coexist because situations that appear to violate the principle of competitive exclusion are, in fact, frequent in nature.

## ■ SUMMARY

1. The rate of growth of any population is defined as the difference between the birth rate and the death rate per individual per unit of time. The actual rate is affected both by emigration from the population and by immigration into it.

2. Most populations exhibit a sigmoid growth curve, which implies a relatively slow start in growth, a rapid increase, and then a leveling off when the carrying capacity of the species' environment is reached. Populations can be harvested most effectively when they are in the rapid growth phase.

3. *r* strategists have large broods and rapid rates of population growth; the populations of *K* strategists are limited in size by the carrying capacity of their environments. They tend to have fewer offspring, with slower rates of population growth.

4. Survivorship curves are used to describe the characteristics of growth in different populations. Type I populations are those in which a large proportion of the individuals approach their physiologically determined age limits. Type II populations have a constant mortality throughout their lives. Type III populations have very high mortality in their early stages of growth, but an individual surviving beyond that point is likely to live a very long time.

5. Gause's principle of competitive exclusion states that if two species compete with one another for the same limited resources, then one will be able to utilize them more efficiently than will the other and will drive its less efficient competitor to extinction. This principle can be restated in terms of the niche concept: No two species can occupy the same niche indefinitely. Organisms studied under field conditions display many ways of avoiding direct competition for the same scarce resources.

6. Each species plays a specific role in its ecosystem; this role is called its niche.

7. An organism's theoretical niche is the total niche that the organism would occupy in the absence of competition. The actual niche of an organism is the actual niche that it occupies in nature.

8. Predator-prey relationships are of crucial importance in limiting population size in nature. They appear to maintain a balance because if the prey is completely eliminated, the predator will die also.

# REVIEW

1. Populations with higher proportions of children are expected to grow _____ rapidly than populations with lower proportions of children.

2. The biotic potential of a population depends on its size. True or false?

3. With respect to its size, a _____ population is more likely to become extinct.

4. Oysters exhibit a type _____ growth curve, whereas human beings exhibit a type _____ curve.

5. The percentage of a population that is living at a given age is defined as the _____ of that population.

# SELF-QUIZ

1. Which one of the following is *least* characteristic of a species that is an *r* reproductive strategist?
   (a) Small size of offspring
   (b) Large numbers of offspring per litter
   (c) Rapid maturation
   (d) Extensive parental care
   (e) All are characteristic

2. What is the difference between a type II and a type III survivorship curve?
   (a) In type II, mortality is higher in the early stages than it is in type III.
   (b) In type III, mortality is the same in early stages as it is in type II.
   (c) In type III, mortality is higher in the early stages than it is in type II.
   (d) In type II, a large proportion of individuals survive to the maximum age, whereas in type III they do not.
   (e) None of the above.

3. Populations of different species that live together are collectively called
   (a) an ecosystem.           (c) an ecotype.           (e) Gaia.
   (b) a community.            (d) a biome.

4. The size of a population after it has completed its period of rapid growth is called
   (a) a density-dependent state.
   (b) a density-independent state.
   (c) the carrying capacity of that place.
   (d) its biotic potential.
   (e) its intrinsic growth rate.

5. The optimal yield that can be derived from a population
   (a) occurs when it has become stable.
   (b) occurs when it has reached its biotic potential.
   (c) occurs when it is growing rapidly.
   (d) is density-independent.
   (e) occurs when it is a *K* strategist.

6. Survivorship
   (a) is the percentage of the original population that lives beyond a given age.
   (b) is characteristic of type II populations.
   (c) brings the sigmoid growth curve to a stable condition.
   (d) determines the carrying capacity of all populations.
   (e) becomes more important as a population grows.

7. The principle of competitive exclusion states that
   (a) *K* strategists exclude *r* strategists.
   (b) *r* stragegists exclude *K* strategists.
   (c) species are excluded primarily by intraspecific competition.
   (d) the number of species in a community will become progressively lower through time.
   (e) if two species are competing for a limited resource, one will tend to eliminate the other.

8. Grazing and other forms of predation (choose two)
   (a) tend to limit the number of species in a given community.
   (b) tend to increase the number of species in a given community.
   (c) determine the distribution of barnacles.
   (d) determined the relationships between the species of warblers that Robert MacArthur studied.
   (e) restrict the innate capacity for growth.

# *THOUGHT QUESTIONS*

1. How is it possible for ecological generalists—those which feed on a wide range of foods—to coexist with specialists in natural communities, as they do throughout the world? Why do all species not simply become generalists or specialists? Which of these habits is a better strategy for survival? Why?

2. How do you think the intensity of competition would differ between an Arctic tundra and a tropical rain forest? How would you test your answer?

3. Which kinds of ecosystems are most resistant to introduced species? Why?

# FOR FURTHER READING

BEDDINGTON, J.R., and R.M. MAY: "The Harvesting of Interacting Species in a Natural Ecosystem," *Scientific American*, November 1982, pages 62-69. An analysis of the changing populations of whales and other animals that feed on the krill (shrimp) populations in the Antarctic Ocean as an example of the problems of using a biological resource without destroying it.

CASE, T.J., and M.L. CODY: "Testing Theories of Island Biogeography," *American Scientist*, vol. 75, 1987, pages 402-411. On islands in the Gulf of California, a natural laboratory for testing theories of island biogeography exists.

CONNOR, E.F., and D. SIMBERLOFF: "Competition, Scientific Method, and Null Models in Ecology," *American Scientist*, vol. 74, 1986, pages 155-162. A central article in the continuing debate about the role of competition in structuring natural communities.

RYAN, M.J.: "Signals, Species, and Sexual Selection," *American Scientist*, vol. 78(1), 1990, pages 46-52. Studies of mate recognition in frogs and fishes reveal preferences for individuals, populations, and even members of closely related species.

VERRELL, P.: "When Males are Choosy," *New Scientist*, January 20, 1990, pages 46-50. Under certain circumstances, males can be choosy, too.

WOLFF, J.O.: "Getting Along in Appalachia," *Natural History*, vol. 95, no. 9, 1986, pages 44-49. Fine account of the author's investigations of two species of mice that occur together in the forests of the southern Appalachians and of the different ways in which they relate to the habitat.

# *H*ow Species Interact within Communities

Competition within communities for survival can be fierce. This young whitetail deer fawn hides while its mother forages for food—lest it itself become food for some hungry predator.

# HOW SPECIES INTERACT WITHIN COMMUNITIES

## Overview

Ecosystems function because of the interactions of the organisms that compose them; such a set of interacting organisms is, in turn, called a community. A community has an extent in space and time, and can, in some respects, be defined objectively. Within a community, the herbivores avoid feeding on certain plants and selectively use others; their actions are mediated by various characteristics of the plants. As a result, the herbivores themselves have become specialized. In some cases, herbivores use chemicals obtained from the plants that they eat to protect themselves from their own predators; in others, they break down the chemicals and thus avoid their toxic effects. Many other phenomena, such as warning or protective coloration and mimicry, are related to the nature of the chemicals that animals obtain from their food or manufacture. Symbiotic relationships generally act to order the flow of energy through ecosystems and to regulate the cycling of nutrients. Learning about these relationships individually is the key to learning about the functioning of ecosystems.

## For Review

*Here are some important terms and concepts that you will encounter in this chapter. If you are not familiar with them, you should review them before proceeding.*

**Energy flow** (Chapter 7)

**Demography** (Chapter 19)

**Competition** (Chapter 19)

The magnificent redwood forest that extends along the coast of central and northern California and into the southwestern corner of Oregon is an example of a community. Within it, the most obvious organisms are redwood trees, *Sequoia sempervirens* (Figure 20-1). These trees are the sole survivors of a genus that once was distributed throughout much of the Northern Hemisphere. With them are regularly associated a number of other plants and animals (Figure 20-2); their coexistence is partly made possible because of the special conditions that are created by the redwoods themselves: shade, water (dripping from the branches), and relatively cool temperatures. This particular, distinctive assemblage of organisms is called the redwood community.

We recognize this community largely because of the redwoods themselves. The overall distributions of the other kinds of organisms that occur together with the redwoods differ greatly from one another—some are smaller than that of the redwoods, and some are larger. In the redwood community, their niches overlap; that is why, in this community, they occur together, a recognizable group. However, it would be incorrect to view the community as a kind of "superorganism," the term that some ecologists in the past used to refer to communities; rather it is a collection of distinct organisms with overlapping requirements in close proximity with one another.

The redwood community also has a historical dimension; the organisms that are now characteristic of this community have each had a complex and unique evolutionary history. At different times in the past, they evolved and then came to be associated with the redwoods. Certainly the present-day redwood community has come into existence only during the past few million years. During this period, the range of the redwood was reduced as the tree was eliminated by climatic change from many of the areas in which it grew previously. In a historical sense, then, the redwood community that we recognize today represents the concordance of the individual histories of the plants, animals, and microorganisms that compose it.

*We recognize a redwood community largely because of the presence of the redwood trees themselves, but many other kinds of organisms exist in this community. The community exists because the ranges of these species overlap in it.*

## Figure 20-1

**Most ecosystems contain many species.** The redwood forest of coastal California and southwestern Oregon is dominated by redwoods *(Sequoia sempervirens)*, although many other kinds of plants live beneath them.

## Figure 20-2

**Plants that occur regularly in the redwood community.**
A  The redwood sorrel *(Oxalis oregana)*.
B  On this sword fern *(Polystichum munitum)*, a ground beetle is shown feeding on a slug.

Although they are not superorganisms, many communities are very similar in their species composition and appearance over wide areas. For example, the open savanna grassland that stretches across much of Africa includes many plant and animal species that co-exist over thousands of square kilometers. Similar interactions between these organisms, some of which have evolved over millions of years, occur throughout these communities. We will now consider some of these interactions and then expand our discussion to a consideration of the ways in which the distributions of organisms are limited—the reasons that particular kinds of organisms occur in particular places, and not in others.

## COEVOLUTION

The interactions between organisms that characterize particular communities and ecosystems have arisen as a result of their evolutionary history. Thus the **primary producers**—photosynthetic organisms, including plants, algae, and photosynthetic bacteria; **herbivores**—organisms that consume photosynthetic organisms; and second- and higher-level predators—organisms that consume other predators—have changed and adjusted to one another continually over millions of years. For example, many of the features of flowering plants have evolved in relation to the dispersal of their pollen to other members of the same species by animals, especially insects (discussed further in Chapter 30). These animals, in turn, have evolved a number of special traits that enable them to obtain food or other resources efficiently from the plants they visit, often from their flowers. In addition, the seeds of many flowering plants have features that make them more likely to be dispersed to new areas of favorable habitat.

Such interactions, which involve the long-term, mutual evolutionary adjustment of the characteristics of the members of biological communities in relation to one another, are examples of **coevolution.** Many of the interspecific interactions we have covered earlier, as in Chapter 19, fall into this heading. Here we consider some additional examples, grouping them under the general headings of predator-prey interactions and symbiosis.

*Figure 20-3*

**Many herbivores eat only particular food plants.** The Gerenuk is an antelope that grazes successfully on some of the spiny shrubs and trees of the African savanna that other browsers do not use as food.

## PREDATOR-PREY INTERACTIONS

### Plant-Herbivore Relationships

The plant-herbivore interface is the point at which the greatest transfer of energy takes place in the world's ecosystems, with the exception of the interface between organic matter and decomposers. In plant-herbivore interactions generally, some 300,000 species of autotrophic organisms, including about 250,000 species of plants, fix about 1% of the sunlight that falls on their leaves and green stems. Once these plants have carried out this process, at least 2 million species of herbivores feed on them. These herbivores integrate the materials in the plants, algae, or photosynthetic bacteria into their own bodies, actually converting about 10% of the stored energy in these organisms. The nature of life on earth has been largely determined by the ways in which plants avoid being eaten and by which herbivores succeed in eating them.

### Plant Defenses

The most obvious ways in which plants limit the activities of herbivores are morphological defenses. Thorns, spines, and prickles play an important role in discouraging browsers, although some of the animals have learned to overcome these obstacles (Figure 20-3). The defensive role of plant hairs, especially those which have a glandular, sticky tip (Figure 20-4), is less obvious. The enormous abundance of insects and the scale at which they operate as herbivores provide some indication of why hairs of various kinds are so important to plants. Simple strengthening of the plant parts, as by the deposition of silica in the leaves, is also an element in the protective system of some plants, such as grasses. If enough silica is present in their cells, the plants may simply be too tough to eat (Figure 20-5).

Significant as these morphological adaptations are, the chemical defenses that occur so widely in plants are even more crucial. As research in this area proceeds, we continue to be astonished by the number and diversity of these features. First, recall that most animals can manufacture only 8 of the 20 required amino acids; they must obtain all of the others from plants. Through their possessing limited quantities of 1 or more of the 12 amino acids that animals require from them, plants can limit their nutritional suitability for animals. As a result, they may be protected from herbivores.

Perhaps the most important element in the defenses of plants against herbivores are the **secondary chemical compounds,** so called to distinguish them from the **primary compounds,** or compounds that are regularly formed as components of the major metabolic pathways, such as respiration. Virtually all plants, and apparently many algae as well, contain secondary compounds that play a defensive role, and these are structurally very diverse. Estimates suggest that perhaps 50,000 to 100,000 secondary plant compounds may exist; some 15,000 of these have already been characterized chemically.

*Figure 20-4*

**Glandular hairs of the wild potato *Solanum berthaultii*.** The hairs are defenses against aphids, which are the most important herbivores on these plants. The hairs are of two types, type A and type B, both seen in this scanning electron micrograph. The shorter type A hairs have a four-lobed head that ruptures on contact, entrapping the aphid or other insect in a quick-setting sticky liquid. The type B hairs play an even more fascinating role in the plant's defenses. The sticky liquid that they secrete contains a chemical that is the main ingredient of the alarm pheromone in aphids. An extract prepared from these hairs greatly disturbed aphids, driving many of them away, when it was tested experimentally by scientists at the Rothamsted Experimental Station in England. Together, the two types of hairs constitute a formidable barrier against the aphids. Breeding experiments have now succeeded in transferring these hairs to the cultivated tomato and potato; the new strains are much more resistant to pests than are the nonhybrid ones.

We now know a great deal about the roles of secondary compounds in nature. As a rule, different compounds are characteristic of particular groups of plants. For example, the mustard family (Brassicaceae) is characterized by a group of chemicals known as mustard oils, the substances that give the pungent aromas and tastes to such plants as mustard, cabbage, watercress, radish, and horseradish. The same tastes that we enjoy signal the presence of chemicals that are toxic to many groups of insects.

Another group of plants, the potato and tomato family (Solanaceae), is rich in alkaloids and steroids, complex ring-forming molecules that exhibit an enormous range of structural diversity. Such compounds occur very widely among the flowering plants, but the particular kinds found in different groups vary greatly. Sometimes the compounds are involved in more complex defensive systems; for example, plants of the milkweed family (Asclepiadaceae) and the closely related dogbane family (Apocynaceae) tend to produce a milky sap that deters herbivores from eating them. In addition, these plants usually also contain cardiac glycosides, molecules named for their drastic effect on heart function in vertebrates.

One of the best known plant groups with toxic effects comprises the related poison ivy, poison oak, and poison sumac (*Toxicodendron* species). All contain a gummy oil called *urushiol*, which contains a substance that causes a severe rash in susceptible people; urushiol can persist for years on clothes or other objects with which it has come into contact. At least 140,000 cases that cause absence from work occur in the United States annually, and many more are not reported. Approximately one out of two people are at least moderately sensitive to these plants. Immediate washing with soap will remove the excess urushiol and prevent its

further spread, but the reaction of exposed body parts seems to be immediate. Urushiol almost certainly functions to protect the plants in which it occurs from herbivores, but this relationship has not been demonstrated experimentally.

To mention one further example, the seeds of castor beans *(Ricinus communis)* produce a protein, ricin, that attacks ribosomes. Ricin is an enzyme that is similar to a toxin produced by the bacterium that causes bacterial dysentery *(Shigella dysenteriae)*. Ricin acts by lopping off an adenine nucleotide from a specific position in the RNA chain within a ribosome; this area of the ribosome is the same in all animals, which explains

*Figure 20-5*

**Plants can defend themselves.** These zebras grazing on the East African savanna dislike eating grasses with leaves reinforced with silica. The more silica in the leaf cells, the less likely zebras and other grazing animals are to eat that particular kind of grass.

the wide toxicity of castor bean seeds. Why such an unusual enzyme in a bacterium should resemble so closely one found in the seeds of a plant is unclear; perhaps the bacterium acquired the gene from the plant in the course of its evolution.

Many plant parts other than castor bean seeds are toxic to humans. One very familiar example is the houseplant called dumb cane, *Dieffenbachia,* a member of the same plant family as philodendrons. Touching a leaf of this plant with the tongue will result in intense pain and eventually produce ulcers and corrosive burns. These effects occur because the leaves of dumb cane have special ejector cells that expel needlelike crystals of calcium oxalate that penetrate the skin and cause the release of histamines. Particularly in families with children, people should take care to know the properties of the plants they grow.

These few examples demonstrate that many angiosperms are protected by a rich and varied chemical arsenal. Mustard oils, alkaloids, steroids, and other classes of secondary compounds are either toxic to most herbivores or disturb their metabolism so greatly that they are unable to complete their development normally. As a consequence, most herbivores tend to avoid the plants that possess these compounds. Therefore the pattern of occurrence of such chemicals has had major consequences on the evolution both of the plants themselves and of the herbivores, especially the insects, that feed on them.

## Producing Defenses When They are Needed

Some of the secondary compounds that plants use to defend themselves from herbivores are not normally present in their tissues. Rather, the plants produce them only when they are needed, in much the same way that immune systems operate in animals. When a leaf of a tomato, potato, or alfalfa plant is injured, a chemical message travels rapidly through the plant that induces the synthesis and accumulation of proteins that inhibit digestion in an animal's gut. The signal is a fragment of the plant's cell wall, and the inhibitory enzymes currently are being identified in several laboratories. Similarly, the infection of a plant by a fungus or bacterium may cause the plant to synthesize a molecule that retards the spread of the infection. A familiar example of such a process is provided by the chemicals that color apples brown when they are cut; interacting with proteins, these chemicals make the apple less attractive to caterpillars, fungi, and other organisms. Such defensive strategies are metabolically efficient: the plants are able to avoid the necessity of producing the chemical until it is actually needed. At other times, they are able to use the energy available to them for their growth and maintenance. It may not surprise you to learn that parasites, such as fungi, that regularly infect particular kinds of plants are often tolerant of the plants' particular defensive substances.

## The Evolution of Herbivores

Associated with each family or other group of plants that is protected by a particular kind of secondary compound are certain groups of herbivores that are able to feed on these plants, often as their exclusive food source. How do these animals manage to avoid the chemical defenses of the plants, and what are the evolutionary antecedents and ecological consequences of such patterns of specialization?

As a starting point, we can offer the observation that some herbivore groups feed on plants of many families, whereas others feed on plants of just a few families. In general, the herbivores that feed on a restricted array of plant families, perhaps only one, feed on plant groups in which certain kinds of secondary compounds are well represented. For example, the larvae of cabbage butterflies (subfamily Pierinae) feed almost exclusively on plants of the mustard and caper families, as well as on a few other small families of plants that are characterized by the presence of mustard oils (Figure 20-6). Similarly, the caterpillars of the monarch butterflies and their relatives (subfamily Danainae) feed on plants of the milkweed and dogbane families (Figure 20-7). No other groups of butterflies have larvae that feed on these particular plants, and the members of other groups of insects that feed on them usually also belong to specialized subgroups whose feeding habits are associated exclusively with the members of those particular plant families.

We can explain the evolution of these particular patterns as follows. Once the ancestors of the caper and mustard families acquired the ability to manufacture mustard oils, they were protected for a time against most or all herbivores that were feeding on other plants in their area. The plants that contained the mustard oils apparently succeeded very well as a result of this particular defense, and they and their descendants thus evolved and migrated, eventually giving rise to the thousands of species of mustards and capers that now grow worldwide.

At some point, certain groups of insects—for example, the cabbage butterflies—developed the ability to break down the mustard oils and thus feed on these mustards and capers without harming themselves. Having acquired this ability, the butterflies themselves had registered an important evolutionary "breakthrough." They were able to use a new resource without having to face competition for it from other herbivores. Often, then, in groups of insects such as the cabbage butterflies, sense organs particularly sensitive to the secondary compounds that their food plants produce have evolved. Clearly, the relationship that has

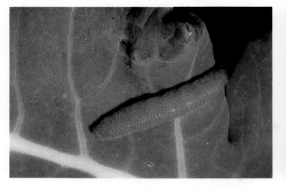

**Figure 20-6**

**Insect herbivores are well suited to their hosts.** The green caterpillars of the cabbage butterfly, *Pieris rapae*, are camouflaged on the leaves of cabbage and other plants on which they feed. The pieris caterpillars are able to break down the toxic mustard oils that prevent most insects from eating cabbage.

formed between the cabbage butterflies and the plants of the mustard and caper families is an example of co-evolution.

*The members of many groups of plants are protected from most herbivores by their secondary compounds. But once the members of a particular herbivore group acquire the ability to feed on them, these herbivores have gained access to a new resource that they can exploit without competition from other herbivores.*

Although the role of secondary compounds in protecting plants from herbivores was first discovered in flowering plants, other groups of photosynthetic organisms have similar defenses. Thus many red, brown, and green algae produce terpenoids and other compounds that deter feeding by herbivores such as fishes and that also inhibit the growth of bacteria in culture. Some evidence suggests that the group of shrimplike crustaceans known as amphipods, which are often abundant on algae, may play a role in marine environments similar to that of the insects on land. Both marine and terrestrial animals may also manufacture toxic chemicals that deter their predators. Therefore similar systems seem to operate both in the sea and on land that deter the dominant herbivores from consuming the primary producers of each community and that deter carnivores from consuming herbivores.

## Chemical Defenses in Animals

Some groups of animals that feed on plants rich in secondary compounds receive an extra benefit, one of great ecological importance. When the caterpillars of monarch butterflies feed on plants of the milkweed family, they do not break down the cardiac glycosides that protect these plants from most herbivores. Instead, they store them in their fat bodies. As a result, these butterflies are themselves protected from predators! The cardiac glycosides in the leaves that the caterpillars eat are concentrated, stored, and passed through the chrysalis stage to the adult and even to the eggs; all stages are protected from predators. A bird that eats a monarch butterfly quickly regurgitates it and thenceforth avoids the conspicuous orange and black pattern that characterizes the adult monarch (Figure 20-8). Locally, however, some birds have acquired the ability to tolerate the protective chemicals and eat the monarchs.

Insects that do feed regularly on plants of the milkweed family are generally brightly colored (Figure 20-9). Among them are brightly colored cerambycid beetles, whose larvae feed on the roots of the milkweed plants; bright blue or green chrysomelid beetles; and bright red bugs (order Hemiptera). In some parts of the

**Figure 20-7**

**Monarch butterflies make themselves poisonous.** All stages of the life cycle of the monarch butterfly *(Danaus plexippus)* are protected from birds and other predators by the poisonous chemicals that occur in the milkweeds and dogbanes on which they feed as larvae. Both caterpillars and adult butterflies "advertise" their poisonous nature with warning coloration.

A

B

## Figure 20-8

**A bluejay learns that monarch butterflies taste bad.**
**A** A cage-reared bluejay, which had never seen a monarch butterfly before, eating one.
**B** The same bird a few minutes later, regurgitating the butterfly. Such a bird is not likely to attempt to feed on orange-and-black insects again.

world, there are also bright red grasshoppers and other very obvious insects. These herbivores clearly are "advertising" their poisonous nature by their bright colors, using an ecological strategy known as **warning coloration.** Similar relationships occur in marine communities; a recent investigation revealed that, of the exposed common coral reef invertebrates at Lizard Island on the Great Barrier Reef off the northeast coast of Australia, three-fourths were toxic to fish; of the protectively camouflaged ones, only one-fourth were toxic.

The kinds of insects that eat plants whose chemical defenses are less obvious than those of the milkweeds are seldom brightly colored. In fact, many of these insects are **cryptically** colored—colored so as to blend in with their surroundings and thus be hidden from predators (Figure 20-10). This is also true of insects such as the larvae of cabbage butterflies, which, although they feed on plants with well-marked chemical defenses, are able to do so because of their ability to break down the molecules involved, rather than store them.

Some marine animals, such as certain nudibranchs (sea slugs; see Chapter 28), acquire defensive chemicals or defensive cells from their prey. Hydroids often pro-

vide such stinging cells to animals that graze on them. Off the coast of Panama, the large nudibranch *Aplysia* grazes selectively on red algae of the genus *Laurencia*, which is protected by elatol, a powerful inhibitor of cell division. Perhaps this is the reason that few fish feed on *Aplysia*. An intensive investigation of marine animals, algae, and flowering plants for new drugs against cancer and other diseases, or as sources of antibiotics, is now underway. It holds great promise because of the enormous diversity of chemical compounds that occur in these organisms.

Animals also manufacture many of the chemicals that they use in their defense. In fact, animals manufacture and use a startling array of substances to perform a quite incredible variety of defensive functions. Bees, wasps, predatory bugs, scorpions, spiders, and many other arthropods have chemicals that they use to defend themselves and to hunt for prey.

A variety of chemical defenses is also found among the vertebrates. The dart-poison frogs of the family Dendrobatidae produce toxic alkaloids in the mucus that covers their skin (Figure 20-11). Some of these toxins are so powerful that a few micrograms will kill a person if injected into the bloodstream. Most

A

B

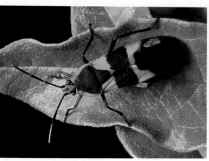

## Figure 20-9

**Feeding as defense.** These insects feed on milkweeds that contain toxic cardiac glycosides. They are not harmed, as they keep the dangerous chemicals in special cell compartments, but any predator that eats the insect will be poisoned, a fact they advertise by displaying warning coloration to their potential predators.
**A** Longhorn beetle *Tetraopes*.
**B** Milkweed bug, *Oncopeltus fasciatus*.

**Figure 20-10**

**A few striking examples of cryptic coloration.**
A  A young jackrabbit in the desert near Tucson displays both adaptive stillness and camouflage.
B  Protective coloration in a flatfish on the sea floor.
C  A tropical hawkmoth at rest, blending in perfectly with the moss- and lichen-covered tree bark.
D  The brown-green-brown coloring of the unicorn caterpillar gives it the appearance of a dried leaf edge.
E  An inchworm caterpillar *(Necophora quernaria)*, which closely resembles a twig.

species of these frogs, which number more than 100 species, are brightly colored; they are favorites in zoos and aquaria for that reason.

## Aposematic Coloration

Warning coloration, as defined above, is also known as **aposematic** coloration. Such coloration is characteristic of animals that have effective defense systems, including not only poisons, but also stings, bites, and other means of repelling predators. An organism possessing such a system will benefit by advertising the fact clearly—for example, by showy colors that are not normally found in that particular habitat. Otherwise, the distasteful or poisonous individual runs the risk of being killed while protecting itself.

*Aposematic, or warning, coloration serves to keep potential predators away from poisonous or otherwise dangerous prey.*

Of course, the animals that display aposematic coloration must also remain together if the system is to be effective. A lone individual that exhibits its warning coloration but is eaten will not deliver a message useful for the survival of other animals with a similar appearance. But if genetically related individuals are similarly colored and live in the same vicinity, the selective advantage is obvious.

Some examples of animals with aposematic coloration are shown in Figure 20-12. Such animals tend to live together in family groups, unlike those which are cryptically colored. If camouflaged animals lived together in groups, one might be discovered by a potential predator, offering a valuable clue to the presence of others.

## Mimicry

Many unprotected species have come, during the course of their evolution, to resemble distasteful ones that exhibit aposematic coloration. Provided that the unprotected animals are present in numbers that are low relative to those of the species which they resemble, they too will be avoided by predators. Such a pat-

of resemblance is called **Batesian mimicry,** after the British naturalist H.W. Bates, who first brought it to general attention in the 1860s. Bates also carried out the first scientific studies of this phenomenon, which has been of great interest to naturalists ever since.

*In Batesian mimicry, unprotected species resemble others that are distasteful. Both species exhibit aposematic coloration. The unprotected mimics will be avoided by predators if they are relatively scarce.*

**Figure 20-11**

**The bright color is a warning.** Frogs of the family Dendrobatidae, such as this individual of *Dendrobates reticulatus,* abundant in the forests of Latin America, are extremely poisonous to vertebrates. More than 200 different alkaloids have been isolated and identified in these frogs, and some of them are playing important roles in neuromuscular research. Two groups of Indians in western Colombia obtain a potent poison for blow-gun darts from extremely toxic species of these frogs that occur in their region.

Many of the best-known examples of Batesian mimicry occur among butterflies and moths. Obviously, predators in systems of this kind must use visual cues to hunt for their prey. Otherwise similar patterns of coloration would not matter to potential predators.

The groups of butterflies that provide the **models** in Batesian mimicry are, not surprisingly, members of groups whose larvae feed on only one or a few closely

**Figure 20-12**

**Warning coloration.** The spotted skunk, *Spilogale putorius,* which can eject a foul-smelling liquid when threatened; the poisonous gila monster, *Heleoderma suspectus,* a member of the only genus of poisonous lizards in the world; the red-and-black African grasshopper *Phymateus morbillosus,* which feeds on highly poisonous *Euphorbia* plants.

related plant families; the plants on which they feed are strongly protected chemically. The model butterflies take poisonous molecules from these plants and retain them in their own bodies. The mimic butterflies, in contrast, belong to groups in which the feeding habits of the larvae are not so restricted. As caterpillars, these butterflies feed on a number of different plant families, but not including those which are protected by toxic chemicals.

One well-known mimic among North American butterflies is the viceroy, *Limenitis archippus* (Figure 20-13). This butterfly, which resembles the poisonous monarch, ranges from central Canada south through much of the United States east of the Sierra Nevada and Cascade Range into Mexico. The larvae feed on willows and cottonwoods, and neither they nor the adults are distasteful to birds. Interestingly, the larvae of the viceroy are hidden from predators on the leaves of their host plants because they resemble bird droppings, whereas the distasteful larvae of the monarch are very conspicuous.

Another kind of mimicry, **Muellerian mimicry,** was named for the German biologist Fritz Mueller, who first described it in 1878. In Muellerian mimicry, several unrelated but protected animal species come to resemble one another. Thus a number of different kinds of stinging wasps have black-and-yellow striped abdomens, but they may not all be descended from a common ancestor that had similar coloration. In general, yellow-and-black and bright red tend to be common color patterns that presumably warn predators relying on vision that animals with such coloration are to be avoided. It is more difficult to prove that a resemblance between two protected animals actually repre-

Models

Monarch

A

Mimics

Pipevine swallowtail

B

Viceroy

Red-spotted purple

*Figure 20-13*

**Mimicry.**
**A** The viceroy butterfly, *Limenitis archippus,* is a North American mimic of the poisonous monarch. Although the viceroy is not related to the monarch, it looks a lot like it, and so predators that have learned how distasteful monarchs are avoid viceroys, too.
**B** The red-spotted purple, *Limenitis arthemis astyanax,* is another member of the same genus as the viceroy, and thus much more closely related to it than is the monarch. However, it does not look at all like the viceroy. Instead, the red-spotted purple is a Batesian mimic of another poisonous butterfly, the pipevine swallowtail, *Battus philenor.* The very different appearances of these two closely related species illustrate vividly the way in which selection can drastically change the appearances of mimics.

***Figure 20-14***

**Batesian mimics.** The familiar yellow-and-black stripes of wasps such as *Vespula arenaria (top)* form the basis for large Batesian and Muellerian mimicry complexes. Here we see Batesian mimics representing three separate orders of insects, all rarer than wasps and with patterns of behavior similar to those of the dangerous wasps they resemble. None of these mimics sting, yet all are conspicuous members of the communities where they occur, flying about actively and remaining in full view at all times. They would be easy prey if they were not protected by their resemblance to wasps.

sents Muellerian mimicry than it is to demonstrate Batesian mimicry experimentally.

In both Batesian and Muellerian mimicry, mimic and model must not only look alike but also act alike if predators are to be deceived. For example, the members of several families of beetles that resemble wasps (Figure 20-14) behave surprisingly like the wasps they mimic, flying often and actively from place to place. Mimics must also spend most of their time in the same habitats as do their models. If they did not, predators would discover that all of those conspicuous animals in one area are not only easily seen but also quite tasty! If the animals that resemble one another are all poisonous, or dangerous, they still gain an advantage by resembling one another, thus achieving collective protection.

*Muellerian mimicry is a phenomenon in which two or more unrelated but protected species resemble one another, thus achieving a kind of group defense.*

## SYMBIOSIS

Symbiotic relationships are those in which two kinds of organisms live together. All symbiotic relationships provide the potential for coevolution between the organisms involved in them, and in many instances the results of this coevolution are fascinating. The major kinds of symbiotic relationships include (1) **commensalism,** in which one species benefits while the other neither benefits nor is harmed; (2) **mutualism,** in which both participating species benefit; and (3) **parasitism,** in which one species benefits but the other is harmed. Parasitism, as mentioned in Chapter 19, can also be viewed as a form of predation.

Examples of symbiosis include lichens, which are associations of certain fungi and green algae or cyanobacteria (discussed in more detail in Chapter 27); mycorrhizae, associations of fungi and the roots of plants; and the associations of bacteria and root nodules that occur in legumes and certain other plants, which enable the host plants to fix atmospheric nitrogen and the bacteria to obtain carbohydrates. A coral reef is a highly complex symbiotic system, involving not only the coral animals but also coralline algae and other autotrophic organisms that are intermingled with the coral animals and contribute greatly to the net productivity of the reef. More broadly, a coral reef is an entire biological community, one in which symbiotic relationships are especially prominent.

*Symbiotic relationships are those in which two or more kinds of organisms live together in often elaborate, more or less permanent relationships.*

## Figure 20-16

**Symbiosis.** Fishes, such as this individual of *Amphiprion perideraion* in Guam, often form symbiotic associations with sea anemones, gaining protection by remaining among their tentacles and gleaning scraps from their food.

### Commensalism

In nature, the individuals of one species often grow attached to those of another. For example, birds nest in trees, or epiphytes grow on the branches of other plants. In general, the host plant is unharmed, while the organism that grows or nests on it benefits. Similarly, various marine animals, such as barnacles, grow on other, often actively moving sea animals and thus are carried passively from place to place (Figure 20-15). These "passengers" presumably gain more protection from predation than they would if they were fixed in one place, and they also reach new sources of food. The increased water circulation that such animals receive as their host moves around may be of great importance, particularly if the passengers are filter feed-

## Figure 20-15

**Some animals live in unexpected places.** These barnacles live on the back of a gray whale. The barnacles have chosen their home wisely, for the whale carries them from place to place so that they have continuous access to fresh sources of the small, free-floating organisms on which they feed.

ers. The gametes of the passenger are also more widely dispersed than would be the case otherwise.

The best-known examples of commensalism—those which virtually define the concept—involve the relationships between certain small tropical fishes and sea anemones, marine aninals that have stinging tentacles (see Chapter 28). These fishes have evolved the ability to live among the tentacles of the sea anemones, even though the tentacles would quickly paralyze other fishes that touched them (Figure 20-16). The anemone fishes feed on the detritus left from the meals of the host anemone, remaining uninjured under remarkable circumstances.

On land, an analogous relationship exists between certain birds and grazing animals. The birds spend most of their time clinging to the animals, picking off insects and other small bits of food, and carry out their entire life cycles in close association with the host animal.

In each of these instances, it is difficult to be certain whether the second partner receives a benefit or not, and there is no clear-cut boundary between commensalism and mutualism. For instance, it may be advantageous to the sea anemone to have particles of food removed from its tentacles; it may then be better able to catch other prey. The association of the grazing mammals and the birds, on the other hand, is quite clearly an example of mutualism. The mammal benefits by having parasites and other insects removed from its body, and the birds benefit by having access to a dependable source of food. The factors that are involved in a particular association can be learned only by patient observation and experimentation.

## Mutualism

Examples of mutualism are of fundamental importance in determining the structure of biological communities. In the tropics, leafcutter ants (Figure 20-17) are often so abundant that they can remove a quarter or more of the total leaf surface of the plants in a given area. They do not eat these leaves directly; rather, they take them to their underground nests, where they chew them up and inoculate them with the spores of particular fungi. These fungi are cultivated by the ants and brought from one specially prepared bed to another, where they grow and reproduce. In turn, the fungi constitute the primary food of the ants and their larvae. The relationship between the leafcutter ants and these fungi, fueled by material cut from the leaves of plants, is an excellent example of mutualism.

Another relationship of this kind involves ants and aphids. Aphids, also called greenflies, are small insects that suck fluids from the phloem of living plants with their piercing mouthparts. They extract a certain amount of the sucrose and other nutrients from this fluid, but much runs out, in an altered form, through their anus. Certain ants have taken advantage of this habit—in effect domesticating the aphids—by carrying the aphids to new plants and using the "honeydew" that they excrete as food (Figure 20-18).

A particularly striking example of mutualism involving ants concerns certain Latin American species of the plant genus *Acacia*. In these species, the leaf parts called stipules are modified as paired, hollow thorns; consequently, these particular species are called "bull's

**Figure 20-18**

**Cattle-raising, ant style.** These ants *(Crematogaster)* are tending willow aphids, obtaining the "honeydew" that the aphids excrete continuously, moving them from place to place, and protecting them from potential predators.

horn acacias." The thorns are inhabited by stinging ants of the genus *Pseudomyrmex*, which do not nest anywhere else.

At the tip of the leaflets of these acacias are unique, protein-rich bodies called Beltian bodies—after Thomas Belt, a nineteenth-century British naturalist who first wrote about them based on his experiences in Nicaragua. Beltian bodies do not occur in species of *Acacia* that are not inhabited by ants, and their role is clear: they serve as a primary food for the ants. In addition, the plants secrete nectar from glands near the bases of their leaves. The ants consume this nectar also, and feed it and the Beltian bodies to their larvae as well (Figure 20-19).

Apparently this association is beneficial to the ants, and one can readily see why they inhabit acacias of this group. The ants and their larvae are protected within the swollen thorns, and the trees provide a ready source of a balanced diet, including the sugar-rich nectar and the protein-rich Beltian bodies. What, if anything, do the ants do for the plants? This had been the question that had fascinated observers for nearly a century, until it was answered by Daniel Janzen, then a graduate student at the University of California, Berkeley, in a beautifully conceived and executed series of field experiments.

Whenever any herbivore lands on the branches or leaves of an acacia that is inhabited by such ants, the ants immediately attack and devour it. Thus the ants protect the acacias from being eaten, and the herbivore also provides additional food for the ants, which continually patrol the branches. Related species of acacias that do not have the special features of the bull's horn acacias and are not protected by ants often have bitter-tasting substances in their leaves that the bull's horn

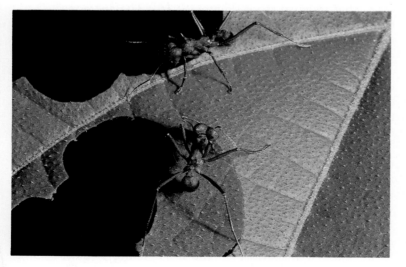

**Figure 20-17**

**Leafcutter ants demolishing a leaf.** Leafcutter ants are enormously abundant in the tropics and often tear up a high proportion of the available leaves at a particular spot to build their "fungus gardens."

# Dynamics of Ecosystems

This bald eagle swooping down on its prey is a carnivore, an organism that feeds on other animals. Carnivores, along with herbivores, detritovores, and decomposers, are all different levels of consumers that constitute the food chains in ecosystems.

# DYNAMICS OF ECOSYSTEMS

## Overview

The earth provides living organisms with much more than a place to stand or swim. Many chemicals cycle between our bodies and the physical environment around us. We live in a delicate balance with our physical environment, one easily disturbed by human activities. The collection of organisms that live in a place, and all the physical aspects of the environment that affect how they live, operate as a fundamental biological unit, the ecosystem. Much of the effort of ecologists focuses on improving our understanding of how ecosystems function.

## For Review    *Here are some important terms and concepts that you will encounter in this chapter. If you are not familiar with them, you should review them before proceeding.*

**Nitrogen** (Chapter 3)

**Metabolic energy** (Chapter 7)

**Respiration** (Chapter 8)

Ecology is concerned with the most complex level of biological integration. It attempts to tell us why particular kinds of organisms can be found living in one place and not in another. In previous chapters we have dealt with the physical and biological variables that govern the distribution of organisms and the factors that maintain the numbers of particular kinds of organisms at certain levels. In this chapter we focus on principles that govern the functioning of assemblages of organisms.

In ecological terms, populations of different organisms that live together in a particular place are called **communities.** A community, together with the nonliving factors with which it interacts, is called an **ecosystem.** An ecosystem regulates the flow of energy and the cycling of the essential elements on which the lives of its constituent plants, animals, and other organisms depend. Communities and ecosystems exist both on land and in the sea (Figure 21-1).

*A community is the interacting set of different kinds of organisms that occur together at a particular place. An ecosystem is that set of organisms, together with the nonliving factors with which it interacts.*

*Figure 21-1*

**A desert at night.** Unlike the moon, all of the earth's surface, even its deserts, is teeming with life, although it may not always appear so. The same ecological principles apply to the organization of all of the earth's communities, both on land and in the sea, although the details differ greatly.

# ECOSYSTEMS

Ecosystems are the most complex level of biological organization. They are systems in which there is a regulated transfer of energy and an orderly, controlled cycling of nutrients. The individual organisms and populations of organisms in an ecosystem act as part of an integrated whole, adjust over time to their role in the ecosystem, and relate to one another in complex ways that we only partly understand. Despite their differences, all ecosystems are governed by the same principles and restricted by the same limitations. The earth is a closed system with respect to chemicals, but an open one in terms of energy. That is, no new chemicals are being added from outside, but energy is constantly being added from the sun. Ecosystems function to regulate the capture and expenditure of that energy and the cycling of those chemicals. As you will see in this chapter, all organisms, including human beings, depend on the activities of a few other organisms to recycle the basic components of life: plants, algae, some bacteria in the case of carbon, and other bacteria in the case of nitrogen.

## THE CYCLING OF NUTRIENTS IN ECOSYSTEMS

All of the substances that occur in organisms, including water, carbon, nitrogen, and oxygen, as well as a number of others that are ultimately derived from the weathering of rocks, cycle through ecosystems. These cycles are geological ones that involve the biologically controlled cycling of chemicals and therefore are called **biogeochemical cycles.** Among the elements that are cycled are some that come from rocks, including phosphorus, potassium, sulfur, magnesium, calcium, sodium, iron, cobalt, and all the other elements that are essential for plant growth (see Chapter 31). All organisms require carbon, hydrogen, oxygen, nitrogen, phosphorus, and sulfur in relatively large quantities; the other elements are required in smaller amounts.

We speak of the **cycling** of materials in ecosystems because these materials are incorporated from the atmosphere or from weathered rock first into the bodies of organisms; they then sometimes pass from these organisms into the bodies of other organisms that feed on these primary ones, and ultimately are returned to the nonliving world. When this occurs, the nutrients may possibly be incorporated again into the bodies of other organisms. Some examples will help to clarify the ways in which different cycles function.

### The Water Cycle

All life directly depends on the presence of water. In fact, the bodies of most organisms consist mainly of this substance. Water is the source of the hydrogen ions whose movements generate ATP in organisms, and for that reason alone, it is indispensable to their functioning. Thus the **water cycle** (Figure 21-2) is the most familiar of all biogeochemical cycles. The oceans cover nearly three-fourths of the earth's surface. From their surface, water evaporates into the atmosphere, a process that is powered by energy from the sun. This water eventually precipitates back to earth and passes into surface and subsurface bodies of fresh water. Most

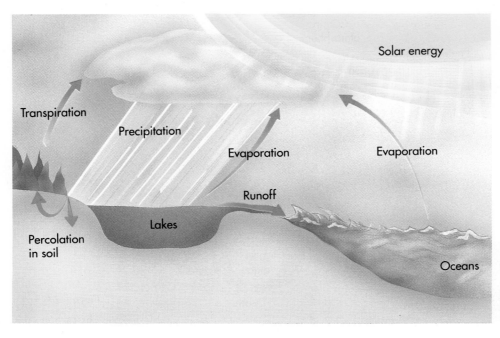

*Figure 21-2*

The water cycle.

of it falls directly into the oceans, but some falls onto the earth. Only about 2% of all the water on earth is fixed in any form—frozen, held in the soil, or incorporated into the bodies of organisms. The rest is free water, circulating between the atmosphere and earth. But regardless of where this water is held temporarily, it eventually returns to the atmosphere and the oceans.

Much less obvious than the surface waters, which we see in streams, lakes, and ponds, is the groundwater, which occurs in permeable, saturated, underground layers of rock, sand, and gravel called **aquifers.** In many areas, groundwater is the most important reservoir of water; it amounts to more than 96% of all fresh water in the United States. It flows much more slowly than does surface water, anywhere from a few millimeters to as much as a meter or so per day. In the United States, groundwater provides about a fourth of the water used for all purposes and provides about half of the population with drinking water. Three-fourths of the country's cities and most of its rural areas rely, at least in part, on groundwater reserves. The use of groundwater throughout the world is growing much more rapidly than is the use of surface water.

*Some 96% of the fresh water in the United States consists of groundwater. This groundwater, which already provides a fourth of all the water used in this country, will become even more important in the future.*

## The Carbon Cycle

The **carbon cycle** is based on carbon dioxide, which makes up only about 0.03% of the atmosphere. The worldwide synthesis of organic compounds from carbon dioxide and water results in the fixation of about a tenth of the roughly 700 billion metric tonnes of carbon dioxide in the atmosphere each year (Figure 21-3). This enormous amount of biological activity takes place as a result of the combined activities of photosynthetic bacteria, algae, and plants. All heterotrophic organisms—including the nonphotosynthetic bacteria and protists, the animals, and the relatively few plants that have lost the ability to photosynthesize—obtain their carbon indirectly from the organisms that fix it. As a result of oxidative respiration and the ultimate decomposition of their bodies, organisms release carbon dioxide to the atmosphere again. Once there, it can be reincorporated into the bodies of other organisms.

*About a tenth of the estimated 700 billion metric tonnes of carbon dioxide in the atmosphere is fixed annually by the process of photosynthesis.*

In addition to the carbon dioxide in the atmosphere, approximately 1000 billion metric tonnes are dissolved in the ocean; more than half of this quantity is in the upper layers, where photosynthesis takes

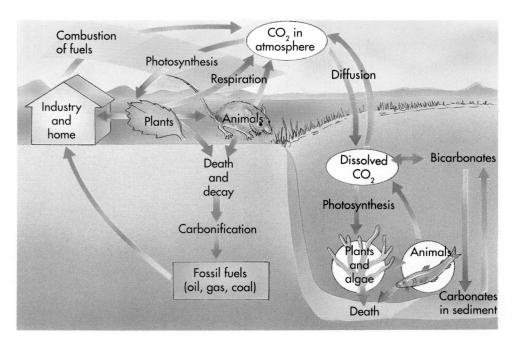

**Figure 21-3**

The carbon cycle.

place. The fossil fuels, primarily oil and coal, contain more than 5 trillion additional metric tonnes of carbon, and between 600 billion and 1 trillion metric tonnes are locked up in living organisms at any one time.

Carbon dioxide and certain other gasses, such as methane, that occur in the earth's atmosphere transmit radiant energy from the sun but trap the longer wavelengths of infrared light, or heat, and prevent them from radiating into space. This creates what is known as the **greenhouse effect.** Planets which lack an atmosphere are much colder than those which possess one; it is estimated that the average temperature of the earth would be about $-20°$ C, instead of the actual $+15°$ C, if the earth did not have an atmosphere.

The earth's greenhouse effect is being made more intense now as a result of human activities, which are increasing the amount of carbon dioxide, chlorofluorocarbons (CFCs), nitrogen oxides, and methane—all "greenhouse gasses"—in the atmosphere. Before widespread industrialization, the concentration of carbon dioxide in the atmosphere was about 260 to 280 parts per million (ppm). During the 25-year period starting in 1958, this concentration increased from 315 ppm to more than 340 ppm and is continuing to rise rapidly. The rise in average global temperatures by about 0.5° C during the past century is consistent with these increased concentrations of carbon dioxide in the atmosphere, but it cannot be demonstrated with certainty that the increase in temperature was caused by the atmospheric changes. But projected increases in the greenhouse gasses over the next 50 years are estimated to cause increases in average global temperatures from 1° to 4° C, a matter of serious concern because of such associated effects as shifts in prime agricultural lands, changes in sea levels, and alterations in rainfall patterns. Remember that the projected changes concern *average* temperatures, and as such cannot be connected with particular hot or cold days or seasons. The earth is warming now more rapidly than at any period in the past, including the times when the glaciers were melting during the ice ages.

## The Nitrogen Cycle

Nitrogen gas constitutes nearly 80% of the earth's atmosphere by volume, but the total amount of fixed nitrogen in the soil, oceans, and the bodies of organisms is only about 0.03% of that figure. The nitrogen in the atmosphere cannot be used by most organisms because few of them possess the special enzyme system necessary to break the very strong triple bond of nitrogen gas ($N_2$). Those which do have the enzymes, including several genera of bacteria, carry out a process called **nitrogen fixation.** Once nitrogen has been fixed, it cycles within biological systems; all living organisms depend on nitrogen fixation. Without it, they would be unable to synthesize proteins, nucleic acids, and other necessary nitrogen-containing compounds. Nitrogen fixation is the means by which a very small fraction of the enormous reservoir of nitrogen that exists in the earth's atmosphere is made available for biological processes (Figure 21-4).

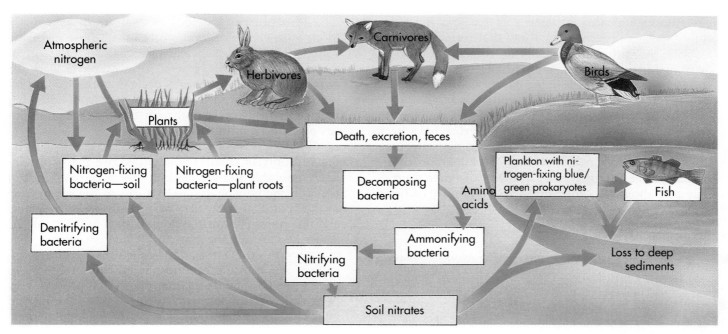

*Figure 21-4*

**The nitrogen cycle.**

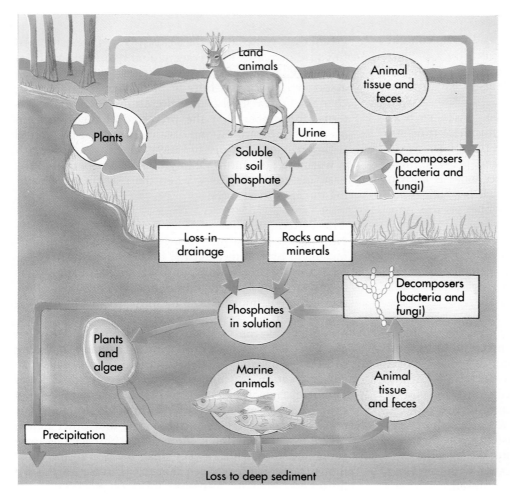

**Figure 21-5**

The phosphorus cycle.

Although dozens of genera of bacteria have the ability to fix nitrogen, only a few that form symbiotic associations with plants usually fix enough nitrogen to be of major significance. Plants that form symbiotic associations with nitrogen-fixing bacteria, such as the legumes (members of the plant family Fabaceae, which includes peas, beans, alfalfa, mesquite, and many other well-known plants), can grow in soils that have very low amounts of nitrogen. It is estimated that a legume crop may add as much as 300 to 350 kilograms of nitrogen per hectare per year, whereas all other sources may add about 15 kilograms. The importance of these symbiotic associations is obvious.

*Although nitrogen gas constitutes nearly 80% of the earth's atmosphere, it becomes available to organisms only through the activities of a few genera of bacteria.*

## The Phosphorus Cycle

In all biogeochemical cycles other than those involving water, carbon, oxygen, and nitrogen, the reservoir of the nutrient exists in mineral form rather than in the atmosphere. The **phosphorus cycle** (Figure 21-5) is presented here as a representative example of all other mineral cycles because of the critical role that phosphorus plays in plant nutrition worldwide.

Phosphates—charged phosphorus ions—exist in the soil only in small amounts because they are relatively insoluble and are present only in certain kinds of rocks. If the phosphates are not lost to the sea by way of rivers and streams, they may be absorbed by plants and incorporated into compounds such as ATP, nucleic acids, and membrane proteins. Fungi associated with plant roots—the associations called **mycorrhizae**—facilitate the transfer of phosphates from the soil to plants (see Chapter 26). Other plants that grow in very poor soils do not form mycorrhizae, but may instead have clusters of fine roots that perform a similar function.

When animals or plants die, shed parts, or, in the case of animals, excrete waste products, the phospho-

**Figure 21-6**

**Living the phosphorus cycle.** Pelicans roosting on a small island in the Gulf of California, Mexico. Sea birds bring up phosphorus from the deepest layers of the sea by eating fish and other marine animals and depositing their remains as guano on the rookeries.

rus may be returned to the soil or water and then cycle again through other organisms. When phosphates are lost to the deep sea, they may be recycled through the activities of sea birds that eat fishes and other animals that feed in deep waters (Figure 21-6).

## THE FLOW OF ENERGY IN ECOSYSTEMS

An ecosystem includes two different kinds of living components, autotrophic and heterotrophic ones. The autotrophic components, consisting of plants, algae, and some bacteria, are able to capture light energy and manufacture their own food. To support themselves, the heterotrophic ones, including animals, fungi, most protists and bacteria, and nongreen plants, must obtain organic molecules that have been synthesized by autotrophs.

Once energy enters an ecosystem, mainly after it is captured as a result of photosynthesis, it is slowly released as metabolic processes proceed. The autotrophs that first acquire this energy provide all of the energy that heterotrophs use. Ecosystems, as well as the organisms that make them up, can in one sense be viewed as systems that have become adapted through time to delay the release of the energy obtained from the sun back into space. Energy flows one way through an ecosystem, whereas the nutrients are continually recycled.

Looking at an ecosystem as a whole, we can speak of its **primary productivity,** which we define as the total amount of light energy that is converted to organic compounds in a given area per unit of time. The **net productivity** of the ecosystem is the total amount of energy fixed by photosynthesis per unit of time, minus that which is expended by the metabolic activities of the organisms in the community. The total weight of all of the organisms living in the ecosystem, called its

**biomass,** increases as a result of its net production. Some ecosystems, such as a cornfield or a cattail swamp, have a very high net primary productivity. Others, such as tropical rain forests (Figure 21-7), also have a relatively high net primary productivity, but a rain forest has a much larger biomass than a cornfield; consequently, the net primary productivity of a rain forest is much lower in relation to its total biomass. In communities such as sugar cane fields, coral reefs, and estuaries, the net primary productivity per square meter per year may range from roughly 3500 to 9000 grams. The productivity of marshlands and tropical forests is somewhat less, and that of deserts is about 200 grams.

### Trophic Levels

Green plants, the primary producers of an ecosystem, generally capture about 1% of the energy that falls on their leaves, converting it to food energy. In especially productive systems, this percentage may be a little higher. When these green plants are consumed by other organisms, usually only about 10% of the plant's accumulated energy is actually converted into the bodies of the organisms that consume them.

Among these consumers, several levels may be recognized. The **primary consumers,** or herbivores, feed directly on the green plants. **Secondary consumers,** carnivores and the parasites of animals, feed directly or indirectly on the herbivores. **Decomposers** break down the organic matter accumulated in the bodies of other

**Figure 21-7**

**The highly specialized plants of the tropics are able to make poor soils very productive.** The tropical rain forest has a large biomass and a high net productivity, even though it grows on soils that are often low in some of the nutrients that are essential for plant growth.

**A Herbivore**

**B Detritivore**

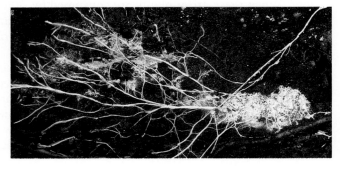

**C Decomposer**

***Figure 21-8***

**A** The grass in the East African savannas is a primary producer, capturing energy from the sun. Grazing mammals obtain their food from the grass, and may in turn be consumed by predators.
**B** This crab, *Gecarcinus quadratus,* photographed on the beach at Mazatln, Mexico, is a detritivore, playing the same role that vultures and similar animals do in other ecosystems.
**C** Fungi, such as the one shown here growing through the soil in Costa Rica, are, along with the bacteria, the primary decomposers of terrestrial ecosystems.

organisms. Another, more general, term that includes decomposers is **detritivores.** Detritivores are organisms that live on the refuse of an ecosystem—not only on dead organisms but on the cast-off parts of organisms. They include large scavengers, such as crabs, vultures, and jackals, as well as decomposers such as bacteria and fungi (Figure 21-8).

All of these levels, and usually additional ones, are represented in any fairly complicated ecosystem. They are called **trophic levels,** from the Greek word *trophos,* which means "feeder." Organisms from each of these levels, feeding on one another, make up a series that is called a **food chain.** The length and complexity of food chains vary greatly. In real life, it is rather rare for a given kind of organism to feed on only one other kind of organism. Usually, each will feed on two or more other kinds and in turn will be fed on by several other kinds of organisms. When diagrammed, the relationship appears as a series of branching lines rather than as one straight line; it is called a **food web** (Figure 21-9).

A certain amount of the energy that is ingested and retained by the organisms at a given trophic level goes toward heat production. A great deal of the energy is used for digestion and work, and usually 40% or less goes toward growth and reproduction. An invertebrate such as a worm or insect typically uses about a quarter of this 40% for growth; in other words, about 10% of the food that an invertebrate eats is turned into its own body and thus into potential food for its predators. Although the comparable figure varies from approximately 5% in carnivores to nearly 20% for herbivores generally, 10% is a good average value for the amount of organic matter that is present at each step in a food chain, or each successive trophic level, and that reaches the next level.

> *A plant fixes about 1% of the sun's energy that falls on its green parts. The successive members of a food chain, in turn, process about 10% of the available energy in the organisms on which they feed into their own bodies.*

Lamont Cole of Cornell University studied the flow of energy in a freshwater ecosystem in Lake Cayuga in upstate New York. He calculated that about 150 calories of each 1000 calories of energy that were fixed by algae and cyanobacteria were transferred into the bodies of small heterotrophs (Figure 21-10). Of these, about 30 calories were incorporated into the bodies of smelt, the principal secondary consumers of the system. If we eat the smelt, we gain about 6 calories from each 1000 calories that originally entered the system. If, on the other hand, trout eat the smelt and we eat the trout, we gain only about 1.2 calories from each original 1000.

Relationships of this kind make it clear that organisms, including people, that subsist on an all-plant diet obviously have more food and energy available to them than do carnivores. Such considerations will become increasingly important in the future, not only for the efficient management of fisheries, but also in an ef-

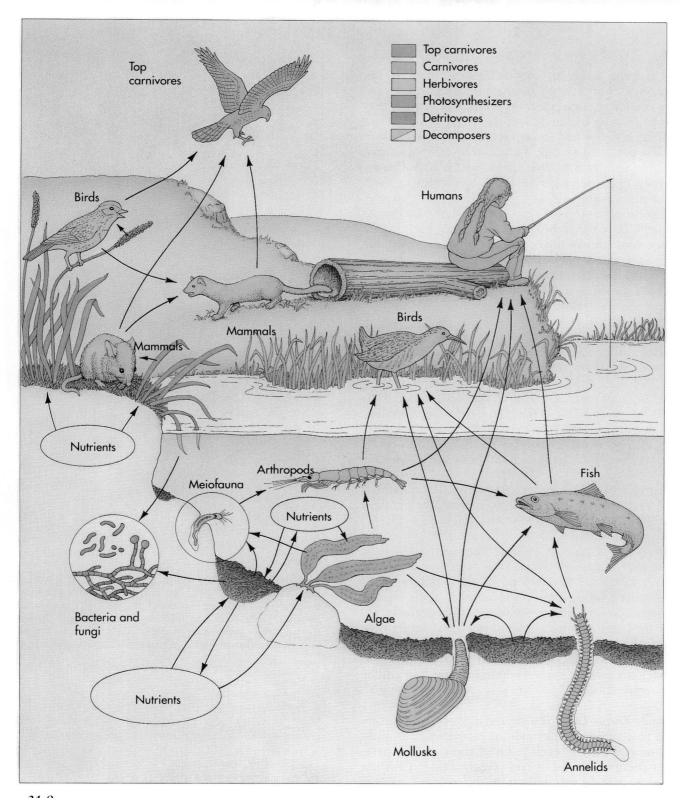

**Figure 21-9**

**The food web in a salt marsh.** Each level of rectangles represents a trophic level feeding on, or gaining energy from, the layer below.

fort to maximize the yield of food for a hungry and increasingly overcrowded world.

Food chains generally consist of only three or four steps. The loss of energy at each step is so great that very little of the original energy remains in the system as usable energy after it has been incorporated successively into the bodies of organisms at four trophic levels. There are generally far more individuals at the lower trophic levels of any ecosystem than at the higher ones. Similarly, the biomass of the primary producers present in a given ecosystem is greater than that of the primary consumers, with successive trophic levels having a lower and lower biomass and correspondingly less potential energy. Larger animals characteristically are members of the higher levels; to some extent, they *must* be larger in order to capture enough prey to support themselves.

The trophic structure of an ecosystem determines the energy flow through it. It is influenced both by the numbers and sizes of the organisms (that is, how much energy they can supply and how much they consume) and by the amount of energy lost at each transfer. These relationships, if shown diagrammatically, appear as **ecological pyramids.**

*Counting Individuals.* If you were to count all the individuals in an ecosystem and assign each to a trophic level, you would obtain a **pyramid of numbers** (Figure 21-11, *A*). Such a pyramid is not a good representation of the flow of energy through the ecosystem, as it doesn't take into account that some organisms are bigger than others—oak trees contribute more than grass plants, and bears more than fleas. Such pyramids rarely count the decomposers, who number in the billions.

*Weighing Biomass.* If you were to weigh all the individuals at each trophic level of the ecosystem, the pyramid you obtain, a **pyramid of biomass** (Figure 21-11, *B*), would not be influenced by differences in size. Such pyramids give a good if indirect representation of the flow of energy through the ecosystem. For most ecosystems on land, the pyramid of biomass points "up," with smaller trophic levels on top. But in some aquatic ecosystems, an inverted pyramid is obtained, the biomass of the heterotrophic zooplankton being larger than that of the autotrophic phytoplankton. This is only possible because the phytoplankton are reproducing at a prodigious rate; it is a very unusual situation among ecosystems generally.

*Measuring Energy Loss.* To get a really accurate picture of the flow of energy through an ecosystem, you must measure it directly at each point of transfer. The **pyramid of energy** that you obtain (Figure 21-11, *C*) is not skewed by size or metabolic rate differences, and it is always "right side up"—pyramids of energy cannot be inverted because of the necessary loss of energy at each step. It is not easy to construct such a pyramid, as you must measure the actual amounts of energy individuals take in, how much they burn up during metabolism, how much they invest in growth, and how much they lose as waste. In the few cases where this has been done carefully, an important point has emerged: *less*

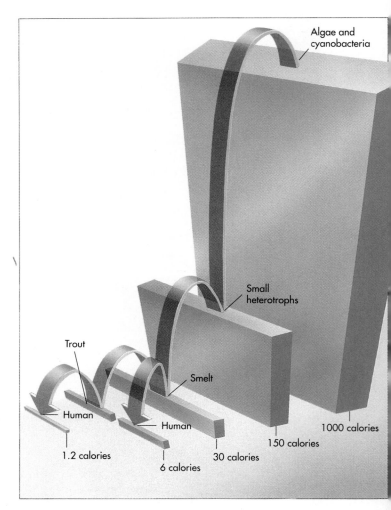

**Figure 21-10**

**The food web in Lake Cayuga.** In Lake Cayuga, autrotrophic plankton fix the energy of the sun, heterotrophic plankton feed on them, and both are consumed by smelt. The smelt are eaten by trout, with about a tenfold loss in fixed energy; for humans, the amount of biomass available in smelt is at least ten times greater than that available in trout, which they prefer to eat.

*than ⅙ of the energy entering one trophic level becomes available to organisms at the next level.*

## ECOLOGICAL SUCCESSION

Even when the climate of a given area remains stable year after year, ecosystems have a tendency to change from simple to complex in a process that is known as **succession.** This process is familiar to anyone who has seen a vacant lot or cleared woods slowly but surely become occupied with larger and larger plants and more and more different kinds of them, or seen a pond

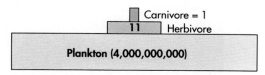

Pyramid of numbers

Carnivore = 1
**11** Herbivore
**Plankton (4,000,000,000)**

Pyramids of biomass

Decomposer 5 grams/
square meter

Second-level carnivore
(1.5 grams/square meter)
First-level carnivore
(11 grams/square meter)
Herbivore
(37 grams/square meter)
**Plankton
(807 grams/square meter)**

Zooplankton
and bottom fauna **21**
Phytoplankton **4**

Pyramid of energy

Decomposer
3890 kilocalories

First-level carnivore (48 kilocalories)
Herbivore (596 kilocalories)
**Plankton
(36,380 kilocalories/square meter/year)**

*Figure 21-11*

Pyramids of numbers, biomass, and energy.

become filled with vegetation that encroaches from the sides and gradually turns it into dry land (Figure 21-12).

Succession is continuous and worldwide in scope. If a wooded area is cleared, and the clearing is left alone, plants will slowly reclaim the area. Eventually, the traces of the clearing will disappear, and the whole area will again be woods. This kind of succession, which occurs in areas that have been disturbed and that were originally occupied by living organisms, is called **secondary succession.** Human beings are often responsible for initiating secondary succession throughout the regions of the world that they inhabit. Secondary succession may also take place after fire has burned off an area, for example, or after the eruption of a volcano (Figure 21-13).

**Primary succession,** in contrast to secondary succession, occurs on some bare, lifeless substrate, such as

A

B

C

*Figure 21-12*

**Succession in a pond.** This pond near Tallahassee, Florida, rapidly filling with aquatic vegetation (**A**). As the process of succession proceeds, the pond is slowly filled (**B**). Gradually the area where it once existed will become indistinguishable from the surrounding vegetation (**C**).

**Figure 21-13**

**Primary succession.** Mount St. Helens in the state of Washington erupted violently on May 18, 1980; the lateral blast devastated more than 600 square kilometers of forest and recreation lands within 15 minutes. **A** shows an area near Clearwater Creek 4 months after the eruption. Five years later, **B**, succession was underway at the same spot, with shrubs, blueberries, and dogwoods following the first plants that became established immediately after the blast.

rocks or open water, where organisms gradually occupy the area and change its nature. On bare rocks, cyanobacteria, algae, or lichens (associations of fungi and algae discussed in Chapter 26) may grow first. Acidic secretions from the lichens help to break down the stone and form small pockets of soil. Mosses may then colonize these pockets of soil, eventually followed by ferns and the seedlings of other plants. Over many thousands of years, or even longer, the rocks may be completely broken down, and the vegetation over an

area where there was once a rock outcrop may be just like that of the surrounding grassland or forest.

Similarly, a lake may gradually accumulate organic matter and fill in (see Figure 22-12). Plants standing along the edges of the lake, such as cattails and rushes, and those growing submerged, such as pondweeds, together with other organisms may contribute to the formation of a rich organic soil. As this process continues, the lake may increasingly be filled in with terrestrial vegetation. Eventually, the area where the lake once stood, like the rock outcrop we just described, may become an indistinguishable part of the surrounding vegetation.

**Figure 21-14**

**Agriculture stops succession.** Intensively cultivated areas are kept permanently at an early successional stage, as shown here in this newly cut hay field. Agriculture expends energy to arrest succession at an early stage and thus to enhance productivity.

*Primary succession takes place in areas that are originally open, such as dry rock faces or new ponds. Secondary succession, in contrast, takes place in areas that have been disturbed after having been occupied by living things earlier.*

Ponds and bare rocks in the same region often come to feature the same kind of vegetation as one another—the vegetation that is characteristic of the region as a whole. This relationship led the American ecologist F.E. Clements, at about the turn of the century, to propose the concept of **climax vegetation**. The term refers to the fact that this characteristic vegetation type is thought to be controlled by the *climate* of the region. However, with an increasing realization that the climate keeps changing, the process of succession is so slow, and the nature of a region's vegetation is being determined to a greater and greater extent by human activities, ecologists no longer consider the term "climax vegetation" to be as useful as they once did.

One of the most interesting features of succession is that the organisms involved in its early stages often are participants in symbiotic systems. Lichens, so important in the early colonization of rocks, are symbiotic associations of cyanobacteria or algae and fungi. This association appears to enable the lichens to withstand harsh conditions of moisture and temperature. Legumes and other plants that harbor nitrogen-fixing bacteria in nodules on their roots are often among the first colonists of relatively infertile soils. Trees with mycorrhizae are apparently more resistant to conditions of moisture stress and perhaps other environmental extremes than are those trees which have no mycorrhizae. It is believed that the first land plants had mycorrhizae and that this association was highly beneficial to them in the sterile, open habitats that they would have encountered.

*Symbiotic associations such as lichens, legumes with their associated nitrogen-fixing bacteria, and plants with mycorrhizal fungi appear to play an unusually important role in early successional communities.*

As any ecosystem matures, there is an increase in total biomass but a decrease in net productivity. The earlier successional stages are more productive than are the later ones. Agricultural systems are examples of early successional stages in which the process is intentionally not allowed to go to completion and the net productivity is high (Figure 21-14). There are more kinds of species in mature ecosystems than in immature ones, and the number of heterotrophic species increases even more rapidly than does the number of autotrophic species. This progression is related to the decreasing net productivity of increasingly mature ecosystems and to the fact that mature ecosystems have a greater ability to regulate the cycling of nutrients.

*Communities at early successional stages have a lower total biomass, higher net productivity, fewer species, many fewer heterotrophic species, and less capacity to regulate the cycling of nutrients than do communities at later successional stages.*

It seems that the plants and animals which appear in the later stages of succession may be more specialized, in general, than those which exist in the earlier stages. The late-successional species fit together into more complex communities and have much narrower ecological requirements, or niches.

## ■ SUMMARY

1. Populations of different organisms that live together in a particular place are called communities. A community, together with the nonliving components of its environment, is called an ecosystem. Ecosystems regulate the flow of energy, ultimately derived from the sun and the cycling of nutrients.

2. Only 2% of the water on earth is fixed in any way; the rest is free. In the United States, 96% of the fresh water is groundwater.

3. About 10% of the roughly 700 billion metric tonnes of free carbon dioxide in the atmosphere is fixed each year through photosynthesis. An additional trillion metric tonnes of carbon dioxide is dissolved in the ocean, and five times as much carbon as is represented in the ocean is locked up as coal, oil, and gas. About as much carbon exists in living organisms at any one time as is in the atmosphere.

4. Plants convert about 1% of the energy that falls on their leaves to food energy. The herbivores that eat the plants and the other animals that eat the herbivores constitute a set of trophic levels. At each of these levels, only about 10% of the energy fixed in the food is fixed in the body of the animal that eats that food. For this reason, food chains are always relatively short.

5. Primary succession takes place in areas that are originally bare, such as rocks or open water. Secondary succession takes place in areas where the communities of organisms that existed initially have been disturbed.

6. Both sorts of succession lead ultimately to the formation of climax communities, whose nature is controlled primarily by the climate of the area concerned, although the human influence on many of these communities is increasing. Such communities have more total biomass, less net productivity, more species, many more heterotrophic species, and a higher capacity of regulating the cycling of nutrients within them than do the earlier successional stages.

## REVIEW

1. The gas _____ makes up nearly four-fifths of the earth's atmosphere.

2. The total weight of all the organisms living in an ecosystem is called the _____ of that ecosystem.

3. The total amount of energy that an area converts to organic compounds over a period of time is called its _____.

4. List four elements that are required for life that are **not** derived from the weathering of rocks.

## SELF-QUIZ

1. In which of the following cycles does the reservoir of the nutrient exist in a mineral form?
   (a) Carbon cycle          (c) Nitrogen cycle          (e) All of the above
   (b) Phosphorus cycle       (d) Two of the above

2. Herbivores are
   (a) primary consumers.     (c) decomposers.            (e) parasites.
   (b) secondary consumers.   (d) detritivores.

3. Communities at early successional stages have _____ relative to the more advanced communities of later stages.
   (a) higher biomass                    (d) less regulated nutrient cycling
   (b) lower productivity                 (e) none of the above
   (c) more species

4. Of the 700 billion metric tonnes of carbon dioxide in the atmosphere, how much of the carbon is fixed into organic compounds by photosynthesis each year?
   (a) 0.03%                 (c) Less than 0.1%          (e) All of it
   (b) 10%                    (d) 5%

5. When a green plant is consumed by another organism, how much of the energy of the plant is converted into the body of the animal that consumes it?
   (a) None of it            (c) 1% of it                (e) ⅓ of it
   (b) All of it             (d) 10% of it

6. The greenhouse effect
   (a) may have begun already, but experts disagree.
   (b) is responsible for the carbon dioxide in the atmosphere.
   (c) keeps the earth about 35° C warmer than it would be without an atmosphere.
   (d) is being decreased as a result of human activities.
   (e) has not been demonstrated.

7. Mycorrhizae are
   (a) fungi.
   (b) symbiotic associations between plants and fungi.
   (c) a form of lichen.
   (d) nitrogen-forming structures on the roots of legumes.
   (e) formed only during early succession.

8. Arrange the following communities in descending order of their net productivity.
   (a) Surface of the Greenland ice cap
   (b) Tropical rain forest
   (c) Temperate woodland
   (d) Desert scrub
   (e) Cornfield

# THOUGHT QUESTIONS

1. How could you increase the net primary productivity of a desert?

2. Why does the productivity of an ecosystem increase as it becomes more mature?

3. At what successional stage would you characterize a field of wheat? What does this imply as to its stability and productivity?

# FOR FURTHER READING

COLINVAUX, P.A.: "The Past and Future Amazon," *Scientific American*, May 1989, pages 102-108. A very interesting article about historical changes in the Amazon rain forest ecosystem, based on current research.

HOUGHTON, R.A., and G.M. WOODWELL: "Global Climatic Change," *Scientific American*, April 1989, pages 36-44. A good summary of recent research on global warming.

SISSON, R.: "Tide Pools: Windows Between Land and Sea," *National Geographic*, vol. 169(2), 1986, pages 252-259. A beautifully illustrated tour of a California tide pool, alive with organisms.

TURNER, M.H.: "Building an Ecosystem from Scratch," *BioScience*, vol. 39, 1989, pages 147-150. In the Arizona desert, scientists and engineers are constructing a self-contained ecosystem as a model of the earth, and they investigate some aspects of what planetary travel will be like.

# Atmosphere, Oceans, and Biomes

These cactus stand lonely in Organ Pipe National Monument, New Mexico. Although other plants grow in profusion in the desert after rain, rains are infrequent. Most plants flower quickly while there is water and then are gone from view. The cactus is able to persist because it is well adapted to a life of scarce water.

# ATMOSPHERE, OCEANS, AND BIOMES

## Overview

There are great differences in climate across the face of the globe that, over billions of years, have resulted in the evolution of diverse terrestrial biomes and comparable associations of organisms in the sea. Biomes are quite distinct, their differences caused ultimately by the major circulation patterns of the atmosphere and the oceans, driven by the unequal distribution of heat from the sun. The biomes and marine communities of the tropics, where at least two-thirds of the species of organisms occur, are by far the richest biologically.

*For Review*     *Here are some important terms and concepts that you will encounter in this chapter. If you are not familiar with them, you should review them before proceeding.*

**Major features of evolution** (Chapter 16)

**Communities** (Chapter 20)

**Symbiosis** (Chapter 20)

**Biogeochemical cycles** (Chapter 21)

---

The major biomes, which are terrestrial communities that occur over wide areas, are easily recognized by their overall appearance and characteristic climates. Each biome is similar in its structure and appearance wherever it occurs on earth and differs significantly from other kinds of biomes. Biomes could be classified in a number of ways; for this book, we have selected some examples as a convenient means for discussing the properties of life on earth from an ecological perspective.

The distribution of biomes results from the interaction of the features of the earth itself, such as the different soil types or the occurrence of mountains and valleys, with two key physical factors: (1) the amounts of heat from the sun that reach different parts of the earth and the seasonal variation in that heat; and (2) global atmospheric circulation and the resulting patterns of atmospheric circulation. Together these factors determine the local climate and, particularly, the amounts and distribution of precipitation.

## THE GENERAL CIRCULATION OF THE ATMOSPHERE

The world contains a great diversity of ecosystems because its climate varies a great deal from place to place. On any given day, Miami, Florida, and Bangor, Maine, often have very different weather. There is no mystery about this. The tropics are warmer than the temperate regions because the sun's rays arrive almost perpendicular at regions near the equator, whereas their angle of incidence spreads them out over a much greater area near the poles, providing less energy per unit area. This simple fact, that some parts of the earth receive more energy from the sun than other parts because the earth is a sphere, is responsible for many of the major climatic differences that occur over the earth's surface and thus indirectly for much of the diversity of ecosystems (Figure 22-1).

The earth's annual orbit around the sun and its daily rotation on its own axis are both important in determining world climate (Figure 22-2). Because of the daily cycle, the climate at a given latitude is relatively constant, with a constant mixing of climates and temperatures at that latitude. Because of the annual cycle and the inclination of the earth's axis at approximately 23.5 degrees from its plane of revolution around the sun, there is a progression of seasons away from the equator in all parts of the earth. One of the poles is closer to the sun than the other at all times because the angle and direction of the earth's inclination are maintained as it rotates around the sun.

### Major Circulation Patterns

Near the equator, warm air rises and flows toward the poles (Figure 22-3, *A*). As it rises, this warm air loses most of its moisture, which is why it rains so much in

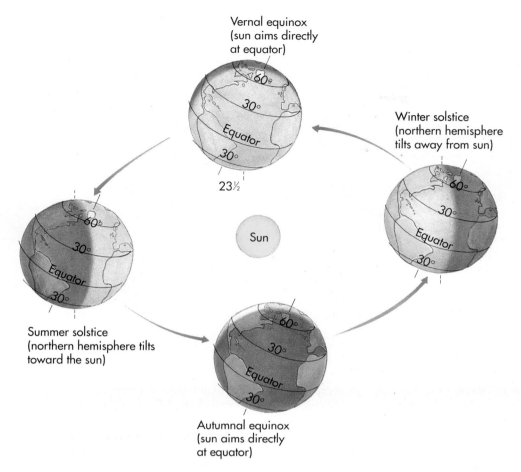

*Figure 22-1*

**The rotation of the earth around the sun has a profound effect on climate.** In the northern and southern hemispheres, temperatures change in an annual cycle because the earth is slightly tilted on its axis in relation to its pathway around the sun.

the tropics. This region of rising air is one of low pressure, the doldrums, which draws air from both north and south of the equator. When the air masses that have risen reach about 20 to 30 degrees north and south latitude, the air, now cooler, sinks and becomes reheated, producing a zone of decreased precipitation—warm air holds more moisture than does cooler air. Consequently, all of the great deserts of the world lie near 20 to 30 degrees north or 20 to 30 degrees south of the equator. Air at these latitudes is still warmer than it is in the polar regions, and it continues to flow toward the poles. It rises again at about 60 degrees north and south latitude and flows back toward the equator. Another air mass that rises here descends near the poles, producing a zone of very low temperature. These interactions, together with the rotation of the earth, result in the major patterns of atmospheric circulation (Figure 22-4).

## ATMOSPHERIC CIRCULATION, PRECIPITATION, AND CLIMATE

Because the moisture-holding capacity of air increases when it is warmed and decreases when it is cooled, precipitation is generally low near 30 degrees north and south latitude, where air is falling and being warmed, and relatively high near 60 degrees north and south latitude, where it is rising and being cooled. Partly as a result of these factors, all of the great deserts of the world lie near 30 degrees north or 30 degrees south latitude, and some of the great temperate forests are near 60 degrees north and south latitude. Other major deserts are formed in the interiors of the large continents; these areas have limited precipitation because of their distance from the sea and sometimes because mountain ranges intercept the moisture-laden winds from the sea. When the latter occurs, the air's mois-

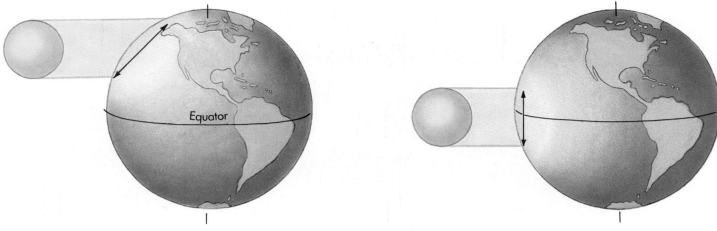

**Figure 22-2**

**Relationship between the earth and the sun are critical in determining the nature and distribution of life on earth.** A beam of solar energy striking the earth in middle latitudes is spread over a wider area of the earth's surface than a similar beam striking the earth near the equator.

ture-holding capacity decreases, resulting in increased precipitation on the windward side of the mountains—the side from which the wind is blowing. As the air descends the other side of the mountains, the lee side, it is warmed, and its moisture-holding capacity increases, tending to block precipitation. The eastern sides of the mountains are much drier than their western sides, and

the vegetation is often very different; this phenomenon is called the **rain shadow effect** (Figure 22-5).

> *The great deserts and associated arid areas of the world lie along the western sides of the continents at about 30 degrees north and south latitude. Other major deserts occur in the interiors of the large continents.*

### Patterns of Circulation in the Ocean

Patterns of circulation in the ocean are determined by the major patterns of atmospheric circulation just discussed, but they are modified by the location of the land masses around which and against which the ocean currents must flow. The circulation is dominated by huge surface **gyrals** (Figure 22-6), circular patterns that move around the subtropical oceans at about 30 degrees north latitude and 30 degrees south latitude. These gyrals move clockwise in the Northern Hemisphere and counterclockwise in the Southern Hemisphere. They profoundly affect life not only in the oceans but also on coastal lands by the ways in which they redistribute heat. For example, the Gulf Stream, in the North Atlantic, swings away from North America near Cape Hatteras, North Carolina, and reaches Europe near the southern British Isles. Because of the Gulf Stream, western Europe is much warmer and thus

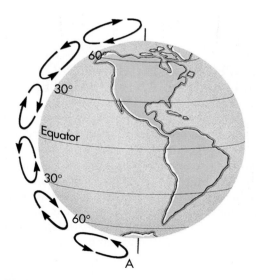

**Figure 22-3**

**The pattern of air movement out from and back to the earth's surface.**

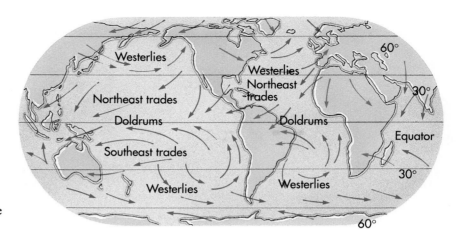

*Figure 22-4*

The major wind currents across the face of the earth.

more temperate than is eastern North America at the same latitudes.

In South America the Humboldt Current carries cold water northward up the west coast and helps to make possible an abundance of marine life that supports the fisheries of Peru and northern Chile. Marine birds, which feed on these organisms, are responsible for the commercially important guano deposits of these countries. These deposits, like those shown in Figure 21-6, are rich in phosphorus, which is brought up from the ocean depths by the upwelling of cold water that occurs from the generally mountainous slopes that border the Pacific. When the coastal waters are not as cold as usual, a devastating phenomenon called "El Nino" occurs, which affects the vitality of both marine and terrestrial populations of animals and plants worldwide.

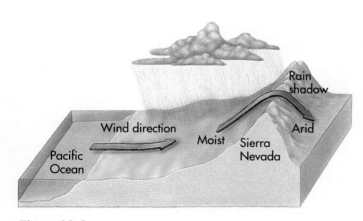

*Figure 22-5*

**The rain shadow effect.** Moisture-laden winds from the Pacific Ocean rise and are cooled when they encounter the Sierra Nevada Mountains. As they cool, their moisture-holding capacity decreases and precipitation occurs. As the air descends on the east side of the range, it warms, its moisture-holding capacity increases, and the air picks up moisture from its surroundings. As a result, desert conditions prevail on the east side of the mountains.

*In the ocean, huge surface gyrals move around the subtropical zones of high pressure between approximately 30 degrees north and south latitude, clockwise in the Northern Hemisphere and counterclockwise in the Southern Hemisphere.*

## THE OCEANS

Nearly three-fourths of the earth's surface is covered by ocean. The seas have an *average* depth of more than 3 kilometers, and they are, for the most part, cold and dark. Heterotrophic organisms are found even at the greatest ocean depths, which reach nearly 11 kilometers in the Marianas Trench of the western Pacific Ocean, but photosynthetic organisms are confined to the upper few hundred meters of water (Figure 22-7). Organisms that live below this level obtain almost all

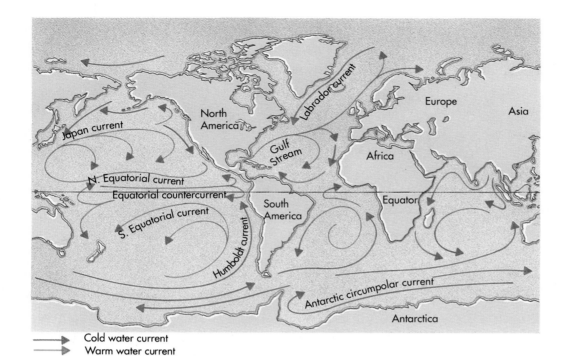

**Figure 22-6**

**Ocean circulation.** The circulation in the oceans moves in great surface spiral patterns called gyrals and profoundly affects the climate on adjacent lands.

of their food indirectly, as a result of photosynthetic activities that occur above; these activities result in organic detritus that drifts downward.

Many fewer species live in the sea than on land. Probably more than 90% of all species of organisms occur on land, including the great majority of members of a few large groups—especially insects, mites, fungi, and plants. Each of these groups has marine representatives, but these constitute only a very small fraction of the total number of species. On land the barriers between habitats are sharper, and variations in elevation, parent rock, degree of exposure, and other factors have all been crucial to the evolution of the millions of species of terrestrial organisms. On the other hand, most phyla originated in the sea, and almost all are represented there now, whereas only a few phyla occur on land.

*Although representatives of almost every phylum occur in the sea, only a relatively few phyla occur on land. However, the few phyla that are terrestrial have many species: an estimated 90% of living species of organisms are terrestrial. This is because the boundaries between different habitats are sharper on land than they are in the sea.*

**Figure 22-7**

**Ocean ecosystems are often diverse in coastal regions.** Fishes and many other kinds of animals find food and shelter among the kelp beds that occur in the coastal water of temperate regions.

The marine environment consists of three major kinds of habitats: (1) the **neritic zone,** the zone of shallow waters along the coasts of the continents; (2) the **surface layers** of the open sea; and (3) the **abyssal zone,** the deep-water areas of the oceans.

## The Neritic Zone

The neritic zone of shallow water is small in area, but it is inhabited by very large numbers of species compared with other parts of the ocean (Figure 22-8). The intense and sometimes violent interaction between sea and land in this zone gives a selective advantage to well-secured organisms that can withstand being washed away by the continual beating of the waves. Part of this zone, the **intertidal,** or **littoral,** region is exposed to the air whenever the tides recede.

Because of the way in which it gives access to the land, the intertidal zone must have been home for the ancestors of the first land organisms. The complex structures necessary to anchor them and protect them from drying out in this turbulent zone seem in some cases to have made possible their success on land. Some of the organisms that occur in intertidal areas, such as fiddler crabs, have internal biological clocks that allow them to attune their feeding activities to the daily ebb and flow of the tides. The world's great fisheries also occur on banks in the coastal zones (Figure 22-9), where nutrients, derived from the land, are often more abundant than in the open ocean.

Partly enclosed bodies of water, such as those which often form at river mouths and in coastal bays, where the salinity is intermediate between that of salt and fresh water, are called **estuaries.** Estuaries are among the most naturally fertile areas in the world, often with rich stands of submerged and emergent plants, algae, and microscopic organisms. They provide the breeding grounds for most coastal shellfish and fish that are harvested both in the estuaries and in open water.

## The Surface Zone

Drifting freely in the upper, better-illuminated waters of the ocean is a diverse biological community, primarily consisting of microscopic organisms called the **plankton.** Fishes and other larger organisms that swim in these same waters constitute the **nekton,** whose members feed mainly on plankton. Together, the organisms that make up the plankton and the nekton provide all of the food for those which live below. Some of the members of the plankton, including the algae and some bacteria, are photosynthetic. Collectively these organisms account for about 40% of all the photosynthesis that takes place on earth, and even more by

**Figure 22-8**

**Diverse communities occur in intertidal zones.** Many different habitats are created by the pounding of the waves and the periodic drying and flooding as the tides move in and out. Tide pools often form among the rocks when the tides recede.

some calculations. Most of the plankton occurs in the top 100 meters of the sea, the zone into which light from the surface penetrates freely. Perhaps half of the total photosynthesis in this zone is carried out by organisms less than 10 micrometers in diameter, including cyanobacteria and the smallest algae.

## The Abyssal Zone

In the deep waters of the sea, below the top 300 meters, occur some of the most bizarre organisms found on earth. Many of these animals have some form of bioluminescence (Figure 22-10), by means of which they communicate with one another or attract their prey. In the mud of the ocean floor, or along rifts from which warm water issues, live similar assemblages of peculiar creatures (Figure 22-11). Bacteria are apparently rather frequent in the deeper layers of the sea and as decomposers are as important in this zone as they are on land and in freshwater habitats.

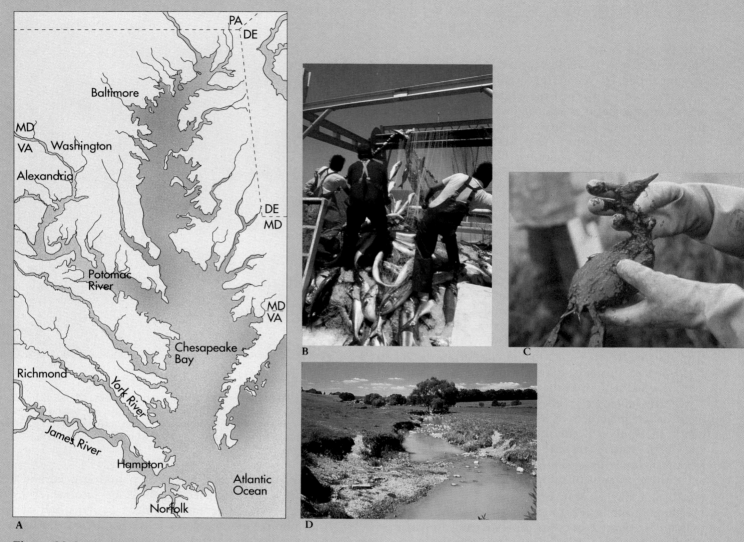

## Figure 22-9

**Chesapeake Bay, which has more than 11,300 kilometers of shoreline, drains more than 166,000 square kilometers in one of the most densely populated and heavily industrialized areas in North America.** The body of open water is about 320 kilometers long and, at some points, nearly 50 kilometers wide.

**A** Large metropolitan areas and shipping facilities make the Bay one of the busiest natural harbors anywhere.

**B** One of the most biologically productive bodies of water in the world, the Bay yielded an annual average of about 275,000 kilograms of fish in the 1960s but only a tenth as much in the 1980s. The human population of the area grew 50% during the same period.

**C** More than 290 oil spills were reported in the Bay in 1983, with oil transport and commercial shipping expected to double by the year 2020. This grebe is coated from an oil spill off the mouth of the Potomac River.

**D** Uncontrolled erosion from certain agricultural practices, pesticides, and increases in nutrients block the light needed for photosynthesis and upset the delicate ecological balance on which the productivity of the Bay depends. The states that border the Bay are cooperating, with the assistance of the Environmental Protection Agency, to try to bring back its former productivity.

*Figure 22-10*

**Symbiosis at sea.** The luminous spot below the eye of this deep-sea fish results from the presence of a symbiotic colony of luminous bacteria. Similar luminous signals are a common feature of deep-sea animals that move about.

*Figure 22-11*

**Deep-sea vents.** These giant beardworms are members of a small phylum of animals, the Pogonophora. They are living along warm-water vents in fissures along the Galapagos Trench in the Pacific Ocean; similar colonies occur at depths of up to 3000 meters. Water jets from these fissures at a scalding 350° C, but it soon cools to the 2° C temperature of the surrounding water. Hydrogen sulfide also emerges from these vents in abundance. Bacteria use the hydrogen sulfide as a source of energy; these bacteria in turn make possible the existence of a diverse community of animals—including these remarkable beardworms—that was discovered in 1977. These ecosystems have proved to be of extraordinary interest because they are among the most important on earth that do not depend in any way on photosynthesis or on energy from the sun.

# FRESH WATER

Freshwater habitats are distinct from both marine and terrestrial ones, but they are very limited in area. Inland lakes cover about 1.8% of the earth's surface, and running water covers about 0.3%. All freshwater habitats are strongly connected with terrestrial ones, with marshes and swamps constituting intermediate habitats. In addition, a large amount of organic and inorganic material continuously enters bodies of fresh water from communities growing on the land nearby (Figure 22-12). Many kinds of organisms are restricted to freshwater habitats (Figure 22-13); when they occur in rivers and streams, they must be able to attach themselves in such a way as to resist or avoid the effects of current, or risk being swept away. In bodies of standing water such matters are of much less importance.

Ponds and lakes, like the ocean, have three zones in which organisms occur: (1) a **littoral** zone; (2) a **limnetic** zone, inhabited by plankton and other organisms that live in open water; and (3) a **profundal** zone, below the limits of effective light penetration (Figure 22-14, *A*). **Thermal stratification** is characteristic of the larger lakes in temperate regions (Figure 22-14, *B*). Because water is densest at about 4° C, water at that temperature sinks beneath water that is either warmer or cooler. In winter, water at 4° C sinks beneath cooler water, and that cooler water at the surface freezes at 0°

*Figure 22-12*

**A nutrient-rich stream in the North Coast Ranges of California in summer.** As in all streams, much organic material falls or seeps into the water from the communities along the edges; this input is responsible for much of the biological productivity of the stream.

**Figure 22-13**

**Freshwater organisms.**
A Speckled darter *(Etheostoma stigmaeum)*.
B Green frog *(Rana clamitans)*.
C Freshwater snail.
D Giant waterbug with eggs on its back.
E Damselfly nymph.
F Bladderwort.

C. Below the ice, the water remains between 0° and 4° C, and plants and animals survive there. In spring, as the ice melts, the surface water is warmed to 4° C and sinks below the cooler water, bringing that water to the top with nutrients from the lower regions of the lake. This process is known as the **spring overturn.**

In summer, warmer water forms a layer over the cooler waters (about 4° C) that lie below. There is an abrupt change in temperature, the **thermocline**, between these two layers. Depending on the climate of the particular area, the warm upper layer may become as much as 20 meters thick during the summer. In the autumn its temperature drops until it is the same as that of the cooler layer underneath, 4° C. When this occurs, the upper and lower layers mix—a process called the **fall overturn.** Therefore colder waters reach the surfaces of lakes in the spring and fall, bringing up fresh supplies of dissolved nutrients.

Lakes can be divided into two categories, based on their production of organic matter. In **eutrophic** lakes

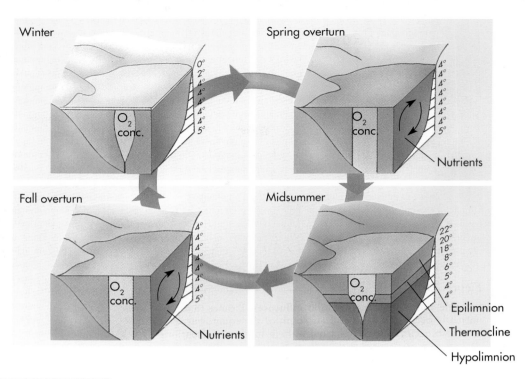

**Figure 22-14**

**Stratification in a lake.** If you have dived into a lake in the summer, you have experienced the existence of these layers directly.

Winter

Spring overturn

Fall overturn

Midsummer

Nutrients

Nutrients

Epilimnion

Thermocline

Hypolimnion

**Figure 22-15**

**A eutrophic farm pond.** The surface bloom of green algae reflects the plentiful supply of nutrients in the water.

there is an abundant supply of minerals and organic matter (Figure 22-15). Oxygen depletion occurs below the thermocline in the summer because of the abundant organic material and the high rate at which aerobic decomposers in the lower layer use oxygen. These stagnant waters again reach the surface after the fall overturn. In **oligotrophic** lakes, on the other hand, organic matter and nutrients are relatively scarce; such lakes are often deeper than eutrophic ones (Figure 22-16), and their deep waters are always rich in oxygen. Oligotrophic lakes are highly susceptible to pollution, as through the addition of excess phosphorus from sources such as fertilizer runoff, sewage, and detergents.

## BIOMES

**Biomes** are climatically defined assemblages of organisms that have a characteristic appearance and are distributed over a wide area on land (Figure 22-17). Biomes are classified in several ways, but, for our purposes, we shall group them into seven categories: (1) tropical forests; (2) savannas; (3) deserts; (4) grasslands; (5) temperate deciduous forests; (6) taiga; and (7) tundra. These biomes differ remarkably from one another because they have evolved in regions with very different climates. These and some of the related communities are shown in Figure 22-18. We shall now examine some of the distinctive features of the seven biomes we have chosen to discuss in more detail.

**Figure 22-16**

**Lake Tahoe, a deep-water lake that is not yet eutrophic.** The drainage of fertilizers applied to the plantings around residences, businesses, and recreational facilities bordering the lake poses an ever-present threat to the maintenance of the deep blue color of its water.

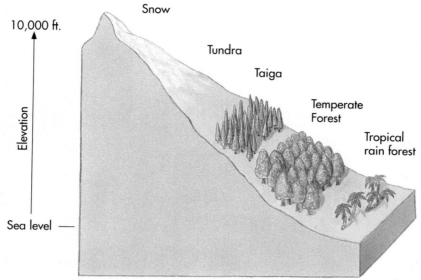

**Figure 22-17**

**Latitude and altitude have similar influences.** Biomes that normally occur far north and far south of the equator at sea level also occur in the tropics but at high mountain elevations.

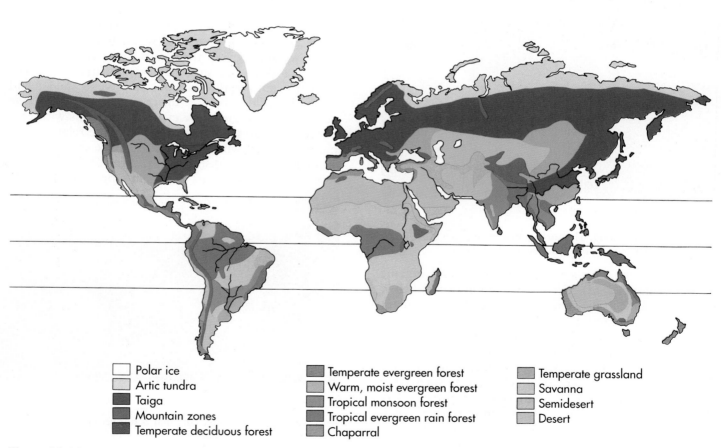

Polar ice
Artic tundra
Taiga
Mountain zones
Temperate deciduous forest

Temperate evergreen forest
Warm, moist evergreen forest
Tropical monsoon forest
Tropical evergreen rain forest
Chaparral

Temperate grassland
Savanna
Semidesert
Desert

**Figure 22-18**

The distribution of biomes.

# TROPICAL RAIN FORESTS: LUSH EQUATORIAL FORESTS

The **tropical rain forests** are the richest biome in terms of number of species, probably containing at least half of the species of terrestrial organisms—more than 2 million species! The rainfall in areas where tropical rain forests occur is generally 200 to 450 centimeters per year, with little difference in its distribution from season to season. The communities that make up tropical forests are rich in species and diverse, so each kind of animal, plant, or microorganism is often represented in a given area by very few individuals. For example, there are seldom fewer than 40 species of tree per hectare; this is 4 or 5 times as many as are typical of temperate forests. In a single square mile of tropical forest in Peru or Brazil, there may be 1400 or more species of butterflies—more than twice the total number found in the United States and Canada combined! The ways of life of tropical organisms are often specialized and highly unusual (Figure 22-19).

Tropical rain forests have a high net productivity, even though they exist mainly on quite infertile soils. Most of the nutrients are held within the plants themselves and are rapidly recycled when the plants die or when parts, such as leaves, are lost. Most of the roots of the tall trees spread through the top 1 or 2 centimeters of the soil, extracting the nutrients promptly from fallen leaves and other plant parts and recycling them back to the trees and other plants from which they have fallen. When tropical rain forests are cut and burned, an abundant supply of nutrients runs out into the soil. This temporary fertilization makes possible the cultivation of crops for a few years, but then, as nutrients are depleted, the people must move on; if there are too many people in a given region, the forest is destroyed (see Chapter 23).

There are extensive tropical rain forests in South America, particularly in and around the Amazon Basin; in Africa, particularly in portions of West Africa; and in Southeast Asia. Human populations in the countries that contain the world's rain forests are destroying them rapidly. Consequently, few of these forests will be left in an undisturbed condition anywhere in the world by the first part of the next century. The destruction of tropical rain forests as an untouched biome will be accompanied by the extinction of as much as a quarter of the total species on earth during the lifetime of many of us.

**Figure 22-19**

Characteristic animals and plants of the tropical rain forest.
A three-toed sloth with young. Sloths use their toes to hang onto the branches of trees.
The emerald tree boa can be found in the tropical rain forests of South America.
A young jaguar in the rain forest of Panama.
Background photo: huge trees dominate tropical rain forests.

531

# SAVANNAS: DRY TROPICAL GRASSLANDS

In areas of reduced annual precipitation or prolonged annual dry seasons, open tropical and subtropical **deciduous forests**—forests in which most of the trees and shrubs lose their leaves at some season of the year— give way to the kind of open grassland called **savannas** (Figure 22-20). The huge herds of grazing mammals, with their associated predators, that inhabit the savannas of Africa are well known and spectacular. Such animal communities occurred in North America during the Pleistocene Epoch but have persisted mainly in Africa. On a global scale, the savanna biome is transitional between tropical rain forest and desert. Generally 90 to 150 centimeters of rain fall each year in savannas. There is a wider fluctuation in temperature here during the year than in the tropical rain forests, and there is seasonal drought. These factors have led to the evolution of an open landscape, often with widely spaced trees; many of the animals and plants are active only during the rainy season. Savannas have often been converted to agricultural purposes throughout the world and provide most of the agricultural products for many tropical and subtropical countries.

*Figure 22-20*

**A view of the savanna.**
Nubian vultures in Niger. Vultures feed on dead animals and do not attack living ones.
A notorious predator of the savanna, the lion.
African buffalo in Kenya.
Background photo: a savanna in Tanzania showing grazing zebras.

# DESERT: ARID AND BURNING HOT

Less than 25 centimeters of annual precipitation usually falls in the world's desert areas—such a low amount that water is the predominant controlling variable for most biological processes. In desert regions the vegetation is characteristically sparse (Figure 22-21). Recall that such regions occur around 20° to 30° north and south latitude, where the warm air that rises near the equator falls and precipitation is limited. Deserts are most extensive in the interiors of continents, especially in Africa (the Sahara Desert), Eurasia, and Australia. Less than 5% of North America is desert. The organisms that occur in deserts are often bizarre in appearance or way of life.

Desert survival depends on water conservation by structural, behavioral, or physiological adaptations (Figure 22-22). Plants and animals may restrict their activity to favorable times of the year, when water is present; they must also avoid high temperatures. Most desert vertebrates live in deep, cool, and sometimes even somewhat moist burrows, and some of them that are active over a greater portion of the year emerge from these burrows only at night, when temperatures are relatively cool. Some, such as camels, can drink large quantities of water when it is available and can then safely withstand the loss of much of it. Many animals simply migrate to or through the desert, where they exploit food that may be abundant seasonally; when the food disappears, the animals move on to more favorable areas.

**Figure 22-21**

**Scenes from the desert.**
Cactus flower, an annual, in Tucson, Arizona. Annuals appear seasonally in this desert region.
Burrowing owls in a California desert. These owls burrow into the ground to conserve water.
A collared lizard, a common vertebrate of the American desert.
Background photo: flats dominated by cholla cactus and shrubs, Borego Valley, California.

**Figure 22-22**

**Strategies by which desert animals conserve water.**
A  Spadefoot toads, which live in the deserts of North America, can burrow nearly a meter below the surface and remain there for as much as 9 months of each year. Underground, their metabolic rate is greatly reduced.
B  On the very dry sand dunes of the Namib Desert in southwestern Africa, the beetle *Onymacris unguicularis* collects fog water by holding up its abdomen at the crest of a dune, thus gathering condensed water on its body.

# GRASSLANDS: SEAS OF GRASS

**Temperate grasslands** once covered much of the interior of North America (Figure 22-23), and they were widespread in Eurasia and South America as well. Such grasslands are often highly productive when they are converted to agriculture, and many of the rich agricultural lands in the United States and southern Canada were originally occupied by **prairies**, another name for temperate grasslands. In eastern Europe and Central Asia, they are called **steppes**. The roots of perennial grasses characteristically penetrate far into the soil, and

grassland soils tend to be deep and fertile. In North America the prairies were once inhabited by huge herds of buffalo and pronghorns, which had wolves, bears, and other predators, as well as various groups of American Indians, hunting them for food and clothing. These are almost all gone now, with most of the prairies having been converted to the richest agricultural region on earth, stretching across a wide region of the north-central United States and adjacent Canada.

**Figure 22-23**

Characteristic plants and animals of the grasslands.
A pronghorn antelope. The "horns" are actually made of matted hair, not horn.
A black-tailed prairie dog.
Lazy susans, a typical grassland annual.
Background photo: bison roam the grassland of North America.

# TEMPERATE DECIDUOUS FORESTS: RICH HARDWOOD FORESTS

In areas of the Northern Hemisphere with relatively warm summers, relatively cold winters, and sufficient precipitation, **temperate deciduous forest** occurs (Figure 22-24). This biome covers very large areas, particularly over much of the eastern United States and Canada and an extensive region in Eurasia. It is the region of deer, bears, beavers, and raccoons—the familiar animals of the temperate regions. Many of the plants flower before the trees form their leaves in the early spring. In temperate deciduous forests the annual precipitation generally ranges from about 75 to about 250 centimeters, well distributed throughout the year but generally unavailable to animals and plants in the winter—because it is usually frozen. Because the deciduous forests represent the remnants of more extensive forests that stretched across North America and Eurasia a few million years ago, these remaining areas— especially those in eastern Asia and eastern North America—share animals and plants that were once more widespread. Alligators, for example, are found only in China and in the southeastern United States (Figure 22-25). The temperate deciduous forest is much richer in species in eastern Asia than in either North America or Europe because climatic conditions have remained more constant and favorable for survival there than in the other two regions.

The **chaparral** of California and adjacent regions is historically derived from deciduous forests. It consists of evergreen, often spiny shrubs and low trees that form extensive communities in dry-summer regions. The forests of conifers that are so impressive in western North America also are remnants of mixed forests that existed in the past, related to both the present-day deciduous forests and the taiga.

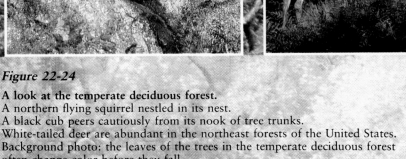

**Figure 22-24**

A look at the temperate deciduous forest.
A northern flying squirrel nestled in its nest.
A black cub peers cautiously from its nook of tree trunks.
White-tailed deer are abundant in the northeast forests of the United States.
Background photo: the leaves of the trees in the temperate deciduous forest often change color before they fall.

**Figure 22-25**

**An American alligator.** There are only two species of the genus *Alligator*; the Chinese species has become very rare and is regarded as endangered.

# TAIGA: GREAT EVERGREEN FORESTS OF THE NORTH

**Taiga** is the northern forest of coniferous trees, primarily spruce, hemlock, and fir, that extends across vast areas of Eurasia and North America (Figure 22-26). Here the winters are long and cold, and most of the limited amount of precipitation falls in the summer. Many large mammals live in the taiga, including elk, moose, deer, and carnivores such as wolves, bears, lynx, and wolverines. Traditionally, much fur trapping has gone on in this region, which is also important in lumber production. Because of the latitude where taiga occurs, the days are short in winter (as little as 6 hours) and correspondingly long in summer. During the summer, plants may grow rapidly, and crops often attain a large size in a surprisingly short time. Marshes, lakes, and ponds are common here and are often fringed by willows or birches. Most of the trees in the taiga tend to occur in dense stands of one or a few species.

*Figure 22-26*

Scenes from the taiga.
This Rocky Mountain elk is greeting the sun in Wyoming.
Moose calf in Wyoming. Moose are one of the characteristic animals of the taiga.
Grizzly bear in Alaska.
Background photo: taiga. A forest dominated by spruces and alders, in Alaska.

# TUNDRA: COLD BOGGY PLAINS OF THE FAR NORTH

Farthest north in Eurasia, North America, and their associated islands, between the taiga and the permanent ice, occurs the open, often boggy grassland community known as the **tundra** (Figure 22-27). This is an enormous biome, extremely uniform in appearance, that covers a fifth of the earth's land surface. Trees are small and are mostly confined to the margins of streams and lakes; in general the tundra's appearance is like that of some parts of the prairies.

Annual precipitation in the tundra is very low, usually less than 25 centimeters, and the water is unavailable for most of the year because it is frozen. During the brief Arctic summers, water sits on frozen ground, and the surface of the tundra is often extremely boggy then. **Permafrost**, or permanent ice, usually exists within a meter of the surface.

**Figure 22-27**

**The tundra.**
Lichens grow during the brief summer months in the tundra.
Squirrel foraging for food.
Timber wolves are very social animals, hunting and raising their young in packs of a dozen or more adults.
Caribou, or reindeer, live in large herds that migrate across the tundra.
Background photo: tundra in Mt. McKinley National Park, Alaska.

## THE FATE OF THE EARTH

Now that you have finished your survey of the biomes and the corresponding communities that occur in the sea, you have gained some appreciation of the different kinds of climate under which these areas have evolved over tens or hundreds of millions of years. In just the last several hundred years, in contrast, human populations have grown over most of the land surface of the globe. Having already converted many biomes to agriculture, urban areas, or other uses, human populations are attacking those biomes—such as the tropical rain forests—which are less suitable for exploitation and about which we know much less. The human population doubled from 1950 to 1987, and most of that growth has occurred in the warmer portions of the globe. There are now so many of us that we must manage the whole earth as a single system if those who follow us are to find a stable ecological situation, one in which they can live out their lives in peace and relative prosperity. In the next chapter, we examine some of the factors that will determine our success or failure in this great enterprise.

## ■ SUMMARY

1. The distribution of major ecosystems on the earth's surface is largely determined by two key physical factors: (1) how much heat from the sun reaches a particular place, and (2) how much moisture it receives, depending in turn on global patterns of atmospheric circulation.

2. Warm air rises near the equator and flows toward the poles, descending at about 30 degrees north and south latitude. Because the air is falling in these regions, it is warmed, and its moisture-holding capacity is therefore increased. The great deserts of the world are formed in these latitudes.

3. Huge circulation patterns occur in the world's oceans, driven by surface "gyrals" that move around subtropical high-pressure zones that occur at 30 degrees north and south latitude. These gyrals, and the hurricanes and typhoons they spawn, spin clockwise in the Northern Hemisphere and counterclockwise in the Southern Hemisphere.

4. Ocean communities occur in three major kinds of environments: (1) the neritic zone, (2) the surface layers, and (3) the abyssal zone. The neritic zone, which lies along the coasts, is small in area but very productive and rich in species. The surface layers are the habitat of plankton (drifting organisms), and nekton (actively swimming ones).

5. Freshwater habitats make up only about 2.1% of the earth's surface; most of them are ponds and lakes. As autumn changes to winter, cooler water forms and sinks because water is most dense at 4° C, mixing the water of the lake. When winter changes to spring, warmer water sinks, and a similar mixing occurs.

6. Biomes are major terrestrial assemblages of plants, animals, and microorganisms that occur together in similar habitats; biomes are defined largely by climate. We discuss seven major biomes, or terrestrial communities, in this book: (1) tropical forests; (2) savannas; (3) deserts; (4) grasslands; (5) temperate deciduous forests; (6) taiga; and (7) tundra.

# REVIEW

1. The biome that is transitional between tropical rain forest and desert is _____ .

2. What does an ecosystem possess that a community does not?

3. Is there a greater number of species on land or in the ocean? Why?

4. Why are the majority of great deserts located near 30 degrees north and south latitude?

5. What are the key characteristics of the desert biome with respect to rainfall?

# SELF-QUIZ

1. Seasons occur because
   (a) the sun revolves slowly around the earth.
   (b) the earth revolves slowly around the sun.
   (c) warm air is concentrated near the equator.
   (d) the earth is tilted relative to the sun's rays.
   (e) the moon creates a pull on the earth's atmosphere.

2. Which of the following phenomena are related to the fact that warm air holds more moisture than colder air?
   (a) Tides
   (b) The formation of glaciers
   (c) Deserts form in the interior of continents
   (d) It rains more in the tropics than elsewhere
   (e) None of the above

3. Lakes, ponds, rivers, and streams cover about what percentage of the earth's surface?
   (a) 20%        (c) 5%        (e) 0.1%
   (b) 10%        (d) 2%

4. During the spring overturn
   (a) cool surface water sinks.
   (b) cool bottom water freezes, and the ice floats to the top.
   (c) ice melts.
   (d) warm surface water sinks.
   (e) nutrients are lost from the surface waters.

5. Which of the following are characteristics of eutrophic lakes?
   (a) Clear water
   (b) Highly susceptible to pollution
   (c) Poor in nutrients
   (d) Decomposing aerobic bacteria abundant
   (e) Little organic matter

6. List the following biomes in order of annual precipitation, with the highest first.
   (a) Temperate deciduous forest
   (b) Grasslands
   (c) Tundra
   (d) Taiga
   (e) Tropical rain forest

7. Which of the following features are characteristic of some desert organisms? (More than one answer possible)
   (a) Loss of leaves in the summer
   (b) Seed germination related to moisture availability
   (c) Deep burrows
   (d) Ability to retain large quantities of water
   (e) Nocturnal activity

8. List the following biomes in order from north to south as they occur in the northern hemisphere:
   (a) Taiga
   (b) Savanna
   (c) Temperate deciduous forest
   (d) Tundra
   (e) Desert

# THOUGHT QUESTIONS

1. What kinds of biological communities would you expect to find on the windward and leeward sides of a mountain range in an area where the annual precipitation ranged between 20 and 100 centimeters per year and was distributed mainly in one rainy season? How would these differences affect human existence in each area?

2. Near the coast in southern California there are two major plant communities. Right along the ocean occurs the coastal sage community, which is dominated by low shrubs that often wither or lose their leaves in the summer. Higher up, on the ridges, occurs the chaparral, a community dominated by evergreen shrubs. Which of these communities would you think grows in the area that receives more rainfall? Why?

# FOR FURTHER READING

ATTENBOROUGH, D.: *The Living Planet: A Portrait of the Earth*, William Collins Sons & Co., Ltd., and British Broadcasting Corporation, London, 1984. A beautifully written and illustrated account of the biomes.

GORE, R.: "Between Monterey Tides," *National Geographic*, February 1990, pages 2-43. Beautifully illustrated account of life along the California coast.

LAWS, R.: "Antarctica: A Convergence of Life," *New Scientist*, September 1983, pages 608-616. A dynamic and beautifully illustrated view of one of the most productive oceans on earth.

McNAUGHTON, S.J.: "Grazing Lawns: Animals in Herds, Plant Form, and Coevolution," *American Naturalist*, vol. 124, 1984, pages 863-886. The fascinating story of the ways in which differences in grasses and in the populations of animals that depend on them help to determine the structure of the savanna biome.

PERRY, D.R.: "The Canopy of the Tropical Rain Forest," *Scientific American*, November 1984, pages 138-147. This article describes efforts to explore the richest and least understood biological community on earth.

WHITEHEAD, J.A.: "Giant Ocean Cataracts," *Scientific American*, February 1989, pages 50-57. Enormous undersea cataracts play a crucial role in determining the chemistry and climate of the deep ocean.

# *O*ur Changing Environment

Human activity is placing a great stress on the biosphere. This sea otter is matted with oil from an oil spill in Alaska, and without help it would not survive the tragedy. No one knows if the Earth will survive.

# OUR CHANGING ENVIRONMENT

## Overview

In 1990 the world population surpassed 5.3 billion individuals, having doubled in less than 40 years. At this unprecedented level, the human population is putting a nearly unbearable strain on the earth's capacity to support us all: 1 out of 4 people lives in extreme poverty, 1 out of 10 is severely malnourished, and other species are being driven to extinction at a rate that has not been achieved since the end of the Cretaceous Period, when the dinosaurs disappeared forever. The loss of a quarter of our topsoil since 1950 is only one of the signs that the world cannot support us indefinitely if we continue to live as we are—and yet our numbers are growing by some 95 million people a year, towards a projected 13 billion a century from now. To improve the situation and achieve global stability, it will be necessary to use the very best knowledge available from the natural and social sciences, to alter our behavior drastically, to improve our agricultural systems, and to control the pollution that we are spreading throughout the world and to the upper layers of the atmosphere. Understanding biology will be one essential requirement for intelligent citizens who wish to carry out their lives successfully in the future.

## For Review    *Here are some important terms and concepts that you will encounter in this chapter. If you are not familiar with them, you should review them before proceeding.*

**Overpopulation** (Chapter 19)

**Population growth characteristics** (Chapter 19)

**Cycling of minerals and flow of energy** (Chapter 21)

**Characteristics of biomes** (Chapter 22)

The view in Figure 23-1 is a portrait of 15.3 million people. It is the city of New York, photographed by a satellite in the spring of 1985. The wakes of ships can be seen in the harbor, wharves lining the shores of the Hudson River. The runways of J.F. Kennedy International Airport are visible at the far right, and the Verrazano Bridge can be seen in the lower center, joining Brooklyn to New Jersey where the Hudson River empties into the Atlantic Ocean. Individual buildings cannot be seen—the scale is too large for that—and from the satellite it is not obvious that the ground below teems with people. At the moment this picture was taken, millions of people within its view were talking, hundreds of thousands of cars struggled through traffic, hearts were broken, babies born, and dead people buried. Were our lens but sharp enough, we would see frozen in time all of this and more, 15,300,000 people busy at life, a panorama of modern industrial society. All of human history has led to this photograph, and in it the future of humanity can be seen.

Our futures, and those of all of the people on earth, are linked to the unseen millions in this photograph, for we share an earth with them. The wisdom with which they manage their environment has deep significance for us all, for their environment is also ours. As human numbers increase, so does their impact on our common environment, posing new challenges to us all.

Now look again at this photograph—15.3 million individuals is a very large number. Holding hands, the people of the city of New York would form a chain that would stretch half way around the world. And the earth is peppered with cities like New York, growing rapidly. More than a dozen other cities contain more than 10 million inhabitants; the three cities that were larger than New York City in 1991 (Mexico City, Tokyo, and Sao Paulo) together have over 50 million inhabitants. In 1987 the human population of the earth reached—and passed—a significant milestone: 5 billion individuals. And it is in precisely this way that the 15.3 million people in this photograph represent the central challenge to the ingenuity of environmental scientists, politicians, and every citizen: there are a lot of people on our earth.

**Figure 23-1**

New York City, as seen from a satellite in 1985.

withdrew and agriculture first developed, there were about 5 million people on earth, distributed over all the continents except Antarctica. With the new and much more dependable sources of food that became available as a result of agriculture (Figure 23-2), the human population began to grow more rapidly. Towns and then cities developed in the areas where agriculture was practiced, and food was sufficiently abundant; such settlements had become widespread by 3000 BC. At the time of Christ, there were an estimated 130 million people on earth—half the population of the United States and Canada today.

The ability to live together on a long-term basis in relatively large settlements made possible, in turn, the specialization of professions in these centers. This was the necessary condition for the development of modern culture. Such advances as the development of metal tools and utensils could not originate until after towns had been formed.

The blip you see at the bend of Figure 23-3 reflects the outbreak of bubonic plague in 1348. Within a generation, 75 million people, over 92% of Europe's population, died. But the world's population recovered quickly from this disaster and rapidly replaced its lost numbers. By 1650 the world population had doubled, and doubled again, reaching 500 million, with many people living in substantial urban centers in which many important innovations were occurring. The Renaissance in Europe, with its renewed interest in science, ultimately led to the establishment of industry in the seventeenth century, and to the Industrial Revolution of the late eighteenth and early nineteenth centuries.

A lot of people consume a lot of food and water, use a great deal of energy and raw materials, and produce a great deal of waste. They also have the potential to solve the problems that arise in an increasingly crowded world, to do what can be done to meet the challenge. In this chapter we delve into the details of how today's living affects the environment within which all future humans must live and into the efforts being mounted to lessen the adverse impact and increase potential benefits. Let us first consider human population growth in a historical context.

## A Growing Population

The earliest fossils that are clearly *Homo sapiens*, our species, come from western Europe; they are about 500,000 years old. Human beings reached North America at least 12,000 to 13,000 years ago, crossing the narrow straits between Siberia and Alaska and moving swiftly to the southern tip of South America. By 10,000 years ago, when the continental ice sheets

**Figure 23-2**

**Much of today's agriculture is still primitive.** Here, in southern India, the people still thresh sorghum using primitive tools whose designs are probably thousands of years old.

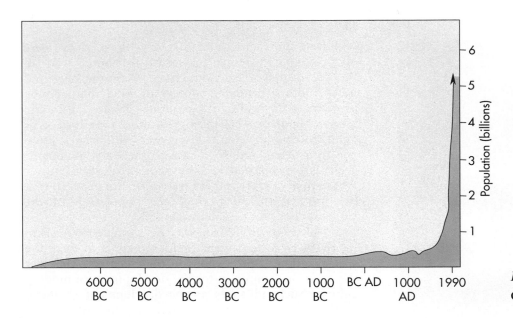

**Figure 23-3**

Growth curve of the human population.

## The Present Situation

For the past 300 years, and probably for much longer, the human birth rate (as a global average) has remained nearly constant, at about 30 births per year per 1000 people. It may be lower now than it has been historically, at about 27 births per year per 1000 people, but this difference is probably not significant. However, with the spread of better sanitation and improved medical techniques, the death rate has fallen steadily, to an estimated 1990 level of about 10 deaths per 1000 people per year. The difference between these two figures amounts to an annual worldwide increase in human population of approximately 1.8%. Such a rate of increase may seem relatively small, but it would lead to a doubling of the world population from its present level of more than 5.3 billion people in only 39 years!

The world population reached an unprecedented 5 billion people by early 1987, and the *annual* increase in population amounts to about 95 million people, a number substantially larger than the total population of Britain, Germany, or Mexico. At this rate, more than 260,000 people are added to the world population each day, or more than 180 every minute! The world population is expected to rise much higher, to well over 6 billion people by the end of the century, and then perhaps to a stable level as high as 13 billion by 2090, according to United Nations estimates (Figure 23-4).

In view of the limited resources available to the human population and the necessity of our learning how to manage these resources well, the first and most necessary step toward global prosperity is to stabilize the human population. One of the surest signs of the pressure we are putting on the environment is our utilization of about 40% of the total global photosynthetic productivity on land. Given that statistic, doubling our population in 39 years certainly poses extraordinarily severe problems. The facts virtually demand restraint in population growth: if and when we develop the technology that would allow greater numbers of people to inhabit the earth in a stable condition, it will be possible to increase our numbers to whatever level might be appropriate at that time.

**Figure 23-4**

**India faces an uncertain future.** India's human population growth is outstripping its resources. At present rates of growth, India will pull even with China in population in about 2020, when each country will have about 1.2 billion people. This scene was photographed at a camp for starving refugees in New Delhi.

*In the early 1990s, the global human population of more than 5.3 billion people was growing at a rate of approximately 1.8% annually. At that rate, it would reach about 6.3 billion people by 2000 and about 8.2 billion by 2020.*

By the year 2000, about 60% of the people in the world will be living in countries that are at least partly tropical or subtropical. An additional 20% will be living in China, and the remaining 20%—1 in 5—in the so-called developed, or industrialized, countries: Europe, the Soviet Union, Japan, the United States, Canada, Australia, and New Zealand. Whereas the populations of the developed countries are growing at an annual rate of only about 0.5%, those of the less developed, mostly tropical countries (excluding China) are growing at an annual rate estimated in 1990 to be about 2.4%. For every person living in an industrialized country in 1950, there were 2 people living elsewhere; by 2020, just 70 years later, there will be 5.

Because the people living in industrialized countries control about 85% of the world's wealth and material goods and enjoy a standard of living perhaps 20 times higher than those found in many developing countries, the changing ratios of total population in each of these sectors should be a matter of special concern for everyone. To provide a few examples, the infant mortality rate in industrial countries in 1990 was 16 per thousand, whereas in developing countries (excluding China) it was 91 per thousand; the respective life expectancies at birth were 74 years versus 59 years, the latter number being even lower in those developing countries where many people are malnourished.

As you learned in Chapter 20, the age structure of a population determines how fast the population will grow (Figure 23-5). For this reason, it is essential in

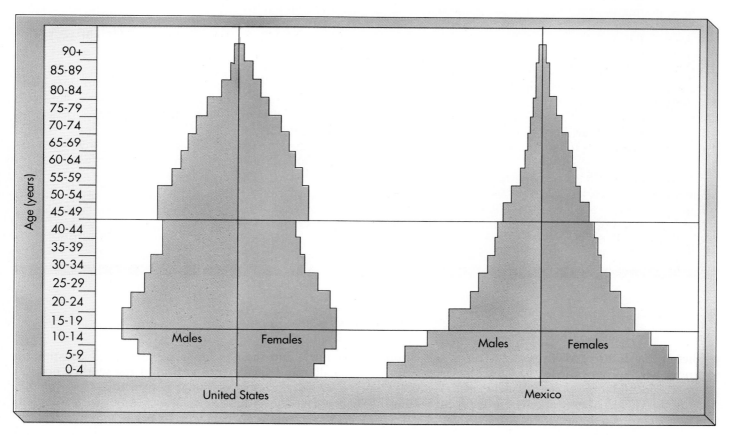

**Figure 23-5**

The age structure of the United States and Mexico in 1977. The rate of population growth in Mexico, already rapid, will increase considerably in the future because so much of its population will be entering into their reproductive years. The "bulge" in the population structure in the United States indicates the "Baby Boomer" generation. Males are indicated on the left in each graph, females on the right.

# The 1990s: A Decisive Decade

As we enter the 1990s, we note that many less developed countries have begun the process of fertility decline and some—such as Thailand and Mexico—are well along the path. The onset of these declines is a watershed in world demographic history. But the question nonetheless remains: will declines continue smoothly or will country birth rates decrease in fits and starts? This is not purely academic. The course of fertility decline over the next decade will largely determine whether the world's ultimate population reaches 10 billion, 15 billion, or some other plateau. The 1990s is truly a decisive decade.

Population projections do not simply extrapolate the effect of present rates into the future. Demographers assume that as countries industrialize, their fertility will decline and that by supporting family planning, these declines can be accelerated. Such an assumption is based on the demographic experience of today's more developed countries. Projections also must make assumptions about precisely when a country's fertility will fall to the "replacement level" of about two children per woman. At this level, couples just "replace" themselves and thus do not increase the size of successive generations. Replacement level fertility ultimately leads to a zero population growth rate and a stationary population size.

To date, world projections have enjoyed a reasonably good track record. In 1958, The United Nations projected a total world population of 6.3 billion for the end of the century. Today, the projection is still 6.3 billion. Such a performance may be striking, but as more and more demographic changes are observed in more and more countries, predicting future size will become less and less certain.

The timing and pace of fertility declines have a large impact on the future size of a country's population. For example, if India resumes a period of fertility decline, ending a "stall" of many years, and reaches replacement level in 2010, its 1990 population of 853 million would grow to 1.3 billion by 2025, and the annual growth rate would fall to half the current rate of 2.1%. If India does not resume its fertility decline, the 2025 population will exceed 2 billion and will be growing at an even faster rate than it is today.

Different fertility declines effect "ultimate" world population size. If replacement were achieved as soon as 2000, a highly unlikely scenario, the world population size would not reach 9 billion. If it were delayed to 2080, world population would be close to triple today's size.

There can be little doubt that the world's birth rate has entered a period of decline, but its future path will be pivotal to humanity's future numbers. What happens now is of central importance. It is highly likely that world population growth will slow to a near halt during the next century. But we are far less certain of its ultimate size.

Modified from Haub, C., M.M. Kent, and M. Yanagishita: *1990 World Population Data Sheet*, Population Reference Bureau, Inc., Washington, D.C., 1990.

predicting the future growth patterns of a population to know what proportion of its individuals have not yet reached childbearing age. In developed countries, such as the United States, about a fifth of the population consists of people under 15 years of age; in developing countries, such as Mexico, the proportion is typically about twice as high. Thus, even if the policies that most tropical and subtropical countries have established to limit their own population growth are carried out consistently for decades, the populations of these countries will continue to grow well into the next century, and the developed countries will constitute a smaller and smaller proportion of the world's population. For example, if India, with a mid-1990 population level of about 853 million people (with 39% under 15 years old), managed to reach a simple replacement reproductive rate by the year 2000, its population would still not stop growing until the middle of the next century. At present rates of growth, India will have a population of nearly 1.4 billion people by 2020 and will still be growing rapidly, judged from the country's age structure and current patterns of growth.

Most countries are devoting considerable attention to slowing the growth rate of their populations, and there are genuine signs of progress. It is estimated that the world population may stabilize by the close of the next century at the level of about 13 billion people. No one knows whether the world can support so many people indefinitely. Finding a way to do so is the greatest task facing humanity's future. The quality of life that will be available for our children in the next century will depend to a large extent on our success.

## FOOD AND POPULATION

Even though experts estimate that enough food is produced in the world to provide an adequate diet for everyone in it, the distribution of this food is so unequal that large numbers of people live in hunger. In the United States, the Soviet Union, Japan, and Europe there is, on the average, a quarter to a third *more* food available per person than the United Nations Food and Agriculture Organization (FAO) regards as the calorie intake necessary to maintain moderate physical activity. In contrast, countries such as Bangladesh, Ecuador, and Kenya have 10% to 15% *less* than this minimum available per person and, for the most part, little cash with which to purchase more. Worldwide, 300 million to 400 million people each consume less than 80% of the U.N.–recommended minimum standards for caloric intake, a diet insufficient to prevent stunted growth and serious health risks. Only the United States, Canada, Argentina, and a few European countries have emerged as consistent exporters of food.

Of the approximately 4.1 billion people living in the tropics and China in 1990, the World Bank estimated that some 1.2 billion—nearly a quarter of the total world population—were living in extreme poverty. These people cannot expect to provide adequate food, clothing, and shelter for themselves and their families from day to day. Among the malnourished people mentioned above, UNICEF estimates that about 13 million children under the age of 5 starve to death every year or die of diseases complicated by malnutrition (Figure 23-6). Worse, many millions of additional children exist only in a state of lethargy, their mental capacities often permanently damaged by their lack of access to adequate quantities of food.

*About 1.2 billion people live in a state of extreme poverty; many of them are malnourished.*

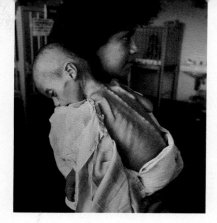

**Figure 23-6**

**Starvation.** Consuelo Andena, 16, holding her 1-year-old baby, Carolina, in the malnutrition ward of the Women and Children's Hospital in Tegucigalpa, Honduras, in 1983. Such tragic scenes are common throughout the tropics, where more than 40,000 babies perish every day from starvation.

One of the most alarming trends taking place in developing countries is the massive movement to urban centers. Mexico City, the largest city in the world, is plagued by smog, traffic, inadequate waste disposal, and other problems—and it will have a population of more than 30 million by the end of the century (Figure 23-7). The prospects of supplying adequate food, water, and sanitation to these people, in a country whose 1990 population (about 90 million) is growing at a rate at which it would double in only 29 years, are almost unimaginable. The lot of the rural poor, mainly farmers, in Mexico is even worse, and it is no wonder that nearly a quarter of all Mexicans indicated in a 1989 poll that they were either "likely" or "very likely" to move to the United States over the next 2 years.

*The proportion of poor, hungry people in the less developed countries, now amounting to a third of their populations, will grow rapidly unless we are remarkably successful in developing solutions to their problems in the near future.*

## THE FUTURE OF AGRICULTURE

One of the greatest and most immediate challenges facing today's world is to produce enough food to feed its expanding population. Even though it is estimated that world food production has expanded 2.6 times since 1950—more rapidly than the human population—virtually all the lands that can be cultivated are already

**Figure 23-7**
Mexico City, the world's largest city.

under cultivation. A quarter of the world's topsoil has been lost from agricultural lands during this period, and the human population is continuing to grow explosively. Much of the world is populated by very large numbers of hungry people who are rapidly destroying the sustainable productivity of the lands they inhabit; at the same time, consumption in industrialized countries such as the United States, running at 20 to 30 times the rate in developing countries, is having an even greater adverse effect on the future of humanity. In the face of these massive problems, we need to consider what the prospects are for increased agricultural productivity in the future, and what we can do to improve these prospects.

For a start, we are going to have to identify more kinds of crops. Nearly all of the major crops now grown in the world have been cultivated for hundreds or even thousands of years. Only a few, including rubber and oil palms (Figure 23-8), have entered widespread cultivation since 1800. One key feature for which nearly all of our important crops were first selected was ease of growth by relatively simple methods.

How many plants do we use at present? Just three species—rice, wheat, and corn—supply more than

half of all human energy requirements; only about 150 kinds of plants are used extensively; and only about 5000 have ever been used for food. It is estimated that there may be tens of thousands of additional kinds of plants, among the world total of some 250,000 species, that could be used for human food if their properties were fully explored and they were brought into cultivation. There are also many uses for plants other than for food. Oral contraceptives were produced from Mexican yams for many years; the muscle relaxants used in surgery worldwide came from curare, an Amazonian vine used traditionally to poison darts used in hunting; and the cure for Hodgkin's disease was developed in the early 1970s from the rosy periwinkle, a widely cultivated plant native to Madagascar.

The reasons for which we select new crops now are often very different from those which appealed to our ancestors who developed agriculture, living as they did in small groups around the foothills of the Near East or on the temperate slopes of the mountains in Mexico. Standards of cultivation have changed, and we use many products from plants—for example, oils, drugs, and other chemicals—that often would not have led to their cultivation earlier. In a time when human activities threaten to drive many of the world's plants and other organisms to extinction during our lifetimes and those of our children, we need to search systematically for new and useful crops that fit the multiple needs of modern society in ways that would not have been considered earlier. We must begin more diligent efforts to find and bring into cultivation more kinds of useful plants before they are gone forever.

*Only about 150 kinds of plants, out of the roughly 250,000 known, play an important role in international trade at present. Many more could be developed by a careful search for new crops.*

## The Prospects for More Food

In improving the world food supply, the most promising strategy is to improve the productivity of crops that are already being grown on lands that are under cultivation. Much of the improvement in food production must take place in the tropics and subtropics, where a rapidly growing majority of the world's people already live, including most of those enduring a life of extreme poverty. These people cannot be fed by exports from the industrial nations, which contribute only about 8% of their total food at present and where agricultural lands are already heavily exploited.

A                                    B

**Figure 23-8**

**Crops of the future.**
**A** Oil palms *(Eleais guineensis)* are grown throughout the
tropical regions of the world as a source of oil. On infertile
tropical soils, trees often are much more successful than cereal
crops.
**B** The roots of *Casuarina*, a fast-growing tree, are inhabited by
nitrogen-fixing actinomycetes, which form nodules there.
*Casuarina* is receiving increasing attention as a possible source
of firewood and for watershed protection in the tropics.

*The Green Revolution has resulted in increased food production
in many parts of the world through the use of improved crop
strains. These strains often depend on increased inputs of
fertilizers, water, pesticides, and herbicides, as well as the
greater use of machinery—means that are not normally
available to the poor.*

During the 1950s and 1960s, the so-called Green
Revolution took place. This revolution depended on
the development of new, improved strains of wheat
and rice at international centers that were organized by
groups such as the Rockefeller and Ford foundations,
with the assistance of many other agencies and of governments. As a result of the efforts made at these centers, the production of wheat in Mexico increased
nearly 10-fold between 1950 and 1970, and Mexico
temporarily became an exporter of wheat rather than
an importer (Figure 23-9). During the same decades,
food production in India was largely able to outstrip
even a population growth of approximately 2.3% annually, and China became self-sufficient in food.

Despite the apparent success of the Green Revolution, the improvements were limited. The agricultural
techniques that were employed required the expenditure of large amounts of energy and abundant supplies
of fertilizers, pesticides, and herbicides, as well as adequate machinery. For example, in the United States it
requires about 1000 times as much energy to produce
the same amount of wheat that results from traditional
farming methods in India. Introducing industrial-coun-
try methods obviously increases the yield, but who will
pay the bill? Energy prices are often held at artificially
low levels in developing countries, so it may actually be
more expensive for a poor rural farmer to grow an
equivalent amount of grain than for a large-scale
farmer using developed-world technology. In this respect, the introduction of Green Revolution methods in
some regions has actually worsened poverty for many
of the people and lessened their access to food, fuel,
and other commodities on which they depend.

Certain nutritional problems have also arisen in
connection with the Green Revolution. The overcon-
centration on cereal crops has tended to lower the production of other nutritionally important plants, including legumes, oilseeds, and vegetables of all kinds. The
reason that legumes and cereals are often nutritionally
combined is that they provide a balanced set of amino
acids required by human beings for proper growth.
The varied strains of crops that are grown on small
farms may also be driven out by fewer kinds of modern
strains and fewer crops, which produce a better yield if
large-scale inputs of chemicals and the use of machinery are possible (Figure 23-10). Despite these short-

**Figure 23-9**

**Dwarf wheat.** Development of improved strains such as this
helped to make Mexico an exporter of wheat during some
years of the 1960s and 1970s.

**Figure 23-10**

**A traditional farm in Thailand.** On it grow coconuts, sugar palms, litchi nuts, and the nitrogen-fixing water fern *Azolla* growing in ponds. Land crabs are frequent and are harvested as a source of protein. Farms of this type are extremely suitable for many sites in the tropics.

term advantages, the loss of the unique, traditional strains of crops presently cultivated by small, rural farmers throughout the world may ultimately prevent the particular crop plants from being able to grow in less favorable habitats or to withstand important diseases in the long run (Figure 23-11). **Monoculture**—the exclusive cultivation of a single crop over wide areas—is an efficient way to use certain kinds of soils, but it carries the risk of an entire crop being destroyed with the appearance of a single pest species or disease.

In improving the world food supply, the most promising strategy is to improve the productivity of crops that are already being grown on lands that are under cultivation. There are relatively few parts of the world where additional land exists that can be brought into cultivation using currently available technology. Biologists have a crucial role to play in the improvement of crops and in the development of new ones. Traditional methods of plant breeding and selection (Figure 23-12) must be applied fully to many crops of importance in the tropics and subtropics (Figure 23-13) in addition to wheat, corn, and rice. Genetic engineering techniques (see Chapter 14) are being applied extensively to crop plants and, to some extent, to animals, and these techniques offer great promise for the

A

D

B

C

**Figure 23-11**

**Who cares if a tropic plant goes extinct?** You should.
A  Most corn *(Zea mays)* is genetically uniform and thus difficult to improve by selection.
B  This field in Mexico is dominated by a wild perennial relative, *Zea diploperennis*, discovered in 1977. The two species can be crossed to produce fertile hybrids, and *Z. diploperennis* is resistant to five of the seven main types of virus diseases that damage corn.
C  Spikelets of *Z. diploperennis*.
D  This great improvement was almost lost. *Z. diploperennis* might never have been found at all because it occurs in only one field. Its single population could easily have been destroyed by the rapidly increasing human population pressures of the area.

# The Potential for New Crops

With their vast and diverse arsenal of natural insecticides, plants offer a reservoir of useful products that we have barely tapped. A number of these plants have proved useful already, and many more will do so in the future as we explore further.

One of the most fertile areas for biological exploration is the study of plant use by people who live in direct contact with natural plant communities. These people know a great deal about the plants with which they come into contact, and the screening that they have performed over many generations can, if its results are understood and catalogued, help us to recognize new useful plants. For example, the herbal curer in the jungles of southern Surinam has a profound knowledge of uses of the plants that grow in his region, but that knowledge is being lost as civilization and modern medicine move into the area.

Some recent triumphs in the search for economically important plants are the following:

Guayule *(Parthenium argenteum)* (Figure 23-A, 1). This widespread shrub of northern Mexico and the southwestern United States is a rich source of rubber. The yield from some cultivated strains is up to 20% rubber. Guayule is now being cultivated in more than 30 countries, and efforts are being made to bring it into even wider cultivation.

Periwinkle *(Catharanthus roseus)* (Figure 23-A, 2). Periwinkle is a garden plant, now widespread in cultivation; as a native plant, it is found only in Madagascar. Two drugs, vinblastine and vincristine, which have been developed from periwinkle by Eli Lilly & Company, are effective in the treatment of certain forms of leukemia. For example, a child who develops Hodgkin's disease, a form of leukemia, now has a 90% chance of survival if treated with these drugs. In 1950, such a child would have had only a 20% chance to live.

Grain amaranths *(Amaranthus* species) (Figure 23-A, 3). Important grain crops of the Latin American highlands in the days of the Incas and Aztecs, the grain amaranths are now minor crops in Latin America; their use was suppressed because they played a role in pagan ceremonies of which the Spaniards disapproved. The grain amaranths, fast-growing plants that produce abundant grain rich in lysine—an amino acid that is rare in most plant proteins but essential for animal nutrition—are now being investigated seriously for widespread development.

Winged bean *(Psophocarpus tetragonolobus)* (Figure 23-A, 4). The winged bean is a nitrogen-fixing tropical vine that produces highly nutritious seeds, pods, and leaves. Its tubers are eaten like potatoes, and its seeds produce large quantities of edible oil. First cultivated locally in New Guinea and Southeast Asia, the winged bean has spread since the 1970s throughout the tropics, where it holds great promise as a productive source of food.

1

2

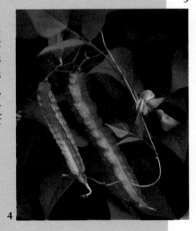

3

4

*Figure 23-12*

**A greenhouse laboratory.** In these experiments at the University of Arizona, Tucson, scientists are attempting to select salt-tolerant crops that will be suited for the naturally saline soils of the world, as well as for those which have become saline through continued irrigation.

future. For example, it will be possible to produce plants that are resistant to specific herbicides, which can therefore be applied much more effectively to crops for weed control. It will soon be possible, through the methods of genetic engineering, to produce new strains of plants that will grow successfully in areas where the crop could not grow before. Eventually it will also be possible to introduce desirable characteristics into important crop plants, such as the ability to fix nitrogen, carry out $C_4$ photosynthesis, or produce substances that deter pests and diseases in abundance. The ability to transfer genes between organisms will be of great importance in the improvement of crop plants before the twentieth century draws to a close.

The oceans were once regarded as an inexhaustible source of food, but overexploitation of their resources is actually limiting the world catch from year to year, while these catches are costing more in terms of energy. It has been estimated that the mismanagement of fisheries, mainly through overfishing, local pollution, and the destruction of fish breeding and feeding grounds, has already lowered the catch of fish in the sea by about 20% from its maximum levels. The decline in the numbers of whales in the oceans of the world is a tragic and well-known example of the way fisheries have been and are being destroyed.

The development of new kinds of food, such as microorganisms grown in culture in nutrient solutions, should definitely be pursued. For example, the photosynthetic, nitrogen-fixing cyanobacterium *Spirulina* is being used in this way, with experimentation under way in several countries to see whether it can be developed into a commercial food source (Figure 23-14). Masses of this bacterium are used traditionally as food

A

*Figure 23-13*

**Plants as energy factories of the future.**
A Water hyacinth, despite its beautiful flowers, is generally regarded as a noxious weed, clogging waterways throughout the tropics.
B But the same rapid growth that makes it such a harmful weed allows it to produce large amounts of biomass rapidly. Here it is being harvested in India as animal feed; it can readily be dried and burned as a source of energy also.

B

A

B

C

**Figure 23-14**

*Spirulina,* **a cyanobacterium that is being grown increasingly as a source of protein.**
**A** Individuals of *Spirulina,* each about 250 micrometers long.
**B** A mat of *Spirulina* on a conveyor belt in a processing plant in Mexico.
**C** Ponds for the production of *Spirulina* in Thailand.

in Africa, Mexico, and other regions. *Spirulina* thrives in very alkaline water and has a higher protein content than do soybeans; the ponds in which it grows are 10 times more productive, on the average, than are wheat fields. Ultimately the enrichment of human food through protein-rich concentrates of microorganisms will provide important nutritional supplements to our diets. But there are psychological barriers that must be overcome to persuade people to eat such foods, and the processes required to produce them tend to be energy-expensive.

## THE TROPICS

More than half of the world's people live in the tropics, and this percentage is increasing rapidly. For global stability and for the sustainable management of the world ecosystem, it will be necessary to solve the problems of food production and regional stability in the areas where most people live. World trade, political and economic stability, and the future of most species

of plants, animals, fungi, and microorganisms depend on our addressing this problem in an adequate fashion.

Many people in the tropics engage in **shifting agriculture**—clearing and cultivating a patch of forest, growing crops for a few years, and then moving on (Figure 23-15). The fertility of the soil has by then returned to its original, very low level—the temporary enrichment that resulted from cutting and burning the forest has been exhausted—and the cultivator must move on and clear another patch of forest. Such agricultural systems work well where human populations are relatively low, but as these numbers grow, there is little opportunity even for the cultivation of traditional crops such as manioc (tapioca, cassava) (Figure 23-16). Firewood gathering is also hastening the demise of many tropical forests; about 1.5 billion people worldwide—a third of the global population—depend on firewood as their major source of fuel and are cutting the local supplies faster than the trees can regenerate themselves.

Shifting agriculture is one major factor in the destruction of tropical forests, but a number of other fac-

*Figure 23-15*

**Slash and burn.** A family has cleared a small patch of forest in the northern Amazon Basin of brazil; they will be able to grow crops on it for a few years and then will need to move on to another part of the original forest. Such activities can be sustained indefinitely if population levels are low enough, but that will rarely be the case in the closing years of the twentieth century.

tors are also significant. We can illustrate them by reference to the tropical rain forest, biologically the richest of the world's biomes. Most other kinds of tropical forest have already been largely destroyed; they tend to grow in more fertile soils and were exploited by human beings earlier than were the tropical rain forests now under such massive attack. It is estimated that about 6 million square kilometers of tropical rain forest still exist in a relatively undisturbed form. This area, about three-fourths the size of the United States exclusive of Alaska, represents about half of the original extent of tropical rain forest. From it, about 160,000 square kilometers are being clear-cut per year (the rate now exceeds 1.5 acres per second), with perhaps an equivalent amount severely disturbed by shifting cultivation, firewood gathering, and allied practices. The total area of tropical rain forest destroyed—and therefore permanently removed from the world total—amounts to an area greater than the size of Indiana each year. At such a rate, all of the tropical rain forests in the world will be gone in about 30 years, but in many regions, the rate of destruction is much more rapid.

As a result of such overexploitation, experts predict there will be little undisturbed tropical forest left anywhere in the world by early in the next century. Many areas now occupied by forest will still be tree-covered, but those trees will represent only a small percentage of those which now grow in these areas. Many species of plants, animals, fungi, and microorganisms can reproduce only under the conditions in which they live in undisturbed forest. Consequently, they are threatened with extinction or, at the very least, exclusion from large areas. In fact, a fifth or even more of the species on earth may become extinct during the next 30 or 40 years, amounting to a million or more

species. Viewing the situation in another way, several species per day are probably becoming extinct now. Many of these species which are becoming extinct inhabit ecologically devastated islands such as St. Helena in the south Atlantic and Rodrigues (Figure 23-17) in the western Indian Ocean, or similarly devastated areas on continents, such as the Atlantic forests of Brazil or the lowlands of western Ecuador. By early in the next century, the rate could easily reach several species per *hour,* and it would continue to climb for at least another 50 years. Overall, this would amount to an extinction event that has been unparalleled for at least 65 million years, since the end of the age of dinosaurs. The number of species in danger of extinction during our lives is far greater than the number that became extinct at that distant time.

The prospect of such an extinction event is of major concern. Only one out of every six tropical organisms has even been given a scientific name, so many species that are about to become extinct will never have been seen by any human being. As these species disappear, so does our opportunity to learn about them, not only scientifically, but also in terms of their possible benefits for ourselves. We have an intrinsic curiosity about the plants, animals, and microorganisms with which we share this planet, and we would like to understand them better individually. Each is unique, and with its loss we lose forever the chance to use it for any purpose whatever. The fact that our entire supply of food is based primarily on 20 kinds of plants, out of the quarter million kinds that are available, should give us pause to consider what it means to be living in a generation during which a high proportion of the remainder are being lost permanently. Many of them would surely be of great use to us if we knew about their properties.

A

B

C

**Figure 23-16**

**Manioc, a key tropical crop.**
A   Manioc is a widespread and very important tropical root crop, comparable in its use to potatoes in temperate regions.
B   A Brazilian woman is preparing the starchy tubers to eat.
C   In Brazil, manioc is being used for the production of ethyl alcohol, which is mixed with gasoline to make a fuel (gasohol) for motors. However, tying up large tracts of land for the production of gasohol takes away land that might be used for food production for the poor.

What biologists can do is help design plans for finding those organisms most likely to be of use and saving them from extinction. They must also participate in sound, globally based schemes to preserve as much as possible of the biological diversity of life so that the options for our descendants may be as numerous as possible. With the loss of tropical forest, and of biological communities throughout the world, we are permanently losing many opportunities not only for knowledge, but also for increased prosperity. It is biologists who must understand this message and inform their fellow citizens of its importance to them.

Viewing the consequences of uncontrolled deforestation in the tropics and subtropics in another way, tropical forests are complex, productive ecosystems that work well in the areas where they have evolved. The sad truth is that we generally do not know how to replace them with other productive ecosystems that

A

B

**Figure 23-17**

**Escape from extinction.**
A This palm is the last individual of its species, *Hypphorbe verschaffeltii,* known in nature. It is shown here on the island of Rodrigues, in the Indian Ocean, the same island where that famous flightless pigeon, the dodo, became extinct more than 300 years ago.
B But *H. verschaffeltii* is not likely to become extinct because it is preserved on a university campus on the island of Mauritius. The palms are the second most important family of plants in the world economically, being surpassed only by the grasses.

**Figure 23-18**

**Destroying the tropical forests.**
**A** These fires are destroying rain forest in Brazil, which is being cleared for cattle pasture.
**B** The consequences of deforestation can be seen on these elevation slopes in Ecuador, which now support only low-grade pastures where highly productive forest, which protected the watersheds of the area, grew in the 1970s.
**C** As time goes by, the consequences of tropical deforestation may become even more severe, as shown here by the extensive erosion in this area of Tanzania from which forest has been removed.

will support human beings. When we cut a forest or open a prairie in the North Temperate Zone, we provide the basis for a farm that we know can be worked for generations. In the tropics, for the most part, we simply do not know how to engage in continuous agriculture in most areas that are not now under cultivation. When we clear a tropical forest, we engage in a one-time consumption of natural resources that will not be available again (Figure 23-18). The complex ecosystems that have been built up over millions of years are now being dismantled, in almost complete ignorance, by the human species.

What biologists must do is to learn more about the construction of sustainable agricultural ecosystems that will meet human needs in tropical and subtropical regions. The ecological principles that we have been reviewing in the last three chapters are universal principles. The undisturbed tropical rain forest has one of the highest rates of net primary productivity of any plant community on earth, and it is therefore logical to assume that it can be harvested for human purposes in a sustainable, intelligent way. Simply passively allowing it to be consumed is something that biologists and concerned individuals should attempt to avoid. Sound development of the tropics, an urgent matter in view of the very large numbers of humans who live in tropical countries, must be based on sound biology as well as on achieving stable human population levels and alleviating the problems of poverty and widespread malnutrition. Biologists must address themselves increasingly to problems that have traditionally been the concern of agronomists, animal breeders, and other agriculturists and must apply the results of their research to the creation of a stable global ecosystem.

## OUR IMPACT ON THE ENVIRONMENT

The simplest way to gain a feeling for the some of the other dimensions of the problem before us is simply to scan the front pages of any newspaper or news magazine or to watch television. Although they are only a sampling, the five "front page" issues that follow teach us a great deal about the scale and complexity of the challenge we face.

### Nuclear Power

At 1:24 AM on April 26, 1986, one of the four reactors of the Chernobyl nuclear power plant blew up. Located in central Russia 100 kilometers north of Kiev, Chernobyl was one of largest nuclear power plants in Europe, producing 1000 megawatts of electricity (enough to light a medium-sized city). Before dawn on April 26, workers at the plant hurried to complete a series of tests of how the generator of reactor #4 performed during a power reduction and took a foolish short-cut: they shut off all the safety systems. The reactors at Chernobyl were graphite reactors designed with

**Figure 23-21**
**The Rhine.**

Six months later, Swiss and German environmental scientists monitoring the effects of the accident were able to report that the blow to the Rhine was not mortal. Enough small aquatic invertebrates and plants had survived to provide a basis for the eventual return of fish and other water life, and the river was rapidly washing out the remaining residues from the spill. A lesson difficult to ignore, the spill on the Rhine has caused the governments of Germany and Switzerland to intensify efforts to protect the river from future industrial accidents and to regulate the creation of chemical and industrial plants on its shores.

## The Threat of Pollution

The pollution of the Rhine is a story that can be told countless time in different places in the industrial world, from Love Canal in New York to the James River in Virginia to Times Beach in Missouri. Nor are all of the pollutants that threaten the sustainability of life immediately toxic. Many forms of pollution arise as by-products of industry. For example, the polymers known as plastics, which we produce in abundance, break down slowly if at all in nature. Even though scientists are attempting to develop strains of bacteria that can decompose plastics, their efforts have been largely unsuccessful. Consequently, virtually all of the plastic items that have ever been produced are still with us. Collectively, they constitute a new form of pollution for which there is, as yet, no solution.

Water pollution is another very serious problem that exists on a global scale. There is simply not enough water available to dispose of the diverse substances that today's enormous human population produces continuously. Despite the implementation of ever-improving methods of sewage treatment throughout the world, our lakes, streams, and groundwater are becoming increasingly polluted. Household detergents, which contain phosphates, may flow into oligotrophic lakes and lead to their eutrophication, as we discussed in Chapter 22. This leads to an overgrowth of algae and a rapid deterioration of water quality.

Widespread agriculture, increasingly carried out by modern methods, also causes very large amounts of many new kinds of chemicals to be introduced into the global ecosystem. These include pesticides, herbicides, and fertilizers (Figure 23-22). Developed countries such as the United States now attempt to carefully monitor the side effects of these chemicals. But unfortunately, large quantities of many toxic chemicals that were manufactured in the past still circulate in the ecosystems of these nations.

For example, the chlorinated hydrocarbons, a class of compounds that includes DDT, chlordane, lindane, and dieldrin, have all been banned for normal use in the United States, where they were once used very widely. These molecules break down very slowly and accumulate in animal fat. Furthermore, as they pass through a food chain, they are increasingly concentrated. DDT caused serious problems by leading to the production of very thin, fragile eggshells in many bird species in the United States and elsewhere until the late 1960s, when it was banned in time to save the

**Figure 23-22**

**Huge quantities of pesticides and herbicides are used in connection with agriculture each year.** One of the most important tasks facing agriculturists and other biologists is the development of integrated pest management and systems involving genetically engineered organisms that will avoid the necessity for such ecological excesses.

birds from extinction. Chlorinated hydrocarbons have many other undesirable side effects, several of which are still poorly understood.

Obviously, a "back to nature" approach—one that ignores the important contributions made to our standard of living by the intelligent use of chemicals—will not allow us to care adequately for the needs of the current world population. Also, such a backward approach would not allow us to feed the additional billions of people who will join us during the next few decades. On the other hand, it is essential that we use our technology as intelligently as possible and with due regard for the protection of the productive capacity of all parts of the earth, on which we all depend.

## Acid Rain

The three smokestacks you see in Figure 23-23 are those of the Four Corners power plant in New Mexico. This facility burns coal, sending the smoke up high into the atmosphere through these stacks, each of which is over 65 meters tall. The smoke that the stacks belch out contains high concentrations of sulfur, which smells bad (like rotten eggs) and produces acid when it combines with the water vapor in air. The intent of those who designed the plant was to release the sulfur-rich smoke high up in the atmosphere, where the winds would disperse and dilute it. This sort of solution to the problem posed by burning high-sulfur coal was first introduced in Britain in the mid-1950s and rapidly became popular in the United States and Europe—there are now about 800 such stacks in the United States alone. Basically they function to carry the acid produced by the fuels away from the areas where they are produced: London no longer suffers from acid fogs,

but the forests and lakes of Sweden are being destroyed.

The environmental effects of this acidity are serious. The sulfur introduced into the upper atmosphere combines with water vapor to produce sulfuric acid, and when the water later falls as rain or snow, the precipitation is acid. Natural rainwater rarely has a pH lower than 5.6; in the northeastern United States, rain and snow now have a pH of about 3.8, about a hundred times as acid as the usual limit. Because the prevailing winds in the temperate latitudes, where most industries are concentrated, are westerlies, the sulfur emissions released by plants in the midwestern United States primarily return to the earth in rain and snow that falls in the eastern United States and Canada, and similar patterns occur in Europe.

Acid precipitation destroys life. Thousands of the lakes of nothern Sweden and Norway no longer support fish; these lakes are now eerily clear. In the northestern United States and eastern Canada, tens of thousands of lakes are dying biologically as a result of acid precipitation (Figure 23-24). At pH levels below 5.0, many fish species and other aquatic animals die, unable to reproduce under these condition. In southern Sweden and elsewhere, groundwater is now regularly found to have a pH between 4.0 and 6.0, its acidity resulting from the acid precipitation that is slowly filtering down into the underground reservoirs, thus threatening the water supplies of future generations.

Trees also suffer. There has been enormous forest damage in the Black Forest in Germany and in the forests of the eastern United States and Canada. It has been estimated that at least 3.5 million hectares of forest in the Northern Hemisphere are being affected by

**Figure 23-23**

Four Corners power plant in New Mexico.

**Figure 23-24**

**Acid rain kills.** Twin Pond in the Adirondacks of upstate New York is one of the many lakes in the region in which the levels of acidity in the lake have killed the fishes, amphibians, and most of the other kinds of animals and plants that once occurred there.

acid precipitation (Figure 23-25), and the problem is clearly growing.

The solution at first seems obvious: capture and remove the emissions instead of releasing them into the atmosphere. But there are serious difficulties in executing this solution. First, it is expensive. Reliable esti-

mates of the cost of installing and maintaining the necessary "scrubbers" in the United States are on the order of $4 to $5 billion a year. Although this is no more than a hundredth of the amount that will ultimately be spent to "bail out" failed savings and loans associations, our national priorities evidently do not yet clearly focus on a healthy environment. An additional difficulty is that the polluter and the recipient of the pollution are far from one another, and neither wants to pay so much for what they view as someone else's problem. The Clean Air Act revisions of 1990 addressed this problem in the United States significantly for the first time.

*Acid rain is a blanket term for the process whereby industrial pollutants such as nitric and sulfuric acids—introduced into the upper atmosphere by factory smokestacks often over 50 meters tall—are spread over wide areas by the prevailing winds and then fall to earth with the precipitation, lowering the pH of groundwater and killing life.*

A

B

C

**Figure 23-25**

**Damage to trees at Camel's Hump Mountain in Vermont.**
A The healthy forest as it was in 1963.
B A view from the same spot 20 years later, in 1983, showing many trees diseased and dying.
C A closer view of some of the dying trees in this area.

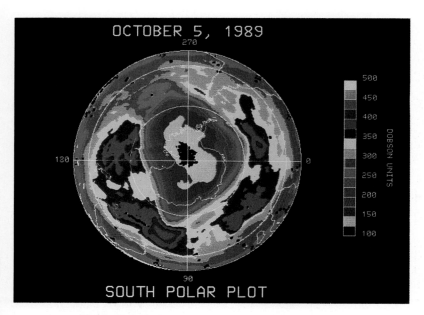

OCTOBER 5, 1989

SOUTH POLAR PLOT

**Figure 23-26**

Satellite picture of the Antarctic ozone hole taken on October 5, 1989.

## The Ozone Hole

The swirling colors of Figure 23-26 are a view of the South Pole in 1986 as viewed from a satellite. This is not the view your eye would see, but rather a computer reconstruction in which the colors represent different concentrations of ozone ($O_3$, a form of oxygen gas, $O_2$). As you can easily see, there is an "ozone hole" over Antarctica, an area about the size of the United States within which the ozone concentration is very much less than it is elsewhere. The hole is not a permanent feature, but rather one that becomes evident each year for a few months at the onset of the Antarctic spring. Every September from 1975 onward, the ozone hole has reappeared—and each year the layer of ozone is thinner and the hole is larger. In 1985 the minimum ozone concentration in the hole was 30% lower than it was 5 years earlier.

The major cause of the ozone depletion was suggested in the early 1970s but was generally accepted slowly. Chlorofluorocarbons (CFCs), chemicals that have been manufactured in large amounts since they were invented in the 1920s, largely for use in cooling systems, fire extinguishers, and styrofoam containers, were percolating up through the atmosphere and reducing $O_3$ molecules to $O_2$ molecules. Although other factors have also been implicated in ozone depletion, the role of these CFCs is so predominant that worldwide agreements to phase out their production within the next decade have been signed. Nonetheless, large amounts of CFCs that were manufactured earlier, and which are still being manufactured, are moving slowly upward through the atmosphere so that the problem will grow worse before the ozone layer that protects all life is stabilized once again.

The thinning of the ozone layer in the stratosphere, 25 to 40 kilometers above the surface of the earth, is a matter of serious concern biologically. This layer protects key biological molecules, especially proteins and nucleic acids, from the harmful ultraviolet rays that bombard the earth continuously in huge amounts from the sun. Life on land, in fact, may have become possible only when the oxygen layer was sufficiently thick that it generated enough ozone, in chemical equilibrium with the oxygen, that the surface of the earth was sufficiently shielded from these destructive rays. This factor may account for the three billion years in which all life was aquatic.

Ultraviolet radiation is a serious human health concern. Thus every 1% drop in the atmospheric ozone content is estimated to lead to a 6% increase in the incidence of skin cancers. The drop of approximately 3% that has occurred worldwide, therefore, is estimated to have led to an increase of perhaps as much as 20% in skin cancers, which are one of the more lethal diseases afflicting human beings. Even so, human beings are relatively resistant to increased ultraviolet radiation, but other organisms, such as the highly susceptible photosynthetic plankton species that are so important to global productivity, are apparently much more highly susceptible.

## ENVIRONMENTAL SCIENCE

**Environmental scientists** attempt to find solutions to environmental problems, considering them in a broad context. Unlike biology or ecology, sciences that seek to learn general principles about how life functions, environmental science is an applied science, dedicated to solving problems. Its basic tools are derived from the sciences of ecology, geology, meteorology (the study of climate), the social sciences, and the many other areas of knowledge that bear on the functioning of the environment and our management of it. Environmental science addresses the problems created by rapid human population growth: an increasing need for energy, a depletion of resources, and a growing level of pollution. These problems are both unavoidable and obvious. Environmental science focuses on solving them.

The problems faced by our severely stressed planet (Figure 23-27) are not insurmountable. A combination of scientific investigation and public action, when brought to bear effectively, can solve environmental

# A World at Risk

**The World at risk.**
1 Acid rain and fog.
2 Drought.
3 Pollution.
4 Overpopulation.
5 Destruction of the rain forest.

# Endangered Species

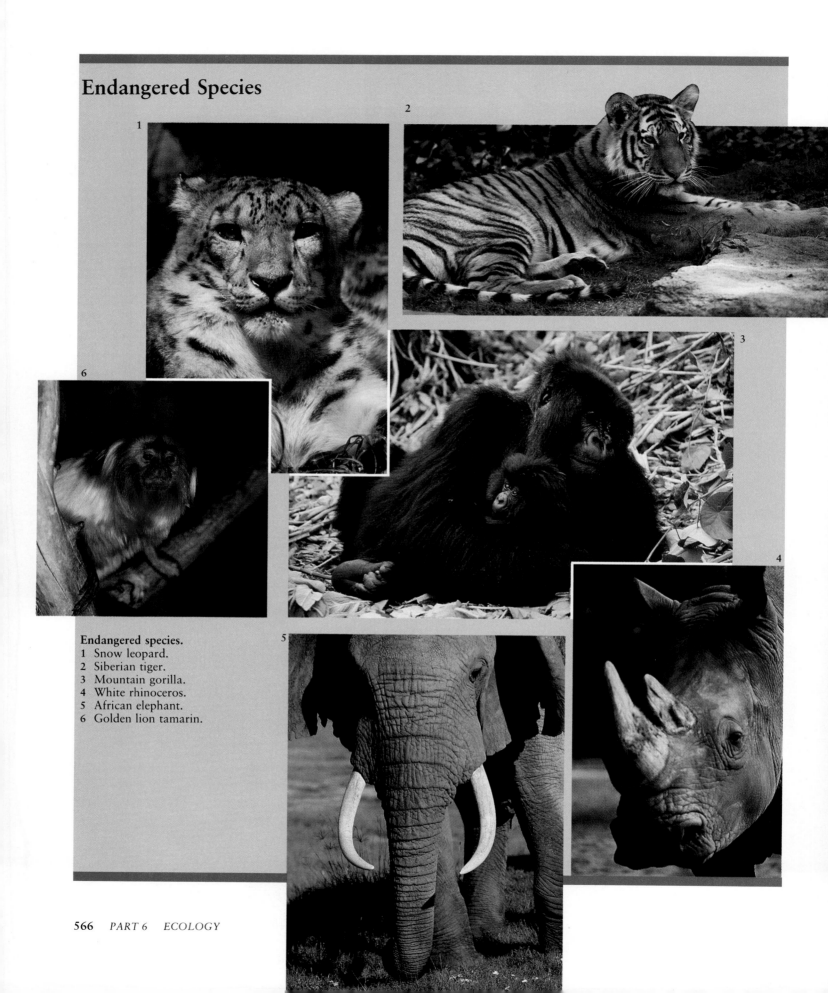

**Endangered species.**
1 Snow leopard.
2 Siberian tiger.
3 Mountain gorilla.
4 White rhinoceros.
5 African elephant.
6 Golden lion tamarin.

**Figure 23-27**

**Air pollution.** The smog evident in this photograph of the Los Angeles basin is kept within the basin by a temperature inversion.

problems that seem intractable. How is success to be achieved? Viewed simply, there are five components to solving any environmental problem:

1. **Assessment.** The first stage in addressing any environmental problem is scientific analysis, the gathering of information. Data must be collected and experiments performed in order to construct a "model" that describes the situation. Such a model can be used to make predictions about the future course of events.
2. **Risk analysis.** Using the results of scientific analysis as a tool, it is then possible to analyze the **environmental impact,** the potential effects of environmental intervention—what could be expected to happen if a particular course of action were followed. It is necessary to evaluate not only the potential for solving the environmental problem, but also any adverse effects that a plan of action might create.
3. **Public education.** When a clear choice can be made among alternative courses of action, the public must be informed. This involves explaining the problem in terms the public can understand, presenting the alternative actions available, and explaining the probable costs and results of the different choices.
4. **Political action.** The public, through its elected officials, makes a choice, selecting a course of action and implementing it. The choices are particularly difficult to implement when environmental problems transcend national boundaries.

5. **Follow-through.** The result of any action taken should be carefully monitored, both to see if the environmental problem is being solved and, more basically, to evaluate and improve the initial evaluation and modeling of the problem. Every environmental intervention is an experiment, and we need the knowledge gained from each one for future applications.

## WHAT BIOLOGISTS HAVE TO CONTRIBUTE

The development of appropriate solutions to the world's environmental problems must rest partly on the shoulders of politicians, economists, bankers, engineers—many different kinds of people. But a large proportion of the basis for solving these problems permanently and achieving a stable, productive world is biological. We have reviewed a number of biological solutions to environmental problems in this chapter.

The energy that bombards the earth comes in essentially inexhaustible amounts from the sun; living systems capture that energy and fix it in molecules that can be utilized by all organisms to sustain their life processes. We and all other living beings depend on the proper utilization of that renewable energy and of the other materials on which our civilization depends. We must come to understand better the processes involved and to develop new systems that will work efficiently in areas where they are not available now. That is our challenge for the future. It is a challenge that must be

met soon. In many parts of the world, the future is happening right now (Figure 23-28).

It is clear that a scientific education has become necessary for everyone so that we may understand the basis for our continued existence on earth and the steps that we will need to take to improve the quality of our lives. Biology should play a major part in that education and is of critical importance in improving the standard of living for our fellow human beings. Biological literacy is no longer a luxury for intelligent human beings who want to play a constructive role in improving the world. It has become a necessity for all of us.

**Figure 23-28**

**The future is now.** The girl gazing from this page faces an uncertain future. She is a refugee, an Afghani. The whims of war have destroyed her home, her family, all that is familiar to her. Her expression carries a message about our own future. The message is that our future, like hers, is now on us, and that the problems humanity faces in living on an increasingly unstable, overcrowded, and polluted earth are no longer hypothetical, a dilemma for our children, but are with us today and demanding a solution.

## ■ SUMMARY

1. When the 1990 college class started school in 1977, there were 4 billion people in the world. There are over 5.3 billion people today, and the world's population is increasing by 1.8% a year. At this rate, there will be over 8 billion people living when your children attend college, double the number living in 1977.

2. An explosively growing human population is placing considerable stress on the environment, consuming ever-increasing quantities of food and water, using a great deal of energy and raw materials, and producing enormous amounts of wastes and pollution.

3. Among the many challenges to the environment posed by human activities are releases of materials into the environment that may harm it: escape of radioactive materials into the atmosphere may increase the incidence of cancer, release of pollutants into rivers may make the water unfit for human consumption, release of industrial smoke into the upper atmosphere leads to acid rain that kills forests and lakes, release of chemicals such as chlorofluorocarbons may destroy the atmosphere's ozone and so expose the world to more ultraviolet radiation, and release of carbon dioxide by burning fossil fuels may increase the world's temperature and so alter weather and ocean levels.

4. Other challenges are created by our attempts to "develop" natural resources: the cutting and burning of the tropical forests of the world to make pasture and cropland is producing a massive wave of extinction.

5. All of these challenges to our future can and must be addressed. Today, environmental scientists and concerned citizens are actively searching for constructuve solutions to these problems.

6. The application of the principles of biology to the human condition has never been more necessary than it is now. Only the full attention of society and many talented individuals to the solution of these grave problems will make it possible for our children and grandchildren to enjoy the same benefits that we enjoy.

# REVIEW

1. Will a country with a greater proportion of its population below childbearing age grow faster or slower in the future than a country that has a greater proportion of its population within the childbearing ages?

2. During the next 30 to 40 years, what percentage of the world's species are expected to become extinct?

3. When agriculture first developed 11,000 years ago, there were approximately _____ people in the world.

4. Three major factors leading to destruction of the tropical forests are _____, _____, and _____.

5. At what annual rate is the world population growing?

◆

# SELF-QUIZ

1. The present population of the world is about
   - (a) 500 million.
   - (b) 100 billion.
   - (c) 10 billion.
   - (d) 5 billion.
   - (e) 50 billion.

2. Which of the following have entered cultivation on a global scale since 1800?
   - (a) Wheat
   - (b) Manioc
   - (c) Rubber
   - (d) Peanuts
   - (e) Soybeans

3. We obtain more than half of our food energy from how many different kinds of plants?
   - (a) 200
   - (b) 1800
   - (c) 3
   - (d) 12
   - (e) 20

4. The population of developing countries is growing at a rate that, if sustained, would lead to a doubling within approximately _____ years.
   - (a) 10
   - (b) 20
   - (c) 30
   - (d) 40
   - (e) 50

5. Of the roughly 70% of total net photosynthetic output produced on land, humans consume approximately
   - (a) 5%.
   - (b) 10%.
   - (c) 20%.
   - (d) 30%.
   - (e) 40%.

6. The most important way to alleviate global warming in the near future is
   - (a) nuclear energy.
   - (b) solar energy.
   - (c) wind energy.
   - (d) energy conservation.
   - (e) to install scrubbers in coal-burning plants.

7. The most promising source of additional food for the near future is
   - (a) culturing microorganisms.
   - (b) hydroponic culture.
   - (c) improving existing cultivation.
   - (d) developing new farmlands.
   - (e) farming the sea.

8. The rapidly growing 75% of the people in the world who live in developing countries have approximately what proportion of the world's wealth?
   - (a) 75%
   - (b) 50%
   - (c) 25%
   - (d) 15%
   - (e) 5%

# THOUGHT QUESTIONS

1. If the AIDS epidemic were to kill as many people as did the Black Plague in the fourteenth century (some 75 million people), what effect would this have on the world's population in the year 2020? How many people would have to die in order for the world's population in the year 2020 to be no larger than it is now (for simplicity, assume they all died in 1990)?

2. Decreased birth rates have ultimately followed decreased death rates in many countries; for example, Germany and Great Britain. Do you think this will eventually occur worldwide and solve our population problems?

3. Some have argued that attempts by the United States to promote lowering of the birth rate in the underdeveloped countries of the tropics is no more than economic imperialism and that it is in the best interests of these countries that their populations grow as rapidly as possible. How would you respond to this argument?

# FOR FURTHER READING

BROWN, L., editor: *State of the World, 1990,* W.W Norton & Co., New York, 1990. A highly recommended, easily read summary of the ecological problems faced by an overcrowded and hungry world, with an excellent chapter on suggested solutions. A new edition appears every year.

ELLIS, W.S.: "The Aral. A Soviet Sea Lies Dying," *National Geographic,* February 1990, pages 73-92. Overproduction of cotton led to overuse of water and pollution by fertilizers and pesticides; the Aral Sea has been almost completely destroyed.

MATTHEWS, S. and SUGAR, J.: "Is our world warming?," *National Geographic,* October 1990, pages 66-99. A vivid, well-illustrated discussion of the greenhouse effect, its causes, and the possible consequences of global warming.

NATIONAL WILDLIFE: "What on Earth Are We Doing," vol. 28(2), February-March 1990. An excellent issue of this outstanding magazine, entirely devoted to the state of the environment, that presents many actions that you can take to help solve the problems discussed in this chapter.

REGANOLD, J., and OTHERS: "Sustainable Agriculture," *National Geographic,* June 1990, pages 112-120. Traditional conservation methods combined with modern technology can reduce farmers' dependence on possibly dangerous chemicals.

REPETTO, R.: "Deforestation in the Tropics," *National Geographic,* April 1990, pages 36-42. Government policies that encourage excessive logging and clearing for ranches and farms are to blame for accelerating the destruction of tropical forests.

SCIENTIFIC AMERICAN: "Managing Planet Earth," September 1989. An entire issue devoted to many aspects of the management of our earth, with a number of excellent articles.

VIETMEYER, N.D.: "Lesser-Known Plants of Potential Use in Agriculture and Forestry," *Science,* vol. 232, 1986, pages 1379-1384. An excellent, brief survey of little-known but extremely useful plants that humans could use to improve food and firewood supplies in the future.

WHITE, R.: "The Great Climate Debate," *National Geographic,* July 1990, pages 36-43. The greenhouse effect and the prospect of global warming are the subjects of scientific and political controversy. This thoughtful article asks whether we should take steps now to avoid consequences we cannot foresee.

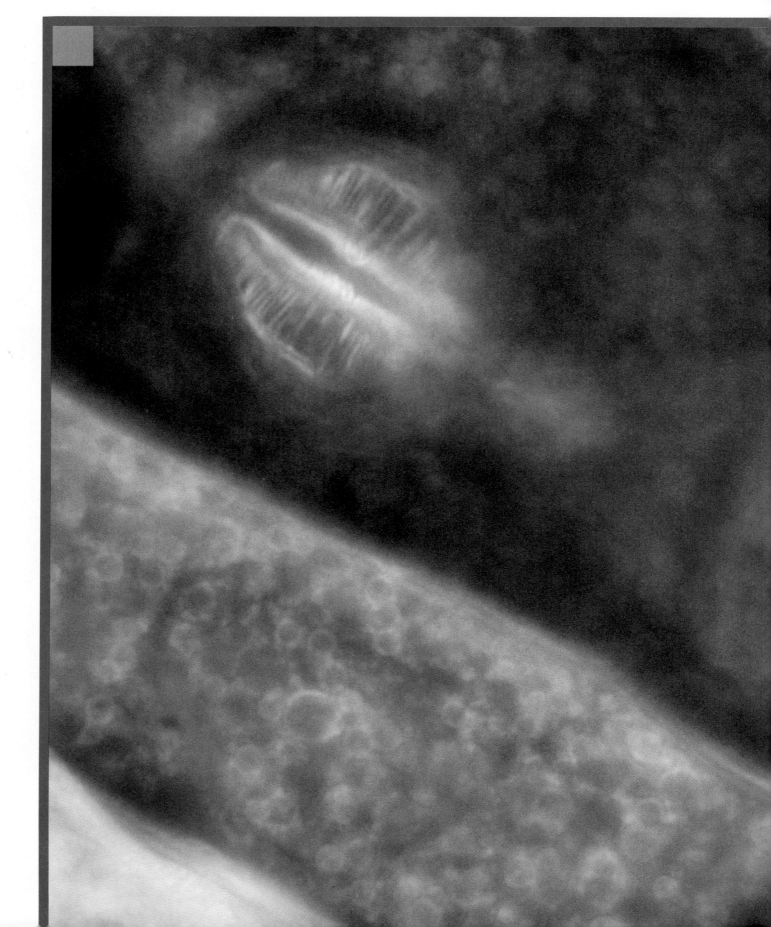

# The Five Kingdoms of Life

The embryo of a tiny rotifer within its parent.
Not all creatures are large or have backbones.
A single drop of water contains a universe of
microscopic life.

# THE FIVE KINGDOMS OF LIFE

## Overview

There are probably at least 10 million different kinds of organisms living on earth, of which only about 1.4 million have been named and catalogued. In order to be able to discuss and study the different kinds of organisms, scientists group them in categories and give those categories particular names. Every organism is assigned to a species, and the species are grouped in genera, which consist of one or more species. The names of organisms are written with the name of the genus first, followed by a distinctive word that, together with the name of the genus, makes up the name of that particular species. In turn, genera are grouped into families, families into orders, orders into classes, classes into phyla (called divisions in plants, fungi, and algae), and phyla into kingdoms. Each of these more inclusive groups has its own distinctive set of properties, so the name of an organism, which indicates its position in the system, tells us a great deal about that organism.

## For Review

*Here are some important terms and concepts that you will encounter in this chapter. If you are not familiar with them, you should review them before proceeding.*

**The five kingdoms of life** (Chapter 2)

**Prokaryotic versus eukaryotic cells** (Chapter 4)

**Definition of a species** (Chapter 17)

**How species form** (Chapter 17)

**Biological classification** (Chapter 18)

All living organisms have a great deal in common. They are all composed of one or more cells, use enzymes to carry out metabolism based on energy transfer via ATP, encode hereditary information in DNA, and have evolved from simpler ancestral forms. In addition, they continue to evolve, and become progressively more complex, and they occur together in populations which have certain properties of their own. These populations make up communities and ecosystems, which provide the overall structure of life on earth. So far in this text, we have stressed these common themes, considering the family of general principles that apply to all organisms. Now it is time to focus on the *differences* between organisms, to consider the diversity of the biological world and the properties of the individual organisms that make it up. For the rest of this text, we examine the parade of life on earth, from the simplest microbes to jellyfish and insects, elephants and bears and redwood trees. Like going to the zoo, examining biological diversity is interesting, challenging, and a great deal of fun (Figure 24-1).

**Figure 24-1**

**Life and death on the forest floor.** This weevil, lying on a leaf in the tropical rain forest of Costa Rica, is a kind of beetle. It has been killed by the fungus whose fruiting structure now rises above it, scattering spores to the four winds. The filaments of the fungus grew through the body of the weevil, killed it, and converted its body into the body of the fungus. There are more than 100,000 named species of fungi and 40,000 named species of beetles, by far the largest family of the animal kingdom.

# THE CLASSIFICATION OF ORGANISMS

Millions of different kinds of organisms exist; to talk about them and study them, it is necessary that they have names. Going back as early as we can trace, organisms have been grouped in basic units, such as oaks, cats, and horses. Eventually, these units, which were often given the same names used by the Romans and Greeks, began to be called **genera** (singular, **genus**). Starting in the Middle Ages, these names were written in Latin, the language of scholarship at that time, or given a Latin form. Thus oaks were assigned to the genus *Quercus*, cats to *Felis*, and horses to *Equus*—names that the Romans applied to these groups of organisms. For genera that were not known to the Romans, of course, new names had to be invented.

*The names of genera were the basic points of reference in classification systems, and these names eventually came to be written in Latin.*

## The Polynomial System

Before the 1750s, scholars usually added a series of additional descriptive terms to the name of the genus when they wanted to designate a particular species. (Both the singular and plural of the word *species* are the same.) These phrases, starting with the name of the genus, made up what came to be known as **polynomials,** strings of Latin words and phrases consisting of up to 12, 15, or even more words. Not only were such polynomials cumbersome, but they also could be altered at will by later authors. As a result, a given organism really did not have a single name that was its alone. Instead, polynomial names were a series of different descriptive phrases that could be related to one another only by scholars. This was a burdensome and imprecise system of naming.

## The Binomial System

The simplified system of naming plants, animals, and microorganisms that has been standard for more than two centuries stems from the works of the Swedish biologist Carl Linnaeus (1707-1778; Figure 24-2). Linnaeus' ambition, like that of some of his predecessors, was to catalog all the kinds of organisms and minerals. In the 1750s he produced several major works that, like his earlier books, employed the polynomial system, which had been well established for nearly a century as the basic system for naming organisms.

**Figure 24-2**

**Carl Linnaeus,** (1707-1778). This Swedish biologist devised the system of naming organisms that is still in use today.

But as a kind of shorthand, Linnaeus also included a two-part name for each species. It is these two-part names, or **binomials,** that have become our standard way of designating them. For example, he designated the willow oak *Quercus phellos* and the red oak *Quercus rubra* (Figure 24-3), even though he also included

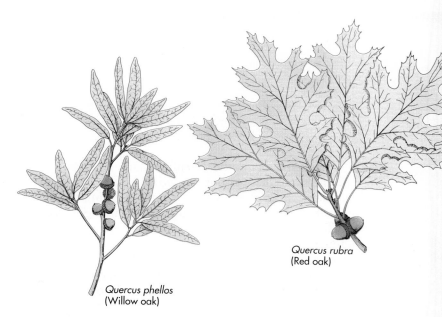

*Quercus rubra*
(Red oak)

*Quercus phellos*
(Willow oak)

**Figure 24-3**

**Two species of oak.** Although they are clearly oaks *(Quercus),* these two species differ sharply in the shapes and sizes of their leaves and in many other features, including their overall geographical distributions.

## A Species Fighting for Survival

The black-footed ferret, *Mustela nigripes,* is one of the most elegant members of the weasel family. It inhabits the plains of the north-central United States, where it preys on prairie dogs, colonial rodents that have been exterminated by poisoning over much of their former extensive range. As a result, black-footed ferrets became so rare that for a time they were feared extinct. One small population, shown here, was discovered by ranchers in September 1981 near Meeteetse, Wyoming. The maximum size of this population included about 129 ferrets in the summer of 1984, but an outbreak of canine distemper killed most of the population in 1985. The 18 ferrets that survived were captured the next year, and there are now over 200 in captivity. When the population has been built up further, attempts will be made to reintroduce the species in nature.

the polynomial names for these species. The convenience of the short, new names proposed by Linnaeus was so great that they were immediately accepted generally.

*In Linnaeus' binomial system the name of a species consists of two parts, the first of which is the generic name.*

By agreement, the scientific names of organisms are the same throughout the world and thus provide a standard and precise way of communicating about organisms, whether the language of a particular scholar is Chinese, Arabic, Russian, or English. In contrast, the names that are given locally often differ greatly from place to place; thus a bear in Australia may be either a koala bear (a marsupial) or one of the large carnivores (placental mammals) more usually known by this name. For general communication, it is important to have one standard set of names. The system of naming plants, animals, and microorganisms established by Carl Linnaeus has served the science of biology well for nearly 250 years.

## WHAT IS A SPECIES?

In Chapter 16 we reviewed the nature of species and saw that there are no absolute criteria that can be applied to the definition of this category. Individuals that belong to a given species—for example, dogs (see Figure 16-2)—may look very unlike one another. Nevertheless, they are generally capable of hybridizing with one another, and the different forms can appear in the progeny of a single mated pair. Species remain relatively constant in their characteristics, can be distinguished from other species, and do not normally interbreed when they come together with other species in nature. For example, dogs are not capable of interbreeding with foxes, which, although they are generally similar, are members of another, completely distinct, group of mammals. In contrast, dogs can and do form fully or partly fertile hybrids with related species such as wolves and coyotes, which are also members of the genus *Canis* (Figure 24-4). The transfer of characteristics between these species has, in some areas, changed the characteristics of both of the interbreeding units.

The criteria just mentioned for species apply primarily to those that regularly **outcross**—interbreed with individuals other than those like themselves. In some groups of organisms, including bacteria and many eukaryotes, **asexual reproduction**—reproduction without sex—predominates. The species of these organisms clearly cannot be characterized in the same

A            B            C

*Figure 24-4*

**A** All dogs, *Canis familiaris,* are capable of interbreeding with one another, regardless of how different they appear.
**B** Coyotes, *Canis latrans,* occur widely in North America.
**C** The grey wolf, *Canis lupus,* inhabits all areas of the Northern Hemisphere. Coyotes and wolves belong to the same genus as dogs, *Canis,* and often produce fertile hybrids with them when their ranges come together. Together with weasels, bears, cats, and other related mammals, all canines belong to the order Carnivora, the carnivores.

way as are the species of outcrossing animals and plants; they do not interbreed with one another, much less with the individuals of other species. Despite these difficulties, biologists generally agree on the kinds of units that they classify as species, although these units share no biological characteristics uniformly.

*Species differ from one another in at least one characteristic and generally do not interbreed freely with one another where their ranges overlap in nature.*

## HOW MANY SPECIES ARE THERE?

Since the time of Linnaeus, about 1.4 million species have been named. This is a far greater number of organisms than Linnaeus suspected to exist when he was developing his system of classification in the eighteenth century. But the actual number of species in the world is undoubtedly much greater, judging from the very large numbers that are still being discovered. **Taxonomists**—scientists who study and classify different kinds of organisms—estimate that there are at least 10 million species of organisms on earth and that at least

two-thirds of these—more than 6 million species—occur in the tropics. Considering that no more than 500,000 tropical species have been named, our knowledge of these organisms is obviously very limited.

## THE TAXONOMIC HIERARCHY

Biological systems of classification are **hierarchical,** with the basic units arranged like boxes within boxes (Figure 24-5). In such a system, grouping genera into the larger, more inclusive categories known as **families** is an effective way to remember and study their characteristics. Families of organisms such as "finches," "legumes," and "mammals" include many genera and have characteristics of their own. Thus the oaks *(Quercus),* beeches *(Fagus),* and chestnuts *(Castanea)* were grouped, along with other genera, into the beech family, Fagaceae, because of the many features they had in common. Similarly, the tree squirrels *(Sciurus),* Siberian and western North American chipmunks *(Eutamias),* and marmots *(Marmota)* were grouped with other related mammals in the family, Sciuridae (Figure 24-6). If one knows that a genus belongs to a particular family, one immediately knows a great many more of its features.

The **taxonomic system** also includes several more inclusive units than families, as mentioned in connection with our survey of the history of life on earth in

# What are the Relatives of the Giant Panda? A Molecular Solution

Ever since the giant panda, a unique mammal of the mountain forests of southwestern China, was brought to the attention of European scientists, scientists worldwide have speculated about its true relationships. Although it was widespread throughout most of central China as recently as 150 to 200 years ago, the giant panda is now confined to 6 widely separated areas amounting in total to 29,500 square kilometers. These areas are mostly in central Sichuan Province, with an outlying area in the Qinling Mountains of Shaanxi Province, to the northeast. The total wild population of giant pandas at present probably numbers as few as 1150 to 1200 individuals. In addition, 13 giant pandas are still alive in zoos outside of China, and 60 to 70 are in zoos within China.

Although Pere David, who provided the first scientific description of the giant panda for European scientists (in 1869), considered it to be a bear, others pointed out that its bones and teeth resembled those of the racoons, a family of mammals from the New World. The argument has raged ever since and, although there is general acceptance of the fact that pandas resemble both bears and raccoons, it has not been possible to reach general agreement about which group of mammals they resemble more closely.

The lesser panda, or red panda *(Ailurus fulgens)*, which is another fascinating Asian mammal, seems similar to giant pandas in many respects, yet it is also similar to the raccoons. Partly because of these relationships, the red panda has often been grouped with the giant panda in a subfamily of the raccoon family, as if they both were kinds of raccoons. While the red panda lacks the remarkable false thumb of the giant panda, an "extra" structure with which it grasps bamboo stalks while eating them, the two are similar in behavior—both feed on bamboos, and they occur in the same general area. We now know the resemblance to be misleading.

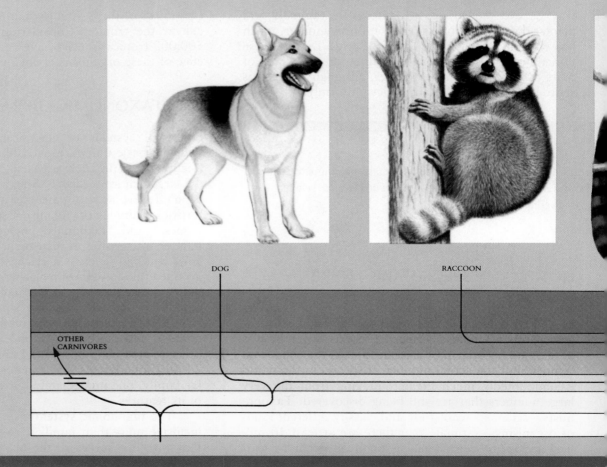

DOG

RACCOON

OTHER CARNIVORES

By matching reactions between blood serum samples from these animals, a comparison that was first made in the 1950s, scientists soon came to the conclusion that the giant panda was indeed related to the bears. By the same methods, the red panda showed up as a separate evolutionary line, independent from, but related to, both bears and raccoons. Zoologist George Schaller, of the New York Zoological Society's Wildlife Conservation International Division, who has probably spent more time studying the behavior of pandas in the wild than anyone else, has concluded on the basis of their behavioral traits that red pandas and giant pandas are closely related to one another, and that they are both more closely related to bears than to raccoons.

Recently, Stephen O'Brien of the U.S. National Cancer Institute and his coworkers at the U.S. National Zoological Park have neatly solved this problem with molecular methods of analysis—DNA hybridization, isozyme similarities, immunological comparisons, and the study of chromosomal morphology after the application of special staining techniques that reveal satellite DNA and other differentiated regions. Their conclusions are illustrated in the diagram. The conclusions clearly indicated that the red panda represents an evolutionary line that diverged from the raccoons fairly soon after they had separated from the bears and that the giant panda diverged from the bears much more recently. As to timing, the original split between raccoons and bears probably occurred 30 to 50 million years ago; that between the red panda and the raccoons about 10 million years later, or about 10 to 20 million years ago. Although the red panda and the giant panda share a number of structural and behavior features, they do not share a common ancestor closer than the common ancestor of the bears and the raccoons. In these studies, modern molecular information has led to the solution of the longstanding evolutionary problem.

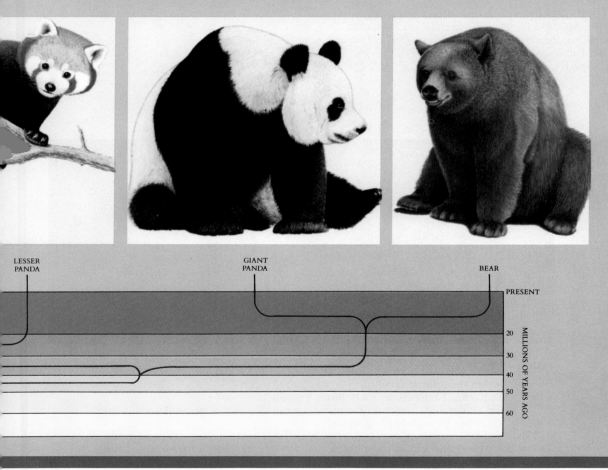

LESSER PANDA

GIANT PANDA

BEAR

PRESENT

20

30

40

50

60

MILLIONS OF YEARS AGO

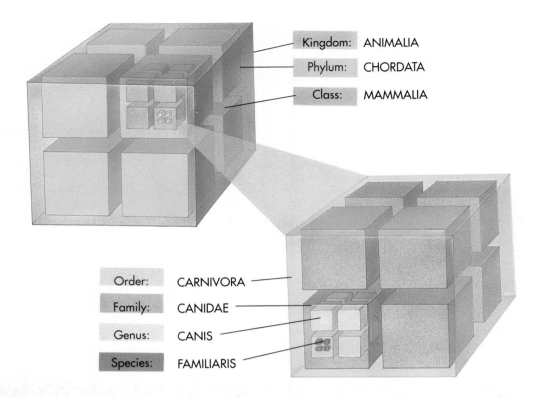

**Figure 24-5**

**The hierarchical system used in classifying an organism, in this case a dog.** Kingdom: Monera, Plantae, Fungi, Protisa, ANIMALIA. Phylum: Anthophyta, Mollusca, CHORDATA, etc. Class: Amphibia, Reptilia, MAMMALIA, etc. Order: Primates, Insectivora, CARNIVORA, etc. Family: Mustelidae, Ursidae, CANIDAE, etc. Genus: *Vulpes, Urocyon, CANIS,* etc. Species: *Canis latrans, C. lupus, C. FAMILIARIS,* etc.

**Figure 24-6**

**Diversity within the squirrel family.**
A   The red squirrel, *Tamiasciurus hudsonicus,* an agile tree dweller.
B   Townsend's chipmunk, *Eutamias townsendii,* a member of a genus of diurnal, brightly colored, very active ground-dwelling squirrels.
C   Yellow-bellied marmot, *Marmota flaviventris,* a large squirrel that lives in burrows. Yellow-bellied marmots are social, tolerant, and playful; they live in harems that consist of a single territorial male together with numerous females and their offspring.

Chapter 18. They function in the same way as do the names of families, making possible more efficient communication about organisms. Families are grouped into **orders,** orders into **classes,** and classes into **phyla** (singular, **phylum**), which are called **divisions** among plants, fungi, and algae. The phyla or divisions are in turn grouped into **kingdoms,** the most inclusive units of classification. To remember the order of the taxonomic categories that are used to classify organisms, you may wish to memorize this phrase: "Kindly Pay Cash Or Furnish Good Security" (kingdom—phylum—class—order—family—genus—species).

*Species are grouped into genera, genera into families, families into orders, orders into classes, and classes into phyla (or divisions). Phyla are the basic units within kingdoms. Such a system is hierarchical.*

Table 24-1 shows, respectively, how the human species, the honey bee, and the red oak are placed in taxonomic categories at the seven different hierarchical levels. The categories at the different levels may include many, a few, or only one item, depending on the nature of the relationships in the particular groups involved. Thus there is only one living genus of the family Hominidae, but, as we have mentioned, several living genera of Fagaceae. Each of the categories at every level implies to someone familiar with the system, or with access to the appropriate books, both a series of characteristics that pertain to that group and also a series of organisms that belong to it.

## THE FIVE KINGDOMS OF ORGANISMS

During the past 50 years, biologists have recognized that the most fundamental distinction among groups of organisms is not that between animals and plants (Figure 24-7), but rather that which separates prokaryotes and eukaryotes. The prokaryotes, or bacteria (Figure 24-8), differ from all other organisms in their lack of membrane-bound organelles and microtubules, as well as in their possession of simple flagella. Their DNA is not associated with proteins and is not located in a nucleus. Bacteria do not undergo sexual recombination in the same sense that eukaryotic organisms do, although forms of genetic recombination occur occasionally in some bacteria. The cell walls of most bacteria contain muramic acid, which is only one of the many biochemical peculiarities by which they differ from all eukaryotes. Bacteria existed for at least 2.5 billion years before the appearance of eukaryotes. In recognition of their highly distinctive features, the bacteria have been assigned to a kingdom of their own, **Monera,** which will be treated in detail in the next chapter.

The features of eukaryotes contrast sharply with those of prokaryotes, as we stressed in Chapter 5 and will now review. All eukaryotes have microtubules in their cytoplasm, and their cells are characterized by classes of membrane-bound organelles. They have a clearly defined nucleus that is enclosed within a double membrane, the nuclear envelope. Their chromosomes are complex structures in which the DNA is characteristically associated with proteins; the chromosomes divide and are distributed in a regular manner to the daughter cells as a result of the process of mitosis. When flagella and cilia are present in eukaryotes, they have a characteristic 9 + 2 structure that consists of microtubules. Mitochondria are among the complex organelles that occur in the cells of all eukaryotes. In

**TABLE 24-1   SAMPLE CLASSIFICATIONS OF THREE REPRESENTATIVE ORGANISMS**

|  | HUMAN BEINGS | HONEY BEE | RED OAK |
|---|---|---|---|
| Kingdom | Animalia | Animalia | Plantae |
| Phylum (division) | Chordata | Arthropoda | Anthophyta |
| Class | Mammalia | Insecta | Dicotyledones |
| Order | Primates | Hymenoptera | Fagales |
| Family | Hominidae | Apidae | Fagaceae |
| Genus | *Homo* | *Apis* | *Quercus* |
| Species | *Homo sapiens* | *Apid mellifera* | *Quercus rubra* |

A     **Figure 24-7**            B

**Kingdoms are the highest level of biological diversity.** For obvious reasons, people have always considered plants and animals to be very different from one another. The southern flying squirrel, *Glaucomys volans,* shown here feeding on berries, appears completely unlike the ladyslipper orchid, *Cypripedium caleolus.* Animals ingest their food, and most move about from place to place; plants manufacture their food through the process of photosynthesis and are stationary. These and other major differences have caused plants and animals to be regarded as distinct kingdoms, with their similarities being appreciated only in the past hundred years or so.

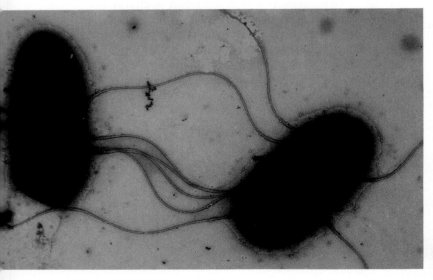

**Figure 24-8**

**The kingdom monera.** Bacteria, such as these *Escherichia coli,* constitute the kingdom Monera. They are prokaryotes, a term that describes a pattern of cellular organization completely different from that of all other organisms. The slender projections extending from the surfaces of these bacteria are called pili.

addition, eukaryotes exhibit integrated multicellularity of a kind that is not found in bacteria, as well as true sexual reproduction.

Despite their often bewildering diversity of form, eukaryotes are much less diverse metabolically than are bacteria. All eukaryotes are believed to have evolved from a common ancestor. Eukaryotes consist primarily of groups of organisms that are single-celled, together with several multicellular groups that were derived from single-celled ancestral eukaryotes (Figure 24-9).

## THE KINGDOMS OF EUKARYOTIC ORGANISMS

The classification of organisms, especially eukaryotes, into kingdoms is clearly somewhat arbitrary, and no scheme has been universally accepted yet. Here we have adopted what is called the **five-kingdom system,** in which four of the kingdoms are eukaryotic organisms. The fifth kingdom, Monera, contains the bacteria, as we mentioned earlier. In the five-kingdom system of classification, the array of diverse eukaryotic phyla consisting predominantly of single-celled organisms is assigned to the kingdom Protista (see Chapter 26). Three major multicellular lines that have originated from this kingdom, each with a great many species, are assigned to separate kingdoms: Animalia, Plantae, and Fungi (Figure 24-10). Each of the three multicellular kingdoms—plants, animals, and fungi—has evolved from a different ancestor among the Protista. Although the fungi include a few single-celled forms, the yeasts, these have been derived from multicellular ancestral forms.

The members of the three multicellular kingdoms derived from the Protista differ in their mode of nutrition. Thus animals ingest their food, plants manufacture it, and fungi digest it by means of enzymes that they secrete and then absorb it. Fungi are filamentous and have no moveable cells; plants are mainly stationary, but some of them have moveable sperm; and many animals are highly motile.

*We recognize five kingdoms of organisms. Two of them, Monera and Protista, are predominantly unicellular; the other three kingdoms, Fungi, Animalia, and Plantae, are multicellular and are derived from individual groups of protists.*

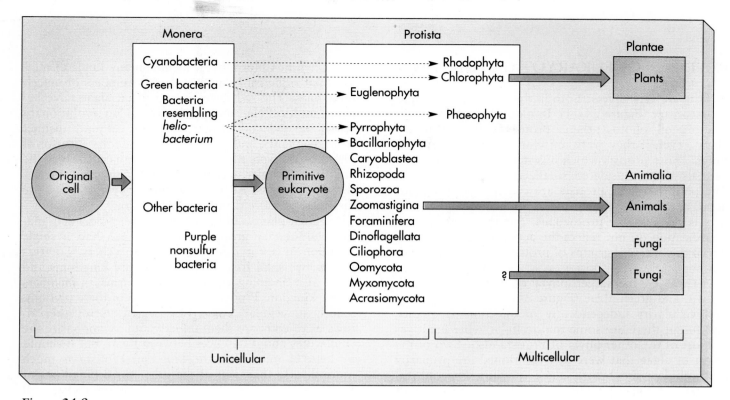

*Figure 24-9*

**Evolutionary relationships of the five kingdoms.** The *solid lines* indicate evolutionary relationships, the *dotted lines* indicate symbiotic events.

*Figure 24-10*

**The multicellular kingdoms.** Animals, fungi, and plants are the three multicellular kingdoms derived from members of the kingdom Protista.
**A** Resplendent quetzal, *Pharomachrus mocinno,* eating the fruits of *Ocotea,* a plant in the laurel family, in Costa Rica.
**B** *Aphelandra tridentata,* a plant of the acanthus family; Costa Rica. Its red flowers are attractive to hummingbirds.
**C** Stinkhorn fungus, *Dictyophora,* on the forest floor in Cameroon, Central Africa.

# FEATURES OF EUKARYOTIC EVOLUTION

Protists are very diverse both in their form and in their biochemistry (Figure 24-11). In older systems of classification, all photosynthetic protists were considered plants, even if they were unicellular, and so were the fungi. Those protists which ingested their food like the animals were considered small, very simple animals. The photosynthetic protists are called **algae** (singular, **alga**), and the traditional name for heterotrophic protists is **protozoa** or **protozoans**. But these schemes of classification do not reflect the actual relationships between the different groups of protists, as do more modern ones.

Three of the larger phyla of protists—the red, brown, and green algae (Figure 24-12)—have attained multicellularity independently of the animals, plants, and fungi; there are some multicellular representatives in most of the other phyla of Protista also. Two of the phyla of algae that were just mentioned are primarily multicellular; the brown algae (phylum Phaeophyta) and the red algae (phylum Rhodophyta). The green al-

gae (phylum Chlorophyta) include many kinds of multicellular organisms and even larger numbers of unicellular ones. The red, brown, and green algae have become multicellular independently. In an evolutionary sense, the red and brown algae are certainly distinct from the plants, and green algae include the ancestor of plants. However, the green algae are basically aquatic and much simpler in form than are plants, so we treat them as protists. All three of these phyla of algae are discussed in more detail in Chapter 26.

Fungi, plants, and animals are clearly defined, major evolutionary groups that have each had a single common ancestor different from the others. For this reason, we shall treat them as separate kingdoms and leave all remaining eukaryotic organisms as members of the kingdom Protista. Grouping all of these predominantly unicellular, eukaryotic organisms together allows us to compare them directly and to emphasize the similarities and differences between them. But it should be kept in mind that the kingdom Protista is much more diverse than any of the other three kingdoms of eukaryotic organisms because this kingdom includes all

A
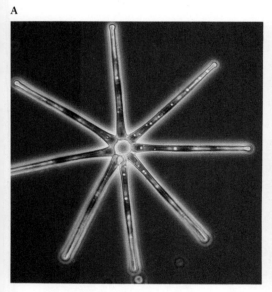

B
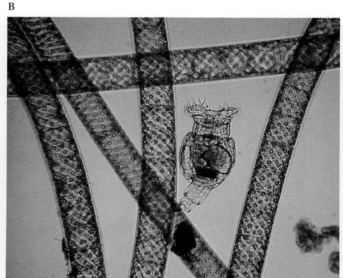

C
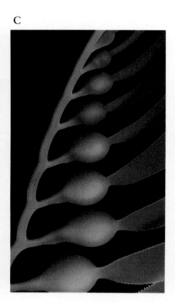

*Figure 24-11*

**Protista, the most diverse kingdom.** The members of the kingdom Protista are very diverse, as suggested by these photographs.
**A** A freshwater diatom, member of a mostly photosynthetic, unicellular group of organisms that occurs widely in fresh and salt water and is characterized by silica shells.
**B** A green alga, *Spirogyra*, a threadlike organism with a single ribbon-shaped spiral chloroplast, by means of which it manufactures its own food, with a rotifer, a small, motile animal that captures and ingests its food.
**C** A kelp, one of the brown algae, a member of a completely multicellular division (phylum) of the kingdom Protista. The individual kelp may be many meters in length. Brown, red, and green algae are not classified as plants, even though they are all photosynthetic, because they were derived independently from different ancestors among the Protista.

## Figure 24-12

**The multicellular protists.** Three phyla of algae have become multicellular independently and differ greatly in their photosynthetic pigments and structure. Different groups of bacteria seem to have given rise to the chloroplasts of each of these phyla of algae as a result of ancient symbiotic events.
**A** The brown algae are large, complex algae that often dominate northern, temperate seacoasts.
**B** *Ulva* is a green alga, a group that includes many unicellular and multicellular organisms.
**C** Some red algae are shown here growing on sponges. Members of this phylum occur at greater depths in the sea than do the members of either of the other phyla.

of the evolutionary lines of eukaryotes except for the three multicellular ones that are recognized as distinct kingdoms. The characteristics of the five kingdoms are outlined in Table 24-2.

*Of the four kingdoms of eukaryotes, the most diverse is Protista, the protists. Probably every phylum of Protista has at least some multicellular representatives included in it, and three of the groups—the red, brown, and green algae—are completely or largely multicellular.*

## SYMBIOSIS AND THE ORIGIN OF THE EUKARYOTIC PHYLA

As discussed in Chapter 4, two of the principal organelles of eukaryotic cells, mitochondria and chloroplasts, share a number of unusual characteristics with each other and with the bacteria from which they were apparently derived.

As you saw in Chapter 17, the first eukaryotes appeared about 1.5 billion years ago; their descendants soon become abundant and diverse. With a very few exceptions, all modern eukaryotes possess mitochondria, and it is therefore clear that these organelles were acquired early in the history of the group. Mitochondria are most similar to nonsulfur purple bacteria, the group that probably gave rise to them originally (see Chapter 4).

In contrast to mitochondria, which are relatively uniform from group to group, chloroplasts fall into three classes that are distinct in their biochemistry. Each of these three classes seems to have been derived from a different bacterial ancestor (see Chapter 25). Even today, so many bacteria and unicellular protists are symbiotic that the incorporation of smaller organisms with desirable features into cells does not appear to be a difficult process. The symbiotic events that gave rise to the chloroplasts of different groups of protists seem to have taken place independently, and the same groups of bacteria appear to have been involved more than once.

*Eukaryotic cells acquired chloroplasts by symbiosis not once, but at least several times during the evolution of different groups of protists.*

## THE EVOLUTION OF MULTICELLULARITY AND SEXUALITY

### Multicellularity

True **multicellularity,** a condition in which the activities of the individual cells are coordinated and the cells themselves are in contact, is a property of eukaryotes alone and one of their major characteristics. The cell walls of bacteria occasionally adhere to one another,

**TABLE 24-2    CHARACTERISTICS OF THE FIVE KINGDOMS**

| KINGDOM | CELL TYPE | NUCLEAR ENVELOPE | MITOCHONDRIA | CHLOROPLASTS | CELL WALL |
|---|---|---|---|---|---|
| Monera | Prokaryotic | Absent | Absent | None (photosynthetic membranes in some types) | Noncellulose (polysaccharide plus amino acids) |
| Protista | Eukaryotic | Present | Present | Present (some forms) | Present in some forms, various types |
| Fungi | Eukaryotic | Present | Present | Absent | Chitin and other noncellulose polysaccharides |
| Plantae | Eukaryotic | Present | Present | Present | Cellulose and other polysaccharides |
| Animalia | Eukaryotic | Present | Present | Absent | Absent |

*Figure 24-13*

**On the road to multicellularity: a colonial protist.** Individual motile, unicellular green algae are united in *Volvox* as a hollow bowl of cells that moves by means of the beating of the flagella of its individual cells. A few cells near the posterior (rear when moving) end of the colony are reproductive cells, but most are relatively undifferentiated. Some species of *Volvox* have cytoplasmic connections between the cells that function in the coordination of the activities of the colony. *Volvox* represents an unusual form of multicellularity that evolved independently of all other multicellular organisms.

and bacterial cells may also be held together within a common sheath. Consequently, some bacteria form filaments, sheets, or three-dimensional bodies, but the degree of integration of individual bacteria within these groups of cells is much more limited than is the case in multicellular eukaryotes.

Multicellularity evolved numerous times among the protists. The brown, green, and red algae all evolved multicellularity independently, as did animals, plants, and fungi. The evolution of multicellularity allowed organisms to deal with their environments in novel ways (Figure 24-13). Distinct types of cells, tissues, and organs can be differentiated within the complex bodies of multicellular organisms. With such functional division within its body, a multicellular organism can protect itself, move about, seek mates and prey, and carry out other activities on a scale and with a complexity that would have been impossible for its unicellular ancestors. With all of these advantages, it is not surprising that multicellularity has arisen independently so many times.

*Multicellularity evolved numerous times among the protists. All of the complex differentiations that we associate with advanced life forms depend on multicellularity.*

| MEANS OF GENETIC RECOMBINATION, IF PRESENT | MODE OF NUTRITION | MOTILITY | MULTICELLULARITY | NERVOUS SYSTEM |
|---|---|---|---|---|
| Conjugation, transduction, transformation | Autotrophic (chemosynthetic or photosynthetic or heterotrophic) | Bacterial flagella, gliding or nonmotile | Absent | None |
| Fertilization and meiosis | Photosynthetic or heterotrophic, or combination of these | 9 + 2 cilia and flagella ameboid, contractile fibrils | Absent in most forms | Primitive mechanisms for conducting stimuli in some forms |
| Fertilization and meiosis | Absorption | Nonmotile | Present in most forms | None |
| Fertilization and meiosis | Photosynthetic, chlorophylls *a* and *b* | 9 + 2 cilia and flagella in gametes of some forms, none in most forms | Present in all forms | None |
| Fertilization and meiosis | Digestion | 9 + 2 cilia and flagella, contractile fibrils | Present in all forms | Present, often complex |

## Sexuality

The second major characteristic of eukaryotic organisms as a group is sexuality. Although some interchange of genetic material does occur in bacteria (see Chapters 4 and 25), it is certainly not a regular, predictable mechanism in the same sense that sex is in eukaryotes. The regular alternation between **syngamy**—the union of male and female gametes—and meiosis constitutes the **sexual cycle** that is characteristic of eukaryotes. It differs sharply from anything found in bacteria.

Some eukaryotes are haploid all their lives, but, with few exceptions, animals and plants are diploid at some stage of their lives. In the cells of animals and plants there are two sets of chromosomes, derived respectively from their male and female parents. These chromosomes segregate regularly by the process of meiosis; because meiosis involves crossing-over, no two products of a single meiotic event are ever identical. As a result, the offspring of sexual, eukaryotic organisms vary widely, thus providing the raw material for evolution.

In many of the unicellular phyla of protists, sexual reproduction occurs only during times of stress. Meiosis may have evolved originally as a means of producing new, well-adapted forms that would increase the chances of survival during such times. The first eukaryotes were probably haploid, and diploids seem to have arisen on a number of separate occasions by the fusion of haploid cells, which then eventually divided by meiosis.

*Sexual organisms are able to evolve rapidly in relation to the demands of their environment because they produce variable progeny. Sexual reproduction, involving the regular alternation between syngamy and meiosis, is the process that makes their evolutionary adjustment possible.*

Eukaryotes are characterized by three major types of life cycles. In the simplest of these, the zygote is the only diploid cell. Such a life cycle is said to be characterized by **zygotic meiosis** because the zygote immediately undergoes meiosis. In most animals the gametes are the only haploid cells. They exhibit **gametic meiosis;** meiosis produces the gametes, which fuse, giving rise to a zygote. In plants there is a regular alternation between a multicellular haploid phase and a multicellular diploid phase. The diploid phase produces spores that give rise to the haploid phase, and the haploid phase produces gametes that fuse to give the zygote. The zygote is the first cell of the multicellular diploid phase. This kind of life cycle is characterized by **alternation of generations** and has **sporic meiosis.** These three major types of eukaryotic life cycles are diagrammed in Figure 24-14.

*In a life cycle characterized by zygotic meiosis, the zygote is the only diploid cell; in one characterized by gametic meiosis, the gametes are the only haploid cells. In plants there is an alternation of generations, in which both diploid and haploid cells divide by mitosis; their type of life cycle is characterized by sporic meiosis.*

## ORGANISMS AND EVOLUTION

All of the groups we have discussed so far are clearly organisms, but there is another group that lies on the borderline between living and nonliving: the viruses. Viruses present a special classification problem. They are nonliving but capable of replication. They are basically fragments of nucleic acids probably derived from ancient cells. They occur in both eukaryotes and bacteria. Most of them have the capacity to organize protein coats around themselves (Figure 24-15). Viruses are able to direct the machinery of their host cells to manufacture more virus material, but they cannot exist on their own. They cannot logically be placed in any of the kingdoms because they are not organisms. Viruses are discussed in Chapter 25.

The remaining chapters of this section present a detailed discussion of the diversity of the major groups of organisms, stressing their evolutionary relationships

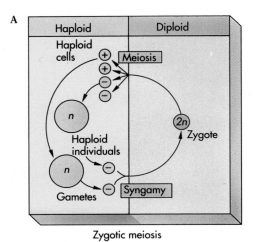

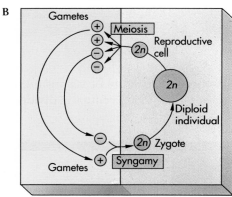

Zygotic meiosis

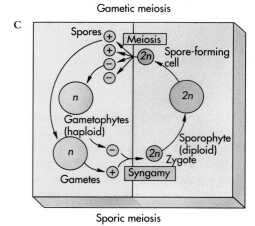

Sporic meiosis

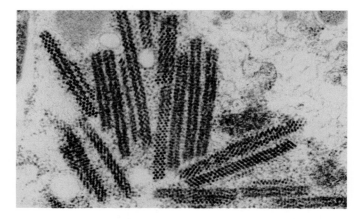

***Figure 24-15***

**Viruses are not organisms.** A kind of virus called GD7 replicating itself in a human white blood cell, or macrophage. Viruses are nonliving segments of DNA or RNA, ultimately derived from the genomes of either bacteria or eukaryotes; they have the ability to take over the protein-synthesizing machinery of their host cells. Many viruses manufacture protein coats around their nucleic acid cores.

***Figure 24-14***

**Diagram of the three major kinds of life cycles in eukaryotes.** Top to bottom: (**A**) zygotic meiosis, (**B**) gametic meiosis, and (**C**) sporic meiosis.

with one another. Multicellularity and sexuality, which we have just discussed, are major themes that underlie the evolutionary advances we will examine. The colonization of the land, which was carried out by multicellular organisms complex enough to function in this harsh habitat, is perhaps the feature that contributed most to the definition of the major groups of organisms that we see today.

## ■ SUMMARY

1. Carl Linnaeus, an eighteenth-century Swedish scientist, developed the system of naming organisms that is still used today. In it, every species of organism has a binomial name, which consists of two words, the first of which is the name of the genus and the second of which, when combined with the name of the genus, designates the species.

2. In the hierarchical system of classification used to describe organisms, genera are grouped into families, families into orders, orders into classes, classes into phyla (called divisions in some groups), and phyla into kingdoms.

3. There are probably at least 10 million species of plants, animals, and microorganisms in the world. We have named only about 1.4 million of them thus far, and only about 500,000 of the estimated 6 million or more species that occur in the tropics.

4. The fundamental distinction among living organisms is that which separates the bacteria (kingdom Monera), which have a prokaryotic form of cellular organization, from the eukaryotes.

5. Most phyla of eukaryotes are placed in the diverse kingdom Protista. Among the phyla of Protista, virtually all have at least some multicellular representatives, and three—red algae (Rhodophyta), brown algae (Phaeophyta), and green algae (Chlorophyta)—are exclusively or largely multicellular.

6. Three groups of multicellular eukaryotes are so large and differ so sharply from their ancestors among the Protista that each is classified as a different kingdom. These are the animals (Animalia), plants (Plantae), and fungi (Fungi).

7. Viruses are not organisms and are not included in the classification of organisms. They are portions of the genomes of organisms.

8. True multicellularity and sexuality are both exclusive properties of the eukaryotes. Multicellularity confers a degree of protection from the environment and the ability to carry out a wider range of activities that is not available to unicellular organisms. Sexuality permits extensive, orderly, genetic reproduction.

## REVIEW

1. _____ _____ devised the binomial system of nomenclature.

2. About how many species of organisms have been named? About how many organisms are estimated to live on the earth?

3. In what way are eukaryotes less diverse than prokaryotes?

4. The three kingdoms that consist exclusively or nearly so of multicellular organisms have evolved from different members of which eukaryotic kingdom?

5. _____ present a special classification problem because they are nonliving but capable of replication.

## SELF-QUIZ

1. List the taxonomic categories from most inclusive to least inclusive.
   (a) Order
   (b) Kingdom
   (c) Species
   (d) Phylum or division
   (e) Family
   (f) Genus
   (g) Class

2. To which (A) kingdom, (B) phylum, and (C) class do you belong?
   (a) Mammalia
   (b) Arthropoda
   (c) Anthophyta
   (d) Animalia
   (e) Chordata

3. Which kingdom contains only unicellular organisms?
   (a) Monera
   (b) Protista
   (c) Fungi
   (d) Plantae
   (e) Animalia

4. Which three of the following are characteristics shared by all eukaryotes?
   (a) Their DNA is not associated with protein
   (b) They have microtubules in their cytoplasm
   (c) They possess membrane-bound organelles in their cytoplasm
   (d) They have a nuclear envelope
   (e) The cell walls of most of them contain muramic acid

5. Which of the following are not included in any kingdom?
   (a) Viruses
   (b) Bacteria
   (c) Algae
   (d) Oak trees
   (e) Human beings

6. Which of the following taxonomic categories is most directly related to the names given to organisms in Greek and Roman times?
   (a) Family
   (b) Genus
   (c) Species
   (d) Kingdom
   (e) Phylum

7. Which of the following groups do *not* contain multicellular representatives?
   (a) Fungi
   (b) Animals
   (c) Plants
   (d) Bacteria
   (e) Arthropods

8. The kingdom Protista is more diverse than the other kingdoms of eukaryotic organisms because
   (a) they are single units derived from individual groups of Protista.
   (b) symbiosis distinguishes Protista from bacteria.
   (c) bacteria are metabolically adapted to Protista.
   (d) evolution has proceeded longer for Protista.
   (e) Protista inhabit a wider array of natural habitats.

# THOUGHT QUESTIONS

1. Why did Carl Linnaeus and his contemporaries divide all kinds of organisms into plants and animals, when we now believe that the most fundamental distinction is between bacteria and eukaryotic organisms? What have we learned since the time of Linnaeus, and how have we learned it?

2. Is photosynthesis, in an evolutionary sense, a characteristic of bacteria only? Explain your answer.

3. Are viruses organisms? What is the evidence for your answer?

# FOR FURTHER READING

JONES, S.B., and A.E. LUCHSINGER: *Plant Systematics,* ed. 2, McGraw-Hill Book Company, New York, 1986. A good overall account of the field.

MARGULIS, L., and K.V. SCHWARTZ: *Five Kingdoms: An Illustrated Guide to the Phyla of Life on Earth,* W.H. Freeman & Company, San Francisco, 1982. A marvelous account of the diversity of organisms, beautifully illustrated and presented in a logical fashion throughout.

ROSS, H.H.: *Biological Systematics,* Addison-Wesley Publishing Co., Inc., Reading, Mass., 1974. A nicely balanced overview of taxonomy.

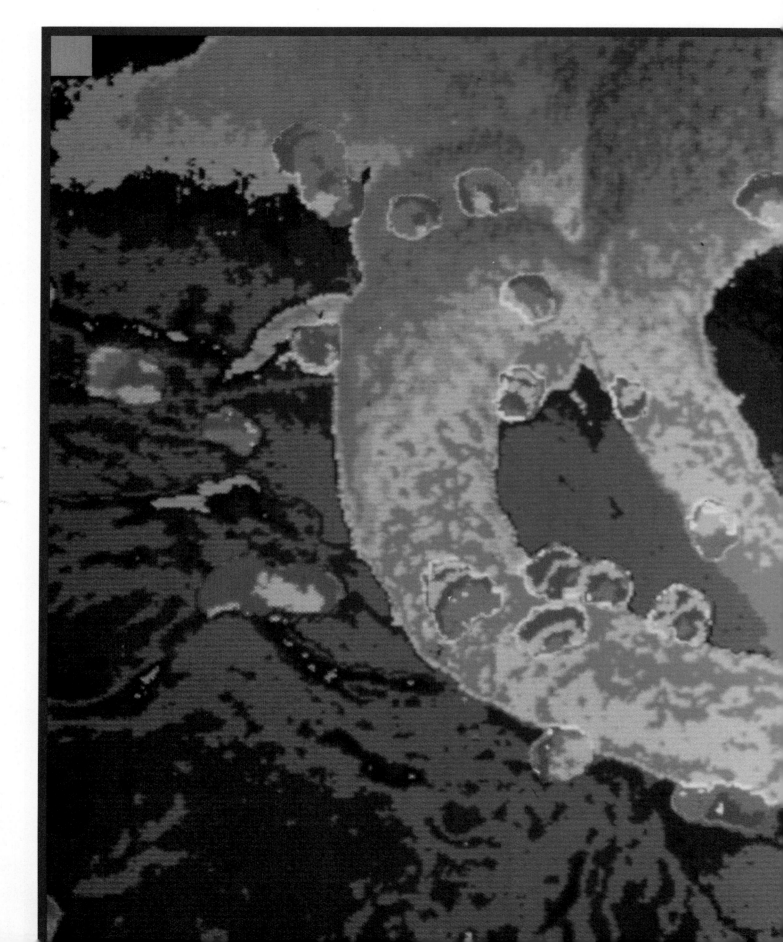

## CHAPTER 25

# The Invisible World: Bacteria and Viruses

The round particles are AIDS viruses, shown here attacking a lymphocyte. The AIDS virus is so deadly because it attacks lymphocytes, the primary cells in the immune system.

# THE INVISIBLE WORLD: BACTERIA AND VIRUSES

## Overview

Invisible to us because they are so small, bacteria and viruses are known to everyone as agents of disease. But the role of bacteria is much broader; our life on earth would not be possible without them because they perform many of the essential functions of ecosystems. Bacteria are the simplest, most ancient, and most abundant of organisms. A virus is a fragment of a nucleic acid, either DNA or RNA, that is able to reproduce within the cells of organisms. Viruses are not alive and are not organisms, but because of their disease-producing potential they are very important biological entities.

## For Review    *Here are some important terms and concepts that you will encounter in this chapter. If you are not familiar with them, you should review them before proceeding.*

**Definition of organism** (Chapter 1)

**Structure of bacterial cell** (Chapter 4)

**Bacterial photosynthesis** (Chapter 8)

**Bacterial cell division** (Chapter 9)

**Nitrogen cycle** (Chapter 21)

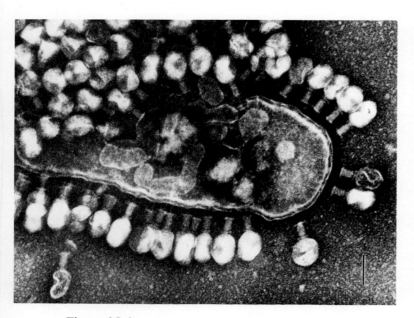

### Figure 25-1

**Viruses invading a bacterium.** T4 bacteriophage (virus) particles surround the end of a cell of a common colon bacterium, *Escherichia coli*. Each particle injects its DNA, a long, coiled molecule within the viral coat, into the bacterium, where the viral DNA directs the production of more virus material by using the machinery of the bacterial cell.

Bacteria are the oldest, the structurally simplest, and the most abundant form of life on earth; they are the only organisms with prokaryotic cellular organization (Figure 25-1). Represented in the oldest rocks from which fossils have been obtained—rocks about 3.5 billion years old—bacteria were abundant for well over 2 billion years before eukaryotes appeared in the world. They were largely responsible for creating the properties of the atmosphere and the soil during the long ages in which they were the only form of life on earth. Bacteria are the only members of the kingdom Monera.

Bacteria are exceedingly abundant: in a single gram of fertile agricultural soil there may be 2.5 billion bacteria. In one hectare of wheat land in England, the weight of bacteria in the soil is approximately equal to that of 100 sheep! It is not surprising, then, that bacteria play a very important part in the web of life on earth. From a human perspective, they are both very useful and very worrisome. On the one hand, bacteria play key roles in the cycling of minerals within our ecosystems, whereas on the other hand, bacteria are responsible for many of the most deadly of human diseases and the most devastating diseases of plants and other animals. Our constant companions, bacteria are present in everything we eat and on everything we touch. No other organisms except plants have as great an impact on human life, for better and for worse.

Although bacteria are the smallest and simplest organisms, they are not the smallest or simplest agents of disease. Viruses, which are segments of nucleic acid, are also often serious agents of disease. Like bacteria, viruses have had a major impact on human history and still do. AIDS, a viral disease that became prominent during the 1980s, is but one familiar example. In this chapter we consider first the bacteria and then viruses (see Figure 25-1).

## BACTERIAL STRUCTURE

We discussed bacterial cell structure in Chapters 3 and 4. The most distinctive characteristic of bacteria is that they lack the internal organization that is the hallmark of the eukaryotic cell. Bacterial cells lack both a nucleus and a system of endoplasmic reticulum. Although the interior of bacterial cells may sometimes contain membranes, such interior membranes do not subdivide the cells into separate compartments, as the endoplasmic reticulum does the interior cytoplasm of eukaryotic cells.

Bacterial cell walls consist of a network of polysaccharide molecules (a polymer of sugars) connected by short peptide cross-links (see Figure 4-6). Bacteria are commonly classified by differences in their cell walls as gram positive and gram negative. The name refers to the Danish microbiologist Hans Christian Gram, who developed a staining process that distinguishes the two classes of bacteria as a way to detect the presence of certain disease-causing bacteria. **Gram-positive** bacteria have a cell wall about 15 to 80 nanometers thick that retains the **gram stain** within the cell, causing the stained cells to appear purple under the microscope. More complex cell walls have evolved in other groups of bacteria. In them, the cell wall is thinner and it does not retain the gram stain; such bacteria are called **gram-negative**. In them, large molecules of **lipopolysaccharide**—a polysaccharide chain with lipids attached to it—are deposited in a lipid bilayer membrane over a thinner wall. Gram-negative bacteria are resistant to many antibiotics to which gram-positive ones, lacking the lipopolysaccharide layer, are susceptible. Beyond these layers, a gelatinous layer, the **capsule,** often surrounds bacterial cells. Bacteria are susceptible to different kinds of antibiotics depending on the structure of their cell walls.

Bacteria are mostly simple in form, varying mainly from straight and rod-shaped (**bacilli**) or spherical (**cocci**) to long and spirally coiled (**spirilla**) (see Figure 4-5). Some bacteria change into stalked structures; grow long, branched filaments; or form erect structures that release the **spores,** single-celled bodies that grow into new bacteria (Figure 25-2). Among the rod-shaped bacteria, some adhere end-to-end after they have divided, forming chains. Still other bacteria form different kinds of cell groups.

Bacteria, like all other living cells, divide. In the case of the bacteria, the mode of division is **binary fission.** In this process, reviewed in Chapter 9, an individual cell simply increases in size and divides in two. The cell membrane and cell wall grow inward and eventu-

### Figure 25-2

**Bacteria are not multicellular, but. . . .**
Although no bacteria are truly multicellular, some adhere to one another.
**A** *Chondromyces crocatus,* one of the gliding bacteria. The rod-shaped individuals move together, forming the composite spore-bearing structures shown here; millions of spores, which are basically individual bacteria, eventually are released from these structures.
**B** *Chroococcus,* a cyanobacterium in which the individuals adhere within a gelatinous capsule in groups of four.

A

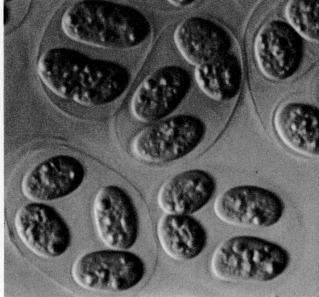

B

ally divide the cell by forming a new wall from the outside toward the center of the old cell (see Figure 9-3).

*Bacterial cells have a very simple structure. The cytoplasm contains no internal compartments or organelles and is bounded by a membrane encased within a cell wall composed of one or more layers of polysaccharide.*

## BACTERIAL ECOLOGY AND METABOLIC DIVERSITY

Bacteria occur in the widest possible range of habitats and play key ecological roles in virtually all of them. Some thrive in hot springs, where the usual temperatures may range as high as 78° C; others have been recovered living beneath 430 meters of ice in Antarctica. Bacteria are abundant in groundwater, where they were once thought to be absent. Still other bacteria, capable of dividing only under high pressures, exist around deep-sea vents, where the water is at temperatures as high as 360° C. These bacteria use the energy they get from the conversion of hydrogen sulfide to sulfur to power their own growth and reproduction, and all other organisms in the vicinity of the deep-sea vents live directly or indirectly by consuming these bacteria (see Figure 22-10).

The reason that bacteria are able to play such a varied ecological role is their metabolic diversity. Different kinds of bacteria vary extensively from one another in almost all aspects of their metabolism, in marked contrast to eukaryotes, which are relatively uniform. All eukaryotes follow the same general patterns of respiration, glucose breakdown, photosynthesis, and synthesis of nucleic acids and proteins. Bacteria are far more diverse in their ways of carrying out each of these vital functions. In view of their abundance and metabolic versatility, it is not surprising that bacteria play a key ecological role in virtually all habitats on earth.

### Photosynthetic Bacteria

Like plants and algae, the photosynthetic bacteria contain chlorophyll, but it is not held within plastids. The photosynthetic processes carried out by different kinds of bacteria are diverse and reflect the long evolutionary path to photosynthesis as it occurs in plants. One group of photosynthetic bacteria, the cyanobacteria (Figure 25-3) (also called *blue-green algae*), was dis-

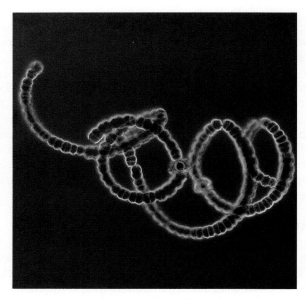

**Figure 25-3**

**Individual cells sometimes adhere in filaments.** The large, clear cell of the cyanobacterium *Anabaena* (the bright object in the middle of the filament) is a heterocyst, an enlarged, specialized cell in which nitrogen fixation occurs. Cyanobacteria such as *Anabaena* exhibit some of the closest approaches to multicellularity among the bacteria.

cussed in Chapters 3 and 17 in connection with the evolution of life on earth. The plastids of the eukaryotic red algae are almost certainly derived from cyanobacterial ancestors. The structure of cyanobacteria is relatively simple. In contrast to those of the red algae, the plastids of the green algae, and therefore those of their descendants the plants, are more similar to bacteria of the genus *Prochloron* (see Figure 4-6). Still other groups of algae, including the brown algae, dinoflagellates, and diatoms, have plastids with different characteristics, suggesting an origin from a different kind of bacterium.

### Chemoautotrophic Bacteria

Chemoautotrophic bacteria derive the energy they use for their metabolism from the oxidation of chemical sources of energy such as the reduced gases ammonia ($NH_3$), methane ($CH_4$), or hydrogen sulfide ($H_2S$). Chemoautotrophs do not require sunlight; in the presence of one of the chemicals just mentioned, they can manufacture all of their own amino acids and proteins. The chemoautotrophs appear to be a very ancient and primitive phylum of bacteria.

### Heterotrophic Bacteria

Most bacteria are heterotrophs, obtaining their energy from organic material formed by other organisms. Once organic material is formed, it must be broken down if it is to be used again by organisms. Bacteria and fungi are the principal organisms that occupy this trophic level, playing the leading role in breaking down the organic molecules that have been formed by biological processes. By doing so, bacteria and fungi make the nutrients in these molecules available once more for recycling. Decomposition is just as indispensible to the continuation of life on earth as is photosynthesis. Obviously, the disease-causing bacteria are also heterotrophic, breaking down living tissue to obtain their nutrients.

## Nitrogen-Fixing Bacteria

Another group of bacteria of great ecological importance is the one that includes the nitrogen-fixing bacteria, such as *Rhizobium,* which lives in nodules on the roots of legumes, and *Frankia,* which forms nitrogen-fixing nodules on the roots of a few kinds of nonleguminous plants. A few other genera of nitrogen-fixing bacteria live free in the soil, but *Rhizobium* is by far the most important of all nitrogen-fixing organisms. It and the other nitrogen-fixing bacteria convert atmospheric nitrogen to a form in which it can be used by living organisms as part of the nitrogen cycle (see Figure 21-4). *Rhizobium* is a member of a group of bacteria that are **aerobic** (require oxygen for growth), gram negative, and mostly motile by means of flagella.

In addition to these heterotropic bacteria, many genera of a photosynthetic group, the cyanobacteria, discussed above in the section on photosynthetic bacteria, have the ability to fix nitrogen. In Asia, rice can often be grown continuously in the same fields without the addition of nitrogen fertilizer because of the presence of nitrogen-fixing cyanobacteria in the paddies. One of the most important nitrogen-fixing genera is *Anabaena,* illustrated in Figure 25-3.

## BACTERIA AS PATHOGENS

Many costly diseases of plants are associated with particular bacteria; almost all kinds of plants are susceptible to one or more kinds of bacterial disease. The symptoms of these plant diseases vary, but they are commonly manifested as spots of various sizes on the stems, flowers, or fruits, or by wilting or local rotting. A familiar example of a bacterial disease is citrus canker, which broke out in Florida in August, 1984, and led to the destruction of over 4 million citrus seedlings in the first 4 months. The ongoing effects of this disease, combined with recent freezes, are threatening the future of the $2.5 billion citrus industry in Florida.

Many human diseases are also caused by bacteria, including typhoid fever, dysentery, plague, cholera, typhus, tetanus, bacterial pneumonia, whooping cough,

and diphtheria. Enormous sums are spent annually in the effort to reduce the likelihood of these infections and their devastating effects. In industrialized countries, it is usually possible to control bacterial infections through the use of antibiotics, but most people in the world do not have access to such drugs and are often the victims of bacterial diseases.

Several genera of **pathogenic** (disease-causing) bacteria are of particular importance for humans. Members of the genus *Streptococcus* (see Figure 4-5, *B*) are associated with strep throat, scarlet fever, rheumatic fever, and other infections (Figure 25-4). Interestingly, the scarlet fever bacterium produces its characteristic and deadly **toxin** (poison) only if it is infected with the appropriate bacterial virus. Tuberculosis, an-

***Figure 25-4***

**Some bacteria are killers.** Jim Henson, creator of the popular television creatures, The Muppets, is shown here with one of them, Kermit the Frog. Henson died suddenly in May, 1990 at the age of 53, the victim of an infection of group A *Streptococcus* bacteria, the same organisms that are responsible for rheumatic fever, scarlet fever, and life-threatening blood diseases. Although pneumonia caused by this particular bacterial strain is rare, other strains of *Streptococcus* afflict about 2 million Americans and claims the lives of 40,000 to 70,000 per year. Once group A *Streptococcus* gets in the blood, it is extremely virulent and very difficult to control with antibiotics.

other bacterial disease, is still a leading cause of death in humans. These diseases are mainly spread through the air.

Another important air-borne microbe is the pathogenic genus *Staphylococcus,* which causes widespread hospital infections. Toxic shock syndrome, caused by bacteria of this genus, is characterized by fever, lowered blood pressure, vomiting, diarrhea, and a rash in which the skin peels. It is caused by certain strains of *S. aureus.* About 85% of the cases of toxic shock syndrome have occurred in menstruating women who are using tampons at the same time the disease occurs, but both men and women can contract the disease. Superabsorbent tampons of a kind that are no longer manufactured contained fibers that provided an environment that enhanced the production of the disease-causing toxin by the bacterium.

One of the more recently detected bacterial diseases of humans is legionellosis (Legionnaire's disease), which is believed to affect about 125,000 people in the United States annually. The disease develops into a severe form of pneumonia that proves fatal to about 15% to 20% of its victims if it is untreated. Discovered in 1976, when a mysterious lung ailment proved fatal to 35 American Legion members attending a convention in Philadelphia, legionellosis is caused by small, flagellated, rod-shaped bacteria with pointed ends that have been given the name *Legionella* (Figure 25-5). These bacteria are common in water, preferring warm water at about 40° to 50° C; they can be spread through air-conditioning units that are not cleaned regularly enough. In the human body, they destroy the monocytes, a type of white blood cell that normally plays a major defensive role against most microorgan-

## The Eradication of Smallpox

One of the greatest triumphs of modern medicine has been the eradication of smallpox everywhere in the world. This dread disease killed millions of people, and when first introduced to new populations, as in the Spanish conquest of Mexico, it sometimes eliminated half or more of the total population. Fortunately there are no known hosts of the smallpox virus other than human beings, so when all of the people who were susceptible in all of the areas where smallpox still occurred were inoculated in the late 1970s, the disease was eliminated completely.

Officials of the World Health Organization, established in 1948 as a specialized agency of the United Nations, noted that smallpox had already been eliminated in North America and Europe. By 1959 the disease had been eliminated throughout the Western Hemisphere, except for five South American countries, and an intensive worldwide campaign was initiated. As late as 1967, smallpox was still endemic in 33 countries, and the campaign appeared

to be faltering. Countries throughout the world, including the Soviet Union and the United States, manufactured and donated large quantities of vaccine, and improved methods of vaccination were developed.

Although reporting was poor, there were probably 10 to 15 million cases of smallpox worldwide in 1967.

Attention was focused on areas in which cases had actually occurred. The last case of smallpox on the Indian subcontinent was contracted by a 3-year-old girl on October 16, 1975. By 1978, no more cases were reported in Somalia, the last country of the world in which the scourge of smallpox persisted, and none have been reported since, anywhere in the world (Figure 25-A).

### Figure 25-A

**The last smallpox victim.** Ali Maow Maalin, of Merka, Somalia, contracted the last known case of smallpox reported anywhere in the world in 1977. At the time, he was 23 years old.

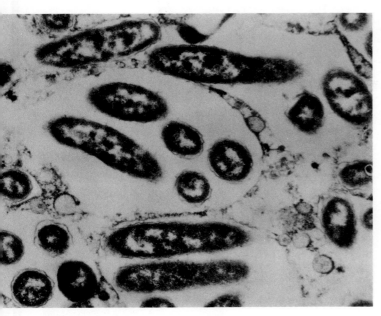

**Figure 25-5**

**Bacteria that causes the disease legionellosis.** The dark, oval objects in this photograph are bacteria of the genus *Legionella*.

isms. The bacteria are gram negative and can be destroyed by treatment with the antibiotic erythromycin.

A number of important bacterial diseases are sexually transmitted. Among the most common are gonorrhea, caused by the bacterium *Neisseria gonorrhoeae*, and syphilis, caused by *Treponema pallidum*, a corkscrew–shaped bacterium of the group called the spirochaetes (see Figure 4-5, C). As you saw in Chapter 4, spirochaetes are thought by some scientists to have given rise to centrioles following their incorporation into eukaryotic cells. At any rate, both gonorrhea and syphilis were easily controlled until recently by antibiotics such as penicillin. But the appearance of penicillin-resistant strains of gonorrhea has made the treatment of this disease much more difficult than was the case formerly. Gonorrhea is much more common and less serious than syphilis, which can be fatal; gonorrhea infected about 500 people per 100,000 in the United States each year during the 1980s, syphilis fewer than 20. Gonorrhea can be detected in men easily because of the discharge of pus from the penis and burning sensations during urination; in women it is more difficult to detect, producing at most very mild symptoms. However, if untreated, it can lead to infection and inflammation of the oviducts, which can cause their blockage and consequently sterility. Syphilis generally produces a hard, painful ulcer called a **chancre** within 3 weeks of infection that soon disappears. From

2 to 4 months later, there is a generalized skin rash, following which the disease may become inactive or may ultimately produce damage to the nervous system or circulatory system. In syphilis, as in gonorrhea, antibiotic resistance has recently become a serious problem.

More common than either syphilis or gonorrhea are **chlamydial infections** caused by the bacterium *Chlamydia trachomatis*. These infections, which are usually relatively mild, are controllable with the antibiotic tetracycline. But if they are left untreated, they can also cause serious complications.

A particularly troublesome bacterial disease that is now spreading throughout the United States is Lyme disease, caused by the spirochaete *Borrelia burgdorferi*. The disease, named for the village of Old Lyme, Connecticut, where it was first recognized, is an inflammatory ailment of humans that is spread by ticks. The symptoms of Lyme disease somewhat resemble those of arthritis. The juvenile stages of the tick *Ixodes dammini*, which is the primary vector of the disease, attach themselves to white-footed mice and occasionally to humans. For many victims, the first sign of infection is a rash that resembles a bull's-eye, which may appear up to a month after the tick's bite. At the end of the summer, when the ticks are adults, they move to different mammals, especially (in the eastern United States) white-tailed deer, and mate on the host. The ticks also infect birds and are presumably spread over wide distances in this way. Fortunately, Lyme disease, if properly diagnosed, can be treated effectively with penicillin or tetracycline.

One human bacterial disease—dental caries, or cavities—affects almost everyone. The disease arises in the film on our teeth, which is called **dental plaque**; this film consists largely of bacterial cells surrounded by a polysaccharide layer. Most of the bacteria are filamentous cells, classified as *Leptotrichia buccalis*, which extend out perpendicular to the surface of the tooth, but many other bacterial species are also present in the plaque. Tooth decay is caused by the bacteria that are present in the plaque, which persists especially in places that are difficult to reach with a toothbrush. Diets that are high in sugars are especially harmful to the teeth because lactic-acid bacteria (especially *Streptococcus sanguis* and *S. mutans*) ferment the sugars to lactic acid, a substance that causes the local loss of calcium from the teeth. Once the hard tissue has started to break down, the breakdown of proteins in the tooth enamel starts, and tooth decay begins in earnest. Fluoride makes the teeth more resistant to decay because it retards the loss of calcium. Because tooth decay is an infectuous disease caused by bacteria, it can, in principle, be controlled by antibiotics; a number of studies are now underway that bear on the problem.

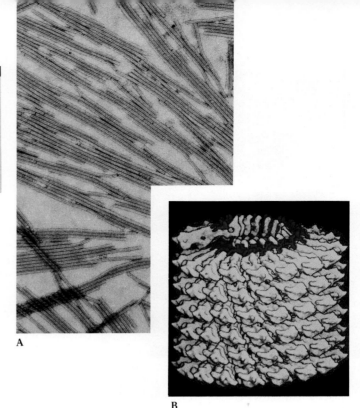

*Bacteria are important disease-causing organisms in plants and animals, including humans. Among the many human diseases that they cause are scarlet fever, rheumatic fever, tuberculosis, typhoid, and legionellosis.*

## VIRUSES

The simplest organisms living on earth today are bacteria, and we think that they closely resemble the first living organisms that evolved on earth. Even simpler than the bacteria, however, are the viruses. The earliest indirect observations of viruses, other than simply observations of their effects, were made near the end of the nineteenth century. At that time, several groups of European scientists, working independently, concluded that the infectious agents associated with a plant disease known as tobacco mosaic and those associated with hoof-and-mouth disease in cattle were not bacteria. They reached this conclusion because these infectious agents were not filtered out of solutions by the kinds of fine-pored porcelain filters that were routinely used to remove bacteria from various media. Investigating the properties of the filtered material, the scientists found that not only were viruses much smaller than any known bacteria, but also that viruses could reproduce themselves only within living host cells and therefore lacked some of the critical machinery by which living cells are able to reproduce themselves.

## ARE VIRUSES ALIVE?

For many years after their discovery, viruses were regarded as very primitive forms of life, perhaps the ancestors of bacteria. We now know this view to be incorrect. The true nature of viruses first became evident in 1933 when Wendell Stanley, then of the Rockefeller Institute, prepared an extract of tobacco mosaic virus and purified it. Surprisingly, the purified virus precipitated (separated out of solution) in the form of crystals (Figure 25-6, A). Stanley was able to show by this method that viruses can better be regarded as chemical matter than as living organisms, at least in any normal sense of the word "living." The purified crystals still retained the ability to infect healthy tobacco plants and so clearly *were* the virus itself, not merely a chemical derived from it. With Stanley's experiments, scientists began to understand the nature of viruses for the first time.

*Figure 25-6*

**Tobacco mosaic virus.**
**A** Virus particles in a crystalline array.
**B** Computer-generated model of a portion of tobacco mosaic virus. An entire virus consists of 2130 identical protein molecules—the yellow knobs, which form a cylindrical coat around a single strand of RNA, colored red in this model. This model is based on x-ray analyses of the virus structure.

Within a few years, other scientists were able to follow up on Stanley's discovery and demonstrate that tobacco mosaic virus consisted simply of an RNA molecule surrounded by a coat of protein molecules (Figure 25-6, B; see also Figure 12-7). Many plant viruses have a similar composition, but most other viruses have DNA in place of RNA. Nearly all viruses form a protein sheath, or **capsid**, around their nucleic acid core. In addition, many viruses form an **envelope**, rich in proteins, lipids, and glycoprotein molecules, around the capsid. The only exceptions are **viroids**, small, naked RNA molecules that affect some plant cells.

*Most viruses form a protein sheath, or capsid, around their nucleic acid core and a lipid-rich protein envelope around the capsid.*

No viruses have the ability to grow or replicate on their own. Viruses reproduce only when they enter cells and utilize the cellular machinery of their hosts (see Figure 25-1). They are able to reproduce themselves in this way because they carry genes that are translated into proteins by the cell's genetic machinery, leading to the production of more viruses. Outside of its host cell, a virus is simply a fragment of nucleic acid encased in protein, a helpless portion of a cellular genome. The fact that individual kinds of viruses contain only a single type of nucleic acid, either DNA or RNA, is one of the major reasons that they can reproduce only within living cells. All true organisms contain both DNA and RNA, and both are essential components of their genetic machinery. Viruses also lack ribosomes and all of the enzymes necessary for protein synthesis and energy production.

In light of their simple chemical nature and total inability to exist independently from other organisms, earlier theories that viruses represent a kind of halfway house between life and nonlife have now largely been abandoned. Instead, viruses are now viewed as fragments of the genomes of other organisms; they could not have existed independently of preexisting organisms.

As an analogy to a virus, consider a computer whose operation is directed by a set of instructions in a program, just as a cell is directed by DNA-encoded instructions. A new program can be introduced into the computer that will cause the computer to cease what it is doing and instead devote all of its energies to making copies of the introduced program. But the new pro-gram is not itself a computer and cannot make copies of itself when outside the computer, lying on the desk. The introduced program, like a virus, is simply a set of instructions.

*Viruses are fragments of DNA or RNA that have become detached from the genomes of bacteria or eukaryotes and have the ability to replicate themselves within cells. Viruses are nonliving and are not organisms. Their genetic material consists of DNA or RNA, but not both.*

## THE DIVERSITY OF VIRUSES

Viruses occur in virtually every kind of organism that has been investigated for their presence. Viruses are almost always highly specific in the hosts they infect and do not reproduce anywhere else. In light of this observation, there should be nearly as many kinds of viruses as there are kinds of organisms—perhaps millions of them. Because there often is more than one kind of virus in a given organism, the actual number of kinds of viruses might even be much greater.

The smallest viruses are only about 17 nanometers in diameter, the largest ones up to 1000 nanometers (1 micrometer) in their greatest dimension; they vary greatly in appearance (Figure 25-7). The largest vi-

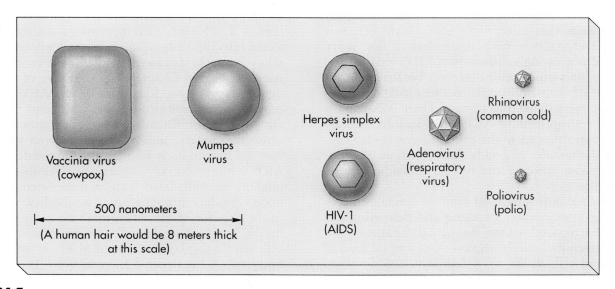

**Figure 25-7**

**The diversity of viruses.** At the same scale, a human hair would be more than 8 meters thick.

ruses, therefore, are just visible with a light microscope. Most viruses can be detected only by using the higher resolution of an electron microscope. Viruses are directly comparable with molecules in size, a hydrogen atom being about 0.1 nanometer in diameter and a large protein molecule being several hundred nanometers in its greatest dimension.

## VIRUSES AND DISEASE

Depending on which genes a virus genomic fragment carries, it can often seriously disrupt the normal functioning of the cells that it infects. For thousands of years, diseases caused by viruses have been known and feared. Among them are smallpox, chickenpox, measles, German measles (rubella), mumps, influenza, colds, infectious hepatitis, yellow fever, polio, rabies, and AIDS, as well as many other diseases not as well known. One series of viral diseases that has been discussed a great deal in recent years is herpes, which includes one category of viruses (herpesvirus 1) associated with cold sores and fever blisters, another (herpesvirus 2) that causes venereal disease, a third (herpesvirus 3) that is responsible for chickenpox and sometimes shingles, and several others.

## AIDS: THE VIRAL SCOURGE OF OUR TIME

A new and particularly vicious viral disease, acquired immunodeficiency syndrome (AIDS), was first reported in the United States in 1981. Affected individuals have no resistance to infection, and all of them eventually die of diseases that most of us easily ward off. Few who contract AIDS survive more than a few years. There is also growing evidence that AIDS victims may likewise exhibit progressive dementia because of a deterioration of their brain cells. Because AIDS is infectious, it is a very dangerous disease. There is essentially no risk of transmission of AIDS from an infected individual to a healthy one in the course of day-to-day contact, but the transfer of bodily fluids, such as blood or semen, between infected and healthy individuals poses a severe risk.

In normal individuals, an army of specialized cells patrols the bloodstream, attacking and destroying any invading bacteria or viruses. But in AIDS patients, this army of defenders stands helpless. One special kind of white blood cell, called a T4 cell, which is discussed further in Chapter 40, is required to rouse the defending cells to action, and in AIDS patients these T4 cells are silent. Like a football team without a quarterback,

defender cells mill about doing nothing to prevent the spread of infection, as if unaware of the problem.

Biologists all over the world began working to determine the cause of AIDS. It was not long before the infectious agent, human immunodeficiency virus (HIV) (Figure 25-8), was identified by laboratories in France and the United States. Study of HIV revealed it to be closely related to an African vervet, or green monkey, virus, suggesting that it might have been introduced to humans in central Africa by a monkey bite. The virus homes in on T4 cells, infecting and killing them until none are left. HIV is transmitted from one person to another with the transfer of body fluids—sexually in semen and vaginal fluid or as a result of small wounds, and in blood transfusions or as a result of the use of nonsterile needles. As a result, the incidence of AIDS is growing very rapidly in the United States, and it is already very high in many African countries. The implications of the AIDS epidemic for college students are discussed in Chapter 42.

In the United States, up to 2 million people were estimated to have been infected by HIV by the late 1980s. Many—perhaps all—of them will eventually come down with AIDS. There is a long latency period after infection before clinical symptoms begin to develop, typically from 2 to 5 years. During this long interval, carriers of HIV have no clinical symptoms but are apparently fully infectious, which makes the spread of AIDS very difficult to control.

Many different kinds of efforts are being made to combat AIDS. Sexual abstinence by infected individuals, the use of condoms, and strictly hygienic procedures in the use of needles and in certain medical practices are all important. In addition, a major attempt is being made to produce a vaccine against AIDS. A **vaccine** is a substance that prevents a disease when injected into the body of an individual who does not already have the disease. The action of vaccines is discussed further in Chapter 40. In the case of AIDS, biologists are inserting fragments of the AIDS virus into otherwise harmless viruses; such fragments of the AIDS virus may be sufficient to prevent infection but not to cause the disease. Ideally, the engineered virus would fool the human body into developing a defense against AIDS without actually experiencing the disease. When humans are inoculated with an engineered virus, their bodies produce proteins called **antibodies,** which bind to the exterior of the virus, marking it for destruction by the body's defender cells. In this case, the infected cells would display proteins on their surfaces specified by genes on the inserted fragment of HIV. There they would be recognized by the body's immune defenses, and antibodies would be made against them. After inoculation, the bloodstream would build up high levels

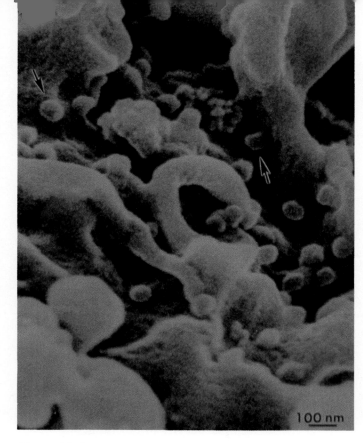

**Figure 25-8**

**AIDS virus in action.** AIDS viruses released from infected T4 cells, signified by arrows, soon spread over neighboring T4 cells, infecting them in turn. The individual AIDS particles are very small; more than 200 million would fit on the period at the end of this sentence.

## Attachment

When an AIDS virus is introduced into the human bloodstream, the virus particle circulates throughout the body but does not infect most of the cells that it encounters. But when it encounters a T4 cell, it infects that cell. Most other animal viruses are similarly narrow in their requirements. Polio goes only to certain spinal nerve cells, hepatitis to the liver, and rabies to the brain. How does a virus such as AIDS recognize a specific kind of target cell such as a T4 cell? Recall from Chapter 5 that every kind of cell in your body has a specific array of cell surface receptors, which are designed to bind particular hormones and growth factors. Cells also possess one or more kinds of cell surface markers that they use to identify themselves to other, similar cells. The AIDS virus recognizes T4 cells because each AIDS particle possesses a glycoprotein on its surface that precisely fits a protein called CD4 on the T4 cell surface. Other viruses possess other glycoproteins that key in on other cell types. Because most human cells lack the CD4 surface marker, they are immune to AIDS infection. But when the AIDS virus encounters a T4 cell, it is able to attach to the T4 cell surface, as these cells contain the proper CD4 marker on their surfaces.

## Entry

After docking with a T4 cell, the AIDS virus penetrates the cell membrane. Like other animal viruses, the AIDS virus enters the cell by endocytosis, with the cell membrane folding inward to form a deep cavity around the virus particle. Plant viruses normally enter the cells of their hosts at points of injury, whereas bacterial viruses (called **bacteriophages,** or "bacteria eaters;" see Figure 25-1) shed their coats outside their host cell, injecting their nucleic acids through the host cell wall and making new coats as they reproduce themselves within.

## Replication

Once within the host cell, the AIDS virus particle sheds its protective coat. This leaves a single strand of virus RNA floating in the cytoplasm, along with a virus enzyme that was also within the virus shell. This enzyme, called **reverse transcriptase,** synthesizes a double strand of DNA complementary to the virus RNA. This double-stranded DNA then inserts itself into the chromosomes of the host T4 cell, where one of two things then happens; either the copy remains quiet ("latent"), its genes not transcribed into virus proteins, or the copy becomes active, taking over part of the host cell machinery and directing it to produce many copies of the virus. In the latter case, the cell eventually dies, releasing thousands of new virus particles, which infect other T4 cells (see Figure 25-9).

of antibodies directed against the virus surface. If any real AIDS viruses come along, they would immediately be bound by these antibodies and not attacked. Engineered viruses such as these are not yet available for use, and it is not clear that they would be effective. There may be many strains of HIV, each requiring a different vaccine for immunity. Work on them, and on other substances that might mitigate the harmful effects of the virus within the body, continues at a high pitch.

## THE INFECTION CYCLE OF AN AIDS VIRUS

Viruses infect bacteria, animals, plants, and probably all other organisms. As an example of how a viral infection proceeds, we shall examine the way in which the AIDS virus infects humans. Most other viral infections follow a similar course, although the details of entry and replication differ in individual cases. The infection cycle of the AIDS virus is outlined in Figure 25-9.

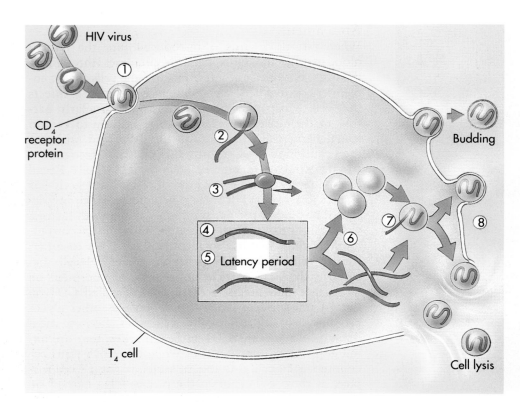

**Figure 25-9**

**The infection of an AIDS virus.**
**1,** The HIV virus responsible for AIDS attaches to a CD4 receptor protein on the surface of a T4 lymphocyte cell and enters the cell by endocytosis. **2,** The viral RNA is released into the cell's cytoplasm. **3,** A DNA copy is made of the virus RNA. **4,** The DNA copy enters the cell's chromosomal DNA. **5,** After a long period (typically 5 years), the virus genes initiate active transcription. **6,** Both HIV RNA and HIV proteins are made. **7,** Complete HIV virus particles are assembled. **8,** Some cells are lysed, releasing free HIV, whereas others bud out HIV by exocytosis.

*Some viruses have the ability to incorporate themselves into the chromosomes of their hosts; in this form, they are said to be latent if they do not begin replicating themselves immediately.*

The conditions that determine when a latent AIDS virus will become active are not well known. Infections by other microbes seem to trigger AIDS outbreaks, perhaps because the infected T4 cells are involved in the immune response. Many latent human viruses are known to be activated by external stimuli, such as ultraviolet radiation. This is precisely what happens when a fever blister develops on your lip because of the activation of a latent herpesvirus, or when a cell bearing a latent retrovirus is suddenly converted into a rapidly dividing cancerous cell by a carcinogen, a process described in Chapter 13.

One of the most important viruses in human history has been the yellow fever virus (Figure 25-10). This virus is one of the arboviruses, or viruses that are spread by arthropods—*Aedes* mosquitoes in the case of yellow fever. In their modes of dispersal, as in their manner of infection, viruses are amazingly versatile. As shown by the recent history of AIDS, and by our increasing understanding of the role of viruses and virus-like particles in causing cancer, it is evident that many new viral diseases will appear among the increasingly crowded human populations of the future. The very best efforts of biological scientists will be necessary to deal with them effectively.

## SIMPLE BUT VERSATILE ORGANISMS

For at least 2 billion years—probably more than half of the history of life on earth—bacteria were the only organisms in existence. They have survived to the present in rich diversity by exploiting an amazingly diverse set of habitats, some of them unchanged since the beginnings of the evolution of the world as we know it. Ecologically, their metabolic activities are of fundamental importance. In a broader sense, bacteria also contribute directly to the functioning of all but a very few eukaryotes because they are the ancestors of mitochondria. Considering that all chloroplasts originated as symbiotic bacteria, it may be said that even photosynthesis, like chemosynthesis and nitrogen fixation, is exclusively a property of bacteria.

Although bacteria are clearly very simple living organisms, viruses are best thought of as segments of

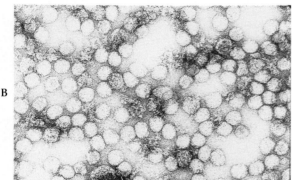

genomes. We have learned a great deal about the structure of viruses, but much remains to be learned, probably including the existence of kinds of particles that are unsuspected at present. The nature of viruses suggests that new forms are evolving constantly. While this process continues, we clearly have a great deal to learn about the existing viruses.

**Figure 25-10**

**Virus diseases are often linked to animal carriers.** The reservoir of yellow fever virus is in monkeys, from which this *Aedes* mosquito spreads it to human beings. The building of the Panama Canal became possible only after the mosquito was eradicated.
**A** *Aedes* mosquito.
**B** Yellow fever virus.

## ■ SUMMARY

1. Bacteria are the only organisms with prokaryotic cellular organization. They are the oldest and simplest organisms but are metabolically much more diverse than all of the other forms of life on earth combined.

2. Bacteria have cell walls consisting of a network of polysaccharide molecules connected by short peptide cross-links. In the gram-negative bacteria, large molecules of lipopolysaccharide are deposited over this layer.

3. Bacteria are rod-shaped (bacilli), spherical (cocci), or spiral (spirilla). Bacilli or cocci may adhere in small groups or chains.

4. Bacteria and fungi break down organic compounds and make the substances in them available for other living organisms once again. Different groups of bacteria fix the energy from the sun in different ways, and some bacterial groups also use chemical processes as a source of energy.

5. Only bacteria are capable of fixing atmospheric nitrogen, thus making it available for their own metabolic activities and those of other organisms. Some of these bacteria live in nodules on the roots of plants, whereas others are free living, found in the soil or in freshwater aquatic or marine habitats.

6. The most frequent form of reproduction by bacteria is binary fission.

7. Viruses are fragments of bacterial or eukaryotic genomes that are able to replicate within cells by using the genetic machinery of those cells. They are not alive and are not organisms.

8. The cells to which a virus will attach are determined by proteins that make up the coats and envelopes of the virus.

9. AIDS, or acquired immunodeficiency syndrome, is a disease of the immune system that is transferred between individuals through bodily fluids such as blood or semen. It can be controlled by changes in human behavior and, in the future, by the development of appropriate vaccines.

## REVIEW

1. Are viruses alive?

2. Chemoautotrophic bacteria derive their energy from the oxidation of chemicals such as ammonia, methane, or hydrogen sulfide. Do they need light energy for this process?

3. *Treponema pallidum*, the cause of syphilis, belongs to what group of bacteria?

4. Can viruses replicate outside of living cells?

5. To what kind of human cells does the AIDS virus attach?

## SELF-QUIZ

1. How is the virus responsible for AIDS transmitted? (Choose two.)
   (a) Sexually   (c) Toilet seats   (e) Coughs or sneezes
   (b) Insect bites   (d) Blood

2. Which of the following bacteria is most similar to the chloroplasts found in plants?
   (a) *Prochloron*   (c) *Heliobacterium*   (e) *Streptococcus*
   (b) *Pseudomonas*   (d) *Rhizobium*

3. Which of the following human diseases are caused by bacteria?
   (a) Tetanus   (c) Toxic shock syndrome   (e) Herpes
   (b) AIDS   (d) Cholera

4. Polio viruses attack nerve cells, hepatitis attacks the liver, and AIDS attacks a specific kind of white blood cell. How does each virus know which kind of cell to attack?
   (a) They enter all cell types, but only certain ones are vulnerable to each kind of virus.
   (b) They key in on certain cell surface molecules characteristic of each cell type.
   (c) The virus in each case was originally derived from that kind of cell.
   (d) The cell types practice virus-specific phagocytosis.
   (e) None of the above.

5. The genetic information of viruses is encoded in a molecule of
   (a) RNA.   (d) either RNA or DNA, but not both.
   (b) DNA.   (e) either DNA or proteins, but not both.
   (c) proteins.

6. How do viruses that attack bacteria enter the bacterial cells?
   (a) They enter them while the bacteria are dividing and naked.
   (b) They enter through conjugation tubes.
   (c) The viruses dissolve small holes in the cell wall and enter whole.
   (d) The viruses shed their coats outside the bacterial cell wall; only the nucleic acid component enters.
   (e) The viruses are so small they can enter through the spaces between the wall components.

# THOUGHT QUESTIONS

1. Why do we say that bacteria are alive and viruses are not?

2. Did the different kinds of viruses originate at different times from bacterial or eukaryotic genomes, or did they have a common ancestor and become more diverse at they attacked many different kinds of hosts? Give the evidence for your answer.

# FOR FURTHER READING

CHILDRESS, J.J.: "Symbiosis in the Deep Sea," *Scientific American,* May 1987, pages 114-120. Bacteria colonize the tube worms and clams at hot-water vents in the deep ocean and supply them with nutrients.

DONOGHUE, H.: "A Mouthful of Microbial Ecology," *New Scientist,* February 5, 1987, pages 61-65. Excellent and well-illustrated account of the communities of bacteria that inhabit our mouths and their effects.

GOODFIELD, J.: *Quest for the Killers,* Birkhauser Press, Boston, 1985. An illuminating account of how smallpox and four other diseases are being combated in the field by public health scientists.

HABICHT, G.S., G. BECK, and J.L. BENACH: "Lyme Disease," *Scientific American,* July 1987, pages 78-83. This rapidly spreading disease has many fascinating properties.

MEE, C.L., JR.: "How a Mysterious Disease Laid Low Europe's Masses," *Smithsonian,* February 1990, pages 66-79. The story of bubonic plague, the Black Death that changed the character of Europe in the 14th century.

RADETSKY, P. "Taming the Wily Rhinovirus," *Discover,* April 1989, pages 38-43. Contemporary approaches to preventing colds, which affect about 1 in 20 people at any given time.

RESENBERG, Z., AND A. FAUCI: "Inside the AIDS Virus," *New Scientist,* February 10, 1990, pages 51-54. Understanding the mechanisms by which the AIDS virus destroys the immune system is of central importance in developing a cure.

SHAPIRO, J.A.: "Bacteria as Multicellular Organisms," *Scientific American,* June 1988, pages 82-89. Explores the sophisticated temporal and spatial control systems that govern colonies of bacteria.

# The Origins of Multicellularity: Protists and Fungi

Looking a bit like sliced watermelons, these marine diatoms are actually single-celled protists, members of the most diverse of the five kingdoms of life.

# THE ORIGINS OF MULTICELLULARITY: PROTISTS AND FUNGI

## Overview

Most kinds of eukaryotes are protists, members of the very diverse kingdom known as Protista. Protists are typically unicellular organisms, but two phyla of algae, the red and brown algae, consist almost entirely of multicellular ones. A third phylum, the green algae, has many multicellular representatives. Multicellularity originated in each of these groups independently. Other protists gave rise to three of the five kingdoms we recognize as distinct: the plants, the animals, and the fungi. Plants manufacture their own food, animals ingest it, and fungi absorb it after releasing enzymes into their surroundings. Along with the bacteria, fungi are the major decomposers of the biosphere, breaking down organic molecules and making the material in them available for recycling. This process is as necessary for the continuation of life on earth as is photosynthesis.

## For Review *Here are some important terms and concepts that you will encounter in this chapter. If you are not familiar with them, you should review them before proceeding.*

**Symbiotic origin of mitochondria and chloroplasts** (Chapters 4 and 25)

**Mitosis** (Chapter 9)

**Major features of evolutionary history** (Chapter 17)

**The five kingdoms of life** (Chapter 24)

**Life cycle patterns** (Chapter 24)

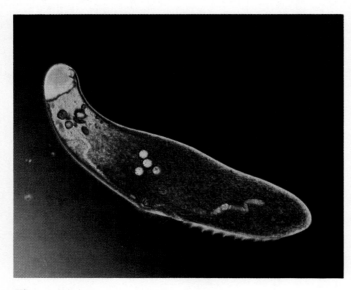

**Figure 26-1**

**Protists.** Although multicellularity has evolved many times among the protists, most protists are microscopic, single-celled organisms. This one is typical in having a complex interior organization. Protists are far more diverse than any other kingdom of life.

The Protista comprise phyla that are primarily unicellular, although every group has at least a few multicellular representatives. Multicellularity probably has evolved independently in each of them. Two of the protist phyla, the red algae (phylum Rhodophyta) and the brown algae (phylum Phaeophyta), consist largely of multicellular organisms that may attain great size. Another group, the green algae (phylum Chlorophyta), are of special interest as the ancestors of plants. Many green algae are multicellular (see Figures 24-12, *B,* and 24-13). But most of the protists are unicellular, and these members exhibit the greatest diversity (Figure 26-1).

No transitional forms between the bacteria and eukaryotes survive. But because of their simple form, we are sure that the first eukaryotes were unicellular organisms similar to some protists. They rapidly differentiated into many distinct evolutionary lines and in turn gave rise to the major groups of eukaryotes—plants, animals, and fungi (Figure 26-2). Later chapters discuss the plants and animals; this chapter includes the fungi. The fungi are a kingdom of diverse organisms that have a great impact on the functioning of the biosphere, decomposing organic materials and return-

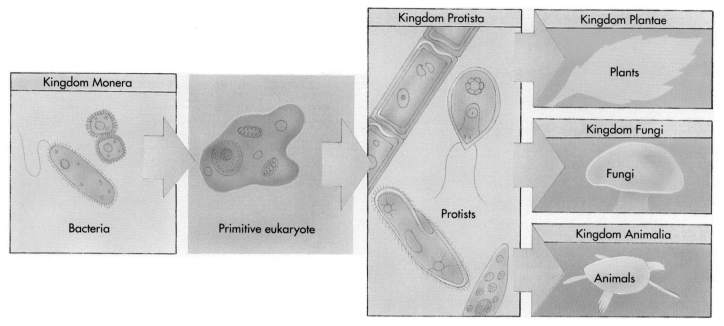

**Figure 26-2**

The evolutionary radiation of the eukaryotes.

ing them to ecosystems. The fungi and those bacteria which perform similar functions are as indispensible as plants to the continuation of life on earth, playing a critical role in the carbon cycle.

## THE PROTISTS

Of the four kingdoms of eukaryotes, the most diverse kingdom is Protista, the protists (Figure 26-3). Among the protists are the simplest eukaryotes, but even single-celled protists have a structure far more complex than that of bacteria. Virtually the full range of diversity of the eukaryotes is found in the protists; in fact, this group includes all the eukaryotes that are not separated as plants, animals, and fungi. Some protists have chloroplasts and manufacture their own food; these are called the algae. Other phyla of protists ingest their food, as do the animals, and still others absorb their food, as do the fungi.

Despite their diversity, grouping all the protists together allows us to compare them directly to emphasize the similarities and differences between different phyla. But in reviewing them, keep in mind that they are by far the most diverse of the eukaryotes—even though metabolically they are far less diverse than are the bacteria.

## SYMBIOSIS AND THE ORIGIN OF EUKARYOTES

To unravel the evolutionary relationships of the protists, one must understand the history of the major symbiotic organelles, mitochondria and chloroplasts (see Chapter 4). Because these were derived in early protists from different groups of bacteria, the relationships of these organelles are different from the overall relationships of the cells within which they occur. One must keep both kinds of relationships in mind to appreciate the protists properly.

Mitochondria are thought to have originated from bacterial ancestors that had characteristics similar to the purple nonsulfur bacteria. Mitochondria are characteristic of all eukaryotes except for several zoomastigotes (see the following section) and *Pelomyxa palustris* (Figure 26-3, *A*), an unusual amoeba-like organism found on the muddy bottoms of freshwater ponds. *Pelomyxa*, a unique organism, is placed in its own phylum. *Pelomyxa* lacks mitosis, and its nuclei divide somewhat like those of bacteria do. The existence of a few protists that lack mitochondria probably indicates that the initial stages of this group's evolution took place before mitochondria were acquired. But this event must have taken place at about the same time that eukaryotes originated—approximately 1.5 billion years ago.

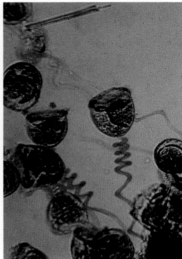

**Figure 26-3**

**Single-celled protists.**
A *Pelomyxa palustris,* a unique, amoeba-like protist, which lacks mitochondria and mitosis. *Pelomyxa* may represent a very early stage in the evolution of the eukaryotic cell.
B The multinucleate plasmodium of a plasmodial slime mold. Individual cells cannot be distinguished in such a plasmodium, but they retain their distinctness.
C *Vorticella,* which is heterotrophic, feeds largely on bacteria and has a retractable stalk. This is an amazing degree of complexity for a single cell; it goes far beyond the complexity found in any bacterium.

*Mitochondria, which probably originated from symbiotic purple nonsulfur bacteria, are absent in two unusual groups of protists. These organisms may represent a stage of evolution before ancestral eukaryotes acquired mitochondria.*

The mitochondria that occur in all but a very few eukaryotes generally possess similar features. On the basis of their characteristics, it is difficult to be sure whether these mitochondria all originated from a single symbiotic event, or whether similar bacteria became symbiotic independently in different groups of early eukaryotes. For chloroplasts, however, the story is very different. As you learned in Chapter 25, three biochemically distinct classes of chloroplasts exist, each resembling a different bacterial ancestor. Thus the chloroplasts of red algae possess chlorophyll *a,* carotenoids, and an unusual class of accessory pigments called phycobilins. These chloroplasts were almost certainly derived from symbiotic cyanobacteria. In contrast, the green chloroplasts of plants and green algae seem to have been derived from bacterial ancestors similar to *Prochloron* (see Figure 4-7). Finally, the chloroplasts of several other groups of algae (brown algae, diatoms, and dinoflagellates), which have chlorophylls *a* and *c,* carotenoids, and distinctive yellowish-brown pigments, probably were derived from a third group of bacteria. These relationships are outlined in Figure 24-9, which you should use as a guide to the relationships between the different groups discussed in this chapter.

*Eukaryotic cells acquired chloroplasts by symbiosis not once, but at least three times during the evolution of different groups of protists.*

## MAJOR GROUPS OF PROTISTS

### Photosynthetic Protists

The photosynthetic protists acquired their chloroplasts as a result of separate symbiotic events, and they are not directly related to one another. Nevertheless, they play a similar ecological role, in which they resemble the plants and photosynthetic bacteria, and it is convenient to discuss them here together.

**Green Algae.** Three phyla of photosynthetic protists consist largely or entirely of multicellular organisms (see Figure 24-12). Among these, the green algae (phylum Chlorophyta) consist of about 7000 species; most are aquatic, but some are semiterrestrial in damp places. They are biochemically similar to the plants, which we suppose were derived from multicellular green algae. The growth form of the marine genus *Ulva* (see Figure 24-12, *B*) is platelike (the individuals are two cells thick), and that of the freshwater genus *Volvox* (see Figure 24-13) is a globular sphere of flagellated cells. Other green algae are filamentous, such as *Spirogyra* (see Figure 24-11, *B*), or unicellular and sometimes flagellated. Flagellated unicellular organisms such as *Chlamydomonas* (Figure 26-4) apparently gave rise to colonial algae such as *Volvox,* in which the individual cells are so integrated with one another that *Volvox* can be considered to be truly multicellular. *Acetabularia,* which we considered in Chapter 12 as a key experimental organism for the study of development, is a member of this phylum (see Figures 12-2 and 12-3). In *Acetabularia,* the individual cells are very large. Many of the green algae have cell walls composed totally or partly of cellulose.

**Red and Brown Algae.** The other two groups of multicellular photosynthetic algae are primarily marine. The red algae (phylum Rhodophyta) comprise about 4000 species. Their chloroplasts are similar to those of the cyanobacteria. An unusual feature of the red algae, some of which grow at greater depths in the sea than any other photosynthetic organism, is that they completely lack flagellated cells at any stage in their life cycle. (A representative red alga is shown in Figure 24-12, *C*.) Some members of this phylum, as well as some green and brown algae, are harvested as important sources of food in the Orient.

The brown algae (phylum Phaeophyta) are unrelated to either the red or green algae. They have flagellated reproductive cells and biochemically distinct chloroplasts. About 1500 species of brown algae exist, some of which attain great size (up to 100 meters long) and complexity. The kelps (see Figures 24-11, *C*, and 24-12, *A*) and rockweeds are one large group of brown algae, often commerically harvested for fertilizer or for alginates, a group of substances widely used as thickening agents and colloid stabilizers in the food, textile, cosmetic, and pharmaceutical industries. Agar, a mucilaginous material extracted from the cell walls of red algae, is used as a culture medium for bacteria and other microorganisms and for a wide variety of commerical applications, including making the capsules that contain vitamins and drugs.

**Primarily Unicellular Photosynthetic Protists.** Several other groups of protists possess chloroplasts. In two of them,

**Figure 26-4**

Life cycle of *Chlamydomonas.* Individual cells of this microscopic, biflagellated alga, which are haploid, divide asexually, producing identical copies of themselves. At times, such haploid cells act as gametes, fusing to produce a zygote, as shown at the lower right-hand side of the diagram. The zygote develops a thick, resistant wall, becoming a zygospore. Within this diploid zygospore, meiosis takes place, ultimately resulting in the release of four haploid individuals. Because of the segregation during meiosis, two of these individuals are what is called the + strain; the other two, the − strain. Only + and − individuals are capable of mating with one another when union does take place, although both may divide asexually and reproduce themselves in that way also.

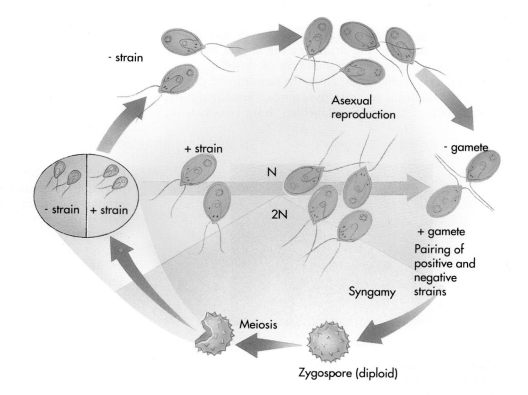

- strain

Asexual reproduction

- gamete

+ strain

N

2N

+ gamete
Pairing of positive and negative strains

Syngamy

Meiosis

Zygospore (diploid)

- strain | + strain

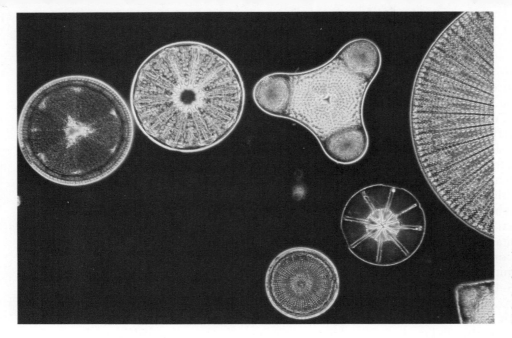

**Figure 26-5**

**Several different kinds of diatoms.** The diverse shapes of the silica shells of some members of this phylum are only hinted at in the forms shown here.

the diatoms (phylum Bacillariophyta) and the dinoflagellates (phylum Dinoflagellata), the chloroplasts are biochemically and structurally similar to one another and to those of the brown algae as well. Diatoms, of which there are about 11,500 living species and many others known only as fossils, have a very characteristic "shell" made of two parts, like the halves of a box. Chemically, the shell is opaline silica (Figure 26-5). Huge masses of diatoms have been deposited in certain locations, such as near Lompoc, California, where the strata are hundreds of meters thick.

Although only about 1000 species of dinoflagellates exist, they are important as the causal agent of "red tides," during which millions of dinoflagellates, secreting toxic substances, color the sea red and poison fish and shellfish by the thousands. Free-living dinoflagellates (phylum Dinoflagellata) have two flagella,

one lying in a shallow girdle around the center of the cell, which is armed with stiff cellulose plates, and the other directed backward. Other dinoflagellates are symbiotic with marine animals such as coral (Figure 26-6). The presence of these dinoflagellates allows the coral colony to produce its own food by photosynthesis, and "bleached" coral—coral from which the dinoflagellates have been lost—is unhealthy and perhaps dying. Because of the presence of symbiotic dinoflagellates, corals are able to flourish in nutrient-poor tropical waters.

## Heterotrophic Protists

Most protists lack chloroplasts or acquire them only occasionally. They are exceedingly diverse, and the relationships between them are poorly understood. Several of the more important groups are presented here as examples. They are predominantly unicellular, but each group also contains multicellular members.

**Figure 26-6**

**Symbiosis is common among the protists.** These tentacles of a coral animal are packed with symbiotic dinoflagellates, the abundant, tiny, brownish objects.

*Ciliates.* The ciliates (phylum Ciliophora) are a large phylum, conprising at least 8000 structurally complex, primarily aquatic species (see Figures 26-1 and 4-1). Perhaps the best-known member of the phylum is *Paramecium* (see Figures 1-4 and 4-2), which has been the subject of comprehensive genetic and life-history studies, as has its relative *Tetrahymena*. Many ciliates move about by the beating of cilia (fine flagella), which occur in lines on their cells. *Vorticella* (see Figure 26-3) remains anchored in one place, using its cilia to sweep food into its gullet. An unusual mode of reproduction known as **conjugation** characterizes the ciliates. In it, nuclei are exchanged between individuals through tubes that connect them during the period of conjugation.

# Hiker's Diarrhea

*Giardia lamblia,* a ciliate, is found throughout the world, including all parts of the United States and Canada. It occurs in water, including the clear water of mountain streams and the water supplies of some cities, infecting at least 40 species of wild and domesticated animals in addition to humans. In 1984 in Philadelphia, 175,000 people had to boil their drinking water for several days following the appearance of *Giardia* in the city's water system. Although most individuals exhibit no symptoms if they drink water infested with *Giardia,* many suffer nausea, cramps, bloating, vomiting, and diarrhea. Only 35 years ago, *Giardia* was thought to be harmless; today, it is estimated that at least 16 million residents of the United States alone are infected by it.

*Giardia* lives in colonies in the upper small intestine of its host; it occurs there in a motile form that cannot survive outside the host's body for long. It is spread in the feces of infected individuals in the form of dormant, football-shaped cysts—sometimes at levels as high as 300 million cysts per gram of stool. These cysts can survive at least 2 months, and they last longer in cool water, such as that of mountain streams. They are relatively resistant to the usual water-treatment agents such as chlorine and iodine but are killed at temperatures greater than about 65° C. Apparently, infected wild animals, such as beavers, can release *Giardia* cysts that lead to infection in humans; pollution by humans or domestic animals therefore is not necessary for stream water to be dangerous, although pollution can be an important contributing factor. There are at least three species of *Giardia* and many distinct strains; how many of them attack humans and under what circumstances are not known with certainty.

In the wilderness, good sanitation is important in preventing the spread of *Giardia.* Dogs, which readily contract and spread the disease, should not be taken into pristine wilderness areas. Drinking water should be filtered—the filter must be capable of eliminating particles as small as 1 micrometer in diameter—or boiled for at least 1 minute; water from natural streams or lakes should never be consumed directly, regardless of how clean it looks. In other regions, good sanitation methods are important to prevent not only *Giardia* but also other diseases.

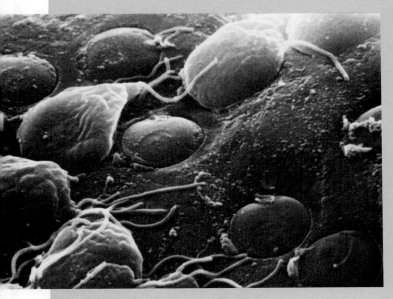

*Figure 26-A*
*Giardia lamblia.*

**Zoomastigotes.** In the zoomastigotes (phylum Zoomastigina), the cells are propelled by the beating of one or more long flagella. There are thousands of species of zoomastigotes, and many of them are parasitic. The serious and widespread diseases trypanosomiasis, or "sleeping sickness" (Figure 26-7), East Coast fever, and Chagas' disease, which are especially prevalent in the tropics, are caused by trypanosomes, one of the groups of zoomastigotes. Sleeping sickness has kept cattle out of much of Africa, thus complicating the problems of feeding the people.

Trypanosomes have complicated life cycles. When they are ready to spread to a mammal from the flies that carry them around, the trypanosomes acquire a thick coat of glycoprotein antigens that protects them from the host's antibodies. Trypanosomes can change this coat so quickly that the immune systems of cattle and humans cannot mount a defense against it. A ma-

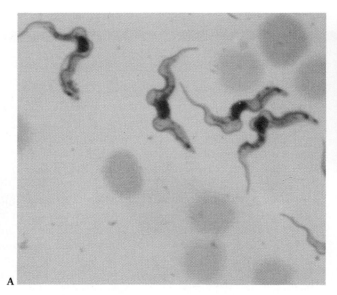

A

B

**Figure 26-7**

**Sleeping sickness is a protist's disease.**
A *Trypanosoma*, the protists that cause sleeping sickness, among red blood cells. The nuclei (dark-staining bodies); the anterior (forward-projecting) flagella; and the undulating, changeable shape of the trypanosomes may be seen in this photograph.
B A tsetse fly, shown here sucking blood from a human arm, in Tanzania, East Africa.

jor effort is now being made to develop vaccines against trypanosomes, and some of the prototype vaccines are being tested in the field.

Other groups of zoomastigotes are equally interesting. Some that live in the guts of wood roaches and termites possess cellulase enzymes that allow these insects to digest cellulose and thus live on a diet of wood. The insects are unable to manufacture these enzymes themselves, but the presence of the zoomastigotes, which are always in their guts, enables them to live as if they did have this ability. Members of this phylum include the choanoflagellates, which have a characteristic collar at one end of the cell from which the single flagellum protrudes (Figure 26-8). The choanoflagellates are clearly the ancestors of the sponges (see Figure 28-3), and probably the ancestors of all other animals as well. Some genera of a satellite group of the zoomastigotes, the euglenoids (sometimes considered to be a separate phylum, Euglenophyta), possess chloroplasts with characteristics resembling those of green algae and plants. They clearly acquired these chloroplasts independently from the other groups, either directly or by ingesting algae that had acquired them earlier.

*Amoeba-Like Protists.* The amoebas (phylum Rhizopoda) are heterotrophic protists that are abundant throughout the world in freshwater and saltwater, as well as in the soil. Many species are parasites of animals. It is estimated that more than 10 million people in the United States are infected by amoebas, about 2 million of whom show symptoms of amoebic dysentery. Amoebas lack cell walls, flagella, meiosis, and any form of sexuality, but they do carry out mitosis and possess mitochondria. They move from place to place

by means of their **pseudopodia,** from the Greek words for "false" and "feet" (Figure 26-9).

The well-known disease malaria is caused by the members of another phylum of amoeba-like protists, the sporozoans (phylum Sporozoa). Sporozoans are

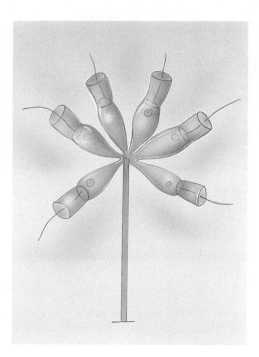

**Figure 26-8**

**Our protist ancestor.** This flagellated protist is a member of the phylum *Zoomastigina*, which gave rise to the sponges. The unique structure of their cells is shared only with that group. The sponges are the most primitive members of the animal kingdom.

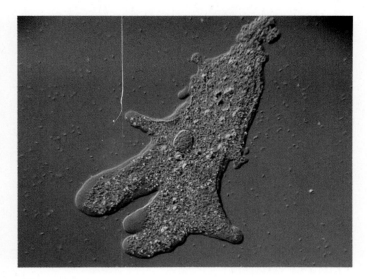

**Figure 26-9**

**An amoeba.** *Amoeba proteus* is a relatively large amoeba that is commonly used in teaching and for research in cell biology. The projections are pseudopodia; an amoeba moves simply by flowing into them. The nucleus of the amoeba is plainly visible.

nonmotile, spore-forming parasites of animals. Nearly 4000 species are known. Approximately 250 million people are afflicted by malaria at any one time, and 2 to 4 million of them die each year. The symptoms include chills, fever, sweating, an enlarged and tender spleen, confusion, and great thirst. Malaria kills most children under 5 years of age who contract it. The disease is caused by sporozoans of the genus *Plasmodium*, which are carried by mosquitoes of the genus *Anopheles* (Figure 26-10). Once the *Plasmodium* organisms reach the bloodstreams of humans or other mammals, they move to the liver, where they begin to divide. They then pass back into the bloodstream and invade the red blood cells, dividing rapidly within them and causing them to become enlarged and ultimately to rupture. This releases toxic substances throughout the body of the host, bringing about the well-known cycle of fever and chills that is characteristic of malaria. Malaria has proved difficult to control both because the mosquitoes have become resistant to insecticides and because the parasites have developed resistance to the chemicals, such as quinine, that are used to kill them.

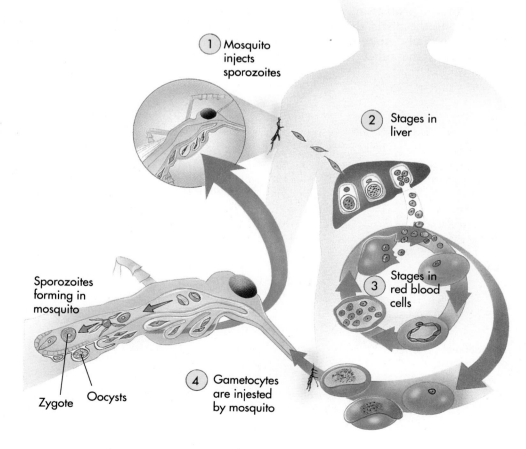

**Figure 26-10**

**Life cycle of *Plasmodium*, the sporozoan that causes malaria.** The stages in the life cycle are called sporozoites, merozoites, and gametocytes, and they play the roles indicated in this diagram.

1 Mosquito injects sporozoites

2 Stages in liver

3 Stages in red blood cells

4 Gametocytes are injested by mosquito

Sporozoites forming in mosquito

Zygote    Oocysts

*Slime Molds.* Among the unrelated phyla of protists that have historically, and incorrectly, been considered to be fungi are two that have amoeba-like stages in their life cycles. The plasmodial slime molds (phylum Myxomycota) exist mainly as flowing, multinucleate masses (see Figure 26-3, *B*) that feed on bacteria and other small bits of organic matter. The more than 500 species of this phylum differentiate into spore-forming bodies, often very ornate, when reproduction occurs. In contrast, the cellular slime molds (phylum Acrasiomycota), of which there are only a few dozen species, exist chiefly as single "amoebas" for most of their life cycle, but swarm together at certain stages to form resistant spores that are dispersed (Figure 26-11). Despite the similarity of their names, these two phyla are clearly unrelated to one another.

## THE MOST DIVERSE KINGDOM OF EUKARYOTES

You are probably convinced now of the diversity of the protists. The phyla we have considered in this chapter consist mainly of microscopic, unicellular organisms, yet they also include massive, multicellular kelps that may exceed 100 meters in length. In their ways of life, the habitats where they occur, and the details of their life cycles, the protists are extraordinarily diverse.

Among the phyla that make up this great group, we are especially interested in those which apparently gave rise to the other exclusively or predominantly multicellular evolutionary lines that we call animals, plants, and fungi. The choanoflagellates, one of the groups of zoomastigotes (see Figure 26-8), appear certainly to be the ancestors of the sponges and probably of all animals; the fundamental similarity between their unique structure and that of the collar cells of sponges (see Chapter 28) is simply too great to be explained in any other way. Green algae, which share with plants the same photosynthetic pigments, cell wall structure, and chief storage product (starch), certainly include the ancestors of plants, a kingdom that achieved its distinctive features when it invaded the land. The ancestors of the fungi are not known because their features are distinct from those of all living protists.

## THE FUNGI

Fungi are an ancient group of organisms at least 400 million, and perhaps 800 million, years old. This distinct kingdom of organisms comprises approximately 63,500 described species (Figure 26-12), and many more await discovery. Scientists who study fungi are called **mycologists.** Although fungi have traditionally been included in the plant kingdom, they have no chlorophyll and resemble plants only in lacking mobility and in growing from the ends of somewhat linear bodies. But even these similarities prove to be misleading when fungi are examined closely. Some plants have motile sperm with flagella; therefore plants certainly originated from ancestors that had flagella. In con-

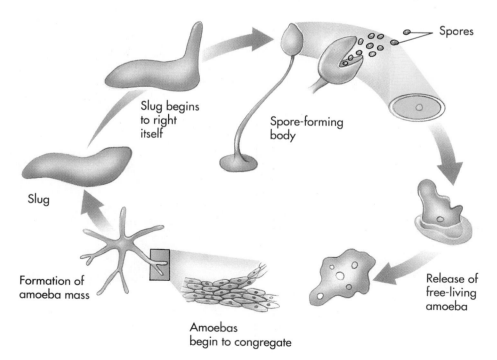

Slug begins to right itself

Spore-forming body

Spores

Slug

Formation of amoeba mass

Amoebas begin to congregate

Release of free-living amoeba

*Figure 26-11*

**Life cycle of a cellular slime mold.** The amoebas emerge from spores, congregate at certain stages of their life cycle, and then differentiate into stalks and spore-forming bodies. Individuals of this group feed on bacteria and other small organisms.

# Ascomycetes

# Basidiomycetes

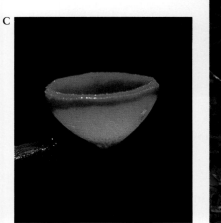

**Figure 26-12**

**The two divisions (phyla) of fungi.** All visible structures of fleshy fungi such as the ones shown here arise from an extensive network of filaments (**hyphae**) that penetrate and interweave with the substrate on which they grow.
**A** A morel, *Morchella esculenta,* a delicious, edible ascomycete (division Ascomycota) that appears in early spring in the northern, temperate woods (especially under oaks).
**B** *Amanita muscaria,* the fly agaric, a poisonous basidiomycete (division Basidiomycota).
**C** A cup fungus (division Ascomycota) in the rain forest of the Amazon Basin.
**D** A shelf fungus, *Coriolus versicolor* (division Basidiomycota), growing on a tree trunk. Basidia, the reproductive structures, line the underside of this fungus.
**E** A puffball (division Basidiomycota). Puffballs release their basidiospores by rupturing when mature.

strast, no fungi ever have flagella, and no evidence indicates that their ancestors possessed flagella. Fungi are basically filamentous in their growth form, consisting of slender filaments, whereas plants are multicellular in three dimensions. Unlike plants, fungi obtain their food by secreting enzymes out of their bodies and onto or into the substrate. They then absorb into their bodies the materials that these enzymes make available.

Many fungi are harmful because they decay, rot, and spoil many different materials as they obtain food. They also can cause serious diseases in plants and animals, including human beings. But other fungi are extremely useful. The manufacture of both bread and beer depends on the biochemical activities of yeasts, single-celled fungi that produce abundant quantities of

ethanol and carbon dioxide. Both cheese and wine achieve their delicate flavors because of the metabolic processes of certain fungi, and other fungi make possible the manufacture of such Oriental delicacies as soy sauce and tofu. Vast industries depend on the biochemical manufacture of organic substances such as citric acid by using fungi in culture, and yeasts are now employed on a large scale to produce protein for the enrichment of animal food. Many antibiotics, including the first one widely used, penicillin, are derived from fungi. Other fungi are used to convert complex organic molecules into other molecules, such as in the synthesis of many commercially important steroids. Because so many are harmful and because so many are beneficial, fungi hold great importance for all of us.

**Figure 26-14**

**Mycelia.** A fungal mycelium growing through leaves on the forest floor in Maryland.

## Fungal Ecology

Fungi, along with bacteria, play an essential role as decomposers in the biosphere. They break down organic materials and return the substances locked up in these molecules to circulation in the ecosystem. In this way, critical biological building blocks, such as compounds of carbon, nitrogen, and phosphorus, that have been incorporated into the bodies of living organisms are released and made available for other organisms. For the same reason, fungi often cause serious damage to materials that are used by human beings and are responsible for major human, animal, and plant diseases.

## Fungal Structure

Fungi exist mainly as slender filaments, or **hyphae** (singular, **hypha**), which are barely visible to the naked eye. These hyphae may be divided into cells by cross walls called **septa** (singular, **septum**). But the septa rarely form a complete barrier, except for those separating the reproductive cells. Cytoplasm characteristically flows freely throughout the hyphae, passing

through the major pores in the septa (Figure 26-13). Because of this cytoplasmic streaming, proteins, which are synthesized throughout the hyphae, may be carried to the actively growing tips of the hyphae. As a result, the growth of fungal hyphae may be rapid when abundant food and water are available and the temperature is high enough.

A mass of hyphae is called a **mycelium** (plural, **mycelia**). This term, like *mycologist,* is derived from the Greek word for fungus, *myketos.* The mycelia of fungi (Figure 26-14) constitute a system that may be many kilometers long, although it is concentrated in a much smaller area. This system grows through and penetrates the environment of the fungus, resulting in a unique relationship between a fungus and its environment. All parts of a fungus are metabolically active, continually interacting with the soil, wood, or other material in which the mycelia are growing.

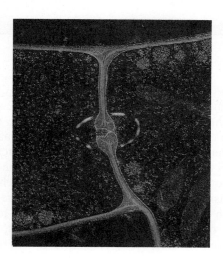

**Figure 26-13**

**A septum.** Transmission electron micrograph of a section through a hypha of the basidiomycete *Laetisaria arvalis* showing a septum. Septa of this kind, lined with a thick, barrel-shaped structure, are characteristic of one phase of the life cycle of basidiomycetes.

In two of the three divisions of fungi, structures composed of interwoven hyphae, such as mushrooms, puffballs, and morels, are formed at certain stages of the life cycle. These structures may expand rapidly because of cytoplasmic streaming and growth in the kilometers of hyphae from which they arise. For this reason, mushrooms can force their way through tennis court surfaces or appear suddenly in your lawn.

The cell walls of fungi are not formed of cellulose, as are those of plants and some groups of protists.

Other polysaccharides are typical constituents of fungal cell walls, such as chitin, which occurs especially frequently. Chitin is the same material that makes up the major portion of the hard shells, or **exoskeletons,** of arthropods, a phylum of animals that includes insects and crustaceans (see Chapter 28). Chitin is far more resistant to microbial degradation than is cellulose.

Mitosis in fungi differs from that found in other organisms. The nuclear envelope does not break down and re-form, and the spindle apparatus is formed within it. In addition, centrioles are lacking in all fungi. Overall, fungal features suggest that the kingdom originated from some unknown group of single-celled eukaryotes that lacked flagella. Certainly, the fungi differ sharply from all other groups of living organisms.

**Spores,** always nonmotile, constitute a common means of reproduction among the fungi. They may be formed through either sexual or asexual processes. When the spores land in a suitable place, they germinate, giving rise to a new fungal hypha. Because the spores are very small, they may remain suspended in the air for long periods. Because of this property, fungal spores may be blown great distances from their place of origin, a factor explaining the extremely wide distributions of many kinds of fungi. Unfortunately, many fungi that cause diseases of plants and animals are spread rapidly and widely by such means.

## MAJOR GROUPS OF FUNGI

Fungi comprise three divisions (phyla), representatives of which are shown in Figure 26-12. The evolutionary relationships between these groups are not clear, although the zygomycetes, which we discuss first, are the simplest in structure.

### Zygomycetes

**Zygomycetes** (division Zygomycota) comprise about 665 described species. The name of this division refers to its chief characteristic, the production of sexual structures called **zygosporangia**[*] (singular, **zygosporangium**) (Figure 26-15). A zygosporangium is formed following fusion of two of the simple reproductive organs of these fungi, which are called **gametangia** (singular, **gametangium**). Within a zygosporangium, the gametes—which are simply nuclei—fuse, forming one or more diploid nuclei, or zygotes.

The zygosporangia of members of this division differ greatly and are often highly ornate. Spores are also formed without sexual reproduction taking place; these **asexual** spores are produced in bodies called **sporangia** (singular, **sporangium**). The life cycle of the black bread mold, *Rhizopus,* is shown in Figure 25-15.

[*]In these fungi, the zygosporangium often has been called a zygospore, but because it contains several to many zygotes, we prefer to consider it a zygosporangium.

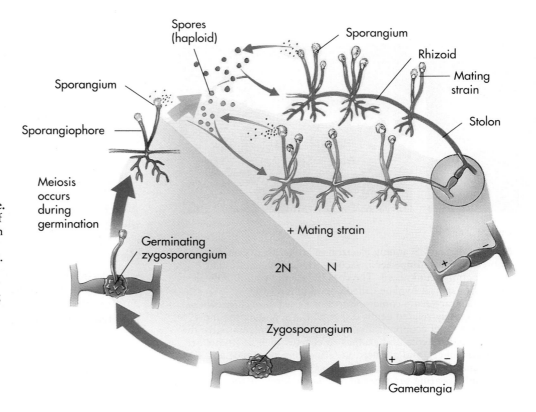

**Figure 26-15**

**Life cycle of *Rhizopus,* a zygomycete.** The hyphae grow over the surface of the bread or other material on which the fungus feeds, producing erect, sporangium-bearing stalks in clumps. If both + and − strains are present in a colony, they may grow together and their nuclei may fuse, producing a zygote. This zygote, which is the only diploid cell of the life cycle, acquires a thick, black coat, and is then called a zygosporangium. Meiosis occurs during its germination, and normal, haploid hyphae grow from the resulting haploid cells.

## Ascomycetes

The ascomycetes (division Ascomycota) comprise about 30,000 described species of fungi. Among them are such familiar and economically important fungi as cup fungi and morels (see Figure 26-12, *A* and *C*), truffles, and yeasts. Ascomycetes also include the organisms that cause many of the most serious plant diseases. The chestnut blight, *Cryptonectria parasitica,* has almost exterminated the American chestnut throughout its native range, throughout its native range, and Dutch elm disease, *Ceratocystis ulmi,* has decimated elms around the world.

The characteristic reproductive structure of the ascomycetes is the **ascus** (plural, **asci**), a club-shaped element that is formed within a structure consisting of densely interwoven hyphae, the **ascocarp**. The hyphae of ascomycetes are haploid. Syngamy, which occurs in the young asci, is immediately followed by meiosis, producing four, eight, or more spores. Asexual reproduction by means of **conidia** (singular, **conidium**), which are multinucleate spores in fungi cut off at the ends of the hyphae, is characteristic of most ascomycetes. Asexual reproduction is also a common feature among the approximately 17,000 species of **Fungi Imperfecti,** a group of fungi in which sexual reproduction is not known (Figure 26-16). Most Fungi Imperfecti would be classified as ascomycetes if their sexual structures were found, judged from the features of their hyphae. Yeasts, the only single-celled fungi, are also primarily ascomycetes.

A                                                           B

### Figure 26-16

**Fungi Imperfecti.** Scanning electron micrograph of conidia (spores) of Fungi Imperfecti, ascomycetes in which sexual reproduction is unknown. The conidia are the round balls at the end of special hyphae called **conidiophores.**

**A** The characteristic conidiophores of *Penicillium. Penicillium* and the closely related genus *Aspergillus* are among the most important fungi economically. They produce penicillin, give the characteristic flavors and aromas to cheeses such as Roquefort and Camembert, and are used in fermenting soy sauce and soy paste, among other applications.

**B** Conidia-bearing branches of *Tolypocladium inflatum.*

### Figure 26-17

**Rusts.** Wheat rust, *Puccinia graminis,* one of about 7000 species of rusts, all of them plant pathogens. This species causes enormous economic losses to wheat wherever it grows, and it is combatted largely by breeding resistant wheat varieties. Mutation and recombination in wheat rust constantly produce new virulent strains and make it necessary to replace the existing wheat varieties constantly. Wheat rust alternates between two different hosts, wheat and barberries, and needs both to complete its life cycle. The sexual stages of wheat rust take place only on barberries, and the eradication of these plants helps to control this disease.

## Basidiomycetes

**Basidiomycetes,** the third division of fungi (division Basidiomycota), with about 16,000 described species, are the most familiar fungi. These include not only the mushrooms, toadstools, puffballs (Figure 26-12, *E*), jelly fungi, and shelf fungi (Figure 26-12, *D*), but also many important plant pathogens among the groups called rusts and smuts (Figure 26-17). In place of the asci of ascomycetes, the basidiomycetes form **basidia** (singular, **basidium**). At the apex of the basidia, the spores that result from meiosis are elevated; otherwise, the details of the life cycle (Figure 26-18) are generally similar to those of the ascomycetes. Asexual reproduction, however, is relatively uncommon.

## LICHENS

A lichen is a symbiotic association between an ascomycete and a photosynthetic partner.*

There are about 13,500 described species of lichens (Figure 26-19). Most of the visible body of a lichen consists of its fungus, but within the tissues of that fungus are found either cyanobacteria or green algae, or both (Figure 26-20). Specialized fungal hyphae penetrate the photosynthetic cells held within them and

---

*About a dozen species of basidiomycetes also form associations with algae, but they are closely related to free-living basidiomycetes and do not resemble any other lichens.

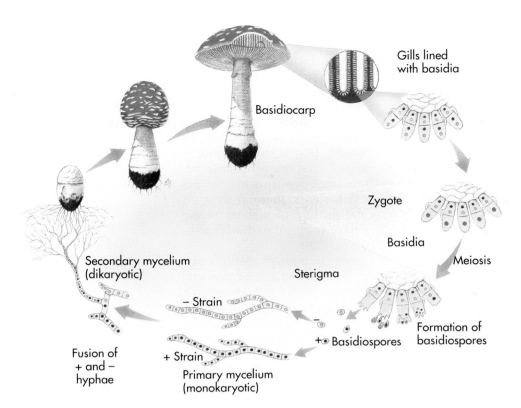

Gills lined with basidia

Basidiocarp

Zygote

Basidia

Meiosis

Sterigma

Secondary mycelium (dikaryotic)

– Strain

Formation of basidiospores

+● Basidiospores

+ Strain

Fusion of + and – hyphae

Primary mycelium (monokaryotic)

**Figure 26-18**

**Life cycle of a basidiomycete.** In primary mycelia, there is only one nucleus in each cell; in the secondary mycelia, which are formed by the fusion of primary mycelia (plasmogamy), there are two nuclei, one derived from each of the strains that gave rise to the secondary mycelia, within each cell. Secondary mycelia ultimately may become massed and interwoven, forming the basidiocarp.

transfer nutrients directly to the fungal partner, which in turn protects its bacterial or algal cells from drying out.

The durable construction of the fungus, linked with the photosynthetic properties of its partner, has enabled lichens to invade the harshest of habitats: the tops of mountains, the farthest north and south latitudes, and dry, bare rock faces in the desert. In such desolate, exposed areas, lichens are often the first colonists, breaking down the rocks and setting the stage for

**Figure 26-19**

**Lichens on a fog-swept rock in coastal California.**

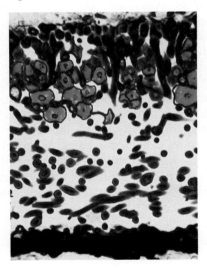

**Figure 26-20**

**Structure of a lichen.** This transverse section of a lichen shows the fungal hyphae (more densely packed into a protective layer on the top) and, especially, the bottom layer of the lichen. The green cells near the upper surface of the lichen are those of a green alga. Penetrated by fungal hyphae, they supply carbohydrates to the fungus.

the invasion of other organisms. Lichens with a cyanobacterium for a photosynthetic partner have a particular advantage because they are able to fix atmospheric nitrogen for themselves; they also contribute it to their habitat where it is used by other pioneering organisms.

Lichens are often strikingly colored because of their pigments, which probably play a role in protecting the photosynthetic partner from the destructive action of the sun's rays. These same pigments may be extracted from the lichens and used as natural dyes, as they were in the traditional method of manufacturing Harris tweed (now, however, it is colored with synthetic dyes).

Lichens are able to survive in inhospitable habitats partly by being able to dry or freeze to a condition one might call suspended animation. Once the drought or cold has passed, the lichens recover quickly and resume their normal metabolic activities, including photosynthesis. The growth of lichens may be extremely slow in harsh environments; many relatively small ones appear to be thousands of years old and therefore are among the oldest living things on earth. Lichens are extremely sensitive to pollutants in the atmosphere and thus can be used as bioindicators of air quality. Many species are characteristically absent near major cities and other sources of pollution.

## MYCORRHIZAE

The roots of about 80% of all kinds of plants normally are involved in symbiotic relationships with certain specific kinds of fungi; it has been estimated that these fungi probably account for as much as 15% of the weight of the plant's roots in many cases. Associations of this kind are termed **mycorrhizae,** from the Greek words for "fungus" and "roots." To a certain extent, the fungi involved in mycorrhizal associations replace and have the same function as the fine projections from the epidermis, or outermost cell layer, of the terminal portions of roots called **root hairs.** When mycorrhizae are present, they aid in the direct transfer of phosphorus, zinc, copper, and probably other nutrients from the soil to the roots of plants. The plant, on the other

**Figure 26-21**

**Mycorrhizae aid plant growth.** Soybeans without endomycorrhizae (left) and with different strains of mycorrhizae (center and right).

hand, supplies organic carbon to the symbiotic fungus (Figure 26-21).

*Mycorrhizae are symbiotic associations between plants and fungi.*

The earliest fossil plants often are found to have mycorrhizal roots. Such associations, which were common during the initial invasion of the land by plants, may have played an important role in allowing that invasion to occur. The soils available at such time would have been relatively infertile and completely lacking in organic matter. Plants that form mycorrhizal associations are particularly successful in similar situations today. When one considers this fact, and the fossil evidence, the suggestion that mycorrhizal associations were characteristic of the earliest plants seems reasonable.

## ■ SUMMARY

1. The kingdom Protista consists of the exclusively or predominantly unicellular phyla of eukaryotes, together with three phyla that include many multicellular organisms: the red algae, brown algae, and green algae. The phyla of protists are highly diverse.

2. *Pelomyxa palustris*, an amoeba-like organism, lacks both meiosis and mitochondria and apparently represents a very early stage in the evolution of the eukaryotes.

3. Chloroplasts originated independently several times in the protists. The process apparently involved the members of at least three different groups of bacteria: cyanobacteria (in the red algae); *Prochloron*-like organisms (in the green algae and euglenoids); and bacteria of a third, different group (in the brown algae, diatoms, and dinoflagellates).

4. The three major multicellular groups of eukaryotes, plants, animals, and fungi, all originated from protists. But because the three groups are so large and important, each is considered a separate kingdom. The three are not related directly to one another. The plants originated from green algae; the sponges, and probably all animals, originated from the choanoflagellates, a group of zoomastigotes; and the ancestors of fungi are unknown.

5. About 10 other phyla of protists are mentioned, including Rhizopoda, the amoebas; Ciliophora, the ciliates; Bacillariophyta, the diatoms; and Dinoflagellata, the dinoflagellates. Two phyla that have stages in their life histories in which individuals move like amoebas, and which have sometimes been grouped with the fungi, are Myxomycota, the plasmodial slime molds, and Acrasiomycota, the cellular slime molds.

6. Among the zoomastigotes are the organisms that cause sleeping sickness and several other primarily tropical diseases. Another group of Zoomastigina, the choanoflagellates, are the ancestors of sponges and probably all animals. A third group, the euglenoids (sometimes separated as the division Euglenophyta), includes a number of genera in which chloroplasts occur.

7. The malarial parasite, *Plasmodium,* is a member of the phylum Sporozoa. Carried by mosquitoes, it multiplies rapidly in the liver of humans and other primates. This results in the cyclical fevers characteristic of malaria because of the release of toxins into the bloodstream of the host.

8. The fungi are a distinct kingdom of eukaryotic organisms characterized by their filamentous growth form, lack of chlorophyll and motile cells, chitin-rich cell walls, characteristic form of mitosis, and external digestion of food by the secretion of enzymes. Along with the bacteria, fungi are the decomposers of the biosphere.

9. Fungal filaments, called hyphae, collectively make up a mass called the mycelium. Mitosis in fungi occurs within the nuclear envelope.

10. Symbiotic systems involving fungi include lichens and mycorrhizae. The fungal partners in lichens are ascomycetes, which derive their nutrients from green algae, cyanobacteria, or both. Mycorrhizae are symbiotic associations between plants and fungi, which are characteristic of about 80% of all plant species.

## *REVIEW*

1. Four characteristics that differentiate plants from fungi are _____, _____, _____, and _____.

2. The _____ algae, a division (phylum) of protists, lack flagellated cells at all stages of their life cycle.

3. What group of protists gave rise to the sponges and probably to all other groups of animals?

4. The slender filaments that make up the body of a fungus are called _____.

5. The three classes of fungi are _____, _____, and _____.

◆

# SELF-QUIZ

1. Which kingdom of eukaryotes is most diverse?
   (a) Monera                (c) Fungi                (e) Animalia
   (b) Protista              (d) Plantae

2. Which of the following groups of algae have chloroplasts that are closely similar to the ones found in plants?
   (a) Red algae             (c) Green algae          (e) Euglenoids
   (b) Brown algae           (d) *Prochloron*

3. Which of the following features distinguish *Pelomyxa* from nearly all other eukaryotes? (Choose two.)
   (a) Mitosis takes place within the nuclear envelope
   (b) Mitochondria absent
   (c) No motile cells
   (d) Mitosis absent
   (e) Meiosis absent

4. Which of the following are chiefly ascomycetes?
   (a) Yeasts
   (b) The fungal partner in lichens
   (c) The fungal partner in mycorrhizae
   (d) Fungi Imperfecti
   (e) Rust fungi

5. Which of the following organisms do not carry out meiosis?
   (a) Amoebas               (c) Red algae            (e) Fungi Imperfecti
   (b) Fungi                 (d) *Prochloron*

6. The largest protists belong to which of the following groups?
   (a) Red algae             (c) Plants               (e) Choanoflagellates
   (b) Trypanosomes          (d) Brown algae

7. Two phyla of protists mentioned in this text have historically, and incorrectly, been placed in the kingdom Fungi. They are
   (a) Plasmodial slime molds    (c) Cellular slime molds    (e) Smuts
   (b) Bread molds               (d) Yeasts

8. Mycorrhizae are
   (a) symbiotic associations between plants and fungi.
   (b) an association found in about 80% of all plants.
   (c) an aid in mineral uptake by plants.
   (d) all of the above.
   (e) none of the above.

# THOUGHT QUESTIONS

1. If fungi have been so successful as to become an entire kingdom with tens of thousands of species, why do you suppose that it has proved so difficult to identify the unicellar protist that gave rise to the group?

2. Both plants and animals have individuals that are very large. Whales can weigh many tons, and a mature redwood tree even more. Why do you suppose that there are no similarly large fungi?

# FOR FURTHER READING

AHMADJIAN, V., and S. PARACER: *Symbiosis: An Introduction to Biological Associations,* University Press of New England, Hanover, NH, 1986. An outstanding account of lichens and other symbiotic systems.

COCHRAN, M.F.: "Back from the Brink. Chestnuts," *National Geographic,* February 1990, pages 128-140. Nearly destroyed by the chestnut blight, the American chestnut is being rescued by dedicated scientists and volunteers.

KOSIKOWSKI, F.V.: "Cheese," *Scientific American,* May 1985, pages 52-59. A fascinating account of how more than 2000 varieties of cheese are made and how bacteria and fungi participate in the process.

LAMBRECHT, F.: "Trypanosomes and Hominid Evolution," *BioScience,* vol. 35, 1985, pages 640-646. This fascinating article charts the probable effects of sleeping sickness in determining the course of human history.

McKNIGHT, K.H., and V. McKNIGHT: *A Field Guide to Mushrooms of North America,* Peterson Field Guide Series, Houghton Mifflin, New York, 1987. Excellent guide that includes most of the common and edible species of North America.

SAFFO, M.B.: "New Light on Seaweeds," *BioScience,* vol. 37, 1987, pages 654-664. Seaweeds occur at different depths in the ocean, their photosynthetic pigments being efficient in harvesting energy for the particular light waves that reach that depth.

# Plants

Fields of wheat and lavender in France. Our earth is blanketed by plants, the sole source of the energy that fuels our lives. Some are lavishly colored, but most are green, colored by the pigments with which they absorb sunlight.

# PLANTS

## Overview

Over a relatively short time, plants evolved into a very diverse group, creating the great forests and other plant communities where the vertebrates, insects, and fungi evolved and diversified. About 266,000 species of plants are now living. Of the 12 divisions (phyla) of plants that have living representatives, the liverworts are the simplest and most distinct, having no trace of vascular tissue and no stomata; the hornworts have stomata but no vascular tissue. Seeds evolved as a means to protect the embryos in the common ancestor of the gymnosperms (4 divisions with living representatives) and angiosperms, or flowering plants. Flowers occur only in the angiosperms; they guide the activities of insects and other pollinators in such a way that pollen is carried efficiently from one flower to another of the same species. Flowering plants dominate every spot on earth except the great northern forests, the polar regions, the highest mountains, and the driest deserts. Despite their overwhelming success, this division originated relatively recently and became dominant throughout the world no more than 100 million years ago.

## For Review
*Here are some important terms and concepts that you will encounter in this chapter. If you are not familiar with them, you should review them before proceeding.*

**History of plants** (Chapter 17)

**Classification of organisms** (Chapter 23)

**Mycorrhizae** (Chapter 26)

Of the five kingdoms of living organisms, the three we have discussed so far consist of organisms that are mostly small relative to us. Bacteria and most protists are unicellular, and fungi, although multicellular, are rarely as large as your fist. But in the plant kingdom, we encounter many species that consist of large individuals. A tree can be 30 meters or more tall, with a mass of many tonnes. Plants are the dominant organisms of the terrestrial landscape, and we believe that they were the first organisms to invade the land successfully from the sea, where life first evolved and flourished (Figure 27-1).

Despite all the diversity of life on earth, only three groups are responsible for the fixation of carbon; by doing so, they provide the sustenance of not only themselves, but that of all other living organisms. The algae and photosynthetic bacteria carry out most photosynthesis in the sea, whereas plants are the dominant photosynthetic organisms on land. Plants are multicellular eukaryotic organisms that have (1) cellulose-rich cell walls; (2) chloroplasts that contain chlorophylls *a* and *b*, together with carotenoids; and (3) starch as their primary carbohydrate food reserve. This chapter discusses the characteristics of the major plant groups (Figure 27-2). Chapters 29 to 31 will treat specific aspects of the biology of plants in more detail.

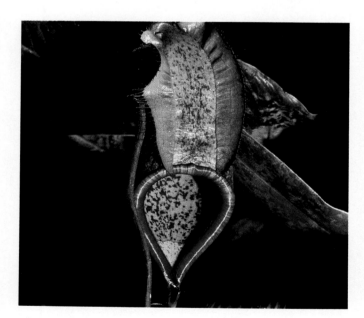

**Figure 27-1**

**Diversity among the plants.** The extraordinary evolutionary radiation of the angiosperms, which has resulted in such strange structures as the insect-trapping leaves of these Malaysian pitcher plants *(Nepenthes albomarginata)*, has made the angiosperms the dominant plant group on land for the past 100 million years.

**Figure 27-2**

**Representatives of four divisions of plants.**
**A** A moss gametophyte.
**B** A branch of Norway spruce, *Picea nigra*, in the Alps. Seeds are produced in the large cones, and pollen is produced in the smaller ones.
**C** The white flowers of this *Bougainvillea* are surrounded by bright purple bracts—modified leaves.
**D** Maidenhair fern, *Adientum pedatum*.

## THE GREEN INVASION OF THE LAND

Plants, fungi, and insects are the only major groups of organisms that occur almost exclusively on land; several other phyla, including chordates and mollusks (Chapter 28), are also well represented on land. The groups that occur almost exclusively in terrestrial habitats—the plants, fungi, and insects—all probably evolved there, whereas the chordates and mollusks clearly originated in the water. Of the groups that did evolve on land, the ancestors of the plants were almost certainly the first to become terrestrial. As you have seen with fungi (Chapter 26) and will see with insects (Chapter 28), a major evolutionary challenge in the transition from an aquatic to a terrestrial habitat is **desiccation,** the tendency of organisms to lose water to the air. In the organisms that were most successful on land, therefore, various structures evolved to conserve water, and other structures evolved to supply water to the different parts of the plant.

Relatively efficient plumbing systems evolved in some of the earliest plants, and only two divisions that have living representatives, the liverworts (division Hepatophyta) and hornworts (division Anthocerophyta) lack them completely. These two divisions, along with the mosses (division Bryophyta), which have reduced vascular systems—strands of specialized cells that conduct water and carbohydrates—have been grouped as **bryophytes.** But most scientists have concluded that the "bryophytes," like the "algae," are a group of organisms that are not directly related to one another. The remaining nine divisions of plants, which have evident and efficient conducting systems, are called the **vascular plants;** they probably share a common ancestor with the mosses.

Vascular plants are named because of their possession of *vascular* tissue, a name that comes from the Latin *vasculum,* meaning a vessel or duct. Vascular tissue consists of specialized strands of elongated cells that run from near the tip of a plant's roots through its stems and into its leaves (Figure 27-3). Different kinds of vascular tissue conducts water with dissolved minerals (nutrients), which come in mainly through the roots, and carbohydrates, which are manufactured in the green parts of the plant (Figure 27-4). Therefore water and nutrients reach all parts of the plant, as do the carbohydrates that provide energy for the synthesis of its different structures. The vascular systems of those mosses which possess them may be less efficient than those of the vascular plants, but perhaps mosses, being much smaller on the average than vascular plants, have less need for efficient conducting systems.

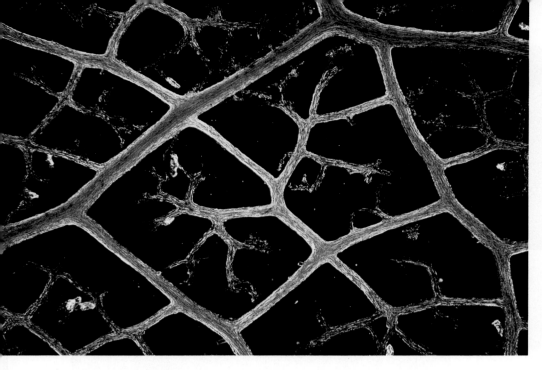

Figure 27-3

**The veins of a vascular plant.** Veins contain strands of specialized cells for conducting water with its dissolved minerals and carbohydrates, respectively. They run from the tips of the roots to the tips of the shoots and throughout the leaves, as shown here in this greatly enlarged photomicrograph of a cleared leaf.

**Figure 27-4**

**The architecture of a plant.** A vascular plant's body design is determined largely by the need to transport water through the plant. The conducting tissues of a vascular plant are shown diagrammatically. Water passes continuously in through the roots, up through the stems, and out through the leaves; at the same time, sugar molecules, manufactured as a result of photosynthesis in the leaves and green stems, pass through a parallel conducting system to all parts of the plant, where they are used for growth.

Most plants also are well protected from drying out by the **cuticle,** an outer covering formed from a waxy substance called **cutin.** The cuticles that cover the exposed surfaces of plants are impermeable to water and thus provide a key barrier to water loss. Passages do exist through the cuticle, in the form of specialized pores called **stomata** (singular, **stoma;** Figure 27-5) in the leaves and sometimes the green portions of the stems. Stomata, which occur on at least some portions of all plants except liverworts, allow carbon dioxide to pass into the plant bodies and allow water and oxygen to pass out of them. The cells that border stomata expand and contract, thus controlling the loss of water while allowing the entrance of carbon dioxide. Most plants depend on the constant flow of water in through the roots or lower portions and out through the stomata.

*Vascular plants and some mosses have a specialized plumbing system, the vascular system. The vascular system involves strands of elongated, specialized cells that transport water and dissolved nutrients and other strands that transport carbohydrate molecules.*

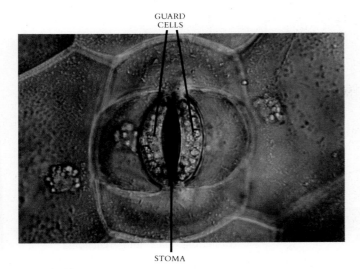

**Figure 27-5**

**A stoma.** The guard cells flanking the stoma contain chloroplasts, unlike the other epidermal cells. Water passes out through the stomata, and carbon dioxide enters by the same portals. The mechanism of opening and closing of stomata is described in Chapter 31.

In addition to their structural features, plants also developed a special relationship with fungi that evidently was a key factor in their successful occupation of terrestrial habitats. Mycorrhizae are characteristic of some 80% of all plants and are frequently seen in fossils of the earliest plants (see Chapter 26). They were probably critical to the success of these early plants in the harsh habitats available on land at the time of the first invasion.

Many other features developed gradually and aided the evolutionary success of plants on land. For example, in the first plants there was no fundamental difference between the above-ground and below-ground parts. Later, roots and shoots with specialized structures evolved, each suited to its particular environment (Chapter 29). **Leaves,** expanded areas of photosynthetically active tissue, evolved and diversified in relationship to the varied habitats that existed on land. Specializations in key reproductive features improved the methods by which plants protected their embryos and were dispersed from place to place. Flowers and seeds were of particular importance.

The more specialized roots, stems, leaves, and reproductive features that evolved in plants were important factors in the overwhelming success of the group on land. An estimated 266,000 species of this kingdom are now in existence, dominating every part of the terrestrial landscape, except the extreme polar regions and the highest mountaintops.

## THE PLANT LIFE CYCLE

### Alternation of Generations

Understanding the different kinds of life cycles that occur among plants provides an important key to understanding their evolutionary relationships. All plants exhibit **alternation of generations,** in which a diploid generation, or **sporophyte,** alternates with a haploid generation, or **gametophyte** (see Figure 24-24, C). *Sporophyte* literally means "spore plant," and *gametophyte* means "gamete plant"; these terms indicate the kinds of reproductive structures that the respective generations produce (Figure 27-6).

Most animals are diploid, and in this respect they resemble the sporophyte generation of a plant. But such animals produce eggs and sperm, which fuse directly to form a zygote. In contrast, the sporophyte generation of a plant does not produce gametes as a result of meiosis. Instead, meiosis takes place in specialized cells, called **spore mother cells,** and results in the production of haploid **spores,** the first cells of the gametophyte generation. Spores do not fuse with one another as gametes do; instead, they divide by meiosis, producing a multicellular haploid individual, the gametophyte (Figure 27-7).

In turn, the gametes—eggs and sperm—eventually are produced by the gametophyte as a result of meiosis. They are haploid, as is the gametophyte that produces them. When they fuse to form a zygote, the first cell of the next sporophyte generation has come into existence. The zygote grows into a sporophyte in which meiosis ultimately occurs.

*Plant life cycles are marked by an alternation of generations of diploid sporophytes with haploid gametophytes. Sporophytes produce spores, which are haploid and result from meiosis; the spores grow into gametophytes. Gametophytes produce gametes, which result from mitosis. The fusion of gametes produces a zygote, the first cell of the sporophyte generation.*

In ferns, mosses, and liverworts, the gametophyte is green and free living; in most other plants, it is not green and is nutritionally dependent on the sporophyte. When you look at a moss or liverwort, you see largely gametophyte tissue; the sporophytes are smaller brown or yellow structures attached to or enclosed within the tissues of the gametophyte. In most other plants, the gametophytes are always much smaller than the sporo-

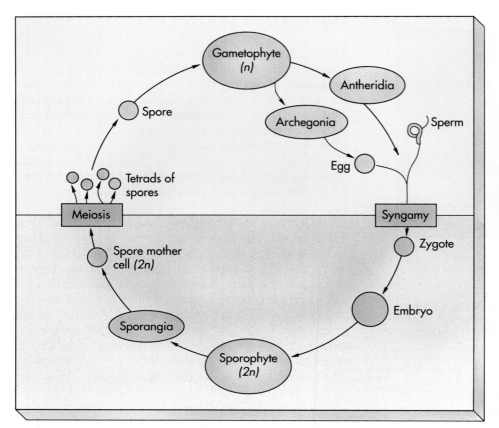

**Figure 27-6**

**A generalized plant life cycle.** In a life cycle of this kind, gametophytes, which are haploid *(n)*, alternate with sporophytes, which are diploid *(2n)*. Antheridia (male) and archegonia (female), which are the sex organs (gametangia), are produced by the gametophyte, and they in turn produce sperm and eggs, respectively. These ultimately come together in the process of syngamy to produce the first diploid cell of the sporophyte generation, the zygote. Meiosis takes place within the sporangia, the spore-producing organs of the sporophyte, resulting in the production of the spores, which are haploid and are the first cells of the gametophyte generation.

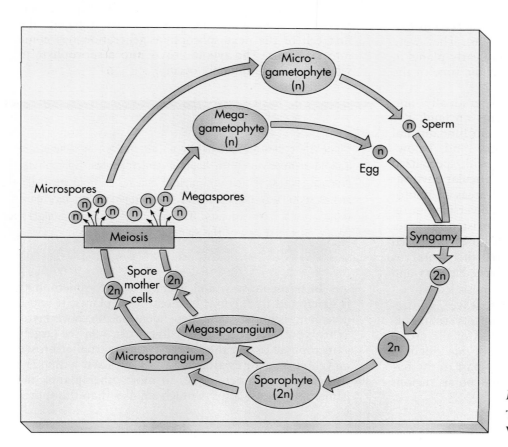

**Figure 27-7**

The life cycle of a heterosporous vascular plant.

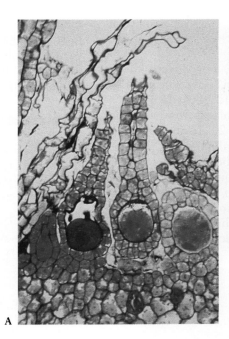
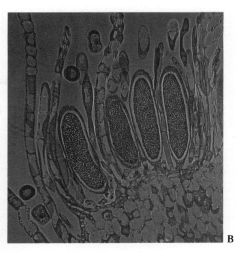

**Figure 27-8**

**Gametangia in plants.**
**A** Transection through the
archegonium of the liverwort
*Marchantia*. A single egg
differentiates within the lower,
swollen portion of the archegonium.
**B** Transections through a group of
moss antheridia. The smaller cells in
each of these elongate structures will
give rise to sperm which, when
liberated by the rupturing
antheridium, swim through free water
to the mouth of the archegonium.

A

B

phytes; in many, they are nutritionally dependent on the sporophytes and enclosed within their tissues. When you look at a vascular plant, what you see is a sporophyte, with very rare exceptions.

## The Specialization of Gametophytes

The gametes of the first plants differentiated within specialized organs called **gametangia** (singular, **gametangium**), in which the eggs and sperm were surrounded by a jacket of cells. Complex multicellular gametangia of this sort are still found in all of the members of the plants kingdom that do not form seeds and in some seed-forming plants as well. Eggs and sperm are formed within different kinds of gametangia. Those in which eggs are formed are called **archegonia** (singular, **archegonium**), and those in which sperm are formed are called **antheridia** (singular, **antheridium**). An archegonium produces only one egg; an antheridium produces many sperm. These structures usually look very different from one another (Figure 27-8). In some plants, including the ferns, antheridia and archegonia occur together on the same gametophyte. In others, and typically in mosses, the two kinds of gametangia are on separate gametophytes. In the more advanced vascular plants, including all but a very few of the vascular plants that form seeds, the gametangia have been lost during the course of evolution. The eggs or sperm develop directly from individual cells of the respective gametophyte. Some of the gametophytes bear only eggs, and others bear only sperm.

When one kind of gametophyte bears antheridia and another kind bears archegonia, the two kinds of gametophytes may look different from one another. If they do, the gametophytes which form antheridia are called **microgametophytes,** and those which form archegonia are called **megagametophytes.** These two formidable-looking terms literally mean no more than "small gametophytes" and "large gametophytes." In nearly all plants that do have two kinds of gametophytes, the gametophytes arise from two kinds of spores, **microspores** and **megaspores.** Plants which produce two different-looking spores are called **heterosporous;** those which produce only one kind of spore are called **homosporous.**

Spores are formed as a result of meiosis in the sporophyte generation. Their differentiation occurs within specialized multicellular structures called **sporangia** (singular, **sporangium**). If a plant forms both megaspores and microspores, each of these will be formed in a different kind of sporangium called, respectively, **megasporangia** and **microsporangia.**

As we describe the principal divisions of plants, you will see a progressive reduction of the gametophyte from group to group and increasing specialization for life on the land, culminating with the remarkable structural adaptations of the flowering plants.

## MOSSES, LIVERWORTS, AND HORNWORTS

Now regarded as separate divisions, not directly related to one another, the mosses (division Bryophyta), liverworts (division Hepatophyta), and hornworts (division Anthocerophyta) are simple plants that retain many primitive features (Figures 27-2, A, and 27-9). They were formerly collectively called the "bryophytes," but that term is no longer believed to reflect direct relationship between these groups. Mosses and liverworts are especially common in relatively moist places, both in the tropics and in temperate regions. In the Arctic and Antarctic, mosses are the most abundant plants. Not only do they boast the most individuals in these harsh regions, but also the greatest number of species. Regardless of where they grow, mosses, liverworts, and hornworts require free water at some time of the year to reproduce sexually because their sperm swim through it to the mouth of the archegonium.

*The term "bryophyte" refers to the mosses, liverworts, and hornworts, which are now understood to be divisions of relatively simple plants that are not directly related to one another.*

### Mosses

Mosses are a diverse group of about 9500 species of small plants. Many other small, tufted plants are mistakenly called "mosses"; for example, Spanish moss is actually a flowering plant, a relative of the pineapple. In many mosses, the stems of the gametophytes have a central strand of water-conducting cells. Unlike the similar conducting cells that occur in vascular plants, the conducting cells of mosses lack specialized wall thickenings. In some mosses, these conducting cells are surrounded by carbohydrate-conducting cells resembling those of the vascular plants. Although these structures are less complex than those of the vascular plants, they probably had a common origin with them.

In mosses, as in liverworts and hornworts, the sporophytes are borne on the gametophytes, from which they derive their food. Moss sporophytes take 6 to 18 months to develop and are generally elevated on a stalk. The life cycle of a moss is summarized in Figure 27-10.

### Liverworts and Hornworts

Liverworts constitute a division of about 6000 species of plants that are generally inconspicuous individually but which may form masses in moist places. Their name dates from the Middle Ages and relates to the liver-shaped outline of the gametophyte in some genera. Liverworts lack conducting tissue, cuticle, and stomata and are the simplest of all living plants. Their sporophytes often remain enclosed in gametophyte tissue until they are mature; in outline, their life cycle resembles that of the mosses. The six genera and approximately 100 species of hornworts do possess stomata, together with a number of unique features that separate them from both mosses and liverworts and suggest strongly that there is no direct evolutionary connection between the members of these groups.

**Figure 27-9**

**Simple plants.**
**A** A liverwort, *Marchantia*. The sporophytes are borne within the tissues of the umbrella-shaped structures that arise from the surface of the flat, green, creeping gametophyte. These particular structures develop archegonia within their tissues; another kind of similar structure, borne on different plants of *Marchantia*, produces the antheridia.
**B** Hair-cup moss, *Polytrichum*. The leaves below belong to the gametophyte. Each of the yellowish-brown stalks, with the capsule at its summit, is a sporophyte. Although moss sporophytes may be green and carry out a limited amount of photosynthesis when they are immature, they are soon completely dependent on the gametophyte in a nutritional sense.

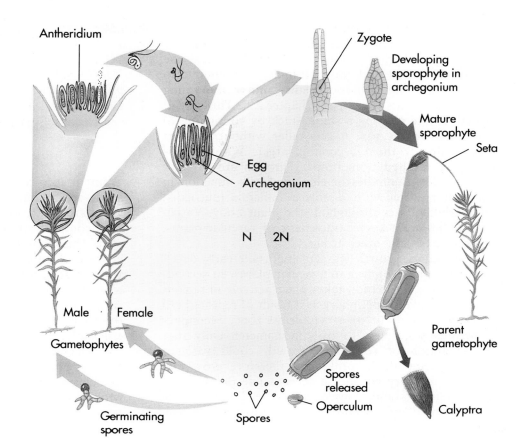

**Figure 27-10**

**Moss life-cycle.** On the gametophytes, which are haploid, the sperm are released from antheridia. They then swim through free water to the archegonia and down their neck to the egg. Fertilization takes place there; the resulting zygote develops into a sporophyte, which is diploid. The sporophyte grows on the gametophyte and eventually produces spores as a result of meiosis. These spores are released and germinate, giving rise to gametophytes. The gametophytes initially are threadlike; they grow along the ground. Ultimately, buds form on them, from which leafy gametophytes arise.

*Mosses often have specialized conducting strands, a cuticle, and stomata, whereas liverworts, which are the most primitive living plants, lack all three features. In the course of the life cycles of both groups, the sporophyte remains attached to, and nutritionally dependent on, the gametophyte.*

The vascular plants are characterized by (1) their large, dominant, and nutritionally independent sporophytes; (2) efficient conducting tissues (see Figures 27-3 and 27-4); (3) specialized leaves, stems, and roots; (4) cuticles and stomata (see Figure 27-5); and (5) the evolution of seeds in some. The vascular plants certainly seem to be derived from an immediate common ancestor, which is apparently closely related to the ancestor of the mosses.

## VASCULAR PLANTS

The first vascular plants appeared approximately 430 million years ago. Fossils of these early plants are rare. Interestingly, fungi (mycorrhizae) are found associated with the roots of many of these fossils, suggesting that this symbiosis may have played a key role in the successful invasion of the land.

*The vascular plants are characterized by their dominant sporophytes; efficient conducting tissues; specialized stems, leaves, and roots; cuticles and stomata; and, in some of them, seeds.*

## Growth of the Vascular Plants

Early vascular plants were characterized by **primary growth,** growth that results from cell division at the tips of the stems and roots. Primary growth is also characteristic of herbaceous plants and of the young stems of all plants today (see Chapter 29). This sort of growth is like erecting a smokestack by continuing to add bricks to the top; it gets taller without the part that is already there becoming larger. Within stems that have resulted from primary growth are **vascular bundles.** These bundles of elongated cells function in conducting water with dissolved minerals (nutrients) and carbohydrates throughout the plant body. Vascular bundles are a key characteristic of vascular plants, not present in any other group.

**Secondary growth** was an important early development in the evolution of vascular plants. In secondary growth, cell division takes place actively in regions around the plant's periphery as a result of repeated cell divisions that occur in a cylindrical zone. Secondary growth causes a plant to grow in diameter. Only after the evolution of secondary growth could vascular plants develop thick trunks and therefore grow tall. This evolutionary advance made possible the development of forests and, consequently, the domination of the land by plants. According to the fossil record, secondary growth had evolved independently in several different groups of vascular plants by the middle of the Devonian Period, approximately 380 million years ago.

*Primary growth results from cell division at the tips of stems and roots, whereas secondary growth results from the division of a cylinder of cells around the plant's periphery.*

## Conducting Systems of Vascular Plants

Two types of conducting elements in the earliest plants have become characteristic of the vascular plants as a group. **Sieve elements** are soft-walled cells that conduct carbohydrates away from the areas where they are manufactured. **Tracheary elements** are hard-walled cells that transport water and dissolved minerals up from the roots. Both kinds of cells are elongated, and both occur in strands. More specialized versions of each occur in the flowering plants, or angiosperms, than in other vascular plants. Sieve elements are the characteristic cell types of a tissue called **phloem;** tracheary elements are characteristic of a tissue called **xylem.** In primary tissues, which result from primary growth, these two types of tissue are often associated with one another in the same vascular strands.

*Water and nutrients are carried in the xylem, which consists primarily of hard-walled cells called tracheary elements. Carbohydrates, in contrast, are carried in the phloem, which consists of soft-walled cells called sieve elements.*

## What is a Seed?

Seeds are a characteristic feature of some groups of vascular plants, collectively known as **seed plants.** A **seed** contains an embryo surrounded by a protective coat. This embryo's development has been temporarily arrested (Figure 27-11). Seeds are units by means of which plants, being rooted in the ground, are dispersed to new places. Many seeds have devices, such as the wings on the seeds of pines or maples or the plumes on the seeds of a dandelion, that help them to travel very efficiently.

The seed is a crucial adaptation to life on land because it protects the embryonic plant from drying out or being eaten when it is at its most vulnerable stage. Most kinds of seeds have abundant food stored in them, either inside the embryo or in specialized storage

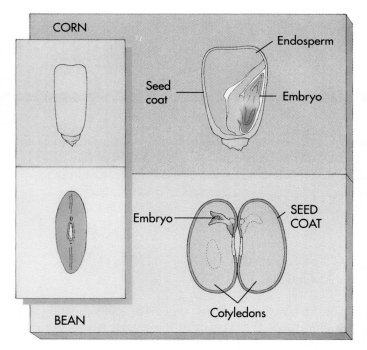

*Figure 27-11*

**Basic structure of seeds.** A seed in which the endosperm is still present at maturity (corn) and one in which it has been incorporated in the embryo cotyledons (bean).

tissue. The rapidly growing young plant uses the seed as a ready source of energy; the seed thus plays the same role as the yolk of an egg. The evolution of the seed was clearly a critical step in the domination of the land by plants.

## SEEDLESS VASCULAR PLANTS

The members of four divisions of vascular plants with living representatives do not form seeds. The ferns (division Pterophyta) are members of the most familiar division of seedless vascular plants (Figure 27-2, C), which includes about 12,000 living species. The club mosses and their relatives (division Lycophyta) include four living genera with a total of about 1000 species. The horsetails, with a single genus (*Equisetum*) and about 15 species, make up the division Sphenophyta. The fourth division, the Psilophyta, consists of two genera, the whisk ferns.

The life cycle of a fern (Figure 27-12) differs from that of a moss primarily in the much more complex development, independence, and dominance of the fern's sporophyte.

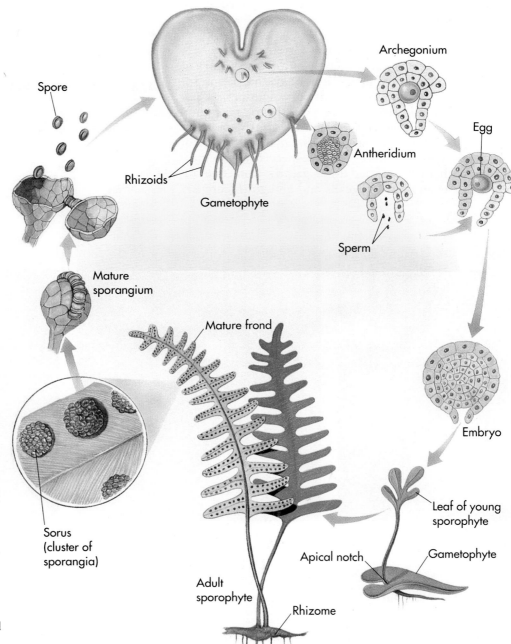

**Figure 27-12**

**Fern life-cycle.** The gametophytes, which are haploid, grow in moist places. Eggs and sperm develop in archegonia and antheridia, respectively, on their lower surface. The sperm, when released, swim through free water to the mouth of the archegonium, entering and fertilizing the single egg. Following the fusion of egg and sperm to form a zygote—the first cell of the diploid sporophyte generation—the zygote starts to grow within the archegonium. Eventually, the sporophyte become much larger than the gametophyte—it is what we know as a fern plant. On its leaves occur clusters of sporangia, within which meiosis occurs and spores are formed. The release of these spores, which is explosive in many ferns, and their germination, lead to the development of new gametophytes.

Labels in figure: Spore · Archegonium · Rhizoids · Antheridium · Egg · Gametophyte · Sperm · Mature sporangium · Embryo · Mature frond · Sorus (cluster of sporangia) · Leaf of young sporophyte · Apical notch · Gametophyte · Adult sporophyte · Rhizome

# GYMNOSPERMS

In four of the divisions of living seed plants, the **ovules**—the structures that eventually become the seeds after fertilization—are not completely enclosed by the tissues of the parent individual on which they are carried at the time of pollination. The members of these four divisions, the conifers, cycads, ginkgo, and gnetophytes, are called **gymnosperms** (Greek *gymnos*, naked, and *sperma*, seed; in other words, naked-seeded plants). In fact, the seeds of gymnosperms often become enclosed by the tissues of their parents by the time they are mature, but their ovules are naked at the time of pollination.

## The Cycads, Ginkgo, and the Gnetophytes

True seeds apparently evolved only once in the history of plants, and the four living divisions of gymnosperms, together with the flowering plants, or angiosperms, are thought to have descended from a seed-bearing common ancestor. But the "gymnosperms" are clearly an artificial group, including all seed-bearing plants that do not possess the special features of angiosperms. The four divisions with living species differ greatly from one another. For example, the cycads and ginkgo (Figure 27-13, *A* and *B*) still have motile sperm, as do all seedless plants. The cycads (division Cycadophyta) comprise 10 genera and about 100 species, widespread throughout the warmer portions of the world; one of the most familiar kinds is the sago palm. But this is not a true palm, despite the palmlike appearance of its leaves; palms are flowering plants. The ginkgo (*Ginkgo biloba*) is the only living species of the division Ginkgophyta. Its seeds resemble small plums, with a fleshy, unpleasantly scented outer covering. The ginkgo is not known anywhere as a wild plant, but it was preserved in cultivation around the temples and in the gardens of Japan and China. The division Gnetophyta comprise 3 genera, including *Ephedra*, or Mormon tea, and about 70 species. It is of special interest as the group most closely related to the flowering plants.

*Seed plants are apparently derived from a common ancestor and consist of four divisions of "gymnosperms," together with the angiosperms, or flowering plants.*

A

B

C

**Figure 27-13**

Seed plants.
A An African cycad, *Encephalartos kosiensis*. The cyads are seed plants with fernlike leaves.
B Maidenhair tree, *Ginkgo biloba,* the only living representative of a group of plants that was abundant 200 million years ago. Among living seed plants, only the cycads and *Ginkgo* have swimming sperm.
C Engelmann spruce, *Picea engelmannii,* in the Rocky Mountains.

## The Conifers

The most familiar division of seed plants, other than the angiosperms, is the division Coniferophyta, the conifers (Figures 27-2, *B*, and 27-13, *C*). With about 550 living species, the conifers include such tree groups as pines, spruces, firs, yews, redwoods, bald cypress, junipers, cedars, and others. Most conifers are evergreens, and they form vast forests in many of the temperate and cooler regions of the globe. Many are very important commercially as major sources of timber and pulp. In the conifer life cycle (Figure 27-14), pollen grains, which contain the male gametes, are carried by the wind to the vicinity of the ovules, the structures holding the female gametes. The ovules are carried on cone scales (in pines and their relatives) or similar structures. Within the seeds of conifers and other gymnosperms, stored food is provided for the embryo within tissue derived from the megagametophyte.

## THE FLOWERING PLANTS

Flowering plants, or angiosperms (division Anthophyta), are the dominant photosynthetic organisms nearly everywhere on land. This great group of some 235,000 species includes our familiar trees (except conifers), shrubs, herbs, grasses, vegetables, and grains—in short, nearly all of the plants that we see every day. Virtually all of our food is derived, directly or indirectly, from the flowering plants; in fact, more than half of the calories we consume come from just three species: rice, corn (maize), and wheat (see Chapter 23). In addition, angiosperms are valuable sources of timber, pulp, textiles, medicines, waxes, resins—a wide array of important products, with many more awaiting discovery.

In a sense, the remarkable evolutionary success of the angiosperms is the culmination of the plant line of

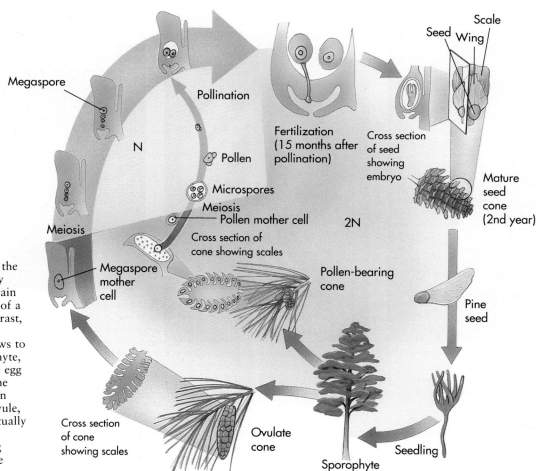

**Figure 27-14**

**Pine life-cycle.** In all seed plants, the gametophyte generation is greatly reduced. A germinating pollen grain is the mature microgametophyte of a pine; megagametophytes, in contrast, develop within the tissues of the ovule. When the pollen tube grows to the vicinity of the megagametophyte, sperm are released, fertilizing the egg and producing a zygote there. The development of the zygote into an embryo takes place within the ovule, which matures into a seed. Eventually the seed falls from the cone and germinates, the embryo resuming growth and becoming a new pine tree.

evolution. Our earliest record of flowering plants is from approximately 123 million years ago, in the early Cretaceous. Angiosperms, therefore, are by far the youngest of all the divisions of plants, despite their world dominance for most of the last hundred million years.

Angiosperms differ from other seed plants in their possession of flowers and fruits, the characteristic structures of the division Anthophyta. The basic structure of a flower consists of four **whorls;** a whorl is a circle of parts present at a single level along an axis. The two outer whorls of the flower are mainly concerned with protecting the flower and attracting insects and other visitors to it; the inner whorls contain the male and/or female gametophytes. The third whorl consists of the **stamens,** which produce pollen grains.

The fourth and innermost flower whorl, the **gynoecium,** consists of the **carpels.** In angiosperms, the ovules are enclosed within the carpels, and pollination is indirect. This contrasts with the other divisions of seed plants, the gymnosperms, in which the ovules are carried on the surface of scales, and the pollen is often blown to them by the wind (Figure 27-15).

In many angiosperms the pollen grains are carried from flower to flower by insects; in others the pollen is transported passively in the wind, as it is in most other seed plants. Chapter 30 discusses the details of angiosperm reproduction. Briefly, however, the pollen reaches a specialized area of the carpel, called the **stigma,** where each pollen grain then forms a **pollen tube,** which grows through the tissue of the carpel and may ultimately reach the egg. Within each pollen grain are two sperm, which travel down the tube. One sperm fuses with the egg to form a zygote.

During meiosis, most angiosperms produce eight haploid nuclei within an **embryo sac.** One of these nuclei becomes an egg. Two others, held together in a large cell at the center of the embryo sac, are called **polar nuclei.** The second sperm from the pollen grain fuses with these two polar nuclei, forming the **primary endosperm nucleus.** Because the primary endosperm nucleus is formed from the fusion of three haploid ($1n$) nuclei, it is triploid ($3n$), whereas the zygote is diploid ($2n$). The primary endosperm nucleus divides rapidly by mitosis, giving rise to a specialized kind of triploid

A

B

C

*Figure 27-15*

**The remarkable diversity of angiosperm flowers.**
**A** Wild geranium. *Geranium,* a woodland plant, showing the 5 free petals, 10 stamens, and fused carpel.
**B** Tiger lily, *Lilium canadense,* with six free, colored, attractive flower parts (they cannot be separated in lilies into sepals and petals) and six free anthers. The carpels are fused together.
**C** Fragrant water lily, *Nymphaea odorata,* a flower with numerous free, spirally arranged parts that intergrade with one another in form.

## Figure 27-16

**Angiosperm life-cycle.** As in the pine, the sporophyte is the dominant generation. Eggs form within the megagametophyte, or embryo sac, inside the ovules, which in turn are enclosed in the carpels—members of the inner whorl of the flower. The carpel is differentiated in most angiosperms into a slender portion, or style, ending in a stigma, the surface on which the pollen grains germinate. The pollen grains, meanwhile, are formed within the sporangia of the anthers and complete their differentiation to their mature, three-celled stage either before or after grains are shed. Fertilization is distinctive in angiosperms, being a double process. A sperm and an egg come together, producing a zygote; at the same time, another sperm fuses with the two polar nuclei, producing the primary endosperm nucleus, which is triploid. Both the zygote and the primary endosperm nucleus divide mitotically, giving rise, respectively, to the embryo and the endosperm. The endosperm is the tissue, unique to angiosperms, that nourishes the embryo and young plant.

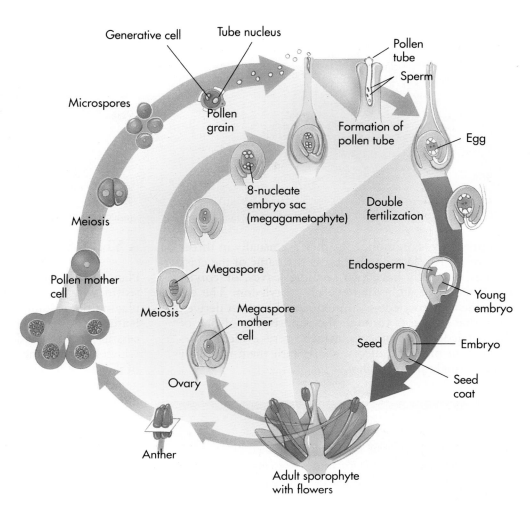

nutritive tissue called the **endosperm,** one of the distinctive features of angiosperms. In different groups of angiosperms, the endosperm either is digested by the growing embryo or is retained in the mature seed to nourish the germinating seedling. The unique process whereby one sperm fuses with the egg while the other fuses with the polar nuclei is called **double fertilization.** This process is characteristic of angiosperms and is found nowhere else. Figure 27-16 outlines the angiosperm life cycle.

*Double fertilization, a process that is unique to the angiosperms, occurs when one sperm nucleus fertilizes the egg and the second one fuses with the polar nuclei. These two events result in the formation of the zygote and the primary endosperm nucleus, respectively. The latter divides to produce the endosperm, the nutritive tissue that occurs in the seeds of angiosperms.*

The two outer whorls of the angiosperm flower are (1) the outer **calyx,** the individual parts of which, the **sepals,** are often green and leaflike, surrounding the flower, and (2) the inner **corolla,** the individual parts of which are called **petals** (Figure 27-17). The petals are often colored and attractive to insects and other animals that visit the flowers and spread their pollen from flower to flower. Chapter 30 presents more detail about how the features of flowers have evolved.

### Monocots and Dicots

The two classes of angiosperms, division Anthophyta, are (1) the Monocotyledones, or **monocots** (about 65,000 species) and (2) the Dicotyledones, or **dicots** (about 170,000 species). The monocots include lilies, grasses, cattails, palms, agaves, yuccas, pondweeds, orchids, and irises. The dicots include the great majority of familiar angiosperms of all kinds.

Monocots and dicots differ from one another in several features. For example, the venation of monocot leaves usually consists of parallel veins, whereas dicot leaves generally have netlike (**reticulate**) veins. In dicot flowers, the members of a given whorl generally occur

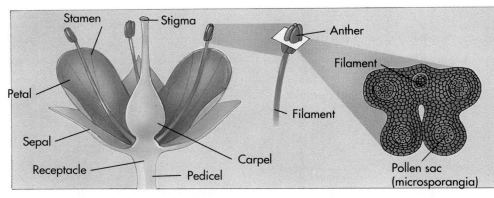

Stamen — Stigma — Anther

Petal — Filament

Sepal — Filament

Receptacle — Carpel

Pedicel

Pollen sac
(microsporangia)

**Figure 27-17**

**Simplified diagram of an angiosperm flower.** The members of the outermost whorl of the flower, the sepals, make up the calyx, and the other whorls, working inward, are the corolla (made up of petals), and androecium (anthers), and gynoecium (carpels). The microsporangia differentiate within the anthers, as do the megasporangia within the carpels.

in fours or fives; in monocots, they are usually in threes (Figure 27-18). Dicot embryos, as the name of the class implies, generally have two seedling leaves, or **cotyledons**; monocot embryos have one cotyledon. The arrangement of cotyledons in monocots has been derived from that in dicots through the suppression of one of the cotyledons. Monocots, which were evidently derived from primitive dicots, share with them a similar kind of single-pored pollen.

*The two classes of angiosperms are Monocotyledones (monocots) and Dicotyledones (dicots). Monocots have one cotyledon, or seedling leaf; usually parallel venation in their leaves; and flower parts often in threes. Dicots have two cotyledons; usually netlike (reticulate) venation; and flower parts in fours or fives.*

## A VERY SUCCESSFUL GROUP

Plants have come to dominate the land all over the world, with vascular plants, and especially angiosperms, playing the leading role in most habitats. Many of the features of angiosperms seem to be correlated with successful growth under arid and semiarid conditions, which have become more widespread during the history of this division. Flowers effectively ensure the transfer of gametes over substantial distances and therefore promote outcrossing because insects, birds, and other animals are more precise in carrying pollen from one plant to another than is the wind. Fruits, often carried from place to place by animals, would have been especially effective in assisting the spread of some of these early angiosperms from one patch of suitable habitat to another. The tough, often leathery leaves of the flowering plants, their efficient cuticles and stomata, and their specialized conducting elements were all important factors in survival and growth under arid conditions, just as they are in stressful conditions at present. The many natural insecticides produced by these plants (Chapter 20) were essential for their survival as well. As the early angiosperms evolved, all of the features that contributed to their success continued to evolve rapidly.

A                                                                                          B

**Figure 27-18**

**Monocots and dicots have different flowers.** The bright red flowers of fire-pink, *Silene virginica*, with their flower parts in fives, are typical of the dicots (**A**); those of *Trillium ozarkanum* variety *pusillum*, with flower parts in three, are typical of the monocots (**B**).

# SUMMARY

1. The evolution of plants' conducting tissues, cuticle, stomata, and seeds has made them less dependent on free water.

2. Plants all have an alternation of generations, in which gametophytes, which are haploid, alternate with sporophytes, which are diploid. The spores that sporophytes form as a result of meiosis grow into gametophytes, which produce gametes (sperm and eggs) as a result of meiosis.

3. There are 12 divisions of plants with living representatives, of which 7 are seedless and 5 form seeds. Liverworts (division Hepatophyta) are the simplest living plants. Traditionally, they have been grouped with other simple plants, the mosses (division Bryophyta) and hornworts (division Anthocerophyta) as the "bryophytes."

4. Vascular plants, consisting of nine living divisions, have well-defined conducting strands, some specialized to conduct water and dissolved materials (xylem), and others specialized to conduct carbohydrate molecules that the plants manufacture (phloem).

5. Liverworts, mosses, hornworts, and ferns have green, nutritionally independent gametophytes. In the first three divisions, the sporophytes are nutritionally dependent on the gametophytes.

6. In all seed plants and a few ferns, the gametophytes are either female (megagametophytes) or male (microgametophytes). Megagametophytes produce only eggs; microgametophytes produce only sperm. The megagametophytes and microgametophytes, in turn, are produced from distinct kinds of spores, megaspores and microspores.

7. In angiosperms, the ovules are enclosed at the time of pollination, which is indirect and often mediated by animals; the other divisions of seed plants, the gymnosperms, have exposed ovules at the time of pollination, and pollination usually does not involve animals.

8. Angiosperm seeds have a unique kind of nutritive tissue, the endosperm, derived as part of the process of double fertilization, which is unique to the division. In other seed plants, the role of the endosperm is played by the tissue of the megagametophyte.

9. Angiosperms evolved at least 123 million years ago, in the early Cretaceous Period, and have been the dominant plants on land for about 100 million years.

10. Angiosperm flowers typically consist of four whorls. From the inside outward, these whorls are called the gynoecium (consisting of the carpels), androecium (stamens), corolla (petals), and calyx (sepals).

11. There are two classes of angiosperms. Monocotyledones, or monocots, include about 65,000 species, largely with parallel venation, flower parts in threes, and one cotyledon. Dicotyledones, or dicots, include approximately 170,000 species, usually with reticulate venation, flower parts in fours or fives, and two cotyledons.

12. Angiosperms were successful on land because of their relatively drought-resistant vegetative features, including their specialized vascular systems, cuticles, and stomata. In addition, flowers facilitate the precise transfer of pollen and therefore outcrossing, even when the stationary individual plants are widely separated. Fruits, with their complex adaptations, enable angiosperms to disperse widely from one favorable patch of habitat to another.

## REVIEW

1. In plants, gametes fuse to form a diploid zygote that grows into the _____ generation.

2. How do moss sperm get to the archegonium?

3. A sperm is haploid. Endosperm is _____.

4. Angiosperms differ from other divisions of seed plants in that their ovules are _____ at the time of pollination.

5. Name the two classes of angiosperms.

## SELF-QUIZ

1. Which of the following were probably the key features in allowing the earliest plants to invade the land?
   (a) Leaves
   (b) Roots
   (c) Mycorrhizae
   (d) A vascular system
   (e) A cuticle

2. Which of the following are found *only* in vascular plants?
   (a) Cuticle
   (b) Carbohydrate molecules
   (c) A vascular system
   (d) Seeds
   (e) Stomata

3. Which plants *lack* antheridia and archegonia?
   (a) Gametophytes
   (b) Bryophytes
   (c) Ferns
   (d) Angiosperms
   (e) None of the above

4. Which of the following features distinguish all mosses from ferns?
   (a) Lack of a vascular system
   (b) Sporophyte nutritionally dependent on gametophyte
   (c) Presence of a nutritionally independent gametophyte
   (d) Alternation of generations
   (e) Free water needed for fertilization

5. Arrange the following whorls of a flower in order, from the inside out:
   (a) Androecium
   (b) Gynoecium
   (c) Calyx
   (d) Corolla

6. An example of a seedless vascular plant is
   (a) a hornwort.
   (b) a fern.
   (c) a pine tree.
   (d) a fruit tree.
   (e) a lily.

7. Which of the following are predominantly characteristics of the angiosperms?
    (a) Double fertilization
    (b) A vascular system
    (c) Flowers
    (d) Fruits
    (e) Insect pollination

# THOUGHT QUESTION

1. Why do both mosses and ferns require free water to complete their reproductive cycles? At which stage is the water required? Do angiosperms also require free water to complete their reproductive cycle? What are the reasons for the difference, if any?

# FOR FURTHER READING

GENSEL, P.G., and H.N. ANDREWS: "Evolution of Early Land Plants," *American Scientist,* vol. 75, 1987, pages 478-489. Excellent account of what we know about the first plants to invade the land.

HEYWOOD, V.H. (editor): *Flowering Plants of the World,* Mayflower Books, Inc., New York, 1978. An excellent account of the diversity of angiosperms.

NICKLAS, K.J.: "Aerodynamics of Wind Pollination," *Scientific American,* July 1987, pages 90-95. The capture of pollen blown about by the wind is not as random as it seems.

NORSTOG, K.: "Cycads and the Origin of Insect Pollination," *American Scientist,* vol. 75, 1987, pages 270-279. Angiosperms are not the only insect-pollinated plants.

PAOLILLO, D.J.: "The Swimming Sperms of Land Plants," *BioScience,* vol. 31, 1981, pages 367-373. Excellent account of the sperm in those groups of plants which retain them.

SCHOFIELD, W.B.: *Introduction to Bryology,* Macmillan Publishing Company, New York, 1985. An outstanding analysis of the liverworts, mosses, and hornworts, with many general chapters.

# CHAPTER ■ 28

# Animals

Most of the world's vertebrate carnivores are becoming rare as humans encroach on natural habitats. Timber wolves such as these used to roam over much of the United States.

# ANIMALS

## Overview

In this chapter we shall trace the long evolutionary history of the animals. All of the major animal phyla first evolved in the sea during the Cambrian Period. All but two of them are still water-dwellers—only the phyla containing vertebrates and the arthropods (spiders and insects) are fully terrestrial. The earliest animals to evolve in the sea had no distinct tissues or organs. Later there evolved a succession of animals with well-defined tissues—first solid worms and then worms with progressively more complex body cavities. The next innovation in body design was segmentation, in which bodies were assembled from similar subunits, like the cars of a train. In the arthropods, jointed appendages evolved. A radical change in the organization of the embryo accompanied the evolution of the sea stars and their relatives, the echinoderms, and of our own phylum, the chordates.

## For Review    *Here are some important terms and concepts that you will encounter in this chapter. If you are not familiar with them, you should review them before proceeding.*

**Major features of evolutionary history** (Chapter 17)

**Vertebrate evolution** (Chapter 18)

**Classification** (Chapter 23)

**The five kingdoms of life** (Chapter 23)

**Choanoflagellates** (Chapter 26)

**Figure 28-1**

**A land dweller.** This spider belongs to the arthropod phylum. The arthropod phylum and our own phylum, the chordates, are the two most successful land-dwelling phyla.

One kingdom of life remains for us to discuss: the animal kingdom. Animals are the most familiar organisms to most people. We are animals, and so are fleas and worms and jellyfish. Animals are multicellular heterotrophs; there are no photosynthetic animals and no unicellular ones. The animal kingdom evolved from protists in water, and much of its evolution since then has been in the sea. Of the 35 phyla in the animal kingdom, only 2 have been overwhelmingly successful as land-dwellers: the arthropod phylum of spiders and insects (Figure 28-1) and our own chordate phylum, which includes the vertebrates. In this chapter we follow that evolutionary journey, tracing the path that has led from the protists to sponges and spiders and bears.

We start with the simplest animals, the sponges, and then talk about jellyfish, snails, and a variety of different worms. We then consider insects, the most successful of animal groups, followed by the sea stars and other echinoderms, which constitute the phylum most closely related to our own, and finish with chordates such as ourselves. The evolutionary relationships between the major groups of animals are discussed throughout this chapter and are presented visually in Figure 28-2.

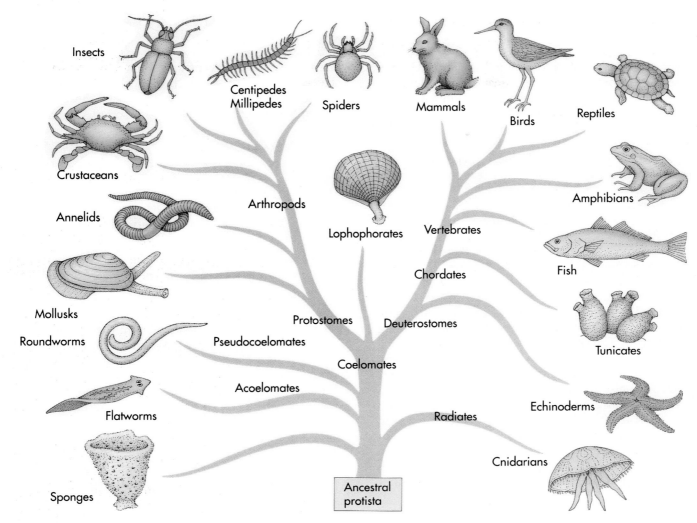

**Figure 28-2**
The animal ancestral tree.

## SOME GENERAL FEATURES OF ANIMALS

Animals are extraordinarily diverse in form. They range in size from a few that are smaller than many protists to others like the truly enormous whales and giant squids. The cells in animals are exceedingly diverse in form and function and are of fundamental importance in making up their complex bodies, which function under a wide variety of circumstances. Except for the sponges, animal cells are organized into **tissues,** which are groups of cells combined into a structural and functional unit. In most animals the tissues are organized into **organs,** which are complex structures that are made up of two or more kinds of tissues.

Most animals reproduce sexually. Their gametes—eggs and sperm—do not divide by mitosis. With few exceptions, animals are diploid; their ga-

metes are the only haploid cells in their life cycles. The complex form of a given animal develops from a zygote formed from the union of male and female gametes. In a characteristic process of embryonic development, discussed in Chapter 43, the zygote first undergoes a series of mitotic divisions and becomes a hollow ball of cells called the blastula. This developmental stage occurs in all animals. In most animals the blastula folds inward at one point to form a hollow sac with an opening at one end called the **blastopore.** An embryo with a blastopore is called a **gastrula.** The subsequent growth and movement of the cells of the gastrula produce the digestive system. The details of early embryonic development differ widely from one phylum of animals to another, as you will see. These details often provide important clues to the evolutionary relationships among the phyla.

# THE SPONGES—ANIMALS WITHOUT TISSUES

Sponges are the simplest of animals. The cells of a sponge are not organized into tissues. Sponges lack organs, and most sponges completely lack symmetry. The bodies of sponges consist of little more than masses of cells embedded in a gelatinous matrix. There is relatively little coordination among the cells—a sponge can pass through a fine silk mesh, with individual clumps of cells separating, and then reaggregate on the other side.

The body of a young sponge is shaped like a sac or vase. The body wall is covered on the outside by a layer of flattened cells called the **epithelial wall.** Facing into the internal cavity are specialized, flagellated cells called **choanocytes,** or collar cells (Figure 28-3). Between the choanocytes and epithelial wall is a gelatinous, protein-rich matrix in which occur various types of amoeboid cells, minute needles of calcium carbonate or silica called **spicules,** and fibers of a tough protein called **spongin.** The spicules and spongin may occur together, or only one of the two may be present. These elements not only strengthen the body of the sponge, but they may also deter predators.

The body of a sponge is perforated by tiny holes. The name of the phylum, Porifera, refers to this system of pores. The beating of the flagella of the many choanocytes that line the body cavity draws water in through the pores and drives it through the sponge, thus providing the means by which the sponge acquires food and oxygen and expels wastes. The flagellum of each choanocyte beats independently. In some sponges, 1 cubic centimeter of tissue can propel more than 20 liters of water a day in and out of the sponge body! The movement of water through the pores and channels of sponges is a primitive form of the circulatory systems that occur in other, more complex animals.

*Sponges are unique in the animal kingdom because they possess choanocytes, which are special flagellated cells whose beating drives water through the body cavity.*

The microscopic examination of individual choanocytes (see Figure 28-3, *B*) reveals an important substructure: the base of each flagellum is surrounded by a collar of small, hairlike projections that resemble a picket fence. The beating flagellum of the choanocyte draws water through the openings in the collar. Any food particles in the water are also drawn in and are trapped. The trapped particles pass directly through the plasma membrane and are later digested either by the choanocyte itself or by a neighboring amoeboid cell.

Each choanocyte closely resembles a protist with a single flagellum and is exactly like the group of unicel-

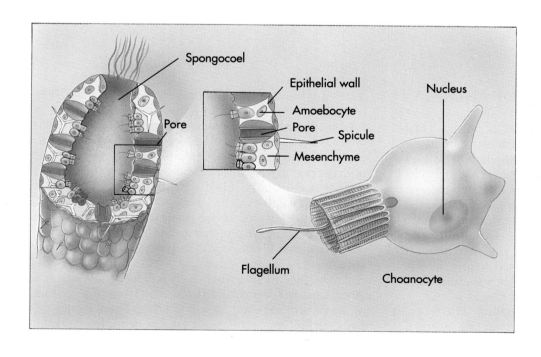

**Figure 28-3**

Anatomy of a sponge.

**Figure 28-4**

**Diversity in sponges.**
**A** A barrel sponge, a large sponge in which the form is somewhat organized.
**B** Red boring sponge *(Cliona delitrix)*, a crustose (encrusting) sponge.

lular zoomastigotes called choanoflagellates (see Chapter 26, Figure 26-8). Based on their unique features and the close resemblance between choanoflagellates and the choanocytes of sponges, it seems almost certain that choanoflagellates are the ancestors of the sponges. They may also be the ancestors of the other animals as well, but it is difficult to be certain of a direct relationship between sponges and other groups.

There are about 5000 species of marine sponges (Figure 28-4) and about 150 additional species that live in fresh water. In the sea, sponges are abundant at all depths. Although some sponges are tiny, no more than a few millimeters across, others, such as the loggerhead sponges, may reach 2 meters or more in diameter. Although larval sponges are free-swimming, the adults are **sessile,** or anchored in place.

## CNIDARIANS: THE RADIALLY SYMMETRICAL ANIMALS

All animals other than sponges are called **eumetazoans,** or "true animals"; they have a definite shape and symmetry and nearly always distinct tissues. Three distinct cell layers form in the embryos of all eumetazoans: an outer **ectoderm,** an inner **endoderm,** and an in-between **mesoderm.** These layers ultimately differentiate into the tissues of the adult animal. In general the nervous system and outer covering layers called **integuments** develop from the ectoderm, the muscles and skeletal elements develop from the mesoderm, and the intestine and digestive organs develop from the endoderm.

*All eumetazoans possess three embryonic tissues: ectoderm, mesoderm, and endoderm.*

Only the two most primitive of the eumetazoan phyla are **radially symmetrical,** with parts arranged around a central axis like the petals of a daisy. These two phyla are Cnidaria (pronounced ni-DAH-ree-ah), which includes jellyfish, hydra, sea anemones, and corals, and Ctenophora (pronounced tea-NO-fo-rah), a minor phylum that includes the comb jellies. The bodies of all other eumetazoans are marked by a fundamental bilateral symmetry (see Figure 28-8).

All cnidarians (Figure 28-5) are carnivores that capture their prey, such as fishes and crustaceans, with the tentacles that ring their mouth. Their bodies, like those of all eumetazoans, consist of three layers and differ greatly from those of sponges. Cnidarians are basically gelatinous in construction. More than 9000 species are known.

There are two basic body forms among the cnidarians, **polyps** and **medusae** (Figure 28-6). Polyps are cylindrical, pipe-shaped animals that are usually attached to a rock. In polyps the mouth faces away from the rock on which the animal is growing and therefore is often directed upward. Many polyps build up a hard shell, an internal skeleton, or both. In contrast, most medusae are free-floating and are often umbrella-

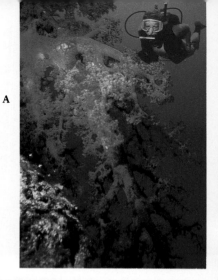

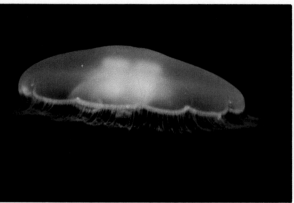

**A**

**B**

**E**

**C**

**D**

*Figure 28-5*

**Representative cnidarians.**
A  A soft coral.
B  Yellow cup coral.
C  A jellyfish, *Aurelia aurita.*
D  A sea anemone.
E  Hydra.

Mouth — Tentacle

Gastrovascular cavity

Epidermis

Gastrodermis
Mesoglea

Polyp type

Gastrovascular cavity

Epidermis

Gastrodermis
Mesoglea

Mouth

Tentacle

Medusa type

*Figure 28-6*

**Body forms of cnidarians.** The two kinds of cnidarians, the medusa *(above)* and the polyp *(below)*. These two phases alternate in the life cycles of many cnidarians, but a number—including the corals and sea anemones, for example—exist only as polyps.

shaped. Their mouths usually point downward, and the tentacles hang down around the mouth. Medusae are commonly known as jellyfish because of their thick, gelatinous interior. Many cnidarians occur only as polyps, whereas others exist only as medusae; still others alternate between these two phases during the course of their life cycles.

*Individual cnidarian species may either be medusae, which are floating, bell-shaped animals with the mouth directed downward, or polyps, which are anchored animals with the mouth directed upward. In some cnidarians these two forms alternate during the life cycle of the organism.*

## Nematocysts

The tentacles of the cnidarians bear stinging cells called **cnidocytes.** The name of the phylum Cnidaria refers to these cells, which are highly distinctive and occur in no other group of organisms. Within each cnidocyte there is a **nematocyst** (Figure 28-7), which is best thought of as a small but very powerful harpoon. Cnidarians use nematocysts to spear their prey and then draw the har-

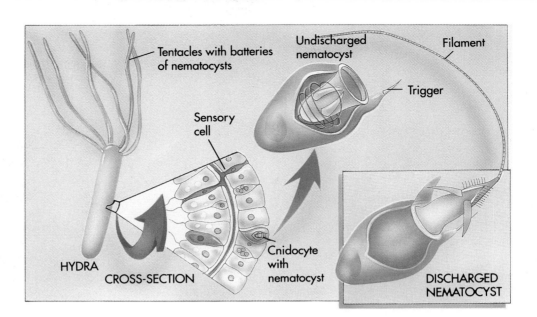

**Figure 28-7**
Structure of *Hydra*.

pooned prey back to the tentacle containing the cnidocyte. The cnidocyte uses water pressure to propel the harpoon. Here's how it works: using transmembrane channels such as those described in Chapter 6, each cnidocyte builds up a very high internal concentration of ions. Because the cnidocyte's membrane is not permeable to water, this creates an intense osmotic pressure. If a flagellum-like trigger on the cnidocyte is touched, other transmembrane channels open, permitting water to rush in. The resulting hydrostatic pressure pushes the barbed filament of the nematocyst violently outward. Because the filament is shot from the nematocyst so forcefully, the barb can penetrate even the hard shell of crustaceans. Nematocyst discharge is one of the fastest cellular processes in nature. The entire process takes place in about 3 milliseconds, with a maximum velocity of 2 meters per second.

walls of the cavity and partially break down the food. But unlike the process in more advanced invertebrates, digestion is not completely extracellular: it only fragments food into small bits that are then engulfed by the cells lining the gut by the process of phagocytosis.

What is important about the evolution of extracellular digestion that precedes the phagocytosis and intracellular digestion in cnidarians is that for the first time it became possible to digest an animal larger than oneself. Cnidarians tackle the job a little at a time.

*Cnidarians are the most primitive animals that exhibit extracellular digestion.*

*Cnidarians characteristically possess a specialized kind of cell called a cnidocyte. Each cnidocyte contains a nematocyst, which is like a harpoon and is used to attack prey. Nematocysts are found in no phylum other than Cnidaria.*

### Extracellular Digestion

A major evolutionary innovation in the cnidarians, as compared with sponges, is the extracellular digestion of food—that is, digestion within a gut cavity rather than within individual cells. This evolutionary advance has been retained by all of the more advanced groups of animals. Cnidarians have a digestive cavity with only one opening (see Figure 28-6). Digestive enzymes, primarily proteases, are released from cells lining the

## THE EVOLUTION OF BILATERAL SYMMETRY

Unlike a radially symmetrical animal, a **bilaterally symmetrical** animal (Figure 28-8) has a right half and a left half that are mirror images of each other. There are a top and a bottom, better known as the **dorsal** and **ventral** portions of the animal, respectively. There is also a front, or **anterior,** end and a back, or **posterior,** end and therefore right and left sides. All eumetazoans other than cnidarians and ctenophores are bilaterally symmetrical (even sea stars, as you will see). The bilaterally symmetrical animals constitute a major advance because this symmetry allows different parts of the body to become specialized in different ways.

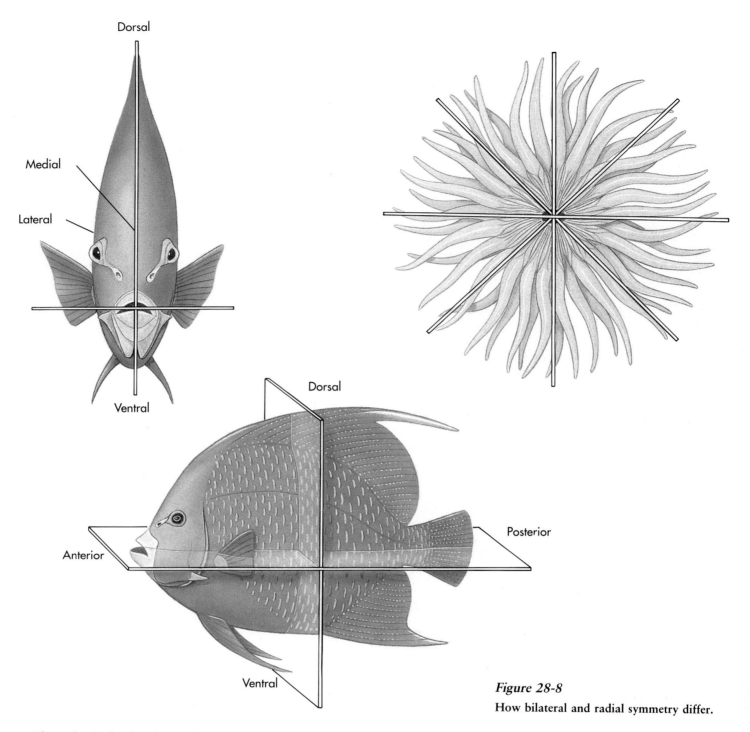

**Figure 28-8**

How bilateral and radial symmetry differ.

Three basic body plans occur in bilaterally symmetrical animals (Figure 28-9):

1. Some bilaterally symmetrical animals have no body cavity at all, other than the digestive system. They are called **acoelomates.**
2. In another group of phyla, the body cavity develops between the mesoderm and the endoderm rather than within the mesoderm. Because of the way it originates, the body cavity of these animals lacks the characteristic lining derived from mesoderm that is found in a true coelom. The kind of cavity described is called a **pseudocoel,** and the animals in which it occurs are called **pseudocoelomates.**
3. In the more advanced phyla, the mesoderm opens during development, forming a particular kind of body cavity called the **coelom.** The digestive, reproductive, and other internal organs develop within or around the margins of this coelom and are suspended within it by double

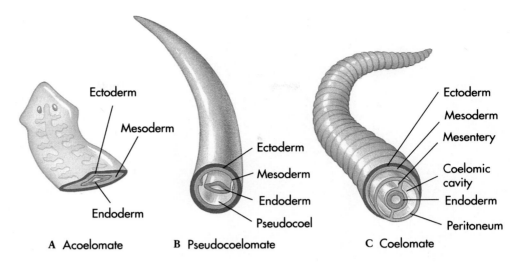

**A** Acoelomate      **B** Pseudocoelomate      **C** Coelomate

**Figure 28-9**

Three plans for construction of animal bodies.

layers of mesoderm known as **mesenteries.** Animals in which a coelom develops are called **coelomates.**

*The body architecture of bilaterally symmetrical animals follows one of three patterns:*
*1. Acoelomate, possessing no body cavity.*
*2. Pseudocoelomate, possessing a cavity between endoderm and mesoderm.*
*3. Coelomate, possessing a cavity bounded by mesoderm.*
*All vertebrates are coelomates.*

## SOLID WORMS: THE ACOELOMATE PHYLA

Among bilaterally symmetrical animals, those with the simplest body plan are the acoelomates; they are solid worms that lack any internal cavity other than the digestive tract (see Figure 28-9, *A*). By far the largest phylum of acoelomates, with about 15,000 species, is Platyhelminthes, which includes the flatworms (Figure 28-10). These ribbon-shaped, soft-bodied animals are flattened from top to bottom, like a piece of tape or ribbon.

Although in structure they are among the simplest of all the bilaterally symmetrical animals, flatworms exhibit traces of many of the evolutionary trends that are so highly developed among the members of more advanced phyla. It is important not to confuse flatworms with earthworms. The members of this phylum are the simplest animals in which organs occur. Flatworms have distinct bilateral symmetry and a definite head at the anterior end. The presence of organs, bilateral symmetry, and a distinct head is characteristic of all the more advanced phyla of animals.

Most species of flatworms are parasitic and occur within the bodies of members of almost every other animal phylum (Figures 28-10, *B*, and 28-11). Flatworms range in size from 1 millimeter or less to many meters long, as in some of the tapeworms.

*The acoelomates, typified by the flatworms, are the most primitive bilaterally symmetrical animals and are the simplest animals in which true organs occur.*

Flatworms lack circulatory systems, and most of them have a gut with only one opening. Therefore they excrete wastes directly into the gut and out through the mouth. To a lesser extent they also excrete wastes by means of specialized, bulblike cells lined with cilia that function primarily to regulate the water balance of the organism. The nervous systems of flatworms are sim-

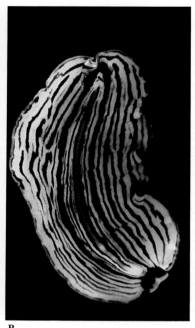

A                                          B

*Figure 28-10*

Flatworms (pylum Platyhelminthes).
A  A common flatworm, *Planaria.*
B  A free-living marine flatworm.

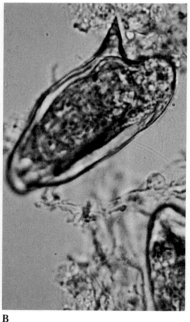

A                                          B

*Figure 28-11*

*Schistosoma.*
A  Copulating male and female individuals. The thicker male has a canal that runs the length of its body, within which the female is held during insemination and egg-laying. The worms mate in the human bloodstream.
B  An egg containing a mature larva.

ple; only the tiny swellings that occur near the leading (anterior) end of some flatworms resemble brains.

The free-living flatworms belong to the class Turbellaria, a large group with about 121 families. There are two classes of parasitic flatworms: flukes (class Trematoda) and tapeworms (class Cestoda). Among the best-known flukes are the blood flukes of the genus *Schistosoma* (see Figure 28-11), which afflict some 200 to 300 million people (about 1 in 20 of the world's population) throughout tropical Asia, Africa, Latin America, and the Middle East. The eggs of *Schistosoma* leave the human body through the urine and the feces, and their larvae develop within the bodies of freshwater snails. Eventually these snails release a highly infectious stage of the worm's life cycle; the individuals of this stage burrow into human skin and eventually reach the intestine or the bladder, where they may live for up to 30 years. causing inflammation, discomfort, and sometimes death.

Tapeworms also alternate between various hosts; the juvenile beef tapeworm occurs in cattle, but the mature form of the tapeworm occurs in humans (Figure 28-12). In a tapeworm, the attachment organ, or **scolex,** attaches to the intestinal wall; the body consists mainly of a series of repetitive segments, the **proglottids.** A mature beef tapeworm may reach 10 meters or more in length. The proglottids that are shed pass out of humans in the feces, scattering embryos that may be ingested by cattle; about 1% of all cattle in the United States are infected.

## THE EVOLUTION OF A BODY CAVITY

The body organization of the other bilaterally symmetrical animals differs from that of the solid worms in an important way: all the other bilaterally symmetrical animals possess an internal body cavity. The evolution of an internal body cavity made possible a significant advance in animal architecture. Consider for a moment the limitations of a solid body: a solid worm has no internal circulatory or digestive system, and all of its internal organs are pressed on by muscles and are thus deformed by muscular activity. Even though flatworms do have digestive systems, they are subject to the same type of problems.

An internal body cavity circumvents these limitations, and its development was an important step in animal evolution. Perhaps the most important advantage of an internal body cavity is that the body's organs are located within a fluid-filled enclosure where they can function without having to resist pressures from the surrounding muscles. In addition, the fluid that fills the cavity may act as a circulatory system,

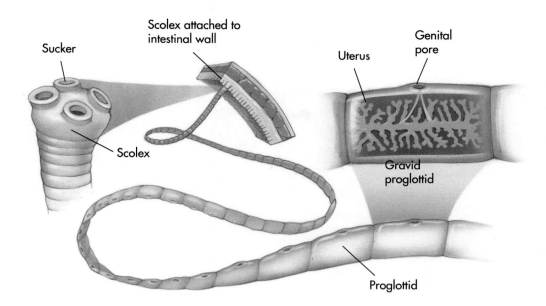

**Figure 28-12**

Structure of the beef tapeworm, *Taenia saginata*.

Labels in figure: Sucker; Scolex attached to intestinal wall; Uterus; Genital pore; Scolex; Gravid proglottid; Proglottid

freely transporting food, water, waste, and gasses throughout the body. Without the free circulation made possible by such a system, every cell of an animal must be within a short distance of oxygen, water, and all of the other substances that it requires.

The digestive system of an animal, like its circulatory system, also functions much more efficiently within an internal body cavity such as a coelom than it can when embedded in other tissues. When the gut is suspended within such a cavity, food can pass through the gut freely and at a rate controlled by the animal because the gut does not open and close when the animal moves. By controlling the rate of passage of the food, the body allows it to be digested much more efficiently than would be possible otherwise. Waste removal is also carried out much more efficiently under such circumstances.

In addition, an internal body cavity provides space within which the gonads (ovaries and testes) can expand, allowing the accumulation of large numbers of eggs and sperm. Such accumulation helps to make possible all of the diverse modifications of breeding strategy that characterize the phyla of higher animals. Furthermore, large numbers of gametes can be released when the conditions are as favorable as possible for the survival of the young animals.

## Nematodes

Nematodes (phylum Nematoda) are bilaterally symmetrical, cylindrical, unsegmented worms that possess a pseudocoelomate body plan (see Figure 28-9, *B*). They lack a defined circulatory system; the circulatory role is performed by the fluids that move within the pseudocoel. Nematodes, like all coelomates, have a complete, one-way digestive tract, which functions like an assembly line, the food being acted on in different ways in each section. First the food is broken down, then absorbed, then the wastes are treated and stored, and so on. The body of a nematode is covered by a flexible, thick cuticle, which is shed periodically as the individual grows. A layer of muscles extends beneath the epidermis along the length of the worm. These longitudinal muscles push both against the cuticle and against the pseudocoel, whipping the body from side to side as the nematode moves.

There are 12,000 recognized species in the phylum Nematoda, most of them microscopic animals that live in soil (Figure 28-13). Nematodes occur practically everywhere. It has been estimated that a spadeful of fertile soil contains, on the average, 1 million nematodes. So many new kinds of nematodes are found when environments are sampled that some scientists think there actually may be 500,000 or more species of this phylum, the great majority of which have never been collected or studied. Almost every species of plant and animal that has been studied has been found to have at least one parasitic species of nematode living on or in it, and nematodes are one of the most serious groups of pests on agricultural and horticultural crops throughout the world. Roundworms and hookworms are common nematode parasites of humans in the United States.

## Rotifers

The second phylum consisting of animals with a pseudocoelomate body plan that we shall consider is the phylum Rotifera, the rotifers (Figure 28-14). Rotifers are common, small, basically aquatic animals that have a crown of cilia at their heads; they range from 0.04 to 2 millimeters long. There are about 2000 spe-

**Figure 28-13**

**Nematodes.** One square meter of ordinary garden lawn or forest soil teems with 2 to 4 million nematodes. Although most are similar in form, they range from about 0.2 millimeter to about 6 millimeters long. A trained nematologist (student of nematodes) must examine slide mounts with a compound microscope to determine which species are present.

cies throughout the world. Bilaterally symmetrical and covered with chitin, rotifers depend on their cilia for both feeding and locomotion, ingesting bacteria, protists, and small animals. They are often called "wheel animals" because the cilia, when they are beating together, resemble spokes radiating from a wheel. Each female rotifer lays between 8 and 20 large eggs during the course of her life.

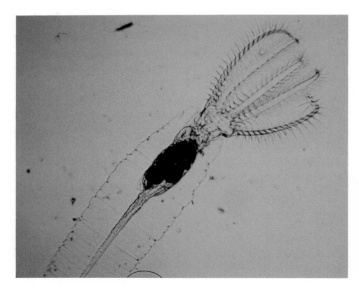

**Figure 28-14**

Rotifers—common, small aquatic animals.

## ADVENT OF THE COELOMATES

We have met two of the three bilaterally symmetrical body plans: the acoelomates, represented by the solid flatworms, and the pseudocoelomates, which are worms with a body cavity that develops between their mesoderm and endoderm. Even though both of these body plans have proven very successful, a third way of organizing the body has evolved and occurs in the bulk of the animal kingdom—the "higher" invertebrates and vertebrates. It involves the development of a coelom, which is a body cavity that originates within the mesoderm (see Figure 28-9, C).

### The Advantage of the Coelom

Both pseudocoelomate and coelomate animals possess a fluid-filled body cavity, a great improvement in body design when compared with the less advanced solid worms. What then is the functional difference between a pseudocoel and a coelom, and why has the latter kind of body cavity been so much more overwhelmingly successful in evolutionary terms? In coelomates the body cavity develops, not between endoderm and mesoderm, but entirely within the mesoderm. This makes it easier for complex organ systems to develop. The evolutionary specialization of the internal organs of the coelomates has far exceeded that of the pseudocoelomates. For example, very few pseudocoelomates possess a true circulatory system, whereas many coelomates have such a system.

In addition, the presence of a coelom allows the digestive tract, by its coiling or folding within the coelom, to be longer than the animal itself. The longer passage allows for storage organs for undigested food, longer exposure to the enzymes for more complete digestion, and even storage and final processing of food remnants. Such an arrangement allows an animal to eat a great deal when it is safe to do so and then to hide during the digestive process, thus limiting the animal's exposure to predators. The tube within the coelom architecture is also more flexible, thus allowing the animal greater freedom to move.

*The evolution of the coelom was a major improvement in animal body architecture. It permitted the development of a closed circulatory system, provided a fluid environment within which digestive, sexual, and other organs could be suspended, and facilitated muscle-driven body movement.*

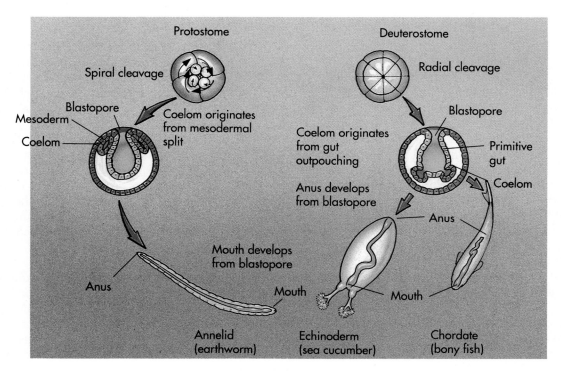

**Figure 28-15**

**The two basic patterns of embryonic development.** The progressive division of cells during embryonic growth is called **cleavage.** In **spiral cleavage,** each new cell buds off at an angle oblique to the axis of the embryo; this kind of cleavage is characteristic of nearly all protostomes. In **radial cleavage,** which is characteristic of deuterostomes, the cells divide parallel to and at right angles to the polar axis. As a result, the pairs of cells that result from each division are positioned directly above and below one another. These differences are correlated with the other features of embryonic development illustrated here.

## An Embryonic Revolution

There are two major branches of coelomate animals, representing two distinct evolutionary lines. In the first, which includes the mollusks, annelids, and arthropods as well as some smaller phyla, the mouth (stoma) develops from or near the blastopore (Figure 28-15). This pattern of embryonic development also occurs in all noncoelomate animals. An animal whose mouth develops in this way is called a **protostome.** If the animal has a distinct anus or anal pore, it develops later in another region of the embryo. The fact that this kind of developmental pattern is so widespread virtually ensures that it is the original one for animals as a whole and that it was characteristic of the common ancestor of all eumetazoan animals.

A second pattern of embryological development occurs in the echinoderms, the chordates, and a few other small, related phyla. In these animals the anus forms from or near the blastopore, and the mouth forms later on another part of the blastula (see Figure 28-15). Animals in this group of phyla are called the **deuterostomes.** They are clearly related to one another by their shared pattern of embryonic development.

In protostomes the developmental fate of each cell in the embryo is fixed when that cell first appears. Even at the four-celled stage, each cell is different, and not one of them, if separated from the others, can develop into a complete animal because the chemicals that act as developmental signals are localized in different parts of the egg. Consequently, the cleavage divisions that occur after fertilization separate different signals into different daughter cells. On the other hand, in deuterostomes the first cleavage divisions of the fertilized embryo result in identical daughter cells, any one of which can, if separated, develop into a complete organism. The commitment of individual cells to developmental pathways occurs later.

We begin our discussion of the coelomate animals with the three major phyla of protostomes—the mollusks, annelids, and arthropods. These three phyla include such familiar animals as clams, snails, octopuses, earthworms, lobsters, spiders, and insects. In the members of these phyla we can observe all of the major advances that are associated with the evolution of the coelom. We shall then consider the two largest phyla of deuterostomes—the echinoderms and the chordates.

# MOLLUSKS

The **mollusks** (phylum Mollusca) include the snails, clams, scallops, oysters, cuttlefish, octopuses, and slugs. Three classes of mollusks are representative of the phylum: (1) Gastropoda—snails, slugs, limpets, and their relatives; (2) Bivalvia—clams, oysters, scallops, and their relatives; and (3) Cephalopoda—squids, octopuses, cuttlefishes, and nautilus (Figure 28-16). The mollusks are the largest animal phylum, except for the arthropods, in terms of named species; there are at least 110,000, and probably at least that many more still to be discovered. Mollusks are abundant in marine, freshwater, and terrestrial habitats. With about 35,000 species, the terrestrial mollusks far outnumber the roughly 20,000 species of terrestrial vertebrates.

The body of a mollusk is composed of a **visceral mass** that contains the body's organs, a muscular foot that is used in locomotion, and a heavy fold of tissue called a **mantle** wrapped around the visceral mass like a cape. Within the visceral mass are found the organs of digestion, excretion, and reproduction. The folds of the mantle enclose a cavity between the dorsal wall and the visceral mass; within this mantle cavity are the mollusk's gills, which consist of a specialized respiratory system of filamentous projections, rich in blood vessels, that capture oxygen and release carbon dioxide. Mollusk gills are very efficient. Many gilled mollusks extract 50% or more of the dissolved oxygen from the water that passes through the mantle cavity. In most members of this phylum the other surface of the mantle also secretes a protective **shell.** Many mollusks can withdraw for protection into their mantle cavity, which lies within the shell. In the squids and octopuses the mantle cavity has been modified to create the jet-propulsion system that enables the animals to move rapidly through the water.

The foot of a mollusk is muscular and may be adapted for locomotion, for attachment, for food capture (in squids and octopuses), or for various combinations of these functions. Some mollusks secrete mucus, forming a path that they glide along on their foot. In the cephalopods (squids and octopuses) the foot is divided into arms, which are called tentacles.

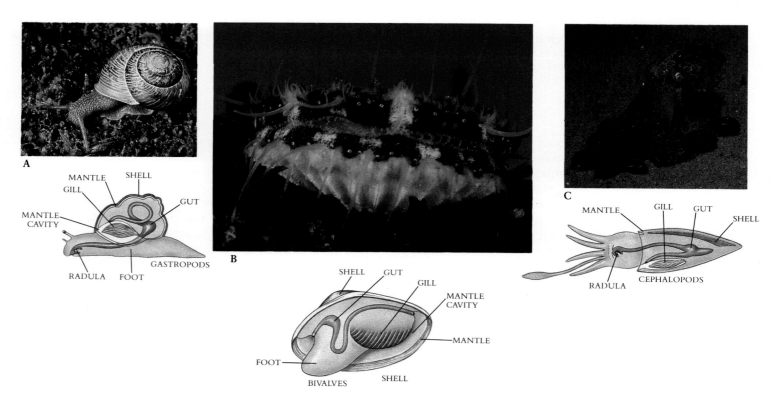

*Figure 28-16*

**The three kinds of mollusks.**
A Gastropod.
B Bivalve.
C Cephalopod.

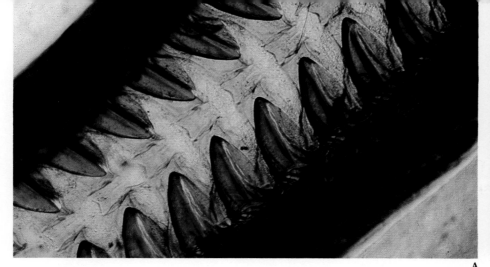

A                                                                                          B

**Figure 28-17**

**Radula.** The radula, or rasping tongue, is unique to mollusks.
A The radula of a snail.
B Enlargement of the rasping teeth on a radula.

One of the most characteristic features of mollusks is the **radula,** a rasping tonguelike organ. All members of the phylum have radula except bivalves, in which it has almost certainly been lost during evolution. The radula consists primarily of chitin and is covered with rows of pointed, backward-curving teeth (Figure 28-17). It is used by some of the snails and their relatives—members of a group known as the gastropods—to scrape algae and other food materials off their substrates and then convey this food to the digestive tract. Other gastropods are active predators; they use their radula to puncture their prey and extract food from it.

vanced animals. In some adult arthropods the segments are fused, making it difficult to perceive the underlying segmentation, but it is usually apparent in embryological development. Even among vertebrates, the backbone and muscular areas are segmented.

*Segmentation is a feature of the advanced coelomate phyla, notably the annelids and the arthropods, although it is not obvious in some of them.*

*Mollusks are the second largest phylum of animals in terms of named species. They characteristically have bodies with three distinct sections: head, visceral mass, and foot. All mollusks except the bivalves also possess a unique rasping tongue called a radula.*

## THE RISE OF SEGMENTATION

Unlike mollusks, the other major phyla of protostomes, annelids and arthropods, have segmented bodies. Just as it is efficient for workers to construct a tunnel from a series of identical prefabricated parts, so these advanced protostome coelomates are "assembled" as a chain of identical segments, like the boxcars of a train. Segmentation underlies the organization of all ad-

## ANNELIDS

The **annelids** (phylum Annelida), one of the major animal phyla, are segmented worms (Figure 28-18). They are abundant in marine, freshwater, and terrestrial habitats throughout the world. Their segments are visible externally as a series of ringlike structures running the length of the body. Internally the segments are divided from one another by partitions. In each of the cylindrical segments of these animals, the digestive and excretory organs are repeated in tandem. Leeches are one of three classes of annelid worms (class Hirudinea). They are freshwater predators or bloodsuckers and were formerly used in medicine for bloodletting, which was thought to be beneficial. Currently the enzymes that leeches use to prevent blood from clotting are extracted and used to treat certain conditions in which blood clots are a serious risk. The two other classes of annelids are Polychaeta, a largely marine

D

**Figure 28-18**

**Annelids.**
**A** Earthworms mating.
**B** Shiny bristleworm, *Oenone fulgida,* a polychaete.
**C** Fan worm, *Sabella melanostigma,* a polychaete.
**D** A freshwater leech with young leeches visible inside its body.

group of about 8000 species, and Oligochaeta, the earthworms, with about 3100 species. Earthworms are often present in very large numbers and are extremely important in aerating and enriching the soil.

The basic body plan of the annelids (Figure 28-19) is a tube within a tube: the internal digestive tract is a tube suspended within the coelom, which is a tube running from mouth to anus. The anterior segments have become modified, and a well-developed **cerebral gan-glion,** or **brain,** is contained in one of them. The sensory organs are mainly concentrated near the anterior end of the worm's body. Some of these are sensitive to light, and elaborate eyes with lenses and retinas have evolved in certain members of the phylum. Separate nerve centers, or ganglia, are located in each segment but are interconnected by nerve cords. These nerve cords are responsible for the coordination of the worm's activities.

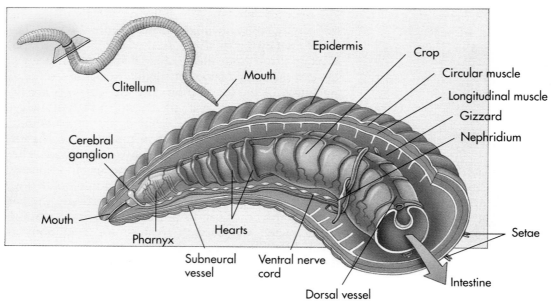

**Figure 28-19**

Body of an earthworm.

Annelids use their muscles to crawl, burrow, and swim. In each segment the muscles play against the fluid in the coelom. This fluid creates a hydrostatic (liquid-supported) skeleton that gives the segment rigidity, like an inflated balloon. Because each segment is separate, each is able to contract or expand independently. Therefore a long body can move in ways that are quite complex. When an earthworm crawls on a flat surface, it lengthens some parts of its body while shortening others.

Each segment of an annelid typically possesses **setae** (singular, **seta**), which are bristles of chitin that help to anchor the worms during locomotion or when it is in its burrow. Because of the setae, annelids are often called "bristle worms." The setae are absent in all but one species of the leeches.

*The annelids are characterized by serial segmentation. The body is composed of numerous similar segments, each with its own circulatory, excretory, and neural elements and each with its own array of setae.*

## ARTHROPODS

With the evolution of the first annelids, many of the major innovations of animal structure had already appeared: the division of tissues into three primary types (endoderm, mesoderm, and ectoderm), bilateral symmetry, coelomic body architecture, and segmentation. However, one further innovation remained. It marks the origin of the body plan characteristic of the most successful of all animal groups, the arthropods. This innovation was the development of jointed appendages.

The name "arthropod" comes from two Greek words, *arthros,* jointed, and *podes,* feet. All arthropods have jointed appendages. The numbers of these appendages are progressively reduced in the more advanced members of the phylum, and the nature of the appendages differs greatly in different subgroups. Thus individual appendages may be modified into antennae, mouthparts of various kinds, or legs.

Arthropod bodies are segmented like those of annelids, a phylum to which at least some of the arthropods are clearly related. The members of some classes of arthropods have many body segments. In others the segments have become fused together into functional groups such as the head or thorax of an insect (Figure 28-20). But even in these the original segments can be distinguished during development of the larvae.

*Arthropods, like annelid worms, are segmented, but in many arthropods the individual segments are fused into functional groups, such as the head or thorax of an insect.*

The arthropods have a rigid external skeleton, or **exoskeleton,** in which chitin is an important element (Figure 28-21). The exoskeleton provides places for muscle attachment, protects the animal from predators and injury, and, most importantly, impedes water loss. As an individual outgrows its exoskeleton, that exoskeleton splits open and is shed, allowing the animal to increase in size. This process, which is known as **ecdysis,** is controlled by hormones. The eggs of arthropods develop into immature forms that may bear little

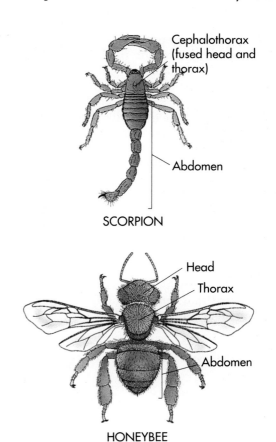

**SCORPION**

**HONEYBEE**

*Figure 28-20*

**Evolution of arthropod segmentation.** Evolution of body segments proceeded from many to few segments among the arthropods; the scorpion and the honeybee are examples of arthropods with different numbers of body segments, as you can see by counting them.

A

B

*Figure 28-21*

**Arthropod exoskeletons.** Some arthropods have a very tough exoskeleton, like this beetle (**A**); others have a fragile exoskeleton, like the widow dragonfly (**B**).

or no resemblance to the adults of the same species; most members of this phylum change their characteristics as they develop from stage to stage during a process called **metamorphosis.**

Arthropods are by far the most successful of all life forms. Approximately a million animal species—about two-thirds of all the named species on earth—are members of this gigantic phylum, and it is estimated that there are many millions more awaiting discovery. There are several times as many species of arthropods as there are of all other plants, animals, and microorganisms put together. A hectare of lowland tropical forest is estimated to be inhabited, on the average, by 41,000 species of insects. Many suburban gardens may have 1500 or more species of this gigantic class of organisms. The insects and other arthropods are abundant in all of the habitats on this planet, but they especially dominate the land, where along with the flowering plants and the vertebrates they determine the very structure of life. In terms of individuals, it has been estimated that approximately a billion billion insects are alive at any one time!

## DIVERSITY OF THE ARTHROPODS

Many arthropods have jaws, or **mandibles,** formed by the modification of one of the pairs of anterior appendages (but not the one nearest the anterior end). These arthropods are called **mandibulates;** they include the crustaceans, insects, centipedes, millipedes, and a few other small groups. The remaining arthropods, which include the spiders, mites, scorpions, and a few other groups, lack mandibles. These animals are called the **chelicerates.** Their mouthparts, called **chelicerae** (Fig-

ure 28-22), evolved from the appendages nearest the anterior end of the animal. They often take the form of pincers or fangs. The crustaceans seem to have evolved mandibles separately and are not thought to be directly related to the insects and other mandibulates.

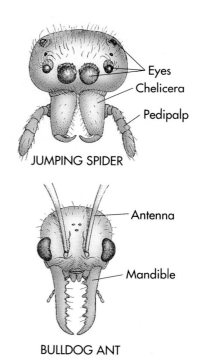

JUMPING SPIDER — Eyes, Chelicera, Pedipalp

BULLDOG ANT — Antenna, Mandible

*Figure 28-22*

**How to tell the difference between chelicerates and mandibulates.** In the chelicerates, such as the jumping spider, the chelicerae are the foremost appendages of the body. Mandibulates, in contrast, such as the bulldog ant, have the antennae at the anterior (front) end of the body.

## Chelicerates

The fossil record of chelicerates goes back as far as that of any multicellular animals, about 630 million years. A major group of extinct arthropods, the trilobites, was also abundant then, and the living horseshoe crabs seem to be directly descended from them. By far the largest of the three classes of chelicerates is the largely terrestrial class Arachnida, with some 57,000 named species, including the spiders, ticks, mites, scorpions, and daddy longlegs (Figure 28-23). Scorpions are probably the most ancient group of terrestrial arthropods; they are known from as early as the Silurian Period, some 425 million years ago. There are about 35,000 named species of spiders (order Araneae). The order Acari, the mites, is the largest in terms of number of species and the most diverse of the arachnids.

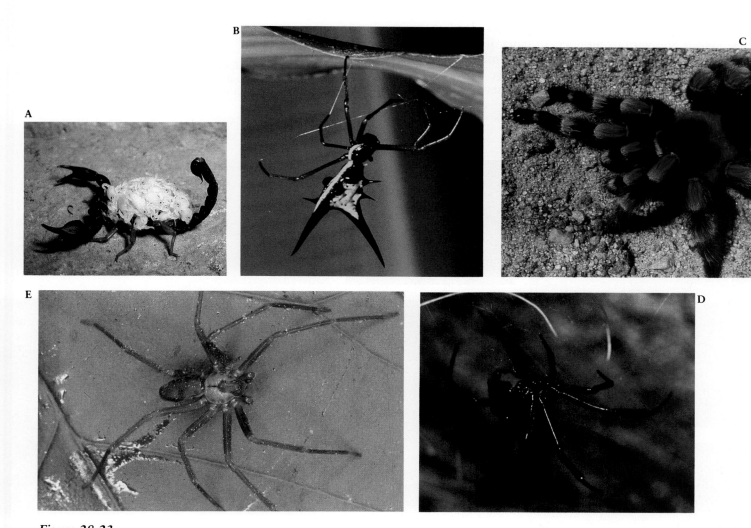

### Figure 28-23

**Arachnids.**

**A** The scorpion *Uroctonus morda*, showing the characteristic pincers and segmented abdomen, ending in a sting, raised over the animals' back. White young scorpions cluster on this individual's back.

**B** An orb-weaving spider, the arrowhead spider *Micranthena*, in Peru.

**C** Tarantulas are large hunting spiders, powerful enough to prey on small lizards, mammals, and birds in addition to insects. They do not spin webs.

**D** One of the two poisonous spiders in the United States and Canada, the black widow spider, *Latrodectus mactans*.

**E** The other genus of poisonous spiders of this area, the brown recluse, *Loxosceles reclusa*. Both species are common throughout temperate and subtropical North America, but bites are rare in humans.

## Mandibulates: Crustaceans

The crustaceans (subphylum Crustacea, with a single class) are a large, diverse group of primarily aquatic organisms, including some 35,000 species of crabs, shrimps, lobsters (Figure 28-24), crayfish, barnacles, water fleas, pillbugs, and related groups (Figure 28-25). Often incredibly abundant in marine and freshwater habitats and playing a role of critical importance in virtually all aquatic ecosystems, crustaceans have been called "the insects of the water." Most crustaceans have two pairs of antennae, three pairs of chewing appendages, and various numbers of pairs of legs. Crustaceans differ from the insects—but resemble the centipedes and millipedes—in that they have legs on their abdomen as well as on their thorax. They are the only arthropods with two pairs of antennae.

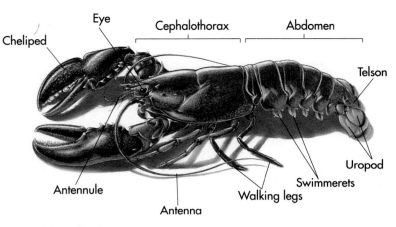

### Figure 28-24

**The body of a lobster, *Homarus americanus*.** Some of the specialized terms used to describe crustaceans are indicated; for example, the head and thorax are fused together into a cephalothorax.

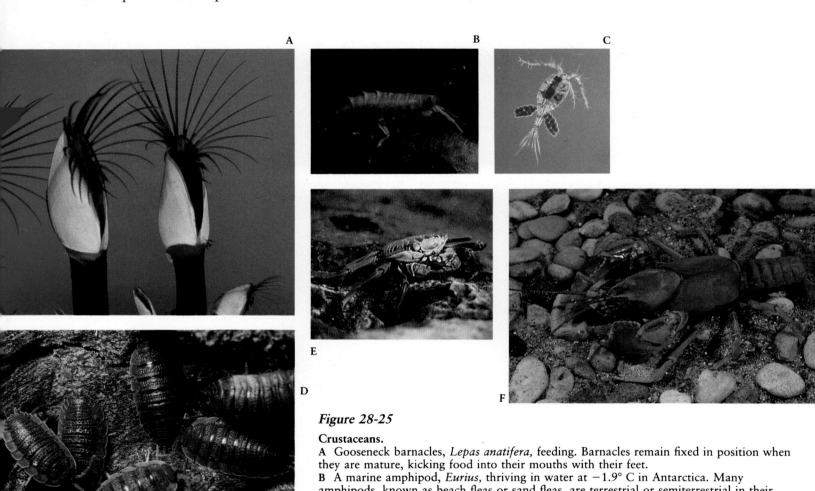

### Figure 28-25

**Crustaceans.**

**A** Gooseneck barnacles, *Lepas anatifera*, feeding. Barnacles remain fixed in position when they are mature, kicking food into their mouths with their feet.

**B** A marine amphipod, *Eurius*, thriving in water at −1.9° C in Antarctica. Many amphipods, known as beach fleas or sand fleas, are terrestrial or semiterrestrial in their habits.

**C** A copepod, member of an abundant group of marine and freshwater crustaceans that are important components of the plankton.

**D** Sowbugs, *Porcellio scaber*, representatives of the terrestrial isopods.

**E** Sally lightfoot crab, *Grapsus grapsus*. This crab species is semiterrestrial, ranging widely over land in search of food.

**F** A freshwater crayfish, *Procambarus*.

## Mandibulates: Insects and Their Relatives

The insects, class Insecta, are by far the largest group of arthropods, whether measured in terms of numbers of species or numbers of individuals; as such, they are the most abundant group of eukaryotes on earth. Insects live in every conceivable habitat on land and in fresh water, and a few have even invaded the sea. One scientist has calculated that there are probably about 300 million insects alive at any one time for each person on earth! More than 70% of all the named animal species are insects, and the actual proportion is undoubtedly much higher because millions of additional forms await detection, classification, and naming. There are about 90,000 described species in the United States and Canada, and the actual number of species in this region probably approaches 125,000. A glimpse at the enormous diversity of insects is presented in Figure 28-26.

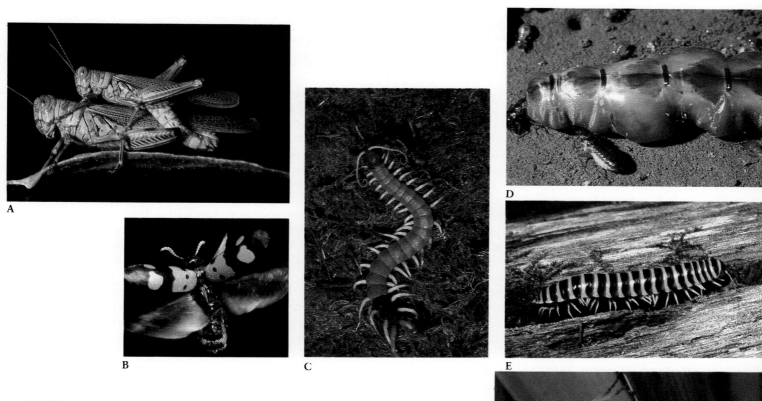

### Figure 28-26

**Insects and some relatives.**
A Copulating grasshoppers.
B An African blister beetle, which has spread its two tough, leathery forewings, exposing its delicate flying wings. More than 350,000 species of beetles—a fifth of all species of animals—have been described.
C A centipede.
D Termites, *Macrotermes bellicosus.* The large, sausage-shaped individual is a queen, specialized for laying eggs; the smaller individuals around it are nonreproductive workers. Only the reproductive males and females (queens) have wings, which are soon lost in the queens.
E A millipede. Centipedes are active predators, whereas millipedes are sedentary herbivores. Millipedes and centipedes are closely related to insects; all of the other animals illustrated here are insects.
F A mayfly, *Hexagenia limbata,* immediately after molting to reach the adult form. Adult mayflies, the reproductive stage, do not feed and live for only a few hours or a few days; their nymphs, which are aquatic, live underwater for months or even years.

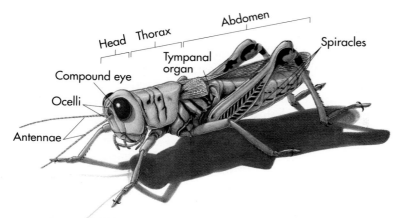

**Figure 28-27**

**The body of a grasshopper.** The diagram illustrates the major structural features of insects, the most numerous group of arthropods. The left wing has been removed to show the features of the abdomen.

Insects have three body sections—the head, thorax, and abdomen; three pairs of legs, all attached to the thorax; and one pair of antennae (Figure 28-27). Most insects have compound eyes, which are composed of many independent visual units (Figure 28-28, *A*). The insect thorax consists of three segments, each of which has a pair of legs, although occasionally one or more of these pairs of legs is absent. The thorax is almost entirely filled with muscles that operate the legs and wings. Insects have two pairs of wings, which are attached to segments of the thorax, although in the flies and some other groups one or both of these pairs has been lost during the course of evolution. The wings of adult insects are solid sheets of chitin with strengthening veins made of chitin tubules. Like many other arthropods, insects have no single major respiratory organ. Their respiratory system consists of small, branched, cuticle-lined air ducts called **tracheae** (singular, **trachea;** Figure 28-28, *B*), which are a series of tubes that transmit oxygen throughout the body. The tracheae ultimately branch into very small **tracheoles.** Air passes into the tracheae through specialized openings called **spiracles,** which occur on many of the segments of the thorax and abdomen.

Also characteristic of insects are complex patterns of metamorphosis, in which certain stages give way to others. In some insects, such as grasshoppers, cockroaches, and termites, the stages grade from one to the other without abrupt changes; in others, such as butterflies and moths, flies, and beetles, the changes are abrupt (Figure 28-29).

## DEUTEROSTOMES

Two outwardly dissimilar large phyla, Echinodermata and Chordata, have a series of key embryological features that are very different from those of the other animal phyla (see Figure 28-15). Because it is extremely unlikely that these features evolved more than once, it is believed that both phyla share a common ancestry. They are the members of a group that we call the deuterostomes. Deuterostomes diverged from protostome ancestors more than 630 million years ago.

Deuterostomes, like protostomes, are coelomates. But they differ fundamentally from protostomes in the way the embryo grows. The blastopore of a deuterostome becomes the animal's anus, and the mouth develops at the other end; for a brief period of development, they have identical daughter cells, each of which, if separated from the others at an early enough stage, can develop into a complete organism; and whole groups of cells move around during the course of embryonic development to form new tissue associations.

A

B

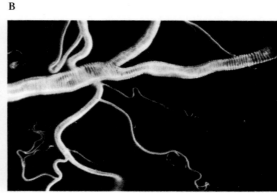

**Figure 28-28**

**Two characteristic features of insects.**
**A** The compound eyes found in insects are complex structures. Three ocelli—simple eyes—can be seen between the compound eyes of the robberfly.
**B** A portion of the tracheal system of a cockroach.

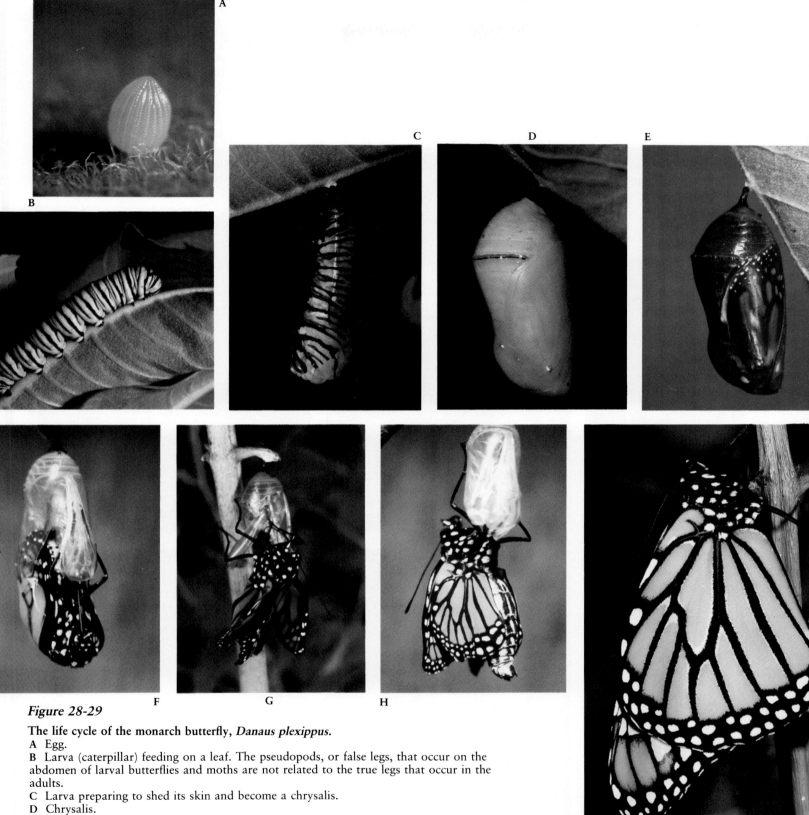

**Figure 28-29**

**The life cycle of the monarch butterfly, *Danaus plexippus*.**

**A** Egg.

**B** Larva (caterpillar) feeding on a leaf. The pseudopods, or false legs, that occur on the abdomen of larval butterflies and moths are not related to the true legs that occur in the adults.

**C** Larva preparing to shed its skin and become a chrysalis.

**D** Chrysalis.

**E** The colors of the wings of the adult butterfly can be seen through the thin outer skin of the chrysalis; the adult is nearly ready to emerge.

**F** to **H** Adult butterfly emerging.

**I** Adult monarch butterfly.

# ECHINODERMS

Echinoderms (phylum Echinodermata) are an ancient group of marine animals that are very well represented in the fossil record. The term *echinoderm* means "spine-skin," an apt description of many members of this phylum, which comprises about 6000 living species. Many of the most familiar animals seen along the seashore, such as the sea stars (starfish), brittle stars, sea urchins, sand dollars, and sea cucumbers, are echinoderms (Figure 28-30). All are bilaterally symmetrical as larvae but radially symmetrical as adults; thus they are said to be fundamentally bilaterally symmetrical. They certainly evolved from bilaterally symmetrical ancestors. Echinoderms are well represented not only in the shallow waters of the sea but also in its abyssal depths. All but a few of them are bottom-dwellers. The adults range from a few millimeters to more than 1 meter in diameter or length.

**Figure 28-30**

**Echinoderms.**
A  A sand dollar, *Echinarachnius parma.*
B  Giant red sea urchin, *Strongylocentrotus franciscanus.*
C  A sea star, *Oreaster occidentalis.*
D  A sea cucumber, *Stichopus.*
E  A feather star.
F  A brittle star, *Ophiothrix.*

## Basic Features of Echinoderms

Echinoderms have a five-part body plan corresponding to the arms of a sea star. As adults, these animals have no head or brain. Their nervous systems consist of central nerve rings from which branches arise. The animals are capable of complex response patterns, but there is no centralization of function. Apparently the centralization of the nervous system is not feasible in animals with radial symmetry.

The **water vascular system** of an echinoderm radiates from a ring canal that encircles the animal's esophagus. Five radial canals extend into each of the five parts of the body and determine its basic symmetry (Figure 28-31). Water enters the water vascular system through a sievelike plate on the animal's surface, from which it flows through a tube to the ring canal. The five radial canals, in turn, extend out of the animal through short side branches into the many hollow tube feet. In some echinoderms, each tube foot has a sucker at its end; in others, suckers are absent. At the base of each tube foot is a muscular sac, the **ampulla,** that contains fluid. When the sac contracts, the fluid is prevented from entering the radial canal and is forced into the tube foot, thus extending it. When extended, the foot attaches itself to the substrate. Longitudinal muscles can then shorten the foot, and the animal is pulled forward to a new position as water is forced back into the foot by the muscular sac. By such action repeated in many tube feet, sea stars, sand dollars, and sea urchins slowly move along.

*As adults, echinoderms are radially symmetrical animals whose bodies typically have a pattern that repeats itself five times. By means of a unique water vascular system connected to their tube feet, they are able to move about.*

## CHORDATES

The chordates (phylum Chordata) are the best understood and most familiar group of animals. There are some 42,500 species of chordates, a phylum that includes the birds, reptiles, amphibians, fishes, and mammals (and therefore our own species).

The three principal features of the chordates were presented in Chapter 18, where the evolution of vertebrates was discussed, but are reviewed here briefly. All chordates have a single, hollow nerve cord, a notochord, and pharyngeal slits (see Figure 18-3), and each of these features has played an important role in the evolution of the phylum. In the more advanced vertebrates the dorsal nerve cord becomes differentiated into the brain and spinal cord. The notochord, which persists throughout the life cycle of some of the invertebrate chordates, becomes surrounded and then is replaced by the vertebral column during the embryological development of vertebrates. Pharyngeal slits are

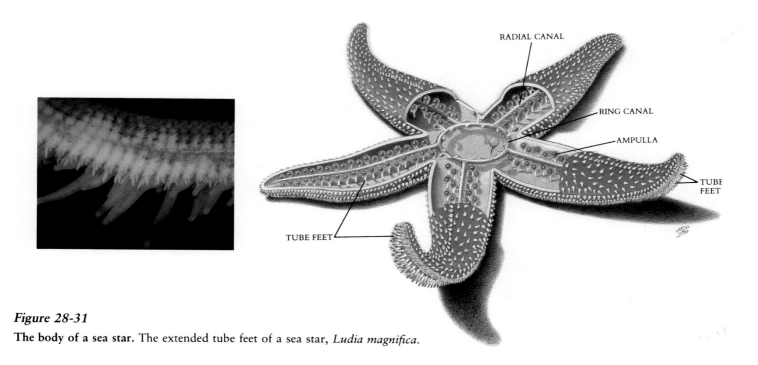

RADIAL CANAL

RING CANAL

AMPULLA

TUBE FEET

TUBE FEET

*Figure 28-31*

**The body of a sea star.** The extended tube feet of a sea star, *Ludia magnifica.*

present in the embryos of all vertebrates but are lost later in the development of the terrestrial vertebrates. However, the presence of these structures in all vertebrate embryos provides a clue to the aquatic ancestry of the group.

*Chordates are characterized by a single, hollow dorsal nerve cord. At some point in the embryonic development of all chordates, the notochord, which is a flexible rod, forms dorsal to the gut, and slits are present in the pharynx. The notochord persists into the adult stage in the less advanced chordates.*

In addition to these three principal features, many other characteristics distinguish the chordates. In their body plan, chordates are more or less segmented, and distinct blocks of muscles can be seen clearly in many less specialized forms (Figure 28-32). Chordates have an internal skeleton against which the muscles work, and either this skeleton or the notochord makes possible the extraordinary powers of locomotion that characterize chordates. Finally, chordates have a tail that extends beyond the anus, at least during their embryonic development; nearly all other animals have a terminal anus.

This great phylum is divided into three major groups. Two of them, the tunicates and the lancelets, are **acraniates**; that is, they lack a brain. The **craniate** chordates are the **vertebrates,** which we have already examined in detail in Chapter 18. We review the two other subphyla of chordates in more detail here.

## Tunicates

The tunicates (subphylum Tunicata) are a group of about 1250 species of marine animals. Most of them are sessile (fixed in one spot) as adults (Figure 28-33).

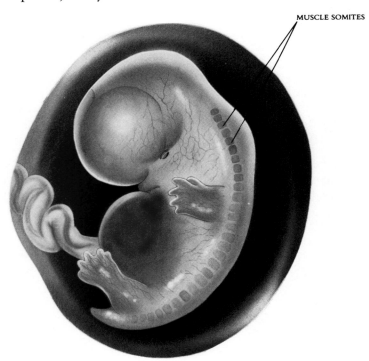

MUSCLE SOMITES

**Figure 28-32**
A human embryo showing somites.

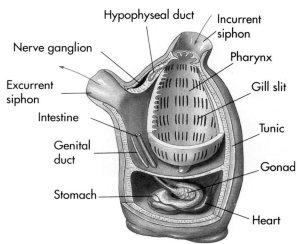

Hypophyseal duct
Incurrent siphon
Nerve ganglion
Pharynx
Excurrent siphon
Gill slit
Intestine
Tunic
Genital duct
Gonad
Stomach
Heart

**Figure 28-33**
Body of a tunicate.

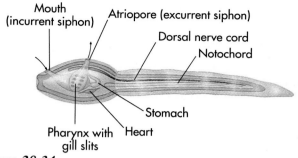

**Figure 28-34**

**Our distant cousins.** The structure of a larval tunicate is much like that of the postulated common ancestor of the chordates.

When they are adults, these animals lack visible signs of segmentation or of a body cavity. It would be very difficult to discern the evolutionary relationships of an adult tunicate by examining its features.

But the tadpolelike larvae of tunicates plainly exhibit all of the basic characteristics of chordates (notochord and nerve cord) and mark the tunicates as having the most primitive combination of features found in any chordate (Figure 28-34). They do not feed, and they have a poorly developed gut. The larvae remain free-swimming for no more than a few days; then they settle to the bottom and attach themselves to a suitable substrate by means of a sucker.

## Lancelets

The lancelets (subphylum Cephalochordata; Figure 28-35) are scaleless, fishlike marine chordates a few centimeters long. They occur widely in shallow waters throughout the oceans of the world. There are about 23 species of lancelets, which were given their English name because of their similarity to a lancet, a small, two-edged surgical knife. In the lancelets the notochord persists through the animal's life and runs the entire length of the dorsal nerve cord.

## Vertebrates

Vertebrates (subphylum Vertebrata) differ from other chordates in that they usually possess a vertebral column, which replaces the notochord to a greater or lesser extent in adult individuals of different members of the subphylum. In addition, the vertebrates, or craniate chordates, have a distinct head with a skull and brain. The hollow dorsal nerve cord of most vertebrates is protected with a U-shaped groove formed by paired projections from the vertebral column (Figure 28-36). The seven classes of living vertebrates were discussed and illustrated in Chapter 18; three of them are

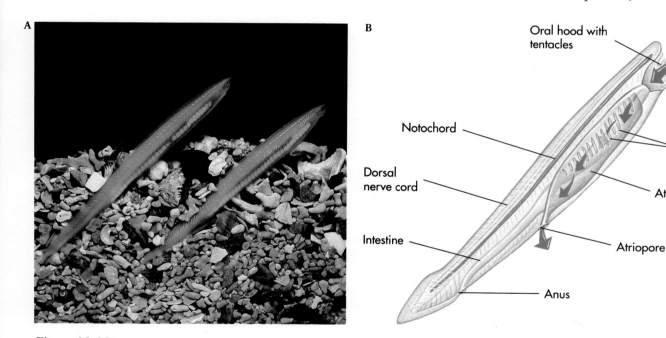

**Figure 28-35**

**Body of a lancelet.**
**A** Two lancelets, *Branchiostoma lanceolatum*, partly buried in shell gravel, with their anterior ends protruding. The muscle segments are clearly visible in this photograph; the numerous square, pale yellow objects along the side of the body are gonads, indicating that these are male lancelets.
**B** The structure of a lancelet, showing the path by which the water is pulled through the animal.

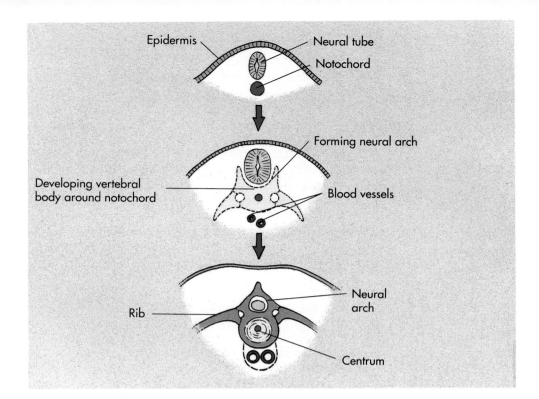

**Figure 28-36**

**Embryonic development of a vertebrate.** During the course of development, the flexible notochord is surrounded and eventually replaced by a cartilaginous or bony covering, the centrum. The neural tube is protected by an arch above the centrum, and the vertebrae may also have a hemal arch, which protects the major blood vessels below the centrum. The vertebral column is a strong, flexible rod against which the muscles pull when the animal swims or moves.

fishes and four are tetrapods. The classes of fishes are Agnatha, the lampreys and hagfishes; Chondrichthyes, the cartilaginous fishes, sharks, skates, and rays; and Osteichthyes, the bony fishes, the dominant group of fishes today. The four classes of tetrapods are Amphibia, the amphibians, including salamanders, frogs, and toads; Reptilia, the reptiles; Aves, the birds; and Mammalia, the mammals.

## ■ SUMMARY

1. The animals, kingdom Animalia, comprise some 35 phyla and at least 4 million species. Animals are heterotrophic, multicellular organisms that ingest their food.

2. The sponges (phylum Porifera) are characterized by specialized, flagellated cells called choanocytes; by a lack of symmetry in bodily organization; and by a lack of tissues and organs.

3. The remaining animals have radially or bilaterally symmetrical body plans and distinct tissues. The cnidarians (phylum Cnidaria) are predominantly marine, radially symmetrical animals with unique stinging cells called cnidocytes, each of which contains a specialized harpoon apparatus, or nematocyst.

4. Bilaterally symmetrical animals include three major evolutionary lines: acoelomates, which lack a body cavity; pseudocoelomates, which develop a body cavity between the mesoderm and the endoderm; and coelomates, which develop a body cavity (coelom) within the mesoderm.

5. Acoelomates are the most primitive bilaterally symmetrical animals. They lack an internal cavity, except for the digestive system, and are the simplest animals that have organs, which are structures made up of two or more tissues. The most prominent phylum of acoelomates is the phylum Platyhelminthes, which includes the free-living flatworms and the parasitic flukes and tapeworms.

6. Pseudocoelomates, exemplified by the nematodes (phylum Nematoda) and rotifers (phylum Rotifera), have a body cavity that develops between the mesoderm and the endoderm. This pseudocoel provides a place to which the muscles can attach, giving these worms enhanced powers of movement.

7. The two major evolutionary lines of coelomate animals—the protostomes and the deuterostomes—are represented among the oldest known fossils of multicellular animals, dating back some 630 million years.

8. In the protostomes the mouth develops from or near the blastopore, and the early divisions of the embryo are spiral. At early stages of development the fate of the individual cells is already determined, and they cannot develop individually into a whole animal. The major phyla of protostomes are Mollusca, Annelida, and Arthropoda.

9. In the deuterostomes the anus develops from or near the blastopore, and the mouth forms later on another part of the blastula. At early stages of development, each cell of the embryo can differentiate into a whole animal. The major phyla of deuterostomes are Echinodermata and Chordata.

10. Arthropods are the most successful of all animals in terms of numbers of individuals and numbers of species, as well as in terms of ecological diversification. Like the annelids, arthropods have segmented bodies; but in arthropods some of the segments have become fused during the course of evolution.

11. Insects, which have six legs, are the largest class of organisms, with at least 2 million and perhaps as many as 30 million species.

12. Echinoderms are marine deuterostomes that are radially symmetrical as adults. They have a unique water vascular system that includes tube feet, by means of which some echinoderms move.

13. Chordates are characterized by a single, hollow dorsal nerve cord; a notochord; and pharyngeal slits, at least as embryos. There are three subphyla: tunicates, lancelets, and vertebrates.

# REVIEW

1. Of the 35 phyla of animals, name the two that are represented by large numbers of terrestrial species.

2. In all eumetazoans, which embryonic layer gives rise to the digestive organs?

3. Name the two groups of animals that have fundamentally distinct patterns of embryonic development.

4. Which animal phylum has the most species?

5. Animals that develop a body cavity between the mesoderm and the endoderm are called _____.

# SELF-QUIZ

1. Choanocytes of sponges bear a striking resemblance to the _____, members of one group of zoomastigotes.
   (a) acoelomates
   (b) choanoflagellates
   (c) cnidarians
   (d) collar cells
   (e) none of the above

2. Where does a coelom originate?
   (a) In the endoderm
   (b) Between the endoderm and the mesoderm
   (c) In the mesoderm
   (d) Between the mesoderm and the ectoderm
   (e) In the ectoderm

3. Some coelomates are
   (a) protostomes.
   (b) deuterostomes.
   (c) spiders.
   (d) human beings.
   (e) all of the above.

4. Which two phyla listed here are deuterostomes?
   (a) Platyhelminthes
   (b) Nematoda
   (c) Arthropoda
   (d) Echinodermata
   (e) Chordata

5. The following are characteristic features of deuterostomes:
   (a) more primitive than protostomes.
   (b) spiral cleavage during embryonic growth.
   (c) coelom originating from split of mesoderm.
   (d) mouth develops from blastopore.
   (e) none of the above.

6. Three characteristics that distinguish the chordates from other animals are
   (a) a single, hollow, dorsal nerve cord.
   (b) a notochord at some time in development.
   (c) pharyngeal slits at some time in development.
   (d) segmentation.
   (e) a nervous system.

# THOUGHT QUESTIONS

1. Why do we believe that the ancestor of the deuterostomes was a protostome?

2. Why is it believed that echinoderms and chordates, which are so dissimilar, are members of the same evolutionary line?

# FOR FURTHER READING

BAVENDAM, F.: "Even for Ethereal Phantasms, It's a Dog-Eat-Dog World," *Smithsonian,* August 1989, pages 94-101. Aspects of the lives of some of the incredibly beautiful sea slugs, or nudibranchs.

CAMERON, J.N.: "Molting in the Blue Crab," *Scientific American,* May 1985, pages 102-109. An interesting discussion of the way the blue crab molts; soft-shelled crabs are individuals caught right after molting.

GOSLINE, J.M., and M.E. DEMONT: "Jet-Propelled Swimming in Squids," *Scientific American,* January 1985, pages 97-103. For short distances a squid can jet through the water as rapidly as a fish by contracting radial and circular muscles in its boneless mantle wall.

HADLEY, N.: "The Arthropod Cuticle," *Scientific American,* July 1986, pages 104-112. A major innovation aiding the success of insects as they conquered the terrestrial environment, the cuticle is far more than a simple waxy covering.

MARTIN, J.: "The Engaging Habits of Chameleons Suggest Mirth more than Menace," *Smithsonian,* June 1990, pages 44-53. Wonderful variety in a single group of tropical lizards.

RICKETTS, E.F., J. CALVIN, and J.W. HEDGPETH: *Between Pacific Tides,* 5th ed., Stanford University Press, Stanford, Calif., 1985. An outstanding, ecologically oriented treatment of the marine organisms of the Pacific Coast of the United States.

WILSON, E.O.: "Empire of the Ants," *Discover,* March 1990, pages 45-50. Outstanding essay on the fascinating ways of ants.

YONGE, C.M.: "Giant Clams," *Scientific American,* April 1975, pages 96-105. These enormous undersea bivalves, although they are filter-feeders, probably owe their large size to the fact that photosynthetic dinoflagellates live within their tissues and nourish them.

8

P L A N T

B I O L O G Y

# The Structure of Plants

The roots of these trees in Ecuador extend directly into the water and are rooted in the soil below the surface. The roots do not drown because they are able to obtain oxygen drawn in from the leaves and piped down to them through internal plumbing.

# THE STRUCTURE OF PLANTS

## Overview

Plants, which cover the surface of the earth in bewildering variety, all possess the same fundamental architecture. Although roots and shoots differ in their basic structure, growth at the tips throughout the life of the individual is characteristic of both. All parts of plants have an outer covering, called dermal tissue, and ground tissue, within which is embedded vascular tissue that conducts water, nutrients, and food throughout the plant. Plants cannot move, but they adjust to their environment by growing and changing their form. Their light-gathering leaves are supported on a stem, which is connected to the water-gathering roots penetrating the soil below.

## For Review
*Here are some important terms and concepts that you will encounter in this chapter. If you are not familiar with them, you should review them before proceeding.*

**Cell structure** (Chapter 4)

**How cells divide** (Chapter 9)

**History of plants** (Chapter 17)

**Major groups of plants** (Chapter 27)

**Monocots and dicots** (Chapter 27)

**Figure 29-1**

**Plants.** In this barrel cactus *(Ferocactus diguetii)*, the stem is greatly swollen and stores water efficiently; the photosynthetic cells lie mainly just beneath the surface of the thick, tough epidermal layers of the stem. Leaves are present only in seedlings.

Plants dominate and give color to the living world, providing the basic structure for the terrestrial communities in which other organisms live (Figure 29-1). Only plants, algae, and a few genera of bacteria can carry out photosynthesis, capturing the energy of the sun and making it available for their own metabolic activities and those of all other organisms. The world is populated with millions of kinds of organisms, nearly all of them dependent on the process of photosynthesis. We began our study of plants in Chapter 27; in this section we consider their structure and function in more detail.

## THE ORGANIZATION OF A PLANT

A plant never becomes an "adult" in the way an animal does. Rather, plants simply keep growing, adding new cells, tissues, and organs at the ends of their shoots and roots, becoming ever larger. Some large patches of prairie grasses are thought to be single individuals that have been growing in one place since the glaciers receded more than 10,000 years ago. Trees attain great ages, and potatoes and many other crops are simply propagated over and over again as parts of a single clone plant, producing generation after generation of genetically identical individuals.

Compared with the other two main evolutionary lines, or kingdoms, of organisms that are fundamentally multicellular—animals and fungi—plants differ because they are green: they photosynthesize. Photosynthesis takes place largely in leaves, which are flattened organs that are arranged to capture the sun's light, and in young green stems. There are no leaves on the underground portions of a plant, obviously, and the growth patterns in the roots and shoots differ fundamentally. This chapter is devoted primarily to outlining these differences and to analyzing the way these organs, together with their specialized cell and tissue types and their appendages, form the plant body. Although the similarities between a cactus, an orchid, and a pine tree might not be obvious at first sight, plants have a fundamental unity of structure. This unity is reflected in the construction plan of their respective bodies; in the way they grow, produce, and transport their food; and in how they regulate their development.

A vascular plant is organized along a vertical axis, like a pipe (Figure 29-2). The part below ground is called the **root**; the part above ground is called the shoot. The root penetrates the soil and absorbs water and various ions, which are crucial for plant nutrition. It also anchors the plant. The shoot consists of stem and leaves. The **stem** serves as a framework for the positioning of the **leaves,** the structure in which most photosynthesis takes place. The arrangement, size, and other characteristics of the leaves are of critical importance in the plant's production of food. Flowers, other reproductive organs, and ultimately fruits and seeds are formed on the shoot as well.

When an embryo is formed within a seed, it remains dormant for a time. Such an embryo consists of an axis, usually with one or two cotyledons, or embryonic leaves (Figure 29-3). As we have seen, monocot embryos usually have a single cotyledon, whereas dicot embryos usually have two. The food in a mature seed may be stored either in endosperm (a common condi-

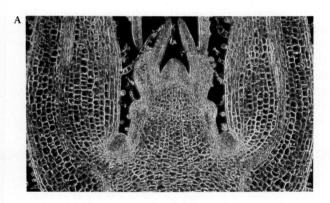

A

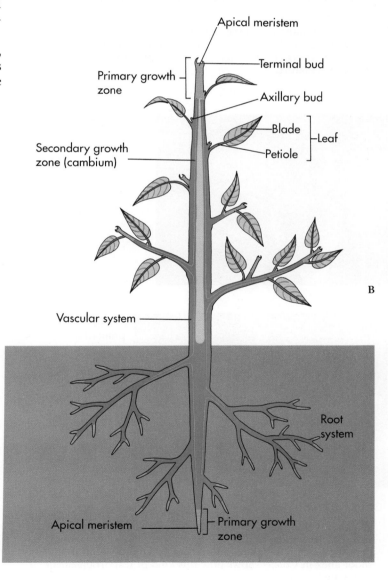

B

Apical meristem
Terminal bud
Primary growth zone
Axillary bud
Blade
Petiole
Leaf
Secondary growth zone (cambium)
Vascular system
Root system
Apical meristem
Primary growth zone

**Figure 29-2**

**The body of a plant.** The terms in this illustration are explained as we discuss different parts of the plant body, which consists of a shoot (stems and leaves) and root. Primary growth takes place as a result of the division of clusters of cells, the **apical meristems,** which are located at the ends of the roots and the stems; secondary growth takes place laterally, allowing the plant to enlarge in girth.
**A** Transection of a shoot apex in *Coleus.*
**B** Diagram of a plant body.

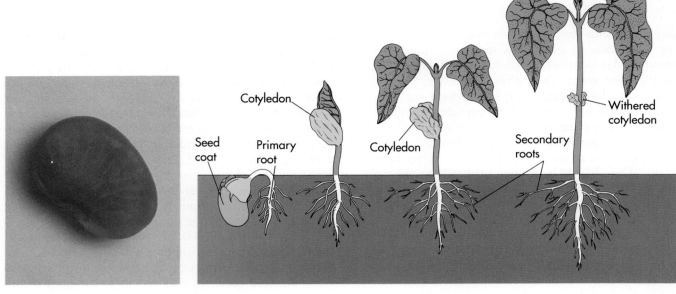

**Figure 29-3**

**Stages of germination in the common bean, *Phaseolus vulgaris*.** In the bean, as in most dicots, the food that is initially produced in the endosperm is absorbed by the embryo during the course of its development and primarily located in the cotyledons by the time the seed is mature.

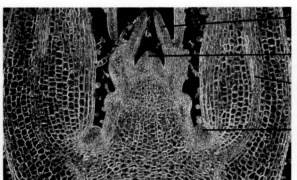

Young leaf primordium

Apical meristem

Older leaf primordium

Lateral bud

**Figure 29-4**

**Where leaves originate.** The leaves of *Coleus* are a familiar sight in greenhouses and as bedding plants. They are borne in opposite pairs, alternating in direction. The transection of a shoot apex in *Coleus* shows how this kind of leaf arrangement is initiated.

tion in monocots) or in the cotyledons (as in many dicots; see Figure 27-11).

In an embryo, the **apical meristems**—regions of active cell division that occur at the tips of roots and shoots—differentiate early, thus establishing the growth pattern that will persist throughout the life of the plant (Figure 29-4). In the germination of a seed, the embryonic root may emerge first, anchoring the seedling in the soil, or the shoot may emerge at the same time or even earlier.

*An apical meristem is a region of active cell division that occurs at or near the tips of the roots and shoots of plants.*

## TISSUE TYPES IN PLANTS

The organs of a plant, the leaves, roots, and stem, are composed of different combinations of tissues, just as your legs are composed of bone, muscle, and connective tissue. A tissue is a group of similar cells organized into a structural and functional unit, cells that are specialized in the same way. In plants there are three major tissue types: **vascular tissue**, which conducts water and dissolved minerals up the plant and conducts the products of photosynthesis throughout; **ground tissue**, in which the vascular tissue is embedded; and **dermal** tissue, the outer protective covering of the plant. Each major tissue type is composed of distinctive kinds of cells whose structures are related to the functions of the tissues in which they occur. For example, as we saw in Chapter 27, vascular tissue is composed of xylem, which conducts water and dissolved minerals, and phloem, which conducts carbohydrates, mostly sucrose, that the plant uses as food. Together, xylem and phloem are the two principal types of conducting tissue in vascular plants.

*The three major types of tissues in plants are (1) dermal tissue, which covers the outside of the plant; (2) ground tissue, which fills its interior; and (3) vascular tissue, which conducts water, minerals, carbohydrates, and other substances throughout the plant.*

## TYPES OF MERISTEMS

Animals grow all over. When you are growing from a child into an adult, your torso grows at the same time your legs do. Imagine if instead you grew in only one place, your legs simply getting longer and longer. This is similar to the way a plant grows. Plants contain zones of unspecialized cells called **meristems,** whose only function is to divide. Every time one of these cells divides, one of its two offspring remains in the meristem, whereas the other goes on to differentiate into one of the three kinds of plant tissue and ultimately become part of the plant body.

Primary growth in plants is initiated by the apical meristems. The growth of these meristems results primarily in the extension of the plant body. As it elongates, it forms what is known as the **primary plant body,** which is made up of the **primary tissues.** The primary plant body comprises the young, soft shoots of a tree or shrub, or the entire plant in some short-lived plants.

Secondary growth involves the activity of the **lateral meristems.** Lateral meristems are cylinders of meristematic tissue. The continued division of their cells results primarily in the thickening of the plant body. There are two kinds of lateral meristems: the **vascular cambium,** which gives rise to ultimately thick accumulations of secondary xylem and phloem, and the **cork cambium,** from which arise the outer layers of the bark on both roots and shoots. The tissues formed from the lateral meristems, constituting most of the bulk of trees

and shrubs, are known collectively as the **secondary plant body,** and its tissues are known as **secondary tissues.** Figure 29-2 compares primary and secondary growth zones.

The ways in which meristems function in the production of a mature plant determine its nature. A **woody plant,** such as a tree or shrub, is one in which secondary growth has been extensive. An **herbaceous plant** is one in which secondary growth has been limited. Herbaceous plants produce new shoots each year, either from underground portions of the plant or from seeds. If they complete their entire life cycle within a year, they are called **annuals;** if shoots are produced year after year, they are called **perennials.** Members of a less common class of plants, **biennials,** form a leafy shoot the first year and then go on to flower the second year. Two extremes are shown in Figure 29-5, a desert annual and a perennial.

*The primary plant body, which includes the young, soft shoots and roots, arises from the apical meristems. Once the lateral meristems begin to function, they produce the secondary plant body, which is characterized by thick accumulations of conducting tissue and the other cell types associated with it.*

## PLANT CELL TYPES

### Ground Tissue

**Parenchyma cells** are the least specialized and the most common of all plant cell types (Figures 29-7 and 29-14 show many parenchyma cells); they form masses in leaves, stems, and roots. Parenchyma cells, unlike some of the other cell types, are characteristically alive at maturity, with fully functional cytoplasm and a nucleus. Therefore they are capable of further division. Most parenchyma cells have only **primary cell walls,** which are mostly cellulose that is laid down while the cells are still growing. **Secondary cell walls,** in contrast, are deposited between the cytoplasm and primary wall of a fully expanded cell.

**Collenchyma cells,** which are also living at maturity, form strands or continuous cylinders beneath the epidermis of stems or leaf stalks and along veins in leaves. They are usually elongated, with unevenly thickened primary walls (Figure 29-6), which are their distinguishing feature. Strands of collenchyma provide much of the support for plant organs in which secondary growth has not taken place.

**Figure 29-5**

**Plants live for very different lengths of time.** Desert annuals such as those shown in **A** complete their entire life span in a few weeks, whereas trees—such as the giant redwood, *Sequoiadendron giganteum* (**B**), which occurs in scattered groves along the west slope of the Sierra Nevada in California—live 2000 years or more. When plants do live for more than a year, they accumulate wood; even in such plants, however, the stems that are produced during the current year are herbaceous.

*Parenchyma cells, which are usually living at maturity, are the most common type of cells in the primary plant body. They lack secondary cell walls. Collenchyma cells, which are also living at maturity, are elongated cells with unevenly thickened primary walls. They provide much of the support for plants in which secondary growth has not taken place.*

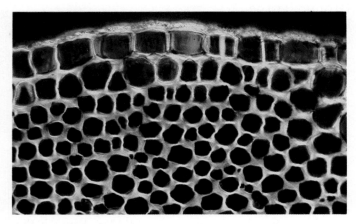

**Figure 29-6**

**Collenchyma.** Cross section of collenchyma cells, with thickened side walls, from a young branch of elderberry *(Sambucus)*. In other kinds of collenchyma cells, the thickened areas may occur at the corners of the cells or in other kinds of strips.

In contrast to parenchyma and collenchyma cells, **sclerenchyma** cells have tough, thick secondary walls; they usually do not contain living protoplasts when they are mature. There are two types of sclerenchyma: **fibers,** which are long, slender cells that usually form strands, and **sclereids,** which are variable in shape but often branched. Sclereids are sometimes called stone cells because they make up the bulk of the stones of peaches and other "stone" fruits, as well as that of nut shells (Figure 29-7). Both fibers and sclereids are tough and thick-walled; they serve to strengthen the tissues in which they occur.

## Dermal Tissue

Flattened **epidermal cells,** which are often covered with a thick, waxy layer called the **cuticle,** cover all parts of the primary plant body. These cells are the most abundant kind in the epidermis, or skin, of a plant. They protect the plant and provide an effective barrier preventing water loss. A number of kinds of specialized cells occur among the epidermal cells, including guard cells, trichomes, and root hairs.

**Guard cells** are paired cells. With the opening that lies between them, they make up the **stomata** (singular, **stoma**) (see Figures 27-4 and 29-14), which occur frequently in the epidermis of leaves and occasionally on the outer parts of the shoot, such as on stems or fruits. The passage of oxygen and carbon dioxide into and out of the leaves, as well as the loss of water from them, takes place almost exclusively through the stomata, which open and shut in response to external factors such as the supply of moisture (see Chapter 31).

**Figure 29-7**

**Sclereids.** Clusters of sclereids ("stone cells"), stained blue in this preparation, in the pulp of a pear. Such clusters of sclereids give pears their gritty texture.

**Trichomes** are outgrowths of the epidermis, superficially like the hairs of mammals, that vary greatly in form in different kinds of plants (Figure 29-8). They play a major role in controlling the loss of water from leaves and other plant parts and in regulating the temperature of plant parts. Similar to trichomes, but actually extensions of single epidermal cells, are **root hairs,** which occur in masses just behind the very ends of the roots (Figure 29-9). They keep the roots in intimate contact with the particles of soil and are soon worn off as the root continues to grow.

## Vascular Tissue

Vascular plants contain two kinds of conducting or vascular tissue. These are the xylem and the phloem.

*Xylem.* **Xylem** is the principal water-conducting tissue of plants. It forms a continuous system running throughout the plant body. Within this system, water passes from the roots up through the shoot in an unbroken stream; Chapter 31 discusses the way in which

A                          B                          C

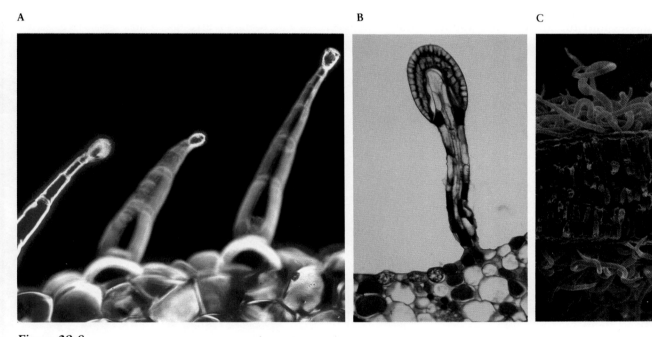

**Figure 29-8**

**Trichomes.**
**A** Three of the multicellular trichomes that cover the leaves of African violets, *Saintpaulia.* A covering of trichomes creates a layer of more humid air near the surface of a leaf, enabling the plant to conserve available supplies of water.
**B** A complex multicellular trichome from the leaves of the sundew, *Drosera.* Such hairs secrete the enzymes that the plant uses to digest the bodies of its insect prey.
**C** The dense mats of trichomes on both sides of the silvery leaves of the desert composite *Encelia farinosa,* seen in this scanning electron micrograph, cause the leaf absorption of visible sunlight to drop from about 85%, as in a green leaf, to about 30%; as a consequence, the leaf temperatures are about 8° to 10° C cooler than if they were green. This lower leaf temperature reduces water loss by some 30% to 50%.

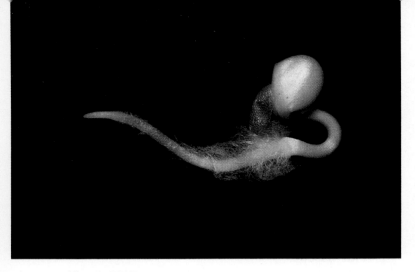

**Figure 29-9**

**Root hairs.** A germinating seedling of radish, *Raphanus sativus*, showing the abundant fine root hairs, which are the small structures just visible in the back of the root apex.

The two principal types of conducting elements in the xylem are **tracheids** and **vessel elements** (Figure 29-10), both of which have thick secondary walls, are elongated, and have no living protoplast at maturity. In conducting elements composed of tracheids, water flows from tracheid to tracheid through openings called **pits** in the secondary walls. In contrast, vessel elements have not only pits but also definite openings, or **perforations,** in their end walls by which they are linked together and through which water flows. A linked row of vessel elements forms a **vessel.** Primitive angiosperms have only tracheids, but the majority of living angiosperms have vessels. Vessels conduct water much more efficiently than do strands of tracheids. In addition to the conducting cells, xylem likewise includes fibers and parenchyma cells.

this stream is maintained. Dissolved minerals also are taken into plants through their roots, as a part of this stream of water. When the water reaches the leaves, much of it passes into the air as water vapor, mainly through the stomata.

*The major types of conducting cells of the xylem are tracheids and vessel elements. Incorporated in the xylem are also fibers and parenchyma cells.*

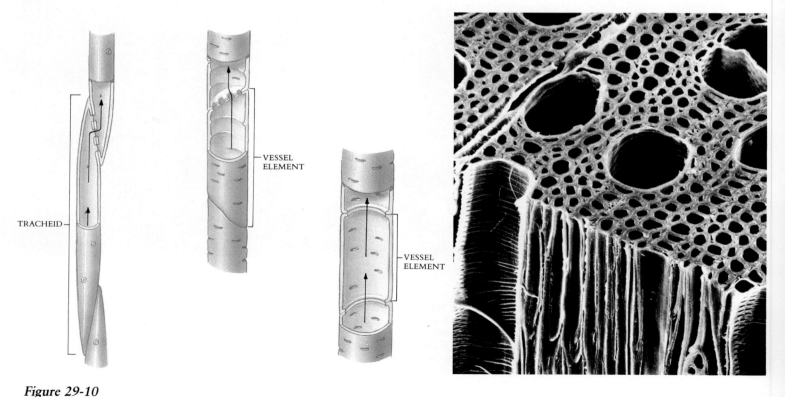

**Figure 29-10**

**Comparison between vessel elements and tracheids.** In tracheids the water passes from cell to cell by means of pits, whereas in vessel elements it moves by way of pores, which may be simple or interrupted by bars.

*Phloem.* Phloem is the principal food-conducting tissue in the vascular plants. Different kinds of plants have one of two different kinds of phloem cells: **sieve cells** or **sieve-tube members.** Clusters of pores known as **sieve areas** occur on both kinds of cells and connect the protoplasts of adjoining sieve cells and sieve-tube members. Both cell types are living, but their nuclei are lost during the process of their maturation.

Angiosperms contain sieve-tube members, phloem cells in which the pores in some of the sieve areas are larger than those in others; such sieve areas are called **sieve plates.** Sieve-tube members occur end-to-end, forming longitudinal series called **sieve tubes.** Other vascular plants contain sieve cells, phloem cells in which the pores in all of their sieve areas are roughly the same diameter. Specialized parenchyma cells known as **companion cells** occur regularly in association with sieve-tube members (Figure 29-11), but are lacking in plants with sieve cells. In an evolutionary sense, sieve-tube members clearly are advanced over sieve cells; they are more specialized and presumably more efficient.

*The principal cell types in phloem are sieve cells or sieve-tube members. Sieve cells and sieve-tube members lose their nucleus in the course of becoming mature.*

# SHOOTS

The shoot of a plant is technically that portion that lies above the cotyledon, and the root is the portion below the place where they are attached. In most plants, all of the above-ground parts are portions of the shoot. Leaves are usually the most prominent organs of the shoot and determine its appearance.

## Leaves

Leaves, outgrowths of the shoot apex, are the light-capturing organs of most plants. The only exceptions to this are found in some plants, such as cacti, in which the stems are green and have largely taken over the function of photosynthesis for the plant (see Figure 29-1).

The apical meristems of stems and roots are capable of growing indefinitely under appropriate conditions. Leaves, in contrast, grow by means of **marginal meristems,** which flank their thick central portions. These marginal meristems grow outward and ulti-

mately form the blade of the leaf, while the central portion becomes the midrib. Once a leaf is fully expanded, its marginal meristems cease to function: their growth is called **determinate,** whereas that of apical meristems is **indeterminate.**

Most leaves have a flattened portion, the **blade,** and a slender stalk, the **petiole** (see Figures 29-2 and 29-12). In addition, there may be two leaflike organs, the **stipules,** that flank the base of the petiole where it

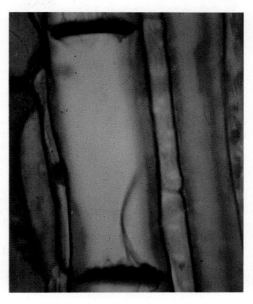

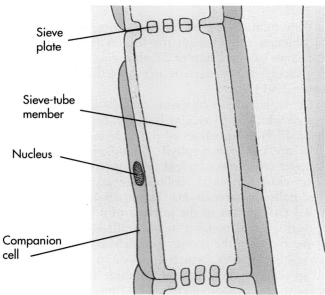

Sieve
plate

Sieve-tube
member

Nucleus

Companion
cell

**Figure 29-11**

**Sieve tubes.** Sieve-tube member from the phloem of squash *(Cucurbita),* connected with the cells above and below to form a sieve tube. Note the thickened end walls, which are at right angles to the sieve tube. The narrow cell with the nucleus at the left of the sieve-tube member is a companion cell.

*Figure 29-12*

**Leaves.** Angiosperm leaves are stunningly variable: they are the primary factor in controlling the specific ways plants capture light and regulate their water loss.
**A** Diverse leaves in the herb layer of a Costa Rican rain forest.
**B** A compound leaf: marijuana *(Cannabis sativa)*. Such a compound leaf, in which the leaflets join the petiole at one point, is said to be palmately compound. A compound leaf is associated with a single lateral bud, located where the petiole is attached to the stem.
**C** A simple leaf, its margin deeply lobed, from the tulip tree *(Liriodendron tulipifera)*.
**D** Another compound leaf from a member of the legume family in the lowland forest of Peru. Such a compound leaf, in which the leaflets are attached all along the main axis, is said to be pinnately compound.
**E** Many unusual arrangements of leaves occur in different kinds of plants. For example, in miner's lettuce *(Claytonia perfoliata),* an herb of the Pacific states, two leaves are completely fused below each of the clusters of flowers, which seem, therefore, to arise from the center of a single leaf.

joins the stem. Veins, usually consisting of both xylem and phloem, run through the leaves. In many monocots, as we saw in Chapter 27, the veins are parallel; in most dicots, the pattern is net or reticulate venation (Figure 29-13).

A typical leaf contains masses of parenchyma through which the vascular bundles, or veins, run (see Figure 27-3). The masses of parenchyma that occur in leaves are called **mesophyll** ("middle-leaf"). Beneath the upper epidermis of a leaf there are one or more layers of closely packed, columnlike parenchyma cells called **palisade parenchyma** (Figure 29-14; see also Figure 8-14). The rest of the interior of a leaf, except for the veins, consists of a tissue called **spongy parenchyma**. Between the spongy parenchyma cells are large intercellular spaces, which function in gas exchange and in the passage of carbon dioxide from the atmosphere to the mesophyll cells. These intercellular spaces are connected, directly or indirectly, with the stomata.

The cells of the mesophyll, especially those near the leaf surface, are packed with chloroplasts (see Figure 8-13). These cells constitute the plant's primary site of photosynthesis. Water and minerals are brought from the roots to the leaves in the xylem strands of the veins. Once the water reaches the ends of the veins, it passes into the photosynthetic mesophyll cells. Because the surfaces of these cells border on the intercellular spaces of the interior of the leaf, and because water can pass through cell membranes and cell walls easily, much of the water evaporates into the intercellular space and can then escape from the leaf through the open stomata. Thus a structure that allows absorption of the essential carbon dioxide also allows the escape of water. At the same time that water is leaving through the stomata, the products of photosynthesis are being transported from the leaves to all other parts of the plant through the phloem.

*The mesophyll in a leaf consists of two types of parenchyma cells, both packed with chloroplasts. Palisade parenchyma cells are columnar and closely packed together, whereas spongy parenchyma cells are loosely packed and separated by large intercellular spaces.*

**Figure 29-13**

**A comparison of dicot and monocot leaves.** The leaves of dicots, such as this African violet relative from Sri Lanka, have net, or reticulate, venation; those of monocots, such as this Latin American palm, have parallel venation. Both leaves have been cleared with chemicals, and the dicot one has been stained with a red dye to make its veins show up more clearly.

## Stems

*Primary Growth.* In the primary growth of a shoot, leaves first appear as **leaf primordia** (singular, **primordium**), or rudimentary young leaves, which cluster around the apical meristem, unfolding and growing as the stem itself elongates (Figure 29-15; see also Figure 29-4). The places on the stem at which leaves form are called **nodes** (see Figure 29-2). The portions of the stem between these attachment points are called the **internodes**. As the leaves expand to maturity, a **bud,** a tiny, undeveloped side-shoot, develops in the **axil** of each leaf, the angle between a branch or leaf and the stem

from which it arises (see Figure 29-4, where the buds are visible as densely staining masses at the base of the larger pair of leaf primordia). These buds, which have their own leaves, may elongate and form lateral branches, or they may remain small and dormant. A hormone diffusing downward from the terminal bud of the shoot continuously suppresses the expansion of the lateral buds (see Chapter 31 where plant hormones are discussed in more detail). These buds begin to expand when the terminal bud is removed. Therefore gardeners who wish to produce bushy plants or dense hedges crop off the tops of the plants, thus removing their terminal buds.

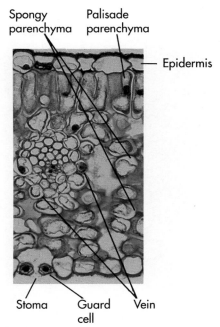

Spongy parenchyma  Palisade parenchyma

Epidermis

Stoma  Guard cell  Vein

**Figure 29-14**

**A leaf in cross section.** Transection of a lily leaf, showing palisade and spongy parenchyma; a vascular bundle, or vein; and the epidermis, with paired guard cells flanking the stoma, which is visible on the lower surface of the leaf and the substomatal space.

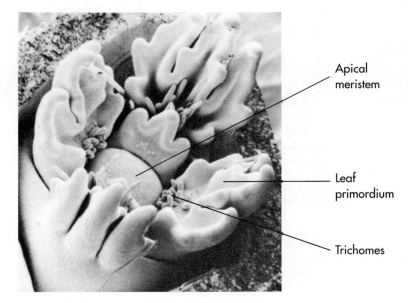

Apical meristem

Leaf primordium

Trichomes

**Figure 29-15**

**The shoot apex.** Scanning electron micrograph of shoot apex of silver maple, *Acer saccharinum,* showing a developing shoot during summer, the season of active growth. The apical meristem, leaf primordia, and trichomes are plainly visible at this stage.

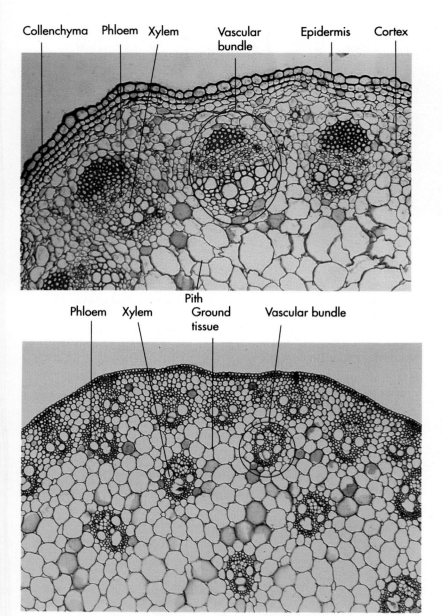

Collenchyma  Phloem  Xylem  Vascular bundle  Epidermis  Cortex

Phloem  Xylem  Pith Ground tissue  Vascular bundle

**Figure 29-16**

**A comparison of dicot and monocot stems.** Transections of a young stem in a dicot, the common sunflower, *Helianthus annuus*, in which the vascular bundles are arranged around the outside of the stem; and a monocot, corn, *Zea mays*, with the scattered vascular bundles characteristic of the class.

Within the soft, young stems, the strands of vascular tissue, xylem and phloem, either occur as a cylinder in the outer portion of the stem, as is common in dicots (Figure 29-16), or are scattered through it, as is common in monocots. The vascular bundles contain both primary xylem and primary phloem (Figure 29-17). At the stage when only primary growth has occurred, the inner portion of the ground tissue of a stem is called the **pith**, and the outer portion is the **cortex**.

*Secondary Growth.* In stems, secondary growth is initiated by the differentiation of the **vascular cambium**, which consists of a thin cylinder of actively dividing cells that is located between the bark and the main stem in mature woody plants. The vascular cambium differentiates from parenchyma cells within the vascular bundles of the stem, between the primary xylem and the primary phloem (Figure 29-18). These begin to divide actively first, and then the vascular cambium grows into a cylinder when some of the parenchyma cells between the bundles also begin to divide. When it has been established, the vascular cambium consists of elongated, somewhat flattened cells with large vacuoles. The cells which divide from the vascular cambium outwardly, toward the bark, become secondary phloem; those which divide from it inwardly become secondary xylem. The cells of the vascular cambium also divide side-to-side, allowing the stem to get larger and larger as the tree or shrub becomes more mature.

While the vascular cambium is becoming established, a second kind of lateral cambium, the **cork cambium,** normally also develops in the outer layers of the stem. The cork cambium usually consists of plates of dividing cells that move deeper and deeper into the stem as they divide. Outwardly, the cork cambium splits off densely packed **cork cells;** they contain a fatty substance and are nearly impermeable to water. Cork cells are dead at maturity. Inwardly, the cork cambium

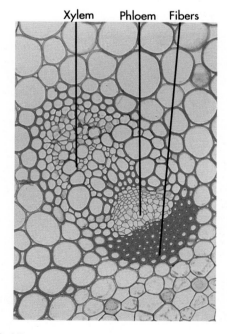

Xylem  Phloem  Fibers

**Figure 29-17**

**A vascular bundle.** Transection of a vascular bundle from a buttercup, *Ranunculus acris*, showing the xylem and phloem.

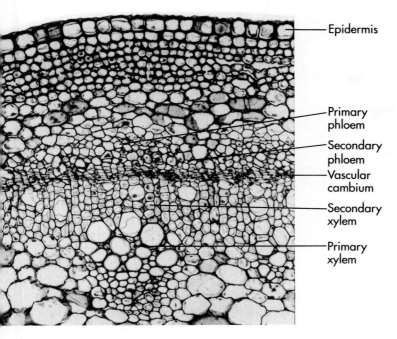

Figure 29-18

**Vascular cambium.** Early stage in differentiation of vascular cambium in an elderberry *(Sambucus canadensis)* stem. The outer part of the cortex consists of collenchyma, and the inner part of parenchyma.

divides to produce a dense layer of parenchyma cells. The cork, the cork cambium that produces it, and this dense layer of parenchyma cells make up a layer called the **periderm,** which is the outer protective covering of the plant. Oxygen reaches the layers of living cells under the bark through areas of loosely organized cells called **lenticels,** which are often easily identifiable on the outer surface of bark (Figure 29-19).

*The periderm consists of cork to the outside, the cork cambium in the middle, and layers of parenchyma cells to the inside of a stem or root. Penetrated by areas of more loosely packed tissue known as lenticels, the periderm retards water loss from the secondary plant body.*

Cork, which covers the surfaces of mature stems or roots, takes the place of the epidermis, which performs a similar function in the younger parts of the plant. **Bark** is a term used to refer to all of the tissues of a mature stem or root outside of the vascular cam-

bium (Figure 29-20). Because the vascular cambium has the thinnest-walled cells that occur anywhere in a secondary plant body, it is the layer at which bark breaks away from the accumulated secondary xylem. The inner layers of the bark are primarily secondary phloem, with the remains of the primary phloem crushed among them. Its outer layers consist of the periderm, and the very outermost ones are cork.

*Wood.* Wood is one of the most useful, economically important, and beautiful products that we obtain from plants. Anatomically, **wood** is accumulated secondary xylem. As the secondary xylem ages, its cells become infiltrated with gums and resins, and the wood becomes darker. For this reason, the wood located nearer the central regions of a given trunk, called **heartwood,** is often darker in color and denser than the wood nearer the vascular cambium, called **sapwood,** which is still actively involved in transport within the plant (Figure 29-21). Commercially, wood is divided into hardwoods and softwoods. **Hardwoods** are the woods of dicots, regardless of how hard or soft they actually may be; **softwoods** are the woods of conifers.

Because of the way it is accumulated, wood often displays rings. In temperate regions, these rings are **annual rings** (see Figure 29-21): they reflect the fact that

A           B

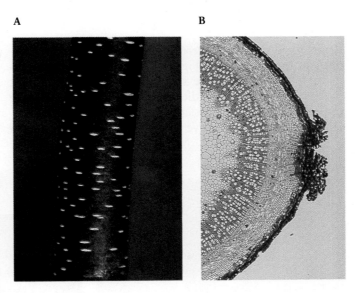

Figure 29-19

**Lenticels.**
**A** Lenticels are the numerous, small, pale, raised areas visible on the bark of paper birch, *Betula papyrifera.* Lenticels allow oxygen to readily diffuse into the living tissues immediately below the bark of woody plants. Highly variable in form in different species, lenticels are a valuable aid to the identification of deciduous trees and shrubs in winter.
**B** Transection through a lenticel in a stem of elderberry, *Sambucus canadensis.*

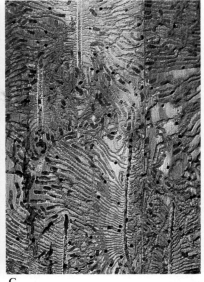

A              B                                C

*Figure 29-20*

**Bark.**

**A** Bark of valley oak, *Quercus lobata,* in California. Bark is highly characteristic of individual trees and shrubs, which often can be recognized by the bark's characteristics even when the trees are leafless (**B**).

**C** Galleries constructed by elm bark beetles, which spread Dutch elm disease from tree to tree, in the cambium of an elm tree. Such beetles bore through the thin layer of living cells that separates bark from the trunk of a tree, thus gaining access to the carbohydrates passing through the phloem.

the vascular cambium divides more actively when water is plentiful and temperatures are suitable for growth than when water is scarce and the weather is cold. The abrupt discontinuity between the layers of larger cells, with proportionately thinner walls, which form in the growing season (in most temperate regions, during the spring and early summer) and those which form later is often very evident. For this reason, the annual rings in a tree trunk can be used to calculate the age of the tree.

## Roots

Roots have simpler patterns of organization and development than do stems (Figure 29-22, *A*). Although different patterns exist, we shall describe a kind of root that is found in many dicots. There is no pith in the center of the vascular tissue of the root in most dicots. Instead, these roots have a central column of xylem with radiating arms. Between these arms are located strands of primary phloem (Figure 29-23). Around the column of vascular tissue, and forming its outer boundary, is a cylinder of cells one or more cell layers thick called the **pericycle**. Branch, or lateral, roots are formed from cells of the pericycle, as we shall discuss further. The outer layer of the root, as in the shoot, is the **epidermis**. The mass of parenchyma in which the vascular tissue of the root is located is the cortex. Its innermost layer, the **endodermis,** consists of specialized cells that regulate the flow of water between the vascular tissues and the outer portion of the root. The endodermis lies just outside of the pericycle. Its cells are surrounded by a thickened waxy band called the **Casparian strip.** It is by the differential passage of minerals and nutrients through these cells that the plant regulates its supply of minerals.

A

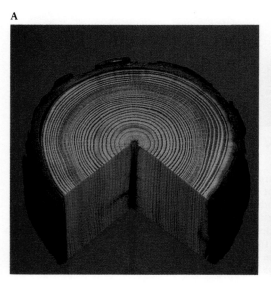

B

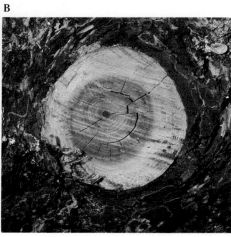

*Figure 29-21*

**Two features of wood.**

**A** Annual rings in a section of pine *(Pinus).*

**B** The distinction between heartwood (dark central portion) and sapwood (light outer portion) is evident in this sawed-off limb of ponderosa pine *(Pinus ponderosa)* in the mountains of California.

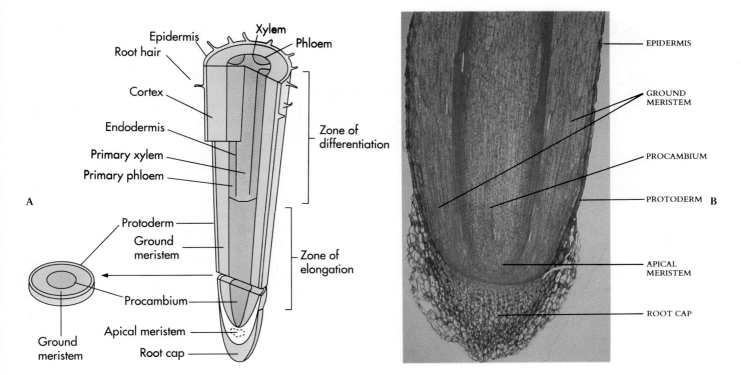

**Figure 29-22**

**A** Root structure. Diagram of primary meristems in the root, showing their relation to the apical meristem.
**B** A root tip. Median longitudinal section of a root tip in corn, *Zea mays*, showing the differentiation of epidermis, cortex, and column of vascular tissues.

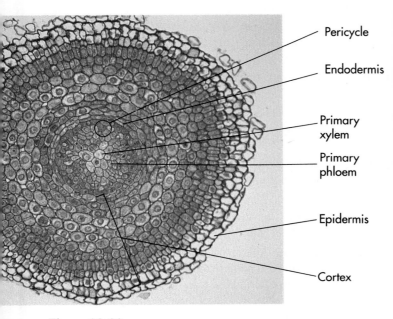

**Figure 29-23**

**A root in cross section.** Cross section through a root of a buttercup, *Ranunculus californicus*. The central xylem and phloem, cortex, and epidermis are visible.

Most dicot roots have a central column of primary xylem with radiating arms and strands of primary phloem between these arms, surrounded by a layer one or more cells thick called the pericycle. The innermost layer of the cortex, or endodermis, consists of cells surrounded by a thickened waxy band called the Casparian strip.

The apical meristem of the root divides and produces cells both inwardly, back toward the body of the plant, and outwardly. Outward cell division results in the formation of a thimblelike mass of relatively unorganized cells, the **root cap,** which covers and protects the root's apical meristem as it grows through the soil.

The root elongates relatively rapidly just behind its tip. Above that zone are formed abundant root hairs (see Figure 29-9). Virtually all absorption of water and minerals from the soil takes place through the root hairs, which greatly increase the surface area and there-

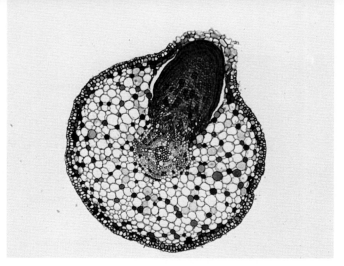

***Figure 29-24***

**Lateral roots.** A lateral root growing out through the cortex of black willow, *Salix nigra*. The origin of lateral roots occurs beneath the surface of the main root, whereas the origin of lateral stems occurs at the surface.

fore the absorptive powers of the root. In plants that have mycorrhizae (Chapter 26), the root hairs are often greatly reduced in number, and the fungal filaments of the mycorrhizae play a role similar to that of the root hairs.

One of the fundamental differences between roots and shoots has to do with the nature of their branching. In stems, branching occurs from buds on the surface of the stem; in roots, branching is initiated well back of the root apex as a result of cell divisions in the pericycle. The lateral root primordia grow out through the cortex toward the surface of the root (Figure 29-24), eventually breaking through and becoming established as lateral roots. Secondary growth in roots, both main roots and laterals, is similar to that in stems, with the vascular cambium being initiated after the division of cells located between the primary xylem and the primary phloem.

## ■ SUMMARY

1. A plant body is basically an axis that has two parts, a root and a shoot. Within it, there are three principal tissue types: vascular tissue, ground tissue, and dermal tissue. Dermal tissue covers the outside of the plant; vascular tissue conducts substances through it; and ground tissue is the matrix in which the vascular tissue is embedded.

2. Plants grow in length by means of their apical meristems, zones of active cell division at the ends of the roots and the shoots.

3. Plants which complete their entire life cycle within a year are called annuals; those which require 2 years to reach maturity and then flower just once are called biennials; and those which flower year after year once they have reached maturity are known as perennials. Perennial plants may be soft, as are herbs, or woody, as are trees and shrubs.

4. Parenchyma cells are the most abundant type of ground tissue in the plant body. They usually have only primary walls, which are composed mainly of cellulose and are laid down while the cells are still growing. Parenchyma cells are living at maturity.

5. Collenchyma cells often form strands or continuous cylinders in the plant body, for which they provide the chief source of strength. They may be recognized by the uneven thickenings of their primary cell walls.

6. Sclerenchyma cells have tough, thick walls with secondary thickening laid down after the cells have reached full size. Fibers, which are elongated strengthening cells, are sclerenchyma cells.

7. Tracheids and vessel elements are the principal conducting elements of the xylem. They have thick cell walls and lack protoplasts at maturity. Water reaches the leaves after entering the plant through the roots and passing upward via the xylem. Water vapor passes out of leaves by entering the intercellular spaces, evaporating, and then moving out through the stomata; carbon dioxide enters the plant by the same route.

8. Carbohydrates are conducted through the plant primarily in the phloem, whose elongated conducting cells are living but lack a nucleus. These conducting cells are called sieve cells and sieve-tube members.

9. The growth of leaves is determinate, and that of shoots and roots is indeterminate. Leaves, which are mainly flattened organs of the shoot specialized for photosynthesis, expand in size by means of marginal meristems.

10. Stems branch because of the growth of buds that form externally at the point where the leaves join the stem; roots branch by forming centers of cell division within their cortex. Young roots grow out through the cortex, eventually breaking through the surface of the root.

11. Secondary growth in both stems and roots takes place following the formation of lateral meristems known as vascular cambia. These cylinders of dividing cells form xylem internally and phloem externally. As a result of their activity, the girth of a plant increases.

12. Wood is accumulated secondary xylem; it often displays rings because it exhibits different rates of growth at different seasons.

13. The cork cambium forms in both roots and stems during the initial stages of establishment of the vascular cambium. It produces cork externally and a dense layer of parenchyma internally. The cork, cork cambium, and underlying parenchyma layers collectively are called the periderm. This layer, the outermost of bark, is perforated, in the stem, by areas of loosely organized tissue called lenticels.

## REVIEW

1. Areas of active cell division that occur at the shoot and root tips are called _____.

2. Primary growth in plants is initiated by the _____, whereas secondary growth involves activity of the _____.

3. Epidermal cells are covered with a waxy _____, which retards water loss from the plant.

4. Which leaf cells are the primary site of photosynthesis?

5. The term _____ refers to all of the tissue of a woody stem outside the vascular cambium.

## SELF-QUIZ

1. The vegetative organs of plants are (pick three)
   (a) flowers.
   (b) stems.
   (c) roots.
   (d) fruits.
   (e) leaves.

2. Parenchyma cells
   (a) are alive at maturity.
   (b) have only primary cell walls.
   (c) are capable of further division.
   (d) are found in both the primary and secondary plant body.
   (e) all of the above.

3. Xylem
   (a) is the principal water-conducting tissue of plants.
   (b) conducts the products of photosynthesis.
   (c) is made of a number of different cell types.
   (d) is not found in roots.
   (e) a and c.

4. What goes on at the lenticels?
   (a) Water absorption
   (b) Gas exchange
   (c) Photosynthesis
   (d) Production of the periderm
   (e) a and c

5. Cells of the _____ regulate the flow of water laterally between the vascular tissues and the cell layers in the outer portions of the root.
   (a) periderm
   (b) endodermis
   (c) pericycle
   (d) pith
   (e) xylem

6. Branch roots arise
   (a) in the axils of other branch roots.
   (b) in the axils of root hairs.
   (c) from the endodermis.
   (d) from the pericycle.
   (e) from the cork cambium.

7. Branch stems arise
   (a) from lateral meristems.
   (b) from the axils of leaf primordia.
   (c) internally.
   (d) from the endodermis.
   (e) from primary phloem strands.

8. Roots become thicker
   (a) because of the accumulation of branch roots.
   (b) as they take in water.
   (c) when they form lateral meristems that become active.
   (d) because they lack a cork cambium.
   (e) because of the activity of the periderm.

# THOUGHT QUESTIONS

1. If you hammer a nail into the trunk of a tree 2 meters above the ground when the tree is 6 meters tall, how far above the ground will the nail be when the tree is 12 meters tall?

2. What is the difference between primary and secondary growth?

# FOR FURTHER READING

GALSTON, A.W., P.J. DAVIS, and R.L. SATTER: *The Life of the Green Plant*, ed. 3, Prentice-Hall, Inc., Englewood Cliffs, N.J., 1980. Very well-written, physiologically oriented treatment of the structure and functioning of plants.

RAVEN, P.H., R.F. EVERT, and H. CURTIS: *The Biology of Plants*, 4th ed., Worth Publishers, Inc., New York, 1981. A comprehensive treatment of general botany, emphasizing structural botany.

SANDVED, K.B., and G.T. PRANCE: *Leaves*, Crown Publishers, Inc., New York, 1984. This book provides a beautiful introduction to the diversity of leaves.

SWAIN, R.B.: "Notes from the Radical Underground," *Discover*, November 1988, pages 16-18. Cooperation and competition between tree roots—a new view of the relationships between plants.

WOODWARD, I.: "Plants, Water and Climate. Part One," *New Scientist*, February 18, 1989, supplement, pages 1-4. Excellent account of the way water moves through plants.

# *F*lowering Plant Reproduction

Called "Yellow Bird," the Columnea flower is an essential element in the reproduction of the plant, for it attracts the insects that carry its pollen to other plants and so ensures the next generation.

# FLOWERING PLANT REPRODUCTION

## Overview

The flowering plants, or angiosperms, dominate every spot on earth except the great northern forests, the polar regions, the high mountains, and the driest deserts. Among the features that have contributed to their success are their unique reproductive structures, the flower and the fruit. Flowers bring about the precise transfer of pollen by insects and other animals. Because they can "control" the activities of these animals, plants can exchange gametes with one another even though each plant is rooted in one place. Fruits play a role of extraordinary importance in the dispersal of angiosperms from place to place. Not only were both flowers and fruits important in making possible the early success of the flowering plants, but their evolution has produced most of the striking differences that we see among different members of the class.

## For Review    *Here are some important terms and concepts that you will encounter in this chapter. If you are not familiar with them, you should review them before proceeding.*

**Meiosis and mitosis** (Chapter 9)

**Evolution of major groups** (Chapter 17)

**Coevolution of insects and flowering plants** (Chapter 21)

**Angiosperm life cycle** (Chapter 27)

**Figure 30-1**

**The heart of a flower.** The rich colors and textures of this flower head of a South African species of *Gazania*, a member of the sunflower family (Asteraceae), have evolved in response to the sensory perceptions and activities of insects. There are two types of flowers in the head; the outermost are extended into rays, and the inner ones, called disc flowers, are symmetrical.

Plants are, with insects, one of the few major groups of organisms to evolve entirely on land. As you learned in Chapter 17, insects and plants coevolved in conquering terrestrial environments, insects forming a key link in the reproduction of many of the flowering plants that now cover the face of our earth (Figure 30-1). Each one of us deals with plant reproduction every day without thinking about it. The bread we eat is the ground-up seed of a grass, wheat; the roses that a boy gives his girlfriend evolved as structures attractive to insects; and the honey that we put on bread is produced by bees from nectar, the bribe a flower uses to induce the bees to carry the flower's gametes to another plant. In this chapter we examine how flowering plants, or angiosperms, reproduce.

## FLOWERS AND POLLINATION

As you saw in Chapter 27, all plant life cycles are characterized by an alternation of generations, which involves diploid sporophytes and haploid gametophytes. In plants, the haploid cells that result from meiosis *always* divide by mitosis to produce a gametophyte generation. In flowering plants, this gametophyte generation is very reduced—the size of a fleck of dust—and completely enclosed within the tissues of its parent sporophyte. The ovules of flowering plants, the structures that become the seeds when mature, are enclosed within carpels, and pollination is indirect. In contrast, gymnosperms have direct pollination, in which the ovules are exposed at the time pollen reaches them.

Unlike the reproductive organs of animals, the reproductive structures of plants are not permanent parts of the adult individual. Flowers and the reproductive organs of other plant groups develop seasonally, being produced at times of the year that are favorable for pollination. In a flower, the corolla and sometimes other floral parts are often brightly colored and attractive to insects and other animals that may visit the flower and thus carry its pollen passively from place to place. Those flowering plants in which the pollen is carried from place to place by wind or in which self-pollination predominates often have dull-colored or green corollas and other flower parts.

## Pollination by Animals

In many angiosperms, the pollen grains are carried from flower to flower by insects and other animals that visit the flowers for food or other rewards (Figure 30-2), or are deceived into doing so because the characteristics of the flower suggest that they may offer such rewards. Successful pollination depends on the plants attracting insects and other animals regularly enough that these animals will carry pollen from one flower of that particular species to another. Flowers that are regularly visited by animals normally provide some kind of food reward, often in the form of a liquid called **nectar.** Nectar is rich not only in sugars but also in amino acids and other substances.

The relationship between such animals, known as **pollinators,** and the flowering plants has been one of the most important features of the evolution of both groups. By using insects to transfer pollen, the flowering plants can disperse their gametes on a regular and more or less controlled basis, despite the fact that they are anchored to their substrate.

For pollination by animals to be effective, it is necessary that a particular insect or other animal visit numerous plant individuals of the same species. The color and form of flowers have been shaped by evolution to promote such specialization. Yellow flowers are particularly attractive to bees, whereas red flowers attract birds but are not particularly noticed by insects (Figure 30-3). Some flowers have very long floral tubes with the nectar produced deep within them; only the long, slender beaks of hummingbirds or the long, coiled tongues of moths or butterflies can reach nectar supplies of this kind. Specialized flowers that attract certain kinds of animals consistently take advantage of these visits in spreading their pollen accurately from individual to individual. The animals that visit flowers in this way perform the same functions for the flowering plants that they do for themselves when they actively search out mates.

The most numerous insect-pollinated angiosperms are those pollinated by bees (see Figure 30-2), a large group of insects consisting of some 20,000 species. Bees are the most frequent, characteristic, and constant visitors to particular kinds of flowers today. On the other hand, they certainly did not pollinate the most primitive angiosperms, in which beetles (Figure 30-4) may have played a similar role; bees were not abundant and may not even have existed when angiosperms first appeared. The diversity of flowering plants that we see today is closely related to later specialization of angiosperms in relation to bees. The coevolutionary relationships between the members of these two groups are often tightly linked.

## Pollination by Wind

In certain angiosperms, pollen is blown about by the wind, as it is in most gymnosperms, and reaches the stigmas passively. For such a system to operate efficiently, the individuals of a given plant species must grow relatively close together, because wind does not carry pollen very far or very precisely as compared with insects or other animals. Because gymnosperms such as spruces or pines grow in dense forests, they are

**Figure 30-2**

**A bumblebee, *Bombus,* covered with pollen while visiting a flower.** This bee will transfer large quantities of pollen to the next flower that it visits.

A       B       C

**Figure 30-3**

**Hummingbirds and flowers.**
**A** Long-tailed hermit hummingbird extracting nectar from the flowers of *Heliconia imbricata* in the forests of Costa Rica. Note the pollen on the bird's beak. Hummingbirds of this group obtain nectar primarily at long, curved flowers that more or less match the length and shape of their beak.
**B** Poinsettia *(Euphorbia pulcherrima)* flowers. The individual flowers each have a large nectary at one side. The clusters of yellowish flowers are made more attractive to birds because of the large red leaves that surround them.
**C** *Penstemon centranthifolius* in the deserts of southern California, showing how attractive the plants that regularly attract hummingbirds may be, even at a distance.

very effectively pollinated by wind. Wind-pollinated angiosperms such as birches, alders, and ragweed (Figure 30-5) also tend to grow in dense stands. The flowers of wind-pollinated angiosperms are usually small, greenish, and odorless, and their petals are reduced in size or absent. They typically produce large quantities of pollen. Certainly the ancestors of the angiosperms were wind-pollinated, but whether the very first members of the group were pollinated by insects or by the wind cannot be determined with certainty. But the association with insects is an ancient one for the flowering plants. Wind pollination seems certainly to have evolved secondarily in all angiosperms in which it occurs today; in other words, they all had insect-pollinated ancestors.

**Figure 30-5**

**The flowers of a birch, *Betula.*** Birches have two different kinds of flowers, pollen-producing (staminate) ones, which hang down in long, yellowish tassels, and ovule-bearing (pistillate) ones, which mature into characteristic conelike structures and occur in small, reddish-brown clusters. Both are shown here.

**Figure 30-4**

**Some insects eat pollen.** Scarab beetles eating pollen at the head of flowers of a relative of the sunflower in Greece.

## Self-Pollination

In some angiosperms the pollen does not reach other individuals at all: instead, it is shed directly onto the stigma of the same flower, sometimes in bud. This results in self-pollination and inbreeding, with evolutionary consequences that are discussed in Chapter 15.

## SEED FORMATION

The entire series of events that occurs between fertilization and maturity is called **development.** During development, cells become progressively more specialized, or **differentiated.** Development in seed plants results first in the production of an embryo, which remains dormant within a seed until the seed germinates. The first stage in the development of a plant zygote is active cell division. The **zygote** formed by union of sperm and egg divides repeatedly to form an organized mass of cells, the **embryo.** In angiosperms the differentiation of cell types within the embryo begins almost immediately after fertilization. By day 5 the principal tissue systems can be detected within the embryo mass, and within another day, the root and shoot apical meristems can be detected. Cell movement does not occur during the process of embryonic development in plants, as it does in animals. Instead, cells differentiate where they are formed, their positions determining in large measure their future developmental fates. In animals, specific cell movements play an important role in development.

Early in the development of an angiosperm embryo, a profoundly significant event occurs. The embryo simply stops developing. In many plants the development of the embryo is arrested soon after apical meristems and the first leaves, or cotyledons, are differentiated. The **integuments,** the coats surrounding the embryo, develop into a relatively impermeable seed coat, which encloses the quiescent embryo within the seed, together with a source of stored food.

Once a seed coat forms around the embryo, most of the embryo's metabolic activities cease; a mature seed contains only about 10% water. Under these conditions, the seed and the young plant within it are very stable. Germination (Figure 30-6) cannot take place until water and oxygen reach the embryo, a process that sometimes involves cracking the seed. Seeds of some plants have been known to remain viable for hundreds of years (Figure 30-7).

The role of environmental factors helps to ensure that the plant will germinate only under appropriate conditions. Sometimes the seeds are held within tough fruits that will not crack until they are exposed to the heat of a fire, a strategy that clearly results in the germination of a plant in an open habitat. The seeds of

**Figure 30-6**

**How a seed germinates.** The germination of a seed, such as this garden pea *(Pisum sativum)*, involves the fracture of the seed coat, from which the young shoot and root emerge.

other plants will germinate only when inhibitory chemicals have been washed out of their seed coats, guaranteeing germination when sufficient water is available.

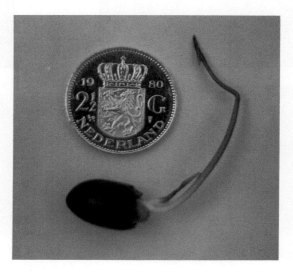

**Figure 30-7**

**Seeds can remain viable for long periods.** A seedling grown from seeds of lotus, *Nelumbo nucifera,* recovered from the mud of a dry lake bed in Manchuria, northern China. The radiocarbon age of this seed indicates that it was formed in about 1515; another seed that was germinated was estimated to be at least a century older. The coin is included to give some idea of the size.

Still other plants will germinate only after they pass through the intestines of birds or mammals or are regurgitated by them. This process both weakens the seed coats and ensures dispersal of the plants involved.

While a seed's embryo is embedded in its coat, water and oxygen are largely excluded and the metabolism in the seed is slowed down greatly. To reactivate its metabolism, the embryo has only to be provided with water, oxygen, and a source of metabolic energy. The final release of arrested development, germination, is then cued to specific signals from the environment.

Seeds are clearly important adaptively in at least three respects:

1. Seeds permit plants to postpone development when conditions are unfavorable and to remain dormant until more advantageous conditions arise. Under conditions in which young plants might or might not become established, a plant can "afford" to have some seeds germinate because others remain dormant.
2. By tying the reinitiation of development to environmental factors, seeds permit the course of embryo development to be synchronized with critical aspects of the plant's habitat.
3. Perhaps most importantly, the dispersal of seeds facilitates the migration and dispersal of genotypes into new habitats. The seed also offers maximum protection to the young plant at its most vulnerable stage of development.

## FRUITS

Paralleling the evolution of flowers of the angiosperms, and of equal importance for their success, has been the evolution of their fruits. Besides the many ways that fruits can be formed, they exhibit a wide array of modes of specialization in relation to their dispersal.

Fruits that have fleshy coverings, often black, or bright blue, or red, are normally dispersed by birds and other vertebrates. Like the red flowers that we discussed in relation to pollination by birds, the red fruits signal an abundant food supply (Figure 30-8, *B*). By feeding on

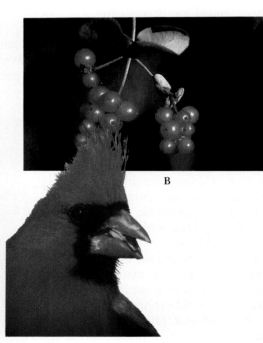

*Figure 30-8*

**Animal-dispersed fruits.**
A  The spiny fruits of this burgrass, *Cenchrus incertus,* adhere readily to any passing animal, as you will know if you have stepped on them.
B  The bright red berries of this honeysuckle, *Lonicera hispidula,* are highly attractive to birds, just as are red flowers. Birds may carry the seeds they contain for great distances stuck on their feet or, internally.
C  The fruits of figs, *Ficus carica,* are flower clusters turned inside out. They are pollinated by tiny, specialized wasps, which require figs to complete their life cycle. The wasps enter through the hole in the end of the fig. When mature, figs are consumed and the seeds of the individual tiny flowers are scattered about by birds and mammals.

A                              B                              C

**Figure 30-9**

**Wind-dispersed fruits.**
**A** False dandelion, *Pyropappus caroliniana*. The "parachutes" disperse the fruits of
dandelions widely in the wind, much to the gardener's despair.
**B** The double fruits of maples, *Acer*, when mature, are blown considerable distances from
their parent trees.
**C** Tumbleweed, *Salsola*, in which the whole dead plant becomes a light, windblown
structure that rolls about scattering seeds.

these fruits, the birds and other animals may carry seeds from place to place and thus transfer the plants from one suitable habitat to another. Other fruits—you can find many examples on your socks in summer—are dispersed by attaching themselves to the fur of mammals or the feathers of birds (Figure 30-8, *A*).

Other kinds of specialized fruit dispersal have evolved many different times in the flowering plants. For instance, mammals such as squirrels disperse and bury seeds and often do not find them again. Other fruits or seeds have wings and are blown about by the wind; the wings on the seeds of pines and those on the fruits of ashes or maples (Figure 30-9, *B*) play identical ecological roles. Dandelions and related plants (Figure 30-9, *A*) provide a familiar example of a kind of fruit

that is dispersed by the wind; the dispersal of seeds of such plants as milkweeds, willows, and cottonwoods is similar. In tumbleweeds, the whole plant is blown about by the wind, scattering seeds as it goes (Figure 30-9, *C*).

Still other fruits, such as those of mangroves, coconuts, and certain other plants that characteristically occur on or near beaches, swamps, or other bodies of water, are regularly spread from place to place in the water (Figure 30-10). Dispersal of this sort is especially important in the colonization of distant island groups, such as the Hawaiian Islands. It has been calculated that the seeds of about 280 original flowering plant species must have reached Hawaii to have evolved into the minimum of 956 native species found there today.

A                              B

**Figure 30-10**

**Water-dispersed fruits.**
**A** Mangrove, *Rhizophora mangle*. The embryos begin to grow on the parent plants and soon become established when they wash away to another silty shore.
**B** A fruit of the coconut, *Cocos nucifera*, sprouting on a sandy beach. One of the most useful plants for humans in the tropics, coconuts have become established even on the most distant islands by drifting in the waves.

# GERMINATION

What happens to a seed when, as a result of factors such as those we have been discussing, it encounters conditions suitable for its germination? First, it imbibes water. Because the tissues of the seed are so dry at the start of germination, the seed takes up water with great force. Once this has occurred, metabolism resumes. Initially, the metabolism may be anaerobic, but when the seed coat ruptures, aerobic metabolism takes over. At this point, it is important that oxygen be available to the developing embryo, because plants, which drown for the same reason people do, require oxygen for active growth. Few plants produce seeds that germinate successfully under water, although some, such as rice, have evolved a tolerance of anaerobic conditions.

Germination and early seedling growth require the mobilization of metabolic reserves stored in the starch grains of **amyloplasts** (plastids that are specialized to store starch) and protein bodies. Fats and oils are also important food reserves in some kinds of seeds. They can readily be digested during germination to produce glycerol and fatty acids, which yield energy through aerobic respiration and can also be converted to glucose. Any of these reserves may be stored in the embryo itself or in the endosperm, depending on the kind of plant.

How are the genes that transcribe the enzymes involved in the mobilization of food resources activated? Experimental studies have shown that, in the endosperm of the cereal grains at least, this occurs when hormones called **gibberellins** are synthesized by the embryo. These hormones initiate a burst of mRNA and protein synthesis. It is not known whether the gibberellins act directly on the DNA or through chemical intermediates in the cytoplasm. DNA synthesis apparently does not occur during the early stages of seed germination, but becomes important when the **radicle**, or embryonic root, has grown out of the seed coats (see Figure 29-3).

*During early germination and seed establishment, the vital mobilization of the food reserves stored in the embryo or the endosperm is mediated by hormones.*

# GROWTH AND DIFFERENTIATION

Once a seed has germinated, the plant's further development depends on the activities of the meristematic tissues, which interact with the environment. As described in Chapter 29, the shoot and root apical meristems give rise to all of the other cells of the adult plant.

Differentiation, the formation of specialized tissues, can be considered to occur in five stages in plants (Figure 30-11):

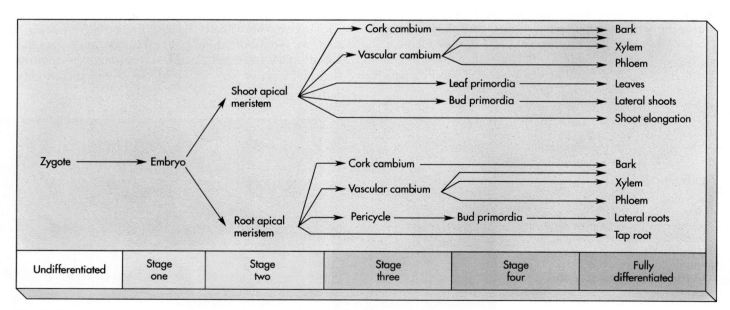

**Figure 30-11**
Stages in plant differentiation.

*Stage 1* is the formation of the embryo by cell division of the zygote formed by the sperm's fertilization of the egg within the ovule.

*Stage 2* is the differentiation within the embryo of the apical meristems, which begins almost immediately in angiosperms. The apical meristems can be detected when there are as few as 40 cells in the growing embryo. Apical meristems are largely responsible for primary growth.

*Stage 3* is the differentiation from apical meristems of the cambia, which are largely responsible for secondary growth.

*Stage 4* is the production of **primordia,** which are cells fully committed to becoming leaves, shoots, or roots. Leaf and shoot primordia develop directly from apical meristem cells, whereas root primordia develop from the root cambium, called the pericycle.

*Stage 5* is the production of fully differentiated tissues and structures, including xylem, phloem, leaves, shoots, and roots.

After a seed germinates, the pattern of growth and differentiation that was established in the embryo is repeated indefinitely until the plant dies. But differentia-tion in plants, unlike that in animals, is largely reversible. Botanists first demonstrated in the 1950s that individual differentiated cells isolated from mature individuals could give rise to entire individuals. F.C. Steward was able to induce isolated bits of phloem taken from carrots to form new plants, plants that were normal in appearance and fully fertile (Figure 30-12). Regeneration of entire plants from differentiated tissue has since been carried out in many plants, including cotton, tomatoes, and cherries. These experiments clearly demonstrate that the original differentiated phloem tissue still contains all of the genetic potential needed for the differentiation of entire plants. No information is lost during plant tissue differentiation, and no irreversible steps are taken.

In plants, regeneration is not confined to the laboratory. Nature does it too. Asexual reproduction is a regeneration process in which differentiated root or stem cells form a new shoot with its own set of roots. As a result, entire plants may form from horizontal roots, as well as above-ground and underground runners such as **stolons** and **rhizomes.** Colonies of aspen, *Populus tremuloides,* often consist of a single individual that has given rise to a colony of genetically identical trees by producing new stems from its horizontal roots (Figure 30-13).

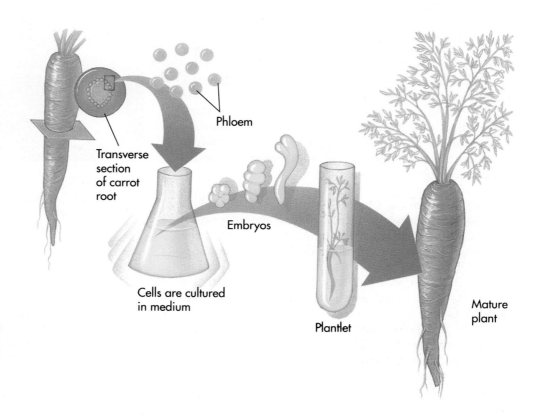

*Figure 30-12*

**Growing a new plant from a bit of adult tissue.** Phloem tissue was isolated from carrots *(Daucus carota),* in the laboratory of F. C. Steward at Cornell University. The discs of tissue were grown in a flask in which the medium was constantly agitated so as to bring a fresh supply of nutrients to the masses of **callus** (undifferentiated cells) that soon formed.

Phloem

Transverse section of carrot root

Embryos

Cells are cultured in medium

Plantlet

Mature plant

**Figure 30-13**

**Clones of aspen.** The contrast between the golden quaking aspens *(Populus tremuloides)* and the dark green Engelmann spruces *(Picea engelmannii)*, evident in this autumn scene near Durango, Colorado, makes it possible to see their clones, large colonies produced by single individuals spreading underground and sending up new shoots periodically.

## REPRODUCTIVE STRATEGIES OF PLANTS

Flowers, seeds, fruits, and asexual reproduction in plants can all be viewed as ways in which the plants increase or decrease the rate of recombination, a genetic process that has been one of the most powerful forces in the evolution of life on earth. By rapidly generating new combinations of alleles, recombination gives natural selection the raw material from which to select new and better-adapted phenotypes. But recombination is not always advantageous. Species that are well adapted to their environments have little to gain by shuffling their genes about because most new combinations are less well-attuned to the plants' present environments. In plants, reproductive strategies that favor **outcrossing,** hybridization with another individual of the same species that is not a close relative and thus promote recombination, occur in some species, whereas different strategies that favor inbreeding and thus minimize recombination occur in other species.

### Factors Promoting Outcrossing

Outcrossing is critically important for the adaptation and evolution of all eukaryotic organisms. One of the ways that outcrossing is enhanced in certain plant species is by the possession of separate male and female flowers. A flower that has only ovules, not pollen (see Figure 30-5), is called **pistillate**; functionally it is female. A flower that produces only pollen is functionally male; it is called **staminate.**

Staminate and pistillate flowers may occur on separate individuals in a given plant species, as in willows. Such plants, whose sporophytes produce either only ovules or only pollen, are called **dioecious** (from the Greek words for "two houses"). In other kinds of

plants—oaks, birches, and ragweed (see Figure 30-5, *A* and *B*)—the two types of flowers may be produced on the same plant; such plants are called *monoecious* ("one house"). The separation of ovule-producing and pollen-producing flowers that occurs in dioecious and monoecious plants makes outcrossing even more likely than it would be otherwise. But most angiosperms are neither dioecious nor monoecious; each of their flowers includes both pollen-producing and ovule-producing structures—that is, stamens and carpels.

An even simpler way that many flowers promote outcrossing is through the physical separation of the anthers and the stigmas. If the flower is constructed in such a way that these organs do not come into contact with one another, there will naturally be a tendency for the pollen to be transferred to the stigma of another flower, rather than to the stigma of its own flower (Figure 30-14, *A* and *B*).

**Figure 30-14**

**Sex in the genus *Epilobium*. A** *Epilobium ciliatum*. In this species, of a kind that is more common in the genus, the anthers shed their pollen directly on the large, creamy, undivided stigmas of the same flower, which are mature at the same time, thus bringing about self-pollination. **B** *Epilobium angustifolium* is strongly outcrossing and is one of the first plants in which the process of pollination was studied (in the 1790s). In it, the anthers first shed pollen, then the style swings up into a similar position in the flower and its four lobes open, **C**; these flowers are functionally staminate at first and then become pistillate about 2 days later. The flowers open progressively up the stem, so that the lowest ones are visited by bees first. Working up the stem, the bees encounter pollen-shedding, staminate-phase flowers, become covered with pollen, and carry it passively to the lower, functionally pistillate flowers of another plant. The stigmas are the white, cross-shaped structures.

Another device that occurs widely in flowering plants and increases outcrossing is **genetic self-incompatibility.** In self-incompatible plants, the pollen from a given individual does not function on the stigmas of that individual, or the embryos resulting from self-fertilization do not function. Self-incompatibility is a mechanism that increases outcrossing even though the flowers of such plants may produce both pollen and ovules, and their stamens and stigmas may be mature at the same time.

## The Flowering of Bamboos and the Starvation of Pandas

In many Asian bamboo species, all the individuals of the species flower and set seed simultaneously. These cycles tend to occur at very long intervals, ranging from 3 years upward. Most of the cycles are between 15 and 60 years long. The most extreme example is the Chinese species *Phyllostachys bambusoides,* which seeded massively and throughout its range in 919, 1114, between 1716 and 1735, in 1833 to 1847, and in the late 1960s. The last event involved cultivated plants in widely separated areas, including England, Russia, and Alabama. *P. bambusoides,* therefore, has a flowering cycle of about 120 years, set by an internal clock that runs more or less independently of environmental circumstances.

The bamboos with cycles of this kind spread mainly by the production of rhizomes, horizontal underground stems. When they do flower, nearly all individuals flower at once, set large quantities of seed, and die. Huge numbers of animals, including rats, pigs, and pheasants, often migrate into areas where bamboos are fruiting to feed on the seeds; humans also gather them for food. Apparently the plants put all of their energy into producing such vast amounts of seeds that large numbers of seedlings survive, even though most of the seeds are eaten.

A particular conservation problem that attracted widespread notice in the 1980s is the relationship between the flowering of bamboos (primarily of the genus *Fargesia*) and the survival of the giant panda, a spectacular animal that still survives in a few mountain ranges in southwestern China. Humans have occupied so much of the former range of the panda that the wild population now consists of only a few thousand individuals, which live only in several widely separated areas. In these places, only one or a few species of bamboo provide virtually all of the pandas' food. The mass flowering of some of these bamboo species in 1984, which led to their death over large areas, drove many of the pandas to the brink of starvation, and only a massive international effort made it possible to rescue many of the animals. The predictable episodes of mass flowering in bamboos will need to be considered carefully in planning for the pandas' survival.

*Figure 30-A*

**1,** A panda in a large enclosure in a stand of the bamboo *Fargesia spathacea*. Bamboos of the genus *Fargesia* are the most important panda food in the Min Mountains of northern Sichuan Province, China, where this photograph was taken. **2,** A Chinese scientist examining a stand of the bamboo *Fargesia nitida* that has flowered and dropped its leaves; the whole plant will soon be dead. This bamboo covers large areas of the Min Mountains and other mountain areas of southwestern China where pandas occur. It flowered extensively in the Min Mountains between 1974 and 1976, and again in 1982 and 1983; as a result, many pandas needed to be rescued and fed in captivity.

## Self-Pollination

Self-pollination is also very frequent among angiosperms. In fact, probably more than half of the angiosperms that occur in temperate regions self-pollinate regularly. Most of these have small, relatively inconspicuous flowers in which the pollen is shed directly onto the stigma, sometimes even before the bud opens (Figure 30-14, C). You might logically ask why there are so many self-pollinated plant species, if outcrossing is just as important for plants, in genetic terms, as it is for animals. There are two basic reasons for the frequent occurrence of self-pollinated angiosperms:

1. Self-pollination is advantageous in harsh climates, and as a result the plants in which it occurs do not need to be visited by animals to produce seed. As a consequence, self-pollinated plants can grow in areas where the kinds of insects or other animals that might visit them are absent or very scarce, as in the Arctic or at high elevations.

2. In genetic terms, self-pollination produces progenies that are more uniform than those which result from outcrossing. Such progenies may contain high proportions of individuals well adapted to particular habitats. Where the habitat occurs regularly, inbreeding is advantageous because it produces a greater proportion of well-adapted progeny than does outcrossing. Many successful weeds are self-pollinating because the habitat of weeds has been made uniform and spread all over the world by human beings.

## ■ SUMMARY

1. The flowers of angiosperms make possible the precise transfer of pollen and thus enable even widely separated plants to outcross effectively. Pollen is transferred by insects, particularly bees, by birds and bats, and by the wind.

2. The change of a zygote into a mature individual, initiated immediately after fertilization, is called development. Development is a process of progressive specialization that results in differentiation, the production of highly individual tissues and structures. In plants, differentiation is fully reversible. Whole plants can be regenerated from cultures of single cells, as long as they retain a living protoplast and a nucleus.

3. Differentiation of angiosperm embryos ceases after the major organs have been laid down, and each now-dormant embryo is encased within a dry casing, becoming a seed that maintains a state of suspended development until it is broken or moistened.

4. The seeds of angiosperms remain within the carpel, which develops into a fruit. Many fruits are fleshy and often sweet; animals that consume fruit may carry the seeds for long distances before excreting them as solid waste. The seeds, not harmed by the animal digestive system, can then germinate in the new location.

5. Seeds, because of their rigid, relatively impermeable seed coat, germinate only when they receive water and appropriate environmental cues. In the germination of seeds, mobilization of the food reserves stored in the cotyledons and in the endosperm is critical. In cereal grains, this process is mediated by hormones known as gibberellins.

6. Outcrossing in different angiosperms is promoted by the separation of the pollen-producing and ovule-producing structures into different flowers, or even onto different individuals; by genetic self-incompatibility; and by the separation of the pollen and the stigmas within a given flower with respect to position or time of maturation.

7. Self-pollination occurs in more than half of the angiosperm species of temperate regions. It is most common among plants in harsh climates or those which occur in widespread, uniform habitats, such as weeds.

# REVIEW

1. Which type of cell division, mitosis or meiosis, produces gametes in plants?

2. _____ are the most numerous and constant pollinators of insect-pollinated flowers.

3. The first step in seed germination occurs when the seed _____.

4. During the early germination of cereal grains, hormones called _____ are synthesized by the embryo; they mediate the mobilization of food reserves.

5. _____ is the hybridization of a plant with another individual of the same species that is not a close relative.

# SELF-QUIZ

1. The flowers of angiosperms
   (a) are often colorful so that insects are attracted to them.
   (b) usually offer nectar or other rewards to insects that visit them.
   (c) are usually attractive only to a specific set of pollinators.
   (d) are often green or dull-colored if self-pollination or wind-pollination predominates.
   (e) all of the above.

2. Why are seeds an important evolutionary improvement over spores?
   (a) Because they contain little water
   (b) Because they can remain dormant until conditions are right for germination
   (c) Because they enhance dispersal and thus migration of genotypes
   (d) Because seeds such as beans, corn, and rice are important food sources
   (e) Both b and c

3. Fleshy fruits that are brightly colored are often dispersed by
   (a) insects.
   (b) the wind.
   (c) water.
   (d) birds.
   (e) attaching to the fur of mammals.

4. Leaf and shoot primordia develop from
   (a) the xylem.
   (b) the epidermis.
   (c) the apical meristem.
   (d) the vascular cambium.
   (e) none of the above.

5. Which three are factors that promote outcrossing in flowering plants?
   (a) Having dioecious plants
   (b) Physical separation of stamens and carpels
   (c) Inbreeding
   (d) Genetic self-incompatability
   (e) Asexual reproduction

6. The most numerous insect-pollinated angiosperms are those pollinated by
   (a) hummingbirds.
   (b) beetles.
   (c) bees.
   (d) butterflies.
   (e) the wind.

7. The wind-pollinated flowers are characteristically *not*
   (a) unisexual.
   (b) dull green or colorless.
   (c) visited by insects of several different kinds.
   (d) odorless.
   (e) clumped; that is, growing in patches or colonies.

8. The regeneration of entire individuals from tissue that has already become differentiated is a property of
   (a) animals.
   (b) plants.
   (c) all organisms.
   (d) mammals.
   (e) carrots only.

## THOUGHT QUESTIONS

1. Angiosperms usually produce both pollen and ovules within a single flower, or at least on a single plant. Why don't all angiosperms simply self-pollinate?

2. Self-pollination does occur within at least some species of a wide variety of different kinds of plants, usually those living in harsh environments or among "weedy" species. Why do you suppose that animals coping with similar environments never developed similar reproductive strategies? Can you think of any animals that do self-fertilize?

3. Why is plant development so much more closely linked with environmental cues than is animal development?

4. When you eat a tossed salad, what plant parts are you consuming?

# FOR FURTHER READING

BATRA, S.W.T.: "Solitary Bees," *Scientific American,* February 1984, pages 120-127. Excellent article on these diverse and fascinating pollinators, some of which are of great commercial importance.

COOK, R.E.: "Clonal Plant Populations," *American Scientist,* vol. 71, 1983, pages 244-253. Excellent discussion on the role of asexual reproduction among plants in natural populations.

JANZEN, D.H.: "Why Bamboos Wait So Long to Flower," *Annual Reviews of Ecology and Systematics,* vol. 7, 1976, pages 347-391. A fascinating essay about the natural history of bamboos.

PETTITT, J., S. DUCKER, and B. KNOX: "Submarine Pollination," *Scientific American,* March 1981, pages 134-145. Sea grasses (marine angiosperms) flower underwater, shedding pollen that is carried from plant to plant by waves.

SCHALLER, G.B., H. JINCHU, P. WENSHI, and Z. JING: *The Giant Pandas of Wolong,* University of Chicago Press, Chicago, 1985. A fascinating account of the mutual adaptations of pandas and bamboo; this book offers the first glimpse of the life of pandas in their remote mountain home in western China.

# How Plants Function

Many plants lose their leaves every year. These Michigan sugar maples signal the approach of winter with a brilliant display of color. Within a month all of these leaves will have fallen to the ground.

# HOW PLANTS FUNCTION

## Overview

The body of a plant is basically a tube embedded in the ground and extending up into the light, where expanded surfaces, the leaves, capture the sun's energy and participate in gas exchange. The warming of the leaves by the sunlight increases evaporation from them, creating a suction that draws water into the plant through the roots and up the plant through the xylem to the leaves. Transport from the leaves and other photosynthetically active structures to the rest of the plant occurs through the phloem. Most of the nutrients critical to plant metabolism are accumulated by the roots and are then carried in the water stream throughout the plant. The growth of the plant itself is regulated by plant hormones. Partly through the mediation of such hormones, plants respond in complex ways to external stimuli such as touch, light, gravity, and day length.

## For Review
*Here are some important terms and concepts that you will encounter in this chapter. If you are not familiar with them, you should review them before proceeding.*

**Osmotic pressure** (Chapter 5)

**Transport of ions across membranes** (Chapter 5)

**Photosynthesis** (Chapter 8)

**Conducting tissues** (Chapter 29)

**Seed germination** (Chapter 30)

### Figure 31-1

**A sunflower.** This sunflower has within it a complex system that transports fluids, as well as specialized organs for reproduction and capturing energy. Although plants do not talk or move from place to place, they are as much alive as you are.

When you first look at a plant, it may not appear vibrantly alive (Figure 31-1). The plant does not bound from one place to another like a gazelle, or growl, or respond to a caress. But the tree's stolid appearance is deceiving; its internal structure is more complex than you might suspect. It has a plumbing system, as you do, that sends fluids from one part to another and special organs for reproduction and the gathering of energy. Like you, it regulates its growth and the functioning of its organs with hormones, chemicals that act as messengers to coordinate the many activities of the body. In this chapter we focus on these activities, first discussing the movement of fluids, and the many substances that they transport, within plants. We then examine one class of these substances in more detail, the hormones that control how the plant grows.

# WATER MOVEMENT

Functionally, a plant is essentially a tube with its base embedded in the ground (Figure 31-2). At the base of the tube are roots, and at its top are leaves. For a plant to function, two kinds of transport processes must occur. First, the carbohydrate molecules that are produced in the leaves by photosynthesis must be carried to all of the other living cells of the plant. To accomplish this, liquid with these carbohydrate molecules dissolved in it must move both up and down the tube. Second, nutrients and water in the ground must be taken up by the roots and ferried to the leaves and other cells of the plant. In this process, liquid moves up the tube.

It is not unusual for many of the leaves of a large tree to be more than 10 stories off the ground. Did you ever wonder how a tree manages to raise water so high? To understand how this happens, imagine filling with water a long, hollow tube closed at one end and placing the tube, open end down, in a full bucket of water. Gravity acts (pushes) on the column of air over the bucket: the weight of the air (at sea level) exerts an amount of pressure that is defined as 1 atmosphere downward on the water in the bucket and thus presses the water up into the tube. But gravity also acts to pull the water down within the tube. The interaction of these two forces, both dependent on gravity, determines the level of water in the tube. At sea level the water rises to about 10.4 meters. If the tube is any higher, a vacuum will form in the upper, closed end of the tube and will fill with water vapor.

To illustrate how water rises higher than 10.4 meters in a plant, open the tube and blow across the upper end. The stream of relatively dry air will cause water molecules to evaporate from the surface of the water in the tube. Does the level of water in the tube fall? No. As water molecules are drawn from the top, they are replenished by new ones that are forced up from the bottom. This, in essence, is what happens in plants. The passage of air across leaf surfaces results in the loss of water by evaporation, creating a suction force at the open upper end of the "tube" while water is pushed up from below by atmospheric pressure. In addition to this principal factor, the **adhesion** of water molecules to the walls of the very narrow tubes that occur in plants also helps to maintain the flow of water to the tops of plants.

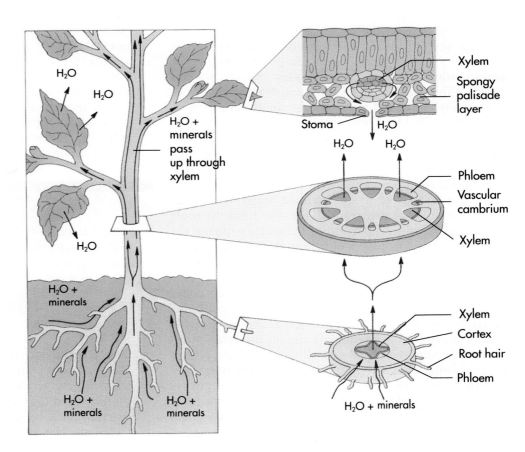

**Figure 31-2**

**The flow of materials into, out of, and within a plant.** Water with minerals passes in through the root hairs and upward in the plant by way of the vascular system; most is eventually lost through the leaves. The minerals are used where they are required for metabolic purposes, and the manufactured carbohydrates pass throughout the plant in the phloem from their places of manufacture.

Why does a column of water in a tall tree not collapse simply because of its weight? The answer is that water has an inherent strength that arises from the cohesion of its molecules; that is, from their tendency to form hydrogen bonds with one another. Because of these hydrogen bonds, a column of water resists separation. This resistance, called tensile strength, varies inversely with the diameter of the column; that is, the smaller the diameter of the column, the greater the tensile strength. Therefore plants must have very narrow transporting vessels to take advantage of the tensile strength.

### Transpiration

More than 90% of the water that is taken in by the roots of a plant is ultimately lost to the atmosphere, almost all of it from the leaves. It passes out primarily through the stomata in the form of water vapor. The process by which water leaves the plant is known as **transpiration.** First, the water passes into the pockets of air within the leaf by evaporating from the walls of the spongy mesophyll that lines the intercellular spaces (see Figure 29-14). These intercellular spaces open to the outside of the leaf by way of the stomata. The water that evaporates from these surfaces of the spongy mesophyll cells is continuously replenished from the tips of the veinlets in the leaves (see Figure 31-2). Because the strands of xylem conduct water within the plant in an unbroken stream all the way from the roots to the leaves, when a portion of the water vapor in the intercellular spaces passes out through the stomata, the supply of water vapor in these spaces is continually renewed.

Because they are constantly losing so much water to the atmosphere, and because the presence of this water is essential to their metabolic activities, growing plants depend on a continuous stream of water entering and leaving their bodies at all times. Photosynthesis and the other vital activities of plants cannot be carried out in its absence. For this reason, water must always be available to plant roots. Such structural features as the stomata, the cuticle, and the substomatal spaces have evolved in response to two contradictory requirements: to maintain this stream of water and to admit

carbon dioxide, essential for photosynthesis, into the plant. But before considering how plants resolve this problem, we must consider the absorption of water by the roots.

### The Absorption of Water by Roots

Most of the water absorbed by the plant comes in through the root hairs, which collectively have an enormous surface area (see Figure 29-9). The root hairs are always **turgid**—plump and swollen with water—because they have a higher concentration of dissolved minerals than does the water in the soil solution; water, therefore, tends to move into them steadily. Once it is inside the roots, the water passes inward to the conducting elements of the xylem.

Water is not the only substance that enters the roots by passing into the cells of the root hairs. The membranes of these root hair cells contain a variety of ion transport channels that actively pump specific ions into them even against large concentration gradients. These ions, many of which are plant nutrients, are then transported throughout the plant as a component of the water flowing through the xylem.

At night, when the relative humidity may approach 100% at the leaf surface, there may be no transpiration from the leaves. Under these circumstances, the negative pressure component of the water potential (suction caused by evaporation) becomes very small or nonexistent. At such times water does not travel upward in the xylem. But active transport of ions into the roots continues to take place under these circumstances. It results in an increasingly high ion concentration within the cells, a concentration that causes water to be drawn into the root hair cells by osmosis. The result is the movement of water into the plant and up the xylem columns, which we call **root pressure.**

### The Regulation of Transpiration Rate

The only way plants can control water loss on a short-term basis is to close their stomata. Many plants can do this when they are subjected to water stress. But the stomata must be open at least part of the time so that $CO_2$, which is necessary for photosynthesis, can enter the plant. In its pattern of opening or closing its stomata, a plant must respond to both the need to conserve water and the need to admit $CO_2$.

The stomata open and close because of changes in the water pressure of their guard cells (Figure 31-3). The guard cells of stomata are the only cells of the epidermis to possess chloroplasts. They stand out for this reason and because of their distinctive shape—they are thicker on the side next to the stomatal opening and thinner on their other sides and ends. When the guard cells are turgid (plump and swollen with water), they become bowed in shape, as do their thick inner walls, thus opening the stomata as wide as possible.

The guard cells use ATP-powered ion transport channels through their membranes to concentrate ions actively. This concentration causes water to enter the guard cells osmotically. As a result, these cells accumulate water and become turgid, opening the stomata. The guard cells remain turgid only so long as the active transport channels pump ions, chiefly potassium ($K^+$), into the cells and so maintain the higher solute concentration there. The surrounding cells maintain the reservoir of potassium ions. Keeping the stomata open requires a constant expenditure of ATP. When the active transport of ions into the guard cells ceases, the higher concentration of ions within the guard cells causes the ions to move out into the surrounding cells by diffusion. Ultimately, water leaves the guard cells also, which then become somewhat "limp" or deflated, and the stomata between them close.

*Stomata open when their guard cells become turgid. Their inner surfaces are thickest, and they bow inwardly when the pressure within the cells is high. Keeping the guard cells turgid requires a constant expenditure of ATP.*

A number of different environmental factors affect the opening and closing of stomata. The most important is water loss; in plants that are wilted because of a lack of water, the stomata tend to close. Also, an increase in $CO_2$ concentration causes the stomata of most species to close. In most plant species, the stomata open in the light and close in the dark; also, very high temperatures (above 30° to 35° C) tend to cause stomata to close. Finally, stomata also exhibit daily rhythms of opening and closing that appear to be controlled within the plant by factors that are poorly understood.

## CARBOHYDRATE TRANSPORT

Most of the carbohydrates manufactured in the leaves and other green parts of the plant are moved through the phloem to other parts of the plant. This process, known as **translocation,** is responsible for the availability of suitable carbohydrate building blocks at the actively growing regions of the plant. The carbohydrates that are concentrated in storage organs such as tubers, often in the form of starch, are also converted into transportable molecules such as sucrose and are moved through the phloem. The liquid in the phloem contains 10% to 25% dissolved solid matter, almost all of which is sucrose.

The movement of substances in the phloem can be remarkably fast; rates of 50 to 100 centimeters per hour have been measured. The movement of water and dissolved nutrients within the sieve tubes is a passive process that does not require the expenditure of energy. The movement of materials transported in the phloem occurs because of hydrostatic pressure, which develops as a result of osmosis (Figure 31-4). The overall process by which this occurs is called **mass flow.** First, sucrose produced as a result of photosynthesis is actively loaded into the phloem tubes of the veinlets. This increases the solute concentration of the sieve

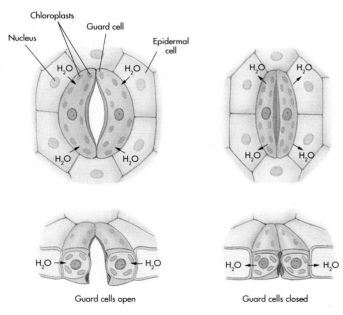

**Figure 31-3**

**Structure of a stomate.**
**A** Open. The thick inner side of each of the two guard cells that make up a stoma bow outwardly, opening the stoma, when solute pressure is high (water pressure is high) within the guard cells.
**B** Closed. When the solute pressure is low (water pressure is low) within the guard cells, they become flaccid and close the stoma.

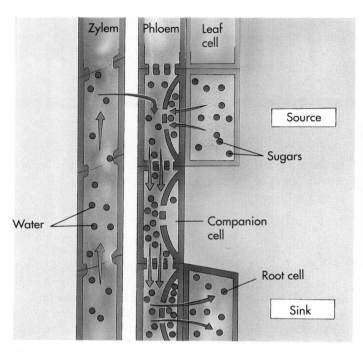

**Figure 31-4**

**Mass flow.** In this diagram, sucrose molecules are indicated by *red dots*, water molecules by *black dots*. Moving from the parenchyma cells of a leaf or another part of the plant into the conducting cells of the phloem, the sucrose molecules are then transported to other parts of the plant by mass flow and unloaded where they are required.

tubes, so water passes into them by osmosis. An area where the sucrose is made is called a **source;** an area where it is being taken from the sieve tubes is called a **sink.** Sinks include the roots and other regions where the sucrose is being unloaded. There the solute concentration of the sieve tubes is decreased as the sucrose is removed. As a result of these processes, water moves in the sieve tubes from the areas where sucrose is being taken in to those areas where it is being withdrawn, and the sucrose moves passively with it.

## PLANT NUTRIENTS

### Essential Nutrients
Plants require a number of inorganic nutrients. Some of these are **macronutrients,** which the plants need in relatively large amounts, and others are **micronutrients,** those required in trace amounts. There are nine macronutrients: carbon, hydrogen, and oxygen (the three elements found in all organic compounds); and nitrogen, potassium, calcium, phosphorus, magnesium, and sul-

fur. Each of these nutrients approaches or exceeds 1% of the dry weight of a healthy plant. The seven micronutrient elements, which in most plants constitute from less than one to several hundred parts per million by dry weight, are iron, chlorine, copper, manganese, zinc, molybdenum, and boron.

*The macronutrients are substances required by plants in relatively large amounts. Each of them approaches or exceeds 1% of a plant's dry weight. Micronutrients are substances that plants require in only trace amounts of one to several hundred parts per million.*

The six macronutrients in addition to carbon, hydrogen, and oxygen are involved in plant metabolism in many ways. Potassium ions regulate the turgor pressure of guard cells and therefore the rate at which the plant loses water and takes in carbon dioxide. Calcium is an essential component of the **middle lamellae,** structural elements laid down between plant cell walls, and it also helps to maintain the physical integrity of membranes. Magnesium is a part of the chlorophyll molecule. The presence of phosphorous in many key biological molecules such as nucleic acids and ATP has been explored in detail in the earlier chapters of this book. Nitrogen is an essential part of amino acids and proteins, chlorophyll, and the nucleotides that make up nucleic acids.

A number of ions are components of enzyme systems and serve as cofactors in essential biochemical reactions. Potassium, which reaches higher concentrations in plants than any other elements except carbon and oxygen, affects the conformation of many proteins and probably affects at least 60 different enzymes while they are functioning. Zinc appears to play a similar role in the synthesis of the important plant growth hormone, auxin. Plants with an inadequate supply of zinc display symptoms that derive mainly from a lack of cell elongation, apparently reflecting a shortage of auxin. When an essential nutrient is present in short supply, a plant will display deficiency symptoms characteristic of a shortage of that nutrient. For this reason, a trained observer can often tell what chemicals should be supplied simply by observing the appearance of a plant (Figure 31-5).

Some kinds of plants have specific nutritional requirements that are not shared by others. For example, silica is essential for the growth of many grasses be-

**Figure 31-5**

**Mineral deficiencies in plants.** Here we see individuals of Marglobe tomatoes *(Lycopersicon esulentum)* grown in hydroponic solution—water to which nutrients have been added.

**A** The (healthy) plant at left was grown in a complete nutrient solution (one that contains all of the essential minerals). Calcium was not added to the nutrient solution of the plant at right, resulting in its unhealthy, dwarfed appearance. Without calcium, the young shoots and roots die.

**B** Leaf of a healthy control plant.

**C** Chlorine-deficient plant, with curled leaves that have patches of dead tissue.

**D** Copper-deficient plant, with blue-green, curled leaves. Copper is required only in very small amounts for normal plant growth.

**E** Zinc-deficient plant, with small, necrotic leaves.

**F** Manganese-deficient plant, with chlorosis (yellowing) between the veins. This chlorosis is caused by a drastic reduction in chlorophyll production.

cause it helps to retard their complete destruction by herbivores (see Figure 20-5), but it is not required by plants in general. Cobalt is necessary for the normal growth of the nitrogen-fixing bacteria associated with the nodules of legumes and is therefore an essential element for the normal growth of these plants. Nickel seems to be essential for soybeans, and its role in the nutrition of other plants needs further investigation.

In general, the elements that animals require reach them through plants, which therefore form an indispensable link between animals and the reservoirs of chemicals in nature. Some elements that animals require, such as iodine, come by way of plants but are not required by the plants. Iodine is very rare in soils; a shortage of it in the human diet can lead to the condition known as goiter.

## REGULATING PLANT GROWTH: PLANT HORMONES

**Hormones** are chemical substances produced in small, often minute, quantities in one part of an organism and then transported to another part of the organism, where they bring about physiological responses. The activity of hormones results from their ability to stimulate certain physiological processes and inhibit others. How they act in a particular instance is influenced both

# Carnivorous Plants

Some plants are able to use other organisms directly as sources of nitrogen, and some require minerals, just as animals normally do. These are the carnivorous plants. Carnivorous plants often grow in acidic soils such as bogs—habitats that are not favorable for the growth of most legumes or of nitrifying bacteria. By capturing and digesting small animals directly, such plants obtain adequate supplies of nitrogen and thus are able to grow in these seemingly unfavorable environments.

The carnivorous plants have adaptations that are used to lure and trap insects and other small animals. The plants digest their prey with enzymes secreted from various kinds of glands.

The Venus flytrap *(Dionea muscipula)* (Figure 31-A, *1* and *2*), which grows in the bogs of coastal North and South Carolina, has three sensitive hairs on each side of each leaf, which, when touched, trigger the two halves of the leaf to snap together. Once enfolded by a leaf, the prey of a Venus flytrap is digested by enzymes secreted from the leaf surfaces.

Pitcher plants (Figure 31-A, *3* and *4*) attract insects by the bright, flowerlike colors within their pitcher-shaped leaves and perhaps also by sugar-rich secretions. Once inside the pitchers, the insects may slide down into the cavity of the leaf, which is filled with water, digestive enzymes, and half-digested prey.

Bladderworts, *Utricularia,* are aquatic; they sweep small animals into their bladderlike leaves by the rapid action of a springlike trapdoor and then digest these animals. In the sundews (Figure 31-A, *5*), the glandular trichomes secrete both sticky mucilage, which traps small animals, and digestive enzymes.

3

4

1

**Figure 31-A**

**Carnivorous plants. 1,** Venus flytrap, *Dionea muscipula,* which inhabits low, boggy ground in North and South Carolina. **2,** A Venus flytrap leaf has snapped together, imprisoning a fly. **3,** A tropical Asian pitcher plant, *Nepenthes.* Insects enter the pitchers, which are modified leaves, seeking nectar, and are trapped and digested. **4,** Yellow pitcher plant, *Sarracenia flava,* which grows in bogs in the southeastern United States. Its pitchers, with their yellow borders, resemble flowers and secrete a sweet-smelling nectar that aids the plants in trapping insects. **5,** Sundew, *Drosera.* A small fly has been trapped by the glandular hairs.

5

by what the hormones themselves are and by how they affect the particular tissue that receives their message.

In animals, hormones are usually produced at definite sites, normally in organs that are solely concerned with their production. In plants, on the other hand, hormones are produced in tissues that are not specialized for that purpose but that carry out other, usually more obvious, functions. There are at least five major kinds of hormones in plants: auxin, cytokinins, gibberellins, ethylene, and abscisic acid. Other kinds of plant hormones certainly exist, but they are less well understood. The study of plant hormones, especially our attempts to understand how they produce their effects, is an active and important field of current research.

*Five major kinds of plant hormones are reasonably well understood: auxins, cytokinins, gibberellins, ethylene, and abscisic acid.*

## The Discovery of the First Plant Hormone

In his later years, the great evolutionist Charles Darwin became increasingly devoted to the study of plants. In 1881 he and his son Francis published a book called *The Power of Movement in Plants*. In this book, the Darwins reported their systematic experiments concerning the way in which growing plants bend toward light, a phenomenon known as **phototropism.** They made many observations in this field and also conducted experiments using young grass seedlings.

Charles and Francis Darwin found that the seedlings they were studying normally bent strongly toward a source of light if the light came primarily from one side. However, if they covered the upper part of a seedling with a cylinder of metal foil so that no light reached its tip, the shoot would not bend (Figure 31-6). The Darwins obtained this result even though the region where the bending normally occurred was still exposed. Light reached this part of the seedling directly, but bending did not occur. But if they covered the end of the shoot with a gelatin cap, which transmitted light, the shoot would bend as if it were not covered at all.

To explain this unexpected finding, the Darwins hypothesized that when the shoots were illuminated from one side, an "influence" arose in the uppermost part of the shoot, was then transmitted downward, and caused the shoot to bend. For some 30 years, the Darwins' perceptive experiments remained the sole source of information about this interesting phenomenon. Then a series of experiments was performed by several

botanists, who demonstrated that the substance that caused the shoots to bend was a chemical. They cut off the tip of a grass seedling and then replaced it, but separated it from the rest of the seedling by a block of agar, a gelatinous medium often used in biological ex-

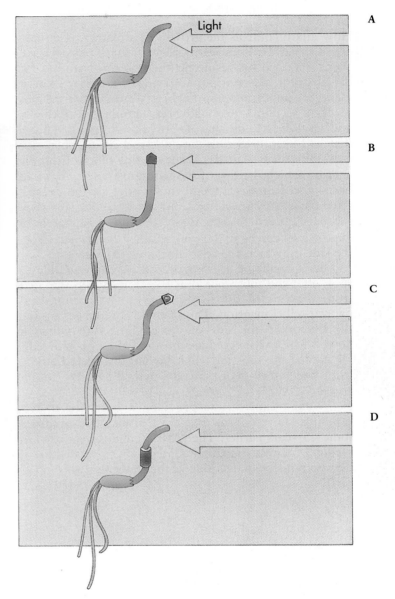

*Figure 31-6*

**The Darwins' experiment.** Young grass seedlings normally bend toward the light (**A**). When the tip of a seedling was covered by a lightproof collar (**B**) (but not when it was covered by a transparent one [**C**]), this bending did not occur. When the collar was placed below the tip (**D**), the characteristic light response still took place. From these experiments, the Darwins concluded that, in response to light, an "influence" that causes bending was transmitted from the tip of the seedling to the area below the tip, where bending normally occurs.

periments. The seedling reacted as if there had been no change. Something was evidently passing from the tip of the seedling through the agar into the region where the bending occurred. The "something" was a plant hormone called **auxin**. We now know that auxin acts to regulate cell growth in plants.

## How the Shoot Apex Uses Auxin to Control Growth

How auxin controls the growth of a plant was discovered in 1926 by Frits Went, a Dutch plant physiologist. Went obtained his results in the course of studies for his doctoral dissertation. Carrying the earlier experiments an important step farther, Went cut off the tips of grass seedlings that had been illuminated normally and set these tips on agar. He then took grass seedlings that had been grown in the dark and cut off their tips in a similar way. Finally, Went cut tiny blocks from the agar on which the tips of the light-grown seedlings had been placed and put them on the tops of the decapitated dark-grown seedlings, but set off to one side (Figure 31-7). Even though these seedlings had not been exposed to the light themselves, they bent *away* from the side on which the agar blocks were placed.

As a result of his experiments, Went was able to show that the substance that had flowed into the agar from the tips of the light-grown grass seedlings enhanced cell elongation. This chemical messenger caused the tissues on the side of the seedling into which it flowed to grow more than those on the opposite side. He named the substance that he had discovered auxin, from the Greek word *auxein,* which means "to increase."

Went's experiments provided a basis for understanding the responses that the Darwins had obtained some 45 years earlier. The grass seedlings bent toward the light because the auxin contents on the two sides of the shoot differed. The side of the shoot that was in the shade had more auxin; therefore its cells elongated more than those on the lighted side, bending the plant toward the light. Later experiments by other investigators showed that auxin in normal plants migrates from the illuminated side to the dark side in response to light and thus causes the plant to bend toward the light.

## THE MAJOR PLANT HORMONES

### Auxin

Only one form of auxin occurs in nature (Figure 31-8), indoleacetic acid (IAA). IAA is produced at the shoot apex in the region of the apical meristem and diffuses continuously downward. The term *auxin* is now used to refer both to the naturally occurring substance and to those related synthetic molecules which produce similar effects.

Auxin acts by increasing the plasticity of the plant cell wall. A more plastic wall will stretch more during active cell growth, which is while its protoplast is swelling. Because very low concentrations of auxin promote cell wall plasticity, it must be broken down rapidly to prevent its accumulation. Plants break auxin down by means of the enzyme **indoleacetic acid oxidase**. By controlling the levels of both IAA and IAA oxidase, plants can regulate their growth very precisely.

Auxin controls various plant responses in addition to those involved in phototropism. One of these is the suppression of lateral bud growth. How can auxin, a growth promoter, also inhibit growth? Apparently the cells around lateral buds produce the chemical ethylene under the influence of auxin. The ethylene, in turn, inhibits the growth of the lateral buds. When the terminal bud is removed, removing the source of auxin, the lateral buds grow, producing bushy plants. The number of flowers on an individual plant is also increased in this situation.

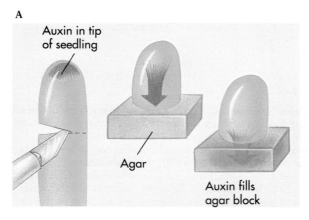
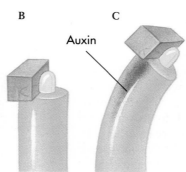

**Figure 31-7**

**Frits Went's experiments.**
**A** Went removed the tips of the grass seedlings and put them on agar.
**B** Blocks of the agar were then put on one side of the ends of other grass seedlings grown in the dark from which the tips had been removed.
**C** The seedlings bent away from the side on which the agar block was placed. Went concluded that the substance that he named auxin promoted the elongation of the cells, and that it accumulated on the side of a grass seedling away from the light.

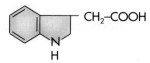

**Figure 31-8**

**Auxin.** Indoleacetic acid (IAA), the only known naturally occurring auxin.

*The only known naturally occurring auxin, indoleacetic acid (IAA), is produced in the apical meristems of shoots and diffuses downward, suppressing the growth of lateral buds. In young grass seedlings and other herbs, it plays a major role in stem elongation, migrating from the illuminated portions of the stem to the dark portions and thus causing the stems to grow toward the light.*

## Synthetic Auxins

Synthetic auxins are routinely used to control weeds. When they are used as herbicides, they are applied in higher concentrations than those at which IAA normally occurs in plants. One of the most important of the synthetic auxins used in this way is 2,4-dichlorophenoxyacetic acid, usually known as 2,4-D. It kills weeds in lawns without harming the grass because 2,4-D affects only broad-leaved dicots. When treated, the weeds literally "grow to death," rapidly depleting all metabolic reserves so that there is no source of energy to carry out transport or other essential functions.

Closely related to 2,4-D is the herbicide 2,4,5-trichlorophenoxyacetic acid (2,4,5-T), which is widely used to kill woody seedlings and weeds. Notorious as the "Agent Orange" of the Vietnam War, 2,4,5-T is easily contaminated with a by-product of its manufacture, **dioxin.** Dioxin, which is believed to be harmful to people, is the subject of great environmental concern.

## Cytokinins

A **cytokinin** is a plant hormone that, in combination with auxin, stimulates cell division in plants and determines the course of differentiation. Substances that have these properties are widespread both in bacteria and in eukaryotes. In vascular plants, most cytokinins seem to be produced in the roots and transported from there throughout the rest of the plant. Cytokinins seem to be necessary for mitosis and cell division to take place. They apparently work by influencing the synthesis or activation of proteins that are specifically required for mitosis.

*Cytokinins are plant hormones that, in combination with auxin, stimulate cell division and determine the course of differentiation. Most are produced in the roots and transported to the rest of the plant.*

The naturally occurring cytokinins all appear to be derivatives of the purine base, adenine. In contrast to auxins, cytokinins *promote* the growth of lateral branches. Similarly, auxins promote the formation of lateral roots, whereas cytokinins inhibit it. As a consequence of these relationships, the balance between cytokinins and auxin determines, along with other factors, the appearance of a mature plant.

## Gibberellins

Gibberellins are named for the fungus genus *Gibberella*, which causes a disease of rice in which the plants grow to be abnormally tall. This "foolish seedling disease" of rice was investigated in the 1920s by Japanese scientists, who found that if they grew the fungus in culture, they could obtain a chemical completely free of the fungus itself that would affect the rice plants in a way similar to the fungus. This substance, isolated in 1939 and chemically characterized in 1954, was the first of what proved to be a large class of naturally occurring plant hormones called the **gibberellins** (Figure 31-9).

Synthesized in the apical portions of both stems and roots, gibberellins have important effects on stem elongation in plants and play the leading role in controlling this process in mature trees and shrubs. In these plants the application of gibberellins characteristically promotes internode elongation, and this effect is enhanced if auxins are also present. Gibberellins are also involved with many other aspects of plant growth, such as inducing flowering and hastening seed germination.

*Gibberellins are a very common and important class of plant hormones. They are produced in the apical regions of shoots and roots and play the major role in controlling stem elongation for most plants, acting in concert with auxin and other plant hormones.*

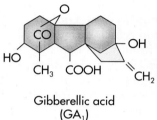

Gibberellic acid
(GA₁)

**Figure 31-9**

**Gibberellin.** Although more than 60 gibberellins have been isolated from natural sources, apparently only GA₁ is active in shoot elongation. All of the other gibberellins have a similar structure. Cabbage *(Brassica oleracea)*, a biennial that is native to the seacoasts of Europe, will "bolt"—flower—when the heads are treated with gibberellin. Most biennials exhibit similar behavior.

## Ethylene

Long before its role as a plant hormone was appreciated, the simple hydrocarbon ethylene ($H_2C$—$CH_2$) was known to defoliate plants when it leaked from gas lights in street lamps. But **ethylene** is a natural product of plant metabolism and appears to be the main factor in the formation of specialized layers of cells that precedes the dropping off of leaves from plants. We have already mentioned the way in which auxin, diffusing down from the apical meristem of the stem, may stimulate the production of ethylene in the tissues around the lateral buds and thus retard their growth. Ethylene also suppresses stem and root elongation, probably for similar reasons.

Ethylene is also produced in large quantities during a certain phase of the ripening of fruits, when their respiration is proceeding at its most rapid rate. At this phase of ripening, complex carbohydrates are broken down into simple sugars, cell walls become soft, and the volatile compounds associated with flavor and scent in the ripe fruits are produced. When ethylene is applied to fruits, it hastens their ripening. One of the first lines of evidence that led to the recognition of ethylene as a plant hormone was the observation that gases that came from oranges caused premature ripening in bananas. Such relationships have led to major commercial uses. Tomatoes are often picked green and then artificially ripened as desired by the application of ethylene. Ethylene is widely used to speed the ripening of lemons and oranges as well. Carbon dioxide produces effects in fruits opposite to those of ethylene, and fruits that are being shipped are often kept in an atmosphere of carbon dioxide if they are not intended to ripen yet.

*Ethylene, a simple gaseous hydrocarbon, is a naturally occurring plant hormone. It plays the key role in controlling the falling off of leaves, flowers, and fruits from the plants on which they form; auxin retards their tendency to fall.*

## Abscisic Acid

**Abscisic acid** is a naturally occurring plant hormone that is synthesized mainly in mature green leaves, fruits, and root caps. The hormone was given its name because applications of it stimulate leaves to age rapidly and fall off (the process of *abscission*), but there is little evidence that it plays an important natural role in this process. Abscisic acid suppresses the growth and elongation of buds and promotes aging, counteracting some of the effects of the gibberellins (which stimulates the growth and elongation of buds) and auxin (which tends to retard aging).

## TROPISMS

**Tropisms,** or responses to external stimuli, control the patterns of growth of plants and thus their appearance. Plants adjust to the conditions of their environment by growth responses. **Positive tropisms** are those in which the movement or reaction is in the direction of the source of the stimulus, whereas **negative tropisms** are those which result in movement or growth in the opposite direction. We consider three major classes of plant tropisms: phototropism, gravitropism, and thigmotropism.

## Phototropism

We introduced **phototropism,** the bending of plants toward unidirectional sources of light, in our discussion of the action of auxin. In general, stems are positively phototropic, growing toward the light, whereas roots are negatively phototropic, growing away from it. The phototropic reactions of stems are clearly of adaptive value because they allow plants to capture greater amounts of light than would otherwise be possible. Auxin is involved in most, if not all, of the phototropic growth responses of plants.

*Phototropisms are growth responses of plants to a unidirectional source of light. They are mostly, if not entirely, mediated by auxin.*

## Gravitropism

Another familiar plant response is **gravitropism,** formerly known as geotropism. This tropism causes stems to tend to grow upward and roots downward (Figure 31-10); both of these responses are clearly of adaptive significance. Stems which grow upward are apt to re-

ceive more light than those which do not; roots which grow downward are more apt to encounter a more favorable environment than those which do not. The phenomenon is now called gravitropism because it is clearly a response to gravity and not to the earth (prefix "geo") as such.

It seems likely that **amyloplasts,** starch-containing plastids, play an important role in the perception of gravity by plants. The amyloplasts are heavy capsules that contain large amounts of calcium and starch. In roots, the cells in which amyloplasts occur are apparently located in the central cells of the root cap; removing the root cap stops the root's responses to gravity in most cases. In shoots, on the other hand, gravity is clearly sensed along the whole length of the stem, probably by the functioning of similar amyloplasts in certain cells. Gravity causes the amyloplasts to fall to the lower side of a given cell. There they apparently set in motion a series of reactions that eventually causes the shoots and roots to bend. The amyloplasts reach the lower side of their cells within a minute, and the bending of the root or shoot may occur within as little as 10 minutes.

*Gravitropism, the response of a plant to gravity, generally causes shoots to grow up and roots to grow down. The force of gravity apparently is sensed in special cells with amyloplasts, starch-containing plastids.*

## Thigmotropism

Still another response of plants that is commonly observed is **thigmotropism,** a name derived from the Greek root *thigma,* meaning "touch." Thigmotropism is defined as the response of plants to touch. The responses by which tendrils curl around and cling to stems or other objects are surprisingly rapid and clear (Figure 31-11); the ways in which twining plants such as bindweed coil around objects are analogous. This behavior is the result of rapid growth responses to touch. Specialized groups of cells in the epidermis appear to be concerned with thigmotropic reactions, but again, their exact mode of action is not well understood.

*Thigmotropisms are growth responses of plants to touch.*

**Figure 31-10**

**Tropism guides plant growth.** The branches of this fallen tree are growing straight up because they are negatively gravitropic and also because they are positively phototropic.

**Figure 31-11**

**Thigmotropism.** These tendrils coil around the stem because of their positive thigmotropism.

## PHOTOPERIODISM

Essentially all eukaryotic organisms are affected by the cycle of night and day, and many features of plant growth and development are keyed to the changes in the proportions of light and dark in the daily 24-hour cycle. Such responses constitute **photoperiodism,** a mechanism by which organisms measure seasonal changes in relative day and night length. One of the most obvious of these photoperiodic reactions concerns the production of flowers by angiosperms.

Day length changes with the seasons; the farther from the equator one is, the greater the variation. The flowering responses of plants fall into three basic categories in relation to day length. **Short-day plants** begin to form flowers when the days become shorter than a critical length (Figure 31-13). **Long-day plants,** on the other hand, initiate flowers when the days become longer than a certain length. Thus many fall flowers are short-day plants, and many spring and early-summer flowers are long-day plants. Commercial plant growers use these responses to day length to bring plants into bloom when they are wanted for sale. In addition to the long-day and short-day plants, a number of plants are described as **day neutral;** they produce flowers whenever environmental conditions are suitable, without regard to day length. The complex chemical basis of these responses is now reasonably well understood.

## TURGOR MOVEMENTS

Some kinds of plant movements are based on reversible changes in the **turgor pressure** (defined as the pressure within a cell resulting from the movement of water into the cell) of specific cells, rather than on differential growth or cell enlargement. One of the most familiar of these has to do with the changing patterns of leaf position that certain plants exhibit at night and in the day. For example, the attractively spotted leaves of the prayer plant *(Maranta)* spread horizontally during the day but become more or less vertical at night (Figure 31-12).

**A**

**B**

**Figure 31-12**

**Prayer plant.** A prayer plant *(Maranta)* in the day with leaflets held horizontally (**A**) and at night with leaflets held down (**B**).

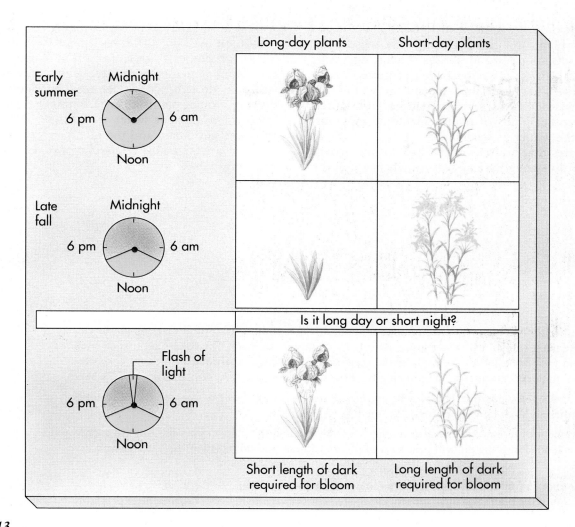

**Figure 31-13**

**How flowering responds to daylength.** The goldenrod is a short-day plant that blooms in the fall throughout the Northern hemisphere, stimulated to do so by the long nights. The iris is a long-day plant that blooms in the spring, stimulated to do so by the short nights that occur in the spring. If the long night of winter is artificially interrupted by a flash of light, the goldenrod will not bloom, and the iris will. In each case, it is the duration of uninterrupted darkness that determines when flowering will occur.

*Short-day plants form flowers when days become shorter than a certain critical day length; long-day plants form flowers when days become longer than a certain length. Day-neutral plants do not have specific day-length requirements for flowering.*

## DORMANCY

Plants respond to their external environment largely by changes in growth rate. As you might imagine, the ability to stop growing altogether when conditions are not favorable is a critical factor in their survival.

In temperate regions, we generally associate dormancy with winter, when low temperatures and the unavailability of water because of freezing make it impossible for plants to grow. During this season the buds of deciduous trees and shrubs remain dormant, and the apical meristems remain well protected inside enfolding scales. Perennial herbs spend the winter underground as stout stems or roots packed with stored food. Many other kinds of plants, including most annuals, pass the winter as seeds.

In climates that are seasonally dry, dormancy will occur primarily during the dry season, whenever in the year it falls. In dry conditions, plants stay in a dormant condition by using strategies similar to those which the plants of temperate areas rely on in winter.

Annual plants occur frequently only in areas of seasonal drought. Seeds are ideal mechanisms for allowing annual plants to bypass the dry season, when there is insufficient water for growth. When it rains, they can germinate and the plants can grow rapidly to take advantage of the relatively short period when water is available.

## ■ SUMMARY

1. Water flows through plants in a continuous column, driven mainly by transpiration through the stomata. The plant can control water loss primarily by closing its stomata. The cohesion of water molecules and their adhesion to the walls of the very narrow cell columns through which they pass are additional important factors in maintaining the flow of water to the tops of plants.

2. The stomata open when their guard cells are turgid and bow out, thus causing the thickened inner walls of these cells to bow away from the opening.

3. The movement of dissolved sucrose and other carbohydrates in the phloem does not require energy. Sucrose is loaded into the phloem near sites of photosynthesis and unloaded at the places where it is required.

4. Nutrients move primarily through plants as solutes in the water column in the xylem. Their selective admission into the plant and their subsequent movement between living cells require the expenditure of energy.

5. The nine macronutrients, substances that are each present at concentrations of 1% or more of a plant's dry weight, are carbon, hydrogen, oxygen, nitrogen, potassium, calcium, phosphorus, magnesium, and sulfur. In addition, there are seven known micronutrients, each present at concentrations of from 1 to several hundred parts per million of dry weight.

6. Hormones are chemical substances produced in small quantities in one part of an organism and transported to another part of the organism, where they bring about physiological responses. The tissues in which plant hormones are produced are not specialized particularly for that purpose, nor are there usually clearly defined receptor tissues or organs.

7. There are five major classes of naturally occurring plant hormones: auxins, cytokinins, gibberellins, ethylene, and abscisic acid. They often interact with one another in bringing about growth responses.

8. Auxins are produced at the tips of shoots and diffuse downward, suppressing the growth of lateral buds. In young grass seedlings and other herbs, they play a major role in promoting stem elongation.

9. Cytokinins are necessary for mitosis and cell division in plants. They promote the growth of lateral buds and inhibit the formation of lateral roots.

10. Gibberellins play the major role in stem elongation in most plants.

11. Tropisms in plants are growth responses to external stimuli. A phototropism is a response to light, gravitropism a response to gravity, and thigmotropism a response to touch.

12. The flowering responses of plants fall into two basic categories in relation to day length. Short-day plants begin to form flowers when the days become shorter than a given critical length; long-day plants do so when the days become longer than a certain length.

13. Dormancy is a necessary part of plant adaptation that allows a plant to bypass unfavorable seasons such as winter, when the water may be frozen, or periods of drought. Dormancy also allows plants to survive in many areas where they would be unable to grow otherwise.

## REVIEW

1. _____ are substances produced in minute quantities in one part of the plant and transported to other parts of the plant where they are physiologically active.

2. The active transport of ions into the root hairs results in a high concentration of ions in these cells, so that water is drawn into the cells by _____.

3. Name the six macronutrients other than carbon, hydrogen, and oxygen that are required for plant growth.

4. In contrast to auxins, cytokinins _____ the growth of lateral branches.

5. _____ is the response of roots to gravity, which causes them to grow downward.

## SELF-QUIZ

1. Some 90% of the water taken in by the roots of plants is lost to the atmosphere. Which leaf structure accounts for the greatest portion of this water loss?
   (a) The cuticle
   (b) Veinlets
   (c) Transpiration
   (d) Stomata
   (e) Mesophyll

2. Water can rise to the top of a 20-meter tree primarily because of
   (a) gravity.
   (b) evaporation.
   (c) adhesion.
   (d) capillarity.
   (e) hydrogen bonding.

3. Keeping guard cells turgid and thus keeping the stomata open requires a constant expenditure of energy. What is the location at which the necessary ATPs are produced?
   (a) The roots
   (b) Photosynthesis carried on in the guard cells themselves
   (c) Mesophyll cells
   (d) Rhizomes
   (e) All the cells of the plant

4. Plants bend toward the light because (pick two)
   (a) cells on the shaded side of the stem elongate more than those on the sunny side.
   (b) cells on the sunny side of the stem elongate more than those on the shaded side.
   (c) cells on the shaded side of the stem accumulate auxin.
   (d) cells on the sunny side of the stem accumulate auxin.
   (e) auxin suppresses lateral bud growth.

5. Dormancy in plants is commonly associated with which two of the following?
   (a) Warm temperatures
   (b) Cold temperatures
   (c) Lack of carbon dioxide
   (d) Lack of nutrients
   (e) Lack of available water

6. Among the environmental factors regulating the opening of stomata are (more than one answer possible)
   (a) temperature.
   (b) $CO_2$ concentration.
   (c) available water.
   (d) $O_2$ concentration.
   (e) intensity of grazing by herbivores.

7. Gibberellins are associated with which of the following growth reactions of plants (more than one answer possible)?
   (a) Bending toward the light
   (b) Growing upright
   (c) The branching of stems
   (d) Flowering (bolting) in biennials
   (e) Resistance to fungi

8. Which of the following classes of plant hormones appears to be necessary for mitosis and cell division to take place?
   (a) Auxins
   (b) Gibberellins
   (c) Abscisic acid
   (d) Cytokinins
   (e) All of the above

# THOUGHT QUESTIONS

1. Why do gardeners often remove many of a plant's leaves after transplanting it?

2. If you grew a plant that initially weighed 200 grams but eventually weighed 50 kilograms in a pot, would you expect the soil in the pot to change weight? If so, how much, and why?

3. When poinsettias are kept inside a house following the holiday season, they rarely bloom again. Why do you think this might be, and what might you do to get them to produce flowers a second time?

# FOR FURTHER READING

EVANS, M., R. MOORE, and K. HASENSTEIN: "How Roots Respond to Gravity," *Scientific American*, December 1986, pages 112-120. Did you ever wonder why roots grow down and not up? These authors speculate that calcium ions settle in roots, where they activate proteins that promote growth.

MANSFIELD, T.A., and W.J. DAVIES: "Mechanisms for Leaf Control of Gas Exchange," *BioScience*, vol. 35, 1985, pages 158-168. Excellent review of some of the factors involved in stomatal opening and the ways in which they are integrated.

MOONEY, H.A., and OTHERS: "Plant Physiological Ecology Today," *BioScience*, vol. 37, 1988, pages 18-67. Nearly an entire issue of *BioScience* devoted to the ways in which plants cope with their environments and in which scientists study them. Highly recommended.

SISLER, E.C., and S.F. YANG: "Ethylene, The Gaseous Plant Hormone," *BioScience*, vol. 33, 1984, pages 233-238. An up-to-date review of this important plant hormone.

STEWART, D.: "Green Giants," *Discover*, April 1990, pages 61-64. More than majestic trees, sequoias are triumphs of hydraulics, wind resistance, and architecture.

TREWAVAS, A.: "How Do Plant Growth Substances Work?" *Plant, Cell, and Environment*, vol. 4, 1981, pages 203-228. A concise introduction to plant hormones.

# PART 9

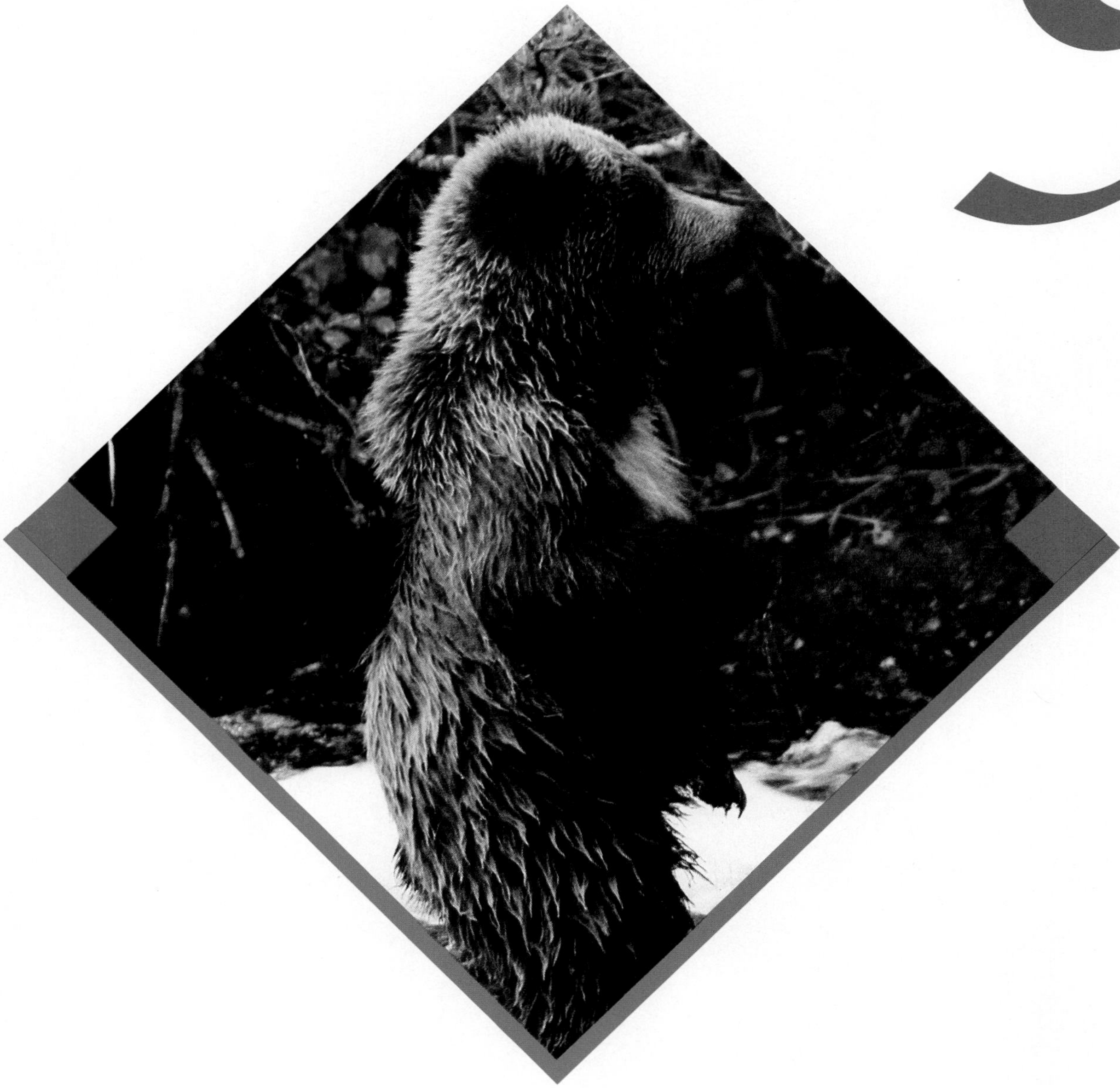

# ANIMAL BIOLOGY

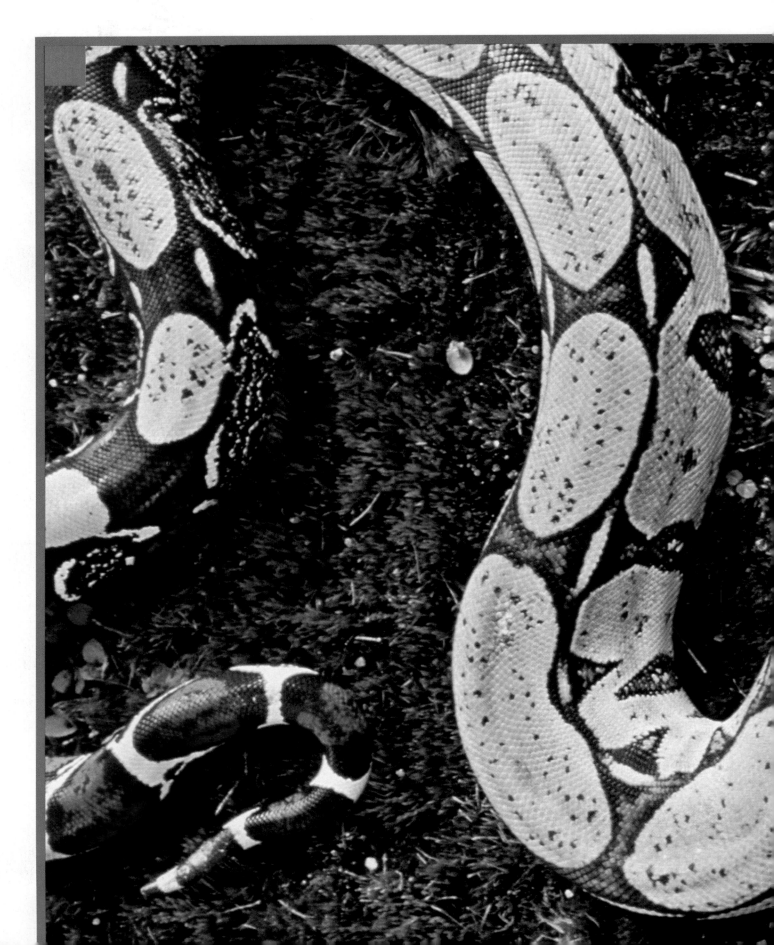

# The Vertebrate Body

All vertebrates share the same basic body plan.
Like you, the body of this red-tailed boa
constrictor is supported by a skeletal scaffold
of jointed bones and a column of vertebrae
that surrounds the dorsal nerve chord.

# THE VERTEBRATE BODY

## Overview

The bodies of vertebrates are composed of between fifty and several hundred kinds of cells, depending on the kind of animal. These cells are of four general kinds, called tissue types: epithelium, connective tissue, muscle, and nerve. Each of these tissue types can be recognized by its structure, function, and origin. The organs of the body are often composed of several different tissue types. The study of the function of cells, tissues, and organs is called physiology. In order for the many cell types to function properly, extracellular conditions must be kept relatively constant. This constancy is referred to as homeostasis, and the body employs a variety of means to prevent significant disturbances.

## For Review

*Here are some important terms and concepts that you will encounter in this chapter. If you are not familiar with them, you should review them before proceeding.*

**The sodium-potassium pump** (Chapter 5)

**The evolution of vertebrates** (Chapter 18)

**Basic structure of chordates** (Chapters 18 and 28)

**Coelom** (Chapter 28)

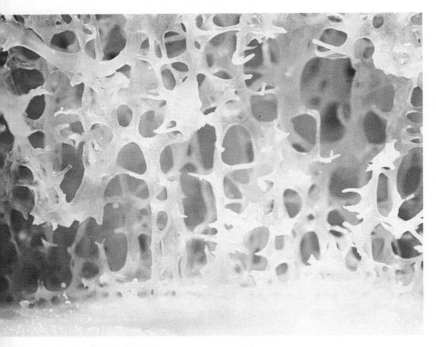

### Figure 32-1

**The interior of bone.** While we think of bone as hard and solid, the interiors of many bones are composed of a delicate and surprising latticework. Bone, like most tissues present in your body, is a dynamic structure, constantly renewing itself.

Most of us have been to a zoo and have seen the many different kinds of animals there—long-necked giraffes, armor-plated rhinoceroses, fish that swim, and birds that fly and sing. Monkeys look at us, perhaps wondering why we are looking at them; snakes slither, squirrels dart about, and peacocks strut. All these animals are vertebrates, and although they may seem very different from each other, they are in fact very much alike "under the skin." All vertebrates share the same basic body plan, with the same sorts of organs operating in much the same way. In this chapter we begin a detailed consideration of the biology of vertebrates and of the often intricate and fascinating structure of their bodies (Figure 32-1). We focus on human biology because the architecture of the human body provides a good focal point for discussing vertebrate bodies in general and because human biology is of particular importance to all of us. We each want to know how our body works and why it functions the way it does.

Not all vertebrates are the same, of course, and some of the differences are important. For example, a fish doesn't breathe the same way you do. Thus we do not limit our discussion to human beings, but rather describe the human animal in the context of vertebrate diversity. The differences reflect our evolutionary history and offer important lessons in why we function the way we do.

# THE HUMAN ANIMAL

An incredible machine of great beauty, the human body has the same general architecture that all vertebrates have (Figure 32-2). It includes a long tube that travels from one end of the body to the other, from mouth to anus; this tube is suspended within an internal body cavity called the coelom (see Chapter 28). In human beings the coelom is divided into two parts: (1) the thoracic cavity, which contains the heart and lungs; and (2) the abdominal cavity, which contains the stomach, intestines, and liver. The bodies of all vertebrates are supported by an internal scaffold or skeleton made up of jointed bones that grow as the body grows. A bony skull surrounds the brain; and a column of bones, the vertebrae, surrounds the dorsal nerve cord, or spinal cord.

Human beings are mammals and, like all other mammals, are warm blooded, regulating their internal temperature at a relatively constant value. Humans keep their temperature at about 98° F (37° C). Like all other mammals, human beings have hair rather than scales or feathers; and like all other mammals except monotremes (for example, the duck-billed platypus), humans do not lay eggs but give birth to young who must be nurtured by a parent. Human development is a lengthy process compared with other mammals; infants are nursed for a long time and mature slowly.

## How the Body is Organized

The bodies of vertebrates, like those of all other multicellular animals, are composed of different cell types. The bodies of adult vertebrates contain between 50 and several hundred different kinds of cells, depending on the kind of vertebrate and how finely you differentiate between cell types. Groups of similar cells are organized into **tissues**, which are structural and functional units.

**Organs** are body structures composed of several different tissues grouped together into a larger structural and functional unit. Your heart is an organ. It contains cardiac muscle tissue wrapped in connective tissue and is the target of many nerves. All these tissues work together to pump blood through your body. An **organ system** is a group of organs that function together to carry out the principal activities of the body. For example, the digestive organ system is composed of individual organs concerned with the breaking up of food (teeth), the passage of food to the stomach (esophagus), the storage of food (stomach), the digestion and absorption of food (intestine), and the expulsion of solid residue (rectum). The liver, a chemical reprocessing plant, is also a component of the digestive system. The human body contains 11 principal organ systems (Table 32-1).

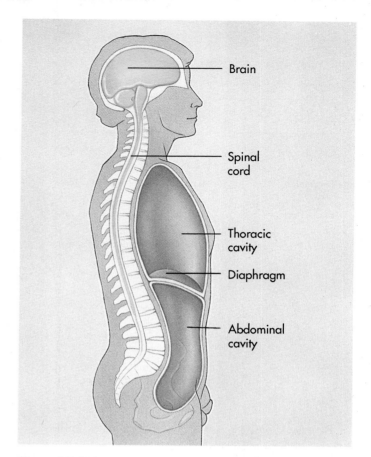

**Figure 32-2**

**Architecture of the human body.** Humans, like all vertebrates, have a dorsal central nervous system, consisting of a spinal cord and brain, enclosed in vertebrae and the skull. In mammals, a muscular diaphragm divides the coelom into the thoracic cavity and the abdominal cavity.

## TISSUES

Early in the development of any vertebrate, the growing mass of cells differentiates into three fundamental layers of cells: **endoderm, mesoderm,** and **ectoderm.** These three kinds of embryonic cell layers in turn differentiate into the hundreds of different cell types that are characteristic of the adult vertebrate body. As you saw in Chapter 28, where they were reviewed in an evolutionary context, these diverse cell types are traditionally grouped on a functional basis into four basic types of tissues: **epithelium, connective tissue, muscle,** and **nerve** (Figure 32-3). Of these, connective tissues are particularly diverse. Blood cells, for example, are classified as one kind of connective tissue, whereas bone is another.

## Epithelium is Protective Tissue

*Epithelial cells are the guards and protectors of the body.* They cover its surface and determine which substances enter it and which do not. The organization of the vertebrate body is fundamentally tubular, with de-

**TABLE 32-1    THE MAJOR VERTEBRATE ORGAN SYSTEMS**

| SYSTEM | FUNCTIONS | COMPONENTS | DETAILED TREATMENT |
|---|---|---|---|
| Circulatory | Transports cells and materials throughout the body | Heart, blood vessels, blood, lymph, and lymph structures | Chapter 39 |
| Digestive | Captures soluble nutrients from ingested food | Mouth, esophagus, stomach, intestines, liver, and pancreas | Chapter 37 |
| Endocrine | Coordinates and integrates the activities of the body | Pituitary, adrenal, thyroid, and other ductless glands | Chapter 36 |
| Urinary | Removes metabolic wastes from the bloodstream | Kidney, bladder, and associated ducts | Chapter 41 |
| Immune | Removes foreign bodies from the bloodstream | Lymphocytes, macrophages, and antibodies | Chapter 40 |
| Integumentary | Covers the body and protects it | Skin, hair, nails, and sweat glands | Chapter 32 |
| Muscular | Produces body movement | Skeletal muscle, cardiac muscle, and smooth muscle | Chapter 35 |
| Nervous | Receives stimuli, integrates information, and directs the body | Nerves, sense organs, brain, and spinal cord | Chapters 33 and 34 |
| Reproductive | Carries out reproduction | Testes, ovaries, and associated reproductive structures | Chapter 42 |
| Respiratory | Captures oxygen and exchanges gases | Lungs, trachea, and other air passageways | Chapter 38 |
| Skeletal | Protects the body and provides support for locomotion and movement | Bones, cartilage, and ligaments | Chapter 35 |

velopmental derivatives of ectodermal cells covering the outside (skin), those of endodermal cells lining the hollow inner core (alimentary canal and gut), and those of mesodermal cells lining the body cavity (coelom). All these kinds of **epidermal,** or "skin," cells are broadly similar in form and function, and they are collectively called the **epithelium.** The epithelial layers of the body function in four different ways:

1. They protect the tissues beneath them from dehydration and mechanical damage, a particularly important function in land-dwelling vertebrates.
2. They provide a selectively permeable barrier that can facilitate or impede the passage of materials into the tissue beneath. Because epithelium encases all the body's surfaces, every substance that enters or leaves the body must cross an epithelial layer.
3. They provide sensory surfaces. Many of the body's sense organs are in fact modified epithelial cells.
4. They secrete materials. Most secretory glands are derived from invaginations of layers of epithelial cells that occur during embryonic development.

Layers of epithelial tissue are usually only one or a few cells thick. Individual epithelial cells possess a small amount of cytoplasm and have a relatively low metabolic rate. Few blood vessels pass through the epithelium; instead, the circulation of nutrients, gases, and wastes in epithelial tissue occurs by diffusion from the capillaries of neighboring tissue.

Epithelial tissues possess remarkable regenerative abilities. The cells of epithelial layers are constantly being replaced throughout the life of the organism. For example, the liver, which is a gland formed of epithelial tissue, can readily regenerate substantial portions of tissue that have been surgically removed from it. The cells lining the digestive tract are continuously replaced every few days.

There are three general classes of epithelial tissue (Figure 32-4): **simple epithelium, stratified epithelium,**

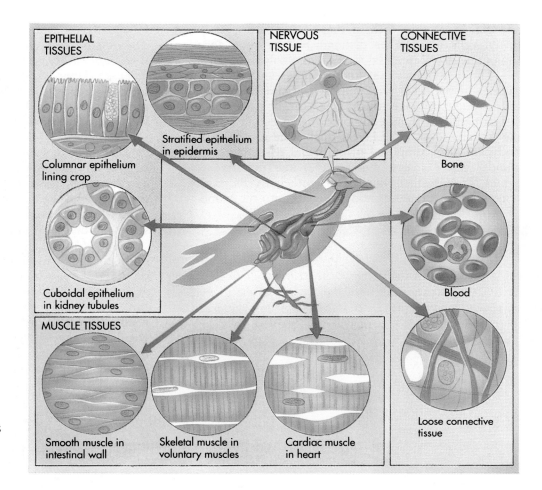

**Figure 32-3**

**Vertebrate tissue types.** Epithelial tissues are indicated by blue arrows; connective tissue by green arrows; muscle tissue by red arrows; and nervous tissue by a yellow arrow.

Within the figure 32-3:

EPITHELIAL TISSUES

Columnar epithelium lining crop

Cuboidal epithelium in kidney tubules

Stratified epithelium in epidermis

NERVOUS TISSUE

CONNECTIVE TISSUES

Bone

Blood

Loose connective tissue

MUSCLE TISSUES

Smooth muscle in intestinal wall

Skeletal muscle in voluntary muscles

Cardiac muscle in heart

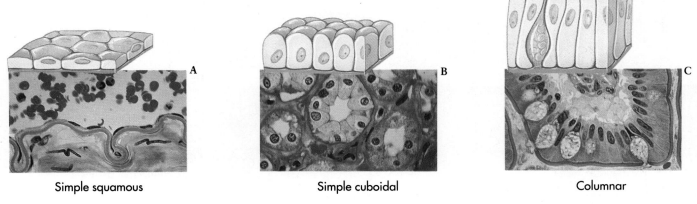

Simple squamous

Simple cuboidal

Columnar

**Figure 32-4**

**Three kinds of epithelial tissue.**
**A** Simple. The cells seen here are lining an artery. The nuclei are flattened, like plates (the round cells above are blood cells in the artery).
**B** Stratified. The cells forming the walls of these kidney tubules are cube-shaped, with round nuclei.
**C** Glandular. Elongated columnar cells from the outer layer of the human ileum. Nuclei are lenghthened, like sausages. Interspersed are round goblet cells, which secrete mucus.

and **glands.** Within each of these general classes, epithelial cells are further distinguished by their form: squamous (flat), cuboidal (cubic), and columnar (tall).

The membranes that line the lungs and the major cavities of the body are composed of simple epithelial cells a single cell layer thick. The skin, or epidermis, is a stratified epithelium composed of more complex epithelial cells several cell layers thick. The glands of the body are also composed of epithelial tissue. Exocrine glands secrete sweat, milk, saliva, and digestive enzymes through ducts lined with epithelial cells. Endocrine glands secrete hormones directly into the blood.

*The skin of the body and the lining of the respiratory and digestive tracts are composed of epithelium. Much of the internal lining of respiratory and digestive tracts is simple epithelium only one cell thick, whereas external skin is composed of epithelium many cell layers thick.*

## Connective Tissue Supports the Body

*The cells of connective tissue provide the body with its structural building blocks and also with its most potent defenses* (Figure 32-5). Beneath the skin and in many internal organs, the cells that make up connective tissue are not stacked tightly; instead, they are spaced well apart from one another. This is called loose connective tissue. In other cases, connective tissue cells are densely packed. Connective tissue cells, which are derived from the mesoderm, also fall into three functional categories: (l) the cells of the immune system, which acts to defend the body; (2) the cells of the skeletal system, which supports the body; and (3) the cells that store and distribute substances throughout the body.

The cells of the immune system are small and roam the body within the bloodstream (Figure 32-6); they are mobile hunters of invading microorganisms and foreign substances (see Chapter 40). There are three principal kinds of immune system cells: (l) **macrophages,** which are mobile, phagocytic cells able to engulf and digest invading bacteria, fungi, and other microorganisms, as well as cellular debris; (2) **lymphocytes,** or white blood cells, which either synthesize an-

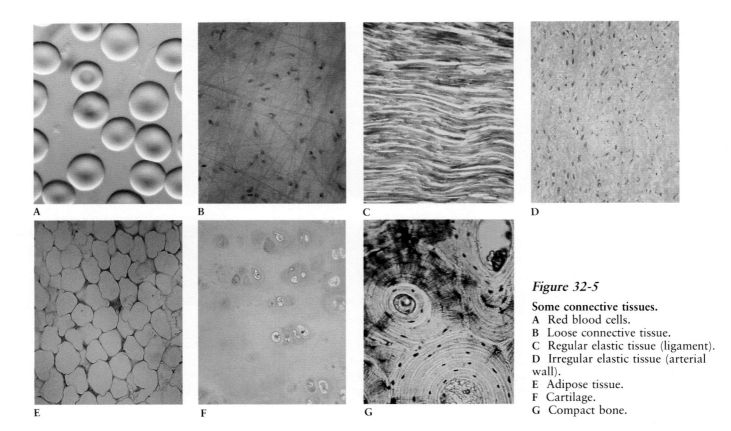

*Figure 32-5*

**Some connective tissues.**
A  Red blood cells.
B  Loose connective tissue.
C  Regular elastic tissue (ligament).
D  Irregular elastic tissue (arterial wall).
E  Adipose tissue.
F  Cartilage.
G  Compact bone.

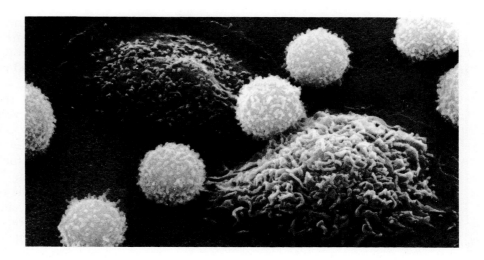

**Figure 32-6**

**Lymphocytes and macrophages.** The lymphocytes are small and spherical; the macrophages are larger and more irregular in form.

tibodies or attack virus-infected cells; and (3) **mast cells,** which synthesize the molecules involved in the body's response to **trauma** (injury), including histamine, which produces the inflammation response, and heparin, which prevents blood clotting. Allergies are the result of the stimulation of mast cells by other cells of the immune system.

Three principal kinds of connective tissues occur in the skeletal system: fibroblasts, cartilage, and bone. The three are distinguished principally by the nature of the matrix that is laid down between the individual cells.

*Fibroblasts.* The most common kind of connective tissue in the vertebrate body consists of **fibroblasts,** which are flat, irregularly branching cells (Figure 32-7) that secrete structurally strong proteins into the matrix between the cells. The most commonly secreted protein is **collagen** (Figure 32-8), which is the most abundant protein in vertebrate bodies. One quarter of all animal protein is collagen. Fibroblasts are active in wound healing. They multiply rapidly in wound tissue, forming granulated fibrous scar tissue that possesses a collagen matrix.

*Cartilage.* **Cartilage** is a specialized connective tissue in which the collagen matrix between cells is formed at positions of mechanical stress. In cartilage the fibers are laid down along the lines of stress in long, parallel arrays. The result of this process is a firm and flexible tissue that has great strength. Cartilage covers the ends of bones that come together in joints, such as the knee, ankle, and elbow.

*Bone.* **Bone** is a specialized connective tissue in which the collagen fibers are coated with a calcium phosphate salt. Bone is more rigid than collagen and is strong but not brittle. The structure of bone and the way in which it is formed are discussed in Chapter 35.

The third general class of connective tissue is composed of cells that are specialized for the accumulation and transport of particular molecules. **Sequestering connective tissues** include pigment-containing cells and the fat cells of adipose tissue. The most important tissues of this class, present in the vascular system rather than in the connective tissue proper, are red blood

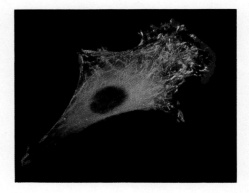

**Figure 32-7**

**Cells of connective tissue are often supported by complex webs of protein fibers.** The different proteins in this fibroblast are made visible by fluorescence microscopy. Actin appears blue here, microtubules appear green, and intermediate fibers appear red.

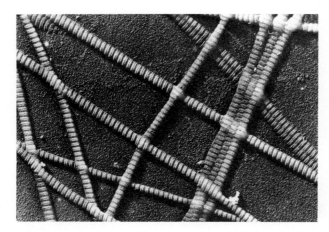

**Figure 32-8**

**Collagen fibers.** Each fiber is composed of many individual collagen strands and can be very strong.

cells, called **erythrocytes** (Figure 32-9). There are about 5 billion erythrocytes in every milliliter of human blood. Erythrocytes act as the transporters of oxygen in the vertebrate body. During the process of their maturation in mammals, they lose their nucleus and mitochondria, and their endoplasmic reticulum is reabsorbed. As a result of these processes, mammalian erythrocytes are relatively inactive metabolically, but they are not empty. Large amounts of the iron-containing protein hemoglobin are produced within the eryth-

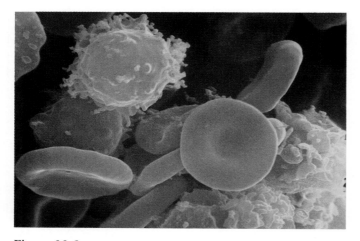

**Figure 32-9**

**Blood cells.** White blood cells, or eukocytes, are roughly spherical and have irregular surfaces with numerous extending pili. Red blood cells, or erythrocytes, are flattened spheres, typically with a depressed center.

rocytes and remain in the mature cell. Hemoglobin is the principal carrier of oxygen in vertebrates, as it is in many other groups of animals. Each erythrocyte contains about 300 million molecules of hemoglobin.

The fluid, or **plasma,** in which the erythrocytes move is both the "banquet table" and the "refuse heap" of the vertebrate body. Practically every substance used by cells is dissolved in plasma. These substances include the sugars, lipids, and amino acids, which are the fuel of the body, as well as the products of metabolism. The plasma also contains inorganic salts such as the calcium used to form bone, fibrinogen from the liver, albumin, which gives the blood viscosity, and antibody proteins produced by lymphocytes. Waste products such as urea, the product of protein breakdown, are also present in the plasma.

## Muscle Tissue Provides for Movement
*Muscle cells, formed from the mesoderm early in development, are the "workhorses" of the vertebrate animal.* The distinguishing characteristic of muscle cells, the one that makes them unique, is the relative abundance of actin and myosin microfilaments within them. These protein microfilaments are present as a fine network in all eukaryotic cells (see Chapter 4), but they are far more abundant in muscle cells. Vertebrates possess three different kinds of muscle cells (Figure 32-10): smooth muscle, skeletal muscle, and cardiac muscle. The last two types are called striated muscle to distinguish them from nonstriated smooth muscle. In nonstriated muscle, microfilaments are only loosely organized. In striated muscle cells, the actin microfilaments are bunched together with thicker filaments of myosin into many thousands of fibers called **myofibrils.** The myofibrils shorten when the actin and myosin filaments slide past each other. Interactions of actin and myosin filaments can cause muscle cells to change shape, as in smooth muscle. Because there are so many filaments all aligned the same way in muscle cells, a considerable force is generated when actin and myosin filaments interact.

## Nerve Tissue Conducts Signals Rapidly
*Nerve cells carry impulses (action potentials) rapidly from one organ to another.* Nerve tissue, the fourth major class of vertebrate tissue, is composed of two kinds of cells: (1) **neurons,** which are specialized for the transmission of nerve impulses; and (2) **supporting glial cells,** which support and insulate the neurons. The supporting cells also assist in the propagation of the nerve impulse by playing an important part in the maintenance of the ionic composition of nerve tissue. They are also believed to supply the neurons with nutrients and other molecules.

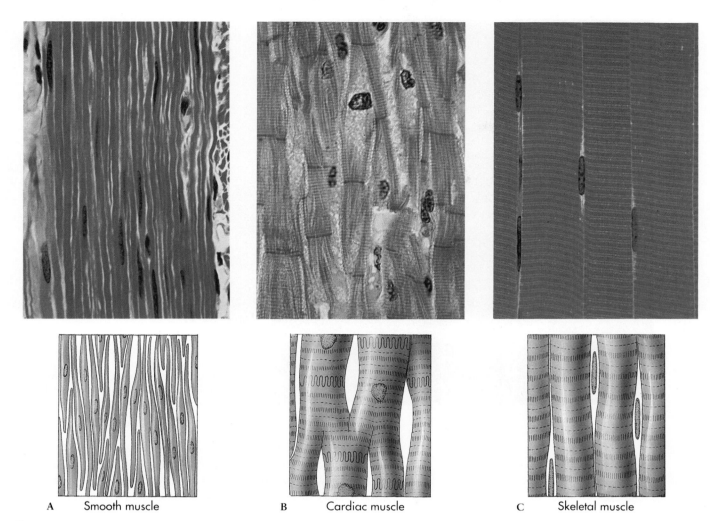

**Figure 32-10**

**Types of muscle.**
**A** Smooth muscle cells are long and spindle shaped, with a single nucleus.
**B** Cardiac muscle cells, such as these from a human heart, also contain a single nucleus and are organized into long branching chains that interconnect, forming a lattice.
**C** Skeletal or striated muscle cells are formed by the fusion of several muscle cells, end to end, to form a long fiber with many nuclei.

Neurons (Figure 32-11) are cells specialized to conduct signals rapidly throughout the body. Their membranes are rich in ion-selective channels, water-filled pores that control the movement of ions into and out of the cell. In doing so, these channels maintain a voltage difference between the interior and exterior of the cell, the equivalent of a battery. When ion channels in a local area of the membrane open, ions reenter from the exterior. This flow of ions is called a current, and it temporarily wipes out the charge difference—a process called **depolarization.** This depolarization in a local area of the membrane tends to open nearby ion channels in the neuron membrane, resulting in a wave of electrical activity that travels, or propagates, down the entire length of the nerve as a **nerve impulse.** The

nature of nerve impulses varies, as described in more detail in Chapter 33, but all nerve impulses propagate along nerves as waves of membrane depolarization.

The cell body of a neuron contains within it the cell nucleus. From the cell body project two kinds of cytoplasmic extensions that carry out the transmission functions of the neuron. The first of these element consists of **dendrites,** which are threadlike protrusions. The dendrites act as antennae for the reception of nerve impulses from other cells or sensory systems. The second kind of cytoplasmic extension that projects from the cell body of a neuron is the **axon.** An axon is a long tubular extension of the cell that carries the nerve impulse away from the cell body, often for considerable distances (Figure 32-12). Because axons can

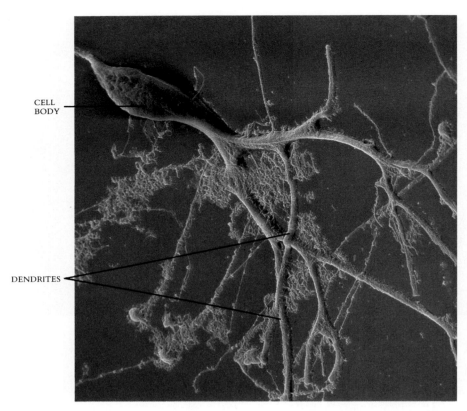

CELL
BODY

DENDRITES

*Figure 32-11*

**A human neuron.** The cell body is at the upper left, with the axon extending up out of view. The branching network of fibers extending down from the cell body is made up of dendrites, which carry signals to the cell body.

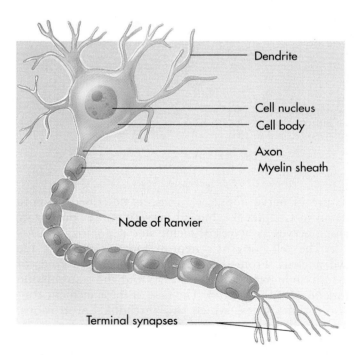

Dendrite

Cell nucleus
Cell body

Axon
Myelin sheath

Node of Ranvier

Terminal synapses

*Figure 32-12*

**Idealized structure of a vertebrate neuron.** Many dendrites lead to the cell body, from which a single long axon extends. In some neurons, specialized for rapid signal conduction, the axons are encased at intervals within myelin sheaths. At its far end, an axon often branches.

be quite long, some nerve cells are very long indeed. A single neuron that activates the muscles in your thumb may have its cell body in the spinal cord and possess an axon that extends all the way across your shoulder and down your arm to your thumb. Single neuron cells more than a meter in length are common.

The nerves of the vertebrate body appear as fine white threads when they are viewed with the naked eye but are actually composed of bundles of axons. Like a telephone trunk cable, these nerves include large numbers of independent communications channels—bundles of hundreds of axons, each connecting a nerve cell to a muscle fiber. In addition, the nerve contains numerous supporting glial cells bunched around the axons. In the **brain** and **spinal cord,** which together make up the **central nervous system,** these supporting glial cells are called **oligodendrocytes cells.** The supporting cells associated with projecting axons, and with all other nerve cells that make up the **peripheral nervous system,** are called **Schwann cells.**

## ORGANS: DIFFERENT TISSUES WORKING TOGETHER

The four classes of vertebrate tissues that we have discussed in this chapter (epithelium, connective tissue, muscle, and nerve) are the building blocks of the vertebrate body (Table 32-2). Each **organ** of the body, a

**TABLE 32-2   MAJOR BODY TISSUES**

| TISSUE | TYPICAL LOCATION | TISSUE FUNCTION | CHARACTERISTIC CELL TYPES |
|---|---|---|---|
| **EPITHELIAL** | | | |
| Simple | Lining of lungs, capillary walls, and blood vessels; lining of some glands | Cells very thin; provides a thin layer across which diffusion can readily occur | Epithelial cells |
| Stratified | Outer layer of skin; lining of mouth | Tough layer of cells; provides protection | Epithelial cells |
| | Lining of parts of respiratory tract | Functions in secretion of mucus; dense with cilia that aid in movement of mucus; provides protection | |
| Glands | Digestive tract; skin | Secretion of enzymes, fluids, and hormones | Gland cells |
| **CONNECTIVE** | | | |
| Connective tissue proper | | | |
| Loose | Beneath skin and other epithelia | Support; provides a fluid reservoir for epithelium | Fibroblasts |
| Dense | Tendons; sheath around muscles; kidney; liver; dermis of skin | Provides flexible strong connections | Fibroblasts |
| Elastic | Ligaments; large arteries; lung tissue; skin | Enables tissues to expand and return to normal size | Fibroblasts |
| Adipose | Fat beneath skin and on surface of heart and other internal organs | Provides insulation, food storage, and support of breasts and kidneys | Fat cells |
| Cartilage | Spinal disks; knees and other joints; ear; nose; tracheal rings | Provides flexible support; functions in shock absorption and reduction of friction on load-bearing surfaces | Chondrocytes |
| Bone | Most of skeleton | Protects internal organs; provides rigid support for muscle attachment | Osteocytes |
| Vascular | | | |
| Blood | Circulatory system | Transports $O_2$, $CO_2$, nutrients, and wastes; functions as highway of immune system, and stabilizer of body temperature | Erythrocytes |
| Cells of immune system | Circulatory system | Defense against bacteria, viruses, and abnormal cells | Lymphocytes; macrophages; mast cells |
| **MUSCLE** | | | |
| Smooth | Walls of blood vessels, stomach, and intestines | Powers rhythmic contractions not under conscious control but commanded by central nervous system | Smooth muscle cells |
| Skeletal | Voluntary muscles of body | Powers walking, lifting, talking, and all other voluntary movement | Skeletal muscle cells |
| Cardiac | Walls of heart | Highly interconnected cells; promotes rapid spread of signal initiating contraction | Heart muscle cells |
| **NERVE** | | | |
| Sensory cells | Eyes; ears; surface of skin | Receives information about body's condition and about exterior world | Rods and cones; muscle stretch receptors |
| Signal-transmitting cells | Nerves | Transmits signals | Neurons |
| Information-processing cells | Brain and spinal cord | Integrates information | Interneurons |
| Glia | Brain and spinal cord; peripheral nerves | Support | Schwann cells; oligodendrocytes |

structure that carries out a specific function, is composed of these tissues, organized and assembled in various ways. Different combinations of tissues are found in different organs.

The many organs that, working together, carry out the principal activities of the body are referred to as **organ systems.** Much of the biology of vertebrates is concerned with the functioning of organ systems. The 11 major organ systems of the human body (see Table 32-1) all work together. The remainder of the text is devoted to describing them in more detail.

# HOMEOSTASIS

As the animal body has evolved, specialization has increased, and nowhere is this more evident than in the many cell types that make up the body of a vertebrate. Each is a sophisticated machine, finely tuned to carry out a precise role within the body. Such specialization of cell function is possible only when extracellular conditions are kept within narrow limits. Temperature, pH, the concentration of glucose and oxygen—all these factors must stay very constant for cells to function efficiently and relate to one another properly.

## Homeostasis Is Basic to Body Function
We call the maintenance of constant extracellular conditions within the body **homeostasis.** Your body employs many mechanisms to ensure that its internal conditions do not vary outside of a narrow range. There are many familiar examples.

***Regulating Levels of Glucose in the Blood.*** Stability of the internal environment is essential for homeostasis. However, the environment within our bodies is continuously being exposed to conditions that, if not regulated, would act to change it. When you eat a large dinner, for example, its digestion introduces a large amount of glucose to your body in a short time. What prevents the level of glucose in your blood from quickly rising to a high level? When glucose levels within the blood exceed normal values, the excess glucose is absorbed by liver cells, which convert it to a storage form, glycogen. When glucose levels in the blood fall below the normal range, the liver breaks down its glycogen to add more glucose to the bloodstream. Thus glucose levels in the fluid surrounding the body's cells change very little over the course of a day, even though the body's intake of glucose many be concentrated within a short period. This is the essence of glucose homeostasis.

***Regulating Body Temperature.*** When the temperature of your blood exceeds 98.6° F (37° C), neurons in the brain detect the temperature change. These neurons provide input to the hypothalamus, which responds by triggering mechanisms for dissipating heat (sweating, dilation of blood vessels, and so on). There are also two sorts of temperature-sensitive nerve endings in your skin, one sensitive to low temperatures, the other to high temperatures. In each case the change in temperature affects ion channels in the membrane of the nerve ending, initiating a nerve impulse in a way we discuss in the next chapter. These signals are also relayed to the control center of a specific part of the brain called the hypothalamus.

## Homeostasis is Maintained by Feedback Loops
To maintain homeostasis, the body must constantly monitor itself and act to correct any deviation. We call such a process of surveillance and response a **feedback loop** (Figure 32-13). Homeostasis, in which changes in the body's condition are detected and reversed, involves **negative feedback** mechanisms, in which a disturbance triggers processes that reduce the disturbance. Instances in which a disturbance is accentuated are called **positive feedback** and are important components of the nerve impulse, labor and delivery, and resistance to disease.

The monitoring of body functions is centered in the brain, much of it in a small, marble-sized region called the **hypothalamus.** Using sensors scattered throughout your body, it monitors temperature, pH, and many other factors, and when a disturbance is detected, it issues orders to combat it. Sometimes these orders alter body functions, such as how fast you breathe; at other times they call for the production of particular hormones.

***Regulating Blood pH and Salt Balance.*** The regulation of your blood's pH and salt balance provides an example of how this feedback loop works. The kidneys play a very important role in regulating the composition of vertebrate blood and thus the internal chemical environment of the body. By selectively removing substances from the blood, the kidneys maintain rigorous control over the concentrations of ions, such as $H^+$, $Na^+$, $K^+$, $Cl^-$, Mg, Ca, and $HCO_3^-$, all of which are maintained within narrow boundaries. This serves to maintain the blood's pH at a constant value, as well as to maintain the proper ion balances for nerve conduction (blocked by even a small excess of Mg) and muscle contraction (a small increase in $K^+$ levels in the blood causes the heart to stop).

When you drink a lot of pure water, you dilute the salts in your blood. That is why athletes drink Gatorade, a properly balanced mixture of water and salts similar to that in plasma. To keep the concentration of ions in your blood within proper limits after your body absorbs a lot of water, the hypothalamus employs simple feedback loops. When uptake of water is excessive, receptors in the hypothalamus detect the resulting decrease in blood salt concentration (Figure 32-14); the hypothalamus responds by lessening its output of a hormone called **antidiuretic hormone (ADH)**. A decrease in ADH inhibits the reabsorption of water by the kidneys and therefore reduces water volume in the blood, compensating for the excessive water intake that triggered the response.

Other feedback loops regulate the levels of specific ions. Thus, when levels of sodium ion in the blood rise, as happens when you eat highly salted food, cells in the kidneys detect the increase and cause a decrease in the production of the hormone **aldosterone**. Aldosterone stimulates the kidneys to absorb sodium; a decrease in aldosterone lessens the reabsorption of sodium by the kidneys and thus lowers the concentration of sodium in the blood, compensating for the excessive sodium intake that triggered the response.

*Regulating Blood Pressure.* The brain maintains a constant blood pressure by another feedback loop that adjusts the rate at which your heart beats, the force of contraction of the heart, and the diameter of some blood vessels to compensate for changes in blood pressure. Blood pressure is measured at special sites in the major arteries where the wall of the artery is thin and contains a highly branched network of nerve endings.

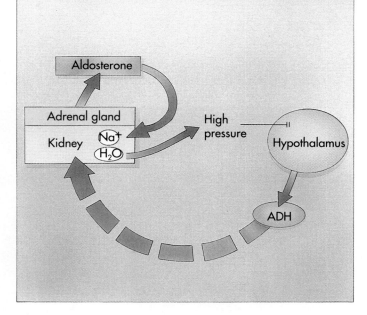

**Figure 32-14**

**Control of water and salt balance within the kidney.** The hypothalamus produces antidiuretic hormone (ADH), which renders the collecting ducts of the kidneys freely permeable to urea and so maximizes water retention. If too much water retention leads to high blood pressure, pressure-sensitive receptors in the hypothalamus detect this and cause the production of ADH to be shut down. If the level of sodium in the blood falls, the adrenal gland initiates production of the hormone aldosterone, which stimulates salt reabsorption by the renal tubules of the kidney.

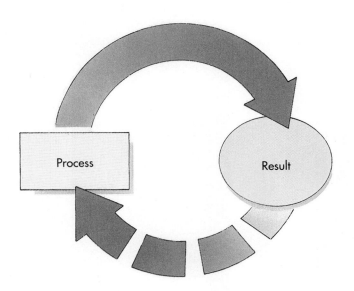

**Figure 32-13**

**A feedback loop.** A body process produces a result such as a change in temperature or an increase in levels of a hormone. In a feedback loop, the magnitude of this result influences the rate of the process. An excessive increase in temperature, for example, may act to slow or halt the body process producing heat.

When blood pressure increases, the wall of the artery is stretched, causing the nerve endings in the wall to send signals to the brainstem. The brain responds by issuing signals that slow down the heart's pacemaker, lowering the rate at which the heart beats and dilating peripheral blood vessels (vasodilation).

When you exercise, your heart becomes stretched by the extra blood flowing through it, and nerve endings within its wall fire signals to the brainstem, which responds by stimulating the heart's pacemaker and thus increasing your heartbeat. Similarly, your heartbeat is increased by the brainstem if levels of carbon dioxide in the blood rise, indicating that the body's cells probably need more oxygen.

## HOMEOSTASIS IS CENTRAL TO THE OPERATION OF THE VERTEBRATE BODY

In the following 11 chapters we will consider the functioning of the vertebrate body's organ systems in detail. These are shown in overview in Figure 32-15. In every case we encounter homeostatic mechanisms that act to coordinate these functions within the narrow limits dictated by the complexity of these highly specialized animals.

# VERTEBRATE BODY ORGAN SYSTEMS

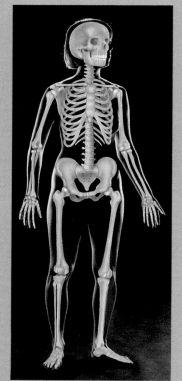

Skeletal system

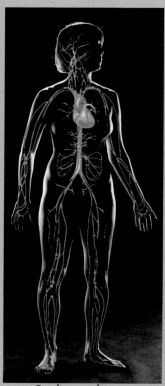

Cardiovascular system

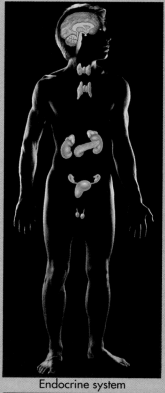

Endocrine system

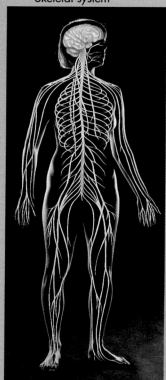

Nervous system

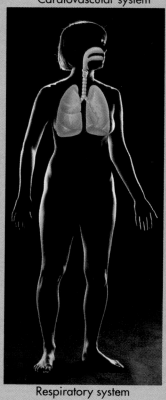

Respiratory system

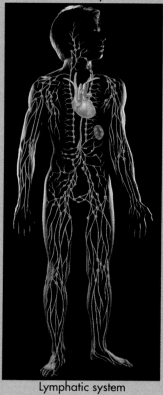

Lymphatic system

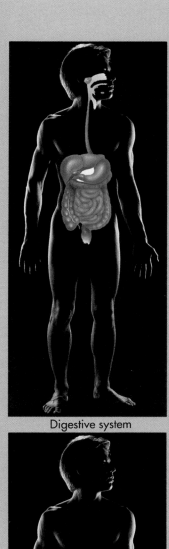

Digestive system

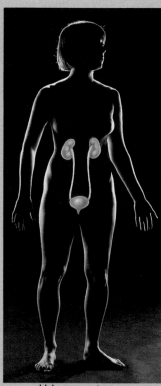

Urinary system

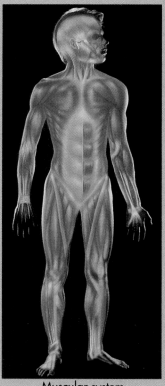

Muscular system

Reproductive system—male

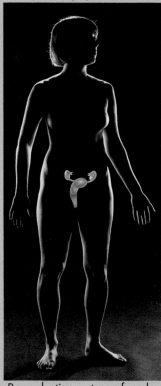

Reproductive system—female

# SUMMARY

1. The four basic types of tissue are epithelium, connective tissue, muscle, and nerve.

2. Epithelium covers the surfaces of the body. The lining of the major body cavities is composed of epithelium, as is the exterior skin.

3. Connective tissue supports the body mechanically and defensively. The structural connective tissues are fibroblasts (which secrete the structural protein collagen), cartilage, and bone. The defensive connective tissues are macrophages and other cells of the immune system that attack and destroy foreign bacteria and virus-infected cells. Other connective tissues are pigment-containing cells, fat cells, and blood cells.

4. The contraction of cells within muscle tissue provides the force for mechanical movement of the body.

5. Nerve cells provide the body with a means of rapid communication. There are two kinds of nerve cells—neurons and supporting glial cells. Neurons are specialized for the conduction of electrical impulses called action potentials.

6. The organs of the body employ feedback loops to maintain constant extracellular conditions around the many cells of the body, a condition called homeostasis.

## REVIEW

1. In human beings the coelom is divided into two parts. What are they?

2. Connective tissue cells are derived from which embryonic tissue?

3. _____ is the most abundant protein in vertebrate bodies.

4. Microfilaments made of actin and myosin are much more abundant in _____ cells than in any other eukaryotic cell type.

5. Neurons have two kinds of projections from the cell body, the _____ and the _____.

## SELF-QUIZ

1. Which of the following is *not* one of the four basic types of tissues of the adult vertebrate body?
   (a) Muscle      (c) Connective      (e) Epithelium
   (b) Nerve      (d) Mesoderm

2. Which of the following is *not* a function of the epithelial layers?
   (a) To secrete material
   (b) To store and distribute substances throughout the body
   (c) To protect the tissues beneath from dehydration and mechanical damage
   (d) To provide a selectively permeable barrier
   (e) To provide sensory surfaces

3. Which of the following is *not* a cell type of the connective tissue?
   - (a) Lymphocytes
   - (b) Fat cells
   - (c) Columnar cells
   - (d) Erythrocytes
   - (e) Macrophages

4. Cartilage functions in many ways. Which of the following is one of them?
   - (a) Protects the body surface
   - (b) Forms the ends of bones in knees and other joints
   - (c) Aids in wound healing
   - (d) Enables limb movement
   - (e) Produces antibodies to bacteria

5. Which of the following is *not* a principal organ system of the human body?
   - (a) Circulatory system
   - (b) Urinary system
   - (c) Lymphatic system
   - (d) Secretory system
   - (e) Muscular system

# THOUGHT QUESTIONS

1. Land was successfully invaded four times—by plants, fungi, arthropods, and vertebrates. Because bodies are far less buoyant in air than in water, each of these four groups evolved a characteristic hard substance to lend mechanical support. Describe and contrast these four substances, discussing their advantages and disadvantages. Can you imagine any other substance (such as plastic) that would have been superior to any of these?

2. Your body contains 206 bones. As you grow, all these bones must increase in size and maintain proper proportions with one another. How is the growth of these bones coordinated?

3. What are the basic elements of feedback control that must be present for an organism to regulate its internal environment? Where would you expect each to be located in the case of the cardiovascular system?

# FOR FURTHER READING

CAPLAN, A.: "Cartilage," *Scientific American*, October 1984, pages 84-97. An interesting account of the many roles played by cartilage in the vertebrate body.

CURREY, J.: *The Mechanical Adaptations of Bones,* Princeton University Press, Princeton, N.J., 1984. A functional analysis of why different bones are structured the way they are, with an unusually well-integrated evolutionary perspective.

HOUK, J.C.: "Control Strategies in Physiological Systems," *FASEB Journal*, February 1988, page 97. General discussion of control systems written for a general audience.

NATIONAL GEOGRAPHIC SOCIETY: *The Incredible Machine*, National Geographic Society, Washington, D.C., 1986. A series of outstanding articles on the human body, focusing on its major organ systems. Beautifully illustrated and fun to read.

# How Animals Transmit Information

How a neuron transmits information may seem an abstract concept, but it has very concrete consequences. These spring moose calves are learning a blizzard of neuronal impulses directed at survival from their mother.

# HOW ANIMALS TRANSMIT INFORMATION

## Overview

All animals except sponges possess special cells called neurons, which are capable of transmitting information in the form of a sequence of stereotyped events called action potentials, or nerve impulses. All nerve impulses are the same. It is in the frequency (pattern) and the point of origin that carries information from one part of the body to another.

## For Review

*Here are some important terms and concepts that you will encounter in this chapter. If you are not familiar with them, you should review them before proceeding.*

**The sodium-potassium pump** (Chapter 5)

**Gated ion channels** (Chapter 5)

**Neuron** (Chapter 32)

**Depolarization** (Chapter 32)

There are several ways in which one cell of your body can communicate with another. One simple way is by direct contact, with an open cytoplasmic connection between two cells that permits the passage of ions and small molecules. This method of communication requires cell contact. It can be understood by imagining that instructions from a manager to his or her employees were limited to those conveyed through direct contact. In the body, communication of this kind is provided by gap junctions, discussed in Chapter 5. It provides a ready means of communication between adjacent cells but is not able to provide rapid and efficient communication between distant tissues.

It would be better, in terms of distant communication, if a manager sent a letter instructing various employees what to do. The body organizes various tissues in just this way. In some cases chemical instructions are sent to these tissues. The instructions are in the form of hormones, small chemical molecules that act as messengers circulating within the body's bloodstream. However, this kind of command system can be too slow, just as the mail may sometimes be when you are waiting for an important letter.

If the message to be delivered to the leg muscles of your body is, "Contract quickly, we are being pursued by a leopard," a quicker means of communication than hormones is desirable. In a city, a person in an emergency does not mail a letter, but rather uses the telephone to shout for help, dialing 911 and requesting assistance. That, in effect, is just what the vertebrate body does. All complex animals possess specialized cells called **neuron** cells (Figure 33-1), which, as you will recall from Chapter 32, are cells specialized for signal transmission. Neurons maintain an electrical charge across their membranes because the membrane is permeable to $K^+$, and there is a concentration difference for $K^+$ across the cell membrane. The sign of this charge difference is such that the interior of the cell is negative with respect to the exterior. This is called **polarization.** Loss of this charge difference is called **depolarization.** An electrical event called an **action potential,** consisting of a depolarization followed by a return to the normal resting voltage difference, can pass down the length of a neuron. The process is similar to the rapid burning of a fuse.

In this chapter we focus on the neuron, the chief functional element of the nervous system. In Chapter 34 we consider the functional organization of the brain and of the nervous system as a whole. We then examine the senses and the ways in which they collect information and transmit it along nerves to the central nervous system. In Chapter 35 we consider the structure and function of muscles and how the brain directs muscle movement, and in Chapter 36 how the brain uses hormones to effect long-term changes in the body's metabolism.

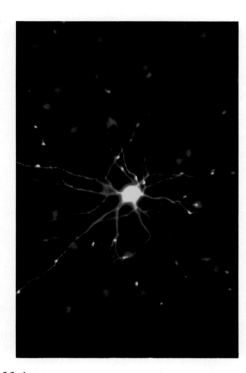

**Figure 33-1**

**A living neuron.** This neuron in the retina of the eye has been injected with a fluorescent dye. Its long, unbranched dendrites are readily apparent.

# THE NEURON

The body contains many different sorts of neurons, some of them tiny with a few projections, others bushy with projections, and still others with extensions several meters long (Figure 33-2). Despite individual neurons often differing in appearance, all neurons have the same functional architecture. Extending from the body of all but the simplest nerve cells are one or more cytoplasmic extensions called **dendrites** (Figure 33-3). Most nerve cells possess a profusion of dendrites, many of which are branched. The dendrites enable the body of the cell to receive inputs from many different sources simultaneously. The surface of the cell body acts to integrate the information arriving from the many different dendrites. If the resulting membrane excitation is large enough, it travels outward from the cell body as an electrical **nerve impulse,** along an axon (see Figure 33-3). Most nerve cells possess a single axon, which may be quite long. The axons controlling muscles in your legs are more than a meter long, and even longer ones occur in larger mammals. In a giraffe (Figure 33-4) a single axon travels from the hoof all the way up to the pelvis.

Neurons are not normally in direct contact with one another. At the far end of most axons, the axonal membrane contains packets of specialized chemicals

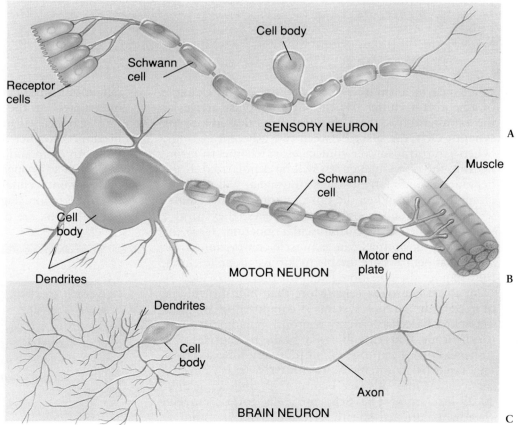

**Figure 33-2**

**Types of vertebrate neurons.**
**A** Sensory neurons (which carry signals from sense organs to the brain) typically have dendrites only in specific receptor cells.
**B** The axons of many motor neurons (which carry commands from the brain to muscles and glands) are encased at intervals by Schwann cells.
**C** Neurons within the brain often possess extensive, highly branched dendrites.

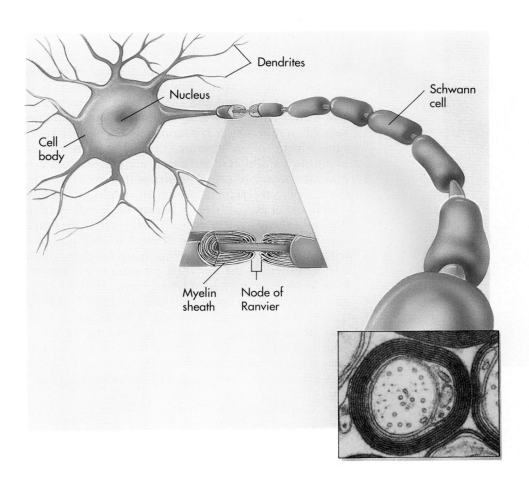

**Figure 33-3**

**Structure of a typical neuron.** Many dendrites carry information to the cell body, which sends information outward along a single long axon. Long axons are encased at intervals by a myelin sheath whose many layers, visible in the electron micrograph, insulate the axon. The electrical signal jumps from uninsulated node to uninsulated node and so moves very rapidly down the axon.

called **neurotransmitters;** acetylcholine is an example of a neurotransmitter. When a nerve impulse arrives at the axon terminus, these chemicals are released into the **synaptic cleft,** the space between two adjacent neurons. When neurotransmitter molecules reach the receiving neuron, the receiving neuron will be depolarized. If the depolarization is large enough, the signal will continue along the nerve. When the tips of axons contact (innervate) a muscle fiber, on the other hand, they form a structure called the **neuromuscular junction.** Here the release of transmitters causes muscle contraction.

Most neurons are unable to survive alone for long; they require the nutritional support that is provided by companion **neuroglia** cells. More than half the volume of vertebrate nervous systems is composed of supporting neuroglia cells. In many neurons, including the motor neurons that extend from brain to muscles, the transmission of impulses along the very long axons is facilitated by neuroglial **Schwann cells,** which envelop the axon at intervals (see Figure 33-3) and act as electrical insulators. The Schwann cells form a **myelin sheath,** a flattened sheath of fatty material found in many, but not all, vertebrate neurons. Such a sheath is interrupted at the spaces between Schwann cells to form gaps called **nodes of Ranvier,** where the axon is in direct contact with the surrounding intercellular fluid. An axon and its associated Schwann cells, or cells with similar properties, form a **myelinated fiber.** Bundles of nonmyelinated and myelinated neurons, which in cross section (Figure 33-5) look somewhat like telephone cables, are called **nerves.**

*Nerve cells specialized for electrical signal transmission are called neurons. Typically a signal is received by a dendrite branch and passes to the cell body, where its influence is merged with that of other incoming signals; the resulting signal is passed outward along a single long axon. When the signal reaches the end of the axon, it is transmitted chemically to another neuron or a target cell.*

**Figure 33-4**

**Some nerve cells are quite large.** From each toe of this giraffe, a single nerve axon extends all the way up to the back—a distance of 3 meters.

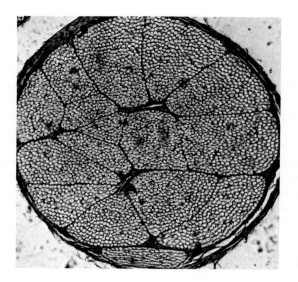

**Figure 33-5**

**A nerve is a bundle of axons bound together by connective tissue.** In this cross section of a nerve, many myelinated neuron fibers are visible, each looking in cross section something like a severed hose.

## THE NERVE IMPULSE

How does the nervous system work? The answer lies in the fact that the nerve membrane contains gated ion channels. They are called gated because they can open or close. Some channels only allow $Na^+$ ions to pass through them, whereas others are selective for $K^+$. Whenever ion channels open or close, there is a change in the movement of ions across the nerve membrane. This current flow then affects the charge separation across the cell, that is, its voltage. Information passes along the nerves of the vertebrate body as electrical currents and associated voltage changes, and the key to understanding how the nervous system works is understanding these nerve impulses.

### The Resting Potential: Result of a $K^+$ Concentration Gradient

Voltage differences are produced by the separation of positive and negative charges; nerve cells generate voltage differences by pumping positive $Na^+$ out of the cell and $K^+$ into it through the active transport channels called Na-K pumps, described in Chapter 5. This generates concentration differences with $Na^+$ high outside the cell and low inside, whereas $K^+$ is high internally and low externally. Subsequent changes in the permeability of the membrane to $Na^+$ and $K^+$ result in ion movements down their concentration gradients, leading to voltage shifts.

*How is the activity of the Na-K pump related to nerve impulses?* The Na-K pump expends ATP to pump $Na^+$ out of the cell and take up $K^+$. $Na^+$ cannot readily move back into the cell once pumped out, despite the low internal $Na^+$ concentration, because there are very few open inward channels through the neuron membrane across which the $Na^+$ can pass (Figure 33-6). Although some $Na^+$ does leak back in across the resting membrane, the cell constantly pumps excess $Na^+$ out. Meanwhile the tendency of $K^+$ to leave the cell results in a charge separation, with the interior polarized negatively with respect to the exterior. To summarize, the Na-K pump maintains normal concentration gradients but is not directly involved in the production of nerve impulses.

*The impermeability of the membrane to negatively charged ions is also important.* $K^+$ ions are pumped inward at the same time that $Na^+$ ions are pumped out, and the $K^+$ ions diffuse back out of the cell through open $K^+$ ion channels. If no other influence were at work, their concentration would soon be the same on both sides of the membrane. However, the membrane is impermeable to proteins and many other negatively charged ions, so that as $K^+$ ions leave, an excess negative charge builds up inside. Because opposite charges attract, the inward attraction of $K^+$ caused by the buildup of inside negative charge tends to counteract the outward push due to the $K^+$ concentration gradient. A steady-state equilibrium is reached at which there is no net movement of $K^+$ across the membrane. At this point, there is a slight excess of negative charge inside the cell.

Both the pumping out of $Na^+$ and the equilibrium of $K^+$ contribute to making the inside of the cell membrane more negatively charged than the outside; in other words, the neuron is polarized. This electrical potential difference across the membrane, which is called the **resting potential,** is the starting point for the transmission of signals by nerves.

*Because of the activity of the $Na^+$-$K^+$ transmembrane pumps and the impermeability of the cell to $Na^+$, concentration gradients for $Na^+$ and $K^+$ are maintained across the cell membrane. As a result, the surface of the neuron carries a positive charge relative to its interior. It is said to be polarized, and the electrical potential difference across the membrane is called the resting potential.*

## The Action Potential: Initiating a Nerve Impulse

The transmission of a signal by a nerve occurs in four phases, which we consider in order: (1) initiation of an action potential by a sufficiently large (threshold) stim-

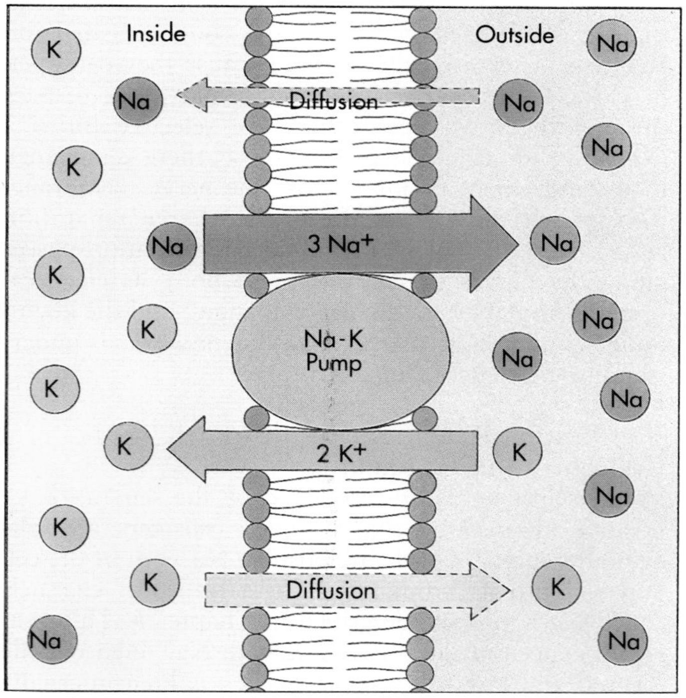

*Figure 33-6*

**The sodium-potassium pump.** The Na-K pump transports $Na^+$ ions to the outside of the cell, creating a high exterior $Na^+$ concentration.

ulus, (2) its transmission along a nerve fiber, (3) its transfer to a target muscle or nerve, and (4) its effect on the target tissue.

A nerve can be stimulated when an electrical current is applied at some site on the neuron membrane. Such events cause structural changes in the ion channels in the membrane, causing them to begin to admit Na$^+$ into the cell; these channels are called **voltage-gated Na$^+$ ion channels.** The result is a sudden flood of Na$^+$ ions down their concentration gradient into the interior of the cell. That is, Na$^+$ ions move from the outside, where the Na$^+$ concentration is high, to the cytoplasm, where the Na$^+$ concentration is low. This movement of positive ions into the cell wipes out the local voltage difference and is therefore called depolarization (Figure 33-7). As Na$^+$ ions pour across the

membrane in response to the large concentration difference, the interior of the cell's membrane in that area actually develops a positive charge relative to the outside (Figure 33-8). The momentary depolarization and return to the resting potential are called the **action potential.** Following the depolarization phase of the action potential, the closing of Na$^+$ channels combines with the leakage of K$^+$ out of the cell to return the voltage difference to the resting potential.

The action potential is an all-or-nothing occurrence. The amount of stimulation required to initiate an action potential is called the **threshold value.** At threshold a regenerative (positive feedback) cycle of Na$^+$ channel opens and depolarization occurs.

## Repolarization: Terminating a Nerve Impulse

Depolarization associated with the action potential causes the voltage-gated Na$^+$ channels in the membrane to close slowly (Figure 33-9) by a process termed Na$^+$ channel inactivation so that no more Na$^+$ ions enter the cell. Movement of K$^+$ ions outward and down their concentration gradient then reestablishes the voltage difference (resting potential) that existed before the stimulus occurred. The whole process of stimulation and recovery is very fast—it takes only about 5 milliseconds. Fully 100 such cycles could oc-

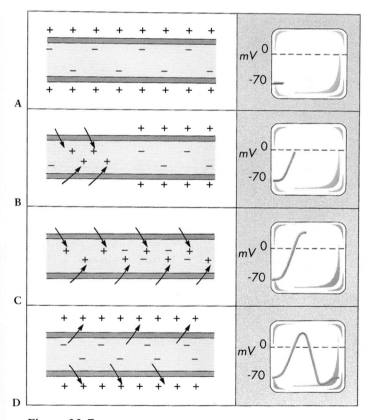

**Figure 33-7**

**The stages of membrane depolarization.**
**A** At rest, there is an excess of exterior positive charge and a net interior negative charge of about −70mV.
**B** When stimulated, Na$^+$ ions enter, abolishing the voltage difference.
**C** Enough positive ions enter to establish a net interior positive charge.
**D** The exit of K$^+$ ions then returns the interior to a net negative charge.

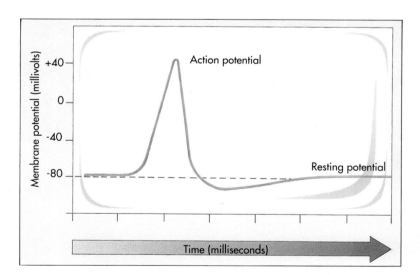

**Figure 33-8**

**The action potential.** The changes in electrical potential associated with depolarization of a nerve membrane can be measured with minute, needle-sharp electrodes inserted into individual neurons. The change is always the same for any given neuron; each impulse alters the membrane potential by the same number of millivolts and takes the same length of time to recover.

cur, one after another, in the time it takes you to say the word "nerve." After many action potentials there may be an accumulation of $Na^+$ inside the cell, but the Na-K pump protects the concentration gradients by using ATP to expel $Na^+$ and take up $K^+$.

## Why Transmission Occurs

During the few milliseconds when the site of stimulation is depolarized by the movement of $Na^+$ inward and before depolarization closes the voltage-gated $Na^+$ channels, the site of stimulation has less charge than the membrane surface surrounding it. This potential difference establishes a small, very localized current in the immediate vicinity, which influences nearby closed voltage-gated $Na^+$ channels to open, permitting $Na^+$ to enter the cell—and depolarizing these sites as well (Figure 33-10).

The action potential thus spreads, with depolarization at one site producing a local current, which induces nearby voltage-gated $Na^+$ channels to open and therefore depolarize the nearby site. In this way the initial depolarization passes outward over the membrane, spreading out in all directions from the site of stimulation. Like a burning fuse, the action potential is usually initiated at one end and travels in one direction, but it would travel out from both directions if it were "lit" in the middle.

For any one neuron the action potential is always the same. Any stimulus that opens enough $Na^+$ channels will propagate an impulse outward as a wave of depolarization with a constant amplitude (the height or strength of the wave) corresponding to wide-open $Na^+$ channels. The stimulus is said to have "fired" the neuron. Every nerve impulse traversing that neuron has the

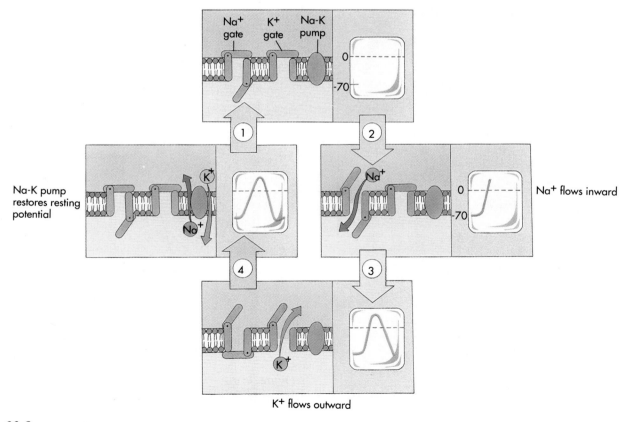

*Figure 33-9*

**The molecular basis of the action potential.**

same amplitude; signals differ from one another only in the frequency of these impulses, *not* in the amplitude of the depolarization.

The time required for voltage-gated Na$^+$ channels to recover from their inactive state, during which another action potential cannot occur, is called the **refractory period**. The nerve impulse moves in one direction only, away from the stimulus site, because during the refractory period the membrane is not sensitive to depolarization. By the time the recovery process has been completed, the action potential has moved too far away for the site at which it was initiated to affect it.

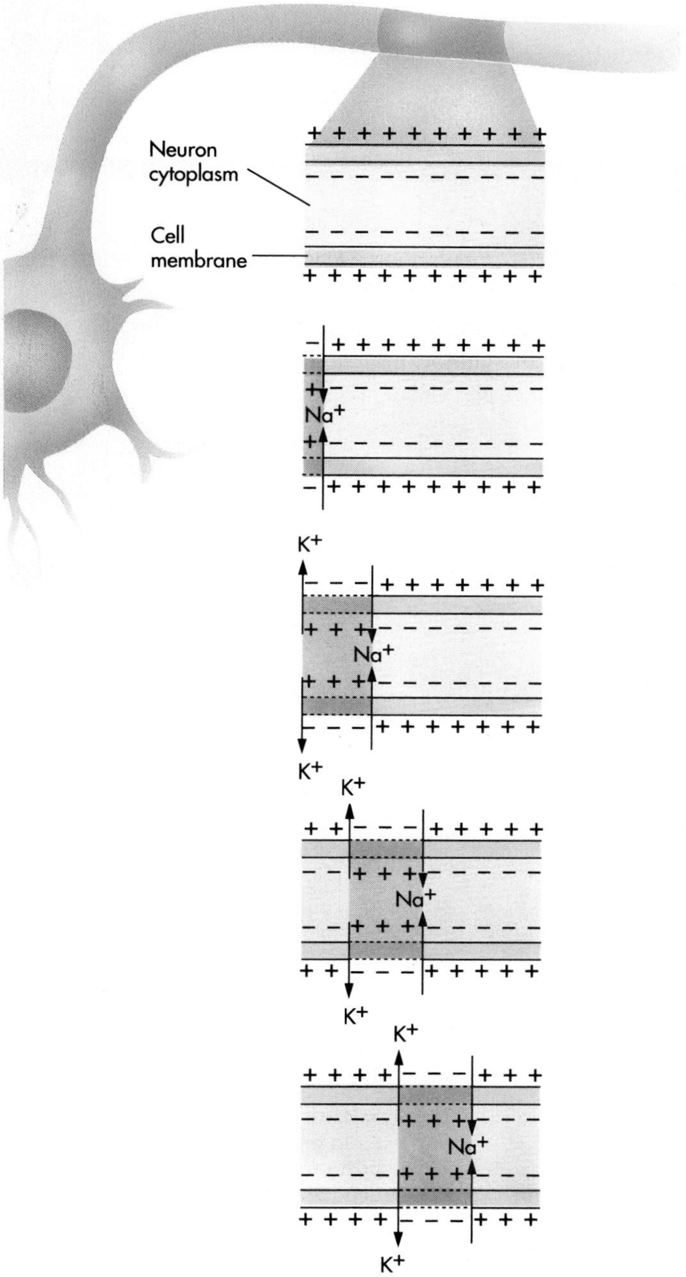

*An action potential arises because of a depolarization of the neuron membrane. This transient voltage shift causes nearby transmembrane ion channel proteins to respond, spreading the action potential.*

## Saltatory Conduction

Many vertebrate neurons possess axons sheathed at intervals by neuroglial Schwann cells. These cells envelop the axon (Figure 33-11), wrapping their cell membrane around it many times to produce a series of layers. This lipid-rich envelope of membrane layers is the myelin sheath. The myelin sheath prevents the transport of ions across the neuron membrane beneath it and thus acts as an electrical insulator, creating a region of high electrical resistance on the axon.

Schwann cells are spaced along such an axon one after the other, with nodes of Ranvier separating each Schwann cell from the next (Figure 33-12). These nodes are critical to the propagation of the nerve impulse in these cells. Within the small gap represented by each node, the surface of the axon is exposed to the fluid surrounding the nerve. Voltage-gated ion channels are concentrated in these zones. The direct fluid contact permits ion transport to occur through these channels and an action potential to be generated. The action potential does not continuously spread down the axon, because current cannot flow across the insulating Schwann cells. Instead, the action potential jumps as an electrical current from one node to the next. When the current reaches a node, it acts as a current that opens voltage-gated Na$^+$ channels; in so doing, it generates a potential difference large enough to create a current that reaches the next node. The arrival of the current at that node opens its voltage-gated Na$^+$ channels, creating another current that passes on to the next node, and so on. This very fast form of nerve impulse conduction is known as **saltatory conduction** (from the Latin *saltare*, "to jump"). An impulse conducted in this fashion moves very fast, up to 120 meters per second

*Figure 33-10*

**Transmission of a nerve impulse.**

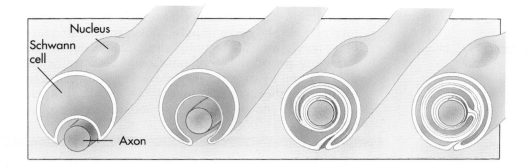

**Figure 33-11**

**How a myelin sheath is made.** Development of the myelin sheath that surrounds many neuron axons involves the envelopment of the axon by a Schwann cell. It is the progressive growth of the Schwann cell membrane around the axon that contributes the many membrane layers characteristic of myelin sheaths. These layers act as an excellent electrical insulator.

for large-diameter neurons. Saltatory conduction is also very cheap metabolically for the cell because the movements of $Na^+$ and $K^+$ are reduced. The activity of the Na-K pump can be much lower when only the nodes are undergoing depolarization than when the entire nerve surface is. Incidentally, the disease multiple sclerosis results in the destruction of large patches of the myelin sheath; the resulting slower transmission of signals in the nervous system results in the characteristic symptoms of the disease.

*Not all nerve impulses propagate as a wave of depolarization spreading along the neuronal membrane. On some vertebrate nerves, impulses travel very much faster by jumping along the membrane in saltatory conduction, leaping from node to node over insulated portions.*

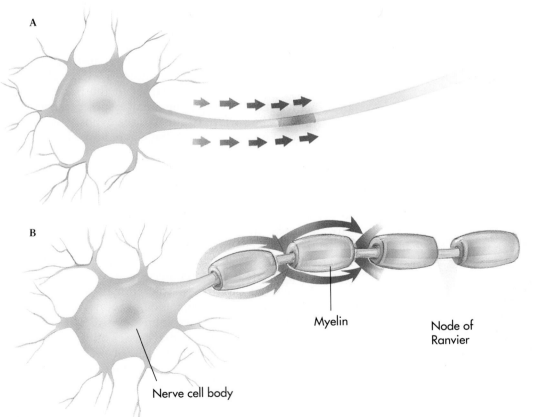

Myelin

Node of Ranvier

Nerve cell body

**Figure 33-12**

**Saltatory conduction.**
A In an unmyelinated fiber, each portion of the membrane becomes depolarized in turn, like a row of falling dominos.
B The signal moves faster along a myelinated fiber because the wave of depolarization jumps electrically from node to node without ever depolarizing the insulated membrane segments between nodes.

**action potential**  A single nerve impulse; a transient all-or-none reversal of the electric potential across a neuron membrane; because it can activate nearby voltage-sensitive channels, an action potential propagates along a nerve cell.

**axon**  A single long process extending out from a neuron that conducts impulses away from the neuron cell body.

**chemically gated ion channel**  A transmembrane pathway for a particular ion that is opened or closed by a chemical such as a neurotransmitter.

**dendrite**  A process extending from a neuron, typically branched, that conducts impulses inward toward the cell body; neurons may have many dendrites.

**depolarization**  The movement of ions across a cell membrane that wipes out locally an electrical potential difference.

**excitatory synapse**  A synapse in which the receptor protein is a chemically gated sodium channel; binding of a neurotransmitter opens the channel and initiates an excitatory electrical potential that increases the ease with which the membrane can be depolarized.

**inhibitory synapse**  A synapse in which the receptor protein is a chemically gated potassium or chloride channel; binding of a neurotransmitter opens the channel and produces an inhibitory electrical potential that reduces the ability of the membrane to depolarize.

**nerve**  A bundle of axons with accompanying supportive cells, held together by connective tissue.

**neuron**  A nerve cell specialized for signal transmission.

**neurotransmitter**  A chemical released at an axon tip that travels across the synapse and binds a specific receptor protein in the membrane on the far side.

**potential difference**  A difference in electrical charge on two sides of a membrane caused by an unequal distribution of ions.

**refractory period**  The recovery period after membrane depolarization during which the membrane is unable to respond to additional stimulation.

**resting potential**  The charge difference that exists across a neuron's membrane at rest (about 70 millivolts).

**synapse**  A junction between a neuron and another neuron or a muscle cell; the two cells do not touch, neurotransmitters crossing the narrow space between them.

**threshold potential**  The minimum change in membrane potential necessary to produce an action potential.

**voltage-gated channel**  A transmembrane pathway for an ion that is opened or closed by a change in the voltage, or charge difference, across the cell membrane.

## TRANSFERRING INFORMATION FROM NERVE TO TARGET TISSUE

An action potential passing down an axon eventually reaches the end of the axon, which is often branched. These terminal branches may be associated either with dendrites of another neuron in the chain or with sites on muscle cells or secretory gland cells. These associations of nerves with other cells are the synapses mentioned earlier. When the tip of a vertebrate axon is examined carefully, it becomes apparent that it does not actually make contact with the target cell it approaches. There is a narrow intercellular gap, 10 to 20 nanometers across, separating the axon tip and the target cell (Figure 33-13), called a synaptic cleft. Synaptic clefts are characteristic of all vertebrate nerve junctions, except for some of the synapses in the brain.

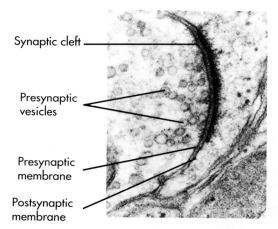

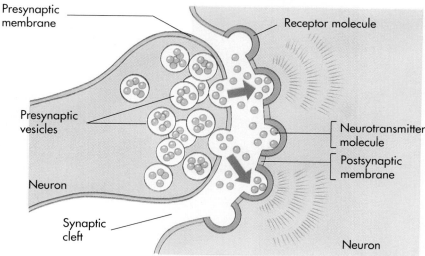

*Figure 33-13*

**A synaptic cleft between two neurons.**

When a nerve signal arrives at a synaptic cleft, it passes across the gap chemically. The membrane on the axonal side of the synaptic cleft is called the **presynaptic membrane.** When a wave of depolarization reaches the presynaptic membrane, it stimulates the release of neurotransmitter chemicals from vesicles at the tip into the cleft. These chemicals rapidly pass to the other side of the gap. Once there, they combine with receptor molecules in the membrane of the target cell, which is called the **postsynaptic membrane.** Binding of a neurotransmitter to a receptor causes ion channels to open. These channels are referred to as chemically gated.

*A synapse is a junction between an axon tip and another cell, almost always including a narrow gap that the impulse cannot bridge. Passage of the impulse across the gap is by a chemical signal from the axon.*

The great advantage of a chemical junction such as this, compared with a direct electrical contact, is that the specific chemical transmitter can be different in different junctions, and the nature of the ion channels activated by a particular transmitter can also be differ-

ent. Over 60 different chemicals have been identified that act as specific neurotransmitters or can modify the activity of neurotransmitters. The result is a great diversity in the response of a postsynaptic cell. The events that occur within the synaptic cleft when a nerve signal arrives depend very much on the identity of the particular neurotransmitter chemical that is released into the cleft. To understand what happens, we first look at the junction between a nerve and a muscle cell, where the situation is a simple one, and then consider nerve-nerve junctions, where the situation is more complex.

## Neuromuscular Junctions

In synapses with muscle cells, called neuromuscular junctions (Figure 33-14), the neurotransmitter is acetylcholine. Passing across the gap, the acetylcholine molecules bind to receptors in the postsynaptic muscle membrane, opening chemically gated $Na^+$ channels different from voltage-gated channels producing action potentials. During the millisecond that these $Na^+$ channels are open, some $10^4$ ions flow inward (Figure 33-15). This ion flow depolarizes the adjacent postsynaptic muscle cell membrane, which contains voltage-gated $Na^+$ channels. In this way acetylcholine initiates a wave of depolarization that passes down the muscle cell. This wave of depolarization permits the entry of calcium, into the muscle, which in turn triggers muscle contraction.

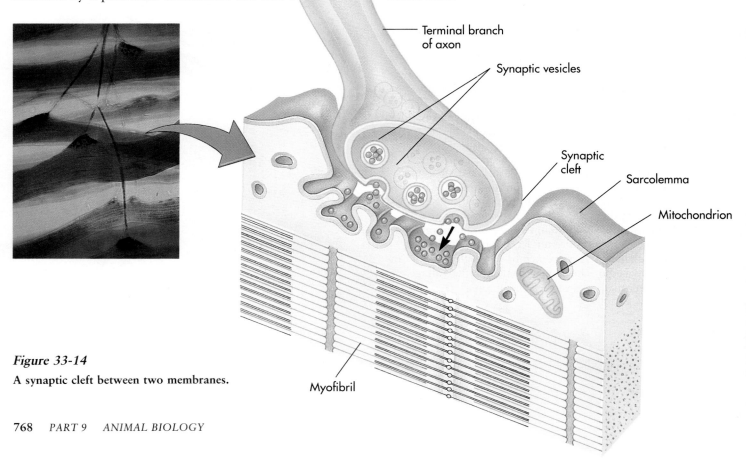

Terminal branch of axon

Synaptic vesicles

Synaptic cleft

Sarcolemma

Mitochondrion

Myofibril

*Figure 33-14*

A synaptic cleft between two membranes.

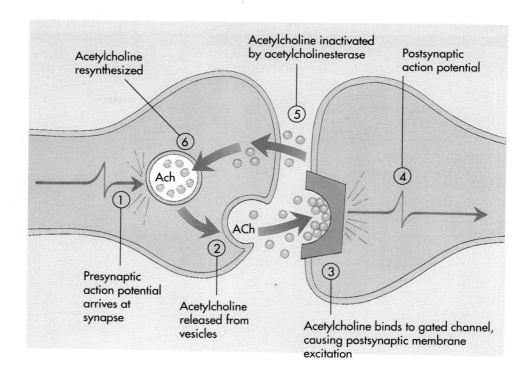

Acetylcholine resynthesized

Acetylcholine inactivated by acetylcholinesterase

Postsynaptic action potential

Presynaptic action potential arrives at synapse

Acetylcholine released from vesicles

Acetylcholine binds to gated channel, causing postsynaptic membrane excitation

**Figure 33-15**

The sequence of events in synaptic transmission.

*At a neuromuscular junction, acetylcholine released from an axon tip depolarizes the muscle cell membrane, permitting the entry of calcium ions, which trigger muscle contraction.*

At the neuromuscular junction it is necessary to destroy the residual neurotransmitter remaining in the synaptic cleft after the last impulse. If this does not occur, the postsynaptic membrane simply remains depolarized because acetylcholine cannot easily diffuse away. This removal of leftover acetylcholine is accomplished by an enzyme, **acetylcholinesterase,** which is present in the synaptic cleft. Acetylcholinesterase is one of the fastest-acting enzymes in the vertebrate body, cleaving one acetylcholine molecule every 40 microseconds. The rapid removal of neurotransmitters by acetylcholinesterase permits as many as 1000 impulses per second to be transmitted across the neuromuscular junction. Many organic phosphate compounds, such as the nerve gases tabun and sarin and the agricultural insecticide parathion, are potent inhibitors of acetylcholinesterase. Because they produce continuous neuromuscular transmission, such compounds can be lethal to vertebrates. Breathing, for example, requires muscular contraction, as does blood circulation.

## Neural Synapses

When the axon connection is with another nerve cell rather than with a muscle, the resulting postsynaptic voltage shift can be depolarizing or hyperpolarizing (increasing the polarization). Vertebrate nervous systems utilize dozens of different kinds of neurotransmitters, each with specific receptors on postsynaptic membranes. In an **excitatory synapse** the receptor protein is a chemically gated channel that is closed at rest. On binding a neurotransmitter that it recognizes (acetylcholine is a typical example), the channel opens and allows movement of all small ions. Because the concentration gradient for $Na^+$ dominates, the result is primarily the entry of $Na^+$, resulting in a depolarization referred to as an excitatory postsynaptic potential. In an **inhibitory synapse** the receptor protein is a chemically gated potassium channel (or in some cases a chloride channel). Binding of its neurotransmitter (**gammaaminobutyric acid,** or **GABA,** is a typical example) opens the channel, leading to the exit of positively charged potassium ions and a more negative interior. The result is termed an inhibitory postsynaptic potential, and this voltage shift reduces the ability of the postsynaptic membrane to depolarize (Figure 33-16).

An individual nerve cell can possess both kinds of synaptic connections to other nerve cells. When signals from both excitatory and inhibitory synapses reach the body of the neuron, the depolarizing effects (which decrease the internal negative charge) and the stabilizing ones (which increase the internal negative charge) in-

# The Postsynaptic Membrane: A Closer Look

Scientists have recently succeeded in reconstructing the membrane channel protein that is the target of the neurotransmitter acetylcholine. Protruding through postsynaptic membranes, these proteins respond to acetylcholine by opening a water-filled channel through the membrane for cations (positively charged ions) to diffuse. The protein has five membrane-spanning subunits arrayed in a circle that opens to create the water channel. Figure 33-A is a cross section through the channel protein, showing the long vertical wall made by the subunits as they traverse the membrane.

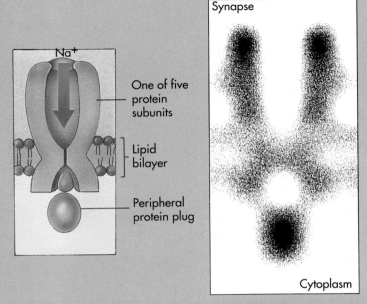

$Na^+$

One of five protein subunits

Lipid bilayer

Peripheral protein plug

Synapse

Cytoplasm

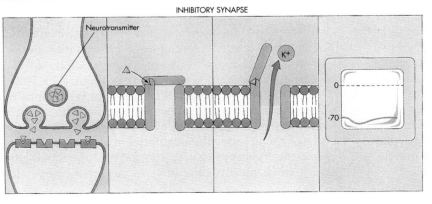

EXCITATORY SYNAPSE

Neurotransmitter

$Na^+$ gate closed

$Na^+$

0

-70

INHIBITORY SYNAPSE

Neurotransmitter

$K^+$

0

-70

### Figure 33-16

**Different kinds of synapses use different ion gates.** Excitatory synapses open $Na^+$ gates, whereas inhibitory synapses open $K^+$ gates.

teract with one another (Figure 33-17). The result is a process of **integration** in which the various excitatory and inhibitory electrical effects tend to cancel or reinforce each other.

*The integration of neural information occurs within individual neurons as the result of the sum of different synapses' electrical effects on a neuron's membrane. Some of these synapses facilitate depolarization, whereas others inhibit it.*

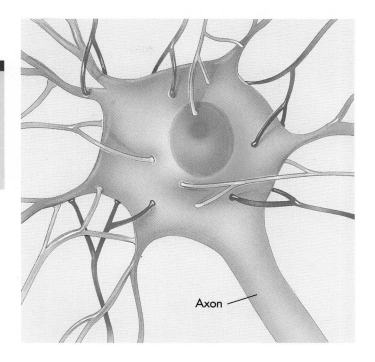

Axon

**Figure 33-17**

**Integration of nerve impulses takes place on the neuron cell body.** The synapses made by some axons are inhibitory, tending to counteract depolarization of the cell body membrane; these are indicated in red. The synapses made by other axons are stimulatory, tending to depolarize the cell body membrane; these are indicated in blue. The summed influences of all these inputs determines whether the axonal membrane will be sufficiently depolarized to initiate a propagating nerve impulse.

## ■ SUMMARY

1. A neuron is a nerve cell with an excitable membrane. It is specialized for the transmission of waves of depolarization along the membranes of processes extending out from the cell body.

2. The resting potential is created by the active pumping by the Na-K pump of $Na^+$ ions out across the membrane and of $K^+$ ions in. Because the membrane is not permeable to $Na^+$ but is permeable to $K^+$, $K^+$ streams back out but $Na^+$ does not flow back in. The result is a net negative charge within the cell.

3. An action potential is initiated by depolarization of an excitable membrane; this depolarization opens voltage-gated $Na^+$ channels. Such an opening allows $Na^+$ entry and thus removes the charge difference.

4. An action potential propagates because the opening of some $Na^+$ ion channels facilitates the opening of other adjacent channels, causing a wave of depolarization to travel down the membrane of the nerve cell. All signals traversing a given neuron have the same amplitude, differing only in the frequency (pattern) of impulses.

5. When an action potential reaches the far end of a nerve cell, the axon tip, it depolarizes the membrane at the tip, causing the release of chemicals from the tip. These chemicals pass across a synaptic gap and interact with chemically-gated ion channels in the membrane of another neuron or of muscle or gland cells. If channel opening results in depolarization, the result is an "excitatory" response. If channel opening hyperpolarizes a cell, the result is termed an "inhibitory" response.

6. The integration of nerve signals occurs on the cell body membranes of individual neurons, which receive both depolarizing waves from excitatory synapses and stabilizing waves from inhibitory synapses. These waves tend to cancel each other out, with the final amount of depolarization depending on the mix of the signals received.

# REVIEW QUESTIONS

1. The action potential in a nerve fiber arises from a sequence of membrane permeability changes. Initially, the permeability to _____ increases.

2. In the _____ refractory period it is impossible to elicit a second action potential.

3. _____ is the depolarization necessary to open a sufficient number of $Na^+$ channels to initiate a nerve impulse.

4. Multiple sclerosis affects impulse conduction by destroying _____.

5. During the falling phase of an action potential the membrane permeability to _____ is increasing.

# SELF-QUIZ

1. Which of the following are *not* properties of the action potential? It
   (a) has a threshold.
   (b) has a refractory period.
   (c) propagates decrementally.
   (d) is an all-or-none occurrence.
   (e) is none of the above; all the statements are correct.

2. The sodium pump
   (a) is responsible for repolarization of the action potential.
   (b) is necessary to maintain ion concentration gradients.
   (c) is fueled by ATP.
   (d) b and c are correct.
   (e) a, b, and c are correct.

3. During the refractory period of an action potential
   (a) the membrane potential is always positive inside.
   (b) $Na^+$ channels are inactivated.
   (c) the $K^+$ permeability is higher than in the resting state.
   (d) b and c are correct.
   (e) a, b, and c are correct.

4. In saltatory conduction along a myelinated nerve fiber
   (a) the nerve impulse is propagated down the entire membrane.
   (b) velocity is lower than in nonmyelinated axons of the same size.
   (c) an action potential skips from one node of Ranvier to another.
   (d) the presence of external $Na^+$ is not required.
   (e) c and d are correct.

5. Transmission at synapses
   (a) is always electrical.
   (b) is chemical.
   (c) is unidirectional.
   (d) b and c are correct.
   (e) none of the above.

# THOUGHT QUESTIONS

1. Why does a small stimulus not produce a small action potential and a large one not produce a large action potential?

2. Local anesthetics used by dentists to deaden pain are blockers of $Na^+$ channels. Why?

# FOR FURTHER READING

AXELROD, J.: "Neurotransmitters," *Scientific American*, June 1974, pages 59-71. An account of the chemicals that lead to the different synaptic responses underlying neuronal integration.

BULLOCK, T.H., R. ORKAND, and A. GRINNELL: *Introduction to Nervous Systems*, W.H. Freeman & Co., San Francisco, 1977. A very good, comprehensive, comparative treatment.

DUNANT, Y., and M. ISRAEL: "The Release of Acetylcholine," *Scientific American*, April 1985, pages 58-66. A challenge to the accepted theory that acetylcholine is emitted by synaptic vesicles.

KATZ, B.: *Nerve, Muscle, and Synapse*, McGraw-Hill Book Co., New York, 1966. A classic description of how a nerve impulse arises and is propagated; concise and unusually lucid.

KEYNES, R.D.: "Ion Channels in the Nerve-Cell Membrane," *Scientific American*, March 1979, pages 126-135. An up-to-date account of the structure of nerve-cell membranes.

KUFFLER, S.W., and J.G. NICHOLLS: *From Neuron to Brain: A Cellular Approach to the Function of the Nervous System*, ed. 2, Sinauer Associates, Inc., Sunderland, Mass., 1984. A superb overview of the mechanisms of nerve excitation and transmission.

MORELL, P., and W.T. NORTON: "Myelin," *Scientific American*, May 1980, pages 88-118. Still reflects current thought. Describes the composition of myelin, its synthesis, and how myelination increases the conduction velocity.

STEVENS, C.F.: "The Neuron," *Scientific American*, September 1979, pages 54-65. A description of the structure and functioning of a typical nerve cell.

## CHAPTER 34

# The Nervous System

This fox is exhibiting a behavior common to all mammals. Although not well understood, yawning is thought to be a reflex, an automatic consequence of nerve stimulation.

# THE NERVOUS SYSTEM

## Overview

In vertebrates the central nervous system (CNS) coordinates and regulates the diverse activities of the body. The activities of the CNS depend on information originating in sensory nerve endings and coded as a pattern of action potentials. From a knowledge of which sensory neurons are sending signals to the CNS and how often they are doing so, the brain builds a picture of the body's internal and external environment. The brain then issues appropriate commands, using one network of motor neurons to direct the voluntary muscles and a second network to control cardiac and smooth muscles. In addition, the brain regulates the release of chemicals called neurohormones to effect long-term changes in physiological activities. Understanding how the immense number of neurons in the brain function to produce consciousness and behavior is one of the most exciting frontiers of modern biology.

## For Review *Here are some important terms and concepts that you will encounter in this chapter. If you are not familiar with them, you should review them before proceeding.*

**Neuron** (Chapters 32 and 33)

**Depolarization** (Chapters 32 and 33)

**Muscle contraction** (Chapter 32)

**Figure 34-1**

**Singing well takes practice.** This baby coyote is greeting the approaching evening. His howling is not as impressive as his dad's — a good performance takes practice. His brain is learning by repetition to control the vocal cords properly.

Vertebrates, like all other animals except sponges, use a network of neurons to gather information about the body's condition and the external environment, to process and integrate that information, and to issue commands to the body's muscles and glands. This network of neurons, the nervous system (Figure 34-1), is the subject of this chapter. The basic architecture of the nervous system is similar throughout the animal kingdom.

## ORGANIZATION OF THE VERTEBRATE NERVOUS SYSTEM

All vertebrate nervous systems can be said to have one underlying mechanism and three basic elements. The underlying mechanism is the nerve impulse, or action potential. The three basic elements are (1) a central processing region, or **brain,** (2) nerves that bring information to the brain, and (3) nerves that transmit commands from the brain.

The vertebrate nervous system consists of two distinct functional components, the central nervous system and the peripheral nervous system. The **central nervous system** is composed of the brain and the spinal cord (Figure 34-2) and is the site of information processing and control within the nervous system. The **peripheral nervous system** includes all the nerve pathways of the body outside the brain and spinal cord. These peripheral pathways are commonly divided into two groups: **sensory**, or **afferent**, **pathways**, which transmit information from the body to the CNS; and **motor**, or **efferent**, **pathways**, which transmit commands to the body from the CNS (Figure 34-3). The motor pathways are further distinguished as the **voluntary**, or **somatic**, **nervous system**, which relays commands to skeletal muscles; and the **involuntary**, or autonomic (Greek *auto*, "self" + *nomos*, "law": self-controlling), **nervous system**, whose nerve fibers stimulate secretion from glands and excite or inhibit the smooth muscles of the body. Along with the voluntary and involuntary motor systems is a **neuroendocrine system**, a network of **endocrine glands** whose hormone secretion is controlled by commands from the CNS. The CNS issues commands through all these systems (Figure 34-4).

Nervous system pathways are composed of individual nerve fibers, axons, and long dendrites, all bundled together like the strands of a telephone cable. Within the CNS these bundles of nerve fibers are called **tracts**. In the peripheral nervous system they are called **nerves**. The cell bodies from which pathways extend are often clustered into groups, called **nuclei** if they are within the CNS and **ganglia** if they are in the peripheral nervous system.

## The Spinal Cord

The spinal cord is composed of a central region containing nerve cell bodies (gray matter) surrounded by ascending (sensory) and descending (motor) tracts consisting of bundles of axons (white matter). Ascending tracts carry sensory information to the brain, whereas descending tracts carry impulses from the motor areas of the brain. These tracts ultimately influence the neurons in the spinal cord controlling the muscles of the body. These neurons are located in the ventral (lower) part of the gray matter of the spinal cord.

In this chapter we begin by considering the brain and CNS, then consider the sensory systems that convey information to the brain, and end with a brief look at the motor pathways.

## EVOLUTION OF THE VERTEBRATE BRAIN

The earliest vertebrates had far more complex brains than their ancestors. Casts of the interior braincases of fossil agnathans, fish that swam 500 million years ago, have revealed a lot about the early evolutionary stages of the vertebrate brain. Although they were very small, these brains already had the three principal divisions that characterize the brains of all contemporary vertebrates: (1) the hindbrain, or **rhombencephalon**; (2) the

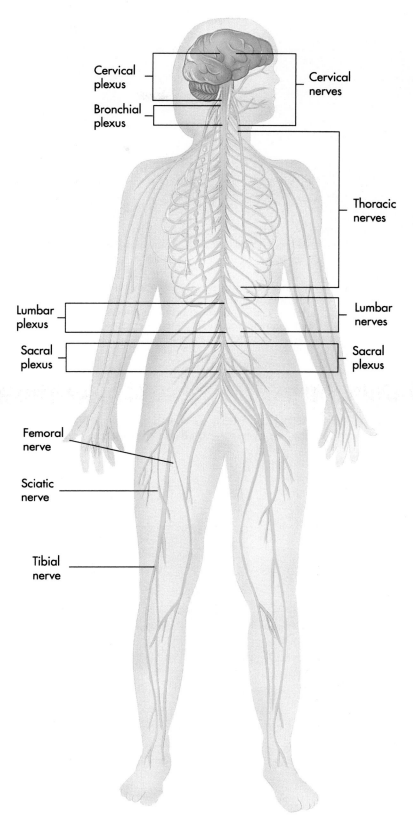

*Figure 34-2*

**The human nervous system.**

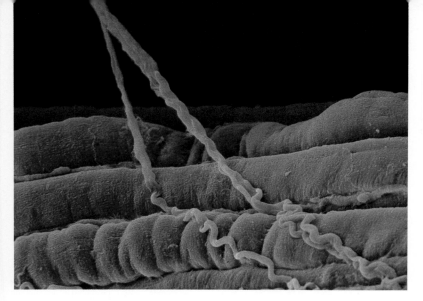

*Figure 34-3*

**Motor nerves transmit commands to individual muscles and glands.** The slender, twisted threads are motor nerves, and the long, thick strands at the bottom are muscle fibers. Often a nerve will carry many independent fibers; these branch off to establish contact with different target muscles.

midbrain, or **mesencephalon;** and (3) the forebrain, or **prosencephalon.**

The hindbrain is the principal component of these early brains (Figure 34-5), as it still is in fishes today. Composed of the **pons,** the **medulla oblongata,** and the **cerebellum,** the hindbrain may be considered an extension of the spinal cord devoted primarily to coordinating motor reflexes. Tracts, cables composed of large numbers of nerve fibers, run up and down the spinal cord to the hindbrain. The hindbrain in turn integrates the afferent signals that come in from the body and determines the pattern of efferent response.

Much of this coordination is carried out by the cerebellum, a small extension of the hindbrain (cerebellum means "little cerebrum"). In the advanced vertebrates the cerebellum plays an increasingly important role as a coordinating center and is correspondingly larger than it is in the fishes. In all vertebrates the cerebellum processes data on the current position and movement of each limb, the state of relaxation or contraction of the muscles affecting that limb, and the general position of the body and its relation to the outside world. These data are gathered in the cerebellum and synthesized, and the resulting orders are issued to efferent pathways.

In fishes the remainder of the brain is devoted to reception and the processing of sensory information. The second major division of the fish brain, the midbrain, is composed primarily of the **optic lobes,** which receive and process visual information. The third major division, the **forebrain,** is devoted to processing olfactory (smell) information.

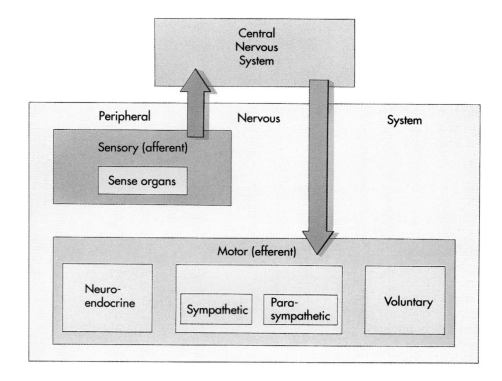

*Figure 34-4*

**The nervous system of vertebrates has a hierarchical architecture.** The central nervous system receives information from the sensory system and issues commands to the motor system; the sensory and motor systems together are called the peripheral nervous system. Three kinds of motor commands are issued: to muscles (voluntary), to organs (autonomic), and to hormone-producing glands (neuroendocrine).

Figure 34-5

**The basic organization of the vertebrate brain can be seen in the brains of primitive fishes.** These brains are divided into the same regions that can be seen in differing proportions in all vertebrate brains: the hindbrain, which is the largest portion of the brain in fishes; the midbrain, which in fishes is a small zone devoted to processing visual information; and the forebrain, which in fishes is devoted primarily to processing olfactory (smell) information. In the brains of terrestrial vertebrates, the forebrain plays a far more dominant role than it does in fishes.

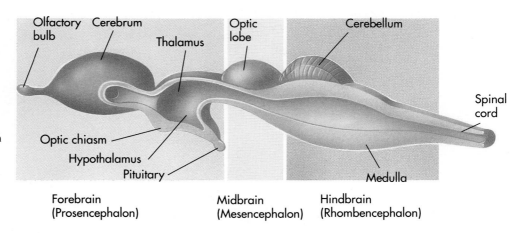

*In early vertebrates the principal component of the brain was the hindbrain, devoted largely to coordinating motor reflexes.*

## The Advent of a Dominant Forebrain

Starting with the amphibians, and much more prominently in the reptiles, one can perceive a pattern that becomes a dominant evolutionary trend in the development of the vertebrate brain: the processing of sensory information becomes increasingly centered in the forebrain (Figure 34-6).

The forebrain in reptiles, amphibians, birds, and mammals is composed of two elements with distinct functions. The **diencephalon** (Greek *dia,* "between") mainly consists of the thalamus and hypothalamus, important centers for sensory information processing and internal homeostasis. The **telencephalon,** or "endbrain" (Greek *telos,* "end"), includes the cerebrum and all those higher functions associated with it.

*In land-dwelling vertebrates the processing of information is increasingly centered in the forebrain.*

*Sensory Integration.* The **thalamus** is one important site of sensory processing in the brain. Auditory, optical, and other information from sensory receptors enters the thalamus and then is passed to the sensory areas of the cerebral cortex, a layer several millimeters thick on the brain's outer surface. Information about posture, derived from the muscles, and information about orientation, derived from sensors within the ear, combines with information from the cerebellum of the hindbrain and passes to the thalamus. The thalamus then processes the information and channels it to the appropriate motor center on the outer surface of the brain.

*Integrating the Body's Responses.* The **hypothalamus** integrates all visceral activities. It controls centers in the medulla that in turn regulate body temperature, respiration, and heartbeat; it also directs the secretions of the brain's major hormone-producing gland, the **pituitary gland.** The hypothalamus is linked by a network of neurons to some areas of the cerebral cortex. This network, along with the hypothalamus, is called the **limbic system.** The operations of the limbic system are responsible for many of the most deep-seated drives and emotions of vertebrates, including pain, anger, sex, hunger, thirst, and pleasure.

*Stereotyped Behavior.* The telencephalon of reptiles and birds consists largely of a layer of nerve tissue called the **corpus striatum,** which controls complicated stereotyped behavior. There is nothing like the corpus striatum in any other group. Much of what we think of as "behavior" in birds and reptiles is really composed of completely determined neural reactions. For example, the complex mating rituals of some birds are often choreographed by the structure of the brain; their pattern is fixed by nerve paths within the corpus striatum.

## The Expansion of the Cerebrum

Fishes and reptiles have brains that are small compared with the size of their bodies; mammals and birds have

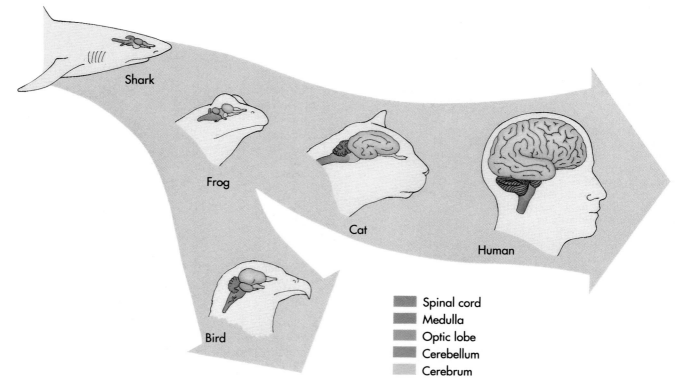

## Figure 34-6

**The evolution of the vertebrate brain.**
In sharks and other fishes, the hindbrain is predominant; the rest of the brain serves primarily to process sensory information. In frogs and other amphibians and reptiles, the forebrain is far larger, and a larger cerebrum devoted to associative activity can be seen. In birds, which evolved from reptiles, the cerebrum is even more pronounced. In cats and other mammals, the cerebrum is the largest portion of the brain. In humans, the cerebrum envelopes much of the rest of the brain.

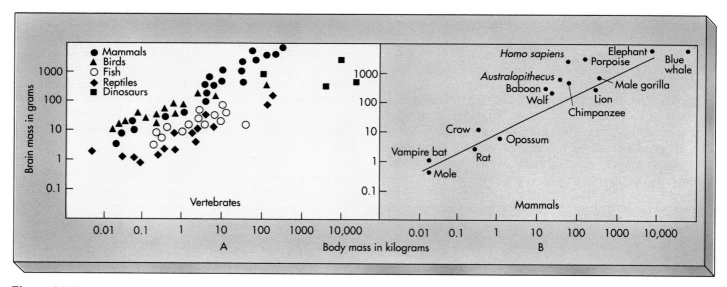

## Figure 34-7

**Brain mass versus body mass.** Among most vertebrates, brain weight is a relatively constant proportion of body weight, so that a plot of brain mass gives a straight line.
**A** However, the proportion of brain mass to body mass is much greater in birds than in reptiles, and even greater in mammals.
**B** Among mammals, humans have the greatest brain mass per unit of body mass. In second place are porpoises.

brains that are proportionately large (Figure 34-7). The increase in brain size in the mammals largely reflects a great enlargement of the cerebrum, the dominant part of the mammalian brain.

The **cerebrum** is the center for correlation, association, and learning in the mammalian brain. It receives sensory data from the thalamus. From the cerebrum, motor nerve fibers extend to the motor nerve columns of the spinal cord, which contain groups of neurons servicing different muscles. Efferent motor neurons extend from these motor columns directly to the muscles. The motor fibers from the motor areas of the cerebrum pass straight through the brain to the voluntary motor regions of the brainstem, and they can be seen as a cable of fibers called the **pyramidal (corticospinal) tract,** a structure that particularly dominates the structure of primate brains.

*The brains of mammals and birds are large relative to their body size. This reflects great enlargement of the cerebrum, the center for correlation, association, and learning.*

# THE HUMAN CNS

The cerebrum, located at the very front of the human brain, is so large compared with the rest of the brain that it appears to envelop it (Figure 34-8). In the brains of humans and other primates the cerebrum is split into two halves, or hemispheres, which are connected only by a nerve tract called the **corpus callosum.** Each hemisphere of the brain is divided further by two deep grooves into four lobes, designated the **frontal, parietal, temporal,** and **occipital lobes** (Figure 34-9).

## The Cerebral Cortex

Much of the neural activity of the cerebrum occurs within a thin, gray layer only a few millimeters thick on its outer surface, overlying a solid white region that consists of myelinated nerve fibers. This layer, the cerebral cortex, mentioned earlier, is densely packed with neuron cell bodies. The human cerebral cortex contains more than 10 billion nerve cells, roughly 10% of all the neurons in the brain. The surface of the cerebral cortex is highly convoluted, particularly in human brains, a property that increases its surface area (and number of cell bodies) threefold.

By examining the effect of injuries to particular sites on the cerebrum, it has been possible to plot

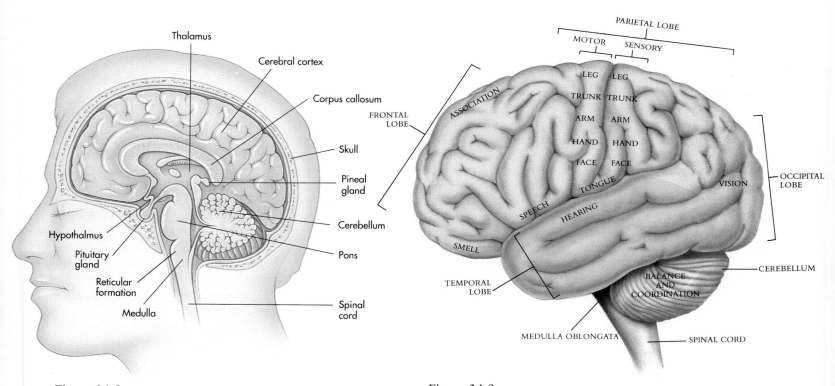

*Figure 34-8*

**A section through the human brain.** The cerebrum envelopes the rest of the brain; only the cerebral cortex, part of the cerebrum, is visible in surface view.

*Figure 34-9*

**The major functional regions of the human brain.**

roughly the location of the major sensory and motor regions. There are four major sensory regions on the cerebral cortex, each of them referred to as a specialized cortex: the motor, somatosensory, auditory, and visual cortices (see Figure 34-9). Specific sensory stimuli cause electrical activity in the corresponding sensory areas of the cortex.

The **motor cortex** straddles the rearmost portion of the frontal lobe. Each point on its surface contains neurons controlling the movement of different muscles of the body. Directly behind the motor cortex, on the leading edge of the parietal lobe, lies the **somatosensory cortex**. Each point on the surface of the somatosensory cortex receives afferent input from sensory receptors from a different part of the body, such as the pressure sensors of the fingertips or the touch receptors on the tongue. This projection forms a kind of map of the body surface on the surface of the cortex. The **auditory cortex** lies within the temporal lobe; different surface regions of this cortex correspond to different sound frequencies. The **visual cortex** lies on the occipital lobe, with different points in the visual field corresponding to different positions on the retina. The visual cortex is also specialized to detect shapes and the movements of objects.

## Associative Organization of the Cerebral Cortex

Only a small portion of the total surface of the cerebral cortex is occupied by the motor and sensory cortices. The remainder of the cerebral cortex is referred to as **associative cortex**. This appears to be the site of sensory information integration from several senses and of higher mental activities, such as planning and contemplation. The associative cortex represents a far greater portion of the total cortex in primates than it does in any other mammals and reaches its greatest extent in human beings. In a mouse, for example, 95% of the surface of the cerebral cortex is occupied by motor and sensory areas. In humans, only 5% of the surface is devoted to motor and sensory functions; the remainder is associative cortex.

The cerebrum consists of two hemispheres, identical in form but differing in function. All four of the specialized sensory and motor regions of the cortex occur on the surface of each hemisphere, with the ones on the *right hemisphere* related to the *left side* of the body, and vice versa. The switch occurs because the descending motor tracts and ascending sensory tracts cross at some level. In addition, the two hemispheres are connected by a mass of nerve fibers called the corpus callosum.

Although the two hemispheres of the brain each contain similar sensory and motor regions, the hemispheres seem to be responsible for different associative activities. Injury to the left hemisphere of the cerebrum often results in the partial or total loss of speech, but a similar injury to the right side does not. There are several **speech centers** controlling different aspects of speech (Figure 34-10). They are almost always located on the left hemisphere of right-handed people; in about one half of left-handed people they are located on the right hemisphere. An injury to one speech center produces halting but correct speech; injury to another speech center produces fluent, grammatical, but meaningless speech; injury to a third center destroys speech altogether.

Injuries to other sites on the surface of the brain's left hemisphere result in impairment of the ability to read, write, or do arithmetic. Comparable injuries to the right hemisphere have very different effects, resulting in impairment of three dimensional vision, musical ability, or the ability to recognize patterns and solve inductive problems.

It has been a popular activity, particularly among journalists, to speculate that the human brain is composed of a "rational" left hemisphere and an "intuitive" right hemisphere because many of the more imaginative associative activities seem to be carried out by the right hemisphere. Some people even argue that the brain actually has two consciousnesses, one dominant over the other, or that the right brain hemisphere is the more ancient, with language and reasoning having evolved much later. Actually, the significance of the clustering of associative activities in different areas of the brain is not at all clear, but it remains a subject of much interest.

## Memory and Learning

One of the great mysteries of the brain involves the basis of memory and learning. If portions of the brain are removed, especially the temporal lobes, memory is impaired but not lost; there is no one part of the brain in which memory appears to reside. Investigators who have tried to prove the physical mechanisms underlying memory often have felt that they were grasping a shadow. Complete understanding of these mechanisms is not available, but one possible mechanism involves long-term modifications of transmission at individual synapses, perhaps by altering the type of ion channel present or the channel's response.

There are two functionally different types of memory. The first, **short-term memory,** is transient, lasting only a few moments. Such memories can readily be removed from the brain by application of an electrical shock. When this is done, short-term memories are wiped from the circuits, but longer-term memories remain. This result suggests that short-term memories are

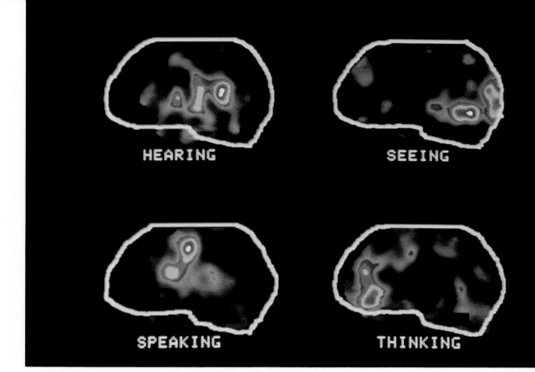

**Figure 34-10**

**Different regions of the brain are active as we do different things.** This illustration shows how the brain reacts in subjects asked to listen to a spoken word, to read that same word silently, to repeat the word out loud, and then to speak a word related to the first.

## The Molecular Basis of Memory

Medical progress sometimes has unsuspected beginnings. About 12 years ago, a retired California postal worker suffered a stroke. He survived, but lost his ability to form memories. When he died in 1983, scientists were allowed to study his brain. They discovered that the amnesia was not the result of massive brain damage, but rather the result of a small, well-defined lesion in the hippocampus, a part of the temporal lobe that is believed to be involved in storing memories.

What is unusual about the hippocampus? The brain cells that had been damaged are now known to be rich in an unusual cell surface receptor, called the NMDA receptor after the chemical (*N*-methyl D-aspartate) that is used to detect it. Since 1986, evidence has been accumulating that NMDA receptors turn on the biochemical activities that lead to the encoding of memories. The NMDA receptor is unusual among the many receptors present on neurons in that it admits calcium ions into the cell. Blocking these receptors in rats blocks the admission of calcium into hippocampal neurons—and also blocks memory, interfering with spatial learning. Most scientists believe that the brain stores memories by linking neurons to form new circuits. Calcium ap-

parently increases synaptic sensitivity and so strengthens synaptic connections, like turning up a volume control.

The NMDA receptor is unique in that it must be stimulated first by a voltage change, which ejects a magnesium ion from the calcium channel, followed by a neurotransmitter, in this case glutamate. The electrical impulse cocks the gun; the second impulse fires it. Memory is particularly effective if, every half second, a firing impulse follows a priming inpulse by about 200 milliseconds. This is, interestingly, the same frequency as the so-called theta rhythm, a brain wave emanating from the hippocampus that has long been thought to be associated with learning. Chemically blocking the NMDA receptors blocks the theta rhythm.

The discovery of the NMDA receptors has taken us much closer to understanding the chemical mechanisms of memory. It is a very active area of research, and new, exciting results are reported almost daily. For example, new evidence suggests that the sensitivity of the NMDA receptor depends on how much of the amino acid glycine it is exposed to. It seems the two-stage switch has a volume control.

**TABLE 34-1    SENSORY TRANSDUCTION AMONG THE VERTEBRATES**

| STIMULUS | RECEPTOR | LOCATION | STRUCTURE | UNDERLYING PROCESS |
|---|---|---|---|---|
| **INTEROCEPTION** | | | | |
| Temperature | Heat receptors and cold receptors | Skin, hypothalamus | Simple nerve endings | Temperature change alters activity of ion channels in membrane |
| Blood chemistry | Carotid bodies | Arterial walls | Simple nerve endings | Change in $O_2$ and $H^+$ concentration alter membrane charge |
| Pain | Nociceptors | Body surfaces | Simple nerve endings | Changes in pressure or temperature open membrane channels |
| Muscle contraction | Stretch receptors | Within muscles | Spiral nerve endings wrapped around muscle spindle | Stretch of spindle deforms nerve |
| Blood pressure | Baroreceptors | Arterial branches | Nerve extends over thin part of arterial wall | Stretch at arterial wall deforms nerve |
| Vibration | Pacinian corpuscles | Deep within skin | Nerve ending within elastic capsule | Severe change in pressure deforms nerve |
| Touch | Meissner's corpuscles Merkel cells | Surface of skin | Nerve ending within elastic capsule | Rapid or extended change in pressure deforms nerve |
| **EXTEROCEPTION** | | | | |
| Gravity | Statocysts | Outer chambers of inner ear | Pebble and cilia | Pebble presses against cilia |
| Motion | Cupula | Semicircular canals of inner ear | Collection of cilia | Deformation of cilia by fluid movement |
| | Lateral-line organ | Within grooves on body surface of fish | Collection of cilia | Deformation of cilia by fluid movement |
| Hearing | Organ of Corti | Cochlea of inner ear | Cilia between membranes | Deformation of membrane by sound waves in fluid |
| Taste | Taste bud cells | Mouth and skin of fish | Chemoreceptors | Binding to specific receptors in membrane |
| Smell | Olfactory neurons | Nasal passage | Chemoreceptors | Binding to specific receptors in membrane |
| Vision | Rod and cone cells | Retina of eye | Array of photosensitive pigment | Light initiates process that closes ion channels |
| Heat | Pit organ | Face of snake | Temperature receptors in two chambers | Temperature of surface and interior chambers compared |
| Electricity | Ampullae of Lorenzini | Within skin of fish | Closed vesicles with asymmetrical ion channel distribution | Electric field alters ion distribution on membranes |
| Magnetism | Unknown | Unknown | Unknown | Deflection at magnetic field initiates nerve impulse? |

stored electrically, in the form of short-term neural excitation. **Long-term memory,** in contrast, appears to involve changes in the way information is processed at neural connections within the brain, such as by the ion channel mechanism referred to above. There may also be changes in the patterns of synapse formation. What is known is that inhibitors of protein synthesis affect the ability to store long-term memories.

## THE SENSORY NERVOUS SYSTEM

All input from sensory neurons to the CNS arrives in the same form, as nerve impulses in afferent sensory neurons (afferent = carrying nerve impulses to the CNS). Every arriving nerve impulse is identical to every other one. The information that the brain derives from sensory input is based on the frequency and pattern with which these impulses arrive and on the identity of the specific neuron that transmits it. A sunset, a symphony, searing pain—to the brain they are all the same, differing only in the source of the impulse and its frequency and pattern. Thus if the auditory nerve is artificially stimulated, the CNS perceives the stimulation as a noise. If the optic nerve is artificially stimulated (for example, by pressing on the eyes), the stimulation is perceived as a flash of light.

Except for visual system photoreceptors, all sensory receptors are able to initiate nerve impulses by opening stimulus-gated $Na^+$ ion channels within sensory neuron membranes and thus depolarizing them.

The receptors differ from one another with respect to the nature of the environmental input that triggers this event. Many kinds of receptors have evolved among vertebrates, with each receptor sensitive to a different aspect of the environment. These receptors, their location, structure, and process are summarized in Table 34-1.

### Sensing Internal Information

Traditionally, the sensing of information that relates to the body itself, its internal condition and its position, is known as **interoception,** or inner perception (Figure 34-11, *A*). In contrast, the sensing of the exterior is called **exteroception** (Figure 34-11, *B*). Many of the neurons and receptors that monitor body functions are simpler than those that monitor the external environment. The simplest sensory receptors are free nerve endings that depolarize in response to direct physical stimulation, to temperature, to chemicals like oxygen diffusing into the nerve cell, or to a binding or stretching of the neuron cell membrane. The vertebrate body uses a variety of such **interoceptors** to obtain information about its internal conditioning.

*Temperature Change.* Two kinds of nerve endings in the skin are sensitive to changes in temperature. One of these is stimulated by a low temperature (cold receptors), whereas the other is stimulated by a high temperature (warm receptors). By comparing information from the two, the receptors can detect absolute temperature and temperature changes.

**Figure 34-11**

By tradition, the senses are grouped into two classes. 1, Interoception. Dancers are able to maintain their proper stance by using their internal sense of balance. 2, Exteroception. The other class senses the external environment. This leaf frog *(Phyllomedusa tarsius)*, photographed in the Amazon rain forest of Ecuador, learns about the world around it by using eyes to perceive patterns of light, just as you are doing in reading this page.

1                    2

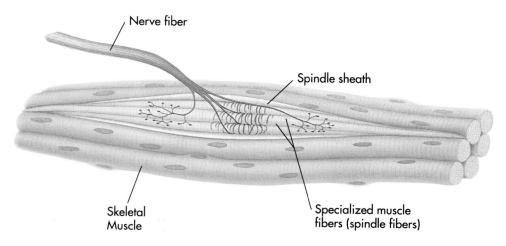

Nerve fiber

Spindle sheath

Skeletal Muscle

Specialized muscle fibers (spindle fibers)

**Figure 34-12**

**A stretch receptor embedded within skeletal muscle.** Stretching of the muscle elongates the spindle fiber, which deforms the nerve endings, causing them to fire and send a nerve impulse out along the nerve fiber.

*Blood Chemistry.* Special receptors called **carotid bodies** are embedded in the walls of arteries at several locations in the circulatory system. By sensing $CO_2$ levels, the carotid bodies provide the sensory input that the body uses to regulate its rate of respiration. When $CO_2$ rises above normal levels, this information is conveyed to the CNS, which reacts by increasing the respiration rate to eliminate the excess $CO_2$.

*Pain.* A stimulus that causes or is about to cause tissue damage is perceived as **pain.** Such a stimulus elicits an array of responses from the CNS, including reflexive withdrawal of a body segment from a source of stimulus and changes in heartbeat and blood pressure. The receptors that produce these effects are called **nociceptors.** They consist of the ends of small, sparsely myelinated nerve fibers located within tissue, usually near the surfaces where damage is most likely to occur. Physical deformation of the ends of the nerve initiates depolarization.

*Muscle Contraction.* Buried deep within the muscles of all vertebrates, except the bony fishes, are specialized muscle fibers called **muscle spindles.** Wrapped around each muscle spindle is the end of an afferent neuron, called a **stretch receptor** (Figure 34-12). When a muscle is stretched, the spindle fiber elongates, stretching the spiral nerve ending of the stretch receptor and stimulating it to fire repeatedly. When the muscle contracts, the tension on the spindle lessens and the stretch receptor ceases to fire. Thus the frequency of stretch receptor discharges along the afferent nerve fiber indicates muscle length and the rate of change of muscle length at any given moment. Stretch receptors sensitive to muscle tension are present in the tendons that connect the muscles to the skeleton. The CNS uses this information to control movements that involve the combined action of several muscles, such as those involved in breathing or walking.

*Blood Pressure.* Blood pressure is sensed by a highly branched network of nerve endings called **baroreceptors,** or **pressure receptors,** within the walls of major arteries in several locations in the circulatory system. When the blood pressure increases, the arterial wall is stretched most where it is thinnest, in the region of the baroreceptor, increasing the rate of firing of the afferent neuron. A decrease in blood pressure causes the wall to move inward and lowers the rate of neuron depolarization. Thus the frequency of impulses arriving from these afferent fibers provides the CNS with a continuous measure of blood pressure. The higher the blood pressure, the higher is the frequency of impulses.

*Touch.* Touch is sensed by pressure receptors below the surface of the skin. There are many different types of receptors. **Meissner's corpuscles** and **Merkel cells** are responsible for the great sensitivity of your fingertips. Meissner's corpuscles fire in response to rapid changes in pressure, and Merkel cells measure the duration and extent to which pressure is applied. Other receptors are sensitive to vibration, and some produce the sensations we refer to as itching.

> *Mechanical stimuli of different types initiate nerve impulses by deforming the sensory neuron membrane. Such deformation opens stimulus-gated $Na^+$ channels in the membrane and initiates depolarization.*

**Balance.** All vertebrates possess gravity receptors known as **statocysts** or **otoliths**. In humans these are located in a series of hollow chambers within the inner ear and provide the information the brain uses to perceive balance. To illustrate how these receptors work, imagine a pencil standing in a glass. No matter which way you tip the glass, the pencil will roll along the rim, applying pressure to the lip of the glass. If you want to know the direction the glass is tipped, you need only inquire where on the rim the pressure is being applied.

The body uses receptors of this sort, broadly called **proprioceptors** (Greek *proprios,* "self"), to sense its position in space. Gravity serves as a reference point; changes in the pressure applied to a nerve ending by a heavy object within the receptor provide the stimulus.

*Motion.* **Motion** is sensed by vertebrates in a way similar to that used to detect vertical position: by employing a receptor in which fluid deflects cilia in a direction opposite to that of the motion. Within the inner ear are three fluid-filled **semicircular canals** (Figure 34-13), each oriented in a different plane at right angles to the other two so that motion in any direction can be detected. Protruding into the canals are sensory cells, which are connected to afferent nerves. Because the three canals are oriented in all three planes, movement in any plane is sensed by at least one of them. Complex movements can be analyzed by comparing the sensory input from each canal.

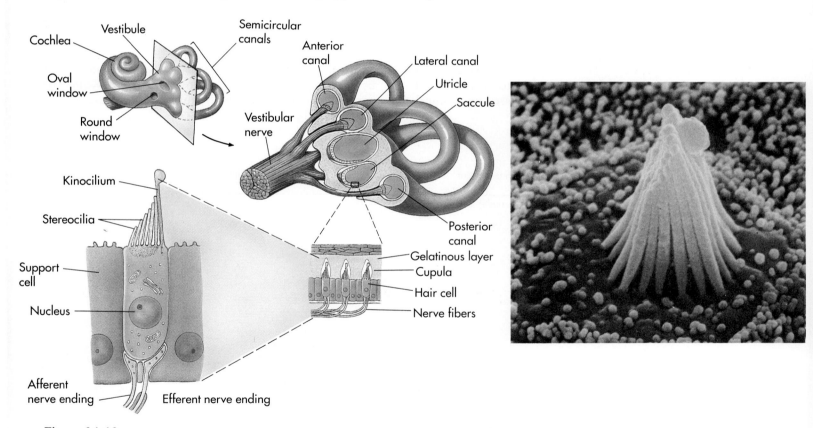

**Figure 34-13**

**Vertebrates sense movement as pressure of a long cilium, the kinocilium, against a battery of surrounding cilia, the stereocilia.** As you can see in the electron micrograph, groups of long cilia form tentlike assemblies called cupulae that project out into passages within the semicircular canals of the ear. Movement causes fluid in the canal to press against the cupula. Because there are three semicircular canals at tight angles to one another, movement in any plane can be detected.

*Complex mechanical receptors can respond to pressure, to gravity, or to motion. In each case the receptors employ mechanical devices like levers to convert the information to a mechanical stimulus, which then initiates the depolarization of a sensory membrane by deformation.*

## Sensing The External Environment

So far, most of the sensory receptors we have examined have been directed at the internal environment of the body or at determining its position in space. Sensing the nature of other objects in the external environment presents quite a different problem. Imagine, for example, the problem of learning about an object standing some distance away. The amount of information that any sensory receptor can obtain about that object is limited both by the nature of the stimulus sensed by that receptor and by the medium, either air or water, through which the stimulus must move to reach the receptor.

The four primary senses that detect objects at a distance use different classes of receptors. Taste and smell use chemical receptors; hearing uses mechanical receptors; and vision uses electromagnetic receptors that sense photons of light. We consider each of these and then examine other sensory systems employed by vertebrates.

*Taste and Smell.* The simplest exteroceptors, like the simplest interoceptors, are chemical ones. Embedded within the membranes of afferent nerve endings or of sensory cells associated with afferent neurons are specific chemical receptors that control chemically gated ion channels and thus induce depolarization when they bind particular molecules. There are two chemical sensory systems that utilize different receptors and process information at different locations in the brain: **taste,** in which the receptors are specialized sensory cells, and **smell,** in which the receptors are neurons. Despite the differences in how stimuli are received and the information processed, there is little difference in the chemical stimuli to which the two chemical senses respond. In taste the receptors are **taste buds** located in the mouth. In smell the receptors are neurons whose cell bodies are embedded in the epithelium of the upper portion of the nasal passage.

*Hearing.* Terrestrial vertebrates detect vibration in air by means of mechanical receptors located within the ear. In the ears of mammals, sound waves beat against

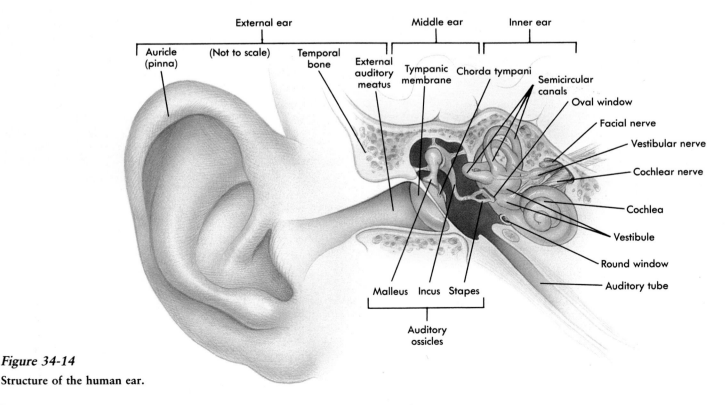

**Figure 34-14**
Structure of the human ear.

a large membrane of the outer ear called the **tympanic membrane** (eardrum), causing corresponding vibrations in three small bones: the hammer, anvil, and stirrup. These bones act together as a lever system, increasing the force of the vibration. The third in line of the levers, the stirrup, pushes against another membrane, the **oval window** (Figure 34-14). Because the oval window is smaller than the tympanic membrane, vibration against it produces more force per unit area of membrane. The combination of the lever system and area difference amplifies sound. The chamber in which all these events occur is called the **middle ear.** It is connected to the pharynx by the eustachian tube in such a way that there is no difference in air pressure between the middle ear and the outer ear. The familiar "ear popping" associated with landing in an aircraft or with the rapid descent of an elevator in a tall building is a result of pressure equalization between the two sides of the eardrum.

The membrane of the oval window is the door to the **inner ear,** where hearing actually takes place. The chamber of the inner ear is shaped like a tightly coiled snail shell and is called the **cochlea,** from the Latin name for "snail."

The auditory receptors are hair cells located in a complex structure called the organ of Corti. These cells rest on the **basilar membrane,** which bisects the cochlea. The hair cells do not project into the fluid filling the cochlea. Instead they are covered with another membrane called the **tectorial membrane.** The bending of the basilar membrane as it vibrates causes the hairs of the receptor cells pressed against the tectorial membrane to bend, depolarizing their associated afferent neurons. Sounds of different frequencies cause different portions of the basilar membrane to vibrate and thus to fire different afferent neurons. The CNS perceives sound intensity in terms of the frequency of discharge in particular nerve fibers. Sound frequency is coded in terms of the pattern of neurons on the basilar membrane that are firing afferent impulses. High frequencies produce the greatest deflection near the oval window. Low frequencies primarily affect the apex of the cochlea.

Our ability to hear depends on the flexibility of the basilar membrane, a flexibility that changes as we grow older. Humans are not able to hear low-pitched sounds, below 20 vibrations or cycles per second, although some other vertebrates can. As children, human beings can hear high-pitched sounds, up to 20,000 cycles per second, but this ability decreases progressively through middle age. Other vertebrates can hear sounds at far higher frequencies (Figure 34-15). Dogs readily detect sounds of 40,000 cycles per second. Thus dogs can hear a high pitched dog whistle when it seems silent to a human observer.

**Figure 34-15**

**The bat-moth game.** This bat is emitting high-frequency "chirps" as it flies. The bat listens for the sound's reflection against the moth and by timing how long it takes for a sound to return, it can effectively catch its prey even in total darkness.

*Vision.* Light is perhaps the most useful stimulus for learning about the external environment. Because light travels in a straight line and arrives virtually instantaneously, visual information can be used to determine both the direction and the distance of an object. No other stimulus provides as much detailed information.

The type of problem associated with perceiving light is one that we have not encountered before in our consideration of sensory systems. All receptors described to this point have been either chemical or mechanical ones. None of them is able to respond to the electromagnetic energy provided by photons of light. **Vision,** the perception of light, is carried out in vertebrates by a specialized sensory apparatus called an eye. Eyes have evolved many times independently in different groups of animals.

Eyes contain sensory receptors that detect photons of light. These receptors are located in the back of the eye, which is organized something like a camera. Light that falls on the eye is focused by a lens on the receptors in the rear, just as the lens of a camera focuses light on film. How does a photon-detecting sensory receptor work? Just as is the case in photosynthesis, the primary photoevent of vision is the absorption of a photon of light by a pigment. The visual pigment is called *cis*-retinal (Figure 34-16), a substance that we encountered as the cleavage product of carotene, a photosynthetic pigment in plants (see Figure 8-3). In

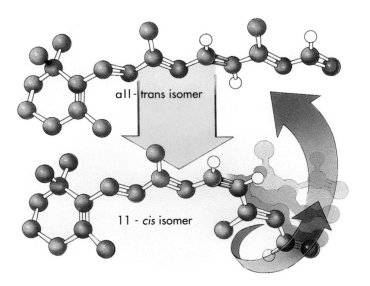

all - trans isomer

11 - cis isomer

**Figure 34-16**

**Light absorption.** When light is absorbed by the 11-*cis* isomer of the visual pigment retinal, the pigment undergoes a change in shape; the linear end of the molecule (right end) rotates about a double bond, indicated here in red. The new isomer is referrred to as all-*trans*. This change in the pigment's shape causes the isomer to detach from the protein opsin to which the pigment is bound, generating a nerve impulse. The all-*trans* isomer of retinal is then converted back to its 11-*cis* form.

# The Evolution of the Eye

In considering the evolution of major groups of organisms, we sometimes find that the differences among them are so great that it is difficult to trace the lines of descent. Especially complex features, such as the eye, appear so unlike any possible earlier structure from which they may have been derived that the question of their origin greatly puzzled early students of evolution.

But when we consider eyes in more detail, we find that they have in fact evolved many times, as shown by the fundamental differences between the eyes found in different groups or organisms. Some eyes consist of a single photoreceptor cell, and others are complex, such as those of the vertebrates,

with focusing lenses and color sensitivity. Zoologists, analyzing these differences, have calculated that eyes have evolved independently, and through intermediate stages, at least 38 different times in various groups of animals (Figure 34-A).

In other words, not all of the structures we call eyes are homologous with one another. By analyzing and understanding their similarities and differences, biologists can more precisely interpret the relationships among the groups in which they occur and more accurately understand the outlines of their evolution. Thus the difficult problem of how eyes evolved, when it is solved, assists us in interpreting broad evolutionary patterns of great interest.

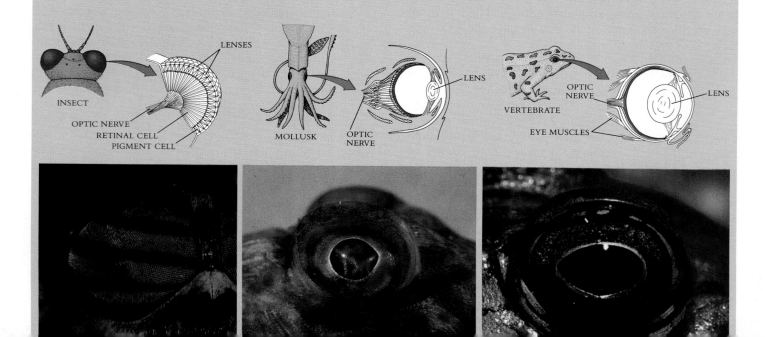

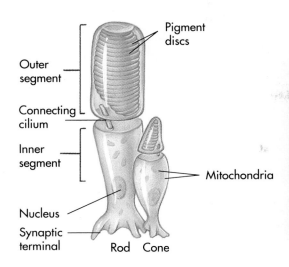

**Figure 34-17**

**Rods and cones.** The broad tubular cell diagrammed on the right is a rod, the shorter, tapered cell next to it, a cone. Although not obvious from the electron micrograph, the pigment-containing outer segments of these cells are separated from the rest of the cells by a partition through which there is only a narrow passage, the "connecting column."

vertebrates the *cis*-retinal pigment is coupled to a transmembrane protein called **opsin** to form **rhodopsin** and **iodopsin.**

The visual pigment in vertebrate eyes is located in the tips of specialized sensory cells called **rod** and **cone cells** (Figure 34-17). Rod cells, which contain rhodopsin, are responsible for black-and-white vision. Cone cells, which contain iodopsin, function in color vision. There are three different kinds of cone cells, each with a different kind of iodopsin molecule that preferentially absorbs light of different wavelengths (Figure 34-18). In each case, photons of light hyperpolarize the photoreceptors via a specific second messenger system.

Vertebrate eyes are lens-focused eyes (Figure 34-19). In them, light first passes through a transparent layer, the **cornea,** which begins to focus the light onto the rear of the eye. Light then passes through the **lens,** a structure that completes the focusing. The lens is a thick disk filled with transparent jelly, somewhat resembling a flattened balloon. In mammals the lens is attached by suspending ligaments to ciliary muscles. When these muscles contract, they change the shape of

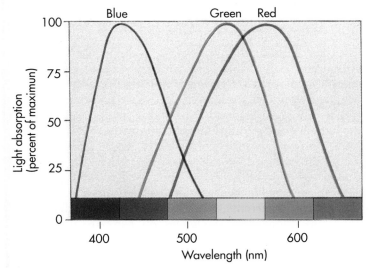

**Figure 34-18**

**The absorption spectrum of human vision.** The wavelength of light absorbed by the visual pigment of cone cells is shifted, depending on the protein to which the pigment is bound. There are three such proteins, producing cones which absorb at 455 nanometers (blue), 530 nanometers (green), and 625 nanometers (red).

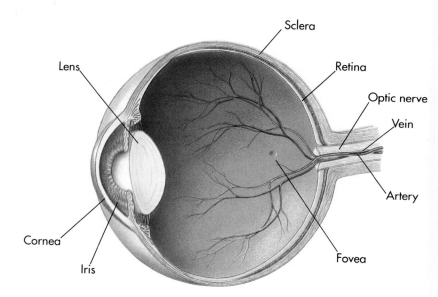

**Figure 34-19**

**Structure of the human eye.** Light passes through the transparent cornea and is focused by the lens on the rear surface of the eye, the retina, at a particular location called the fovea. The retina is rich in rods and cones.

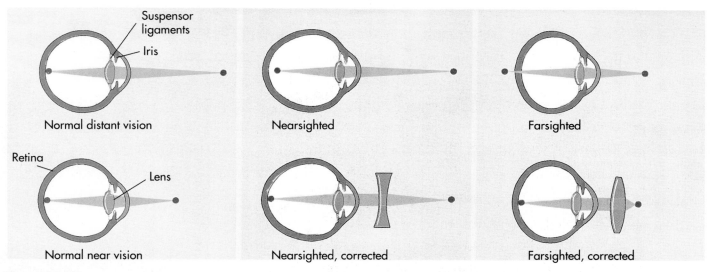

Normal distant vision     Nearsighted     Farsighted

Normal near vision     Nearsighted, corrected     Farsighted, corrected

*Figure 34-20*

**Focusing the human eye.** Contraction of ciliary muscles pulls on suspensor ligaments and changes the shape of the lens, which alters its point of focus forward or backward. In nearsighted people the ciliary muscles place the point of focus in front of the fovea rather than on it. The problem can be corrected with glasses or contact lenses, which extend the focal point back to where it should be. In farsighted people, the ciliary muscles make the opposite error, placing the point of focus behind the retina. Corrective lenses can shorten the focal point.

the lens (Figure 34-20) and thus the point of focus on the rear of the eye. In amphibians and fishes the lens does not change shape. These animals instead focus their images by moving the lens in and out, thus operating the same way that a camera does. In all vertebrates the amount of light entering the eye is controlled by a shutter, called the **iris,** between the cornea and the lens. The iris reduces the size of the transparent zone, or **pupil,** of the eye through which the light passes.

The field of receptor cells, rods and cones, that lines the back of the eye is the **retina.** The retina contains about 3 million cones, most of them located in the central region of the retina called the **fovea,** and approximately 1 billion rods. The eye forms a sharp image in the central fovea region of the retina, which is a region composed almost entirely of cone cells.

Each foveal cone cell makes a one-to-one connection with a special neuron called a **bipolar cell** (Figure 34-21). Each bipolar cell is connected in turn to an individual visual ganglion cell, whose axon is part of the **optic nerve.** The optic nerve transmits visual impulses more or less directly to the brain. The frequency of pulses transmitted by any one receptor provides information about light intensity. The pattern of action potentials present in different foveal axons provides a point-to-point image of the visual field onto the visual cortex. It also contains information related to the shape of an object, its orientation, and its movement. Three different types of cone cells (red, blue, and green) provide information about the color of the image.

Most vertebrates have two eyes, one on each side of the head. When both are trained on the same object, the image that each sees is slightly different because eye views the object from a different angle. The bilateral

location of eyes, with its slight displacement of images (an effect called **parallax**), permits sensitive depth perception. By comparing the differences between the images provided by the two eyes with the physical distance to specific objects, vertebrates learn to interpret different degrees of disparity between the two images as representing different distances—stereoscopic vision (Figure 34-22). We are not born with the ability to perceive distance; we learn it. Stereoscopic vision develops in babies only over a period of months.

## Other Environmental Senses in Vertebrates

Vision is the primary sense used by all vertebrates that live in a light-filled environment, but the wavelength and intensity of visible light are by no means the only stimuli available to vertebrates for assessing this environment. Animals are known that sense polarized light (the octopus), ultrasound (bats), heat (snakes), electricity (catfish), and perhaps magnetism (birds). Most vertebrate sensory systems evolved early, while vertebrates were still confined to the sea. Some, such as sensors that detect electric fields, are not used by land vertebrates, presumably because air is a very poor conductor of electric currents. Others, such as sensors that respond to liquid pressure waves on fish, are used by land vertebrates to detect pressure waves in air, to hear. Because the terrestrial environment is so physically different from the sea, it has presented some new sensory opportunities. Air, for example, transmits heat much better than water does, and snakes have evolved heat sensors that detect heat sources and in doing so provide quite detailed three-dimensional images (Figure 34-23).

No one vertebrate has a perfect repertoire of sensory systems, able to finely discriminate all potential

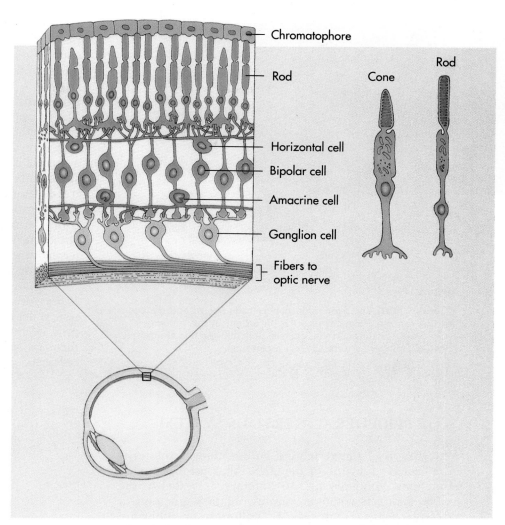

**Figure 34-21**

**Structure of the retina.** Note that the rods and cones are at the rear of the retina, not the front. Light passes through several layers of ganglion and bipolar cells before it reaches the rods and cones.

cues. Rather, vertebrates have evolved sensory nervous systems that meet particular evolutionary challenges. Thus some hunting mammals such as dogs have evolved highly capable olfactory systems, but primates have not. It is of some interest to note, for example, the detailed sensory information in our own environment that we could potentially utilize if we possessed the proper sensory receptors. We cannot utilize thermal radiation, as snakes do, or ultrasound, as bats do, to construct a three-dimensional image.

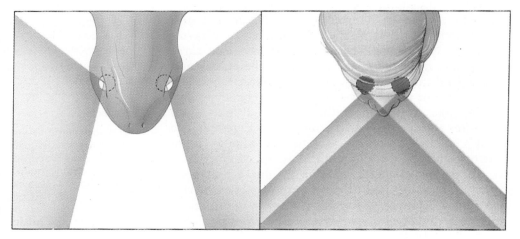

**Figure 34-22**

**Stereoscopic vision.** Vertebrates have eyes located toward the front of the head, so the two fields of vision overlap. When eyes are located on the sides of the head, the two vision fields do not overlap and stereoscopic vision does not occur.

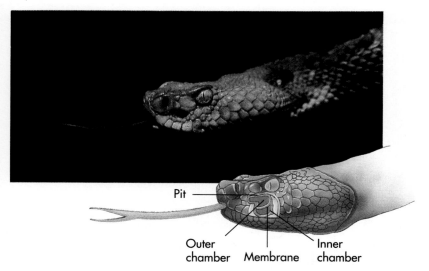

Pit

Outer chamber    Membrane    Inner chamber

**Figure 34-23**

"Seeing" heat. The depression between the nostril and the eye of this rattlesnake opens into the pit organ. In the cutaway portion of the diagram you can see that the organ is composed of two chambers separated by a membrane.

## THE PERIPHERAL NERVOUS SYSTEM

Integration of the vertebrate body's many activities is the primary function of the CNS. All control of bodily functions, voluntary and involuntary, is vested in the CNS. It directs voluntary and involuntary functions in different ways (Figure 34-24). **Voluntary** functions are movements of skeletal (striated) muscle that are directed by somatic motor pathways from the brain and spinal cord (Figure 34-25). These are the pathways that coordinate your fingers when you grasp a pencil, spin your body when you dance, and put one foot ahead of the other when you walk. The direction of these muscular movements by the CNS is subject to conscious control by the motor regions of the cortex.

Many functions directed by the CNS are not subject to conscious control; these are referred to as **involuntary** or **autonomic**. The motor pathways that carry commands from the CNS to regulate the glands and nonskeletal muscles of the body are collectively called the autonomic nervous system. The autonomic nervous system takes your temperature, monitors your blood pressure, and sees to it that your food is digested. The body's *internal condition* is thus maintained within narrow limits by the autonomic nervous system.

Both the voluntary and the involuntary nervous systems are motor pathways. In both cases, sensory input from receptors provides the CNS with information. The CNS processes this information, reaches decisions

about the appropriate responses, and then issues a series of commands to accomplish these responses. These commands travel to the body's muscles and organs along motor pathways (see Figure 34-24).

### Neuromuscular Control

The voluntary nervous system directs the striated muscles of the body and regulates their contraction by means of the stretch receptors described earlier. When a muscle extends or contracts, receptors in the muscle spindle are depolarized, initiating a nerve impulse that travels along afferent nerve fibers to a cell body in the dorsal root ganglion outside the spinal cord. Fibers from the dorsal root ganglion make synaptic contact with neurons in the middle of the spinal cord, usually termed **interneurons**. These interneurons in turn affect the motor neurons controlling the muscles.

*Reflex Arcs.* In animals of relatively simple construction, a nerve impulse passes through the system of neurons, eventually reaching the body muscles and causing them to contract. The motion that results is called a **reflex**, because it is an automatic consequence of the

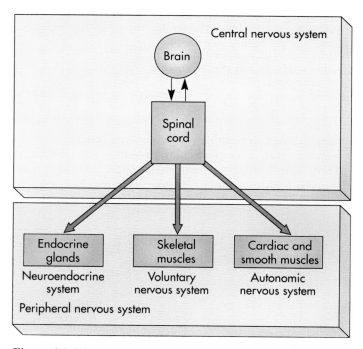

**Figure 34-24**

**How the central nervous system issues commands.** The central nervous system issues commands via three different systems. The neuroendocrine system is a network of endocrine glands whose hormone production is controlled by commands from the central nervous system. The voluntary nervous system is a network of nerves that extends to the striated muscles. The involuntary or autonomic nervous system is a network of nerves that extends to cardiac and smooth muscles and some glands.

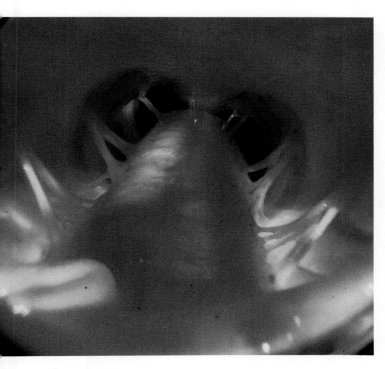

**Figure 34-25**

**A view down the human spinal cord.** Pairs of spinal nerves can be seen extending out from it. It is along these nerves that the central nervous system, consisting of the brain and spinal cord, communicates with the body.

nerve stimulation. The simple chain of events is called a **reflex arc;** in this case the impulse passes from sensory receptor → motor neuron → muscle.

The simplest reflex arcs are **monosynaptic reflex arcs,** in which the afferent nerve cell makes synaptic contacts directly with a motor neuron in the spinal cord whose axon travels directly back to the muscle. A monosynaptic reflex arc is relatively stable but can be influenced by motor efferents from higher levels of the CNS. All the voluntary muscles of your body possess such monosynaptic reflex arcs, although usually in conjunction with other more complex feedback pathways. It is through these more complex paths that voluntary control is established.

In the few cases where the monosynaptic reflex arc is the only feedback loop present, its function can be clearly seen. The well-known "knee jerk" is one such reflex (Figure 34-26). If the patellar ligament just below the kneecap is struck lightly by the edge of your hand or by a doctor's rubber hammer, the sudden pull that results stretches the muscles of the upper leg, which are attached to the ligament. Stretch receptors in these muscles immediately send an impulse along afferent nerve fibers to the spinal cord, where these fibers synapse directly with motor neurons that extend back to upper leg muscles, stimulating them to contract and the leg to jerk upward. Such reflexes play an important role in maintaining posture. Most reflexes are more complex than the knee jerk. Reflexes are important components of many movements (Figure 34-27).

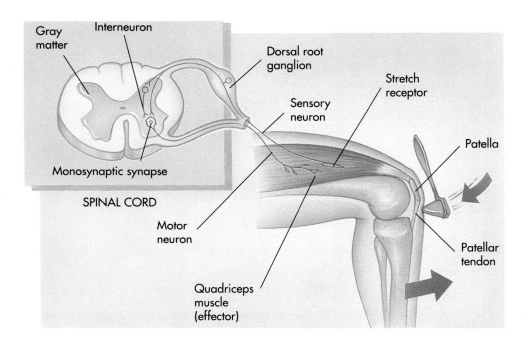

**Figure 34-26**
The knee-jerk reflex.

**Figure 34-27**

**Yawn.** There are some reflexive behaviors that are common to almost all mammals. Do you know any mammal that doesn't yawn?

## Neurovisceral Control

The glands, smooth muscle, and cardiac muscle of the body (the "viscera") respond to a second network of motor neurons, the autonomic nervous system. This system is the other part of the CNS's motor command network. It plays a major role in controlling the body's internal environment and thus homeostasis.

***The Autonomic Nervous System.*** Control of the body's glands and involuntary muscles is carried out by the autonomic nervous system, composed of two elements usually acting antagonistically (in opposition to each other) (Figure 34-28):

1. The **parasympathetic nervous system** consists of a network of (a) long efferent axons that make synaptic contact with ganglia in the immediate vicinity of an organ and (b) short efferent neurons extending from the ganglia to the organ.
2. The **sympathetic nervous system** is composed of a network of (a) short efferent CNS axons that extend to ganglia located near the spine and (b) long efferent neurons extending from the ganglia directly to each target organ.

***Control of the Autonomic Nervous System.*** Autonomic nerve impulses must cross two synapses in traveling from the CNS out to its target organ, whether they travel along sympathetic or parasympathetic nerves (Figure 34-29). The first synapse is in the ganglion, between the axon of a neuron extending from the CNS and the dendrites of the autonomic neuron's cell body. The second synapse is between the autonomic neuron's axon and the target organ. The neurotransmitter in the ganglion is acetylcholine for both sympathetic and parasympathetic nerves. However, the neurotransmitter between the terminal autonomic neuron axon and the target organ is different in the two antagonistic elements of the autonomic nervous systems. In the parasympathetic system the neurotransmitter at the terminal synapse is acetylcholine, just as it is in the ganglion. In the sympathetic system the neurotransmitter at the terminal synapse is either adrenaline (epinephrine) or noradrenaline (norepinephrine), both of which have an effect *opposite* to that of acetylcholine.

Most glands (except the adrenal gland), smooth muscles, and cardiac muscle controlled by the auto-nomic nervous system have inputs from *both* the sympathetic and the parasympathetic systems. Usually these two systems antagonize one another, but either may be excitatory. Thus, depending on which of the two components of the autonomic nervous system is selected by the CNS, an arriving signal will either stimulate or inhibit the organ.

Each gland, smooth muscle, and cardiac muscle constantly receives stimulatory signals through one nerve and inhibitory signals by way of the other nerve. The CNS controls activity in each case by varying the ratio of the two signals.

Thus an organ receiving nerves from both components of the autonomic nervous system will be subject to the effects of two opposing neurotransmitters. If the sympathetic nerve ending excites a particular organ, the parasympathetic synapse usually inhibits it. For example, the sympathetic system speeds up the heart and inhibits gastrointestinal motility and secretion, inhibiting digestion, whereas the parasympathetic system slows down the heart and increases gastrointestinal activity. In many cases the opposing systems are organized so that the parasympathetic system stimulates the activity of normal body functions; for example, the churning of the stomach, the contractions of the intestine, and the secretions of the salivary glands. The sympathetic system, on the other hand, generally mobilizes the body for greater activity, as in increased respiration or a faster heartbeat.

## CHEMICAL TRANSMITTERS IN THE CNS

The effect of a chemical neurotransmitter is not a function of its chemical identity, but rather depends on the nature of the postsynaptic ion channel activated by that neurotransmitter. Therefore a particular compound may be excitatory or inhibitory in its effects. There are many neurotransmitters in the CNS, and they fall into several chemical categories. In most cases a neurotransmitter has been "assigned" a particular role based on studies showing that drugs known to interfere with or augment the effect of a transmitter have consistent behavioral effects.

Acetylcholine was the first neurotransmitter to be

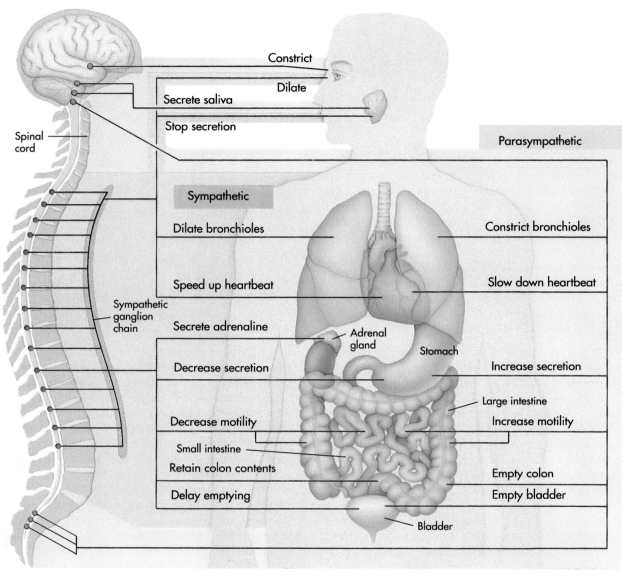

**Figure 34-28**

**The sympathetic and parasympathetic nervous systems.** The ganglia of sympathetic nerves are located near the spine, and the ganglia of parasympathetic nerves are located far from the spine, near the organs they innervate. As you can see, a nerve path runs from both of the systems to every organ indicated, except the adrenal gland.

The following labels appear in the figure: Constrict; Dilate; Secrete saliva; Stop secretion; Spinal cord; Parasympathetic; Sympathetic; Dilate bronchioles; Constrict bronchioles; Speed up heartbeat; Slow down heartbeat; Sympathetic ganglion chain; Secrete adrenaline; Adrenal gland; Stomach; Decrease secretion; Increase secretion; Large intestine; Decrease motility; Increase motility; Small intestine; Retain colon contents; Empty colon; Delay emptying; Empty bladder; Bladder.

discovered in the autonomic nervous system, but it is widely used in the brain as a central neurotransmitter. A deficiency of acetylcholine, probably resulting from degeneration and death of neurons, has been suggested as a cause of the type of dementia seen in the elderly referred to as **Alzheimer's disease.** Norepinephrine, the other major autonomic transmitter, probably affects arousal and some emotional responses.

Both acetylcholine and norepinephrine are chemi-cals known as **monoamines.** Other monoamine trans-mitters are dopamine and serotonin. Dopamine ap-pears to be involved in emotional behavior, with excess dopamine causing schizophrenia. Dopamine inhibitors are used in the treatment of these diseases. Dopamine is also important in the motor systems. Deficiencies in dopamine cause the shaking and wild movements seen in Parkinson's disease, which can be effectively treated by administration of the precursor to dopamine, L-

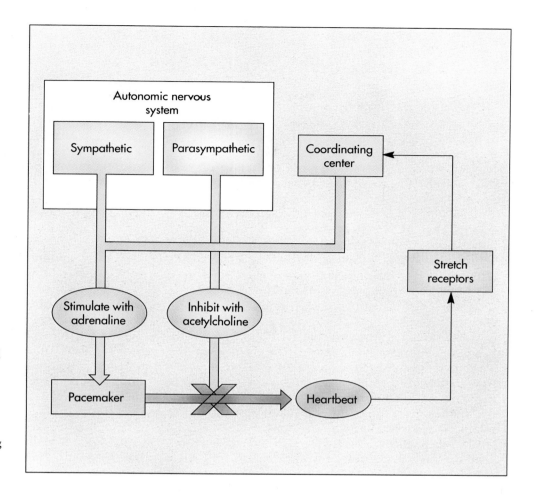

**Figure 34-29**

**The control of heartbeat.** The control of heartbeat occurs at two levels:
1 Each heartbeat is initiated by the pacemaker, which is stimulated or inhibited by signals from the autonomic nervous system.
2 Stretch receptors in heart muscle provide information to a coordinating center in the brain, which modifies the autonomic commands.

dopa. Overproduction of dopamine may be the cause of some depressive illnesses. Serotonin appears to be involved in the regulation of the sleep cycle.

Amino acids are another class of neurotransmitters. Glutamate is probably the main excitatory transmitter in the CNS, whereas glycine is one of the major inhibitory transmitters. GABA is also inhibitory, and GABA deficits seem to cause certain types of anxiety. Antianxiety drugs such as diazepam (Valium) increase GABA levels.

Learning and memory have begun to be understood in terms of long-term synaptic changes, some of which appear to be mediated by a family of chemical transmitters termed **neuromodulators.** The effects of neuromodulators on neuronal activity are more subtle than the brief changes in ion permeability that result in postsynaptic potentials. In many cases these substances seem to cause structural modifications of ion channel proteins, so that their response to a particular transmit-

ter is altered. Many neuromodulators are small proteins called **neuropeptides.** Examples include vasopressin (antidiuretic hormone), somatostatin, oxytocin, the enkephalins, and the endorphins. The distinction between neuromodulators and transmitters is not a matter of chemical identity. The difference lies in the time scale and nature of the effect of the substance on information processing. Transmitters mediate effects that are immediate and quickly over. Neuromodulators mediate effects that are slow and more lasting and that typically involve second messengers within the cell. These messengers modify the behavior of membrane receptors for a neurotransmitter or the operation of voltage-gated ion channels.

In the next chapter we consider how animals move. In animals, movement is achieved through the movement of muscle, whose control rests with the nervous system.

# Drugs and the Central Nervous System

Many of the drugs abused in the United States affect pathways in the CNS related to the experience of pain and pleasure. To understand their effects it is first necessary to discuss the mechanism of pain perception. An injury severe enough to damage tissue results in the release of several neuromodulators at the injury site, including prostaglandins and bradykinin. These substances enhance the activity of nerve endings that initiate pain messages. These sensory impulses pass along pain afferents into the spinal cord, where synaptic connections with ascending interneurons allow the information to be passed on to the brain. The neurotransmitter released by pain-sensitive sensory neurons is a peptide called substance P.

The intensity with which pain is perceived depends on the emotional state of the pain sufferer. Variations in pain perception are attributable in part to the effects of peptide neuromodulators called enkephalins. These substances are released by neurons descending from the brain and making presynaptic inhibitory connections with neurons in the spinal cord. Enkephalin release appears to reduce or even shut off the passage of pain information into the nervous system. Opium derivatives, such as morphine or heroin, have an analgesic, or pain-reducing, effect because they are similar enough in chemical structure to bind to the receptors normally utilized by enkephalins.

In addition to the opiate receptors present in the body's pathways, similar receptors seem to be present in the limbic system of the brain, an area concerned with emotional behavior and the experience of pleasure. Administration of drugs such as heroin and cocaine causes emotional "highs" by directly stimulating these areas of the brain. Cocaine is particularly effective because, when inhaled, it immediately passes into the limbic areas of the brain. The enkephalins have been used in an attempt to treat drug addiction but, unfortunately, they themselves appear to be addictive.

What about other drugs? Alcohol is a general CNS depressant, inhibiting respiration and numbing pain. These effects are believed to result from a general structural alteration of the cell membrane similar to that produced by general anesthetics. The euphoria caused by alcohol is less well understood and is the focus of much current research. Amphetamines cause the release of monoamine neurotransmitters such as norepinephrine, causing an increase in alertness while simultaneously suppressing appetite. They are therefore often abused by individuals attempting to lose weight. The hallucinogenic drug LSD (lysergic acid) interacts with the neurotransmitter serotonin, causing euphoria and hallucinations. The mechanism of action of another hallucinogenic drug, marijuana, has not yet been determined with any degree of certainty. In all these cases it is important to determine the biological mechanism of action, not only to treat abuse, but also to learn more about how the brain performs its functions.

## ■ SUMMARY

1. The vertebrate nervous system is made up of the central nervous system (CNS), consisting of the brain and the spinal cord, and the peripheral nervous system. Within the peripheral nervous system, sensory pathways transmit information to the CNS, and motor pathways transmit commands from it.

2. The motor pathways are divided into somatic, or voluntary, pathways, which relay commands to skeletal muscles, and autonomic, or involuntary, pathways, which stimulate the glands and other muscles of the body.

3. A vertebrate brain consists of a hindbrain, or rhombencephalon; a midbrain, or mesencephalon; and a forebrain, or prosencephalon. The hindbrain was the principal component of the brains of early vertebrates, with the forebrain becoming increasingly dominant in reptiles, amphibians, birds, and mammals.

4. In birds and mammals the brain is much larger in proportion to the body than is the case in other vertebrates, reflecting a great increase in the size of the cerebrum.

5. In humans and other primates the cerebrum is split into two hemispheres, connected by a nerve tract called the corpus callosum. Each hemisphere is divided further by deep grooves into four lobes, the frontal, parietal, temporal, and occipital lobes. The lobes have different functions.

6. Sensory receptors respond to three classes of stimuli: mechanical, chemical, and electromagnetic. No matter what type of stimulus, a response usually takes the same form, depolarization of a sensory neuron membrane.

7. Many of the body's internal receptors are simple receptors in which a nerve ending becomes depolarized as a response to deformation of a nerve membrane or to a chemical- or temperature-induced opening of ion channels in the nerve membrane.

8. Hearing organs in terrestrial vertebrates amplify airborne sound waves and direct them at a fluid-containing chamber, the cochlea, within the ear. Mechanical receptors in the chamber are then deformed by the fluid-borne sound waves.

9. Vision, like photosynthesis, uses a pigment as a primary photoreceptor. The vertebrate eye is designed as a lens-focused camera. The fovea at the center of the retina transmits a point-to-point image to the brain.

10. Other stimuli sensed by vertebrate receptors include polarized light, ultrasound, heat, electricity, and magnetism.

11. The somatic muscles are directed by the motor neurons of the CNS. The activity of the CNS in directing the somatic muscles is modulated by stretch receptors, which are embedded in the muscles linked to afferent fibers returning to the CNS.

12. Smooth muscles, cardiac muscles, and glands are directed by antagonistic command nerve pairs of the autonomic nervous system, one of which stimulates while the other inhibits. In general the parasympathetic nerves stimulate the activity of normal internal body functions and inhibit alarm responses, and the sympathetic nerves do the reverse.

## REVIEW

1. The three principal divisions that characterize the brains of all living vertebrates are _____ , _____ , and _____ .

2. What part of your forebrain is responsible for learning?

3. Short-term memory appears to be stored _____ , whereas long-term memory seems to involve _____ in the neural connections within the brain.

4. The visual pigment is _____ .

5. The neurotransmitter at the terminal synapse in the sympathetic nervous system is either _____ or _____ .

# SELF-QUIZ

1. Bundles of nerve fibers in the CNS are called
   (a) tracts.
   (b) dendrites.
   (c) axons.
   (d) nerves.
   (e) ganglia.

2. The hindbrain is devoted primarily to
   (a) sensory integration.
   (b) processing of olfactory information.
   (c) processing of visual information.
   (d) coordinating motor reflexes.
   (e) integrating visceral activities.

3. The hypothalamus controls
   (a) respiration.
   (b) sensory integration.
   (c) heartbeat.
   (d) body temperature.
   (e) all but b.

4. The ability to hear often decreases with age because
   (a) the cilia degenerate.
   (b) the hair cells stiffen.
   (c) the flexibility of the basilar membrane changes.
   (d) the tympanic membrane breaks.
   (e) the tympanic canal straightens.

5. The autonomic nervous system is composed of which two of the following elements?
   (a) The sympathetic nervous system
   (b) The sympatric nervous system
   (c) The voluntary nervous system
   (d) The peripheral nervous system
   (e) The parasympathetic nervous system

6. The transmitter at autonomic ganglia is
   (a) Norepinephrine
   (b) Acetylcholine
   (c) Serotonin
   (d) Glycine
   (e) Glutamic acid

7. The nature of the action of a chemical neurotransmitter depends primarily on the
   (a) chemical released from the synapse.
   (b) ion channel activated by the neurotransmitter.
   (c) width of the synaptic cleft.
   (d) enzymatic hydrolysis.
   (e) structure of the synapse.

# THOUGHT QUESTIONS

1. Many writers have stated that the hindbrain of the human brain is its most primitive element. How would you respond to this proposition?

2. Most of us have sensed at one time or another an oncoming storm by detecting the increase in humidity in the air. What sort of receptors detect humidity? Why do you suppose that hot days seem so much hotter when it is humid?

3. The literature of science fiction is awash with stories of extrasensory perception, and some research laboratories are actively engaged in attempts to demonstrate it. Can you describe a sensory receptor that might function in "extrasensory" perception?

4. The heat-detecting pit receptor of snakes is a very effective means of "seeing" at night. It is the same sort of sensory system employed by soldiers in "snooper-scopes" and in heat-seeking missiles. Why do you suppose other night-active vertebrates, such as bats, have not evolved this sort of sensory system?

# FOR FURTHER READING

ALLPORT, S.: *Explorers of the Black Box: The Search for the Cellular Basis of Memory,* W.W. Norton & Co., New York, 1986. A vivid account of the pioneering studies of Eric Kandel and others in their efforts to demonstrate how we remember. Easy to read, this book shows scientists in action, gathering data and disputing among themselves about what the data mean.

FINKE, R.: "Mental Imagery and the Visual System," *Scientific American,* May 1986, pages 84-92. Explains the theory that the way in which visual information is perceived and carried to the brain has a strong influence on mental imagery.

GIBBONS, B.: "The Intimate Sense of Smell," *National Geographic,* September 1986, pages 324-361. A detailed account of the human sense of smell; interesting and well illustrated.

MASLAND, R.: "The Functional Architecture of the Retina," *Scientific American,* December 1986, pages 102-111. A close look at how neurobiologists are beginning to study the paths of individual neurons by selectively injecting single cells with fluorescent dyes.

MILLER, J.A.: "A Matter of Taste," *BioScience,* vol. 40(2), 1990, pages 78-82. Innovative methods and materials have made possible new insights into the way our sense of taste works.

WINTER, P., and J. MILLER: "Anesthesiology," *Scientific American,* April 1985, pages 124-131. A clear description of how certain drugs act to block consciousness.

# How Animals Move

This ermine, a short-tailed weasel, is leaping after a mole, a small rodent that may provide its next meal. Of all the five kingdoms, only the animals actively move over the land, using intricate combinations of muscle contraction to propel themselves over the surface.

# HOW ANIMALS MOVE

## Overview

One of the obvious differences between animals and plants is that most animals move about from one place to another and plants don't. Whether it is swimming, flying, or walking, movement almost always requires that an individual shift the position of a limb against a resistance. This is done by contracting muscles attached to the limb. In many invertebrates the muscles are attached to an outer skeleton, which must become thicker as the animal becomes larger to perform its function as an anchor point. In vertebrates, on the other hand, muscles are often attached to bones, and muscle contraction moves one bone relative to another within a flexible skin that stretches to accommodate the movement. The mechanism for all muscle contraction is the interaction between the proteins actin and myosin, the basis for what is known as the sliding filament model of striated muscle contraction.

*For Review*   *Here are some important terms and concepts that you will encounter in this chapter. If you are not familiar with them, you should review them before proceeding.*

**Actin filaments** (Chapter 4)

**ATP** (Chapter 6)

**Vertebrate tissues** (Chapter 32)

**Depolarization** (Chapter 33)

**Figure 35-1**

**Vertebrate motion.** A mountain lion springs into motion, attempting to seize a snowshoe hare. The integration of nerves and muscles and the flexibility of the skin combine to make this leap possible.

Some animals remain in one place all their adult lives. For example, the barnacles that encrust submerged rocks and the bottoms of ships are immobile. Nevertheless, barnacles use muscles to feed and reproduce themselves. Most animals move about from place to place, often with great speed (Figure 35-1). Many animals are creatures of constant activity, and few stay still for long. Of the three multicellular kingdoms that evolved from protists, only animals explore their environment in this active way; plants and fungi move only by growing, or as the passive passengers of wind and water. To move, all animals use the same basic mechanism, the contraction of muscles. In addition, animals use muscles to move blood, breathe, and process foods. In this chapter we examine how animals use muscles to achieve movement, with our focus on vertebrates.

## THE MECHANICAL PROBLEMS POSED BY MOVEMENT

If you've ever tried to lift a boulder or push down a big tree, you are familiar with the basic problem posed by movement, which is that gravity tends to hold objects in one place. The reason you cannot toss a boulder with your little finger is that gravity is pulling down on the boulder far harder than your finger can push up. To move the boulder, you have to lift with a greater force than gravity is exerting. All motion must meet this simple requirement.

Organisms use the chemical energy of ATP to supply that force. When an ATP molecule is split into ADP and $P_i$ (inorganic phosphate), 7.3 kilocalories of energy per mole (the atomic weight of a substance—in this case ATP—expressed in grams) is made available to do the work of movement. Organisms apply this energy to alter the length of structural elements within the cytoskeletons of muscle cells, causing the cells to shorten. When many muscle cells shorten all at once, they can exert great force.

## The Need for a Skeleton

If this were all there is to movement, organisms would not move. Instead they would simply pulsate as their muscles contracted and relaxed, contracted and relaxed, in futile cycles. For a muscle to produce movement, it must direct its force against another object.

Some soft-bodied invertebrates like slugs, which live on land but have no shell or internal skeleton, move by attaching themselves to the material over which they move.

*Hydroskeletons.* Most soft-bodied invertebrates, which have neither an internal nor an external skeleton, solve the problem of what to direct muscle force against by using the relative incompressibility of the water within their bodies as a kind of skeleton. They simply direct the force of their muscles against the water. This sort of skeleton is called a **hydroskeleton.** Earthworms have hydroskeletons, and so do jellyfish. In this case the action of the muscles expelling the water produces an opposite reaction, causing the animal to move.

*Exoskeletons.* Most animals, however, are able to move because the opposite ends of their muscles are attached to hard parts of their bodies, so that, for example, muscle contraction results in the beating of a wing or the lifting of a leg (Figure 35-2). When the hard body part to which the muscles are attached is a shell

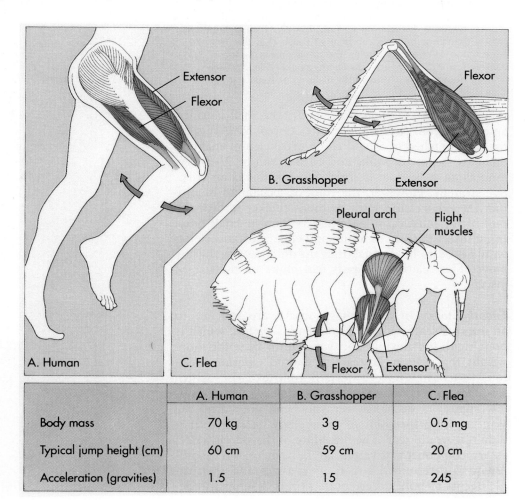

**Figure 35-2**

**An example of the action of muscles in a flea, a grasshopper, and a human being.** Despite differing by several orders of magnitude in size, all three jump to similar heights. In all three, antagonistic muscle pairs control the movement of the legs. Muscles can exert force only by becoming shorter. All three animals shown here have double sets of muscles that work in opposite directions. In the flea and the grasshopper ( **B** and **C**), the muscles are attached to the inside of the skeleton, whereas in the human (**A**), the muscles are attached to the bones. In each case the flexor muscles move the lower leg closer to the body and the extensor muscles move it away. Because the flea muscle is more massive for its size, it is capable of producing much greater acceleration.

|  | A. Human | B. Grasshopper | C. Flea |
|---|---|---|---|
| Body mass | 70 kg | 3 g | 0.5 mg |
| Typical jump height (cm) | 60 cm | 59 cm | 20 cm |
| Acceleration (gravities) | 1.5 | 15 | 245 |

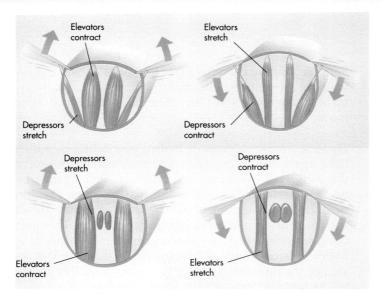

Elevators
contract

Elevators
stretch

Depressors
stretch

Depressors
contract

Depressors
stretch

Depressors
contract

Elevators
contract

Elevators
stretch

*Figure 35-3*

**Muscles in insects.** In most insects, as in birds, the raising and lowering of wings is achieved by the alternate contraction of extensor muscles (elevators) and flexor muscles (depressors). Four insect orders (including flies, mosquitoes, wasps, bees, and beetles), however, beat their wings at frequencies from 100 to more than 1000 times per second, faster than nerves can carry successive impulses! In these insects, the flight muscles are not attached to the wings at all, but rather to the wall of the thorax, which is distorted in and out by their contraction. The reason that these muscles can beat so fast is that the contraction of one set stretches the other, initiating its contraction in turn. Much less energy is expended in such beating than in that of the wings of birds or bats.

that encases or surrounds the body, the shell is called an **exoskeleton.** Arthropods, for example, have muscles that are attached to a rigid chitin exoskeleton, enabling them to swim, to walk, and to fly (Figure 35-3). As long as an individual is small enough, this is an effective strategy. However, the rules of mechanics require that the exoskeleton must be much thicker to bear the pull of the muscles in large insects than in small ones. In an insect the size of a human being, the exoskeleton would need to be so thick that the animal could hardly move. This relationship puts real limits on the size of insects.

*Endoskeletons.* In vertebrates, muscles are attached to an internal scaffold of bone, an **endoskeleton,** which is both rigid and flexible and yet able to bear far more weight than chitin. Instead of a rigid exterior skeleton, vertebrates have a soft, flexible exterior, which stretches to accommodate the movement of their bodies (Figure 35-4). Whenever you bend your arm, the skin covering the joint of your elbow stretches; if it didn't, it would tear. The flexibility of the skin is a necessary component of vertebrate movement, which otherwise is determined largely by muscles and their attachments to bones. Vertebrate skin is described in detail in Chapter 40.

**WALKING**

*Figure 35-4*
On the move.

**CRAWLING**

**RUNNING**

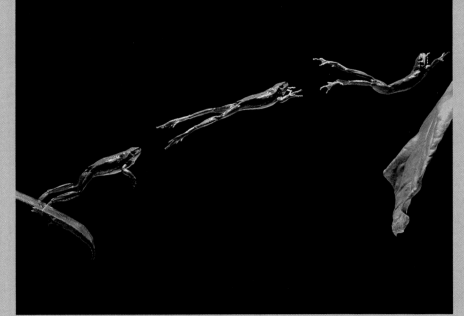

LEAPING

CLIMBING

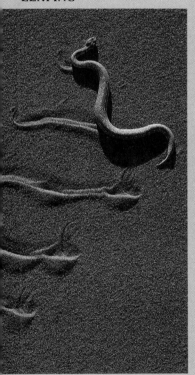

ALONG FOR THE RIDE

SWIMMING

FLYING

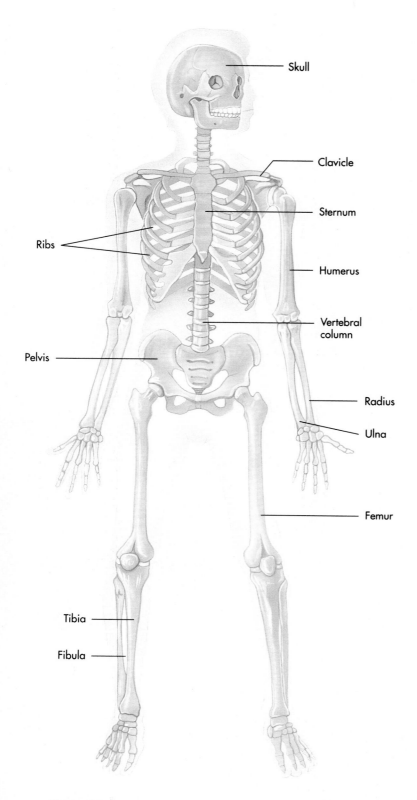

**Figure 35-5**
The human skeletal system.

## THE HUMAN SKELETON

The endoskeleton of humans is made up of 206 individual bones (Figure 35-5). These can be grouped according to their functions. The 80 bones of the **axial skeleton** support the main body axis, and the 126 bones of the **appendicular skeleton** support the arms and legs (Figure 35-6). Interestingly, the motor control systems of the body have evolved to control the muscles of the axial skeleton (postural muscle) and the appendages (manipulatory muscle) more or less independently.

### The Axial Skeleton
The axial skeleton is made up of the skull, backbone, and rib cage. Of the skull's 28 bones, 8 form the cranium that encases the brain; the rest are facial bones

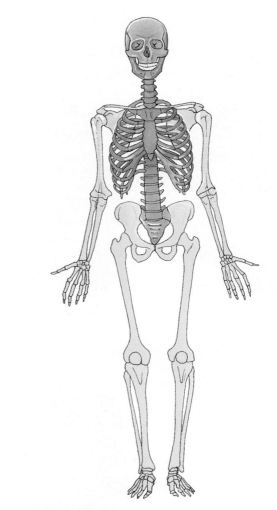

**Figure 35-6**

**Axial and appendicular skeletons.** The axial skeleton is shown in red, and the appendicular skeleton is shown in yellow.

and middle-ear bones. The skull also contains the hyoid bone. It is suspended at the back of the jaw by muscles and a form of connective tissue called a **ligament** and supports the base of the tongue.

The skull is attached to the anterior (upper) end of the backbone, which is also called the **spine** or **vertebral column.** The spine is made up of 33 vertebrae, stacked one on another to provide a flexible column that surrounds and protects the spinal cord. Curving forward from the vertebrae are 12 pairs of ribs, which are attached at the front to the breastbone, or **sternum,** forming a protective cage around the heart and lungs.

## The Appendicular Skeleton

The 126 bones of the appendicular skeleton are attached to the axial skeleton at the shoulders and hips. The shoulder or **pectoral girdle** is composed of two large, flat shoulder blades (scapulae), each connected to the breastbone by a slender, curved collarbone (clavicle). The arms are attached to the pectoral girdle. Each arm and hand contains 32 bones. The clavicle is the most frequently broken bone of the body; if you fall on an outstretched arm, a large component of the force is transmitted to the clavicle.

The **pelvic girdle** forms a bowl that provides strong connections for the legs, which carry the weight of the body. Each leg and foot contains a total of 30 bones.

## BONE: WHAT VERTEBRATE SKELETONS ARE MADE OF

Bone is a special form of connective tissue in which collagen fibers (see Chapter 32) are coated with a calcium phosphate salt. The great advantage of bone over chitin as a structural material is that it is strong without being brittle. To understand the properties of bone, first consider those of fiberglass. Fiberglass is composed of glass fibers embedded in epoxy glue. The individual fibers are rigid, giving great strength, but they are brittle. The epoxy component of fiberglass, on the other hand, is flexible but weak. The composite, fiberglass, is both rigid and strong. When a fiber breaks because of stress and a crack starts to form, the crack runs into glue before it reaches another fiber. The glue distorts and reduces the concentration of the stress, and the adjacent fibers consequently are not exposed to the same high stress. In effect, the glue acts to spread the stress over many fibers.

The construction of bone is similar to that of fiberglass because the collection of the fibrils in bone runs in various directions. Small, needle-shaped crystals of a calcium-containing mineral, hydroxyapatite, surround and impregnate collagen fibrils of bone. The fibrils are placed parallel to the axes of long bones and also parallel to the curved ends of bones in joints. As a result of the way these fibers are placed, no crack can penetrate far into bone without encountering a hard mass of hydroxyapatite crystals embedded in a collagenous matrix. Bone is more rigid than collagen. On the other hand, bone is more flexible and resistant to fracture than is hydroxyapatite—or chitin.

Bone is a dynamic, living tissue that is constantly being reconstructed throughout the life of an individual. New bone is formed by cells called **osteoblasts,** which secrete the collagen fibers on which calcium is later deposited. Bone is laid down in thin, concentric layers called **lamellae,** like so many layers of paint on an old pipe. The lamellae are laid down as a series of tubes around narrow channels called **Haversian canals,** which run parallel to the length of the bone. The Haversian canals are interconnected and contain blood vessels and nerve cells. The blood vessels provide a lifeline to living bone-forming cells, whereas the nerves control the diameter of the blood vessels and thus the flow through them. When bone is first formed in the embryo, osteoblasts use the cartilage skeleton as a template for bone formation. Later, new bone is formed along lines of stress. Mature osteoblasts become trapped by the bone that they lay down, and are then called **osteocytes.** Flat bones, such as the sternum, originate from sheets of dense connective tissue as the osteoblasts do their work.

*Bone is formed in two stages: first, collagen is laid down in a matrix of fibrils along lines of stress, and then calcium minerals impregnate the fibrils. These minerals provide rigidity, whereas the collagen provides flexibility.*

The bones of the vertebrate skeleton are composed of two structural elements. The ends and interiors of long bones are composed of an open lattice of bone called **spongy bone tissue,** or **marrow.** Within this lattice framework, most of the body's red blood cells are formed. Surrounding the spongy bone tissue at the core of bones are concentric layers of **compact bone tissue** (Figure 35-7) in which the collagen fibrils are laid down in a pattern that is far denser than the marrow. The compact bone tissue gives the bone the strength to withstand mechanical stress.

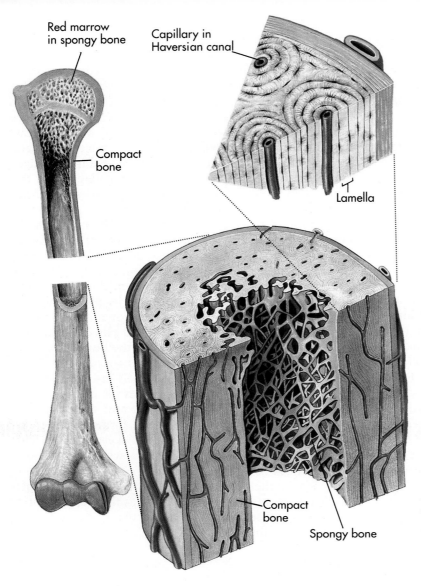

Red marrow in spongy bone

Capillary in Haversian canal

Compact bone

Lamella

Compact bone

Spongy bone

**Figure 35-7**

**The organization of compact bone, shown at three levels of detail.** Some parts of bone are dense and compact, giving the bone strength. Other parts are spongy, with a more open lattice; it is here that most red blood cells are formed.

## JOINTS: HOW BONES OF THE SKELETON ARE ATTACHED TO ONE ANOTHER

The mechanical movement of the body occurs when bones move relative to one another. These interactions occur at **joints,** where one bone meets another. There are three kinds of joints (Figure 35-8):

*Sutures* are nearly immobile joints connected by a thin layer of connective tissue. The cranial bones of your skull are joined by sutures. In a fetus, these bones are not fully formed, and there are open areas of connective tissue between the bones ("soft spots," or fontanels) that allow the bones to slide over one another slightly as the fetus makes its journey down the birth canal during childbirth. Later, bone replaces this connective tissue.

*Cartilaginous joints* are slightly movable joints bridged by cartilage. The vertebral bones of your spine are separated by pads of cartilage called intervertebral disks. The disks allow some movement while acting as efficient shock absorbers.

*Freely movable joints* are swinging joints bridged by pads of cartilage and held together by bands of cartilage called **ligaments.** The pads of cartilage that tip the two bones do not touch. Synovial membranes filled with fluid occur where tendons glide over bones, and this fluid lubricates the gliding of the bones across one another. **Rheumatoid arthritis** is a degenerative and very painful disorder in which the immune system attacks the bursa membrane; white blood cells degrade the cartilage and other connective tissue, and bone is deposited in the joint.

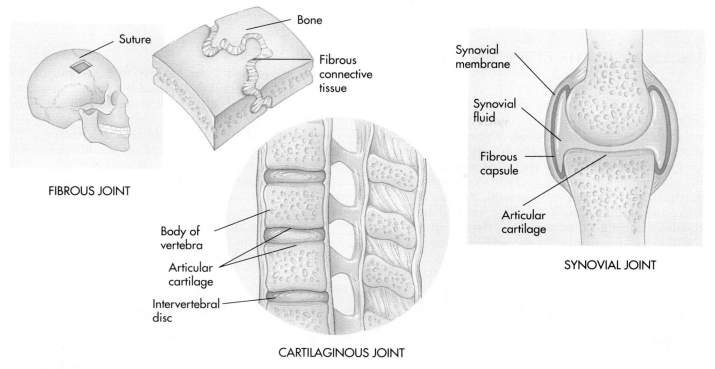

FIBROUS JOINT

Suture

Bone

Fibrous
connective
tissue

Body of
vertebra

Articular
cartilage

Intervertebral
disc

CARTILAGINOUS JOINT

Synovial
membrane

Synovial
fluid

Fibrous
capsule

Articular
cartilage

SYNOVIAL JOINT

*Figure 35-8*

Three types of joints: fibrous, cartilagenous, and synovial.

## Tendons Connect Bone to Muscle

Muscles are attached to bones by straps of dense collagenous connective tissue called **tendons.** Bones pivot about joints because of where the tendons are attached to them. Each muscle acts to pull on a specific bone. One end of the muscle, the **origin,** attaches to a bone that remains stationary during a contraction. This provides an object against which the muscle can pull. The other end of the muscle, the **insertion,** is attached to a bone that moves if the muscle contracts.

In the freely movable joints of vertebrates, muscles are attached in opposing pairs, flexors and extensors (Figure 35-9). When the flexor muscle of your leg contracts, the lower leg is moved closer to the thigh. When the extensor muscle of your leg contracts, the lower leg is moved in the opposing direction, further away. To use the body's energy efficiently, it is necessary to control antagonistic pairs of muscles. It would be futile if, when you lifted an object, the triceps contracted at the same time as the biceps. Your arm would not move.

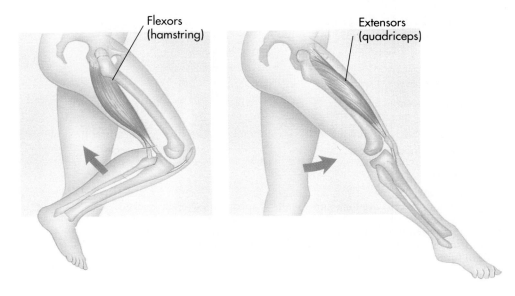

Flexors
(hamstring)

Extensors
(quadriceps)

*Figure 35-9*

Flexor and extensor muscles.

# MUSCLE: HOW THE BODY MOVES

Even though the cells of almost all eukaryotic organisms appear capable of shape changes, many multicellular animals have evolved specialized cells devoted almost exclusively to this purpose. These cells contain numerous filaments of the proteins actin and myosin. Such specialized animal cells are called **muscle cells.** As we mentioned in Chapter 32, vertebrates possess three different kinds of muscle cells: smooth muscle, skeletal muscle, and cardiac muscle. Of these, skeletal and cardiac muscle show clear patterns of striations caused by the geometric arrangement of the actin and myosin filaments.

## Smooth Muscle

Smooth muscle was the earliest form of muscle to evolve, and it is found throughout the animal kingdom. **Smooth muscle** cells are long and spindle shaped, with each cell containing a single nucleus (Figure 35-10, *A*). The interiors of smooth muscle cells are packed with myofibrils composed of actin and myosin filaments, but the individual myofibrils of a cell are not aligned into organized assemblies as they are in skeletal and cardiac muscle. Smooth muscle tissue is organized into sheets of cells. In some tissues smooth muscle cells contract only when they are stimulated by a nerve or hormone. Examples are the muscles found lining the walls of many vertebrate blood vessels and those which make

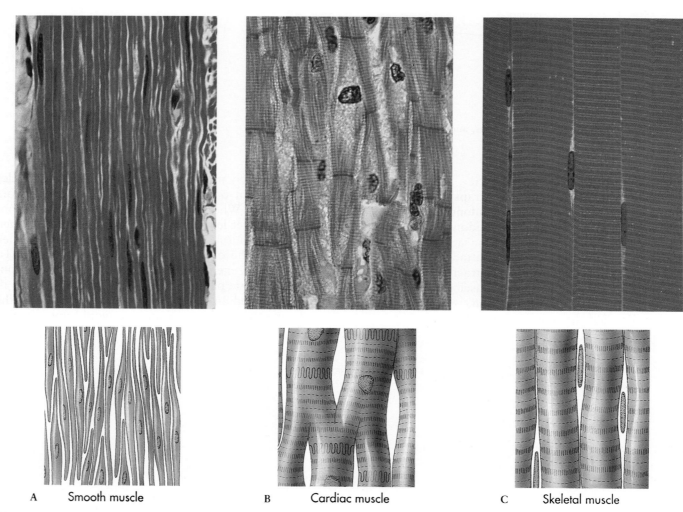

| A | Smooth muscle | B | Cardiac muscle | C | Skeletal muscle |

*Figure 35-10*

**Types of muscle.**
A  Smooth muscle cells are long and spindle shaped, with a single nucleus.
B  Cardiac muscle cells, such as these from a human heart, also contain a single nucleus and are organized into long branching chains that interconnect, forming a lattice.
C  Skeletal or striated muscle cells are formed by the fusion of several muscle cells, end to end, to form a long fiber with many nuclei.

up the iris of the vertebrate eye. In other smooth muscle tissue, such as that found in the wall of the gut, the individual cells contract spontaneously, leading to a slow, steady contraction of the tissue. Smooth muscles are also classified as unitary or multiunit. In unitary smooth muscle, cells are electrically coupled to one another so the muscle contracts as a unit. In multiunit smooth muscle, individual cells are not coupled and must be activated separately.

## Skeletal Muscle

Skeletal muscles are the muscles associated with the skeleton (Figure 35-11). They, along with cardiac muscle, are called **striated muscles** because they are marked with obvious lines (Latin *striae;* see Figure 35-10, C). Skeletal muscle cells are produced during development by the fusion of several cells at their ends to form a very long fiber. Each muscle cell or muscle fiber still contains all the original nuclei pushed out to the periphery of the cytoplasm. Each skeletal muscle is a tissue made up of numerous individual muscle cells that act as a unit. These skeletal muscle cells are specialized for rapid contractions and large forces. In contrast, smooth muscles are specialized for slow, maintained contraction with minimal energy utilization. Imagine a large raft being towed upstream by many small canoes, with each canoe bound to the raft by its own towline. This is analogous to the contraction of smooth muscle, with each smooth muscle cell participating individually in the contraction of the muscle. Now imagine placing all the rowers in one galley where they row in concert, pulling the raft far more effectively. This is analogous to the contraction of skeletal muscle, in which numerous muscle cells pool their resources.

## Cardiac Muscle

The vertebrate heart is composed of striated muscle fibers arranged very differently from the fibers of skeletal muscle. Instead of very long multinucleate cells running the length of the muscle, heart muscle is composed of chains of single cells, each with its own nucleus. Each cell is coupled to its neighbors electrically by gap junctions. These chains of cells are organized into fibers that branch and interconnect, forming a latticework (see Figure 35-10, B). This lattice structure is critical to the way heart muscle functions. Heart contraction is initiated at one location by the opening of transmembrane channels that admit ions into the muscle cells there, altering the voltage difference across their membranes. This change in membrane charge is a depolarization, similar to that discussed in Chapter 33. When two cardiac muscle fibers touch one another, their membranes make an electrical junction, and the electrical depolarization of the initial fiber initiates a

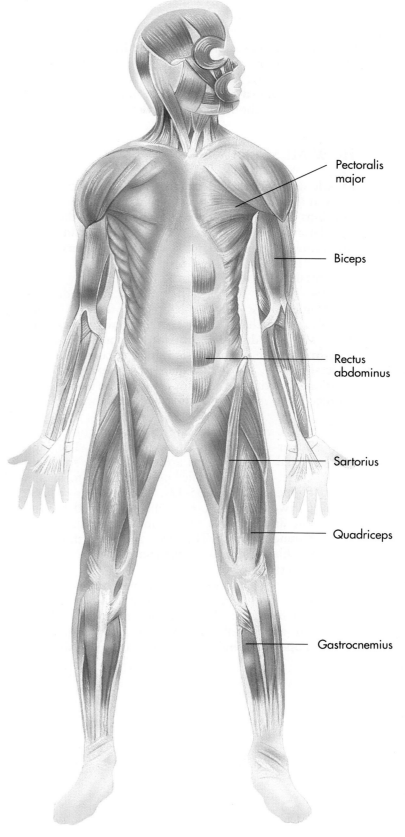

**Figure 35-11**

**The human muscular system.**

wave of contraction throughout the heart. This wave of depolarization rapidly passes from one fiber to another across these junctions. For this reason the heart contracts in an orderly fashion as a unit so as to propel the blood throughout the body.

## HOW MUSCLES WORK

Muscle cells move by expanding and contracting portions of their surfaces. This ability to alter surface relationships arises from dynamic changes in their cytoskeletons. The key elements driving these changes are tiny cables within muscle cells called myofilaments.

### The Structure of Myofilaments

Far too fine to see with the naked eye, individual **myofilaments** are only 6 nanometers thick. Myofilaments are long chains of the proteins **actin** and **myosin.** There are two additional proteins, **troponin** and **tropomyosin,** that are important in initiating contractions in skeletal muscle; we discuss them more later. The organization of actin and myosin is such that muscles are arranged into repeating units called **sarcomeres.**

1. *Actin.* Actin microfilaments are one of the two major components of myofilaments. The individual actin proteins are the size of a small enzyme. Actin molecules polymerize to form thin filaments (Figure 35-12). The filaments consist of two strings of monomers wrapped around one another, like two strands of pearls loosely wound together. The result is a long, thin, helical filament with a diameter of about 6 nanometers.

2. *Myosin.* The other major protein associated with myofilaments, myosin, is a protein molecule more than 10 times longer than an individual actin molecule. Myosin has an unusual shape: one end of the molecule consists of a

very long rod, whereas the other end consists of a double-headed globular region. In electron micrographs a myosin molecule looks like a two-headed snake. Like actin, myosin spontaneously forms into filaments (Figure 35-13). The heads of the myosin molecule are capable of interacting with actin to form what are known as crossbridges.

### How Myofilaments Contract

The contraction of myofilaments occurs when the heads of the myosin filaments change their orientation with respect to their rodlike backbone. The myosin heads tilt in a way similar to flexing the hand at the wrist. As a result of this tilt, the actin filaments are pulled so as to slide past the myosin filaments. In effect, myosin "walks" step by step along actin (Figure 35-14). Each step uses a molecule of ATP.

How does microfilament sliding lead to cell movement? Myofilaments convert the sliding of fibers into motion by anchoring the actin to a particular anchoring protein called **actinin.** Actinin is found widely distributed on the interior surfaces of the plasma membranes of eukaryotic cells. In striated muscle it is the major component of the Z line, which separates individual sarcomeres. Because the actin is not free to move with respect to the actinin to which it is bound, the zones between actinin anchors shorten when the microfilaments slide past myosin. Because the myofilament may be attached at both of its ends to membrane, its overall shortening moves the membranes to which the myofilament is attached.

Myofilaments are composed of a highly ordered complex of actin and myosin. The secret of muscle contraction lies in the way in which the actin and myosin fibers are combined. They *interdigitate.* The arrangement is diagrammed in Figure 35-15, with a myosin filament interposed between two pairs of actin filaments, with the myosin heads (crossbridges) jutting out toward the actin filaments on each side.

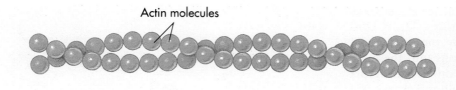

Actin molecules

*Figure 35-12*

**An actin filament.**

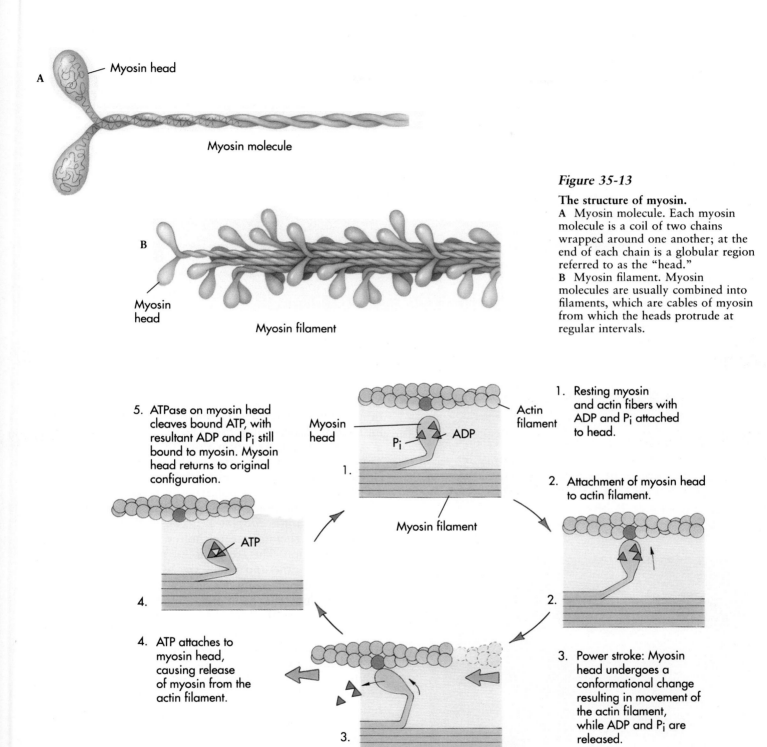

A — Myosin head

Myosin molecule

B

Myosin head

Myosin filament

*Figure 35-13*

**The structure of myosin.**
**A** Myosin molecule. Each myosin molecule is a coil of two chains wrapped around one another; at the end of each chain is a globular region referred to as the "head."
**B** Myosin filament. Myosin molecules are usually combined into filaments, which are cables of myosin from which the heads protrude at regular intervals.

5. ATPase on myosin head cleaves bound ATP, with resultant ADP and $P_i$ still bound to myosin. Mysoin head returns to original configuration.

Myosin head

$P_i$   ADP

Actin filament

1.

Myosin filament

1. Resting myosin and actin fibers with ADP and $P_i$ attached to head.

2. Attachment of myosin head to actin filament.

2.

ATP

4.

4. ATP attaches to myosin head, causing release of myosin from the actin filament.

3.

3. Power stroke: Myosin head undergoes a conformational change resulting in movement of the actin filament, while ADP and $P_i$ are released.

*Figure 35-14*

**The mechanisms of myofilament contraction.** Myosin moves along actin (from left to right in this diagram) by first binding to it and then hunching forward as the result of a change in the shape of the myosin head. The splitting of ATP recocks the mechanism, returning the myosin head to its extended position.

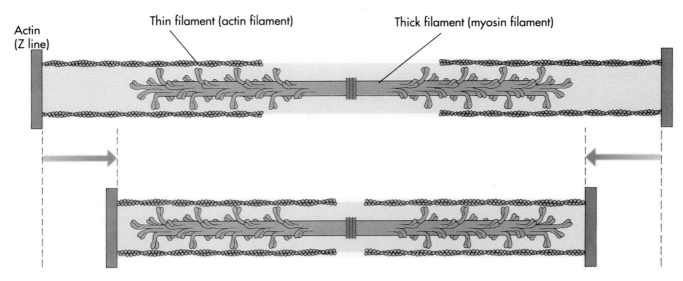

**Figure 35-15**

**The interaction of actin and myosin filaments in vertebrate muscle.** As the right-hand end of the myosin molecule "walks" along the actin filaments, pulling them toward the center, the left-hand end of the same myosin molecule "walks" in a leftward direction, pulling its actin filaments toward the center. The result is that both actins move toward the center—contraction.

*The contraction of vertebrate muscles and many other kinds of cell movement in eukaryotes result from the movements of microfilaments within cells. The microfilaments are composed of long, parallel fibers of actin cross-connected by myosin. Their movement results from a ATP-driven shape change in myosin.*

**Figure 35-16**

**Neuron axons often branch when they connect with skeletal muscle.** The thick central portion is the body of the axon, with small branches each ending in a motor end plate, where the nerve and muscle join to form a neuromuscular junction.

## How Nerves Signal Muscles to Contract

In vertebrate striated skeleton muscle, contraction is initiated by a nerve impulse. The nerve impulse arrives as a wave of depolarization along the nerve membrane. The nerve fiber is embedded in the surface of the muscle fiber, forming a **neuromuscular junction** (Figure 35-16). When a wave of depolarization reaches the end of a neuron, at the point where the neuron attaches to a muscle (the **motor endplate**), it causes that membrane to release the chemical acetylcholine from the nerve cell into the junction. The acetylcholine passes across to the muscle membrane and opens the ion channels of that membrane, depolarizing it.

How does the depolarization of the nerve fiber membrane cause the contraction of the muscle fibers? The endoplasmic reticulum of a striated muscle cell, which is called the **sarcoplasmic reticulum,** wraps around each myofibril like a sleeve (Figure 35-17). As a result, the entire length of every myofibril is very close to the intracellular space bounded by the sarcoplasmic reticulum membrane system. Within the sarcoplasmic reticulum are embedded numerous ion channels that permit calcium to pass through them. In resting muscle, calcium ions are actively pumped into the sarcoplasmic reticulum by an ATP-driven calcium pump, thus concentrating all the calcium ions of that cell within the spaces of the sarcoplasmic reticulum. The depolarization of the muscle fiber membrane opens calcium channels in the sarcoplasmic reticular membrane and thus causes the release of this concentrated calcium into the cytoplasm (sarcoplasm). This calcium acts as a trigger to initiate contraction of the myofibril. It does this in the following way (Figure 35-18):

1. In resting muscle, myosin filaments are not free to interact with actin, because the sites on actin where the myosin heads must make contact are covered by the protein tropomyosin. Another protein, troponin, binds to both actin and tropomyosin.

2. Troponin molecules are also able to bind calcium ions, and when they do, the troponin molecules change their shape. As a result of this change in shape, tropomyosin is repositioned to a new location, where it does not cover the myosin binding sites on actin. Only when this repositioning has occurred can contraction take place. The myosin heads are now free to form crossbridges with actin and, with ATP expenditure, move along the actin in a stepwise fashion to shorten the myofibril.

Thus the release of calcium by the nerve's stimulation of the sarcoplasmic reticulum acts as a "calcium-activated switch," triggering the contraction of the myofibril.

## A VOCABULARY OF MOVEMENT

**actin** One of the two major proteins that make up myofilaments (the other is myosin).

**bone** Hard, flexible material that forms the vertebrate endoskeleton; a fiberglass-like material composed of calcium phosphate salts embedded within a matrix of collagen.

**cartilage** A strong, flexible connective tissue in the skeletons of vertebrates; forms much of the skeleton of embryos, but is largely replaced by bone in adults of most species.

**ligament** A band or sheet of connective tissue that links bone to bone.

**motor endplate** The point where a neuron attaches to a muscle; a neuromuscular synapse.

**myofilaments** A contractile microfilament within muscle cells composed largely of actin and myosin; sometimes called myofibrils.

**myosin** One of the two protein components of myofilaments (the other is actin).

**sarcomere** The fundamental unit of contraction in skeletal muscle; the repeating bands of actin and myosin that appear between two Z lines.

**sarcoplasmic reticulum** The endoplasmic reticulum of a muscle cell; a sleeve of membrane that wraps around each myofilament.

**tendon** A strap of cartilage that attaches muscle to bone.

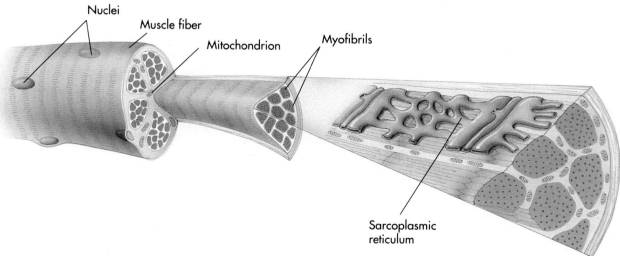

Nuclei
Muscle fiber
Mitochondrion
Myofibrils
Sarcoplasmic reticulum

**Figure 35-17**

**The sarcoplasmic reticulum is a system of membranes that wraps around individual myofibrils of a muscle fiber.** The transverse tubules are continuous with the plasma membrane of the muscle fiber and relay the wave of depolarization initiated at a neuromuscular junction to individual myofibrils.

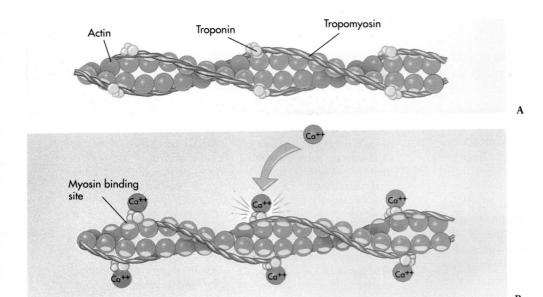

Actin  Troponin  Tropomyosin

A

Myosin binding site

Ca++

Ca++  Ca++  Ca++

Ca++  Ca++  Ca++

B

**Figure 35-18**

**How calcium controls muscle contraction.**
**A** When the muscle is at rest, a long filament composed of the molecule tropomyosin blocks the myosin-binding sites of the actin molecule. Without actin's ability to form links with myosin at these sites, muscle contraction cannot occur.
**B** When calcium ions bind to another protein, troponin, the resulting complex displaces the filament of tropomyosin, exposing the myosin-binding sites of actin, then cross-links form between actin and myosin and contraction occurs.

## How Striated Muscle Contracts

As noted earlier, striated muscle cells are produced during development by the fusion of several cells, end to end, to form a very long fiber (Figure 35-19). A single fiber will typically run the entire length of a verte-brate muscle. Each cell, or muscle fiber, still contains all the original nuclei, pushed out to the periphery of the cytoplasm by a central cable of 4 to 20 myofibrils. The cytoplasm in striated muscle is given a special name, the **sarcoplasm.** The myofibrils that run down

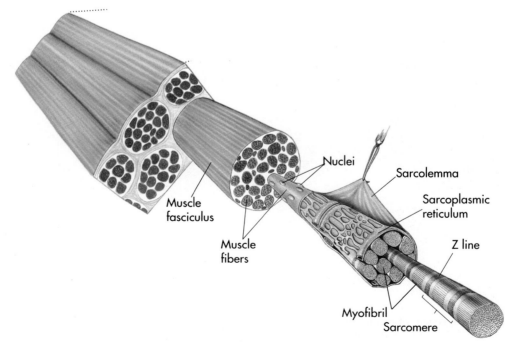

Muscle fasciculus

Muscle fibers

Nuclei

Sarcolemma

Sarcoplasmic reticulum

Z line

Myofibril

Sarcomere

**Figure 35-19**

**The organization of striated muscle.**

the center of a muscle cell are highly organized to promote simultaneous contractions (see Figure 35-19):

1. A myofibril is made up of a long chain of contracting units called **sarcomeres,** lined up like the cars on a train.
2. Each sarcomere is composed of interdigitating filaments of actin and myosin. One collection of actin filaments is attached to the front and another to the back of the sarcomere. These front and back assemblies of actin filaments are not long enough to reach each other in the center of the sarcomere. They are joined to one another by interdigitating myosin filaments.
3. The sarcomere contracts when the heads of the myosin filaments change their shape. Since these heads are bound to the actin filaments, the effect of their contraction is to pull the myosin along the actin.
4. The orientation of myosin within vertebrate muscle is such that it moves along actin toward the Z line. Because both ends of the myosin filaments move in this manner simultaneously, the effect is to pull the two Z lines together, contracting the sarcomere.
5. Simultaneous contraction of all the sarcomeres of a myofibril results in an abrupt and forceful shortening of the myofibril. All the myofibrils of a muscle fiber also usually contract at the same time, producing a very strong contraction in the length of the muscle fiber cell.

The sarcomeres are lined up with one another all along the length of the stacked myofibrils (see Figure 35-10, C). This gives striated muscle, as viewed with a light microscope, the distinctive pattern of bands or striations for which it is named.

*Muscle cells are rich in actin and myosin, which form myofibrils that are capable of contraction. Many muscle cells contracting in concert can exert considerable force.*

## SUMMARY

1. In vertebrates, movement results from the contraction of muscles anchored to bones. When the limb containing the bone is pulled to a new position, the skin stretches to accommodate the change.

2. Bone is a form of connective tissue in which collagen fibers are impregnated with calcium salts. The salts act like the glass fibers in fiberglass, creating a material that is strong without being brittle.

3. There are three kinds of muscle: smooth muscle, which is organized into sheets and contracts spontaneously; skeletal muscle, which is organized into trunks of long fibers and contracts only when stimulated by a nerve; and cardiac muscle, in which the fibers are interconnected and may initiate contraction spontaneously.

4. Muscle cells contract as a result of shortening of myofilaments within the cytoskeleton. The myofilaments are composed of the proteins actin and myosin, together with two other proteins that together control the contraction.

5. In a myofilament the myosin is located between adjacent actin filaments. Changes in the shape of the ends of the myosin molecule, driven by the splitting of ATP molecules, cause the myosin molecule to move along the actin, producing contraction of the myofilament.

6. In vertebrate skeletal muscle, contraction is initiated by a nerve impulse. Acetylcholine passes across the neuromuscular junction from the nerve to the muscle, initiating the process that causes the muscle to contract.

# REVIEW

1. The muscles of a beetle are attached to its _____, whereas your muscles are attached to your _____.

2. Most of the red blood cells in your body are produced in the _____ of long bones.

3. The major components of myofibrils are _____ and _____, both of which spontaneously form into filaments.

4. When troponin molecules bind _____, they change shape and muscle contraction occurs.

# SELF-QUIZ

1. Which is more likely to be found at the core of bones?
   (a) Compact bone tissue
   (b) Spongy bone tissue (marrow)
   (c) Osteoblasts
   (d) Myofilaments
   (e) Nothing; the interior is hollow

2. Haversian canals
   (a) have bone lamellae laid down around them in concentric rings.
   (b) run parallel to the long axes of bones.
   (c) are interconnected.
   (d) have blood vessels and nerves running through them.
   (e) all the above.

3. Which of the following is *not* a major component of muscle?
   (a) Albumin
   (b) Actin
   (c) Myosin
   (d) Troponin
   (e) Tropomyosin

4. When you lift your leg, what types of muscles are doing the lifting?
   (a) Smooth muscles
   (b) Striated muscles
   (c) Cardiac muscles
   (d) a and b
   (e) All the above

5. The source of energy for muscle contraction is
   (a) actin.
   (b) myosin.
   (c) sarcomeres.
   (d) myofibrils.
   (e) ATP.

6. Sarcomeres in skeletal muscle
   (a) contain actin and myosin.
   (b) are joined at the Z line.
   (c) do not change length when a muscle contracts.
   (d) store and release $Ca^+$.
   (e) are as long as the muscle itself.

7. In activating contraction, $Ca^+$ binds to
   (a) actin.
   (b) myosin.
   (c) troponin.
   (d) tropomyosin.
   (e) none of the above.

# THOUGHT QUESTIONS

1. Myofilaments can contract forcefully, pulling membranes attached to the two ends toward one another. Myofilaments cannot expand, however, pushing membranes attached to the two ends of a myofilament apart from one another. Why is it that myofilaments can pull but not push?

2. Among long-distance runners and committed joggers, the long bones of the leg often develop "stress fractures," numerous fine cracks running parallel to one another along the lines of stress. In most instances, stress fractures occur when runners push themselves much farther than they are accustomed to running. Runners who train by gradually increasing the distances they run develop stress fractures rarely, if ever. What protects this second kind of runner?

# FOR FURTHER READING

CARAFOLI, E., and J. PENNISTON: "The Calcium Signal," *Scientific American*, November 1985, pages 70-78. Discusses how the release of calcium ions is the only known way in which the electricity of the nervous system is able to produce changes in the body. Nerves regulate all muscle contractions and hormone secretions by controlling the level of $Ca^{++}$ ions.

COHEN, C.: "The Protein Switch of Muscle Contraction," *Scientific American*, November 1975, pages 36-45. Explains how proteins associated with myofilaments interact with $Ca^{++}$ ions to trigger contraction.

SCHMIDT-NIELSEN, K.: *Animal Physiology: Adaptation and Environment*, ed. 3, Cambridge University Press, New York, 1983. Chapter 11 presents an outstanding treatment of muscles and bones from an evolutionary perspective.

# Hormones

Hormones in action. The sparring of these bull elks occurs only during mating season, when the hormonal cycle of elk females is favorable for mating and ignites fierce competition among potential suitors.

# HORMONES

## Overview

Hormones are chemical messengers the body uses to regulate its activities. Unlike neurotransmitters, whose effects usually last for relatively short periods, the effects of hormones tend to persist for a long time. Hormones usually act at sites far from where they are manufactured, unlike neurotransmitters, which affect a particular postsynaptic cell. Hormones regulate growth and maintain physiological conditions within narrow bounds. Most hormones seem to function similarly in all vertebrates.

## For Review

*Here are some important terms and concepts that you will encounter in this chapter. If you are not familiar with them, you should review them before proceeding.*

**Receptor proteins** (Chapter 5)

**Neurotransmitters** (Chapter 33)

**Operation of neurons** (Chapter 33)

**Central nervous system** (Chapter 34)

**Figure 36-1**

**Our bodies are complex machines whose activities must be coordinated for integration.** The nervous control system and the hormonal control system provide this integration.

The tissues and organs of an adult mammal participate in a multitude of activities, capturing oxygen and digesting food, walking and singing, and seeing far into the distance. All these activities that our bodies perform every day must be coordinated to avoid conflict and to maximize interaction. This is why we think of ourselves as an *organism* (Figure 36-1), rather than as a smoothly functioning collection of organs. Integration of the many activities of the vertebrate body is the primary function of the central nervous system. Some of the control of body functions by the central nervous system is directed through the motor nerves of the voluntary and autonomic components of the peripheral nervous system. In this chapter we learn of a second control system, the endocrine system, which differs in its speed of expression, duration of response, and narrowness of application.

## THE IMPORTANCE OF CHEMICAL MESSENGERS

To control the vertebrate body, the central nervous system employs a battery of specific molecules called hormones as signals. Why use a chemical signal rather than an electrical one? For electrical signals to be effective, they must be transmitted to individual cells. This would require many nerves and would be wasteful when the desired result is to affect the metabolic or other activity of a group of tissues, organs, or organ systems. The advantage of chemical molecules over electrical signals as messengers within the body is twofold. First, chemical molecules can spread to all tissues via the blood. Second, each kind of hormone molecule has a unique shape unlike any other, much as every human face is unique. The mechanism (receptor) coupling a hormone to intracellular processes can also be different in different cells. The human body uses various kinds of signals. In some cases messengers are differently shaped molecules. In other cases the same messenger activates receptors that trigger diverse intercellular events (Figure 36-2).

## The Role of Receptors

How does the body recognize a molecule with a particular shape? It does so by designing a template that exactly matches the shape of a potential signal molecule—a glove to fit the hand. Using such a template, the body can recognize a signal molecule with exquisite precision, selecting one individual molecule from billions of others. These marvelous templates are the **receptor proteins** you encountered in Chapter 5. The AIDS virus, for example, infects certain cells of the immune system, and not the cells of the lungs or foot, because the immune cells possess a particular cell surface receptor that the virus recognizes.

In Chapters 33 and 34 you saw how receptor proteins play a critical role in the nervous system as the targets of neurotransmitters. As you recall, nerve cells have highly specific cell surface receptors embedded in their membranes; each receptor is tuned to respond to a different neurotransmitter molecule. How the neuron responds to stimulation depends on which of its receptors encounter its particular neurotransmitter. *The great advantage of a molecular messenger is that it can be directed at a particular protein receptor on its "target cells" that recognizes only this molecule, ignoring all other molecules.* In each case the operating principle the body uses is the same: only cells whose membranes contain an appropriate receptor protein will respond to a molecular message.

*Chemical communication within the vertebrate body involves two elements, a molecular signal and a protein receptor on target cells. The system is highly specific because each protein receptor has a shape that only its particular signal molecule fits.*

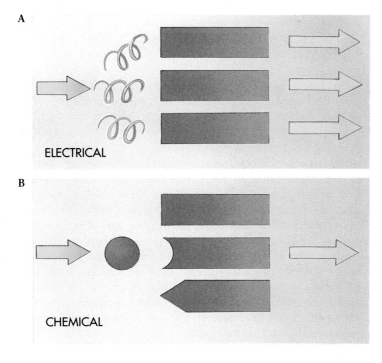

**Figure 36-2**

**A key advantage of chemical signals is specificity.** An electrical signal might excite many adjacent neurons to fire (**A**), while a chemical signal can be targeted to a specific neuron (**B**).

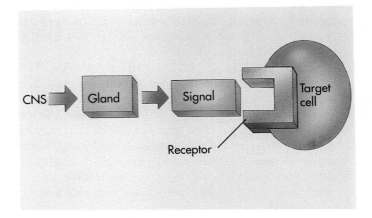

**Figure 36-3**

**The path of hormonal communication in vertebrates.** The central nervous system (CNS) issues a command that reaches the appropriate gland. The gland secretes a hormone that acts as a signal to a particular receptor, or target. After the hormone binds to the receptor, the receptor responds by changing its shape, which triggers a change in cell activity.

## How Molecular Signals are Sent

The path of communication within the vertebrate body can be visualized as a series of simple steps (Figure 36-3):

1. *Issuing the command.* The central nervous system is the most important regulator of the body's activities, although other organs can also independently control the release of chemical messengers. For many hormones an area of the brain called the hypothalamus controls the release of chemical messengers from the pituitary gland. The pituitary gland is discussed later.
2. *Transporting the signal.* Hormones may act on an adjacent cell, be carried throughout the body by the bloodstream, or even pass to a different organism.
3. *Hitting the target.* When a hormone encounters a cell with a matching receptor, called a target cell, the hormone binds to that receptor.
4. *Having an effect.* When the hormone binds it, its receptor protein responds by changing shape, which triggers a change in cell activity.

## Factories for Making Molecular Messengers: Endocrine Glands

There are three classes of molecular messengers: messengers that work only within cells, called second messengers; neurotransmitters released by axons; and stable molecular messengers released from endocrine glands called **hormones.** A hormone is a chemical messenger, often a steroid or peptide, that is stable enough to be transported in active form far from where it is produced and that typically acts at a distant site. Hormones thus are a very different class of molecular messenger than neurotransmitters, which tend to have effects that last only for a short time. Hormones are designed for more lasting effect.

**Neurohormones** are secreted by nerve cells. The most important neurohormones travel through the bloodstream from their release sites in the brain to target cells in closely associated endocrine glands. Some hormones are secreted from ductless glands that may in turn respond to different hormones, including some neurohormones. Ductless glands are called **endocrine glands.** Endocrine glands need to be clearly distinguished from glands with ducts, called exocrine glands, which secrete sweat, milk, digestive enzymes, and other material from the body. Endocrine glands are the hormone-producing factories of the body. By churning out large amounts of hormones, endocrine glands act to amplify greatly the initial neurohormone signal issued by the central nervous system (Figure 36-4).

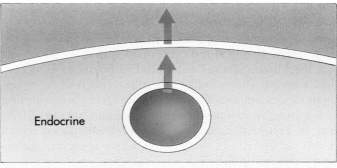

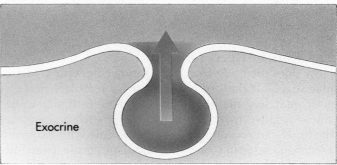

**Figure 36-4**

**Endocrine and exocrine glands.** Endocrine glands do not have ducts and produce hormones, the body's chemical messengers. Exocrine glands do have ducts and secrete sweat, milk, digestive enzymes, and other materials.

## HOW HORMONES WORK: GETTING THE MESSAGE ACROSS

Neurohormones and endocrine hormones act in one of two fundamental ways: either they enter the target cell or they do not.

### Steroid Hormones Enter Cells

Some protein receptors designed to recognize hormones are located in the cytoplasm of the target cell. The hormones in these cases are lipid-soluble molecules, typically steroids, that pass across the cell membrane and bind to receptors within the cytoplasm (Figure 36-5). This complex of receptor and hormone then binds to the DNA in the nucleus and causes a change in the pattern of gene activity. It is the change in gene activity that is responsible for the effect of the hormone. The steroids that weight lifters and other athletes sometimes use turn on genes and thus trick their muscle cells into added growth.

All **steroid hormones** are derived from cholesterol, a complex molecule composed of three six-membered carbon rings and one five-membered carbon ring; it resembles a fragment of chain-link fence. The hormones that promote the development of the secondary sexual

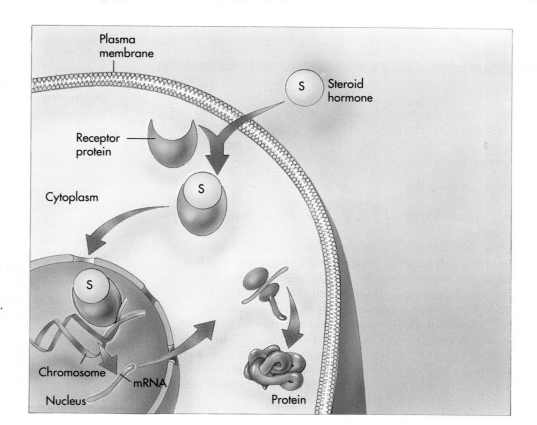

**Figure 36-5**

**How steroid hormones work.** Steroid hormones are lipid soluble and thus readily pass through the plasma membrane of cells into the cytoplasm. There they bind to specific receptor proteins, forming a complex that enters the nucleus and binds to specific regulatory sites on chromosomes. The binding initiates transcription of the genes regulated by the site and thus results in the production of specific proteins.

characteristics are steroids. They include cortisone and testosterone, as well as the hormones estrogen and progesterone, which are used in birth control pills.

*Steroid hormones enter a target cell, bind to a cytoplasmic receptor, and penetrate the nucleus, where they initiate the transcription of some genes while repressing the transcription of others.*

## Peptide Hormones do not Enter Cells

Other hormone receptors are embedded within the cell membrane, with their recognition region directed outward from the cell surface (Figure 36-6). Hormones, typically peptides, bind to these receptors on the cell surface. This binding then triggers events within the cell cytoplasm, usually through intermediates known as second messengers.

Some **peptide hormones,** such as epinephrine (also called adrenaline), are small molecules derived from the amino acid tyrosine; others are short polypeptide chains. Most hormones that circulate within the brain belong to this second class. Still other hormones are large proteins, consisting of very long polypeptide chains, such as insulin.

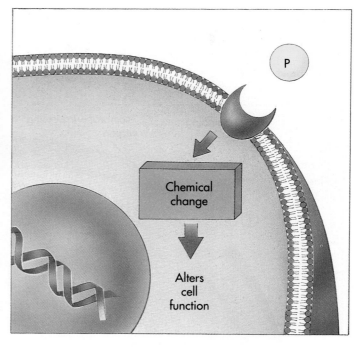

**Figure 36-6**

**How peptide hormones work.** Peptide hormones (P) bind to receptors on the cell surface. This binding triggers chemical changes within the cytoplasm. These changes alter cell function, usually through the agency of chemicals known as second messengers.

## A Peptide Hormone in Action: How Insulin Works

How does the binding of a peptide hormone to the *surface* of a cell produce changes within? The peptide hormone insulin provides a well-studied example of how peptide hormones achieve their effect within target cells (Figure 36-7). Most vertebrate cells have receptors for insulin in their membranes, the number ranging from fewer than 100 to more than 100,000 in some liver cells. The receptors are glycoproteins. One part protrudes from the surface of the cell and binds insulin, whereas another spans the membrane and extends into the cytoplasm.

> *Stage one: activation by insulin.* The binding of insulin causes a change in the shape of the insulin receptor so that within the cell, phosphate groups are actively added to the tyrosine amino

acid side groups of proteins. This phosphorylation of proteins activates the second stage of the insulin response.

*Stage two: amplification of signal.* As a result of insulin receptor–induced enzyme activity, small molecules of cyclic AMP are formed from ATP. Cyclic AMP activates a variety of enzymes in the cell that stimulate the uptake of glucose from the blood and the formation of glycogen, causing levels of glucose in the blood to fall. In addition to cyclic AMP, insulin binding also promotes the production of **inositol phosphates,** which are cleaved from the plasma membrane. A single insulin molecule binding to its receptor releases many mediator molecules into the cell, and each mediator activates many enzyme molecules in an expanding cascade of response.

## Second Messengers

Insulin mediators are one example of **second messengers,** intermediary compounds that couple extracellular signals to intracellular processes and also amplify a hormonal signal. Second messengers set into action a variety of events within the affected cell, depending on the enzymatic profile of the cell. In the early 1960s Earl Sutherland described the first second messenger, a cyclic form of adenosine monophosphate, **cyclic AMP (cAMP).** When hormones such as epinephrine bind receptors on liver cells (Figure 36-8), the receptor changes shape and binds a cell protein called **G protein,** causing it in turn to bind the nucleotide GTP and activate another membrane protein, **adenylate cyclase.** The result of these complex interactions is the production of large amounts of cAMP by the activated adenylate cyclase. *The cAMP is the amplified form of the epinephrine hormonal message.* Another common second messenger is inositol phosphate.

A variety of peptide hormones use cAMP as a second messenger. The cAMP has different effects in various target cells because different enzymes are present in different target cells and tissues. In muscle cells, cAMP is induced by epinephrine and activates the enzyme protein kinase A, which in turn activates the enzyme that breaks down glycogen into glucose. In cells of the ovary, cAMP is induced by a luteinizing hormone and stimulates the cells to produce a specific follicle cell enzyme.

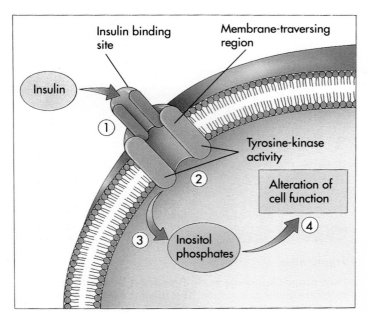

**Figure 36-7**

How insulin works.

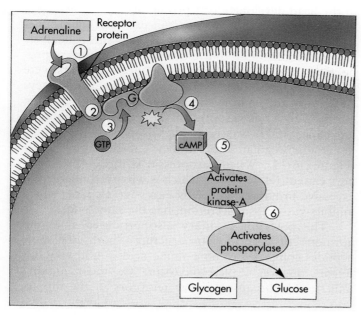

**Figure 36-8**

**Second messengers. 1,** Peptide hormones bind to specific receptor proteins on the surface of target cells. **2,** The hormone enters the cytoplasm. **3,** The receptor changes shape and binds to a protein called G. **4,** The receptor then binds to GTP and activates a membrane protein called adenyl cyclase. Adenyl cyclase produces cAMP. **5,** cAMP activates protein kinase A. **6,** Protein kinase A activates phosphorylase, which breaks glycogen down into glucose.

## Thyroxine is Unique: It Goes Directly into the Nucleus

One important endocrine hormone is neither steroid nor peptide. The hormone thyroxine, produced by the thyroid gland, is a modified form of the amino acid tyrosine and contains four iodine atoms. Thyroxine acts on target cells in a unique way. It diffuses directly into the cell cytoplasm and then into the cell nucleus, where it interacts with a protein receptor attached to DNA to initiate production of particular growth-promoting messenger RNAs. Thyroxine is the only hormone known to go directly into a target cell nucleus without first binding to receptors on the cell membrane or in the cytoplasm.

## THE MAJOR ENDOCRINE GLANDS AND THEIR HORMONES

Vertebrates have about a dozen major endocrine glands that together make up the **endocrine system** (Figure 36-9). In this section we briefly examine the principal endocrine glands and the hormones they produce. These principal endocrine glands and their hormones are summarized in Table 36-1.

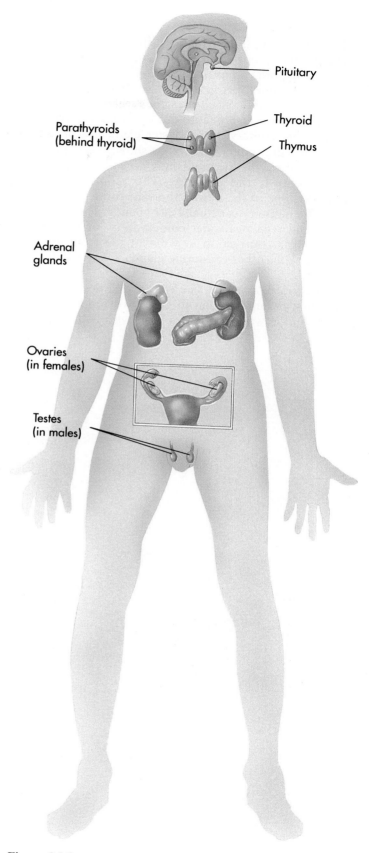

**Figure 36-9**

**The human endocrine system.**

# The Pituitary: The Master Gland

The pituitary, located in a bony recess in the brain below the hypothalamus, is the site of production of nine major hormones. Because many of these hormones act principally to influence other endocrine glands (Figure 36-10), it was fashionable until relatively recently to regard the pituitary as a "master gland" orchestrating the endocrine system. In fact, as shown here, that role is reserved for the hypothalamus because the hypothalamus controls the pituitary. The pituitary remains, however, one of the most important endocrine glands.

The pituitary is actually two glands. The back *posterior* end regulates water conservation, milk letdown, and uterine contraction in women; the front *anterior* end regulates other endocrine glands.

*The Posterior Pituitary.* The role of the posterior pituitary first became evident in 1912, when a remarkable medical case was reported: a man who had been shot in the head developed a surprising disorder—he began to urinate every 30 minutes, unceasingly. The bullet had lodged in the pituitary gland, and subsequent research demonstrated that removal of the pituitary produces these unusual symptoms. Pituitary extracts were shown to contain a substance that makes the kidneys conserve water, and eventually in the early 1950s the peptide hormone **vasopressin** (also called antidiuretic hormone, ADH) was isolated. Vasopressin is the hormone that regulates the kidneys' retention of water (Figure 36-11). When vasopressin is missing, the kidneys cannot retain water, which is why the bullet led to excessive urination (and why excessive alcohol, which inhibits vasopressin secretion, has the same effect).

The posterior pituitary also produces a second hormone of very similar structure—both are short peptides composed of nine amino acids—and very different function, called **oxytocin**. Oxytocin initiates milk release because sensory receptors in the nipples

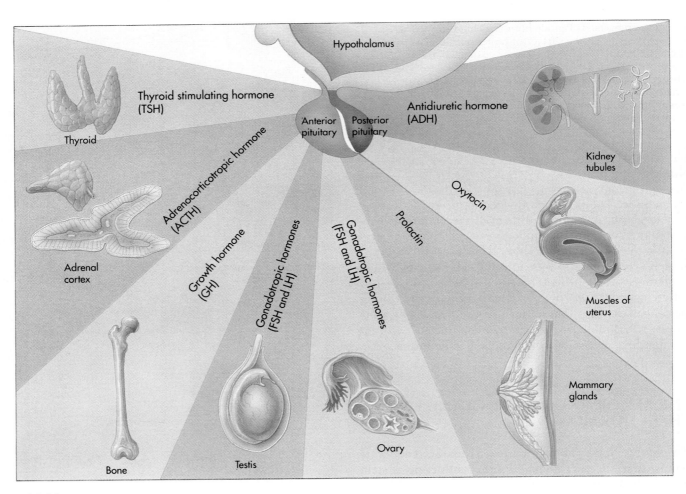

*Figure 36-10*

**The master gland.** Interactions between the anterior lobe and the posterior lobe of the pituitary (two distinct glands) and various organs of the human body.

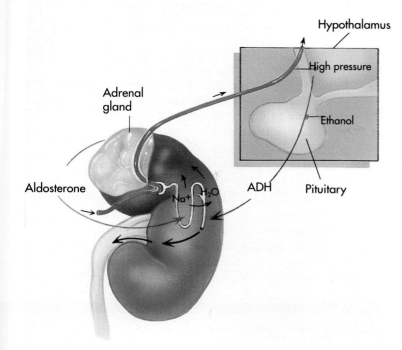

**Figure 36-11**

**How hormones control water chemistry.** Control of water and salt balance within the kidney is centered in the hypothalamus. The hypothalamus produces antidiuretic hormone (ADH), which renders the collecting ducts of the kidneys freely permeable to water and so maximizes water retention. If too much water retention leads to high blood pressure, pressure-sensitive receptors in the hypothalamus detect this and cause the production of ADH to be shut down. If the level of sodium in the blood falls, the adrenal gland initiates production of the hormone aldosterone, which stimulates salt reabsorption by the renal tubules of the kidney. Feedback loops are noted by the colored lines with the double bar.

send messages to the hypothalamus, causing oxytocin release. Oxytocin stimulates the contraction of the muscles around the ducts into which mammary glands secrete milk. Oxytocin also stimulates uterine contractions in women during childbirth, which is why the uterus of a nursing mother returns to normal size after its extension during pregnancy more quickly than does the uterus of a mother who does not nurse her baby.

Vasopressin and oxytocin are synthesized inside nerve cells within the hypothalamus. The two hormones are transported down nerve cell axons from the hypothalamus to synapses located within the posterior pituitary, where they are stored in the axon terminals. The hormones are released into the bloodstream when

*The posterior lobe of the pituitary, a distinct gland, is connected with the hypothalamus by neural connections. It secretes the important hormones vasopressin and oxytocin.*

a nerve impulse reaches the terminals from the hypothalamus.

***The Anterior Pituitary.*** The key role of the anterior pituitary first became understood in 1909, when a 38-year-old South Dakota farmer was cured of the growth disorder acromegaly by the surgical removal of a pituitary tumor. Acromegaly is a form of giantism in which the jaw begins to protrude and features thicken (Figure 36-12). It turned out that giantism is almost always associated with pituitary tumors. Robert Wadlow, born in Alton, Illinois, in 1928, grew to a height of 8 feet, 11 inches and weighed 475 pounds before he died from infection at age 22—the tallest human being ever recorded (Figure 36-13). Skull x-rays showed he had a pituitary tumor. So did the 8-foot, 2-inch Irish giant Charles Byrne, born in 1761; his skeleton, preserved in the Royal College of Surgeons, London, shows the effects of a pituitary tumor.

Why did removal of the pituitary tumor cure the South Dakota farmer? Pituitary tumors produce giants because the tumor cells produce prodigious amounts of a growth-promoting hormone. This **growth hormone (GH)**, a peptide of 191 amino acids, is normally produced in only minute amounts by the anterior pituitary gland and usually only during periods of body growth, such as infancy and puberty.

**Figure 36-12**

**Acromegaly.** This ancient carving of the Egyptian pharaoh Akhenaton, who ruled from 1379-1362 BC, exhibits many characteristics of acromegaly; if so, it is the oldest known case.

**Figure 36-13**

**The Alton giant.** This photo, taken when Robert Wadlow of Alton, Illinois, was 13 years old, shows him towering over his father and 9-year-old brother. Born at a normal size, he developed a growth-hormone–secreting pituitary tumor as a young child and never stopped growing.

We now know that the anterior pituitary gland of vertebrates produces seven major peptide hormones, each controlled by a particular releasing factor secreted from cells in the hypothalamus.

*Thyroid-stimulating hormone (TSH).* TSH stimulates the thyroid gland to produce thyroid hormone, which in turn stimulates oxidative respiration.

*Luteinizing hormone (LH).* LH plays an important role in the female menstrual cycle (see Chapter 42). It also stimulates the male gonads to produce testosterone, which initiates and maintains the development of male secondary sexual characteristics, those external features not involved in reproduction.

*Follicle-stimulating hormone (FSH).* FSH is significant in the female menstrual cycle (see Chapter 42). In males, it stimulates certain cells in the testes to produce a hormone that regulates the development of sperm.

*Adrenocorticotropic hormone (ACTH).* ACTH stimulates the adrenal cortex to produce corticosteroid hormones. Some of these hormones regulate the production of glucose from fat; others regulate the balance of sodium and potassium ions in the blood; and still others contribute to the development of the male secondary sexual characteristics.

*Somatotropin,* or *growth hormone (GH).* GH stimulates the growth of muscle and bone throughout the body.

*Prolactin (PRL).* PRL stimulates the breasts to produce milk.

*Melatonin,* or *melanocyte-stimulating hormone (MSH).* In reptiles and amphibians, MSH stimulates color changes in the epidermis. This hormone has no known function in mammals.

*Seven major peptide hormones are secreted by the anterior lobe of the pituitary. Very similar in structure, they have a wide variety of functions.*

In addition to their endocrine functions, these and many other hormones have been shown also to be associated with particular populations of cells within the central nervous system. This is an area of very active research; biologists are attempting to uncover the as yet unknown role of these hormones in the central nervous system. Whatever their function there, it is clear that the same hormone may play different roles in different parts of the body because cells have different enzymatic profiles. Hormones are signals used in different tissues for different reasons, just as raising your hand is a signal that has different meanings in different contexts: in class it indicates you have a question; in a football game it signals a "fair catch"; and when a policeman does it on a street, it means "stop." The evolving use of hormones by vertebrates has been a conservative process. Rather than produce a new hormone for every use, vertebrates have often adapted a hormone already at hand for a new use, when this new use does not cause confusion.

## The Thyroid: A Metabolic Thermostat

The thyroid gland is shaped like a shield (Greek *thyros*, meaning "shield"). It lies just below the Adam's apple in the front of the neck. It makes several hormones, the two most important of which are **thyroxine,** which increases metabolic rate and promotes growth, and **calcitonin,** which stimulates calcium uptake.

Without adequate thyroxine (also called thyroid hormone), growth is retarded. Children with underactive thyroid glands are not able to carry out carbohydrate breakdown and protein synthesis at normal rates, a condition called **cretinism,** which results in stunted growth. Mental retardation is also seen because thyroxine is needed for normal development of the CNS. Adults with too little of this hormone also have slowed metabolism, affecting their mental performance.

As mentioned earlier, the hormone thyroxine is made from the amino acid tyrosine, with four iodine atoms added to it. If the iodine concentration in a person's diet is too low, the thyroid cannot make adequate amounts of thyroxine and will grow larger in a futile attempt to manufacture more of the hormone. The greatly enlarged thyroid gland that results is called a **goiter** (Figure 36-14). This need for iodine in your diet is why iodine is added to table salt.

## The Parathyroids: Builders of Bones

The parathyroid glands are four small glands attached to the thyroid. Small and unobtrusive, they were ignored by researchers until well into this century. The first suggestions that the parathyroids produce a hormone came from experiments in which they were removed from dogs: the concentration of calcium in the dogs' blood plummeted to less than half the normal level. However, if an extract of parathyroid gland was administered, calcium levels returned to normal. If an excess was administered, calcium levels became *too* high, and the bones of the dogs literally were dismantled by the extract. It was clear that the parathyroid glands were producing a hormone that acted on calcium uptake into, and release from, bone.

The hormone produced by the parathyroids is **parathyroid hormone** (**PTH**). It is one of only two hormones in your body that are absolutely essential for survival (the other, discussed in the next section, is aldosterone, produced by the adrenal glands). PTH acts to regulate levels of calcium in your blood. Recall from Chapter 35 that calcium ions are the key actors in vertebrate muscle contraction; by altering calcium release, nerve impulses cause muscles to contract. You cannot live without the muscles that pump your heart and drive your body, and these muscles cannot function if calcium levels are not kept within narrow limits. Calcium is also important for normal nerve activity.

PTH acts as a fail-safe to make sure calcium levels never fall too low. PTH is released into the bloodstream, where it travels to the bones and acts on the osteoclast cells within bones, stimulating them to dismantle bone tissue and release calcium into the bloodstream. PTH also acts on the kidneys to resorb calcium ions from the urine and leads to the activation of vita-

**Figure 36-14**

**A goiter.** This condition is caused by a lack of iodine in the diet.

min D, necessary for calcium absorption by the intestine. A diet deficient in vitamin D leads to poor bone formation, a condition called rickets. PTH is synthesized by the parathyroids in response to falling levels of calcium ions in the blood; the body essentially sacrifices bone to keep calcium levels within the narrow limits necessary for proper functioning of muscle and nerve.

The thyroid also plays a role in maintaining proper calcium levels (Figure 36-15). If levels of calcium in your blood become too high, the thyroid hormone calcitonin, mentioned earlier, stimulates calcium deposition in bone, thus lowering calcium levels in the blood.

## The Adrenals: Two Glands in One

There are two adrenal glands, one located just above each kidney. Each adrenal gland is composed of two parts: an inner core called the **medulla,** which produces the peptide hormones epinephrine and norepinephrine, and an outer layer called the **cortex,** which produces the steroid hormones **cortisol** and **aldosterone.**

***Adrenal Medulla: Emergency Warning Siren.*** The medulla releases epinephrine and norepinephrine in times of

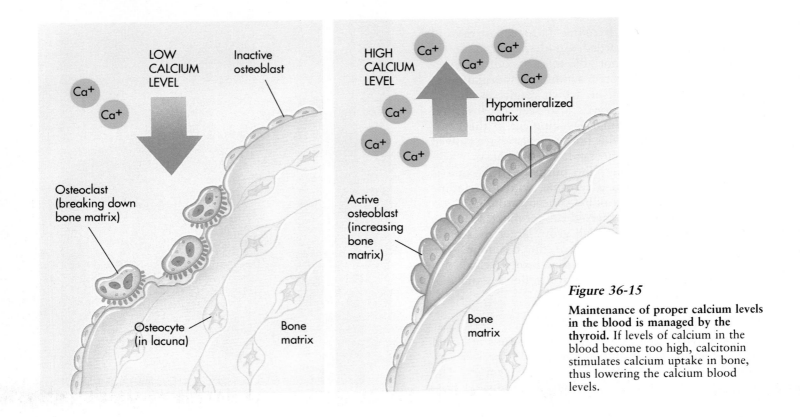

**Figure 36-15**

**Maintenance of proper calcium levels in the blood is managed by the thyroid.** If levels of calcium in the blood become too high, calcitonin stimulates calcium uptake in bone, thus lowering the calcium blood levels.

stress. Epinephrine and norepinephrine act in the body as emergency signals that stimulate rapid deployment of body fuel. The "alarm" response throughout the body is identical to the individual effects achieved by the sympathetic nervous system but is longer lasting. Among the effects of these hormones are an accelerated heartbeat, increased blood pressure, higher levels of blood sugar, dilated blood vessels, and increased blood flow to the heart and lungs. These hormones thus can be thought of as extensions of the sympathetic nervous system.

***Adrenal Cortex: Maintaining the Proper Amount of Salt.*** Cortisol (also called hydrocortisone) acts on many different cells in the body to maintain nutritional well-being. It stimulates carbohydrate metabolism and acts to reduce inflammation. Derivatives of this hormone, such as prednisone, have widespread medical use as anti-inflammatory agents. The ability of many cortisol-derived steroids to stimulate muscle growth has also led the abuse of so-called anabolic steroids by athletes (Figure 36-16).

Aldosterone acts primarily at the kidney to promote the uptake of sodium and other salts from the urine. Sodium ions play critical roles in nerve conduction and many other body functions. Their concentration also has a critical influence on blood pressure. Without aldosterone, sodium ions are not retrieved

**Figure 36-16**

**Many athletes have succumbed to the abuse of steroids.** Although steroids stimulate muscle growth, their abuse carries many serious health risks.

from body fluids and are lost in the urine. The loss of salt in the blood causes water to leave the bloodstream and enter cells; thus blood pressure falls. Aldosterone also acts in the opposite way to promote the export of potassium out of the body, stimulating the kidneys to secrete potassium ions into the urine. When aldosterone levels are too low, potassium levels in the blood may rise to dangerous levels. Aldosterone is, with PTH, one of the two endocrine hormones essential for survival. Removal of the adrenal glands is invariably fatal.

## The Pancreas: The Body's Dietitian

The pancreas gland is located behind the stomach and is connected to the front end of the small intestine by a small tube. It secretes a variety of digestive enzymes into the gut through this tube, and for a long time was thought to be solely an exocrine gland. In 1869, however, a German medical student named Paul Langerhans described some unusual clusters of cells scattered throughout the pancreas (Figure 36-17). At the end of the last century doctors had begun to notice that patients with injuries to the pancreas often develop **diabetes mellitus,** a common and serious disorder in which the affected individuals develop elevated levels of glucose in the blood. They are unable to take up glucose from the blood, even though the level of blood glucose is high. In Type I diabetes mellitus, insulin secretion is abnormally low. In Type II diabetes mellitus, there is an abnormally low number of insulin receptors on the target tissue, but the level of insulin is normal in the blood. Such individuals literally starve; they lose weight and may eventually suffer brain damage and even death if their condition is untreated. In 1893 it was suggested that the clusters of cells in the pancreas, which came to be called **islets of Langerhans,** produced something that prevented diabetes mellitus.

The substance, which we now know to be the peptide hormone **insulin,** was not isolated until 1922, when two young doctors working in a Toronto hospital succeeded where many others had not. On January 11, 1922, they injected an extract purified from beef pancreas glands into a 13-year-old boy, a diabetic whose weight had fallen to 65 pounds and who was not expected to survive. The hospital record note gives no indication of the historic importance of the trial, only stating, "15 cc of MacLeod's serum. 7½ cc into each buttock." With this single injection, the glucose level in the boy's blood fell 25%. A more potent extract soon brought levels down to near normal. This was the first instance of successful insulin therapy. Today cases of diabetes can be treated by supplying insulin to the affected individual daily. Others are treated by a combination of exercise and diet. Active research on the possibility of transplanting islets of Langerhans holds much promise of a lasting treatment for diabetes.

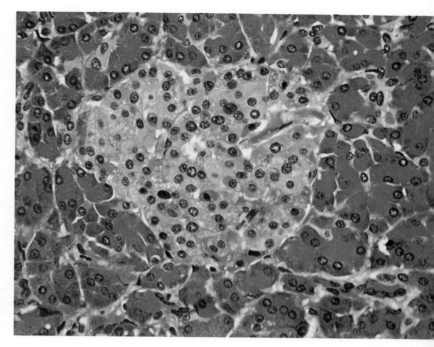

***Figure 36-17***

**Islets of Langerhans.** Glucagon and insulin are produced by clumps of cells within the pancreas called islets of Langerhans, which are stained dark in this preparation.

*Diabetes is a condition in which individuals are unable to obtain glucose from their blood because they either lack insulin (Type I) or have an abnormally low number of insulin receptors (Type II). It is a serious disease and can be fatal if untreated.*

We now know that the islets of Langerhans in the pancreas produce *two* hormones that interact to govern the levels of glucose in the blood (Figure 36-18), insulin and **glucagon.** Insulin is a storage hormone, designed to put away nutrients for leaner times. It promotes the accumulation of glycogen in the liver and triglycerides in fat cells. When you eat, **beta-cells** in the islets of Langerhans secrete insulin, storing away glucose to be used later. Later, when body activity causes the level of glucose in the blood to fall as it is used up as fuel, other cells in the islets of Langerhans called **alpha-cells** secrete glucagon, which causes liver cells to release stored glucose and fat cells to break down triglycerides. The two hormones thus work together to keep glucose levels in the blood within narrow bounds.

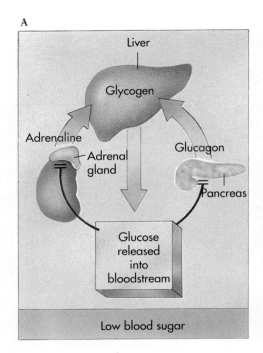

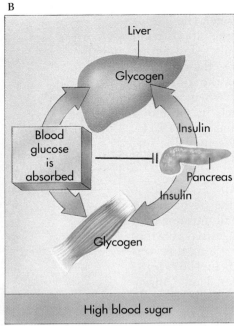

*Figure 36-18*

**Hormonal control of blood glucose levels.**
**A** When blood glucose levels are low, cells within the pancreas release the hormone glucagon into the bloodstream; and other cells within the adrenal gland, situated on top of the kidneys, release the hormone adrenaline into the bloodstream. When they reach the liver, these two hormones both act to increase the liver's breakdown of glycogen to glucose.
**B** When blood glucose levels are high, other cells within the pancreas produce the hormone insulin, which stimulates the liver and muscles to convert blood glucose into glycogen. The glucose level in the blood determines the levels of insulin and glycogen in the blood via feedback loops to the pancreas and adrenal gland.

## Other Endocrine Glands

The ovary and testes are important endocrine glands, producing sex hormones (estrogen, progesterone, and testosterone), which are described in detail in Chapter 42. Surprisingly, the gut is also a major endocrine gland, secreting hormones that regulate the release of secretions, which in turn play a key role in food digestion; they are discussed in Chapter 37. The only other major endocrine gland is the **pineal gland,** which sits in the center of the brain. It is small, about the size of a pea, and is shaped like a pine cone (hence its name). It is the last endocrine gland discovered in the human body, and its function is not known. It was discovered in 1958 that the pineal gland secretes a hormone, **melatonin,** a derivative of the amino acid tryptophan. In hamsters, melatonin regulates reproductive biology, and in frogs it influences pigmentation, but its function in humans is not well understood.

The pineal gland in reptiles is located closer to the surface and is called the "third eye" because it is structurally similar to the retina and responds directly to light. In humans the pineal gland in not connected to the central nervous system directly, but the gland *is* connected, via the sympathetic nervous system, to the eyes. Melatonin seems to be released by the human pineal gland as a response to darkness, in a cyclical biological rhythm keyed to daylight. Perhaps the pineal gland is involved in establishing daily biorhythms. It has also been implicated in mood disorders such as winter depression, also called SADS (seasonal affective disorder syndrome), and in a variety of other roles concerning sexual development.

## HOW THE BRAIN CONTROLS THE ENDOCRINE SYSTEM

The 12 major endocrine glands (Table 36-1) do not function independently. Their activities are coordinated at two levels: (1) the six anterior pituitary hormones regulate many of the activities of the other endocrine glands, and (2) the pituitary gland is controlled by the CNS via the hypothalamus.

### Releasing Hormones: The Brain's Chemical Messengers

How the pituitary gland is regulated by the brain was until recently one of the great mysteries of medicine. It is suspended by a short stalk to the hypothalamus (Figure 36-19), which is part of the diencephalon region of the forebrain, located at the base of the brain. Within the hypothalamus, information about the body's many internal functions (called interoceptive information) is processed and regulatory commands are issued. Some of these commands involve functions such as the regulation of body temperature, the intake of food and water, reproductive behavior, and response to pain and emotion. These commands are issued to the pituitary gland, which in turn sends chemical signals to the various hormone-producing glands of the body. For this reason, injury to the hypothalamus causes a decrease in the anterior pituitary's production of hormones. The hypothalamus thus regulates the body through a chain of command in the same way that a general gives orders to a chief of staff, who relays them to lower-ranking commanders.

from 1 million pigs, however, the first of these hormones, a short peptide called **thyrotropin-releasing hormone (TRH)**, was isolated in 1969 (Figure 36-20). The release of TRH from the hypothalamus triggers secretion of thyrotropin from the anterior pituitary.

Six other hypothalamic regulatory hormones have since been isolated, which together govern all the hormones secreted by the anterior pituitary. For each releasing hormone secreted by the hypothalamus, a corresponding hormone is synthesized by the anterior pituitary. When the pituitary gland receives a releasing hormone from the hypothalamus, the anterior lobe responds by secreting the corresponding pituitary hormone.

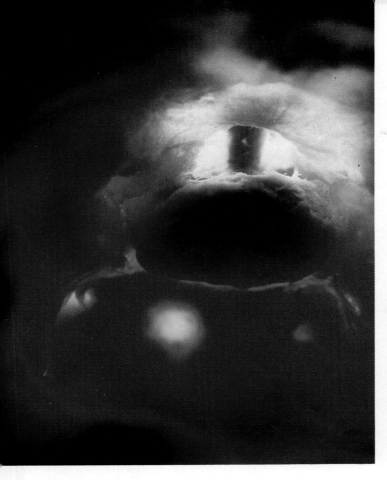

*Figure 36-19*

**The pituitary gland hangs by a short stalk from the hypothalamus.** The pituitary regulates the hormone production of many of the body's endocrine glands. Here it is enlarged 15 times.

The anterior pituitary is connected to the hypothalamus by special blood vessels only a few millimeters long. Through these vessels passes a group of hormones produced in the hypothalamus called releasing hormones, which command the anterior pituitary to initiate the production (release) of specific hormones in distant endocrine glands.

The hypothalamus processes information about the body's internal functions and issues regulatory commands that direct the pituitary gland to send further chemical signals to the various hormone-producing glands of the body.

No nerves connect the anterior pituitary gland to the hypothalamus or to any other part of the brain. How are the commands sent from the hypothalamus to the pituitary? In the 1930s a network of tiny blood vessels was discovered that spans the short distance between the hypothalamus and anterior pituitary; researchers wondered if perhaps the hypothalamus employed chemical messengers. The experimental difficulties in isolating such hypothalamic hormones were great because very little of them are present in any one brain. After concentrating the hypothalamus glands

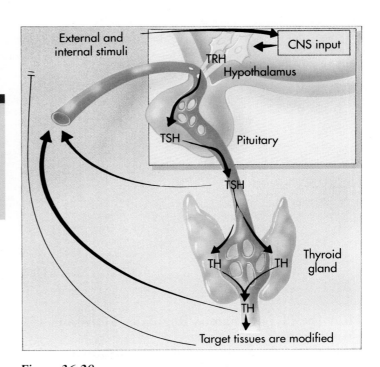

*Figure 36-20*

**The release of thyroid hormone (TH)** from the thyroid gland is the end result of a long series of events mediated by the central nervous system (CNS).

## TABLE 36-1    PRINCIPAL ENDOCRINE GLANDS AND THEIR HORMONES

| ENDOCRINE GLAND AND HORMONE | TARGET TISSUE | PRINCIPAL ACTIONS | CHEMICAL NATURE |
|---|---|---|---|
| **HYPOTHALAMUS** | | | |
| Oxytocin | Uterus | Stimulates contraction of uterus | Peptide (9 amino acids) |
| | Mammary glands | Stimulates ejection of milk | |
| ADH | Kidneys | Stimulates reabsorption of water; conserves water | Peptide (9 amino acids) |
| **ANTERIOR LOBE OF PITUITARY** | | | |
| Somatotropin (growth hormone, GH) | General | Stimulates growth by promoting protein synthesis and breakdown of fatty acids | Protein |
| Prolactin (PRL) | Mammary glands | Stimulates milk production | Protein |
| Thyroid-stimulating hormone (TSH) | Thyroid gland | Stimulates secretion of thyroid hormones | Glycoprotein |
| Adrenocorticotropic hormone (ACTH) | Adrenal cortex | Stimulates secretion of adrenal cortical hormones | Polypeptide |
| Follicle-stimulating hormone (FSH) | Gonads | Stimulates ovarian follicle spermatogenesis | Glycoprotein |
| Luteinizing hormone (LH) | Gonads | Stimulates ovulation and corpus luteum formation in females; stimulates secretion of testosterone in males | Glycoprotein |
| **OVARY** | | | |
| Estrogens | General Female reproductive structures | Stimulate development of secondary sex characteristics in females and growth of sex organs at puberty; prompt monthly preparation of uterus for pregnancy | Steroid |
| Progesterone | Uterus | Completes preparation of uterus for pregnancy | Steroid |
| | Breasts | Stimulates development | |
| **TESTIS** | | | |
| Testosterone | General | Stimulates development of secondary sex characteristics in males and growth spurt at puberty | Steroid |
| | Male reproductive structures | Stimulates development of sex organs; stimulates spermatogenesis | |
| **ADRENAL MEDULLA** | | | |
| Adrenaline and noradrenaline (epinephrine and norepinephrine) | Skeletal muscle, cardiac muscle, blood vessels | Initiate stress responses; increase heart rate, blood pressure, metabolic rate; dilate blood vessels; mobilize fat; raise blood sugar level | Amino acid derivatives |

## How the Hypothalamus Regulates Hormone Production

Control over the production of hormones produced by the anterior pituitary gland is exercised in two ways.

*CNS Control.* The production of the hormones GH, PRL, and MSH is controlled by both releasing and in-hibitory signals produced by the hypothalamus. Consider the growth hormone somatotropin (GH) (Figure 36-21). The releasing signal for GH is the growth-hormone–releasing hormone (GHRH) produced by the hypothalamus; GHRH stimulates the anterior pituitary to produce somatotropin. The inhibiting signal, which is also produced at the same time by the hypothalamus,

TABLE 36-1    PRINCIPAL ENDOCRINE GLANDS AND THEIR HORMONES—cont'd

| ENDOCRINE GLAND AND HORMONE | TARGET TISSUE | PRINCIPAL ACTIONS | CHEMICAL NATURE |
|---|---|---|---|
| **ADRENAL CORTEX** | | | |
| Aldosterone | Kidney tubules | Maintains proper balance of sodium and potassium ions | Steroid |
| Cortisol | General | Adaptation to long-term stress; raises blood glucose level; mobilizes fat | Steroid |
| **PINEAL GLAND** | | | |
| Melatonin | Gonads, pigment cells | Function not well understood; influences pigmentation in some vertebrates; may control biorhythms in some animals; may help control onset of puberty in humans | Amino acid derivative |
| **THYMUS** | | | |
| Thymosin | White blood cells | Promotes production and maturation of white blood cells | Protein |
| **THYROID GLAND** | | | |
| Thyroid hormone (thyroxine) | General | Stimulates metabolic rate; essential to normal growth and development | Iodinated amino acid |
| Calcitonin | Bone | Lowers blood calcium level by inhibiting loss of calcium from bone | Polypeptide (32 amino acids) |
| **PARATHYROID GLANDS** | | | |
| Parathyroid hormone | Bone, kidneys, digestive tract | Increases blood calcium level by stimulating bone breakdown; stimulates calcium reabsorption in kidneys; activates vitamin D | Polypeptide (34 amino acids) |
| **ISLETS OF PANCREAS** | | | |
| Insulin | General | Lowers blood glucose; increases storage of glycogen | Polypeptide (51 amino acids) |
| Glucagon | Liver, adipose tissue | Raises blood glucose level; stimulates breakdown of glycogen in liver | Polypeptide (29 amino acids) |

is somatostatin. Somatostatin inhibits the anterior pituitary from producing somatotropin. The hypothalamus thus regulates growth by mediating the relative rates of production of GHRH and somatostatin. In a similar way the hypothalamus regulates the production of PRL and MSH. Thus the release and inhibition of hormones is controlled by the release of *other* hormones.

*Feedback Control.* The levels of all other hormones produced by the anterior pituitary are controlled by negative feedback from the target glands. For example, when LH stimulates the gonads to release testosterone into the bloodstream, that testosterone in turn inhibits the hypothalamus. The hypothalamus then ceases to transmit LH-releasing hormone to the pituitary.

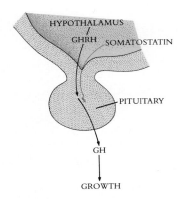

**Figure 36-21**

**Levels of the growth-promoting hormone GH are regulated by both stimulatory and inhibitory signals.** The production of GH by the pituitary is stimulated by a hypothalamic-releasing hormone, GHRH, and inhibited by another pituitary hormone, somatostatin. If levels of GHRH become too high, the excess somatostatin that is induced in the pituitary shuts down some of the production of GH (indicated by the double bar), so that effective levels of GH circulating in the blood do not fluctuate.

*The regulation of pituitary hormone production is achieved by (1) pairs of hypothalamic hormones with opposite effects and (2) feedback loops, through which the hypothalamus is sensitive to hormone levels in the blood.*

## NONENDOCRINE HORMONES

### Neuropeptides

In 1974 Swedish researchers first reported the existence in the brain of **neurohormones**. They isolated two small peptides called **enkephalins** (Figure 36-22), only five amino acid units long, which acted as powerful narcotics. The enkephalins appear to play a role in integrating afferent impulses from the pain receptors. A second type of peptide hormone has since been found that is larger: the brain hormones called **endorphins** are polypeptides 32 amino acid units long. Endorphins appear to regulate emotional responses in the brain. Morphine has such a potent analgesic (pain-relieving) effect on the CNS because it mimics the effects of the endorphins.

More than 20 peptides have now been identified as acting as neurotransmitters in the brain, and some investigators believe the number will eventually exceed 100. Their study is causing a revolution in how scientists view the brain's internal activity. The role of chemical messengers within the brain is one of the most active areas of biological research.

### Prostaglandins

**Prostaglandins** are modified lipids produced from membrane phospholipids by virtually all cells. They do not circulate in the blood, but rather accumulate in regions of tissue disturbance or injury. They stimulate smooth muscle contraction and expansion and contraction of blood vessels. Aspirin relieves headache pain because it inhibits prostaglandin production; overproduction of prostaglandins swells cerebral blood vessels so that their walls press against nerve tracts in the brain, causing pain.

### Atrial Peptides

Another nonendocrine hormone of considerable interest is a small peptide manufactured in the heart, called **atrial natriuretic hormone (ANH)**, that circulates throughout the body. Receptors for ANH have been identified in cells of blood vessels, kidneys, and adrenal glands. The atrial peptides apparently help regulate

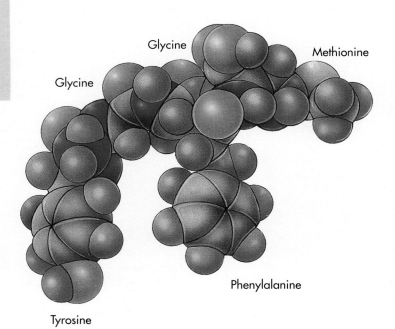

**Figure 36-22**

**Molecular structure of enkephalin.** A potent narcotic, enkephalin is a peptide composed of a linear chain of five amino acids.

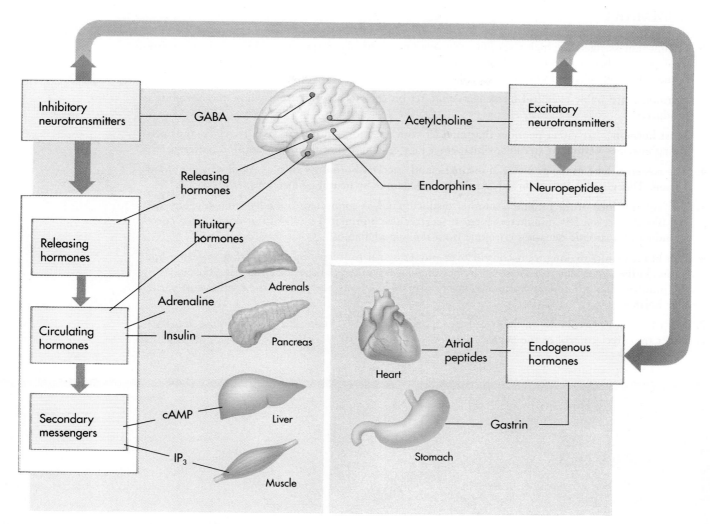

**Figure 36-23**

An overview of neuroendocrine control.

blood pressure and volume. Because they reduce blood volume, atrial peptides are being investigated as a potential treatment for high blood pressure.

## THE BODY'S MESSENGERS

Hormonal communication is essential to the body's proper functioning. The many kinds of hormones we have encountered in this chapter form a network of communication (Figure 36-23) that reaches to all the tissues of the body. The hormones are all ultimately under the control of the central nervous system, of which they are a chemical extension. You will encounter hormones repeatedly in the following chapters. They are one of the principal integrating elements of the vertebrate body.

# SUMMARY

1. Chemical messengers are highly specific, affecting only particular target cells. This exquisite sensitivity is possible because each messenger is keyed to a particular receptor protein present in the membranes of target cells and not in other cells.

2. Hormones are stable chemical messengers that act on cells located far from where the hormone is produced.

3. Most hormones are either peptides that interact with receptors on the target cell surface, thus activating enzymes within, or steroids, which enter into cells and alter their transcription patterns.

4. The posterior lobe of the pituitary, a distinct gland, secretes hormones such as ADH into the bloodstream. This gland is linked directly to the hypothalamus by neural connections.

5. The anterior lobe of the pituitary, another distinct gland, is connected to the hypothalamus by special blood vessels a few millimeters long. It secretes seven principal kinds of hormones, each corresponding to a specific releasing hormone from the hypothalamus.

6. The brain maintains long-term control over physiological processes by synthesizing releasing hormones in the hypothalamus. These hormones direct the synthesis of specific circulating hormones by the pituitary gland. Pituitary hormones travel out into the body and initiate the synthesis of particular hormones in target tissues.

7. Most hormones circulating in the bloodstream are produced by endocrine glands, whose activity is under the direct control of the nervous system.

## REVIEW

1. The neuroendocrine system is a network of _____, whose production is controlled by the central nervous system.

2. The _____ sends signals to the pituitary gland and thus controls the neuroendocrine system.

3. When the hypothalamus releases thyroid-releasing hormone to the pituitary, it causes release of _____ by the anterior pituitary.

4. Inhibitory signals for the production of growth hormone, prolactin, and melanocyte-stimulating hormone are produced by the _____.

5. If blood glucose levels are high, the islets of Langerhans in the pancreas produce the hormone _____, which stimulates the liver to convert blood glucose to glycogen.

## SELF-QUIZ

1. Hormones are
   (a) enzymes.
   (b) regulatory chemicals.
   (c) produced at one place in the body but exert their influence at another.
   (d) a and c.
   (e) b and c.

2. An example of a peptide hormone is
   (a) cortisone.              (c) testosterone.           (e) estrogen.
   (b) progesterone.           (d) oxytocin.

3. Which of the following is *not* one of the seven principal pituitary hormones?
   (a) Thyroid hormone                    (d) Somatotropin
   (b) Follicle-stimulating hormone       (e) Luteinizing hormone
   (c) Melanocyte-stimulating hormone

4. Norepinephrine and epinephrine (also called adrenaline) are hormones that produce an "alarm" response. They are produced by the adrenal gland. Where is the adrenal gland?
   (a) In the brain              (c) In the neck               (e) Near the thymus
   (b) Just above the kidneys    (d) In the testes or ovaries

5. The hypothalamus controls water and salt balance within the kidneys by regulating the production of the hormone
   (a) estrogen.               (c) glucagon.               (e) aldosterone.
   (b) insulin.                (d) antidiuretic hormone.

# THOUGHT QUESTIONS

1. If you are lost in the desert with a case of liquor and are desperately thirsty, should you drink the liquor? Explain your answer.

2. Why do you suppose the brain goes to the trouble of synthesizing releasing hormones rather than simply directing the production of the pituitary hormones immediately?

# FOR FURTHER READING

BERRIDGE, M.: "The Molecular Basis of Communication Within the Cell," *Scientific American*, October 1985, pages 142-152. An up-to-date account of what is known about second messengers in the cell.

BLOOM, F.E.: "Neuropeptides," *Scientific American*, October 1981, pages 148-168. An account of recent advances in the study of endorphins and other brain hormones.

CANTIN, M., and J. GENEST: "The Heart as an Endocrine Gland," *Scientific American*, February 1986, pages 76-81. A good example of how our body self-regulates its activities; in addition to pumping blood, the heart secretes a hormone that fine-tunes the control of blood pressure.

DAVIS, J.: *Endorphins: New Waves in Brain Chemistry*, Doubleday & Co., Inc., New York, 1984. A popular account of current research on brain hormones.

## CHAPTER 37

# How Animals Digest Food

This flying squirrel is eating a moth for dinner. While this may not seem appetizing to you, it provides the squirrel with a nutritious meal.

# HOW ANIMALS DIGEST FOOD

## Overview

Digestion is the conversion of foods that are not directly absorbable into small molecules that can be easily used by cells as energy sources and building blocks. In vertebrates, food in the form of proteins, carbohydrates, and fats is digested as it passes through a long, one-way digestive tract. The food is first pulverized in the mouth and then degraded into molecular fragments in the stomach. The digestion of these fragments to simple molecules is completed in the small intestine. The small molecules that are the products of digestion are then absorbed into the body through the walls of the small intestine, and the residual solids are concentrated in the large intestine and eliminated.

## For Review

*Here are some important terms and concepts that you will encounter in this chapter. If you are not familiar with them, you should review them before proceeding.*

**Acid** (Chapter 2)

**Lysosomes** (Chapter 4)

**Oxidative respiration** (Chapter 7)

**Insulin and glucagon** (Chapter 36)

A vertebrate body is a complex organization of many cells. Like a city, it contains many individuals that carry out specialized functions. It has its own police (macrophages), its own construction workers (fibroblasts), and its own telephone company (the nervous system). Just as in a city, the many individual cells of the vertebrate body need to be provided with food that is trucked in from elsewhere (Figure 37-1). Among the cells of a vertebrate's body there are no farmers; no vertebrate contains photosynthetic cells. Instead, all of an animal's cells are nourished with food that the animal obtains outside itself and transports to the individual cells. Many of the major organ systems of an animal are involved in this acquisition of energy. The digestive system acquires organic foodstuffs; the respiratory system acquires the oxygen necessary to metabolize organic foodstuffs; the circulatory system transports both food and oxygen to the individual cells of the body; and the excretory system rids the body of waste produced by the metabolism of organic compounds. In this chapter we consider the first of these activities, digestion.

**Figure 37-1**

**All vertebrates need to eat to survive.** This lion is eating a zebra, which he caught on the grasslands of Kenya.

## THE NATURE OF DIGESTION

Animals are thermodynamic machines, expending energy to maintain order throughout their bodies. Sources of energy are therefore essential for survival. Animals obtain the metabolic energy needed for growth and activity by degrading the chemical bonds of organic molecules. This breakdown is referred to as catabolism. In Chapters 6 and 7 we considered these degradation processes in detail. They include the breakdown of sugar molecules in glycolysis and the oxidation of pyruvate in the citric acid cycle. What catabolic processes have in common is that they act on amino acids, lipids, sugars, and fragments of these molecules to produce energy in the form of ATP, together with water and $CO_2$ as the waste products. Eating another organism, however, does not in itself provide a rich source of such molecules to an animal, because few organisms contain significant concentrations of free sugars and amino acids. Instead, in foods the simple molecules are incorporated into long chains, into starches, fats, and proteins. Before an animal can obtain the energy from its food, it must degrade these molecules into the simple compounds from which the molecules were built. This process is called **digestion.**

Like most of the body's other systems, digestion has become increasingly complex during the evolution of animals. Their protist ancestors, as you have seen, simply incorporated food particles within their bodies by phagocytosis, combining these with hydrolytic enzymes in a process of intracellular digestion (Figure 37-2). Cnidarians break down the bodies of their prey by secreting enzymes into their central cavity, a cavity in contact with the external environment, and then incorporating the fragments into their cells, where digestion continues. This is a simple example of extracellular digestion. Flatworms, on the other hand, have a highly branched digestive system with only one opening (Figure 37-3, *A*). Because of its branching, such a digestive system has a greatly increased absorptive surface. Although some of the food particles are broken down by the extracellular secretion of enzymes, as in cnidarians, most are simply incorporated into the cells that line the flatworm's digestive tract and digested there.

Other, more complex animals have a more complete digestive tract, with a mouth and an anus. In them, there are various methods for breaking down the food mechanically, as by grinding in the gizzard of an earthworm or bird, or enzymatically. Many such animals (Figure 37-3, *B* and *C*) also have a crop, which can be filled with food when it is available; this stored food can be used later.

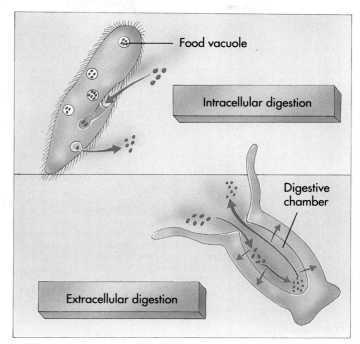

**Figure 37-2**

**Types of digestion.** *Intracellular digestion:* the most ancient animals used intracellular digestion (phagocytosis), which limited the size of the food particle to a fraction of the animal's size. *Extracellular digestion:* among the cnidarians we first see a digestive cavity, in which food particles are broken down extracellularly, outside of cells; final digestion is carried out within the cells that line the cavity.

## Agents of Digestion

Digestion is carried out in two ways: by hydrochloric acid (HCl), which breaks up large proteins into smaller pieces, and by a variety of highly specific enzymes (Table 37-1). Enzymes that break up proteins into amino acids are called **proteases;** enzymes that break up starches and other carbohydrates into sugars are called **amylases;** and enzymes that break up lipids and fats into small segments are called **lipases.** There is also a DNase and an RNase, which break up DNA and RNA. Most digestive enzymes cannot tolerate high acid concentrations, so the vertebrate digestive process is carried out in two phases: acid digestion takes place first, in the stomach, and then the food moves to the small intestine, where the acid is neutralized and a variety of digestive enzymes continues the digestive processes.

*In the stomach, proteins are partially denatured by acids; the fragments are then cleaved into individual amino acids by a variety of enzymes in the stomach and small intestine. Also in the small intestine, starches are digested by amylases and fats are digested by lipases.*

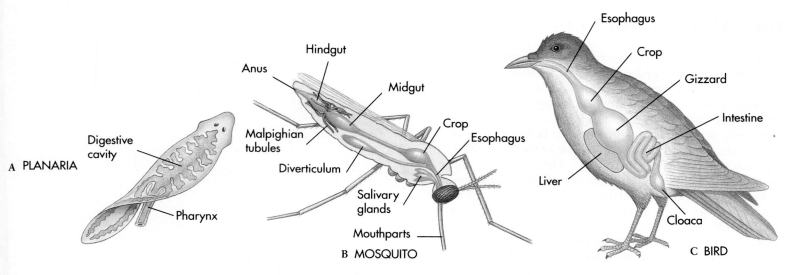

**A PLANARIA**

**B MOSQUITO**

**C BIRD**

*Figure 37-3*

**The digestive systems of various animal phyla.**
**A** The digestive system of a flatworm, in which there is only one opening to the outside. Most of the digestion in flatworms takes place after the food particles are incorporated into the cells that line the digestive cavity.
**B** The digestive system of an adult female mosquito. Most blood that the mosquito sucks from another animal is stored in the diverticulum for later use.
**C** The digestive system of a bird. A bird's crop plays a role similar to that of the diverticulum of the mosquito. In the gizzard, hard food is mixed and crushed with rocks and pebbles, often called grit, and thus prepared for more efficient digestion in the intestine.

**TABLE 37-1    DIGESTIVE ENZYMES**

| LOCATION | ENZYME | SUBSTRATE | DIGESTION PRODUCT |
|---|---|---|---|
| Salivary gland | Amylase | Starch Glycogen | Disaccharides |
| Stomach | Pepsin | Proteins | Short peptides |
| Small intestine | Peptidases | Short peptides | Amino acids |
|  | Nucleases | DNA, RNA | Sugars, nucleic acid bases |
|  | Lactase Maltase Sucrase | Disaccharides | Glucose, monosaccharides |
| Pancreas | Lipase | Triglycerides | Fatty acids, glycerol |
|  | Trypsin Chymotrypsin | Proteins | Peptides |
|  | DNase | DNA | Nucleotides |
|  | RNase | RNA | Nucleotides |

## ORGANIZATION OF THE VERTEBRATE DIGESTIVE SYSTEM

The general organization of the digestive tract is the same in all vertebrates, although different elements are emphasized in different groups. In all vertebrates, acid digestion of proteins takes place in the stomach, after which food passes to the upper part of the small intestine, called the duodenum, where a multitude of digestive enzymes continues the digestive process. The products of digestion then pass across the wall of the small intestine into the bloodstream. Figure 37-4 illustrates the organization of the human digestive system.

Specializations among the digestive systems of different kinds of vertebrates reflect differences in the way these animals live. The initial components of the gastrointestinal tract are the mouth and pharynx. The pharynx is the gateway to the esophagus. Fishes have a large pharynx with gill slits, not unlike that of the lancelets, whereas air-breathing vertebrates have a greatly reduced pharynx. Adult amphibians, which are carnivores, have a short intestine; the food they ingest is readily digested, including the soluble carbohydrate glycogen. Many birds, in contrast, subsist on plant material. The primary structural component of plants is cellulose, a rigid carbohydrate that resists digestion. Birds have a convoluted small intestine, by means of which they prolong the process of digestion and aid absorption of digestion products. Many animals have teeth, and chewing (mastication) breaks up food into small particles and mixes food with fluid secretions. Birds, which lack teeth to masticate their food, break up food in their stomachs, which have two chambers. In one of these chambers—the gizzard—small pebbles that are ingested by the bird are churned together with the food by muscular action; this churning serves to grind up the seeds and other hard plant material into smaller chunks before their digestion in the second chamber of the stomach. Mammals that digest grass and other vegetation often have stomachs with multiple chambers where bacteria aid the digestion of cellulose.

## WHERE IT ALL BEGINS: THE MOUTH

The food of all vertebrates is taken in through their mouths, which in all groups except birds typically contain teeth. The teeth of different kinds of vertebrates are specialized in different ways, depending on whether they usually feed on animals or plants and how they obtain what they eat. Human beings are omnivores, eating both plant and animal food regularly. As a re-

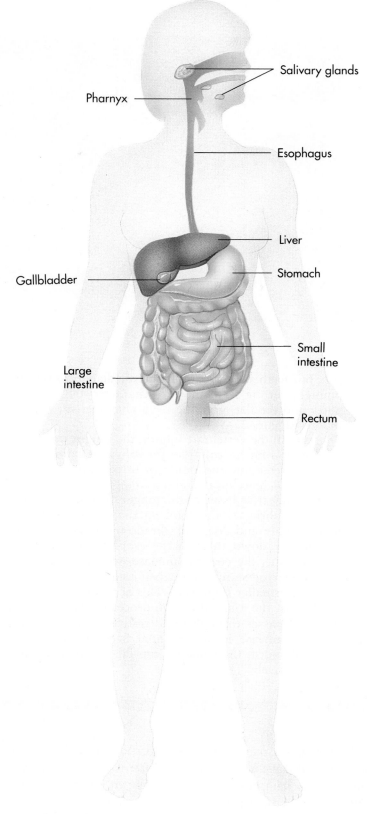

*Figure 37-4*

**The human digestive system.**

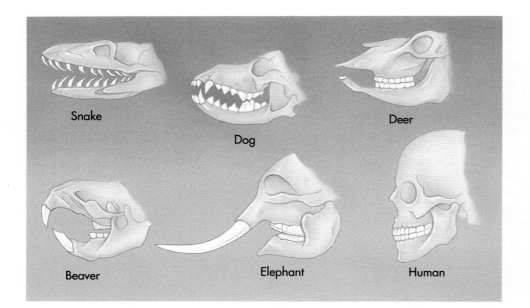

**Figure 37-5**

**Vertebrate teeth—tools to fit many functions.** The teeth of vertebrates are specialized for particular tasks. In the snake, the teeth slope backward to aid in retention of prey during swallowing. In carnivores such as the dog, "canine" teeth specialized for ripping food predominate. In herbivores such as deer, grinding teeth predominate. In the beaver, the foreteeth are specialized as incisors—chisels. In the elephant, two of the upper front teeth are specialized as weapons. Human beings are omnivores—we consume both plant and animal food—and have a relatively broad-function mouth containing canine, chisel, and grinding teeth.

sult, our teeth are structurally intermediate between the pointed, cutting teeth that are characteristic of carnivores and the flat, grinding teeth characteristic of herbivores (Figure 37-5).

Within the vertebrate mouth, the tongue mixes the food with a mucus solution, the **saliva.** In humans, saliva is secreted into the mouth by three pairs of salivary glands, which are located above and below the jaw and in the connective tissue of the tongue. The saliva moistens and lubricates the food so that it is swallowed more readily and does not abrade the tissue it passes on its way down through the esophagus. The saliva also contains the enzyme **amylase,** which initiates the breakdown of starch and other complex polysaccharides into smaller fragments. This action by amylase is the first of the many digestive processes that occur as the food passes through the digestive tract. However, salivary amylase is not essential because similar enzymes found in the pancreas can digest carbohydrates very effectively.

*Vertebrate teeth serve to shred animal tissue and to grind plant material. In humans, the ripping and tearing teeth are in front, the grinding teeth in the rear. Saliva secreted into the mouth moistens the food, which aids its journey into the digestive system and begins the enzyme-catalyzed process of degradation.*

## THE JOURNEY OF FOOD TO THE STOMACH

After passing through the opening at the back of the mouth (Figure 37-6), the food passes through the upper esophageal sphincter and enters a tube called the **esophagus,** which connects the pharynx to the stomach. No further digestion takes place in the esophagus. Its role is to move food down toward the stomach. In adult human beings the esophagus is about 25 centimeters long, and its lower end opens into the stomach proper. The lower two thirds of the esophagus is enveloped in smooth muscle. Successive waves of contraction of these muscles move food down through the esophagus to the stomach. Such rhythmic sequences of waves of muscular contraction in the walls of a tube are called **peristalsis.** Because the movement of food through the esophagus is primarily caused by these peristaltic contractions, humans can swallow even if they are upside down.

The exit of food from the esophagus to the stomach is controlled by the **lower esophageal (cardiac) sphincter.** When this sphincter is contracted, it prevents the food in the stomach from moving back up the esophagus.

If the sphincter does not close properly, the food backs up into the esophagus, and the acid coating it causes a burning sensation called "heartburn" (which, as you can see, actually has nothing to do with the heart). This problem often arises during pregnancy, when the digestive organs are displaced far upward.

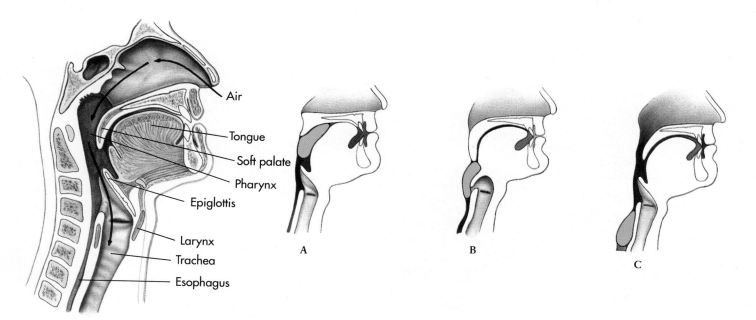

**Figure 37-6**

**How humans swallow.** As food passes back past the rear of the mouth (**A**), it presses the soft palate against the back wall of the pharynx, sealing off the nasal passage. As the food passes on down, a flap of tissue called the epiglottis folds down (**B**), sealing the respiratory passage. After the food enters the esophagus, the soft palate relaxes and the epiglottis is raised (**C**), opening the respiratory passage between the nasal cavity and the trachea.

## PRELIMINARY DIGESTION: THE STOMACH

The stomach (Figure 37-7) is a saclike portion of the digestive tract. The stomach has several functions: temporary storage, mechanical breakdown of food, and the chemical digestion of proteins. In the stomach the digestive process is organized; the stomach collects ingested food, partially hydrolyzes protein, and feeds its contents in a controlled fashion into the primary digestive organ, the small intestine. The interior of the stomach, like that of the rest of the digestive tract, is continuous with the outside of the body. The epithelium overlies a deep layer of connective tissue (Figure 37-8), called **mucosa**, below which is located a complex array of muscles, blood vessels, and nerves.

### How the Stomach Digests Food

The epithelium of the stomach is the source of the "digestive juices," a combination of HCl and digestive enzymes that, when mixed with food, forms a semisolid material called chyme. The upper epithelial surface of the stomach is dotted with deep depressions called **gastric pits** (Figure 37-9). In them the epithelial membrane

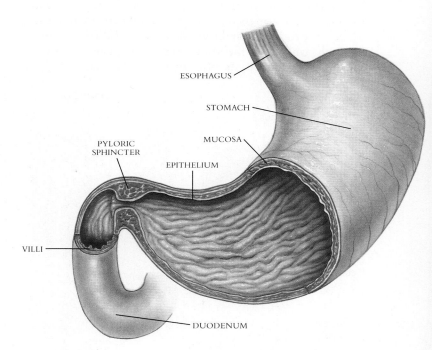

**Figure 37-7**

**The upper digestive tract of a human.**

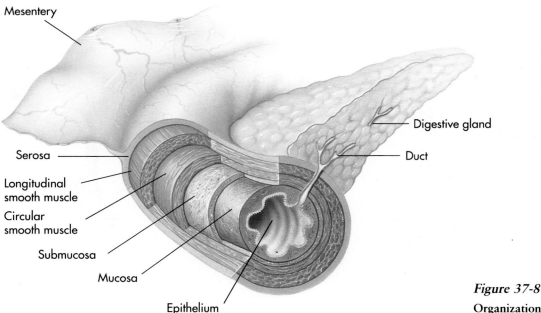

Mesentery

Digestive gland

Serosa

Duct

Longitudinal
smooth muscle

Circular
smooth muscle

Submucosa

Mucosa

Epithelium

**Figure 37-8**

**Organization of the vertebrate digestive tract.**

is invaginated, forming exocrine glands within the mucosa. These exocrine glands contain two kinds of secreting cells, **parietal cells,** which secrete HCl and **chief cells,** which secrete pepsinogen. Within the stomach, HCl cleaves a terminal fragment from the pepsinogen, converting pepsinogen into the protease pepsin.

Many of the epithelial cells that line the stomach are specialized for the secretion of mucus. This mucus, which is produced in large quantities, lubricates the stomach wall and facilitates the movement of food within the stomach. It also protects the cells of the

stomach wall from abrasion by the food and, most importantly, protects the walls of the stomach from its own digestive juices, the **gastric fluid,** which otherwise would eat its lining away (for example, causing ulcers).

The human stomach secretes about 2 liters of HCl every day, creating a very concentrated acid solution. This solution is actually about 150 millimolar, and thus 3 million times more acidic than the blood. The HCl breaks up connective tissue. It does this because the very low pH values, between 1.5 and 2.5, created by the HCl disrupt the attraction between carboxyl

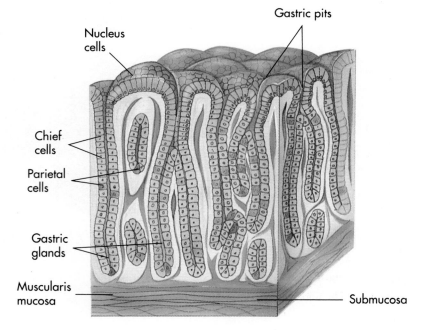

Gastric pits

Nucleus
cells

Chief
cells

Parietal
cells

Gastric
glands

Muscularis
mucosa

Submucosa

**Figure 37-9**

**Gastric pits.** Gastric pits are deep invaginations of the stomach epithelium down into the underlying mucosa, at the base of which parietal and chief cells secrete hydrochloric acid and a protein that is cleaved into the enzyme pepsin.

and amino side groups of proteins that is responsible for the tertiary structure of proteins. This process causes the folded proteins of connective tissue to open out and disrupts their associations with one another.

The action of the acid disintegrates the food into molecular fragments and thus is an essential prelude to digestion. The acid itself does not carry out any further digestive activity. It has a very limited ability to break proteins down into amino acids, or carbohydrates into their constituent sugars, and it does not attack fats at all. The further digestion of protein is accomplished by the enzyme pepsin, which cleaves proteins into short polypeptides. Because the fragments produced by pepsin are large and frequently charged, they cannot pass across the epithelial membrane. Except for water, some vitamins, and alcohol, no absorption takes place through the stomach wall. Although pepsin aids protein digestion, it is not essential because enough proteases are also secreted by the pancreas to digest proteins completely.

## The Stomach Secretes the Hormone Gastrin

It is important that a stomach not produce *too* much acid. If it did, it would be impossible for the body to neutralize the acid later in the small intestine, a step that is essential for the terminal stages of digestion. The stomach controls the production of acid by means of hormones produced by endocrine cells scattered throughout its epithelial layer (Table 37-2) (see Figure 37-9). The hormone **gastrin** regulates the synthesis of HCl by the parietal cells of the gastric pits, permitting such synthesis to occur only when the pH of the stomach contents is higher than about 1.5. Other hormones interact to maintain a constant stomach pH (Figure 37-10).

Some stomachs greatly overproduce gastrin, which results in excessive acid production. The excessive acid may attack and burn holes in the walls of the small intestine. These holes are called **duodenal ulcers.** The contents of the small intestine are not normally acidic, and this organ is much less able to withstand the disruptive actions of stomach acids than is the wall of the stomach. For this reason over 90% of all ulcers are duodenal, although other ulcers sometimes occur in the stomach when its mucous barrier is damaged, such as by aspirin and alcohol.

> In the stomach, concentrated acid breaks up connective tissue and protein into molecular fragments, which are further digested by pepsin into short polypeptides. Carbohydrates and fats are not digested in the stomach.

The inner surface of the stomach is highly convoluted. For this reason it can fold up when empty and open out like an expanding balloon as it fills with food. There is, of course, a limit to how much food a stomach can hold. The human stomach has a volume of about 50 milliliters when empty; when full, it may

| TABLE 37-2 | HORMONES OF DIGESTION | | | | |
|---|---|---|---|---|---|
| HORMONE | CLASS | SOURCE | STIMULUS | ACTION | NOTE |
| Gastrin | Polypeptide | Pyloric portion of stomach | Entry of food into stomach | Secretion of HCl | Unusual in that it acts on same organ that secretes it |
| Cholecystokinin | Polypeptide | Duodenum | Arrival of food in small intestine | Stimulates gallbladder contraction, and so the release of bile salts into intestine<br>Stimulates secretion of digestive enzymes by pancreas | CCK bears a striking structural resemblance to gastrin |
| Secretin | Polypeptide | Duodenum | HCl in duodenum | Stimulates pancreas to secrete bicarbonate, which neutralizes stomach acid | The first hormone to be discovered (1902) |

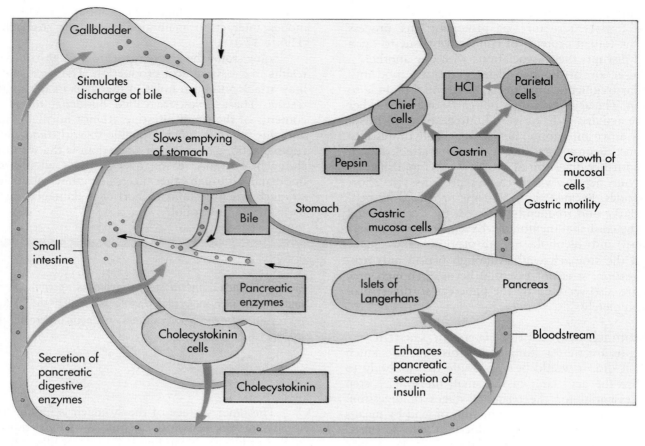

**Figure 37-10**

Hormones of the digestive system.

have a volume 50 times larger, from 2 to 4 liters. Carnivores, often consuming large meals at infrequent intervals, possess stomachs that are able to distend much more than can our stomachs or those of most other mammals.

## Leaving the Stomach

The digestive tract exits from the stomach at a muscular constriction known as the **pyloric sphincter**. The pyloric sphincter is the gate to the small intestine, the organ within which the final stages of digestion occur. The pyloric sphincter therefore is the traffic light of the digestive system. The capacity of the small intestine is limited, and its digestive processes take time. Consequently, efficient digestion requires that only relatively small portions of food be introduced from the stomach into the small intestine at any one time. When a small volume of food, which is by now semisolid, highly acidic chyme, passes into the small intestine, the acid introduced with the food acts as a signal and prompts the closing of the pyloric sphincter. As time passes, the food is digested and the acid that entered the small in-

testine with it is neutralized. At a certain point in the process, the pH of the small intestine reaches a level that signals the pyloric sphincter to open once again. Another small portion of food is introduced from the stomach into the small intestine, and the process continues.

## TERMINAL DIGESTION AND ABSORPTION: THE SMALL INTESTINE

The small intestine is the true digestive vat of the vertebrate body. Within it, carbohydrates, proteins, and fats are broken down into sugars, amino acids, and fatty acids. Once these small molecules have been produced, they all pass across the epithelial wall of the small intestine into the bloodstream. Some of the enzymes necessary for these digestive processes are secreted by the cells of the intestinal wall. Most, however, are introduced into a short initial segment of the small intestine, the **duodenum**, through a duct from a gland called the pancreas.

The small intestine is approximately 6 meters long. The first 25 centimeters, about 4% of the total length, is the duodenum. It is at this point that pancreatic enzymes and bile enter the intestine, initiating digestion. Absorption of water and the products of digestion by the bloodstream occurs in the later sections of the intestine, the jejunum and the ileum. The epithelial wall of the small intestine is covered with fine fingerlike projections called **villi,** which are microscopic (Figure 37-11). In turn, each of the epithelial cells covering the villi is covered on its outer surface by a field of cytoplasmic projections called **microvilli** (Figure 37-12). Both kinds of projections greatly increase the absorptive surface of the epithelium lining the small intestine. The average surface area of the small intestine of an adult human being is about 300 square meters. The membranes of the epithelial cells contain some enzymes that complete digestion, as well as carrier systems that actively transport sugars and amino acids across the membrane; fatty acids cross passively by diffusion. After absorption from the lumen, end-products of digestion enter capillaries (sugars and amino acids) in the villi or enter lymph vessels known as lacteals (triglycerides and fatty acids).

*Most digestion occurs in the first 25 centimeters of the 6-meter length of the small intestine, in a zone called the duodenum. The rest of the small intestine is devoted to the absorption of the products of digestion.*

The amount of material passing through the small intestine is startlingly large. An average human consumes about 800 grams of solid food and 1200 milliliters of water each day, for a total volume of about 2 liters. To this amount is added about 1.5 liters of fluid from the salivary glands, 2 liters from the gastric secretions of the stomach, 1.5 liters from the pancreas, 0.5 liter from the liver, and 1.5 liters of intestinal secretions. The total adds up to a remarkable 9 liters. However, although the flux is great, the *net* passage is small. Almost all these fluids and solids are reabsorbed during their passage through the intestines, with about 8.5 liters passing across the walls of the small intestine and

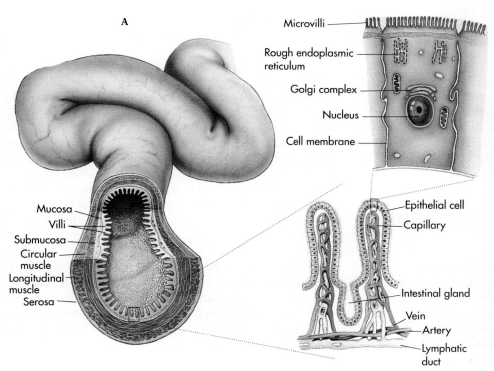

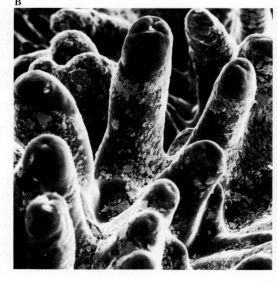

*Figure 37-11*

**The small intestine.**
A  Cross-section of the small intestine.
B  Villi, shown in a scanning electron micrograph, are very densely clustered, giving the small intestine an enormous surface area, which is very important for efficient absorption.

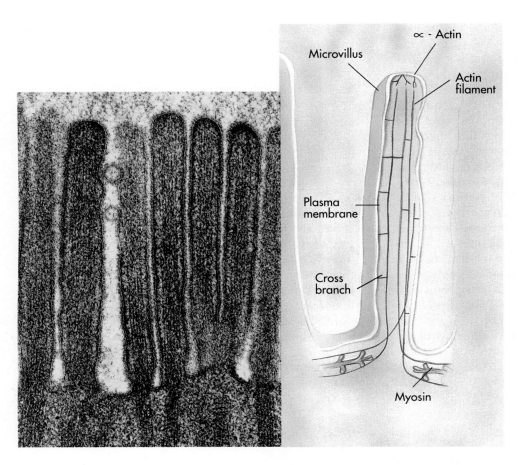

**Figure 37-12**
**Intestinal microvilli.**

an additional 350 milliliters through the wall of the large intestine. Of the 800 grams of solid and 9 liters of liquid that enter the digestive tract in 1 day, only about 50 grams of solid and 100 milliliters of liquid leave the body as feces. The normal fluid absorption efficiency of the digestive tract thus approaches 99%, which is very high indeed.

## ORGANS THAT SUPPORT THE DIGESTIVE SYSTEM

### The Pancreas: Making Digestive Enzymes

The **pancreas** is a large gland situated near the junction of the stomach and the small intestine. It is one of the body's major exocrine glands, secreting a host of different enzymes that act in the duodenum to break down carbohydrates, proteins, and fats. These enzymes include proteases for breaking down proteins, lipases for digesting fats, and enzymes that break down carbohydrates. The pancreas also functions as an endocrine gland.

The pancreas has two types of exocrine cells. The first type functions exactly opposite to the parietal cells of the gastric pits. Instead of secreting an acid, HCl, these specialized cells secrete a base, **bicarbonate.** The alkaline bicarbonate that is secreted by the pancreas is critical to successful digestion, since most of the enzymes secreted by the pancreas will not work in acid solution. The introduction of bicarbonate into the duodenum neutralizes the acid derived from the stomach and thus permits the digestive enzymes to function. Since acid is secreted in the stomach, and bicarbonate is secreted in the intestine, there is no net effect of digestion on the body's acid-base balance (see Chapter 41).

The pancreas has yet a third function critical to metabolism. As we noted in Chapter 36, the islets of Langerhans, distributed throughout the exocrine regions of the pancreas, function as endocrine glands, producing the hormones that act in the liver and elsewhere to regulate the level of sugar in the blood.

### The Liver: Major Component of the Digestive System

Because fats are insoluble in water, they tend to enter the small intestine as small globules that are not attacked readily by the enzymes secreted by the pancreas.

Before fats can be digested by pancreatic lipases, they must be made soluble. This process is carried out by a collection of detergent molecules secreted by a second gland, the **liver**. The liver is the body's principal metabolic factory, turning foodstuffs arriving from the digestive tract in the bloodstream into substances that are utilized by the different cells of the body. It is the largest internal organ of the body. In an adult human being the liver weighs about 1.5 kilograms and is the size of a football.

The liver carries out a wide variety of metabolic functions, many of which we discuss in later chapters. It supplies quick energy, metabolizes alcohol, makes proteins, stores vitamins and minerals, regulates blood clotting, regulates the production of cholesterol, and detoxifies poisons. It also produces the detergent molecules already mentioned, known as **bile salts** and secretes them through a duct into the duodenum. Bile acts as a superdetergent. It combines with fats to form microscopic droplets called micelles in a process known as emulsification. The components of bile include bile salts, cholesterol, and the phospholipid lecithin. All these combine to render fat soluble. Human beings concentrate and store bile manufactured in the liver in the **gallbladder**. When chyme enters the small intestine, a hormone known as cholecystokinin (CCK) stimulates contraction of the gallbladder to release bile into the duodenum.

*How the Liver Regulates Blood Glucose Levels.* It is important that the blood of vertebrates maintain a relatively constant composition, since the different tissues of the body need specific compounds as energy sources. Brain cells, for example, can store very little glucose and lack the enzymes to convert fat or amino acids into glucose. Brain cells are thus very sensitive to the level of available glucose. They are totally dependent on blood plasma for glucose and cease to function if the level of glucose in the blood falls much below normal values.

Maintaining a constant level of metabolites in the blood requires active control by the body's organs. A moment's reflection shows why. Most vertebrates eat sporadically. In the United States, most people eat three meals a day. Food enters the digestive system at intervals separated by long periods of fasting. Much of the food is digested relatively quickly, with the metabolites, such as glucose and amino acids, passing through the lining of the small intestine into the bloodstream. Without active control of the levels of these and other metabolites, they would suddenly become much more abundant in the blood right after a meal and then fall rapidly during a period of starvation, when the metabolites are being removed from the bloodstream by metabolizing cells and are not being replenished.

Control over metabolite levels in the blood is achieved in a very logical way: by establishing a reservoir, or metabolic bank, in the liver. One of the liver's most important functions is regulation of the blood's metabolite levels. This regulation is achieved by means of a special shunt in the circulatory system. Blood returning from the stomach and small intestine flows into the portal vein, which carries it not to the heart but to the liver (Figure 37-13). Within the liver, blood passes through a network of fine passages called sinuses. Only after flowing through the liver is the blood collected into the hepatic vein and delivered to the vena cava and the heart.

When excessive amounts of glucose are present in the blood passing through the liver, a situation that occurs soon after a meal, the liver converts the excess glucose into the starchlike glucose polymer, glycogen, stimulated by the hormone insulin (see Figure 36-7). The liver stores this glycogen. Some glycogen is also stored in the muscles, where it is easily available to fuel muscle contraction. When blood glucose levels decrease, as they do in a period of fasting or between meals, the glucose deficit in the blood plasma is made up from the glycogen reservoir, stimulated by the hormone glucagon. The human liver stores enough glycogen to supply glucose to the bloodstream for about 10 hours of fasting. If fasting continues, the liver begins to convert other molecules, such as amino acids, into glucose to maintain the glucose level in the blood. The liver, the body's metabolic reservoir, thus acts much like a bank, making deposits and withdrawals in the "currency" of glucose molecules.

*In addition to its many other roles, the liver acts to regulate the level of blood glucose, maintaining it within narrow bounds.*

Also like a bank, the liver exchanges currencies, converting other molecules, such as amino acids and fats, to glucose and subsequently glycogen for storage. Excess amino acids that may be present in the blood are converted to glucose by liver enzymes, and the glucose is stored as glycogen. The first step in this conversion is the removal of the amino group ($NH_3$) from the amino acid, a process called deamination. Unlike plants, animals cannot reuse the nitrogen from these amino groups and must excrete it as nitrogenous

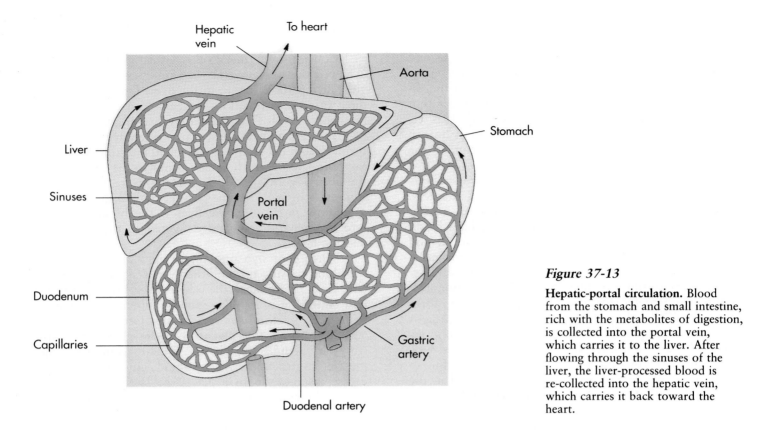

*Figure 37-13*

**Hepatic-portal circulation.** Blood from the stomach and small intestine, rich with the metabolites of digestion, is collected into the portal vein, which carries it to the liver. After flowing through the sinuses of the liver, the liver-processed blood is re-collected into the hepatic vein, which carries it back toward the heart.

waste. The product of amino acid deamination, ammonia ($NH_4$), forms a complex with carbon dioxide to form urea (see Figure 41-10). In some vertebrates, amino acids are not deaminated directly, but instead are converted to uric acid. The uric acid is released by the liver into the bloodstream, where the kidneys subsequently remove it.

The liver has a limited storage capacity. When its glycogen reservoir is full, it continues to remove excess glucose molecules from the blood by converting them to fat, which is stored elsewhere in the body. In humans, for example, long periods of overeating and the resulting chronic oversupply of glucose frequently result in the deposition of fat around the stomach or on the hips.

## CONCENTRATION OF SOLIDS: THE LARGE INTESTINE

The large intestine, or **colon,** is much shorter than the small intestine, occupying approximately the last meter of the intestinal tract. No digestion takes place within the large intestine, and only about 4% of the absorption of fluids by the body occurs there. The large intestine is not convoluted, lying instead in three relatively straight segments, and its inner surface does not possess villi. Consequently the large intestine has less than one-thirtieth the absorptive surface area of the small intestine. Although sodium, vitamin K, and some other products of bacterial metabolism are absorbed across its wall, the primary function of the large intestine is to act as a refuse dump. Within it, undigested material, including large amounts of plant and cellulose (Figure 37-14), is compacted and stored. Many bacteria live and actively divide within the large intestine, where they play a role in the processing of undigested material into the final excretory product, **feces.** Bacterial fermentation produces gas within the colon at a rate of about 500 milliliters per day. This rate increases greatly after the consumption of beans or other vegetable matter because the passage of undigested plant material into the large intestine provides material for fermentation.

*The large intestine serves primarily to compact the solid refuse remaining after digestion of food, thus facilitating its elimination.*

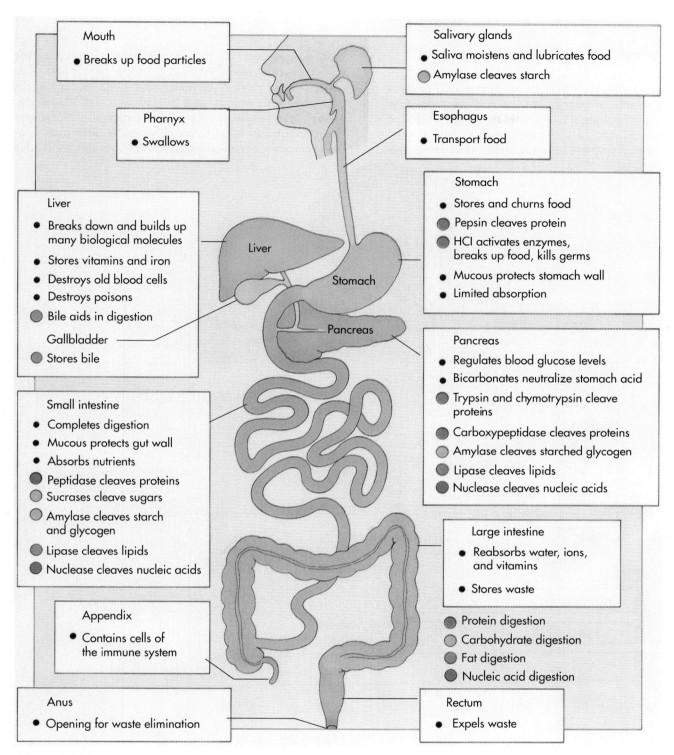

**Mouth**
- Breaks up food particles

**Salivary glands**
- Saliva moistens and lubricates food
- Amylase cleaves starch

**Pharnyx**
- Swallows

**Esophagus**
- Transport food

**Liver**
- Breaks down and builds up many biological molecules
- Stores vitamins and iron
- Destroys old blood cells
- Destroys poisons
- Bile aids in digestion

**Gallbladder**
- Stores bile

**Stomach**
- Stores and churns food
- Pepsin cleaves protein
- HCl activates enzymes, breaks up food, kills germs
- Mucous protects stomach wall
- Limited absorption

**Pancreas**
- Regulates blood glucose levels
- Bicarbonates neutralize stomach acid
- Trypsin and chymotrypsin cleave proteins
- Carboxypeptidase cleaves proteins
- Amylase cleaves starched glycogen
- Lipase cleaves lipids
- Nuclease cleaves nucleic acids

**Small intestine**
- Completes digestion
- Mucous protects gut wall
- Absorbs nutrients
- Peptidase cleaves proteins
- Sucrases cleave sugars
- Amylase cleaves starch and glycogen
- Lipase cleaves lipids
- Nuclease cleaves nucleic acids

**Large intestine**
- Reabsorbs water, ions, and vitamins
- Stores waste

- Protein digestion
- Carbohydrate digestion
- Fat digestion
- Nucleic acid digestion

**Appendix**
- Contains cells of the immune system

**Anus**
- Opening for waste elimination

**Rectum**
- Expels waste

*Figure 37-14*

**The organs of the digestive system, and their functions.**

The final segment of the gastrointestinal tract is a short extension of the large intestine called the **rectum.** Compacted solids within the colon pass into the rectum as a result of the peristaltic contractions of the muscles encasing the large intestine. From the rectum, the solid material passes out of the anus through two **anal sphincters.** The first is composed of smooth muscle; it opens involuntarily in response to a pressure-generated nerve signal from the rectum. The second sphincter, in contrast, is composed of striated muscle. It is subject to voluntary control from the brain, thus permitting a conscious decision to delay defecation.

# NUTRITION

The ingestion of food by vertebrates serves two ends: it provides a source of energy, and it also provides raw materials that the animal is not able to manufacture for itself (Figure 37-15). There are two basic nutritional states: (1) the absorptive state, which immediately fol- lows meals, and (2) the postabsorptive state, during which reserve stores of energy must be tapped. The process of digestion is matched to the needs of the body by both neural and hormonal control systems. Neural control is mediated by external parasympathetic input and local reflexes, whereas hormonal control primarily involves insulin and glucagon (and to a

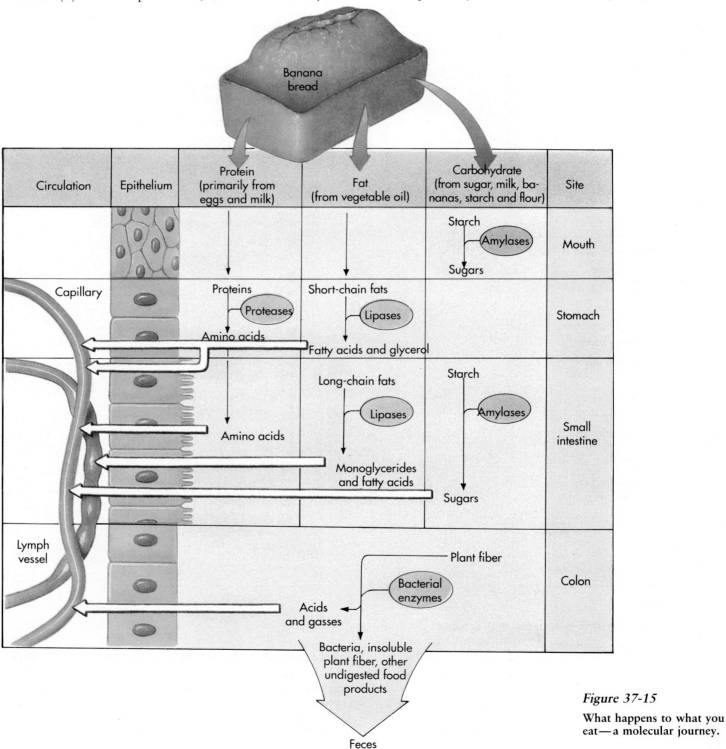

**Figure 37-15**

**What happens to what you eat—a molecular journey.**

lesser degree somatostatin). The liver maintains a very constant level of glucose in the blood and stores several hours' reserve of glucose in the form of glycogen. Any intake of food in excess of that required to maintain the glycogen reserve results in one of two consequences. Either the excess glucose is metabolized by the muscles and other cells of the body, or it is converted to fat and stored within fat cells. Fats have a much higher energy content per gram than carbohydrates and proteins and represent a very efficient way to store energy. Thus we have the simple equation:

$$\text{FOOD} - \text{EXERCISE} = \text{FAT}$$

In wealthy countries such as those of North America and Europe, obesity results from chronic overeating and from unbalanced diets high in fat and is a significant human health problem. In the United States, about 30% of middle-aged women and 15% of middle-aged men are classified as overweight, weighing at least 20% more than the average weight for their height (Figure 37-16). Being overweight is strongly correlated with coronary heart disease and many other disorders.

Over the course of their evolution, many vertebrates have lost the ability to synthesize different substances that nevertheless continue to play critical roles in their metabolism. Substances that an animal cannot manufacture for itself but that are necessary for its health must be obtained in other ways—in its diet. When essential organic substances are used in trace amounts, they are called **vitamins.** Many vitamins are required cofactors for intracellular enzyme systems. Human beings, monkeys, and guinea pigs, for example, have lost the ability to synthesize ascorbic acid (vitamin C) and will develop the disease **scurvy**—characterized by weakness, spongy gums, and bleeding of the skin and mucus membranes, which can ultimately prove fatal—if vitamin C is not supplied in sufficient quantities in their diets. All other mammals, as far as is known, are able to synthesize ascorbic acid. Human beings require at least 13 different vitamins (Table 37-3).

*A vitamin is an organic substance that is required in minute quantities by an organism for growth and activity, but that the organism cannot synthesize.*

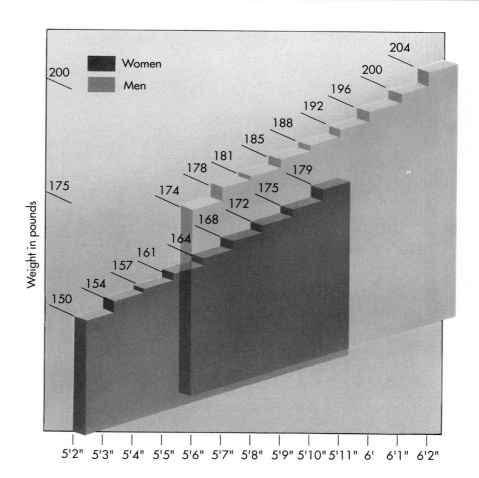

**Figure 37-16**

**How fat are you?** Obesity is usually characterized as the state of being more than 20% heavier than the average person of the same sex and height. These are the weight values at which obesity begins in Americans for a variety of heights, using average weights compiled in 1985.

**TABLE 37-3    MAJOR VITAMINS**

| VITAMIN | FUNCTION | DIETARY SOURCE | RECOMMENDED DAILY ALLOWANCE (milligrams) | DEFICIENCY SYMPTOMS | SOLUBILITY |
|---|---|---|---|---|---|
| Vitamin A (retinol) | Used in making visual pigments, maintenance of epithelial tissues | Green vegetables, milk products, liver | 1 | Night blindness, flaky skin | Fat |
| B-Complex Vitamins | | | | | |
| $B_1$ | Coenzymne in $CO_2$ removal during cellular respiration | Meat, grains, legumes | 1.5 | Beriberi, weakening of heart, edema | Water |
| $B_2$ (riboflavin) | Part of coenzymes FAD and FMN, which play metabolic roles | In many different kinds of foods | 1.8 | Inflammation and breakdown of skin, eye irritation | Water |
| $B_3$ (niacin) | Part of coenzymes $NAD^+$ and $NADP^+$ | Liver, lean meats, grains | 20 | Pellagra, inflammation of nerves, mental disorders | Water |
| $B_5$ (pantothenic acid) | Part of coenzyme-A, a key connection between carbohydrate and fat metabolism | In many different kinds of foods | 5 to 10 | Rare: fatigue, loss of coordination | Water |
| $B_6$ (pyridoxine) | Coenzyme in many phases of amino acid metabolism | Cereals, vegetables, meats | 2 | Anemia, convulsions, irritability | Water |
| $B_{12}$ (cyanocobalamin) | Coenzyme in the production of nucleic acids | Red meat, dairy products | 0.003 | Pernicious anemia | Water |
| Biotin | Coenzyme in fat synthesis and amino acid metabolism | Meat, vegetables | Minute | Rare: depression, nausea | Water |
| Folic acid | Coenzyme in amino acid and nucleic acid metabolism | Green vegetables, whole | 0.4 | Anemia, diarrhea | Water |
| Vitamin C | Important in forming collagen, cement of bone, teeth, connective tissue of blood vessels; may help maintain resistance to infection | Fruit, green leafy vegetables | 45 | Scurvy, breakdown of skin, blood vessels | Water |
| Vitamin D (calciferol) | Increases absorption of calcium and promotes bone formation | Dairy products, cod liver oil | 0.01 | Rickets, bone deformities | Fat |
| Vitamin E (tocopherol) | Protects fatty acids and cell membranes from oxidation | Margarine, seeds, green leafy vegetables | 15 | Rare | Fat |
| Vitamin K | Essential to blood clotting | Green leafy vegetables | 0.03 | Severe bleeding | Fat |

## Dangerous Eating Habits

In the United States, serious eating disorders have become much more common since the mid-1970s. The most frequent of these are **anorexia nervosa,** a condition in which the afflicted person literally starves himself or herself; and **bulimia,** a condition in which individuals gorge themselves and then cause themselves to vomit so that their weight stays constant. For reasons that we do not understand, 90% to 95% of those suffering from these eating disorders are female, and researchers estimate that 5% to 10% of the adolescent girls and young women in the United States suffer from eating disorders. As many as one of five female high-school and college students may suffer from bulimia.

Those who suffer from anorexia are typically shy, well-behaved young women who feel embarrassed about their bodies. Eventually, they often begin to show signs of severe malnutrition and abnormally low temperatures and pulse rates. Although malfunction of the pituitary gland may produce similar symptoms, anorexia nervosa is now understood to be a psychiatric disturbance that has severe psychic and psychological consequences. Its medical consequences may be severe enough to result in death.

Bulimia is usually accompanied by severe feelings of guilt, depression, and a feeling of anxiety and helplessness. The eating binges that are characteristic of bulimia often involve large quantities of carbohydrate-rich junk food. The accompanying medical consequences result from the frequent purging, which reduces the body's supply of potassium. Potassium plays an important role in regulating body fluids; its loss may lead to muscle weakness or paralysis, an irregular heartbeat, and kidney disease. Like anorexia nervosa, bulimia often leads to decreased resistance to infection.

Although both anorexia nervosa and bulimia are now believed to be psychiatric disorders, scientists are continuing to search actively for physiological explanations for these serious conditions. Changes in the levels of certain hormones are associated with these conditions, but whether these changes cause or result from the conditions has not been demonstrated clearly. In treating these conditions, it is essential that the shame often associated with them be recognized as a problem and that professional help be sought promptly. Both a psychiatrist or psychologist and a physician should be involved, and the family as a whole must play a strong, supportive role for the treatment to be effective. Even with effective treatment, both conditions may persist for years; it is therefore essential that they be understood. Parents, teachers, and others should be alert to the existence of anorexia and bulimia and should provide the kind of education about food and diet that, coupled with a supportive environment, will lead to the early detection and cure of these eating disorders if they do occur.

---

Some of the substances that vertebrates are not able to synthesize are required in more than trace amounts. Many vertebrates require one or more of the 20 amino acids necessary to synthesize proteins. Human beings are unable to synthesize 8 of these 20 amino acids: lysine, tryptophan, threonine, methionine, phenylalanine, leucine, isoleucine, and valine. These amino acids, called **essential amino acids,** must be obtained by humans from proteins in the food they eat (Figure 37-17). Complete proteins refer to proteins containing all the essential amino acids, whereas incomplete proteins lack some essential amino acids. All vertebrates have also lost the ability to synthesize certain polyunsaturated fats that provide backbones for fatty acid synthesis. Some essential substances that vertebrates do synthesize for themselves cannot be manufactured by the members of other animal groups. Some carnivorous insects, for example, are unable to synthesize the cholesterol that is required for the synthesis of steroid hormones; they must therefore obtain cholesterol in their diet. Other insects have the ability to convert other steroids, which they consume with their food, to cholesterol.

In addition to energy and those organic compounds that cannot be synthesized, the food that an animal consumes must also supply essential minerals such as calcium and phosphorus. Food must also include a wide variety of **trace elements,** which are min-

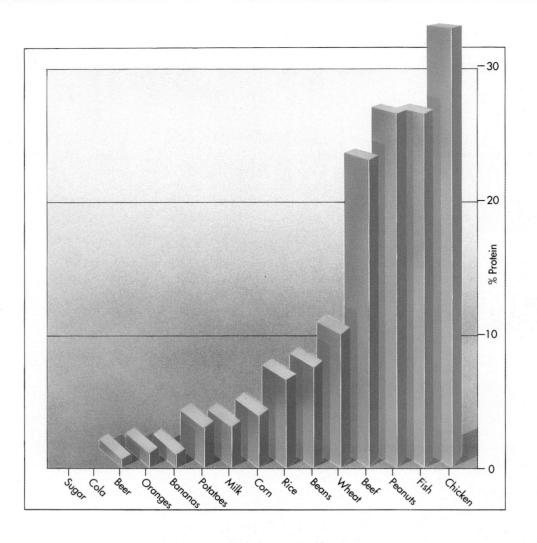

**Figure 37-17**

The protein content of a variety of common foods.

erals that are required in very small amounts. Among the trace elements are iodine (a component of thyroid hormone), cobalt (a component of vitamin $B_{12}$), zinc and molybdenum (components of enzymes), manganese, and selenium. All these, with the possible exception of selenium, are also essential for plant growth; they are obtained by the animals that require them either directly from plants or from animals that have eaten the plants.

Interestingly, one essential characteristic of food is simply bulk, its content of undigested fiber. The large intestine of humans, for example, has evolved as an organ adapted to process food that has a relatively high fiber content. Diets that are low in fiber, which are common in the United States, result in a slower passage of food through the colon than is desirable. This low dietary fiber content is thought to be associated with the levels of colon cancer in the United States, among the highest in the world.

The vertebrate digestive systems that we have described in this chapter are much larger and more complex than those of the invertebrates discussed in Chapter 28. In the vertebrate systems the working surfaces are convoluted and therefore much more extensive, and the different portions of the system are specialized for particular functions in the digestive process. Nutritional specialization in the mammals has likewise become complex and has led to many of the problems that humans face in selecting an appropriate diet. In the next chapter, we shall consider the way in which vertebrates acquire oxygen, another critical component of their systems for energy use and release.

# ■ SUMMARY

1. Digestion is the rendering of parts of organisms into amino acids and sugars, which can be metabolized by organisms.

2. The digestive tract of vertebrates is one-way. The initial portion leads from a mouth through an esophagus to a stomach.

3. In most mammals the stomach juices are concentrated acid, in which the protein-digesting enzyme pepsin is active.

4. Food passes from the stomach to the small intestine, where the pH is neutralized and various enzymes synthesized in the pancreas act to complete digestion. Most digestion occurs in the first 25 centimeters of the small intestine, in a zone called the duodenum.

5. The products of digestion are absorbed across the walls of the small intestine, which possess numerous villi and so achieve a very great surface area. Amino acids and sugars are transported by specific transmembrane channels, whereas fatty acids, which are lipid soluble, passively cross the membranes of the villi. In the process of digestion, fats need to be made soluble by detergent molecules secreted by the liver.

6. The large intestine has little digestive or absorptive activity; it functions principally to compact the refuse that is left over from digestion for easier elimination.

7. Vertebrates lack the enzymes necessary to synthesize many necessary compounds and must obtain these enzymes, which are called vitamins, from their diet. A number of trace elements must also be present in the diet.

## REVIEW

1. _____ is the process of breaking down food into simple compounds.

2. A band of muscle called the _____ controls the entrance to the duodenum from the stomach.

3. The numerous _____ that cover the epithelial wall of the small intestine greatly increase the surface area for absorption of digested foods.

4. If food energy is not metabolized by muscle or other body cells, it is stored as _____.

5. Which vitamin do you get from your symbiotic intestinal bacteria?

## SELF-QUIZ

1. Starchy foods such as potatoes begin being digested in the
   (a) mouth.
   (b) esophagus.
   (c) stomach.
   (d) duodenum.
   (e) small intestine.

2. Protein-rich foods such as steak are broken down in the stomach by the combined action of (pick two)
    (a) low pH.
    (b) high pH.
    (c) amylase.
    (d) pepsin.
    (e) bile.

3. Carbohydrate digestion begins in the
    (a) mouth.
    (b) stomach.
    (c) small intestine.
    (d) large intestine.
    (e) esophagus.

4. Most of the enzymes that complete the breakdown of food items into simple sugars, amino acids, and fatty acids are secreted by which of the following?
    (a) Salivary glands
    (b) Gastric pits
    (c) Liver
    (d) Pancreas
    (e) Large intestine

5. The main function of the large intestine is
    (a) digestion.
    (b) absorption.
    (c) secretion.
    (d) fermentation.
    (e) compaction.

6. Which of the following is an essential amino acid?
    (a) Alanine
    (b) Leucine
    (c) Cystine
    (d) Asparagine
    (e) Proline

# THOUGHT QUESTIONS

1. Human beings obtain vitamin K from symbiotic bacteria living in their gastrointestinal tract. Many bacteria also produce ascorbic acid, vitamin C. Can you suggest a reason why people have not evolved a symbiotic relationship with bacteria that would result in their obtaining bacterial vitamin C?

2. Many digestive enzymes are synthesized in the pancreas and released into the duodenum. Since this is the case, why do mammalian digestive systems go to the trouble of producing pepsin in the stomach? What does pepsin do that the other enzymes do not?

# FOR FURTHER READING

D<small>E</small>GABRIELE, R.: "The Physiology of the Koala," *Scientific American*, July 1980, pages 110-117. These Australian marsupials are adapted to a highly specific and unusual diet.

HARRIS, A.R.: "Why do More Women get Ulcers?" *Chatelaine*, vol. 61, March 1988, page 32. Discusses the link between ulcers in women and risk factors such as smoking, alcohol, and stress.

MOOG, F.: "The Lining of the Small Intestine," *Scientific American*, November 1981, pages 154-176. A clear description of the most important absorptive surface in the human body.

SCRIMSHAW, N.S., and V.R. YOUNG: "The Requirements of Human Nutrition," *Scientific American*, September 1976, pages 50-64. If you want to know what you should eat and what perhaps you should not, this article will point the way. A particularly good treatment of the important roles of trace elements in the human diet.

WINICK, M.: *Control of Appetite, Current Concepts In Nutrition*, John Wiley & Sons, Inc., New York, 1988. Describes appetite regulation by control centers in the brain and receptors in the gastrointestinal tract. Discusses that the CNS pathways involved in satiety involve endogenous opiates and mediate responses to stress and reward.

# How Animals Capture Oxygen

All animals obtain oxygen and dispose of carbon dioxide. The life-sustaining process of respiration is so effortless that we are often unaware of it.

# HOW ANIMALS CAPTURE OXYGEN

## Overview

Oxygen enters the bodies of animals by diffusion into water. The evolution of respiratory mechanisms among the vertebrates has favored changes that maximize the rate of this diffusion. The most efficient aquatic mechanism to evolve is the gill of bony fishes; the most efficient aerial respiration mechanism is the two-cycle lung of birds. Both these structures achieve high efficiency by countercurrent flow. In the vertebrates, hemoglobin plays a critical role in the transport of oxygen in the respiration process.

## For Review

*Here are some important terms and concepts that you will encounter in this chapter. If you are not familiar with them, you should review them before proceeding.*

**Chemistry of carbon dioxide** (Chapter 2)

**Diffusion** (Chapter 5)

**Oxidative respiration** (Chapter 7)

**Adaptation of vertebrates to terrestrial living** (Chapters 18 and 28)

**Red blood cells** (Chapter 32)

**Figure 38-1**

**Respiration.** Most of us think of respiration as breathing. This boy is blowing his horn with air from his lungs. Respiration also refers to the metabolic processes that use oxygen and generate carbon dioxide.

Human beings, like all other animals, obtain carbon compounds by consuming other organisms and then metabolize these carbon compounds to obtain the energy they use to move, to grow, and to think (Figure 38-1). The biochemical mechanism of this **oxidative metabolism** is discussed in Chapter 8. Basically, animals obtain their energy by removing electrons from carbon compounds they eat and then using these electrons to drive a series of proton-pumping channels in the membrane to generate ATP. Afterwards, the electrons are donated to oxygen gas ($O_2$), which combines with hydrogen ($H_2$) to form water ($H_2O$). The carbon atoms that are left over after the electrons have been stripped off combine with oxygen and are released as carbon dioxide ($CO_2$). In most vertebrates the water produced as a result of this metabolism is simply diluted into the much larger volume of the body's internal fluid: metabolism is in effect a process that utilizes oxygen and produces carbon dioxide. This final balance sheet determines one of the principal physiological challenges facing all animals—how to obtain oxygen and dispose of carbon dioxide. The uptake of oxygen and the release of carbon dioxide together are called **respiration.**

## WHERE OUR OXYGEN COMES FROM: THE COMPOSITION OF AIR

Each of the oxygen gas molecules that are the raw material of respiration has been produced by photosynthesis. As we saw in Chapters 3 and 8, the oxygen in the earth's air is the product of photosynthesis. Initially, photosynthetic organisms in the oceans released oxygen gas molecules into the water, which diffused into the atmosphere; later, terrestrial plants began to release additional oxygen directly into the air. Both processes are still occurring. The present atmosphere, which we call **air,** is rich in oxygen.

Dry air has a very constant composition: 78.09% nitrogen, 20.95% oxygen, 0.93% argon and the other inert gases, and 0.03% carbon dioxide. By convention, physiologists group the noble gases present in air (they constitute less than 1% of the total volume) with nitrogen because, like nitrogen, they are inert in most physiological processes. Convection currents cause air to have a constant composition to an altitude of at least 100 kilometers, although there is much less air present at high altitudes.

The amount of air present at a given altitude is usually expressed in terms that depend on its weight. Imagine a column of air standing on the ground and extending up into the sky as far as the atmosphere goes. This column has many gas molecules in it, and they all experience the force of gravity. For this reason the column, which you might consider "as light as air," actually weighs a lot. This weight cannot be sensed because it is equally distributed over the entire body.

How much does all this air weigh? It weighs enough to push down on one end of a U-shaped column of mercury sufficiently to raise the other end of the column 760 millimeters at sea level under a set of specified, standard conditions (Figure 38-2). An apparatus that measures air pressure in this way is called a barometer; 760 mm Hg (millimeters of mercury) is therefore the **barometric pressure** of the air at sea level, on the average. This pressure is also defined as 1 atmosphere of pressure.

Within the column of air, each separate gas exerts a pressure as if it existed alone. The total air pressure is simply made up of the component **partial pressures (P)** of each of the individual gases. Thus at sea level the total of 760 millimeters of air pressure is composed of:

760 × 79.02% = 600.6 millimeters of nitrogen
760 × 20.95% = 159.2 millimeters of oxygen
760 × 0.03% = 0.2 millimeter of carbon dioxide

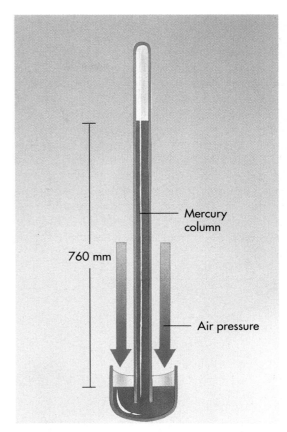

*Figure 38-2*

**A simple mercury barometer.** The weight of air pressing down on the surface of the mercury in the open dish pushes the mercury down into the dish and up the tube. The greater the air pressure pushing down on the mercury surface, the farther up the tube the mercury will be forced. At sea level, air pressure will cause a standard column of mercury to rise 760 millimeters.

At altitudes above 6000 meters, human beings do not survive long. The air still contains 20.95% oxygen, but the total air pressure is lower, as is the partial pressure of oxygen ($P_{O_2}$). The atmospheric pressure at such a height is about 380 millimeters, so the $P_{O_2}$ is only:

380 × 20.95% = 79.6 millimeters of oxygen

This figure is only half of the $P_{O_2}$ at sea level. Since the oxygen content of the blood is related to the partial pressure, the amount of oxygen available to the body tissues is reduced.

# THE EVOLUTION OF RESPIRATION

Animals do not capture oxygen from their environments actively; to do so would require the expenditure of more energy than the animals would obtain by utilizing the captured oxygen. In all animals the capture of oxygen and discharge of carbon dioxide are passive processes; the gases move into and out of cells by diffusion. The force that drives the movement of the gases is the difference in $Po_2$ between the interior of the organism and its exterior environment. Oxygen will diffuse into a cell only if the $Po_2$ in the cell is less than the $Po_2$ in the air or water surrounding the cell. The greater the difference in partial pressures (the more oxygen outside the cell than inside), the more rapid is the diffusion.

Oxygen diffuses slowly. The levels of oxygen required by oxidative metabolism in most organisms cannot be obtained by diffusion alone over distances greater than about 0.5 millimeter. This factor severely limits the size of organisms that obtain their oxygen entirely by diffusion from the environment into the cytoplasm. Single-celled protists are small enough that diffusion distance presents no problem, but as the size of an organism increases, the problem soon becomes significant.

Major changes in the mechanism of respiration have occurred during the evolution of animals (Figure 38-3). In general, these changes have tended to optimize the rate of diffusion by (1) increasing the surface area over which diffusion takes place, (2) decreasing

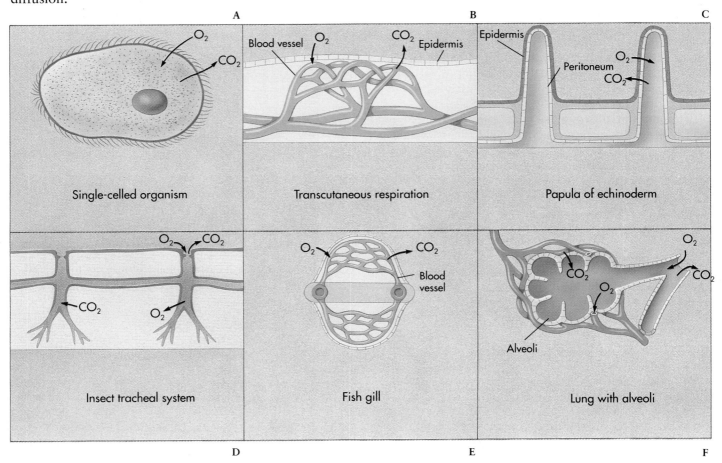

*Figure 38-3*

**Gas exchange in animals may take place in a variety of ways.**
A  Gases diffuse directly into single-celled organisms.
B  Amphibians and many other multicellular organisms respire through their skin.
C  Echinoderms have protruding papulae, which provide an increased respiratory surface.
D  Insect respire through spiracles, openings in their cuticle.
E  The gills of fishes provide a very large respiratory surface and employ countercurrent exchange.
F  Mammalian lungs provide a large respiratory surface but do not permit countercurrent exchange.

the thickness of tissue through which the gas must pass to reach the interior of the organism, and (3) increasing the concentration difference in partial pressure between the organism and its environment.

## Creating a Water Current

In animals, most cells are in direct contact with the surrounding extracellular fluid. The problem, therefore, is how to keep the composition of this fluid constant. Most of the more primitive animal phyla possess no special respiratory organs. The sponges (phylum Porifera), cnidarians (phylum Cnidaria), many flatworms (phylum Platyhelminthes) and roundworms (phylum Nematoda), and some annelids (phylum Annelida) all obtain their oxygen by diffusion directly from surrounding water. How do they overcome the limits imposed by diffusion? By beating with their cilia, these organisms create a water current, by means of which they continuously replace the water over the diffusion surface. Because of this continuous replenishment with water containing fresh oxygen, *the exterior oxygen partial pressure does not decrease as diffusion proceeds.* Although each oxygen gas molecule that passes into the organism has been removed from the surrounding volume of water, the exterior $P_{O_2}$ does not fall, because a new volume of water with a higher $P_{O_2}$ is constantly replacing the one from which the oxygen has been removed.

## Increasing the Diffusion Surface

Most of the more advanced invertebrates (mollusks, arthropods, echinoderms), as well as the vertebrates, possess special respiratory organs that both increase the surface area available for diffusion and reduce the thickness of the tissue separating the internal fluid from the surroundings (see Figure 38-3). These organs are of two kinds: (1) those that facilitate exchange with water and (2) those that facilitate exchange with air. As a rough rule of thumb, aquatic respiratory organs increase the diffusion surface by extensions of tissue, called **gills,** that project from the body out into the water. Atmospheric respiratory organs, on the other hand, involve invaginations into the body; in terrestrial vertebrates the respiratory organs are internal sacs called **lungs.**

Perhaps the simplest of the respiratory organs are the tracheae of arthropods (see Chapter 28). Tracheae are extensive series of passages connecting the surface of the animal to all portions of its body (see Figure 38-3, *D*). Oxygen diffuses from these passages directly to the cells, without the intervention of an active circulatory system. Piping air directly to the cells in this manner works very well in small organisms such as insects, since air must move only a relatively short distance within their bodies.

## THE GILL AS AN AQUEOUS RESPIRATORY MACHINE

By far the most successful aqueous respiratory organ, the gill, evolved among the bony fishes. In these animals, water passes through the mouth into two cavities, which are situated behind the mouth on each side of the head. From these cavities the water passes back out of the body. The gills hang like curtains between the mouth and the entrance to each of the cavities.

Many fishes that swim continuously, such as tuna, have practically immobile covers over their gill cavities. They swim with their mouths partly open, forcing water constantly over the gills. Most bony fishes, however, have flexible gill covers that permit a pumping motion using muscles. In such gills, there is an uninterrupted one-way flow of water over the gills even when the fish is not swimming (Figure 38-4).

In addition to maintaining a high diffusion rate by providing a continuous flow of water, the gills of fishes are constructed in such a way that they actually maximize the $P_{O_2}$ difference between tissue and environment. To understand how this is done, we must look carefully at the structure of a gill. Each gill is composed of thin, membranous gill filaments that project out into the flow of water (Figure 38-5). Within each filament are rows of thin, disklike lamellae arrayed parallel to the direction of water movement. Water flows over these lamellae from front to back. Here is the key: within each lamella, the blood circulation is arranged

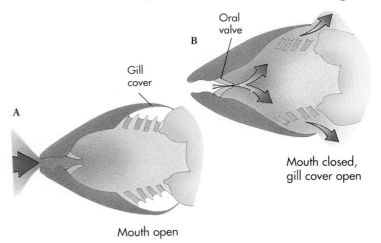

***Figure 38-4***

**How a fish breathes.** The gills are suspended between the mouth cavity and the outside, under a hard cover called the operculum. Breathing occurs in two stages:
**A** When the oral valve of the mouth is open, closing the operculum increases the volume of the mouth cavity so that water is drawn in.
**B** When the oral valve is closed, opening the operculum decreases the volume of the mouth cavity, forcing water out past the gills to the outside.

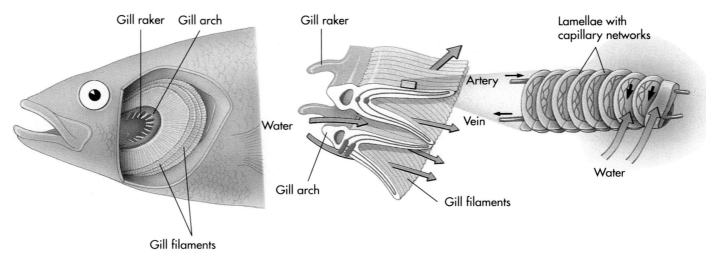

Gill raker | Gill arch | Gill raker | Lamellae with capillary networks

Artery

Water

Vein

Gill arch

Water

Gill filaments

Gill filaments

Gill filaments

*Figure 38-5*

**Structure of a fish gill.** Water passes from the gill arch out over the filaments (from left to right in the diagram). Water always passes the lamellae in the same direction—which is opposite to the direction the blood circulates across the lamellae. The sucess of the gill's operation critically depends on this opposite orientation of blood and water flow.

so that *blood is carried in the direction opposite to the movement of the water,* from the back of the lamella to the front.

Because the water flowing over the lamellae and the blood flowing within lamellae run in opposite directions, the difference in oxygen partial pressures is maximized. At the back of the gill the least oxygenated blood meets the least oxygenated water and is able to remove oxygen from the water. By the time the blood reaches the front of the gill, it has acquired a lot of oxygen, but it is able to acquire still more oxygen by diffusion from the water entering the gill. It can do this because the new water entering the front of the lamellae is richer in oxygen than the water that has already flowed past the gills and lost some of its oxygen. This kind of **countercurrent flow** (Figure 38-6) ensures a continuous $P_{O_2}$ gradient, and diffusion continues to occur all along the gill.

If the flows of water and blood had been in the same direction, the $P_{O_2}$ difference would have been high initially as the oxygen-free blood met the new water that was entering. Its value would have fallen rapidly, however, as the water lost oxygen to the blood. Since the blood $P_{O_2}$ rises as the water oxygen falls, much of the oxygen in the water would remain there when the blood and water partial pressures became equal and diffusion ceased. In countercurrent flow, in contrast, the blood oxygen level encountered by the water becomes lower and lower as the oxygen level in the water falls. The result is that, in countercurrent flow, the blood can attain oxygen partial pressures as high as those which exist in the water entering the gills. Fish gills are the most efficient respiratory machines

that occur among large organisms. Gills are able to maximize the rate of diffusion in an oxygen-poor medium, obtaining as much as 85% of the available oxygen.

*The gill is the most efficient of all respiratory organs. Its great efficiency derives from the countercurrent flow of water past the blood vessels of the gills.*

## FROM AQUATIC TO ATMOSPHERIC BREATHING: THE LUNG

Although aquatic animals were the first to evolve, their descendants have successfully invaded the land many times. On land the respiratory challenge is very different than the one that exists in water.

Water is relatively poor in dissolved oxygen; it contains only 5 to 10 milliliters of oxygen per liter of water. Air, in contrast, is rich in oxygen, containing about 210 milliliters of oxygen per liter of air. Not surprisingly, many members of otherwise aquatic groups utilize atmospheric air as a source of oxygen; these include many mollusks, crustaceans, and fishes.

When organisms first became fully terrestrial, the air became the source of their oxygen. An entirely new respiratory apparatus evolved, one that was based on internal passages rather than on gills. Why were gills

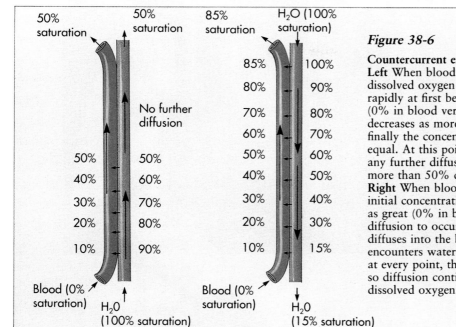

Non countercurrent exchange    Countercurrent exchange

**Figure 38-6**

**Countercurrent exchange.**
**Left** When blood and water flow in the same direction, dissolved oxygen gas can diffuse from the water into the blood rapidly at first because of the large concentration difference (0% in blood versus 100% in water), but the difference decreases as more oxygen diffuses from water into blood, until finally the concentrations of oxygen in water and blood are equal. At this point there is no concentration difference to drive any further diffusion. In this example, blood can obtain no more than 50% dissolved oxygen in this fashion.
**Right** When blood and water flow in opposite directions, the initial concentration difference between water and blood is not as great (0% in blood versus 15% in water) but is sufficient for diffusion to occur from water to blood. As more oxygen diffuses into the blood, raising its oxygen concentration, it encounters water with high and higher oxygen concentrations; at every point, the oxygen concentration is higher in the water, so diffusion continues. In this example, blood obtains 85% dissolved oxygen.

not maintained in terrestrial organisms, since they are such superb oxygen-capturing mechanisms? Gills were lost for two principal reasons:

1. Air is less buoyant than water. Because the fine, membranous lamellae of gills lack structural strength, they must be supported by water to avoid collapsing on one another. A fish out of water, although awash in oxygen, soon suffocates because its gills collapse into a mass of tissue. This collapse greatly reduces the diffusion surface of the gill.
2. Water diffuses into air through the process of **evaporation.** Atmospheric air is rarely saturated with water vapor, except immediately after a rainstorm. Consequently, organisms that live in air are constantly losing water to the atmosphere. Gills would have provided an enormous surface for water loss.

Two main systems of internal oxygen exchange evolved among terrestrial organisms. One was the tracheae of insects mentioned earlier, and the other was the lung. Both systems sacrifice respiratory efficiency to maximize water retention. Insects prevent excessive water loss by closing the external openings of the tracheae whenever possible. They do this whenever body carbon dioxide levels are below a certain level.

Lungs, in contrast, minimize the effects of drying out by eliminating the one-way flow of oxygen that was such an effective means of increasing the efficiency of aquatic respiratory systems. In organisms that respire by means of lungs, the air moves into the lung through a tubular passage and then back out again via the same passage. When each breath is completed, the lung still contains a volume of air, the **residual volume.** In human beings this volume is about 1200 milliliters. Each inhalation adds from 500 milliliters (resting) to 3000 milliliters (exercising) of additional air. Each exhalation removes approximately the same volume as inhalation added, reducing the air volume in the lung once more to about 1200 milliliters. Because the diffusion surfaces of the lungs are not exposed to fully oxygenated air, but rather to a mixture of fresh and partly depleted air, the difference in partial pressures of oxygen is far from maximal, and the respiratory efficiency of lungs is much less than that of gills. Oxygen capture is lessened by this two-way flow of air—but so is water loss.

## Evolution of the Lung

There is so much more oxygen in air than in water that low respiratory efficiency does not appear to have presented a critical problem to early land dwellers. The amphibian lung evolved from the swim bladder of the fish, an important regulator of buoyancy. The amphibian lung is essentially a sac with a convoluted internal membrane (Figure 38-7, *A*). Because the inner membrane surface of the lung is convoluted, the surface

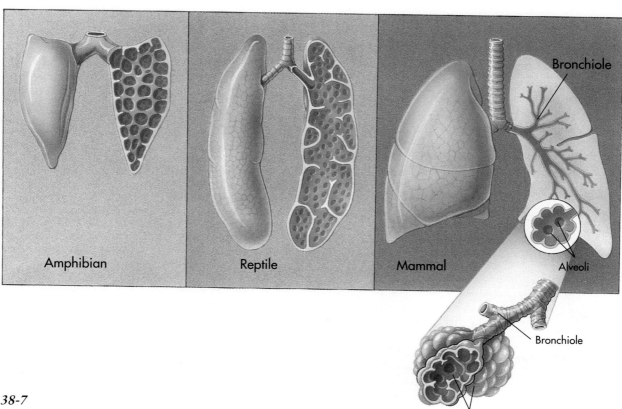

**Figure 38-7**
**Evolution of the vertebrate lung.**

area of amphibian lungs available for diffusion is great, although it is still not large enough to provide all the oxygen necessary for the organism. Consequently, amphibians obtain much of their oxygen by diffusion through their moist skin.

Reptiles are far more active than amphibians, and they have significantly greater metabolic demands for oxygen. The early reptiles could not rely on their skins for respiration. Living fully on land, reptiles are "watertight," avoiding desiccation by possessing a dry, scaly skin. Again, as in aquatic organisms, the respiratory apparatus has changed in ways that tend to optimize respiratory efficiency. The lungs of reptiles possess many small chambers within their surface called **alveoli,** which are clustered together like a bunch of grapes (Figure 38-7, B). The alveoli greatly increase the diffusion surface of the lung.

The metabolic demands for oxygen became even greater with the evolution of birds and mammals, which, unlike reptiles and amphibians, maintain a constant body temperature by heating their bodies metabolically. The lungs of mammals contain many clusters of alveoli (Figure 38-7, C). Humans, for example, have about 300 million alveoli in their two lungs. The in-

crease in the number of alveoli enlarged yet again the total diffusion surface of the lung. In humans the total surface devoted to diffusion can be as much as 80 square meters, an area about 42 times the surface area of the body.

*In amphibians, one-way flow through the respiratory organ was abandoned in favor of a saclike lung. Increases in efficiency among the reptiles and mammals have been achieved by enlargement of the lung's internal surface area.*

There is a limit to the improvements that can be made by increases in the diffusion surface of the lung, a limit that is probably approached by the more active mammals. With the advent of birds, flying introduced respiratory demands that exceeded the capacity of a saclike lung. Many birds rapidly beat their wings for prolonged periods during flight; such rapid wing movement uses up a lot of energy quickly because it depends on the frequent contraction of wing muscles. Flying birds must carry out intensive oxidative respiration within their cells to replenish the ATP expended by contracting flight muscles. They thus require a great deal of oxygen, more oxygen than a saclike lung, even one with a large surface such as a mammalian lung, is capable of delivering. The lungs of birds cope with the demands of flight by employing a new respiratory mechanism, one that produces a significant improvement in respiratory efficiency.

An avian lung works like a two-cycle pump (Figure 38-8). When a bird inhales air, the air passes directly to a nondiffusing chamber called the **posterior air sac.** When the bird exhales, the air flows into a lung. On the following inhalation, the air passes from the lung to a second air sac, the **anterior air sac.** Finally, on the second exhalation, the air flows from the anterior air sac out of the body. What is the advantage of this complicated passage? It creates a unidirectional flow of air through the lungs! Thus there is no dead volume as in the mammalian lung, and the air passing through the lung of a bird is always fully oxygenated. In fish gills, flow of blood and water are in opposite directions, 180 degrees apart, whereas in bird lungs the latticework of capillaries is arranged *across* the air

flow, at a 90-degree angle. This **crosscurrent flow** is not as efficient at extracting oxygen as a fish gill, but the oxygenated blood leaving the lung can still contain more oxygen than exhaled air, a capacity not achievable by mammalian lungs.

## THE STRUCTURE AND MECHANICS OF THE RESPIRATORY SYSTEM

Humans possess a pair of lungs located in the chest, or **thoracic,** cavity. The two lungs hang free within the cavity, being connected to the rest of the body only at the one position where the lung's blood vessels and air tube enter (Figure 38-9). This air tube is called a **bronchus.** It connects each lung to a long tube called the **trachea,** which passes upward in the body, past the voice box, or **larynx,** and opens into the rear of the mouth. Air normally enters through the nostrils, where it is warmed. In addition, the nostrils are lined with hair that filter out dust and other particles. As the air passes through the nasal cavity, an extensive array of cilia on its epithelial lining further filters the air and moistens it (Figure 38-10). Mucous secretory ciliated cells in the bronchi also trap foreign particles and carry them upward, where they can be swallowed. The air

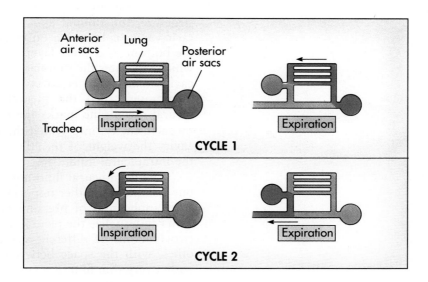

*Figure 38-8*

**How a bird breathes.**
The respiratory system of a bird is composed of anterior air sacs, lungs, and posterior air sacs. Breathing occurs in two cycles. *Cycle 1:* Air is drawn from the trachea into the posterior air sacs and then is exhaled through the lungs. *Cycle 2:* Air is drawn from the lungs into the anterior air sacs and then is exhaled through the trachea. Passage of air through the lungs is always in the same direction, from posterior to anterior (right to left here). Because blood circulates in the lung from anterior to posterior, the lung achieves a type of countercurrent flow and thus is very efficient at picking up oxygen from the air.

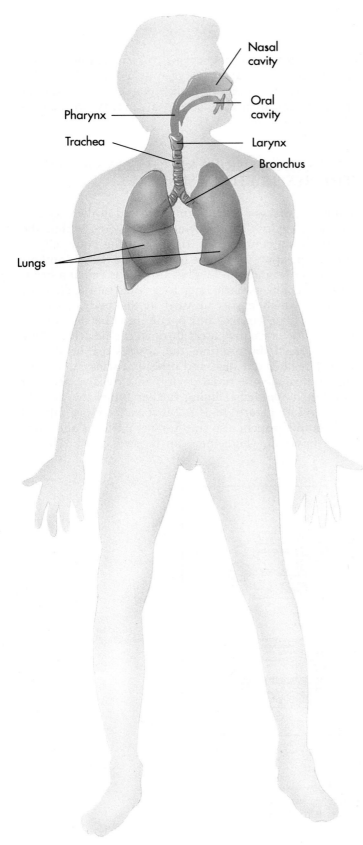

**Figure 38-9**

The human respiratory system.

**Figure 38-10**

Respiratory cilia such as these line the trachea.

then passes through the back of the mouth, crossing the path of food as it enters first the larynx and then the trachea. From there it passes down through the bronchus to the lungs. The epiglottis covers the trachea whenever food is swallowed.

The lungs consist of millions of small sacs called alveoli. These alveoli are surrounded by so many capillaries it is as if blood were flowing over them in a constant sheet. These alveoli are connected to the bronchi by a branching network of tubes called bronchioles, some of which are surrounded by smooth muscle and sensitive to metabolites and oxygen.

The human respiratory apparatus is simple in structure, functioning as a one-cycle pump. The thoracic cavity is bounded on its sides by the ribs and on the bottom by a thick layer of muscle, the **diaphragm,** which separates the thoracic cavity from the abdominal cavity. Each lung is covered by a very thin, smooth membrane called the **pleural membrane.** A second pleural membrane marks the interior boundary of the thoracic cavity, into which the lungs hang. Within the cavity the weight of the lungs is supported by water, the **intrapleural fluid.**

The intrapleural fluid not only supports the lungs, but also plays another important role by permitting an even application of pressure to all parts of the lung. You can visualize the pleural membranes as a system of two balloons of different sizes, one nested inside the other, with the space between them completely filled with a very thin film of water. The inner balloon opens out to the atmosphere (Figure 38-11); thus two forces act on this inner balloon: air pressure from the atmosphere pushes it outward, and water pressure from the intrapleural fluid pushes it inward.

The active pumping of air in and out through the lungs is called **breathing.** During **inhalation,** muscular contraction causes the walls of the chest cavity to expand so that the rib cage moves outward and upward (Figure 38-12). The diaphragm is dome shaped when

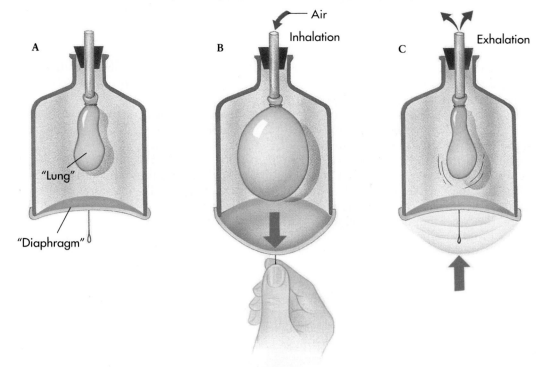

*Figure 38-11*

**A simple experiment that shows how you breathe.** In the jar is a balloon (**A**). When the diaphragm is pulled down, as shown in **B**, the balloon expands; when it is relaxed (**C**), the balloon contracts. In the same way, air is taken into your lungs when your diaphragm pulls down, expanding the volume of your lung cavity. When your diaphragm pushes back, the volume decreases and air is expelled.

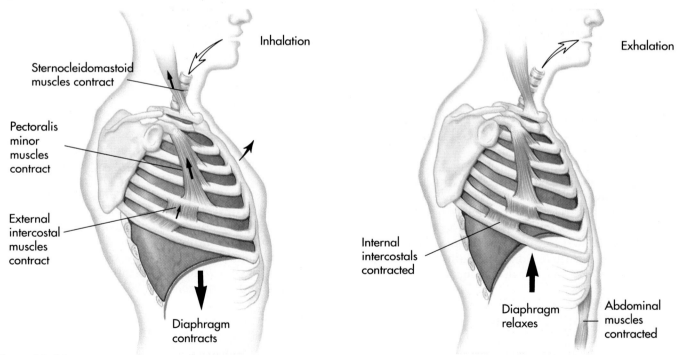

*Figure 38-12*

**How a human breathes.**
Inhalation: The diaphragm pushes up and the chest cavity expands, increasing its volume. As a result of the larger volume, air is sucked in through the trachea.
Exhalation: The diaphragm and chest walls return to their normal positions, reducing the volume of the chest cavity and forcing air outward through the trachea.

relaxed and moves downward when it contracts. In effect, we have enlarged the outer balloon by pulling it in all directions. This expansion causes the fluid pressure to decrease to a level less than that of the air pressure within the inner balloon. Since fluids are incompressible, the wall of the inner balloon is pulled out. If you have difficulty visualizing this, think about how much force you have to apply to pull apart a wet pair of glass slides, compared to moving them horizontally. As the inner balloon expands, its internal air pressure tends to decrease, and since the inner balloon is connected to the atmosphere, air rapidly moves in from the atmosphere to equalize pressure. During **exhalation,** the ribs and diaphragm return to their original resting position. In doing so, they exert pressure on the fluid. This pressure is transmitted uniformly by the fluid over the entire surface of the lung (the inner balloon), forcing air from the inner cavity back out to the atmosphere. Intrapleural pressure is always more negative than atmospheric pressure and pressure within the alveoli.

*The active pumping of air in and out through the lungs is called breathing. The lungs function by suction during inhalation. The expansion of the chest cavity draws air into the lungs, and the return of the ribs to their resting position drives air from the lungs.*

## HOW RESPIRATION WORKS: GAS TRANSPORT AND EXCHANGE

When oxygen has diffused from the air into the moist cells lining the inner surface of the lung, its journey has just begun. Passing from these cells into the bloodstream, the oxygen is carried throughout the body by the circulatory system (to be described in Chapter 39). It has been estimated that it would take a molecule of oxygen 3 years to diffuse from your lung to your toe if transport depended only on diffusion, unassisted by a circulatory system.

Oxygen moves within the circulatory system on carrier proteins that bind dissolved molecules of oxygen. This binding occurs in the capillaries surrounding the alveoli of the lungs. The carrier proteins later release their oxygen molecules to metabolizing cells at distant locations in the body.

The carrier protein that is used by all vertebrates is hemoglobin. **Hemoglobin** is a protein composed of four polypeptide subunits; each of the four polypeptides is combined with iron in such a way that oxygen

can be bound reversibly to the iron. Hemoglobin is synthesized by erythrocytes (red blood cells) and remains within these cells, which circulate in the bloodstream like ships bearing cargo.

### Oxygen Transport

Only a small amount (5%) of oxygen gas is transported from lungs to body tissues and dissolved in blood plasma. The remaining 95% is bound to hemoglobin within red blood cells. The higher the $P_{O_2}$ in the air within the lungs, the more oxygen will dissolve in the blood and combine with hemoglobin. The relationship is not a simple proportion, however. If the $P_{O_2}$ is halved, the oxygen content is reduced only by about 25%. This facilitates oxygen uptake from the lungs into the blood and also aids unloading of oxygen from the blood to the tissues. Hemoglobin molecules act like little oxygen sponges, soaking oxygen up within red blood cells and causing more to diffuse in from the blood plasma. At the $P_{O_2}$ encountered in the blood supply of the lung, most hemoglobin molecules are saturated with oxygen. In the tissues the $P_{O_2}$ is much lower so that hemoglobin gives up its bound oxygen.

In tissue, the presence of carbon dioxide ($CO_2$) causes the hemoglobin molecule to assume a different shape, one that gives up its oxygen more easily (Figure 38-13). This augments the unloading of oxygen from hemoglobin. The effect of $CO_2$ on oxygen binding is called the Bohr effect. It is of real importance, since $CO_2$ is produced by the tissues at the site of cell metabolism. For this reason the blood unloads oxygen more readily to those tissues undergoing metabolism and generating $CO_2$.

### $CO_2$ Transport

At the same time that the red blood cells are unloading oxygen, they are also absorbing $CO_2$ from the tissue. Perhaps one fifth of the $CO_2$ that the blood absorbs is bound to hemoglobin. Another 8% is simply dissolved in plasma. The remaining 72% of the $CO_2$ diffuses from the plasma into the cytoplasm of the red blood cells. There an enzyme, **carbonic anhydrase,** catalyzes the combination of $CO_2$ with water to form carbonic acid ($H_2CO_3$), which dissociates into bicarbonate ($HCO_3^-$) and hydrogen ($H^+$) ions. This process removes large amounts of $CO_2$ from the blood plasma, facilitating the diffusion of more $CO_2$ into it from the surrounding tissue. The facilitation is critical to $CO_2$ removal, since the difference in $CO_2$ concentration between blood and tissue is not large (only 5%).

The red blood cells carry their cargo of bicarbonate ions back to the lungs. The lower $CO_2$ concentration in the air inside the lungs causes the carbonic anhydrase reaction to proceed in the reverse direction, releasing gaseous $CO_2$, which diffuses outward from the

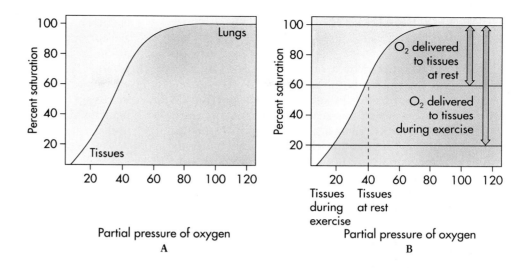

**Figure 38-13**

**Oxygen-hemoglobin dissociation curves for human hemoglobin.** Hemoglobin is saturated with oxygen in blood leaving the lungs (**A**), but little of this oxygen remains after circulation through the tissues. In tissue such as exercising muscle (**B**), the concentration of oxygen, measured as pressure in millimeters of mercury, is low, about 30 millimeters. At such low pressures, oxygen tends to dissociate from hemoglobin. In the lungs, where oxygen concentrations are much higher (typically 10 millimeters), oxygen molecules do not tend to dissociate from hemoglobin; hemoglobin molecules are fully saturated with all the oxygen they can carry.

blood into the alveoli. With the next exhalation, this $CO_2$ leaves the body. The fifth of the $CO_2$ bound to hemoglobin also leaves because hemoglobin has a greater affinity for oxygen than for $CO_2$ at low $CO_2$ concentrations. The diffusion of $CO_2$ outward from the red blood cells causes the hemoglobin within these cells to release its bound $CO_2$ and take up oxygen instead. The red blood cells, with their newly bound oxygen, then start the next respiratory journey (Figure 38-14).

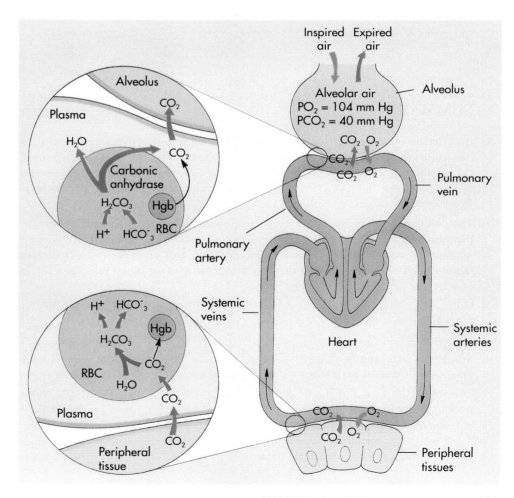

**Figure 38-14**

**The respiratory journey.**

## HOW THE BRAIN CONTROLS BREATHING

Each breath is initiated by a respiratory control center in the brain, which sends the nerve signals to the diaphragm and, for deeper breathing, to the intercostal muscles. These contractions regularly expand the chest cavity. Other neurons then act to inhibit the stimulation of these muscles so that they relax and the body exhales. These signals are not subject to voluntary control; we cannot consciously stop breathing. You breathe more slowly when you sleep, more rapidly when you run. You need less oxygen when asleep because many of your muscles are not being used, and more oxygen when you run because you are using muscles intensively.

The body regulates this change in rate of breathing by a simple feedback mechanism, a respiratory center in the brainstem. Chemoreceptors in the brain (central chemoreceptors) detect changes in $CO_2$ levels and transmit nerve impulses to the respiratory center, which sends appropriate signals to the muscles of the diaphragm and rib cage. There is also a second set of chemoreceptors in the aorta and carotid artery. These are primarily sensitive to pH and greatly reduced amounts of oxygen. When the $CO_2$ level rises, the respiratory center increases the rate of breathing. Thus, when you run and rapid muscle contraction is adding increased amounts of $CO_2$ to your blood, specific brain neurons detect the rise in $CO_2$, leading to more rapid and deeper breathing.

## LOOKING AHEAD TO CIRCULATION

In Chapter 39 we consider in more detail the journey of the red blood cells. Traveling between the lungs, where they acquire oxygen and release $CO_2$, and respiring tissues, where they release oxygen and acquire $CO_2$, the body's red blood cells traverse a complex highway that passes to all parts of the body. The red blood cells are not the only traffic on this highway. Just as the roads and sidewalks transport all the commerce of a city, so the circulatory system of the vertebrate body transports all the material that moves from one part of the body to another.

## ■ SUMMARY

1. All animals obtain oxygen by diffusion. The evolution of respiratory mechanisms among animals has tended to favor changes that improve the rate of diffusion. These include changes that decrease the length of the path over which diffusion occurs, changes that increase the surface area over which diffusion occurs, and changes that maximize the difference in oxygen partial pressures between environment and tissue.

2. The most efficient aquatic respiratory organ is the gill of bony fishes. Fishes take water in through their mouths, move it past their gills, and pass it out of their bodies. This one-way flow is the secret of high respiratory efficiency, since it permits fishes to establish a countercurrent flow of blood: blood vessels are located within the gills in such a way that blood flows in a direction opposite that of water.

3. Gills will not work in air, since air is not buoyant enough to support their fine latticework of passages. That is the reason fishes drown in air even though there is much more oxygen available in air than in water.

4. The amphibian lung is a simple sac with two-way flow in and out. Gaseous diffusion across the lung's small internal surface area is supplemented by diffusion across the moist skin.

5. The further evolution of the lung in reptiles and mammals has involved no fundamental changes. Instead, there has been a progressive increase in the internal surface area of the lung, achieved by partitioning the inner surface into increasingly numerous chambers called alveoli. The lungs of humans and other mammals possess about 300 million alveoli, with a combined surface area that is about 42 times the body's surface area.

6. A fundamental change in the atmospheric lung is characteristic of birds, which utilize a series of air chambers and two-cycle breathing to effect a one-way flow of air. The establishment of a one-way flow of the diffusing medium permits a form of countercurrent flow, which is by far the most efficient diffusion mechanism.

7. Humans, like other terrestrial vertebrates, breathe by expanding the cavity within which the lungs hang. This action expands the lungs, sucking air inward.

8. In the respiration of terrestrial vertebrates, hemoglobin within the red blood cells binds the oxygen, which diffuses across the lung capillaries into the blood plasma. The circulating system carries these red blood cells to the respiring tissues of the body.

9. In respiring tissues, there is much less oxygen than in the blood and much more $CO_2$ because of the consumption of oxygen and generation of $CO_2$ by respiring cells. Hemoglobin responds to the higher $CO_2$ concentration by unloading its oxygen, which diffuses out of the red blood cells into the tissue. The $CO_2$ is absorbed into the red blood cells and carried to the lungs, where it is discharged.

## REVIEW

1. The uptake of oxygen and the release of carbon dioxide together are called _____.

2. An increase in the carbon dioxide concentration of blood causes the same effect on hemoglobin's ability to combine with oxygen as does a decrease in the partial pressure of oxygen within the lungs. True or false?

3. The first lungs seem to have evolved from the _____ of fishes.

4. Although lungs are less efficient respiratory structures than gills, they have the great advantage to terrestrial organisms of minimizing _____.

5. The protein _____ carries oxygen from your lungs throughout your body via the circulatory system.

## SELF-QUIZ

1. Which of the following phyla possess organisms with no special respiratory organs?
   (a) Porifera
   (b) Cnidaria
   (c) Platyhelminthes
   (d) Chordata
   (e) a, b, and c

2. Within the lamellae of gills, the blood circulation is arranged so that blood is carried in the opposite direction to the movement of water. The functional significance of this arrangement is that
   (a) it helps to maintain the temperature of the organism equal to the water temperature, thus enhancing diffusion.
   (b) it results from a developmental constraint.
   (c) it ensures a continuous gradient of concentration difference between the blood and the water, so that diffusion continues to occur all along the gill.
   (d) it increases the surface area for diffusion.
   (e) it allows some kinds of fishes to continue to get oxygen even if they are not moving.

3. Arrange the following in the order in which air contacts them during breathing, starting from the mouth and nose.
   (a) Bronchus
   (b) Alveoli
   (c) Larynx
   (d) Hemoglobin
   (e) Trachea

4. Which type of terrestrial organism has the most efficient type of lung?
   (a) Terrestrial mollusks
   (b) Amphibians
   (c) Reptiles
   (d) Birds
   (e) Mammals

5. Air normally enters your body through the nostrils, which are filled with hairs. What is the function of these hairs?
   (a) They filter out dust and other particles
   (b) They are a noise-making device
   (c) They slow and regulate the passage of air
   (d) They moisten the air
   (e) They allow you to breathe while you eat

6. Most oxygen transported in the blood
   (a) is dissolved in the plasma.
   (b) is carried by lymphocytes.
   (c) is bound to hemoglobin.
   (d) reacts with water to form bicarbonate.
   (e) is carried by myoglobin.

7. Most carbon dioxide transported in the blood
   (a) is dissolved as $CO_2$ in the plasma.
   (b) is carried by lymphocytes.
   (c) is bound to hemoglobin.
   (d) reacts with water to form bicarbonate.
   (e) is carried by myoglobin.

8. Intrapleural pressure is always
   (a) more negative than intraalveolar pressure.
   (b) more negative than atmospheric pressure.
   (c) 760 mm Hg.
   (d) increased on inspiration.
   (e) a and b.

# THOUGHT QUESTIONS

1. Often people who appear to have drowned can be revived, in some cases after being underwater for as long as half an hour. In every case of full recovery after extended submergence, however, the person had been in very cold water. Is this observation consistent with the fact that oxygen is twice as soluble in water at 0° C as it is at 30° C?

2. Can you think of a reason why a respiratory system has not evolved in which oxygen is actively transported across respiratory membranes, in place of the passive process of diffusion across these membranes that is employed universally?

3. If by accident your pleural membrane were punctured, would you be able to breathe?

# FOR FURTHER READING

FEDER, M., and W. BURGGREN: "Skin Breathing in Vertebrates," *Scientific American*, November 1985, pages 126-142. Discusses how many vertebrates do a significant portion of their breathing through their skin.

PERUTZ, M.F.: "Hemoglobin Structure and Respiratory Transport," *Scientific American*, December 1978, pages 92-125. An account of how hemoglobin changes its shape to facilitate oxygen binding and unloading, by the man who won a Nobel Prize for unraveling the structure of hemoglobin.

RALOFF, J.: "New Clues to Smog's Effect on Lungs," *Science News*, vol. 132, August 8, 1987, page 86. Describes how the lung's defense mechanisms can be damaged by environmental contamination.

SCHMIDT-NIELSEN, K.: "How Birds Breathe," *Scientific American*, December 1971, pages 73-79. A fascinating account of the discovery of unidirectional flow in avian lungs, by a great comparative physiologist.

# Circulation

Without the insulation of their feathers, these ptarmigan would soon freeze, as they have a very efficient circulation system that transmits not only oxygen and food throughout the body, but also temperature.

# CIRCULATION

## Overview

Circulatory systems are the highways over which red blood cells carry oxygen to the tissues and remove carbon dioxide. The fluid of the blood also transports glucose and amino acids to the cells and carries away wastes. The composition of the blood is kept constant by the liver and kidneys, which monitor and adjust metabolite and waste product levels in the blood. The key to circulation in vertebrates is the organ that pumps the blood through the system, the heart.

## For Review    *Here are some important terms and concepts that you will encounter in this chapter. If you are not familiar with them, you should review them before proceeding.*

**Glycogen** (Chapter 2)

**Gills** (Chapters 28 and 38)

**Erythrocytes** (Chapter 32)

**How hemoglobin carries oxygen** (Chapter 38)

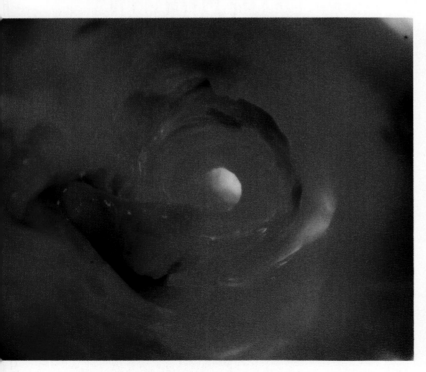

*Figure 39-1*

**A look down an artery.** The interior of an artery caked with fatty deposits. Such deposits retard the circulation of blood. Fragments that break off can block arterial branches and produce strokes or heart attacks.

Of the many tissues of the vertebrate body, few have the emotive impact of blood. In movies and in real life the sight of blood connotes violence and bodily harm. In literature, blood symbolizes the life force, which is not a bad analogy in real life either. This chapter is about blood, its circulation through the body, and the many ways in which it affects the body's functions (Figure 39-1). Although vertebrates possess many other organ systems that are necessary for life, it is the activities of the blood that bind them together into a functioning whole.

## THE EVOLUTION OF CIRCULATORY SYSTEMS

The capture of nutrients and gases from the environment is one of the essential tasks that all living organisms must carry out. In animals, with their relatively large bodies with many elements, this function has become more complex. In less complex animals, such as roundworms, the fluid within the body cavity constitutes a primitive kind of circulatory system, one that permits materials to pass from one cell to another without leaving the organism. Within such simple systems, which are called **open circulatory systems**, there is generally no distinction between circulating fluid and body fluid. Arthropods have open circulatory systems

890    *PART 9    ANIMAL BIOLOGY*

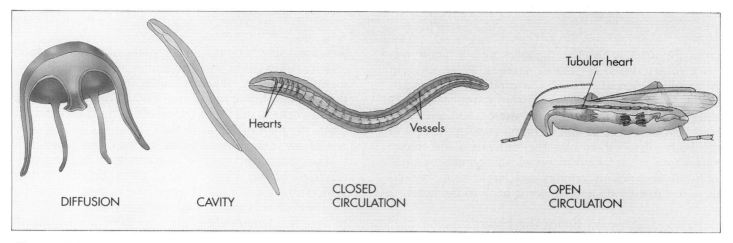

**Figure 39-2**

**Evolution of circulatory systems.**
Simple diffusion—jellyfish. Without any movement of the organism, diffusion reaches all parts of the body.
Circulation in a body cavity—a nematode. As the worm moves, its contracting muscles push fluid back and forth through the body cavity.
A closed circulatory system—annelid. The series of hearts in each segment push blood out through a system of vessels to the body's tissues, then back via a second vessel system.
An open circulatory system—insect. The tubular heart of insects pumps blood out to the tissues of the body, from which it perfuses back rather than traveling in vessels.

(see Chapter 28), in which a muscled tube within the central body cavity forces the cavity fluid out through a network of interior channels and spaces. The fluid then flows back into the central cavity. Most animals, however, have **closed circulatory systems,** in which the circulatory system fluid is separated from the rest of the body's fluids and does not mix with them (Figure 39-2). In annelids, for example, two major tubes or vessels extend the length of the worm, with branches extending out from each tube to the muscles, skin, and digestive organs.

A great advantage of closed circulatory systems is that they permit regulation of fluid flow by means of muscle-driven changes in the diameters of the vessels. In other words, with a closed circulatory system the different parts of the body can maintain different circulation rates.

A closed circulatory system is characteristic of all vertebrates. It has four principal functions: (1) nutrient and waste transport, (2) oxygen and carbon dioxide transport, (3) temperature maintenance, and (4) hormone circulation. The function of temperature maintenance is associated with the passage of the blood throughout the body; it is important in all vertebrates.

## THE CARDIOVASCULAR SYSTEM

The circulatory system of vertebrates is composed of three elements: (1) the **heart,** a muscular pump that we

consider later in this chapter; (2) the blood vessels (Figure 39-3), a network of tubes that pass through the body; and (3) the blood, which circulates within these vessels. The plumbing of the closed circuit, the heart and vessels, is known collectively as the **cardiovascular system.** Blood moves within this cardiovascular system, leaving the heart through vessels known as **arteries.**

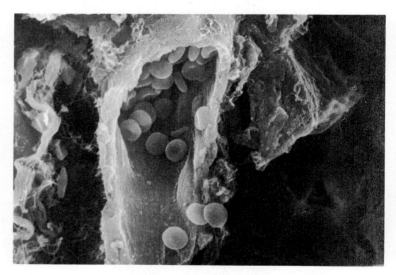

**Figure 39-3**

**A blood vessel.** This ruptured tube is a blood vessel, an element of the human circulatory system. It is full of red blood cells, which move through these blood vessels transporting oxygen and carbon dioxide from one place to another in the body.

From the arteries the blood passes into a larger network of **arterioles,** or smaller arteries. From these, it eventually is forced through the **capillaries,** a fine latticework of very narrow tubes (from Latin *capillus,* "a hair"). While passing through these capillaries, the blood exchanges gases and metabolites with the cells of the body. After traversing the capillaries, the blood passes into a third kind of vessel, the **venules** or small **veins.** A network of venules and larger veins collects the circulated blood and carries it back to the heart.

## Arteries: Highways From the Heart

Arteries carry blood away from the heart. The walls of the arteries are made up of three layers of tissue (Figure 39-4, *A*). The innermost one is composed of a thin layer of **endothelial** cells. Surrounding these cells is a thick layer of smooth muscle and elastic fibers, which in turn is encased within an envelope of protective connective tissue. Because this sheath is elastic, the artery is able to expand its volume considerably in response to a pulse of fluid pressure, much as a tubular balloon might respond to air blown into it. The steady contraction of the muscle layer strengthens the wall of the vessel against overexpansion.

## Arterioles: Little Arteries

The arterioles differ from the arteries in two ways. They are smaller in diameter, and the muscle layer that surrounds an arteriole can be relaxed under the influence of hormones and metabolites to enlarge the diameter. When the diameter increases, the blood flow also increases, an advantage during times of high metabolic activity. Conversely, most arterioles are in contact with many nerve fibers. When stimulated, these nerves cause the muscular lining of the arteriole to contract and thus constrict the diameter of the arteriole. Such contraction limits the flow of blood to the extremities during periods of low temperature or stress. You turn pale when you are scared or need to conserve heat because contractions of this kind constrict the arterioles in your skin. You blush for just the opposite reason. When you overheat or are embarrassed, the nerve fibers connected to muscles surrounding the arterioles are inhibited, which relaxes the smooth muscle and causes the arterioles in the skin to dilate. This brings heat to the surface for escape.

## Capillaries: Where Exchange Takes Place

Blood is a medium of exchange, carrying materials to and from the many cells of the body. All this exchange takes place in the capillaries, which have narrow walls suitable for the passage of gases and metabolites. Capillaries have the simplest structure of any element in the cardiovascular system (Figure 39-4, *C*). They are little more than tubes one cell thick and on the average about 1 millimeter long; they connect the arterioles with the venules. The internal diameter of the capillaries is, on the average, about 8 micrometers. Surprisingly, this is little more than the diameter of a red blood cell (5 to 7 micrometers). However, red blood cells squeeze through these fine tubes without difficulty (Figure 39-5).

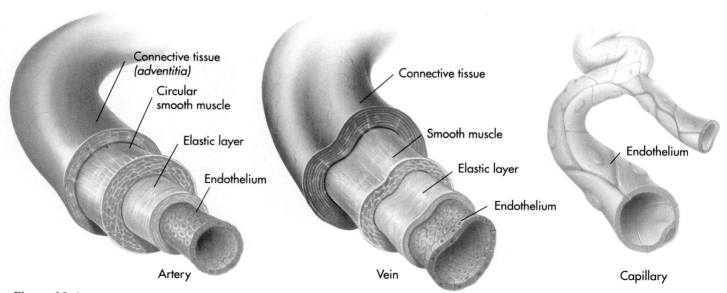

**Figure 39-4**

The structure of some important blood vessels.

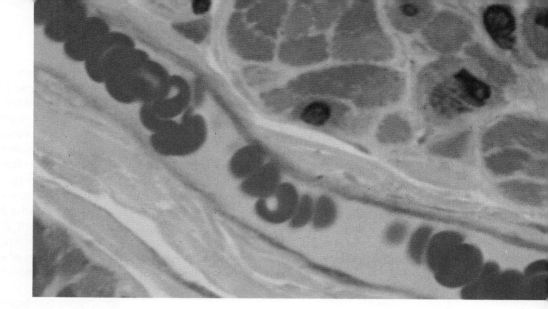

**Figure 39-5**

**Red blood cells in a capillary passing along in single file.** Many capillaries are even smaller than that shown here from the bladder of a monkey. Red blood cells will even pass through capillaries narrower than their own diameter, pushed along by the pressure generated by a pumping heart.

The intimate contact between the walls of the capillaries and the membranes of the red blood cells facilitates the diffusion of gases and metabolites between them. No cell of the body is more than 100 micrometers from a capillary. At any one moment, about 5% of your blood is in your capillaries. Some capillaries, called **thoroughfare channels,** connect arterioles and venules directly (Figure 39-6). From these channels, loops of true capillaries leave and return. It is through these loops that almost all exchange between the blood and the cells of the remainder of the body occurs. The entry to each loop is guarded by a ring of muscle called a **precapillary sphincter,** which, when closed, blocks flow through the capillary. Such restriction of entry to the capillaries in surface tissue is another powerful means of limiting heat loss from an animal's body during periods of cold.

The entire body is permeated with a fine mesh of these capillaries, a network that amounts to several thousand miles in overall length. If all the capillaries in your body were laid end to end, they would extend across the United States. Although individual capillaries have high resistance to flow because of their small diameters, the cross-sectional area of the extensive cap-

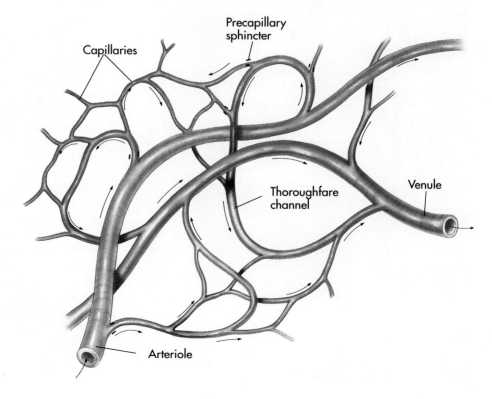

**Figure 39-6**

**The capillary network connects arteries with veins.** The most direct connection is via thoroughfare channels that connect arterioles directly to venules. Branching from these thoroughfare channels is a network of finer channels, the capillary network. Most of the exchange between body and red blood cells occurs while the red blood cells are in this capillary network.

Capillaries

Precapillary sphincter

Thoroughfare channel

Venule

Arteriole

illary network is greater than that of the arteries leading to it, so that the blood pressure is actually lower in the capillaries than in the arteries.

## Veins and Venules: Returning Blood to the Heart

Veins are vessels that return blood to the heart. Veins do not have to accommodate the pulsing pressures that arteries do, because much of the force of the heartbeat is weakened by the high resistance and great cross-sectional area of the capillary network. The walls of veins, although similar in structure to those of the arteries, have much thinner layers of muscle and elastic fiber (see Figures 39-4, *B,* and 39-7). An empty artery is still a hollow tube, like a pipe, but when a vein is empty, its walls collapse like an empty balloon.

The internal passageway of veins is often quite large. The diameter of the largest veins in the human body, the **venae cavae,** which lead into the heart, is fully 3 centimeters. The reason that veins are so much larger than arteries is related to the lower pressure of blood flowing within veins back toward the heart—it is advantageous to minimize any further resistance to its flow, and a larger tube presents much less resistance to flow than does a smaller tube. Thus only a small pressure difference is required to return blood to the heart. Veins also have unidirectional valves that aid the return of blood to the heart.

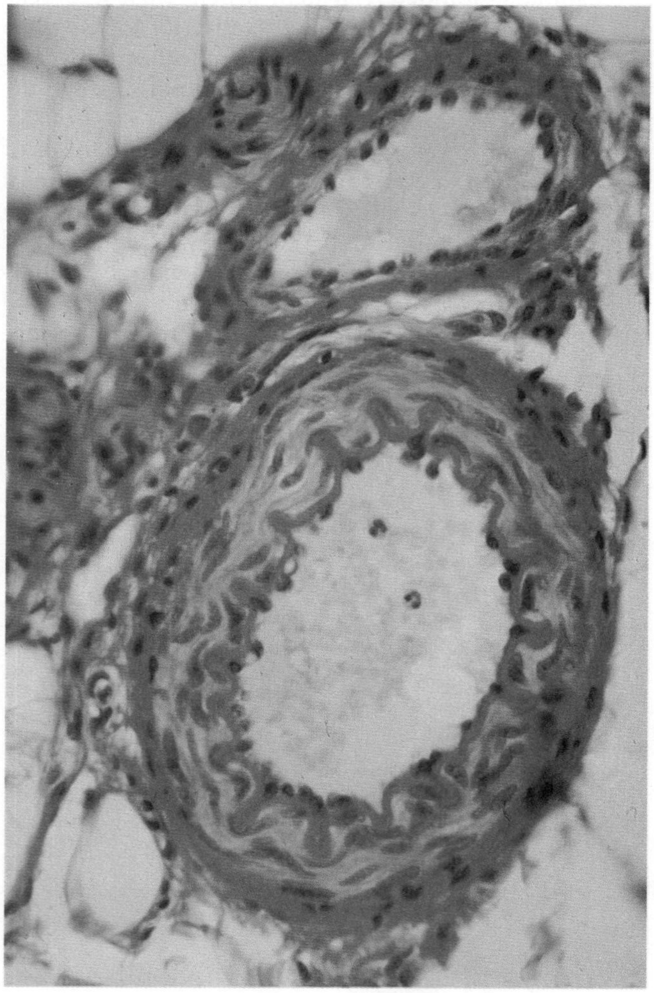

*Figure 39-7*

**A closer look at blood vessels.** The vein *(top)* has the same general structure as an artery *(bottom),* but much thinner layers of muscle and elastic fiber. An artery will retain its shape when empty, but a vein will collapse.

*The major vessels of the circulatory system are tubes of cells encased within three sheaths: (1) a layer of elastic fibers, which renders the diameter of the vessel elastic to accommodate pulses of blood pumped from the heart; (2) a layer of muscle serviced by nerves, which permits the body to control the diameter of the vessel and strengthens the wall against overexpansion; and (3) a layer of connective tissue, which protects the vessel.*

## The Lymphatic System: Recovering Lost Fluid

We speak of the cardiovascular system as closed because all its tubes are connected with one another and none is simply open ended. In another sense, however, the system is open—it is open to passage through the walls of the capillaries. Fluids are forced out across the thin walls of the capillaries on the arteriole side by the hydrostatic pressure that builds up because of the great reduction in vessel diameter. Although this loss is an unavoidable consequence of a closed circulatory system (which could not function properly without tiny vessels with thin walls), it poses difficulties in maintaining the integrity of that system.

Difficulties arise because of how *much* liquid is lost from the cardiovascular system as blood passes through the capillaries. In a human being, about 3 liters of fluid leave the cardiovascular system in this way each day, a quantity amounting to more than half the body's total supply of about 5.6 liters of blood. In counteracting the effects of this process, the body utilizes a second, *open* circulatory system called the **lymphatic system.** The elements of the lymphatic system gather liquid from the interstitial fluid (the spaces surrounding cells) and return it to the cardiovascular circulation (Figure 39-8). Open-ended lymph capillaries gather up fluids by diffusion and carry them through a series of progressively larger vessels to two large lym-

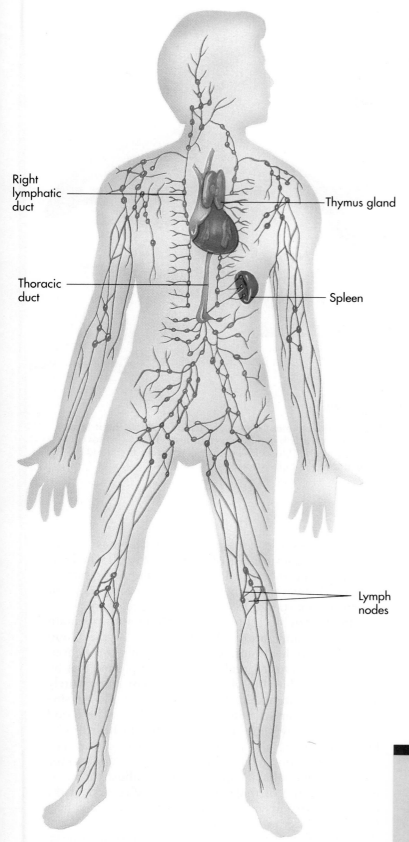

**Figure 39-8**

The human lymphatic system.

Right lymphatic duct

Thymus gland

Thoracic duct

Spleen

Lymph nodes

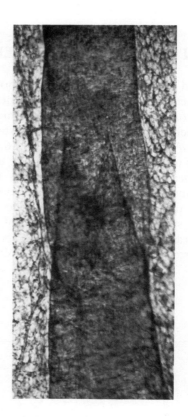

**Figure 39-9**

**A lymphatic vessel valve, magnified 25 times.** Flow from bottom to top is not retarded because such flow tends to force the inner cone open; flow from top to bottom is prevented because such flow tends to force the inner cone closed.

phatic vessels, which resemble veins. These lymphatic vessels drain into veins in the lower part of the neck through one-way valves. Your heart does not pump fluid through the lymphatic system; instead, fluid is driven through it when its vessels are squeezed by the movements of the body's muscles. The lymphatic vessels contain a series of one-way valves (Figure 39-9), which permit movement only in the direction of the neck. The lymphatic system has several other important functions. It returns proteins to the circulation, transports fats absorbed from the intestine, and carries bacteria and dead blood cells to the lymph nodes and spleen for destruction.

*Much of the water within blood plasma is forced out during passage through the capillaries. This water is collected by an open circulatory system, the lymphatic system, and is returned to the bloodstream.*

# THE CONTROL OF BODY TEMPERATURE

Mammals and birds maintain a constant body temperature by expending metabolic energy. They differ in this respect from most other organisms, which typically do not use metabolism to maintain a constant internal temperature. As discussed in Chapter 36, mammals monitor their internal temperature closely, using a battery of temperature-sensitive neurons to inform the central nervous system when body temperature strays beyond normal bounds. If the body overheats, the central nervous system initiates a variety of actions to lower temperature, including sweating, panting, and dilation of arterioles in the skin (Figure 39-10). If the

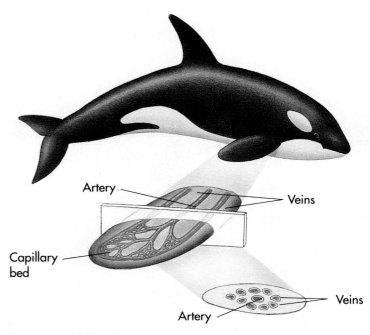

**Figure 39-11**

**Many marine mammals, such as this killer whale, limit heat loss in cold water by countercurrent flow.** Heat is removed from the arterial blood by nearby veins before the blood circulates to the body's surface.

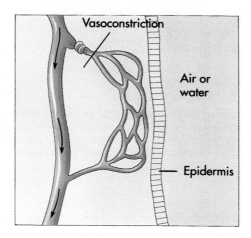

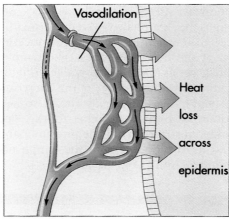

**Figure 39-10**

**Regulation of heat loss.** The amount of heat lost at the body's surface can be regulated by controlling the flow of blood to the surface. Constriction of the surface blood vessels limits flow and lessens heat loss; dilation of them increases blood flow and thus increases heat loss.

body becomes too cool, the central nervous system may initiate constriction of surface arterioles to limit heat loss. Control of heat loss from the body surface through changes in the blood flow in the skin is one of the most important aspects of thermoregulation.

In all vertebrates, regardless of how they maintain their body temperature, the circulation of the blood distributes the heat more or less uniformly throughout the body. This heat circulation is accomplished first by a network of fine blood vessels that pass immediately beneath the external surfaces of the body. These surfaces are in contact with the environment. The blood that circulates within this network absorbs heat from a hot environment and releases heat to a cold one. The further circulation of this heated or cooled blood to the interior of the animal's body tends to adjust the internal temperature of that animal in the direction of the external environment. In many marine mammals, heat loss to cold ocean waters is limited by countercurrent flow in the limbs, which removes heat from arterial blood before it reaches the surface (Figure 39-11).

# THE BLOOD

About 8% of the body mass of most vertebrates is taken up by the blood circulating through their bodies. This blood is composed of a fluid plasma, together with several different kinds of cells that circulate within that fluid.

## Blood Plasma: The Blood's Fluid

**Blood plasma** is a complex solution of water with three very different components:

1. *Metabolites and wastes.* If the circulatory system is thought of as the highway of the vertebrate body, the blood contains the traffic traveling on that highway. Dissolved within the plasma are glucose, lipids, and all the other metabolites, vitamins, hormones, and wastes that circulate between the cells of the body.
2. *Salts and ions.* Like the seas in which life arose, plasma is a dilute salt solution. The chief plasma ions are sodium, chloride, and bicarbonate. In addition, there are trace amounts of other salts, such as calcium and magnesium, as well as of metallic ions, including copper, potassium, and zinc. The composition of the plasma, therefore, is not unlike that of seawater.
3. *Proteins.* Blood plasma is 90% water. Passing by all the cells of the body, blood would soon lose most of its water to them by osmosis if it did not contain as high a concentration of proteins as the cells it passes. Water does not move from the blood vessels into the surrounding cells, because blood plasma contains proteins. Some of these are antibody and globulin proteins that are active in the immune system, as well as a small amount of fibrinogen, a protein that plays a role in blood clotting. Taken together, however, these proteins make up less than half of the amount of protein that is necessary to balance the protein content of the other cells of the body. The rest consists of a protein, **serum albumin,** which circulates in the blood as an osmotic counterforce. Human blood contains 46 grams of serum albumin per liter.

*Blood plasma contains metabolites, wastes, and a variety of ions and salts. It also contains high concentrations of the protein serum albumin, which functions to keep the blood plasma in osmotic equilibrium with the cells of the body.*

## Blood Cells: Cells that Circulate Through the Body

Although blood is liquid, some 40% of its volume is actually occupied by cells. There are three principal types of cells in the blood (Figure 39-12): erythrocytes, leukocytes, and platelets.

***Erythrocytes Carry Hemoglobin.*** Each milliliter of blood contains about 5 billion oxygen-carrying **erythrocytes,** or **red blood cells.** Each erythrocyte is a flat disk with a central depression on both sides (Figure 39-13), something like a doughnut with a hole that doesn't go all the way through. Attached to the outer membranes of the erythrocytes is a collection of polysaccharides, which determines the blood group of the individual. Almost the entire interior of each cell is packed with hemoglobin.

Mature erythrocytes in mammals contain neither a nucleus nor protein-synthesizing machinery. Because they lack a nucleus, these cells are unable to repair themselves, and they therefore have a rather short life; any one erythrocyte lives only about 4 months. New erythrocytes are constantly being synthesized and released into the blood by cells within the soft interior marrow of bones.

***Leukocytes Defend the Body.*** Less than 1% of the cells in human blood are **leukocytes,** or **white blood cells;** there are about 1 or 2 leukocytes for every 1000 red blood cells. Leukocytes are larger than red blood cells; they contain no hemoglobin and are essentially colorless. There are several kinds of leukocytes, each with a different function. All these functions, however, are related to the defense of the body against invading microorganisms and other foreign substances, as you will see in Chapter 40. Leukocytes are not confined to the bloodstream; they also migrate out into the interstitial fluid.

If you prick your skin, some of the injured cells release chemicals that cause the capillaries in the vicinity to expand. The resulting increase in blood flow is one component of the **inflammatory response** making the wound look red and feel warm. In inflammation, **granulocytes,** which are circulating leukocytes, push out through the walls of the distended capillaries to the site of the injury. The granulocytes are classified into three groups by their staining properties. About 50% to 70% of them are **neutrophils,** which begin to stick to the interior walls of the blood vessels at the site of the injury. They then form projections that enable them to push their way into the infected tissues, where they engulf microorganisms and other foreign particles. **Basophils,** a second kind of leukocyte, contain granules that rupture and release chemicals that enhance the inflam-

| Blood cell | Life span in blood | Function |
| --- | --- | --- |
| Erythrocyte | 120 days | O₂ and CO₂ transport |
| Neutrophil 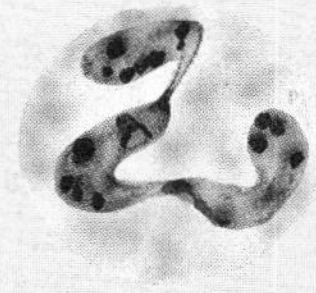 | 7 hours | Immune defenses |
| Eosinophil 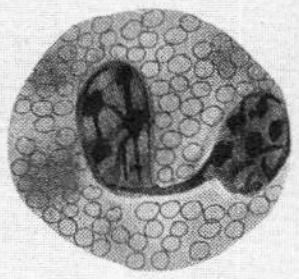 | Unknown | Defense against parasites |
| Basophil 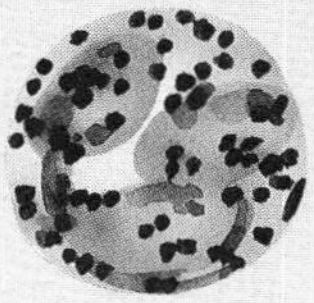 | Unknown | Inflammatory response |
| Monocyte 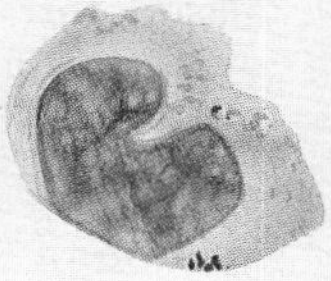 | 3 days | Immune surveillance (precursor of tissue macrophage) |
| B-lymphocyte 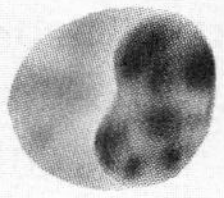 | Unknown | Antibody production (precursor of plasma cells) |
| T-lymphocyte 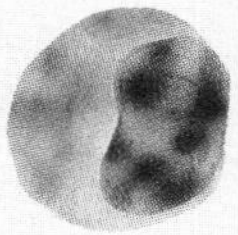 | Unknown | Cellular immune response |
| Platelets  | 7-8 days | Blood clotting |

**Figure 39-12**

Types of blood cells.

matory response; they are important in causing allergic responses. The function of the third kind of leukocyte, the **eosinophils,** is not clear, but they may play a role in defending against parasitic infections.

Another kind of circulating leukocyte, the **monocyte,** is also attracted to the sites of inflammation, where they are converted into **macrophages**—enlarged, amoeba-like cells that entrap microorganisms and particles of foreign matter. They usually arrive at the site of inflammation after the neutrophils, and they are important in the immune response because they play a key role in antibody production (see Chapter 40).

*Platelets Help Blood to Clot.* Certain large cells within the bone marrow, called **megakaryocytes,** regularly

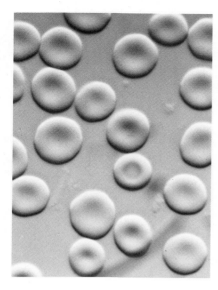

**Human erythrocytes, magnified 1000 times.** Human erythrocytes lack nuclei, which gives them a characteristic collapsed appearance, rather like a pillow on which someone has sat.

pinch off bits of their cytoplasm. These cell fragments, called **platelets,** contain no nuclei; they enter the bloodstream, where they play an important role in controlling blood clotting. A blood clot is a seal of a ruptured blood vessel. The ruptured vessel seals itself by generating a matrix of long fibers and trapped cells that fills the gap from components present in the plasma. In a clot the gluey substance is a protein called fibrin (derived from fibrinogen), which sticks platelets together to form a tight, strong seal.

Recently scientists have discovered that the fibrin that forms blood clots is generated in a spreading cascade of molecular events (Figure 39-14). The clotting process is initiated by injury to blood vessel cells attracting platelets, which releases a protein factor that starts the cascade. At each stage in the cascade that follows, proteins from cells and blood combine in fast-rising waves, involving many more molecules in the progressive steps of the process. Billions of molecules of fibrin can be formed from a single clot-initiating event.

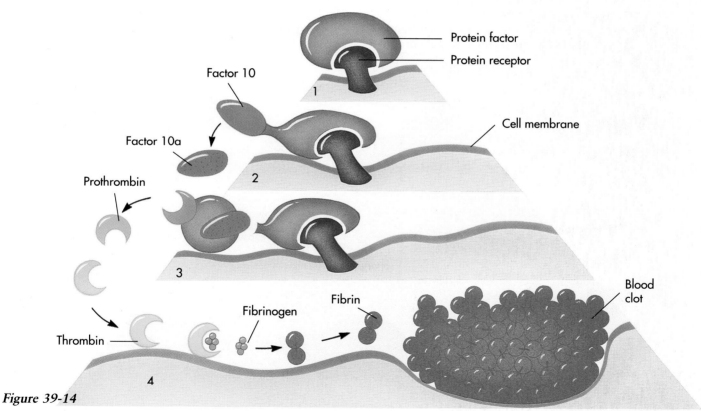

Figure 39-14

**How does blood clot?**
Step 1  A blood clot is initiated by the binding of a membrane receptor by a circulating protein factor.
Step 2  The complex of protein receptor and protein factor then binds with factor 10, changing it to the active form, factor 10a.
Step 3  Each molecule of 10a binds to another protein factor, and this complex catalyzes the conversion of prothrombin to thrombin.
Step 4  Thrombin interacts with fibrinogen, promoting its conversion to fibrin. Finally fibrin traps blood cells that seal the wound.

# THE EVOLUTION OF THE VERTEBRATE HEART

Any closed circulatory system requires both a system of passageways through which fluid can circulate and a pump to force the fluid through them. In the circulation of blood the pump is the heart.

Early chordates, such as the lancelets, seem to have had simple tubular hearts, which amounted to little more than a specialized, muscular zone of the ventral artery, which beats in simple waves of contraction. When gills evolved in the early fishes, a more efficient pump was necessary to force blood through the fine capillary network. The fish heart can be considered a tube that has four chambers arrayed one after the other (Figure 39-15). The first two chambers (**sinus venosus** and **atrium**) are collection chambers; the second two (**ventricle** and **conus arteriosus**) are pumping chambers. When a fish heart beats, contraction starts in the rear chamber (the sinus venosus) and spreads progressively forward to the conus arteriosus. The same heartbeat sequence is characteristic of all vertebrate hearts.

After fish blood leaves the gills, it circulates through the rest of the body; in terrestrial vertebrates,

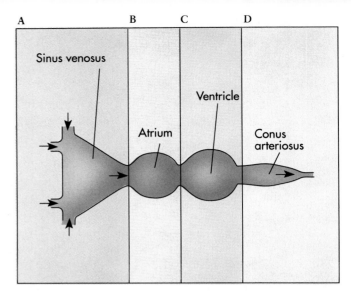

**Figure 39-15**

**Schematic diagram of the fish heart. A,** *Sinus venosus*. A large collection chamber contributes the least possible resistance to venous flow. **B,** *Atrium*. A vestibule about the size of the ventricle is required to deliver blood to the pump quickly in increments of suitable volume. **C,** *Ventricle*. A thick-walled pumping chamber is required to provide the energy to move the blood through the gills. **D,** *Conus arteriosus*. Blood leaving the ventricle passes through a second elongated chamber to smooth the pulsations and add still more thrust.

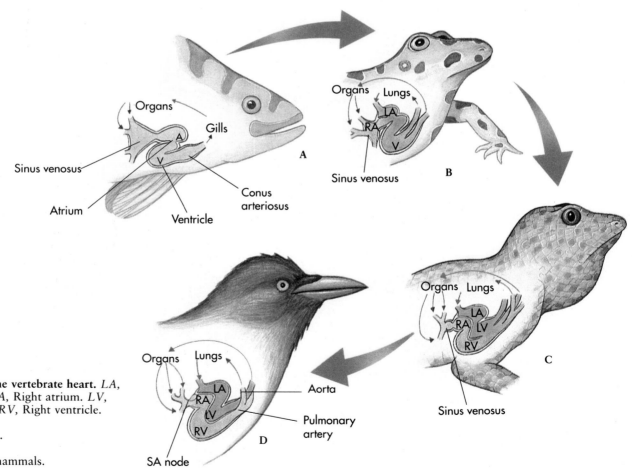

**Figure 39-16**

**Evolution of the vertebrate heart.** *LA,* Left atrium. *RA,* Right atrium. *LV,* Left ventricle. *RV,* Right ventricle.
A Fishes.
B Amphibians.
C Reptiles.
D Birds and mammals.

# How Americans Die

Every year 1.5 million people in the United States die from heart disease, cancer, stroke, accidents, and chronic lung disease. Previously, these were not the primary cause of death. At the turn of the century, the great killers were the infectious diseases.

The top 5 causes of death in 1986 were:

1. Heart disease        763,380
2. Cancer               465,440
3. Stroke               147,390
4. Accidents             93,990
5. Chronic lung disease  75,220

Unlike infectious disease, chronic illness does not strike at random, threatening everyone equally. On the contrary, deaths from chronic illnesses are influenced greatly by individual living habits, so strongly that it is possible to predict in large measure how one will die from how one lives. As many as 70% of the deaths from these "top killers" could be prevented by altered living habits.

Heart disease is the greatest killer in the United States today. It is not, however, an inevitable result of being alive. A few simple changes in lifestyle can drastically lower the risk of heart disease. Strokes, which are caused by stoppage of blood flow to part of the brain, share many of the same risk factors as heart disease. The three principal risk factors are:

*Smoking.* About a fourth of Americans (51 million) smoke, leading to almost a third of heart-disease deaths because smoking elevates cholesterol levels and thus atherosclerosis, according to the U.S. surgeon general. Smokers also run a 20% to 60% greater chance of stroke. It is not easy to quit smoking—nicotine has been declared an addictive drug—but the best way for smokers to avoid death by heart disease and stroke is to throw their cigarettes away.

*High cholesterol levels.* If your blood cholesterol level is high—over 200 for the average adult—it can be lowered through diet. Fewer saturated (animal) fats (eggs and meat) and more plant fats should be eaten. If blood cholesterol is over 240, many physicians recommend cholesterol-lowering drugs.

*High blood pressure.* Hypertension (high blood pressure) is a killer. High blood pressure causes up to one half of all strokes. If your blood pressure exceeds 140 over 90 mm Hg (140 represents the pushing pressure of the heart, 90 the pressure in vessels during the resting portion of the heartbeat), it can be lowered with a low-salt, low-fat diet.

---

however, the blood that receives oxygen and discharges carbon dioxide in the lungs is returned to the heart by two large veins, the **pulmonary veins,** and then pumped forcefully out through the body; this results in much-improved rates of circulation.

To achieve this double circulation, the arrangement of the amphibian heart differs from that found in fishes in two ways (Figure 39-16, *B*). First, the atrium is divided into two chambers: a right atrium, which receives unoxygenated blood from the sinus venosus for circulation to the lungs, and a left atrium, which receives oxygenated blood from the lungs through the pulmonary vein for circulation throughout the body. This division extends partway into the ventricle. Second, the conus arteriosus is partly separated by a dividing wall, which directs oxygenated blood into the aorta and unoxygenated blood into the pulmonary arteries, which lead to the lungs. The aorta leads to the body's network of arteries.

Taken together, these modifications divide the circulatory system into two separate pathways: (1) the **pulmonary circulation,** between the heart and the lungs; and (2) the **systemic circulation,** between the heart and the rest of the body. However, since the divisions in the heart ventricles and conus arteriosus are not complete, oxygenated and unoxygenated blood are mixed, reducing somewhat the oxygen level of the blood pumped to the body.

Further evolution among the vertebrates (Figure 39-17) has resulted in the complete closing of the dividing wall in the ventricle, which results in a total division of the pumping chamber into two parts in birds (Figure 39-16, *C*) and mammals. In these groups the four-chambered heart acts as a double pump, the left side pumping oxygenated blood to the general body circulation and the right side pumping unoxygenated blood from the veins to the lungs. Such efficient hearts function well in helping to maintain constant internal

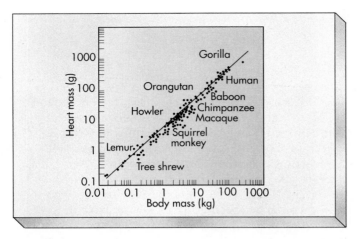

**Figure 39-17**

**Heart size is constant.** The heart is about 0.6% of the total body mass of any mammal, regardless of its size. This is true of a mouse and equally true of an elephant.

temperatures in birds and mammals. They also circulate blood through the lungs much more rapidly and efficiently than in other vertebrates, thus greatly increasing the efficiency with which oxygen is captured by the bloodstream.

## THE HUMAN HEART

The human heart, like that of all mammals and birds, is therefore a double pump; a frontal section through the heart clearly shows its organization (Figure 39-18). The left side has two connected chambers, and so does the right, but the two sides are not connected with one another.

### Circulation Through the Heart

Let's follow the journey of blood through the human heart, starting with the entry of oxygenated blood into the heart from the lungs. Oxygenated blood from the lungs enters the left side of the heart, emptying directly into the **left atrium** through large vessels called the **pulmonary veins.** From the atrium, blood flows through an opening into the adjoining chamber, the **left ventricle.** Most of this flow, roughly 80%, occurs while the heart is relaxed. When the heart starts to contract, the atrium contracts first, pushing the remaining 20% of its blood into the ventricle.

After a slight delay, the ventricle contracts. The walls of the ventricle are far more muscular than those of the atrium, and as a result this contraction is much stronger. It forces most of the blood out of the ventricle in a single strong pulse. The blood is prevented from going back into the atrium by a large one-way

valve, the **mitral valve,** whose flaps are pushed shut as the ventricle contracts. Strong fibers that prevent the flaps from moving too far when closing are attached to their edges. If the flaps did move too far, they would project out into the atrium. The fibers that prevent this operate in much the same way as a chain on a screen door—the door can be opened only as far as the slack in the chain permits.

Prevented from reentering the atrium, the blood within the ventricle takes the only other passage out of the contracting left ventricle. It moves through a second opening that leads into a large vessel called the **aorta.** The aorta is separated from the left ventricle by a one-way valve, the **aortic valve.** Unlike the mitral valve, the aortic valve is oriented to permit the flow of the blood *out* of the ventricle. Once this outward flow has occurred, the aortic valve closes, thus preventing the reentry of blood from the aorta into the heart.

The aorta and all the other blood vessels that carry blood away from the heart are arteries. Many of these arteries branch from the aorta, carrying oxygen-rich blood to all parts of the body. The first to branch are the coronary arteries, which carry freshly oxygenated blood to the heart itself; the muscles of the heart do not obtain their supply of blood from within the heart.

The blood that flows into the arterial system eventually returns to the heart after delivering its cargo of oxygen to the cells of the body. In returning, it passes through a series of veins, eventually entering the right side of the heart. Two large veins collect blood from the systemic circulation. The **superior vena cava** drains the upper body, the **inferior vena cava** the lower body. These veins empty deoxygenated blood into the right atrium. The right side of the heart is similar in organization to the left side. Blood passes from the right atrium into the right ventricle through a one-way valve, the **tricuspid valve.** It passes out of the contracting right ventricle through a second valve, the **pulmonary valve,** into the **pulmonary arteries,** which carry the deoxygenated blood to the lungs. The blood then returns from the lungs to the left side of the heart with a new cargo of oxygen, which is pumped to the rest of the body. The human circulatory system as a whole is outlined in Figure 39-19.

### How the Heart Contracts

The overall contraction of the heart consists of a carefully orchestrated series of muscle contractions (Figure 39-20). First the atria contract together, followed by the ventricles. Contraction is initiated by the **sinoatrial (SA) node,** the small cluster of excitatory cardiac muscle cells derived from the sinus venosus embedded in the upper wall of the right atrium. The cells of the SA node act as a pacemaker for the rest of the heart. Their

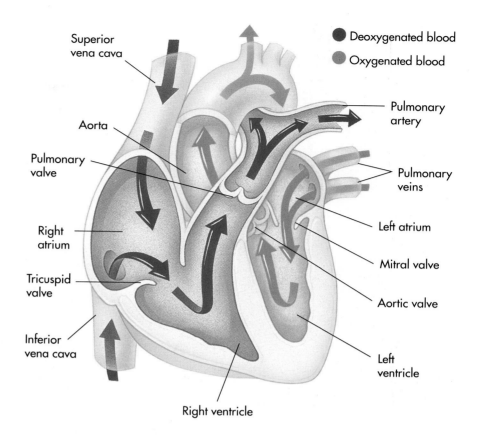

**Figure 39-18**

**The path of blood through the human heart.** Entering the right atrium by way of the superior vena cava, the blood passes into the right ventricle and then through the pulmonary valve to the pulmonary artery and the lungs. Oxygenated blood from the lungs then returns to the heart by way of the pulmonary veins, entering the left atrium and then the left ventricle, from which it enters the general circulatory system of the body by way of the aorta.

# Hypertension and Its Consequences

The central nervous system regulates blood pressure at normal levels in several different ways. A reflex involving pressure receptors in the arteries regulates blood pressure over periods of seconds to minutes. Blood pressure regulation over periods of hours to days involves neural and hormonal mechanisms that regulate the volume of the blood through effects on water and salt reabsorption by the kidneys. When this long-term regulation breaks down, the result may be a chronic elevation of blood pressure, referred to as hypertension.

In one form of hypertension, called high-renin hypertension, there are abnormally high levels of the hormone renin, released by cells in the kidney. Renin causes salt to be reabsorbed, accompanied by water, increasing the blood volume and therefore blood pressure. A second form of hypertension, known as essential hypertension, is not caused by elevated levels of renin. Although it is not fully understood, one possible explanation for this disorder is a decrease in the release of a newly discovered hormone released from cells in the wall of the right atrium, called atrial natriuretic hormone (ANH).

Whenever the blood pressure is chronically elevated, there is an increased chance that blood vessels will rupture. When this occurs in the brain, the result is a stroke that may seriously damage this delicate structure. Hypertension also increases the work needed by the heart to pump blood, requiring an increased oxygen supply. As a result, the heart may fail. Hypertension is also associated with an increased accumulation of fat deposits on the walls of the coronary arteries (atherosclerosis), which ultimately leads to impaired blood flow to regions of the heart. Finally, hypertension damages the nephrons of the kidneys, often leading to a vicious cycle of further increases in salt and water retention and therefore in blood pressure.

membranes spontaneously depolarize with a regular rhythm that determines the rhythm of the heart's beating. Each depolarization initiated within this pacemaker region passes quickly from one cardiac muscle cell to another in a wave that envelops both the left and the right atria almost instantaneously.

*The contraction of the heart is initiated by the periodic spontaneous depolarization of cells of the SA node; the resulting wave of depolarization passes over both the left and the right atria and causes their muscle cells to contract.*

But the wave of depolarization does not immediately spread to the ventricles. Almost 0.1 second passes before the lower half of the heart starts to contract. The reason for the delay is that the atria of the heart are separated from the ventricles by connective tissue, and connective tissue cannot propagate a depolarization wave. The depolarization would not pass to the ventricles at all except for a slender connection of cardiac muscle cells known as the **atrioventricular (AV) node,** which connects to a strand of specialized muscle in the ventricular septum known as the **bundle of His.** Bundle branches divide to the right and left; on reaching the apex of the heart, each branch further divides into **Purkinje fibers,** which initiate contraction there. The passage of the wave triggers the almost simultaneous contraction of all the cells of the right and left ventricles because the bundle branches conduct very rapidly. The cells involved in the passage of the depolarization wave from the atria to the ventricles have small diameters; thus they propagate the depolarization slowly, causing the delay mentioned. This delay permits the atria to finish emptying their contents into the corresponding ventricles before those ventricles start to contract.

*The passage of the wave of depolarization from the AV node via the bundle of His and bundle branches to the Purkinje fibers causes the almost simultaneous contraction of all the cells of the right and left ventricles.*

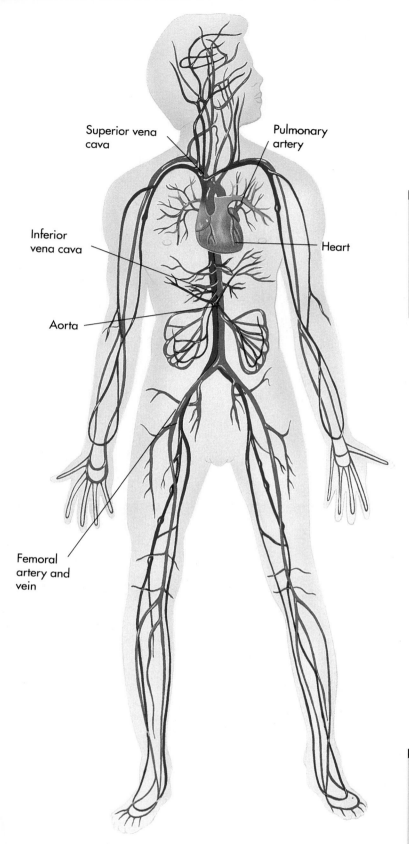

**Figure 39-19**

The human circulatory system.

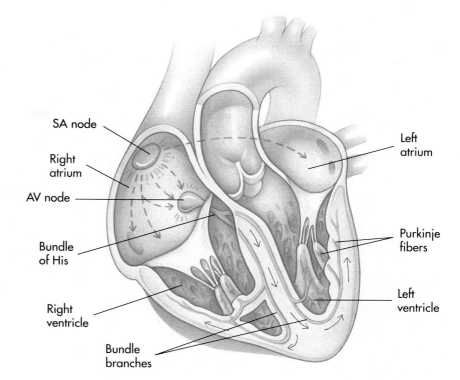

SA node

Right
atrium

AV node

Bundle
of His

Right
ventricle

Bundle
branches

Left
atrium

Purkinje
fibers

Left
ventricle

**Figure 39-20**

**How the heart contracts.** Contraction of the human heart is initiated by a wave of depolarization that begins at the SA node. After passing over the right and left atria and causing their contraction, the wave of depolarization reaches the AV node. From there it passes through the bundle of His, which has a bundle branch to each ventricle. From the tips of the ventricles, the depolarization is conducted rapidly over their surfaces by the branches Purkinje fibers.

## Monitoring the Heart's Performance

As you can see, the heartbeat is not simply a squeeze-release, squeeze-release cycle, but rather a little play in which a series of events occur in a predictable order. You may watch the play in several ways, depending on the events that you are observing. The simplest way to monitor heartbeat is to listen to the heart at work. The first sound you hear, a low-pitched *lub*, is the closing of the mitral and tricuspid valves at the start of ventricular contraction. A little later you hear a higher-pitched *dub*, the closing of the pulmonary and aortic valves at the end of ventricular contraction. If the valves are not closing fully, or if they open too narrowly, turbulence is created within the heart. This turbulence can be heard as a **heart murmur.** It often sounds like liquid sloshing.

A second way to examine the events of the heartbeat is to monitor the blood pressure. During the first part of the heartbeat the atria are filling. At this time the pressure in the arteries leading from the left side of the heart out to the tissues of the body decreases slightly as the blood moves out of the arteries, through the vascular system, and into the atria. This period is referred to as the **diastolic period,** and the lowest arterial pressure is called the diastolic pressure. During the contraction of the left ventricle, a pulse of blood is forced into the systemic arterial system, immediately raising the blood pressure within these vessels. This pushing period, which ends with the closing of the aor-

tic valve, is referred to as the **systolic period.** The highest arterial pressure is termed systolic pressure. Blood pressure values are measured in millimeters of mercury; they reflect the height to which a column of mercury would be raised in a tube by an equivalent pressure. By the conventional system of measurement, normal blood pressure values are 70 to 90 diastolic and 110 to 130 systolic. These would be abbreviated as 110/70 or 130/90, respectively. When the inner walls of the arteries accumulate fats, as they do in the condition known as **atherosclerosis,** the diameters of the passageways are narrowed. If this occurs, the systolic blood pressure is elevated.

A third way to monitor the progress of events during a heartbeat is to measure the waves of depolarization. Because the human body basically consists of water, it conducts electrical currents rather well. A wave of membrane depolarization passing over the surface of the heart generates an electrical current that passes in a wave throughout the body. The magnitude of this electrical pulse is tiny, but it can be detected with sensors placed on the skin. A recording made of these impulses (Figure 39-21) is called an **electrocardiogram.**

In a normal heartbeat, three successive electrical pulses are recorded. First, there is an atrial excitation, caused by the depolarization associated with atrial contraction (P wave). A tenth of a second later, there is a much stronger ventricular excitation, reflecting both the depolarization of the ventricles and the relaxation

# Diseases of the Heart and Blood Vessels

Cardiovascular diseases are the leading cause of death in the United States; more than 42 million people have some form of cardiovascular disease. **Heart attacks** are the main cause of cardiovascular deaths in the United States, accounting for about a fifth of all deaths. They result from an insufficient supply of blood reaching an area of heart muscle, leading to death of some cells. Heart attacks may be caused by a blood clot forming somewhere in the vessels and blocking the passage of blood through those vessels. They may also result if a vessel is blocked sufficiently by atherosclerosis. Recovery from a heart attack is possible if the segment of the heart tissue damaged was small enough that the other blood vessels in the heart can enlarge their capacity and resupply the damaged tissues. **Angina pectoris,** which literally means "chest pain," occurs for reasons similar to those which cause heart attacks, but it is not as severe. The pain may occur in the heart and often also in the left arm and shoulder. It is a warning sign that the blood supply to the heart is inadequate but not enough to cause cell death.

**Strokes** are caused by an interference with the blood supply to the brain. They often occur when a blood vessel bursts in the brain, and they may be associated with a **thrombus,** or coagulation (clotting) of blood cells or other elements in one of the vessels. Such a thrombus may be caused by cancer or other diseases. The effects of a stroke depend on how severe the damage is and where in the brain the stroke occurs.

**Atherosclerosis** is a buildup within the arteries (Figure 39-A, *1, 2,* and *3*). Atherosclerosis contributes both to heart attacks and to strokes. The accumulation within the arteries of fatty materials, of abnormal amounts of smooth muscle cells, of deposits of cholesterol or fibrin, or of cellular debris of various kinds can all impair the arteries' proper functioning. When this condition is severe, the arteries can no longer expand and contract properly, and the blood moves through them with difficulty. The accumulation of cholesterol is thought to be the prime contributor to atherosclerosis, and diets low in cholesterol and unsaturated fats are now prescribed to help to prevent this condition.

**Arteriosclerosis,** or hardening of the arteries, occurs when calcium is deposited in arterial walls. It tends to occur when atherosclerosis is severe. Not only is flow through such arteries restricted, but they also lack the ability to expand as normal arteries do to accommodate the volume of blood pumped out by the heart. This forces the heart to work harder.

1

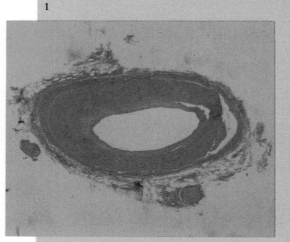

2

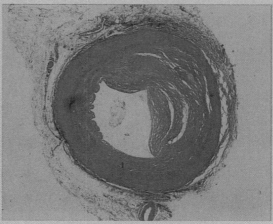

3

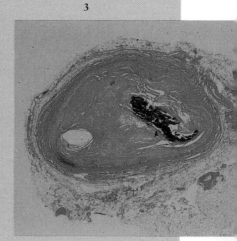

*Figure 39-A*

**The path to a heart attack. 1,** The coronary artery shows only minor blockage. **2,** The artery exhibits severe atherosclerosis—much of the passage is blocked by buildup on the interior walls of the artery. **3,** The coronary artery is essentially completely blocked.

of the atria (QRS wave). Finally, perhaps 0.2 second later, there is a third pulse, caused by the relaxation of the ventricles (T wave).

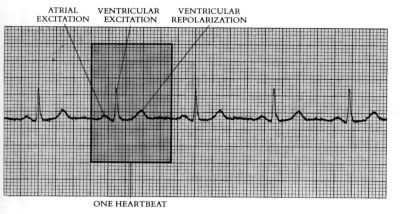

ATRIAL EXCITATION   VENTRICULAR EXCITATION   VENTRICULAR REPOLARIZATION

ONE HEARTBEAT

*Figure 39-21*

**An electrocardiogram.**

## THE CENTRAL IMPORTANCE OF CIRCULATION

The evolution of multicellular organisms has depended critically on the ability to circulate nutrients and other materials to the various cells of the body and to carry metabolic wastes away from them. The digestive processes described in Chapter 37, by which vertebrates obtain metabolizable foodstuffs, and the respiratory processes described in Chapter 38, by which vertebrates obtain the oxygen necessary for aerobic metabolism, both depend critically on the transport of food and oxygen to cells and the removal of the end-products of their metabolism. Vertebrates carefully regulate the operation of their circulatory systems; by doing so, they are able to integrate their bodily activities. This regulation is carried out by the nervous system, the subject of Chapter 34.

## ◼ SUMMARY

1. The circulatory system of vertebrates is closed, permitting a more exact control of pressure and resistance in the system.

2. The plasma of the circulating blood contains the proteins and ions that are necessary to maintain the blood's osmotic equilibrium with the surrounding tissues.

3. The general flow of blood circulation through the body is a circuit starting from the heart, which pumps blood out via muscled arteries to the capillary networks that interlace the tissues of the body; the blood returns to the heart from these capillaries via the veins.

4. A second, open circulatory system, the lymphatic system, gathers liquid from the body that has been lost from the circulatory system by diffusion and returns it via a system of lymph capillaries, lymph vessels, and two large lymphatic ducts to veins in the lower part of the neck.

5. The heart of mammals and birds is a double pump, pushing both pulmonary (lung) circulation and systemic (general body) circulation. Because the two circulations are kept separate within the heart, the systemic circulation receives only fully oxygenated blood.

6. The four-chambered mammalian heart has evolved from two of the chambers of the fish heart, by the creation of dividing walls within its two central chambers. The chambers at the ends of the fish heart were gradually lost, although its pacemaker cells have been retained in their original location in the mammalian heart.

7. The contraction of the heart is initiated at the SA node, or pacemaker, as a periodic spontaneous depolarization of these cells. The wave of depolarization spreads across the surface of the two atrial chambers, causing all these cells to contract.

8. The passage of the wave of depolarization to the ventricles is briefly delayed by tissue that insulates the two segments of the heart from one another. Only a narrow channel of cardiac muscle cells connects the atria and the ventricles. The delay in the passage of the wave of depolarization permits the atria to empty completely into the ventricles before ventricular contraction occurs.

9. The heartbeat can be heard, or monitored, by tracking changes in blood pressure through the period of filling and contracting of the atria (the diastolic period) and the contraction of the left ventricle (the systolic period). The waves of depolarization can also be measured directly; a recording of these pulses is called an electrocardiogram.

## REVIEW

1. The circulatory system of vertebrates is composed of three elements: the heart, the _____, and the blood.

2. No cell of your body is more than _____ away from a capillary.

3. The function of _____ is to keep the blood plasma in osmotic equilibrium with the cell of the body.

4. In what order does blood flow through the four chambers of your heart, starting from the lungs?

5. The sound of turbulence caused by only partial closing of heart valves is called _____.

## SELF-QUIZ

1. Which of the following is *not* part of the cardiovascular system?
   (a) The heart       (c) Arteries       (e) Capillaries
   (b) Veins           (d) The liver

2. Which of the following is a structural component of arteries, veins, *and* capillaries?
   (a) Endothelium     (c) Smooth muscle     (e) All of the above
   (b) An elastic layer (d) Connective tissue

3. In general, veins are larger in diameter than arteries. Why?
   (a) To reduce resistance to flow              (d) Because arteries are more elastic
   (b) To increase resistance to flow            (e) None of the above
   (c) Because cholesterol builds up only in arteries

4. Erythrocytes are packed with
   (a) fibrinogen.     (c) hemoglobin.     (e) platelets.
   (b) serum albumin.  (d) antibody proteins.

5. Which of the following events occurs first in the cardiac cycle?
   (a) Atrial contraction              (d) The QRS wave of the electrocardiogram
   (b) An action potential in the SA node (e) Ventricular contraction
   (c) Arterial systolic pressure

6. Blood pumped out of the left ventricle moves into the
   (a) left atrium.    (c) right ventricle.    (e) superior vena cava.
   (b) pulmonary veins. (d) aorta.

# THOUGHT QUESTIONS

1. Instead of evolving an entire second open circulatory system—the lymphatic system—to collect water lost from the blood plasma during passage through the capillaries, why haven't vertebrates simply increased the level of serum albumin in their blood?

2. Starving animals often exhibit swollen bodies rather than emaciated ones in early stages of their deprivation. Why?

3. The hearts of the more advanced vertebrates pump blood entirely by pushing action. Why do you suppose hearts have not evolved that act like suction pumps, drawing blood into the heart as it expands, rather than pushing it out as the heart contracts?

# FOR FURTHER READING

GOLDSTEIN, G., and A.L. BELZ: "The Blood-Brain Barrier," *Scientific American*, September 1986, pages 74-83. Many chemicals will not pass from the bloodstream to the cells of the brain. This article explains that this "barrier" is the result of the structure of brain capillaries, which possess an unusual array of membrane channels.

ROBINSON, T., S. FACTOR, and E. SONNEBLINK: "The Heart As a Suction Pump," *Scientific American*, June 1986, pages 84-91. Between birth and death our hearts beat millions of times. These authors argue that the heart is aided greatly in this Herculean task by a very clever trick: contraction compresses elastic elements within the heart muscles, which then bounce back to expand the ventricles.

ZUCKER, M.B.: "The Functioning of the Blood Platelets," *Scientific American*, June 1980, pages 86-103. A description of the many roles of platelets in human health, with emphasis on their role in blood clotting.

## CHAPTER 40

# How the Body Defends Itself

Animals defend themselves from predators with a variety of mechanisms. This armadillo is protected from attack by its armor-plated covering. However, many of the organisms harmful to animals can broach the external protection and invade the body. The immune system defends the body against such internal invaders.

# HOW THE BODY DEFENDS ITSELF

## Overview

Most diseases that afflict human beings are in fact microbial infections, invasions of the body by viruses, bacteria, fungi, or protists. To defend against this onslaught, vertebrates possess physical barriers and a variety of rapid-response defenses. However, the most important defense consists of a sophisticated collection of cells and tissues called the immune system. The immune system constantly monitors the bloodstream and tissues for the presence of any foreign cells or molecules. When an infection occurs, the immune systems ensure that the invading microbes are attacked and destroyed. No vertebrate can live for long without these defenses against infection. A dramatic illustration of the consequence of the loss of an effective immune system is seen in AIDS (acquired immunodeficiency syndrome), where individuals die from bacterial and viral infections that would go unnoticed in a healthy person.

## For Review

*Here are some important terms and concepts that you will encounter in this chapter. If you are not familiar with them, you should review them before proceeding.*

**AIDS** (Chapter 25)

**Macrophage** (Chapter 32)

**Lymphocyte** (Chapters 32 and 39)

**The lymphatic system** (Chapter 39)

*Figure 40-1*

**The flu epidemic of 1918 killed 22 million people in 18 months.** With 25 million Americans infected during the influenza epidemic, it was hard to provide care for everyone. The Red Cross often worked around the clock.

When you think of how animals defend themselves, it is natural to think of armor, of dinosaurs covered like tanks with heavy plates, of turtles and clams and armadillos. However, armor offers no protection against the most dangerous enemies that vertebrates face—their distant relatives, the microbes. Every vertebrate body offers a feast in nutrients for single-celled creatures too tiny for you to see with the naked eye, as well as a warm, sheltered environment in which they can grow and reproduce. Like Europeans first discovering the New World, a microbe entering a vertebrate body is faced with a rich ecosystem ripe for plundering. We live in a world awash with microbes, and no vertebrate body could long withstand their onslaught unprotected. We survive because we have evolved a variety of very effective defenses against this constant attack. These defenses are the subject of this chapter. As we review them, it is important to keep in mind that our defenses are far from perfect—microbial infection is still a major cause of death among humans. Some 22 million Americans and Europeans died of flu within 18 months in 1918-1919 (Figure 40-1). More than 3 million people will die of malaria *this year*. Attempts to improve our defenses against infection are among the most active areas of scientific research today.

## THE BODY'S DEFENSES

Your body is defended from infection the same way that knights defended medieval cities. There are walls and moats to make secret entry difficult, roaming patrols that attack strangers, and sentries that challenge anyone wandering about, and call patrols if a proper ID is not presented.

*Walls and moats.* The outermost defense of the vertebrate body is the **skin** and the mucous membranes. In some vertebrates like rhinoceroses the skin is very thick and tough. In all vertebrates it offers a surprisingly efficient barrier to penetration by microbes. The lungs also have important barriers that protect their delicate alveoli from invasion.

*Roaming patrols (nonspecific defenses).* The initial response of the vertebrate body to infection is a battery of **nonspecific defenses,** including chemicals and cells that kill microbes. These defenses act very rapidly after the onset of infection.

*Sentries (specific defenses).* The vertebrate body also employs two types of cells that scan the surfaces of every cell in the body. Together they are called **the immune system.** One kind of cell aggressively attacks and kills any cell identified as foreign, whereas the other type marks the foreign cell or virus for elimination by the roaming patrols.

## SKIN

Skin is the outermost layer of the vertebrate body, and provides its first defense against invasion by microbes. It also serves to keep the body watertight so that it does not loose excessive water to the air by evaporation. Skin is the largest organ of the vertebrate body. In an adult human, 15% of the total weight is skin. Many other specialized cells are crammed in among skin cells; one square centimeter of human skin, about what a dime covers, contains 200 nerve endings, 10 hairs and muscles, 100 sweat glands, 15 oil glands, 3 blood vessels, 12 heat-sensing organs, 2 cold-sensing organs, and 25 pressure-sensing organs.

Vertebrate skin is composed of three layers (Figure 40-2): an outer **epidermis,** a lower **dermis,** and an underlying layer of **subcutaneous tissue.**

### Epidermis is the "Bark" of the Vertebrate Body

The epidermis of skin is from 10 to 30 cells thick, about as thick as this page. The outer layer, called the **stratum corneum,** is the one you see when you look at your arm or face. Cells from this layer are continuously subjected to damage. They are abraded, injured, and worn by friction and stress during the body's many activities. They also lose moisture and dry out. The body deals with this damage not by repairing cells but by replacing them. Cells from the stratum corneum are shed continuously. They are replaced by new cells produced deep within the epidermis. The cells of the innermost layer of the epidermis, called the **stratum basal layer,**

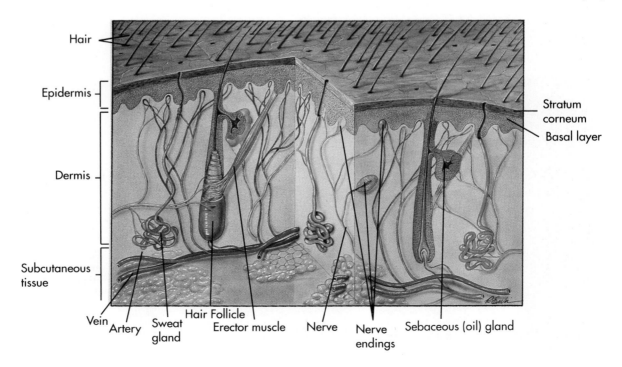

*Figure 40-2*

Human skin.

Hair — Epidermis — Dermis — Subcutaneous tissue — Stratum corneum — Basal layer — Vein — Artery — Sweat gland — Hair Follicle — Erector muscle — Nerve — Nerve endings — Sebaceous (oil) gland

are among the most actively dividing cells of the vertebrate body. New cells formed there migrate upward, and as they move they form keratin protein, which makes the skin tough. Each cell eventually arrives at the outer surface and takes its turn in the stratum corneum, ready to be shed and replaced by a newer cell. A cell normally lives in the stratum corneum for about a month. **Psoriasis**, familiar to some 4 million Americans as persistent dandruff, is a chronic skin disorder in which new cells reach the epidermal surface every 3 or 4 days, about 7 times faster than normal.

## The Lower Skin Layers Provide Support and Insulation

The dermis of skin is from 15 to 40 times thicker than the epidermis. The thick dermis provides structural support for the epidermis and a matrix for the many nerve endings, muscles, and specialized cells residing within skin. The wrinkling that occurs as we grow older takes place here (Figure 40-3). A fine network of blood vessels passes through it. The leather used to manufacture belts is derived from very thick animal dermis.

The layer of subcutaneous tissue below the dermis is composed primarily of fat-rich cells. They act as shock absorbers and provide insulation, which conserves body heat. This tissue varies greatly in thickness in different parts of the body. The eyelids have none of it, whereas the buttocks and thighs may have a lot of it. The subcutaneous tissue of the skin on the soles of your feet may be a quarter-inch thick or more.

## The Battle is on the Surface

The skin not only defends the body by providing a nearly impenetrable barrier, but also reinforces this defense with chemical weapons on the surface. The oil and sweat glands within the human epidermis, for example, lower the pH at the skin's surface to 3 to 5, an acid level that inhibits the growth of many microorganisms. Sweat also contains the enzyme lysozyme, which attacks and digests the cell walls of many bacteria.

Two other routes of entry must also be guarded:

*The digestive tract.* Many bacteria are present in the food that we eat. Most of these are killed by saliva, which also contains lysozyme; by the strong digestive acids present in the stomach; and by protein-digesting enzymes in the intestine.

*The respiratory tract.* Ciliated epithelial cells in the nasal cavity entrap many bacteria before they can enter the airway. Many microbes are present in the air we breathe. The cells lining the smaller bronchi and bronchioles secrete a layer of sticky mucus that traps microorganisms before they can reach the warm, moist lungs, ideal breeding grounds for microbes. Cilia on the cells lining these passages continually sweep the mucus upward, where it can be swallowed, carrying potential invaders out of the lungs like bound prisoners to be destroyed by gastric HCl.

The surface defenses of the vertebrate body are very effective, but they are occasionally breached. By breathing, by eating, by cuts and nicks, bacteria and vi-

*Figure 40-3*

**For the most part, skin ages gradually, but on the face, the changes are more dramatic.** At 19, this woman's face appears youthful and smooth. Forty years later, her production of skin oil is much less; her face appears less smooth and elastic and begins to exhibit wrinkles.

ruses now and then enter our bodies. When these invaders reach deeper tissue, a second line of defense comes into play, the vertebrate body's nonspecific defenses.

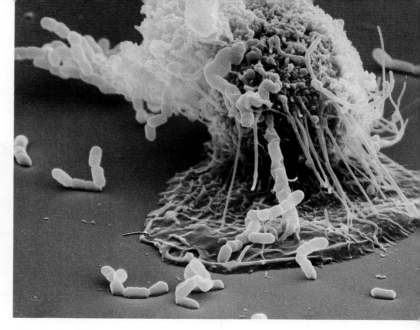

*The surface defenses of the body consist of the skin and mucous membranes, which eliminate many invading organisms before they can enter the body tissues.*

## NONSPECIFIC DEFENSES

The vertebrate body uses a host of nonspecific cellular and chemical devices to defend itself. They all have one property in common: they respond to *any* microbial infection. They do not pause to inquire as to the identity of the invader, but swing into action immediately. Of the many nonspecific defenses, the four that are of most importance are (1) cells that ingest invading microbes; (2) antimicrobial proteins that kill pathogens; (3) the inflammatory response, which speeds defending cells to the point of infection; and (4) the temperature response, which elevates body temperature to slow the growth of invading bacteria.

### Cells that Kill Invading Microbes

Perhaps the most important of the vertebrate body's nonspecific defenses are the cells that attack invading microbes. These cells patrol the bloodstream and await invaders within the tissues. There are three basic kinds: macrophages, neutrophils, and natural killer cells. Each of these cells kills invading organisms differently.

*Foot soldiers.* **Macrophages** ("big eaters") kill bacteria one at a time by ingesting them, much as an amoeba ingests a food particle. Flowing cytoplasmic extensions stick to the invading bacterium (Figure 40-4) and pull it inside the macrophage by means of endocytosis. Once inside the macrophage, the bacterium is killed very efficiently: the membrane-lined vacuole containing the bacterium is fused with a lysosome. This activates lysosomal enzymes that liberate large quantities of oxygen "free radicals" that literally rip the bacteria apart chemically. Although some macrophages are fixed within particular organs, including lungs, liver

**Figure 40-4**

**A macrophage in action.** In this scanning electron micrograph a macrophage is "fishing" with long, sticky cytoplasmic extensions. Bacterial cells unfortunate enough to come in contact with the extensions are drawn back to the macrophage and engulfed.

sinusoids, spleen, and brain, most of the body's macrophages patrol the byways of the body, circulating in the blood, lymph, and interstitial fluid between cells.

*Kamikazes.* **Neutrophils** are white blood cells that ingest bacteria in the same way macrophages do. Both of these defending cells are **phagocytes,** cells that kill invading cells by engulfing them. Neutrophils, however, are kamikazes; they also release chemicals (which are identical to household bleach) to "neutralize" the entire area, killing any other bacteria in the neighborhood and themselves in the process. Macrophages kill only one invading cell at a time, but they live to keep on doing it.

*Internal security patrol.* **Natural killer cells** do not kill invading microbes, but rather the cells that are infected by them. They are particularly effective at detecting and attacking body cells that have been infected with viruses. Natural killer cells are not phagocytes; rather they kill by attacking and puncturing the membrane of the target cell (Figure 40-5). Creation of the hole allows water to rush into the target cell, which swells and bursts (Figure 40-6). Natural killer cells are also able to detect cancer cells, which

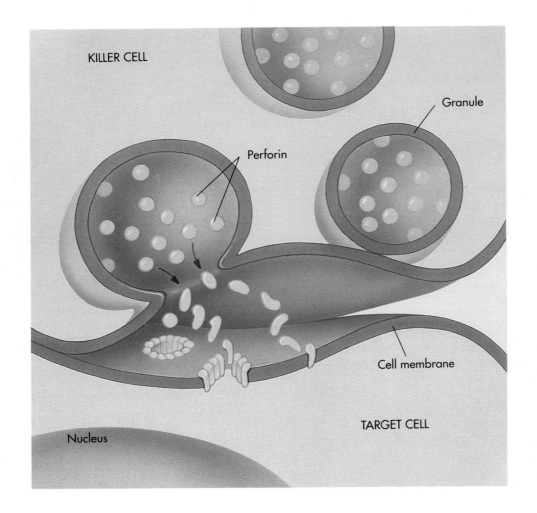

KILLER CELL

Granule

Perforin

Cell membrane

TARGET CELL

Nucleus

*Figure 40-5*

**How cytotoxic T cells kill target cells.** The intial event is the tight binding of the T cell to the target cell. Binding initiates a chain of events within the T cell in which granules loaded with perforin molecules move to the outer cell membrane and disgorge their contents into the intercellular space over the target. The perforin molecules insert into the membrane like staves of a barrel to form a pore that admits water and ruptures the cell.

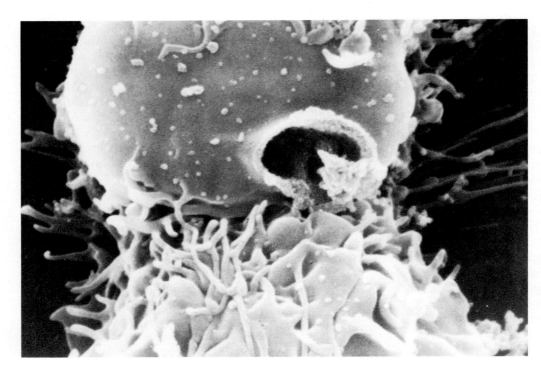

*Figure 40-6*

**Death of a tumor cell.** A natural killer cell has attacked this cancer cell, punching a hole in its cell membrane. Water has rushed in, making it balloon out. Soon it will burst.

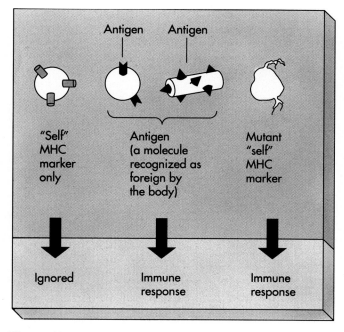

Antigen    Antigen

"Self"
MHC
marker
only

Antigen
(a molecule
recognized as
foreign by
the body)

Mutant
"self"
MHC
marker

Ignored

Immune
response

Immune
response

*Figure 40-7*

**The immune system ignores a cell that has a "self" MHC marker and attacks cells that have a "nonself" MHC marker.**

they kill before the cancer cell has a chance to develop into a tumor. The vigilant surveillance by natural killer cells is one of the body's most potent defenses against cancer.

How do these three kinds of cells distinguish *self* (the body's own cells) from *nonself* (foreign cells)? It is very important not to unleash the destructive power of these cells on the body's own cells. When the body's defenders do turn on the body itself, in **autoimmune diseases,** the results can be fatal. The patrolling cells do not attack their own body because all the cells of their body contain a surface protein marker that identifies them. This protein, called a **MHC (major histocompatibility complex) marker,** acts like a "dogtag"—each person has a different version, although all the cells within a particular person contain the same one (Figure 40-7). The cells of the immune system simply ignore any cells having "self" MHC markers alone; they are called "nonspecific" because they will attack *any* cell that lacks the "self" MHC marker. In autoimmune diseases, antibodies are formed against "self" cells because the immune system fails to distinguish between foreign and host tissue.

## Proteins that Kill Invading Microbes

The vertebrate body employs a very effective chemical defense, the **complement system,** that "complements" its cellular defenses. The complement system consists of a battery of more than a dozen different proteins that normally circulate in the blood plasma in an inactive state. Its defensive activity is triggered by the cell walls of bacteria and fungi (and also by the binding of antibodies to invading microbes, discussed later in this chapter). On detection of a bacterial cell wall, these proteins interact to form a **membrane attack complex (MAC)** like that produced by natural killer cells. The MAC inserts itself into the pathogen's cell membrane, forming a hole (Figure 40-8). Like a dagger through the heart, this wound is fatal to the invading cell—water pulled in by osmosis causes it to swell and burst.

Proteins of the complement system also act to amplify the effect of other body defenses. Some amplify the inflammatory response (discussed next) by stimulating histamine release, others attract phagocytes to the site of infection, and still others coat invading microbes, roughening the microbe's surfaces so that macrophages may more readily stick to them.

Another class of proteins that play a key role in body defenses are **interferons.** Released by virus-infected cells, they diffuse out to other cells and inhibit the ability of viruses to infect them. As you will see later in this chapter, their principal role is to sound the alert for the immune system.

## The Inflammatory Response

One of the most generalized nonspecific responses to infection is the **inflammatory response.** Infected or injured cells release chemical alarm signals, most notably histamines and prostaglandins. These alarm signals promote local expansion of blood vessels, which both increases the flow of blood to the site of infection or injury and, by stretching their thin walls, makes the capillaries more permeable. This is what produces the redness and swelling so often associated with infection. The larger, leakier capillaries promote the migration of phagocytes (macrophages and neutrophils) from the blood to the interstitial fluid surrounding cells, where they can engulf bacteria. Monocytes arrive first, spilling out chemicals that kill the bacteria in the vicinity (as well as tissue cells and themselves), followed by macrophages that clean up the remains of all the dead cells. This counterattack by phagocytes can take a considerable toll; the pus associated with some infections is a mixture of dead or dying neutrophils, broken down tissue cells, and dead pathogens. In some cases (for example, arthritis) inflammation occurs in the absence of infection. This is another example of a misdirected immune response.

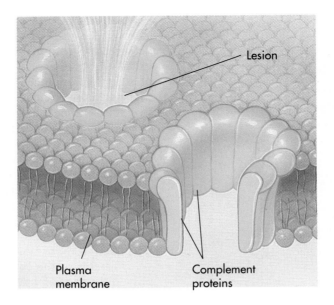

**Figure 40-8**

**How complement creates a hole in a cell.** As the diagram shows, the complement proteins form a complex transmembrane channel resembling the perforin-lined lesion in cytotoxic T cells.

Labels in figure: Lesion; Plasma membrane; Complement proteins

---

*The inflammatory response acts to clear an infected area of infecting microbes and dead tissue cells so that tissue repair can begin.*

### The Temperature Response

When macrophages encounter invading microbes, they release chemical substances called **pyrogens** (Greek *pyr*, fire), which pass through the bloodstream to the brain. When they reach the cluster of neurons in the hypothalamus that serves as the body's thermostat, they act to boost the body's temperature several degrees above the normal value of 37° C (98.6° F). The higher-than-normal temperature that results is called a **fever**. Fever contributes to the body's defense by stimulating phagocytosis, by inhibiting microbial growth, and by causing the body to reduce blood levels of iron, which bacteria need in large amounts to grow. Very high fevers, however, are dangerous because excessive heat may inactivate critical enzymes. In general, temperatures greater than 103° F are considered dangerous; those greater than 105° F are often fatal.

The nonspecific defenses, both chemical and cellular, provide the vertebrate body with a sophisticated defense against microbial infection. Only occasionally do bacterial or viruses overwhelm them. When this happens, they face yet a third line of defense, more difficult to evade than any they have encountered. It is the immune system, the most elaborate of the body's defenses. Unlike other defenses, the immune system remembers previous encounters with potential invaders, and if they reappear, the immune system is ready for them.

## SPECIFIC DEFENSES: THE IMMUNE SYSTEM

Few of us pass through childhood without being infected by a microbe. Measles, chickenpox, mumps—these are rites of passage, childhood illnesses that most of us experience before our teens. They are diseases of childhood because most of us suffer through them as children *and never catch them again.* Once you have had measles, you are immune. The mechanism that provides you with this immunity to such childhood diseases is the **immune system,** your body's most powerful means of resisting infection. Your immune system is the backbone of your health, protecting you not only from measles, but also from many other far more serious diseases. It is only in the last few years, as the result of an explosion of interest and knowledge, that biologists have begun to get a clear idea of how our immune system works.

## Discovery of the Immune Response

In 1796 an English country doctor named Edward Jenner carried out an experiment that marks the beginning of the study of immunology (Figure 40-9). Smallpox was a common and deadly disease in those days, and only those who had previously had the disease and survived it were immune from the infection—except, Jenner observed, milkmaids. Milkmaids who had caught another, much milder form of "the pox," called cowpox (it was caught by people who worked with cows), rarely caught smallpox. It was as if they had already had the disease. Jenner set out to test the idea that cowpox conferred protection against smallpox. He deliberately infected people with material that induced cowpox, causing them to catch this mild illness, and many of them became immune to smallpox, just as he had predicted.

Jenner's work demonstrated that it is possible for the human body to protect itself against disease very effectively when it is able to make suitable preparations. We now know that smallpox is caused by a virus called variola and that cowpox is caused by a different, although similar, virus. Jenner's patients injected with cowpox virus mounted a defense against the cowpox infection, a defense that was also effective against a later infection of the similar smallpox virus. Jenner's procedure of injecting a harmless microbe into a person or animal to confer resistance to a dangerous one is called **vaccination.** Modern attempts to develop resistance to malaria, herpes, and other diseases are focusing on the virus **vaccinia,** which is related to the cowpox virus used by Jenner. Such attempts are using the methods of genetic engineering (see Chapter 14) to incorporate genes encoding the protein-polysaccharide coats of the other viruses into the chromosome of vaccinia. Since vaccinia does not cause disease, it allows the body to be exposed to the protein coats of the other viruses in a harmless way, thus enabling the body to build up resistance to them.

It was a long time before people learned how one microbe can confer resistance to another, however. Further important information was added almost a century after Jenner by Louis Pasteur of France. In 1879 Pasteur was studying fowl cholera, a serious disease in chickens that we now know to be caused by a bacterium. From diseased chickens, Pasteur could isolate a culture of bacteria that would elicit the disease if injected into other healthy birds. One day Pasteur accidentally left his bacterial culture out on a shelf at the end of the day and went on vacation. Two weeks later he returned and injected this extract into healthy birds. The extract had been weakened; the injected birds became only slightly ill and then recovered. Surprisingly, however, the vaccinated birds could not then be in-

*Figure 40-9*

**This famous painting shows Edward Jenner inoculating patients with cowpox in the 1790s and thus protecting them from smallpox.** The underlying principles of vaccination were not understood until more than a century later.

fected with fowl cholera! They stayed healthy even if injected with massive doses of active fowl cholera bacteria, whereas control chickens receiving the same injections all died. Clearly something about the bacteria could elicit immunity, if only the bacteria did not kill the bird first.

We now know what that "something" was: molecules protruding from the surface of the bacterial cells. Every cell has on its surface a tangle of proteins, carbohydrates, and lipids (Figure 40-10), and it was the presence of foreign molecules on the surface of the cholera bacteria to which the chickens were responding. These bacterial cell surface molecules are different from any of the bird's own. "Nonself" molecules such as these are called **antigens.** Chickens injected with heat-killed fowl cholera bacteria are immune to later infection because the bacterial antigens cause the chickens to produce proteins called **antibodies.** These antibodies are

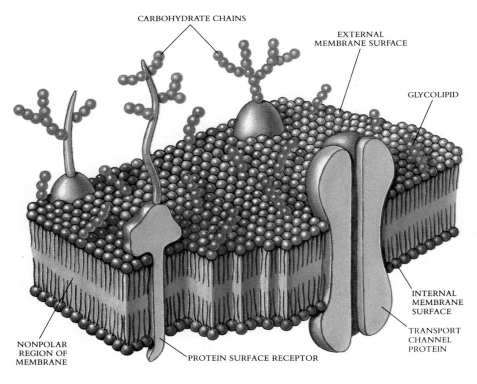

CARBOHYDRATE CHAINS

EXTERNAL
MEMBRANE SURFACE

GLYCOLIPID

INTERNAL
MEMBRANE
SURFACE

TRANSPORT
CHANNEL
PROTEIN

NONPOLAR
REGION OF
MEMBRANE

PROTEIN SURFACE RECEPTOR

**Figure 40-10**

**The outer surface of a cell is not smooth, but rather a tangle of glycolipid and protein imbedded within the membrane.** Glycolipid molecules often serve as highly specific cell surface markers that identify specific cell types. The two major kinds of transmembrane protein are transport channels and receptors. Transport channels import ions, sugars, and other molecules into the cell. Receptors bind hormones, growth factors, neurotransmitters, and in the case of immune receptors, other proteins.

able to recognize any future cholera invaders, treated or normal, and prevent them from causing disease. The production of antibodies directed against a specific antigen is called an **immune response**. The immune response is one component of a complex system of recognition and defense that we call the immune system.

*An immune response takes place when foreign proteins, called antigens, cause the production of other proteins, called antibodies, which recognize any antigens of the same sort with which they may come into contact in the future.*

## THE CELLS OF THE IMMUNE SYSTEM

Our immune system is neither localized to one place in the body nor is it controlled by any central organ such as the brain. Rather, it is composed of a host of individual cells, an army of defenders that rush to the site of an infection to combat invading microorganisms. These cells, the white blood cells mentioned in Chapter 39, arise in the bone marrow and circulate in blood

and lymph. Of the 100 trillion cells in an adult human being, two in every 100 ($2 \times 10^{12}$) are white blood cells. Although not bound together, the body's white blood cells exchange information and act in concert as a functional, integrated system. They are found not only in blood and lymph, but also in lymph nodes, spleen, liver, thymus, and bone marrow (Figure 40-11).

White blood cells are larger than the red blood cells that ferry oxygen to the body's tissues, but they are formed from the same cells in bone marrow, called **hematopoietic stem cells** (Figure 40-12). Unlike mature red blood cells, every white blood cell has a nucleus. Four kinds of white blood cells are involved in the immune system: phagocytes, T cells, B cells, and the nonspecific natural killer cells discussed earlier in this chapter. T and B cells are collectively referred to as **lymphocytes** (Table 40-1).

*The immune system is composed of white blood cells. Four principal classes are involved: phagocytes (including macrophages), natural killer cells, and two kinds of lymphocytes (T cells and B cells).*

**TABLE 40-1    CELLS OF THE IMMUNE SYSTEM**

| CELL TYPE | FUNCTION |
| --- | --- |
| Helper T cells | Commander of the immune responses, the helper T cell detects infection and sounds the alarm, initiating both T cell and B cell responses. |
| Inducer T cells | Not involved in the immediate response to infection, these cells mediate the maturation of T cells that are involved. |
| Cytotoxic T cells | Recruited by helper T cells, these are the foot soldiers of the immune response, detecting and killing bacteria and infected body cells. |
| Suppressor T cells | These cells dampen the activity of T and B cells, scaling back the defense after the infection has been checked. |
| B cells | Precursors of plasma cells, these cells are specialized to recognize particular foreign antigens. |
| Plasma cells | Biochemical factories, these cells are devoted to the production of antibody directed against a particular foreign antigen. |
| Mast cells | Initiators of the inflammatory response, which aids the arrival of white blood cells at a site of infection. |
| Monocytes | Precursors of macrophages. |
| Macrophages | The body's first line of defense, they also serve as antigen-presenting cells to B cells; later they engulf antibody-covered cells. |
| Killer cells | These lymphocytes recognize and kill foreign cells: natural killer (NK) cells detect and kill a broad range of foreign cells; killer (K) cells attack only antibody-coated cells. |

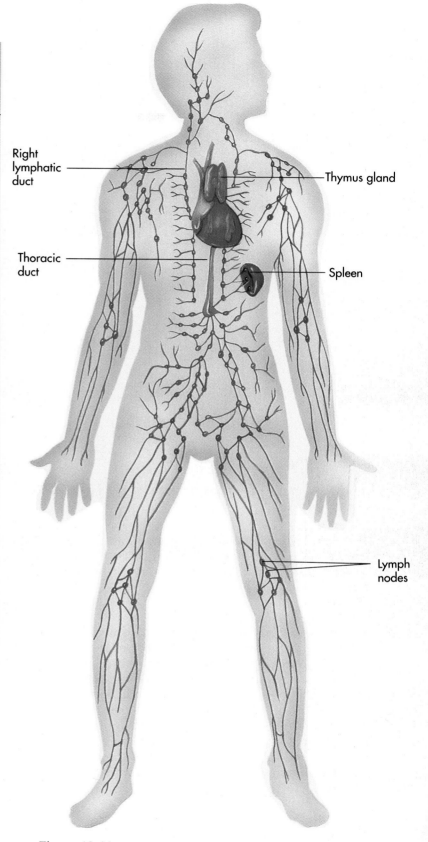

*Figure 40-11*

**The human immune system.**

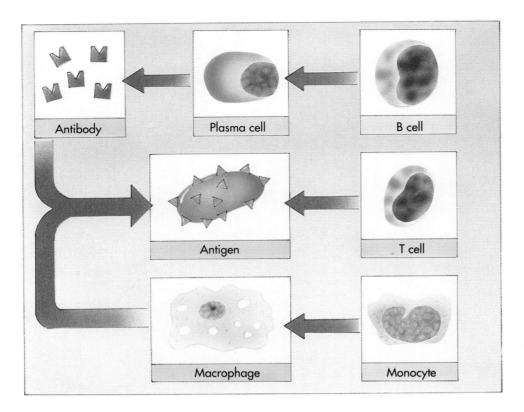

**Figure 40-12**

**Players in the immune response.** Three different kinds of cells mount attacks against the antigen-studded foreign cell in the center of the diagram. As described in the text, each line of attack actually involves the cooperative action of many cell types.

Labels in figure: Antibody · Plasma cell · B cell · Antigen · T cell · Macrophage · Monocyte

# THE ARCHITECTURE OF THE IMMUNE DEFENSE

The vertebrate immune system is a multilayered defense that uses a patrolling army of cells to attack and destroy invading microorganisms and eliminate infected cells.

## Sounding the Alarm

Macrophages respond to a pathogen encounter by secreting an alarm signal—proteins that initiate the immune response (Figure 40-13). Among these alarm signals are **gamma-interferon,** which activates monocytes to mature into macrophages, and **interleukin-1,** which activates a class of white blood cell called **helper T cells.** Helper T cells respond to the alarm being broadcast by macrophages by simultaneously initiating two different parallel immune responses, the cell-mediated immune response and the humoral immune response (Figure 40-14).

## The Cell-Mediated Immune Response

The activation of helper T cells by interleukin-1 unleashes a chain of events known as the **cell-mediated immune response,** in which special **cytotoxic T cells** (*cytotoxic* means "cell poisoning") recognize and destroy infected body cells. Their mechanism of killing is the same as that of natural killer cells—they puncture the membranes of target cells. Helper T cells initiate the response, activating both cytotoxic T cells and other elements of the system. Cell-mediated immunity is essential for the destruction of host cells that have been infected by viruses or have become abnormal (for example, in some cancers).

## The Humoral Immune Response

When a helper T cell is stimulated to respond to a foreign antigen, it not only activates the T cell–mediated immune response, as just described, but also simultaneously activates a second, more longer-range defense, referred to as the **humoral,** or **antibody, immune response.** The key player in this stage of defense against infection is another kind of lymphocyte, the **B cell.** B cells recognize invading pathogens much as T cells do, but unlike T cells they do not attack the pathogen directly; rather they mark them for destruction by the nonspecific body defenses. The humoral immune response is specialized to destroy invading bacteria and viruses and inactivate foreign molecules that otherwise would be toxic.

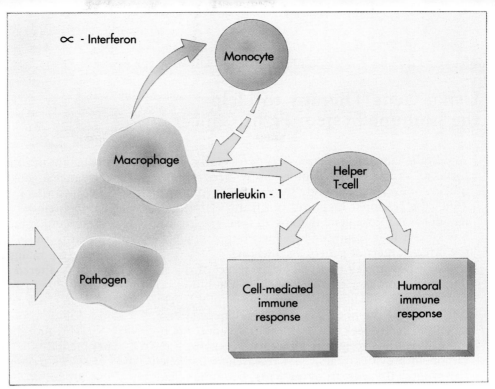

**Figure 40-13**

Overview of the human immune defense.

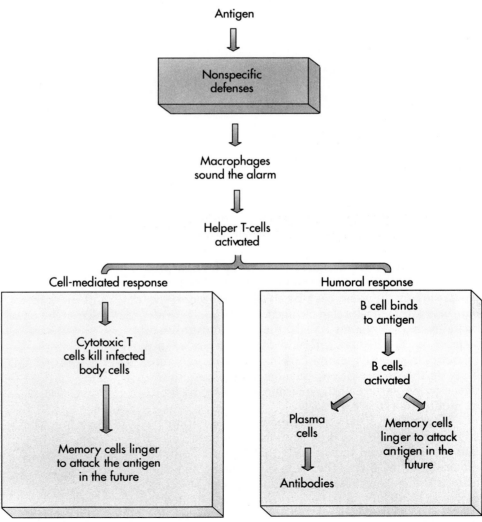

**Figure 40-14**

Overview of the cell-mediated and humoral immune responses.

## Using Gene Therapy to Help the Immune System Fight Cancer

Malignant melanoma is a particularly lethal form of skin cancer. Tumors grow very rapidly, quickly subverting the body's defenses. In August of 1990, the National Institutes of Health (NIH) gave its approval to use gene therapy to treat this cancer, one of the first instances in which humans will be the direct targets of genetic engineering.

The human cell that will be the target of this pioneering effort is a special kind of white blood cell called a tumor-infiltrating lymphocyte (TIL). This cell is part of the body's cancer surveillance system. It will normally seek out and attack a cancerous tumor but is not strong enough by itself to control the tumor. In previous years, NIH researchers had tried removing TIL cells from patients with malignant melanoma, culturing billions of them in test tubes, and then returning the cells to the patients' bloodstreams—but even in greatly increased numbers the cells were not strong enough to cure the cancer, although about half the patients improved.

Here is how the gene therapy will be done. Researchers will remove TIL cells from a patient with malignant melanoma and insert into each cell's chromosomes a gene that will command the cell to produce a protein called tumor necrosis factor (TNF). This protein kills tumor cells by blocking them from establishing a blood supply. The cells will then be returned to the patient's bloodstream to seek out and invade the malignant melanoma tumors. As each genetically altered TIL cell finds and enters a tumor, it will be able to attack using a much stronger weapon, TNF, in effect becoming a factory that makes the tumor-killing protein inside the tumor itself.

This first attempt at human gene therapy is experimental, and the degree to which it will be successful in treating malignant melanoma will not be known for a year or more. Because the approach shows great promise, it has medical researchers very excited. The chairman of the Recombinant DNA Advisory Committee that approved the therapy called the advance historic: "What we're doing today is adding gene therapy to vaccines, antibiotics, and radiation in the medical arsenal. Medicine has been waiting thousands of years for this."

## HOW THE ANTIBODY DEFENSE WORKS

Each B cell has on its surface about 100,000 copies of a protein called an **antibody**. Each B cell bears a different version of the antibody on its surface. Antibodies are proteins that are designed to bind to foreign proteins; because each cell's version of the antibody protein is slightly different, each B cell is specialized to recognize a different foreign antigen. Because the body contains many different B cells, there is almost always at least one B cell that will bind to the surface of *any* microorganism. At the onset of an infection, the antibody on the surface of one or more B cells bind to antigens on the surface of the microbe, inducing the B cell to proliferate (Figure 40-15).

After about 5 days and numerous cell divisions, a large clone of cells has been produced from each B cell that was stimulated by antigen to proliferate. Some of the proliferating B cells then stop reproducing and dedicate all their resources to producing and secreting more copies of the antibody protein that responded to the antigen. The secreting B cells are then called **plasma cells**. They live only a few days but secrete a great deal of antibody during that time. One cell will typically secrete more than 2000 molecules per second. Antibodies constitute about 20% by weight of the total protein in blood plasma, forming a plasma protein fraction known as gamma-globulins. Antibodies cover a virus or bacterium by binding to molecules on its surface and mark it for destruction by macrophages, natural killer cells, or complement.

*In the antibody defense, one or more B cells recognize a foreign antigen and are stimulated to divide repeatedly, creating a large number of cells with the same antibody. Some of them proceed to produce large quantities of antibody molecules directed against the antigen. The antibodies bind to any antigen they encounter and mark for destruction cells or viruses bearing the antigen.*

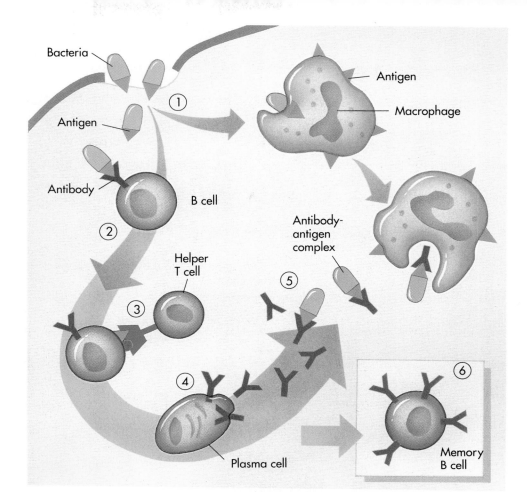

**Figure 40-15**

**The humoral immune response.**
1 Bacterial invasion triggers an immediate response by macrophages, and those macrophages which ingest bacterial cells display bacterial antigens on their surfaces.
2 Other bacterial cells are bound by shape-matching antibody protruding from the surface of a B-cell.
3 The B-cell then interacts with a helper T-cell and is activated to undergo cell division.
4 Some of the B-cell progeny differentiate into plasma cells, which secrete antibody of the kind that binds to the infecting bacteria.
5 Infecting bacteria are bound by free antibody, and in this way marked for destruction by macrophages.
6 Other B-cell progeny differentiate into memory cells, able to respond to the antigen at a future date.

## HOW DO ANTIBODIES RECOGNIZE ANTIGENS?

The cell surface receptors of lymphocytes (T cell receptors and B cell antibodies) can recognize specific antigens with great precision. Even single amino acid differences between proteins can often be discriminated, with a receptor recognizing one form and not the other. This high degree of precision is a necessary property of the immune system, since without it the identification of foreign antigens would not be possible in many cases; the differences between "self" and foreign ("non self") molecules can be very subtle.

### Antibody Structure

Antibody molecules (also called immunoglobulins) consist of four polypeptide chains (Figure 40-16). There are two identical short strands, called **light chains,** and two identical long strands, called **heavy chains.** The amino acid sequences of the two kinds of chains suggest that they evolved from a single ancestral sequence of about 110 amino acids. Modern light chains contain two of these basic 110 amino acid units or domains; and heavy chains contain three, or in some cases four, of them. The four chains are held together

by disulfide (——S—S——) bonds, forming a Y-shaped molecule.

The specificity of antibodies resides in the two arms of the Y. Three small segments at the end of each arm come together to form a cleft that acts as the binding site for the antigen. Both arms always have exactly the same cleft. The specificity of the antibody molecule for an antigen depends on the precise shape of these clefts. An antigen fits into one of the clefts like a hand into a glove; changes in the amino acid sequence of an antibody can alter the shape of its clefts and by doing so change the antigen that can bind to that antibody, just as changing the size of a glove will alter which hand can fit it.

*An antibody molecule recognizes a specific antigen because it possesses two clefts or depressions into which an antigen can fit, much as a substrate fits into an enzyme's active site. Changes in the amino acid sequence at the position of these clefts alter their shape and thus change the identity of the antigen that is able to fit into them.*

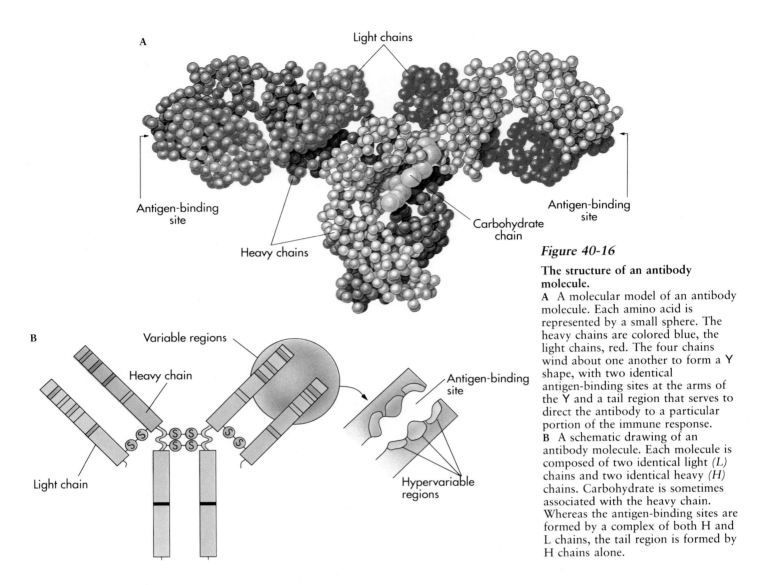

Light chains

A

Antigen-binding site

Heavy chains

Carbohydrate chain

Antigen-binding site

B

Variable regions

Heavy chain

Light chain

Antigen-binding site

Hypervariable regions

*Figure 40-16*

**The structure of an antibody molecule.**
**A** A molecular model of an antibody molecule. Each amino acid is represented by a small sphere. The heavy chains are colored blue, the light chains, red. The four chains wind about one another to form a Y shape, with two identical antigen-binding sites at the arms of the Y and a tail region that serves to direct the antibody to a particular portion of the immune response.
**B** A schematic drawing of an antibody molecule. Each molecule is composed of two identical light *(L)* chains and two identical heavy *(H)* chains. Carbohydrate is sometimes associated with the heavy chain. Whereas the antigen-binding sites are formed by a complex of both H and L chains, the tail region is formed by H chains alone.

## HOW CAN THE IMMUNE SYSTEM RESPOND TO SO MANY DIFFERENT FOREIGN ANTIGENS?

The vertebrate immune response is capable of recognizing as foreign practically any "nonself" molecule presented to it, that is, literally millions of different antigens. It is estimated that a human being is able to make between $10^6$ and $10^9$ different antibody molecules. It has long been a puzzle how this is done, since research has demonstrated that there are only a few hundred antibody-encoding genes on vertebrate chromosomes, not millions. What process, then, is responsible for generating the great diversity of antibodies? Two ideas have been proposed:

1. The **instructional theory** proposed that the antigen elicited the appropriate antibody, like a shopper ordering a custom-made suit.

2. The **clonal theory** proposed that there were indeed millions of different kinds of stem cells in the bone marrow and that an antigen caused those few encoding an appropriate antibody to proliferate, creating a clone of descendants expressing the appropriate antibody.

We now know the clonal theory to be correct. Within our bone marrow the stem cells destined to form B cells express an incredible diversity of antibody-encoding genes; each cell encodes only one form of antibody, but every cell is different from practically every other one.

How do vertebrates generate millions of different stem cells, each producing a unique antibody, when their chromosomes encode only a few hundred copies of such genes? They accomplish this by rearranging parts of the antibody-encoding genes as each B cell matures. Antibody genes do not exist as single sequences

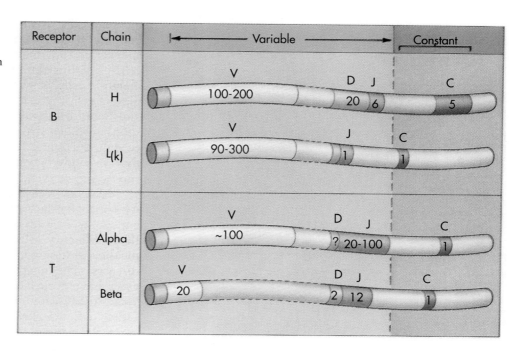

**Figure 40-17**

**The immune response library.**
The gene specifying the variable region of an antibody protein is assembled from DNA segments. Each segment typically exists in several copies—in some cases, hundreds. Thus for the B receptor chain there are several hundred copies of the "variable," or V segment, 20 copies of the "diversity," or D segment, and 6 copies of the "joining," or J segment. One specific heavy chain gene is produced in a particular stem cell by a maturation process that selects one copy of each V, D, and J segment at random and joins them to one copy of the constant region (C). A single gene encoding a specific light chain is similarly assembled from clusters of V and J segments, and the heavy and light chains are joined to form a complete B or T receptor. Millions of different combinations are possible.

of nucleotides, as do the genes encoding all other proteins, but rather are first *assembled* by joining together three or four DNA segments. Each segment, corresponding to a region of the antibody molecule, is encoded at a different site on the chromosome. These chromosomal sites are composed of a cluster of similar sequences (Figure 40-17), each sequence varying from others in its cluster by small degrees. When an antibody is assembled, one sequence is selected at random from each cluster, and the DNA sequences selected from the various clusters are brought together by DNA recombination to form a composite gene. The process is not unlike going into a large department store and picking at random one coat, one shirt or blouse, one pair of pants, one pair of socks, and one pair of shoes—few people would come out of the store wearing the same outfit.

Because a cell may end up with any heavy-chain gene and any light-chain gene during its maturation, the total number of different antibodies possible is staggering. It is:

HEAVY (16,000 combinations) × LIGHT (1200 combinations)
= 19 million different possible antibodies

Every mature stem cell divides to produce a clone of descendant lymphocytes, and each cell of the clone carries the particular rearranged gene assembled earlier, when that stem cell was undergoing maturation.

As a result, all the cells of a clone produce the specific immune receptor encoded by that stem cell, and no other. Because an adult vertebrate contains many millions of stem cells and each stem cell undergoes the maturation process independently, millions of different cells exist in each of us, each one of them specializing in a different antibody.

*Antibodies are encoded by genes that are assembled during stem cell maturation by rearrangement of the DNA. Because each component is selected at random from many possibilities, a vast array of different antibodies is produced.*

## THE NATURE OF IMMUNITY

When a particular B cell is stimulated by an invading microbe to begin dividing, producing a clone of proliferating cells with the same antibody, all these identical cells do not go on to become plasma cells. Instead, many persist as circulating lymphocytes called **memory cells.** These provide an accelerated response to any later encounter with the stimulating antigen, because

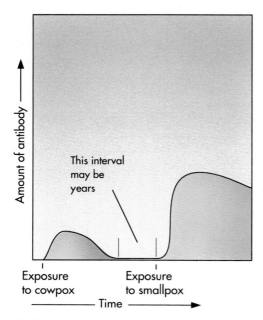

**Figure 40-18**

**Immunity.** Immunity to smallpox in Jenner's patients occurred because their innoculation with cowpox stimulated their bodies to produce antibodies that would recognize either cowpox or smallpox. Some of these antibodies, and the cells that produce them, remain in the bloodstream for a long time. A second exposure, this time to smallpox, stimulates the body to produce large amounts of the antibody much more rapidly than before.

there are now many cells that can respond to it, rather than a few.

## The Primary and Secondary Immune Response

The first time a particular kind of pathogen invades the vertebrate body, there are only a few B cells that by chance may have the antibody that can recognize it. The immune response that this first encounter sets off is called a **primary response**. It takes several days for these few cells to form a clone of cells that will produce antibody. The *next time* the body is invaded by the same pathogen, however, the immune system is ready; as a result of the first infection, there are now a small army of B cells that can recognize that pathogen—the

memory cells. Remember, only some of the clones of dividing B cells became plasma cells in the first response; all the others are still there, a host of memory cells patrolling the bloodstream. Because each of these memory cells is well along the road to becoming a plasma cell, the **secondary immune response** is swifter, and because there are so many more of them, the response is much stronger (Figure 40-18). With each succeeding encounter, the bank of memory cells carrying that antibody becomes larger, so that the immune response grows even quicker and stronger.

Memory cells can survive for several decades, which is why most of us rarely contract mumps, chickenpox, or measles a second time once we have had them. Memory cells are also why vaccination against measles, polio, and smallpox are effective against these diseases. The microbes causing these childhood diseases have a surface that changes little from year to year, so the same antibody is effective decades later. Other diseases like flu are caused by microbes whose surface-specifying genes mutate rapidly; thus new strains appear every year or so that are not recognized by memory cells from previous infections. That is why immunity to flu lasts only a few years—the memory cells persist, but the antigen continually shifts to new forms. The ability of flu to defeat our immune defenses by constantly changing its surface is but one of several strategies pathogens have employed to defeat the vertebrate immune system.

## DEFEAT OF THE IMMUNE SYSTEM

All mammals and birds possess an immune system similar to the human one described in this chapter. During the evolutionary history of the vertebrates, however, several microbes have developed strategies, some of them quite successful, for defeating vertebrate immune defenses (Figure 40-19). As you might expect, these strategies are responsible for very serious diseases.

One of these strategies consists of a direct attack on the immune mechanism itself. If you had to design

A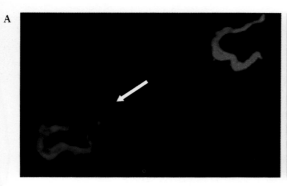

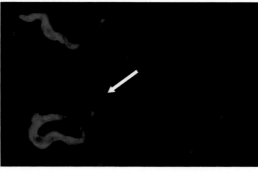

B **Figure 40-19**

**How a trypanosome escapes the immune system.** The trypanosomes in **A** express a particular surface antigen; antibody directed against that antigen causes them to fluoresce red. The trypansomes in **B** express a different surface antigen, to which a green-fluorescing antibody has been bound. The individual in the lower left is in the process of switching its coat antigen and so is labeled with both antibodies.

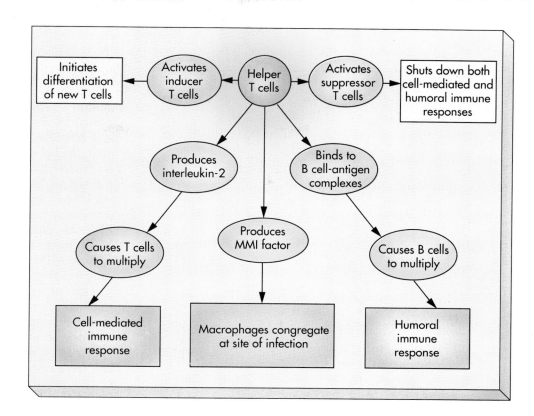

**Figure 40-20**

The many roles of the helper T cell.

such an attack, perhaps the most sensitive target would be the helper T cells. Helper T cells are the key to the entire immune response (Figure 40-20), responsible for initiating proliferation of both T cells and B cells. Without helper T cells, the immune system is unable to mount a response to *any* foreign antigen.

AIDS is a deadly disease for just this reason. The AIDS virus mounts a direct attack on all cells that have a specific protein called **CD4** on their surfaces. It is found on a class of lymphocyte cells called T4 cells, which includes helper T cells, which is why the AIDS virus destroys the body's population of helper T cells. CD4 receptors are also found on the surface of macrophages, and as a result macrophages also become infected with the AIDS virus. Much of the transmission of AIDS from one individual to another is thought to occur within macrophages passed as part of body fluids.

AIDS-infected T4 cells die, but only after releasing progeny viruses (Figure 40-21) that infect other T4 cells, until the entire population of T4 cells is destroyed. In a normal individual, T4 cells make up 60% to 80% of circulating T cells; in AIDS patients T4 cells often become too rare to detect (Figure 40-22).

The effect of T4 cell destruction by AIDS infection is to wipe out the human immune defense. An effective immune response, either a T cell–mediated cellular immune response or a B cell–mediated antibody response, is impossible without helper T cells to initiate the response. With no defense against infection, any of

a variety of otherwise commonplace infections proves fatal. It is for this reason that AIDS is a particularly devastating disease.

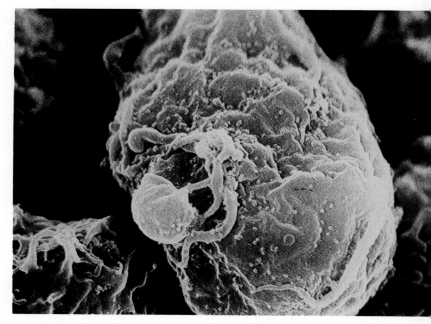

**Figure 40-21**

**AIDS virus.** AIDS viruses released from infected T cells, the tiny balls in this micrograph, soon spread over neighboring helper T cells, infecting them in turn. The individual AIDS particles are very small; over 200 million would fit on the period at the end of this sentence.

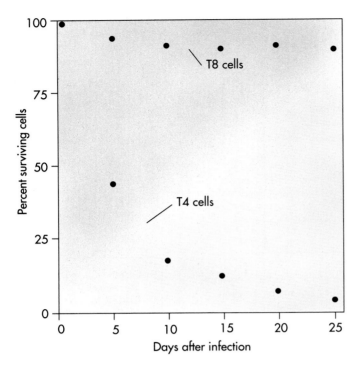

**Figure 40-22**

**Survival of T cells in culture after exposure to AIDS virus.** The virus has little effect on the number of $T_8$ cells, but it causes the number of $T_4$ cells to decline dramatically.

---

*AIDS destroys the ability of the immune system to mount a defense against any infection. The AIDS virus attacks and destroys T4 cells, without which no immune response can be initiated.*

---

Although the AIDS virus became a prominent cause of human disease only recently, possibly transmitted to humans from African green monkeys in Central Africa, it is already clear that AIDS is one of the most serious diseases in human history (Figure 40-23). The fatality rate of AIDS is 100%; no patient exhibiting the symptoms of AIDS has ever been known to survive more than a few years. The disease is *not* highly infectious; it is transmitted from one individual to another during the direct transfer of internal body fluids, typically in semen or vaginal fluid during sex and in blood during transfusions or by contaminated hypodermic needles. Not all individuals exposed to AIDS (as judged by antibodies in their blood directed against AIDS virus) have yet come down with the disease.

There were about 1.5 million to 2 million such individuals in the United States by the mid-1980s, and it is estimated that at least 30% to 50%, and possibly all, of them will eventually contract AIDS. Most will die within 2 years of onset of the symptoms unless additional strategies for the treatment of AIDS are discovered first.

Efforts to develop a vaccine against AIDS continue, both by splicing portions of the AIDS surface protein gene into the vaccinia virus (Figure 40-24) and by attempting to develop a harmless strain of AIDS. These approaches, although promising, have not yet proved successful and are limited by the fact that different strains of

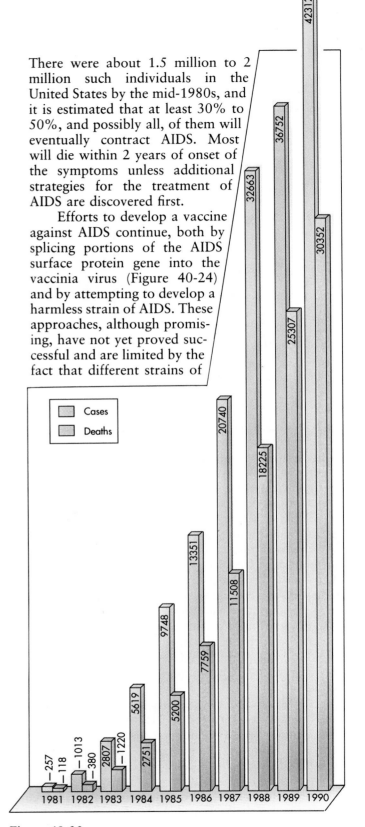

**Figure 40-23**

**The AIDS epidemic.** The U.S. Public Health Service estimates that by 1992, there will have been 179,000 deaths in the United States caused by AIDS.

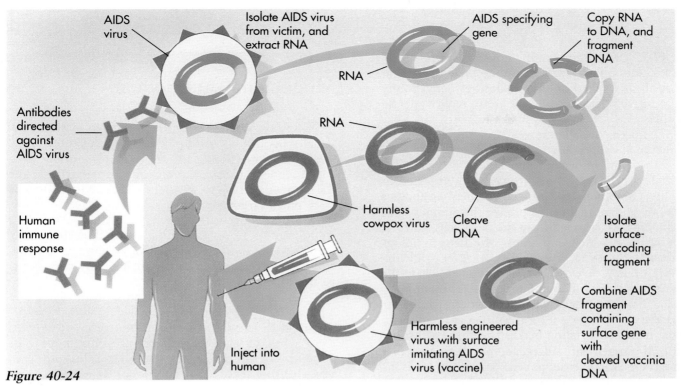

**Figure 40-24**

**How scientists are attempting to construct a vaccine for AIDS.** Of the several genes of the human immunodeficiency virus (HIV), one is selected that encodes a surface feature of the virus. All of the other HIV genes are discarded. This one gene is not in itself harmful to humans; it is simply the shape of one of the HIV surface proteins. This one gene, or fragment of it, is inserted into the DNA of a harmless vaccinia cowpox virus, resulting in a harmless virus whose surface immitates the AIDS virus. Persons injected with the vaccinia virus do not become ill (the vaccinia virus is harmless), but they do develop antibodies directed against the infecting virus surface. Because the surface contains HIV proteins, the new antibodies would serve to protect the infected person against any subsequent exposure to the HIV virus.

AIDS virus seem to possess different surface antigens. Like flu, AIDS indulges in some form of antigen shifting. Recent studies have described a stable coat protein that is a prime candidate for a vaccine. Drugs that inhibit specific genes of the AIDS virus or block the operation of enzymes critical to the synthesis of viral RNA are also being investigated.

## ALLERGY

Although the human immune system provides very effective protection against viruses, bacteria, parasites, and other microorganisms, sometimes it does its job too well, mounting a major defense against a harmless antigen. Such immune responses are called **allergic reactions**. Hay fever, the sensitivity that many people exhibit to proteins released from plant pollen, is a familiar example of an allergy. In response to as little as 20 pollen grains per cubic meter, a sensitive person's immune defense will swiftly mount a defense. Many other people are sensitive to proteins released from the feces of a minute house-dust mite called *Dermatophagoides*

**Figure 40-25**

The house-dust mite *Dermphagoides*.

(Figure 40-25), which lives in the house dust present on mattresses and pillows and consumes the dead skin cells that all of us shed in large quantities daily. Many people sensitive to feather pillows are in reality allergic to the mites that are residents of the feathers.

What makes an allergic reaction uncomfortable, and sometimes dangerous, is the involvement of antibodies with a kind of heavy chain called "E." Antibodies with class E heavy chains are typically attached to mast cells. The binding of antigen to these antibodies initiates an inflammatory response; histamines and prostaglandins are released from the mast cells, causing dilation of blood vessels and a host of other physiological changes: sneezing, runny noses, fever—all the symptoms of hay fever. In some instances when the body possesses substantial amounts of class E antibody directed against an antigen, allergic reactions can be far more dangerous than hay fever, resulting in anaphylactic shock, in which swelling makes breathing difficult.

Not all antigens are **allergens,** initiators of strong immune responses. Nettle pollen, for example, is as abundant in the air as ragweed pollen, but few people are allergic to it. Also, all people do not develop allergies; the sensitivity seems to run in families. It seems that allergies require both a particular kind of antigen and a high level of class E antibody. The antigen must be able to bind simultaneously to two adjacent E antibodies on the surface of the mast cell in order to trigger the mast cell's inflammatory response, and only certain antigens are able to do this. The class E antibodies must be produced in large enough amounts that many mast cells will have antibody molecules spaced close to one another, rather than the few such cells typical of a normal immune response. Only certain people churn out these high levels of E antibody. It is this combination of appropriate antigen on the one hand and inappropriately high levels of particular class E antibodies on the other that produce the allergic response.

Hay fever and other allergies are often treated by injecting sufferers with extracts of the antigen, a process called desensitization. Allergy shots work best for pollen allergies and for allergy to the venom of bee and wasp stings; they are not effective against food or drug allergies. The strategy of desensitization is to produce high levels of normal (class G) antibody in the bloodstream, so that when a particular antigen is encountered, it will be mopped up by the G antibodies before encountering E antibodies on mast cells. Actually, there seems to be little correlation between levels of circulating G antibody and successful desensitization, and it is not clear why the procedure works as well as it does. A more ideal therapy would be to lower the amounts of class E antibody produced during the immune response, an approach that is being actively investigated.

# Monoclonal Antibodies

Only a small portion of an antigen molecule actually fits into an antibody's recognition site. In the case of protein antigens, for example, this portion (sometimes called a **determinant**) is typically in the size range of two to six amino acids. Most antigens are much larger than this; as a result, different portions of the antigen molecule can fit into different antibody sites. A typical antigen will thus elicit many different antibodies, each fitting a different portion of the antigen surface. Such an antibody response is said to be **polyclonal.**

Antibodies offer great promise in medicine and research, because they recognize biological molecules with exquisite precision. However, it is often critical that biological tools be specific in order to be useful, just as a letter must have a specific address to reach its destination. A polyclonal response presents a whole phone book of potential addresses; to find one address, an investigator needs instead an antibody that is directed against only one determinant (a **monoclonal antibody**). In 1984 Cesar Milstein of England and George Khohler of Switzerland were awarded a Nobel Prize for learning how to engineer an antibody response that is monoclonal. They devised an easy procedure for isolating a single clone of plasma cells, all producing the same antibody molecule.

Milstein and Khohler mixed plasma cells that were producing antibody with cancer cells, malignant lymphocytes called myelomas (Figure 40-A). Neither plasma cells nor myeloma cells lived long when Milstein and Khohler mixed them together: plasma cells normally live only a few cell generations anyway, and the cancer cells they employed were mutant ones, unable to survive without a certain metabolite the cells could no longer synthesize. Some cells did live in the mixture, however, growing and dividing—these were a new kind of cell produced by the fusion of a plasma cell with a myeloma cell. These cell hybrids, called **hybridomas** (Figure 40-B), were directed by genes of the plasma cell to make the metabolite necessary for growth and were directed by the genes of the myeloma to grow and divide ceaselessly, as cancer cells. What made the experiment of profound importance was that the hybridoma cells continued to produce the antibody in which the plasma cell had specialized. Isolating sin-

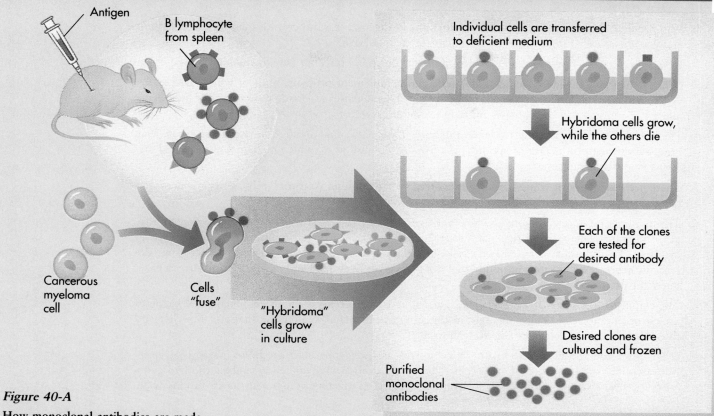

**Figure 40-A**

**How monoclonal antibodies are made.**

gle hybridoma cells from the mixture, Milstein and Khohler obtained rapidly growing cell lines that could be maintained in culture indefinitely, with every cell of the culture producing the same antibody molecule—monoclonal antibodies.

Monoclonal antibodies have proved to be of great importance to industry because they can be used to purify specific molecules from complex mixtures. Interferon, present in only trace amounts in tissue extracts, was first purified in this way. Monoclonal antibodies have also revolutionized many aspects of biological research. It has proved possible, for example, to generate monoclonal antibodies directed against each of the many proteins that protrude from a cell's

surface and in this way to learn a great deal about cell surface receptors. The T receptor, which plays such an important role in this discussion, was first isolated in 1984 by investigators employing a monoclonal antibody. In medicine, monoclonal antibodies offer great promise as vehicles for delivering specific therapies. There is an intensive search underway, for example, for antigens that occur only, or predominantly, on cancer cells, against which radioactive monoclonal antibodies could be targeted to kill cancer cells selectively.

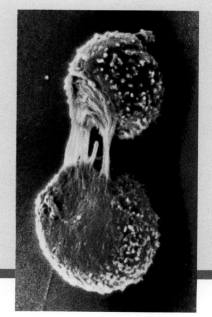

**Figure 40-B**

**A hybridoma dividing.**

# ■ SUMMARY

1. Skin provides the vertebrate body with its first line of defense. It is difficult for microbes to penetrate and presents a very inhospitable environment for their survival.

2. Phagocytes are the vertebrate body's second line of defense, attacking any invading microbe.

3. Other nonspecific defenses include killer cells and complement proteins, both of which destroy invaders by punching holes in their cell membranes. These defenses are made more effective by inflammatory responses, and by increases in body temperature.

4. The immune response is the introduction of foreign proteins, called antigens, into a vertebrate that brings about the production of specific proteins, called antibodies, which recognize the original antigens from that point on.

5. There are four types of white blood cell involved in the immune system: T cells, B cells, macrophages and other phagocytes, and natural killer cells.

6. The immune response is initiated by helper T4 cells. Helper T4 cells, when stimulated by macrophages, simultaneously activate two parallel responses: the cell-mediated immune response, in which cytotoxic T cells attack infected body cells, and the humoral immune response, a longer-range defense in which B cells secrete antibody that binds circulating antigen and marks cells or viruses bearing it for destruction.

7. The exquisite specificity of antibodies for particular antigens reflects the three-dimensional shape of a cleft in the molecule. Slight changes in the amino acid sequence alter the shape of the cleft and thus the identity of molecules able to fit into it.

8. Vertebrates can recognize many different antigens as foreign because the bone marrow of vertebrates contains many different stem cells. During maturation, a stem cell assembles the two genes encoding its particular antibody or T receptor by splicing together component parts, randomly selecting each part from a large library of possibilities.

## REVIEW

1. The procedure, first used by Edward Jenner, of injecting a harmless microbe that is similar to a harmful one so that the body will produce antibodies that will recognize the harmful microbe if it is encountered sometime later, is called _____.

2. The two types of lymphocytes are _____ and _____.

3. An antibody molecule recognizes a specific _____ because the antibody molecule possesses specific binding sites formed by a complex of heavy and light chains.

4. The first kind of immune system cell to respond to an invading virus is either a killer cell or a _____.

5. Allergies require both a particular type of antigen and a high level of class _____ antibody.

## SELF-QUIZ

1. When Pasteur injected a 2-week-old culture of fowl cholera bacteria into chickens, the vaccinated birds did not die. What would have happened if he had heat-killed the 2-week-old culture before injecting it?
   (a) All the vaccinated chickens would have died
   (b) The heat killing would not have changed the result
   (c) Some of the vaccinated chickens would have died
   (d) None of the above

2. Which of the following occurs *last* in an immune response to infection?
   (a) Macrophages ingest and destroy infected cells
   (b) Natural killer cells attack infected cells
   (c) B cells proliferate
   (d) Macrophages secrete chemicals that attract other macrophages
   (e) Helper T cells are mobilized

3. How do vertebrates generate millions of different stem cells that each produce a unique receptor, when their chromosomes encode only a few hundred copies of these genes?
   (a) By mutation
   (b) By rearrangement of segments of DNA
   (c) By antigen-elicited receptor production
   (d) By meiotic recombination
   (e) None of the above

4. The AIDS virus is remarkably effective at short-circuiting the immune response because it infects what cell type?
   (a) B cells
   (b) T4 lymphocytes
   (c) Red blood cells
   (d) Mast cells
   (e) Stem cells

# THOUGHT QUESTIONS

1. Why do you suppose the human immune system encodes only about a hundred basic antibody-encoding genes, when much more diversity could be generated by encoding thousands of copies of each?

2. AIDS is a virus that destroys the human immune system by killing helper T cells, which are necessary to activate the immune response. The African green monkeys from which the AIDS virus is thought to have arisen do not suffer from AIDS. How do you imagine they have escaped this?

# FOR FURTHER READING

GALLO, R.C., and L. MONTAGNIER: "AIDS in 1988," *Scientific American*, October 1987, page 40. In their first collaboration the two investigators who established the cause of AIDS describe how HIV was isolated and linked to AIDS, current status of AIDS research, and the prospects for an AIDS therapy.

JARET, P.: "Our Immune System: The Wars Within," *National Geographic*, June 1986, pages 702-736. A very readable account of current progress in the study of the human immune system, with striking photographs by Lennart Nilsson.

LAWRENCE, J.: "The Immune System in AIDS," *Scientific American*, December 1985, pages 84-93. An excellent overview of how T cells function in the immune response and how AIDS thwarts that response.

SMITH, K.A.: "Interleukin-2," *Scientific American*, March 1990, pages 50-57. The first hormone of the immune system to be recognized, it helps the body to mount a defense against microorganisms by triggering the multiplication of only those cells which attack an invader.

TONEGAWA, S.: "The Molecules of the Immune System," *Scientific American*, October 1985, pages 122-131. An account of what is currently known of the structure of the B and T receptors.

# C H A P T E R ■ 4 1

# *The Control of Water Balance*

All animals need water—some, like this hippo, luxuriate in it. Others, like the kangaroo rat, never drink. All animals, however, need to maintain a proper water balance necessary for life.

# THE CONTROL OF WATER BALANCE

## Overview

Vertebrates live in salt water, in fresh water, and on land, and each of these environments poses different problems for balancing water retention with proper salt concentration. This is known as osmoregulation and is one important function of the kidneys. Vertebrates conserve or excrete water, depending on the environment in which they live, by regulating the passage of water through their excretory system. A second, equally important function of the kidneys is the maintenance of electrolyte and acid-base homeostasis. A third is the elimination of waste products.

## For Review    *Here are some important terms and concepts that you will encounter in this chapter. If you are not familiar with them, you should review them before proceeding.*

**Sodium chloride** (Chapter 2)

**Diffusion** (Chapter 5)

**Hypotonic solutions** (Chapter 5)

**Hormone regulation of water retention** (Chapters 32 and 36)

**Figure 41-1**

**Water affects us all.** This African elephant is knee-deep in water, and enjoys its tromp through the mud every bit as much as you might enjoy going to the beach. For him, and us, water is both a necessity and a pleasure.

The first vertebrates evolved in water, and the physiology of all members of the subphylum still reflects this origin (Figure 41-1). Approximately two thirds of every vertebrate's body is water. If the amount of water in the body of a vertebrate falls much lower than this, the animal will die. In this chapter we discuss the various strategies animals employ to keep from losing or gaining too much water. We might have chosen the evolution of the heart or another system to illustrate the increasing complexity of the adaptation of vertebrates to their varied modes of existence. Instead, we have chosen to consider the ways in which they control their water balance because these strategies are so closely tied to their exploitation of the varied environments where they occur (Figure 41-2).

## OSMOREGULATION

Plasma membranes are freely permeable to water but have very low permeability to salts and ions. This property of differential permeability forms the basis for many life processes, including the nerve conduction that we discussed in Chapter 33. If the concentration of salts and ions dissolved in the water surrounding a vertebrate's body were the same as that within the body, the differential permeability of the cells would present no problem; there would be no tendency for

**Figure 41-2**

**A kangaroo rat never drinks.** The kangaroo rat, *Dipodomys panamintensis*, has very efficient kidneys that can concentrate urine to a high degree by reabsorbing water. As a result, it avoids losing any more moisture than necessary. Because kangaroo rats live in dry or desert habitats, this feature is extremely important to them.

water to leave or enter the body. The osmotic pressure of the body fluids would be essentially the same as that of the surroundings. This is true of most marine invertebrates, but only of sharks and their relatives among the vertebrates. Such animals, called **osmoconformers,** maintain the osmotic concentration of their body fluids at about the same level as that of the medium in which they are living, and they change the osmotic concentrations of their body fluids when the osmotic concentration of the medium changes.

## The Problems Faced by Osmoregulators

Among the aquatic vertebrates other than sharks, mechanisms have evolved by which they control the osmotic concentration of their body fluids more precisely than other aquatic animals. In other words, they are osmoregulators. **Osmoregulators** maintain an internal solute concentration that does not vary, regardless of the environment in which the vertebrate lives. The maintenance of a constant internal solute concentration has permitted the vertebrates to evolve complex patterns of internal metabolism. It does, however, require constant regulation of the animal's internal water concentration.

Freshwater vertebrates must maintain much higher concentrations of salts in their bodies than those in the water surrounding them. In other words, their body fluids are **hypertonic** relative to their environment, and

water tends to enter their bodies. They must therefore exclude water to prevent the fluids within them from being diluted.

Marine vertebrates have only about one-third the osmotic concentration of the surrounding seawater in their body fluids. Their body fluids are therefore said to be **hypotonic** relative to the environment in which the animals live, and water tends to leave their bodies. For this reason, marine animals must retain water to prevent dehydration.

On land, surrounded by air, the bodies of vertebrates have a higher concentration of water than does the air surrounding them. They therefore tend to lose water to the air by evaporation. This situation is faced to some degree by the amphibians, which live on land only part of the time. It is also faced by all terrestrial reptiles, birds, and mammals, which must conserve water to prevent dehydration. Figure 41-3 shows the concentration of ions in the body fluids of some representative vertebrates.

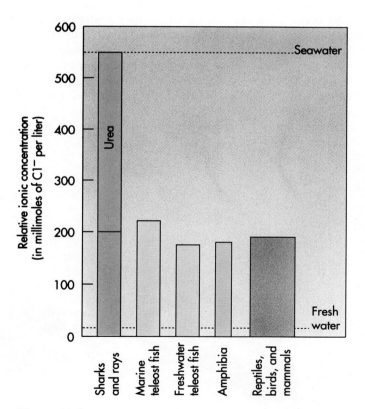

**Figure 41-3**

**The concentration of ions is roughly similar in the bodies of different classes of vertebrates.** Sharks hold the concentration of solutes in their blood at about the level in seawater, or at a slightly higher level, by adding urea to their bloodstream. Terrestrial vertebrates have ion concentrations not unlike those of the fishes from which they evolved.

## How Osmoregulation is Achieved

Animals have evolved a variety of mechanisms to cope with these problems of water balance, all of them based in one way or another on the animal's excretory system. In many animals the removal of water or salts is coupled to the removal of metabolic wastes from the body. Simple organisms, such as many protists and sponges, employ contractile vacuoles for this purpose. Many freshwater invertebrates employ **nephrid organs,** in which water and waste pass from the body across the membrane into a collecting organ, from which they are ultimately expelled to the outside through a pore. The membrane acts as a filter, retaining proteins and sugars within the body, while permitting water and dissolved waste products to leave.

Insects use a similar filtration system, with a significant improvement that helps them guard against water loss. The excretory organs in insects are the **Malpighian tubules.** Malpighian tubules (Figure 41-4) are tubular extensions of the digestive tract that branch off before the hindgut. Potassium ion is secreted into the tubules, causing body water and organic wastes to flow into them from the body's circulatory system because of the osmotic gradient. Blood cells and protein are too large to pass across the membrane into the Malpighian tubules. Because the system of tubules empties into the hindgut, however, the water and potassium can be reabsorbed by the hindgut, and only small molecules and waste products are excreted from the insect. Malpighian tubules provide a very efficient means of water conservation.

Like insects, vertebrates use a strategy that couples water balance and salt concentration with waste excretion. Instead of relying on the secretion of salts into the excretory organ to establish an osmotic gradient, however, vertebrates rely on pressure-driven filtration. Whereas insects use an osmotic gradient to *pull* the blood through the filter, the vertebrates *push* blood through the filter, using the higher blood pressure that a closed circulatory system makes possible. Fluids are forced through a membrane that retains proteins and large molecules within the body but passes the small molecules out. Water is then reabsorbed from the filtrate as it passes through a long tube.

*Terrestrial animals require efficient means of water conservation. In eliminating metabolic wastes, both insects and vertebrates filter the blood to retain blood cells and protein; then the water is collected from the filtrate by passing the filtrate through a long tube across whose walls it is reabsorbed.*

Like the osmotically collected waste fluid of insects, the fluid that is passed through the filter of vertebrates contains many small molecules that are of value to the organism, such as glucose, amino acids, and various salts or ions. Vertebrates have evolved a means of

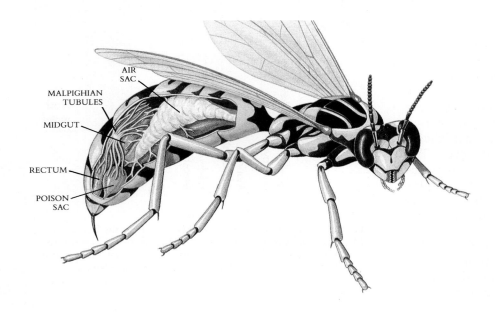

**Figure 41-4**

**Malpighian tubules.** The Malpighian tubules of insects are extensions of the hindgut that collect fluid from the body's circulatory system. Water is later reabsorbed, whereas wastes in the fluid are eliminated.

AIR SAC

MALPIGHIAN TUBULES

MIDGUT

RECTUM

POISON SAC

selectively reabsorbing these valuable small molecules without absorbing the waste molecules that are also dissolved in the filtered waste fluid or **urine.** Selective reabsorption gives the vertebrates great flexibility, since the membranes of different groups of animals can and have evolved different transport channels and thus the ability to reabsorb different molecules. This flexibility is a key factor underlying the ability of different vertebrates to function in many diverse environments. They can reabsorb small molecules that are especially valuable in their particular habitat and not absorb wastes. Among the vertebrates, the apparatus that carries out the processes of filtration, reabsorption, and secretion is the **kidney.** It can function, with modifications, in fresh water, in the sea, and on land.

## THE ORGANIZATION OF THE VERTEBRATE KIDNEY

The kidney is a complex structure of repeating elements called **nephrons,** each of which has a tubular and a vascular component. The outer layer of the kidney is called the cortex, and the inner region, the medulla. From the medulla a series of converging tubes leads to the ureter and, ultimately, to the bladder. All vertebrate kidneys carry out three functions:

1. *Filtration,* in which blood is passed through a filter that retains blood cells and proteins but passes water and small molecules such as amino acids, glucose, and salts.
2. *Reabsorption,* in which desirable ions and metabolites are recaptured from the filtrate, leaving metabolic wastes such as urea and water behind for later elimination. Organisms have expended a great deal of energy to obtain these molecules, and they need to be reclaimed.

3. *Secretion,* in which the kidney first secretes and then excretes $K^+$, $H^+$, $NH_4^+$, and certain drugs and foreign organic materials.

### Filtration
The filtration device of the vertebrate kidney consists of a large number of individual tubular filtration-reabsorption devices called nephrons (Figure 41-5). At the front end of each nephron tube is a filtration apparatus called a **Malpighian corpuscle.** In each Malpighian corpuscle, an arteriole enters and splits into a fine network of vessels called a **glomerulus** (Figure 41-6). It is the walls of these capillaries that act as a filtration device. Blood pressure forces fluid through the capillary walls, which are differentially permeable. These walls withhold the proteins and other large molecules, while passing water and small molecules such as glucose, ions, and ammonia, the primary nitrogenous waste product of metabolism.

### Reabsorption
At the back end of each nephron tube is a reabsorption device that operates like the mammalian small intestine, discussed in Chapter 37. The fluid that passes out of the capillaries of each glomerulus, called the **glomerular filtrate,** enters the tube portion of that nephron within which reabsorption takes place.

### Excretion
A major function of the kidney is the elimination of a variety of potentially harmful substances that animals eat, drink, or inhale. In this process the roughly 2 million nephrons that form the bulk of the two human kidneys receive a flow of approximately 2000 liters of blood per day. The mammalian kidney is able to concentrate urine with a salt concentration well above that of the blood. As a part of the process, the kidney is

*Figure 41-5*

**The basic organization of the vertebrate nephron.** The nephron tube of the freshwater fish is a basic design that has been retained in the kidneys of marine fishes and terrestrial vertebrates, which evolved later. Sugars, small proteins, and ions such as $Ca^{++}$ and $PO_4$ are recovered in the proximal arm; ions such as $Na^+$ and $Cl^-$ are recovered in the distal arm; and water is recovered in the collecting duct.

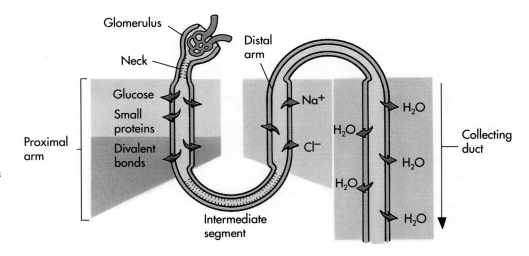

***Figure 41-6***

**Bowman's capsules.** The spherical structures in this micrograph are Bowman's capsules. In each, a fine network of capillaries, the glomerulus, is connected to a nephron tube by an arteriole.

able to build up the concentration of such materials as $H^+$, $K^+$, $NH_4^+$, drugs, and various foreign organic materials in the urine, and thus to excrete them from the body.

## EVOLUTION OF THE VERTEBRATE KIDNEY

The same basic design has been retained in all vertebrate kidneys, although there have been some changes. In the following sections we discuss the structure and function of the kidney in the different major groups of vertebrates.

### Freshwater Fishes

Kidneys are thought to have evolved first among the freshwater fishes. A freshwater fish drinks little and produces large amounts of urine. Because the body fluid of a freshwater fish is hyperosmotic as compared with the water in which the fish lives, water is not reabsorbed in its nephrons. The excess water that enters its body passes instead through the nephron tubes to the bladder, from which it is eliminated as urine. Within the urine is not only the excess water but also all the small molecules that were not reabsorbed while passing through the nephron tubes. Notable among these molecules is ammonia, the principal metabolic

waste product of nitrogen metabolism (present in solution as ammonium ion, $NH_4^+$). Ammonia in higher concentrations is toxic, but because the urine contains so much water, the concentration of ammonia is low enough not to harm the fish.

### Marine Fishes

Although most groups of animals clearly seem to have evolved first in the sea, marine bony fishes probably evolved from freshwater ancestors, as was mentioned in Chapter 19. In making the transition to the sea, they faced a significant new problem of water balance, because their body fluids are hypoosmotic with respect to the water that surrounds them. For this reason, water tends to leave their bodies, in which the fluids are less concentrated osmotically than is seawater. To compensate, marine fishes drink a lot of water, excrete salts instead of reabsorbing them, and reabsorb water. This places radically different demands on their kidneys than those faced by freshwater fishes, thus turning the tables on an organ that had originally evolved to eliminate water and reabsorb salts. As a result, the kidneys of marine fishes have evolved important differences from their freshwater relatives. For example, they possess active ion transport channels that conduct ions with two or three charges—molecules that are particularly abundant in seawater—*out* of the body and *into* the tube. In the sea, the water that the fish drinks is rich in ions such as $Ca^{++}$, $Mg^{++}$, $SO_4^=$, and $PO_4^=$, all of which must be excreted. Because of the functioning of these ion transport channels, the direction of movement is reversed compared with that found in the kidneys of their freshwater ancestors.

### Sharks

Except for one species of shark found in Lake Nicaragua, all sharks, rays, and their relatives live in the sea (Figure 41-7). Some of these members of the class Chondrichthyes have solved the osmotic problem posed by their environment in a different way than have the bony fishes. Instead of actively pumping ions out of their bodies through their kidneys, these fishes use their kidneys to reabsorb the metabolic waste product urea, creating and maintaining urea concentrations in their blood 100 times as high as those that occur among the mammals. As a result, the sharks and their relatives become isotonic with the surrounding sea. They have evolved enzymes and tissues that tolerate these high concentrations of urea. Because they are isotonic with the water in which they swim, they avoid the problem of water loss that other marine fishes face. Since sharks do not need to drink large amounts of seawater, their kidneys do not have to remove large amounts of divalent ions from their bodies.

*Figure 41-7*

**A great white shark.** Although the seawater in which it is swimming contains a far higher concentration of ions than does the shark's body, the shark avoids losing water osmotically by maintaining such a high concentration of urea that its body fluids are osmotically similar to the sea around it.

## Amphibians and Reptiles

The first terrestrial vertebrates were the amphibians; the amphibian kidney is identical to that of the freshwater fishes, their ancestors. This is not surprising, since amphibians spend a significant portion of their time in fresh water, and when on land they generally stay in wet places.

Reptiles, on the other hand, live in diverse habitats, many of them very dry. The reptiles that live mainly in fresh water, like some kinds of crocodiles and alligators, occupy a habitat similar to that of the freshwater fishes and amphibians and have similar kidneys. Marine reptiles, which consist of some crocodiles, some turtles, and a few lizards and snakes, possess kidneys similar to those of their freshwater relatives. They eliminate excess salts not by kidney excretion but rather by means of salt glands located near the nose or eye.

Terrestrial reptiles, which must conserve water to survive, reabsorb much of the water in the kidney filtrate before it leaves the kidneys, and so excrete a concentrated urine. This urine cannot, however, become any more concentrated than the blood plasma; otherwise, the body water of reptiles would simply flow into the urine while it was in the kidneys. Ammonia is toxic and must be excreted in a dilute form. This would cause excessive fluid loss in a terrestrial organism. The solution is to convert ammonia to a less toxic substance. In the relatively concentrated urine of most reptiles, therefore, nitrogenous waste is no longer excreted in the form of ammonia, but either as the solid urea or as uric acid. These metabolic conversions take place in the liver.

## Mammals and Birds

Your body possesses two kidneys, each about the size of a small fist, located in the lower back region (Figure 41-8). These kidneys represent a great increase in efficiency compared with those of reptiles. Mammalian, and to a lesser extent avian, kidneys can remove far more water from the glomerular filtrate than can the kidneys of reptiles and amphibians. Human urine is four times as concentrated as blood plasma. Some desert mammals achieve even greater efficiency: a camel's urine is 8 times as concentrated as its plasma, a gerbil's 14 times as concentrated, and some desert rats and mice have urine more than 20 times as concentrated as their blood plasma (see Figure 41-2). Mammals and birds achieve this remarkable degree of water conservation by using a simple but superbly designed mechanism: they greatly increase the local salt concentration in the tissue through which the nephron tube passes and then use this osmotic gradient to draw the water out of the tube.

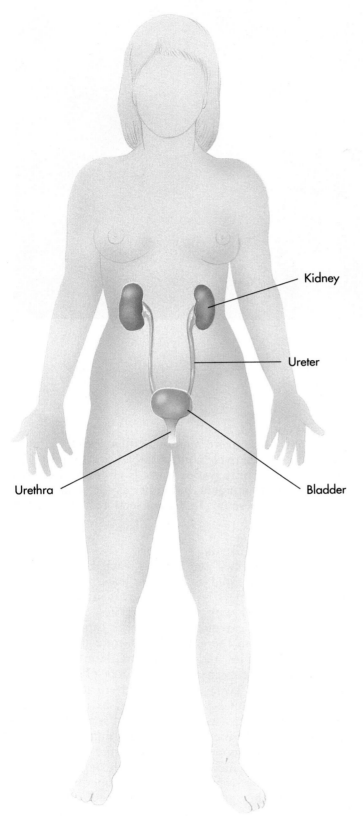

**Figure 41-8**

The urinary system of the human female.

## HOW THE MAMMALIAN KIDNEY WORKS

Remarkably, mammals and birds have brought about this major improvement in efficiency by a very simple change—they bend the nephron tube. A single mammalian kidney contains about a million nephrons. Each of these is composed of a glomerulus that is connected to a nephron tube, which is called a **renal tubule** in mammals. Between its proximal and distal segments, each renal tubule is folded into a hairpin loop called the **loop of Henle** (Figure 41-9).

The kidney uses the hairpin loop of Henle to set up a countercurrent flow. Just as in the gills of a fish, as discussed in Chapter 38, but with water being absorbed instead of oxygen, countercurrent flow enables water reabsorption to occur with high efficiency. In general, the longer the hairpin loop, the more water can be reabsorbed. Animals such as desert rodents that have highly concentrated urine have exceptionally long loops of Henle. The process involves the passage of *two* solutes across the membrane of the loop: salt (NaCl) and urea, the waste product of nitrogen metabolism. It has long been known that animals fed high-protein diets, yielding large amounts of urea as waste products, can concentrate their urine better than animals excreting lower amounts of urea, a clue that urea plays a pivotal role in kidney function. Figure 41-10 illustrates the flow of materials discussed in the following paragraphs.

1. Filtrate from the glomerulus passes down the descending loop. The walls of this portion of the tubule are impermeable to either salt or urea but are freely permeable to water. Because (for reasons we describe soon) the surrounding tissue has a high osmotic concentration of urea, water passes out of the descending loop by osmosis, leaving behind a more concentrated filtrate.
2. At the turn of the loop, the walls of the tubule become permeable to salt, but much less permeable to water. As the concentrated filtrate passes up the ascending arm, salt passes out into the surrounding tissue by diffusion. (The surrounding tissue, although it has much urea, does not contain as much salt as the concentrated filtrate.) This makes salt more concentrated within the bottom of the loop and in the surrounding tissues.
3. Higher in the ascending arm, the walls of the tubule contain active-transport channels that pump out even more salt. This active removal of salt from the ascending loop encourages even more water to diffuse outward from the filtrate.

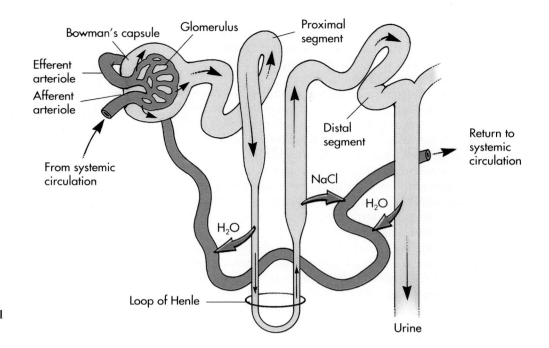

**Figure 41-9**

Organization of a mammalian renal tubule.

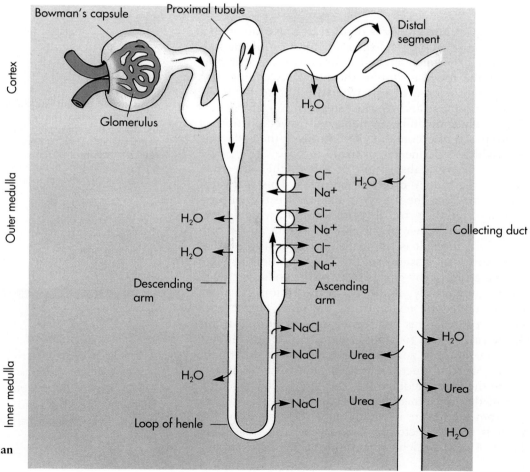

**Figure 41-10**

The flow of materials in the human kidney.

Left behind in the filtrate is the urea that initially passed through the glomerulus as nitrogenous waste; eventually the urea concentration becomes very high in the tubule.

4. Finally, the tubule empties into collecting ducts that pass back through the tissue; unlike the tubule, the lower portion of the collecting ducts is permeable to urea. During this final passage, the concentrated urea in the filtrate diffuses out into the surrounding tissue, which has a lower urea concentration. A high urea concentration in the tissue results, which is what caused water to move out of the filtrate by osmosis when it first passed down the descending arm.

5. As the filtrate passes down the collecting duct, even more water passes outward by osmosis because the osmotic concentration of the surrounding tissue reflects both the urea diffusing out from the duct *and* the salt diffusing out from the ascending arm. The sum of these two is greater than the osmotic concentration of urea in the filtrate (the salt has already been removed).

In effect, the kidney is divided into two functional zones (Figure 41-11): (1) the outer portion of the kidney, called the **outer medulla**, contains the upper portion of the loop, including the upper ascending arm where reabsorption of salt from the filtrate by active transport occurs; and (2) the inner portion of the kidney, or **inner medulla**, contains the lower portion of the loop and also contains the bottom of the collecting duct, which is permeable to urea.

The active reabsorption of salt in the outer medulla of the kidney drives the process. This reabsorption of salt from the filtrate in one arm of the loop establishes a gradient of salt concentration, with the salt concentration higher in the inner medulla at the bottom of the loop. It is this high salt concentration that raises the total tissue osmotic concentration so high that water passes by osmosis out of the collecting duct. Just as importantly, the active reabsorption of salt in the outer medulla concentrates the renal filtrate with respect to urea. This high concentration of urea in the filtrate causes urea to diffuse outward into surrounding tissue in the only zone where it is able to do so, the lower collecting duct, creating a high urea concentration in the inner medulla. It is this high urea concentration in the inner medulla that causes water to diffuse out from the filtrate in the initial descending arm. The water is then collected by blood vessels in the kidney (Figure 41-12), which carry it into the systemic circulation.

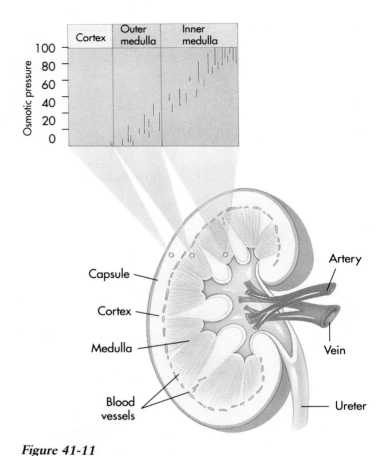

***Figure 41-11***

**Structure of the human kidney.** The cortex has an osmotic concentration like that of the rest of the body. The outer medulla has a somewhat higher osmotic concentration, primarily salt. The osmotic concentration within the inner medulla becomes progressively higher at greater depths; this part of the kidney maintains substantial concentrations of urea, which is a major component of its high osmotic concentration.

*The mammalian kidney achieves a high degree of water reabsorption by using the NaCl and urea in the glomerular filtrate to increase the osmotic concentration of the kidney tissue. This facilitates the movement of the water from the filtrate out into the surrounding tissue, where it is collected by blood vessels impermeable to the high urea concentration but permeable to water.*

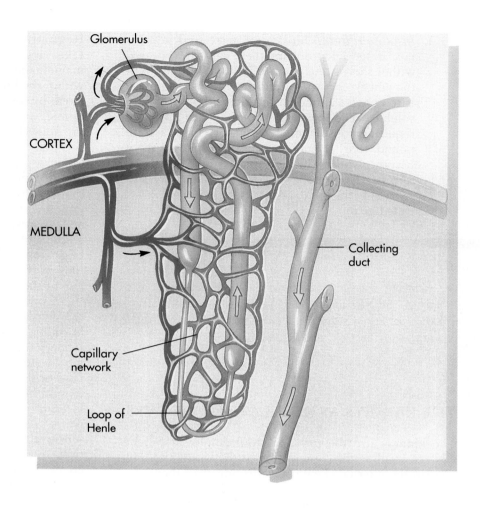

**Figure 41-12**

**How the blood supply collects water from the kidney.**

# EXCRETION OF NITROGENOUS WASTES

All animals contain proteins and nucleic acids, which are both nitrogen-containing molecules. When a heterotroph consumes another animal, the proteins and nucleic acids are broken down. In Chapter 37 we discussed the functioning of the liver. In the liver, enzymes break down amino acids by removing the amino group ($NH_2$), and combine it with $H^+$ ions to form ammonia ($NH_3$); the remainder of the amino acid is then converted to sugar or lipid. Ammonia is quite toxic to all cells, and so it is necessary to dispose of the ammonia produced during the breakdown of protein, nucleic acids, and other nitrogen-containing molecules. Because even very low concentrations of ammonia can kill cells, it is necessary to transport it in very dilute solution.

The need to transport and excrete ammonia in very dilute solution has posed a major evolutionary challenge in situations where water conservation is im-

portant. In freshwater fishes, which have, if anything, too much water, dilution presents no problem, and toxic ammonia is flushed out in highly diluted form. In saltwater fishes and terrestrial animals, however, water is precious, and it is not practical to excrete highly diluted ammonia. Three general solutions to this problem have been adopted:

1. *Flushing.* Both freshwater and saltwater fishes carry out protein breakdown in the gills so that little ammonia actually enters the body; it is simply carried away in the water passing across the gills.
2. *Detoxification.* Most mammals and many other land animals convert ammonia to **urea,** which is far less toxic and so can be transported and excreted at far higher concentrations. The urea is carried by the bloodstream to the kidneys, where it is excreted as a principal component of urine.

3. *Insolubilization.* Birds and terrestrial reptiles face a special problem: their eggs are encased within shells; thus metabolic wastes build up as the embryo grows within the egg. The solution is to convert ammonia to **uric acid,** which is insoluble. The process, although lengthy and requiring considerable energy, produces a compound that crystallizes and precipitates as it becomes more concentrated. Adult birds and terrestrial reptiles carry out the same process, excreting the final result as a semisolid paste called guano.

*The metabolic breakdown of protein produces ammonia as a by-product. Because ammonia is very toxic, animals have devised a variety of strategies for removing it. These include flushing, detoxification, and insolubilization.*

## THE KIDNEYS AS REGULATORY ORGANS

The kidneys are the principal tool employed by the CNS to regulate the composition of the blood and thus control the internal chemical environment of the body. By directing the kidneys to remove selectively particular substances from the blood, the CNS is able to maintain rigorous control over the concentration of ions and other chemicals. Thus, although almost all amino acids are retained by the kidneys, almost half of the urea in entering blood is eliminated. In normal individuals no glucose is eliminated, although in diabetics, in whom glucose cannot be absorbed into cells, excess glucose is excreted. Concentrations of ions such as $H^+$, $Na^+$, $K^+$, $Cl^-$, $Mg^{++}$, $Ca^{++}$, and $HCO_3^-$ are all maintained within narrow boundaries. This serves to maintain the blood's pH at a constant value, as well as maintain the proper ion balances for nerve conduction and muscle contraction.

*The kidneys are as much regulatory organs as they are excretory organs. The concentrations of many ions within the bloodstream, such as $H^+$, $Na^+$, $K^+$, $Cl^-$, $Mg^{++}$, $Ca^{++}$, and $HCO_3^-$, are maintained within narrow boundaries by the kidneys.*

## Regulation of Kidney Function

Like most organs concerned with homeostasis—the maintenance of constant physiological conditions within the body—the operation of the kidneys is regulated by the CNS, which uses the voluntary, autonomic, and hormonal controls we have discussed in previous chapters.

Why is it necessary to regulate the operation of the kidneys? As described in Chapter 32, the proper operation of the many organ systems of the body requires that the osmotic concentration of the blood be maintained within narrow bounds. For this reason, it is not always desirable for your body to retain the same amount of water. If you have consumed an unusually large amount of water, for example, the balance of ions in the blood can only be preserved if you retain less of the water than you would otherwise.

Control of water and salt balance is centered in the hypothalamus. The hypothalamus produces the hormone vasopressin (antidiuretic hormone), which renders the collecting ducts of the kidneys freely permeable to urea, which passes out into the interstitial fluid of the deep medulla and thus maximizes water retention by causing water to pass out of the descending loop into the surrounding tissue. When the uptake of water is excessive, blood pressure rises and pressure-sensitive neurons in the atria detect this rise. In addition, osmoreceptors in the hypothalamus detect a decrease in osmolarity and decrease the hypothalamus output of vasopressin (see Figure 41-12). A decrease in levels of vasopressin renders the collecting ducts of the renal tubules less permeable to water, lowering the osmotic concentration within the medulla, and in that way inhibits the reabsorption of water and increases urine volume. As mentioned in Chapter 36, alcohol consumption is one factor that suppresses vasopressin secretion.

Another example of regulation of the kidneys by the CNS concerns the salt balance in your body. The amount of salt in your diet can vary considerably, and yet it is important to many physiological processes that salt levels in your blood not vary widely. When levels of sodium ion in the blood fall, the adrenal gland increases its production of the hormone aldosterone, described in Chapter 36. Aldosterone stimulates active sodium ion reabsorption across the walls of the ascending arms of your kidneys' renal tubules. If the level of sodium in the blood falls, the adrenal gland increases its production of aldosterone, which acts to increase salt reabsorption and thus decrease the amount of sodium lost in the urine. In the total absence of this hormone, human beings may excrete up to 25 grams of salt a day.

# Replacing the Kidneys: Dialysis and Transplant

The kidneys are essential to life. They not only eliminate toxic waste products such as urea, but they maintain salt and water balance. There are many causes of kidney failure; one is hypertension, a prolonged increase in systemic blood pressure. Other common causes are bacterial or viral infection (nephritis) and diabetes mellitus. Whatever the cause, there are two basic treatment options: kidney transplants and dialysis.

Everyone has two kidneys, and each is capable of meeting the needs of the body, so removal of one would scarcely be noticeable. In addition, all the essential regulatory mechanisms are "built in" to the kidney. Neural input is not needed. The problem in kidney transplant is rejection by the immune system of the recipient. This difficulty can be prevented by matching the HLA antigens (see Chapter 40) that define "self" and "nonself" tissues as closely as possible, combined with drugs that suppress the immune response. There is a national computer network in which the tissue types of individuals needing new kidneys are recorded, so that when a kidney becomes available, a match can be determined quickly. However, in order for there to be a good match, the supply of organs for donation must be large enough. There are more individuals in need of kidneys than available organs, and there is a need to encourage organ donation.

Dialysis is a method of replacing the kidney by using an artificial device to remove toxic wastes from the blood. This may be done in two ways. In hemodialysis an individual has a tube in an artery in the arm that allows arterial blood to pass from the patient's circulation into a dialysis unit and back into a vein. Inside the dialysis unit the blood passes through hollow plastic fibers that allow waste products to diffuse out of the blood. In hemodialysis the patient must be connected to a dialysis unit two or three times a week. Patients must carefully manage their salt and water intake because the dialysis machine, unlike the kidney, does not regulate osmolarity and salt balance.

In a new method, continuous ambulatory peritoneal dialysis, the lining of the abdominal cavity is used instead of a dialysis machine. A patient is provided with a tube that allows the abdominal cavity to be filled with dialysis fluid, which is changed several times daily. This procedure involves fewer diet and fluid restrictions because of the properties of the peritoneum, but great care is needed to prevent infections. In some cases a combination of hemodialysis and peritoneal dialysis is used.

Both methods of dialysis are only short-term solutions because they are expensive, severely restrict the normal activity of an individual, and do not fully replace all the functions of a normal kidney. Successful transplant, on the other hand, effectively cures the recipient's disease, permitting a fully normal lifestyle. In the meantime, dialysis enables a patient to wait for an extended time for a matching kidney to become available.

## ■ SUMMARY

1. Animals that live in fresh water tend to gain water from their surroundings, whereas those that live in salt water or on land tend to lose water.

2. Insects solve the problem of dehydration by conserving water. In particular, they avoid excreting water along with body wastes. They pump ions into Malpighian tubules, so that body fluids are drawn in by osmosis. These body fluids contain the wastes created by metabolism. The wall of the tubule acts as a filter, passing wastes but retaining proteins and blood cells. Insects then reabsorb the water and useful metabolites back out, leaving the wastes behind to be excreted.

3. Instead of sucking body liquid through a filter as insects do, vertebrates push it through. They can do this because they have a closed circulation system that operates under considerable pressure.

4. The vertebrate kidney is composed of many individual units called nephrons, each made up of two segments, the first a filter and the second a resorption tube.

5. Freshwater fishes do not reabsorb water; their bodies already gain too much by direct diffusion from the water. Marine fishes excrete the salts in the water they drink and reabsorb water. Some marine vertebrates, notably sharks, maintain high body levels of urea so that they are isotonic with the sea and do not tend to gain or lose water.

6. Amphibians and reptiles have kidneys much like those of freshwater fishes. Birds and mammals achieve much greater water conservation by bending the nephron tube, producing what is known as the loop of Henle. This creates a countercurrent flow, which greatly increases the efficiency of water reabsorption.

7. The longer the loop of Henle, the greater the osmotic concentration that can be achieved and the more water that can be reclaimed from the urine.

8. The mammalian kidneys achieve a high degree of water reabsorption by using the salts and urea in the glomerular filtrate to increase the osmotic concentration of the kidney tissue. This facilitates the movement of the water from the filtrate out into the surrounding tissue, where it is collected by blood vessels impermeable to the high urea concentration but permeable to the water.

9. The kidneys are regulated by antidiuretic hormone, which affects osmolarity, and aldosterone, which regulates $Na^+$ balance.

## REVIEW

1. Approximately _____ of every vertebrate's body is water.

2. Among vertebrates, the organ that carries out filtration of waste and reabsorption of nonwaste molecules is the _____.

3. To help maintain their water balance, _____ fishes drink lots of water, whereas _____ fishes drink very little water.

4. Which part of the human kidney—the outer cortex, the outer medulla, or the inner medulla—has the highest concentration of urea?

5. The principal function of the kidneys is to maintain homeostatic ion and _____ balance within the body.

## SELF-QUIZ

1. Which of the following is an example of a vertebrate osmoconformer?
   (a) Kangaroo rats      (c) Freshwater fishes      (e) Amphibians
   (b) Whales      (d) Sharks

2. Which of the following is/are recovered in the collecting arm of the vertebrate nephron?
   (a) Small proteins      (c) Bivalent ions      (e) Water
   (b) Glucose      (d) Monovalent ions

3. The urine of terrestrial _____ cannot be more concentrated (hyperosmotic) than their blood.
   - (a) fishes
   - (b) amphibians
   - (c) reptiles
   - (d) birds
   - (e) mammals

4. Mammalian kidneys can concentrate urine to the extent that they do because they have long loops of Henle, where two solutes pass across the membrane of the loop. These solutes are
   - (a) water.
   - (b) salt (NaCl).
   - (c) glucose.
   - (d) protein.
   - (e) urea.

5. An insect conserves water by
   - (a) secreting potassium ions into its Malpighian tubules.
   - (b) pumping sodium ions out of its nephrons.
   - (c) drawing urea out of its Malpighian corpuscles.
   - (d) filtering sodium ions through the walls of its glomerulus.
   - (e) reabsorbing sodium ions into its nephrons.

# THOUGHT QUESTIONS

1. Salt substitutes sold under such names as "light salt" usually consist in part of KCl, which has a salty taste that is not as pleasant as that of NaCl. Why would consumption of KCl as a substitute for NaCl be less likely to increase blood pressure?

2. Atherosclerosis of the renal artery, restricting blood flow to one kidney, results in hypertension. Explain why. HINT: the hypertension is not caused by failure of urine formation by the flow-impaired kidney.

# FOR FURTHER READING

BEEUWKES, R: "Renal Countercurrent Mechanisms, or How to Get Something for (Almost) Nothing," In C.R. TAYLOR, et al., editors: *A Companion to Animal Physiology*, Cambridge University Press, New York, 1982. A clearly presented summary of current ideas about how the human kidney works.

HEATWOLE, H: "Adaptations of Marine Snakes," *American Scientist*, vol. 66, 1978, pages 594-604. Several groups of snakes are able to live in the sea by clever adaptations that modify salt and water balance.

SMITH, HW: *From Fish to Philosopher*, ed. 2, Little, Brown, & Co., Boston, 1961. A broad and well-written account of the evolution of the vertebrate kidney.

# *S*ex and Reproduction

This male lion is guarding his two cubs. Reproduction is evolution's most essential process, allowing the transmission of genes from one generation to the next. The urge to reproduce is shared by all animals, including humans.

# SEX AND REPRODUCTION

## Overview

Almost all vertebrates reproduce sexually. Sex evolved in the sea, and its modification for organisms living on land entailed evolutionary innovations to avoid drying out. Most terrestrial vertebrates, including a few genera of primitive mammals, reproduce by means of eggs. Human beings and other placental mammals have adopted a different solution, nourishing their developing young within the mother's body. Sex in human beings can play an important role in pair bonding as well as in reproduction. Vertebrate development may be divided artificially into several stages, although in reality these stages are parts of a continuous, dynamic process. The developmental process takes longer in mammals than in other vertebrates; it lasts some 266 days from fertilization to birth in human beings and continues for some time thereafter.

## For Review

*Here are some important terms and concepts that you will encounter in this chapter. If you are not familiar with them, you should review them before proceeding.*

**Meiosis** (Chapter 9)

**Vertebrate evolution** (Chapter 18)

**Amniotic egg** (Chapter 20)

**Mammals** (Chapter 28)

**Hormones** (Chapter 36)

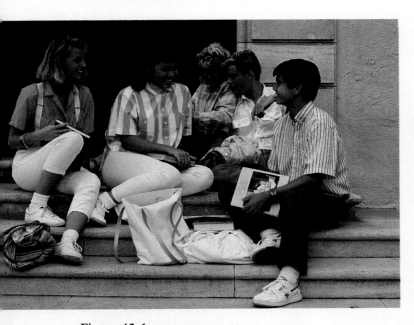

**Figure 42-1**

**Students in the 1990s will have to make informed decisions about sex.** Contraception and the prevention of AIDS and other sexually transmitted diseases are major concerns among students today.

On any dark night you can hear biology happening. The cry of a cat in heat, insects chirping outside the windows, frogs croaking in swamps, wolves howling in a frozen northern scene—all these are the sounds of evolution's essential act, reproduction. Few subjects pervade our everyday thinking more than sex; few urges are more insistent. They are no accident, these strong feelings—they are a natural part of being human. The frog in its swamp has an urgent desire to reproduce itself, a desire that has been patterned within it by a long history of evolution. It is a pattern that we share. The reproduction of our families spontaneously elicits in us a sense of rightness and fulfillment. It is difficult not to return the smile of a new infant, not to feel warmed by it and by the look of wonder and delight to be seen on the faces of the parents. This chapter deals with sex and reproduction among the vertebrates, of which we human beings are one kind. Few subjects are of more direct concern to students than sex. Many students must make important decisions about sex, and all need to be well informed (Figure 42-1). The average age at which American females have sex for the first time is between 16 and 17 years of age. The subject is thus of far more than academic interest.

## SEX EVOLVED IN THE SEA

Sexual reproduction first evolved among marine organisms. The eggs of most marine fishes are produced in batches by the females; when they are ripe, the eggs are simply released into the water. Fertilization is achieved by the release of sperm by the male into the water containing the eggs. Seawater itself is not a hostile environment for gametes or for young organisms.

The effective union of free gametes in the sea, an example of **external fertilization,** poses a significant problem for all marine organisms. Eggs and sperm become diluted rapidly in seawater, so their release by females and males must be almost simultaneous if successful fertilization is to occur. Because of this necessity, most fish vertebrates restrict the release of their eggs and sperm to a few brief and well-defined periods. There are few seasonal cues in the ocean that organisms can use as signals, but one that is all-pervasive is the cycle of the moon. Approximately every 28 days, the moon revolves around the earth; variations in its gravitational attraction cause the differences in ocean level that we call tides. Many different kinds of marine organisms sense the changes in water pressure that accompany the tides, and much of the reproduction that takes place in the sea is timed by the lunar cycle.

The invasion of the land by organisms from the sea meant facing for the first time in the history of life on earth the danger of drying out. This problem was all the more severe for the marine organisms' small and vulnerable reproductive cells. Obviously the gametes could not simply be released near one another on land; they would soon dry up and perish. There are many possible solutions to this problem. We have already considered the seeds of plants, the spores of fungi, and the eggs of arthropods, all successful adaptations to an existence on dry land. Vertebrates have solved the problem in yet other ways.

## VERTEBRATE SEX AND REPRODUCTION: FOUR STRATEGIES

The five major classes of vertebrates have evolved reproductive strategies that are quite different, ranging from external fertilization to bearing live young. In these differences mostly reflect different approaches to protecting gametes from drying out, since four of the five major groups of vertebrates are terrestrial.

### Fishes: External Fertilization Works Well in Water

Some vertebrates, of which the bony fishes are the most abundant, have remained aquatic. All fishes reproduce in the water in essentially the same way as all other aquatic animals. Fertilization in most species is external, with the eggs containing only enough yolk material to sustain the developing zygote for a short time. After the initial dowry of yolk has been exhausted, the growing individual must seek its food from the waters around it. Many thousands of eggs may be fertilized in an individual mating, but few of the resulting zygotes survive their aquatic environment and grow to maturity. Some succumb to bacterial infection, many others to predation. The development of the fertilized eggs is speedy, and the young that survive achieve maturity rapidly.

### Amphibians: Still Tied to Water

A second group of vertebrates, the amphibians (frogs, toads, salamanders), have invaded the land without fully adapting to the terrestrial environment. The life cycle of the amphibians is still inextricably tied to the presence of free water. Among most amphibians, fertilization is still external, just as it is among the fishes and other aquatic animals. Many female amphibians lay their eggs in a puddle or a pond of water. Among the frogs and toads, the male grasps the female and discharges fluid containing sperm onto the eggs as she releases them (Figure 42-2).

*Figure 42-2*

**The eggs of frogs are fertilized externally.** When frogs mate, as these two are doing, the clasp of the male induces the female to release a large mass of mature eggs, over which the male discharges his sperm.

The development time of the amphibians is much longer than that of the fishes, but amphibian eggs do not include a significantly greater amount of yolk. Instead, the process of development consists of two distinct life stages, a larval and an adult stage, like some of the life cycles found among the insects.

The development of the aquatic larval stage of the amphibians is rapid because it uses yolk supplied from the egg. The larvae then function, often for a considerable period of time, as independent food-gathering machines. They scavenge nutrients from their environments and often grow rapidly. Tadpoles, which are the larvae of frogs, can grow in a matter of days from creatures no bigger than the tip of a pencil into individuals as big as goldfish. When an individual larva has grown to a sufficient size, it undergoes a developmental transition, or **metamorphosis**, into the terrestrial adult form.

*The invasion of land by vertebrates was tentative at first, with the amphibians keeping the external fertilization and reproduction by means of eggs that are characteristic of most of the fishes, the group from which they evolved.*

**Figure 42-3**

**Turtles mating.** The introduction of semen containing sperm by the male into the female's body occurs during copulation. Reptiles such as these turtles were the first terrestrial vertebrates to develop this form of reproduction, which is particularly suited to a terrestrial existence.

## Reptiles and Birds: Evolution of the Watertight Egg

The reptiles were the first group of vertebrates to abandon aquatic habitats completely. Unlike the eggs of amphibians, reptile eggs are fertilized internally within the mother before they are laid, the male introducing his **semen**, a fluid containing sperm and fluid secretions, directly into her body (Figure 42-3). By this means, fertilization still occurs in an environment that is protected from drying out, even though the adult animals are fully terrestrial. Most vertebrates that fertilize internally utilize a tube, the **penis**, to inject semen into the female. Birds are the only significant exception to this rule. Composed largely of tissue that can become rigid and erect, the penis penetrates into the female reproductive tract.

Many reptiles are **oviparous**, the eggs being deposited outside the body of the mother; others are **viviparous**, forming eggs that hatch within the body of the mother. The young of viviparous animals, therefore, are born alive.

Most birds lack a penis (swans are an exception) and achieve internal fertilization simply by the male slapping semen against the reproductive opening of the female before the eggs have formed their hard shells. This kind of mating occurs more quickly than that of most reptiles. All birds are oviparous, the young emerging from their eggs outside the body of the mother. Birds encase their eggs in a harder shell than do their reptilian ancestors. They also hasten the development of the embryo within each egg by warming the eggs with their bodies. The young that hatch from the eggs of most bird species are not able to survive unaided, since their development is still incomplete. The young birds therefore are fed and nurtured by their parents, and they grow to maturity gradually.

The shelled eggs of reptiles and birds constitute one of their most important adaptations to life on land, since these eggs can be laid in dry places. Each egg is provided with a large amount of yolk and is encased within a membranous cover. The zygote develops within the egg, eventually achieving the form of a miniature adult before it completely uses up its supply of yolk and leaves the egg to face its fate as an adult.

*The first major evolutionary change in reproductive biology among land vertebrates was that of the reptiles and their descendants, the birds. Although both groups are still oviparous like fishes, they practice internal fertilization, with the zygote encased within a watertight egg.*

## Mammals: Mothers Nourish their Young

The most primitive mammals, the monotremes, are oviparous like the reptiles from which they evolved. The living monotremes consist solely of the duckbilled platypus and the echidna. No other mammals lay eggs. All other members of this class are viviparous.

The young of viviparous mammals are nourished and protected by their mother, an outstanding characteristic of the members of this class. The mammals other than monotremes have approached the problem of nourishing their young in two ways:

1. Marsupials give birth to live embryos at a very early stage of development. The tiny animals crawl to and enter pouches on the mother's body, where they continue their development. They eventually emerge when they are able to function on their own.
2. The placental mammals retain their young for a much longer period within the body of the mother. To nourish them, they have evolved a specialized, massive network of blood vessels called a placenta, through which nutrients are channeled to the embryo from the blood of the mother.

*The second major evolutionary change in reproductive biology among land vertebrates was that of the marsupials and placental mammals. These groups nourish their young within a pouch or inside the body until the young have reached a fairly advanced stage of development.*

## THE HUMAN REPRODUCTIVE SYSTEM

The human reproductive system (Figure 42-4), like those of all other vertebrates, joins **egg cells** and **sperm cells.** The egg is fertilized within the female, and the zygote develops into a mature fetus there.

### Males

The human male gamete, or **sperm,** is highly specialized for its role as a carrier of genetic information. Produced after meiosis, the sperm cells have 23 chromosomes instead of the 46 found in most cells of the human body. Unlike these other cells, sperm do not complete their development successfully at 37° C (98.6° F), the normal human body temperature. The sperm-producing organs, or **testes,** move during the course of fetal development out of the body proper and into a sac called the **scrotum** (Figure 42-5). The scrotum,

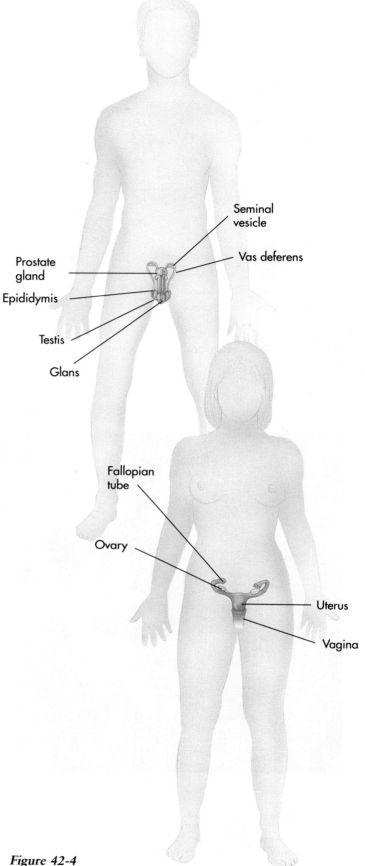

**Figure 42-4**

The human reproductive system.

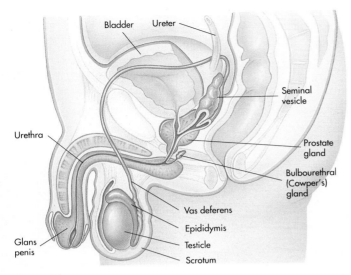

**Figure 42-5**

Organization of the male reproductive organs.

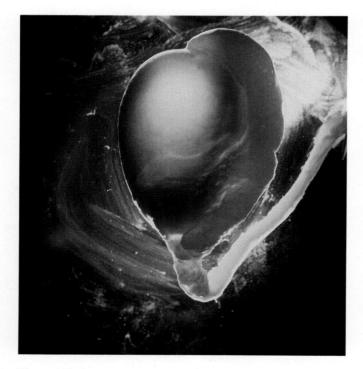

**Figure 42-6**

**Human testes.** The testicle is the darker sphere in the center of the photograph; within it sperm are formed. Cupped above the testicle is the epididymis, a highly coiled passageway within which sperm complete their maturation. Extending away from the epididymis is a long tube, the vas deferens.

which hangs between the legs of the male, maintains the testes at a temperature about 3° C cooler than that of the rest of the body.

*Male Gametes Are Formed in the Testes.* The testes (Figure 42-6) are composed of several hundred compartments, each of which is packed with large numbers of tightly coiled tubes called **seminiferous tubules.** The tubes themselves are the sites of sperm-cell production, or **spermatogenesis** (Figure 42-7). The full process of sperm development takes about 2 months. The number of sperm that are produced is truly incredible. A typical adult male produces several hundred million sperm each day of his life. Those that are not ejaculated from the body are broken down and their constituents reabsorbed, in a continual cycle of renewal.

The testes also contain interstitial cells that secrete the male sex hormone testosterone. All the cells of the testes, gametes and supporting cells, require a combination of the pituitary hormones FSH and LH for their normal function.

After the sperm cells complete their differentiation within the testes, they are delivered to a long coiled tube called the **epididymis,** where they are stored and mature further. The sperm cells are not motile when they arrive in the epididymis, and they must remain there for at least 18 hours before their motility develops. From the epididymis, the sperm are delivered to another long tube, the **vas deferens,** where they are stored. When they are delivered during intercourse, the sperm travel through a tube from the vas deferens to the urethra, where the reproductive and urinary tracts join, emptying through the penis.

*The Male Gametes Are Delivered by the Penis.* The penis is an external tube composed of three cylinders of spongy tissue (Figure 42-8). In cross section the arteries and veins can be seen to run within the spongy tissue nearest the surface, beneath which two of the cylinders sit side by side. Below the pair of cylinders is a third cylinder, which contains in its center the **urethra,** through which both semen (during ejaculation) and urine (during liquid elimination) pass. The spongy tissue that makes up the three cylinders is riddled with small spaces between its cells. When nerve impulses from the CNS cause the dilation of the arterioles leading into this tissue, blood collects within the spaces. This causes the tissue to become distended and the penis to become erect and rigid. Continued stimulation by the CNS is required for this erection to be maintained.

Erection can be achieved without any physical stimulation of the penis. Mental imagery is a common initiating factor. However, the physical stimulation of

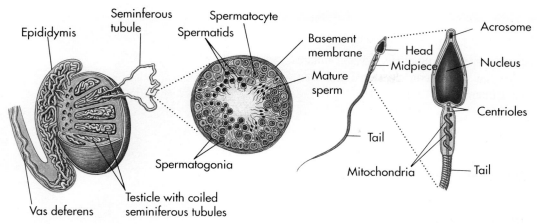

**Figure 42-7**

**The interior of the testes, site of spermatogenesis.** Within the seminiferous tubules of the testes, cells called spermatogonia develop by means of meiosis into sperm, passing through the spermatocyte and spermatid stages. Each sperm possesses a long tail coupled to a head, which contains a haploid nucleus.

the penis usually is required for any delivery of semen to take place. Stimulation of the penis, as by repeated thrusts into the vagina of a female, leads first to the mobilization of the sperm. In this process, muscles encircling the vas deferens contract, moving the sperm along the vas deferens into the urethra. Eventually the stimulation leads to the violent contraction of the muscles at the base of the penis. The result is ejaculation, the ejection of about 5 milliliters of semen out of the penis. Semen is a collection of secretions from the pros-

tate gland, seminal vesicles, and Cowper's glands. Semen provides metabolic energy sources for the sperm. Within this small volume are several hundred million sperm. The odds against any one individual sperm cell successfully completing the long journey to the egg and fertilizing it are extraordinarily high. Successful fertilization requires a high sperm count; males with less than 20 million sperm per milliliter are generally considered sterile (Figure 42-9).

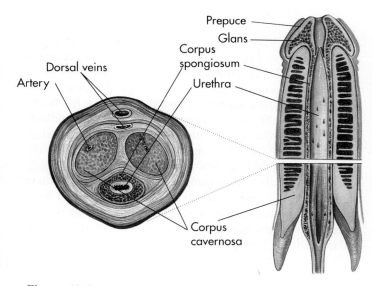

**Figure 42-8**

**A penis in longitudinal section and in cross section.**

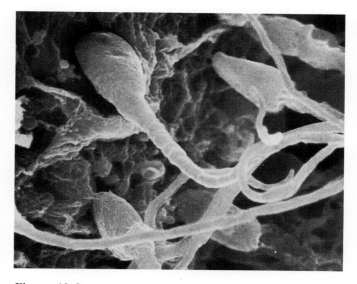

**Figure 42-9**

**Human sperm.** Only the heads and a portion of the long slender tails of these sperm are shown in this scanning electron micrograph.

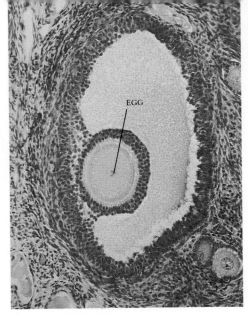

**Figure 42-11**

A mature egg within an ovarian follicle of a cat.

An adult male produces sperm continuously, several hundred million each day of his life. The sperm are stored and then delivered during sexual intercourse.

## Females

Fertilization requires more than insemination. There must be a mature egg to fertilize.

***Female Gametes Are Formed in the Ovaries.*** Eggs are produced within the ovaries of females (Figure 42-10). The ovaries are compact masses of cells, 2 to 3 centimeters long, located within the abdominal cavity. Eggs develop from cells called **oocytes,** which are located in the outer layer of the ovary. Unlike males, whose gamete-producing cells (spermatogonia) are constantly dividing, females have at birth all the oocytes that they will ever produce. At each cycle of ovulation, one or a few of these oocytes initiate development; the others remain in a developmental holding pattern. This long maintenance period is one reason that developmental abnormalities crop up with increasing frequency in pregnancies of women who are over 35 years old. The oocytes are continually exposed to mutation through-

out life, and after 35 years the odds of a harmful mutation having occurred become high enough to increase significantly the incidence of fetal abnormalities.

***Only One Female Gamete Matures Each Month.*** At birth a female's ovaries contain some 2 million oocytes, all of which have begun the first meiotic division. At this stage they are called **primary oocytes.** Meiosis is arrested, however, in prophase of the first meiotic division. Very few oocytes ever proceed to develop further. With the onset of puberty, the female matures sexually. At this time the release of FSH initiates the resumption of the first meiotic division in a few oocytes, but a single oocyte soon becomes dominant, the others regressing (Figure 42-11). Approximately every 28 days after that, another oocyte matures, although the exact timing may vary from month to month. It is rare for more than about 400 out of the approximately 2 million oocytes with which a female is born to mature during her lifetime. When they do mature, the egg cells are called **ova** (singular, **ovum**), the Latin word for "egg."

Unlike male gametogenesis, the process of meiosis in the oocytes does not result in the production of four haploid gametes. Instead, a single haploid ovum is produced, the other meiotic products being discarded as **polar bodies.** The process of meiosis is stop-and-go rather than continuous (Figure 42-12).

***Fertilization Occurs as an Egg Journeys to the Uterus.*** When the ovum is released at ovulation, it is swept by the beating of cilia into one of the **fallopian tubes,** which leads away from the ovary into the uterus

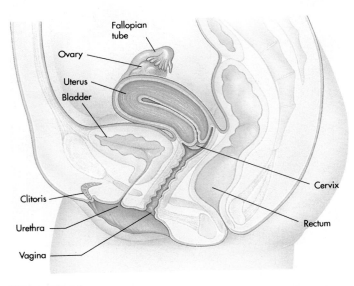

**Figure 42-10**

Organization of the female reproductive organs.

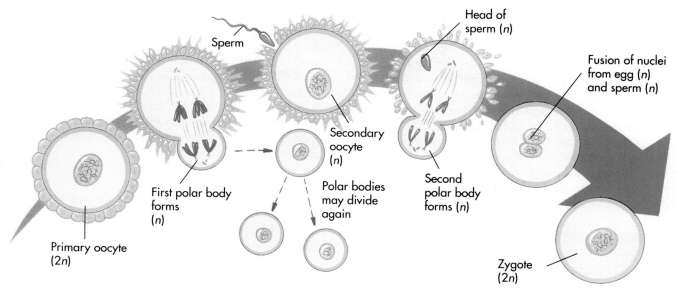

**Figure 42-12**

**The meiotic events of oogenesis.** A primary oocyte is diploid (2n, with 46 chromosomes in human beings). In its maturation the first meiotic division is completed, and one division product is eliminated as a polar body. The other product, the secondary oocyte, is released during ovulation. The second meiotic division does not occur until after fertilization and results in production of a second polar body and a single haploid (n, with 23 chromosomes in humans) egg nucleus. Fusion of the haploid egg nucleus with a haploid sperm nucleus produces a diploid zygote, from which an embryo subsequently forms.

(Figure 42-13). Smooth muscles lining the fallopian tube contract rhythmically; these rhythmic, peristaltic contractions move the egg down the tube to the uterus. The journey is a slow one, taking about 5 to 7 days to complete. If the egg is unfertilized, it loses its capacity to develop in a few days. For this reason the sperm cannot simply lie in wait within the uterus. Any sperm cell that is to fertilize an egg successfully must make its way up the fallopian tube, a long passage that few survive.

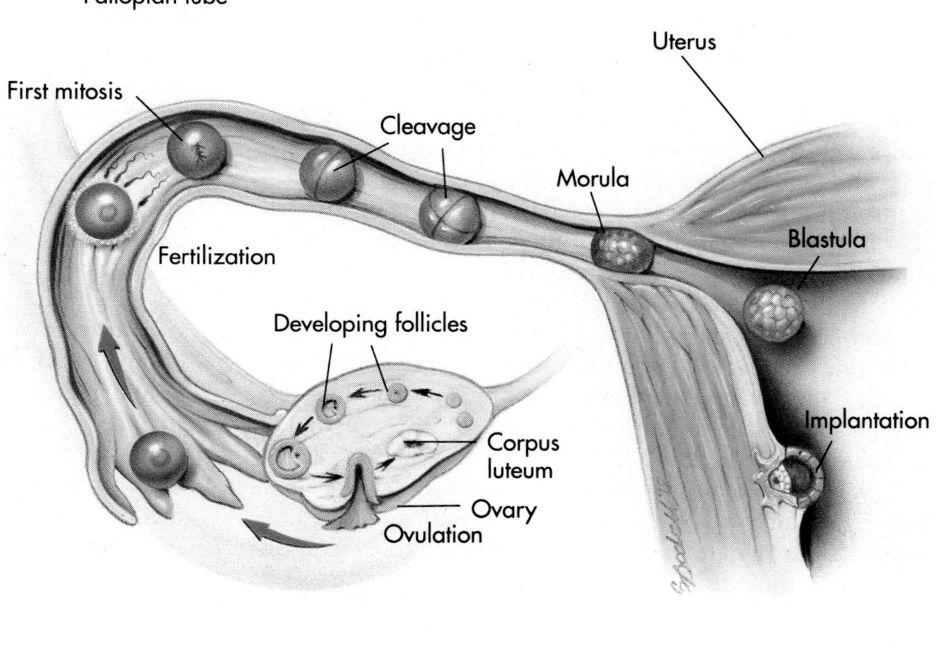

**Figure 42-13**

**The journey of an egg.** Produced within a follicle and released at ovulation, an egg is swept up into a fallopian tube and carried down by waves of contraction of the tube walls. Fertilization occurs within the tube by sperm journeying upward. Several mitotic divisions occur while the fertilized egg continues its journey down the fallopian tube, so that by the time it enters the uterus, it is a hollow sphere of cells. The sphere, a new embryo, implants itself within the wall of the uterus, where it continues its development.

Sperm are deposited within the **vagina,** a muscular tube about 7 centimeters long, which leads to the mouth of the uterus. This opening is bounded by a muscular sphincter called the **cervix.** The **uterus** is a hollow, pear-shaped organ about the size of a small fist. Its inner wall, the **endometrium,** has two layers. The outer of these layers is shed during menstruation while the one beneath it generates the next layer. Sperm entering the uterus swim and are carried upward by waves of motion in the walls of the uterus, enter the fallopian tube, and swim upward against the current generated by peristaltic contractions, which are carrying the ovum downward toward the uterus.

*All the eggs that a woman will produce during her life develop from cells that are already present at her birth. Their development is halted early in meiosis, and one or a few of these cells resume meiosis each 28 days to produce a mature egg. On maturation, eggs travel to the uterus. Fertilization by a sperm cell, if it occurs, happens en route.*

When a successfully fertilized egg reaches the uterus, the new embryo attaches itself to the endometrial lining and thus starts the long developmental journey that eventually leads to the birth of a child.

## SEXUAL CYCLES

For efficient reproduction, mating or the release of sperm must occur as soon as the mature egg or eggs become available. Otherwise, few of the sperm will survive long enough to achieve successful fertilization. Among those vertebrates that practice internal fertilization, the females typically signal the successful development and release of an egg, **ovulation,** by the release of chemical signals called **pheromones.** Female dogs signal their reproductive readiness in this manner, and the male is able to detect their pheromones in the air, even at very low concentrations. In many mammals the female does not actually release a mature egg until mating has occurred. For them, the physical stimulus of copulation causes the pituitary gland to release a signal that triggers ovulation.

When a female vertebrate does not possess a mature egg, she will often reject the sexual advances of males. Most female mammals are sexually receptive, or "in heat," for only a few short periods each year. The period in which the animal is in heat is called **estrus;** the periods of estrus correspond to ovulation events during a periodic cycle, the **estrous cycle.** In general, small mammals have many estrous cycles in rapid succession, whereas larger ones have fewer cycles spaced farther apart. During estrus the female is willing to mate and either has ovulated or will soon ovulate. Meanwhile the endometrium of the uterus has developed into an ideal environment into which an embryo can implant and develop.

Human beings and some apes provide the exception to the rule of cycles of receptivity. Human females are sexually receptive throughout the reproductive cycle. The reproductive cycle of egg production and release takes on average about 28 days. One cycle follows another, continuously, and a human female may mate at any time during a cycle. Successful fertilization, however, is possible only during a period of 3 to 4 days starting the day before ovulation.

## SEX HORMONES

A lot of signaling goes on during sex. Unlike humans, most other viviparous females signal the male when a mature egg has been released at the beginning of an estrous (sexually receptive) cycle; it is also necessary that many different processes within the female and within the male be coordinated during the process of gametogenesis and reproduction. The delayed sexual development that is common in mammals, for example, entails a nonsexual juvenile period, after which changes occur that produce sexual maturity. These changes occur in many parts of the body, a process that requires the simultaneous coordination of further development in many different kinds of tissues. The production of gametes is another carefully orchestrated process, involving a series of carefully timed developmental events. Successful fertilization begins yet another developmental "program," in which the female body prepares itself for the many changes of pregnancy.

All this signaling is carried out by a portion of the brain, the hypothalamus. The signals by which the hypothalamus regulates reproduction are hormones produced by the brain, which are carried by the bloodstream to the various organs of the body. Some reproductive hormones are steroids, complex carbon-ring lipids (Figure 42-14); others are peptides. On the cell surfaces of the target organs of the reproductive hormones are located specific receptors to which a particular hormone binds. In the case of steroid hormones, the receptors are located within the cytoplasm of the receptor cells. Once bound to a receptor, a steroid hormone molecule is transported to the nucleus, where it

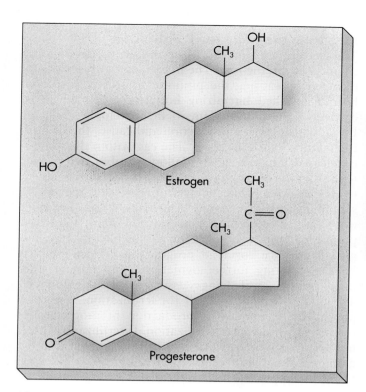

**Figure 42-14**

**The steroid sex hormones estrogen and progesterone.** Estrogen prepares and maintains the uterine lining for pregnancy; progesterone stimulates thickening of the uterine lining during pregnancy.

binds to specific locations on the chromosomes and alters the pattern of gene transcription. By so doing, the molecule initiates the physiological changes associated with that hormone.

*The body uses hormones as signals to control various body functions in the reproductive cycle (Table 42-1). Most reproductive hormones are steroids. Steroid hormones bind to specific receptor sites on target cells, enter the cells, and alter the pattern of gene expression in the target cells.*

## THE HUMAN REPRODUCTIVE CYCLE

The reproductive cycle of female mammals, including that of human beings, is composed of two distinct phases, the follicular phase and the luteal phase.

### Triggering the Maturation of an Egg

The first or **follicular phase** of the reproductive cycle is marked by the hormonally controlled development of eggs within the ovary. The anterior pituitary, after re-

---

**TABLE 42-1    REPRODUCTIVE HORMONES**

| | |
|---|---|
| **MALE** | |
| Follicle-stimulating hormone (FSH) | Stimulates spermatogenesis |
| Luteinizing hormone (LH) | Stimulates secretion of testosterone |
| Testosterone | Stimulates development and maintenance of male secondary sexual characteristics |
| **FEMALE** | |
| Follicle-stimulating hormone (FSH) | Stimulates growth of ovarian follicle |
| Luteinizing hormone (LH) | Stimulates conversion of ovarian follicles into corpus luteum; stimulates secretion of estrogen |
| Estrogen | Stimulates development and maintenance of female secondary sexual characteristics; prompts monthly preparation of uterus for pregnancy |
| Progesterone | Completes preparation of uterus for pregnancy; helps maintain female secondary sexual characteristics |
| Oxytocin | Stimulates contraction of uterus; initiates milk release |
| Prolactin | Stimulates milk production |

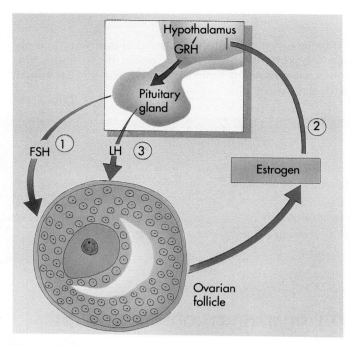

**Figure 42-15**

**Mammalian egg maturation is under hormonal control.**
1 Gonadotrophic releasing hormone (GRH) causes the pituitary to produce the follicle-stimulating hormone (FSH) and release it into the blood circulation; when FSH reaches the ovaries, it initiates final egg development; 2 FSH also causes the ovaries to produce the hormone estrogen; rising levels of estrogen in the blood cause the pituitary to shut down its production of FSH and instead produce the luteinizing hormone LH; 3 When this LH circulates back to the ovaries, it inhibits estrogen production and initiates ovulation.

ceiving a chemical signal from the hypothalamus (Figure 42-15), starts the cycle by secreting **follicle-stimulating hormone (FSH),** which binds to receptors on the surface of the follicles, initiating the final development and maturation of the egg. Normally, only a few eggs at any one time have developed far enough to respond immediately to the FSH. FSH levels are reduced before other eggs reach maturity, so that in every cycle only a few eggs ripen.

The reduction of the levels of FSH is achieved by a feedback command to the pituitary. In addition to starting final egg development, FSH also triggers the production of the female sex hormone **estrogen** by the ovary. Rising estrogen levels in the bloodstream feed back to the hypothalamus and cut off the further production of FSH. In this way, only the few eggs that are already developed far enough for their maturation to be initiated by the FSH are taken into the final stage of development. The rise in estrogen level and the maturation of one or more eggs completes the follicular phase of the estrous cycle.

## Preparing the Body for Fertilization

The second or **luteal phase** of the cycle follows smoothly from the first. The hypothalamus responds to estrogen by causing the pituitary to secrete a second hormone, called **luteinizing hormone (LH),** which is carried in the bloodstream to the developing follicle. LH inhibits estrogen production and causes the wall of the mature follicle to burst. The egg within the follicle is released into one of the **fallopian tubes,** which extend from the ovary to the uterus. This process is called **ovulation.** Meanwhile, the ruptured follicle repairs itself, filling in and becoming yellowish. In this condition it is called the **corpus luteum,** which is simply the Latin phrase for "yellow body." The corpus luteum soon begins to secrete a hormone, **progesterone,** which inhibits FSH, preventing further ovulations, and initiates the many physiological changes associated with pregnancy. The body is preparing itself for fertilization. The corpus luteum continues its production of progesterone for several weeks after fertilization. If fertilization does not occur soon after ovulation, then production of progesterone slows and eventually ceases, marking the end of the luteal phase.

*The reproductive cycle of mammals is composed of two phases, which alternate with one another. During the follicular phase, some of the eggs within the ovary complete their development. During the following luteal phase, the mature eggs are released into the fallopian tubes, a process called ovulation. If fertilization does not occur, ovulation is followed by a new follicular phase, the start of another cycle.*

In the absence of estrogen and progesterone, the pituitary can again initiate production of FSH, thus starting another reproductive cycle. In human beings the next cycle follows immediately after the end of the preceding one. A cycle usually occurs every 28 days, or a little more frequently than once a month, although this varies in individual cases. The Latin word for month is *mens,* which is why the reproductive cycle in humans is called the **menstrual cycle,** or monthly cycle (Figure 42-16). In human beings and some other primates, the hormone progesterone has among its many effects a thickening of the walls of the uterus in preparation for the implantation of the developing embryo. When fertilization does not occur, the decreasing levels of progesterone cause this thickened layer of blood-rich

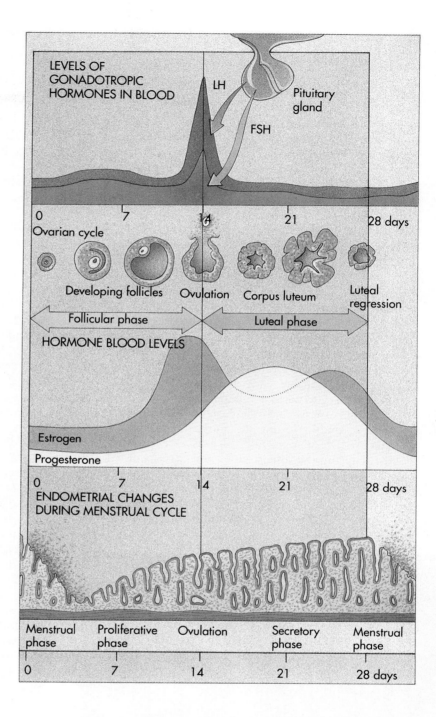

LEVELS OF GONADOTROPIC HORMONES IN BLOOD

LH

FSH

Pituitary gland

0    7    14    21    28 days

Ovarian cycle

Developing follicles    Ovulation    Corpus luteum    Luteal regression

Follicular phase    Luteal phase

HORMONE BLOOD LEVELS

Estrogen

Progesterone

0    7    14    21    28 days

ENDOMETRIAL CHANGES DURING MENSTRUAL CYCLE

Menstrual phase    Proliferative phase    Ovulation    Secretory phase    Menstrual phase

0    7    14    21    28 days

**Figure 42-16**

**The human menstrual cycle.** The growth and thickening of the uterine (endometrial) lining is governed by levels of the hormone progesterone; menstruation, the sloughing off of the blood-rich tissue, is initiated by lower levels of progesterone.

tissue to be sloughed off, a process that results in the bleeding associated with **menstruation.** Menstruation, or "having a period," usually occurs about midway between successive ovulations, or roughly once a month, although its timing varies widely even for individual females.

Two other hormones are important in the female reproductive system. Prolactin is needed for the production of milk and is secreted by the anterior pituitary. Oxytocin, secreted by the posterior pituitary, causes milk release. In combination with uterine prostaglandins, oxytocin initiates labor and delivery.

# AIDS on the College Campus

As most college students now know, AIDS is a serious disease that is rapidly becoming common in the United States and around the world. The disease, first reported in 1981, is transmitted by a virus and is always fatal. Over 130,000 individuals in the United States had contracted AIDS by 1990, and nearly 70 Americans will die each day of AIDS this year. The World Health Organization estimates that 8 million people have contracted the AIDS virus worldwide. Few, if any, of these patients will escape AIDS.

The name **AIDS** is shorthand for **acquired** (transmitted from another infected individual) **immunodeficiency** (a breakdown of the body's ability to defend itself against disease) **syndrome** (a spectrum of symptoms). The disease is fatal because no one can survive for long without an immune system to defend against viral and bacterial infections and to ward off cancer. The virus also infects the CNS, often leading to serious mental disorders.

AIDS is not the only fatal disease to threaten humans, and it is not the most contagious. What makes AIDS an unusually serious threat is that the virus causing the disease does not have its effect immediately on infection. Recently infected people usually show no symptoms of the disease at all. Only much later, typically 5 years, does the virus begin to multiply and attack the immune system. During these years, however, the infected person is an un-knowing carrier, able to transmit the virus to others. It is the large reservoir of undiagnosed, infected individuals that casts such a shadow over our future. Current estimates of the number of individuals in their twenties and thirties who are infected with the virus in the United States vary, but most estimates exceed one per thousand, suggesting both a staggering load of future suffering that will be difficult to avoid and a great danger that the infection will spread further. It would be folly to assume that college campuses will escape infection. Every student should face that fact squarely.

*What causes AIDS?* The virus that causes AIDS is called **human immunodeficiency virus** (**HIV**) (Figure 42-A). It is a fragile virus that does not survive outside body cells. It is present in the body fluids of infected individuals (notably in blood and in semen and vaginal fluid). HIV is transmitted from one individual to another when body fluid is transferred from an infected individual. It is *not* transmitted in the air or by casual contact. You cannot catch AIDS from a bathroom seat, from a hot tub shared with an AIDS victim, from kissing an AIDS carrier, or by being bitten by a mosquito that bit an AIDS victim. In studies of 619 households of AIDS victims, not one family member contracted the AIDS virus.

The *only* way you can become infected with HIV is to come into contact with the body fluids of an infected person. Among college students, there are two important routes of infection.

1. *Sexual intercourse.* Both semen and vaginal fluid of infected individuals have high levels of HIV. This means that vaginal, anal, and oral sex with an infected individual can all transmit the virus successfully—and to either sex. Absorption across the vaginal wall or into the penis offers a ready means for the virus to enter the body; the tiny tears produced during anal sex may facilitate entry even better. The microscopic abrasions that everybody has in their mouth from eating and chewing are a third easy means of entry, making oral sex also dangerous.

The only safe way to have sex with an infected individual is to use a condom, and use it correctly. When used correctly, condoms offer the best available protection from infection. It is important, however, that they be used properly. For some 10% of couples using condoms for birth control, the woman

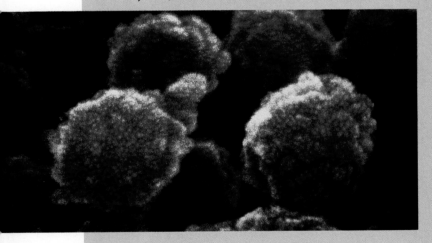

**Figure 42-A**
AIDS viruses infecting a cell of the immune system.

becomes pregnant, almost always as a result of careless use. This suggests that it is not wise to mix alcohol or drugs with sexual encounters; they may cloud your judgment and lead you to do things you wouldn't do with a clearer head, such as forgetting to use a condom or using it carelessly. It is a mistake that could cost you your life.

2. *Drug use.* A needle used more than once by an infected individual typically harbors large quantities of HIV, both in the fluid that remains behind in the needle and in the body of the hypodermic syringe. Anyone else who re-uses the needle will become infected. The use of intravenous drugs is itself dangerous—both illegal and life threatening—but if you engage in such folly, do not compound the damage by employing a used needle or syringe.

*Who is at risk?* You are. AIDS is commonly perceived as a disease of homosexual men because the disease first appeared in the United States among the gay community. Because homosexuals tend to confine their sexual interactions to one another, the disease initially spread among homosexuals without entering the larger heterosexual community. That initial segregation appears to be ending: although only 4% of the AIDS cases diagnosed in 1989 were heterosexual nondrug users, the incidence of HIV among heterosexuals is now expanding. Estimates vary widely. The U.S. Public Health Service estimates that between 1.5 million and 2 million people in the United States now harbor the virus. In Africa, where the virus first infected humans several years before it spread to this country, the epidemic has proceeded further than in the United States, even though homosexuality is rare there. In Africa, sexual transmission of AIDS is almost exclusively heterosexual and occurs in *both* directions, female to male as well as male to female. In some central African countries, as many as 10% of the adult individuals are thought to carry HIV.

*The AIDS antibody test.* Can a person find out if he or she has been infected with HIV? Yes, easily. A simple test identifies infected individuals by detecting antibodies in their blood directed against HIV. Such antibodies are detectable only in the bodies of infected individuals. They are the remnants of the body's attempt to ward off the HIV infection.

*How an AIDS test works.* The standard test for detecting the presence of antibody directed against HIV is called the ELISA antibody test. In the test, about 5 ml of blood (roughly the amount that would fit in a paper drinking straw) is drawn and checked to see whether any antibodies are present that will interact with bits of HIV attached to the surface of a plastic dish. If such antibodies are present, the test is said to be positive.

*How to get an AIDS test.* The Red Cross offers an inexpensive walk-in AIDS antibody test at a variety of locations in most metropolitan areas. The Centers for Disease Control also maintains a national AIDS hotline that you can contact for information and advice by calling 1-800-342-7154.

*Confidentiality.* At the Red Cross the testing is completely confidential. The test is coded with a number that you select—the test is not associated with your name—and you telephone a few days later for the result of the test identified with that number. But maintaining the confidentiality of a positive result is very difficult, even though the test itself is completely confidential. When an individual does have a positive AIDS test, it is necessary to see a doctor frequently to monitor possibilities of disease progression. All this has to be documented in the physician's office, along with billing information. Because these files may be accessed when you sign a medical release form or pay bills with insurance, it is impossible to maintain confidentiality over a long period.

The only way to survive AIDS is not to contract it. The only way (short of contaminated needles) that any student will contract AIDS is by having unprotected sex with someone who has the virus. But the 5-year lag prevents anyone from knowing who is infected and who is not, and the disease continues to spread. Although widespread antibody testing on college campuses has been proposed by some, to identify infected individuals and so dampen the spread of the virus while it is still confined to a relatively few individuals, the proposal is controversial, and the associated dangers of invasion of privacy are a real concern. In the absence of such wide-scale on-campus testing, any student you have sex with might be a carrier and not know it. To avoid becoming part of the epidemic, you have to accept the responsibility of protecting your health.

# THE PHYSIOLOGY OF HUMAN INTERCOURSE

Few physical activities are more pleasurable to humans than sexual intercourse. It is one of the strongest drives directing human behavior, and, as such, it is circumscribed by many rules and customs. Few subjects are at the same time more private and of more general interest.

Until relatively recently, the physiology of human sexual activity was largely unknown. Perhaps because of the prevalence of strong social taboos against the open discussion of sexual matters, research on the subject was not being carried out, and detailed information was lacking. Each of us learned from anecdote, from what our parents or friends told us, and eventually from experience. Largely through the pioneering efforts of William Masters and Virginia Johnson in the last 25 years, and an army of workers who have followed them, this gap in the generally available information about the biological nature of our sexual lives has now largely been filled.

Sexual intercourse is referred to by a variety of names, including copulation and coitus, as well as a host of more informal ones. It is common to partition the physiological events that accompany intercourse into four periods, although the division is somewhat arbitrary, with no clear divisions between the four phases. The four periods are **excitement, plateau, orgasm,** and **resolution** (Figure 42-17).

## Excitement
The sexual response is initiated by commands from the brain that increase the heartbeat, blood pressure, and rate of breathing. These changes are very similar to the ones that the brain induces in response to alarm. Other changes increase the diameter of blood vessels, leading to increased peripheral circulation. The nipples commonly become erect and more sensitive. In the genital area of males this increased circulation leads to the vasocongestion in the penis that produces erection. Similar swelling occurs in the **clitoris** of the female, a small knob of tissue composed of a shaft and glans much like the male penis but without the urethra running through it. The female experiences additional changes that prepare the vagina for sexual intercourse. The increased circulation leads to swelling and parting of the lips of tissue, or **labia,** that cover the opening to the vagina; the vaginal walls become moist; and the muscles encasing the vagina relax.

## Plateau
The penetration of the vagina by the thrusting penis results in the repeated stimulation of nerve endings both in the tip of the penis and in the clitoris. The clitoris, which is now swollen, becomes very sensitive and withdraws up into a sheath or "hood." Once it has withdrawn, the stimulation of the clitoris is indirect, with the thrusting movements of the penis rubbing the clitoral hood against the clitoris. The nerve stimulation that is produced by the repeated movements of the penis within the vagina elicits a continuous sympathetic nervous system response, greatly intensifying the physiological changes that were initiated in the excitement phase. In the female, pelvic thrusts may begin, whereas in the male the penis maintains its rigidity.

## Orgasm
The **climax** of intercourse is reached when the stimulation is sufficient to initiate a series of reflexive muscular contractions. The nerve signals producing these contractions are associated with other nervous activity within the CNS, activity that we experience as intense pleasure. In females the contractions are initiated by impulses in the hypothalamus, which causes the pituitary to release large amounts of the hormone oxytocin. This hormone in turn causes the muscles in the uterus and around the vaginal opening to contract. Or-

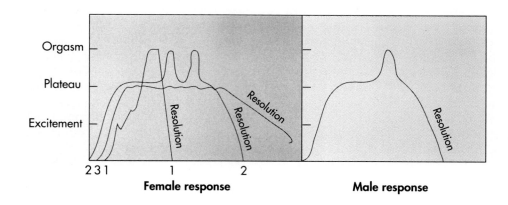

**Female response**

**Male response**

*Figure 42-17*

**The human orgasmic response.** Among females, the response is highly variable. It may be typified by one of the three patterns illustrated here. Among males, the response does not vary as much as it does among females, with a single intense response peak.

gasmic contractions occur about one second apart. There may be one intense peak of contractions (an "orgasm") or several, or the peaks may be more numerous but less intense (see Figure 42-17).

Analogous contractions occur in the male, initiated by nerve signals from the brain. These signals first cause **emission,** in which the rhythmic peristaltic contraction of the vas deferens and of the **prostate gland** causes the sperm and seminal fluid to move to a collecting zone of the urethra. This collecting zone, which is located at the base of the penis, is called the **bulbourethral** area. Shortly after the sperm move into the bulbourethral zone, nerve signals from the brain induce violent contractions of the muscles at the base of the penis, resulting in the ejaculation of the collected semen out through the penis. As in the female orgasm, the contractions are spaced about a second apart, although in the male they continue for a few seconds only. Unlike those of the female, the orgasmic contractions in the male do not vary in their pattern; they are restricted to the single intense wave of contractions that is associated with ejaculation.

### Resolution

After ejaculation, males lose their erection and enter a **refractory period** often lasting 20 minutes or longer, in which sexual arousal is difficult to achieve and ejaculation is almost impossible. After orgasm, the bodies of both men and women return slowly, over a period of several minutes, to their original physiological state.

## CONTRACEPTION AND BIRTH CONTROL

In most vertebrates, sexual intercourse is associated solely with reproduction. In most vertebrates reflexive behavior that is deeply ingrained in the female limits sexual receptivity to those periods of the sexual cycle when she is fertile. In human beings, however, sexual behavior serves a second important function, the reinforcement of pair bonding, the emotional relationship between two individuals. The evolution of strong pair bonding is not unique to human beings, but it was probably a necessary precondition for the evolution of our increased mental capacity. The associative activities that make up human "thinking" are largely based on learning, and learning takes time. Human children are very vulnerable during the extended period of learning that follows their birth, and they require parental nurturing. It is perhaps for this reason that human pair bonding is a continuous process, not restricted to short periods coinciding with ovulation. Among all the vertebrates, human females and a few

species of apes are the only ones in which the characteristic of sexual receptivity throughout the reproductive cycle has evolved and has come to play a role in pair bonding.

Not all human couples want to initiate a pregnancy every time they have sexual intercourse, yet sexual intercourse may be a necessary and important part of their emotional lives together. Among some religious groups, this problem does not arise, or is not recognized because members of these groups believe that sexual intercourse has only a reproductive function and thus should be limited to situations in which pregnancy is acceptable—among married couples wishing to have children. Most couples, however, do not limit sexual relations to procreation, and among them, unwanted pregnancy presents a real problem. The solution to this dilemma is to find a way to avoid reproduction without avoiding sexual intercourse, an approach that is commonly called **birth control** or contraception.

Several different approaches are commonly taken to achieve birth control. These methods (Figure 42-18) differ from one another in their effectiveness and in their acceptability to different couples (Table 42-2).

### Abstinence

The simplest and most reliable way to avoid pregnancy is not to have sex at all. Of all methods of birth control, this is the most certain—and the most limiting, since it denies a couple the emotional support of a sexual relationship.

A variant of this approach is to avoid sexual relations only on the two days preceding and following ovulation because this is the only period during which successful fertilization is likely to occur. The rest of the sexual cycle is relatively "safe" for intercourse. This approach, called the **rhythm method,** is satisfactory in principle but difficult in application because ovulation is not easy to predict and may occur unexpectedly. The effectiveness of the rhythm method is low; the failure rate is estimated to be 12% to 40% (12 to 40 pregnancies per 100 women practicing the rhythm method per year). This is to be compared with a failure rate of 90% for unprotected sex with relations at random.

Another variant of this approach is to have only incomplete sex—withdraw the penis before ejaculation, a procedure known as **coitus interruptus.** This requires considerable willpower, often destroys the emotional bonding of intercourse, and is not as reliable as it might seem. Prematurely released sperm can be secreted by the penis within its lubricating fluid, and a second sexual act may transfer sperm ejaculated earlier. The failure rate of this approach is estimated at 9% to 25%, which is not much better than that of the rhythm method.

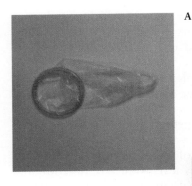

A

B

C

D

E

**Figure 42-18**

Shown here are five of the common means used to achieve birth control.
A  Condom.
B  Diaphragm and spermicidal jelly.
C  Foams.
D  Oral contraceptives.
E  Vaginal sponge.

## Sperm Blockage

If sperm are not delivered to the uterus, fertilization cannot occur. One way to prevent the delivery of sperm is to encase the penis within a thin rubber bag, or **condom.** Many males do not favor the use of condoms because they tend to decrease the sensory pleasure of the male. In principle this method is easy to apply and foolproof, but in practice it proves to be less effective than one might expect, with a failure rate of from 3% to 15%. Nevertheless, it is the most commonly employed form of birth control in the United States. Condoms are also beginning to see widespread use as a means of avoiding AIDS. More than a billion condoms were sold in 1989.

A second way to prevent the entry of sperm into the uterus is to place a cover over the cervix. The cover may be a relatively tight-fitting **cervical cap,** which is worn for days at a time, or a rubber dome called a **diaphragm,** which is inserted immediately before intercourse. Because the dimensions of individual cervices vary, a cervical cap or diaphragm must be fitted by a physician. Failure rates average from 5% to 20% for diaphragms, perhaps because of the propensity to insert them carelessly when in a hurry. Failure rates for cervical caps are somewhat lower, about 8%.

## Sperm Destruction

A third general approach to birth control is to remove or destroy the sperm after ejaculation. This can in principle be achieved by washing out the vagina immediately after intercourse, before the sperm have a chance to travel up into the uterus. Such a procedure is called by the French name for "wash," **douche.** This method turns out to be difficult to apply well because it involves a rapid dash to the bathroom immediately after ejaculation and a very thorough washing. The failure rate has been estimated as high as 40%.

Sperm delivered to the vagina can be destroyed there with spermicidal jellies, sponges, or foams. These require application immediately before intercourse. The failure rate varies widely, from 3% to 22%.

**TABLE 42-2    METHODS OF BIRTH CONTROL\***

| DEVICE | ACTION | FAILURE RATES | ADVANTAGES | DISADVANTAGES |
|---|---|---|---|---|
| Intrauterine devices | Small plastic or metal devices placed in the uterus that somehow prevent fertilization or implantation; some contain copper, others release hormones | 0.5-4 | Convenient, highly effective, need to be replaced infrequently | Can cause excess menstrual bleeding and pain; danger of perforation, infection, and expulsion; not recommended for those who are childless or not monogamous, risk of pelvic inflammatory disease or infertility; dangerous in pregnancy |
| Oral contraceptives | Hormones, either in combination or progestin only, that primarily prevent release of egg | 4-5, depending on type | Convenient and highly effective; provide significant noncontraceptive health benefits, such as protection against ovarian and endometrial cancers | Pills must be taken regularly; possible minor side-effects, which new formulations have reduced; not for women with cardiovascular risks, smokers over 35 |
| Condom | Thin rubber sheath for penis that collects semen | 10 | Easy to use, effective, and inexpensive; protects against some sexually transmitted diseases | Requires male cooperation; may diminish spontaneity; may deteriorate on the shelf |
| Diaphragm with spermicide | Soft rubber cup that covers entrance to uterus, prevents sperm from reaching egg, and holds spermicide | 2-20 | No dangerous side-effects; reliable if used properly; provides some protection against sexually transmitted diseases and cervical cancer | Requires careful fitting; some inconvenience associated with insertion and removal; may be dislodged during sex |
| Cervical cap | Miniature diaphragm that covers cervix closely, prevents sperm from reaching egg, and holds spermicide | Probably comparable to diaphragm | No dangerous side-effects; fairly effective; can remain in place longer than diaphragm | Problems with fitting and insertion; comes in limited number of sizes |
| Foams, creams, jellies, vaginal suppositories | Chemical spermicides inserted in vagina before intercourse that also prevent sperm from entering uterus | 10-25 | Can be used by anyone who is not allergic; protects against some sexually transmitted diseases; no known side effects | Relative unreliable, sometimes messy; must be used 5 to 10 minutes before each act of intercourse |
| Sponge | Acts as sperm barrier and releases spermicide | 13-16 | Safe; easy to insert; provides some protection against sexually transmitted diseases; can be left in place for 24 hours | Relatively unreliable; one size only; some sensitivity and removal problems; cannot be used during menstruation |
| Implant Not available as birth control in U.S. | Capsules surgically implanted under skin that slowly release a hormone that blocks release of eggs | 0.3 | Very safe, convenient, and effective; very long-lasting (5 years); may have nonreproductive health benefits like those of oral conceptives | Irregular or absent periods; necessity of minor surgical procedure to insert and remove |
| Injectable contraceptive Unavailable in U.S. | Injection every 3 months of a hormone slowly released from the muscle that prevents ovulation | 1 | Convenient and highly effective; no serious side-effects other than occasional heavy menstrual bleeding | Animal studies suggest that it may cause cancer, though new studies of women are mostly encouraging |

Source: American College of Obstetricians and Gynecologists: Benefits, Risks, and Effectiveness of Contraception, Washington, D.C., ACOG.
\*Approximate effectiveness of these reversible methods of birth control is measured in pregnancies per 100 actual users per year

## Prevention of Egg Maturation

Since about 1960 a widespread form of birth control in the United States has been the daily ingestion of hormones, or **birth control pills.** These pills contain estrogen and progesterone, either taken together in the same pill or in separate pills taken sequentially. In the normal sexual cycle of a female, these hormones act to shut down the production of the pituitary hormones FSH and LH. The artificial maintenance of high levels of estrogen and progesterone in a woman's bloodstream fools the body into acting as if ovulation had already occurred, when in fact it has not: the ovarian follicles do not ripen in the absence of FSH, and ovulation does not occur in the absence of LH. For these reasons, birth control pills provide a very effective means of birth control, with a failure rate of 0% to 10%. A small number of women using birth control pills experience undesirable side effects, such as blood clotting and nausea. The long-term consequences of the prolonged use of these pills are not yet known, since they have been in widespread use for only 25 years. To date, however, there has been no conclusive evidence of any serious side effects for the great majority of women.

## Surgical Intervention

A completely effective means of birth control, although it is permanent, is the surgical removal of a portion of the tube through which gametes are delivered to the reproductive organs (Figure 42-19). After such an opera-tion, no gametes are delivered during intercourse, which is unaffected in all other respects. The failure rate of such surgical approaches is 0%. The great disadvantage of surgical intervention is that it generally renders the person permanently sterile.

In males, such an operation involves the removal of a portion of the vas deferens, the tube through which sperm travel to the penis. This procedure is called a **vasectomy.** The operation is simple and can be carried out in a physician's office.

In females the comparable operation involves the removal of a section of each of the two fallopian tubes through which the egg travels to the uterus. Since these tubes are located within the abdomen, the operation, called a **tubal ligation,** is more difficult than a vasectomy and is even more difficult to reverse.

The surgical removal of the entire uterus, an operation that is not uncommon but is usually performed for medical reasons other than birth control, is called a **hysterectomy.**

## Prevention of Embryo Implantation

The insertion of a coil or other irregularly shaped object into the uterus is an effective means of birth control, since the irritation in the uterus prevents the implantation of the descending embryo within the uterine wall. Such **intrauterine devices (IUDs)** have a failure rate of only 0.5% to 4%. Their high degree of effectiveness probably reflects their convenience; once they are inserted, they can be forgotten. The great disadvan-

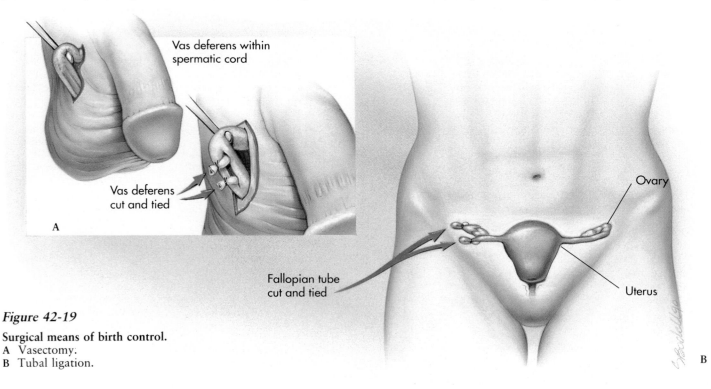

Vas deferens within spermatic cord

Vas deferens cut and tied

A

Fallopian tube cut and tied

Ovary

Uterus

B

*Figure 42-19*

**Surgical means of birth control.**
A  Vasectomy.
B  Tubal ligation.

tage of this method is that almost a third of the women attempting to use IUDs cannot; the devices cause them cramps, pain, and sometimes bleeding.

Another effective way to prevent embryo implantation is the use of the "morning after" pill, which contains 50 times the dose of estrogen present in birth control pills. The failure rate is only 4% to 5%, but many women are uneasy about taking such high hormone doses.

## ABORTION

Reproduction can be prevented after fertilization if the embryo is **aborted** (removed) before its development and birth. During the first trimester this can be accomplished by **vacuum suction** or by **dilation and curettage,** in which the cervix is dilated and the uterine wall is scraped with a spoon-shaped surgical knife called a curette. Chemical methods are also being developed that cause abortion early in the first trimester, apparently with complete safety. One such drug, **RU 486,** is already in use in France and China, although not yet approved for use in the United States. Administration of RU 486 followed by prostaglandins that induce uterine contractions is almost 100% effective when taken within 49 days of the patient's last menstrual period.

In the second trimester the embryo can be removed by injecting a 20% saline solution into the uterus, which induces labor and delivery of the fetus. In general, the more advanced the pregnancy, the more difficult and dangerous the abortion is to the woman.

As a method of birth control, abortion takes a great emotional toll, both on the woman undergoing the abortion and often on others who know and care for her. Abortion also presents serious moral problems for some. Many people believe that the fetus is a living person from the time of conception and that abortion is simply murder. There are many countries in which abortions are defined as a crime. In the United States a fetus is not legally considered a person until birth, and abortions are permitted by law during the first two trimesters. They are illegal in the third trimester, however, except when the mother's life is endangered. The Supreme Court ruling on legalized abortion in the first two trimesters is a relatively recent one, and it is still the subject of intense controversy. The abortion rate in the United States (and most other developed countries) exceeds the birth rate.

## AN OVERVIEW OF VERTEBRATE REPRODUCTION

Vertebrates carry out reproduction in many different ways, reflecting the evolutionary course of their invasion of the land. These differences are reflected in whether fertilization is external or internal, in whether zygotes begin life as eggs or develop internally, in the size of broods and the frequency with which they are produced. The story does not end here. The course of development of the zygote is influenced in many ways by the reproductive strategy of the organism. A zygote nourished by the yolk of an egg develops differently from one nourished by its mother's blood supply. In the next chapter we consider these differences in developmental processes in detail.

Much of the diversity seen among the vertebrates in reproduction and development, however, reflects the transition to the land. As discussed in other chapters of this section, the biology of all vertebrates is similar in most other respects. You employ a different mechanism than a fish does to obtain the oxygen necessary for metabolism, because extracting oxygen from air presents different problems than extracting it from water. However, having obtained oxygen, you circulate it to your tissues in much the same manner that a fish does. The similarities among vertebrates are more striking than the differences between them.

## ■ SUMMARY

1. Sexual reproduction evolved in the sea. Among most fishes and the amphibians, fertilization is external, with gametes being released into water. Amphibians and the great majority of fishes are oviparous, the young being nurtured by the egg rather than by the mother.

2. Successful invasion of the land by vertebrates involved major changes in reproductive strategy. The first change was the watertight egg of the reptiles, which could survive in dry places. Internal fertilization, which is also characteristic of the reptiles, is another important survival strategy on land. Birds, as well as a primitive group of mammals called monotremes, have the same kinds of eggs as reptiles do.

3. The second important change in reproductive adaptation to life on land was that of the marsupials. In them, fertilization is internal. The young are typically born within a few weeks of fertilization, but then they are nourished and protected during the course of their further development within a pouch.

4. A third change was that of the placental mammals. Their young are nourished within the mother's body by means of a large, complex structure, the placenta, which exchanges material from her bloodstream with her offspring via a common circulation of blood.

5. The male gametes, or sperm, of mammals are produced within the testes. In human males hundreds of millions of sperm are produced each day. Sperm mature and become motile in the epididymis, and they are stored in the vas deferens. Stimulation of the penis causes it to become distended and erect and causes the sperm to be delivered from the vas deferens to the urethra at the base of the penis. Further stimulation causes violent muscle contractions, which ejaculate the sperm from the penis.

6. At birth, female mammals contain all the gametes, or oocytes, that they will ever have. A human female has approximately 2 million oocytes. All but a very few of these are arrested in meiotic prophase. At each ovulation the first meiotic division of one or a few eggs is completed. The second meiotic division does not occur until after fertilization.

7. Fertilization occurs within the fallopian tubes. The journey of an egg to the uterus takes 5 to 7 days, and it is viable for only 24 hours unless it is fertilized. Consequently, only those eggs which have been reached within 24 hours by sperm swimming up the fallopian tubes from the uterus can be fertilized successfully. Fertilized eggs continue their journey down the fallopian tubes and attach to the lining of the uterus, where their development proceeds.

8. Reproduction in mammals is regulated by hormones, typically produced by the pituitary on commands from the hypothalamus region of the brain. These chemical signals circulate in the bloodstream. When a hormone molecule reaches its target tissue, it binds to specific receptors of the target cell.

9. The reproductive cycle of mammals is called the estrous cycle. It is composed of two phases: (1) a follicular phase, in which some eggs in the ovary are hormonally signaled to complete their development; and (2) a luteal phase, in which one or more mature eggs are released into the fallopian tube, a process called ovulation. A complete menstrual cycle in a human female takes about 28 days.

10. Human intercourse is marked by four physiological periods: excitement, plateau, orgasm, and resolution. Orgasm in women is highly variable and may be prolonged. Orgasm in men is uniformly abrupt; it coincides with the ejaculation of sperm.

11. A variety of birth control procedures is practiced by humans, of which condoms used by men and birth control pills used by women are perhaps the most common. Intrauterine devices (IUDs) and surgical procedures that block the delivery of gametes are becoming increasingly common. Birth control is not always effective, and some women terminate unwanted pregnancies. In the United States, as in other developed countries, the abortion rate exceeds the birth rate.

## REVIEW

1. In both fishes and amphibians, fertilization is usually internal. True or false?

2. During which phase of the human menstrual cycle is the progesterone level in the blood highest?

3. What goes on in seminiferous tubules?

4. Where are eggs fertilized in the female human reproductive system?

# SELF-QUIZ

1. Humans are
   - (a) viviparous.
   - (b) marsupial mammals.
   - (c) placental mammals.
   - (d) a and b.
   - (e) a and c.

2. Follicle-stimulating hormone (FSH) is produced by the _____ and binds to receptors on the _____.
   - (a) hypothalamus
   - (b) pituitary gland
   - (c) adrenal gland
   - (d) ovary
   - (e) follicle

3. Human male gametogenesis produces _____ gamete(s); human female gametogenesis produces _____ gamete(s).
   - (a) one
   - (b) two
   - (c) three
   - (d) four
   - (e) five

4. Human sexual response generally occurs as four phases. Put them in the order in which they take place.
   - (a) Resolution
   - (b) Orgasm
   - (c) Excitement
   - (d) Plateau

# THOUGHT QUESTIONS

1. Probably the most controversial issue to arise in our consideration of sex and reproduction is abortion. List arguments for and against abortion. Under what conditions would you permit abortions? Forbid them? Do you think the disadvantages of abortion are in any way counterbalanced by the advantages it provides to underdeveloped countries with very high birth rates and swelling populations? Should the United States promote or oppose dissemination of information about abortion in underdeveloped countries?

2. Some fishes and many reptiles are viviparous, retaining fertilized eggs within a mother's body to protect them. Birds, however, although they evolved from reptiles, never employ this means of protecting their eggs. Can you think of a reason why?

3. Relatively few kinds of animals have both male and female sex organs, whereas most plants do. Propose an explanation for this.

# FOR FURTHER READING

LAGERCRANTZ, H., and A. SLOTKIN: "The 'Stress' of Being Born," *Scientific American*, April 1986, pages 100-107. Passage through the narrow birth canal triggers the release of hormones important to the newborn's future survival.

LEIN, A.: *The Cycling Female*, W.H. Freeman & Co., San Francisco, 1979. A short, informal description of the human menstrual cycle and its physical and emotional effects on women.

LEISHMAN, K.: "Heterosexuals and AIDS," *The Atlantic*, February 1987, pages 39-58. A chilling account of the difficulty of modifying sexual behavior, despite the knowledge of the dangers associated with AIDS.

C H A P T E R ■ 4 3

# Development

Adulthood is achieved only after a long journey of growth and development. This young bobcat is almost grown—only months before, it was an embryo.

# DEVELOPMENT

## Overview

Development in humans and other vertebrates is a complex, dynamic process, a symphony of cell movement and change that starts with the formation of a single diploid cell from sperm and egg. From this single cell will arise 100 trillion others as the individual that cell is destined to become develops and grows to adulthood. The developmental journey is begun with a series of rapid cell divisions, resulting in a ball of cells. Some of these cells then migrate to the inside to form a more-or-less spherical structure that possesses the three primary tissue layers. Although the details differ, this process occurs in all deuterostomes. The development of the specific tissues of the body follows. Then organs are formed. Only when this pattern of development is complete (some 3 months in humans) does growth begin in earnest, a process that in humans takes another 6 months before birth. A human baby is 266 days old when it is born.

## For Review

*Here are some important terms and concepts that you will encounter in this chapter. If you are not familiar with them, you should review them before proceeding.*

**Deuterostomes** (Chapter 28)

**Chordates** (Chapter 28)

**The amniote egg** (Chapter 28)

**Terrestrial reproductive strategies** (Chapter 42)

### Figure 43-1

A human fetus at 18 weeks is not yet halfway through the 38 weeks—about 9 months—it will spend within its mother, but already it has developed many distinct behaviors, such as the sucking reflex that is so important to survival after birth.

In the preceding 11 chapters, you have learned what an adult vertebrate body is like and how it functions. Chapters have described the various tissues of the vertebrate body and then considered how vertebrates eat, breathe, and sense the world around them. All the complexity you have encountered, the beauty and cleverness of design, are established during an elaborate developmental process we consider in this chapter. Development is not simply the beginning of life—it is the process that determines life's form and substance.

In almost all vertebrates, development begins with an encounter between two haploid gametes that unite to form a single diploid cell called a zygote. This zygote grows by a process of cell division and differentiation into a complex multicellular animal composed of many different tissues and organs (Figure 43-1). The process of development comprises the events that occur after the union of the two haploid gametes. Although some of its details differ from group to group, development is fundamentally the same in all vertebrates.

In this chapter we discuss the stages of vertebrate development (Table 43-1), then consider the mechanisms governing developmental changes, and conclude with a detailed description of the events that occur during the course of human development.

**TABLE 43-1 STAGES OF DEVELOPMENT**

| | |
|---|---|
| 1. Fertilization | The male and female gametes form a zygote. |
| 2. Cleavage | The zygote rapidly divides into many cells, with no overall increase in size. These divisions set the stage for development, since different cells receive different portions of the egg cytoplasm and hence different regulatory signals. |
| 3. Gastrulation | The cells of the zygote move, forming three cell layers. These layers are the primary cell types: ectoderm, mesoderm, and endoderm. |
| 4. Neurulation | In all chordates the first organ to form is the notochord, followed by formation of the dorsal nerve cord. |
| 5. Neural crest formation | The first uniquely vertebrate event is the formation of the neural crest. From it develop many of the uniquely vertebrate structures. |
| 6. Organogenesis | The three primary cell types then proceed to combine in various ways to produce the organs of the body. |

## FERTILIZATION: BEGINNING THE ACTION

In vertebrates, as in all sexual animals, the first step in reproduction is the union of male and female gametes, a process called **fertilization.** Fertilization consists of three stages: (1) penetration, (2) activation, and (3) pronuclear fusion. The male gametes of vertebrates, like those of other animals, are small, motile sperm. Each sperm is shaped like a tadpole, with a head containing a haploid nucleus and a long tail. Sperm are among the smallest cells in the body. The female gametes, called eggs, or **oocytes,** are large cells. In many vertebrates the eggs contain significant amounts of yolk.

### Penetration: A Sperm Must Work to Enter an Egg

In fishes and amphibians, fertilization is typically external, whereas in all other vertebrates it occurs internally. Internal fertilization is achieved by the release of a mature egg into a body cavity into which sperm can be introduced. There the egg can be fertilized by one of the many sperm introduced into the female reproductive tract during mating. The actively swimming sperm migrate up the oviduct until they encounter a mature egg.

Like a traveling princess, the mammalian egg is surrounded by a great deal of baggage (Figure 43-2). The egg cell itself is encased within an outer membrane called the **zona pellucida,** which is in turn surrounded

*Figure 43-2*

**Some eggs attract more sperm than do others.** Eggs such as this human ovum (the large cell in the center of the photograph) are often surrounded by numerous nutritional cells. Often many sperm crowd around such a cell, whereas in other vertebrates a solitary sperm may effect penetration, as the hamster sperm has done.

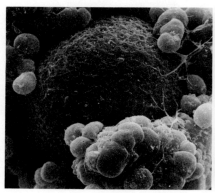
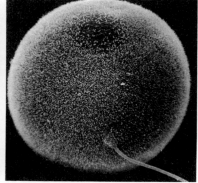

by a protective layer of follicle cells. The first sperm to make its way through the zona pellucida adheres to the egg membrane by the tip of the sperm cell head, the acrosome. From its acrosome, the sperm releases enzymes that cause the plasma membranes of the sperm and egg cell to fuse. Egg cytoplasm bulges out at this point, engulfing the head of the sperm and permitting the sperm nucleus to enter the cytoplasm of the egg (Figure 43-3).

## Activation: The Egg Responds Quickly to the Sperm's Entry

Entry of the sperm nucleus into the egg has three effects:

1. The series of events initiated by sperm penetration is collectively called **egg activation.** In general, the penetration of the first sperm initiates changes in the egg cell membrane that prevents the entry of other sperm.

2. In mammals, eggs are produced by meiotic divisions that occur early in a female's life, but in most cases meiosis stops at an intermediate stage. The penetration of the sperm triggers the resumption of meiosis. On sperm penetration, the chromosomes in the egg nucleus complete meiosis, producing two egg nuclei. One of these two newly formed nuclei is extruded from the egg cell as a polar body, leaving a single haploid egg nucleus within the egg.

3. In vertebrates, a third effect of sperm penetration is the rearrangement of the egg cytoplasm and its metabolic activation. Around the point of sperm entry a series of cytoplasmic movements is initiated within the egg. These movements ultimately establish the bilateral symmetry of the developing organism. In frogs, for example, sperm penetration causes an outer pigmented cap of egg cytoplasm to rotate toward the point of entry, uncovering a **gray crescent** of

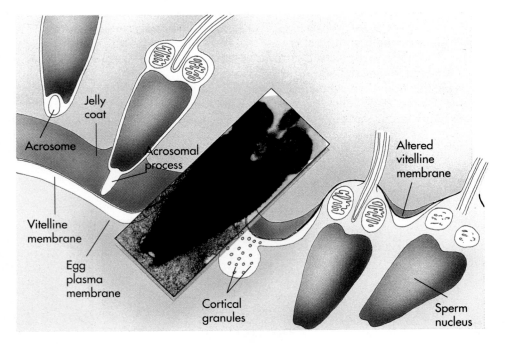

*Figure 43-3*

**Sperm penetration of a sea urchin egg.** The electron micrograph in the middle of the diagram shows a sperm imbedded in the egg, magnified 50,000 times.

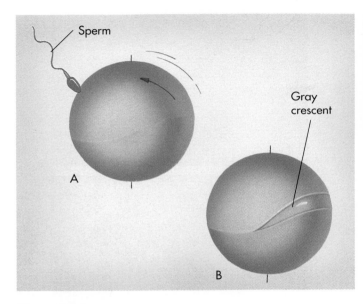

**Figure 43-4**

**Gray crescent formation in frogs.** Appearance of the gray crescent opposite the point of sperm penetration.

interior cytoplasm opposite the point of penetration (Figure 43-4). The position of the gray crescent determines the orientation of initial cell division. A line drawn between the point of sperm entry and the gray crescent would bisect the right and left halves of the future adult.

In some vertebrates it is possible to activate an egg without the entry of a sperm by simply pricking the egg membrane. If the development of any egg is stimulated in this manner, the egg may go on to develop parthenogenetically. A few kinds of amphibians, fishes, and reptiles rely entirely on parthenogenetic reproduction. The eggs of some domestic animals, such as turkeys, can be made to reproduce by parthenogenesis.

### Pronuclear Fusion: Joining of Egg and Sperm Creates a Diploid Nucleus

The third stage of fertilization is the pronuclear fusion of the entering sperm nucleus with the haploid nucleus of the egg to form a diploid zygotic nucleus. This pronuclear fusion is triggered by the activation of the egg. If a sperm nucleus is introduced by microinjection without activation of the egg, pronuclear fusion of the two nuclei will not take place. The nature of the signals that are exchanged between the two nuclei, or sent from one to the other, is not known.

> The three stages of fertilization are penetration, activation, and pronuclear fusion. Penetration initiates a complex series of developmental events, including major movements of cytoplasm, that eventually lead to the fusion of the egg and sperm nuclei.

## CLEAVAGE: SETTING THE STAGE FOR DEVELOPMENT

The second major event in vertebrate reproduction is the rapid division of the zygote into a larger and larger number of smaller and smaller cells. This period of division, called **cleavage,** is not accompanied by any increase in the overall size of the embryo. The resulting tightly packed mass of about 32 cells is called a **morula,** and each individual cell in the morula is referred to as a **blastomere.** Blastomeres are by no means equivalent to one another. Different blastomeres may contain different components of the egg cytoplasm, particularly at later stages of division. When these components differ, they dictate different developmental fates for the cells in which they are present. The cells of the morula continue to divide without an overall increase in size, each cell secreting a fluid into the center of the cell mass. Eventually a hollow mass of 500 to 2000 cells is formed, called a **blastula.** This hollow mass of cells surrounds a fluid-filled cavity called the **blastocoel.**

### The Presence of Yolk Determines the Pattern of Cleavage

The pattern of cleavage division is greatly influenced by the yolk (Figure 43-5). As we discussed in the last chapter, vertebrates have embraced a variety of reproductive strategies that involve different patterns of yolk utilization.

***Primitive Aquatic Vertebrates.*** When eggs contain little or no yolk, cleavage occurs throughout the whole egg (Figure 43-6). This pattern is called **holoblastic cleavage.** Holoblastic cleavage was characteristic of the ancestors of the vertebrates and is still seen in groups such as the lancelets and agnathans. It results in the formation of a symmetrical blastula composed of cells of approximately equal size.

***Amphibians and Advanced Fishes.*** The eggs of bony fishes and frogs contain much more yolk in one hemi-

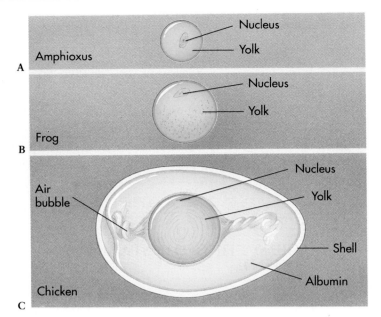

Figure 43-5

**Three kinds of eggs.**
A In the primitive vertebrate *Amphioxus* the organization of the egg is simple, with a central nucleus surrounded by yolk.
B In a frog egg there is much more yolk, and the nucleus is displaced toward one pole.
C Bird eggs have complex organization, with the nucleus astride the surface by a large central yolk like a spot painted on a balloon.

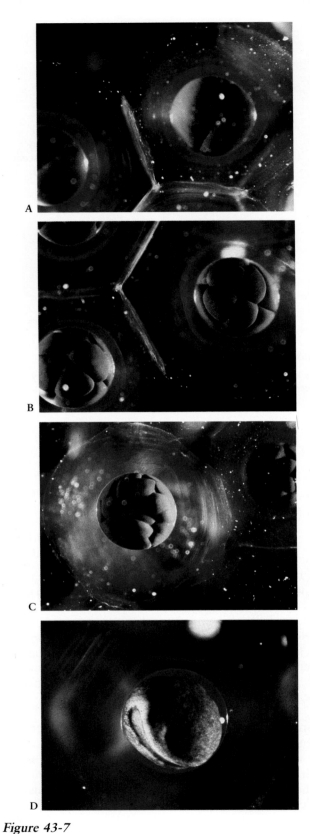

Figure 43-7

**Creation of a frog embryo.** The initial cleavage divisions (**A**), on the side of the zygote facing you, produce a cluster of cells (**B**), which soon expands to become a compact mass of cells (**C**). This mass eventually invaginates into the interior of the embryo (**D**), forming a gastrula-stage embryo.

sphere than in the other. Because yolk-rich cells divide much more slowly than do those which are poor in yolk, unequal holoblastic cleavage of these eggs results in a very asymmetrical blastula (Figure 43-7), with large cells containing a lot of yolk at one pole and a concentrated mass of small cells containing very little yolk at the other.

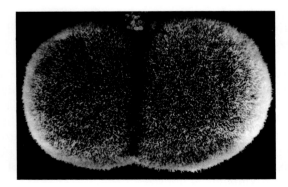

Figure 43-6

**Holoblastic cleavage.** Holoblastic cleavage is symmetrical, dividing an egg into equal portions. The egg dividing here is that of a mouse.

*Reptiles and Birds.* Some eggs are composed almost entirely of yolk, with a small amount of cytoplasm concentrated at one pole. Such eggs are typical of reptiles, birds, and some fishes. In such eggs, cleavage occurs only in the tiny disk of polar cytoplasm, called the **blastodisc,** which lies astride the large ball of yolk material (Figure 43-8). Such a pattern of cleavage is called **meroblastic cleavage.** The resulting blastoderm is not spherical, but rather has the form of a hollow cap perched on the yolk.

*Mammals.* Mammalian eggs are in many ways similar to the reptilian eggs from which they evolved, except that they contain very little yolk. Because there is no mass of yolk to impede cleavage in mammalian eggs, the cleavage of the developing zygote is holoblastic. Such cleavage forms a ball of cells surrounding a blastocoel. In mammalian eggs, there is an inner cell mass concentrated at one pole (Figure 43-9). This interior plate of cells is analogous to the blastodisc of reptiles. It goes on to form the developing embryo. The outer sphere of cells is called a **trophoblast.** It is analogous to the cells that form the membrane that functions as a watertight covering and conserves water within the reptilian egg. These cells have changed during the course of mammalian evolution to carry out a very different function. The trophoblast develops into the chorion and a complex series of membranes known as the placenta, which connects the developing embryo to the blood supply of the mother.

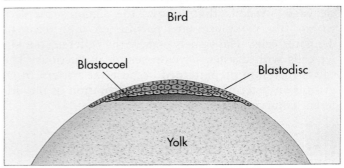

*Figure 43-9*

**Mammals and birds are more similar than they seem.** A mammalian blastula is composed of a sphere of cells, the trophoblast, surrounding a cavity, the blastocoel, and an inner cell mass. An avian (bird) blastula is a disk rather than a sphere, a blastodisc resting astride a large yolk mass; in bird eggs the blastocoel is a cavity between the blastodisc and the yolk.

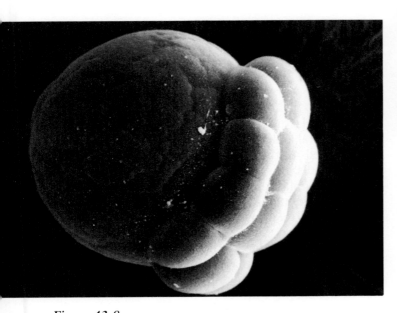

*Figure 43-8*

**Meroblastic cleavage.** Meroblastic cleavage is asymmetrical, with only a portion of a fertilized egg actively dividing to form a cell mass. The dividing cells of this fish embryo are magnified 400 times.

Development is initiated in the zygote by a series of rapid cell divisions called cleavage divisions, producing a ball of cells called a blastula. The evolution of the amniotic egg in reptiles caused an alteration in the pattern of cleavage, which was related to the presence of a yolk. This kind of cleavage pattern is carried on by mammals, a reflection of their ancestry.

## A Blastula's Cells are Already Different from One Another

Viewed from the outside, the blastula looks like a simple ball of cells that all resemble one another. But the apparently close similarity of these cells is misleading. In fact, they differ from one another in three essential respects:

1. Each cell contains a different portion of cytoplasm derived from the egg.
2. Some cells are larger than others, containing more yolk and dividing more slowly.
3. Each cell is in contact with a different set of neighboring cells.

Egg cells contain many substances that act as genetic signals during the early stages of zygote development. But these signal substances are not distributed uniformly within the egg cytoplasm. Instead, they are clustered at specific sites within the egg. The location of each site is genetically determined by information encoded on the mother's chromosomes. When the egg is activated during fertilization, its cytoplasm reorients itself with respect to the site of sperm entry. During the cleavage divisions that follow, the signal substances within this cytoplasm are partitioned into different daughter cells. The signals endow the different daughter cells with distinct developmental instructions. The egg is therefore **prepatterned** because the pattern of its cytoplasm determines the future orientation of the different embryonic cells.

## GASTRULATION: THE ONSET OF DEVELOPMENTAL CHANGE

The first visible results of prepatterning and of the cell orientation within the blastula can be seen immediately after completion of the cleavage divisions. Certain groups of cells *move* inward from the surface of the sphere in a carefully orchestrated migration called **gastrulation.**

How can cells move within a cell mass? Recall from Chapter 4 that cell shape can readily be changed by microfilament contraction. Apparently the migrating cells creep over the stationary ones by means of a series of microfilament contractions. The migrating cells move as a single mass because they adhere to one another. How do cells "know" to which other cell to adhere? Within an adhering cell, genes have been expressed that cause the synthesis of specific adhesive molecules on the cell surface that adhere to similar adhesive molecules on the surfaces of the other adhering cells.

### The Form of the Blastula Shapes the Movements of Gastrulation

During gastrulation, about half of the cells of the blastula move into the interior of the hollow ball of cells. Just as the pattern of cleavage divisions in different groups of vertebrates depends heavily on the amount and distribution of yolk in the egg, so the pattern of gastrulation varies among the vertebrates, depending on the shape of the blastulas produced by the earlier cleavage divisions.

*Aquatic Vertebrates.* In fishes and other aquatic vertebrates with asymmetrical yolk distribution in their eggs, the blastula produced by the cleavage divisions has two distinct poles, one more rich in yolk than the other. The hemisphere of the blastula comprising cells rich in yolk is called the **vegetal pole;** the opposite hemisphere, comprising cells relatively poor in yolk, is called the **animal pole.** In primitive chordates such as lancelets, the animal hemisphere bulges inward, invaginating into the blastocoel cavity. Eventually the inward-moving wall of cells pushes up against the opposite side of the blastula, and then it ceases to move. The resulting two-layered, cup-shaped embryo is the **gastrula.** The hollow crater resulting from the invagination is called the **archenteron,** and it becomes the progenitor of the gut. The opening of the archenteron, the future anus of the lancelet, is the **blastopore.**

Gastrulation in the lancelets produces an embryo with two cell layers, an outer ectoderm and an inner endoderm. A third cell layer, the mesoderm, forms soon afterward between these two layers from pouches pinched off of the endoderm. The formation of these three primary cell types sets the stage for all subsequent tissue and organ differentiation because the descendants of each cell type are destined to have very different developmental fates (Table 43-2).

| TABLE 43-2 | DEVELOPMENTAL FATES OF THE PRIMARY TISSUES |
|---|---|
| Ectoderm | Skin, central nervous system, sense organs, neural crest |
| Mesoderm | Skeleton, muscles, blood vessels, heart, gonads |
| Endoderm | Digestive tract, lungs, many glands |

In the blastula of amphibians the yolk-laden cells of the vegetal pole are fewer and far larger than the yolk-free cells of the animal pole. Because of this distribution of cells, it is mechanically not feasible to invaginate the blastula at the vegetal pole. Instead, a layer of cells from the animal pole folds down over the yolk-rich cells and then invaginates inward (Figure 43-10). The place where the invagination begins is called the **dorsal lip.** As in the lancelets, the invaginating cell layer eventually eliminates the blastocoel cavity, its cells pressing against the inner surface of the opposite side of the embryo. In both fishes and amphibians, the opening of the cavity produced by the invagination is called the blastopore. In this case the blastopore is filled with yolk-rich cells, the **yolk plug.** The outer layer of cells in the gastrula, which is formed as a result

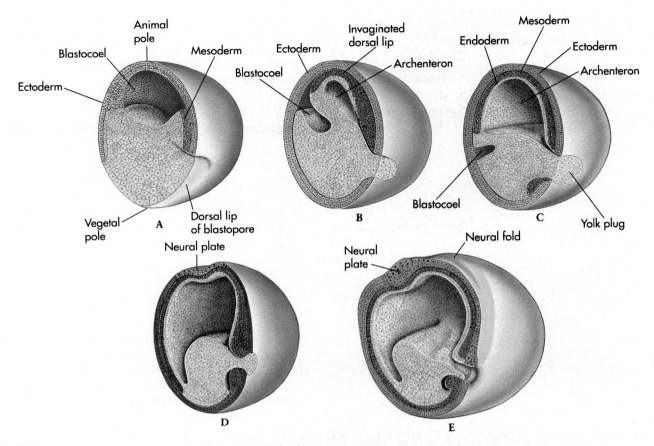

**Figure 43-10**

**Frog gastrulation.**
**A** A layer of cells from the animal pole folds down over the yolk cells, forming the dorsal lip.
**B** The dorsal lip zone then invaginates into the hollow interior, or blastocoel, eventually pressing against the far wall. The three principal tissues (ectoderm, mesoderm, endoderm) become distinguished here.
**C** The inward movement of the dorsal lip creates a new internal cavity, the archenteron, which opens to the outside through the plug of yolk remaining at the point of invagination.
**D** The neural plate later forms from ectoderm and folds down over the surface of the embryo.
**E** The neural plate moves to the interior, completing gastrulation.

of these cell movements, is the ectoderm, and the inner layer is the endoderm. Some ectodermal cells migrate between the ectoderm and endoderm to form the mesoderm layer.

*Reptiles, Birds, and Mammals.* In the blastodisc of a chick the developing embryo is not shaped like a sphere. Instead, it is a hollow cap of cells situated over the animal pole of the large yolk mass. Despite this seemingly great difference between the developing embryos of birds, reptiles, and mammals on the one hand and amphibians on the other hand, the pattern of es-

tablishing the three primary cell layers is basically similar in all of these groups.

There is no yolk separating the two sides of the blastodisc in reptiles, birds, and mammals (Figure 43-11). Consequently and without cell movement, the lower cell layer is able to differentiate into endoderm, the upper layer into ectoderm. Just after this differentiation, much of the mesoderm and endoderm arises by the invagination of cells from the upper layer inward, along the edges of a furrow that appears at the longitudinal midline of the embryo. The site of this invagination, which is analogous to an elongated blastopore,

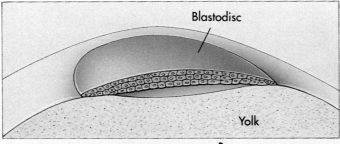

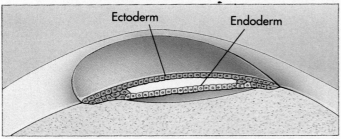

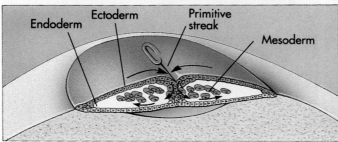

**Figure 43-11**

**Gastrulation of the chick blastodisc.** The upper layer of the blastodisc differentiates into ectoderm, the lower layer into endoderm. Among the cells that migrate into the interior through the dorsal primitive streak are future mesodermal cells.

appears as a slit on the surface of the gastrula. Because of its appearance, it is called the **primitive streak.** Gastrulation occurs at the site of formation of a primitive streak in the reptiles and their descendants, the birds and mammals (Figure 43-12).

*The many cells of the blastula gain unequal portions of egg cytoplasm during cleavage. This asymmetry results in the activation of different genes and a repositioning of cells with respect to one another, which establishes the three primary cell types: ectoderm, mesoderm, and endoderm.*

The events of gastrulation determine the basic developmental pattern of the vertebrate embryo. By the end of gastrulation, distribution of cells into the three primary cell types has been completed. Although the position of the yolk mass dictates changes in the details of gastrulation, the end result of the process is fundamentally the same in all deuterostomes: the ectoderm is destined to form the epidermis and neural tissue; the mesoderm to form the connective tissue, muscle, and vascular elements; and the endoderm the lining of the gut and its derivatives.

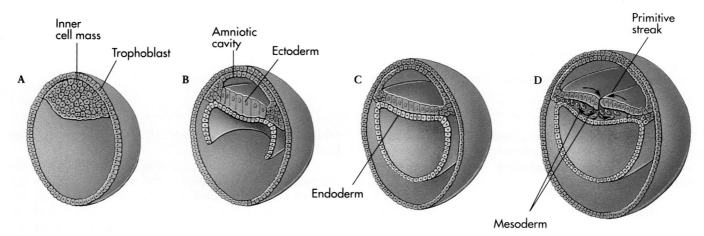

**Figure 43-12**

**Mammalian gastrulation.** The amniotic cavity forms within the inner cell mass (**A**), and in its base, layers of ectoderm and endoderm differentiate (**B and C**), as in the chick blastodisc. A primitive streak develops, through which cells destined to become mesoderm migrate into the interior (**D**), again reminiscent of gastrulation in the chick.

# NEURULATION: THE DETERMINATION OF BODY ARCHITECTURE

In the next step in vertebrate development the three primary cell types begin their development into the tissues and organs of the body. In all chordates, tissue differentiation begins with the formation of two characteristic morphological features, the notochord and the hollow dorsal nerve cord. This stage in development, called **neurulation,** occurs only in the chordates.

The first of these two structures to form is the notochord (see Figure 28-34). It is first visible soon after gastrulation is complete, forming from mesoderm tissue along the midline of the embryo, below its dorsal surface.

After the notochord has been laid down, the dorsal nerve cord forms from the endoderm. The region of the ectoderm that is located above the notochord later differentiates into the spinal cord and brain. The process is illustrated in Figure 43-13. First, a layer of ecto-dermal cells situated above the notochord invaginates inward, forming a long groove called the **neural groove** along the long axis of the embryo. Then the edges of this groove move toward each other and fuse, creating a long, hollow tube, the **neural tube,** which runs beneath the surface of the embryo's back.

> *The key developmental event that marks the evolution of the chordates is neurulation, the elaboration of a notochord and a dorsal nerve cord.*

While the neural tube is forming from ectoderm, the rest of the basic architecture of the body is being rapidly determined by changes in the mesoderm (Figure 43-14). On either side of the developing notochord, segmented blocks of tissue form. Ultimately, these blocks, or **somites,** give rise to the muscles, vertebrae, and connective tissue. As the process of development continues, more somites are formed progressively. Many of the significant glands of the body, including the kidneys, adrenal glands, and gonads, develop within another strip of mesoderm that runs alongside the somites. The remainder of the mesoderm layer moves out and around the inner endoderm layer of cells and eventually surrounds it entirely. As a result of this movement, the mesoderm forms a hollow tube within the ectoderm. The space within this tube is the coelom; it contains the endoderm layers that ultimately form the lining of the stomach and gut.

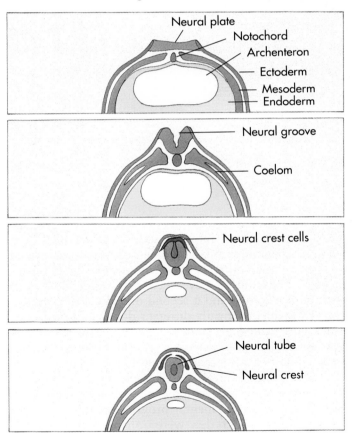

**Figure 43-13**

**Neural tube formation.** The neural tube forms above the notochord (**A**) when cells of the neural plate fold together to form the neural groove (**B**), which eventually closes (**C**) to form a hollow tube (**D**). As this is happening, some of the cells from the dorsal margin of the neural tube differentiate into the neural crest, which is characteristic of vertebrates.

## SEX DETERMINATION

In humans and all other mammals, sex is determined by the Y chromosome. Any individual that possesses a Y chromosome develops into a male, whereas any individual that lacks a Y chromosome does not, becoming instead a female. Thus males are XY and females are XX. In the summer of 1990, researchers found the gene on the Y chromosome that nudges a human embryo towards maleness. By studying the rare exceptions when an XX individual is male and an XY is female, they identified a smidgen of DNA that is absent from an XY female's Y chromosome but present on the X chromosomes of an XX male. A single gene was isolated within this short segment that seems to be present in all male mammals but not in females. Tests are now underway to see if insertion of a DNA fragment containing this gene into an XX mouse embryo will turn a "her" into "him."

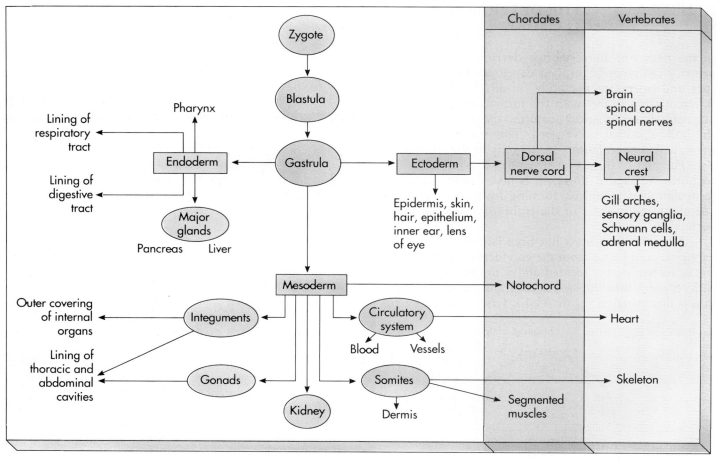

**Figure 43-14**

**Derivation of the major tissue types.** The key role of the neural crest is evident from the many characteristically vertebrate features that derive from it.

## HOW CELLS COMMUNICATE DURING DEVELOPMENT

In the process of vertebrate development (Figure 43-15) the relative position of particular cell layers determines, to a large extent, the organs that develop from them. By now you may have wondered how these cell layers know where they are. For example, when cells of the ectoderm situated above the developing notochord give rise to the neural groove, how do these cells know they are above the notochord?

The solution to this puzzle is one of the outstanding accomplishments of **experimental embryology,** the study of how embryos form. It was worked out by German biologist Hans Spemann and his student Hilde Mangold early in this century. They removed cells from the dorsal lip of an amphibian blastula (Figure 43-16) and transplanted them to a different location on another blastula. The dorsal lip region of amphibian blastulas develops from the gray crescent zone and is the site of origin of those mesoderm cells which later produce the notochord. The new location corresponded to that of the future belly of the animal. What happened? The embryo developed *two* notochords, one normal dorsal one and a second one along its belly!

By using genetically different donor and host blastulas, Spemann and Mangold were able to show that the notochord produced by transplanting dorsal lip cells contained host cells as well as transplanted ones. The transplanted dorsal lip cells had acted as **organizers** of notochord development. As such, these cells stimulated a developmental program in the belly cells of embryos to which they were transplanted: the development of a notochord. The belly cells clearly contained this developmental program but would not have expressed it in the normal course of their development. The transplantation of the dorsal lip cells caused them to do so. These cells had *induced* the ectoderm cells of the belly to form a notochord. This phenomenon as a whole is known as **induction.**

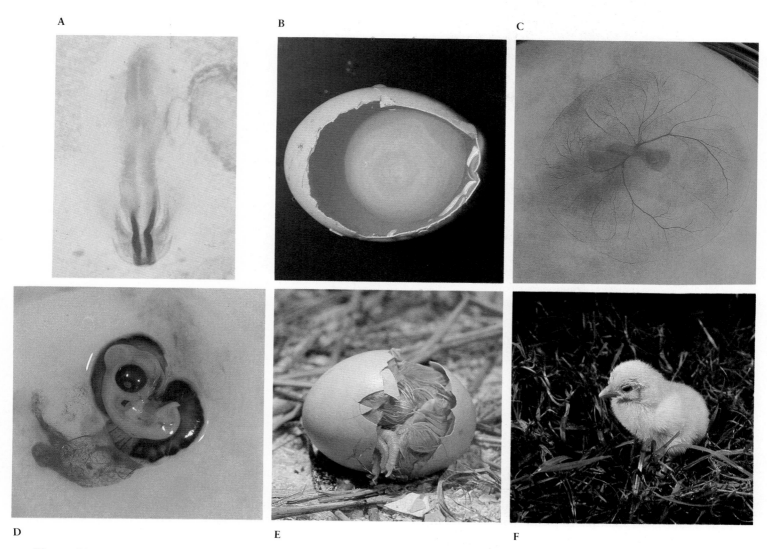

A

B

C

D

E

F

**Figure 43-15**

**Development in the chick.**
**A** After 24 hours, development of the neural groove is well advanced.
**B** After 36 hours, the embryo has grown much larger.
**C** After 72 hours (3 days), the eyes have already formed. In this photograph the shell has been removed to show the extensive blood circulation around the developing embryo.
**D** After 7 days, most of the body's internal organs are present.
**E** After 21 days, the chick pecks its way out of the egg. Note that it does not have much room to maneuver.
**F** This pensive chick is 1 day old. Although not yet an adult, it is fully able to live on its own.

*Induction is the determination of the course of development of one tissue by another tissue.*

The process of induction that Spemann and Mangold discovered appears to be the basic mode of development in vertebrates. Inductions between the three primary tissue types—ectoderm, mesoderm, and endoderm—are referred to as **primary inductions.** Inductions between tissues that have already differenti-

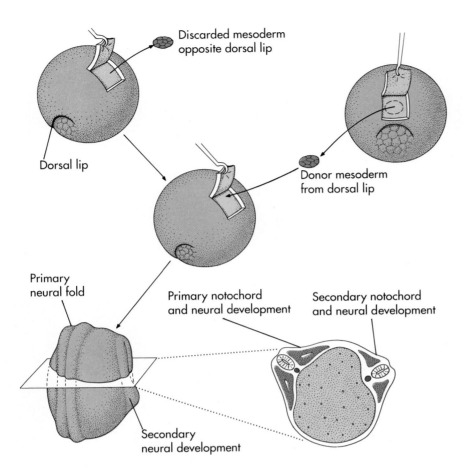

Discarded mesoderm
opposite dorsal lip

Dorsal lip

Donor mesoderm
from dorsal lip

Primary
neural fold

Primary notochord
and neural development

Secondary notochord
and neural development

Secondary
neural development

**Figure 43-16**

Spermann and Mangold's dorsal lip transplant experiment.

ated are called **secondary inductions.** The differentiation of the CNS during neurulation by the interaction of dorsal ectoderm and dorsal mesoderm to form the neural tube is an example of primary induction. In contrast, the differentiation of the lens of the vertebrate eye from ectoderm by interaction with tissue from the CNS is an example of secondary induction.

The eye develops as an extension of the forebrain, a stalk that grows outward until it comes into contact with the epidermis (Figure 43-17). At a point directly above the growing stalk, a layer of the epidermis pinches off, forming a transparent lens. When the optic stalks of the two eyes have just started to project from the brain and the lenses have not yet formed, a person

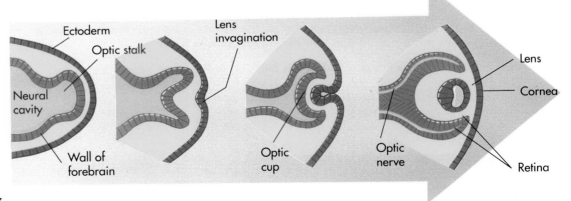

Ectoderm

Optic stalk

Lens invagination

Lens

Cornea

Neural cavity

Wall of forebrain

Optic cup

Optic nerve

Retina

**Figure 43-17**

**Development of the vertebrate eye proceeds by induction.** The eye develops as an extension of the forebrain called the optic stalk that grows out until it contacts the ectoderm. This contact induces the formation of a lens from the ectodermal tissue.

can remove one of the budding stalks and transplant it to a region underneath a different epidermis, such as that of the belly. When this critical experiment was performed by Spemann, a lens still formed, this time from belly epidermis cells in the region above where the budding bulge had been transplanted.

The chemical nature of the induction process is not known in detail. If one interposes a nonporous barrier, such as a layer of cellophane, between the inducer and the target tissue, no induction takes place. A porous filter, in contrast, does permit induction to occur. It is thought that the inducer cells produce a protein factor that binds to the cells of the target tissue, stimulating mitosis in them and initiating changes in gene expression.

## THE NATURE OF DEVELOPMENTAL DECISIONS

An adult vertebrate contains hundreds of different cell types, each expressing different aspects of the total genetic information. What factors determine which genes are expressed in a particular cell and which are not? In a liver cell, what mechanism keeps the genetic information that specifies nerve cell characteristics turned off? Does the differentiation of that particular cell into a liver cell entail the physical loss of the information specifying other cell types? No. But cells progressively lose the capacity to *express* ever-larger portions of their genomes. *Development is a process of progressive restriction of gene expression.*

Some cells become **determined** very early. For example, all of the egg cells of a human female are set aside very early in the life of the embryo, yet some of these cells will not achieve differentiation into functional oocytes for more than 40 years. To a large degree, the fate of a particular cell is determined by the place at which it is located in the developing embryo. By changing a cell's location, an experimenter can alter its developmental destiny. However, this is only true up to a certain point in a cell's development. At some stage, the ultimate fate of every cell becomes fixed and irreversible, a process referred to as **commitment.**

When a cell is "determined," it is possible to predict its developmental fate; when a cell is "committed," that developmental fate cannot be altered. Determination often occurs very early in development, commitment somewhat later.

## THE COURSE OF HUMAN DEVELOPMENT

Vertebrates seem to have evolved largely by the addition of new instructions to the developmental program. The development of the human embryo shows its evolutionary origins. It proceeds through a series of stages, the earlier stages unchanged from those which occur in the development of more primitive vertebrates.

If we did not have an evolutionary perspective, we would be unable to account for the fact that human development proceeds in much the same way as development in a chick. In both embryonic chickens and embryonic human beings, the blastodisc is flattened. In a chick egg the blastodisc is pressed against a yolk mass; in a human embryo the blastodisc is similarly flat despite the absence of a yolk mass. In human blastodiscs a primitive streak forms and gives rise to the three primary cell types, just as it does in the chick blastodisc.

Human development takes much longer than chicken development, an average of 266 days from fertilization to birth, the familiar 9 months of pregnancy. But what may not be so readily apparent is how very early the critical stages of development outlined in this chapter occur during the course of human pregnancy.

## First Trimester: Development is Completed Early

*The First Month.* In the first week after fertilization, the fertilized egg undergoes cleavage divisions. The first of these divisions occurs about 30 hours after the fusion of the egg and the sperm and the second, 30 hours later. Cell divisions continue until a blastocyst forms. During this period the embryo continues the journey down the mother's oviduct, a journey that the egg initiated. On about the sixth day the embryo reaches the uterus, attaches to the uterine lining, or **endometrium,** and penetrates into the tissue of the lining. The trophoblast begins to grow rapidly, initiating the formation of membranes. One of these membranes, the **amnion,** will enclose the developing embryo, whereas another, the **chorion,** will interact with uterine tissue to form the placenta that will nourish the growing embryo.

Ten to eleven days after fertilization, gastrulation takes place. The primitive streak can be seen on the surface of the embryo, and the three primary tissue types are differentiated. Around the developing embryo the placenta starts to form from the chorion.

In the third week, neurulation occurs. It is marked by the formation of the neural tube along the dorsal axis of the embryo, as well as by the appearance of the first somites, from which the muscles, vertebrae, and connective tissue develop. By the end of the week, over a dozen somites are evident, and the blood vessels and

# In Vitro Fertilization

Half a million American women cannot have children because their oviducts are blocked or nonexistent. Even though they have normal ovaries, produce normal eggs, and have a normal uterus, there is simply no place for their eggs to become fertilized, no highway for sperm to reach the uterus. Thirteen years ago, in 1978, a child was born to such a mother. An egg was removed from the mother's ovary and fertilized on a glass plate with the father's sperm. The resulting zygote was allowed to grow and divide for 2 days, and then the embryo was placed in the mother's uterus. It embedded itself in the uterine wall, just as if it had arrived through her oviducts, and proceeded to develop. Louise Brown was born 264 days later, the world's first "test-tube baby."

The process which gave rise to Louise Brown is called "in vitro fertilization" (Latin *vitro,* glass). In the 13 years since she was born, the procedure has become a standard medical procedure performed in hundreds of clinics around the world. Many clinics handle more than 250 patients a year. Not all attempts are successful—indeed, less than a quarter of fertilization attempts "take." Still, thousands of babies now live that without this technique would not have been born.

In vitro fertilization has given rise to an entirely new realm of ethical questions. For example, there is nothing in the procedure that requires the egg and sperm to be donated by a married couple. What if the egg is donated by another woman, who later demands "her" child? Nor is there anything in the procedure that requires the uterus to be that of the egg donor. What if another woman carries the baby to term, and later demands "her" child? It is not clear what the answers to these ethical questions should be—the situations have never arisen before—but we are going to have to decide what the answers are going to be because the questions will continue to arise with increasing frequency.

Other avenues of in vitro fertilization present even more troubling ethical issues. An in vitro fertilized embryo need not be reimplanted immediately—the embryo can be frozen, stored, and used later by an entirely different couple. Should it be legal to charge a fee for this process—can embryos be sold? Or how about this: work with cattle has led to in vitro fertilization procedures that allow the fertilized egg to be dissected after several cleavage divisions in such a manner that all four or eight division products go on to form embryos that develop normally. All four or eight individuals are genetically identical clones. What if a couple wants to use this approach to make identical twins—should humans be cloned?

---

gut have begun to develop. At this point the embryo is about 2 millimeters long.

In the fourth week, **organogenesis** (the formation of body organs) occurs (Figure 43-18, *A*). The eyes form, and the tubular heart begins to pulsate, develops four chambers, and begins a rhythmic beating that stops only with death. At 70 beats per minute, the little heart is destined to beat more than 2.5 billion times during a lifetime of about 70 years before it ceases. More than 30 pairs of somites are visible by the end of the fourth week, and the arm and leg buds have begun to form. The embryo more than doubles in length during this week to about 5 millimeters.

All of the major organs of the body have begun their formation by the end of the fourth week of devel-opment. Although the developmental scenario is now far advanced, many women are not aware that they are pregnant at this stage.

Early pregnancy is a very critical time in development because the proper course of events can be interrupted easily. In the 1960s, many pregnant women took the tranquilizer **thalidomide** to minimize discomforts associated with early pregnancy. Unfortunately, this drug interferes with fetus limb bud development, and its widespread use outside the United States resulted in many deformed babies. Also during the first and second months of pregnancy, contraction of rubella (German measles) by the mother can upset organogenesis in the developing embryo. Most spontaneous abortions occur in this period.

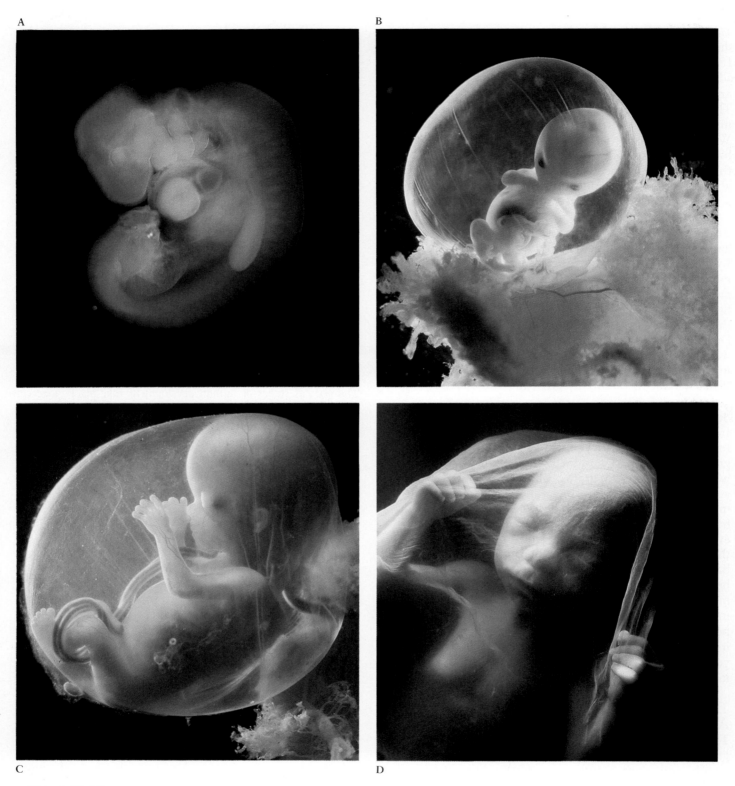

*Figure 43-18*

**The developing human.**
A  4 weeks.
B  7 weeks.
C  3 months.
D  4 months.

**The Second Month.** **Morphogenesis** (the formation of shape) takes place during the second month (Figure 43-18, *B*). The miniature limbs of the embryo assume their adult shapes. The arms, legs, knees, elbows, fingers, and toes can all be seen—as well as a short, bony tail. The bones of the embryonic tail, an evolutionary reminder of our past, later fuse to form the **coccyx.** Within the body cavity, the major organs, including the liver, pancreas, and gallbladder, become evident. By the end of the second month, the embryo has grown to about 25 millimeters in length, weighs perhaps a gram, and begins to look distinctly human.

**The Third Month.** The nervous system and sense organs develop during the third month (Figure 43-18, *C*). By the end of the month, the arms and legs begin to move. The embryo begins to show facial expressions and carries out primitive reflexes such as the startle reflex and sucking. By the end of the third month, all of the major organs of the body have been established. Development of the embryo is essentially complete at 8 weeks. From this point on, the developing human being is referred to as a **fetus** rather than an embryo. What remains is essentially growth.

## Second Trimester: The Fetus Begins to Grow in Earnest

In the fourth and fifth months of pregnancy, the fetus grows to about 175 millimeters in length, with a body weight of about 225 grams. Bone enlargement occurs actively during the fourth month. During the fifth month the head and body become covered with fine hair. This downy body hair, called **lanugo,** is another evolutionary relict and is lost later in development. By the end of the fourth month, the mother can feel the baby kicking. By the end of the fifth month, she can hear its rapid heartbeat with a stethoscope. In the sixth month, growth begins in earnest. By the end of that month, the baby weighs 0.6 kilogram (about 1½ pounds) and is over 0.3 meter (1 foot) long, but most of its prebirth growth is still to come. The baby cannot yet survive outside the uterus without special medical intervention.

## Third Trimester: The Pace of Growth Accelerates

The third trimester is predominantly a period of growth rather than one of development. In the seventh, eighth, and ninth months of pregnancy the weight of the fetus doubles several times. This increase in bulk is not the only kind of growth that occurs. Most of the major nerve tracks are formed within the brain during this period, as are new brain cells. All of this growth is fueled by nutrients provided by the mother's bloodstream. Within the placenta these nutrients pass into the fetal blood supply (Figure 43-19). The undernourishment of the fetus by a malnourished mother can adversely affect this growth and result in severe retardation of the infant. Retardation resulting from fetal malnourishment is a severe problem in many underdeveloped countries where poverty is common.

By the end of the third trimester, the neurological growth of the fetus is far from complete and, in fact, continues long after birth. But by this time the fetus is able to exist on its own. Why doesn't development continue within the uterus until neurological development is complete? Because physical growth would continue as well, and the fetus is probably as large as it can get and still be delivered through the pelvis (Figure 43-20) without damage to mother or child. Birth takes place as soon as the probability of survival is high. For better or worse, the infant is then a person.

*The critical stages of human development take place quite early. All the major organs of the body have been established by the end of the third month. The following 6 months are essentially a period of growth.*

## POSTNATAL DEVELOPMENT

Growth continues rapidly after birth. Babies typically double their birth weight within 2 months (Figure 43-21). But different organs grow at different rates. The reason that adult body proportions are different from those of infants is that different parts of the body grow at different rates or stop growing at different times. The head, for example, which is disproportionately large in infants, grows more slowly in infants than does the rest of the body. Such a pattern of growth, in which different components grow at different rates, is referred to as **allometric growth.**

In most mammals, brain growth is entirely a fetal phenomenon. In chimpanzees the growth rate of the brain and that of the cerebral portion of the skull rapidly decelerate after birth, whereas the bones of the jaw continue to grow. The skull of an adult chimpanzee therefore looks very different from that of a fetal chimpanzee. In human beings, on the other hand, the brain and cerebral skull continue to grow at the same rate after birth as before. During gestation and after birth the

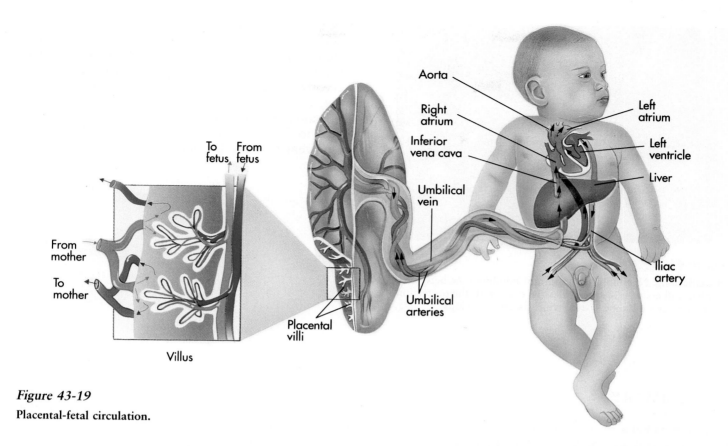

**Figure 43-19**

Placental-fetal circulation.

Aorta

Right atrium

Inferior vena cava

Umbilical vein

Left atrium

Left ventricle

Liver

Iliac artery

Umbilical arteries

To fetus From fetus

From mother

To mother

Placental villi

Villus

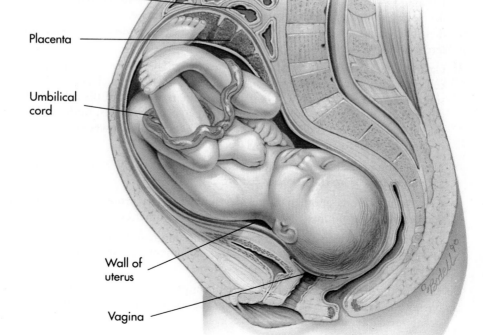

**Figure 43-20**

**Position of the fetus just before birth.** A developing fetus is a major addition to a woman's anatomy. The stomach and intestines are pushed far up, and there is often considerable discomfort from pressure on the lower back. In a natural delivery the fetus exits through the vagina, which must dilate (expand) considerably to permit passage.

Intestine

Placenta

Umbilical cord

Wall of uterus

Vagina

developing human brain generates neurons (nerve cells) at an average rate estimated at more than 250,000 per minute; it is not until about 6 months after birth that this astonishing production of new neurons ceases permanently. Because both brain and jaw continue to grow, the jaw-skull proportions do not change after birth, and the skull of an adult human being looks very similar to that of a human fetus. It is primarily for this reason that a young human fetus seems so incredibly adultlike.

**Figure 43-21**

**Breastfeeding.** Nursing is an important bonding time between mother and child. It also provides the child with the protection of its mother's immune system while its own develops.

# ■ SUMMARY

1. Fertilization is the union of an egg and a sperm to form a zygote. Fertilization is external in fishes and amphibians, internal in all more advanced vertebrates. The three stages of fertilization are (1) penetration, in which the sperm cell moves past the cells surrounding the egg and penetrates the egg membrane; (2) activation, in which a series of cytoplasmic movements are initiated by penetration; and (3) pronuclear fusion, in which the sperm and egg nuclei fuse.

2. Cleavage is the rapid division of the newly formed zygote into a mass of perhaps a thousand cells, without any increase in overall size. Because the egg is structured with respect to the location of developmentally important regulating signals, the future embryo becomes structured by the cleavage divisions. These divisions, in effect, partition the egg cytoplasm into small portions that contain different regulatory elements.

3. Gastrulation is the mechanical movement of portions of the blastosphere, forming the three basic cell types: ectoderm, endoderm, and mesoderm. In eggs that lack a yolk the movement is one of simple invagination. When a yolk is present, the movement of the cells is affected by it. In amphibians the cell layers move down and around the yolk. In reptiles, birds, and mammals, cells establish the three primary cell types as an upper layer (ectoderm), a lower layer (endoderm), and a layer that invaginates inward from the upper layer (mesoderm).

4. Neurulation in chordates is the formation of the first tissues, particularly the notochord and the dorsal nerve cord, from the primary cell types.

5. Cells influence one another during development by a process of induction. In this process, substances that exist on the surface of one cell induce other cells to divide.

6. At some point during animal development the ultimate developmental fate of cells becomes fixed and unalterable. The cells are then said to be committed, even though they may not exhibit any of the characteristics they will eventually assume.

7. Most of the critical events in the development of a human occur in the first month. Cleavage occurs during the first week, gastrulation during the second week, neurulation during the third week, and organ formation during the fourth week.

8. The second and third months of the first trimester are devoted to morphogenesis and to the elaboration of the nervous system and sensory organs. By the end of this period, the development of the embryo is essentially complete.

9. The last 6 months of human pregnancy are essentially a period of growth, devoted to increase in size and formation of nerve tracks within the brain. Most of the weight of a fetus is added in the final 3 months of pregnancy.

# REVIEW QUESTIONS

1. The union of a haploid egg and a sperm forms a diploid cell called a _____.

2. The outer membrane surrounding an egg is called the zona _____.

3. _____ refers to the activation of an egg in the absence of a sperm.

4. Rapid division of the zygote into masses of cells is referred to as _____ .

5. The three primary cell layers arising during development are the ectoderm, _____, and endoderm.

6. The placenta is formed by a combination of cells from the _____ and the uterine endometrium.

7. The ability of one cell type to produce changes in the development of a second cell type is referred to as _____.

# SELF-QUIZ

1. Which of the following is the first step in fertilization?
   (a) Flagellation
   (b) Capacitation
   (c) Pronuclear fusion
   (d) Penetration
   (e) Activation

2. For a sperm to be capable of fertilizing an egg, it must have undergone
   (a) flagellation.
   (b) capacitation.
   (c) mitosis.
   (d) penetration.
   (e) activation.

3. The ball of 32 cells resulting from early cleavage is a
   (a) morula.
   (b) blastocyst.
   (c) blastula.
   (d) chorion.
   (e) acrosome.

4. The placenta is derived from the
   (a) corpus luteum.
   (b) zona pellucida.
   (c) trophoblast.
   (d) chorion.
   (e) acrosome.

5. The migration of cells following completion of cleavage is called
   (a) tropism.
   (b) gastrulation.
   (c) placentation.
   (d) activation.
   (e) induction.

6. The notochord is derived from
   (a) ectoderm.
   (b) ectoplasm.
   (c) endoderm.
   (d) mesoderm.
   (e) a combination of c and d.

7. The formation of the neural tube is an example of
   (a) gastrulation.
   (b) invagination.
   (c) primary induction.
   (d) secondary induction.
   (e) capacitation.

8. Which of the following occurs earliest in the process of embryonic development?
   (a) Organogenesis
   (b) Neurulation
   (c) Morphogenesis
   (d) Formation of limb buds
   (e) Formation of endocrine glands

# THOUGHT QUESTIONS

1. In reptiles and birds the fetus is basically masculine and fetal estrogen hormones are necessary to induce the development of female characteristics. In mammals the reverse is true, the fetus being basically female, with fetal hormones acting to induce the development of male characteristics. Can you suggest a reason why the pattern that occurs in reptiles and birds would not work in mammals?

2. Female armadillos always give birth to four offspring of the same sex. Can you suggest a mechanism that would account for this?

# FOR FURTHER READING

BROWDER, L.: *Developmental Biology*, ed. 2, Saunders College Publishing, Philadelphia, 1984. An exceptionally well-illustrated undergraduate text with many striking photographs.

DALE, B.: *Fertilization in Animals*, Edward Arnold, Publishers, London, 1983. A brief text devoted entirely to the process of animal fertilization, with good illustrations and up-to-date discussions of physical mechanisms.

DRYDEN, R.: *Before Birth*, Heinemann Educational Books, Inc., London, 1978. A detailed description of human development.

NILSSON, L., and J. LINDBERG: *Behold Man*, Little, Brown & Co., Boston, 1974. A wonderful collection of color photographs of the developing human fetus.

RUGH, R., and L.B. SHETTLES: *From Conception to Birth: The Drama of Life's Beginnings*, Harper & Row, Publishers, New York, 1971. A detailed treatment of human development, from fertilization to birth, with excellent photographs.

SAUNDERS, J.: *Developmental Biology*, Macmillan Publishing Co., New York, 1982. A very clear undergraduate text, easily understood by students with little background. A particularly strong point of this text is its emphasis on how experiments have established the basic information.

TRINKAUS, J.: *Cells into Organs: The Forces that Shape the Embryo*, ed. 2, Prentice-Hall, Inc., Englewood Cliffs, N.J., 1984. An advanced but easily understood discussion of the physical mechanisms underlying developmental change.

WASSARMAN, P.M.: "Fertilization in Mammals," *Scientific American*, December 1988, page 78. Describes the cellular mechanisms of sperm penetration and fertilization.

WESSELLS, N.K.: *Tissue Interactions and Development*, Benjamin-Cummings Publishing Co., Menlo Park, Cal., 1977. A brief and lucid account of the experiments that led to our current understanding of the mechanisms of development.

# Animal Behavior

These geckos are having a fight. Without actually hurting each other, they threaten and posture; and the most convincing of them wins. In nature, success often depends on fine nuances of behavior.

# ANIMAL BEHAVIOR

## Overview

Behavior allows an animal to adapt to its environment, and natural selection has played an important role in shaping the behavior patterns of animals. Behavior is crucial to survival and reproduction because it serves as a mechanism of response to the environment; it is often the way by which animals find food or a place to live, avoid predators, and locate mates. The response an animal shows to an environmental stimulus such as food or a mate depends on its internal state and sensory system. Genetics and learning influence behavior in different ways in animals with simple and complex nervous systems; instinct and experience often interact during development to form behavior.

## For Review    *Here are some important terms and concepts that you will encounter in this chapter. If you are not familiar with them, you should review them before proceeding.*

**Neurons and interneurons** (Chapter 8)

**Adaptation** (Chapter 15)

**Reproductive isolation** (Chapter 17)

**Memory and learning** (Chapter 34)

**Sensing the environment** (Chapter 39)

**Hormonal control of physiological processes** (Chapter 40)

We can all relate easily to the study of animal behavior. Each of us has heard birds sing, watched a honeybee extract nectar from a flower, or housebroken a pet. As humans, we interact socially with others. Behavior can be defined as the way an organism responds to a stimulus in its environment; the stimulus might be as simple as the odor of food. In this sense, a bacterial cell "behaves" by swimming toward higher concentrations of sugar.

Obviously this is a very simple behavioral response, but it is just one of a variety of simple responses that are suited to the life of bacteria and allow these organisms to live and reproduce. During the course of the evolution of multicellular animals, the nervous system became more complex and behavior became more complex as well, with more elaborate forms of behavior evolving to meet the demands of the environment. Peripheral and central nervous systems involve networks of sensory and command fibers to perceive and process the information of stimuli in the environment and trigger an adaptive motor response, which we see as a pattern of behavior.

When we see an animal behave in a certain way, we can explain it in two different ways. First, we might ask *how* it all works; that is, how the animal's senses, nerve networks, or internal state provide a physiological basis for the behavior. In this way we would be asking a question of **proximate causation.** To analyze the proximate cause, or mechanism, of behavior, we would measure hormone levels or record the firing patterns of nerve cells. We could also ask *why* the behavior evolved; that is, what was its adaptive value? This is a question concerning **ultimate causation.** To study the ultimate, or evolutionary, cause of a behavior, we would measure how it influenced the animal's survival or reproductive success. If you are attending carefully to your dog when he wants to mate, you may explain his frenetic activity as being caused by hormones, internal messengers released in the spring that cause him to seek out females. Evolutionarily speaking, your dog shows this behavior to pass on his genes. In effect, genes guide the dog to make more genes.

In this chapter we consider the mechanisms by which an animal responds to its environment, as well as the adaptiveness of behavior. We see that scientists have taken different approaches to the study of behavior, some focusing on instinct, others emphasizing learning. The picture that will emerge is one of behavioral biology as a diverse science that draws strongly from allied disciplines such as neurobiology, physiology, and ecology. Today, the overarching theme of the study of behavior is evolution. In our discussion of human behavior, we see how such an evolutionary perspective can be controversial when applied to the social behavior of humans.

## APPROACHES TO THE STUDY OF BEHAVIOR

The study of behavior has had a long history of controversy. One theme of the controversy concerns the importance of genetics and learning in the shaping of behavior. Do genes determine behavior, or do animals behave the way they do because they have learned to do so from experience? Is behavior the result of **nature** (instinct) or **nurture** (experience)? Although in the past it has been argued as an "either/or" situation, we see that both instinct and learning play significant roles, often interacting to produce the final behavioral product. The scientific study of instinct and learning, as well as their relationship, has led to the growth of several disciplines such as ethology, behavioral genetics, behavioral neuroscience, and psychology.

## Ethology

Ethology is the study of the natural history of behavior. Because of their training as zoologists and evolutionary biologists and their emphasis on the study of animal behavior under natural conditions (that is, in the field), ethologists (Figure 44-1) believe that behavior is largely instinctive, or innate, and results from the programming of behavior by natural selection. Ethologists emphasize that because behavior is **stereotyped** (appearing in the same form in different individuals of a species), it is based on programmed neural circuits. These circuits are structured from genetic blueprints and cause an animal to show a relatively complete behavior the first time it is produced.

For example, geese incubate their eggs in nests scooped out of the ground. If a goose notices an egg that has been knocked out of the nest accidentally, the goose will extend its neck towards the egg, get up, and roll the egg back into the nest with its bill. Because this behavior seems so reasonable to us, it is tempting to believe that goose saw the problem and figured out what to do. In fact, the behavior is entirely stereotyped and largely instinctive. According to ethologists, egg retrieval behavior is triggered by detecting a **sign stimulus,** the egg out of the nest. A component of the goose's nervous system, the **innate releasing mechanism,** provides the neural instructions for the motor program, or **fixed-action pattern** (Figure 44-2). More generally, the sign stimulus is the "signal" in the environment that triggers the behavior. The innate releasing mechanism

### Figure 44-1

**The founding fathers of ethology.** Karl von Frisch, Konrad Lorenz, and Niko Tinbergen were the pioneers of ethology. In 1973, they received the Nobel Prize for their pathbreaking contributions to behavioral science. von Frisch lead the study of honeybee communication and sensory biology. Lorenz focused on social development (imprinting) and the natural history of aggression. Tinbergen was the first behavioral ecologist.

*Figure 44-2*

The complex series of muscular contractions this chameleon uses to capture an insect is always the same once the capture behavior has begun.

refers to the sensory mechanism detecting the "signal," and the fixed-action pattern is the stereotyped act. Similarly, a frog unfolds its long, sticky tongue at the sight of an insect, and a male stickleback fish will attack another male showing a bright red underside. Such responses certainly appear to be programmed, but what evidence supports the underlying ethological view that behavior has a genetic and neural basis?

*Behavioral Genetics.* In a famous experiment carried out in the 1940s, Robert Tryon studied the ability of rats to find their way through a maze with many blind alleys and only one exit, where a reward of food awaited (Figure 44-3). It took a while, as false avenues were tried and rejected, but eventually some individuals learned to zip right through the maze to the food, making few incorrect turns. Other rats never seemed to learn the correct path. Tryon bred the "maze-bright" rats with one another, establishing a colony from the fast learners, and similarly established a second "maze-dull" colony by breeding the slowest learning rats with each other. He then tested the offspring in each colony to see how quickly they learned the maze. The offspring of maze-bright rats learned even more quickly than their parents had, whereas the offspring of maze-dull parents were even poorer at maze learning. After repeating this procedure over several generations, Tryon was able to produce two behaviorally distinct types of rat with very different maze-learning ability. Clearly the ability to learn the maze was to some degree hereditary, governed by genes that were being passed from parent to offspring. And the genes are specific for this behavior, rather than being general ones

that influence many behaviors—the abilities of these two groups of rats to perform other behavioral tasks such as running a completely different kind of maze did not differ.

Tryon's research is an example of how a study can illustrate the genetic component of a behavior. Under natural conditions, rats with an ability to learn quickly may have had some advantage that led to increased survival and reproductive success. Working with species hybrids, other studies have shown that cricket and treefrog courtship song and lovebird nest building behavior has a genetic basis. At a greater level of detail, research on the mating behavior of mutant fruitflies has shown that even single allele differences produce behaviorally different individuals.

*The genetic basis of behavior can be shown by artificial selection and hybridization studies. Some studies have identified the alleles that control behavior.*

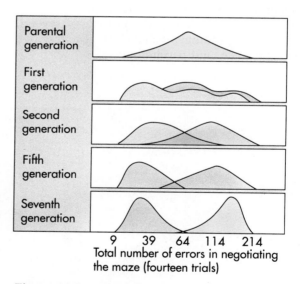

*Figure 44-3*

**Tryon was able to select among rats for ability to negotiate a maze, demonstrating that this ability is directly influenced by genes.** What he did was test a large group of rats, select the few that ran the maze in the shortest time, and let them breed with one another; he then tested their progenies and again selected those with the quickest maze-running times for breeding. By seven generations he succeeded in halving the average time an inexperienced rat required to negotiate the maze. Parallel selection for slow running time was also successful—it more than doubled the average running time.

***The Neural Basis of Behavior.*** The way we perceive the world around us depends on the structure of our sensory systems. We hear sounds of certain frequencies and see light of certain wavelengths. Our sense of smell decodes chemical information in the air. These sensory channels of **audition, vision,** and **olfaction** govern our responses to stimuli in the environment. Research on the neurobiology of stimulus detection has revealed the way in which sensory and other aspects of the nervous system are organized to govern perception and behavior, thus providing support for the ethologist's concept of the innate releasing mechanism. The study of the neural basis of behavior is called **neuroethology.**

Nerve cells are often specialized for the detection of certain stimuli. Frogs and toads prey on insects by flipping out a sticky tongue. To do this, a toad must be able to "track" prey in the environment; this is accomplished with light-sensitive nerve cells located in the retina of each eye. These retinal cells enable the toad to identify prey such as insects or worms. As an insect moves in front of a toad, its image enters the lens of the eye and is focused on the retina. The image thus

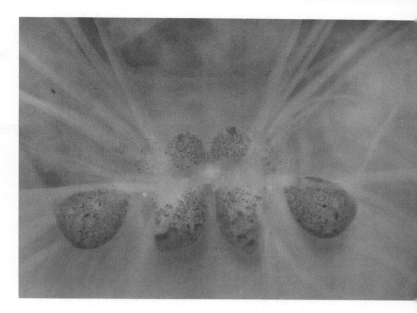

***Figure 44-5***

**Small brain, large nerve cells.** The brain of this sea hare has relatively few large neurons, making this animal easy to study for neurobiologists.

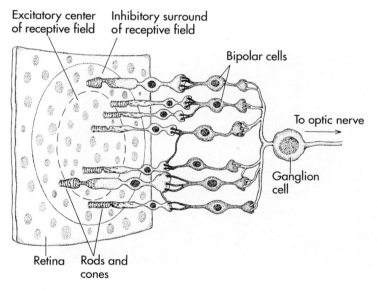

***Figure 44-4***

**Prey detectors in the eye of the toad.** Light-sensitive cells (rods and cones) embedded in the retina monitor the object motion. Ganglion cells receive input from bipolar cells, which in turn process signals from a group of light-receptor cells. Each ganglion cell has a receptive field. It monitors receptor activity in a portion of the retina's surface. Large objects cast an image on more of the surface of the retina than does the movement of small objects such as insects. Large objects inhibit ganglion cells from relaying impulses to the optic nerve, whereas small objects cause the ganglion cells to fire action potentials via the optic nerve to the brain.

moves over groups of retinal receptor cells that fire several nerve impulses and relay information via ganglion cells (which process information from several retinal cells) to the brain (Figure 44-4). A toad attempts to capture what it sees depending on the size of the object on which it has focused and whether the image is moving horizontally. Thus the innate releasing mechanism that responds to the sign stimuli provided by an insect and turns on the fixed-action pattern of flipping out the tongue at the "target" may be thought of as a group of specialized "insect detector" nerve cells.

Some animals have been extremely useful as models for research on the neural basis of behavior because they have relatively simple nervous systems and show behaviors that are easy to record. A sea hare has a brain that features a small number of very large individual nerve cells (Figure 44-5). A simple, yet important, behavior shown by the sea hare is its **escape response,** which is made up of a series of contractions and relaxations of muscles that causes its body to flex dorsally and ventrally and thus move. The response occurs when the sea hare detects the presence of one of its predators, a sea star. The motor patterns that compose the escape behavior result from impulses fired from three groups of neurons in the brain.

The neurobiology of escape behavior has also been examined in the cockroach. In this case, **mechanoreceptor hairs,** sensory hairs located on appendages at the end of the abdomen, detect microcurrents of air and fire nerve impulses that are transmitted rapidly along a giant interneuron to a thoracic ganglion, where they cause motor neurons to fire to initiate locomotion.

The response time is extremely fast—it takes only about 60 milliseconds from the time the air movement is perceived.

*Studies of the nerve networks that are the basis of response to stimuli support the concept of the innate nature of behavior. Components of the nervous system evolved in response to ecological pressures such as predation and the need to capture food.*

## Psychology and the Study of Animal Behavior

In contrast to the instinct theory of behavior developed by ethologists, other students of animal behavior focused on learning as the major element that shapes behavior. These **comparative psychologists** were not interested in naturalistic studies or evolutionary theory and worked primarily in laboratory settings on rats. The main contribution of comparative psychology has been to identify the ways in which animals learn.

**Learning** is formally defined as the creation of changes in behavior that arise as a result of experience, rather than as a result of maturation. There are generally considered to be two broad categories of learning. The simplest learned behaviors are **nonassociative,** ones that do not require the animal to form an association between two stimuli, or between a stimulus and a response. Learned behaviors that require associative activity within the CNS are termed **associative.**

The two major forms of nonassociative learning are **habituation** and **sensitization.** Habituation can be regarded as learning *not* to respond to a stimulus. Learning to ignore unimportant stimuli is a critical ability to an animal confronting a barrage of stimuli in a complex environment. In many cases, the stimulus evokes a strong response when it is first encountered, but then the magnitude of response gradually declines after repeated exposure. When you sit down in a chair, you feel your own weight at first, but soon you are not conscious of it. Your brain "tunes out" this information. Sensitization, by contrast, is learning to be hypersensitive to a stimulus. After encountering an intense stimulus, such as an electric shock, an animal may often react vigorously to a mild stimulus that it would previously have ignored.

Associative behaviors are more complex. Thinking is associative (Figure 44-6), as are other **cognitive** behaviors. Study of nonassociative behaviors in inverte-

A

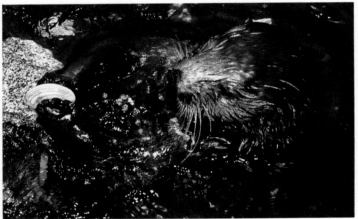

B

*Figure 44-6*

**Associative behaviors.**

**A** This chimpanzee is fashioning a tool. It is stripping the leaves from a twig, which it will then use to dig into a termite nest. Advance preparations such as this strongly suggest that the chimpanzee is consciously planning ahead, with full knowledge of what it intends to do.

**B** This sea otter is having dinner while swimming on its back. It is using the rock as a tool to break open a clam, bashing the clam on the rock "anvil." Often a sea otter will keep a favorite rock for a long time, suggesting that it has a clear idea of what it is going to use the rock for. The sea otter may learn this pattern of eating behavior from others while young, but the capacity to use tools and consciously to foresee their future use certainly depends on inherited abilities.

brates such as the sea hare *Aplysia* suggest that many if not all associative behaviors may be built up from simpler nonassociative elements.

Associative learning is the alteration of behavior by experience leading to the formation of an association between two stimuli or between a stimulus and a response. Nonassociative learning involves no such associations.

**Figure 44-7**

**Skinner box.** The rat rapidly learns that pressing the lever results in the appearance of a food pellet. This kind of learning, trial and error with a reward for success, can also work for far more complex tasks.

Learning mechanisms can be grouped according to the way stimuli become associated when behavior is modified, or **conditioned.**

The repeated presentation of a stimulus in association with a response can cause the brain to form an association between them, even if the stimulus and the response have never been associated before. If you present meat powder (stimulus) to a dog, it will salivate (response). In his famous study of **classical conditioning,** the Russian psychologist Ivan Pavlov also presented to a dog a second, unrelated stimulus (a light) at the same time that meat powder was blown into its mouth. As expected, the dog salivated. After repeated trials, the dog eventually would salivate in response to the light alone. The dog had learned to associate the unrelated light stimulus with the meat stimulus. Early experimenters believed that *any* stimulus could be linked in this way to any response; however, as you will see below, we now know this is not true.

In classical conditioning, learning does not influence whether the reinforcing stimulus is received. In **operant conditioning,** by contrast, the reward follows only after the correct behavioral response, so the animal must make the proper association before it receives the reinforcing stimulus. American psychologist B.F. Skinner studied such conditioning in rats by placing them in a box of a type that came to be called a "Skinner box" (Figure 44-7). Once inside, the rat would explore the box; occasionally it would accidentally press a lever, and a pellet of food would appear. At first the rat would ignore the lever and continue to move about, but soon it learned to press the lever to obtain food. When it was hungry, it would spend all of its time pushing the bar. This sort of trial-and-error learning is of major importance to most vertebrates. Comparative psychologists used to believe that animals could be conditioned to perform any learnable behavior in response to any stimulus by operant conditioning, but as in the case of classical conditioning, we now know this

is not so. Today instinct is considered to play a major role in guiding what type of information can be learned, as we will see.

***The Development of Behavior.*** Roughly 20 years ago, most behavioral biologists came to believe that behavior has both genetic and learned components, and the polarization of the schools of ethology and psychology drew to an end. It has become clear that certain types of learning are not always possible. Some animals have innate predispositions towards forming certain associations. For example, rats easily learn to associate smells with foods that make them ill but are unable to associate sounds or colors with these foods, no matter how many trials they undergo. Similarly, pigeons associate colors with food but cannot make associations between sounds and food; they *can* associate sounds and danger—but cannot associate colors with danger. The sorts of associations that are possible are genetically determined. That is, classical conditioning is possible only within boundaries set by instinct.

Nor is trial-and-error learning free of inherent limits imposed by an animal's genetic makeup. Rats in a Skinner box that learn to press a lever for food cannot learn to press the same lever to avoid an electric shock; they can learn to jump to avoid the shock, but not to obtain food. Animals are innately programmed to learn some things more readily than others. Instinct determines the boundaries of learning.

It now seems clear that animals are innately programmed to respond to specific clues in particular behavioral situations. These innate programs have evolved because they represent adaptive responses. Rats, which forage at night and have a highly developed sense of smell, are better able to identify dangerous food by odor than by color. The seed a pigeon eats may have a distinctive color that it can see, but it makes no sound that a pigeon can hear. Evolution has biased behavior with instincts that make adaptive response more likely.

The current view that learning is guided genetically was developed from a study by Peter Marler on how white-crowned sparrows acquire their courtship song. The song is sung by mature males and is characteristic only of the white-crowned sparrow species. By rearing male birds in soundproof incubators provided with a speaker and a microphone, Marler could completely control what a bird heard as it matured and record on tape the song it produced as an adult. He found that birds that heard no song at all during growth sang a poorly developed song as adults. When played the song of the song sparrow, a related bird species, males also sang a poorly developed song as adults. But when both the white-crowned and song-sparrow songs were played to developed, mature males, they sang a fully developed, white-crowned sparrow song (Figure 44-8). Marler's research suggests that birds have a **genetic template,** or instinctive program, to guide learning the appropriate song. Song acquisition is based on learning, but only the song of the *correct* species can be learned. The genetic template for learning is *selective*.

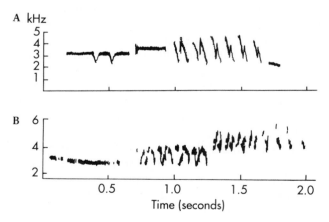

### Figure 44-8

**Song development in birds.** Sonograms of songs produced by white-crowned sparrow males exposed to their own species' song during development (**A**) or which heard no song during rearing (**B**). This illustrates that the genetic program itself is insufficient to produce a normal song.

> *Single genes are not responsible for complex behaviors, and complex behavior is not entirely governed by learning. Genes influence what can be learned; the way that genes and experience interact is adaptive. In different species, instinct and learning vary in importance, but in many species, genes seem to set limits on the extent to which behavior can be modified.*

### The Physiology of Behavior

Psychologists criticized ethology's emphasis on instinct because it ignored the study of internal factors that control behavior. Why does a male bird defend a territory and sing only during the breeding season? Ethologists would answer that birds sang when they were in the right *motivational state*, or *mood*, and had the appropriate *drive*. But what do these terms mean? They are simply "black-box" concepts that give a name to some internal control mechanism that remains unknown.

Today we understand the internal control of behavior from the study of physiology. One focus has been on the physiological control of reproductive behavior. Studies of lizards, birds, rats, and other animals have shown that hormones play an important role in the control of behavior and provide a chemical basis for motivation. Animals show reproductive behaviors such as courtship only during the breeding season. They are able to monitor changes in day length to trigger a series of events that involve the release of hormones from the endocrine glands, hypothalamus, pituitary, and ovaries and testes. Ultimately, the steroid hormones estrogen and testosterone released from the ovaries and testes travel to the brain and cause animals to show behaviors associated with reproduction (Figure 44-9). Bird song and territorial behavior depend on the level of testosterone in the male, and the receptivity of females to male courtship depends on estrogen.

Hormones are therefore a proximate cause of behavior. To control reproductive behavior, they are released at the time of the year during which conditions for the growth of young are favorable. Environmental stimuli that trigger hormone release and thus behavior include the courtship activities of males as well as changes in the physical environment, such as in temperature and day length.

*Behavioral Rhythms.* Many animals exhibit behaviors that vary in a regular fashion. Geese migrate south in the fall, birds sing in the early morning, bats fly at night rather than in daylight hours, and we humans get

**Figure 44-9**

**Dewlap display of a male yarrow spiny lizard.** Under hormonal stimulation, males extend the fleshy, colorful dewlap to court females. This behavior also stimulates hormone release and egg-laying in the female.

mans have lived for months underground in apartments where all light is artificial and there are no external cues of any kind; left to set their own schedules, most people adopt daily activity patterns (one period of activity plus one period of sleep) of about 25 hours, although there is considerable variation. Some individuals exhibited 50-hour clocks and were active for as long as 36 hours each period! In the real world, the day/night cycle resets the free-running clock every day to a cycle of 24 hours.

*Circadian rhythms are endogenous cycles of about 24 hours that occur in the absence of external clues.*

sleepy at night and are active in the daytime. Why do regularly repeating patterns of behavior occur, and what determines when they occur? Why do we sleep at night, and what determines how long we sleep? The study of questions such as these has revealed that rhythmic animal behaviors are based on both **exogenous** (external) timers and **endogenous** (internal) rhythms.

Much of the study of endogenous rhythms has focused on behaviors that seem keyed to a daily cycle, such as sleeping. Many of these behaviors have a strong endogenous component, as if they were driven by a **biological clock.** In the absence of any clues from the environment, the behaviors continue on a regular cycle. Endogenous rhythms of about 24 hours that occur even in the absence of external cues are called **circadian** ("about a day") **rhythms.** Almost all fruitfly pupae hatch in the early morning, even if kept in total darkness throughout their week-long development—they keep track of time with an internal clock whose pattern is determined by a single gene.

It turns out that the rhythm of most biological clocks does not exactly match that of the environment, so an exogenous cue is required to keep the behavior properly in time with the changing real-world environment. Because the duration of each individual cycle of the behavior deviates slightly from 24 hours, an individual kept under constant conditions gradually drifts out of phase with the outside world. Exposure to an environmental cue resets the clock. Light is the most common cue for resetting circadian rhythms.

The most obvious circadian rhythm in humans is the sleep/activity cycle. In controlled experiments, hu-

Many important biological rhythms occur on cycles longer than 24 hours. Annual cycles of breeding, hibernation, and migration are examples of behaviors that occur on a yearly cycle, so-called **circannual behaviors.** These behaviors seem to be largely timed by hormonal and other physiological changes that are keyed to exogenous factors such as day length. The degree to which endogenous biological clocks underlie circannual rhythms is not known—it is very difficult to perform constant-environment experiments of several years' duration.

The physiological mechanism of endogenous biological clocks is unknown, although there has been a great deal of study and speculation, most of it centered on regularly occurring molecular interactions. The mechanism of the biological clock remains one of the most tantalizing puzzles in biology today.

## ANIMAL COMMUNICATION

The sign stimuli we discussed earlier are used by predators to locate prey or are used by prey to avoid predators. However, other stimuli are **social releasers,** or signals produced by one individual to communicate with another individual, usually of the same species. In this case it is adaptive for both the sender and receiver of the signal to exchange information, perhaps concerning the readiness to mate or the location of a food source. Communication may occur through a number of sensory channels; signals may be visual, acoustical, chemical, tactile, or electric. Much of the research in animal behavior concerns identifying the nature of the signal and determining how it is perceived.

## Courtship

Depending on their reproductive condition, animals produce signals to communicate with potential mates. Courtship behaviors usually consist of a series of fixed-action patterns (Figure 44-10), each released by some action of the partner, and releasing in turn a fixed-action pattern by the partner. Courtship signals are often **species specific**—they limit communication to members of the same species. The flash patterns of fireflies (which are actually beetles) are coded for species identity and thus play an important role in reproductive isolation (Figure 44-11). The chemical sex attractant, or **pheromone,** of the female silk moth conveys information only to males of the species. The antennae of the male are covered with **chemoreceptors,** which are nerve cells modified to detect extraordinarily small quantities of the sex pheromone. Many insects, amphibians, and birds produce species-specific sound signals to attract mates (Figure 44-12). Some fish use electric currents for the same purpose.

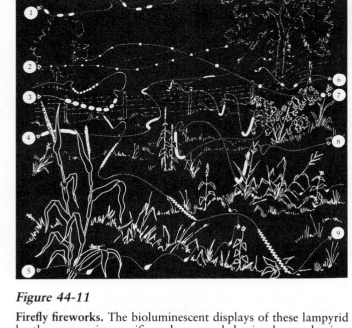

**Figure 44-11**

**Firefly fireworks.** The bioluminescent displays of these lampyrid beetles are species-specific and serve as behavioral reproductive isolating mechanisms. Each number represents the flash pattern of a male of a different species.

① Female gives head-up display to male

② Male swims zigzag to female and then leads her to nest

③ Male shows female entrance to nest

④ Female enters nest and spawns while male stimulates tail

⑤ Male enters nest and fertilizes eggs

**Figure 44-10**

**A stimulus/response chain.** Stickleback courtship involves a sequence of behaviors leading to the fertilization of eggs.

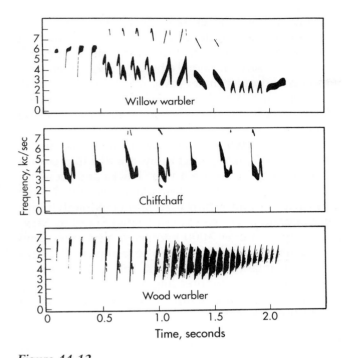

**Figure 44-12**

**Singing a different tune.** These sonograms illustrate how birds announce their species' names. At one time, these three species were considered to be one.

infants are programmed to recognize the consonant sounds characteristic of human speech (including those not present in the particular language they learn) while ignoring a world full of other sounds; they learn by trial and error (the "babbling" phase) how to make these sounds.

Although human languages appear on the surface to be very different, in fact they share many basic structural similarities. Researchers believe these similarities reflect how our brains handle abstract information, a genetically determined characteristic all humans share. The differences among English, French, Japanese, and Swahili are learned—but any human can learn them. All human languages draw from the same set of 40 consonant sounds (English uses two dozen of them), and every normal human baby can distinguish among all 40 of them.

Learning a particular language is something most humans do while young. Children who have not heard certain consonant sounds as infants can only rarely distinguish or produce them as adults. That is why Americans never master the throaty French /r/, whereas speakers of French typically replace the English /th/ with /z/, and why native Japanese substitute /l/ for the unfamiliar English /r/. Children quickly and effortlessly

## Communication in Social Groups

Many insects, fish, birds, and mammals live in social groups in which information is communicated between group members. For example, in mammalian societies some individuals serve as "guards" and keep watch for predators (Figure 44-13). When danger occurs, an **alarm call** is given; group members respond by absconding and seeking shelter. Social insects such as ants and honeybees produce chemicals called **alarm pheromones** that trigger attack behavior. Ants deposit **trail pheromones** between the nest and a food source to induce cooperation during foraging. Honeybees have an extremely complex **dance language** behavior that directs nestmates to rich nectar sources.

## Human Language

Some primates have a "vocabulary" that allows individuals to communicate the identity of specific predators such as eagles, leopards, and snakes. Chimps and gorillas can learn to recognize and use a large number of symbols and have no trouble using them to communicate abstract concepts. However, they cannot assemble symbols into sentences. This requires very complex "wiring" of the brain, a complexity that only humans have achieved.

Language develops at an early age in man. Human

**Figure 44-13**

**Sentry duty.** A meerkat sentinel on duty. Meerkats, *Suricata suricata*, are a species of highly social mongoose living in the semiarid sands of the Kalahari Desert. A recent field study of Oxford University zoologist Dr. David MacDonald has revealed that theirs is an astonishingly complex and cooperative society. This meerkat is taking his turn to act as a lookout for predators; under the security of his vigilance the other group members can focus their attention on foraging.

learn a vocabulary of thousands of words. This rapid-learning ability seems to be genetically programmed and disappears in most of us as we grow older.

Although language is the primary channel of human communication, much evidence exists to suggest that odor and nonverbal signals ("body language") may also be important. In an animal as socially complex and intelligent as the human, it is difficult to sort out the relative contribution of the composite (that is, multichannel) signals humans produce.

*The study of animal communication involves studies of the specificity of a signal, its information content, and the methods used to produce and receive it. Communication plays an important ecological role.*

## ECOLOGY AND BEHAVIOR

In this chapter we have often referred to behavior as being adaptive. Nobel laureate Niko Tinbergen was the first ethologist to study the **survival value** and **adaptive significance** of behavior. Survival value refers to the way a behavior may allow an animal to avoid a predator, and the adaptive significance of a behavior is the way in which it contributes to reproductive success. He noted that an animal's environment, or ecology, presents certain "problems," such as locating a nest or food, avoiding predators, or finding mates. Behavior can be seen as a trait that evolves to "solve" such problems. Currently, **behavioral ecologists** study the ways in which behavior serves as adaptation and allows an animal to increase or even maximize its success in reproduction. In this section we examine several categories of behavior that have been studied with respect to animal ecology.

### Orientation and Migration

Animals may travel to and from a nest to feed or move regularly from one place to another. To do this they must orient themselves by tracking stimuli in the environment.

Movement toward or away from some stimulus is called a **taxis**. The crowding of flying insects about outdoor lights is a familiar example, insects being attracted to the light. They are said to be positively phototactic. Other insects, such as the common cockroach, avoid light (are negatively phototactic). Nor is light the only stimulus. Trout orient in a stream so as to face

upstream, against the current. Not all responses involve a specific orientation. In some cases an individual simply becomes more active under certain conditions. If an animal moves randomly but is active under poor conditions and quiet under favorable ones, then it will tend to stay in favorable areas. These changes in activity level are dependent on stimulus intensity and are called **kineses**.

Long-range two-way movements are called **migrations**. In animals, many migrations are tied to a circannual clock, occurring once a year. Ducks and geese migrate down flyways from Canada across the United States each fall and return each spring. Monarch butterflies migrate from the eastern United States to Mexico, a journey of over 3000 kilometers that takes from 2 to 5 generations of butterflies to complete. Perhaps the longest migration is that of the golden plover, which flies from Arctic breeding grounds to wintering areas in southeastern South America, a distance of some 13,000 kilometers.

Migration patterns can be genetically determined. When colonies of bobolinks became established in the western United States, far from their normal range in the Midwest and East, these birds did not migrate directly to their winter range in South America; rather they migrated east to their ancestral range and then south along the old flyway (Figure 44-14). The old pattern was not changed, but rather an additional pattern was added.

*Migrations are long-range two-way movements by animals, often occurring once a year with the change of seasons.*

Biologists have studied migration with great interest, and we now have a good idea of how these feats of navigation are achieved. Birds and other animals navigate by looking at the sun and the stars. The indigo bunting, for example, which flies during the day and uses the sun as a guide, compensates for the movement of the sun in the sky as the day progresses by reference to the North Star, which does not move in the sky. Other birds such as starlings compensate for the sun's apparent movement by use of an internal clock. If captive birds are shown an experimental sun in a fixed position, they will change their orientation to it at a constant rate of about 15 degrees per hour.

It is important to note the distinction between **orientation** (the ability to follow a bearing) and **naviga-**

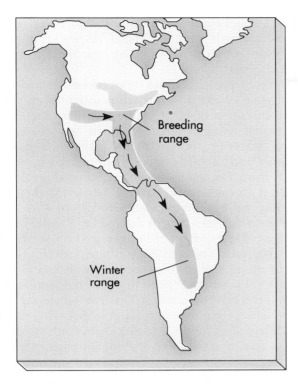

**Figure 44-14**

**Birds on the move.** The migratory path of California bobolinks. These birds came recently to the far West from their more established range in the Midwest. When they migrate to South America in the winter, they do not fly directly, but rather fly to the Midwest first and then use the ancestral flyway.

tion (the ability to set or adjust a bearing, and then follow it)—"compass" versus "map and compass." Orientation mechanisms such as those of the indigo bunting and the starling are well understood; navigation map senses such as that of the bobolink are very poorly understood.

Many migrating bird species appear to use a compass. They have the ability to detect the earth's magnetic field and orient themselves with respect to it. If such birds are studied in a closed indoor cage, they will attempt to move in the correct geographical direction, even though there are no visible external cues. However, if an investigator places a powerful electromagnet near the cage to change the magnetic field, he or she can alter at will the direction in which the birds attempt to move. Little is known about the sensory receptors that birds employ to detect magnetic fields. Magnetite, a magnetized iron ore, has been found in the heads of some birds, but the exact nature of the receptor is unknown.

Although bird migration has been well studied, we know relatively little about how other migrating animals navigate. The green sea turtles that introduce

Chapter 1 of this text migrate from Brazil halfway across the Atlantic Ocean to Ascension Island—how do they find this tiny island, half an ocean away? How do the young that hatch on Ascension Island know how to find Brazil, thousands of miles away over the open sea? How do they find their way back, as adults, when they breed, perhaps 30 more years later? We don't know.

## Foraging Behavior

Feeding is one of the most basic of all animal behaviors. Because all animals are heterotrophs, all of them must eat to survive, and many complex behaviors have evolved that influence what an animal eats and how it obtains its food. These behaviors are collectively called **foraging behaviors.**

Some foraging behaviors involve actively hunting prey. A chameleon will snap up an insect that comes close, a fixed-action pattern released by sight of the insect. Lions will hunt in groups, some driving wildebeests or other game past others waiting in ambush. Some animals, such as spiders, set a trap, spiders by building a web.

If you consider *what* animals eat, foraging behaviors fall into two broad categories, with considerable variation in each: (1) some animals are specialists and feed on only one kind of organism, whereas (2) others are generalists, and feed on many different kinds of organisms. Oystercatchers, for example, are shore birds that feed only on mussels. They have highly specialized foraging behavior (Figure 44-15), stabbing repeatedly

**Figure 44-15**

**Foraging behavior in oystercatchers.** An oystercatcher forages by stabbing the sand in search of buried oysters.

into the mollusk to open its shell. This unusual behavior is of little use in catching fish or other prey, but is a very efficient way of opening mussels. Generalists rarely exhibit such complex foraging behaviors and so are probably not as efficient at catching any one item of food—but they can take advantage of more then one kind of prey.

How general are generalists? Not very, it turns out, when their foraging behavior is studied in detail. Usually an animal will concentrate on one kind of organism and ignore other potential sources of food, then switch to a second favored kind when the first becomes rare. An animal foraging in this way may have a **search image,** or "mental idea," for the food it is selecting. An animal may visualize in its brain the appearance of its prey. Use of a search image is advantageous, as it permits an individual to focus on particular food items and to modify its foraging accordingly.

Because foraging behavior is of obvious importance to growth and reproduction, one would expect it to be the target of strong evolutionary pressures. One aspect of the evolution of foraging behavior has held a particular fascination for students of behavior: does natural selection favor animals that are more efficient foragers, animals that maximize energy intake over energy expenditure? We would certainly expect evolution to favor efficiency—but does it? To investigate this, biologists have studied situations in which foraging behavior involves a variety of trade-offs. For fish, food items come in a variety of sizes, larger ones containing more food energy but being harder to catch and less abundant. If one can measure the food value of different prey items, the energy costs of pursuing prey of different sizes and handling prey of different sizes, it is, in principle, possible to calculate the optimal foraging behavior. But in practice it is difficult to measure all variables, so precise predictions are usually impossible. But it has become clear from a variety of studies that the foraging behaviors of most animals are far from random, that individuals select prey in a highly efficient fashion. Animals seem to maximize their energy gain per unit of time spent foraging.

## Territoriality

**Territoriality** is a form of behavior in which individual members of a species will exclusively use an area holding some limiting resource, such as a foraging ground or a group of females. The critical aspect of territorial behavior is defense against intrusion by other individuals. Territories are defended by advertising through displays that they are occupied and by overt aggression. A bird sings from its perch within a territory to prevent a takeover by a neighboring bird. If an intruder persists, it will be attacked. But singing is energetically expensive, and attacks can lead to injury. Moreover, advertisement through song or visual display can reveal one's position to a predator. Why bear these costs and take such risks?

Over the past 2 decades it has become increasingly clear that an *economic* approach is insightful in studying the evolution of territoriality. As there are energy costs to defending a territory, there are also energy benefits. Studies of nectar-feeding birds such as hummingbirds and sunbirds make this point clear. A bird benefits from having exclusive use of a patch of flowers because it can efficiently harvest the nectar they produce. But the benefits of exclusive use of such a resource outweigh the costs of defense only under certain conditions. For example, if flowers are very scarce, the nectar they yield does not have enough calories to balance the number of calories used up in defense. In this case it is not advantageous to be territorial. Similarly, if flowers are very abundant, a bird can efficiently meet its daily energy requirements *without* behaving territorially. Defending abundant resources isn't worth it from an energy standpoint. But for intermediate levels of flower availability, the benefits do outweigh the costs, and territorial behavior will be favored.

*The "economic" approach has been very useful in examining the evolution and ecology of territoriality and other behaviors, such as foraging. It is assumed that animals that gain more energy from a behavior than they expend will have an advantage in survival and reproduction over animals that behave in less efficient ways. Thus estimating energy benefits and costs of a behavior is one way the adaptiveness of behavior can be determined.*

## The Ecology of Reproduction

During the breeding season, animals make several important "decisions" concerning whom to mate with, how many mates to have, and how much time and energy should be devoted to rearing offspring. Males and females usually differ in their **reproductive strategies,** or sets of behaviors that have evolved to maximize reproductive success or fitness. The reproductive behaviors shown by males and females have evolved in response to ecology, which in this case refers to how food resources, nest sites, or members of the opposite sex are spatially distributed in the environment.

Darwin was the first to observe that females do not simply mate with the first male encountered, but seem to somehow "evaluate" a male's quality and then decide whether to mate. **Mate choice** occurs because individuals that select superior-quality mates will leave behind more offspring. The benefits of choosing a mate may lie in acquiring "good genes" for offspring or resources such as favorable nest sites and energetically valuable food.

If members of one sex are selective in mating, then this will cause members of the opposite sex to compete with one another to "demonstrate" that they are high-quality mates. This may result in aggressive competition among males (or females) to hold territories, if territory quality is considered in mate evaluation. Darwin referred to this competition for mates as a process of **sexual selection.** Because of sexual selection, males and females may look very different and show **sexual dimorphism.** Differences between males and females may be in size, or a trait may occur in one sex but not in the other. For example, stag deer have elaborate antlers; these are absent in female deer. A peacock has a large, colorful tail that he struts in front of a female, but a peahen is drab in color and has no elaborate feathers (Figure 44-16).

Individuals may mate with only one individual during the breeding season and form a long-lasting pair bond or may have more than one mate. **Mating systems** such as **monogamy** (a male mates with one female), **polygyny** (a male mates with more than one female), and **polyandry** (a female mates with more than

**Figure 44-17**

**Pink flamingos.** These birds are monogamous and show almost no size or plumage differences.

one male) are other aspects of male and female reproductive strategy (Figure 44-17). These patterns of reproduction have also evolved to maximize male and female fitness. Much research has shown that mating systems have a strong ecological component. For example, a male may defend a territory that holds nest sites or food sources necessary for a female to reproduce, and the territory might have resources sufficient for more than one female. If males differ in the quality of territories they hold, a female will do her best to mate with a male in a high-quality territory. Such a male may already have a mate, but it is still to the advantage of the female to breed with a mated male in a high-quality territory rather than with an unmated male in a low-quality territory. Thus polygyny may evolve. Decisions about mating are also constrained by the needs of offspring. If the presence of both parents is necessary to successfully rear young, then monogamy may be favored. This is generally the case in birds, in which over 90% of all species form monogamous pair bonds.

**Figure 44-16**

**Sexual dimorphism in deer.** Stags have a rack of antlers used in defending a harem of females from other males. Antlers are absent in females.

Natural selection has favored the evolution of behaviors which maximize the reproductive success of males and females. By evaluating and selecting a mate with superior qualities, it is possible for an animal to increase the growth and survival chance of offspring.

## The Evolution of Animal Societies

*Society* is a household word; we tend to think of it in terms of the groups in which humans live and the cultural environment of man. But species of organisms as diverse as the slime mold, cnidarians, insects, fish, birds, prairie dogs, lions, whales, and chimpanzees exist in social groups. To encompass many types of social phenomena we can broadly define a **society** as a group of organisms of the same species that are organized in a cooperative manner. Why have individuals in some species given up a solitary existence to become members of a group? Recent research has focused on the advantages and disadvantages of group living, and much attention has been given to the evolution of the trait that defines social life: cooperation.

*Sociobiology: The Biological Basis of Social Behavior.* In the early 1970s, E.O. Wilson of Harvard University, Richard Alexander of the University of Michigan, and Robert Trivers, now at the University of California, initiated what has become a major movement within biology, the attempt to study and understand animal social behavior as a *biological* process, a process with a genetic basis that is shaped by evolution.

A biological view of social behavior being the result of evolution would predict that the behaviors characteristic of particular animals would, by and large, be suited to their mode of living—that is, in the sense of Darwin, be adaptive. You would expect that natural selection would have favored those gene combinations which allow animals to adapt more completely to particular habitats. The study of the biological basis of social behavior of animal societies is called **sociobiology**. We first describe the insights into the origin of social behavior that sociobiology has given us. We then consider one of the most contentious topics in the history of biology: can biology tell us anything about human nature?

*Group Living.* Living as a member of a group can actually be viewed as selfish behavior. A bird that joins a flock does so because it may have greater protection from predators. In fact, as flock size increases, the risk of predation decreases because there are more individuals present to vigilantly scan the environment. A member of a flock may also increase its feeding rate if it can acquire information from other flock members about the location of new, rich food sources. But there are costs to group living. For example, parasites and disease are more easily spread within groups, and the advantages of increased group size may be balanced by the disadvantages of loss of young to blood-sucking insect parasites.

*Altruism.* You may think of **altruism** as a heroic human behavior such as jumping into a river to save a drowning person. But altruism, or self-sacrificing behavior, is not limited to humans and occurs in extreme forms in other animals. In many species it is an important aspect of cooperation. A honeybee worker has a barbed sting that remains in the skin of a vertebrate. As the bee flies off, the anchored sting causes the worker to eviscerate itself and thus commit suicide. Vampire bats may share blood meals at the roost. One of the most important aspects of altruism concerns assisting another individual in reproducing. In some species, nonreproductive or even sterile individuals exist to help other group members rear offspring. How can such behavior evolve?

One of the great misconceptions about social behavior is that altruism has evolved because a certain act benefits the group as a whole, or even the entire species. This **group selection** argument has been used (incorrectly!) to explain how animals regulate the size of their populations. For example, in some bird species, males display and compete with each other on the mating grounds to try to secure centrally located territories. Some males are able to hold territories, others are not. Females choose males with territories as mates, and males unable to hold territories may never mate. V.C. Wynne-Edwards, an animal ecologist, explained that nonterritorial males were sacrificing their own reproduction to limit population growth. In other words, some males did not reproduce because it was "good for the species." That is to say, too large a population might exhaust limited resources, with the result that the whole group or species might become extinct.

There is a very important flaw in this explanation of the evolution of altruism. How could the trait of altruism be passed from generation to generation if males that have the trait never leave any offspring? There is simply no Darwinian logic to this argument. We must look for other evolutionary explanations of altruistic behavior.

*Reciprocity.* Robert Trivers proposed that individuals may form "partnerships" in which continuous exchanges of altruistic acts occur. Altruistic acts are mutually reciprocated, hence Trivers' theory is called **reciprocal altruism**. It is important to note that this theory requires that the individuals of an altruistic pair be **unrelated**; that is, they share no genes in common. Studies of alliances among male baboons show that a male usually solicits the help of a particular unrelated male and that "favorite partners" occur. In the evolution of altruism through reciprocity, "cheaters" (nonreciprocators) are discriminated against and are cut off from receiving future aid. According to Trivers, cheating should not occur if the cost of not reciprocating exceeds the benefits of receiving future aid.

*Kin Selection.* The most influential theory of the origin of altruism was presented by sociobiologist William D. Hamilton in 1964. This theory is perhaps best introduced by quoting a passing remark made in a pub in 1932 by the great population geneticist J.B.S. Haldane. Haldane had said that he would willingly lay down his life for *two brothers* or *eight first cousins.* Do you see what Haldane was getting at? Each brother and Haldane share half of their genes in common—Haldane's brothers each had a 50% chance of receiving any given allele that Haldane had obtained. Consequently, it is statistically true that two of his brothers would carry as many of Haldane's particular combinations of alleles to the next generation as would Haldane himself. Similarly, Haldane and a first cousin would share an eighth of their alleles: their sibling parents would each share half of their alleles, and each of their children would receive half of these, of which half on the average would be in common—0.5 × 0.5 × 0.5 = 0.125, or one-eighth. Eight first cousins would therefore pass on as many of those genes to the next generation as would Haldane himself. Hamilton saw Haldane's point clearly: *evolution will favor any strategy that increases the net flow of a combination of genes to the next generation.*

Altruism has costs and benefits. Hamilton showed that by directing aid toward kin, or close genetic relatives, the reduction in an altruist's (personal) fitness may be outweighed by the increased reproductive success of relatives. From an evolutionary perspective, selection will favor the behavior that maximizes the propagation of alleles. If giving up one's own reproduction to help relatives reproduce accomplishes this, then even sterility can be favored by natural selection. Selection acting to favor the propagation of genes by directing altruism toward relatives is called **kin selection.**

In Hamilton's theory of the evolution of altruism by kin selection, he coined the term **inclusive fitness** to describe the sum of genes propagated by personal reproduction and the effect of help on reproduction by relatives. It is important to note that inclusive fitness does not simply result from adding the number of genes passed on directly via an individual's own offspring and the number of genes passed on via relatives other than offspring. Rather, inclusive fitness is the sum of the number of genes directly passed on in an individual's offspring and those genes passed on indirectly by kin (other than offspring) whose existence results from the benefit of the individual's altruism.

The theory of kin selection proposed by W.D. Hamilton predicts that altruism is likely to be directed toward close relatives. The closer the degree of relatedness, the greater the potential genetic payoff in inclusive fitness.

## INSECT SOCIETIES

We have already mentioned the striking forms altruism can take in the social insects. The evolution of the honeybee's suicidal sting and the sterility of worker bees was an enigma to Darwin, who considered the insect societies to present a fatal blow to his theory. Before we try to solve Darwin's problem, it is useful to refresh our memories as to how natural selection works. Evolution acts on *individuals,* not on populations—selection favors the genes borne by those individuals which leave the most offspring. But what constitutes an individual in an insect society?

In truly social insects (some bees and wasps, all ants, and all termites), natural selection has acted on the *colony;* the society itself is the individual and the unit acted on by evolution. Each colony is made up of reproductive and sterile members (Figure 44-18). A honeybee hive, for example, has a single queen who is the sole egg-layer and tens of thousands of her off-

*Figure 44-18*

**Reproductive division of labor.** A bee colony. The queen, with a red spot on her thorax, is the sole egg-layer. Her daughters (workers) are sterile.

spring, who are female workers having generally non-functional ovaries. The sterility of workers is altruistic: during the course of evolution these offspring gave up their personal reproduction to help their mother rear more of their sisters.

Hamilton explained the origin of this trait with the theory of kin selection. He notes that because of an unusual system of sex determination present in bees, wasps, and ants, workers share a very high proportion of genes, theoretically as many as 75%. Because of this close genetic relatedness, *workers propagate more genes by giving up their own reproduction to assist their mother in rearing more of their sisters, some of which will start new colonies and thus reproduce.* This is how workers maximize their inclusive fitness.

*The colony is the evolving entity in insect societies. Because workers share large fractions of genes, they propagate more genes by not directly reproducing but helping their mother reproduce.*

Social insect colonies are composed of highly integrated groups of individuals, called **castes,** each of which performs a set of tasks. Their specialization is so extreme and their organization so rigid that these insect societies as a whole exhibit many of the properties of an individual organism and are sometimes considered as a *superorganism.* A human body relies on millions of individual cells that are specialized to perform many different tasks. In a similar way, a beehive or ant colony is a cohesively organized group of individuals in which certain individuals perform specialized tasks on which the survival and reproduction of the entire colony depend. Only one component of a human body, the gonads, is responsible for reproduction. In a similar way, only one component of the beehive or ants' nest, the queen, is involved in the reproduction of that colony. All the cells of a human body are related to one another by descent from one fertilized zygote. Similarly, all the members of a nest or hive are descended from an individual queen.

A honeybee colony may have up to 50,000 sterile females and a single female queen who lays all the eggs. The queen maintains her dominance by secreting a pheromone, called "queen substance," that suppresses development of the ovaries in other females, turning them into sterile workers. Drones (male bees) are produced in a hive only for purposes of mating.

When the colony becomes too large, some members do not receive a sufficient quantity of queen sub-stance, and the colony begins preparations for swarming. Workers make several new queen cells, in which new queens begin to develop. As these mature, the old queen acknowledges their presence by making a pulsating sound known as "quacking"; the most mature of the developing queens responds with a "tooting" sound. A scout worker returns with directions to a new hive site, and the old queen and a swarm of female workers leave to find the new hive. Left behind, the new queen emerges, kills the other candidate queens, flies out to mate, and returns to assume rule of the hive.

The lifestyles of social insects can be so unusual as to be bizarre, none more so than the leafcutter ants (Figure 44-19). Leafcutters are farmers. They live in colonies of up to several million individuals, growing crops of fungi beneath the ground. Their moundlike nests are underground "cities" covering more than 100 square yards, with hundreds of entrances and chambers as deep as 16 feet underground. Long lines of leafcutters march daily from the mound to a tree or bush, cut its leaves into small pieces, and carry the pieces back in another long line to the mound. Small worker ants chew the leaf fragments into a mulch, which they spread like a carpet in underground chambers. Soon a luxuriant garden of fungi is growing. Nurse ants carry the larvae of the nest around to browse on choice spots. Other workers weed out undesirable kinds of fungi. The **division of labor** among workers is related to worker size.

*Figure 44-19*

Leafcutter ants carrying leaf fragments back to the nest. The ants do not live on the leaf material—they feed it to fungi that they raise in underground gardens.

## VERTEBRATE SOCIETIES

In contrast to the highly structured and integrated insect societies and their remarkable forms of altruism, vertebrates form far less rigidly organized social groups. This seems paradoxical because vertebrates have larger brains and are capable of more complex behavior, yet they show generally lower degrees of altruism. Why? Apparently this is because of the lower amount of gene sharing among group members (maximally 50% in nearly all vertebrates). Nevertheless, many complex vertebrate social systems exist in which individuals show both reciprocity and kin-selected altruism. But vertebrate societies are also characterized by a greater degree of conflict and aggression among group members (although conflict does at times occur in insect societies). In vertebrates, conflict generally centers around access to food resources and mates. Features of vertebrate societies that have been studied in detail in several species are **cooperative breeding, alarm calling,** and **mating systems ecology.**

In some species of birds such as the African pied kingfisher and the Florida scrubjay, cooperative breeding systems have evolved. For example, a pair of scrubjays may have several other birds present in their territory that serve as **helpers at the nest** (Figure 44-20), assisting in feeding the pair's offspring, keeping watch for predators, and defending the territory. Helpers are fully capable of breeding on their own, but they remain as nonreproductive altruists for a time. Nests with helpers have more offspring than those which do not. Helpers are most often the fledged offspring of the pair they assist, so the situation resembles that of a family. The evolution of this type of cooperative breeding in

---

# Naked Mole Rats—A Rigidly Organized Vertebrate Society

One exception to the general rule that vertebrate societies are not rigidly organized is the naked mole rat, a tiny, hairless rodent that lives in East Africa. Adult naked mole rats are about 8 to 13 centimeters long and weigh up to 60 grams—about the size of a sausage (Figure 44-A). Unlike other kinds of mole rats, which live alone or in small family groups, naked mole rats congregate in large underground colonies with a far-ranging system of tunnels and a central nesting area. It is not unusual for a colony to contain 80 or more animals.

Naked mole rats live by eating large underground roots and tubers, which they locate by tunneling constantly. Naked mole rats tunnel in teams. Each mole rat has protruding front teeth, which make it look something like a pocket-sized walrus; it uses these teeth to chisel away the earth from the blind face at the end of a tunnel. When the leading mole rat has loosened a pile of earth, it pushes the pile between its feet and then scuttles backward through the tunnel, moving the pile of earth with its legs. When this animal finally reaches the opening, it gives the pile to another, who kicks the dirt out of the tunnel. Then, free of its pile of dirt, the tunneler returns to the end of the tunnel to dig again, crawling on tiptoe over the backs of a long train of other tunnelers that are moving backwards with their own dirt piles.

Naked mole rat colonies are unusual vertebrate societies not only because they are large and well organized. They also have a breeding structure that one might normally associate with truly social insects such as ants and termites. All of the breeding is done by a single "queen," who has one or two male consorts. The worker group, composed of both sexes, keeps the tunnel clear and forages for food. This is very unusual—few mammals surrender breeding rights without contest or challenge. But if the queen is removed from the colony, havoc breaks out among the workers: individuals attack each other, and they all compete to become part of the new power structure. When another female becomes dominant and starts breeding, things settle down once again and discord disappears.

Recent DNA fingerprinting studies have shown that colony members may share 80% of their genes. Kin selection may have been important in the evolution of this insectlike mammal society.

**Figure 44-A**

A naked mole rat.

**Figure 44-20**

**A scrubjay family.** Helpers cooperate with parents to rear young.

birds has been explained by using the inclusive fitness concept discussed earlier.

From the previous example it should be clear that some aspects of vertebrate behavior are self-sacrificing and present a puzzle to evolutionists. Particularly puzzling is the fact that vertebrates are often organized into social groups in such a way that the activities of certain individuals benefit the group at the potential expense of those individuals themselves. The meerkat in Figure 44-13, for example, is maintaining a lookout on a termite mound deep in the Kalahari desert of Southern Africa, exposed to predators and the full glare of the sun. In doing this, the lookout draws attention to itself and thus exposes itself to greater danger than if it were not a sentry. Behaviors such as this seem to be contrary to an individual's self interest.

Individuals in some vertebrate species such as the meerkat not only serve as sentries for the group by keeping watch, but may also give an alarm call to communicate the presence of a predator when one is sighted. An alarm call causes individuals to seek shelter. Why would an individual place itself in jeopardy by revealing its location by calling? Who benefits from hearing an alarm call?

Paul Sherman of Cornell University has answered this question through years of field observations of alarm calling in Belding's ground squirrel. Alarm calls are given when a predator such as a coyote or badger is spotted. Such predators may attack a calling squirrel, so giving a signal places the caller at risk. The social unit of a ground squirrel colony is female-based; the group tends to comprise a female and her daughters, sisters, aunts, and nieces. Males in the colony are not genetically related to these females. By marking all squirrels in a colony with an individual dye pattern on their fur (using Lady Clairol hair color!) and recording which individuals gave calls and the social circumstances of calling, Sherman found that females with relatives living nearby were more likely to give alarm calls than were females without kin nearby. Males tend to call much less frequently. Alarm calling therefore seems to represent **nepotism**; that is, it favors relatives.

Vertebrate societies, like insect societies, have a particular type of organization. A social group of vertebrates has a certain size, stability of members, and may vary in the number of breeding males and females and type of mating system. Sociobiologists have learned that the way in which a group is organized is related to the species' ecology. The features of the environment that influence the evolution of a particular type of social organization are often food type and predation.

African weaver birds provide an excellent example. There are roughly 90 species of these finchlike birds, which construct nests of woven vegetation. These species can be divided according to the type of social group they form. One group builds camouflaged solitary nests in the forest (Figure 44-21). Males and females have dull plumage, look alike, and are monogamous; they forage for insects to feed their young. The second group nests in colonies in trees on the African savanna. These species feed in flocks on seeds and are polygynous. The feeding and nesting ecology is correlated with the type of breeding relationships. In the forest, insects are hard to find and both parents must cooperate in feeding the young. Because they make repeated trips to collect food, their drab feather coloration does not call the attention of predators to their nest. The cryptically camouflaged nests further reduce predation. On the open savanna, building a hidden nest is not an option. Rather, they are protected from predators by nesting in spiny trees, which are not very abundant. This shortage of safe nest sites means that birds must nest together in a communal fashion. Because seeds occur abundantly, a female can acquire all the food needed to rear young without a male's help. The male, free from the duties of parenting, spends his time competing with other males for the best nest sites in the tree and courting many females. Bright male plumage has evolved because it is attractive to females, and a polygynous mating system is favored.

*Social behavior in vertebrates is often characterized by kin-selected altruism. Altruistic behavior is involved in cooperative breeding in birds and alarm calling in mammals. The organization of a vertebrate society represents an adaptive response to ecological conditions.*

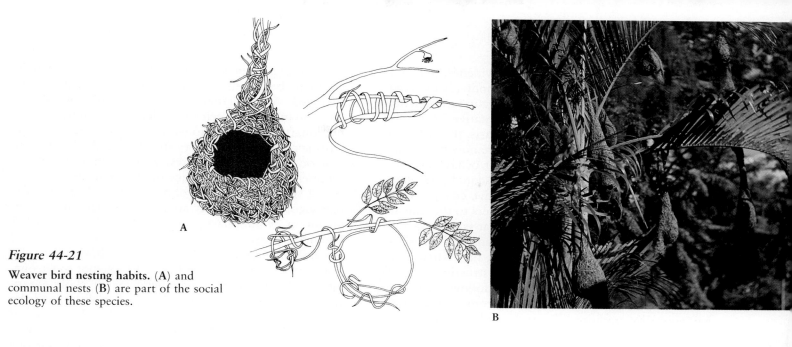

*Figure 44-21*

**Weaver bird nesting habits. (A)** and communal nests **(B)** are part of the social ecology of these species.

## HUMAN SOCIOBIOLOGY

Sociobiology is a comparative science. The same theory is applied to study the origin of social behavior in very different species. As we saw in our discussion of insect and vertebrate societies, altruism in bees, birds, and mammals can be explained with the same concept of kin selection. Almost all biologists agree that social behavior has a biological basis, and that one unifying theory, evolution, explains it. But is man just another social animal whose behavior can be explained with Darwinian concepts and fully understood?

As a social species, humans have an unparalleled complexity. Indeed, we are the only species with the intelligence to contemplate the social behavior of other animals. Intelligence and the ability to learn complex matters are just two human traits (Figure 44-22). If an ethologist took an inventory of human behavior, he or she would record kin-selected altruism and reciprocity, other elaborate social contracts, extensive parental care, conflicts between parents and offspring, violence, and warfare. A variety of mating systems such as monogamy, polygyny, and polyandry would be described, along with a number of sexual behaviors such as adultery and homosexuality. Behaviors such as adoption that appear to defy evolutionary explanation would also be a part of the ethologist's catalog. And remember, this incredible variation in behavior all occurs *in one species* and any of these traits can change within any *individual*. Are these behaviors rooted in our biology?

During the course of human evolution and the emergence of civilization, two processes led to adaptive change. One is **biological evolution.** We have a primate heritage, reflected in the extensive sharing of genetic material between humans and our closest relatives, chimpanzees. Our upright posture, bipedal locomotion,

power, and precision hand grips are adaptations whose origins are traceable through our primate ancestors. Kin-selected and reciprocal altruism can also be seen in nonhuman primates, as well as other shared traits such as aggression and different types of mating systems. In nonhuman primates we can demonstrate through careful study that these social traits are adaptive. We may

*Figure 44-22*

**The young of primates and birds go through long periods of learning before they can survive on their own.** This young lady, not yet 2 years old, has much yet to learn before she can wear Daddy's hat.

*speculate,* based on various lines of evidence, that similar traits evolved in early humans. If individuals showing certain behaviors had an advantage in reproduction over other individuals showing alternative behaviors, and these social traits had a genetic basis, then the alleles for the expression of these traits are now part of the human genome and may influence our behavior.

The second process that underscored the emergence of civilization and led to adaptive change is **cultural evolution.** Cultural evolution refers to the transfer of information across generations that is necessary to survival. It is a nongenetic mode of adaptation. Many human adaptations—the use of tools, the formation of cooperative hunting groups, the construction of shelters, and marriage practices—do not follow Mendelian rules of inheritance and are passed from generation to generation by tradition. To anthropologists interested in the origin of human behavior, a culture is as real a way of conveying adaptations across generations as is the gene. Human cultures are also extraordinarily diverse. The way in which children are socialized among Trobriand Islanders, Pygmies, and Yanomamo Indians are very different. Again we must remember that this fantastic variation occurs within one species, and that individual behavior is very flexible.

Given this great flexibility, how can biological components of human behavior be identified? One way is to study behaviors that are **cross cultural.** In spite of cultural variation, there are some traits that characterize all human societies. For example, all cultures have an **incest taboo** forbidding marriages between close relatives. Incestuous matings lead to a greater chance of exposing disorders such as mental retardation and hemophilia. Natural selection may have acted to create a cultural norm to avoid a serious biological problem. Genes that were responsible for guiding this behavior might have been fixed in human populations because of their adaptive effects.

Although human mating systems vary, polygyny is found to be the most common when all cultures are surveyed. Most mammalian species are polygynous; the human pattern seems to reflect our mammalian evolutionary heritage. Nonverbal communication patterns such as smiling and raising the hand as a greeting also occur in many cultures. Perhaps these behaviors represent a common heritage of communication.

A significant number of biologists and social scientists vigorously resist any attempt to explain human behavior in evolutionary terms. In E.O. Wilson's 1975 landmark treatise *Sociobiology: The New Synthesis,* the one chapter devoted to human behavior was denounced vehemently by critics as an attempt to encourage thinking about human behavior in evolutionary terms. Biology has not been free from politics, and the critics of human sociobiology believe that there is still a potential for abuse. Moreover, if we consider human behavior to be the product of evolution and that our actions are influenced by genes and are at least in part "hardwired," doesn't this suggest that unpleasant aspects of human behavior such as aggression and violence cannot easily be modified? Doesn't such a view affect how we perceive the prospects for positive social change?

Darwinian theory can provide us with an overarching evolutionary *perspective* on human nature, one which stresses the unity of the human species and the origin of some important, adaptive behaviors. But, because human behavior is affected by *both* innate and learned components and because many human activities such as art, music, and religion are strongly influenced by culture and are not easy to study as adaptation, it is unlikely to give us any *resolution* of the fine details of human nature. The study of the biology of human behavior will always be a provocative one.

## ■ SUMMARY

1. Behavior is an adaptive response to stimuli in the environment. An animal's sensory system monitors the environment and has specialized neural elements to detect and process environmental information.

2. Behavior is both instinctive and controlled by genes and is learned through experience. Genes are thought to limit the extent to which behavior can be modified and the types of associations that can be made.

3. The simplest forms of learning involve sensitization and habituation. More complex associative learning may also occur in this way, by the strengthening and weakening of existing synapses, although learning may also involve the formation of entirely new synapses.

4. An animal's internal state influences when and how a response will occur. Hormones cause an animal's behavior and perception of stimuli to change in an adaptive way.

5. Animals communicate by producing visual, acoustical, chemical, and electric signals. These signals are involved in mating, food finding, predator defense, and other social situations.

6. Many behaviors are important ecologically and serve as adaptations. Foraging and territorial behaviors have evolved because they allow animals to use resources efficiently.

7. Male and female animals show different reproductive behaviors that maximize fitness. Usually, males are competitive and females show mate choice. Mating systems are related to a species' ecology.

8. Insects, vertebrates, and other animals show altruistic behavior. Altruism may evolve through reciprocity or be directed toward genetic relatives. Cooperative behavior often increases an individual's inclusive fitness.

9. Individuals live in social groups because it is advantageous to do so. Animal societies are characterized by cooperation and conflict. The organization of a society is related to the ecology of a species.

10. Human behavior is extremely rich and varied and may result from both biology and culture. Evolutionary theory can give us limited but important insight into human nature.

## REVIEW

1. Evolutionary questions about the adaptiveness of behavior concern _____ causation.

2. The genetic basis of behavior can be demonstrated in the laboratory from studies of _____ selection and _____.

3. The simplest form of learning is _____.

4. Male birds sing to defend a territory when the hormone _____ is secreted by the testes.

5. Endogenous rhythms of about 24 hours are called _____ rhythms.

6. Honeybee workers have given up their own reproduction to help their mothers rear sisters to maximize their _____.

7. The mating system in which a female mates with more than one male is _____.

8. Alarm calls given by Belding's ground squirrel benefit _____.

## SELF-QUIZ

1. The giant nerve cells that control escape behavior in the sea hare are an example of which concept in ethology?
   (a) Sign stimuli
   (b) Innate releasing mechanism
   (c) Fixed-action pattern
   (d) Motivation
   (e) Associative learning

2. Marler's study of song acquisition in the white-crowned sparrow shows that bird song
   (a) is entirely innate.
   (b) is entirely learned.
   (c) develops normally when a bird hears no song.
   (d) develops normally when a bird hears another's species song.
   (e) is learned, but guided by a genetic program.

3. The species specificity of courtship signals implies that they play a role in
   (a) reproductive isolation.
   (b) alarm calling.
   (c) kin selection.
   (d) controlling a rhythm.
   (e) olfaction.

4. Behavioral ecologists study
   (a) neural circuits.
   (b) circadian rhythms.
   (c) the adaptiveness of behavior.
   (d) motivation.
   (e) habituation.

5. Territorial behavior should evolve when
   (a) resources are very scarce.
   (b) resources are very abundant.
   (c) species are monogamous.
   (d) the cost of defense is less than the benefits of the energy gained from defense.
   (e) the benefits of being territorial are less than the risks of injury during defense.

6. The genetic basis for altruism in bees, wasps, and ants is caused by
   (a) an unusual type of sex determination.
   (b) inclusive fitness.
   (c) kin selection.
   (d) cooperative breeding.
   (e) reciprocity.

7. Which factor(s) have influenced the evolution of weaver bird social organization?
   (a) Competition
   (b) Temperature and rainfall
   (c) Predation and food type
   (d) Only predation
   (e) Kin selection

8. Most bird species are
   (a) monogamous.
   (b) polygynous.
   (c) polyandrous.
   (d) cooperative breeders.
   (e) strongly sexually dimorphic.

# THOUGHT QUESTIONS

1. A recent study of Yanomamo Indians has shown that high-ranking men have more wives, more children, and commit murder more often than other Yanomamo men. Does this mean that there is a gene for violence in humans?

2. There are many speculations about the evolutionary origin of human behavior, but it is usually not possible to rigorously test such hypotheses and critically evaluate alternative, nonevolutionary explanations. Does this mean that the evolutionary study of human behavior is not scientific?

3. Swallows often hunt in groups, whereas hawks and other predatory birds usually are solitary hunters. Can you suggest an explanation for this difference?

4. Can you suggest an evolutionary reason why many vertebrate reproductive groups are composed of one male and numerous females, rather than the reverse?

# FOR FURTHER READING

BORGIA, G: "Sexual Selection in Bowers Birds," *Scientific American,* June 1986, pages 92-100. Behavior among birds is often bizarre, but none is more so than that of the Australasian bower birds, in which the females choose their mates depending on how well they adorn their bowers or ritualized nests.

DAWKINS, R.: *The Selfish Gene,* Oxford University Press, New York, 2nd edition, 1989. An entertaining account of the sociobiologist's view of behavior.

GOULD, J., and P. MARLER: "Learning by Instinct," *Scientific American,* January 1987, pages 74-85. A clear and interesting account of the relative roles of instinct and learning in behavior. The authors argue that learning is often limited or controlled by instinct.

GRIFFIN, D.: "Animal Thinking," *American Scientist,* vol. 72, 1984, pages 456-463. An exciting discussion of the possibility that animals have consciousness, this article describes many examples of what appears to be "awareness" by animals.

HEINRICH, B.: *Ravens in Winter: A Zoological Detective Story,* Summit Books, New York, 1989. What do ravens do in winter, and how are their social systems organized? Vermont zoologist Bernd Heinrich tells us about his outstanding field investigations of these intelligent animals.

HUBER, F., and J. THORSON: "Cricket Auditory Communication," *Scientific American,* December 1985, pages 60-68. An unusually clear example of how nervous system activity underlies animal behavior.

JOLLY, A.: "The Evolution of Primate Behavior," *American Scientist,* vol. 73, 1985, pages 230-239. A fascinating survey of behavior among primates that indicates a progressive development of intelligence rather than a sudden, full-blown appearance when humans evolved.

VERRELL, P.: "When Males are Choosy," *New Scientist,* January 20, 1990, pages 46-50. Under certain circumstances, males can be choosy, too.

WILLIAMS, A.O.D.: "Giant Brain Cells in Mollusks," *Scientific American,* vol. 244, 1981, pages 68-75. A good introduction to neural aspects of behavior and the use of the sea hare as a model system.

# CLASSIFICATION OF ORGANISMS

The classification used in this book is explained in Chapter 24. It recognizes a separate kingdom, Monera, for the bacteria (prokaryotes) and divides the eukaryotes into four kindgoms: the diverse and predominantly unicellular Protista, and three large, characteristic multicellular groups derived from them: Fungi, Plantae, and Animalia. Viruses, which are considered nonliving, are not included in this appendix, but are treated in Chapter 25.

## KINGDOM MONERA

Bacteria; the prokaryotes. Single-celled, sometimes forming filaments or other forms of colonies. Bacteria lack a membrane-bound nucleus and chromosomes, sexual recombination, and internal compartmentalization of the cells; their flagella are simple, composed of a single fiber of protein. They are much more diverse metabolically than are the eukaryotes. Their reproduction is predominantly asexual. About 2500 species are currently recognized, but many times that many probably exist.

## KINGDOM PROTISTA

Eukaryotic organisms, including many evolutionary lines of primarily single-celled organisms. Eukaryotes have a membrane-bound nucleus and chromosomes, sexual recombination, and extensive internal compartmentalization of the cells; their flagella are complex, with 9 +2 internal organization. They are diverse metabolically, but much less so than are bacteria; protists are heterotrophic or autotrophic and may capture prey, absorb their food, or photosynthesize. Reproduction in protists is either sexual, involving meiosis and syngamy, or asexual.

### Phylum Caryoblastea
One species of primitive amoebalike organism, Pelomyxa palustris, which lacks mitosis, mitochondria, and chloroplasts.

### Phylum Dinoflagellata
Dinoflagellates; unicellular, photosynthetic organisms, most of which are clad in stiff, cellulose plates and have two unequal flagella that beat in grooves encircling the body at right angles. About 1000 species.

### Phylum Rhizopoda
Amoebas; heterotrophic, unicellular organisms that move from place to place by cellular extensions called pseudopods and reproduce only asexually, by fission. Hundreds of species.

### Phylum Sporozoa
Sporozoans: unicellular, heterotrophic, nonmotile, spore-forming parasites of animals. About 3900 species.

### Phylum Acrasiomycota
Cellular slime molds; unicellular, amoeba-like, heterotrophic organisms that aggregate in masses at certain stages of their cycle and form compound sporangia. About 65 species.

### Phylum Myxomycota
Plasmodial slime molds; heterotrophic organisms that move from place to place as a multicellular, gelatinous mass, forming sporangia at times. About 450 species.

### Phylum Zoomastigina
Zoomastigotes and euglenoids; a highly diverse phylum of mostly unicellular, heterotrophic or autotrophic, flagellated free-living or parasitic protists (flagella one to thousands) Thousands of species.

### Phylum Phaeophyta
Brown algae; multicellular, photosynthetic, mostly marine protists with chlorophylls $a$ and $c$ and an abundant carotenoid (fucoxanthin) that colors the organisms brownish. About 1500 species.

### Phylum Chrysophyta
Diatoms and related groups; mostly unicellular, photosynthetic organisms with chlorophylls $a$ and $c$ and fucoxanthin. About 11,500 living species.

## Phylum Chlorophyta

Green algae; a large and diverse phylum of unicellular or multicellular, mostly aquatic organisms with chlorophylls *a* and *b*, carotenoids, and starch, accumulated within the plastids (as it also is in plants) as the food storage product. About 7000 species.

## Phylum Ciliophora

Ciliates; diverse, mostly unicellular, heterotrophic protists, characteristically with large numbers of cilia. About 8000 species.

## Phylum Oomycota

Oomycetes; water molds, white rusts, and downy mildews. Aquatic or terrestrial unicellular or multicellular parasites or saprobes that feed on dead organic matter. About 475 species.

## Phylum Rhodophyta

Red algae; mostly marine, mostly multicellular protists with chloroplasts containing chlorophyll *a* and physobilins. About 4000 species.

# KINGDOM FUNGI

Filamentous, multinucleate, heterotrophic eukaryotes with cell walls rich in chitin; no flagellated cells present. Mitosis in fungi takes place within the nuclei, the nuclear envelope never breaking down. The filaments of fungi grow through the substrate, secreting enzymes and digesting the products of their activity. Septa between the nuclei in the hyphae normally complete only when sexual or asexual reproductive structures are being cut off. Asexual reproduction frequent in some groups. The nuclei of fungi are haploid, with the zygote the only diploid stage in life cycle. About 100,000 named species.

## Division Zygomycota

Zygomycetes; bread molds and other microscopic fungi that occur on decaying organic matter. Hyphae aseptate except when forming sporangia or gametangia. About 600 species.

## Division Ascomycota

Ascomycetes; yeasts, molds, many important plant pathogens, morels, cup fungi, and truffles. Hyphae divided by incomplete septa except when asci, the structures characteristic of sexual reproduction, are formed. Meiosis takes place within asci. About 300,000 named species.

## Division Basidiomycota

Basidiomycetes; mushrooms, toadstools, bracket and shelf fungi, rusts, and smuts. Meiosis takes place within basidia. About 25,000 named species.

## Fungi Inperfecti

An artificial group of about 25,000 named species; in them, the reproductive structures are not known.

## Lichens

Lichens are symbiotic associations between an ascomycete (a few basidiomycetes are also involved) and either a green alga or a cyanobacterium. At least 25,000 species.

# KINGDOM PLANTAE

Multicellular, photosynthetic, primarily terrestrial eukaryotes derived from the green algae (phylum Chlorophyta) and, like them, containing chlorophylls *a* and *b*, together with carotenoids, in chloroplasts and storing starch in chloroplasts. The cell walls of plants have a cellulose matrix and sometimes become lignified; cell division is by means of a cell plate that forms across the mitotic spindle. The vascular plants have an elaborate system of conducting cells consisting of xylem (in which water and minerals are transported) and phloem (in which carbohydrates are transported); the mosses have a reduced vascular system, which the liverworts and hornworts, which may not be directly related to the mosses, lack. Plants have a waxy cuticle that helps them to retain water, and most have stomata, flanked by specialized guard cells, which allow water to escape and carbon dioxide to reach the chloroplast-containing cells within their leaves and stems. All plants have an alternation of generations with reduced gametophytes and multicellular gametangia. About 270,000 species.

## Division Bryophyta

Mosses, hornworts, and liverworts. Bryophytes have green photosynthetic gametophytes and usually brownish or yellowish sporophytes with little or no chlorophyll. About 16,600 species.

## Division Psilophyta

Whisk ferns, a group of vascular plants. Two genera and several species.

## Division Lycophyta

Lycopods (including clubmosses and quillworts). Vascular plants. Five genera and about 1000 species.

## Division Sphenophyta
Horsetails. Vascular plants. One genus *(Equisetum)*, 15 species.

## Division Pterophyta
Ferns. Vascular plants, often with characteristically divided, feathery leaves (fronds). About 12,000 species.

## Division Coniferophyta
Conifers. Seed-forming vascular plants; mainly trees and shrubs. About 550 species.

## Division Cycadophyta
Cycads; tropical and subtropical palmlike gymnosperms. Ten genera, about 100 species.

## Division Ginkgophyta
One species, the ginkgo or maidenhair tree.

## Division Anthophyta
Flowering plants, or angiosperms, the dominant group of plants; characterized by a specialized reproductive system involving flowers and fruits. About 235,000 species.

# KINGDOM ANIMALIA

Animals are multicellular eukaryotes that characteristically ingest their food. Their cells are usually flexible; in all of the approximately 35 phyla except sponges, these cells are organized into structural and functional units called tissue, which in turn makes up organs in most animals. In animals, the cells move extensively during the development of the embryos; the blastula, a hollow ball of cells, forms early in this process and is characteristic of the group. Most animals reproduce sexually; their nonmotile eggs are much larger than their small, flagellated sperm. The gametes fuse directly to produce a zygote and do not divide by mitosis as in plants. More than a million species of animals have been described, and at least several times that many await discovery.

## Phylum Porifera
Sponges. Animals that mostly lack definite symmetry and possess neither tissues nor organs. About 10,000 marine species, mostly marine.

## Phylum Cnidaria
Corals, jellyfish, hydras. Mostly marine, radially symmetrical animals that mostly have distinct tissues; two basically different body forms, polyps and medusae. About 10,100 species.

## Phylum Platyhelminthes
Flatworms; bilaterally symmetrical acoelomates; the simplest animals that have organs. About 13,000 species.

## Phylum Nematoda
Nematodes, eelworms, and roundworms; ubiquitous, bilaterally symmetrical, cylindrical, unsegmented, pseudocoelomate worms, including many important parasites of plants and animals. More than 12,000 described species, but the actual number is probably 500,000 or more species.

## Phylum Mollusca
Mollusks; bilaterally symetrical, protostome coelomate animals that occur in marine, freshwater, and terrestrial habitats. Many mollusks possess a shell. At least 110,000 species.

## Phylum Annelida
Annelids; segmented, bilaterally symmetrical, protostome coelomates; the segments are divided internally by septa. About 12,000 species.

## Phylum Arthropoda
Arthropods; bilaterally symmetrical protostome coelomates with a segmented body, chitonous exoskeleton, complete digestive tract, dorsal brain and paired nerve cord, and jointed appendages. Arthropods are the largest phylum of animals, with nearly a million species described and many more to be found.

## Phylum Echinodermata
Echinoderms; sea stars, brittle stars, sand dollars, sea cucumbers, and sea urchins. Complex deuterostome, coelomate, marine animals that are more or less radially symmetrical as adults. About 6000 living species.

## Phylum Chordata
Chordates; bilaterally symmetrical, deuterostome, coelomate animals that have at some stage of their development a notochord, pharyngeal slits, a hollow nerve cord on their dorsal side, and a tail. The best-know group of animals; about 45,000 species.

# A P P E N D I X ■ B

## ANSWERS

### Chapter 1
*Review*
1. Cellular organization, growth, metabolism, reproduction, homeostasis, heredity.
2. DNA
3. theory
4. 4.5 billion
5. homologous
6. 4—Protista, Animalia, Plantae, Fungi

*Self-Quiz*
1. a, b, d
2. c
3. b
4. b
5. a
6. d

### Chapter 2
*Review*
1. water
2. neutrons, protons, electrons
3. starch, glycogen
4. DNA
5. covalent

*Self-Quiz*
1. c
2. a
3. c, d, e
4. a, b, c, d
5. e

### Chapter 3
*Review*
1. evolution
2. 3.5
3. electrical sparks, energetic UV light or heat
4. coacervates
5. 1.5 billion

*Self-Quiz*
1. e
2. c
3. c
4. b, c, d
5. d

### Chapter 4
*Review*
1. cell membrane
2. channels, receptors, markers
3. lysosomes, golgi bodies
4. mitochondria, chloroplasts, centrioles
5. nucleus, mitochondria, centriole

*Self-Quiz*
1. a
2. a, b, c
3. d
4. c, d, a, b

### Chapter 5
*Review*
1. glycerol, fatty acid, alcohol
2. major histocompatibility complex proteins
3. diffusion
4. proton pump
5. hormones

*Self-Quiz*
1. e
2. a
3. d, e
4. d, e
5. e

### Chapter 6
*Review*
1. work
2. no
3. the sun
4. cellular respiration
5. enzymes, proteins

*Self-Quiz*
1. a, b, e
2. a *or* b
3. c
4. c
5. a
6. d
7. b, c
8. a, d, e

# Chapter 7
*Review*
1. ATP
2. glycolysis, oxidation of pyruvate, citric acid cycle
3. pyruvate
4. mitochondrion
5. 36 molecules of ATP

*Self-Quiz*

| | |
|---|---|
| 1. a | 4. a |
| 2. b | 5. b |
| 3. e | |

*Problem $2.168 \times 10^{25}$*

# Chapter 8
*Review*
1. pigments
2. fix carbon
3. $NADP^+$, $NAD^{++}$
4. photosynthesis
5. thylakoid

*Self-Quiz*

| | |
|---|---|
| 1. a | 4. b |
| 2. d | 5. a |
| 3. b, e | |

# Chapter 9
*Review*
1. fission
2. mitosis
3. metaphase plate
4. synapsis
5. recombinants

*Self-Quiz*

| | |
|---|---|
| 1. a | 4. e, b |
| 2. b | 5. b |
| 3. e | |

# Chapter 10
*Review*
1. genotype, phenotype
2. *w*
3. a test cross
4. Law of Independent Assortment
5. crossing over
6. 3:1
7. 25%
8. crossing over
9. false
10. true
11. 25%
12. linkage

*Self-Quiz*
1. c
2. b
3. d
4. a
5. c
6. b
7. d
8. c
9. e

*Genetics Problems*
1. He only chose to study two pure-breeding varieties of any given trait, he could have chosen to study more. In any given cross, the maximum number of alleles that he could have observed is four, if the two parents were each heterozygous for a different pair of alleles.
2. Alleles segregate in meiosis, and the products of that segregation is incorrectly shown as being between pods, each pod shown as uniformly wrinkled or round.
3. The probability of getting two genes on the same chromosome is $(1/n)^2$.
4. Somewhere in your herd you have cows and bulls that are not homozygous for the dominant gene "polled." Since you have many cows and probably only one or some small number of bulls, it would make sense to concentrate on the bulls. If you have only homozygous "polled" bulls, you could never produce a horned offspring regardless of the genotype of the mother. The most expedient thing to do would be to keep track of the matings and the phenotype of the offspring resulting from these matings and render ineffective any bull found to produce horned offspring.

5. It would not be possible on the basis of the information presented to substantiate a claim of infidelity. You do not know if the woolly trait is the result of a single gene product, or even if the trait is dominant or recessive. Assuming for the moment that it was the effect of a single dominant allele *W*, the man could still be a heterozygote for the gene and, when mated to a recessive homozygous female, would expect to produce woolly-headed offspring only half the time.

6. ½

7. Albinism, *a*, is a recessive gene. If heterozygotes mated, you would have the following:

|   | A | a |
|---|---|---|
| A | AA | Aa |
| a | Aa | aa |

Clearly one-fourth would be expected to be albinos.

8. The best thing to do would be to mate Dingleberry to several dames homozygous for the recessive gene that causes the brittle bones. Half of the offspring would be expected to have brittle bones if Dingleberry were a heterozygous carrier of the disease gene. Although you could never be 100% certain Dingleberry was not a carrier, you could reduce the probability to a reasonable level.

9. Your mating of *DDWw* and *Ddww* individuals would look like the following:

|    | Dw | Dw | dw | dw |
|----|------|------|------|------|
| DW | DDWw | DDWw | DdWw | DdWw |
| Dw | DDww | DDWw | Ddww | Ddww |
| DW | DDWw | DDWw | DdWw | DdWw |
| Dw | DDww | DDww | Ddww | Ddww |

Long-wing, red-eyed individuals would result from 8 of the possible 16 combinations, and dumpy, white-eyed individuals would never be produced.

10. Breed Oscar to Heidi. If half of the offspring are white eyed, then Oscar is a heterozygote.

11. Both parents carry at least one of the recessive genes. Since it is recessive, the trait is not manifested until they produce an offspring who is homozygous.

12. To solve this problem, let's first look at the second cross, where the individuals were crossed with the homozygous-recessive sepia flies *se/se*. In one case, all the flies were red-eyed:

Unknown genotype

|       |    | Se    | Se    |
|-------|----|-------|-------|
| Sepia | Se | Se/se | Se/se |
|       | se | Se/se | Se/se |

The only way to have all red-eyed flies when bred to homozygous sepia flies is to mate the sepia fly with a homozygous red-eyed fly. In the other case, half of the offspring were black eyed and the other half red-eyed.

Unknown genotype

|       |    | Se    | se    |
|-------|----|-------|-------|
| Sepia | se | Se/se | se/se |
|       | se | Se/se | se/se |

The unknown genotype in this case must have been *Se/se,* since this is the only mating that will produce the proper ratio of sepia-eyed flies to red-eyed flies. Since the ratio of this unknown genotype and the one previously determined was 1:1, we must deduce the genotype of the original flies, which, when mated, will produce a 1:1 ratio of *Se/se* to *Se/Se* flies.

Unknown original 1

|            |    | Se    | Se    |
|------------|----|-------|-------|
| Original 2 | Se | Se/Se | Se/Se |
|            | se | Se/se | Se/se |

You can see from this diagram that if one of the original flies was homozygous for red eyes and the other was a heterozygous individual, the proper ratio of heterozygous and a homozygous offspring would be obtained.

13. (a) It could have originated as a mutation in his germ cell line. (b) Since their son Alex was a hemophiliac, the disease almost certainly originated with Alexandra; Nicholas II would have contributed only a silent Y chromosome to Alex's genome. There is a 50% chance that Anastasia was a carrier.

# Chapter 11
## Review
1. (a) 46, (b) 22
2. Barr body
3. 5%
4. Klinefelter's syndrome
5. amniocentesis or chorionic villi sampling

## Self-Quiz
1. c
2. a
3. b, c
4. c
5. b
6. a
7. e
8. c

## Genetics Problems
1.

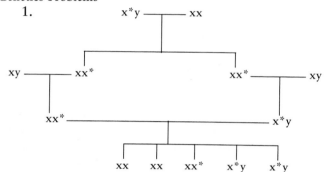

2. $\qquad I^A I^O \times I^B I^O \rightarrow I^O I^O$

3.

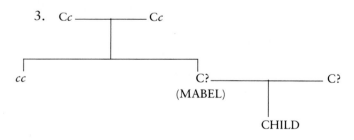

c is a rare allele, and Mabel's child can develop cystic fibrosis only if her husband also carries the allele:

$\left[\begin{array}{c}\text{probability that husband}\\ \text{has } c \text{ allele}\end{array}\right] = \text{frequency of } c = 1/20$

$\left[\begin{array}{c}\text{probability that Mabel}\\ \text{has } c \text{ allele}\end{array}\right] = 2/3$

In addition, the probability of transmission even if both parents carry the allele is only 25%; therefore the overall probability is $1/20 \times 2/3 \times 1/4 = 1/120$

4. 45 (44 autosomes + one X)
5. AO
6. dominant
7. dominant
8.

9. Let $a$ = albino. The genotype of a father is $Aa$.
10. PARENTS BABY
    O and O  O
    A and B  AB
    B and B  B
    AB and O  A
11. b—autosomal dominant
12. sex-linked recessive
13. dominant

# Chapter 12
## Review
1. in the nucleus
2. DNA
3. thymine, cytosine
4. enzymes
5. nucleic acid

## Self-Quiz
1. d
2. a
3. a
4. c
5. c
6. e
7. d

# Chapter 13
## Review
1. genes
2. translated
3. three
4. mRNA, tRNA
5. repressor protein
6. nuclease
7. RNA
8. oncogenes

*Self-Quiz*
1. a, b, c
2. b
3. a
4. c
5. e
6. b, d
7. a

## Chapter 14
*Review*
1. crossing over, assortment during meiosis, transposition
2. a replication origin
3. conjugation
4. prophase
5. satellite DNA
6. tandem cluster

*Self-Quiz*
1. b    5. b
2. e    6. c
3. a    7. a
4. c    8. a

## Chapter 15
*Review*
1. microevolution
2. genetic drift, selection
3. selection
4. industrial melanism
5. parallel adaptation

*Self-Quiz*
1. d    4. a
2. d    5. a
3. e

## Chapter 16
*Review*
1. species
2. microevolution, macroevolution or different species
3. geographical isolation *now* or ecological evolution before 150 years ago
4. prezygotic, postzygotic
5. macroevolution

*Self-Quiz*
1. e    5. a
2. d    6. c
3. a    7. b
4. c

## Chapter 17
*Review*
1. Pangaea
2. phyla
3. 3.5 billion
4. fungi, insects, vertebrates, plants
5. Permian

*Self-Quiz*
1. b, e, a, c, d
2. c
3. a
4. a, b, d, e
5. c
6. c
7. c
8. a

## Chapter 18
*Review*
1. amniotic
2. false
3. fishes
4. *Australopithecus*
5. three

*Self-Quiz*
1. b, d, e
2. c, e, b, a, d
3. a, d, c, e, b
4. c, d, a, e, b
5. c, b, e, d, a
6. b
7. a

## Chapter 19
*Review*
1. more
2. false
3. small
4. III, I
5. survivorship

*Self-Quiz*
1. d
2. c
3. b
4. c
5. c
6. a
7. e
8. a, d

## Chapter 20

### Review

1. autotrophs, heterotrophs
2. brightly
3. toxic
4. lowered
5. mutualism, commensalism, parasitism, mutualism, commensalism

### Self-Quiz

| | | |
|---|---|---|
| 1. b | 4. d | 7. e |
| 2. c | 5. b | |
| 3. a | 6. b | |

## Chapter 21

### Review

1. nitrogen
2. biomass
3. net primary productivity
4. oxygen, nitrogen, carbon, hydrogen

### Self-Quiz

1. b
2. a
3. d
4. b
5. d
6. c
7. b
8. e, b, c, d, a

## Chapter 22

### Review

1. savannas or chaparral
2. land (the physical environment)
3. land—a more diverse habitat
4. warm air at the equator rises and cools, losing its moisture; it falls back at 30-degrees latitude, warming and absorbing moisture, and so creating deserts.
5. less than 25 cm per year of rainfall

### Self-Quiz

1. d
2. c, d
3. d
4. d
5. d
6. e, a, b, d, c
7. b, c, d, e
8. d, a, c, b

## Chapter 23

### Review

1. faster
2. 20%
3. 5 million
4. grazing, firewood gathering, lumbering
5. 1.7% annually

### Self-Quiz

| | |
|---|---|
| 1. d | 5. e |
| 2. c | 6. d |
| 3. c | 7. c |
| 4. c | 8. c |

## Chapter 24

### Review

1. Carl Linnaeus
2. 1.5 million, 5 million
3. metabolically
4. Protista
5. viruses

### Self-Quiz

1. b, d, g, a, e, f, c
2. A (d), B (e), C (a)
3. a
4. b, c, d
5. a
6. b
7. d
8. e

## Chapter 25

### Review

1. no
2. no
3. spirochetes
4. no
5. white blood cells (T4)

### Self-Quiz

1. a, d
2. a
3. a, c, d
4. b
5. d
6. d

## Chapter 26

*Review*
1. lack of chloroplasts, external digestion, filamentous growth, complete lack of flagella
2. rhodophyta or red
3. choanoflagellates (one of the groups of zoomastigotes)
4. hyphae
5. zygomycetes, ascomycetes, basidiomycetes

*Self-Quiz*
1. b
2. c
3. b, d
4. a, d
5. d
6. d
7. a, c
8. d

## Chapter 27

*Review*
1. sporophyte
2. they swim
3. triploid
4. enclosed
5. monocots, dicots

*Self-Quiz*
1. c, e
2. c
3. d
4. a, e
5. b, a, d, c
6. b
7. a, c, d, e

## Chapter 28

*Review*
1. Arthropoda, Chordata
2. endoderm
3. protostomes, deuterostomes
4. Arthropoda
5. Pseudocoelomates

*Self-Quiz*
1. b
2. c
3. e
4. d, e
5. e
6. a, b, c

## Chapter 29

*Review*
1. meristems
2. apical meristem, lateral meristem
3. cuticle
4. mesophyll
5. bark

*Self-Quiz*
1. b, c, e
2. e
3. e
4. b
5. b
6. d
7. b
8. c

## Chapter 30

*Review*
1. meiosis
2. bees
3. imbibes water
4. gibberellins
5. outcrossing

*Self-Quiz*
1. e
2. e
3. d
4. c
5. a, b, d
6. c
7. c
8. b

## Chapter 31

*Review*
1. hormones
2. osmosis
3. nitrogen, phosphorus, calcium, potassium, magnesium, sulfur
4. promote
5. gravitropism

*Self-Quiz*
1. d
2. b
3. b
4. a, c
5. b, e
6. a, b, c
7. a, b, d
8. d

## Chapter 32

### Review

1. (a) thoracic or respiratory cavity, (b) abdominal or digestive cavity
2. mesoderm
3. collagen
4. muscle
5. (a) axon, (b) dendrites

### Self-Quiz

1. d    4. b
2. b    5. d
3. c

## Chapter 33

### Review

1. sodium
2. absolute
3. action potential
4. myelin
5. potassium

### Self-Quiz

1. e    4. c
2. c    5. d
3. b

## Chapter 34

### Review

1. (a) hindbrain or rhombencephalon, (b) midbrain or mesencephalon, (c) forebrain or prosencephalon
2. cerebrum
3. (a) electrically, (b) structural changes
4. *cis*-retinal
5. (a) adrenaline, (b) noradrenaline

### Self-Quiz

1. a
2. d
3. e
4. c
5. a, e
6. b
7. b

## Chapter 35

### Review

1. (a) exoskeleton (chitin), (b) bones
2. spongy bone tissue or marrow
3. (a) actin, (b) myosin
4. calcium ions

### Self-Quiz

1. b    5. e
2. e    6. a
3. a    7. c
4. b

## Chapter 36

### Review

1. hormones
2. hypothalamus
3. thyroid-stimulating hormone
4. hypothalamus
5. insulin

### Self-Quiz

1. e    4. b
2. d    5. d
3. a

## Chapter 37

### Review

1. digestion
2. pyloric sphincter
3. villi
4. fat
5. K

### Self-Quiz

1. a
2. a, d
3. a
4. d
5. e
6. b

## Chapter 38

### Review

1. respiration
2. true
3. pharnyx, swim bladder
4. water loss
5. hemoglobin

### Self-Quiz

1. e
2. c
3. c, e, a, b, d
4. d
5. a
6. c
7. d
8. e

## Chapter 39

*Review*

1. blood vessels
2. 100 micrometers
3. serum albumin
4. left atrium, left ventricle, right atrium, right ventricle
5. a heart murmur

*Self-Quiz*

1. d          4. c
2. a          5. b
3. a          6. d

## Chapter 40

*Review*

1. vaccination
2. (a) T cells, (b) B cells
3. antigen
4. macrophage
5. E

*Self-Quiz*

1. b          3. b
2. a          4. b

## Chapter 41

*Review*

1. two-thirds
2. kidney
3. (a) marine, (b) freshwater
4. inner medulla
5. water

*Self-Quiz*

1. d
2. e
3. c
4. b, e
5. a

## Chapter 42

*Review*

1. false
2. secretory phase
3. sperm production
4. fallopian tubes

*Self-Quiz*

1. e
2. b, e
3. d, a
4. c, d, b, a

## Chapter 43

*Review*

1. zygote
2. pellucida
3. parthenogenesis
4. cleavage
5. mesoderm
6. chorion
7. induction

*Self-Quiz*

1. d          5. b
2. b          6. a
3. a          7. c
4. d          8. b

## Chapter 44

*Review*

1. ultimate
2. hybridization
3. habituation (or sensitization)
4. testosterone
5. circadian
6. inclusive fitness
7. polyandry
8. close relatives (kin)

*Self-Quiz*

1. b          5. d
2. e          6. a
3. a          7. c
4. c          8. a

# *Glossary*

**abortion** (L. *abortum*, miscarried) The termination of pregnancy before the fetus reaches the stage of viability, which is approximately 20 to 28 weeks of gestation.

**abscission** (L. *ab*, away, off + *scisso*, dividing) In vascular plants, the dropping of leaves, flowers, fruits, or stems at the end of a growing season, as the result of formation of a corky layer of young cells at the base.

**absorption** (L. *absorbere*, to swallow down) The movement of water and of substances dissolved in water into a cell, tissue, or organism.

**absorption spectrum** The range of photons that a given atom or molecule is capable of absorbing, depending on the electron energy levels that are available in it.

**abyssal zone** (Gr. *abyssos*, bottomless) The marine environment of the deep-water areas of the ocean.

**accessory pigment** A pigment, such as a carotenoid or chlorophyll *b*, that increases the percentage of the photons of sunlight that are harvested.

**acetylcholine** The most important of the numerous chemical neurotransmitters responsible for the passing of nerve impulses across synaptic junctions; the neurotransmitter in neuromuscular (nerve-muscle) junctions.

**acetylcholinesterase** An enzyme that removes the leftover acetylcholine from the synaptic cleft at the neuromuscular junction after the last impulse; one of the fastest-acting enzymes in the vertebrate body.

**acid** Any substance that dissociates to form H⁺ ions when dissolved in water.

**acid rain** A blanket term for the process whereby industrial pollutants such as nitric and sulfuric acids—introduced into the upper atmosphere by factory smokestacks—are spread over wide areas by the prevailing winds and then fall to earth with the precipitation, lowering the pH of groundwater and killing life.

**acoelomate** (Gr. *a*, not + *koiloma*, cavity) A bilateral symmetrical animal not possessing a body cavity, such as a flatworm.

**acquired immune deficiency syndrome (AIDS)** An infectious and usually fatal human disease caused by a retrovirus, HIV, which attacks T cells. The virus multiplies within and kills individual T cells, releasing thousands of progeny that infect and kill other T cells, until no T cells remain, leaving the afffected individual helpless in the face of microbial infections because his or her immune system is now incapable of marshaling a defense against them. *See* T cell.

**actin** (Gr. *actis*, ray) One of the two major proteins that make up myofilaments (the other is myosin); it provides the cell with mechanical support and plays major roles in determining the shape of cells and in cell movement.

**actinin** (Gr. *actis*, ray) A protein anchored to actin that enables myofilaments to convert the sliding of muscle fibers into cell movement.

**action potential** A single nerve impulse; a transient all-or-none reversal of the electric potential across a neuron membrane; because it can activate nearby voltage-sensitive channels, an action potential propagates along a nerve cell.

**activating enzyme** Any of a battery of 20 enzymes, one or more of which recognizes a particular three-base anticodon sequence in a tRNA molecule.

**activation** The process by which the regulatory protein binds to DNA and turns on the transcription of specific genes.

**activation energy** The energy that a molecule must possess in order for it to undergo a specific chemical reaction.

**active transport** The transport of a solute across a membrane by protein carrier molecules to a region of higher concentration by the expenditure of chemical energy. One of the most important functions of any cell.

**actual rate of population increase** The difference between the birth rate and the death rate per given number of individuals per unit of time.

**adaptation** (L. *adaptare*, to fit) Any peculiarity of structure, physiology, or behavior that promotes the likelihood of an organism's survival and reproduction in a particular environment.

**adenine** An organic molecule composed of two carbon-nitrogen rings.

**adenosine diphosphate (ADP)** The molecule resulting from the breaking off of a phosphate group from adenosine thriphosphate.

**adenosine triphosphate (ATP)** A molecule composed of ribose, adenine, and a triphosphate group. ATP is the chief energy currency of all cells. Cells focus all of their energy resources on the manufacture of ATP from ADP and phosphate, which requires the cell to supply 7 kilocalories of energy obtained from photosynthesis or from electrons stripped from foodstuffs to form 1 mole of ATP. Cells then use this ATP to drive endergonic reactions.

**adhesion** (L. *adhaerere*, to stick to) The molecular attraction exerted between the surfaces of unlike bodies in contact, as water molecules to the walls of the narrow tubes that occur in plants.

**aerobic** (Gr. *aer*, air + *bios*, life) Oxygen-requiring.

**aerobic pathway** A metabolic pathway at least one step of which is an oxidation/reduction reaction that depends on oxygen gas as an electron acceptor; includes the citric acid cycle and pyruvate oxidation.

**afferent pathway** See *sensory* pathway.

**AIDS** See acquired immunodeficiency syndrome.

**albumen** (L. white) The white of an egg, which provides additional nutrients and water for the embryo.

**aldosterone** (from ald[ehyde] + ster[ol] + [horm]one) A steroid hormone of the adrenal cortex that controls the salt and water balance in the body.

**alga,** pl. **algae** (L.) A protosynthetic protist cell; algae contain chloroplasts.

**alkaptonuria** (alkali + Gr. *haptein*, to possess + *ouron*, urine) A genetic disorder caused by the lack of an enzyme necessary to break down homogentisic acid (alkapton) in urine.

**allantois** (Gr. *allas*, sausage + *eidos*, form) A membrane of the amniotic egg that functions in respiration and excretion in birds and reptiles and plays an important role in the development of the placenta in most mammals.

**allele** (Gr. *allelon*, of one another) One or two or more alternative forms of a gene.

**allele frequency** The relative proportion of a particular allele among individuals of a population. Not equivalent to gene frequency, although the two terms are sometimes confused.

**allergen** Any substance that initiates a strong immune response in a particular individual.

**allergy** (Gr. *allos*, different + *ergon*, action) An unusual sensitivity to a certain substance that does not invoke a response in most people. Hay fever is a common manifestation of allergy.

**allometric growth** (Gr. *allos*, other + *meros*, part) A pattern of growth in which different components grow at different rates.

**allosteric interaction** (Gr. *allos*, other + *stereos*, shape) The change in shape that occurs when an activator or inhibitor binds to an enzyme; these changes result when specific, small molecules bind to the enzyme, molecules that are not substrates of that enzyme.

**alpha-cell** A cell in the islets of Langerhans that produces the hormone glucagon, which causes liver cells to release stored glucose and fat cells to break down triglycerides.

**alternation of generations** A reproductive life cycle in which the diploid phase produces spores that give rise to the haploid phase, and the haploid phase produces gametes that fuse to give rise to the zygote. The zygote is the first cell of the multicellular diploid phase.

**altruism** (L. *alter*, other) Self-sacrificing behavior.

**alveolus,** pl. **alveoli** (L. a small cavity) One of the many small, thin-walled air sacs within the lungs in which the bronchioles terminate.

**Alzheimer's disease** A type of dementia seen in the elderly, probably resulting from a deficiency of acetylcholine, which leads to degeneration and death of neurons.

**amino acids** Gr. *Ammon*, referring to the Egyptian sun god, near whose temple ammonium salts were first prepared from camel dung) Molecules containing an amino group ($^-NH_2$), a carboxyl group ($^-COOH$), a hydrogen atom, and a functional group designated R, all bonded to a central carbon atom; the 20 different amino acids that occur in proteins are grouped into five chemical classes.

**amnion** (Gr. membrane around the fetus) The innermost of the extraembryonic membranes; the amnion forms a fluid-filled sac around the embryo in amniotic eggs.

**amniotic egg** An egg that is isolated and protected from the environment by a more or less impervious shell; the shell protects the embryo from drying out, nourishes it, and enables it to develop outside of water.

**ampulla,** pl. **ampullae** (L. pear-shaped bottle) A flask-shaped organ or bladder on an aquatic animal.

**amylase** (from amyl + -ase, enzyme suffix) An enzyme that breaks up starches and other carbohydrates into sugars.

**amyloplast** (Gr. *amylon*, starch + *plastos*, formed) A plant cell called a plastid that specializes in storing starch.

**anabolism** (Gr. *ana*, up + *bolein*, to throw) A process in which more complex molecules are built up.

**anaerobic** (Gr. *an*, without + *aer*, air + *bios*, life) Any process that can occur without oxygen; includes glycolysis and fermentation; anaerobic organisms can live without free oxygen.

**anaphase** (Gr. *ana*, up + *phasis*, form) The third and shortest stage of mitosis during which the daughter chromosomes move rapidly to the opposite poles of the cell.

**animal pole** In fishes and other aquatic vertebrates with asymmetrical yolk distribution in their eggs, the hemisphere of the blastula comprising cells relatively poor in yolk.

**anion** (Gr. *anion*, to go up) A negatively charged ion.

**annelid** (L. *annulus*, ring) Any of a phylum (Annelida) of worms or wormlike animals characterized by a soft, elongated body composed of a series of similar ring-like segments. Earthworms and leeches are annelids.

**annual** (L. *annus*, year) A plant that germinates, flowers, produces seed, and dies within the same year or growing season.

**annual ring** Any of the concentric rings of wood seen when the stem of a tree or shrub is cut across; each ring shows one year's growth.

**anterior** (L. *ante*, before) Located before or toward the front; in animals, the head end of an organism.

**anther** (Gr. *anthos*, flower) The part of the stamen of a flower that bears the pollen.

**antheridium,** pl. **antheridia** (Gr. *anthos*, flower) The male or sperm-producing organ in plants such as ferns and mosses.

**antibody** (Gr. *anti*, against) A protein substance produced in the blood by a B cell lymphocyte in response to a foreign substance (antigen) and released into the bloodstream. Binding to the antigen, antibodies mark them for destruction by other elements of the immune system.

**anticodon** (Gr. *against* + L. *code*) The three-nucleotide sequence at the end of a tRNA molecule that is complementary to, and base pairs with, an amino-acid–specifying codon in mRNA.

**antidiuretic hormone (ADH)** See vasopressin.

**antigen** (Gr. *anti*, against + *genos*, origin) A foreign substance, usually a protein, that stimulates lymphocyte B cells to proliferate and secrete specific antibodies that bind to the foreign substance, labeling it as foreign and destined for destruction.

**aorta** (Gr. *aeirein*, to lift) The main artery of vertebrate systemic blood circulation; it carries the blood from the left side of the heart to all regions of the body except the lungs.

**aortic valve** A one-way valve that permits the flow of blood out of the ventricle, then closes, thus preventing the reentry of blood from the aorta into the heart.

**apical meristem** (L. *apex*, top + Gr. *meristos*, divided) A region of active cell division that occurs at or near the tips of the roots and shoots of plants.

**aposematic** See Warning coloration.

**appendicular skeleton** (L. *appendicula*, a small appendage) The skeleton of the limbs of the human body containing 126 bones.

**archegonium,** pl. **archegonia** (Gr. *archegonos*, first of a race) The multicellular female reproductive organ in plants such as ferns and mosses.

**archenteron** (Gr. *arch*, early + *enteron*, intestine) The primitive intestinal or alimentary cavity of a gastrula.

**arteriole** A smaller artery, leading from the arteries to the capillaries.

**artery,** pl. **arteries** (Gr. *arteria*, artery) One of the tubular branching vessels that carries the blood from the heart through the body.

**arthropod** (Gr. *arthron*, joint + *pous*, foot) A phylum of hard-shelled animals with jointed legs and a segmented body.

**artificial selection** The differential reproduction of genotypes in response to demands imposed by human intervention.

**asexual** An organism that reproduces without forming gametes—that is, without sex. Asexual reproduction, therefore, does not involve sex; its outstanding characteristic is that an individual offspring is genetically identical to its parent.

**associative** Learning behaviors that require associative activity within the central nervous system.

**associative cortex** The major portion of the cerebral cortex that appears to be the site of sensory information integration and of higher mental activities such as planning and contemplation.

**aster** The array of microtubules which radiate from the centrioles when the latter reach the poles of the cells.

**atom** (Gr. *atomos*, indivisible) A core (nucleus) of protons and neutrons surrounded by an orbiting cloud of electrons; the chemical behavior of an atom is largely determined by the distribution of its electrons, particularly the number of electrons in its outermost level.

**atomic mass** The atomic mass of an atom consists of the combined weight of all of its protons and neutrons.

**atomic number** The number of protons in the nucleus of an atom; in an atom that does not bear an electric charge (i.e., one that is not an ion), the atomic number is also equal to the number of electrons.

**ATP** *See* adenosine triphosphate (ATP).

**atrial natriuretic hormone (ANH)** A nonendocrine hormone manufactured in the heart that circulates throughout the body and apparently helps regulate blood pressure and volume.

**atrioventricular (AV) node** A slender connection of cardiac muscle cells that receives the heartbeat impulses from the sinoatrial node and conducts them by way of the bundle of His.

**atrium** (L. main room) Either of the chambers of the heart that receives blood from the veins and forces it into a ventricle. The heart of mammals, birds, and reptiles has two atria; that of fishes and amphibians has one atrium.

**auditory cortex** A major sensory region on the cerebral cortex lying within the temporal lobe; different surface regions of this cortex correspond to different sound frequencies.

**autoimmune disease** A disease in which antibodies are formed against the body's own cells because the immune system fails to distinguish between foreign and host tissue.

**autonomic nervous system** (Gr. *autos*, self + *nomos*, law) The motor pathways that carry commands from the central nervous system to regulate the glands and nonskeletal muscles of the body; also called the involuntary nervous system.

**autosome** (Gr. *autos*, self + *soma*, body) Any of the 22 pairs of human chromosomes that are similar in size and morphology in both males and females.

**autotroph** (Gr. *autos*, self + *trophos*, feeder) Self-feeder; an organism that can harvest light energy from the sun by the process of photosynthesis, thus converting it to chemical energy. Contrasts with heterotroph.

**auxin** (Gr. *auxein*, to increase) A plant hormone that controls cell elongation, among other effects.

**axial skeleton** The skeleton of the head and trunk of the human body containing 80 bones.

**axil** (L. *axilla*, armpit) The angle between a branch or leaf and the stem from which it arises.

**axon** (Gr. *axle*) A single, long process extending out from a neuron that conducts impulses away from the neuron cell body.

**B cell** A lymphocyte that recognizes invading pathogens much as T cells do, but instead of attacking the pathogen directly, they mark them for destruction by the nonspecific body defenses.

**bacillus**, pl. **bacilli** (L. *bacullus*, rod) A straight or rod-shaped bacterium.

**backcross** The crossing of a hybrid individual with one of its parents or with a genetically equivalent individual; the most common type of backcross is a testcross, a cross between an individual with a dominant phenotype (which might in principle be either homozygous for a dominant allele, heterozygous for that allele, or heterozygous for that allele and a recessive one) with an individual that is homozygous for the recessive allele.

**bacteriophage** (Gr. *bakterion*, little rod + *phagein*, to eat) A virus that infects bacterial cells; also called a phage.

**bacterium**, pl. **bacteria** (Gr. *bakterion*, dim. of *baktron*, a staff) The simplest cellular organism; its cells are smaller, prokaryotic in structure, and lack interior organization.

**bark** A term used to refer to all of the tissues of a mature stem or root outside of the vascular cambium.

**barometric pressure** (Gr. *atmos*, vapor + *sphaira*, globe) The weight of the earth's atmosphere over a unit area of the earth's surface; measured with mercury barometer at sea level, this corresponds to the pressure required to lift a column of mercury 760 millimeters.

**baroreceptor** (Gr. *baros*, weight) A nerve cell or group of cells sensitive to changes in pressure, such as blood pressure.

**Barr body** (after Murray L. Barr, Canadian anatomist) The inactivated X-chromosome in female mammals which can be seen as a deeply staining body, which remains attached to the nuclear membrane.

**basal body** In cells that contain flagella or cilia, a form of centriole that anchors each flagellum.

**base** Any substance that combines with $H^+$ ions; having a pH value above 7.

**basidium**, pl. **basidia** (Gr. basis, base) A small, club-shaped structure on basidiomycetes that produces spores.

**basilar membrane** A membranous part of the cochlea that forms the fibrous base supporting the organ of Corti.

**basophil** A leukocyte containing granules that rupture and release chemicals that enhance the inflammatory response; important in causing allergic responses.

**Batesian mimicry** (after Henry W. Bates, English naturalist) A situation in which a palatable or nontoxic organism resembles another kind of organism that is distasteful or toxic; both species exhibit warning coloration.

**behavior** A coordinated neuromotor response to changes in external or internal conditions; a product of the integration of sensory, neural, and hormonal factors.

**beta-cell** A cell in the islets of Langerhans that secretes insulin when a person eats, storing away glucose to be used later.

**biennial** (L. *biennium*, a two-year period) A plant that normally requires two growing seasons to complete its life cycle. Biennials flower in the second year of their lives.

**bilateral symmetry** (L. *bi*, two + *lateris*, side; Gr. *symmetria*, symmetry) A body form in which the right and left halves of an organism are approximate mirror images of each other.

**bile salt** A molecule produced by the liver that acts as a superdetergent combining with fats in emulsification, a process that renders fats soluble.

**binary fission** (L. *binarius*, consisting of two things or parts + *fissus*, split) Asexual reproduction of a cell by division into two equal, or nearly equal, parts. Bacteria divide by fission.

**binocular** (L. *bi*, two + *ocularis*, the eye) Referring to vision in which both eyes are located at the front of the head and function together to provide a three-dimensional view of the object being examined.

**binomial system** (L. *bi*, twice, two + Gr. *nomos*, usage, law) A system of nomenclature that uses two words; the first names the genus and the second designates the species.

**biogeochemical cycle** (Gr. *bios*, life + *geo*, earth + *chem*, chemical) A geological cycle that involves the controlled cycling of chemicals; includes both substances derived from weathering of rocks and those that occur in organisms.

**biological evolution** The extensive behavior patterns of humans rooted in our evolutionary biology which has led to adaptive change.

**biomass** (Gr. *bios*, life + *maza*, lump or mass) The total weight of all of the organisms living in the ecosystem being measured.

**biome** (Gr. *bios*, life + *-oma*, mass, group) A major terrestrial assemblage of plants, animals, and microorganisms that occur over wide geographical areas and have distinct characteristics; the largest ecological unit.

**biotic potential** *See* innate capacity for increase.

**blade** The wide, flat part of a leaf.

**blastocoel** (Gr. *blastos*, sprout + *koilia*, belly) The central cavity of a blastula.

**blastodisc** A disklike aggregation of formative protoplasm at one pole of the yolk of a fertilized egg, containing the nucleus.

**blastomere** (Gr. *blastos*, sprout + *meros*, part) An individual cell of the morula.

**blastopore** (Gr. *blastos*, sprout + *poros*, a path or passage) In vertebrate development, the opening that connects the archenteron cavity of a gastrula stage embryo with the outside; represents the future mouth in some animals (protostomes), the future anus in others (deuterostomes).

**blastula** (Gr. a little sprout) In vertebrates, an early embryonic stage consisting of a hollow, fluid-filled ball of cells one layer thick; a vertebrate embryo after cleavage and before gastrulation.

**blood type** In humans, the type of cell surface antigens present on the red blood cells of an individual; genetically determined, alternative alleles yield different surface antigens. When two different blood types are mixed, the cell surfaces often interact, leading to agglutination. One genetic locus encodes the ABO blood group, another the Rh blood group, and still others encode other surface antigens.

**bronchus**, pl. **bronchi** (Gr. *bronchos*, windpipe) One of a pair of respiratory tubes branching from the lower end of the trachea (windpipe) into either lung.

**calcitonin** (L. *calcem*, lime) A thyroid hormone that stimulates calcium uptake.

**calorie** (L. *calor*, heat) The amount of energy in the form of heat required to raise the temperature of 1 gram of water 1 degree C.

**Calvin cycle** (after Melvin Calvin, American chemist) The series of dark reactions in which ATP and NADPH produced by the light reactions are used to fix carbon. In this process ribulose 1,5-bisphosphate is carboxylated and the products run backward through a series of reactions also found in the glycolytic sequence to form fructose 6-phosphate molecules, some of which are used to reconstitute RuBP. The remainder enters the cell's metabolism as newly fixed carbon in glucose.

**calyx** (Gr. *kalyx*, a husk, cup) The sepals collectively; the outermost flower whorl.

**cambium, pl. cambia** (L. *combiare*, to exchange ) In vascular plants, embryonic tissue zones (meristems) that run parallel to the sides of roots and stems; consists of the cork cambium and the vascular cambium.

**cancer** (L. crab) Unrestrained invasive cell growth; a tumor or cell mass resulting from uncontrollable cell division.

**capillary** (L. *capillaris*, hairlike) A blood vessel with a very slender, hairlike opening; the blood exchanges gases and metabolites within the capillaries; they join the end of an artery to the beginning of a vein.

**capillary action** The movement of a liquid along a surface as a result of the combined effects of cohesion and adhesion.

**carbohydrate** (L. *carbo*, charcoal + *hydro*, water) An organic compound consisting of a chain or ring of carbon atoms to which hydrogen and oxygen atoms are attached in a ratio of approximately 1:2:1; a compound of carbon, hydrogen, and oxygen having the generalized formula $(CH_2O)n$.

**carbon cycle** The worldwide circulation and reutilization of carbon atoms.

**carbon fixation** A process in which atmospheric $CO_2$ is incorporated into carbon-containing molecules.

**carcinogen** (Gr. *karkinos*, cancer + *-gen*) Any cancer-causing agent.

**cardiovascular system** (Gr. *kardia*, heart + L. *vasculum*, vessel) The blood circulatory system and the heart that pumps it; collectively, the blood, heart, and blood vessels.

**carpel** (Gr. *karpos*, fruit) A leaflike organ in angiosperms that encloses one or more ovules; one of the members of the gynoecium.

**carrying capacity** The size at which a population stabilizes in a particular place.

**catabolism** (Gr. *katabole*, throwing down) A process in which complex molecules are broken down into simpler ones.

**catalysis** (Gr. *katalysis*, dissolution + *lyein* to loosen) The process carried out by enzymes in which the subunits of polymers are held together and their bonds are stressed.

**catalyst** (Gr. *kata*, down + *lysis*, a loosening) A general term for a substance that speeds up specific chemical reactions by lowering the energy required to activate or start the reaction; an enzyme is a biological catalyst.

**cell** (L. *cella*, a chamber or small room) The smallest unit of life; the basic unit of organization of all organisms; composed of a nuclear region containing the hereditary apparatus within a larger volume called the cytoplasm bounded by a lipid membrane.

**cell biology** The study of how cells are constructed and how they grow, divide, and communicate at the molecular, subcellular, and cellular levels of organization.

**cell surface antigen** A sugar molecule that has been encoded by an enzyme to a lipid on the surface of a blood cell and that acts as a recognition marker in our immune system.

**cell surface marker** A specific set of proteins unique to a particular cell that enables each cell to signal to the environment what type of cell it is; other cell surface markers are glycolipids.

**cell surface receptor** An information-transmitting protein that extends across a plasma membrane; among the information it transmits are the presence of hormones and the signals that pass from one nerve to another.

**cellular respiration** The process in which the energy stored in a glucose molecule is released by oxidation; hydrogen atoms are lost by glucose and gained by oxygen.

**central nervous system** The portion of the nervous system in vertebrates composed of the brain and spinal cord that is the site of information processing and control within the nervous system.

**centriole** (Gr. *kentron*, center of a circle + L. *olus*, little one) An organelle associated with the assembly and organization of microtubules; an endosymbiont which occurs in animals and most protists, but not in plants or fungi.

**centromere** (Gr. *kentron*, center + *meros*, a part) A constricted region of the chromosome joining two sister chromatids, to which the kinetochore is attached.

About 220 nucleotides in length, it is composed of highly repeated DNA sequences (satellite DNA).

**cerebellum** (L. *little brain*) The hindbrain region of the vertebrate brain; it integrates information about body position and motion, coordinates muscular activities, and maintains equilibrium.

**cerebral cortex** The layer of gray matter covering the cerebrum; the seat of conscious sensations and voluntary muscular activity.

**cerebrum** (L. brain) The portion of the vertebrate brain that occupies the upper part of the skull, consisting of two cerebral hemispheres united by the corpus callosum. It is the primary association center of the brain, coordinating and processing sensory input and coordinating motor responses.

**chaparral** (Sp. *chaparro*, evergreen oak) Extensive communities of evergreen, often spiny shrubs and low trees in dry-summer areas of California and adjacent regions.

**Chargaff's rule** (after Erwin Chargaff, American biochemist) The observation that in all natural DNA molecules, the amount of adenine is always equal to the amount of thymine, and the amount of guanine is always equal to the amount of cytosine.

**chemical bond** The force holding two atoms together; the force can result from the attraction of opposite charges (ionic bond), or from the sharing of one or more pairs of electrons (a covalent bond).

**chemically gated ion channel** A transmembrane pathway for a particular ion that is opened or closed by a chemical such as a neurotransmitter.

**chemiosmosis** (chem, chemical action + Gr. *osmos*, act of pushing, thrust) The mechanism responsible for almost all of the adenosine triphosphate (ATP) harvested from eaten food and for all the ATP produced by photosynthesis.

**chemoautotroph** An autotrophic bacterium that uses chemical energy released by specific inorganic reactions to power its life processes, including the synthesis of organic molecules.

**chemoreceptor** In a male insect, a nerve cell modified to detect tiny quantities of the sex pheromone.

**chiasma, pl. chiasmata** (Gr. a cross) In meiosis, the points of crossing-over where portions of chromosomes have been exchanged during synapsis; a chi-

asma appears as an X-shaped structure under a light microscope.

**chief cell** An exocrine gland within the mucosa of the stomach which secretes pepsinogen.

**chloroplast** (Gr. *chloros*, green + *plastos*, molded) An energy-producing organelle containing bacteria-like elements with vesicles containing chlorophyll; in plant cells it is the site of photosynthesis.

**chromatid** Gr. *chroma*, color + L. *-id*, daughters of) One of two daughter strands of a duplicated chromosome which is joined by a single centromere.

**chromatin** (Gr. *chroma*, color) The complex of DNA and proteins of which eukaryotic chromosomes are composed.

**chromosomal rearrangement** In eukaryotes, when large segments of chromosomes change their relative location or undergo duplication, often with drastic effects on the expression of the genetic message.

**chromosomal theory of inheritance** The theory that reproduction involves the initial union of only two cells, egg and sperm; that the two homologous chromosomes of each pair segregate during meiosis; that gametes have a copy of one member of each pair of homologous chromosomes while diploid individuals have a copy of both members of each pair; and that during meiosis, each pair of homologous chromosomes orients on the metaphase plate independent of any other pair.

**chromosome** (Gr. *chroma*, color + *soma*, body) In an eukaryotic cell long threads of DNA associated with protein which contain hereditary information.

**cilium, pl. cilia** (L. eyelash) Used to refer to flagella, which are numerous and organized in dense rows; cilia propel cells through water; in human tissue, they move water over the tissue surface.

**citric acid cycle** The cyclic series of reactions in which pyruvate, the product of glycolysis, enters the cycle to form citric acid and is then oxidized to carbon dioxide. Also called the Krebs cycle (after its discoverer) and the tricarboxylic acid (TCA) cycle (citric acid possesses three carboxyl groups).

**class** A taxonomic category ranking below a phylum (division), and above an order.

**classical conditioning** The repeated presentation of a stimulus in association with a response that causes the brain to form an association between them, even if the stimulus and the response have never been associated before.

**cleavage** The second stage of the 10 reactions of glycolysis in which the six-carbon product of the first stage is split into two three-carbon molecules. One is G3P, and the other is converted to G3P by another reaction; (2) the progressive division of cells during embryonic growth.

**cleavage furrow** During cytokinesis, the area where the cytoplasm is progressively pinched inward by the decreasing diameter of the microfilament belt.

**climax community** A self-perpetuating community in which populations remain stable and exist in balance with each other and the environment; the final stage of a succession.

**clone** (Gr. *klon*, twig) A line of cells, all of which have arisen from the same single cell by mitotic division; one of a population of individuals derived by asexual reproduction from a single ancestor; one of a population of genetically identical individuals.

**cloning** Producing a cell line or culture all of whose members contain identical copies of a particular nucleotide sequence; an essential element in genetic engineering, cloning is usually carried out by inserting the desired gene into a virus or plasmid, infecting a cell culture or plasmid, infecting a cell culture with the hybrid virus, and selecting for culture a cell that has taken up the gene.

**closed circulatory system** A system in most animals in which the circulatory system fluid is separated from the rest of the body's fluids and does not mix with them.

**codominance** Referring to a situation in which both the 1A and 1B alleles in the heterozygote are expressed simultaneously.

**codon** (L. code) The basic unit of the genetic code; a sequence of three adjacent nucleotides in DNA or mRNA that code for one amino acid or for polypeptide termination.

**coelom** (Gr. *koilos*, a hollow) A body cavity formed between layers of mesoderm and in which the digestive tract and other internal organs are suspended.

**coenzyme** A cofactor that is a nonprotein organic molecule.

**coevolution** (L. *co-*, together + *e-*, out + *volvere*, to fill) A term that describes the long-term evolutionary adjustment of one group of organisms to another.

**cognitive** (L. *cognoscere*, to know) Thinking; using the mind.

**collenchyma cell** (Gr. *killa*, glue + *en-*, in + *chyma*, what is poured) Living plant tissue with cells whose walls are thickened and usually elongated.

**commensalism** (L. *cum*, together with + *mensa*, table) A symbiotic relationship in which one species benefits while the other neither benefits nor is harmed.

**community** L. *communitas*, community, fellowship) The population of different species that live together in a particular place.

**companion cell** A specialized parenchyma cell with a large nuclei adjacent to the sieve tube in the phloem of vascular plants.

**competition** Interaction between individuals of two or more species for the same scarce resources; intraspecific competition, interaction for the same scarce resources between individuals of a single species.

**competitive exclusion** The hypothesis that if two species are competing with one another for the same limited resource in the same place, one will be able to use that resource more efficiently than the other and eventually will drive that second species to extinction locally.

**complement system** The chemical defense of a vertebrate body that consists of a battery of proteins that become activated by the walls of bacteria and fungi; complements the cellular defenses.

**concentration gradient** The concentration difference of a substance as a function of distance; in a cell, a greater concentration of its molecules in one region than in another.

**condensation** The coiling of the chromosomes into more and more tightly compacted bodies begun during the G2 phase of the cell cycle.

**cone cell** A specialized sensory cell of the retina of the eye that contains iodopsin and functions in color vision.

**conjugation** (L. *conjugare*, to yoke together) An unusual mode of reproduction that characterizes the ciliates, in which nuclei are exchanged between individuals through tubes connecting them during conjugation.

**consumer** In ecology a heterotroph that derives its energy from living or freshly killed organisms or parts thereof. Primary consumers are herbivores; secondary consumers are carnivores or parasites.

**conus arteriosus** The anterior-most chamber of the embryonic heart in vertebrate animals; a conical structure in the upper left portion of the right ventrical in humans, supplying the pulmonary artery.

**cooperative breeding** A feature of some vertebrate bird societies in which several birds of the same species will not breed but rather assist as helpers for a breeding pair.

**cork cell** A cell split off by the cork cambium, which contains a fatty substance and is nearly impermeable to water; cork cells are dead at maturity.

**corolla** (L. *cornea*, crown) The petals, collectively; usually the conspicuously colored flower whorl.

**corpus callosum** (N.L. callous body) The band of nerve fibers that connect the two hemispheres of the cerebrum in humans and other primates.

**corpus luteum** (N.L. yellow body) A structure that develops from a ruptured follicle in the ovary after ovulation; it secretes the hormone progesterone, which maintains the uterus during pregnancy.

**cortex** (L. bark) In vascular plants the primary ground tissue of a stem or root, bounded externally by the epidermis and internally by the central cylinder of vascular tissue; in animals the outer, as opposed to the inner, part of an organ, as in the adrenal, kidney, and cerebral cortexes.

**cotyledon** (Gr. *kotyledon*, a cup-shaped hollow) Seed leaf; monocot embryos have one cotyledon and dicots have two.

**countercurrent exchange** In organisms the passage of heat or of molecules (such as oxygen, water, or sodium ions) from one circulation path to another moving in the opposite direction; because the flow of the two paths is in opposite directions, a concentration difference always exists between the two channels, facilitating transfer.

**covalent bond** (L. *co-*, together + *valare*, to be strong) A chemical bond formed by the sharing of one or more pairs of electrons.

**crista**, pl. **cristae** (L. crest) In mitochondria the enfoldings of the inner mitochondrial membrane, which form a series of "shelves" containing the electron-transport chains involved in adenosine triphosphate formation.

**cross** A mating between two different strains of a plant.

**crosscurrent flow** In bird lungs the latticework of capillaries arranged across the air flow, at a 90-degree angle.

**crossing-over** An essential element of meiosis occurring during prophase when the four chromatids exchange portions of DNA strands.

**cryptic coloration** An organism colored to blend in with its surroundings and thus be hidden from predators; camouflage.

**cuticle** (L. *cutis*, skin) A very thin film covering the outer skin of many plants.

**cutin** (L. *cutis*, skin) A waxy, waterproof substance that is the chief ingredient of the cuticle of a plant.

**cyanobacteria** (Gr. *kyanos*, dark-blue + *bakterion*, dim. of *baktron*, a staff.) Sometimes called "blue-green algae"; a very important group of photosynthetic bacteria in the history of life on earth; producing oxygen, they played the decisive role in increasing the concentration of free oxygen in the earth's atmosphere from below 1% to the current level of 21%.

**cyclic AMP (cAMP)** A cyclic form of adenosine phosphate that acts a chemical messenger and amplifies a hormonal signal.

**cytokinesis** (Gr. *kytos*, hollow vessel + *kinesis*, movement) The C phase of cell division in which the cell itself divides, creating two daughter cells.

**cytoplasm** (Gr. *kytos*, hollow vessel + *plasma*, anything molded) A semifluid matrix which occupies the volume between the nuclear region and the cell membrane. It contains the sugars, amino acids, and proteins with which the cell carries out its everyday activities of growth and reproduction.

**cytoskeleton** (Gr. *kytos*, hollow vessel + *skeleton*, a dried body) In the cytoplasm of all eukaryotic cells, a network of protein fibers that support the shape of the cell and anchor organelles, such as the nucleus, to fixed locations.

**cytotoxic T cell** (Gr. *kytos*, hollow vessel + toxin) A special T cell activated during cell-mediated immune response that recognizes and destroys infected body cells.

**dark reactions** Chemical reactions during the second phase of photosynthesis; so-called because as long as adenosine triphosphate generated during the first phase of photosynthesis is available, they occur as readily in the absence of light as in its presence.

**deciduous** (L. *decidere*, to fall off) In vascular plants, shedding all the leaves at a certain season.

**decomposers** Organisms that break down the organic matter accumulated in the bodies of other organisms.

**dehydration reaction** Water-losing; the process in which a hydroxyl (OH) group is removed from one subunit of a polymer and a hydrogen (H) group is removed from the other subunit.

**demography** (Gr. *demos*, people + *graphein*, to draw) The statistical study of population; the measurement of people, or, by extension, of the characteristics of people.

**dendrite** (Gr. *dendron*, tree) A process extending from a neuron, typically branched, that conducts impulses inward toward the cell body; neurons have many dendrites.

**density** The number of individuals in a population in a given area.

**density-dependent effect** An effect controlled by a factor that comes into play particularly when the population size is larger.

**density-independent effect** An effect controlled by a factor that operates regardless of population size.

**dental plaque** The film on teeth, composed largely of bacterial cells surrounded by a polysaccharide layer, which gives rise to dental caries, or cavities; (2) plaque, clear area in a sheet of bacterial cells growing in culture, resulting from the killing (lysis) of contiguous cells by viruses.

**deoxyribonucleic acid (DNA)** The basic storage vehicle or master plan of heredity information; it is stored as a sequence of nucleotides in a linear nucleotide polymer. Two of the polymers wind around each other like the outside and inside rails of a circular staircase.

**depolarization** The movement of ions across a cell membrane that wipes out locally an electrical potential difference.

**desert** A hot, dry, barren region, usually sandy or rocky and without trees; less than 25 centimeters of annual precipitation is the norm in such a region.

**desmosome** In adhering junctions, buttonlike welds that hold cells tightly together.

**determinate** Having flowers that arise from terminal buds and thus terminate a stem or branch.

**determined** Referring to a cell in which predicting its development fate is possible.

**detritivores** (L. *detritus*, worn down + *vorare*, to devour) Organisms that live on dead organic matter; included are large scavengers like vultures, smaller animals like crabs, and decomposers such as bacteria and fungi.

**deuterostome** (Gr. *deuteros*, second + *stoma*, mouth) An animal in whose embryonic development the anus forms from or near the blastopore, and the mouth forms later on another part of the blastula; Also characterized by radial cleavage.

**development** In a flowering plant the entire series of events that occur between fertilization and maturity.

**dicot** Short for dicotyledon; a class of flowering plants generally characterized by having two cotyledons, netlike veins, and flower parts in fours or fives.

**diffusion** (L. *diffundere*, to pour out) The net movement of molecules to regions of lower concentration as a result of random, spontaneous molecular motions; the process tends to distribute molecules uniformly.

**digestion** (L. *digestio*, separating out, dividing) The process by which food is changed chemically into materials that the cells can assimilate, store, oxidize, or use as nourishment.

**dihybrid** (Gr. *dis*, twice + L. *hibrida*, mixed offspring) An individual heterozygous for two genes.

**dioecious** (Gr. *di*, two + *eikos*, house) Having male and female flowers on separate plants of the same species.

**diploid** (Gr. *diploos*, double + *eidos*, form) A cell, tissue, or individual with a double set of chromosomes.

**directional selection** A form of selection in which selection acts to eliminate one extreme from an array of phenotypes, thus the genes promoting this extreme become less frequent in the population.

**disaccharide** (Gr. *dis*, twice + *sakcharon*, sugar) A sugar formed by linking two monosaccharide molecules together; sucrose (table sugar) is a disaccharide formed by linking a molecule of glucose to a molecule of fructose.

**disruptive selection** A form of selection in which selection acts to eliminate rather than favor the intermediate type.

**diurnal** (L. *diurnalis*, day) Active during the day.

**divergence** Increasing separation. Species that become progressively more different from one another as the result of each accumulating a different set of DNA mutations are said to diverge.

**division** A major taxonomic group of the plant kingdom, comparable to a phylum of the animal kingdom; divisions are divided into classes.

**DNA** *See* deoxyribonucleic acid.

**DNA polymerase** An enzyme that catalyzes DNA replication only at the OH ends of DNA strands.

**dominant allele** An allele that dictates the appearance of heterozygotes. One allele is said to be dominant over another if an individual heterozygous for that allele has the same appearance as an individual homozygous for it.

**dormancy** (L. *dormire*, to sleep) A period during which growth ceases and is resumed only if certain requirements, as of temperature or day length, have been fulfilled.

**dorsal** (L. *dorsum*, the back) Toward the back, or upper surface; opposite of ventral.

**double bond** A covalent bond sharing two pairs of electrons.

**double fertilization** A process unique to the angiosperms, in which one sperm nucleus fertilizes the egg and the second one fuses with the polar nuclei. These two events result in the formation of the zygote and the primary endosperm nucleus, respectively.

**double helix** A helix composed of two molecules winding around each other, as in DNA.

**duodenum** (L. *duodeni*, twelve each; with reference to its length, about twelve finger breadths) The initial, short segment of the small intestine, about 25 centimeters long, in which most digestion occurs.

**duplex** The DNA of a chromosome that exists as one very long, double-stranded fiber, which extends unbroken through the entire length of the chromosome.

**ecdysis** (Gr. *ekdysis*, stripping off) The shedding of the outer covering or skin of certain animals, especially the shedding of the exoskeleton by arthropods.

**ecological pyramid** The relationships of the trophic structure of an ecosystem shown diagrammatically.

**ecology** (Gr. *oikos*, house + *logos*, word) The study of the relationships of organisms with one another and with their environment.

**ecosystem** (Gr. *oikos*, house + *systema*, that which is put together) A community, together with the nonliving factors with which it interacts.

**ecotype** (Gr. *oikos*, house + L. *typus*, image) A locally adapted variant of an organism, differing genetically from other ecotypes.

**ectoderm** (Gr. *ectos*, outside + *derma*, skin) The outer layer of cells formed during the development of the embryos of animals. Skin, hair, nails, tooth enamel, and the essential parts of the nervous system grow from the ectoderm.

**efferent pathway** *See* motor pathway.

**egg activation** A collective term for the series of events initiated by sperm penetration.

**egg cell** A female reproductive cell; ovum.

**electron** A subatomic particle with a negative electric charge; the negative charge of one electron exactly balances the positive charge of one proton; electrons orbit the atom's positively charged nucleus and determine its chemical properties.

**electron transport** A collective term describing the series of membrane-associated electron carriers generated by the citric acid cycle. It puts the electrons harvested from the oxidation of glucose to work driving proton-pumping channels.

**element** A substance that cannot be separated into different substances by ordinary chemical methods.

**embryo** (Gr. *en*, in + *bryein*, to swell) The early developmental stage of an organism produced from a fertilized egg; in plants, a young sporophyte, in animals, a young organism before it emerges from the egg or from the body of its mother; in humans, the first 2 months of intrauterine life.

**endergonic** (Gr. *endon*, within + *ergon*, work) Describing reactions in which the products contain more energy than the reactants and require an input of usable energy from an outside source before they can proceed; these reactions are not spontaneous.

**endocrine gland** (Gr. *endon*, within + *krinein*, to separate) A ductless gland producing hormonal secretions that pass directly into the bloodstream or lymph.

**endocrine system** The dozen or so major endocrine glands of a vertebrate.

**endocytosis** (Gr. *endon*, within + *kytos*, cell) The process by which the edges of plasma membranes fuse together and form an enclosed chamber called a vesicle; it involves the incorporation of a portion of an exterior medium into the cytoplasm of the cell by capturing it within the vesicle.

**endoderm** (Gr. *endon*, within + *derma*, skin) The inner layer of cells formed during development of early vertebrate embryos, destined to give rise to the epithelium that lines certain internal structures, such as most of the digestive tract and its outgrowths, most of the respiratory tract, and the urinary bladder, liver, pancreas, and some endocrine glands.

**endometrium** (Gr. *endon*, within + *metrios*, of the womb) The two-layered lining of the uterus; the outer layer is shed during menstruation while the one beneath it generates the next layer.

**endoplasmic reticulum** (Gr. *endon*, within + *plasma*, from cytoplasm; L. *reticulum*, network) A series of membranes that subdivides the interior of eukaryotic cells into separate compartments. It is the most distinctive feature of eukaryotic cells and one of the most important because it enables the cells to carry out different metabolic functions in separate compartments; those portions containing a dense array of ribosomes are called "rough ER," and other portions with fewer ribosomes are called "smooth ER."

**endoskeleton** (Gr. *endon*, within + *skeletos*, hard) In vertebrates an internal scaffold of bone to which muscles are attached.

**endosperm** (Gr. *endon*, within + *sperma*, seed) A nutritive tissue characteristic of the seeds of angiosperms, which develops from the union of a male nucleus and the polar nuclei of the embryo sac. The endosperm is either digested by the growing embryo or retained in the mature seed to nourish the germinating seedling.

**endosymbiont** (Gr. *endon*, within + *bios*, life) An organism that is symbiotic within another; the major endosymbionts that occur in eukaryotic cells are mitochondria, chloroplasts, and centrioles.

**endothelial** (Gr. *endo*, within + *thele*, nipple) Describing the innermost layer of tissue that lines the arteries.

**energy** The capacity to bring about change, to do work.

**energy levels** The placement in separate concentric rings of electrons in two orbitals that are different distances from the nucleus.

**energy shells** *See* energy levels.

**entropy** (Gr. *en*, in + *tropos*, change in manner) A measure of the disorder of a system; a measure of energy that has become so randomized and uniform in a system that it is no longer available to do the work.

**environmental science** An applied science dedicated to finding solutions to environmental problems.

**enzyme** (Gr. *enzymos*, leavened, from *en*, in + *zyme*, leaven) A protein capable of speeding up specific chemical reactions by lowering the energy required to activate or start the reaction but remains unaltered in the process.

**epidermis** (Gr. *epi*, on or over + *derma*, skin) The outermost layer of cells; in vertebrates, the nonvascular external layer of skin of ectodermal origin; in invertebrates, a single layer of ectodermal epithelium; in plants, the flattened, skinlike outer layer of cells.

**epistasis** (Gr. *epistasis*, a standing still) An interaction between the products of two genes in which one modifies the phenotypic expression produced by the other.

**epithelium** (Gr. *epi*, on + *thele*, nipple) A thin layer of cells forming a tissue that covers the internal and external surfaces of the body; simple epithelium, the membranes that line the lungs and the major cavities of the body which are a single cell layer thick; stratified epithelium, the skin or epidermis, composed of more complex epithelial cells which are several cell layers thick.

**erythrocyte** (Gr. *erythros*, red + *kytos*, hollow vessel) A red blood cell, the carrier of hemoglobin; erythrocytes act as the transporters of oxygen in the vertebrate body. During the process of their maturation in mammals, they lose their nucleus and mitochondria, and their endoplasmic reticulum is reabsorbed.

**estrogen** (Gr. *oestros*, frenzy + *genos*, origin) Any of various hormones that induce a series of physiological changes in females, especially in the reproductive or sexual organs.

**estrous cycle** The periodic cycle in which periods of estrus correspond to ovulation events.

**estrus** (L. *oestrus*, frenzy) The period of maximum female sexual receptivity, associated with ovulation of the egg; being "in heat."

**estuary** (L. *aestus*, tide) A partly enclosed body of water, such as those that often form at river mouths and in coastal bays, where the salinity is intermediate between that of salt and fresh water.

**ethology** (Gr. *ethos*, habit or custom + *logos*, discourse) The study of patterns of animal behavior in nature.

**euchromatin** (Gr. *eu*, good + *chroma*, color) Chromatin that is extended except during cell division, from which RNA is transcribed.

**eukaryote** (Gr. *eu*, good + *karyon*, kernel) A small, membrane-bound structure that possesses an internal chamber called the cell nucleus. The appearance of eukaryotes marks a major event in the evolution of life, since all organisms on earth other than bacteria are eukaryotes.

**eumetazoan** (Gr. *eus*, good + *meta*, with + *zoion*, animal) A "true animal"; an animal with a definite shape and symmetry and nearly always distinct tissues.

**eutrophic** (Gr. *eutrophos*, thriving) Referring to a lake in which an abundant supply of minerals and organic matter exists.

**evaporation** The escape of water molecules from the liquid to the gas phase at the surface of a body of water.

**evolution** (L. *evolvere*, to unfold) Genetic change in a population of organisms over time (generations). Darwin proposed that natural selection was the mechanism of evolution.

**exergonic** (L. *ex*, out + Gr. *ergon*, work) Used to describe any reaction producing products that contain less free energy than that possessed by the original reactants and which tend to proceed spontaneously.

**exocytosis** (Gr. *ex*, out of + *kytos*, cell) The extrusion of material from a cell by discharging it from vesicles at the cell surface; the reverse of endocytosis.

**exoskeleton** (Gr. *exo*, outside + *skeletos*, hard) An external hard shell that encases a body; in arthropods, comprised mainly of chitin; in vertebrates, comprised of bone.

**experiment** The test of a hypothesis; a successful experiment is one in which one or more alternative hypotheses are demonstrated to be inconsistent with experimental observation and are thus rejected.

**external fertilization** The fertilization of an egg outside the body of the female, as in many aquatic vertebrates such as fish and amphibians.

**exteroception** (L. *exter*, outside + Eng. *(re)ceptive*) The sensing of information that relates to the external environment of the body.

**F₁** (**first filial generation**) The offspring resulting from a cross.

**F₂** (**second filial generation**) The offspring resulting from a cross between members of the F₁ generation.

**facilitated diffusion** The transport of molecules across a membrane by a carrier protein in the direction of lowest concentration.

**factor** The term Mendel used to describe the bits of encoded information that parents transmit to their offspring, which act in the offspring to produce the trait; now called a gene.

**fall overturn** A process in which the warm upper layer of water in a lake drops its temperature until it is the same as the cooler layer underneath and then the upper and lower layers mix, bringing up free supplies of dissolved nutrients.

**fallopian tube** (after Gabriel Fallopius, Italian anatomist) Either of a pair of slender tubes through which ova from the ovaries pass to the uterus.

**family** A taxonomic group ranking below an order and above a genus.

**fat** A molecule containing many more C—H bonds than carbohydrates contain, thus providing more efficient energy storage.

**fatty acid** A long hydrocarbon chain ending with a —COOH group; fatty acids are components of fats, oils, phospholipids, and waxes; saturated, a fatty acid with all internal carbon atoms having two hydrogen side groups; unsaturated, a fatty acid with a double bond; polyunsaturated, a fatty acid with more than one double bond.

**feedback inhibition** A regulatory mechanism in which a biochemical pathway is regulated by the amount of the product that the pathway produces.

**fermentation** (L. *fermentum*, ferment) Respiration in which the final electron acceptor is an organic molecule.

**fertilization** (L. *ferre*, to bear) The union of male and female gametes to form a zygote.

**fetus** (L. *pregnant*) An animal embryo during the later stages of its development in the womb; in humans, a developing individual is referred to as a fetus from the end of the second month of gestation until birth.

**fever** (A.S. *fefer*) A higher-than-normal elevation of body temperature.

**fiber** (L. *fibra*) One of the narrow, elongated cells in the sclerenchyma of plants.

**fibroblast** (L. *fibra,* fiber + Gr. *blastos,* sprout) A flat, irregularly branching cell of connective tissue that secretes structurally strong proteins into the matrix between the cells.

**fitness** The genetic contribution of an individual to succeeding generations, relative to the the contributions of other individuals in the population.

**fixed-action pattern** A stereotyped animal behavior response, thought by ethologists to be based on programmed neural circuits.

**flagellum,** pl. **flagella** (L. *flagellum,* whip) A fine, long, threadlike organelle protruding from the surface of a cell; in bacteria, a single protein fiber capable of rotary motion that propels the cell through the water; in eukaryotes, an array of microtubules with a characteristic internal 9 + 2 microtubule structure, capable of vibratory but not rotary motion; used in locomotion and feeding; common in protists and motile gametes. A cilium is a small flagellum.

**flavin adenine dinucleotide (FADH$_2$)** A coenzyme used to carry less energetic electrons during the oxidation of glucose.

**follicle** (L. *folliculus,* small ball) In a mammalian ovary one of the spherical chambers containing an oocyte.

**follicle stimulating hormone (FSH)** A hormone secreted by the anterior lobe of the pituitary gland that stimulates the growth of the follicles of the ovaries.

**food chain** A series of organisms from each trophic level that feed on one another.

**food web** The food relationships within a community. A diagram of who eats whom.

**foraging behaviors** A collective term for the many complex, evolved behaviors that influence what an animal eats and how it is obtained.

**forebrain** The third major division of a fish brain devoted to processing olfactory (smell) information.

**founder principle** The effect by which rare alleles and combinations of alleles may be enhanced in the new populations.

**free energy change** The total change in useable energy that results from a chemical reaction or other process; equal to the change in total energy (the heat content or enthalpy) minus the change in unavailable energy (the disorder or entropy times temperature). *See* entropy.

**frequency** In statistics, defined as the proportion of individuals in a certain category, relative to the total number of individuals being considered.

**frontal lobe** The anterior part of either of the two lobes of the cerebrum.

**fruit** In angiosperms, a mature, ripened ovary (or group of ovaries), containing the seeds; also applied informally to the reproductive structures of some other kinds of organisms.

**functional group** The special group of atoms attached to an organic molecule; most chemical reactions that occur within organisms involve the transfer of a functional group from one molecule to another or the breaking of a carbon-carbon bond.

**Fungi Imperfecti** A large group of fungi in which sexual reproduction is not known.

**gallbladder** A sac attached to the liver, in which human beings concentrate and store bile manufactured in the liver.

**gamete** (Gr. wife) A haploid reproductive cell; upon fertilization, its nucleus fuses with that of another gamete of the opposite sex; the resulting diploid cell (zygote) may develop into a new diploid individual, or, in some protists and fungi, may undergo meiosis to form haploid somatic cells.

**gametophyte** (Gr. *gamete,* wife + *phyton,* plant) In plants, the haploid (1n), gamete-producing generation, which alternates with the diploid (2n) sporophyte.

**gamma-aminobutyric acid (GABA)** One of many neurotransmitters that vertebrate nervous systems use, each with specific receptors on postsynaptic membranes: GABA opens the channel, which leads to the exit of positively charged potassium ions and a more negative interior.

**ganglion,** pl. **ganglia** (Gr. a swelling) A group of nerve cells forming a nerve center in the peripheral nervous system.

**gap junction** In communicating junctions, channels or pores through the two cell membranes and across the intercellular space which provide for electrical communication between cells and for flow of ions and small molecules.

**gastric fluid** (Gr. *gastros,* stomach) The digestive juice of the stomach.

**gastric pit** A deep depression in the upper epithelial surface of the stomach where exocrine glands are formed within the mucosa.

**gastrin** A polypeptide hormone that regulates the synthesis of HCl by the parietal cells of the gastric pits.

**gastrula** (Gr. little stomach) In vertebrates, the embryonic stage in which the blastula with its single layer of cells turns into a three-layered embryo made up of ectoderm, mesoderm, and endoderm, surrounding a cavity (archenteron) with one opening (blastopore).

**gastrulation** The inward movement of certain cell groups from the surface of the blastula.

**gene** (Gr. *genos,* birth, race) The basic unit of heredity. A sequence of DNA nucleotides on a chromosome that encodes a polypeptide or RNA molecule and so determines the nature of an individual's inherited traits.

**gene expression** The process in which an RNA copy of each active gene is made, and the RNA copy directs the sequential assembly of a chain of amino acids at a ribosome.

**gene frequency** The frequency with which individuals in a population possess a particular gene. Often confused with allele frequency.

**genetic code** The "language" of the genes; the mRNA codons specific for the 20 common amino acids constitute the genetic code.

**genetic counseling** The process of identifying parents at risk for producing children with genetic defects and of assessing the genetic state of early embryos.

**genetic disorder** The harmful effect produced when a detrimental allele occurs at a significant frequency in human populations.

**genetic drift** Random fluctuations in allele frequencies over time.

**genetic engineering** A collective term for the techniques of transferring genes from one kind of organism to another and multiplying them.

**genetic map** A diagram showing the relative positions of genes.

**genetics** (Gr. *genos,* birth, race) The study of the way in which the traits of an individual are transmitted from one generation to the next.

**genome** (Gr. *genos,* offspring + L. *oma,* abstract group) The genetic information of an organism.

**genotype** (Gr. *genos,* offspring + *typos,* form) The total set of genes present in the cells of an organism. Also used to refer to the set of alleles at a single gene locus.

**genus,** pl. **genera** (L. race) A taxonomic group that ranks below a family and above a species.

**germination** (L. *germinare,* to sprout) The resumption of growth and development by a spore or seed.

**gibberellin** (*Gibberella,* a genus of fungi) A common and important class of plant hormones produced in the apical regions of shoots and roots and play the major role in controlling stem elongation for most plants.

**gill** Part of the body of a fish or crab, for example, by which it breathes in water; oxygen passes in and carbon dioxide passes out through the thin membranous walls of the gills.

**gland** (L. *glandis,* acorn) Any of several organs in the body, such as exocrine or endocrine, which secrete substances for use in the body; glands are composed of epithelial tissue.

**glomerular filtrate** The fluid that passes out of the capillaries of each glomerulus.

**glomerulus** (L. a little ball) A network of capillaries in a vertebrate kidney, whose walls act as a filtration device.

**glucagon** A hormone produced by the alpha-cells in the islets of Langerhans that raises blood sugar level by breaking down glycogen to glucose.

**glucose** A common six-carbon sugar; the most common monosaccharide in most organisms.

**glycogen** (Gr. *glykys,* sweet + *gen,* of a kind) Animal starch; a storage polymer occurring frequently in animals and characterized by complex branching.

**glycolysis** (Gr. *glykys,* sweet + *lyein,* to loosen) The harvesting of chemical energy by rearranging the chemical bonds of glucose to form two molecules of pyruvate and two molecules of ATP.

**goiter** (L. *guttur,* throat) An enlargement of the thyroid gland resulting from a deficiency of iodine in the diet.

**Golgi body** (after Camillo Golgi, Italian physician) Flattened stacks of membranes in the cytoplasm that function in the collection, packaging, and distribution of molecules synthesized in the eukaryotic cell.

**Golgi complex** (after Camillo Golgi, Italian physician) A collective term for Golgi bodies.

**gradualism** The gradualism model of evolution, which assumes that evolution proceeds gradually with progressive change in a given evolutionary line.

**grana,** sing. **granum** (L. grain or seed) In chloroplasts, stacks of membrane-bound disks (thylakoids); the thylakoids contain the chlorophylls and carotenoids and are the sites of the light reactions of photosynthesis.

**gravitropism** (L. *gravis*, heavy + *tropes*, turning) The response of a plant to gravity, which generally causes shoots to grow up and roots to grow down.

**greenhouse effect** The process in which carbon dioxide and certain other gases, such as methane, that occur in the earth's atmosphere transmit radiant energy from the sun but trap the longer wavelengths of infrared light, or heat, and prevent them from radiating into space.

**ground tissue** A type of tissue in which the vascular tissue of a plant is embedded.

**group selection** The argument about social behavior that altruism has evolved because a certain act benefits the group as a whole or even the entire species.

**growth hormone (GH)** A hormone secreted by the anterior pituitary, which regulates the growth of the body.

**guanine** (Sp. from *Quechua*, huanu, dung) A purine base found in DNA and RNA; its name derives from the fact that it occurs in high concentration as a white crystalline base, $C_5H_5H_5O$, in guano and other animal excrements.

**guard cells** Pairs of specialized epidermal cells that surround a stoma. When the guard cells are turgid, the stoma is open; when they are flaccid, it is closed.

**gymnosperm** (Gr. *gymnos*, naked + *sperma*, seed) A seed plant with seeds not enclosed in an ovary; the conifers are the most familiar group.

**habitat** (L. *habitare*, to inhabit) The place where individuals of a species live.

**habituation** (L. *habitus*, condition) A major form of nonassociate learning in which an individual learns not to respond to a stimulus.

**half-life** The length of time it takes for half of the $^{14}C$ present in a sample to be converted to $^{12}C$.

**haploid** (Gr. *haploos*, single + *eidos*, form) The gametes of a cell, tissue, or individual with only one set of chromosomes.

**Hardy-Weinberg equilibrium** (after G.H. Hardy, English mathematician and G. Weinberg, German physician) A mathematical description of the fact that the relative frequencies of two or more alleles in a population do not change because of Mendelian segregation; allele and genotype frequencies remain constant in a random-mating population in the absence of inbreeding, selection or other evolutionary forces; usually stated: if the frequency of allele $A$ is $p$ and the frequency of allele $a$ is $q$, then the genotype frequencies after one generation of random mating will always be $(p + q)2 = p2 + 2pq + q2$.

**Haversian canal** (after Clopton Havers, English anatomist) Narrow channels that run parallel to the length of a bone and contain blood vessels and nerve cells.

**heart** The muscular organ that pumps the blood throughout the body of a vertebrate by contracting and relaxing.

**helix** (L. anything of spiral shape, fr. Gr. helix) In DNA, describing the shape of two of the polymers that wind around each other like the outside and inside rails of a circular staircase.

**helper T cell** A class of white blood cells that initiates both the cell-mediated immune response and the humoral immune response.

**hematopoietic stem cell** (Gr. *haimatos*, blood + *poiesis*, a making) The cells in bone marrow where blood cells are formed.

**hemoglobin** (Gr. *haima*, blood + L. *globus*, a ball) A globular protein in vertebrate red blood cells and in the plasma of many invertebrates that carries oxygen and carbon dioxide; an essential part of each molecule is an iron-containing heme group, which both binds $O_2$ and $CO_2$ and gives blood its red color.

**herbaceous plant** (L. *herba*, herb) A plant in which secondary growth has been limited; herbaceous plants produce new shoots each year.

**herbivore** (L. *herba*, grass + *vorare*, to devour) Any organism that eats plants.

**heredity** (L. *heredis*, heir) The transmission of characteristics from parent to offspring.

**heterochromatin** (Gr. *heteros*, different + *chroma*, color) That portion of a eukaryotic chromosome that remains permanently condensed and therefore is not transcribed into RNA. Most centromere regions are heterochromatic.

**heterotroph** (Gr. *heteros*, other + *trophos*, feeder) An organism that does not have the ability to produce its own food. *See also* autotroph.

**heterozygote** (Gr. *heteros*, other + *zygotos*, a pair) Referring to a diploid individual carrying two different alleles of a gene on its two homologous chromosomes.

**hierarchical** (Gr. *hieros*, sacred + *archos*, leader) Referring to a system of classification, in which successively smaller units of classification are included within one another.

**high-energy bond** A chemical bond that has a low activation energy and is broken easily, which releases its energy.

**histone** (Gr. *histos*, tissue) A complex of small, very basic polypeptides rich in the amino acids arginine and lysine; histones form the core of nucleosomes around which DNA is wrapped.

**holoblastic cleavage** (Gr. *holos*, whole + *blastos*, germ) The pattern of cleavage that occurs throughout the whole egg when eggs contain little or no yolk.

**homeostasis** (Gr. *homeos*, similar + *stasis*, standing) The maintaining of a relatively stable internal physiological environment in an organism, or steady-state equilibrium in a population or ecosystem, usually involves some form of feedback self-regulation.

**homeotherm** (Gr. *homoios*, similar + *therme*, heat) An organism, such as a bird or mammal, capable of maintaining a stable body temperature independent of the environmental temperature; "warm-blooded." *See* endothermic.

**hominid** (L. *homo*, man) Human beings and their direct ancestors; a member of the family Hominidae; Homo sapiens is the only living member.

**homologous chromosome** (Gr. *homologia*, agreement) One of the two nearly identical versions of each chromosome; chromosomes that associate in pairs in the first stage of meiosis. In diploid cells, one chromosome of a pair that carry equivalent genes.

**homology** (Gr. *homologia*, agreement) A condition in which the similarity between two structures or functions is indicative of a common evolutionary origin.

**homozygote** (Gr. *homos*, same or similar + *zygotos*, a pair) A diploid individual whose two copies of a gene are the same. An individual carrying identical alleles on both homologous chromosomes is said to be homozygous for that gene.

**hormone** (Gr. *hormaein*, to excite) A chemical messenger, often a steroid or peptide, produced in a small quantity in one part of an organism and then transported to another part of the organism, where it brings about a physiological response.

**human immunodeficiency virus (HIV)** The virus responsible for acquired immunodeficiency syndrome (AIDS), a deadly disease that destroys the human immune system. HIV is a retrovirus (its genetic material is RNA) that is thought to have been introduced to humans from African green monkeys.

**hybrid** (L. *hybrida*, the offspring of a tame sow and a wild boar) A plant that results from the crossing of dissimilar parents.

**hybridization** The mating of unlike parents.

**hydrogen bond** A molecule formed by the attraction of the partial positive charge of one hydrogen atom of a water molecule with the partial negative charge of the oxygen atom of another.

**hydrolysis reaction** (Gr. *hydro*, water, + *lyse*, break) The process of tearing down a polymer by adding a molecule of water; a hydrogen is attached to one subunit and a hydroxyl to the other, which breaks the covalent bond; essentially the reverse of a dehydration reaction.

**hydrophobic** (Gr. *hydro*, water + *phobos*, hating) Refers to nonpolar molecules, which do not form hydrogen bonds with water and therefore are not soluble in water.

**hydroskeleton** (Gr. *hydro*, water + *skeletos*, hard) The skeleton of most soft-bodied invertebrates that have neither an internal nor an external skeleton; they use the relative incompressibility of the water within their bodies as a kind of skeleton.

**hypertonic** (Gr. *hyper*, above + *tonos*, tension) Refers to a cell that contains a higher concentration of solutes than its surrounding solution.

**hypha,** pl. **hyphae** (Gr. *hyphe*, web) A filament of a fungus; a mass of hyphae comprises a mycelium.

**hypothalamus** (Gr. *hypo*, under + *thalamos*, inner room ) The region of the brain under the thalamus, controlling temperature, hunger, thirst, and producing hormones that influence the pituitary gland.

**hypothesis** (Gr. *hypo*, under + *tithenai*, to put) A proposal that might be true; no hypothesis is ever proved to be correct—all hypotheses are provisional, proposals that are retained for the time being as being useful but may be rejected in the future if found to be inconsistent with new information; a hypothesis that stands the test of time, often tested and never rejected, is called a theory.

**hypotonic** (Gr. *hypo*, under + *tonos*, tension) Refers to the solution surrounding a cell which has a lower concentration of solutes than does the cell.

**immune response** The production of antibodies directed against a specific antigen.

**immune system** A vertebrate body defense composed of white blood cells; one kind attacks and kills cells identified as foreign while the other type marks the foreign invaders for elimination by the roaming patrols (nonspecific defenses).

**inbreeding** The breeding of genetically related plants or animals. In plants, inbreeding results from self-pollination; in animals, inbreeding results from matings between relatives; inbreeding tends to increase homozygosity.

**inclusive fitness** A term that describes the sum of the number of genes directly passed on in an individual's offspring and those genes passed on indirectly by kin (other than offspring) whose existence results from the benefit of the individual's altruism.

**incomplete dominance** The ability of two alleles to produce a heterozygous phenotype that is different from either homozygous phenotype.

**independent assortment** Mendel's second law: the principle that segregation of alternative alleles at one locus into gametes is independent of the segregation of alleles at other loci; only true for gene loci located on different chromosomes or those so far apart on one chromosome that crossing-over is very frequent between the loci. *See* Mendel's second law.

**induced** Describing how the lac operon is transcribed when lactose binding to the repressor protein changes its shape so that it can no longer sit on the operator site and block polymerase binding.

**induction** The determination of the course of development of one tissue by another tissue.

**industrial melanism** (Gr. *melas*, black) A phrase used to describe the evolutionary process in which initially light-colored organisms become dark as a result of natural selection.

**inflammatory response** (L. *inflammare*, to flame) A generalized nonspecific response to infection that acts to clear an infected area of infecting microbes and dead tissue cells so that tissue repair can begin.

**inhalation** (L. *in*, in + *halare*, to breathe) The act of breathing or drawing air into the lungs.

**inhibitor** A chemical whose binding alters the shape of a protein and shuts off enzyme activity.

**initiation complex** A complex consisting of a ribosome, mRNA, and a tRNA molecule, the formation of which begins polypeptide synthesis.

**innate** (L. *innatus*, born) Describing a characteristic based partly or wholly on inherited gene differences.

**inner ear** The innermost part of the ear, behind the middle ear, containing the essential organs of hearing and equilibrium.

**inner medulla** The inner portion of the kidney, which contains the lower portion of the loop of Henle and the bottom of the collecting duct, which is permeable to urea.

**inositol phosphate** A mediator molecule produced during insulin receptor-induced enzyme activity.

**insertion** The end of a muscle that is attached to a bone that moves if the muscle contracts.

**instinct** (L. *instinctus*, impelled) Stereotyped, predictable, genetically programmed behavior.

**instructional theory** A statement attempting to answer the question of how the human body is able to make such a great diversity of antibodies by proposing that the antigen elicited the appropriate antibody, like a shopper ordering a custom-made suit.

**insulin** (L. *insula*, island) A peptide hormone secreted by the islets of Langerhans that acts as a storage hormone; it enables the body to use sugar and other carbohydrates by regulating the sugar metabolism of the body.

**integration, neural** The summation of the repolarizing and repolarizing effects contributed by all excitatory and inhibitory synapses acting on a neuron.

**integument** (L. *integumentum*, covering) The natural outer covering layers of an animal; develops from the ectoderm.

**interferon** In vertebrates, a protein produced in virus-infected cells that inhibits viral multiplication.

**interneuron** A nerve cell found only in the middle of the spinal cord which acts as a functional link between sensory neurons and motor neurons.

**interoception** (L. *interus*, inner + Eng. *(re)ceptive*) The sensing of information that relates to the body itself, its internal condition, and its position.

**interphase** That portion of the cell cycle preceding mitosis; it includes the G1 phase when cells grow, the S phase when a replica of the genome is synthesized, and a G2 phase when preparations are made for genomic separation.

**intron** (L. *intra*, within) A segment of DNA transcribed into mRNA but removed before translation. These untranscribed regions make up the bulk of most eukaryotic genes.

**involuntary nervous system** *See* autonomic nervous system.

**ion** An atom in which the number of electrons does not equal the number of protons; an ion does carry an electrical charge.

**ionic bond** A chemical bond formed between ions as a result of the attraction of opposite electrical charges.

**ionization** The process of spontaneous ion formation; as when the covalent bonds of water sometimes break spontaneously, one of the protons dissociates from the molecule. Because the dissociate proton lacks the negatively charged electron that it had shared in the covalent bond with oxygen, its own positive charge is not counterbalanced; it is a positively charged hydrogen ion, $H^+$. The remaining bit of the water molecule retains the shared electron from the covalent bond and has one less proton to counterbalance it; it is a negatively charged hydroxyl ion $(OH^-)$.

**ionizing radiation** High energy radiation such as x-rays and gamma rays.

**iris** (L. rainbow) A contractile disk, or shutter, between the cornea and the lens that controls the amount of light entering the eye.

**islets of Langerhans** (after Paul Langerhans, German anatomist) The small, scattered endocrine glands in the pancreas that secrete insulin.

**isolating mechanisms** Mechanisms that prevent genetic exchange between individuals of different populations or species; may be behavioral, morphological, or physiological.

**isotonic** (Gr. *isos*, equal + *tonos*, tension) Refers to a cell with the same concentration of solutes as its environment.

**isotope** (Gr. *isos*, equal + *topos*, place) An atom that has the same number of protons but different numbers of neutrons.

**joint** The part of a vertebrate where one bone meets and moves on another.

**karyotype** (Gr. *karyon*, kernel + *typos*, stamp or print) The particular array of chromosomes that an individual possesses.

**kidney** In vertebrates, one of the pair of organs that carries out the processes of filtration, reabsorption, and secretion.

**kin selection** Selection that acts to favor the propagation of genes by directing altruism towards relatives.

**kinetic energy** The energy of motion.

**kinetochore** (Gr. *kinetikos*, putting in motion + *choros*, chorus) A disk of protein bound to the centromere to which microtubules attach during mitosis, linking each chromatid to the spindle.

**kingdom** The chief taxonomic category; in this book we recognize five kingdoms: Monera, Protista, Fungi, Animalia, and Plantae.

**Krebs cycle** (after Hans A. Krebs, German-born English biochemist) The citric acid cycle; also called the tricarboxylic acid (TCA) cycle.

**K-selection** (from the *K* term in the logistic equation) Natural selection under conditions that favor survival when populations are controlled primarily by density-dependent factors.

**lac system** A cluster of genes encoding three proteins that bacteria use to obtain energy from the sugar lactose.

**lamella**, pl. **lamellae** (L. a little plate) A thin, platelike structure; in chloroplasts, a layer of chlorophyll-containing membranes; in bivalve mollusks, one of the two plates forming a gill; in vertebrates, one of the thin layers of bone laid concentrically around the Haversian canals.

**larva**, pl. **larvae** (L. a ghost) Immature form of an animal that is quite different from the adult and undergoes metamorphosis in reaching the adult form; examples are caterpillars and tadpoles.

**larynx** (Gr.) The voice box; the upper end of the human windpipe, lying between the pharynx and trachea, that contains the vocal cords and acts as an organ of voice.

**lateral meristems** (L. *latus*, side + Gr. *meristos*, divided) In vascular plants, the meristems that give rise to secondary tissue; the vascular cambium and cork cambium.

**Law of Independent Assortment** *See* Mendel's Second Law.

**Law of Segregation** *See* Mendel's First Law.

**leaf** One of the thin, usually flat, green parts of a tree or other plant, that grows on the stem or up from the roots; an expanded area of photosynthetically active tissue on a plant.

**leaf primordium** (L. *primordium*, beginning) A lateral outgrowth from the apical meristem that eventually becomes a leaf.

**learning** The creation of changes in behavior that arise as a result of experience, rather than as a result of maturation.

**lens** A transparent oval body in the eye directly behind the iris, that focuses light rays upon the retina.

**lenticels** (L. *lenticella*, a small window) Spongy areas in the cork surfaces of stem, roots, and other plant parts that allow interchange of gases between internal tissues and the atmosphere through the periderm.

**life cycle** The sequence of phases in the growth and development of an organism, from zygote formation to gamete formation.

**ligament** (L. *ligare*, to bind) A band or sheet of connective tissue that links bone to bone.

**light chains** The two identical short strands of the four polypeptide chains of an antibody molecule.

**light reactions** The resultant synthesis of adenosine triphosphate, which takes place in the presence of light during the first process of photosynthesis.

**limbic system** The network of neurons linking the hypothalamus to some areas of the cerebral cortex together with the hypothalamus; responsible for many of the most deep-seated drives and emotions of vertebrates, including pain, anger, sex, hunger, thirst, and pleasure.

**linkage** The patterns of assortment of genes that are located on the same chromosome; important because if the genes are located relatively far apart, crossing-over is more likely to occur between them than if they are located close together.

**lipase** (Gr. *lipos*, fat + -ase, suffix used for enzymes) An enzyme that breaks up lipids and fats into small segments.

**lipid** (Gr. *lipos*, fat) A loosely defined group of molecules that are insoluble in water but soluble in oil; oils such as olive, corn, and coconut are lipids, as well as waxes such as bee's wax and ear wax.

**lipid bilayer** The basic foundation of all biological membranes; in such a layer, the nonpolar tails of phospholipid molecules point inward, forming a nonpolar zone in the interior of the bilayer. Lipid bilayers are selectively permeable and do not permit the diffusion of water-soluble molecules into the cell.

**littoral** (L. *litus*, shore) Referring to the shoreline zone of a lake or pond that is exposed to the air whenever water recedes.

**liver** The largest internal organ of the human body that is the body's principal metabolic factory, turning foodstuffs arriving from the digestive tract in the bloodstream into substances that are used by the different cells of the body.

**locus**, pl. **loci** (L. place) The position on a chromosome where a gene is located.

**long-term memory** A functional type of memory that appears to involve changes in the way information is processed at neural connections within the brain.

**loop of Henle** (after F.G.J. Henle, German anatomist) A hairpin loop formed by a tubule conveying urine when it enters the inner layer of the kidney and then turns around to pass up again into the outer layer of the kidney.

**luteal phase** The second phase of the reproductive cycle during which the mature eggs are released into the fallopian tubes, a process called ovulation.

**luteinizing hormone (LH)** A hormone produced by the anterior lobe of the pituitary gland, which in the female stimulates the development of the corpus luteum.

**lymph** (L. *lympha*, clear water) In animals, a colorless fluid derived from blood by filtration through capillary walls in the tissues.

**lymph node** Located throughout the lymph system, lymph nodes remove dead cells, debris, and foreign particles from the circulation.

**lymphatic system** An open circulatory system composed of a network of vessels that function to collect the water within blood plasma forced out during passage through the capillaries and to return it to the bloodstream; the lymphatic system also returns proteins to the circulation, transports fats absorbed from the intestine, and carries bacteria and dead blood cells to the lymph nodes and spleen for destruction.

**lymphocyte** (Gr. *lympha*, water + Gr. *kytos*, hollow vessel) A white blood cell; a cell of the immune system which either synthesizes antibodies (B cells) or attacks virus-infected cells (T cells).

**Lyonization** (after Mary F. Lyon, English geneticist) The inactivation of one X-chromosome in female mammals.

**lyse** (Gr. *lysis*, loosening) To disintegrate a cell by rupturing its cell membrane.

**lysosome** (Gr. *lysis*, a loosening + *soma*, body) A membrane-bound organelle, formed by the Golgi complex, which contains digestive enzymes; important for digesting worn-out cellular components, making way for newly formed ones while recycling the materials locked up in the old ones; a primary lysosome is one that is not actively functioning; a secondary lysosome is one that fuses with a food vacuole or other organelle so that its pH falls and hydrolytic enzymes are activated.

**macroevolution** (Gr. *mackros*, large + L. *evolvere*, to unfold) The creation of new species and the extinction of old ones.

**macromolecule** (Gr. *makros*, large + L. *moliculus*, a little mass) An extremely large molecule; refers specifically to carbohydrates, lipids, proteins, and nucleic acids.

**macrophage** (Gr. *makros*, long + -*phage*, eat) A phagocytic cell of the immune system able to engulf and digest invading bacteria, fungi, and other microorganisms, as well as cellular debris.

**major histocompatibility complex (MHC)** A protein cell surface marker anchored in plasma membrane, which the immune system uses to identify "self." All the cells of a given individual have the same "self" marker called a MHC protein.

**Malpighian corpuscle** (after Marcello Malpighi, Italian anatomist) A renal corpuscle; a filtration apparatus at the front end of each nephron tube of a vertebrate kidney.

**Malpighian tubule** (after Marcello Malpighi, Italian anatomist) A tubular extension of the digestive tract opening into the hindgut of insects that functions as an excretory organ.

**mantle** In the body of a mollusk, a heavy fold of tissue that is wrapped around the visceral mass like a cape.

**marginal meristem** The method by which leaves grow; marginal meristems grow outward and ultimately form the blade of the leaf, while the central portion becomes the midrib.

**marrow** (A.S. *mearg*) The soft tissue that fills the cavities of most bones and is the source of red blood cells.

**marsupial** (L. *marsupium*, pouch) A mammal in which the young are born early in their development, sometimes as soon as 8 days after fertilization, and are retained in a pouch; opossums and kangaroos are marsupials.

**marsupium** (L. pouch) A pouch on the abdomen of a female marsupial for carrying its young.

**mass** In chemistry, the total number of protons and neutrons in the nucleus of an atom; approximately equal to the atomic weight.

**mass flow** The overall process by which the movement of materials takes place in the phloem of plants.

**mast cell** A cell of the immune system that synthesizes the molecules involved in the body's response to trauma, including histamine and heparin.

**medulla** (L. marrow) The inner portion of an organ, in contrast to the cortex or outer portion, as in the kidney or adrenal gland; (2) the part of the brain that controls breathing and other involuntary functions, located at the top end of the spinal cord; also called medulla oblongata.

**megagametophyte** (Gr. *megas*, large + *gamos*, marriage + *phyton*, plant) In heterosporus plants, the female gametophyte, located within the ovule of seed plants.

**megaspore** (Gr. *megas*, large + *sporos*, seed) A spore of comparatively large size from which a female gametophyte develops.

**meiosis** (Gr. *meioun*, to make smaller) A special form of nuclear division that precedes gamete formation in sexually reproducing eukaryotes.

**Meissner's corpuscles** Receptors below the skin surface that fire in response to rapid changes in pressure.

**melatonin** (from mela[nin] + [sero]tonin) A hormone secreted by the pineal gland, whose function in humans is not well understood.

**membrane attack complex (MAC)** The battery of complement proteins that interact like natural killer cells by inserting itself into the pathogen's cell membrane, forming a hole, and killing the invading cell.

**memory cell** A large group of cells produced by the cloning of proliferating cells with the same antibody by a particular B cell that persist as circulating lymphocytes.

**Mendelian ratio** (after Gregor Mendel, Austrian monk) Referring to the characteristic 3:1 segregation ratio that Mendel observed, in which pairs of alternative traits were expressed in the $F_2$ generation in the ratio of three-fourths dominant to one-fourth recessive.

**Mendel's First Law** (after Gregor Mendel, Austrian monk) The Law of Segregation: central premises that state (1) that alleles do not blend in heterozygotes; (2) that alleles segregate in heterozygous individuals; and (3) that alleles have an equal probability of being included in either gamete.

**Mendel's Second Law** (after Gregor Mendel, Austrian monk) The Law of Independent Assortment. The statement that genes located on different chromosomes assort independently of one another.

**menstrual cycle** (L. *mens,* month) Monthly cycle; the term used to describe the reproductive cycle, usually occurring every 28 days, in humans.

**menstruation** (L. *mens,* month) Periodic sloughing off of the blood-enriched lining of the uterus when pregnancy does not occur. The menstrual cycle in primates is the cycle of hormone-regulated changes in the condition of the uterine lining, which is marked by the periodic discharge of blood and disintegrated uterine lining through the vagina (menstruation).

**meristem** (Gr. *merizein,* to divide) In plants, a zone of unspecialized cells whose only function is to divide.

**meroblastic cleavage** (Gr. *meros,* part + *blastos,* sprout) A type of cleavage in the eggs of reptiles, birds, and some fishes, which occurs only in the blastodisc.

**mesentery** (Gr. *mesos,* middle + *enteron,* intestine) A double layer of mesoderm within the coelom.

**mesoderm** (Gr. *mesos,* middle + *derma,* skin) One of the three embryonic germ layers that form in the gastrula; gives rise to muscle, bone and other connective tissue, the peritoneum, the circulatory system, and most of the excretory and reproductive systems.

**mesophyll** (Gr. *mesos,* middle + *phyllon,* leaf) The photosynthetic parenchyma of a leaf, located within the epidermis. The vascular strands (veins) run through the mesophyll.

**messenger RNA (mRNA)** A class of RNA in which each molecule is a long, single strand of RNA that passes from the nucleus to the cytoplasm; during polypeptide synthesis, mRNA molecules bring information from the chromosomes to the ribosomes to direct which polypeptide is assembled.

**metabolism** (Gr. *metabole,* change) The process by which all living things assimilate energy and use it to grow.

**metamorphosis** (Gr. *meta,* after +*morphe,* form + *osis,* state of) Process in which a marked change in form occurs during postembryonic development, for example, tadpole to frog or larval insect to adult.

**metaphase** (Gr. *meta,* middle + *phasis,* form) The stage of mitosis characterized by the alignment of the chromosomes on a plane in the center of the cell.

**metaphase plate** In metaphase, an imaginary plane passing through the circle around the spindle midpoint where the chromosomes array themselves.

**metastasis,** pl. **metastases** (Gr. to place in another way) The spread of cancerous cells to other parts of the body, forming new tumors at distant cites.

**microbody** A cellular organelle bounded by a single membrane and containing a variety of enzymes; generally derived from endoplasmic reticulum.

**microevolution** (Gr. *mikros,* small + L. *evolvere,* to unfold) Refers to the evolutionary process itself; evolution within a species; also called adaptation.

**microfilament** (Gr. *mikros,* small + L. *filum,* a thread) In cells, a protein thread composed of parallel fibers of actin cross-connected by myosin; their movement results from an ATP-driven shape change in myosin. The contraction of vertebrate muscles and many other kinds of cell movement in eukaryotes result from the movements of microfilaments within cells.

**microgametophyte** (Gr. *mikros,* small + *gamos,* marriage + *phyton,* plant) In heterosporus plants, the male gametophyte.

**microsporangium,** pl. **microsporangia** (Gr. *mikros,* small + *sporus,* seed + *angeion,* a vessel) A sporangium containing microspores, homologous with the sac containing the pollen in flowering plants.

**microspore** (Gr. *mikros,* small + *sporus,* seed) In plants, a spore that develops into a male gametophyte; in seed plants, it develops into a pollen grain.

**microtubule** (Gr. *mikros,* small + L. *tubulus,* little pipe) In eukaryotic cells, a long, hollow cylinder about 25 nanometers in diameter, composed of the protein tubulin. Microtubules influence cell shape, move the chromosomes in cell division, and provide the functional internal structure of cilia and flagella.

**microvillus,** pl. **microvilli** (Gr. *mikros,* small + L. *villus,* tuft of hair) A microscopic hairlike projection growing on the surface of the epithelial cells that cover the villi.

**middle ear** The hollow space between the eardrum and the inner ear; in humans it contains three small bones that transmit sound waves from the eardrum to the inner ear.

**middle lamella** The space, impregnated with pectins, between two new plant cells.

**migration** Long-range, two-way movements by animals, often occurring once a year with the change of seasons.

**mimicry** (Gr. *mimos,* mime) The resemblance in form, color, or behavior of certain organisms (mimics) to other more powerful or more protected ones (models), which results in the mimics being protected in some way.

**mitochondrion,** pl. **mitochondria** (Gr. *mitos,* thread + *chondrion,* small grain) A tubular or sausage-shaped organelle 1 to 3 micrometers long. Bounded by two membranes, they closely resemble the aerobic bacteria from which they were originally derived. As chemical furnaces of the cell, they carry out its oxidative metabolism.

**mitosis** (Gr. *mitos,* thread) The M phase of cell division in which the microtubular apparatus is assembled, binds to the chromosomes, and moves them apart; this phase is the essential step in the separation of the two daughter genomes.

**model** The groups of butterflies and moths that provide many of the best-known examples of Batesian mimicry.

**modified ratio** A modified Mendelian ratio of 9:7 instead of the usual 9:3:3:1 which illustrates the effects of epistasis.

**mole** (L. *moles,* mass) The atomic weight of a substance, expressed in grams; 1 mole is defined as the mass of $6.0222 \times 10^{23}$ atoms.

**molecule** (L. *moliculus,* a small mass) The smallest unit of a compound that displays the properties of that compound.

**mollusk** Any of a large phylum (Mollusca) of invertebrate animals having a soft, unsegmented body, usually covered with a hard shell secreted by a covering mantle, and a muscular foot. Snails, clams, scallops, and oysters belong to the phylum.

**Monera** (Gr. *moneres,* individual, solitary) The kingdom to which bacteria have been assigned.

**monoamine** An amine containing one amino group, especially one that functions as a neurotransmitter; acetylcholine and norepinephrine are monoamines.

**monoclonal antibody** An antibody produced in the laboratory by fusing genetically distinct cells and cloning the resulting hybrids so that each hybrid cell produces the same antibody.

**monocot** Short for monocotyledon: a flowering plant in which the embryos have only one cotyledon, the flower parts are often in threes, and the leaves typically are parallel-veined. Compare dicot.

**monoculture** (Gr. *monos,* one + L. *cultivare,* to cultivate) The exclusive cultivation of a single crop over a wide area.

**monocyte** (Gr. *monos,* single + *kytos,* hollow vessel) A type of circulating leukocyte that becomes a phagocytic cell (macrophage) after moving into tissues.

**monosaccharide** (Gr. *monos,* one + *sakcharon,* sugar) A simple sugar.

**monosynaptic reflex arc** A simple reflex arc in which the afferent nerve cell makes synaptic contacts directly with a motor neuron in the spinal cord, whose axon travels directly back to the muscle.

**monotreme** (Gr. *mono*, single + *treme*, hole) An egg-laying mammal; the only two are the duck-billed platypus and the echidna, or spiny anteater.

**morphogenesis** (Gr. *morphe*, form + *genesis*, origin) The formation of shape; the growth and differentiation of cells and tissues during development.

**morphology** (Gr. *morphe*, form + *logos*, discourse) The study of form and its development; includes cytology (the study of cell structure), histology (the study of tissue structure), and anatomy (the study of gross structure).

**morula** (L. *morum*, mulberry) The mass of blastomeres forming the embryo of many animals, just after the segmentation of the ovum and before the formation of a blastula.

**motor cortex** A sensory region on the cerebral cortex containing neurons that control the movement of different body muscles.

**motor endplate** The point where a neuron attaches to a muscle; a neuromuscular synapse.

**motor pathways** (L. mover) The nerve pathways that transmit commands to the body from the central nervous system.

**mRNA** *See* messenger RNA.

**mucosa,** pl. **mucosae** A deep layer of connective tissue overlaid by epithelium, containing glands that secrete mucus.

**Muellerian mimicry** (after Fritz Mueller, German biologist) A phenomenon in which two or more unrelated but protected species resemble one another, thus achieving a kind of group defense.

**multicellularity** A condition in which the activities of the individual cells are coordinated and the cells themselves are in contact; a property of eukaryotes alone and one of their major characteristics.

**muscle** (L. *musculus*, mouse) The tissue in the body of people and animals that can be tightened or loosened to make the body move.

**muscle cell** *See* muscle fiber.

**muscle fiber** Muscle cell; a long, cylindrical, multinucleated cell containing numerous myofibrils, which is capable of contraction when stimulated.

**muscle spindle** A sensory end organ that is attached to a muscle and sensitive to stretching.

**mutagen** (L. *mutare*, to change) A chemical capable of damaging DNA.

**mutant** (L. *mutare*, to change) A mutated gene; an organism carrying a gene that has undergone a mutation.

**mutation** (L. *mutare*, to change) A change in the genetic message of a cell.

**mutational repair** Repairs undertaken by the cells, such as excising altered nucleotides or re-forming single ruptured bonds; not always accurate, and some of the mistakes are incorporated into the genetic message.

**mutualism** (L. *mutuus*, lent, borrowed) A symbiotic relationship in which both participating species benefit.

**mycelium,** pl. **mycelia** (Gr. *mykes*, fungus) In fungi, a mass of hyphae.

**mycology** (Gr. *mykes*, fungus) The study of fungi. One who studies fungi is called a mycologist.

**mycorrhiza,** pl. **mycorrhizae** (Gr. *mykes*, fungus + *rhiza*, root) A symbiotic association between fungi and plant roots.

**myelin sheath** (Gr. *myelinos*, full of marrow) A flattened sheath of fatty material found in many, but not all, vertebrate neurons; made up of the membranes of Schwann cells.

**myelinated fiber** The structure formed by an axon and its associated Schwann cells, or cells with similar properties.

**myofibril** (Gr. *myos*, muscle + L. *fibrilla*, little fiber) A contractile microfilament within muscle, composed of myosin and actin.

**myofilament** (Gr. *mys*, muscle + L. *filare*, to spin) A contractile microfilament within muscle cells composed largely of actin and myosin; sometimes called myofibril.

**myosin** (Gr. *mys*, muscle + in, belonging to) One of two protein components of myofilaments (the other is actin).

**NADP (nicotinamide adenine dinucleotide phosphate)** A coenzyme that functions as an electron donor in many of the reduction reactions of biosynthesis. NADP4 is the oxidized NADPH2, the reduced form of NADP.

**natural killer cell** A cell that does not kill invading microbes but rather the cells that are infected by them.

**natural selection** The differential reproduction of genotypes caused by factors in the environment; leads to evolutionary change.

**nature or nurture** Instinct or behavior; the question of which factor plays what role in shaping learning and behavior.

**navigation** (L. *naviage*, to move, direct) The ability to set or adjust a bearing, and then follow it.

**nectar** (Gr. *nektar*) A sweet liquid found in many flowers, rich in sugar and amino acids, which attracts insects and birds that carry out pollination.

**negative control** The process of shutting off transcription by use of a regulatory site.

**nematocyst** (Gr. *nema*, thread + *kystos*, bladder) A coiled, threadlike stinging process of cnidarians, discharged to capture prey and for defense.

**nephrid organ** A filtration system of many freshwater invertebrates in which water and waste pass from the body across the membrane into a collecting organ, from which they are expelled to the outside through a pore.

**nephron** (Gr. *nephros*, kidney) The functional unit of the vertebrate kidney; a human kidney has more than 1 million nephrons that filter waste matter from the blood; each nephron consists of a Bowman's capsule, glomerulus, and tubule.

**neritic zone** (L. *nerita*, a sea mussel) The marine environment of shallow waters along the coasts of the continents.

**nerve** A bundle of axons with accompanying supportive cells, held together by connective tissue.

**nerve cord** The main trunk along the back to which the nerves that reach the different parts of the body are connected; a major characteristic of chordates.

**nerve fiber** An axon.

**nerve impulse** A rapid, transient, self-propagating reversal in electric potential that travels along the membrane of a neuron.

**neural groove** The long groove formed along the long axis of the embryo by a layer of ectodermal cells.

**neural tube** The dorsal tube, formed from the neural plate, that differentiates into the brain and spinal cord.

**neuralation** (Gr. *neuron*, nerve) The elaboration of a notochord and a dorsal nerve cord that marks the evolution of the chordates.

**neuroendocrine system** A network of endocrine glands whose hormone secretion is controlled by commands from the central nervous system.

**neuroethology** (Gr. *neuron*, nerve + *ethos*, habit or custom + *logos*, discourse) The study of the neural basis of behavior.

**neuroglia** (Gr. *neuron*, nerve + *glia*, glue) The delicate connective tissue forming a supporting network for the conducting elements of nervous tissue in the brain and the spinal cord.

**neurohormone** A hormone, such as an enkephalin or endorphin, secreted by nerve cells.

**neuromodulator** A chemical transmitter that mediates effects that are slow, longer lasting, and typically involve second messengers within the cell.

**neuromuscular junction** A structure formed when the tips of axons contact (innervate) a muscle fiber.

**neuron** (Gr. nerve) A nerve cell specialized for signal transmission.

**neurotransmitter** (Gr. *neuron*, nerve + L. *trans*, across + *mitere*, to send) A chemical released at an axon tip that travels across the synapse and binds a specific receptor protein in the membrane on the far side.

**neutron** (L. *neuter*, neither) A subatomic particle located within the nucleus of an atom similar to a proton in mass, but, as its name implies, is neutral and possesses no charge.

**neutrophil** An abundant type of granulocyte capable of engulfing microorganisms and other foreign particles; neutrophils comprise about 50% to 70% of the total number of white blood cells.

**niche** (L. *nidus*, nest) The role an organism plays in the environment; actual niche is the niche that an organism occupies under natural circumstances; theoretical niche is the niche an organism would occupy if competitors were not present.

**nicotinamide adenine dinucleotide** (**NAD$^+$**) A composite molecule consisting of two nucleotides bound together; an important coenzyme that functions as an electron acceptor in many oxidative reactions.

**nitrogen fixation** The incorporation of atmospheric nitrogen into nitrogen compounds, a process that can be carried out only by certain microorganisms.

**nociceptor** A receptor response from the central nervous system in response to a pain stimulus.

**nocturnal** (L. *nocturnus*, night) Active primarily at night.

**node** (L. *nodus*, knot) The place on the stem where a leaf is formed; *see* internode.

**node of Ranvier** (after L.A. Ranvier, French histologist) A gap formed at the point where two Schwann cells meet and where the axon is in direct contact with the surrounding intercellular fluid.

**nonassociative** A learned behavior that does not require an animal to form an association between two stimuli, or between a stimulus and a response.

**nonrandom mating** A phenomenon in which individuals with certain genotypes sometimes mate with one another more commonly than would be expected on a random basis.

**nonsense codon** A chain-terminating codon; a codon for which there is no tRNA with a complementary anticodon. There are three: UAA, UAG, and UGA.

**notochord** (Gr. *noto*, back + L. *chorda*, cord) In chordates, a dorsal rod of cartilage that forms between the nerve cord and the developing gut in the early embryo.

**nuclear envelope** The double membrane (outer and inner) surrounding the surface of the nucleus of eukaryotes.

**nuclear pore** Shallow depressions, like the craters of the moon, that are scattered over the surface of the nuclear envelope; such a pore contains many embedded proteins that act as molecular channels, permitting certain molecules to pass into and out of the nucleus.

**nucleic acid** A nucleotide polymer; a long chain of nucelotides; chief types are deoxyribonucleic acid (DNA), which is double stranded, and ribonucleic acid (RNA), which is typically single stranded.

**nucleolus, pl. nucleoli** (L. a small nucleus) Aggregations of rRNA and some ribosomal proteins that are transported into the nucleus from the rough ER and accumulate at those regions on the chromosomes where active synthesis of rRNA is taking place.

**nucleosome** (L. *nucleus*, kernel + *soma*, body) The basic packaging unit of eukaryotic chromosomes, in which the DNA molecule is wound around a ball of histone proteins. Chromatin is composed of long strings of nucleosomes, like beads on a string.

**nucleotide** A single unit of nucleic acid, composed of a phosphate, a five-carbon sugar (either ribose or deoxyribose), and a purine or a pyrimidine.

**nucleus** (L. *a kernel*, dim. fr. *nux*, nut) A spherical organelle (structure) characteristic of eukaryotic cells; the repository of the genetic information that directs all activities of a living cell; in at-

oms, the central core, containing positively charged protons and (in all but hydrogen) electrically neutral neutrons.

**occipital lobe** The posterior lobe of each cerebral hemisphere.

**olfaction** (L. *olfactum*, smelled) The process or function of smelling.

**oncogene** (Gr. *oncos*, tumor) A cancer-causing gene.

**oncogene theory** (Gr. *oncos*, cancer.) The hypothesis that cancer results from the action of a specific tumor inducing one gene.

**one gene-one enzyme hypothesis** The relationship that genes produce their effects by specifying the structure of enzymes and that each gene encodes the structure of a single enzyme.

**oocyte** (Gr. *oion*, egg + *kytos*, vessel) A cell in the outer layer of the ovary that gives rise to an ovum; primary oocyte, any of the 2 million oocytes a female is born with, all of which have begun the first meiotic division.

**open circulatory system** A system within an organism that permits materials to pass from one cell to another without leaving the organism and in which generally no distinction exists between circulating fluid and body fluid.

**operant conditioning** A learning mechanism in which the reward follows only after the correct behavioral response.

**operator** A site of negative gene regulation; a sequence of nucleotides that may overlap the promoter, which is recognized by a repressor protein. Binding of the repressor protein to the operator prevents binding of the polymerase to the promoter and so blocks transcription of the structural genes of an operon.

**operon** (L. *operis*, work) A cluster of functionally related genes transcribed onto a single mRNA molecule. A common mode of gene regulation in prokaryotes, it is rare in eukaryotes other than fungi.

**optic nerve** The nerve of sight, which goes from the brain to the eyeball and terminates in the retina.

**optimal yield** Exploitation at the early, most productive part of the rising portion of the sigmoid growth curve to maximize harvesting of the population; used by humans in agriculture and fisheries.

**order** A taxonomic category ranking below a class and above a family.

**organ** (L. *organon*, tool) A complex body structure composed of several different kinds of tissue grouped together in a structural and functional unit.

**organ system** A group of organs that function together to carry out the principal activities of the body.

**organelle** (Gr. *organella*, little tool) A specialized part of a cell; mitochondria are organelles.

**organism** Any individual living creature, either unicellular or multicellular.

**origin** (L. *oriri*, to rise) The end of a muscle attached to a bone that remains stationary during a contraction.

**osmoconformer** An animal that maintains the osmotic concentration of their body fluids at about the same level as that of the medium in which they are living.

**osmoregulation** The maintenance of a constant internal solute concentration by an organism, regardless of the environment in which the vertebrate lives.

**osmosis** (Gr. *osmos*, act of pushing, thrust) The diffusion of water across a membrane that permits the free passage of water but not that of one or more solutes.

**osmotic pressure** The increase of hydrostatic water pressure within a cell as a result of water molecules that continue to diffuse inward toward the area of lower water concentration (the concentration of the water is lower inside than outside the cell because of the dissolved solutes in the cell).

**osteoblast** (Gr. *osteon*, bone + *blastos*, bud) A bone-forming cell.

**osteocyte** (Gr. *osteon*, bone + *kytos*, hollow vessel) A mature osteoblast.

**otolith** *See* statocyst.

**outcross** A term used to describe species that interbreed with individuals other than those like themselves.

**outer medulla** The outer portion of the kidney, which contains the upper portion of the loop of Henle, including the upper ascending arm where reabsorption of salt from the filtrate by active transport occurs.

**oval window** A membrane in the ear of a mammal or other vertebrate connecting the middle ear with the inner ear.

**ovary** (L. *ovum*, egg) In animals, the organ that produces eggs. (2) In flowering plants, the enlarged basal portion of a carpel, which contains the ovule(s); the ovary matures to become the fruit.

**oviparous** (L. *ovum*, egg + *parere*, to bring forth) Referring to reproduction in which the eggs are developed after leaving the body of the mother, as in reptiles.

**ovulation** The successful development and release of an egg by the ovary.

**ovule** (L. *ovulum*, a little egg) A structure in a seed plant that becomes a seed when mature.

**ovum, pl. ova** (L. egg) A mature egg cell; a female gamete.

**oxidation** (Fr. *oxider*, to oxidize) The loss of an electron during a chemical reaction from one atom to another; takes places simultaneously with reduction; the second stage of the 10 reactions of glycolysis.

**oxidative metabolism** A collective term for metabolic reactions requiring oxygen.

**oxidative respiration** Respiration in which the final electron acceptor is molecular oxygen.

**oxytocin** (Gr. *oxys*, sharp + *tokos*, birth) A hormone of the posterior pituitary gland affecting contraction of the uterus in childbirth and stimulating lactation.

**pain** A stimulus the body receives that causes or is about to cause tissue damage.

**palisade parenchyma** (L. *palus*, stake + Gr. *para*, beside + *en*, in + *chein*, to pour) A type of parenchyma cells, which are columnar, closely packed together, and located between the upper epidermis of a leaf.

**pancreas** (Gr. *pan*, all + *kreas*, flesh) In vertebrates, the principle digestive gland; a large gland situated between the stomach and the small intestine that secretes a host of digestive enzymes and the hormones insulin and glucagon.

**Pangaea** Name given to the giant land mass that comprised the major continents during the early Jurassic Period.

**parasite** (Gr. *para*, beside + *sitos*, food) A symbiotic relationship in which one organism benefits and the other is harmed.

**parasympathetic nervous system** (Gr. *para*, beside + *syn*, with + *pathos*, feeling) One of two subdivisions of the autonomic nervous system (the other is the sympathetic). The two subdivisions operate antagonistically: the parasympathetic system stimulates resting activities such as digestion and restoration of the body to normal after emergencies by inhibiting alarm functions initiated by the sympathetic system.

**parathyroid hormone (PTH)** (Gr. *para*, beside + thyroid + hormone) A hormone produced by the parathyroid glands that regulates the way the body uses calcium.

**parenchyma** (Gr. *para*, beside + *en*, in + *chein*, to pour) The least specialized and most common cell of all plant cells; alive at maturity and capable of further division; usually photosynthetic or storage tissue.

**parthenogenesis** (Gr. *parthenos*, virgin + Eng. genesis) The development of an adult from an unfertilized egg; a common form of reproduction in insects.

**partial pressures (P)** The component of each individual gas, nitrogen, oxygen, and carbon, which together comprise the total air pressure.

**pathogen** (Gr. *pathos*, suffering + Eng. *genesis*, beginning) A disease-causing organism.

**pectin** (Gr. *pektos*, curdled, congealed, fr. *pegnynai*, to make fast or stiff) A complex form of plant starch having short, linear amylose branches consisting of 20 to 30 glucose subunits.

**pedigree** (L. *pes*, foot + *grus*, crane) A family tree; the patterns of inheritance observed in family histories; used to determine the mode of inheritance of a particular trait.

**pelvic girdle** The bony arch that provides strong connections and support for the legs.

**penis** (L. *tail*) The male organ of copulation; in mammals, it is also the male urinary organ.

**peptide** (Gr. *peptein*, to soften, digest) Two or more amino acids linked by peptide bonds.

**peptide bond** A covalent bond linking two amino acids; formed when the positive (amino, or $NH_3$) group at one end and a negative (carboxyl, or COO) group at the other end undergo a chemical reaction and lose a molecule of water.

**peptide hormone** A hormone that interacts with a receptor on the cell surface and initiates a chain of events within the cell by increasing the levels of secondary messengers.

**perennial** (L. *per*, through + *annus*, a year) A plant that lives for more than a year and produces flowers on more than one occasion.

**peripheral nervous system** (Gr. *peripherein*, to carry around) All of the neurons and nerve fibers outside the central nervous system, including motor neurons, sensory neurons, and the autonomic nervous system.

**peristalsis** (Gr. *peri*, around + *stellein*, to wrap) The rhythmic sequences of waves of muscular contraction in the walls of a tube.

**peroxisome** A membrane-bound spherical body, apparently derived from smooth endoplasmic reticulum (ER), that carries one set of enzymes active in converting fats to carbohydrates and another set that detoxifies various potentially harmful molecules—strong oxidants—that form in cells.

**petal** (Gr. *petalon*, leaf) One of the parts of a flower that is usually colored; one of the leaves of a corolla.

**petiole** (L. *petiolus*, a little foot) The stalk of a leaf.

**pH** Refers to the relative concentration of $H^+$ ions in a solution. The numerical value of the pH is the negative of the exponent of the molar concentration. Low pH values indicate high concentrations of $H^+$ ions (acids), and high pH values indicate low concentrations.

**phage** *See* bacteriophage.

**phagocyte** (Gr. *phagein*, to eat + *kytos*, hollow vessel) A cell that kills invading cells by engulfing them; includes neutrophils and macrophages.

**phagocytosis** (Gr. *phagein*, to eat + *kytos*, hollow vessel) A form of endocytosis in which cells engulf organisms or fragments of organisms.

**pharynx** (Gr. gullet) In vertebrates, a muscular tube that connects the mouth cavity and the esophagus; it serves as the gateway to the digestive tract and to the windpipe, or trachea.

**phenotype** (Gr. *phainein*, to show + *typos*, stamp or print) The realized expression of the genotype; it is the observable expression of a trait (affecting an individual's structure, physiology, or behavior), which results from the biological activity of proteins or RNA molecules transcribed from the DNA.

**pheromone** (Gr. *pherein*, to carry + [hor]mone) A chemical sex signal secreted by certain female animals, which signals their reproductive readiness.

**phloem** (Gr. *phloos*, bark) In vascular plants, a food-conducting tissue basically composed of sieve elements, various kinds of parenchyma cells, fibers, and sclereids.

**phosphate group**—$PO_4$; a chemical group commonly involved in high-energy bonds.

**phosphodiester** The name of the bond resulting from the formation of a nucleic acid chain in which individual sugars are linked together in a line by the phosphate groups. The phosphate group of one sugar binds to the hydroxyl group of another, forming an—O—P—O bond.

**phospholipid** (Gr. *phosphoros*, light-bearer, + *lipos*, fat) The lipid molecule that forms the foundation of a plasma membrane; similar to a fat molecule, but having only two fatty acids attached to its glycerol backbone, the third position being occupied by a highly polar organic alcohol that readily forms hydrogen bonds with water. One end of a phospholipid molecule is therefore strongly nonpolar (water insoluble), whereas the other end is extremely polar (water soluble). The two nonpolar fatty acids extend in one direction, roughly parallel to each other, and the polar alcohol group points in the other direction.

**phosphorus cycle** A critical mineral cycle important in worldwide plant nutrition in which phosphates are transferred from the soil to the plants and then recycled when the plants die.

**photocenter** An array of pigment molecules that act as a light antenna, directing photon energy captured by any of its members toward a single pigment molecule and thus amplifying the light-gathering powers of the individual's pigment molecules.

**photon** (Gr. *photos*, light) The unit of energy of light.

**photoperiodism** (Gr. *photos*, light + *periodos*, a period) A mechanism that organisms use to measure seasonal changes in relative day and night length.

**photophosphorylation** (Gr. *photos*, light + *phosphoros*, bringing light) A fundamental form of photosynthetic light reaction in which organisms use a network of chlorophyll molecules (a photocenter) to channel photon excitation energy to one pigment molecule, referred to as P700. P700 then donates an electron to an electron transport chain, which drives a proton pump and returns the electron to P700.

**photorespiration** A process in which $CO_2$ is released without the production of ATP or NADPH; because it produces neither ATP nor NADPH, photorespiration acts to undo the work of photosynthesis.

**photosynthesis** (Gr. *photos*, light -*syn*, together + *tithenai*, to place) The process by which plants, algae, and some bacteria use the energy of sunlight to create the more complicated molecules that make up living organisms from carbon dioxide ($CO_2$) and water ($H_2O$).

**photosynthetic membrane** The location where light reactions take place.

**photosystem I** A term to describe the more ancient bacterial photosystem of algae and plants.

**photosystem II** A photocenter system in which molecules of chlorophyll *a* are arranged with a different geometry in the photocenter so that more of the shorter-wavelength photons of higher energy are absorbed than in the more ancient bacterial photosystem.

**phototropism** (Gr. *photos*, light + *trope*, turning to light) A growth response of a plant to a unidirectional source of light.

**phylogeny** (Gr. *phylon*, race, tribe) The evolutionary relationships among any group of organisms.

**phylum, pl. phyla** (Gr. *phylon*, race, tribe) A major taxonomic category, ranking above a class.

**physiology** (Gr. *physis*, nature + *logos*, a discourse) The study of the function of cells, tissues, and organs.

**pigment** (L. *pigmentum*, paint) A molecule that absorbs light.

**pineal gland** (L. *pinus*, pine tree) A small endocrine gland in the center of the brain that secretes the hormone melatonin; it appears to function in humans as a light-sensing organ and in a variety of other roles concerning sexual development.

**pinocytosis** (Gr. *pinein*, to drink + *kytos*, cell) A form of endocytosis in which the material brought into the cell is a liquid containing dissolved molecules.

**pistil** (L. *pistillum*, pestle) The part of a flower that produces seeds, typically consisting of an ovary, a style, and a stigma; such organs taken collectively is known as the gynoecium. A flower that has only ovules and no pollen is called pistillate; functionally it is female.

**pith** The ground tissue occupying the center of the stem or root within the vascular cylinder, usually consisting of parenchyma.

**pituitary** (L. *pituita*, phlegm) The major hormone-producing gland of the brain, under the control of the hypothalamus; it secretes hormones that promote growth, stimulate glands, and regulate many other bodily functions.

**placenta, placentae** (L. a flat cake) A specialized organ, held within the womb in the mother, across which she supplies the offspring with food, water, and oxygen and through which she removes wastes.

**plankton** (Gr. *planktos*, wandering) The small organisms that float or drift in water, especially at or near the surface.

**plasma** (Gr. form) The fluid of vertebrate blood; contains dissolved salts, metabolic wastes, hormones, and a variety of proteins, including antibodies and albumin; blood minus the blood cells.

**plasma membrane** A lipid bilayer with embedded proteins that control the permeability of the cell to water and dissolved substances.

**plasmid** (Gr. *plasma*, a form or form) A small fragment of DNA that replicates independently outside the main polymer; a carbohydrate composed of many monosaccharide sugar subunits linked together in a long chain.

**plasmodesmata** (Gr. *plasma*, something molded + *desma*, band) The cytoplasmic connections that extend through pairs of holes in the cell walls in plants.

**plate tectonics** The theory that the earth's crust is divided into a series of vast, platelike parts, which explains the movement of continental drift.

**platelet** (Gr. dim. of *plattus*, flat) A fragment of a megakaryocyte that floats in the blood and plays an important role in controlling blood clotting.

**point mutation** A mutation that changes one or a few nucleotides; may result from physical or chemical damage to the DNA or from spontaneous errors during DNA replication.

**polar bodies** The products resulting from the process of meiosis in the oocytes, with only a single haploid ovum being produced and the other meiotic products (called polar bodies) being discarded.

**polar molecule** A molecule with positively and negatively charged ends; one portion of a polar molecule attracts electrons more strongly than another portion, with the result that the molecule has electron-rich (−) and electron-poor (+) regions, giving it magnetlike positive and negative poles. Water is one of the most polar molecules known.

**polar nuclei** In flowering plants, two nuclei (usually), one derived from each end (pole) of the em-

bryo sac, which become centrally located; they fuse with a male nucleus to form the primary (3n) endosperm nucleus.

**polarization** The charge difference of a neuron so that the interior of the cell is negative with respect to the exterior.

**pollen** (L. fine dust) A fine, yellowish powder consisting of grains or microspores, each of which contains a mature or immature male gametophyte. In flowering plants, pollen is released from the anthers of flowers and fertilizes the pistils.

**pollen tube** A tube that grows from a pollen grain. Male reproductive cells move through the pollen tube into the ovule.

**pollination** The transfer of pollen from the anthers to the stigmas of flowers for fertilization, as by insects or the wind.

**polyclonal** An antibody response in which an antigen elicits many different antibodies, each fitting a different portion of the antigen surface.

**polygyny** (Gr. *poly*, many + *gyne*, woman, wife) A mating choice in which a male mates with more than one female.

**polymer** (Gr. *polus*, many + *meris*, part) A large molecule formed of long chains of similar molecules.

**polymerization** (Gr. *polus*, many + *meris*, part + *izein*, to combine with) A process in which identical protein subunits are attracted to one another chemically and assemble spontaneously into long chains.

**polymorphism** (Gr. *polys*, many + *morphe*, form) The presence in a population of more than one allele of a gene at a frequency greater than that of newly arising mutations.

**polynomial system** (Gr. *polys*, many + (bi)nomial) Before Linnaeus, naming a genus by use of a cumbersome string of Latin words and phrases.

**polyp** A cylindrical, pipe-shaped cnidarian usually attached to a rock with the mouth facing away from the rock on which it is growing. Coral is made up of polyps.

**polypeptide** (Gr. *polys*, many + *peptein*, to digest) A general term for a long chain of amino acids linked end to end by peptide bonds; a protein is a long, complex polypeptide.

**polysaccharide** (Gr. *polys*, many + *sakcharon*, sugar) A sugar polymer; a carbohydrate composed of many monosaccharide sugar subunits linked together in a long chain.

**population** (L. *populus*, the people) Any group of individuals, usually of a single species, occupying a given area at the same time.

**population genetics** The branch of genetics that deals with the behavior of genes in populations.

**population pyramid** A bar graph using 5-year age categories by which the characteristics of a population can be illustrated graphically.

**positive control** The process of turning on transcription by use of a regulatory site.

**posterior** (L. *post*, after) Situated behind or farther back.

**postsynaptic membrane** The membrane of the target cell of the synaptic cleft.

**potential difference** A difference in electrical charge on two sides of a membrane caused by an unequal distribution of ions.

**potential energy** Energy having the potential to do work; stored energy.

**prairie** A temperate grassland of the United States and southern Canada.

**precapillary sphincter** A ring of muscle which guards each capillary loop and which, when closed, blocks flow through the capillary.

**predation** (L. *praeda*, prey) The eating of other organisms. The one doing the eating is called a predator, and the one being consumed is called the prey.

**pressure receptor** *See* baroreceptor.

**presynaptic membrane** The membrane on the axonal side of the synaptic cleft.

**prey** (L. *prehendere*, to grasp, seize) An organism eaten by another organism.

**primary induction** Inductions between the three primary tissue types—ectoderm, mesoderm, and endoderm.

**primary plant body** The part of a plant that includes the young, soft shoots, and roots, and which arises from the apical meristems.

**primary mRNA transcript** *See* RNA transcript.

**primary nondisjunction** The failure of homologous chromosomes to separate in meiosis I; the cause of Down syndrome.

**primary producers** Photosynthetic organisms, including plants, algae, and photosynthetic bacteria.

**primary structure of a protein** The sequence of amino acids that make up a particular polypeptide chain.

**primary tissue** Tissue that comprises the primary plant body.

**primitive streak** The site of invagination on a blastodisc, which appears as a slit on the surface of the gastrula.

**primordium,** pl. **primordia** (L. *primus*, first + *ordiri*, begin) The first cells in the earliest stages of the development of an organ or structure.

**productivity** The total amount of energy of an ecosystem fixed by photosynthesis per unit of time: net productivity, productivity minus that which is expended by the metabolic activity of the organisms in the community.

**progesterone** A steroid hormone secreted by the corpus luteum that makes the lining of the uterus more receptive to a fertilized ovum.

**proglottid** (Gr. *proglottis*, tip of the tongue) One of the segments or joints of a tapeworm, containing both male and female sexual organs.

**prokaryote** (Gr. *pro*, before + *karyon*, kernel) A simple, ancient, bacterial organism that was small, single-celled, lacked external appendages, and had little evidence of internal structure.

**promoter** An RNA polymerase binding site; the nucleotide sequence at the end of a gene to which RNA polymerase attaches to initiate transcription of mRNA.

**prophase** (Gr. *pro*, before + *phasis*, form) The first stage of mitosis during which the chromosomes become more condensed, the nuclear envelope is reabsorbed, and a network of microtubules (called the spindle) forms between opposite poles of the cell.

**prostaglandin** (from prosta[te] gland + -in) A modified lipid produced from membrane phospholipids by virtually all cells. It stimulates smooth muscle contraction and expansion and contraction of blood vessels.

**prostate gland** (Gr. *prostates*, one standing in front) A large gland surrounding the male urethra just below the bladder. Its secretions, which transport sperm cells, make up a large part of the semen.

**protease** (Gr. *proteios*, primary + -ase, enzyme ending) An enzyme that breaks up proteins into amino acids.

**protein** (Gr. *proteios*, primary) A long chain of amino acids linked end to end by peptide bonds; because the 20 amino acids that occur in proteins have side groups with very different chemical properties, the function and shape of a protein is critically affected by its particular sequence of amino acids.

**protist** (Gr. *protos*, first) A member of the kingdom Protista, which includes unicellular eukaryotic organisms and some multicellular lines derived from them.

**proton** A subatomic particle in the nucleus of an atom that carries a positive charge; the number of charged protons determines the chemical character of the atom, because it dictates the number of electrons orbiting the nucleus and available for chemical activity; there is one electron for each proton.

**proton pump** A channel in the cell, along with the sodium-potassium pump, that transports molecules against a concentration gradient by expending energy.

**protozoa** (Gr. *protos*, first + *zoon*, animal) The traditional name given to heterotrophic protists.

**proximate causation** The question of how an animal behaves in a certain way, the mechanism of behavior.

**pseudocoel** (Gr. *pseudos*, false + *koiloma*, cavity) A body cavity similar to the coelom except that it is unlined.

**pseudopod, pl. pseudopodia** (Gr. *pseudos*, false + *pous*, foot) "False foot;" a temporary protrusion of a one-celled organism that serves as a means of locomotion.

**pulmonary circulation** The part of the human circulatory system that carries the blood from the heart to the lungs and back again.

**punctuated equilibrium** A hypothesis of the mechanism of evolutionary change, which proposes that long periods of little or no change are punctuated by periods of rapid evolution.

**Punnett square** (after Reginald C. Punnett, English geneticist) A diagram useful in analyzing a Mendelian model of the outcome of an $F_2$ generation derived from a heterozygous individual.

**pupil** (L. *pupilla*, little doll) The opening in the center of the iris of the eye where light enters the eye.

**Purkinje fiber** (after Johannes E. Purkinje, Bohemian physiologist) Any of the modified muscle fibers of the heart that make up a network by which muscle impulses are conducted through the heart.

**pyloric sphincter** (Gr. *pyloros*, gatekeeper) A muscular constriction between the stomach and the small intestine that functions as a kind of traffic light of the digestive system.

**pyramid of biomass** The structure obtained from weighing all the individuals at each trophic level of the ecosystem.

**pyramid of energy** The structure obtained from measuring the flow of energy through an ecosystem directly at each point of transfer.

**pyramid of numbers** The structure obtained from counting all the individuals in an ecosystem and assigning each to a trophic level.

**pyruvate** The three-carbon compound that is the endproduct of glycolysis and the starting material of the citric acid cycle.

**pyruvate dehydrogenase** The complex of enzymes that removes the $CO_2$ from pyruvate; one of the largest known enzymes, containing 48 polypeptide chains.

**quaternary structure of a protein** A term to describe the way the subunits of protein are assembled into a whole.

**radial symmetry** (L. *radius*, a spoke of a wheel + Gr. summetros, symmetry) The regular arrangement of parts around a central axis, so that any plane passing through the central axis divides the organism into halves that are approximate mirror images.

**radicle** (L. *radicula*, root) The part of the plant embryo that develops into the root.

**radioactive** An isotope, such as carbon-14, in which the nucleus tends to break up into elements with lower atomic numbers in a process called radioactive decay.

**radioactivity** The emission of nuclear particles and rays by unstable atoms as they decay into more stable forms; measured in curies, one curie equals 37 billion disintegrations a second.

**radula** (L. scraper) A rasping tonguelike organ characteristic of most mollusks.

**rain shadow effect** The phenomenon in which the eastern sides of mountains are much drier than their western sides, and vegetation is often very different.

**receptor-mediated endocytosis** The process in which a protein called a receptor, embedded within a network of proteins called coated pits located on regions of plasma membrane, detects the presence of a particular target

molecule and reacts by initiating endocytosis of the coated pit. By doing so, it traps the target molecule within the new vesicle. Transport across the membrane occurs only when the correct molecule is present and positioned for transport.

**receptor protein** A highly specific cell surface receptor embedded in a cell membrane that responds only to a specific messenger molecule.

**recessive allele** An allele, whose phenotype effects are masked in heterozygotes by the presence of a dominant allele.

**recombinant DNA** A DNA molecule created in the laboratory by molecular geneticists who join together bits of several genomes into a novel combination.

**recombination** The formation of new gene combinations; in bacteria, it is accomplished by the transfer of genes into cells, often in association with viruses; in eukaryotes, it is accomplished by reassortment of chromosomes during meiosis, and by crossing-over.

**recombination map** The construction of a genetic map by using the frequencies of crossing-over events in crosses.

**rectum** The short and final segment of the large intestine extending from the colon to the anus.

**reducing power** The use of the energy of light to extract hydrogen atoms from water.

**reduction** (L. *reductio*, a bringing back: originally "bringing back" a metal from its oxide) The gain of an electron during a chemical reaction from one atom to another; takes place simultaneously with oxidation.

**reflex** (L. *reflectere*, to bend back) An automatic consequence of a nerve stimulation; the motion that results from a nerve impulse passing through the system of neurons, eventually reaching the body muscles and causing them to contract.

**reflex arc** The nerve path in the body leading from stimulus to reflex action.

**refractory period** The recovery period after membrane depolarization during which the membrane is unable to respond to additional stimulation; (2) the period after ejaculation, lasting 20 minutes or longer, during which males lose their erection, arousal is difficult and ejaculation almost impossible.

**regulatory site** A special nucleotide sequence of a gene that acts as a point of control when the tran-

scription of individual genes begin.

**release factor** A special protein that releases a newly made polypeptide when a stop codon is encountered during protein synthesis.

**releasing hormone** A peptide hormone produced by the hypothalamus that stimulates the secretion of specific hormones by the anterior pituitary.

**renal** (L. *renes*, kidneys) Pertaining to the kidney.

**replication fork** The structure formed at the end where the double-stranded DNA molecule separates during DNA replication.

**replication origin** The sequence of DNA necessary for replication.

**replication unit** An individual zone of a eukaryotic chromosome which replicates as a discrete unit.

**repression** (L. *reprimere*, to press back, keep back) The process of blocking transcription by the placement of the regulatory protein between the polymerase and the gene, thus blocking movement of the polymerase to the gene.

**repressor** (L. *reprimere*, to press back, keep back) A protein that regulates transcription of mRNA from DNA by binding to the operator and so preventing RNA polymerase from attaching to the promoter.

**resolving power** The ability of a microscope to distinguish two lines as separate.

**respiration** (L. *respirare*, to breathe) The utilization of oxygen; in terrestrial vertebrates, the inhalation of oxygen and the exhalation of carbon dioxide.

**resting membrane potential** The charge difference that exists across a neuron's membrane at rest (about 70 millivolts).

**restriction endonuclease** A special kind of enzyme that can recognize and cleave DNA molecules into fragments; the basic tools of genetic engineering.

**restriction enzyme** An enzyme that cuts a DNA strand at a particular place.

**restriction fragment-length polymorphism (RFLP)** An associated genetic mutation marker detected because the mutation alters the length of DNA segments.

**retina** (L. a small net) The structure in the eye that is sensitive to light and receives optical images; contains a field of receptor cells, rods, and cones.

**retrovirus** (L. *retro,* turning back) A virus whose genetic material is RNA rather than DNA; when a retrovirus infects a cell, it makes a DNA copy of itself, which it can then insert into the cellular DNA as if it were a cellular gene.

**reverse transcriptase** An enzyme that synthesizes a double strand of DNA complementary to viral RNA; found only in association with retroviruses such as the AIDS virus.

**Rh blood group** A set of cell surface markers on human red blood cells; named for the Rhesus monkey in which they were first described.

**rhizome** (Gr. *rhizoma,* mass of roots) In vascular plants, a usually more or less horizontal underground stem; may be enlarged for storage or may function in vegetative reproduction.

**ribonucleic acid (RNA)** The other principal form of nucleic acid which is similar in structure and is made as a template copy of portions of the DNA. This copy passes out into the rest of the cell, where it provides a blueprint specifying the amino acid sequence of proteins.

**ribose** A five-carbon sugar.

**ribosomal RNA (rRNA)** A class of RNA molecules found, together with characteristic proteins, in ribosomes; during polypeptide synthesis, they provide the site on the ribosome where the polypeptide is assembled.

**ribosome** An organelle composed of protein and RNA, which translate RNA copies of genes into protein.

**RNA** *See* ribonucleic acid.

**RNA polymerase** The enzyme that transcribes RNA from DNA.

**RNA transcript** The newly assembled mRNA chain, which is called the primary mRNA transcript, of the nucleotide sequence of the gene.

**rod** A specialized sensory cell in the retina of the eye responsible for black-and-white vision.

**root** The part of a plant that grows downward usually into the ground, to hold the plant in place, to absorb water and mineral foods from the soil, and often to store food material.

**root cap** A thimblelike mass of relatively unorganized cells, which covers and protects the root's apical meristem as it grows through the soil.

**root hairs** Fine projections from the epidermis, or outermost cell layer, of the terminal portion of roots.

**root pressure** In vascular plants, the pressure that develops in roots, primarily at night, caused by the continued, active accumulation of ions by the roots of a plant at times when transpiration from the leaves is very low or absent.

**r-selection** (from the *r* term in the logistic equation) Natural selection under conditions that favor survival when populations are controlled primarily by density-independent factors; contrasts with K-selection.

**saltatory conduction** A very fast form of nerve impulse conduction in which the impulses leap from node to node over insulation portions.

**sarcoma** (Gr. *flesh*) A cancerous tumor that involves connective or hard tissue, such as muscle.

**sarcomere** (Gr. *sarx,* flesh + *meris,* part of) The fundamental unit of contraction in skeletal muscle; the repeating bands of actin and myosin that appear between two Z lines.

**sarcoplasm** (Gr. *sarx,* flesh + *plassein,* to form, mold) A special name for the cytoplasm in striated muscle.

**sarcoplasmic reticulum** (Gr. *sarx,* flesh + *plassein,* to form, mold; L. *reticulum,* network) The endoplasmic reticulum of a muscle cell; a sleeve of membrane that wraps around each myofilament.

**savannah** (Sp. zavana) A tropical or subtropical region of open grassland that is a transitional biome between tropical rain forest and desert.

**Schwann cells** (after Theodor Schwann, German anatomist) The supporting cells associated with projecting axons along with all the other nerve cells that make up the peripheral nervous system.

**scientific creationism** A view that the biblical account of the origin of the earth is literally true, that the earth is much younger than most scientists believe, and that all species of organisms were individually created just as they are today.

**sclereid** (Gr. *skleros,* hard) In vascular plants, a type of sclerenchyma cell that is thick-walled or lignifed, usually pitted and not elongated; also called stone cell.

**sclerenchyma cell** (Gr. *skleros,* hard + *en,* in + *chymein,* to pour) A cell of variable form and size with more or less thick, often lignified, secondary walls; may or may not be living at maturity; includes fibers and sclereids. Col-

lectively, sclerenchyma cells may make up a kind of tissue called sclerenchyma.

**scolex** (Gr. worm) The organ in a tapeworm that attaches to the intestinal wall.

**scrotum** The sac containing the testicles which hangs between the legs of the male.

**scurvy** (A.S. *scurf, sceorf*) A disease characterized by weakness, spongy gums, and bleeding of the skin and mucous membranes, resulting from insufficient quantities of vitamin C in the diet.

**second messenger** An intermediary compound that couples extracellular signals to intracellular processes and also amplifies a hormonal signal.

**secondary cell wall** A parenchymal cell that is deposited between the cytoplasm and primary wall of a fully expanded cell.

**secondary chemical compound** A chemical that is not involved in primary metabolic processes and that plays the dominant role in protecting plants from being eaten by herbivores or predators.

**secondary growth** In vascular plants, growth that results from the division of a cylinder of cells around the plant's periphery; secondary growth causes a plant to grow in diameter.

**secondary immune response** The swifter response of the body the second time it is invaded by the same pathogen because of the presence of memory cells, which quickly become plasma cells.

**secondary induction** An induction between tissues that have already differentiated.

**secondary plant body** The part of a plant characterized by thick accumulations of conducting tissue and the other cell types associated with it.

**secondary sex characteristics** External differences between male and female animals; not directly involved in reproduction.

**secondary structure of a protein** The twisting or folding of a polypeptide chain; results from the formation of hydrogen bonds between different amino acid side groups of a chain; the most common structures that form are a single-stranded helix, an extended sheet, or a cable containing three (as in collagen) or more strands.

**secondary tissues** The tissues that comprise the secondary plant body.

**seed** A structure that develops from the mature ovule of a seed plant; contains an embryo surrounded by a protective coat.

**seed coat** The outer layer of a seed, developed from the integuments of the ovule.

**seed plants** A collective term for the groups of vascular plants that produce seeds.

**segregation of alleles** *See* Mendel's first law.

**segregation of alternative traits** The finding that alternative traits segregate in crosses and may mask each other's appearance.

**selection** The process by which some organisms leave more offspring than competing ones, and their genetic traits tend to appear in greater proportions among members of succeeding generations than the traits of those individuals that leave fewer offspring.

**selectively permeable membrane** (L. *seligere,* to gather apart + *permeare,* to go through) Refers to the ability of a cell membrane to allow passage across the membrane of some solutes but not others, the result of specific protein channels extending across the membrane; some molecules can pass through a specific kind of channel, others cannot.

**self-fertilization** The process in which a plant that has both male and female gametes can fertilize itself.

**self-pollination** The transfer of pollen from another to a stigma in the same flower or to another flower of the same plant, leading to self-fertilization.

**semen** (L. seed) The fluid produced in the male reproductive organs that contains sperm and fluid secretions.

**semicircular canal** Any of three fluid-filled canals in the inner ear that help to maintain balance.

**semipermeable membrane** *See* selectively permeable membrane.

**sensory pathway** The nerve pathway of the peripheral nervous system that transmits commands to the the body from the central nervous system.

**sepal** (L. *sepalum,* a covering) A member of the outermost whorl of a flowering plant; collectively, the sepals constitute the calyx.

**septum, pl. septa** (L. *saeptum,* a fence) A partition or cross-wall that divides hyphae into cells.

**seta, pl. setae** (L. bristle) In an annelid, bristles of chitin that help to anchor the worm during locomotion or when it is in its burrow.

**sex chromosomes** The X and Y chromosomes that are different in the two sexes and that are involved in sex determination.

**sex-linked characteristic** A genetic characteristic that is determined by genes located on the sex chromosomes.

**sexual reproduction** Reproduction that involves the regular alternation between syngamy and meiosis; its outstanding characteristic is that an individual offspring inherits genes from two parent individuals.

**sexual selection** Natural selection perpetuating certain characteristics that attract one sex to the other, such as bright feathers in birds.

**shell** The hard outer covering of a mollusk.

**shifting agriculture** A form of agriculture common in tropical areas in which people clear and cultivate a patch of forest, grow crops for a few years, and then move on.

**shoot** In vascular plants, the above ground parts such as the stem and leaves.

**short-term memory** A functional type of memory that is transient, lasting only a few moments.

**sieve cell** In the phloem (food-conducting tissue) of vascular plants, a long, slender sieve element with relatively unspecialized sieve areas and with tapering end walls that lack sieve plates; found in all vascular plants except angiosperms, which have sieve-tube members.

**sieve tube** In the phloem of angiosperms, a series of sieve-tube members arranged end-to-end and interconnected by sieve plates.

**sigmoid growth curve** An S-shaped population growth curve that implies a relatively slow start in growth, a rapid increase, and then a leveling off when the carrying capacity of the species' environment is reached.

**sign stimulus** According to ethologists, the "signal" in the environment that triggers the instinctive behavior of an animal.

**single bond** A covalent bond that shares only one electron pair.

**sink** An area in a vascular plant where sucrose is taken from a sieve tube.

**sinus** (L. curve) A reservoir or channel within the liver containing venous blood.

**skin** The outer layer of tissue of the human or animal body.

**smell** A chemical sensory system in which the receptors are neurons whose cell bodies are embedded in the epithelium of the upper portion of the nasal passage.

**smooth muscle** Nonstriated muscle; lines the walls of internal organs and arteries and is under involuntary control.

**sociobiology** The study of the biological basis of social behavior of animal societies.

**society** A group of organisms of the same species that are organized in a cooperative manner.

**soluble** Referring to polar molecules that dissolve in water and are surrounded by a hydration shell.

**solute** The other kinds of molecules dissolved in water. *See* solution, solvent.

**solution** A mixture of molecules such as sugars, amino acids, and ions dissolved in water.

**solvent** The most common of the molecules dissolved in a solution, usually a liquid, commonly water.

**somatic cells** (Gr. *soma*, body) All the other diploid body cells of an animal.

**somatic nervous system** (Gr. *soma*, body) A motor pathway of the peripheral nervous system whose nerve fibers stimulate secretion from glands and excrete or inhibit the smooth muscles of the body; the voluntary system, as contrasted with the involuntary, or autonomic nervous system.

**somatosensory cortex** A sensory region on the leading edge of the parietal lobe that receives afferent input from sensory receptors of many different parts of the body.

**somite** A segmented block of tissue on either side of a developing notochord.

**source** An area in a vascular plant where sucrose is made.

**species**, pl. **species** (L. kind, sort) A kind of organism; a species ranks next below a genus.

**species specific** Limited in reaction or effect to one species.

**speech centers** Regions in both hemispheres of the brain that are responsible for different associative speech activities.

**sperm** (Gr. *sperma*, sperm, seed) A sperm cell; the human male gamete.

**spermatogenesis** (Gr. *sperma*, sperm, seed + *gignesthai*, to be born) The formation and development of spermatozoa.

**sphincter** (Gr. *sphinkter*, band, from sphingein, to bind tight) In vertebrate animals, a ringlike muscle that surrounds an opening or passage of the body and can contract to close it.

**spinal cord** The thick, whitish cord of nerve tissue extending from the medulla oblongata down through most of the spinal column and from which nerves to various parts of the body branch off.

**spindle** The motive assembly that carries out the separation of chromosomes during cell division; composed of microtubules and assembled during prophase at the equator of the dividing cell.

**spindle fibers** An axis of microtubules formed by separating pairs of centrioles.

**spine** The flexible supporting column of bone along the middle of the back in the body of vertebrates.

**spirellum**, pl. **spirella** (L. *spira*, spire, coil) A spirally coiled bacterium.

**spongin** A tough protein that helps to strengthen the body of a sponge either by itself or together with spicules.

**spongy bone tissue** *See* marrow.

**spongy parenchyma** A leaf tissue composed of loosely arranged chloroplast-bearing cells. *See* palisade parenchyma.

**sporangium**, pl. **sporangia** (Gr. *spora*, seed + *angeion*, a vessel) A structure in which asexual spores are produced in zygomycetes.

**spore** (Gr. *spora*, seed) A haploid reproductive cell, usually unicellular, capable of developing into an adult without fusion with another cell. Spores result from meiosis, as do gametes, but gametes fuse immediately to produce a new diploid cell.

**sporic meiosis** The type of life cycle of a plant in which both diploid and haploid cells divide by mitosis.

**sporophyte** (Gr. *spora*, seed + *phyton*, plant) The spore-producing, diploid (2n) phase in the life cycle of a plant having alternation of generations.

**stabilizing selection** A form of selection in which selection acts to eliminate both extremes from an array of phenotypes, the result of which is the increase in the frequency of the intermediate type, which is already the most common.

**stable population** A population whose size remains the same through time.

**stamen** (L. thread) The part of the flower that contains the pollen, consisting of a slender filament that supports the anther. A flower that produces only pollen is called staminate; it is functionally male.

**stasis** (Gr. a standing still) Referring to a lack of evolutionary change.

**statocyst** (Gr. *statos*, standing + *kystis*, bladder, sac) A gravity receptor located in a series of

hollow chambers within the human inner ear that provides information the brain uses to perceive balance.

**stem** The above-ground axis of vascular plants; stems are sometimes below ground (as in rhizomes and corns).

**steppe** (Russ. step) A temperate grassland of eastern Europe and central Asia.

**stereotyped** Animal behavior that appears in the same form in different individuals of a species.

**steroid** (Gr. *stereos*, solid + L. *ol*, from oleum, oil) A kind of lipid; many of the molecules that function as messengers and pass across cell membranes are steroids, such as the male and female sex hormones and cholesterol.

**steroid hormone** A hormone derived from cholesterol; those that promote the development of the secondary sexual characteristics are steroids.

**stigma** (Gr. mark) A specialized area of the carpel of a flowering plant that receives the pollen.

**stoma**, pl. **stomata** (Gr. mouth) A specialized opening in the leaves of some plants that allows carbon dioxide to pass into the plant body and allows water and oxygen to pass out of them.

**stratum basale layer** The cells of the innermost layer of the epidermis of the skin of the vertebrate body.

**stratum corneum** The outer layer of the epidermis of the skin of the vertebrate body.

**stretch receptor** A sense organ that is sensitive to any stretching of the tissue in which it is found; the muscle spindle is a stretch receptor.

**striated muscle** (L. *striare*, to groove) A type of muscle with fibers of cross bands usually contracted by voluntary action.

**stroma** (Gr. anything spread out) The fluid matrix inside the chloroplast within which the thylakoids are embedded.

**stromatolite** (Gr. *stromatos*, a spread + Eng. -lite) A massive limestone deposit produced by cyanobacteria.

**subcutaneous tissue** The underlying layer of vertebrate skin.

**subspecies** A subdivision of a species, often a geographically distinct race.

**substrate** (L. *substratus*, strewn under) A molecule on which an enzyme acts.

**substrate-level phosphorylation** The generation of ATP by coupling its synthesis to a strongly exergonic (energy-yielding) reaction.

**succession** In ecology, the slow, orderly progression of changes in community composition that takes place through time. Primary succession occurs in nature over long periods of time; secondary succession occurs when a climax community has been disturbed.

**sucrose** Table sugar; a common disaccharide found in many plants; a molecule of glucose linked to a molecule of fructose.

**sugar** Any monosaccharide or disaccharide.

**supercoil** An advanced coil of the DNA that occurs when the string of nucleosomes wraps up into higher-order coils.

**supporting glial cell** A nerve cell that supports and insulates the neurons.

**surface layers** The marine environment of the top layers of the open sea.

**surface tension** A tautness of the surface of a liquid, caused by the cohesion of the molecules of liquid. Water has an extremely high surface tension.

**surface-to-volume ratio** A term describing cell size increases, as the volume of a cell grows much more rapidly than does the surface area of a cell.

**survivorship** The percentage of an original population that is living at a given age.

**sympathetic nervous system** A subdivision of the autonomic nervous system that functions as an alarm response; increases heartbeat while slowing down everyday functions; produces responses opposite to those of the parasympathetic nervous system.

**synapse** (Gr. *synapsis*, a union) A junction between a neuron and another neuron or a muscle cell; the two cells do not touch, neurotransmitters cross the narrow space between them.

**synapse, excitatory** A synapse in which the receptor protein is a chemically gated sodium channel; binding of a neurotransmitter opens the channel and initiates an excitatory electrical potential that increases the ease with which the membrane can be depolarized.

**synapse, inhibitory** A synapse in which the receptor protein is a chemically gated potassium or chloride channel; binding of a neurotransmitter opens the channel and produces an inhibitory electrical potential that reduces the ability of the membrane to depolarize.

**synapsis** (Gr. *contact*, union) The close pairing of homologous chromosomes that occurs early in prophase I of meiosis; with the genes of the chromosomes thus aligned, a DNA strand of one homologue can pair with the complimentary DNA strand of the other.

**synaptic cleft** The space between two adjacent neurons.

**syngamy** (Gr. *syn*, together with + *gamos*, marriage) Fertilization; the union of male and female gametes.

**systemic circulation** The circulation of blood between the heart and the rest of the body except the lungs.

**systolic period** (Gr. *systole*, contraction) The pushing period of heart contraction, which ends with the closing of the aortic valve, during which a pulse of blood is forced into the systemic arterial system, immediately raising the blood pressure within these vessels.

**T cell** A type of lymphocyte involved in cell-mediated immune response and interactions with B cells; also called a T lymphocyte.

**taiga** (Russ.) A swampy, coniferous evergreen forest land of subarctic Siberia between the tundra and the steppes.

**taste** The specialized sensory receptors located in the mouth by which the flavor of a substance is perceived by the taste buds.

**taste bud** One of the groups of receptor cells in the mouth that are organs of taste.

**taxonomy** (Gr. *taxis*, arrangement + *nomos*, law) The science of the classification of organisms.

**telophase** (Gr. *telos*, end + *phasis*, form) The fourth and final stage of mitosis during which the mitotic apparatus is disassembled, the nuclear envelope re-forms, and the chromosomes uncoil.

**temperate grassland** A region of vegetation consisting mainly of grasses or grasslike plants whose soil tends to be deep and fertile.

**temporal lobe** (L. *temporalis*, the temples) The part of each cerebral hemisphere, in front of the occipital lobe, that contains the center of hearing in the brain.

**tendon** (Gr. *tenon*, stretch) A strap of cartilage that attaches muscle to bone.

**territoriality** A form of behavior in which an animal establishes an area as its own and defends it from encroachments by others, usually of its own species.

**tertiary structure of a protein** The three-dimensional shape of a protein; primarily the result of hydrophobic interactions of amino acid side groups and, to a lesser extent, of hydrogen bonds between them; forms spontaneously.

**test cross** A cross between a heterozygote and a recessive homozygote; a procedure Mendel used to further test his hypotheses.

**testis**, pl. **testes** (L. witness) In mammals, the sperm-producing organ.

**testosterone** (Gr. *testis*, testicle + *steiras*, barren) A hormone secreted by the testes that is responsible for the secondary sex characteristics of males.

**thalamus** (Gr. *thalamos*, chamber) That part of the vertebrate forebrain just posterior to the cerebrum; governs the flow of information from all other parts of the nervous system to the cerebral cortex.

**theory** (Gr. *theorein*, to look at) A well-tested hypothesis supported by a great deal of evidence.

**thermal stratification** A process characteristic of larger lakes in temperate regions in which water at a temperature of 4° C sinks beneath water that is either warmer or cooler.

**thermodynamics** (Gr. *therme*, heat + *dynamis*, power) The study of transformations of energy, using heat as the most convenient form of measurement of energy. The first law of thermodynamics states that the total energy of the universe remains constant. The second law of thermodynamics states that the entropy, or degree of disorder, tends to increase.

**thigmotropism** (Gr. *thigma*, touch + *trope*, a turning) The growth response of a plant to touch.

**thorax** (Gr. a breastplate) The part of the body between the neck and the abdomen.

**thoroughfare channel** A capillary that connects arterioles and venules directly.

**threshold value** The minimal change in membrane potential necessary to produce an action potential.

**thrombus** (Gr. clot) A coagulation (clotting) that forms in a blood vessel and obstructs circulation.

**thylakoid** (Gr. *thylakos*, sac + *-oides*, like) A flattened saclike membrane in the chloroplast of a eukaryote; thylakoids are stacked on top of one another in arrangements called grana.

**thymine** A pyrimidine occurring in DNA but not in RNA; *see also* uracil.

**thyroxine** (Gr. *thyros*, shield) A hormone secreted by the thyroid that increases metabolic rate and promotes growth.

**tight junction** In organizing junctions, belts of protein that isolate parts of plasma membrane and form a barrier separating surfaces of cell.

**tissue** (L. *texere*, to weave) A group of similar cells organized into a structural and functional unit.

**trachea**, pl. **tracheae** (L. windpipe) In vertebrates, the windpipe.

**tracheid** (Gr. *tracheia*, rough) An elongated cell with thick, perforated walls, that serves to carry water and dissolved minerals through a plant and provides support. Tracheids form an essential element of the xylem of vascular plants.

**tracts** The bundles of nerve fibers within the central nervous system.

**transcription** (L. *trans*, across + *scribere*, to write) The first stage of gene expression in which a polymerase enzyme assembles an mRNA molecule whose sequence is complementary to the DNA.

**transcription unit** The portion of a gene segment that is transcribed into mRNA; it consists of the elements that are involved in the translation of the mRNA—the ribosome-binding site and the coding sequences.

**transfection** (L. *trans*, across + [in]fect) A technique used in studying tumors which consists of the isolation of nuclear DNA from human cells isolated from tumors, its cleavage into random fragments by using enzymes, and testing the fragments individually for the ability of any particular fragment to induce cancer in the cells that assimilate it.

**transfer RNA (tRNA)** (L. *trans*, across + *ferre*, to bear or carry) A second class of RNA that floats free in the cytoplasm; during polypeptide synthesis, tRNA molecules transport amino acids to the ribosome and position each amino acid at the correct place on the polypeptide chain.

**transformation** (L. *trans*, across + *formare*, to shape) Referring to the transfer of naked DNA from one organism to another; first observed as uptake of DNA fragments among pneumococcal bacteria.

**translation** (L. *trans*, across + *locare*, to put or place) The second stage of gene expression in which a ribosome assembles a polypeptide, using the mRNA to specify the amino acids.

**translocation** (L. *trans*, across + *locare*, to put or place) In plants, the process in which most of the carbohydrates manufactured in the leaves and other green parts of the plant are moved through the phloem to other parts of the plant.

**transmission electron microscope** A microscope that uses a beam of electrons rather than a light beam; scanning electron microscope—a microscope that beams electrons on the surface of a specimen as a fine probe, which passes back and forth rapidly.

**transpiration** (L. *trans*, across + *spirare*, to breathe) The loss of water vapor by plant parts, primarily through the stomata.

**transport form** Refers to sugars that are converted before they are moved within an organism; in many organisms, glucose is converted before it is moved and is less readily consumed (metabolized) while being moved.

**transposon** (L. *transponere*, to change the position of) A DNA sequence carrying one or more genes and flanked by insertion sequences that confer the ability to move from one DNA molecule to another; an element capable of transposition, the changing of chromosomal location.

**triphosphate group** Referring to three phosphate groups linked in a chain.

**trisomic** (Gr. *tri*, three + *soma*, body) A human individual that has received an extra autosome.

**trophic level** (Gr. *trophos*, feeder) A step in the flow of energy through an ecosystem.

**trophoblast** (Gr. *trope*, nourishment + *blastos*, germ, sprout) An outer sphere of cells external to the embryo in many mammals, having the function of supplying it with nourishment.

**tropical rain forest** A forest in a region near the equator, characterized by year-round warmth and very heavy rainfall; the largest is in the Amazon Basin of South America; they are the richest of all biomes in terms of the number of species.

**tropism** (Gr. *trope*, turning) The response to an external stimuli by a plant; a positive tropism is one in which the movement or reaction is in the direction of the source of the stimulus; a negative tropism is one in which the movement or growth is in the opposite direction.

**tropomyosin** (Gr. *tropos*, turn + *myos*, muscle) A major protein element of muscle tissue; it binds to actin and helps regulate the interaction of actin and myosin.

**troponin** (Gr. *tropos*, turn) A protein in muscle tissue regulated by calcium ions, important in muscle contraction.

**true-breeding** Strains of plants that produce offspring resembling their parents in all respects.

**tumor** (L. swollen) A ball of cells, constantly expanding in size.

**tundra** (Russ.) A vast, level, treeless plain in the arctic regions of Siberia and North America; the ground beneath is frozen even in summer; tundra covers one fifth of the earth's land surface.

**turgor** (L. *turgere*, to swell) The pressure exerted on the inside of a plant cell wall by the fluid contents of the cell; the interior of the cell is hypertonic in relation to the fluids surrounding it and so gains water by osmosis.

**turgor pressure** (L. *turgor*, a swelling) The pressure within a cell resulting from the movement of water into the cell. A cell with high turgor pressure is said to be turgid.

**twofold rotational symmetry** Describing the nucleotides at one end of the recognition sequence that are complementary to those at the other end so that the two strands of the DNA duplex have the same nucleotide sequence running in opposite directions for the length of the recognition sequence.

**tympanic membrane** The large membrane of the outer ear; the eardrum.

**ultimate causation** The question of why a certain animal behavior evolved or what its adaptive value was.

**ultrasound** A noninvasive procedure that uses sound waves to produce an image of the fetus but which harms neither the mother nor the fetus.

**undulipodia** (L. *undulatus*, wavy + Gr. *pous*, foot ) A term that describes the flagella of eukaryotic cells to underscore their complete distinctiveness from bacterial flagella, which they resemble only to a limited degree in external form and function.

**unicellular** Composed of a single cell.

**universal donor** A type O blood donor whose blood is compatible with any individual's immune system.

**universal recipient** An individual with an AB blood type who may receive any type of blood.

**uracil** A pyrimidine found in RNA but not in DNA; *see also* thymine.

**urea** (Gr. *ouron*, urine) An organic molecule formed in the vertebrate liver; the principal form of disposal of nitrogenous wastes by mammals.

**urethra** (Gr. *ourein*, to urinate) A duct in a male, through which both semen (during ejaculation) and urine (during liquid elimination) pass.

**uric acid** A waste product derived from ammonia found in the urine of reptiles and birds.

**urine** (Gr. *ouron*, urine) The liquid waste filtered from the blood by the kidney.

**useful energy** The amount of energy remaining to do work as progressively more energy is degraded to heat.

**uterus** (L. womb) In mammals, a chamber in which the developing embryo is contained and nurtured during pregnancy.

**vaccination** The injection of a harmless microbe into a person or animal to confer resistance to a dangerous one.

**vaccine** (L. *vacca*, cow) A substance that prevents a disease when injected into the body of an individual who does not already have the disease.

**vaccinia** A virus related to cowpox, which is the focus of modern researchers in their attempt to develop resistance to malaria, herpes, and other diseases.

**vacuole** (L. *vacuus*, empty) A cavity in the cytoplasm of a cell, bound by a single membrane and containing water and waste products of cell metabolism. Typically found in plant cells.

**vacuum suction** The removal of uterine contents by using a hollow curette or catheter to which a suction apparatus is attached. Used prior to 12th week of pregnancy.

**vagina** (L. sheath) The membranous passage in female animals that leads to the mouth of the uterus.

**variable** Any factor that influences a process; in evaluating alternative hypotheses about one variable, all other variables are held constant so that the investigator is not misled or confused by other influences.

**vas deferens** (L. *vas*, a vessel + *deferre*, to carry down) In mammals, the tube carrying sperm from the testes to the urethra.

**vascular bundle** In vascular plants, a strand of tissue containing primary xylem and primary phloem; these bundles of elongated cells function in conducting water with dissolved minerals and carbohydrates throughout the plant body.

**vascular cambium** In vascular plants, the cambium that gives rise to secondary phloem and secondary xylem; the activity of the vascular cambium increases stem or root diameter.

**vascular tissue** In vascular plants, a major tissue type that conducts water and dissolved minerals up through the plant and conducts the products of photosynthesis throughout.

**vasopressin** A posterior pituitary hormone that regulates the kidney's retention of water.

**vector** The infected genome that harbors the foreign DNA and carries it into bacterial targets; an organism that transmits disease-causing microorganisms.

**vegetal pole** The hemisphere of the blastula comprising cells rich in yolk.

**vein** (L. *vena*, a blood vessel) In plants, a vascular bundle forming a part of the framework of the conducting and supporting tissue of a stem or leaf; (2) in animals, a blood vessel carrying blood from the tissues to the heart.

**ventral** (L. *venter*, belly) Referring to the bottom portion of an animal.

**ventrical** (L. *ventriculus*, belly) Either of the two lower chambers of the heart that receives blood from the atria and forces it into the arteries.

**venule** (L. *vena*, vein) A small vein, especially one that begins at the capillaries and connects them with the larger veins.

**vertebral column** *See* spine.

**vertebrate** An animal having a backbone made of bony segments called vertebrae.

**vesicle** (L. *vesicula*, a little bladder) Membrane-enclosed sacs within eukaryotic organisms created by the weaving in sheets of endoplasmic reticulum through the interior of the cell.

**vessel** (L. *vas*, a vessel) A tubelike element in the xylem of angiosperms, composed of dead cells (vessel elements) arranged end to end. Its function is to conduct water and minerals from the soil.

**vessel element** In vascular plants, a typically elongated cell, dead at maturity, which conducts water and solutes in the xylem; vessel elements make up vessels. Tracheids are less specialized conducting cells.

**villus**, pl. **villi** (L. a tuft of hair) In vertebrates, fine, microscopic, fingerlike projections lining the small intestine that serve to increase the absorptive surface area of the intestine.

**visible light** The range of colors from violet (380 nanometers) to red (750 nanometers) that a human can see.

**vision** The perception of light, carried out in vertebrates by a specialized sensory apparatus called an eye.

**visual cortex** A major sensory region of the cerebral cortex lying on the occipital lobe, with different points in the visual field corresponding to different positions on the retina.

**vitamin** (L. *vita*, life + *amine*, of chemical origin) An organic substance required in minute quantities by an organism for growth and activity, but that the organism cannot synthesize.

**viviparous** (L. *vivus*, alive + *parere*, to bring forth) Referring to reproduction in which eggs develop within the mother's body.

**voltage-gated channel** A transmembrane pathway for an ion that is opened or closed by a change in the voltage, or charge difference, across the cell membrane.

**voluntary nervous system** *See* somatic nervous system.

**warning coloration** An ecological strategy of some organisms that "advertise" their poisonous nature by the use of bright colors.

**water cycle** The most familiar of the biogeochemical cycles in which free water circulates between the atmosphere and earth.

**water vascular system** The system of water-filled canals connecting the tube feet of echinoderms.

**whorl** A circle of parts present at a single level along an axis.

**wood** Accumulated secondary xylem; heartwood, the central, nonliving wood in the trunk of a tree; hardwood, the wood of dicots, regardless of how hard or soft it actually is; softwood, the wood of conifers.

**woody plant** A plant, such as a tree or shrub, in which secondary growth has been extensive.

**xylem** (Gr. *xylon*, wood) In vascular plants, a specialized tissue, composed primarily of elongate, thick-walled conducting cells, that transports water and solutes through the plant body.

**yolk** (O.E. *geolu*, yellow) The stored food in egg cells that provides the primary food supply for the embryo.

**zona pellucida** An outer membrane that encases a mammalian egg.

**zygomycete** (Gr. *zygon*, yoke + *mykes*, fungus) A type of fungus whose chief characteristic is the production of sexual structures called zygosporangia, which result from the fusion of two of its simple reproductive organs.

**zygosporangium,** pl. **zygosporangia** (Gr. *zygon*, yoke + *spora*, seed + *angeion*, a vessel) A sexual structure formed by the fusion of two of its simple reproductive organs, called gametangia, in a zygomycete.

**zygote** (Gr. *zygotos*, paired together) The diploid (2n) cell resulting from the fusion of male and female gametes (fertilization). A zygote may either develop into a diploid individual by mitotic divisions or undergo meiosis to form haploid (n) individuals that divide mitotically to form a population of cells.

# Illustration Credits

9-14 Illustration by Nadine Sokol
   Photo by B.A. Palevitz &
   E.H. Newcomb; BPS/Tom
   Stack and Assoc.
9-15 Raychel Ciemma
9-16 Raychel Ciemma
9-17 Barbara Cousins
9-18 Barbara Cousins
9-19 Illustrations by Barbara
   Cousins
   Photo by C.A.
   Hasenkampf/BPS
9-20 Kevin Somerville
9-21 Kevin Somerville
9-22 Raychel Ciemma

Ch 10 Alan Carey/Image Works
10-1 Michel Tcherevkoff/The
   Image Bank
10-2 H. Oscar/Visuals Unlimited
10-3 Richard Gross
10-4 National Library of
   Medicine/Picture Research
10-5 R.W. Van Olman
10-6, A Gail Hudson
   B Bill Ober
10-7 Nadine Sokol
10-8 Pagecrafters/Nadine Sokol
10-9 Globe Photos
10-10 Nadine Sokol
10-11 Nadine Sokol
10-12 Nadine Sokol
10-13 Nadine Sokol
10-14 Kevin Somerville
10-15 Carolina Biological Supply
10-16 Raychel Ciemma
10-17 Pagecrafters/Nadine Sokol
10-18 Barbara Cousins
10-19 Pagecrafters/Nadine Sokol
10-20, A Albert Blakeslee "Corn
   and Man", *Journal of
   Heredity*, vol. 5, pg. 511,
   1914
   B Pagecrafters/Nadine
   Sokol
10-21 Nadine Sokol
10-22 Fred Bruemmer
Illustrations for Genetics Problems
by Pagecrafters/Nadine Sokol

Ch 11 Gary Brady
11-1 The Bettmann Archive
11-2 CNRI/Science Photo Library
11-3, A Louis McGavran/Denver
   Children's Hospital
   B Richard Hutchings/Photo
   Researchers
11-4 Pagecrafters/Nadine Sokol
11-5 Raychel Ciemma
11-6 Pagecrafters/Nadine Sokol
11-7 Pagecrafters/Nadine Sokol
11-8, A Frank B. Sloop Jr., M.D.
   B Pagecrafters/Nadine Sokol

11-9 Pagecrafters/Nadine Sokol
11-10 Murayama 1981/BPS
11-11 Pagecrafters/Nadine Sokol
11-12 Pagecrafters/Nadine Sokol
11-13 David M. Phillips/Visuals
   Unlimited
11-14 Pagecrafters/Nadine Sokol
11-15 Illustration by
   Pagecrafters/Nadine Sokol
   Photo courtesy of Michael
   Ochs Archives, Venice, CA
11-16 Kevin Somerville
11-17 Washington University
   Medical School
11-18 Pagecrafters/Nadine Sokol
Illustrations for Genetics Problems
by Pagecrafters/Nadine Sokol

Ch 12 Science VU-UCLBL/Visuals
   Unlimited
12-1 John Eastcott/YVA
   Momatiuk/The Image Works
12-2 BPS
12-3 Raychel Ciemma
12-4 Ann Crossway/Calgene, Inc.
12-5 Bill Ober
12-6, A A.K. Kleinschmidt
   B Lee D. Simon/Photo
   Researchers, Inc.
   C Lee D. Simon/Photo
   Researchers, Inc.
12-7 Barbara Cousins
12-8 Pagecrafters/Nadine Sokol
12-9 Pagecrafters/Nadine Sokol
12-10 J.D. Watson/Cold Spring
   Harbor Laboratory Archive
12-11 A.C. Barrington Brown from
   J.D. Watson, *The Double
   Helix*, Atheneum, New York
   1968
12-12 Reprinted by permission
   from *Nature*, vol. 171, pp
   737-738, 1953, Macmillan
   Journals Ltd
12-13 Richard J. Feldman/NIH
12-14 George Klatt/Nadine Sokol
12-15 Barbara Cousins
12-16 Nadine Sokol
12-17 Barbara Cousins
12-18 Raychel Ciemma
12-19 Barbara Cousins
12-20 Barbara Cousins
12-21 Barbara Cousins/Raychel
   Ciemma
12-22 Bill Ober

Ch 13 Sally Elgin/Washington
   University
13-1 Sally Elgin/Washington
   University
13-2 Barbara Cousins
13-3 Pagecrafters/Nadine Sokol
13-4 Kevin Somerville

13-5 Kevin Somerville
13-6 Bill Ober
13-7 Nadine Sokol
13-8 Nadine Sokol
13-9 Nadine Sokol
13-10 Nadine Sokol
13-11 Nadine Sokol
13-12 Nadine Sokol
13-13 S. Cohen and J. Shapiro,
   *Scientific American*, pp.
   40-49, February 1980
13-14 Barbara Cousins
13-15 Claus Pelling
13-16 Illustration by Nadine Sokol
   Photo by Jack Griffith
13-17 Raychel Ciemma
13-18 Raychel Ciemma
13-19 Pagecrafters/Nadine Sokol
13-20 David Scharf/Discover 1987,
   Family Media, Inc.
13-21 Nadine Sokol
13-22 del Regato, et al: *Ackerman
   and del Regato's Cancer:
   Diagnosis, Treatment, and
   Prognosis*
13-23 Guenter Albrecht-Buehler
13-24 Raychel Ciemma
13-25, A Pagecrafters/Nadine
   Sokol
   B Lili Robins
   C Lili Robins
13-26 Barbara Cousins
13-27 Raychel Ciemma
13-28 Raychel Ciemma
13-29 Pagecrafters/Nadine Sokol
13-30 Lili Robins
Genetic Shorthand chart by Liz
Rudder/Kathleen Scogna
13-A American Cancer Society
13-B Pagecrafters/Nadine Sokol

14-1 R.Z. Brinster and R.E.
   Hammer
14-2 Kevin Somerville
14-3 Kevin Somerville
14-4 Stanley N. Cohen
14-5 Barbara Cousins
14-6 Photo by BioProducts,
   Rockland, ME
   Illustration by Barbara
   Cousins
14-7 Nadine Sokol
14-8 Nadine Sokol
14-9 Nadine Sokol
14-10 Courtesy of Monsanto
   Company
14-11 Barbara Cousins
14-12 Courtesy of Monsanto
   Company
14-13 Courtesy of Monsanto
   Company

14-14 Courtesy of Monsanto
   Company
14-15 Courtesy of Monsanto
   Company
14-16 Courtesy of Monsanto
   Company
14-17 Nadine Sokol
14-18 Animals Animals
14-19 Barbara Cousins
14-20 Barbara Cousins
14-A Lifecodes Corp./1988
   Discover Publications

Ch 15 P. & G. Bowater/The Image
   Bank
15-1 Nadine Sokol
15-2 Raychel Ciemma
15-3 E.S. Ross
15-4 E.S. Ross
15-5 E.S. Ross
15-6 Michael Coyne/Image Bank
15-7 Nadine Sokol
15-8, A Kevin Somerville
   B Lili Robins
15-9 Lili Robins
15-10 Pagecrafters/Nadine Sokol
15-11 Breck Kent/Animals Animals
15-12 Pagecrafters/Nadine Sokol
15-13 Bill Ober
15-14 Raychel Ciemma
15-15 Pagecrafters/Nadine Sokol
15-16 Raychel Ciemma
15-17 Nadine Sokol
15-18 Bill Ober
15-19 Bill Ober
15-20 Gary Milburn/Tom Stack &
   Assoc.
15-21 Nadine Sokol
15-22 Pagecrafters/Nadine Sokol

Ch 16 Mark Newman/Tom Stack
   & Assoc.
16-1 The Image Bank
16-2 Frank Oberle/Photographic
   Resources
16-3 E.S. Ross
16-5 Photos by Paul Ehrlich
   Illustrations by Bill
   Ober/Pagecrafters/Nadine
   Sokol
16-6 Map by Pagecrafters/Nadine
   Sokol
   Illustrations by Bill Ober
16-7 E.S. Ross
16-8 Kenneth Y. Kaneshiro
16-9 Map by Pagecrafters/Nadine
   Sokol
   Photo by Ken Lucas and
   Breck Kent
16-10 Bill Ober
16-A Walt Disney Productions

Ch 17 A.J. Copley/Visuals Unlimited
17-1 E.R. Degginger/Earth Scenes
17-2, A John Gurche
B John D. Cunningham/Visuals Unlimited
C James Amos/National Geographic Society
17-3 Courtesy of La Brea Tarpits Museum, LaBrea, CA
17-4 Pagecrafters/Nadine Sokol
17-5 Bill Ober
17-6 S. Giordano, III
17-7 Illustrations by Nadine Sokol Photo by M.R. Walter
17-8 Biological Photo Service
17-9, A John Gurche
B S. Conway Morris, *Phil Trans R Soc London*, vol. B285, pp. 272-274, 1971
17-10, A Karl Nicklas
B William A. Shear, *Science*, vol. 224, pp. 492-494, 1984 by the American Association for the Advancement of Science
17-11, A Frank M. Carpenter
B John Gerlach, Animals Animals
17-12 Pagecrafters/Nadine Sokol Illustrations by Bill Ober
17-13 Ron Ervin
17-14 Peabody Museum of Natural History, Yale University. Painted by Rudolph Zallinger
17-15 E.S. Ross
17-16 A. Montanari
17-17 Discover Magazine, Inc.
17-18 Kjell Sandved

Ch 18 Michael Nichols/Magnum Photos, Inc.
18-1 John Reader/Science Photo Library
18-2 Bill Ober
18-3 Bill Ober
18-4 Raychel Ciemma
18-5 Steve Martin/Tom Stack & Assoc.
18-6 Kevin Somerville
18-7, A Norbert Wu
B G.J. James/Biological Photo Service
C Marty Snyderman
18-8 Raychel Ciemma
18-9, A Alex Kerstitch
B Marty Snyderman
C Rob McKenzie
D John Cunningham/Visuals Unlimited
18-10 Bill Ober

18-11 Bill Ober
18-12, A John Shaw/Tom Stack and Assoc.
B John Shaw/Tom Stack and Assoc.
C Cleveland P. Hickman
D John Cunningham/Visuals Unlimited
E David M. Dennis/Tom Stack
18-13, A Brian Parker/Tom Stack and Assoc.
B William J. Weber/Visuals Unlimited
C John Cancalosi/Tom Stack and Assoc.
D Rod Planck/Tom Stack and Assoc.
18-14 Rudolph Zallinger/Yale Peabody Museum
18-15 Bill Ober
18-16 Michael W. Nickell
18-17 Ron Ervin
18-18, A E.S. Ross
B E.S. Ross
C John Cunningham, Visuals Unlimited
D Alan Nelson/Tom Stack & Assoc.
E John Cunningham/Visuals Unlimited
18-19 Raychel Ciemma
18-20, A J.E. Wapstra, National Photo Index of Australian Wildlife
B J. Adcock/Visuals Unlimited
18-21 Animals Animals
18-22, A Brian Parker/Tom Stack & Assoc.
B Rod Planck/Tom Stack & Assoc.
C E.S. Ross
D Cleveland P. Hickman
18-23 Ron Ervin
18-24, A Russell Mittermeier
B Gary Milburn/Tom Stack & Assoc.
18-25, A Russell Mittermeier and Andy Young
B Russell Mittermeier
C Russell Mittermeier
D Russell Mittermeier and Andy Young
18-26 Bill Ober/Pagecrafters
18-27, A Courtesy of The Living World, St. Louis Zoo, St. Louis, MO
B John Reader
18-28 Barbara Cousins
18-29 Allen Carroll and Jay Maternes/National Geographic, November 1985

18-30 Pagecrafters/Nadine Sokol
18-31 Douglas Waugh/Peter Arnold, Inc.

Ch 19 Fred Whitehead
19-1 Brian Parker/Tom Stack and Assoc.
19-2, A World Wildlife Fund/Helmut Diller
B World Wildlife Fund/Helmut Diller
C World Wildlife Fund/Paul Barruel
19-3 John Shaw
19-4 Pagecrafters/Nadine Sokol
19-5 Pagecrafters/Nadine Sokol
19-6 Gianni Tortoli, 1969 National Geographic Society
19-7 Peter Arnold, Inc.
19-8 Courtesy of National Museum of Natural History, Smithsonian Institution
19-9 Mrs. Rudolph A. Peterson
19-10 Pagecrafters/Nadine Sokol
19-11 Pagecrafters/Nadine Sokol
19-12 Heather Angel/Biofotos
19-13, A Anne Westheim/Animals Animals
B Barbara Cousins
19-14 Bill Ober
19-15 Pagecrafters/Nadine Sokol
19-16 Biological Branch, Department of Lands, Queensland, Australia
19-17 Biological Branch, Department of Lands, Queensland, Australia
19-18 Promotion Australia
19-19 Rolf O. Peterson

Ch 20 Steve Maslowski/Visuals Unlimited
20-1 David Swanlund, courtesy of Save-The-Redwood League
20-2 E.S. Ross
20-3 Gerald Corsi/Tom Stack & Assoc.
20-4 R.W. Gulson
20-5 Joe McDonald/Animals Animals
20-6 E.S. Ross
20-7 E.S. Ross
20-8, A Lincoln P. Brower, University of Florida
B Lincoln P. Brower, University of Florida
20-9 E.S. Ross
20-10 E.S. Ross
20-11 James L. Castner
20-12, A William J. Weber/Visuals Unlimited
B Alex Kerstitch
C E.S. Ross

20-13 Paul Opler
20-14, A E.S. Ross
B E.S. Ross
C E.S. Ross
D E.S. Ross
20-15 Jeff Foott
20-16 Larry Tackett/Tom Stack & Assoc.
20-17 E.S. Ross
20-18 E.S. Ross
20-19, A Daniel H. Janzen
B Daniel H. Janzen
C C.W. Rettenmeyer
D Daniel H. Janzen
20-20 E.S. Ross

Ch 21 Photographic Resources
21-1 Steve McCurry
21-2 Nadine Sokol
21-3 Nadine Sokol
21-4 Nadine Sokol
21-5 Nadine Sokol
21-6 E.S. Ross
21-7 John D. Cunningham/Visuals Unlimited
21-8, A T.W. Ransom/Biological Photo Service
B D. Wrobel/Biological Photo Service
C Kjell B. Sandved
21-9 Raychel Ciemma
21-10 Lili Robins
21-11 Pagecrafters/Nadine Sokol
21-12 Dwight Kuhn
21-13 Peter Frenzen
21-14 Rod Planck/Tom Stack

Ch 22 E.S. Ross
22-1 Bill Ober
22-2 Bill Ober
22-3 Barbara Cousins
22-4 Barbara Cousins
22-5 Bill Ober
22-6 Pagecrafters/Nadine Sokol
22-7 Marty Snyderman
22-8 Ann Wertheim/Animals Animals
22-9, A Map by Pagecrafters/Nadine Sokol
B Photos by Chesapeake Bay Foundation
C Photos by Chesapeake Bay Foundation
D Photos by Chesapeake Bay Foundation
22-10 Jim and Cathy Church
22-11 J. Frederick Grassle, Woods Hole Oceanographic Institute
22-12 E.S. Ross
22-13 Fred Rhode/Visuals Unlimited

22-14 Nadine Sokol
22-15 John D. Cunningham/ Visuals Unlimited
22-16 Daniel W. Gotshall
22-17 Bill Ober
22-18 Bill Ober
22-19 Background photo: Breck P. Kent/Earth Scenes
22-19, **A** Zig Leszcyznski/Animals Animals
   **B** Michael Fogden/Animals Animals
   **C** The Image Bank
22-20 Background photo: W. Perry Conway/Tom Stack and Assoc.
22-20, **A** P & W Ward/Animals Animals
   **B** The Image Bank
   **C** Ken Cole/Animals Animals
22-21 Background photo: Doug Sokell/Visuals Unlimited
22-21, **A** John Gerlach/Earth Scenes
   **B** John Gerlach/Animals Animals
   **C** Breck P. Kent/Animals Animals
22-22, **A** J. Cancalosi/Tom Stack & Assoc.
   **B** William H. Hamilton
22-23 Background photo: J. Weber/Visuals Unlimited
22-23, **A** Brian Parker/Tom Stack and Assoc.
   **B** Margot Conte/Animals Animals
   **C** Charlie Palek/Earth Scenes
22-24 Background photo: Dwight Kuhn
22-24, **A** Dwight Kuhn
   **B** T. Kitchin/Tom Stack and Assoc.
   **C** Gary W. Griffin/Animals Animals
22-25 E.S. Ross
22-26 Background photo: Will Troyer/Visuals Unlimited
22-26, **A** Roger and Donna Aitkenhead/Animals Animals
   **B** Dianna Stratton/Tom Stack and Assoc.
   **C** Johnny Johnson/Animals Animals
22-27 Background photo: W.J. Weber/Visuals Unlimited
22-27, **A** John Gerlach/Tom Stack and Assoc.
   **B** Rick McIntyre/Tom Stack & Assoc.
   **C** M & C Ederegger/Peter Arnold, Inc.
   **D** Cleveland P. Hickman

**Ch 23** Norbert Wu
23-1 Joe McDonald/Earth Scenes
23-2 E.S. Ross
23-3 Pagecrafters/Nadine Sokol
23-4 R. Koch/Contrasto Picture Group
23-5 Pagecrafters/Nadine Sokol
23-6 J.B. Forbes, *St. Louis Post-Dispatch*
23-7 Stephanie Maze/Woodfin Camp, Inc.
23-8 Ghillean Prance
23-9 Nigel Smith/Earth Scenes
23-10 Hugh Iltis
23-11 Hugh Iltis
23-12 Alex Kerstitch
23-13, **A** James L. Castner
   **B** D.O. Hall, King's College, University of London
23-14 D.O. Hall, King's College, University of London
23-15 E.S. Ross
23-16, **A** E.S. Ross
   **B** E.S. Ross
   **C** D.O. Hall, King's College, University of London
23-17 Harold More
23-18 E.S. Ross
23-19 Byron Augustin/Tom Stack & Assoc.
23-20 Pagecrafters/Nadine Sokol. Courtesy of Nuclear Regulatory Commission
23-21 Holt Studios/Earth Scenes
23-22 Fred Ward
23-23 Grant Heilman Photography
23-24 John D. Cunningham/ Visuals Unlimited
23-25, **A** Richard Klein
   **B** Richard Klein
   **C** Richard Klein
23-26 NASA
23-A, **1** Eloy Rodriquez
   **2** L. Mellichamp/Visuals Unlimited
   **3** George H. Huey/Earth Scenes
   **4** Rodale Press
   **5** L. Mellichamp/Visuals Unlimited
**unn 23-3** USFWS Photo/Gary Zahm
**unn 23-4** J. Gebhardt/The Image Bank
**unn 23-5** World Wildlife Fund
**unn 23-6** John Chellman/Animals Animals
**unn 23-7** Terry G. Murphy/Animals Animals

**unn 23-8** A. Spalteholz/The Image Bank
**unn 23-9** John Chellman/Animals Animals
**unn 23-10** Zig Leszczynski/ Animals Animals
**unn 23-11** Andrew Plumptre/Oxford Scientific Films
23-27 Mickey Gibson/Tom Stack & Assoc.
23-28 Steve McCurry

**Ch 24** Peter Parks/Oxford Scientific Films
24-1 *Exxon*, vol 3, 3rd quarter, 1985
24-2 Library of Congress/Picture Research
24-3 Bill Ober
24-4, **A** Andre Gallant/The Image Bank
   **B** Brian Parker/Tom Stack & Assoc.
24-5 Barbara Cousins
24-6, **A** Rod Planck/Tom Stack & Assoc.
   **B** E.S. Ross
   **C** Brian Parker/Tom Stack & Assoc.
24-7, **A** John Cunningham/Visuals Unlimited
   **B** Rod Planck/Tom Stack & Assoc.
24-8 Fred Hossler/Visuals Unlimited
24-9 Pagecrafters/Nadine Sokol
24-10, **A** Michael Fogden
   **B** Michael Fogden
   **C** E.S. Ross
24-11, **A** J. Robert Waaland; BPS/Tom Stack & Assoc.
   **B** T.E. Adams/Visuals Unlimited
   **C** Marty Snyderman
24-12, **A** Marty Snyderman
   **B** John Cunningham/Visuals Unlimited
   **C** Jeff Rotman
24-13 John D. Cunningham/ Visuals Unlimited
24-14 Pagecrafters/Nadine Sokol
24-15 A. Friedmann/Visuals Unlimited

**Ch 25** Peter Arnold, Inc.
25-1 Lee D. Simon/Photo Researchers, Inc.
24-2, **A** J.J. Cadamone, Jr. & B.K. Pugashetti; BPS/Tom Stack & Assoc.
   **B** David M. Phillips/Visuals Unlimited
25-3 Dwight Kuhn
25-4 Globe Photos

25-5 Centers for Disease Control, Atlanta
25-6, **A** K.G. Murti/Visuals Unlimited
   **B** Donald L.D. Caspar, *Science,* vol. 227, pg. 773-776, February 1985
25-7 Pagecrafters/Bill Ober
25-8 Nadine Sokol
25-9 Nadine Sokol
25-10, **A** Dwight Kuhn
   **B** Centers for Disease Control, Atlanta

**Ch 26** Manfred Kage/Peter Arnold, Inc.
26-1 Manfred Kage/Peter Arnold, Inc.
26-2 Nadine Sokol
26-3, **A** E.W. Daniels
   **B** E.S. Ross
   **C** John Cunningham/Visuals Unlimited
26-4 Nadine Sokol
26-5 William Patterson/Tom Stack & Assoc.
26-6 Robert Sisson/National Geographic Society
26-7, **A** Ed Reschke
   **B** E.S. Ross
26-8 Nadine Sokol
26-9 Phillip Harrington/Peter Arnold, Inc.
26-10 Nadine Sokol
26-11 Nadine Sokol
26-12, **A** Kjell Sandved
   **B** Ed Pembleton
   **C** Walt Anderson/Tom Stack
   **D** John Cunningham/Visuals Unlimited
   **E** Robert Simpson/Tom Stack & Assoc.
26-13 Harvey C. Hoch
26-14 Kjell Sandved
26-15 Nadine Sokol
26-16 David M. Phillips/Visuals Unlimited
26-17 Terry Ashley/Tom Stack & Assoc.
26-18 Bill Ober
26-19 E.S. Ross
26-20 Ed Reschke
26-21 Nadine Sokol

**Ch 27** Marcel Isy-Schwart/The Image Bank
27-1 Hoi-Sen
27-2, **A** E.S. Ross
   **B** Rod Planck/Tom Stack & Assoc.
   **C** Rod Planck/Tom Stack & Assoc.
   **D** James Castner

27-3 E.J. Cable/Tom Stack & Assoc.
27-4 Raychel Ciemma
27-5 Terry Ashley/Tom Stack & Assoc.
27-6 Pagecrafters/Nadine Sokol
27-7 Pagecrafters/Nadine Sokol
27-9, **A** Kirtlye Perkins/Visuals Unlimited
  **B** E.S. Ross
27-10 Nadine Sokol
27-11 Raychel Ciemma
27-12, **A** Kjell Sandved
  **B** Runk Schoenberger/Grant Heilman Photography
  **C** Brian Parker/Tom Stack & Assoc.
27-13, **A** Kjell Sandved
  **B** Runk Schoenberger/Grant Heilman Photography
  **C** Jeff Rotman
27-14 Nadine Sokol
27-15 E.S. Ross
27-16 Nadine Sokol
27-17 Nadine Sokol
27-18 Ed Pembleton

**Ch 28** Johnny Johnson/Animals Animals
28-1 Susan Glascock
28-2 Raychel Ciemma
28-3 Nadine Sokol
28-4, **A** David J. Wrobel/Biological Photo Service
  **B** Mickey Gibson/Animals Animals
28-5, **A** Marty Snyderman
  **B** Bill Ober
  **C** Neil G. McDaniel/Tom Stack & Assoc.
  **D** Rick Harbo
  **E** Gwen Fidler/Tom Stack & Assoc.
28-6 Bill Ober
28-7 Nadine Sokol
28-10, **A** T.E. Adams/Visuals Unlimited
  **B** Stan Elems/Visuals Unlimited
28-11, **A** John Cunningham/Visuals Unlimited
  **B** R. Calentine/Visuals Unlimited
28-12 Scott Bodell
28-13 V.R. Ferris
28-14 Visuals Unlimited
28-15 Scott Bodell
28-16, **A** Kjell Sandved
  **B** Milton Rand/Tom Stack & Assoc.
  **C** Kjell Sandved
    Illustrations by Bill Ober

28-17 Kjell Sandved
28-18, **A** David M. Dennis/Tom Stack & Assoc.
  **B** Kjell Sandved
  **C** Kjell Sandved
  **D** Gwen Fidler/Tom Stack & Assoc.
28-19 Kevin Somerville
28-20 M.K. Ryan
28-21, **A** Hans PfetSchinger/Peter Arnold, Inc.
  **B** John Gerlach
28-22 M.K. Ryan
28-23, **A** E.S. Ross
  **B** Kjell Sandved
  **C** John Cunningham/Visuals Unlimited
  **D** Anne Mareton/Tom Stack & Assoc.
  **E** Rod Planck/Tom Stack & Assoc.
28-24 Bill Ober
28-25, **A** Kjell Sandved
  **B** Kjell Sandved
  **C** T.E. Adams/Visuals Unlimited
  **D** E.S. Ross
  **E** Cleveland P. Hickman
  **F** J. Mac Gregor/Peter Arnold, Inc.
28-26, **A** J.A. Adcock/Visuals Unlimited
  **B** Kjell Sandved
  **C** Alex Kerstitch
  **D** Kjell Sandved
  **E** E.S. Ross
  **F** E.S. Ross
28-27 George Venable
28-28, **A** Kjell Sandved
  **B** Thomas Eisner
28-29, **A** John Shaw/Tom Stack & Assoc.
  **B** Gwen Fidler/Tom Stack & Assoc.
  **C** Gwen Fidler/Tom Stack & Assoc.
  **D** Gwen Fidler/Tom Stack & Assoc.
  **E** John Shaw/Tom Stack & Assoc.
  **F** K.A. Blanchard/Visuals Unlimited
  **G** K.A. Blanchard/Visuals Unlimited
  **H** Sal Giordino Hall
  **I** Sal Giordino Hall
28-30, **A** Jeff Rottman
  **B** Daniel Gotshall
  **C** Alex Kirstitch
  **D** Carl Roessler/Tom Stack
  **F** Bill Ober

28-31, **A** Kjell Sandved
  **B** George Venable
28-32 M. England: *A Color Atlas of Life Before Birth-Normal Fetal Development*, ed. 1, Mosby-Yearbook, 1990
28-33, **A** Jim and Cathy Church
  **B** Bill Ober
28-34 Bill Ober
28-35, **A** Heather Angel
  **B** Barbara Cousins

**Ch 29** Thomas Kitchen/Tom Stack & Assoc.
29-1 E.S. Ross
29-2 Photo by Terry Ashley/Tom Stack & Assoc. Illustration by Bill Ober
29-3, **A** Don Sindelac
  **B** Bill Ober
29-4, **A** E.S. Ross
  **B** Terry Ashley/Tom Stack & Assoc.
29-5 Randy Moore/Visuals Unlimited
29-6 George Wilder/Visuals Unlimited
29-7 Richard H. Gross
28-8, **A** Jeff Rotman
  **B** Danial Gotshall
  **C** Alex Kerstitch
29-9 Doug Wechsler/Earth Scenes
29-10, **A** Bill Ober
  **B** Bill Ober
  **C** Wilfred A. Cote/Spectrum Images
29-11, **A** Randy Moore/Visuals Unlimited
  **B** Bill Boer
29-12 Michael Fogden
29-13 Kjell Sandved
29-14 R.F. Evert
29-15 R.A. Gregory/Visuals Unlimited
29-16, **A** Dwight Kuhn
  **B** John Cunningham/Visuals Unlimited
29-17 E.J. Cable/Tom Stack & Assoc.
29-18 R.F. Evert
29-19, **A** Ed Reschke
  **B** Carolina Biological Supply
29-20 E.S. Ross
29-21, **A** Carolina Biological Supply
  **B** E.S. Ross
29-22, **A** Nadine Sokol
  **B** Terry Ashley/Tom Stack & Assoc.
29-23 Jim Solliday/BPS
29-24 E.J. Cable/Tom Stack & Assoc.

**Ch 30** Harry Haralambou/Peter Arnold, Inc.
30-1 Keith Murakami/Tom Stack & Assoc.
30-2 Gary Milburn/Tom Stack & Assoc.
30-3, **A** Michael Fogden
  **B** E.S. Ross
  **C** E.S. Ross
30-4 Kjell Sandved
30-5 John Cunningham/Visuals Unlimited
30-6 Barry L. Runk/Grant Heilman Photography
30-7 David Priestley
38-8, **A** James Castner
  **B** E.S. Ross
  **C** E.S. Ross
30-9, **A** James Castner
  **B** John Cunningham/Visuals Unlimited
  **C** John Cunningham/Visuals Unlimited
30-10, **A** Ed Pembleton
  **B** James Castner
30-11 Nadine Sokol
30-12 Barbara Cousins
30-13 Peter Hoch
30-14 Peter Hoch
30-A George Schaller

**Ch 31** J. Robert Stottlemyer/Biological Photo Service
31-1 E.S. Ross
31-2 Raychel Ciemma
31-3 Bill Ober
31-4 Nadine Sokol
31-5 Emmanuel Epstein
31-6 Nadine Sokol
31-7 Barbara Cousins
31-8 Pagecrafters/Nadine Sokol
31-9, **A** Nadine Sokol
  **B** Silvan H. Wattwer
31-10 John Cunningham/Visuals Unlimited
31-11 John D. Cunningham/Visuals Unlimited
31-12 E.S. Ross
31-13 Pagecrafters/Nadine Sokol/Bill Ober
31-A, 1 J.A.L. Cooke/Earth Scenes
  2 R. Mitchell/Tom Stack & Assoc.
  3 Kjell Sandved
  4 H.A. Miller/Visuals Unlimited
  5 John Cunningham/Visuals Unlimited

**Ch 32** J. Carmichael/The Image Bank
32-1 Lennart Nilsson
32-2 Barbara Cousins

32-3 Nadine Sokol
32-4 Barbara Cousins
32-5 M.K. Ryan
32-6 Emma Shelton
32-7 J.V. Small
32-8 *St. Louis Globe Democrat*
32-9 David M. Phillips/Visuals Unlimited
32-10 Photos by John Cunningham/Visuals Unlimited Illustrations by Kevin Somerville
32-11 John Cunningham/Visuals Unlimited
32-12 Nadine Sokol
32-13 Nadine Sokol
32-14 Nadine Sokol
unn 32-1 Cynthia Turner Alexander/Terry Cockerham, Synapse Media Production

Ch 33 Steve McCutcheon/Visuals Unlimited
33-1 David I. Vaney, from John Nolte: *The Human Brain*, Mosby-Year Book, St. Louis, MO, 1988
33-2 Nadine Sokol
33-3 Nadine Sokol
33-4 E.S. Ross
33-5 E.R. Lewis/BPS/Tom Stack & Assoc.
33-6 Kevin Somerville
33-7 Kevin Somerville
33-8 Kevin Somerville
33-9 Kevin Somerville
33-10 Pagecrafters/Nadine Sokol
33-11 Nadine Sokol
33-12 Nadine Sokol
33-13 Barbara Cousins
33-14 Barbara Cousins
33-15 Barbara Cousins
33-16 Kevin Somerville
33-17 Nadine Sokol
33-A, 1 Chikashi Toyoshima and Nigel Unwin, "Ion channel of acetycholine receptor reconstructed from images of postsynaptic membranes," *Nature*, November 1988
   2 Nadine Sokol

Ch 34 Stephen J. Krasemann/DRK Photo
34-1 E. Wilkinson/Animals Animals
34-2 Nadine Sokol
34-3 Lennart Nilsson
34-4 Pagecrafters/Nadine Sokol
34-5 Nadine Sokol
34-6 Raychel Ciemma

34-7 Pagecrafters/Nadine Sokol
34-8 Barbara Cousins
34-9 Bill Ober
34-10 Courtesy of *Washington University Magazine*, vol. 58, no. 2, Summer 1988
34-11, A Martha Swope
   B Michael Fogden
34-12 Raychel Ciemma
34-13 Bill Ober
34-14 Marsha J. Dohrmann
34-15 Stephen Dalton/Oxford Scientific Films
34-16 Kevin Somerville
34-17, A Barbara Cousins
   B Scott Mittman
34-18 Kevin Somerville
34-19 Bill Ober
34-20 Barbara Cousins
34-21 Raychel Ciemma
34-22 Kevin Somerville
34-23, A Leonard Lee Rue III
   B Nadine Sokol
34-24 Pagecrafters/Nadine Sokol
34-25 Lennart Nilsson
34-26 Barbara Cousins
34-27, A Stephen J. Krasemann/DRK Photo
   B M.P. Kahl/DRK Photo
   C Wayne Lynch/DRK Photo
   D C.C. Lockwood/DRK Photo
   E M.P. Kahl/DRK Photo
   F Stephen J. Krasemann/DRK Photo
   G Don and Pat Valenti/DRK Photo
   H Stephen J. Krasemann/DRK Photo
   I Stephen J. Krasemann/DRK Photo
   J FPG International
   K Blumebild/H. Armstrong Roberts
   L Kimbenn/FPG International
34-28 Raychel Ciemma
34-29 Raychel Ciemma
34-A, 1 Dwight Kuhn
   2 Kjell Sandved
   3 Kjell Sandved Illustrations by Bill Ober

Ch 35 Dwight Kuhn
35-1 Leonard Rue III/Visuals Unlimited
35-2 Raychel Ciemma
35-3 Barbara Cousins
35-4 Walking: Jen and Des Bartlett/NHPA Crawling: Anthony Bannister/Animals Animals

Running: G. Ziesler/Peter Arnold, Inc.
Leaping: Stephen Dalton/NHPA
Climbing: Charles G. Sommers, Jr./Tom Stack & Assoc.
Along for the ride: Frans Lanting
Swimming: Frans Lanting
Flying: Dwight Kuhn
35-5 Nadine Sokol
35-6 Nadine Sokol
35-7 Bill Ober
35-8 Barbara Cousins
35-9 Barbara Cousins
35-10 Kevin Somerville
35-11 Nadine Sokol
35-12 Barbara Cousins
35-13 Raychel Ciemma
35-14 Raychel Ciemma
35-15 Raychel Ciemma
35-16 John Cunningham/Visuals Unlimited
35-17 Kevin Somerville
35-18 Barbara Cousins
35-19 Bill Ober

Ch 36 Michael Quinton/Visuals Unlimited
36-1 Janeart Ltd./The Image Bank
36-2 Nadine Sokol
36-3 Nadine Sokol
36-4 Nadine Sokol
36-5 Nadine Sokol
36-6 Nadine Sokol
36-7 Kevin Somerville
36-8 Kevin Somerville
36-9 Nadine Sokol
36-10 Kevin Somerville
36-11 Nadine Sokol
36-12 *Akhenaten*. The National Museum at Berlin, DDR, Egyptian Museum, number 14 512.
36-13 Robert Wadlow in L.H. Behrens and D.P. Barr, "Hyperpituitarism Beginning in Infancy," *Endocrinology* vol. 16, pg. 125, January 1932
36-15 Barbara Cousins
36-16 Mieke Maas/The Image Bank
36-17 Ed Reschke
36-18 Nadine Sokol
36-19 Lennart Nilsson
36-20 Nadine Sokol
36-21 Bill Ober
36-22 Kevin Somerville
36-23 Raychel Ciemma

Ch 37 Michael Quinton/Visuals Unlimited

37-1 Visuals Unlimited
37-2 Nadine Sokol
37-3 Bill Ober
37-4 Nadine Sokol
37-5 Barbara Cousins
37-6 Bill Ober
37-7 Bill Ober
37-8 Christy Krames
37-9 Nadine Sokol
37-10 Raychel Ciemma
37-11 Raychel Ciemma
37-12 Nadine Sokol
37-13 Raychel Ciemma
37-14 Raychel Ciemma
37-15 Bill Ober
37-16 Lili Robins
37-17 Lili Robins

Ch 38 Michael Quinton/Visuals Unlimited
38-1 Frank Oberle/Photographic Resources
38-2 Barbara Cousins
38-3 Kevin Somerville
38-4 Kevin Somerville
38-5 Kevin Somerville
38-6 Kevin Somerville
38-7 Kevin Somerville
38-8 Painting by Bill Ober; Diagrams by Pagecrafters/Nadine Sokol
38-9 Nadine Sokol
38-10 Ellen Dirkson/Visuals Unlimited
38-11 Barbara Cousins
38-12 Christy Krames
38-13 Pagecrafters/Nadine Sokol
38-14 Raychel Ciemma

Ch 39 Steve McCutcheon/Visuals Unlimited
39-1 Lennart Nilsson
39-2 Nadine Sokol
39-3 W. Rosenberg/BPS/Tom Stack & Assoc.
39-4 Kevin Somerville
39-5 Ed Reschke
39-6 Bill Ober
39-7 Ed Reschke
39-8 Nadine Sokol
39-9 Ed Reschke
39-10 Nadine Sokol
39-11 Bill Ober
39-12 Nadine Sokol
39-13 Makio Murayama/BPS/Tom Stack & Assoc.
39-14 Barbara Cousins
39-15 Pagecrafters/Nadine Sokol
39-16 Bill Ober
39-17 Pagecracfters/Nadine Sokol
39-18 Barbara Cousins

39-19 Nadine Sokol
39-20 Barbara Cousins
39-A, 1 Gernsheim Collection, Harry Ranson Humanities Research Center, Univeristy of Texas at Austin
2 Barbara Cousins

Ch 40 Joe McDonald/Animals Animals
40-1 *Smithsonian Magazine,* p. 142, January 1989
40-2 Bill Ober
40-3 George Johnson
40-4 Manfred Kage/Peter Arnold, Inc.
40-5 Barbara Cousins
40-6 *Scientific American,* January, 1988
40-7 Pagecrafters/Nadine Sokol
40-8 Barbara Cousins
40-9 John Cunningham/Visuals Unlimited
40-10 Bill Ober
40-11 Nadine Sokol
40-12 Nadine Sokol
40-13 Nadine Sokol
40-14 Pagecrafters/Nadine Sokol/Kathleen Scogna
40-15 Barbara Cousins
40-16, A Bill Ober
B Pagecrafters/Nadine Sokol
40-17 Raychel Ciemma
40-18 Pagecrafters/Nadine Sokol
40-20 Pagecrafters/Nadine Sokol
40-21 Lennart Nilsson
40-22 Pagecrafters/Nadine Sokol
40-23 Pagecrafters/Nadine Sokol

40-24 Barbara Cousins
40-25 Larry Arlian, Wright State University
40-A Barbara Cousins

Ch 41 Frans Lanting
41-1 Cleveland P. Hickman
41-2 Larry Brock/Tom Stack & Assoc.
41-3 Pagecrafters/Nadine Sokol
41-4 Bill Ober
41-5 Scott Bodell
41-6 Cleveland P. Hickman
41-7 Marty Snyderman
41-8 Nadine Sokol
41-9 Scott Bodell
41-10 Scott Bodell
41-11 Barbara Cousins
41-12 Barbara Cousins

Ch 42 Jeff Rutman
42-1 Kaz Mori/The Image Bank
42-2 Hans Pfletschinger/Peter Arnold, Inc.
42-3 Cleveland P. Hickman
42-4 Nadine Sokol
42-5 Barbara Cousins
42-6 Lennart Nilsson
42-7 Kevin Somerville
42-8 Ron Ervin
42-9 Lennart Nilsson
42-10 Barbara Cousins
42-11 Ed Reschke
42-12 Barbara Cousins
42-13 Scott Bodell
42-14 Pagecrafters/Nadine Sokol
42-15 Nadine Sokol
42-16 Kevin Somerville
42-17 Pagecrafters/Nadine Sokol

42-18 Trent Stephens
42-19 Scott Bodell

Ch 43 Dwight Kuhn
43-1 Lennart Nilsson
43-2 David M. Phillips/Visuals Unlimited
43-3 Illustration by Nadine Sokol
43-4 Barbara Cousins
43-5 Barbara Cousins
43-6 David M. Phillips/Visuals Unlimited
43-7 Carolina Biological Supply
43-8 David M. Phillips/Visuals Unlimited
43-9 Kevin Somerville
43-10 Kevin Somerville
43-11 Kevin Somerville
43-12 Bill Ober
43-13 Scott Bodell
43-14 Pagecrafters/Nadine Sokol
43-15 Heather Angel
43-16 Bill Ober
43-17 Barbara Cousins
43-18 Lennart Nilsson
43-19 Nadine Sokol
43-20 Scott Bodell
43-21 Claire Perry/The Image Works

Ch 44 Dwight R. Kuhn
44-1 Bettman Archive
44-2 Stephen Dalton/National Audobon Society Collection/Photo Researchers
44-3 Nadine Sokol/Pagecrafters
44-4 John Alcock: *Animal Behavior* 4th ed., Sinauer, Sunderland, MA, 1988
44-5 A.O.D. Willows

44-6, A Linda Koebner/Bruce Coleman, Inc.
B Jeff Foote/Bruce Coleman, Inc.
44-7 Will Rapport/Photo Researchers, Inc.
44-8 John Alcock: *Animal Behavior* 4th ed., Sinauer, Sunderland, MA, 1988
44-9 Dan Kline/Visuals Unlimited
44-10 Illustration by Bill Ober Photo by Animals Animals
44-11 A. Thornall and J. Alcock: *Insect Mating Systems,* Sinauer, Sunderland, MA, 1984
44-12 P. Marler and W. Hamilton: *Mechanisms of Animal Behavior,* John Wiley and Sons, New York, NY, 1984
44-13 David W. MacDonald
44-14 Bill Ober
44-15 Joe McDonald/Tom Stack & Assoc.
44-16 William J. Weber/Visuals Unlimited
44-17 Z. Leszcynski/Animals Animals
44-18 E.S. Ross
44-19 Paulette Brunner/Tom Stack and Assoc.
44-20 E.O. Wilson: *Sociobiology: The New Synthesis,* Harvard University Press, Cambridge, MA, 1975
44-21, A John Cunningham, Visuals Unlimited
B Jim Grier: *Biology of Animal Behavior,* Mosby-Year Book, St. Louis, MO, 1984
44-22 George Johnson

# Index

Antibody, 919-920, 924, 927
acquired immunodeficiency
syndrome virus, 602-603
heavy chain, 925
light chain, 925
molecular structure of, 926
monoclonal, 932-933
polyclonal, 932
Antibody immune response; *see*
Humoral immune response
Antibody test
in acquired immunodeficiency
syndrome, 967
ELISA, 967
Anticodon, 316, 325
Antidiuretic hormone, 751, 832-833,
840
Antigen, 919, 926-927
antibody recognition of, 925
cell surface, 268
Rh blood group, 270
Anura, 441-443
Aorta, 902
Aortic valve, 902
Ape, 449-451, 452
*Aphelandra tridentata,* 583
Aphid, mutualism and, 496
Apical meristem, 683, 684, 689, 691,
695
*Aplysia,* 490, 1007
Apocynaceae, 487
Aposematic coloration, 491
Appearance, diversity in, 234
Appendicitis, 382
Appendicular skeleton, 810, 811
Appendix, vermiform, 382
Aquatic vertebrate
cleavage of primitive, 981
gastrulation and, 984-985
Aquifers, 506
$^{40}$Ar dating of fossils; *see* Argon-40
dating of fossils
Arachnida, 667
Araneae, 667
Arbovirus, 604
Archaen Period, 74
*Archaeopteryx,* 20, 445, 446
Archegonium, 634, 635, 637, 639
Archenteron, 984
Arctic fox, 253
Arginine, 54
biosynthesis of, 305
Argon-40 dating of fossils, 414
Arithmetic progression of food supply,
18
Arizona, greenhouse laboratory in,
554
Armadillo, 911
South American, 16, 17
Armoured fish evolution, 437
Aromatic amino acids, 54
Arrowhead spider, 667
*Artemisia tridentata,* 533
Arterial wall tissue, 745
Arteriole, 891, 892
Arteriosclerosis, 906
Artery, 890, 891, 892, 894
pulmonary, 902
Arthritis, rheumatoid, 812
Arthropod, 650, 663, 665-670, A3
land invasion of, 421
tracheae, respiration, 875
Artificial selection, 372
dogs and, 395
Asclepiadaceae, 487
Ascocarp, 622
Ascomycete, 622, A2
Ascorbic acid, 864
Ascus, 622
Asexual reproduction, 220-221
plants and, 709

Asian pitcher plant, 724
Asiatic fungus, 476-477
Asparagine, 54
Asparagus, 692
*Asparagus officinalis,* 692
Aspartic acid, 54, 69, 70
Aspen, 709, 710
clones of, 710
*Aspergillus,* 622
Associative behavior, 1006
Associative cortex, 782
Associative learning, 1006-1007
Assortment, independent, 244-245,
246
Aster, 216
Asteraceae, 702
Atherosclerosis, 903, 905, 906
deaths from, 901
Atmosphere, 520-523
carbon dioxide in, 506-507
Atmospheric climate, 521-523
Atmospheric precipitation, 521-523
Atmospheric pressure, 720
Atom, 32-35
in functional groups, 44
functional groups of, 44, 45
seeing, 59
Atomic energy levels, 34
Atomic number, 32
ATP; *see* Adenosine triphosphate
ATPase; *see* Adenosine triphosphatase
Atrial natriuretic hormone, 842-843,
903
Atrial peptide, 842-843
Atrioventricular node, 904
Atrium
fish, 900
left, 902
Audition, 1005
Auditory cortex, 782
*Aurelia aurita,* 654
Australian skink, 443
*Australopithecus,* 434, 451-453, 454,
455
Autoimmune disease, 917
Autonomic function, 794
Autonomic nervous system, 777, 794,
797
Autosome, 263
Autotroph, 160
metabolic efficiency of, 166
Autotrophic plankton, 512
Auxins, 726, 727
AV node; *see* Atrioventricular node
Avery, Oswald, 291
Aves, 445-446, 676
Avian blastula, 983
Avian sarcoma virus, 329-331, 332
Axial skeleton, 810-811
Axil, 691
Axon, 748, 759, 766, 767
in giraffe, 761
neuronal, 747-748, 759
skeletal muscles and, 818
*Azolla,* 552

**B**

B allele of ABO blood group, 270
B cell, 921, 924
immune system and, 921, 922, 924
receptors for, 927
Bacillariophyta, 614
Bacilli, 595
*Bacillus thuringiensis,* 354
Backbone, 573, 810-811
vertebrate, 435-436
Background radiation, 559
Bacteria, 25, 72, 582, 593-599, A1;
*see also* Bacterial cell
aerobic, 99-100
anaerobic photosynthetic, 101

Bacteria—cont'd
chromosome replication in, 301
development of, 6
DNA-containing organelles
resembling, 99-102
DNA replication in, 99-102, 210
earliest, 72, 73
exponential growth in population
of, 467
features of, 581
flagella of, 83, 105
fossils of, 72, 418
genetic alterations in, 355
luminous, 527
methane-producing, 72-73
origin of modern, 73-74
photosynthetic, 73, 89, 596
anaerobic, 101
electron in, 192
symbiotic, 101
population growth of, 469
restriction enzymes and, 345-346
rod-shaped, 85
spherical, 85
spiral, 85
spirochaete, 102
structure of, 85-88, 595-596
toxins of, 354
viruses invading, 594
heredity and, 291-292
Bacterial capsule, 595
Bacterial cell, 85-88
deoxyribonucleic acid in, 88
replication of, 300, 301
division of, 210-211
structure of, 88
Bacterial toxins, 354
Bacteriophage, 345
acquired immunodeficiency
syndrome and, 603
T$_4$, 594
Balance, 727
water; *see* Water balance
*Balanus,* 474
*Balanus balanoides,* 473
Bald eagle, 503
Baldness, 241
Baleen whale, 384
Bamboo, flowering, 711
Barbule of feather, 446, 447
Bark, 693, 694
Barnacle, 495
competition and, 473
gooseneck, 668
niche, 474
Barometer, mercury, 873
Barometric pressure, 873
Baroreceptor, 786
Barr body, 265
Barrel cactus, 682
Barrel sponge, 653
Basal body, 102
Base, 44
organic nitrogen-containing, 58
Base pairing, 298
Basidiomycete, 622, 623, A2
Basidiomycota, 619
Basidium, 622
Basilar membrane, 789
Basophil, 897-898
Bat, 449, 531
flight muscles of, 808
sound frequencies and, 789
Bates, H.W., 492
Batesian mimicry, 492-493, 494
*Battus philenor,* 493
Beach fleas, 668
Beachy experiment, 353
Beadle, George, 301
Beadle and Tatum's procedure for
isolating mutations in
*Neurospora,* 303, 304

*Beagle,* voyage of, 14, 15
Bean
germination in, 684
winged, 553
Bear, grizzly, 393, 536
Beardworm, 527
Beaver, teeth of, 852
Bee
colony of, 1017
honey, 581
muscles in, 808
in pollination, 703
wild, 372
Bee-eater, 447
Beef tapeworm, 658, 659
Beetle, 533, 574
African blister, 669
blister, 425, 669
elm bark, 694
flour, 473
ground, 484, 485
lampyrid, 1010
longhorn, 490
muscles in, 808
scarab
pollen, 704
South American, 666
Behavior, 1001-1025
animal; *see* Animal behavior
associative, 1006
circannual, 1009
conditioned, 1007
development, 1007-1008
foraging, 1013-1014
neural basis of, 1005
nonassociative, 1006
and physiology, 1008-1009
stereotyped, 1003
Behavioral genetics, 1004
Behavioral isolation, 400-401
Behavioral rhythm, 1008-1009
Belt, Thomas, 496
Beltian bodies, 496, 497
Beriberi, 864
Beta-carotene, 188, 189
Beta-cell, 837
Beta-galactosidase, 326
*Betula,* 704
*Betula papyrifera,* 693
Bicarbonate, pancreas and, 858
Biennial plant, 685
Bilateral symmetry in evolution,
655-657
Bilayer membrane permeability, 115
Bile salts, 859
Binary fission, 211
in bacteria, 595
Binocular vision, 449
Binomial system in classification of
organisms, 575-576
Biochemical pathway, 149
metabolism and, 152, 160-161
Biogeochemical cycle, 505
Biogeography, 417
Biological clock, 76, 1009
Biological diversity, 571-678
in animals, 649-678
of bacteria, 593-599
of fungi, 618-627
of kingdoms of life, 573-591
in plants, 629-647
of protists, 611-618
of viruses, 600-607
Biological theory, history of, 14-16
Biologist and environmental problems,
567-568
Biology
animal; *see* Animal biology
cell; *see* Cell biology
levels-of-organization approach in,
24, 25

*Homo erectus*, 454
*Homo habilis*, 453, 455
*Homo sapiens*, 455-457
Homologue, 222, 226
Homology, 381
Homosporous plant, 635
Homozygote, 240, 254, 372
   test plant, 242
Honduras, 549
Honeybee, 665, 1018
   classification of, 581
   dance language of, 1011
Honeysuckle berries, 706
Hoofed mammal, evolution of, 381
Hooke, R., 83
Hormone, 825-845
   adrenocorticotropic, 834, 840
   antidiuretic, 832-833, 840
      salt balance and, 751
   atrial natriuretic, 842-843, 903
   behavior, 1008
   as body messenger, 843
   in brain control of endocrine
      system, 838-841
   in communication, 843
   digestion and, 855, 856
   endocrine gland, 828, 831-838,
      840-841
   follicle-stimulating, 834, 840, 963,
      964
   growth, 357, 833, 834, 840
   growth-hormone–releasing,
      840-841, 842
   hypothalamus production of,
      840-841, 842
   importance of, 826-828
   insulin, 830
   in integration of body activity, 826
   luteinizing, 834, 840, 963, 964
   melanocyte-stimulating, 834
   nonendocrine, 842-843
   parathyroid, 835, 841
   peptide, 829, 830
   plant, 723-728
      discovery of first, 725
   releasing, 838-839
   as second messenger, 830
   sex, 838, 962-963
   steroid, 828-829, 962, 963
   thyroid, 834-835, 839, 841
   thyroid-stimulating, 834, 840
   thyrotropin-releasing, 839
   vertebrate communication and, 827
   in water and salt balance in kidney,
      833
Hornwort, 631, 636
Horse
   evolution of, 21
   race, 374
Horsetail, 639
House-dust mite, 931
HRV; *see* Holmes ribgrass virus
H$_2$S; *see* Hydrogen sulfide
H$_2$SO$_4$; *see* Sulfuric acid
Human
   abdominal cavity of, 741
   appearance of, 232
   birth weight in, 376
   body of, 741
   brain of, 780, 781
   cell of, 741
      size of, 75
   chromosome of, 212, 213, 262-266
      deoxyribonucleic acid in, 303
   classification of, 581
   development of, 993
   digestive system of, 851, 853
   ear of, 788
   egg of, 979
      journey to uterus of, 960-962
      size of, 75

Human—cont'd
   embryo of, 674
   evolution of, 455-457
   eye of, 791, 792
   genetic engineering and genome of,
      355
   heart of, 902-907
      path of blood flow through, 903
   immune system of, 921, 923
   kidney of, 945
   liver of, 75
   menstrual cycle in, 965
   muscles in, 807, 815
   nervous system of, 777
   orgasmic response in, 968
   red blood cell of, 75
   reproductive cycle in, 963-965
   reproductive system of, 957
      egg response to sperm in, 979,
         980-981
      male, 957
   respiratory system of, 880, 881
   similarity of, to gorillas and
      chimpanzees, 433
   skeleton of, 741, 810-811
   spinal cord of, 741, 795
   testes of, 958
   thoracic cavity of, 741
   traits of, 235
Human fetus, 978
Human genetics; *see* Genetics, human
Human growth hormone, 357, 833,
   834, 840
Human immunodeficiency virus, 604,
   931
Human population, 470
   growth of, 546
Humboldt Current, 523
Hummingbird, 704
Humoral immune response, 922, 923,
   925
Humpbacked whale, 470
Huntington's disease, 272, 275-276,
   277
   phenotype of, 241
Hyacinth, water, 554
Hybrid, 235
   sterile, 396
Hybrid offspring, 237-238
Hybridization, 395-396
   barriers to, 398-405
Hybridoma, 932-933
*Hydra*, 654, 655
Hydration shell, 43
Hydrocarbon, 51
   chlorinated, 561-562
Hydrocarbon ethylene, 728
Hydrochloric acid, 43, 854
   in digestion, 849
   ionization of, 43-44
   in stomach, 854-855
Hydrogen, 36, 37, 38
   atom of, 33
   covalent bonds of, 37
   enzyme activity and, 149
   primitive earth and, 67
   size of molecule of, 75
Hydrogen bond, 37, 41-42
   deoxyribonucleic acid and, 61
   in enzyme shape, 148
Hydrogen gas, 36, 37, 38
Hydrogen sulfide, 527
   chemoautotrophic bacteria and, 596
   in primitive earth, 67
Hydrolysis, 45, 46
   lysosomes and, 99
Hydrophobic bond, 43
Hydrophobic compound, 43
   enzyme activity and, 148
Hydroskeleton, 807
Hydroxyl group, 44, 45

*Hylobates muelleri*, 452
Hylobatidae, 451
*Hyophorbe verschaffeltii*, 557
Hypercholesterolemia, 272
   phenotype of, 241
Hypertension, 901, 903
   essential, 903
   high-renin, 903
Hypertonic concentration, 120, 939
Hyphae of fungi, 619, 620
Hypothalamus
   homeostasis and, 750
   hormone production and, 840-841,
      842
   in integrating body responses, 979
   kidney function and, 948
   in release of hormones, 828,
      832-834, 838-839
Hypothesis, 9-11, 12
   one gene–one enzyme, 301-305
Hypotonic concentration, 120, 939
*Hyracotherium*, 21
Hysterectomy, 972

**I**

IAA; *see* Indoleacetic acid
Ice formation, 42
Ichthyosaurs, 427
*Ichthyostega*, 423
Immune response, 919-920
   cell-mediated, 922
   humoral, 922
   primary, 928
   secondary, 928
Immune response library, 927
Immune system, 918-922
   acquired immunodeficiency disease
      syndrome and, 966
   antigens and; *see* Antigen
   as body defense, 911, 913
   breast feeding and, 996
   cells of, 744, 920, 921
   human, 921, 923
   major histocompatibility marker
      and, 917
   as major vertebrate organ system,
      742
   trypanosome escape and, 928
Immunity, 927-928
Immunoglobulin, 925; *see also*
   Antibody
Impulse, nerve; *see* Nerve impulse
In vitro fertilization, 992
Inbreeding, 372
Inchworm caterpillar, 491
Incisors, 852
Incompatibility, genetic-self
   autoimmune disease and, 917
   plants and, 711
Incomplete dominance, 252
Independent assortment, 244-245, 246
Indeterminate apical meristem, 689
India
   agriculture in, 545
   population of, 546
Individual in ecosystem, 511
Indoleacetic acid, 726
Inducer T cell, 921
Induction in animal development,
   988-991
Inductive reasoning, 9
Industrial country, population growth
   of, 547
Industrial melanism, 378
   Kettlewell's experiment in, 379
Inferior vena cava, 902
Inflammatory response, 897, 917
Information
   reception of
      cells in, 130-132
      plasma membrane in, 119

Information—cont'd
   sensing of internal, 785-787
   storage of hereditary, 288-293
   transmission of, 757-773
      nerve impulse in, 761-766
      nerve to target tissue, 767-771
      neuron in, 759-760
Inhalation, 880-882
Inheritance; *see also* Genetics;
      Heredity
   chromosomal theory of, 246
   of human traits, 235
   patterns of, 266-268
Inhibition
   allosteric, 153
   end-product, 149
   feedback, 152, 153
Inhibitor, 149
Inhibitory synapse, 767, 769, 770
Initiation complex, 319
Initiation factor in protein synthesis,
   319
Innate capacity increase, 466-468
Innate releasing mechanism,
   1003-1004
Inner ear, 789
Inner medulla, 946
Inositol phosphate, 830
Insect, 669-670
   characteristic features of, 670
   circulation in, 891
   compound eye of, 670
   gas exchange in, 874
   genetic engineering and, 353-354,
      355
   invasion of land by, 421
   Malpighian tubules of, 940
   muscles in, 808
   osmoregulation in, 940
   plant development and, 702
   pollen eating, 704
   society behavior of, 1017-1018
Insectivores, ancestor of, 450
Insulin, 830, 837, 841
   action of, 830
   blood glucose and, 838
Insulin hormone, 830
Integration, 771
Integument, 653
   of seed plants, 705
Integumentary system, 742
Interbreeding, 395-396
Intercellular connections, 132-133
Intercourse, sexual; *see* Sexual
   intercourse
Interferon, 917
Interleukin-1, 922
Intermediate fiber, 103, 104
Interneuron, 794
Internode, stem, 691
Interoception, 784, 785-787
Interoceptor, 785
Interphase, 214-216
Intertidal region, 525
Intestinal cell
   microvilli and, 858
   smooth endoplasmic reticulum and,
      94
Intestine
   large, 860-861
   small, 856-858
      digestive enzymes and, 850
Intrapleural fluid, 880
Intraspecific competition, 472
Intrauterine device, 971, 972
Intron, 321, 325
Invertebrate
   evolutionary history of marine, 422
   osmoregulation in, 940
Involuntary nervous system, 777, 794,
   797

Lion tamarin, 566
Lipase, 849, 850
Lipid, 46, 47, 51-52
  complex carbon-ring, 962
  endoplasmic reticulum and, 93, 94
  plasma membrane and, 118
  synthesis of, 94
Lipid bilayer, 114-116
Lipid membrane, 112, 113-116
Lipopolysaccharides, 595
*Liriodendron tulipifera,* 690
Litchi nuts, 552
Littoral region, 525
Littoral zone, 527
Liver, 858-860
  metabolites and, 859
  sinuses of, 859
Liver cell
  of human, 75
  of rat, 97
  smooth endoplasmic reticulum and,
    94
Liverwort, 631, 635, 636
Living fossils, 72
Lizard, 427
  yarrow spiny, 1009
Lobe-finned fish, 415
  evolution of, 437
  primitive amphibian compared to,
    442
Lobster, 668
Locus, 240, 254
Locust, migratory, 468
*Locusta migratoria,* 468
Long-day plant, 730
Long-tailed hermit hummingbird, 704
Long-term memory, 785
Longhorn beetle, 490
*Lonicera hispidula,* 706
Loop of Henle, 944
Loose connective tissue, 744
Loostrife, European purple, 467
Lorenz, Konrad, 1003
Los Angeles air pollution, 567
Lotus, 705
Lower esophageal sphincter, 852
*Loxosceles reclusa,* 667
Lucy, *Australopithecus,* 453
*Ludia magnifica,* 673
Luminous bacteria, 527
Lung, 875, 876-879
  amphibian, 877-878
  bird, 878-879
    air sac in, 879
  cancer of, 329, 330
  evolution of, 878-879
  mammalian, 874, 878
    gas exchange in, 874
  in reptile, 878
  residual volume of, 877
  in vertebrate adaptation, 440
Lungfish, 439
  evolution of, 386
Luteal phase of egg maturation, 964
Luteinizing hormone, 834, 840, 963,
    964
  egg maturation and, 964
*Lycopersicon esculentum,* 723
Lycophyta, 639, A2
Lyell, C., 16
Lyme disease, 599
Lymphatic system, 752, 894-895
Lymphatic vessel valve, 895
Lymphocyte, 745
  AIDS virus and, 593
  tumor-infiltrating, 924
Lyonization, 265
Lysergic acid, 800
Lysine, 54
Lysosomal cell, 98-99

Lysosome, 98-99
  comparison of bacterial, animal, and
    plant cell, 106
  as digestive enzyme, 98-99
  eukaryotes and, 92
  size of, 75
Lysozyme, 147
  amino acid sequence of, 56
  in body defense, 914
*Lythrum salicaria,* 466, 467

# M

MAC; *see* Membrane attack complex
MacArthur, R., 473
MacDonald, D., 1011
Macroevolution, 379-386
  defined, 366, 367, 373
  pace of, 387
  patterns of distribution and,
    384-386
  of titanotheres, 381
Macromolecule, 45-46
Macronutrient, plant, 722
Macrophage, 745, 898
  in body defense, 915, 921, 922
*Macrotermes bellicosus,* 669
Magnesium, 36
  active transport and, 127
  plants and, 722
Magnetism, 784
Maidenhair fern, 631
Maidenhair tree, 640
Major histocompatibility complex,
    132
Major histocompatibility complex
    marker, 917
Malaria, 616-617
  sickle-cell trait and, 273
  vaccine for, 359
Malaysian pitcher plants, 630
Male reproductive system in human,
    957-959
  illustration of, 753
  as major vertebrate organ system,
    742
Malnourishment, 549
Malpighian corpuscle, 941
Malpighian tubule, 940
Maltase, 850
Malthus, Thomas, 18
Maltose, 48
Mammal, 448-449
  brain mass in, 780
  cleavage in, 983
  egg of
    maturation of, 964
    response to sperm of, 980
  and evolution, 422
    of heart, 900
    of hoof, 381
  gastrulation in, 985-986
  insect eating, 450
  kidneys of, 943
    renal tubule organization in, 945
  lung in, 874, 878
  marine, 896
  placental, 449, 450, 957
    parallel adaptation in, 385
  reproduction of, 957
Manchurian dry lake bed, 705
Mandible of arthropod, 666
Mandibulate, 666, 668-670
Manganese, 36
Mangold, H., 988, 990
Mangold and Spermann's dorsal lip
    transplant experiment, 990
Mangrove, 707
*Manihot esculenta,* 557
Manioc, 557
Manta ray, 439
Mantle, 662

Maple
  silver, 691
  sugar, 717
  wind dispersion of, 707
*Maranta,* 730
*Marchantia,* 635, 636
Marginal meristem, 689
Margulis, L., 102
Marijuana, 690
Marine amphipod, 668
Marine fish kidney, 942
Marine green algae, 84, 289
Marine invertebrate evolutionary
    history, 422
Marine mammals, 896
Marine organisms in Cambrian Period,
    413
Marler, P. 1008
Marmot, yellow-bellied, 580
*Marmota flaviventris,* 580
Marrow, bone, 811, 812
Marsh, salt, 510, 511
Marsh weed, 466, 467
Marsupial, 418, 448, 449, 957
  parallel adaptation in, 385
Mass extinction, 423-424
Mast cell, 745, 921
Master gland, 832-834
Mating, 1015
  nonrandom, 370, 371-372
  in turtles, 956
Matrix, 176
  mitochondrial, 100
*Mawsonites,* 420
Maximum efficiency, 180
Mayfly, 669
Mayr, Ernst, 395-396
Measles, German, 992
Meat diet, 330
Mechanoreceptor hair, 1005
Mediterranean monk seal, 466
Medium
  complete, 301
  defined, 301
  minimal, 303
Medulla
  adrenal, 835-836, 840
  renal, 946
Medulla oblongata, 778
Medusa, 654
  cnidarian, 653, 654
Meerkat, 1011
Megachannel Extra-Terrestrial Assay,
    77
Megagametophyte, 635, 641, 643
Megakaryocyte, 899
Megasporangia, 635, 644
Megaspore, 635
Meiosis, 219-220, 224
  in eukaryotes, 587
  in fern, 639
  first division of, 223-227
  gametic, 588
  mammalian egg and, 980
  mitosis compared to, 228
  in plant, 633, 634
  second division of, 227
  sporic, 588
  stages of, 222-227
  zygotic, 588
Meischer, F., 293
Meissner's corpuscle, 784, 786
Melanism, industrial, 378, 379
Melanocyte-stimulating hormone, 834
Melatonin, 834, 838, 841
Membrane
  basilar, 789
  bilayer, 115
  cell; *see* Plasma membrane
  eukaryotic, 107
  lipid, 112, 113-116

Membrane—cont'd
  photosynthetic, 89, 190
  plasma; *see* Plasma membrane
  pleural, 880
  postsynaptic, 768, 770
  presynaptic, 768
  prokaryotic, 107
  protein movement, 115
  protein secretion, 98
  selectively permeable, 125
  tectorial, 789
  tympanic, 789
Membrane attack complex, 917
Membrane support, 117
Memory, 782-785
  molecular basis of, 783
Mendel, Gregor Johann, 236-245
  experiments of; *see* Mendelian
    genetics
  first law of, 244
  second law of, 245
Mendelian genetics, 233-259; *see also*
    Genetics
  chromosomes and, 245-247
  crossing-over in, 248-250
  early ideas about heredity in,
    235-236
  experiment results in, 237-239
  experimental design in, 237
  garden pea in, 236-237
    cross of, 242
  gene interaction in, 251-253
  independent assortment in, 244-245
  interpretation in, 239-242
  results of experiments in, 237-239
  science of, 253-254
  test cross and, 243-244
Mendelian ratio, 240, 244-245
  modified, 252-253
Mendelian trait, 236
Menstrual cycle, 964, 965
Menstruation, 965
Mercury barometer, 873
Meristem
  apical, 683, 684, 685, 689, 691,
    695
    indeterminate, 689
  lateral, 685
  marginal, 689
    determinate, 689
Merkel cell, 784, 786
Meroblastic cleavage, 983
*Merops,* 447
*Merychippus,* 21
Meselson and Stahl experiment, 297,
    299
Mesencephalon, 778
Mesenteries, 657
Mesoderm, 653, 741
  animal development and, 984
  neurulation and, 987
*Mesohippus,* 21
Mesophyll, 201, 690
Mesozoic Era, 417, 419, 422, 424
  mammals in, 448
  reptiles in, 444
Messenger
  chemical, 826-828, 838-839
  second, 830
Messenger ribonucleic acid, 313, 317
Messenger ribonucleic acid transcript,
    primary, 315
META; *see* Megachannel
    Extra-Terrestrial Assay
Metabolic circle, 169-170
Metabolic efficiency in food chains,
    166
Metabolic machine, 203
Metabolism, 139-157
  activation of, 145-146